Tools for achieving peak academic performance!

Included with every new copy of *Essentials of Human Anatomy & Physiology,* Eighth Edition:

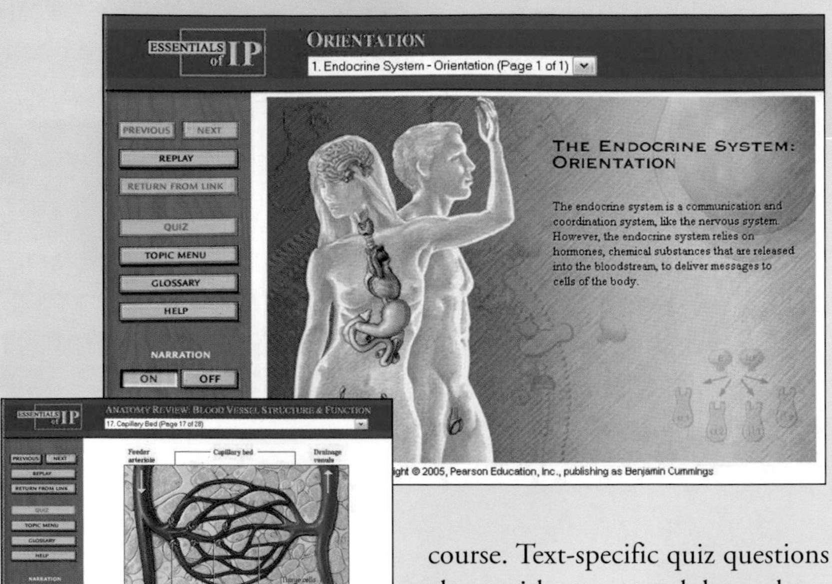

ENHANCED!

Essentials of InterActive Physiology® (IP) CD-ROM

Adapted from the award-winning *InterActive Physiology*® program, this dynamic and highly effective learning tool uses animations, tutorials, and quizzes to teach the toughest concepts in the course. Text-specific quiz questions are included for each chapter, along with a new module on the endocrine system. Icons with references to Essentials of IP tutorials appear in chapter summaries of the text to integrate book coverage and media.

The Anatomy & Physiology Place Open-access Website

www.anatomyandphysiology.com

Students do not need an access code or password to use the wide variety of interactive review exercises, practice quizzes, learning activities, and case studies on the Anatomy & Physiology Place for *Essentials of Human Anatomy & Physiology.* But teachers can use their user name and password (see next page) to register for special, password-protected resources, including an instructor's guide and lecture outlines.

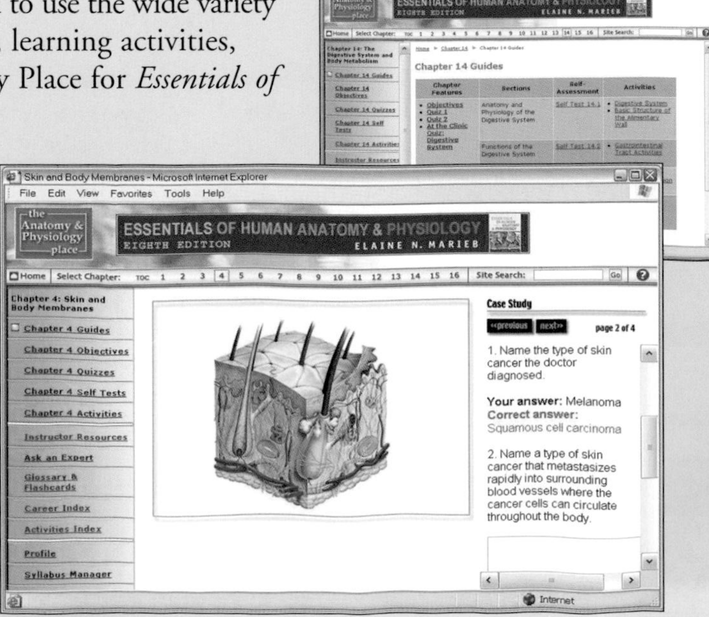

oaching your A&P students is easier than ever!

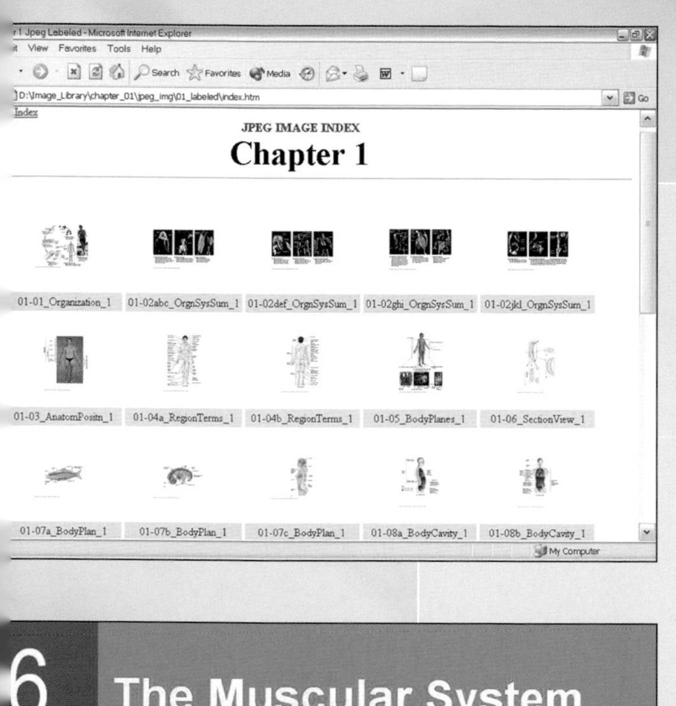

ENHANCED!

The Instructor's Resource CD-ROM

ISBN: 0-8053-7338-1

This teaching resource has been upgraded to include web browser-based access to a library of thumbnail images for easier, more convenient use. Also features lecture outlines with selected images from the text in PowerPoint® format. Also included on this dual-platform CD-ROM are all of the figures and tables from the text, organized by chapter as PowerPoint® files and as JPEGs in three formats: with labels and leaders, with leaders only, and without leaders or labels.

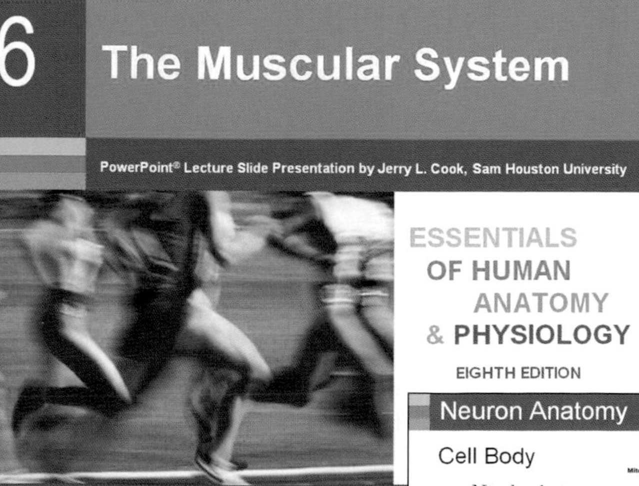

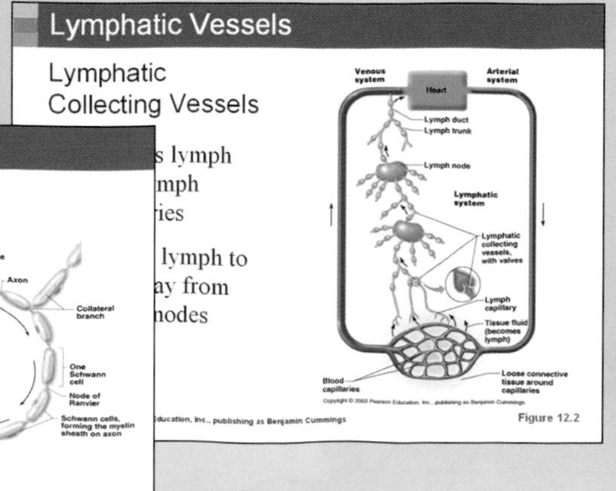

ESSENTIALS OF HUMAN ANATOMY & PHYSIOLOGY

EIGHTH EDITION

ELAINE N. MARIEB, R.N., PH.D.

Holyoke Community College

San Francisco Boston New York
Cape Town Hong Kong London Madrid Mexico City
Montreal Munich Paris Singapore Sydney Tokyo Toronto

Publisher: Daryl Fox
Sponsoring Editor: Serina Beauparlant
Development Manager: Claire Alexander
Project Editor: Mary Ann Murray and Karoliina Tuovinen
Managing Editor: Wendy Earl
Production Supervisor: Janet Vail
Editorial Assistant: Alex Streczyn
Cover and Text Design: Yvo Riezebos
Compositor: GTS Graphics
Proofreader: Martha Ghent
Indexer: Kathy Pitcoff
Senior Manufacturing Buyer: Stacey Weinberger
Executive Marketing Manager: Lauren Harp

Cover Image: Veer
Photo and illustration credits follow the Appendices.

Library of Congress Cataloging-in-Publication Data

Marieb, Elaine Nicpon
Essentials of human anatomy & physiology / Elaine N. Marieb.—8th ed.
p. cm.
Includes index.
ISBN 0-8053-7328-4 (text) — ISBN 0-8053-7327-6 (package)
1. Human physiology. 2. Human anatomy. I. Title: Essentials of human anatomy and physiology.
II. Title.

QP34.5.M264 2006
612—dc22 2004060029

ISBN 0-8053-7327-6 (For sale copy with CD-ROM)

ISBN 0-8053-7342-X (Professional copy with CD-ROM)

ISBN 0-13-193481-3 (High School Edition)

11—DOR—08

www.aw-bc.com

Preface to the Teacher

The eighth edition of *Essentials of Human Anatomy and Physiology,* like its predecessors, introduces students to the structure and function of the human body. Timely topics, such as immunology and clinical advances, have been updated so that explanations are both accurate and current. To ensure that this text remains relevant and meaningful to the students who will use it, the Focus on Careers feature introduced in the sixth edition has been continued. Most importantly, this edition retains the unintimidating writing style that has successfully helped students with a limited background in the sciences grasp the fundamental concepts of human anatomy and the inner workings of the body.

Organization

As before, extensive reviews by anatomy and physiology teachers helped me decide on a particular presentation. My choice was to remain with a sequence of topics that encourages learning. The first chapter orients students to the new world of learning in which they become immersed in anatomical terminology, names of major body areas, and brief overviews of functions that must be performed to sustain life.

Chemistry, cells, tissues, and the first organ system (skin) are treated in succession. Most students find that the skin is an interesting system, and thus the transition to the organ system discussions is easily accomplished. Subsequent chapters include systems requiring a good deal of anatomical terminology (for example, skeletal, muscular, and nervous systems). Each organ system is approached from simple to increasingly complex levels. Building in an understanding of the concepts, rather than rote memorization, is emphasized.

Organizational Flexibility

A good textbook meets the needs of both teachers and students. Because every teacher has a personal style and philosophic approach to teaching and course organization, I have written each chapter to allow flexibility in topic presentation. All chapters are self-contained; hence a teacher may choose a chapter sequence that is personally preferred. Topic omission may be necessary in programs in which less than one semester is devoted to anatomical and physiological concepts.

New to This Edition

Much thought was given to the student learning process and the needs of teachers when undertaking this new edition. Improvements to this new edition include changes in anatomical terms as guided by International Federation of Associations of Anatomists (IFAA) as well as changes in the following areas.

New Features

New "Prove It Yourself" boxes. These active-learning exercises encourage students to verify facts for themselves rather than take them on faith. For example, it's easy to demonstrate quickly the presence of an enzyme in your own saliva. Look for instructions in the Chapter 2 "Prove It Yourself" box on page 49. To calculate your basal metabolic rate to see how much energy your body needs to function, see the "Prove It Yourself" box in Chapter 14, page 491.

New Figure legend questions accompany several selected figures and prod the student to interpret the concepts or illustrated processes or to make predictions about "what's next." Answers can be found upside down on the same page.

Revisions in the Art Program

Several new pieces of art were added in this edition and many figures were reconceptualized. Keeping in mind that automatic learning is very important in studying and learning a subject like A&P, I continued to pay close attention to maintaining a uniform style and consistent use of color in recurring structures across chapters. Special attention was given to the labeling of each illustration to ensure that it is consistent with the discussion in the accompanying text.

An Exceptional Teaching Package

Teaching anatomy and physiology is a challenging and rewarding endeavor. To assist teachers in that task, our new teaching package is designed to help you use and derive the greatest benefit from this text.

- ***New* Instructor's Resource CD-ROM** This powerful lecture presentation resource offers all of the illustrations, photos, and tables from the text in both JPEG and PowerPoint® formats. You can quickly locate any image using directories of thumbnail images that are viewable through your web browser. Also included are customizable PowerPoint® lecture slides with art and lecture notes for each chapter.
- **The Anatomy & Physiology Place Website** This redesigned and updated website includes learning activities, self-tests, quizzes, case studies, and more for every chapter. The password-protected Instructor's Resource section contains all of the media resources from the Instructor's Resource CD-ROM as well as an electronic version of the Instructor's Guide.
- ***Updated and Expanded* Instructor's Guide/ Test Bank** The Instructor's Guide has been revised to support each chapter in the eighth edition. Expanded lecture outlines, updated references, and new teaching tips make this guide a useful tool for teachers. The improved test bank now includes over 1,500 questions (including many new and updated questions) in multiple formats with page references to the text and difficulty ratings. An electronic Test-Gen version of the test bank is also available, and allows teachers to edit and customize the test bank to suit their specific needs.
- ***New* Transparency Acetates** This high-quality acetate package includes all illustrations—approximately 275 images—(excluding photographs) from the text. They are easy to use and to read in any lecture setting.
- ***Essentials of Human Anatomy & Physiology Laboratory Manual,* Third Edition** by Elaine N. Marieb. Designed for use in an introductory one-semester anatomy and physiology laboratory course, this lab manual has been thoughtfully developed to reflect the scope and pace of the essentials-level course. While it complements any one-semester anatomy and physiology text, it will be most effectively used with *Essentials of Anatomy & Physiology,* Eighth Edition. One of the best-selling laboratory manuals for the one-semester course, this lab manual offers 27 concise, activity-based lab exercises that explore basic concepts in anatomy and physiology. It includes a four-color histology atlas, an overview of safety in the lab, and a complete microscope exercise. Each lab is presented with learning objectives, exceptionally cogent summaries of key concepts, and meaningful activities that build students' observational laboratory skills.
- **Instructor's Guide to *Essentials of Human Anatomy & Physiology Laboratory Manual,* Third Edition** This useful guide includes detailed directions for setting up the laboratory, comments on the exercises, and answers to questions that appear in the laboratory manual. It includes an answer key, MultiMedia resources, and hints for lab preparation and presentation.

An Outstanding Student Learning Package

- ***Updated and Expanded Essentials of InterActive Physiology*® (IP)** An award-winning, dynamic, and highly effective learning tool. This CD-ROM features activities specifically selected for your one-semester course from the IP software. Included are animations, tutorials, and quizzes, plus text-specific quiz questions for each chapter. Icons with references to Essentials of IP tutorials appear in the chapter summaries of the text to help integrate book coverage and media.
- ***New* The Anatomy and Physiology Place** (www.anatomyandphysiology.com) Students will be able to access a special edition of the Anatomy and Physiology Place for *Essentials of Human Anatomy and Physiology.* Included are practice quizzes, learning activities, bone review, histology review, and case studies.

Additional Supplements

- ***Revised Anatomy & Physiology Coloring Workbook: A Complete Study Guide,* Eighth Edition** by Elaine Marieb. This excellent stand-alone workbook promotes learning with a variety of visual and written exercises, combining the best features and activities of both a study guide and a coloring book.
- **Anatomy Flashcards** by Glenn Bastian. Always a useful, inexpensive study tool, these flashcards provide students with anatomical and physiological images and concepts for quick review.
- **A.D.A.M.® The Inside Story CD-ROM.** This best-selling software adds tremendous value through dissectible anatomy, animations, high-quality graphics, interactive exercises, and quizzes. Available for Windows or Macintosh.

Special Features Retained from Previous Editions

- **Presentation of anatomy, physiology, and clinically important diseases** The content of this edition continues to present a balance of anatomical and physiological concepts. Discussions of diseases and dysfunctions appear where they enhance and reinforce an understanding of normal human structure and function. Additionally, complex topics are explained consistently by the use of analogies to foster, rather than discourage, student interest. In instances requiring more than the usual amounts of memorization of terminology and facts, the material is presented in both a text discussion and tabular form to reinforce learning.
- **Writing style** The writing style is intentionally informal. New terminology is set in boldface or italic type, followed by phonetic pronunciation. Terms are defined within the text and again in the glossary.
- **"Systems in Sync" feature** This feature summarizes in simple form for the student the "must know" information about the interactions that occur between the system being studied and all other body systems. This approach, provided for each body system, helps to emphasize that the body is a dynamic composite of interdependent parts.
- **Appendices** The appendices present in the previous edition have been retained. These include Appendix A, which lists important word roots, prefixes, and suffixes; Appendix B, the Periodic Table; and Appendix C, which provides key information on vitamins and minerals.
- **Pedagogical devices** Several pedagogical devices are used to ensure that students learn important terminology and concepts. Each chapter begins with student objectives. A summary of text contents and review questions are found at the end of each chapter. Additionally, a list of word roots, prefixes, and suffixes is included in Appendix A.
- **"At the Clinic" question sections** Unique "At the Clinic" sections end each chapter and ask questions that help students apply new knowledge to clinical situations.
- **"A Closer Look" boxes** "A Closer Look" boxes have been updated to capture student interest with new advances in science and intriguing topics. Subjects include timely topics such as AIDS (pp. 418–419) and recent medical advances in treating cancer (pp. 100–101).
- **Clinical examples with an emphasis on homeostasis** Because the normal and most desirable condition of the body is homeostasis, or maintenance of a stable internal environment, the concept of homeostasis is introduced early (Chapter 1) and is then emphasized throughout the book. The understanding that loss of homeostasis leads to some kind of pathology or disease—temporary or permanent—is also stressed. Thus, pathological conditions are introduced and integrated with the text material as appropriate to clarify normal functioning, not as an end in themselves. The normal functional process is always explained first. The clinical examples chosen are those seen most often by people working in the allied health fields and are intended to familiarize students with the possible consequences when body structures are damaged or functional processes diverge from normal pathways. In each case, the disease conditions are indicated visually by the symbol of a red seesaw that has been tipped off balance to

remind students that disease is a loss of homeostasis.

- **Developmental aspects and aging** Each chapter ends with a section on developmental aspects. The section presents the formation of that system in the embryo and follows the changes that occur through old age. Important health problems unique to that system are introduced, and emphasis is placed on problems that accompany the aging process. Because many health workers are intricately involved in the care of elderly people, this emphasis should be valuable to today's student.
- **"Focus on Careers" Feature** This feature, which utilizes interviews with working professionals in allied health fields (such as LPN, phlebotomist, dental hygienist, and others) introduces students to several careers that depend to a substantial extent on a background in human anatomy and physiology. This feature may help students to validate their professional career choices or perhaps suggest new options they haven't yet considered.

Acknowledgements

Many people contributed to my efforts in the creation of the first seven editions of this book. I am especially grateful to those who allowed me to use them as "sounding boards" to try out my analogies and ideas.

I also would like to thank the following reviewers for their thoughtful critiques, which helped me with this revision: Donald Rodd, University of Evansville; Regina Rector, William Rainey, Harper College; Charles Rinehimer, Northampton Community College; Susan Anderson, Central Georgia Technical College; Deborah S. Temperly, Delta College; Anne Keddy-Hector, Austin Community College; Tom Nowak, British Columbia Institute of Technology; Ralph E. Ferges, Palomar Community College; Amy Nunnally, Front Range Community College, Westminster Campus; Jerry Cook, Sam Houston State University; Chris Gan, Rogue Community College; Kathy Gardner, North Dakota State University; Eric Genz-Mould, Shoreline Community College; Laura Hebert, Angelina College; Ann Henninger, Wartburg College; Mark Jonasson, Crafton Hills College; Karla Pouillon, Everett Community College; Susan Robe, Kennebec Valley Technical College; Kimberly Turk, Mitchell Community College.

In addition, I wish to thank past reviewers, whose comments continue to guide my efforts: Dwane Aulner, Indian Hills Community College; Kay Brashear, El Centro College; Yolanda Browne, Del Mar College; Ray Canham, Richland College; Elizabeth Carl, Nassau Technological Center (BOCES); Lydia Chavana, South Texas Vocational Technical College; Barbara Clarke, American University; John Conroy, University of Winnipeg; Jean Cons, College of San Mateo; Luci Constant, W.M.L. Dawson School of Practical Nursing; Paula Curbo, Hill College; Teresa DiFiori, Pasadena City College; Martha Dixon, Diablo Valley College; Grover Ericson, College of West Virginia; Eleanor Shorette Fowler, School of Practical Nursing, Tewksbury Hospital; Mildred Galliher, Cochise College; Anne Genge, St. Lawrence College; Daniel Gong, Seattle Central Community College; Annette Gould, Broome-Delaware-Tioga BOCES; Keith Graham, Lutheran College of Health Professionals; Sue Hall, Rogue Community College; John Harley, Eastern Kentucky University; Gloria Hillert, Triton College; Carol Holley, San Jacinto College; Dawn Holtzmeier, Hocking Technical College; Mary Hopkins, Kiamichi Vocational/Technical College; Nancy James, Space Coast Massage Institute; Murray Jensen, University of Minnesota; Drucilla Jolly, Forsyth Technical College; Dennis Kalichstein, Ocean County College; Steve Kirk, Parker College of Chiropractic; Fred Klaus, East Texas State University; Bonnie Kroemmelbein, Upper Bucks County AVTS Practical Nursing Program; John Lammert, Gustavus Adolphus College; Jeri Lindsey, Tarrant County Junior College; Bennie Marshal, STOP/CETA School of Practical Nursing; William Matthai, Tarrant County Junior College; Walter Matulis, Mid-Michigan Community College; Richard McKeeby, Union County College; Roxine McQuitty, Madison Area Technical College; Lew Milner, North Central Technical College; Lynne Osborn, Middlesex Community College; William Palmore, Triton College; Elspeth Payne, Massage Therapy Institute of Colorado; Bridget Price, Vocational Nursing School of California; Dell Redding, Evergreen State College; Bill Rohrer, Union County College; Jessica Sand, Union County College; Joan Sandell, Gogebic Community College;

Leba Sarkis, Aims Community College; Ann Senisi, Nassau Technological Center (BOCES); Rhonda Shepperd, College of West Virginia; Jan Strandjord, Indiana Vocational/Technical College; Diane Swafford, Fuller School of Massage Therapy; Carlene Tonini, College of San Mateo; Patricia Turner, Howard Community College; Jacqueline Tuttle, Forsyth Technical College; Ann Welch, Indiana Vocational/Technical College; Deborah Wiepz, Madison Area Technical College; Claudia Williams, Campbell University; Eileen Williams, Nassau Technological Center (BOCES); Iris Winkelhake, Southeast Community College; Edward Zalisko, Blackburn College.

The staff of Benjamin Cummings contributed immensely in the form of support and guidance and deserve a hearty round of applause, one and all. I would especially like to thank Mary Ann Murray, my project editor, for her painstaking review and editing of the art manuscript and for her management of the supplements package, which is outstanding. When Mary Ann moved on to another position in the company, Karoliina Tuovinen stepped in and sufficiently tied up loose ends. Thanks also to the production team at Wendy Earl Productions (Janet Vail and Lorna Strutt), whose diligent handling and control of the many details of production made that process as hassle-free as humanly possible.

Elaine Marieb

Elaine N. Marieb

Benjamin Cummings
1301 Sansome Street
San Francisco, CA 94111

Preface to the Student

This book is written with you, the student, in mind. Human anatomy and physiology is more than just interesting—it is fascinating. To help get you involved in the study of this subject, a number of special features are incorporated throughout the book.

The *informal writing style* invites you to learn more about anatomy and physiology without intimidation. We want you to enjoy reading this book.

Topic boxes and *tables* are designed with you in mind. The topic boxes present scientific information that can be applied to your daily life. When reading the topic boxes, you will probably find yourself saying, "I didn't know that," or "Now I understand why. . . ." The tables are summaries of important information in the text. You should be able to use the tables when studying for an exam or reviewing an important topic.

Important terms are defined within the text as they are introduced and are listed at the ends of the chapters. An extensive *glossary* is provided at the back of the book to help you review these terms.

Phonetic spellings are provided for many of these important terms, especially those that are likely to be unfamiliar to you. To read these, you will need to remember the following rules:

1. Accent marks follow stressed syllables. The primary stress is shown by ′, and the secondary stress by ″.
2. Unless otherwise noted, assume that vowels at the ends of syllables are long and vowels followed by consonants are short. Exceptions to this rule are indicated by a bar (¯) over the vowel, which indicates a long vowel, or a breve sign (˘) over the vowel, indicating that the vowel is short.

For example, the phonetic spelling of "thrombophlebitis" is *throm″bo-flĕ-bi′tis*. The next-to-the-last syllable (*bi′*) receives the greatest stress, and the first syllable (*throm″*) gets the secondary stress. The vowel in the second syllable comes at the end of the syllable and is long. The vowel that comes at the end of the third syllable is short because it has a breve sign.

Any exam causes anxiety. Exams in anatomy and physiology are no exception. To help you better prepare for an exam or comprehend the material you have just read, extensive *summaries, review questions,* and *clinical problems* ("At the Clinic") are found at the ends of the chapters. The *Systems in Sync* feature, which accompanies each system, will also be very helpful in your preparations because it lays out very simply the most important ways body systems interact with each other.

The art program is designed to help you learn the different structures and functions of the human body. All figures are referred to within the text discussion. The best way to make use of the extensive art program is to carefully study each illustration when it is discussed in the text.

Be sure to check out The Anatomy & Physiology Place, at www.anatomyandphysiology.com. You'll find an array of quizzes, activities, and other features to help you make the grade in A&P. To help you better understand physiological concepts, combine your reading of the text with your use of the *Essentials of InterActive Physiology®* CD-ROM. And remember—in every chapter summary you'll find handy references to the CD-ROM IP and The Anatomy & Physiology Place **WEB** to help you on specific topics.

I hope that you enjoy *Essentials of Human Anatomy and Physiology* and that this book makes learning A&P a fun and rewarding process. Write to me and let me know how you make out in your A&P course.

Elaine Marieb

Elaine N. Marieb

Benjamin Cummings
1301 Sansome Street
San Francisco, CA 94111

Contents

➤

➤

15 The Urinary System 501

16 The Reproductive System 527

➤

1 The Human Body: An Orientation

YOUR GOALS

After completing this chapter, you will have mastered the objectives listed below.

OBJECTIVE CHECKLIST

AN OVERVIEW OF ANATOMY AND PHYSIOLOGY (p. 2)

- Define *anatomy* and *physiology*.
- Explain how anatomy and physiology are related.

LEVELS OF STRUCTURAL ORGANIZATION (pp. 2–7)

- Name the levels of structural organization that make up the human body, and explain how they are related.
- Name the organ systems of the body, and briefly state the major functions of each system.
- Classify by organ system all organs discussed.
- Identify the organs shown on a diagram or a dissectible torso.

MAINTAINING LIFE (pp. 7–9)

- List functions that humans must perform to maintain life.
- List the survival needs of the human body.

HOMEOSTASIS (pp. 9–11)

- Define *homeostasis*, and explain its importance.
- Define *negative feedback*, and describe its role in maintaining homeostasis and normal body function.

THE LANGUAGE OF ANATOMY (pp. 11–19)

- Verbally describe the anatomical position or demonstrate it.
- Use proper anatomical terminology to describe body directions, surfaces, and body planes.
- Locate the major body cavities, and list the chief organs in each cavity.

An Overview of Anatomy and Physiology

Most of us are naturally curious about our bodies; we want to know what makes us tick. This curiosity is even seen in infants, who can keep themselves happy for a long time staring at their own hands or pulling their mother's nose. Older children wonder where food goes when they swallow it, and some believe that they will grow a watermelon in their belly if they swallow the seeds. They scream loudly when approached by medical personnel (fearing shots that sting), but they like to play doctor. Adults become upset when their hearts pound, when they have uncontrollable hot flashes, or when they cannot keep their weight down.

Anatomy and physiology, subdivisions of biology, explore many of these topics as they describe how our bodies are put together and how they work.

Anatomy

Anatomy (ah-nat′o-me) is the study of the structure and shape of the body and body parts and their relationships to one another. Whenever we look at our own body or study large body structures such as the heart or bones, we are observing *gross anatomy;* that is, we are studying large, easily observable structures. Indeed, the term *anatomy,* derived from the Greek words meaning to cut (*tomy*) apart (*ana*), is related most closely to gross anatomical studies because in such studies preserved animals or their organs are dissected (cut up) to be examined. On the other hand, if a microscope or magnifying instrument is used to see very small structures in the body, we are studying *microscopic anatomy.* The cells and tissues of the body can only be seen through a microscope.

Physiology

Physiology (fiz″e-ol′o-je) is the study of how the body and its parts work or function (*physio* = nature; *ology* = the study of). Like anatomy, physiology has many subdivisions. For example, *neurophysiology* explains the workings of the nervous system, and *cardiac physiology* studies the function of the heart, which acts as a muscular pump to keep blood flowing throughout the body.

Relationship between Anatomy and Physiology

In the real world, anatomy and physiology are always related. The parts of your body form a well-organized unit, and each of those parts has a job to do to make the body operate as a whole. Structure determines what functions can take place. For example, the lungs are not muscular chambers like the heart and cannot pump blood through the body, but because the walls of their air sacs are very thin, they *can* exchange gases and provide oxygen to the body. The intimate relationship between anatomy and physiology is stressed throughout this book to make your learning meaningful.

Levels of Structural Organization

From Atoms to Organisms

The human body exhibits many levels of structural complexity (Figure 1.1). The simplest level of the structural ladder is the *chemical level,* which we will study in Chapter 2. At this level, **atoms,** tiny building blocks of matter, combine to form *molecules* such as water, sugar, and proteins. Molecules, in turn, associate in specific ways to form microscopic **cells,** the smallest units of all living things. The *cellular level* is examined in Chapter 3. Individual cells vary widely in size and shape, reflecting their particular functions in the body.

The simplest living creatures are composed of single cells, but in complex organisms like trees or human beings, the structural ladder continues on to the *tissue level.* **Tissues** consist of groups of similar cells that have a common function. As we will discuss in Chapter 3, each of the four basic tissue types (epithelial, connective, muscular, and neural) plays a definite but different role in the body.

An **organ** is a structure that is composed of two or more tissue types and performs a specific function for the body. At the *organ level* of organization, extremely complex functions become possible. For example, the small intestine, which digests and absorbs food, is composed of all four tissue types. All the body's organs are grouped so that a number of organ systems are formed. An **organ system** is a group of organs that cooperate to accomplish a common purpose. For example,

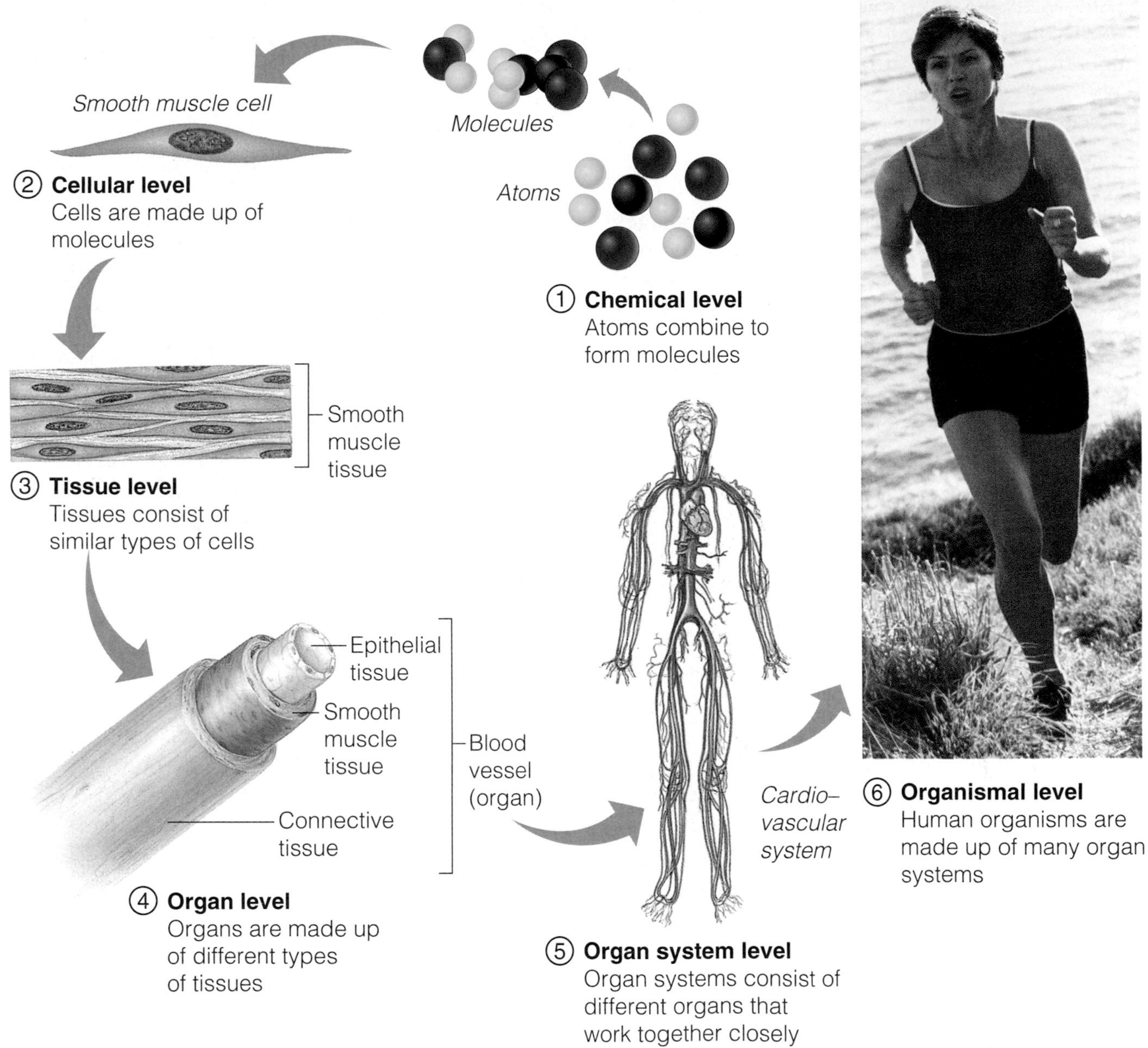

FIGURE 1.1 Levels of structural organization. In this diagram, components of the cardiovascular system are used to illustrate the various levels of structural organization in a human being.

the digestive system includes the esophagus, the stomach, and the small and large intestines (to name a few of its organs). Each organ has its own job to do, and, by working together, they keep food moving through the digestive system so that it is properly broken down and absorbed into the blood, providing fuel for all the body's cells. In all, 11 organ systems make up the living body, or the **organism,** which represents the highest level of structural organization, the *organismal level.* The major organs of each of the systems are shown in Figure 1.2. Refer to the figure as you read through the following descriptions of the organ systems.

Organ System Overview

Integumentary System

The **integumentary** (in-teg″u-men′tar-e) **system** is the external covering of the body, or the skin. It waterproofs the body and cushions and protects the deeper tissues from injury. It also excretes salts and urea in perspiration and helps regulate body temperature. Temperature, pressure, and pain receptors located in the skin alert us to what is happening at the body surface.

Skeletal System

The **skeletal system** consists of bones, cartilages, ligaments, and joints. It supports the body and provides a framework that the skeletal muscles use to cause movement. It also has a protective function (for example, the skull encloses and protects the brain). *Hematopoiesis* (hem″ah-to-poi-e′sis), or formation of blood cells, goes on within the cavities of the skeleton. The hard substance of bones acts as a storehouse for minerals.

Muscular System

The muscles of the body have only one function—to *contract,* or shorten. When this happens, movement occurs. Hence, muscles can be viewed as the "machines" of the body. The mobility of the body as a whole reflects the activity of *skeletal muscles,* the large, fleshy muscles attached to bones. When these contract, you are able to stand erect, walk, leap, grasp, throw a ball, or smile. The skeletal muscles form the **muscular system.** These muscles are distinct from the muscles of the heart and of other hollow organs, which move fluids (blood, urine) or other substances (such as food) along definite pathways within the body.

Nervous System

The **nervous system** is the body's fast-acting control system. It consists of the brain, spinal cord, nerves, and sensory receptors. The body must be able to respond to irritants or stimuli coming from outside the body (such as light, sound, or changes in temperature) and from inside the body (such as decreases in oxygen or stretching of tissue). The sensory receptors detect these changes and send messages (via electrical signals called *nerve impulses*) to the central nervous system (brain and spinal cord) so that it is constantly informed about what is going on. The central nervous system then assesses this information and responds by activating the appropriate body effectors (muscles or glands).

Endocrine System

Like the nervous system, the **endocrine** (en′do-krin) **system** controls body activities, but it acts much more slowly. The endocrine glands produce chemical molecules called *hormones* and release them into the blood to travel to relatively distant target organs.

The endocrine glands include the pituitary, thyroid, parathyroids, adrenals, thymus, pancreas, pineal, ovaries (in the female), and testes (in the male). The endocrine glands are not connected anatomically in the same way that parts of the other organ systems are. What they have in common is that they all secrete hormones, which regulate other structures. The body functions controlled by hormones are many and varied, involving every cell in the body. Growth, reproduction, and food use by cells are all controlled (at least in part) by hormones.

Cardiovascular System

The primary organs of the **cardiovascular system** are the heart and blood vessels. Using blood as the transporting fluid, the cardiovascular system carries oxygen, nutrients, hormones, and other substances to and from the tissue cells where exchanges are made. White blood cells and chemicals in the blood help to protect the body from such foreign invaders as bacteria, toxins, and tumor cells. The heart acts as the blood pump, propelling blood through the blood vessels to all body tissues.

Lymphatic System

The role of the **lymphatic system** is complementary to that of the cardiovascular system. Its organs include lymphatic vessels, lymph nodes, and other lymphoid organs such as the spleen and tonsils. The lymphatic vessels return fluid leaked from the blood to the blood vessels so that blood can be kept continuously circulating through the body. The lymph nodes (and other lymphoid organs) help to cleanse the blood and house the cells involved in immunity.

FIGURE 1.2 The body's organ systems. The structural components of each organ system are illustrated in the diagrammatic views. The major functions of the organ systems are listed beneath each illustration.

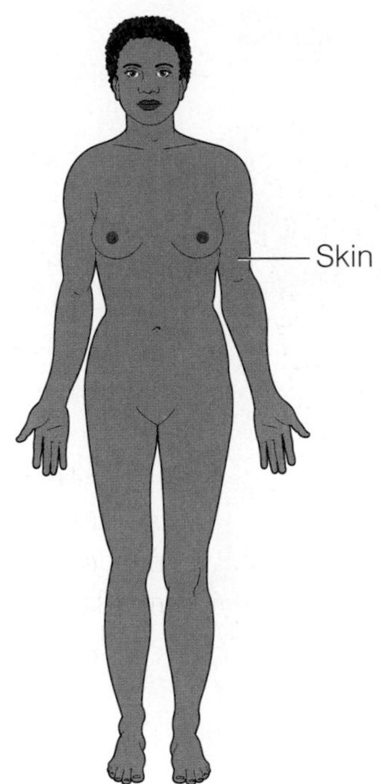

(a) **Integumentary system**
Forms the external body covering; protects deeper tissue from injury; synthesizes vitamin D; location of cutaneous (pain, pressure, etc.) receptors; and sweat and oil glands.

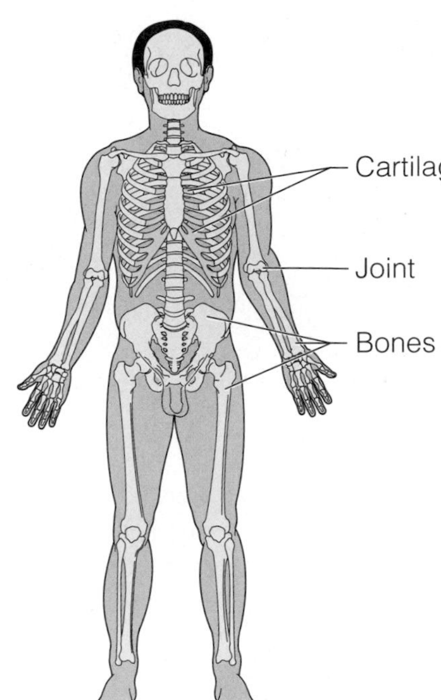

(b) **Skeletal system**
Protects and supports body organs; provides a framework the muscles use to cause movement; blood cells are formed within bones; stores minerals.

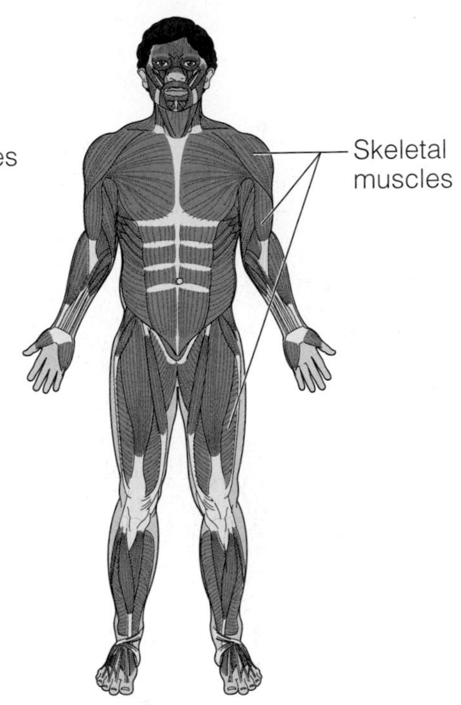

(c) **Muscular system**
Allows manipulation of the environment, locomotion, and facial expression; maintains posture; produces heat.

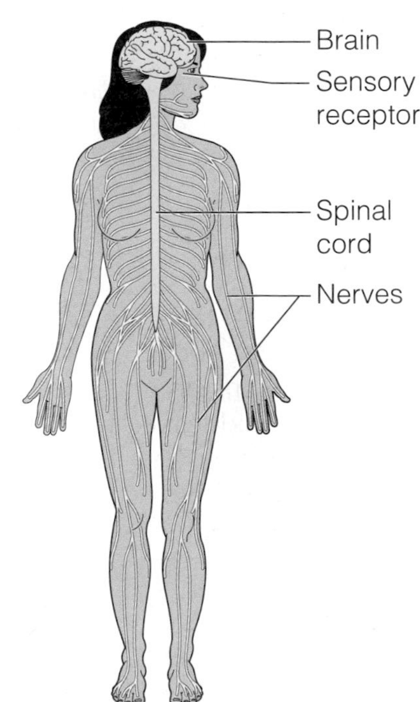

(d) **Nervous system**
Fast-acting control system of the body; responds to internal and external changes by activating appropriate muscles and glands.

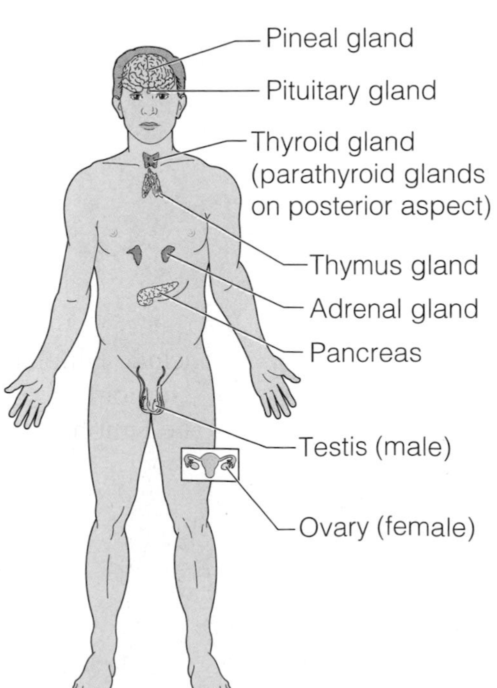

(e) **Endocrine system**
Glands secrete hormones that regulate processes such as growth, reproduction, and nutrient use (metabolism) by body cells.

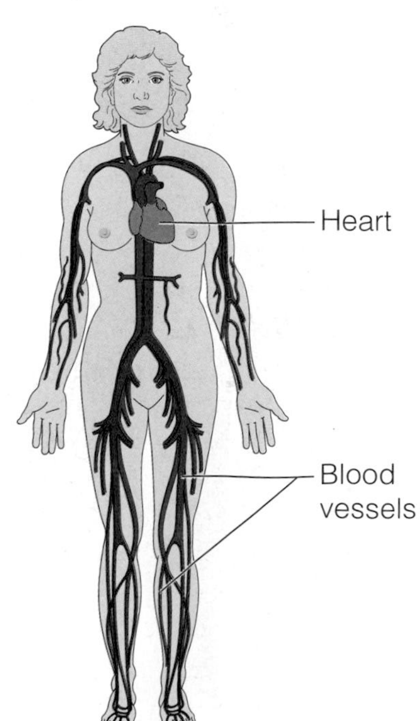

(f) **Cardiovascular system**
Blood vessels transport blood, which carries oxygen, carbon dioxide, nutrients, wastes, etc.; the heart pumps blood.

(Continues on page 6)

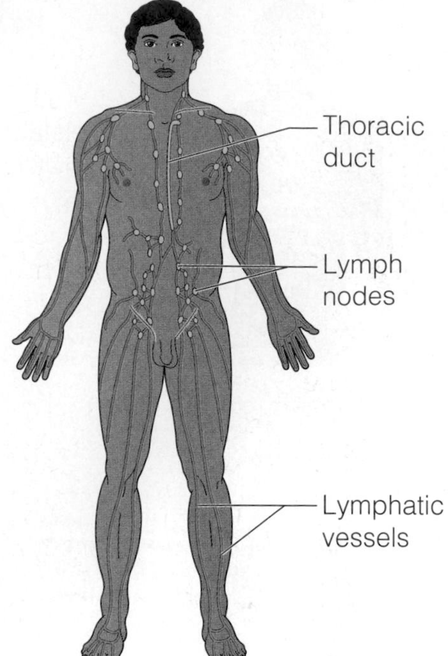

(g) Lymphatic system
Picks up fluid leaked from blood vessels and returns it to blood; disposes of debris in the lymphatic stream; houses white blood cells involved in immunity.

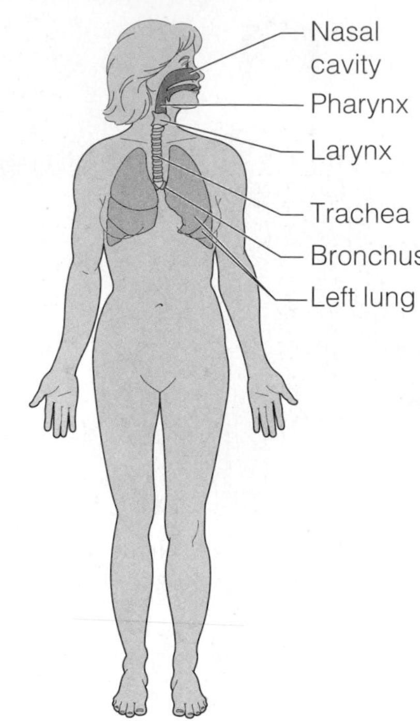

(h) Respiratory system
Keeps blood constantly supplied with oxygen and removes carbon dioxide; the gaseous exchanges occur through the walls of the air sacs of the lungs.

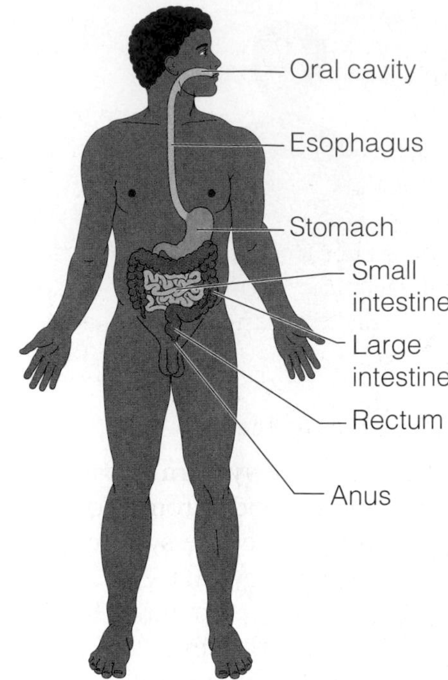

(i) Digestive system
Breaks food down into absorbable units that enter the blood for distribution to body cells; indigestible foodstuffs are eliminated as feces.

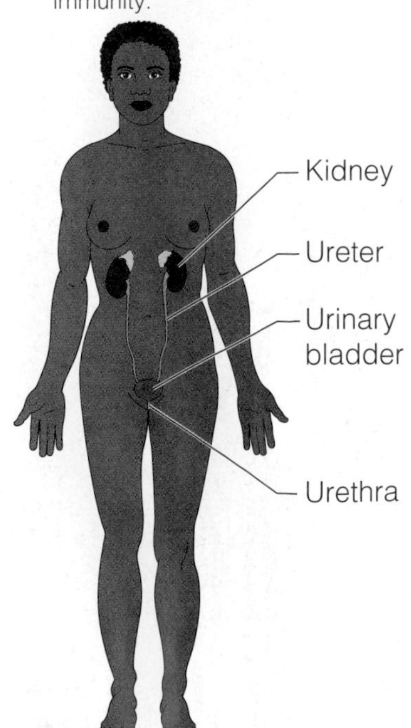

(j) Urinary system
Eliminates nitrogenous wastes from the body; regulates water, electrolyte, and acid-base balance of the blood.

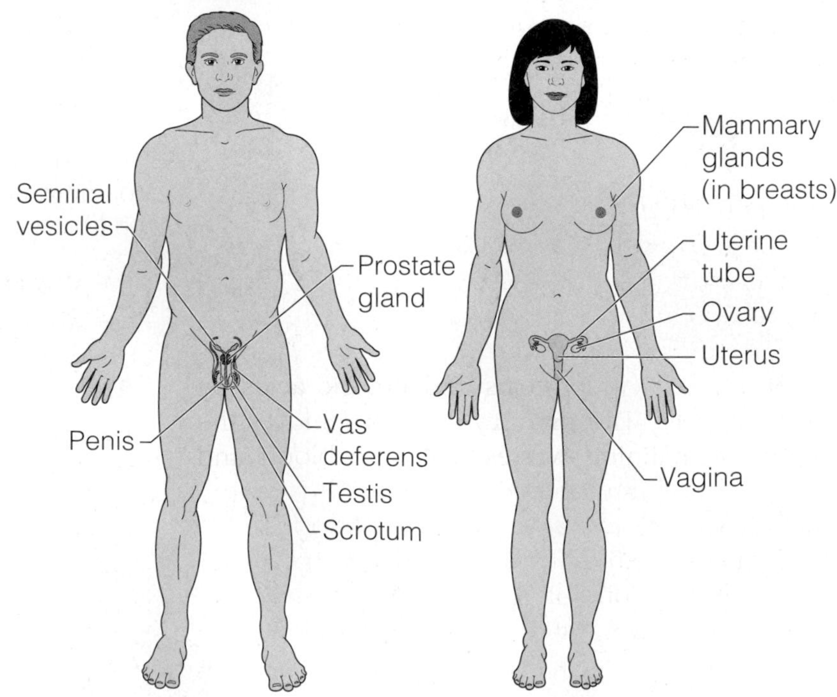

(k) Male reproductive system **(l) Female reproductive system**
Overall function of the reproductive system is production of offspring. Testes produce sperm and male sex hormone; ducts and glands aid in delivery of viable sperm to the female reproductive tract. Ovaries produce eggs and female sex hormones; remaining structures serve as sites for fertilization and development of the fetus. Mammary glands of female breast produce milk to nourish the newborn.

Respiratory System

The job of the **respiratory system** is to keep the body constantly supplied with oxygen and to remove carbon dioxide. The respiratory system consists of the nasal passages, pharynx, larynx, trachea, bronchi, and lungs. Within the lungs are tiny air sacs. It is through the thin walls of these air sacs that gas exchanges are made to and from the blood.

Digestive System

The **digestive system** is basically a tube running through the body from mouth to anus. The organs of the digestive system include the oral cavity (mouth), esophagus, stomach, small and large intestines, and rectum. Their role is to break down food and deliver the products to the blood for dispersal to the body cells. The undigested food that remains in the tract leaves the body through the anus as feces. The breakdown activities that begin in the mouth are completed in the small intestine. From that point on, the major function of the digestive system is to reclaim water. The liver is considered to be a digestive organ because the bile it produces helps to break down fats. The pancreas, which delivers digestive enzymes to the small intestine, also is functionally a digestive organ.

Urinary System

The body produces wastes as by-products of its normal functions, and these wastes must be disposed of. One type of waste contains nitrogen (examples are urea and uric acid), which results from the breakdown of proteins and nucleic acids by the body cells. The **urinary system** removes the nitrogen-containing wastes from the blood and flushes them from the body in urine. This system, often called the *excretory system,* is composed of the kidneys, ureters, bladder, and urethra. Other important functions of this system include maintaining the body's water and salt (electrolyte) balance and regulating the acid-base balance of the blood.

Reproductive System

The **reproductive system** exists primarily to produce offspring. Sperm are produced by the testes of the male. Other male reproductive system structures are the scrotum, penis, accessory glands, and the duct system, which carries sperm to the outside of the body. The ovary of the female produces the eggs, or ova; the female duct system consists of the uterine tubes, uterus, and vagina. The uterus provides the site for the development of the fetus (immature infant) once fertilization has occurred.

Maintaining Life

Necessary Life Functions

Now that we have introduced the structural levels composing the human body, the question that naturally follows is: What does this highly organized human body do? Like all complex animals, human beings maintain their boundaries, move, respond to environmental changes, take in and digest nutrients, carry out metabolism, dispose of wastes, reproduce themselves, and grow. We will discuss each of these necessary life functions briefly here and in more detail in later chapters.

Organ systems do not work in isolation; instead, they work together to promote the well-being of the entire body. Because this theme will be emphasized throughout this book, it is worthwhile to identify the most important organ systems contributing to each of the necessary life functions (Figure 1.3). Also, as you read through this material, you may want to refer back to the more detailed descriptions of the organ systems provided on pp. 3 through 7 and in Figure 1.2.

Maintaining Boundaries

Every living organism must be able to maintain its boundaries so that its "inside" remains distinct from its "outside." Every cell of the human body is surrounded by an external membrane that contains its contents and allows needed substances in while generally preventing the entry of potentially damaging or unnecessary substances. The body as a whole is also enclosed by the integumentary system, or skin. The integumentary system protects internal organs from drying out (which would be fatal), from bacteria, and from the damaging effects of heat, sunlight, and an unbelievable number of chemical substances in the external environment.

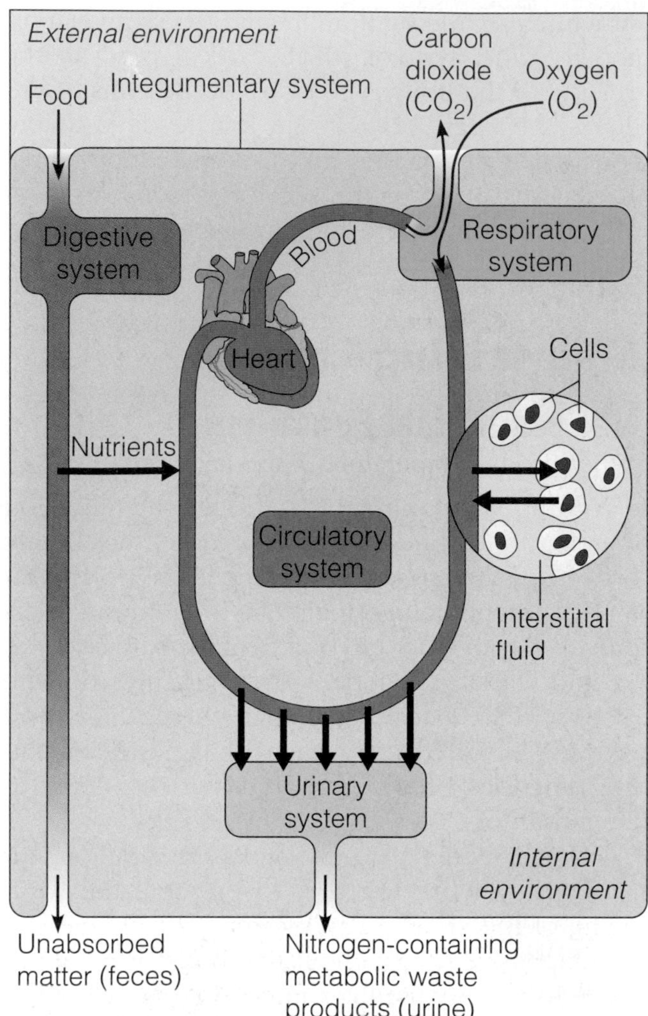

FIGURE 1.3 Examples of selected interrelationships among body organ systems. The integumentary system protects the body as a whole from the external environment. The digestive and respiratory systems, in contact with the external environment, take in nutrients and oxygen, respectively, which are then distributed by the blood to all body cells. Elimination from the body of metabolic wastes is accomplished by the urinary and respiratory systems.

Movement

Movement includes all the activities promoted by the muscular system, such as propelling ourselves from one place to another by walking, swimming, and so forth, and manipulating the external environment with our fingers. The muscular system is aided by the skeletal system, which provides the bones that the muscles pull on as they work. Movement also occurs when substances such as blood, foodstuffs, and urine are propelled through the internal organs of the cardiovascular, digestive, and urinary systems, respectively.

Responsiveness

Responsiveness, or **irritability,** is the ability to sense changes (stimuli) in the environment and then to react to them. For example, if you cut your hand on broken glass, you involuntarily pull your hand away from the painful stimulus (the glass). It is not necessary to think about it—it just happens! Likewise, when the amount of carbon dioxide in your blood rises to dangerously high levels, the response is an increase in your breathing rate to blow off the excess carbon dioxide.

Because nerve cells are highly irritable and can communicate rapidly with each other by conducting electrical impulses, the nervous system bears the major responsibility for responsiveness. However, all body cells exhibit responsiveness to some extent.

Digestion

Digestion is the process of breaking down ingested food into simple molecules that can then be absorbed into the blood for delivery to all body cells by the cardiovascular system. In a simple, one-celled organism like an amoeba, the cell itself is the "digestion factory," but in the complex, multicellular human body, the digestive system performs this function for the entire body.

Metabolism

Metabolism is a broad term that refers to all chemical reactions that occur within body cells. It includes breaking down complex substances into simpler building blocks, making larger structures from smaller ones, and using nutrients and oxygen to produce ATP molecules, the energy-rich molecules that power cellular activities. Metabolism depends on the digestive and respiratory systems to make nutrients and oxygen available to the blood and on the cardiovascular system to distribute these substances throughout the body. Metabolism is regulated chiefly by hormones secreted by the glands of the endocrine system.

Excretion

Excretion is the process of removing *excreta* (ek-skre′tah), or wastes, from the body. If the body is to continue to operate as we expect it to, it must get rid of the nonuseful substances produced during digestion and metabolism. Several organ systems participate in excretion. For example, the digestive system rids the body of indigestible food residues in feces, and the urinary system disposes of nitrogen-containing metabolic wastes in urine.

Reproduction

Reproduction, the production of offspring, can occur on the cellular or organismal level. In cellular reproduction, the original cell divides, producing two identical daughter cells that may then be used for body growth or repair. Reproduction of the human organism, or making a whole new person, is the task of the organs of the reproductive system, which produce sperm and eggs. When a sperm unites with an egg, a fertilized egg forms, which then develops into a bouncing baby within the mother's body. The function of the reproductive system is exquisitely regulated by hormones of the endocrine system.

Growth

Growth is an increase in size, usually accomplished by an increase in the number of cells. For growth to occur, cell-constructing activities must occur at a faster rate than cell-destroying ones.

Survival Needs

The goal of nearly all body systems is to maintain life. However, life is extraordinarily fragile and requires that several factors be available. These factors, which we will call *survival needs,* include nutrients (food), oxygen, water, and appropriate temperature and atmospheric pressure.

Nutrients, taken in via the diet, contain the chemicals used for energy and cell building. Carbohydrates are the major energy-providing fuel for body cells. Proteins and, to a lesser extent, fats are essential for building cell structures. Fats also cushion body organs and provide reserve fuel. Minerals and vitamins are required for the chemical reactions that go on in cells and for oxygen transport in the blood.

All the nutrients in the world are useless unless **oxygen** is also available, because the chemical reactions that release energy from foods require oxygen. Approximately 20 percent of the air we breathe is oxygen. It is made available to the blood and body cells by the cooperative efforts of the respiratory and cardiovascular systems.

Water accounts for 60 to 80 percent of body weight. It is the single most abundant chemical substance in the body and provides the fluid base for body secretions and excretions. Water is obtained chiefly from ingested foods or liquids and is lost from the body by evaporation from the lungs and skin and in body excretions.

For good health, **body temperature** must be maintained at around 37°C (98°F). As body temperature drops below this point, metabolic reactions become slower and slower and finally stop. When body temperature is too high, chemical reactions proceed too rapidly, and body proteins begin to break down. At either extreme, death occurs. Most body heat is generated by the activity of the skeletal muscles.

The force exerted on the surface of the body by the weight of air is referred to as **atmospheric pressure.** Breathing and the exchange of oxygen and carbon dioxide in the lungs depend on appropriate atmospheric pressure. At high altitudes, where the air is thin and atmospheric pressure is lower, gas exchange may be too low to support cellular metabolism.

The mere presence of these survival factors is not sufficient to maintain life. They must be present in appropriate amounts as well; excesses and deficits may be equally harmful. For example, the food ingested must be of high quality and in proper amounts; otherwise, nutritional disease, obesity, or starvation is likely.

Homeostasis

When you really think about the fact that your body contains trillions of cells in nearly constant activity, and that remarkably little usually goes wrong with it, you begin to appreciate what a marvelous machine your body really is. The word **homeostasis** (ho″me-o-sta′sis) describes the body's ability to maintain relatively stable internal conditions even though the outside world is continuously changing. Although the literal translation of *homeostasis* is "unchanging" (*homeo* = the same; *stasis* = standing still), the term does not really mean an unchanging state. Instead, it indicates a *dynamic* state of equilibrium, or a balance in which internal conditions change and vary but always within relatively narrow limits.

Q *If this control system were regulating room temperature, what apparatus would be the effector?*

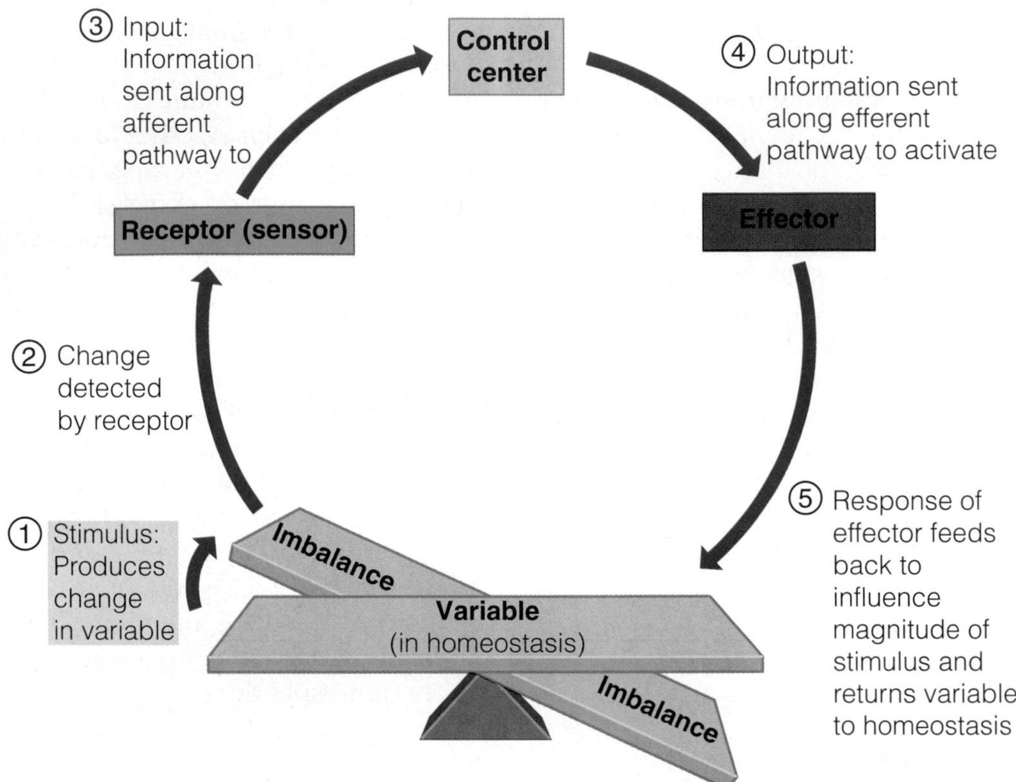

FIGURE 1.4 The elements of a homeostatic control system. Communication between the receptor, control center, and effector is essential for normal operation of the system.

In general, the body demonstrates homeostasis when its needs are being adequately met and it is functioning smoothly. Virtually every organ system plays a role in maintaining the constancy of the internal environment. Adequate blood levels of vital nutrients must be continuously present, and heart activity and blood pressure must be constantly monitored and adjusted so that the blood is propelled with adequate force to reach all body tissues. Additionally, wastes must not be allowed to accumulate, and body temperature must be precisely controlled.

Homeostatic Control Mechanisms

Communication within the body is essential for homeostasis and is accomplished chiefly by the nervous and endocrine systems, which use electrical signals delivered by nerves or bloodborne hormones, respectively, as information carriers. The details of how these two regulating systems operate are the subjects of later chapters, but the basic characteristics of the neural and hormonal control systems that promote homeostasis will be explained here.

Regardless of the factor or event being regulated (this is called the *variable*), all homeostatic control mechanisms have at least three components (Figure 1.4). The first component is a **receptor.** Essentially, it is some type of sensor that monitors and responds to changes in the environment. It responds to such changes, called *stimuli,* by sending information (input) to the second element, the *control center.* Information flows from the receptor to the control center along the *afferent pathway.* (It may help to remember that information traveling along the *afferent* pathway *approaches* the control center.)

A *The heat-generating furnace or oil burner.*

The **control center,** which determines the level (set point) at which a variable is to be maintained, analyzes the information it receives and then determines the appropriate response or course of action.

The third component is the **effector,** which provides the means for the control center's response (output) to the stimulus. Information flows from the control center to the effector along the *efferent pathway.* (*Efferent* information *exits* from the control center.) The results of the response then *feed back* to influence the stimulus, either by depressing it (negative feedback) so that the whole control mechanism is shut off or by enhancing it (positive feedback) so that the reaction continues at an even faster rate.

Most homeostatic control mechanisms are **negative feedback mechanisms.** In such systems, the net effect of the response to the stimulus is to shut off the original stimulus or reduce its intensity. A frequently used example of a negative feedback system is a home heating system connected to a thermostat. In this situation, the thermostat contains both the receptor and the control center. If the thermostat is set at 20°C (68°F), the heating system (effector) will be triggered ON when the house temperature drops below that setting. As the furnace produces heat, the air is warmed. When the temperature reaches 20°C or slightly higher, the thermostat sends a signal to shut off the furnace. Your body "thermostat," located in a part of your brain called the *hypothalamus,* operates in a similar way to regulate body temperature. Other negative feedback mechanisms regulate heart rate, blood pressure, breathing rate, and blood levels of glucose, oxygen, carbon dioxide, and minerals.

Because they tend to increase the original disturbance (stimulus) and to push the variable *farther* from its original value, **positive feedback mechanisms** are rare in the body. Typically these mechanisms control infrequent events that occur explosively and do not require continuous adjustments. Blood clotting and the birth of a baby are the most familiar examples of positive feedback mechanisms.

Homeostatic Imbalance

Homeostasis is so important that most disease can be regarded as a result of its disturbance, a condition called **homeostatic imbalance.** As we age, our body organs become less efficient, and our internal conditions become less and less stable. These events place us at an increasing risk for illness and produce the changes we associate with aging.

Examples of homeostatic imbalance will be provided throughout this book to enhance your understanding of normal physiological mechanisms. These homeostatic imbalance sections are preceded by the symbol ▲ to alert you that an abnormal condition is being described. ▲

The Language of Anatomy

Learning about the body is exciting, but our interest sometimes dwindles when we are confronted with the terminology of anatomy and physiology. Let's face it. You can't just pick up an anatomy and physiology book and read it as though it were a novel. Unfortunately, confusion is inevitable without specialized terminology. For example, if you are looking at a ball, "above" always means the area over the top of the ball. Other directional terms can also be used consistently because the ball is a sphere. All sides and surfaces are equal. The human body, of course, has many protrusions and bends. Thus, the question becomes: Above what? To prevent misunderstanding, anatomists have accepted a set of terms that allow body structures to be located and identified clearly with just a few words. This language of anatomy is presented and explained next.

Anatomical Position

To accurately describe body parts and position, we must have an initial reference point and use directional terms. To avoid confusion, it is always assumed that the body is in a standard position called the **anatomical position.** It is important to understand this position because most body terminology used in this book refers to this body positioning *regardless* of the position the body happens to be in. The face-front diagrams in Figure 1.5 and Table 1.1 illustrate the anatomical position. As you can see, the body is erect with the feet parallel and the arms hanging at the sides with the palms facing forward.

Q *Study this figure for a moment to answer these two questions. Where would you hurt if you (1) pulled a groin muscle or (2) cracked a bone in your olecranal area?*

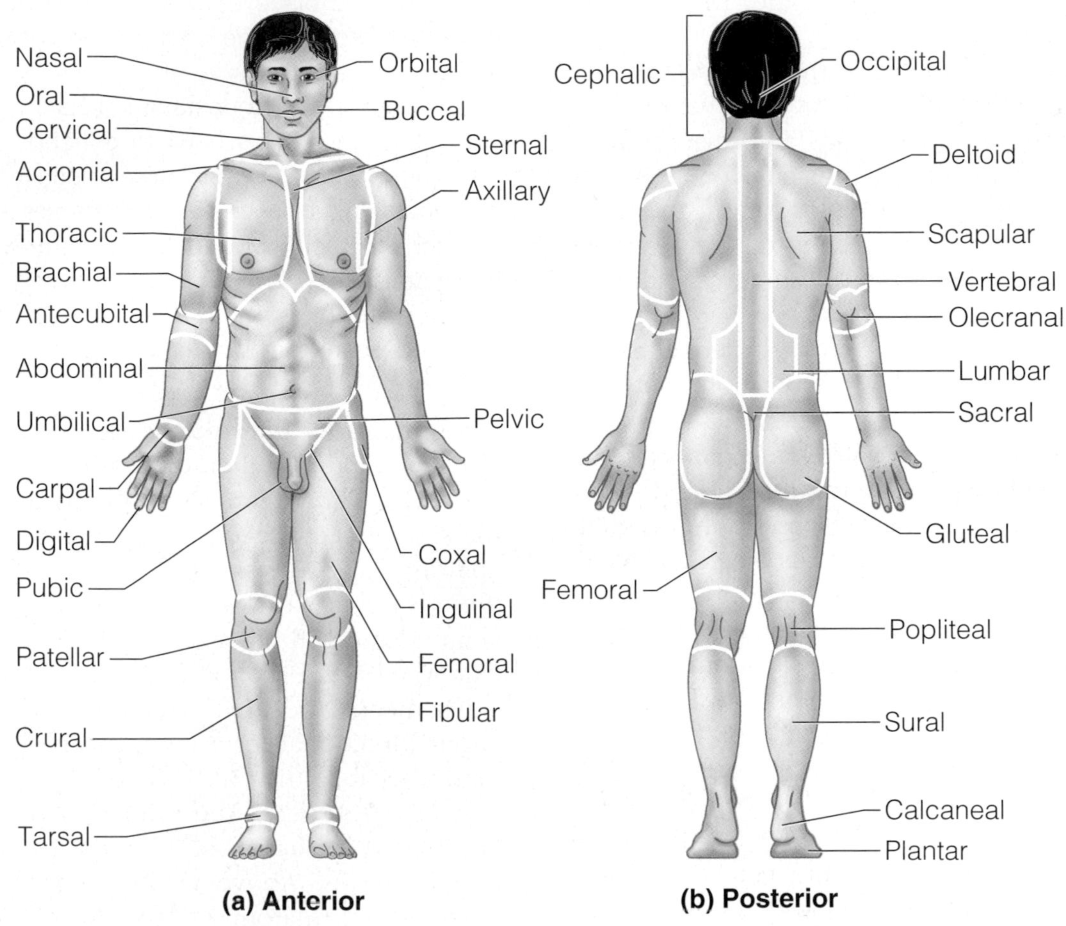

FIGURE 1.5 Surface anatomy: Regional terms. (a) Anterior body landmarks. **(b)** Posterior body landmarks. The heels are raised slightly to show the inferior plantar surface (sole) of the foot

- Stand up and assume the anatomical position. Notice that it is similar to "standing at attention" but is less comfortable because the palms are held unnaturally forward with thumbs pointing away from the body rather than hanging cupped toward the thighs.

Regional Terms

There are many visible landmarks on the surface of the body. Once you know their proper anatomical names, you can be specific in referring to different regions of the body.

Anterior Body Landmarks

Look at Figure 1.5a to find the following body regions. Once you have identified all the anterior body landmarks, cover the labels that describe what the structures are, and again go through the list, pointing out these areas on your own body.

A *(1) Your inguinal area. (2) Your posterior elbow region.*

- **abdominal** (ab-dom′ĭ-nal): anterior body trunk inferior to ribs
- **acromial** (ah-kro′me-ul): point of shoulder
- **antecubital** (an″te-ku′bĭ-tal): anterior surface of elbow
- **axillary** (ak′sĭ-lar″e): armpit
- **brachial** (bra′ke-al): arm
- **buccal** (buk′al): cheek area
- **carpal** (kar′pal): wrist
- **cervical** (ser′vĭ-kal): neck region
- **coxal** (kox′al): hip
- **crural** (kroo′ral): leg
- **digital** (dij′ĭ-tal): fingers, toes
- **femoral** (fem′or-al): thigh
- **fibular** (fib′u-lar): lateral part of leg
- **inguinal** (in′gwĭ-nal): area where thigh meets body trunk; groin
- **nasal** (na′zul): nose area
- **oral** (o′ral): mouth
- **orbital** (or′bĭ-tal): eye area
- **patellar** (pah-tel′er): anterior knee
- **pelvic** (pel′vik): area overlying the pelvis anteriorly
- **pubic** (pu′bik): genital region
- **sternal** (ster′nul): breastbone area
- **tarsal** (tar′sal): ankle region
- **thoracic** (tho-ras′ik): chest
- **umbilical** (um-bil′ĭ-kal): navel

Posterior Body Landmarks

Identify the following body regions in Figure 1.5b, and then locate them on yourself without referring to this book.

- **calcaneal** (kal-ka′neul): heel of foot
- **cephalic** (seh-fă′lik): head
- **deltoid** (del′toyd): curve of shoulder formed by large deltoid muscle
- **femoral** (fem′or-al): thigh
- **gluteal** (gloo′te-al): buttock
- **lumbar** (lum′bar): area of back between ribs and hips
- **occipital** (ok-sip′ĭ-tal): posterior surface of head
- **olecranal** (ol-eh-cra′nel): posterior surface of elbow
- **popliteal** (pop-lit′e-al): posterior knee area
- **sacral** (sa′krul): area between hips
- **scapular** (skap′u-lar): shoulder blade region
- **sural** (soo′ral): the posterior surface of lower leg; the calf
- **vertebral** (ver′tĕ-bral): area of spine

The **plantar** region, or the sole of the foot, actually on the inferior body surface, is illlustrated along with the posterior body landmarks in Figure 1.5b.

Directional Terms

Directional terms used by medical personnel and anatomists allow them to explain exactly where one body structure is in relation to another. For example, we can describe the relationship between the ears and the nose informally by saying, "The ears are located on each side of the head to the right and left of the nose." Using anatomical terminology, this condenses to, "The ears are lateral to the nose." Thus, using anatomical terminology saves a good deal of description and, once learned, is much clearer. Commonly used directional terms are defined and illustrated in Table 1.1. Although most of these terms are also used in everyday conversation, keep in mind that their anatomical meanings are very precise.

Before continuing, take a minute to check your understanding of what you have read in Table 1.1. Give the relationship between the following body parts using the correct anatomical terms.

The wrist is ______________ to the hand.

The breastbone is ______________ to the spine.

The brain is ______________ to the spinal cord.

The lungs are ______________ to the stomach.

The thumb is ______________ to the fingers. (Be careful here. Remember the anatomical position.)

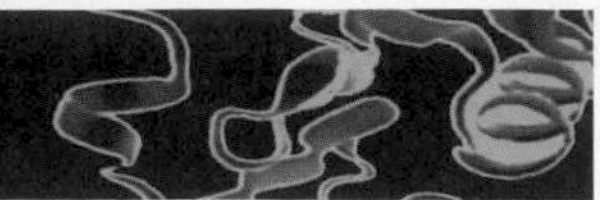

TABLE 1.1 Orientation and Directional Terms

Term	Definition	Illustration	Example
Superior (cranial or cephalad)	Toward the head end or upper part of a structure or the body; above		The forehead is superior to the nose.
Inferior (caudal)†	Away from the head end or toward the lower part of a structure or the body; below		The navel is inferior to the breastbone.
Anterior (ventral)*	Toward or at the front of the body; in front of		The breastbone is anterior to the spine.
Posterior (dorsal)*	Toward or at the backside of the body; behind		The heart is posterior to the breastbone.
Medial	Toward or at the midline of the body; on the inner side of		The heart is medial to the arm.
Lateral	Away from the midline of the body; on the outer side of		The arms are lateral to the chest.
Intermediate	Between a more medial and a more lateral structure		The armpit is intermediate between the breastbone and shoulder.
Proximal	Close to the origin of the body part or the point of attachment of a limb to the body trunk		The elbow is proximal to the wrist (meaning that the elbow is closer to the shoulder or attachment point of the arm than the wrist is).
Distal	Farther from the origin of a body part or the point of attachment of a limb to the body trunk		The knee is distal to the thigh.
Superficial (external)	Toward or at the body surface		The skin is superficial to the skeleton.
Deep (internal)	Away from the body surface; more internal		The lungs are deep to the rib cage.

†The term *caudal*, literally "toward the tail," is synonymous with *inferior* only to the inferior end of the spine.

* *Ventral* and *anterior* are synonymous in humans; this is not the case in four-legged animals. *Ventral* refers to the "belly" of an animal and thus is the inferior surface of four-legged animals. Likewise, although the dorsal and posterior surfaces are the same in humans, the term *dorsal* refers to an animal's back. Thus, the dorsal surface of four-legged animals is their superior surface.

Focus on Careers

Dental Hygienist

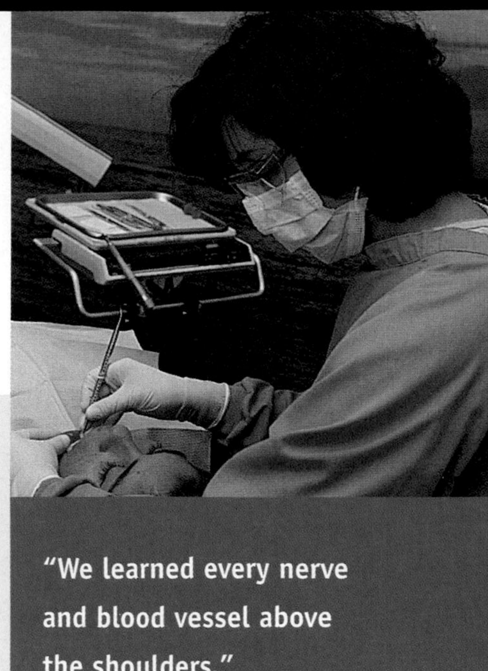

Dental hygienists need thorough training in human anatomy and physiology in order to properly care for their patients.

FEEL a bit nervous when getting your teeth cleaned? Many people do. But probably you'll feel better after you hear how well trained your dental hygienist is.

Dental hygienists are licensed preventive oral health professionals who provide educational, clinical, and therapeutic services to the public. Take Robin Mendica, a hygienist for 15 years, who holds an associate's degree in dental hygiene from a community college. Mendica's course work included extensive training in whole-body anatomy and physiology, plus specialized classes that focused on head and neck anatomy. "We learned every nerve and blood vessel above the shoulders," she recalls. "This is important because dental hygienists can give injections of anesthetic for deep cleaning and root cleaning procedures. You have to know where the nerves are to minimize pain and avoid injuries that could lead to facial paralysis."

When patients come into the office, Mendica conducts a preliminary examination: "I look at their dental X rays and charts to remind myself of any problems or concerns they may have. Then I examine the mouth inside and look at their teeth, tongue, crowns, and fillings. I check for any signs of tooth decay or decay under the gingiva." Then Mendica polishes patients' teeth, removing stains, tartar, and plaque (a soft, sticky deposit of bacteria that leads to periodontal problems) in preparation for the dentist's examination.

Mendica plays an important role in alerting the dentist to symptoms of dental problems and other health conditions. "Untreated dental conditions can lead to trouble in other parts of the body; I've seen tooth abscesses spread into the sinuses and cause bad infections. If people have lost a lot of teeth, they can't chew food properly, and that can cause digestive problems. In school we also learned to recognize nodules, precancerous conditions, and oral cancer."

What does Mendica enjoy most about her work? "I love dealing with people. Every patient is different, and I love teaching them about good oral hygiene and motivating them to take care of their teeth." She tells of many patients she has helped, such as the man who tried to "glue" a rotten molar back in his mouth with bubble gum, or the hapless woman who attempted to brighten stained bridge work with white nail polish. "I consider patient education an important part of my job, because I'm a big advocate of regular dental exams. Getting your teeth cleaned every six months is an investment in your lifelong health."

Dental hygiene education requires a high school diploma or general equivalency diploma (GED) and generally lasts two to four years. Two-year programs offer a diploma, certificate, or associate's degree; four-year programs offer a bachelor's degree. Master's degrees are available for those interested in education, research, or administration.

"We learned every nerve and blood vessel above the shoulders."

An accredited dental hygiene program requires an average of 1,948 clock hours of curriculum. This includes 585 clock hours of supervised clinical instruction and courses in anatomy, physiology, chemistry, and microbiology. Dental hygienists must be licensed by the state in which they practice. State licensing requirements vary but usually include

- Graduation from an accredited dental hygiene program
- Successful completion of the written National Board Dental Hygiene Examination
- Successful completion of a regional or state clinical board examination

Please note that accreditation procedures vary from state to state. To learn more, contact the American Dental Hygienists Association (ADHA) at
444 N. Michigan Avenue
Suite 3400
Chicago, IL 60611
(312) 440-8900 or (800) 735-5121
http://www.adha.org

For additional information on this career, click the Focus on Careers link at www.anatomyandphysiology.com.

Q *Which section type would separate the two eyes?*

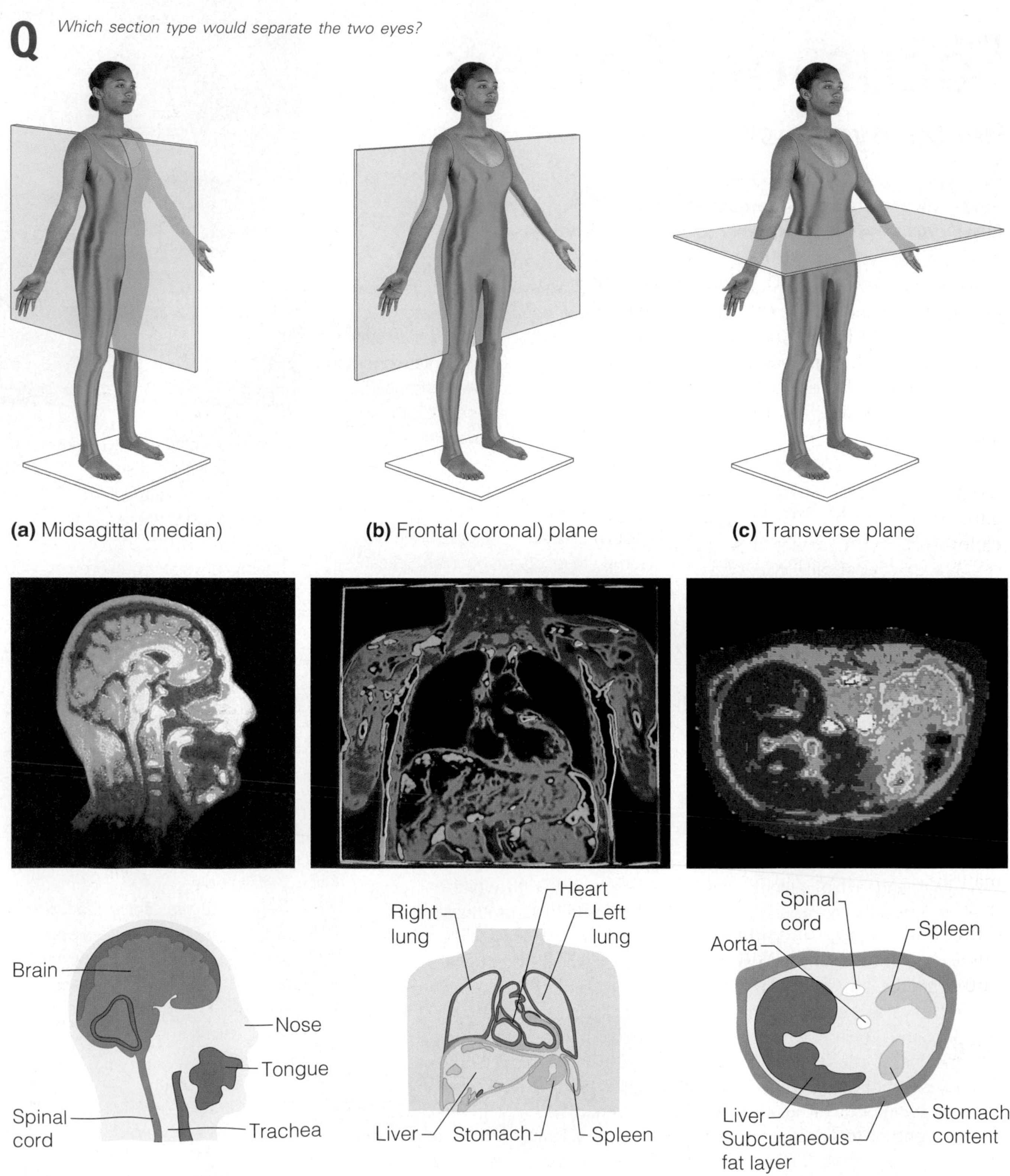

FIGURE 1.6 The anatomical position and planes of the body. The top row of the figure illustrates three major planes of space (midsagittal, frontal, and transverse) relative to humans in the anatomical position. Selected areas of the body, visualized using MRI scans taken at corresponding planes, are illustrated in the center row. Diagrams identifying body organs seen in the MRI scans are at the bottom.

A A midsagittal section would separate the two eyes.

Body Planes and Sections

When preparing to look at the internal structures of the body, medical students find it necessary to make a **section,** or cut. When the section is made through the body wall or through an organ, it is made along an imaginary line called a **plane.** Since the body is three-dimensional, we can refer to three types of planes or sections that lie at right angles to one another (Figure 1.6).

A **sagittal** (saj′ĭ-tal) **section** is a cut made along the lengthwise, or longitudinal, plane of the body, dividing the body into right and left parts. If the cut is made down the median plane of the body and the right and left parts are equal in size, it is called a **midsagittal,** or **median, section.**

A **frontal section** is a cut made along a lengthwise plane that divides the body (or an organ) into anterior and posterior parts. It is also called a **coronal** (ko-ro′nal) **section.**

A **transverse section** is a cut made along a horizontal plane, dividing the body or organ into superior and inferior parts. It is also called a **cross section.**

Sectioning a body or one of its organs along different planes often results in very different views. For example, a transverse section of the body trunk at the level of the kidneys would show kidney structure in cross section very nicely; a frontal section of the body trunk would show a different view of kidney anatomy; and a midsagittal section would miss the kidneys completely. Information on body organ positioning that can be gained by taking magnetic resonance imaging (MRI) scans along different body planes is illustrated in Figure 1.6. (MRI scans are described further in the "A Closer Look" box on pp. 20–21).

Body Cavities

Anatomy and physiology textbooks typically describe two sets of internal cavities that provide different degrees of protection to the organs within them (Figure 1.7). These cavities differ in their mode of embryological development and purpose and in their lining membranes. Consequently, the dorsal, or neural, body cavity is not named as an internal body cavity in many anatomical references. However, the idea of two major sets of internal body cavities is a useful learning concept and will continue to be used here.

Dorsal Body Cavity

The **dorsal body cavity** has two subdivisions, which are continuous with each other. The **cranial cavity** is the space inside the bony skull. The brain is well protected because it occupies the cranial cavity. The **spinal cavity** extends from the cranial cavity nearly to the end of the vertebral column. The spinal cord, which is a continuation of the brain, is protected by the vertebrae, which surround the spinal cavity.

Ventral Body Cavity

The **ventral body cavity** is much larger than the dorsal cavity. It contains all the structures within the chest and abdomen, that is, the visceral organs

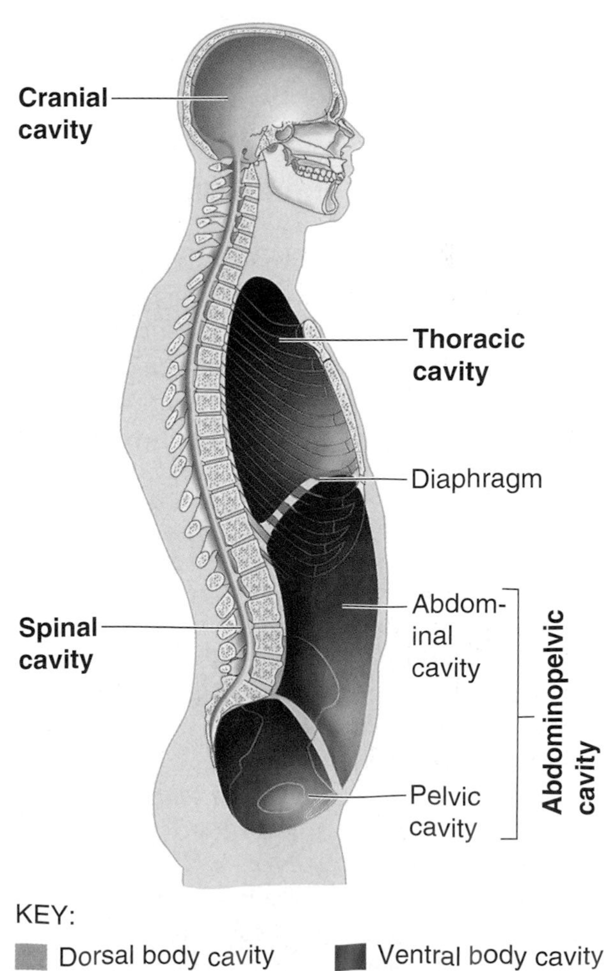

FIGURE 1.7 Body cavities. Notice the angular relationship between the abdominal and pelvic cavities.

in those regions. Like the dorsal cavity, the ventral body cavity is subdivided. The superior **thoracic cavity** is separated from the rest of the ventral cavity by a dome-shaped muscle, the **diaphragm** (di′ah-fram). The organs in the thoracic cavity (lungs, heart, and others) are somewhat protected by the rib cage. A central region called the **mediastinum** separates the lungs into right and left cavities in the thoracic cavity. The mediastinum itself houses the heart, trachea, and other visceral organs.

The cavity inferior to the diaphragm is the **abdominopelvic** (ab-dom″ĭ-no-pel′vik) **cavity.** Some prefer to subdivide it into a superior **abdominal cavity,** containing the stomach, liver, intestines, and other organs, and an inferior **pelvic cavity,** with the reproductive organs, bladder, and rectum. However, there is no actual physical structure dividing the abdominopelvic cavity. If you look carefully at Figure 1.7, you will see that the pelvic cavity is not continuous with the abdominal cavity in a straight plane, but that it tips away from it in the posterior direction.

Homeostatic Imbalance

When the body is subjected to physical trauma (as often happens in an automobile accident, for example), the most vulnerable abdominopelvic organs are those within the abdominal cavity, because the cavity walls of that portion are formed only of trunk muscles and are not reinforced by bone. The pelvic organs receive a somewhat greater degree of protection from the bony pelvis in which they reside. ▲

Because the abdominopelvic cavity is quite large and contains many organs, it is helpful to divide it up into smaller areas for study. A scheme commonly used by medical personnel divides the abdominopelvic cavity into four more or less equal regions called *quadrants.* The quadrants are then simply named according to their relative positions—that is, right upper quadrant, right lower quadrant, left upper quadrant, and left lower quadrant (Figure 1.8a).

Another system, used mainly by anatomists, divides the abdominopelvic cavity into nine separate *regions* by four planes, as shown in Figure 1.8b. Although the names of the nine regions are unfamiliar to you now, with a little patience and study they will become easier to remember. As you locate these regions in the figure, notice the organs they contain by referring to Figure 1.8c.

- The **umbilical region** is the centermost region, deep to and surrounding the umbilicus (navel).
- The **epigastric** (ep″ĭ-gas′trik) **region** is located superior to the umbilical region (*epi* = upon, above; *gastric* = stomach).
- The **hypogastric (pubic) region** is inferior to the umbilical region (*hypo* = below).
- The **right** and **left iliac,** or **inguinal, regions** are lateral to the hypogastric region (*iliac* = superior part of the hip bone).
- The **right** and **left lumbar regions** lie lateral to the umbilical region (*lumbus* = loin).
- The **right** and **left hypochondriac** (hi″po-kon′dre-ak) **regions** flank the epigastric region and contain the lower ribs (*chondro* = cartilage).

Prove It Yourself

Negative Feedback Systems Operate Continuously

Generally, negative feedback systems keep controlled variables within their acceptable range most of the time. Any variations usually are so small that we do not notice them.

You can demonstrate the small, ongoing changes that occur in controlled variables with the following exercise. In this case, the controlled variables are your balance and your ability to control voluntarily the position of your limbs.

First, stand still and look at a distant point. Everything appears stable, as if your body is not moving at all. Then look at the same point through powerful binoculars, and notice that your field of vision appears unsteady.

Why? Your body is constantly making small muscular adjustments to keep you upright and to hold your hands and arms steady. You become aware of these tiny motions only because the binoculars have magnified their effect on your vision.

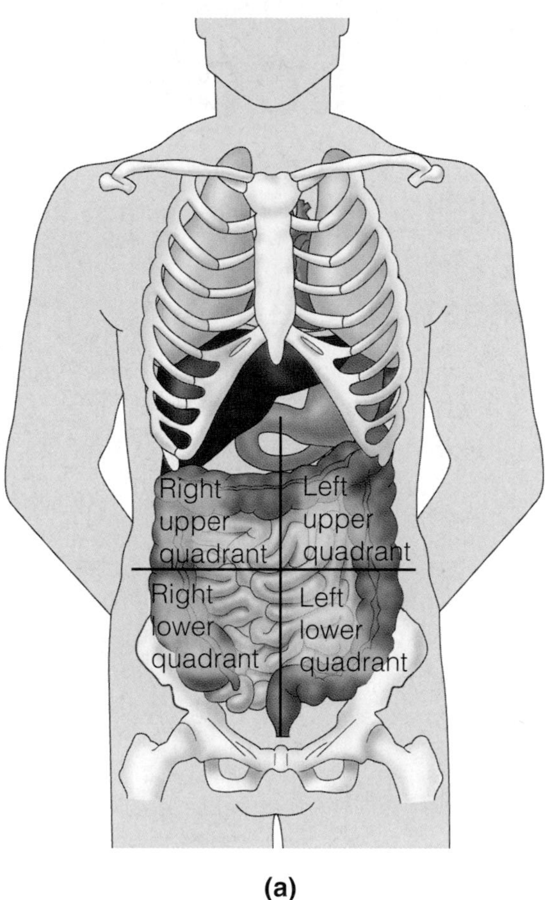

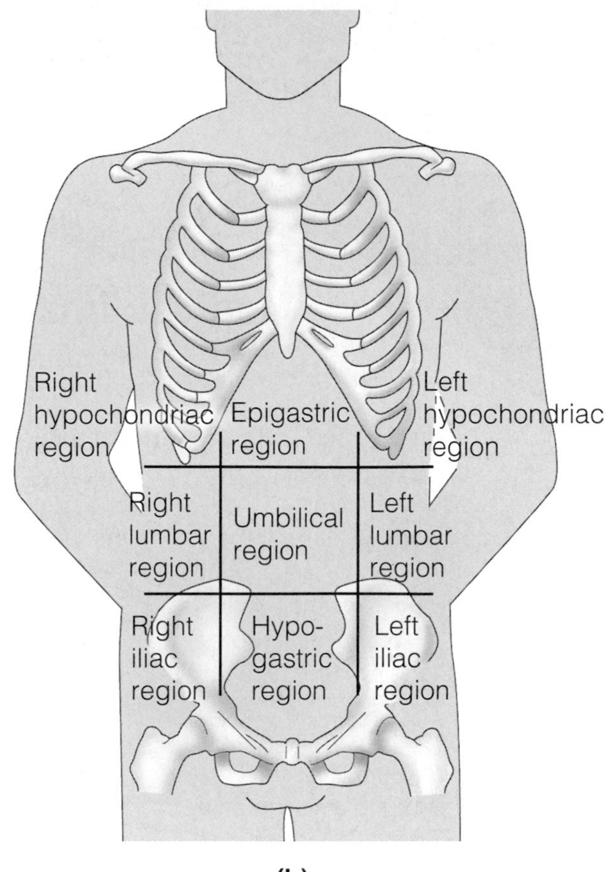

FIGURE 1.8 Abdominopelvic surface and cavity. (a) The four quadrants. **(b)** Nine regions delineated by four planes. The superior horizontal plane is at the inferior aspect of the ribs; the inferior horizontal plane is at the superior aspect of the hip bones, and the vertical planes are just medial to the nipples. **(c)** Anterior view of the ventral body cavity showing superficial organs.

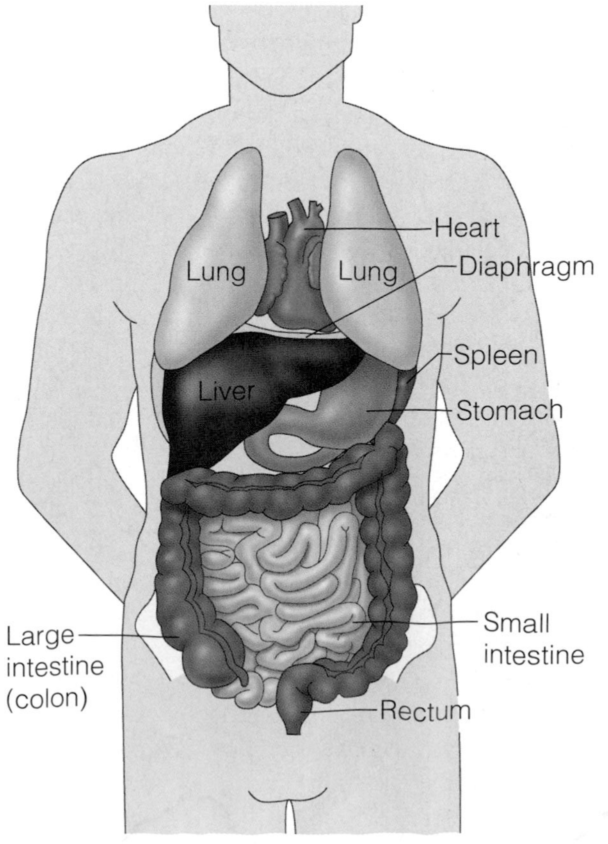

A Closer Look

Medical Imaging: Illuminating the Body

BY bombarding the body with energy, new scanning techniques can reveal the structure of internal organs, and wring out information about the private and, until now, secret working of their molecules. These new imaging techniques are changing the face of medical diagnosis.

Until about 50 years ago, the magical but murky X ray was the only means of peering into a living body. What X rays did and still do best was visualize hard, bony structures and locate abnormally dense structures (tumors, tuberculosis nodules) in the lungs. The 1950s saw the birth of nuclear medicine, which uses radioisotopes to scan the body, and ultrasound techniques. In the 1970s, CT, PET, and MRI scanning techniques were introduced.

The best known of new imaging devices is *computed tomography (CT)* (formerly called *computerized axial tomography [CAT]*), a refined version of X ray. A CT scanner confines its beam to a thin slice of the body and ends the confusion resulting from images of overlapping structures seen in conventional X rays. CT's clarity has all but eliminated exploratory surgery. As the patient is slowly moved through the doughnut-shaped CT machine, its X-ray tube rotates around the body. Different tissues absorb the radiation in varying amounts. The device's computer translates this information into a detailed, cross-sectional picture of the body region scanned. CT scans are at the forefront in evaluating most problems that affect the brain (see photo a), abdomen, and calcification of the coronary arteries in those at an elevated risk for heart disease. Special ultrafast CT scanners have produced a technique called *dynamic spatial reconstruction (DSR),* which provides three-dimensional images of body organs from any angle. It also allows their movements and changes in their internal volumes to be observed at normal speed, in slow motion, and at a specific moment in time. Although DSR can be used to evaluate the lungs and certain other mobile organs, its greatest value has been to visualize the heart beating and blood flowing through blood vessels. This allows heart defects, constricted blood vessels, and the status of coronary bypass grafts to be assessed.

Another computer-assisted X-ray technique is *digital subtraction angiography (DSA)* (*angiography* = vessel pictures). This technique provides an unobstructed view of diseased blood vessels. Conventional radiographs are taken before and after a contrast medium is injected into an artery. Then the computer subtracts the "before" image from the "after" image, eliminating all traces of body structures that obscure the vessel. DSA is often used to identify blockages in the arteries that supply the heart wall and the brain (photo b).

Just as the X ray spawned "new technologies," so did nuclear medicine in the form of *positron emission tomography (PET)*. PET excels in observing metabolic processes. PET's greatest clinical value has been its ability to provide insights into brain activity in those affected by mental illness, Alzheimer's disease, and epilepsy. One of its most exciting uses is to determine which areas of the healthy brain are most active during certain tasks (speaking, listening to music, and so on). The patient is given an injection of short-lived radioisotopes that have been tagged to biological molecules (such as glucose) and then positioned in the PET scanner. As the radioisotopes are absorbed by the most active brain cells, high-energy gamma rays are produced. The computer analyzes the gamma emission and produces a picture of the brain's biochemical activity in vivid colors.

Ultrasound imaging, or *ultrasonography,* has distinct advantages over the approaches described so far: the equipment is inexpensive and it employs high-frequency sound waves (ultrasound) as its energy source. Ultrasound, unlike ionizing forms of radiation, has no harmful effects on living tissues (as far as we know). The body is probed with pulses of sound waves, which cause echos when reflected and scattered by body tissues. The echoes are analyzed by computer to construct visual images of body organs of interest. Because of its safety, ultrasound is the imaging technique of choice for obstetrics, that is, for determining fetal age and position and locating the placenta. Because sound waves have very low penetrating power and are rapidly scattered in air, ultrasonography is of little value for looking at air-filled structures (the lungs) or those surrounded by bone (the brain and spinal cord).

Another technique that depends on nonionizing radiation is *magnetic resonance imaging (MRI),* which uses magnetic fields up to 60,000 times stronger than the Earth's to pry information from the body's tissues. The patient lies in a chamber within a huge magnet. Hydrogen molecules spin like tops in the magnetic field, and their energy is enhanced by radio waves. When the radio waves are turned off, the energy is released and translated by the computer into a visual image. MRI is immensely popular because it can do many things a CT scan cannot. Dense structures do not show up in MRI, so bones of the skull and/or vertebral column do not impair the view of *soft tissues* such as the brain (see Figure 1.6a, p. 16). MRI is also particularly good at detecting degenerative disease of various kinds. Multiple sclerosis plaques, for example, do not show up well in CT scans but are dazzlingly clear in MRI scans. A key issue being investigated by MRI studies is how brain development and behavior change with growth or experience.

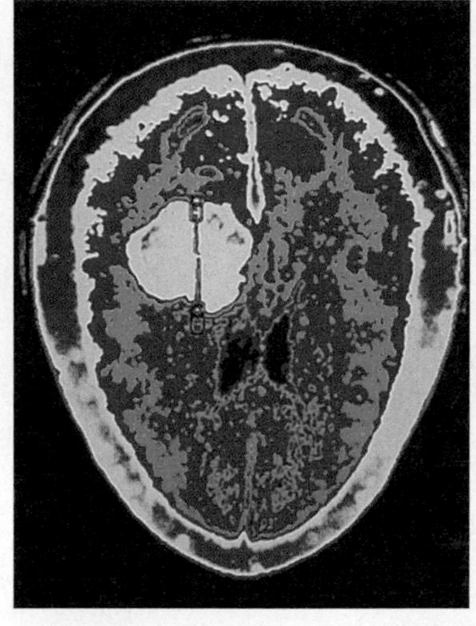

(a)

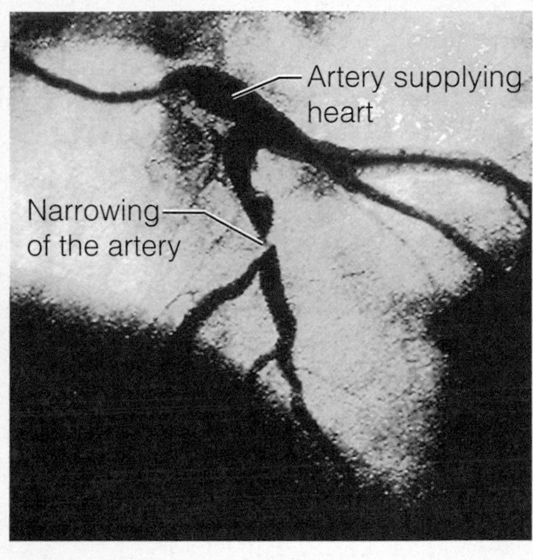

(b)

Two different methods for illuminating the body. **(a)** CT scan showing a brain tumor (oval area on right side of brain). **(b)** DSA image of arteries supplying the heart.

A newer variation of MRI, called *magnetic resonance spectroscopy (MRS),* maps the distribution of elements other than hydrogen to reveal more about how disease changes body chemistry. In 1992, MRI technology leaped forward with the development of the *functional MRI,* which allows tracking of blood flow into the brain in real time. Until then, matching thoughts, deeds, and disease to corresponding brain activity had been the sole domain of PET. Because functional MRI does not require injections of tracer elements, it provides another, perhaps more desirable, alternative to such studies. Despite its advantages, the powerful magnets of the MRI present some problems. For example they can "suck" metal objects, such as implanted pacemakers and loose tooth fillings, through the body. Also, there is no convincing evidence that such magnetic fields are risk free.

As you can see, modern medical science has many remarkable tools at its disposal. CT and PET scans account for about 25 percent of all imaging. Ultrasonography, because of its safety and low cost, is the most widespread of the new techniques. Conventional X rays remain the workhorse of diagnostic imaging techniques and still account for more than half of all imaging currently done.

SUMMARY

Media study tools that could provide you with additional help in reviewing specific key topics of Chapter 1 are referenced below.

IP = Interactive Physiology;

WEB = A&P Place website.

AN OVERVIEW OF ANATOMY AND PHYSIOLOGY (p. 2)

1. Anatomy is the study of structure. Observation is used to see the sizes and relationships of body parts.
2. Physiology is the study of how a structure (which may be a cell, an organ, or an organ system) functions or works.
3. Structure determines what functions can occur; therefore, if the structure changes, the function must also change.

LEVELS OF STRUCTURAL ORGANIZATION (pp. 2–7)

1. There are six levels of structural organization. Atoms (at the chemical level) combine, forming the unit of life, the cell. Cells are grouped into tissues, which in turn are arranged in specific ways to form organs. A number of organs form an organ system that performs a specific function for the body (which no other organ system can do). Together, all of the organ systems form the organism, or living body.

   WEB Exercise: Chapter 1, Levels of Biological Organization.
2. For a description of organ systems naming the major organs and functions, see pp. 3–7.

MAINTAINING LIFE (pp. 7–9)

1. To sustain life, an organism must be able to maintain its boundaries, move, respond to stimuli, digest nutrients and excrete wastes, carry on metabolism, reproduce itself, and grow.
2. Survival needs include food, oxygen, water, appropriate temperature, and normal atmospheric pressure. Extremes of any of these factors can be harmful.

HOMEOSTASIS (pp. 9–11)

1. Body functions interact to maintain homeostasis, or a relatively stable internal environment within the body. Homeostasis is necessary for survival and good health; its loss results in illness or disease.
2. All homeostatic control mechanisms have a receptor that responds to environmental changes and a control center that assesses those changes and produces a response by activating a third element, the effector.
3. Most homeostatic control systems are negative feedback systems, which act to reduce or stop the initial stimulus.

THE LANGUAGE OF ANATOMY (pp. 11–19)

1. Anatomical terminology is relative and assumes that the body is in the anatomical position (erect, palms facing forward).
2. Regional terms. Visible landmarks on the body surface may be used to specifically refer to a body part or area. See pp. 12 and 13 for terms referring to anterior and posterior surface anatomy.

   WEB Exercise: Chapter 1, Anatomical Terminology: Orientation and Directional Terms.
3. Directional terms
   a. Superior (cranial, cephalad): above something else, toward the head.
   b. Inferior (caudal): below something else, toward the tail.
   c. Anterior (ventral): toward the front of the body or structure.
   d. Posterior (dorsal): toward the rear or back of the body or structure.
   e. Medial: toward the midline of the body.
   f. Lateral: away from the midline of the body.
   g. Intermediate: between a more medial and a more lateral structure.
   h. Proximal: closer to the point of attachment.
   i. Distal: farther from the point of attachment.
   j. Superficial (external): at or close to the body surface.
   k. Deep (internal): below or away from the body surface.
4. Body planes and sections
   a. Sagittal section: separates the body longitudinally into right and left parts.

b. Frontal (coronal) section: separates the body on a longitudinal plane into anterior and posterior parts.
c. Transverse (cross) section: separates the body on a horizontal plane into superior and inferior parts.

WEB Exercise: Chapter 1, Body Planes.

5. Body cavities
   a. Dorsal: well protected by bone; has two subdivisions.
      (1) Cranial: contains the brain.
      (2) Spinal: contains the spinal cord.
   b. Ventral: less protected than dorsal cavity; has two subdivisions.
      (1) Thoracic: The superior cavity that extends inferiorly to the diaphragm; contains heart and lungs, which are protected by the rib cage.
      (2) Abdominopelvic: The cavity inferior to the diaphragm that contains the digestive, urinary, and reproductive organs. The abdominal portion is vulnerable because it is protected only by the trunk muscles. There is some protection of the pelvic portion by the bony pelvis. The abdominopelvic cavity is often divided into four quadrants or nine regions (see Figure 1.8).

WEB Exercise: Chapter 1, Dorsal and Ventral Cavities.

REVIEW QUESTIONS

MULTIPLE CHOICE

More than one choice may apply.

1. Consider the following levels: (1) chemical; (2) tissue; (3) organ; (4) cellular; (5) organismal; (6) systemic. Which of the following choices has the levels listed in order of increasing complexity?
   a. 1, 2, 3, 4, 5, 6
   b. 1, 4, 2, 5, 3, 6
   c. 3, 1, 2, 4, 6, 5
   d. 1, 4, 2, 3, 6, 5
   e. 4, 1, 3, 2, 6, 5
2. Which of the following is (are) involved in maintaining homeostasis?
   a. Effector
   b. Control center
   c. Receptor
   d. Feedback
   e. Lack of change
3. Which is not essential to survival?
   a. Water
   b. Oxygen
   c. Gravity
   d. Atmospheric pressure
   e. Nutrients
4. Anatomical terms that apply to the backside of the body in the anatomical position include
   a. ventral and anterior.
   b. back and rear.
   c. posterior and dorsal.
   d. head and lateral.
5. A neurosurgeon orders a spinal tap for a patient. Into what body cavity will the needle be inserted?
   a. Ventral
   b. Thoracic
   c. Dorsal
   d. Cranial
   e. Pelvic
6. Which of the following groupings of the abdominopelvic regions is medial?
   a. Hypochondriac, hypogastric, umbilical
   b. Hypochondriac, lumbar, inguinal
   c. Hypogastric, umbilical, epigastric
   d. Lumbar, umbilical, iliac
   e. Iliac, umbilical, hypochondriac

SHORT ANSWER ESSAY

1. Define anatomy and physiology.
2. Why would you have a hard time trying to learn and understand physiology if you did not also understand anatomy?
3. List the 11 organ systems of the body, briefly describe the function of each, and then name two organs in each system.
4. In addition to being able to metabolize, grow, digest food, and excrete wastes, what functions must an organism perform if it is to survive?
5. Define homeostasis.
6. What is the consequence of loss of homeostasis, or homeostatic imbalance?
7. Describe the anatomical position.
8. On what body surface is each of the following located: nose, calf of leg, ears, umbilicus, fingernails?
9. Several pairs of structures are given next. In each case, choose the one that meets the condition given first.
   a. Distal—the knee/the foot
   b. Lateral—the cheekbone/the nose
   c. Superior—the neck/the chin
   d. Anterior—the heel/the toenails
   e. External—the skin/the skeletal muscles
10. What kind of section would have to be made to cut the brain into anterior and posterior parts?

11. Which of the following organ systems—digestive, respiratory, reproductive, circulatory, urinary, or muscular—are found in *both* subdivisions of the ventral body cavity? Which are found in the thoracic cavity only? In the abdominopelvic cavity only?

AT THE CLINIC

1. A nurse informed John that she was about to take blood from his antecubital region. What part of his body was she referring to? Later, she came back and said that she was going to give him an antibiotic shot in the deltoid region. Did he take off his shirt or drop his pants to get the shot? Before John left the office, the nurse noticed that his left sural region was badly bruised. What part of his body was bruised?
2. How is the concept of homeostasis (or its loss) related to disease and aging? Provide examples to support your reasoning.
3. When we begin to become dehydrated, we usually become thirsty, which causes us to drink fluids. On the basis of what you now know about control systems, decide whether the thirst sensation is part of a negative or positive feedback control system and defend your choice.
4. Jennie Dip fell off her motorcycle and tore a nerve in her axillary region. She also tore ligaments in her cervical and scapular regions and broke the only bone of her right brachial region. Explain where each of her injuries is located.
5. Mr. Petros is behaving abnormally, and doctors strongly suspect he has a brain tumor. Which medical imaging device—conventional X ray, DSA, PET, ultrasound, or MRT—would be best for precisely locating a tumor within the brain? Explain your choice.

Basic Chemistry

YOUR GOALS

After completing this chapter, you will have mastered the objectives listed below.

OBJECTIVE CHECKLIST

CONCEPTS OF MATTER AND ENERGY (pp. 26–27)

- Differentiate clearly between matter and energy.
- List the major energy forms, and provide one example of how each energy form is used in the body.

COMPOSITION OF MATTER (pp. 27–32)

- Define *chemical element,* and list the four elements that form the bulk of body matter.
- Explain how elements and atoms are related.
- List the subatomic particles, and describe their relative masses, charges, and positions in the atom.
- Define *radioisotope,* and describe briefly how radioisotopes are used in the diagnosis and treatment of disease.

MOLECULES AND COMPOUNDS (pp. 32–33)

- Recognize that chemical reactions involve the interaction of electrons to make and break chemical bonds.
- Define *molecule,* and explain how molecules are related to compounds.

CHEMICAL BONDS AND CHEMICAL REACTIONS (pp. 33–39)

- Differentiate between ionic, polar covalent, and nonpolar covalent bonds, and describe the importance of hydrogen bonds.
- Contrast synthesis, decomposition, and exchange reactions.

>>

(Objective Checklist, continued)

BIOCHEMISTRY: THE CHEMICAL COMPOSITION OF LIVING MATTER (pp. 39–52)

- Distinguish between organic and inorganic compounds.
- Differentiate clearly between a salt, an acid, and a base.
- List several salts (or their ions) vitally important to body functioning.
- Explain the importance of water to body homeostasis, and provide several examples of the roles of water.
- Explain the concept of pH, and state the pH of blood.
- Compare and contrast carbohydrates, lipids, proteins, and nucleic acids in terms of their building blocks, structures, and functions in the body.
- Differentiate between fibrous and globular proteins.
- Compare and contrast the structure and functions of DNA and RNA.
- Define *enzyme,* and explain the role of enzymes.
- Explain the importance of ATP in the body.

Many short courses in anatomy and physiology lack the time to consider chemistry as a topic. So why include it here? Very simply, the food you eat and the medicines you take when you are ill are composed of chemicals. Indeed, your entire body is made up of chemicals—thousands of them—continuously interacting with one another at an incredible pace.

Although it is possible to study anatomy without much reference to chemistry, chemical reactions underlie all body processes—movement, digestion, the pumping of your heart, and even your thoughts. This chapter presents the basics of chemistry and biochemistry (the chemistry of living material), providing the background you will need to understand body functions.

Concepts of Matter and Energy

Matter

Matter is the "stuff" of the universe. With some exceptions, it can be seen, smelled, and felt. More precisely, matter is anything that occupies space and has mass (weight). Chemistry studies the nature of matter—how its building blocks are put together and how they interact.

Matter exists in solid, liquid, and gaseous states. Examples of each state are found in the human body. *Solids,* like bones and teeth, have a definite shape and volume. *Liquids* have a definite volume, but they conform to the shape of their container. Examples of body liquids are blood plasma and the interstitial fluid that bathes all body cells. *Gases* have neither a definite shape nor a definite volume. The air we breathe is composed of a mixture of gases.

Matter may be changed both physically and chemically. *Physical changes* do not alter the basic nature of a substance. Examples include changes in state, such as ice melting to become water and cutting food into smaller pieces. *Chemical changes do* alter the composition of the substance—often substantially. Fermenting grapes to make wine and the digestion of food in the body are examples of chemical changes.

Energy

In contrast to matter, **energy** is massless and does not take up space. It can only be measured by its effects on matter. Energy is commonly defined as the ability to do work or to put matter into motion. When energy is actually doing work (moving objects), it is referred to as **kinetic** (kĭ-neh′tik) **energy.** When it is inactive or stored (as in the batteries of an unused toy), it is called **potential energy.** All forms of energy exhibit both kinetic and potential work capacities.

Actually, energy is a physics topic, but it is difficult to separate matter and energy. All living things are built of matter, and, in order to grow and function, they require a continuous supply of energy. Thus, matter is the substance, and energy is the mover of the substance. Because this is so, it is worth taking a brief detour to introduce the forms of energy used by the body as it does its work.

Forms of Energy

- **Chemical energy** is stored in the bonds of chemical substances. When the bonds are broken, the (potential) stored energy is unleashed and becomes kinetic energy (energy in action). For example, when gasoline molecules are broken apart in your automobile engine, the energy released powers your car. In like manner, all body activities are "run" by the chemical energy harvested from the foods we eat.
- **Electrical energy** results from the movement of charged particles. In your house, electrical energy is the flow of electrons along the wiring. In your body, an electrical current is generated when charged particles (called *ions*) move across cell membranes.
- **Mechanical energy** is *directly* involved in moving matter. When you ride a bicycle, your legs provide the mechanical energy that moves the pedals. We can take this example one step further back: As the muscles in your legs shorten, they pull on your bones, causing your limbs to move (so that you can pedal the bike).
- **Radiant energy** travels in waves, that is, energy of the electromagnetic spectrum, which includes X rays and infrared, visible light, radio, and ultraviolet waves. Light energy, which stimulates the retinas of your eyes, is important in vision. Ultraviolet waves are responsible for that suntan we get at the beach but also stimulate our bodies to make vitamin D.

Energy Form Conversions

With a few exceptions, energy is easily converted from one form to another. For example, an electrical current carried to a lamp socket is converted into light energy by the bulb. In the body, chemical energy of foods is trapped in the bonds of a high-energy chemical called ATP (adenosine triphosphate), and ATP's energy may ultimately be transformed into the electrical energy of a nerve impulse or mechanical energy of shortening muscles.

Energy conversions are quite inefficient, and some of the initial energy supply is always "lost" to the environment as heat. (It is not really lost, because energy cannot be created or destroyed, but the part given off as heat is *unusable.*) You can easily demonstrate this principle by touching a lightbulb that has been lit for an hour or so. You will soon discover that some of the electrical energy reaching the bulb is producing heat instead of light. Likewise, all energy conversions that occur in the body liberate heat. It is this heat that makes us warm-blooded animals and contributes to our relatively high body temperature, which has an important influence on body functioning. For example, when matter is heated, its particles begin to move more quickly; that is, their kinetic energy (energy of motion) increases. This is important to the chemical reactions that occur in the body because, up to a point, the higher the temperature, the faster those reactions occur. We will learn more about this later.

Composition of Matter

Elements and Atoms

All matter is composed of a limited number of substances called **elements,** unique substances that cannot be broken down into simpler substances by ordinary chemical methods. Examples of elements include many commonly known substances, such as oxygen, carbon, gold, copper, and iron.

So far, 112 elements are known with certainty, and numbers 113 to 116 are alleged. Ninety-two of these occur naturally; the rest are produced artificially in accelerator devices. Four elements—carbon, oxygen, hydrogen, and nitrogen—make up about 96 percent of body weight, but several others are present in small or trace amounts. A complete listing of the elements appears in the

TABLE 2.1 Common Elements Making Up the Human Body

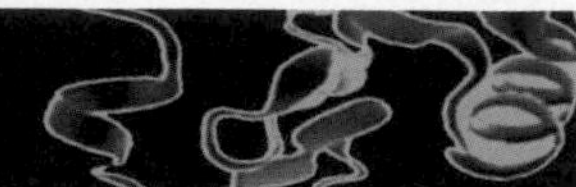

Element	Atomic symbol	Percentage of body mass	Role
Major (96.1%)			
Oxygen	O	65.0	A major component of both organic and inorganic molecules; as a gas, essential to the oxidation of glucose and other food fuels, during which cellular energy (ATP) is produced.
Carbon	C	18.5	The primary elemental component of all organic molecules, including carbohydrates, lipids, proteins, and nucleic acids.
Hydrogen	H	9.5	A component of most organic molecules; in ionic form, influences the pH of body fluids.
Nitrogen	N	3.2	A component of proteins and nucleic acids (genetic material).
Lesser (3.9%)			
Calcium	Ca	1.5	Found as a salt in bones and teeth; in ionic form, required for muscle contraction, neural transmission, and blood clotting.
Phosphorus	P	1.0	Present as a salt, in combination with calcium, in bones and teeth; also present in nucleic acids and many proteins; forms part of the high-energy compound ATP.
Potassium	K	0.4	In its ionic form, the major intracellular cation; necessary for the conduction of nerve impulses and for muscle contraction.
Sulfur	S	0.3	A component of proteins (particularly contractile proteins of muscle).
Sodium	Na	0.2	As an ion, the major extracellular cation; important for water balance, conduction of nerve impulses, and muscle contraction.
Chlorine	Cl	0.2	In ionic form, a major extracellular anion.
Magnesium	Mg	0.1	Present in bone; also an important cofactor for enzyme activity in a number of metabolic reactions.
Iodine	I	0.1	Needed to make functional thyroid hormones.
Iron	Fe	0.1	A component of the functional hemoglobin molecule (which transports oxygen within red blood cells) and some enzymes.

Trace (less than 0.01%)*

Chromium (Cr), Cobalt (Co), Copper (Cu), Fluorine (F), Manganese (Mn), Molybdenum (Mo), Selenium (Se), Silicon (Si), Tin (Sn), Vanadium (V), Zinc (Zn)

*Referred to as the *trace elements* because are required in very minute amounts; many found as part of enzymes or required for enzyme activation.

TABLE 2.2 Subatomic Particles

Particle	Position in atom	Mass (amu)	Charge
Proton (p^+)	Nucleus	1	+
Neutron (n^0)	Nucleus	1	0
Electron (e^-)	Orbitals outside the nucleus	1/1800	−

periodic table, an odd-shaped checkerboard that appears in Appendix B and in chemistry classrooms the world over. The most abundant elements found in the body and their major roles are listed in Table 2.1.

The building block of an element, or the smallest particle that still retains its special properties, is called an **atom.** Because all elements are unique, the atoms of each element differ from those of all other elements. Each element is designated by a one- or two-letter chemical shorthand called an **atomic symbol.** In most cases, the atomic symbol is simply the first (or first two) letter(s) of the element's name. For example, C stands for carbon, O for oxygen, and Ca for calcium. In a few cases, the atomic symbol is taken from the Latin name for the element. For instance, sodium is indicated by Na (from the Latin word *natrium*).

Atomic Structure

The word *atom* comes from the Greek word meaning "incapable of being divided," and historically this idea of an atom was accepted as a scientific truth. According to this notion, you could theoretically divide a pure element, such as a block of gold, into smaller and smaller particles until you got down to the individual atoms, and then could subdivide no further. We now know that atoms, although indescribably small, are clusters of even smaller (subatomic) particles and that, under very special circumstances, atoms can be split into these smaller particles. Even so, the old idea of atomic indivisibility is still very appropriate, because an atom loses the unique properties of its element when it is split into its subparticles.

The atoms representing the 112-plus elements are composed of different numbers and proportions of three basic subatomic particles, which differ in their mass, electrical charge, and location within the atom (Table 2.2). **Protons (p^+)** have a positive charge, whereas **neutrons (n^0)** are uncharged, or neutral. Protons and neutrons are heavy particles and have approximately the same mass (1 atomic mass unit, or 1 amu). The tiny **electrons (e^-)** bear a negative charge equal in strength to the positive charge of the protons, but their mass is so small that it is usually designated as 0 amu.

The electrical charge of a particle is a measure of its ability to attract or repel other charged particles. Particles with the same type of charge (+ to + or − to −) repel each other, but particles with unlike charges (+ to −) attract each other. Neutral particles are neither attracted to nor repelled by charged particles.

Because all atoms are electrically neutral, the number of protons an atom has must be precisely balanced by its number of electrons (the + and − charges will then cancel the effect of each other). Thus, hydrogen has one proton and one electron, and iron has 26 protons and 26 electrons. For any atom, the number of protons and electrons is always equal. Atoms that have gained or lost electrons are called *ions* as discussed shortly.

Planetary and Orbital Models of an Atom

The **planetary model** of an atom portrays the atom as a miniature solar system (Figure 2.1a) in which the protons and neutrons are clustered at the center of the atom in the atomic nucleus. Because the nucleus contains all the heavy particles, it is fantastically dense and positively charged. The tiny

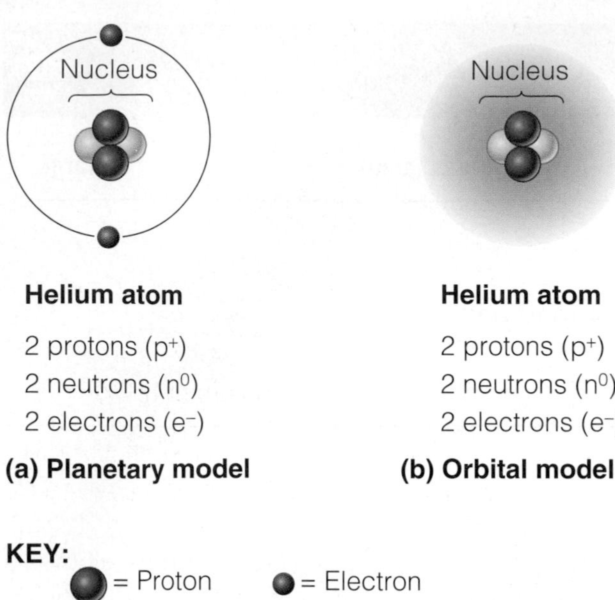

FIGURE 2.1 The structure of an atom. The dense central nucleus contains the protons and neutrons. **(a)** In the planetary model of atomic structure, the electrons move around the nucleus in fixed orbits. **(b)** The orbital model recognizes that we never know exactly where electrons are; therefore electrons are shown as a cloud of negative charge.

electrons orbit around the nucleus in fixed, generally circular orbits, like planets around the sun. But we can never determine the exact location of electrons at a particular time because they jump around following unknown paths. So, instead of speaking of specific orbits, chemists talk about *orbitals*—regions around the nucleus in which a given electron or electron pair is *likely* to be found most of the time. This more modern model of atomic structure, called the **orbital model,** has proved to be more useful in predicting the chemical behavior of atoms. As illustrated in Figure 2.1b, the orbital model depicts the general location of electrons outside the nucleus as a haze of negative charge referred to as the *electron cloud.* Regions where electrons are most likely to be found are shown by denser shading rather than by orbit lines. Regardless of which model is used, notice that the electrons have the run of nearly the entire volume of the atom and determine its chemical behavior (that is, its ability to bond with other atoms). Though now considered outdated, the planetary model is simple and easy to understand and use. Most of the descriptions of atomic structure in this book use that model.

Hydrogen is the simplest atom, with just one proton and one electron. You can visualize the spatial relationships within the hydrogen atom by imagining it enlarged until its diameter equals the length of a football field. In that case, the nucleus could be represented by a lead ball the size of a gumdrop in the exact center of the sphere and the lone electron pictured as a fly buzzing about unpredictably within the sphere. This mental picture should serve to remind you that most of the volume of an atom is empty space, and nearly all of the mass is concentrated in the central nucleus.

Identifying Elements

All protons are alike, regardless of the atom being considered. The same is true of all neutrons and all electrons. So what determines the unique properties of each element? The answer is that atoms of different elements are composed of *different numbers* of protons, neutrons, and electrons.

The simplest and smallest atom, hydrogen, has one proton, one electron, and no neutrons (Figure 2.2). Next is the helium atom, with two protons,

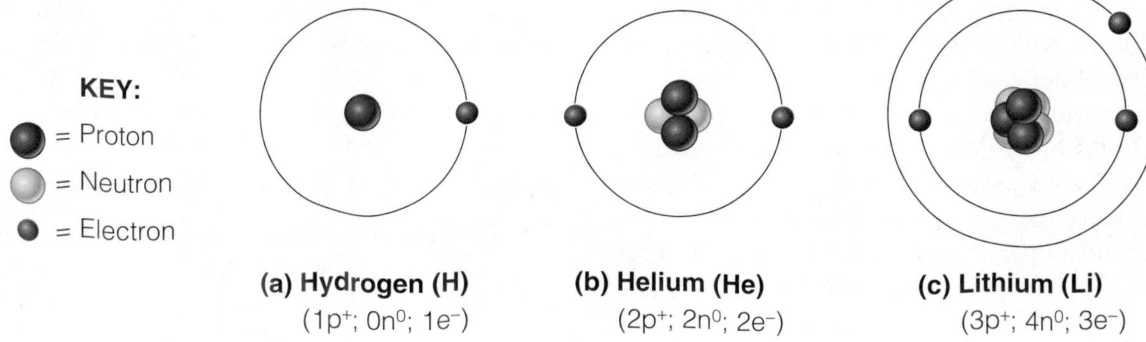

FIGURE 2.2 Atomic structure of the three smallest atoms.

Q *Which of these isotopes is the heaviest?*

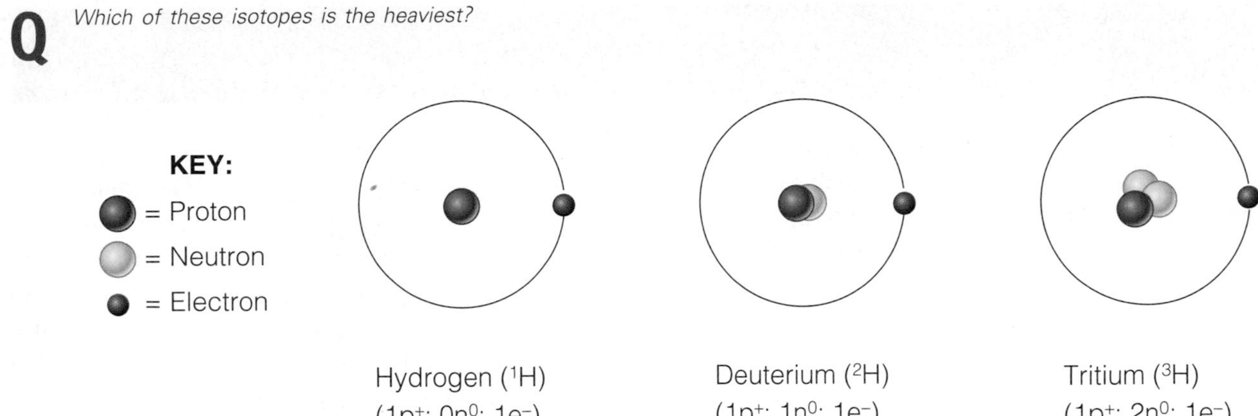

FIGURE 2.3 Isotopes of hydrogen.

two neutrons, and two orbiting electrons. Lithium follows with three protons, four neutrons, and three electrons. If this step-by-step listing of subatomic particles were continued, all known atoms could be described by adding one proton and one electron at each step. The number of neutrons is not as easy to pin down, but light atoms tend to have equal numbers of protons and neutrons, whereas in larger atoms neutrons outnumber protons. However, all we really need to know to identify a particular element is its atomic number, mass number, and atomic weight. Taken together, these indicators provide a fairly complete picture of each element.

Atomic Number

Each element is given a number, called its **atomic number,** that is equal to the number of protons its atoms contain. Atoms of each element contain a different number of protons than the atoms of any other element; hence, its atomic number is unique. Because the number of protons is always equal to the number of electrons, the atomic number *indirectly* also tells us the number of electrons that atom contains.

Atomic Mass Number

The **atomic mass number** of any atom is the sum of the protons and neutrons contained in its nucleus. (The mass of the electrons is so small that it is ignored.) Hydrogen has one bare proton and no neutrons in its nucleus; thus its atomic number and atomic mass number are the same (1). Helium, with 2 protons and 2 neutrons, has a mass number of 4. The atomic mass number is written as a superscript to the left of the atomic symbol (see the examples in Figure 2.3).

Atomic Weight and Isotopes

At first glance, it would seem that the **atomic weight** of an atom should be equal to its atomic mass. This would be so if there were only one type of atom representing each element. However, the atoms of almost all elements exhibit two or more structural variations; these varieties are called **isotopes** (i′so-tōps). Isotopes have the same number of protons and electrons but vary in the number of *neutrons* they contain. Thus, the isotopes of an element have the same atomic number but have different atomic masses. Because all of an element's isotopes have the same number of electrons (and electrons determine bonding properties), their chemical properties are *exactly* the same. As a general rule, the atomic weight of any element is approximately equal to the mass number of its most abundant isotope. For example, as we said before, hydrogen has an atomic number of 1, but it also has isotopes with atomic masses of 1, 2, and 3 (Figure 2.3). Its atomic weight is 1.0079, which reveals that its lightest isotope is present in much greater amounts than its heavier forms. The atomic numbers, mass numbers, and atomic weights for elements commonly found in the body are provided in Table 2.3.

A *Tritium.* (printed upside down)

TABLE 2.3 Atomic Structures of the Most Abundant Elements in the Body

Element	Symbol	Atomic number (# of p)	Mass number (# of p + n)	Atomic weight	Electrons in valence shell
Calcium	Ca	20	40	40.08	2
Carbon	C	6	12	12.011	4
Chlorine	Cl	17	35	35.453	7
Hydrogen	H	1	1	1.008	1
Iodine	I	53	127	126.905	7
Iron	Fe	26	56	55.847	2
Magnesium	Mg	12	24	24.305	2
Nitrogen	N	7	14	14.007	5
Oxygen	O	8	16	15.999	6
Phosphorus	P	15	31	30.974	5
Sodium	Na	11	23	22.99	1
Sulfur	S	16	32	32.064	6

The heavier isotopes of certain atoms are unstable and tend to decompose to become more stable; such isotopes are called **radioisotopes.** The why of this process is very complex, but apparently the "glue" that holds the atomic nuclei together is weaker in the heavier isotopes. This process of spontaneous atomic decay is called **radioactivity** and can be compared to a tiny explosion. All types of radioactive decay involve the ejection of particles (*alpha* or *beta particles*) or electromagnetic energy (*gamma rays*) from the atom's nucleus and are damaging to living cells. Alpha emission has the least penetrating power; gamma radiation has the most. Contrary to what some believe, ionizing radiation does not damage the atoms in its path directly. Instead it sends electrons flying, like a bowling ball through pins, all along its path. It is these electrons that do the damage.

Radioisotopes are used in minute amounts to tag biological molecules so that they can be followed, or traced, through the body and are valuable tools for medical diagnosis and treatment. For example, PET scans, which use radioisotopes, are discussed in the "A Closer Look" box on pp. 20–21. Additionally, a radioisotope of iodine can be used to scan the thyroid gland of a patient suspected of having a thyroid tumor. Radium, cobalt, and certain other radioisotopes are used to destroy localized cancers.

Molecules and Compounds

When two or more atoms combine chemically, **molecules** are formed. If two or more atoms of the same element bond together, a molecule of that element is produced. For example, when two hydrogen atoms bond, the product is a molecule of hydrogen gas:

$$\text{H (atom)} + \text{H (atom)} \rightarrow H_2 \text{ (molecule)*}$$

*Notice that when the number of atoms is written as a subscript, the subscript indicates that the atoms are joined by a chemical bond. Thus, 2H represents two unjoined atoms, but H_2 indicates that the two hydrogen atoms are bonded together to form a molecule.

In the example given, the atoms taking part in the reaction are indicated by their atomic symbols, and the composition of the product is indicated by a *molecular formula* that shows its atomic makeup. The chemical reaction is shown by writing a *chemical equation*.

When two or more *different* atoms bind together to form a molecule, the molecule is more specifically referred to as a molecule of a **compound.** For example, four hydrogen atoms and one carbon atom can interact chemically to form methane:

$$4H + C = CH_4 \text{ (methane)}$$

Thus, a molecule of methane is a compound, but a molecule of hydrogen gas is not—it is instead called molecular hydrogen.

It is important to understand that compounds always have properties quite different from those of the atoms making them up, and it would be next to impossible to determine the atoms making up a compound without analyzing it chemically. Notice that just as an atom is the smallest particle of an element that still retains that element's properties, a molecule is the smallest particle of a compound that still retains the properties of that compound. If you break the bonds between the atoms of the compound, properties of the atoms, rather than those of the compound, will be exhibited.

Chemical Bonds and Chemical Reactions

Chemical reactions occur whenever atoms combine with or dissociate from other atoms. When atoms unite chemically, chemical bonds are formed.

Bond Formation

It is important to understand that a chemical bond is not an actual physical structure, like a pair of handcuffs linking two people together. Instead, it is an energy relationship that involves interactions between the electrons of the reacting atoms. Because this is so, we will devote a few words to the role of electrons in bond formation.

Role of Electrons

As illustrated in Figure 2.2, electrons occupy generally fixed regions of space around the nucleus; these regions are called **electron shells,** or **energy levels.** The maximum number of electron shells in any atom known so far is seven, and these are numbered 1 to 7 from the nucleus outward. The electrons closest to the nucleus are those most strongly attracted to its positive charge, and those farther away are less securely held. As a result, the more distant electrons are likely to interact with other atoms.

Perhaps this situation can be compared to the development of a child. During infancy and the toddler years, the child spends most of its time at home and is shaped and molded by the ideas and demands of its parents. However, when the child goes to school, it is increasingly influenced by friends and other adults, such as teachers and coaches. Thus, just as the child is more likely to become involved with "outsiders" as it roams farther from home, electrons are more influenced by other atoms as they get farther and farther away from the positive influence of the nucleus.

There is an upper limit to the number of electrons that each electron shell can hold. Shell 1, closest to the nucleus, is small and can accommodate only 2 electrons. Shell 2 holds a maximum of 8. Shell 3 can accommodate up to 18 electrons. Subsequent shells hold larger and larger numbers of electrons. In most (but not all) cases the shells tend to be filled consecutively.

The only electrons that are important when considering bonding behavior are those in the atom's outermost shell. This shell is called the **valence shell;** its electrons determine the chemical behavior of the atom. As a general rule, the electrons of inner shells do not take part in bonding.

When the valence shell of an atom contains 8 electrons, the atom is completely stable and is chemically inactive (inert). When the valence shell contains fewer than 8 electrons, an atom will tend to gain, lose, or share electrons with other atoms to reach a stable state. When any of these events occurs, chemical bonds are formed. Examples of chemically inert and reactive elements are shown in Figure 2.4.

The key to chemical reactivity is referred to as the *rule of 8s;* that is, atoms interact in such a way that they will have 8 electrons in their valence shell. The first electron shell represents an exception to this rule, because it is "full" when it has 2 electrons. As you might guess, atoms must approach each other very closely for their electrons to interact—in fact, their outermost electron shells must overlap.

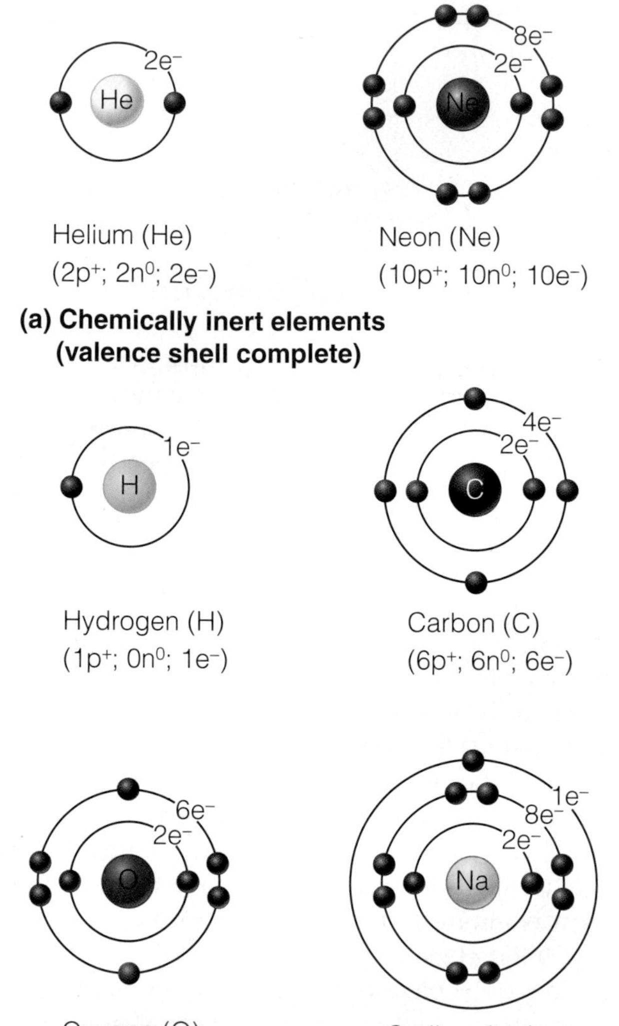

FIGURE 2.4 Chemically inert and reactive elements. (a) Helium and neon are chemically inert because in each case the outermost valence shell (energy level) is fully occupied by electrons. **(b)** Elements in which the valence shell is incomplete are chemically reactive. Such atoms tend to interact with other atoms to gain, lose, or share electrons to fill their valence shells. (To simplify the diagrams, each atomic nucleus is shown as a circle with the atom's symbol in it; individual protons and neutrons are not shown.)

Types of Chemical Bonds

Ionic Bonds **Ionic** (i-on'ik) **bonds** form when electrons are completely transferred from one atom to another. Atoms are electrically neutral, but when they gain or lose electrons during bonding, their positive and negative charges are no longer balanced, and charged particles, called **ions,** result. When an atom gains an electron, it acquires a net negative charge because it now has more electrons than protons. Negatively charged ions are more specifically called *anions.* When an atom loses an electron, it becomes a positively charged ion, a *cation,* because it now possesses more protons than electrons. (It may help you to remember that a cation is a positively charged ion by thinking of its "t" as a plus [+] sign.) Both anions and cations result when an ionic bond is formed. Since opposite charges attract, the newly created ions tend to stay close together.

The formation of sodium chloride (NaCl), common table salt, provides a good example of ionic bonding. As illustrated in Figure 2.5, sodium's valence shell contains only 1 electron and so is incomplete. However, if this single electron is "lost" to another atom, shell 2, which contains 8 electrons, becomes the valence shell; thus sodium becomes a cation (Na^+) and achieves stability. Chlorine needs only 1 electron to fill its valence shell, and it is much easier to gain 1 electron (forming Cl^-) than it is to try to "give away" 7. Thus, the ideal situation is for sodium to donate its valence-shell electron to chlorine, and this is exactly what happens in the interaction between these two atoms. Sodium chloride and most other compounds formed by ionic bonding fall into the general category of chemicals called **salts.**

Covalent Bonds Electrons do not have to be completely lost or gained for atoms to become stable. Instead, they can be shared in such a way that each atom is able to fill its valence shell at least part of the time.

Molecules in which atoms share electrons are called *covalent molecules,* and their bonds are **covalent bonds** (*co* = with; *valent* = having power). For example, hydrogen, with its single electron, can become stable if it fills its valence shell (energy level 1) by sharing a pair of electrons—its own and one from another atom. As shown in Figure 2.6a on p. 36, a hydrogen atom can share an electron pair with another hydrogen atom to form a molecule of hydrogen gas. The shared electron pair orbits the whole molecule and satisfies the stability needs of both hydrogen atoms. Likewise, 2 oxygen atoms, each with 6 valence-shell electrons, can share 2 pairs of electrons (form double bonds) with

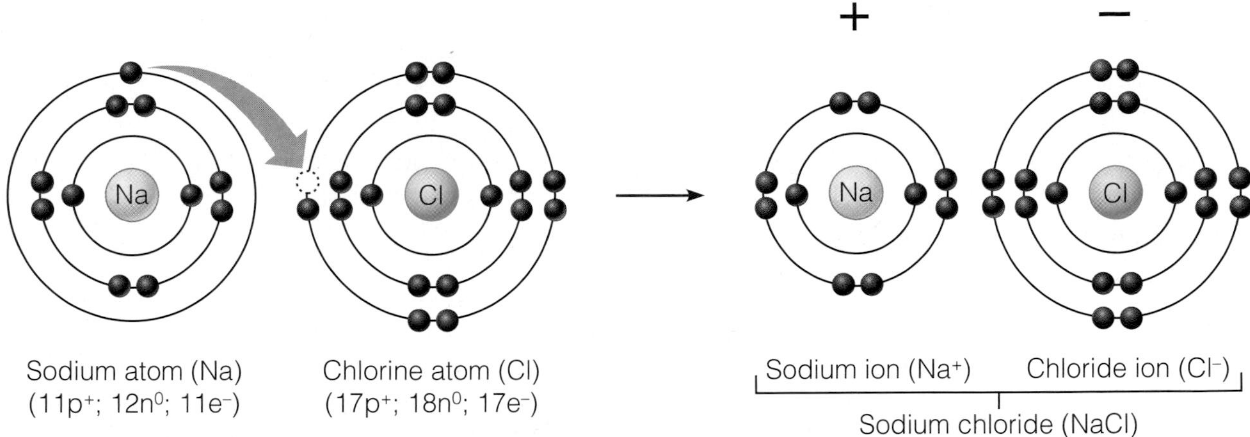

FIGURE 2.5 Formation of an ionic bond. Both sodium and chlorine atoms are chemically reactive because their valence shells are incompletely filled. Sodium gains stability by losing one electron, whereas chlorine becomes stable by gaining one electron. After electron transfer, sodium becomes a sodium ion (Na^+), and chlorine becomes a chloride ion (Cl^-). The oppositely charged ions attract each other.

each other (Figure 2.6b) to form a molecule of oxygen gas (O_2).

A hydrogen atom may also share its electron with an atom of a different element. Carbon has 4 valence-shell electrons but needs 8 to achieve stability. As shown in Figure 2.6c, when methane (CH_4) is formed, carbon shares 4 electron pairs with 4 hydrogen atoms (1 pair with each hydrogen atom). Because the shared electrons orbit and "belong to" the whole molecule, each atom has a full valence shell enough of the time to satisfy its stability needs.

In the covalent molecules described thus far, electrons have been shared *equally* between the atoms of the molecule. Such molecules are called *nonpolar covalently bonded molecules*. However, electrons are not shared equally in all cases. When covalent bonds are made, the molecule formed always has a definite three-dimensional shape. A molecule's shape plays a major role in determining just what other molecules (or atoms) it can interact with; the shape may also result in unequal electron-pair sharing. The following two examples illustrate this principle (Figure 2.7).

Carbon dioxide is formed when a carbon atom shares its 4 valence-shell electrons with 2 oxygen atoms. Oxygen is a very electron-hungry atom and attracts the shared electrons much more strongly than does carbon. However, because the carbon dioxide molecule is linear (O=C=O), the electron-pulling power of one oxygen atom is offset by that of the other, like a tug-of-war at a standoff. As a result, the electron pairs are shared equally and orbit the entire molecule, and carbon dioxide is a nonpolar molecule.

A water molecule is formed when 2 hydrogen atoms bind covalently to a single oxygen atom. Each hydrogen atom shares an electron pair with the oxygen atom, and again the oxygen has the stronger electron-attracting ability. But in this case, the molecule formed is V-shaped (H\O/H). The two hydrogen atoms are located at one end of the molecule, and the oxygen atom is at the other. Consequently, the electron pairs are not shared equally and spend more time in the vicinity of the oxygen atom, causing that end of the molecule to become slightly more negative (indicated by δ^-) and the hydrogen end to become slightly more positive (indicated by δ^+). In other words, a *polar molecule,* a molecule with two charged *poles,* is formed.

Polar molecules orient themselves toward other polar molecules or charged particles (ions, proteins, and others), and they play an important role in chemical reactions that occur in body cells. Because body tissues are 60 to 80 percent water, the fact that water is a polar molecule is particularly significant, as will be described shortly.

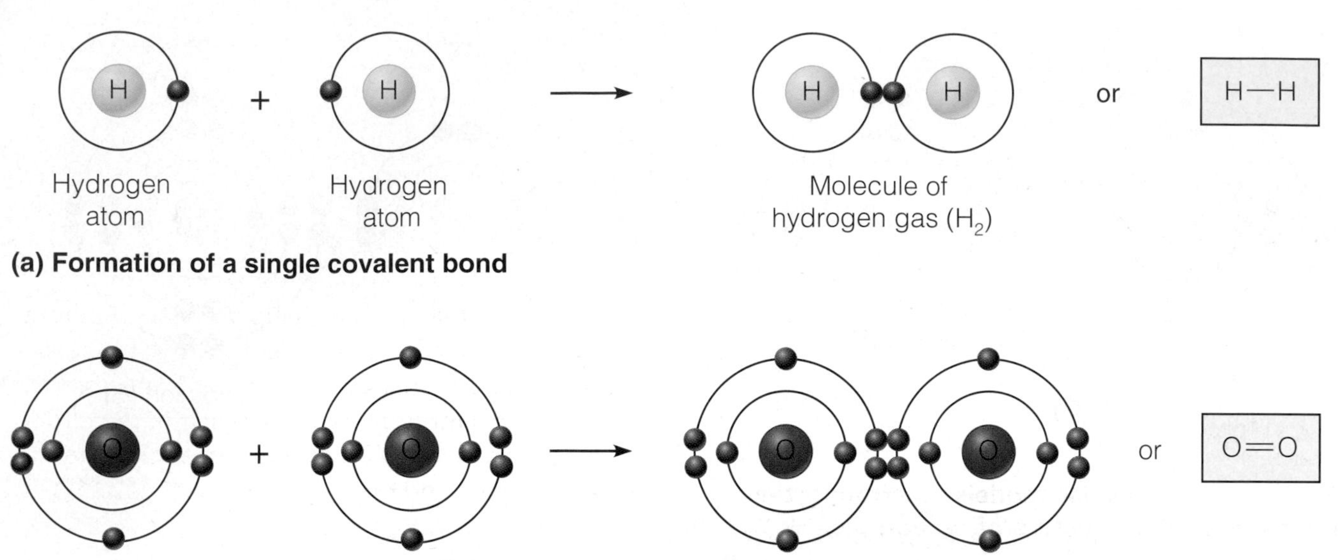

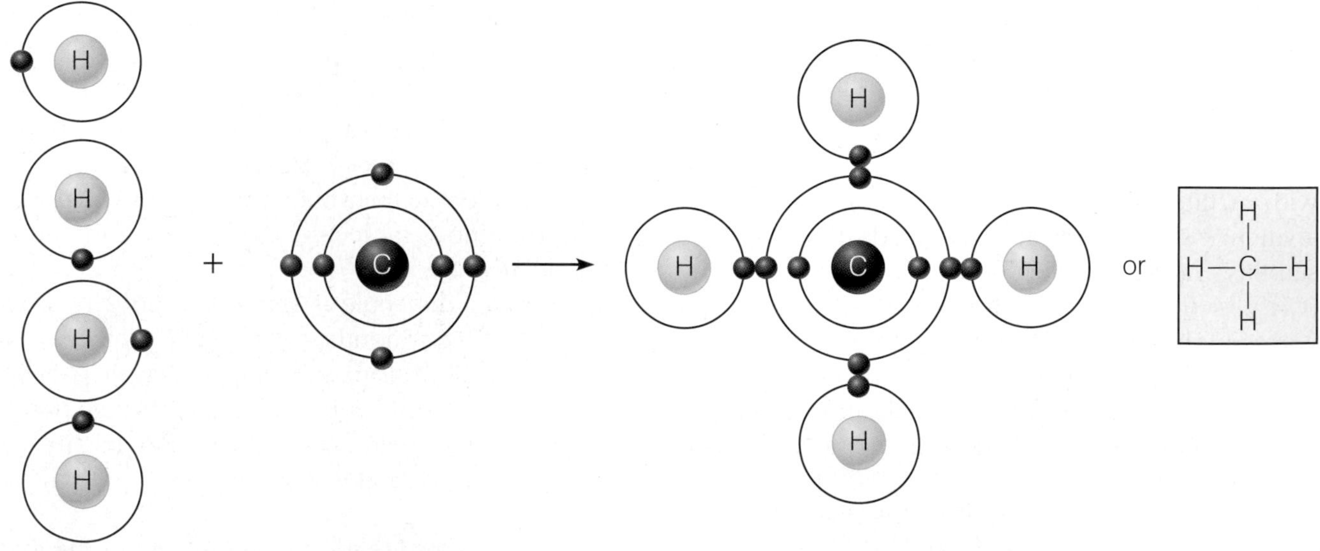

FIGURE 2.6 Formation of covalent bonds. (a) Formation of a single covalent bond between two hydrogen atoms to form a molecule of hydrogen gas. **(b)** Formation of a molecule of oxygen gas. Each oxygen atom shares two electron pairs with its partner; thus a double covalent bond is formed. **(c)** Formation of a molecule of methane. A carbon atom shares four electron pairs with four hydrogen atoms. In the diagrams of molecules shown in the colored boxes at the far right, each pair of shared electrons is indicated by a single line between the sharing atoms.

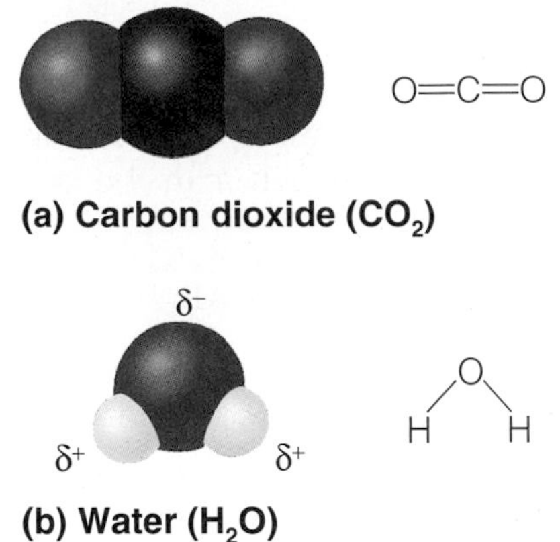

FIGURE 2.7 Molecular models illustrating the three-dimensional structure of carbon dioxide and water molecules.

Hydrogen Bonds **Hydrogen bonds** are extremely weak bonds formed when a hydrogen atom bound to one electron-hungry nitrogen or oxygen atom is attracted by another electron-hungry atom, and the hydrogen atom forms a "bridge" between them. Hydrogen bonding is common between water molecules (Figure 2.8a) and is reflected in water's surface tension. The surface tension of water causes it to "ball up," or form spheres, when it sits on a surface and allows some insects, such as water striders (Figure 2.8b), to walk on water as long as they tread lightly.

Hydrogen bonds are also important *intramolecular bonds;* that is, they help to bind different parts of the *same* molecule together into a special three-dimensional shape. These rather fragile bonds are very important in helping to maintain the structure of protein molecules, which are essential functional molecules and body-building materials.

Patterns of Chemical Reactions

Chemical reactions involve the making or breaking of bonds between atoms. The total number of atoms remains the same, but the atoms appear in new combinations. Most chemical reactions have one of the three recognizable patterns described next.

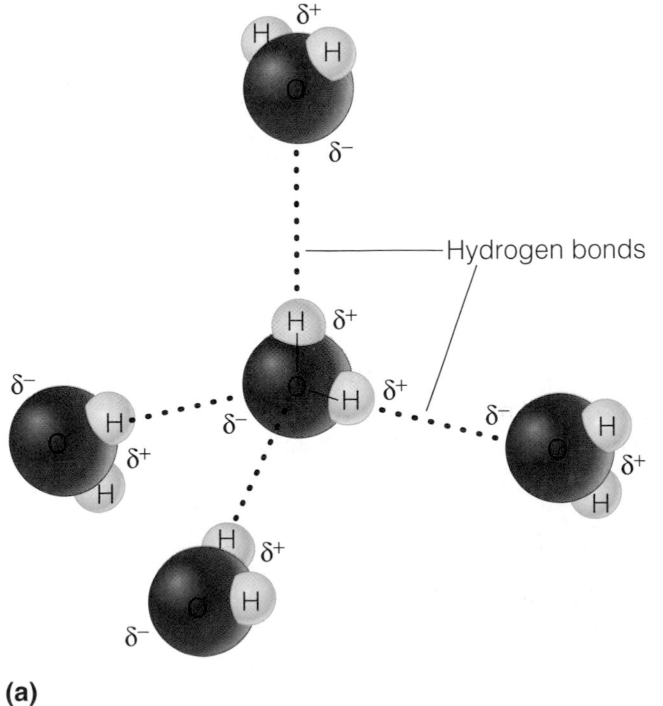

FIGURE 2.8 Hydrogen bonding between polar water molecules. **(a)** The slightly positive ends (indicated by δ^+) of the water molecules become aligned with the slightly negative ends (indicated by δ^-) of other water molecules. **(b)** Water's high surface tension, a result of the combined strength of its hydrogen bonds, allows a water strider to walk on a pond without breaking the surface.

Q *In all reactions shown, chemical bonds are being altered. What atomic subparticle is involved in these alterations?*

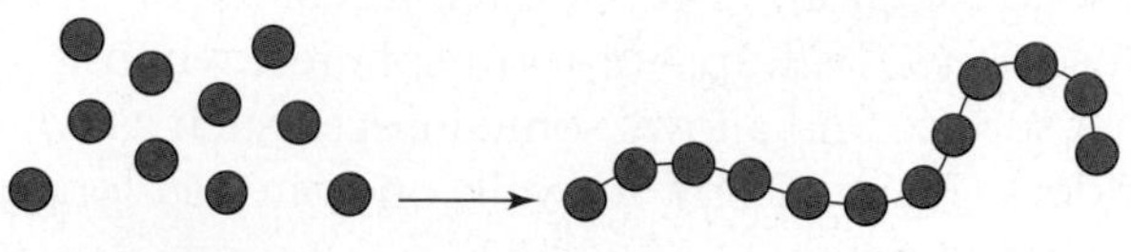

Amino acids Protein molecule

(a) Example of a synthesis reaction: amino acids are joined to form a protein molecule

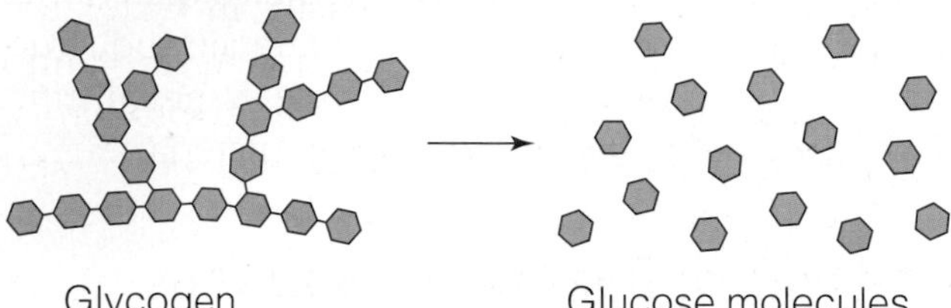

Glycogen Glucose molecules

(b) Example of a decomposition reaction: breakdown of glycogen to release glucose units

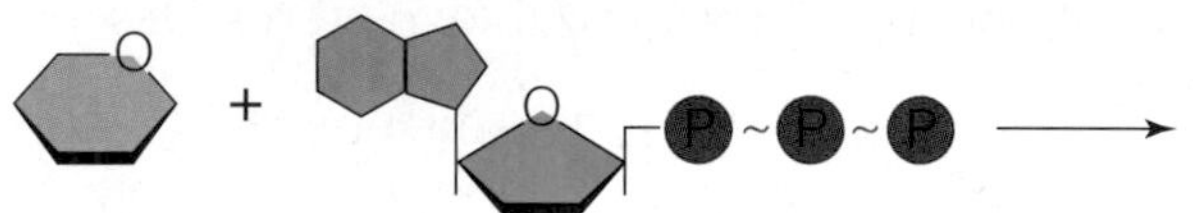

Glucose Adenosine triphosphate (ATP)

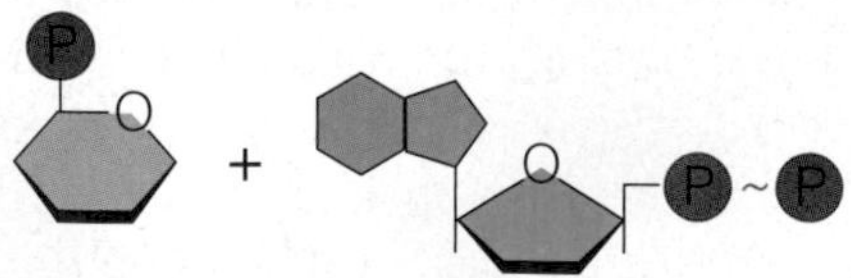

Glucose phosphate Adenosine diphosphate (ADP)

(c) Example of an exchange reaction: ATP transfers its terminal phosphate group to glucose to form glucose phosphate

FIGURE 2.9 Patterns of chemical reactions. **(a)** In synthesis reactions, smaller particles (atoms, ions, or molecules) are bonded together to form larger, more complex molecules. **(b)** Decomposition reactions involve the breaking of bonds. **(c)** In exchange reactions, bonds are both made and broken.

A *Electrons.*

Synthesis Reactions

Synthesis reactions occur when two or more atoms or molecules combine to form a larger, more complex molecule, which can be simply represented as

$$A + B \rightarrow AB$$

Synthesis reactions always involve bond formation. Because energy must be absorbed to make bonds, synthesis reactions are energy-absorbing reactions.

Synthesis reactions underlie all anabolic (constructive) activities that occur in body cells. They are particularly important for growth and for repair of worn-out or damaged tissues. As shown in Figure 2.9a, the formation of a protein molecule by the joining of amino acids into long chains is a synthesis reaction.

Decomposition Reactions

Decomposition reactions occur when a molecule is broken down into smaller molecules, atoms, or ions and can be indicated by

$$AB \rightarrow A + B$$

Essentially, decomposition reactions are synthesis reactions in reverse. Bonds are always broken, and the products of these reactions are smaller and simpler than the original molecules. As bonds are broken, chemical energy is released.

Decomposition reactions underlie all catabolic (destructive) processes that occur in body cells; that is, they are molecule-degrading reactions. Examples of decomposition reactions that occur in the body include the digestion of foods into their building blocks and the breakdown of glycogen (a large carbohydrate molecule stored in the liver) to release glucose (Figure 2.9b) when blood sugar levels start to decline.

Exchange Reactions

Exchange reactions involve both synthesis and decomposition reactions: bonds are both made and broken. During exchange reactions, a switch is made between molecule parts (changing partners, so to speak), and different molecules are made. Thus, an exchange reaction can be generally indicated as

$$AB + C \rightarrow AC + B \text{ and } AB + CD \rightarrow AD + CB$$

An exchange reaction occurs, for example, when ATP reacts with glucose and transfers its end phosphate group to glucose, forming glucose-phosphate (Figure 2.9c). At the same time, the ATP becomes ADP. This important reaction, which occurs whenever glucose enters a body cell, effectively traps the glucose fuel molecule inside the cell.

Biochemistry: The Chemical Composition of Living Matter

All chemicals found in the body fall into one of two major classes of molecules; they are either inorganic or organic compounds. The class of the compound is determined solely by the presence or absence of carbon. With a few so far unexplainable exceptions (such as carbon dioxide gas), **inorganic compounds** lack carbon and tend to be small, simple molecules. Examples of inorganic compounds found in the body are *water, salts,* and many (but not all) *acids* and *bases.* **Organic compounds** are carbon-containing compounds. The important organic compounds in the body are *carbohydrates, lipids, proteins,* and *nucleic acids.* All organic compounds are fairly (or very) large covalently bonded molecules.

Inorganic and organic compounds are equally essential for life. Trying to put a price tag on which is more valuable can be compared to trying to decide whether the ignition system or the engine is more essential to the operation of a car.

Inorganic Compounds

Water

Water is the most abundant inorganic compound in the body. It accounts for about two-thirds of body weight. Among the properties that make water so vital are the following:

1. **High heat capacity.** Water has a *high heat capacity;* that is, it absorbs and releases large amounts of heat before its temperature changes appreciably. Thus, it prevents the sudden changes in body temperature that might otherwise result from intense sun exposure, chilling winter winds, or internal events (such as vigorous muscle activity) that liberate large amounts of heat.
2. **Polarity/solvent properties.** Because of its polarity, water is an excellent solvent; indeed, it is often called the "universal solvent." A *solvent* is a liquid or gas in which smaller amounts of other substances, called *solutes* (which may be gases, liquids, or solids), can be dissolved or suspended. The resulting mixture is called a *solution* when the solute particles are exceedingly minute, and a *suspension* when the solute particles are fairly large. Translucent mixtures with solute particles of intermediate size are called *colloids.*

   Small reactive chemicals such as salts, acids, and bases dissolve easily in water and become evenly distributed. Molecules cannot react chemically unless they are in solution, so virtually all chemical reactions that occur in the body depend upon water's solvent properties.

   Because nutrients, respiratory gases (oxygen and carbon dioxide), and wastes can dissolve in water, water can act as a transport and exchange medium in the body. For example, all these substances are carried from one part of the body to another in blood plasma and are exchanged between the blood and tissue cells by passing through interstitial fluid.

   Specialized molecules that lubricate the body also use water as their solvent. Such lubricants include the mucus that eases feces along the large intestine and saliva, which moistens food and prepares it for digestion. Serous fluids reduce friction between internal organs, and synovial fluids "oil" the ends of bones as they move within joint cavities.
3. **Chemical reactivity.** Water is an important *reactant* in some types of chemical reactions. For example, to digest foods or break down biological molecules, water molecules are added to the bonds of the larger molecules. Such reactions are called *hydrolysis reactions,* a term that specifically recognizes this role of water.
4. **Cushioning.** Water also serves a protective function. In the form of cerebrospinal fluid, water forms a cushion around the brain that helps to protect it from physical trauma. Amniotic fluid, which surrounds a developing fetus within the mother's body, plays a similar role in protecting the fetus.

Salts

The **salts** of many metal elements are commonly found in the body, but the most plentiful salts are those containing calcium and phosphorus, found

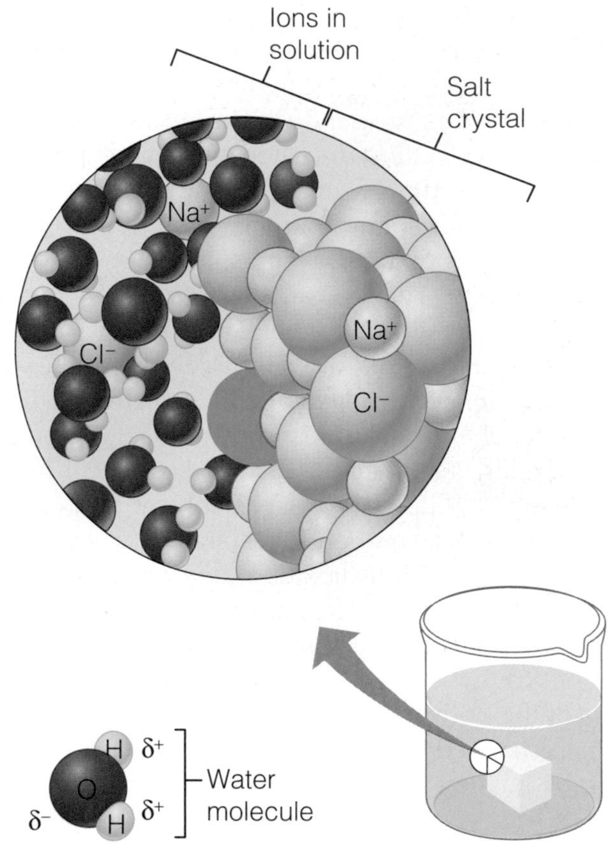

FIGURE 2.10 Dissociation of a salt in water. The slightly negative ends of the water molecules (δ^-) are attracted to Na^+, whereas the slightly positive ends of water molecules (δ^+) orient toward Cl^-, causing the ions of the salt crystal to be pulled apart.

chiefly in bones and teeth. When dissolved in body fluids, salts, which are ionic compounds, easily separate into their ions. This process, called *dissociation,* occurs rather easily because the ions have already been formed. All that remains is to pull the ions apart. This is accomplished by the polar water molecules, which orient themselves with their slightly negative ends toward the cations and their slightly positive ends toward the anions and thereby overcome the attraction between them (Figure 2.10).

Salts, both in their ionic forms and in combination with other elements, are vital to body functioning. For example, sodium and potassium ions are essential for nerve impulses, and iron forms part of the hemoglobin molecule that transports oxygen within red blood cells.

Because ions are charged particles, all salts are **electrolytes**—substances that conduct an electrical current in solution. When ionic (or electrolyte) balance is severely disturbed, virtually nothing in the body works. The functions of the elements found in body salts are summarized in Table 2.1 (on p. 28).

Acids and Bases

Like salts, acids and bases are electrolytes. That is, they ionize and then dissociate in water and can then conduct an electrical current.

Characteristics of Acids **Acids** have a sour taste and can dissolve many metals or "burn" a hole in your rug. But, the most useful definition of an acid is that it is a substance that can release *hydrogen ions (H^+)* in detectable amounts. Because a hydrogen ion is essentially a hydrogen nucleus (a "naked proton"), acids are also defined as **proton donors.**

When acids are dissolved in water, they release hydrogen ions and some anions. The anions are unimportant; it is the release of the protons that determines an acid's effects on the environment. The ionization of hydrochloric acid (an acid produced by stomach cells that aids digestion) is shown in the following equation:

$$\underset{\text{(hydrochloric acid)}}{HCl} \rightarrow \underset{\text{(proton)}}{H^+} + \underset{\text{(anion)}}{Cl^-}$$

Other acids found or produced in the body include acetic acid (the acidic component of vinegar) and carbonic acid.

Acids, like hydrochloric acid, that ionize completely and liberate all their protons are called *strong acids.* Acids that ionize incompletely, as do acetic and carbonic acid, are called *weak acids.* For example, when carbonic acid dissolves in water, only some of its molecules ionize to liberate H^+.

$$\underset{\text{(carbonic acid)}}{H_2CO_3} \rightarrow \underset{\text{(proton)}}{H^+} + \underset{\text{(anion)}}{HCO_3^-} + \underset{\text{(carbonic acid)}}{H_2CO_3}$$

Characteristics of Bases **Bases** have a bitter taste, feel slippery, and are **proton acceptors.** The hydroxides are common inorganic bases. Like acids, the hydroxides ionize and dissociate in water; but in this case, the *hydroxyl ion (OH^-)* and some cations are released. The ionization of sodium hydroxide (NaOH), commonly known as lye, is shown as

NaOH	→	Na^+	+	OH^-
(sodium hydroxide)		(cation)		(hydroxyl ion)

The hydroxyl ion is an avid proton (H^+) seeker, and any base containing this ion is considered a strong base. By contrast, *bicarbonate ion* (HCO_3^-), an important base in blood, is a fairly weak base.

When acids and bases are mixed, they react with each other (in an exchange reaction) to form water and a salt:

HCl	+	NaOH	→	H_2O	+	NaCl
(acid)		(base)		(water)		(salt)

This type of exchange reaction, in which an acid and a base interact, is more specifically called a **neutralization reaction.**

pH: Acid-Base Concentrations The relative concentration of hydrogen (and hydroxyl) ions in various body fluids is measured in concentration units called **pH** (pe-āch) **units.** The idea for a pH scale was devised in 1909 by a Danish biochemist (and part-time beer brewer) named Sørensen and is based on the number of protons in solution expressed in terms of moles per liter. (The *mole* is a concentration unit; its precise definition need not concern us here.) The pH scale runs from 0 to 14 (Figure 2.11), and each successive change of 1 pH unit represents a tenfold change in hydrogen-ion concentration.

At a pH of 7, the scale midpoint, the number of hydrogen ions exactly equals the number of hydroxyl ions, and the solution is neutral; that is, neither acidic nor basic. Solutions with a pH lower than 7 are acidic: the hydrogen ions outnumber the hydroxyl ions. A solution with a pH of 6 has 10 times as many hydrogen ions as a solution with a pH of 7, and a pH of 3 indicates a 10,000-fold ($10 \times 10 \times 10 \times 10$) increase in hydrogen-ion concentration. Solutions with a pH number higher than 7 are alkaline, or basic, and solutions with a pH of 8 and 12 (respectively) have 1/10 and 1/100,000 the number of hydrogen ions present in a solution with a pH of 7.

Living cells are extraordinarily sensitive to even slight changes in pH; and acid-base balance is carefully regulated by the kidneys, lungs, and a number of chemicals called **buffers,** which are

Q *Which ion is responsible for increased acidity?*

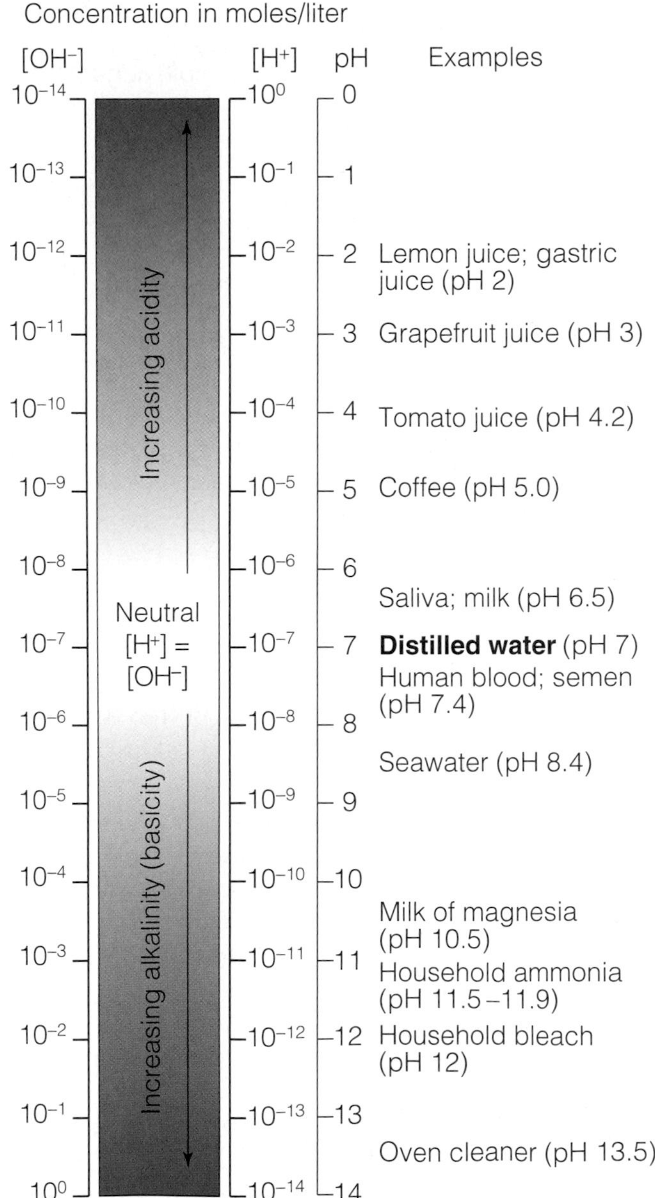

FIGURE 2.11 The pH scale and pH values of representative substances. The pH scale is based on the number of hydrogen ions in solution. The actual concentration (expressed in moles per liter) of hydrogen ions [H^+] and the corresponding hydroxyl ion concentration [OH^-] are indicated for each pH value noted. At a pH of 7, the concentrations of hydrogen and hydroxyl ions are equal, and the solution is neutral. A solution with a pH below 7 is acidic; above 7, basic or alkaline.

A *The hydrogen ion.*

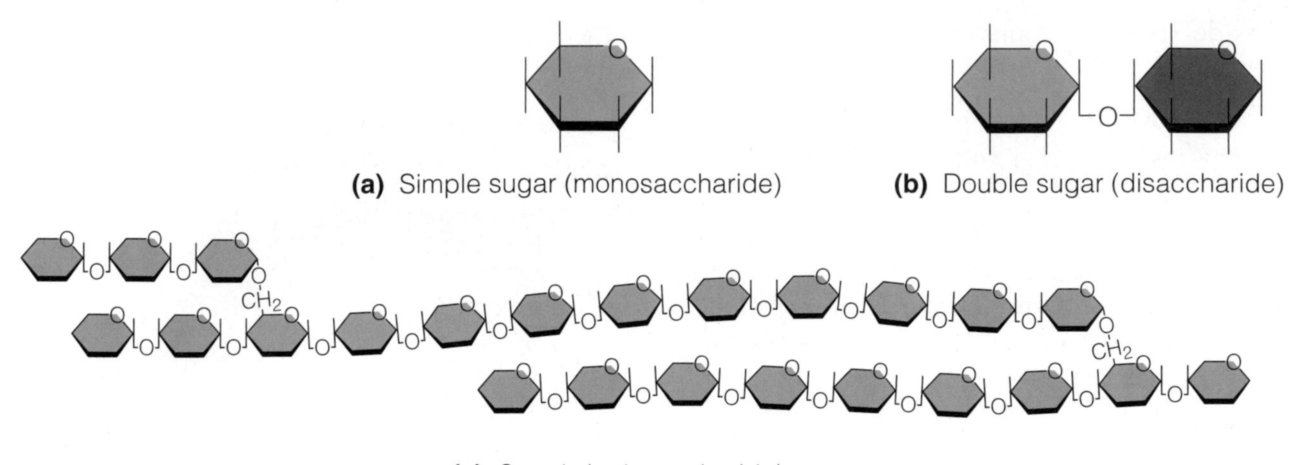

(a) Simple sugar (monosaccharide)

(b) Double sugar (disaccharide)

(c) Starch (polysaccharide)

FIGURE 2.12 Carbohydrates. (a) The generalized structure of a monosaccharide. **(b)** and **(c)** The basic structures of a disaccharide and a polysaccharide, respectively.

present in body fluids. As described in some detail in Chapter 15, weak acids and weak bases are important components of the body's buffer systems, which act to maintain pH stability by taking up excess hydrogen or hydroxyl ions.

Because blood comes into close contact with nearly every body cell, regulation of blood pH is especially critical. Normally, blood pH varies in a narrow range, from 7.35 to 7.45. When blood pH changes more than a few tenths of a pH unit from these limits, death becomes a distinct possibility. Although there are hundreds of examples that could be given to illustrate this point, we will provide just one very important one: when blood pH begins to dip into the acid range, the amount of life-sustaining oxygen that the hemoglobin in blood can carry to body cells begins to fall rapidly to dangerously low levels. The approximate pH values of several body fluids and of a number of commonly ingested substances appear in Figure 2.11.

Organic Compounds

Carbohydrates

Carbohydrates, which include sugars and starches, contain carbon, hydrogen, and oxygen. With slight variations, the hydrogen and oxygen atoms appear in the same ratio as in water; that is, 2 hydrogen atoms to 1 oxygen atom. This is reflected in the word *carbohydrate,* which means "hydrated carbon," and in the molecular formulas of sugars. For example, glucose is $C_6H_{12}O_6$, and ribose is $C_5H_{10}O_5$.

Carbohydrates are classified according to size as monosaccharides, disaccharides, or polysaccharides. Because monosaccharides are joined to form the molecules of the other two groups, they are the structural units, or building blocks, of carbohydrates.

Monosaccharides **Monosaccharide** means one (*mono*) sugar (*saccharide*), and thus monosaccharides are also referred to as *simple sugars.* They are single-chain or single-ring structures, containing from 3 to 7 carbon atoms (Figure 2.12a).

The most important monosaccharides in the body are glucose, fructose, galactose, ribose, and deoxyribose. **Glucose,** also called *blood sugar,* is the universal cellular fuel. *Fructose* and *galactose* are converted to glucose for use by body cells. *Ribose* and *deoxyribose* form part of the structure of nucleic acids, another group of organic molecules.

Disaccharides **Disaccharides,** or *double sugars* (Figure 2.12b), are formed when two simple sugars are joined by a synthesis reaction known as **dehydration synthesis.** In this reaction, a water molecule is lost as the bond forms (Figure 2.13).

Some of the important disaccharides in the diet are *sucrose* (glucose-fructose), which is cane sugar; *lactose* (glucose-galactose), found in milk;

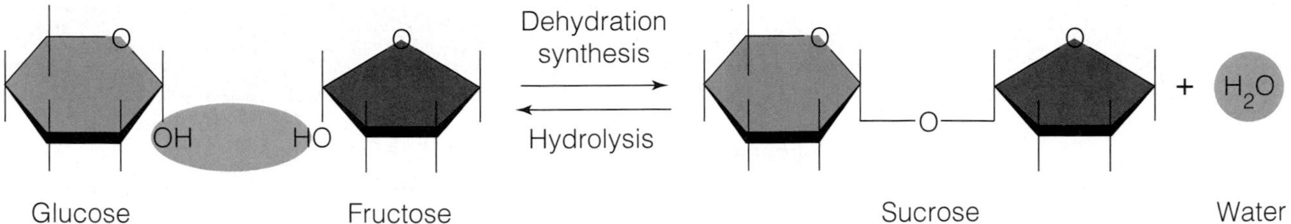

FIGURE 2.13 Dehydration synthesis and hydrolysis of a molecule of sucrose. In the reaction going to the right (the dehydration synthesis reaction), glucose and fructose are joined through a process that involves the removal of a water molecule at the bond site. The resulting disaccharide is sucrose. Sucrose is broken down to its simple sugar units when the reaction is reversed (goes to the left). In this hydrolysis reaction, a water molecule must be added to the bond to release the monosaccharides.

and *maltose* (glucose-glucose), or malt sugar. Because the double sugars are too large to pass through cell membranes, they must be broken down (digested) to their monosaccharide units to be absorbed from the digestive tract into the blood. This is accomplished by **hydrolysis.** As a water molecule is added to each bond, the bond is broken, and the simple sugar units are released (Figure 2.13).

Polysaccharides **Polysaccharides** (literally, "many sugars") are long, branching chains of linked simple sugars (Figure 2.12c). Because they are large, insoluble molecules, they are ideal storage products. Another consequence of their large size is that they lack the sweetness of the simple and double sugars.

Only two polysaccharides, starch and glycogen, are of major importance to the body. *Starch* is the storage polysaccharide formed by plants. We ingest it in the form of "starchy" foods, such as grain products and root vegetables (potatoes and carrots, for example). *Glycogen* is a slightly smaller, but similar, polysaccharide found in animal tissues (largely in the muscles and the liver). Like starch, it is formed of linked glucose units.

Carbohydrates provide a ready, easily used source of food energy for cells, and glucose is at the top of the "cellular menu." When glucose is oxidized (combined with oxygen) in a complex set of chemical reactions, it is broken down into carbon dioxide and water. Some of the energy released as the glucose bonds are broken is trapped in the bonds of high-energy ATP molecules, the energy "currency" of all body cells. If not immediately needed for ATP synthesis, dietary carbohydrates are converted to glycogen or fat and stored. Those of us who have gained weight from eating too many carbohydrate-rich snacks have a firsthand awareness of this conversion process!

Small amounts of carbohydrates are used for structural purposes and represent 1 to 2 percent of cell mass. Some sugars are found in our genes, and others are attached to outer surfaces of cell membranes, where they act as road signs to guide cellular interactions.

Lipids

Lipids are a large and diverse group of organic compounds (Table 2.4). They enter the body in the form of fat-marbled meats, egg yolks, milk products, and oils. The most abundant lipids in the body are triglycerides, phospholipids, and steroids. Like carbohydrates, all lipids contain carbon, hydrogen, and oxygen atoms, but in lipids, carbon and hydrogen atoms far outnumber oxygen atoms, as illustrated by the formula for a typical fat named tristearin: $C_{57}H_{110}O_6$. Most lipids are insoluble in water but readily dissolve in other lipids and in organic solvents such as alcohol and acetone.

Triglycerides The **triglycerides** (tri-glis′er-īdz), or **neutral fats,** are composed of two types of building blocks, **fatty acids** and **glycerol.** Their synthesis

TABLE 2.4 Representative Lipids Found in the Body

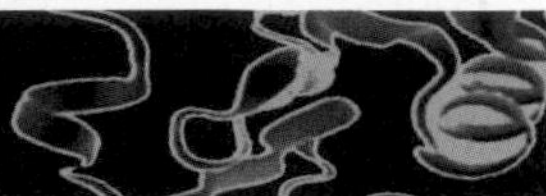

Lipid type	Location/function
Neutral fats (Triglycerides)	Found in fat deposits (subcutaneous tissue and around organs); protect and insulate the body organs; the major source of stored energy in the body.
Phospholipids (Cephalin and others)	Found in cell membranes; participate in the transport of lipids in plasma; abundant in the brain and the nervous tissue in general, where they help to form insulating white matter.
Steroids	
Cholesterol	The basis of all body steroids.
Bile salts	A breakdown product of cholesterol; released by the liver into the digestive tract, where they aid in fat digestion and absorption.
Vitamin D	Produced in the skin, on exposure to UV (ultraviolet) radiation, from a modified cholesterol molecule; necessary for normal bone growth and function.
Sex hormones	Estrogen and progesterone (female hormones) and testosterone (male sex hormone) produced from cholesterol; necessary for normal reproductive function; deficits result in sterility.
Corticosteroids (adrenal cortical hormones)	Cortisol, a glucocorticoid, is a long-term antistress hormone that is necessary for life; aldosterone helps regulate salt and water balance in body fluids by targeting the kidneys.
Other lipoid substances	
Fat-soluble vitamins:	
A	Found in orange-pigmented vegetables (carrots) and fruits (tomatoes); part of the photoreceptor pigment involved in vision.
E	Taken in via plant products such as wheat germ and green leafy vegetables; may promote wound healing and contribute to fertility, but not proven in humans; an antioxidant; may help to neutralize free radicals (highly reactive particles believed to be involved in triggering some types of cancers).
K	Made available largely by the action of intestinal bacteria; also prevalent in a wide variety of foods; necessary for proper clotting of blood.
Prostaglandins	Derivatives of fatty acids found in cell membranes; various functions depending on the specific class, including stimulation of uterine contractions (thus inducing labor and abortions), regulation of blood pressure, and control of motility of the gastrointestinal tract; involved in inflammation.
Lipoproteins	Lipoid and protein-based substances that transport fatty acids and cholesterol in the bloodstream; major varieties are high-density lipoproteins (HDLs) and low-density lipoproteins (LDLs).

involves the attachment of three fatty acids to a single glycerol molecule. The result is an E-shaped molecule that resembles the tines of a fork (Figure 2.14a). Although the glycerol backbone is the same in all neutral fats, the fatty acid chains vary; this results in different kinds of neutral fats. Triglycerides may be solid (typical of animal fats) or liquid (plant oils). In general, animal fats tend to be *saturated,* whereas oils are *unsaturated.* In saturated fats, all carbons have single bonds. The

Q *Triglycerides and phospholipids are similar. What is the major structural difference between them?*

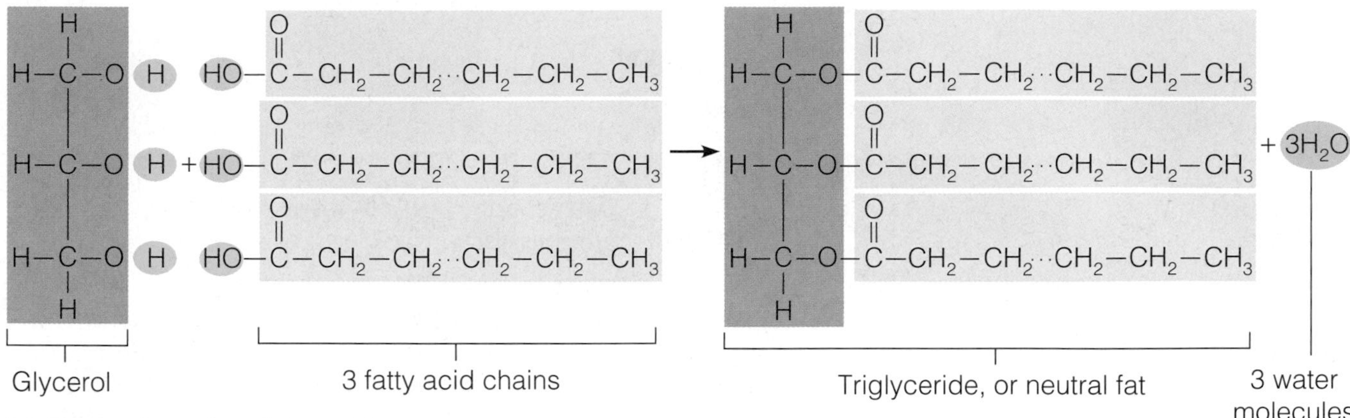

(a) Formation of a triglyceride

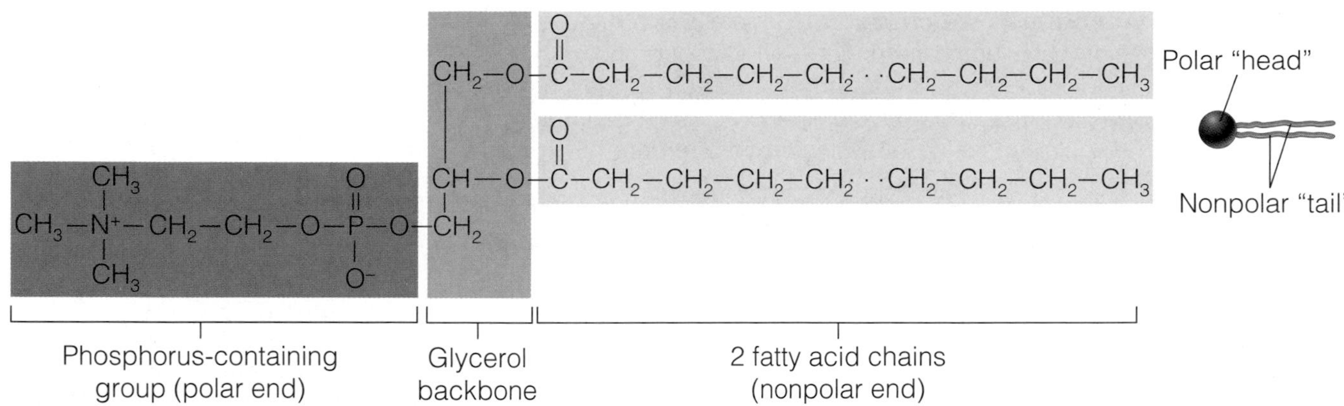

(b) Phospholipid molecule (phosphatidylcholine)

(c) Cholesterol

FIGURE 2.14 Lipids. (a) Triglycerides, or neutral fats, are synthesized by dehydration synthesis. In this process, three fatty acid chains are attached to a single glycerol molecule, and a water molecule is lost at each bond site. **(b)** Structure of a typical phospholipid molecule. Two fatty acid chains and a phosphorus-containing group are attached to the glycerol backbone. **(c)** The generalized structure of cholesterol. Cholesterol is the basis for all steroids made in the body.

They both have a glycerol backbone and fatty acid chains. However, triglycerides have three attached fatty acid chains and phospholipids have only two; the third is replaced by a phosphorus-containing group.

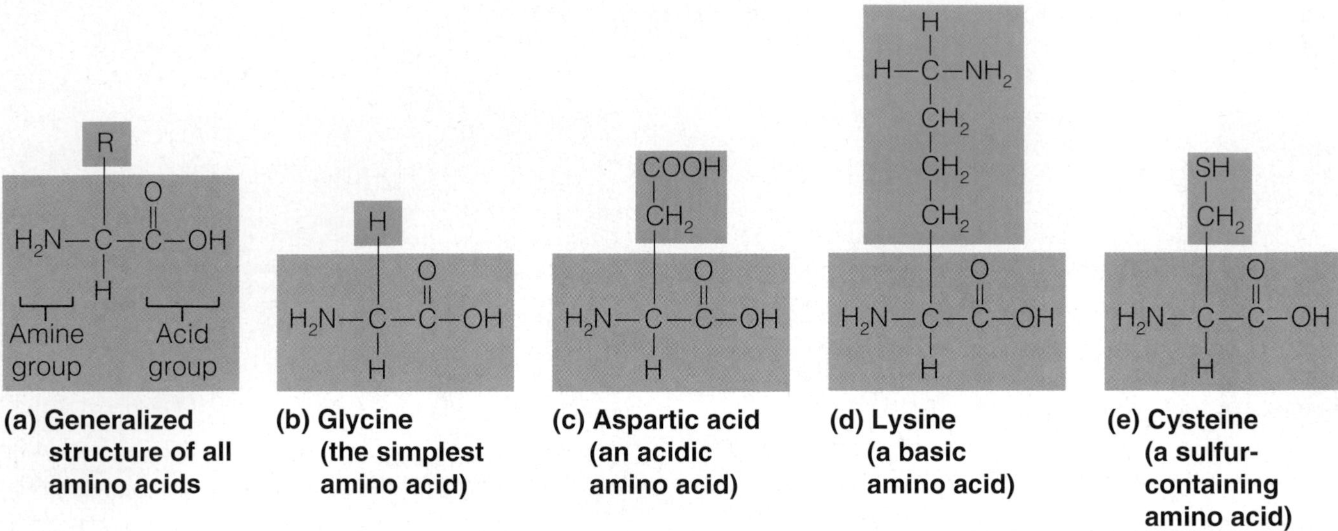

(a) Generalized structure of all amino acids **(b) Glycine (the simplest amino acid)** **(c) Aspartic acid (an acidic amino acid)** **(d) Lysine (a basic amino acid)** **(e) Cysteine (a sulfur-containing amino acid)**

FIGURE 2.15 Amino acid structures. (a) Generalized structure of amino acids. All amino acids have both an amine ($—NH_2$) group and an acid (—COOH) group; they differ only in the atomic makeup of their R-groups (green). **(b)–(e)** Specific structures of four amino acids. The simplest (glycine) has an R-group consisting of a single hydrogen atom. An acid in the R-group makes the amino acid (aspartic acid in this example) more acidic. An amine group in the R-group makes it more basic (as in lysine). The presence of sulfur (—SH) in the R-group of cysteine hints that this is an amino acid likely to take part in intramolecular bonding.

carbons of unsaturated fats have some double (or triple) bonds and thus have the ability to bind with more hydrogen atoms or atoms of a different type.

Triglycerides represent the body's most abundant and concentrated source of usable energy. When they are oxidized, they yield large amounts of energy. They are stored chiefly in fat deposits beneath the skin and around body organs, where they help insulate the body and protect deeper body tissues from heat loss and bumps.

Phospholipids **Phospholipids** (fos′fo-lip″idz) are very similar to the triglycerides. They differ in that a phosphorus-containing group is always part of the molecule and takes the place of one of the fatty acid chains. Thus, phospholipids have two instead of three attached fatty acids (Figure 2.14b).

Because the phosphorus-containing portion (the "head") bears an electrical charge, it gives phospholipids special chemical properties and polarity. For example, the charged region attracts and interacts with water and ions, but the fatty acid chains (the "tail") do not. The presence of phospholipids in cellular boundaries (membranes) allows cells to be selective about what may enter or leave.

Steroids **Steroids** are basically flat molecules formed of four interlocking rings (Figure 2.14c); thus their structure differs quite a bit from that of fats. However, like fats, steroids are made largely of hydrogen and carbon atoms and are fat-soluble.

The single most important steroid molecule is **cholesterol,** which enters the body in animal products such as meat, eggs, and cheese. A certain amount is also made by the liver, regardless of dietary intake. Cholesterol is found in all cell membranes, and it is particularly abundant in the brain. Cholesterol is the raw material used to form vitamin D, some hormones (sex hormones and cortisol), and bile salts.

Homeostatic Imbalance

Saturated fats, along with cholesterol, have been implicated as substances that encourage atherosclerosis (the deposit of fatty substances in artery walls) and eventual arteriosclerosis (hardening of the arteries). As a result, olive oil and liquid spreads made from polyunsaturated fats are being promoted as products that allow us to "have our cake and eat it too"—good-tasting substitutes that (unlike butter) do not damage our arteries. ▲

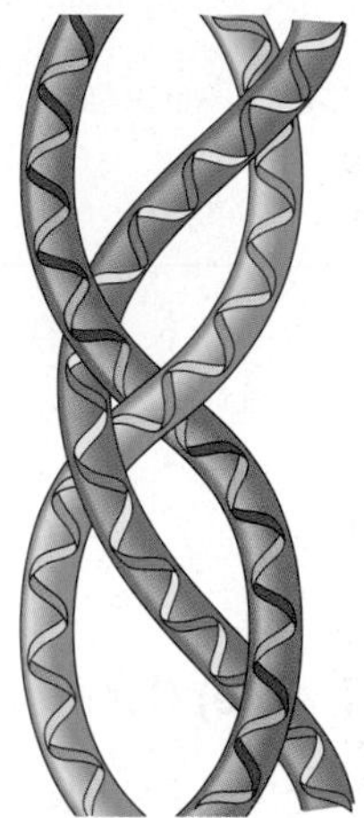

(a) Triple helix of collagen (a fibrous or structural protein).

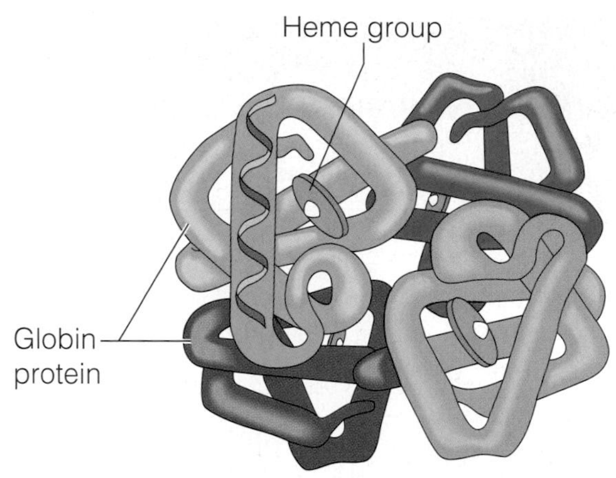

(b) Hemoglobin molecule composed of the protein globin and attached heme groups. (Globin is a globular or functional protein.)

FIGURE 2.16 General structure of (a) a fibrous protein and (b) a globular protein.

Proteins

Proteins account for over 50 percent of the organic matter in the body, and they have the most varied functions of the organic molecules. Some are construction materials; others play vital roles in cell function. Like carbohydrates and lipids, all proteins contain carbon, oxygen, and hydrogen. In addition, they contain nitrogen and sometimes sulfur atoms as well.

The building blocks of proteins are small molecules called **amino** (ah-me′no) **acids.** About 20 common varieties of amino acids are found in proteins. All amino acids have an *amine group* (NH_2), which gives them basic properties, and an *acid group* (COOH), which allows them to act as acids. In fact, all amino acids are identical except for a single group of atoms called their *R-group* (Figure 2.15). Hence, it is differences in the R-groups that make each amino acid chemically unique.

Amino acids are joined together in chains to form large, complex protein molecules that contain from 50 to thousands of amino acids. (Amino acid chains containing fewer than 50 amino acids are called *polypeptides*.) Because each type of amino acid has distinct properties, the sequence in which they are bound together produces proteins that vary widely both in structure and function. Perhaps this can be made more understandable if the 20 amino acids are compared to a 20-letter alphabet. The letters (amino acids) are then used in specific combinations to form words (a protein). Just as a change in one letter of any word can produce a word with an entirely different meaning (flour → floor) or is nonsensical (flour → fluur), changes in kinds of amino acids (letters) or in their positions in the protein allow literally thousands of different protein molecules to be made. The structure of proteins is specified by our genes, as will be described in Chapter 3.

Fibrous and Globular Proteins Based on their overall shape and structure, proteins are classed as either fibrous or globular proteins (Figure 2.16). The strandlike **fibrous proteins,** also called **structural proteins,** appear most often in body structures. They are very important in binding structures together and for providing strength in certain body tissues. For example, *collagen* (kol′ah-jen) is found in bones, cartilage, and tendons and is the most abundant protein in the body. *Keratin* (ker′ah-tin) is the structural protein of hair and nails and the material that makes skin tough.

Globular proteins are mobile, generally spherical molecules that play crucial roles in virtually all biological processes. Because they *do things* rather than just form structures, they are also called **functional proteins.** As noted in Table 2.5, the scope of their activities is remarkable. Some (antibodies) help to provide immunity; others (hormones) help to regulate growth and development. Still others, called *enzymes* (en′zīmz), are biological

TABLE 2.5 Representative Groups of Functional Proteins

Functional group	Role(s) in the body
Antibodies (immunoglobulins)	Highly specialized proteins that recognize, bind with, and inactivate bacteria, toxins, and some viruses; function in the immune response, which helps protect the body from "invading" foreign substances.
Hormones	Help to regulate growth and development. Examples include • Growth hormone—an anabolic hormone necessary for optimal growth. • Insulin—helps regulate blood sugar levels. • Nerve growth factor—guides the growth of neurons in the development of the nervous system.
Transport proteins	Hemoglobin transports oxygen in the blood; other transport proteins in the blood carry iron, cholesterol, or other substances.
Catalysts (enzymes)	Essential to virtually every biochemical reaction in the body; increase the rates of chemical reactions by at least a millionfold; in their absence (or destruction), biochemical reactions cease.

catalysts that regulate essentially every chemical reaction that goes on within the body.

The fibrous structural proteins are exceptionally stable; the globular functional proteins are quite the opposite. Hydrogen bonds are critically important in maintaining their structure, but hydrogen bonds are fragile and are easily broken by heat and excesses of pH. When their three-dimensional structures are destroyed, the proteins are said to be *denatured* and can no longer perform their physiological roles. This is because their function depends on their specific structure—most importantly, on the presence of particular collections of atoms called **active sites** on their surface that "fit" and interact chemically with other molecules of complementary shape and charge (Figure 2.17). As hinted earlier, hemoglobin becomes totally unable to bind and transport oxygen when blood pH becomes too acidic, and pepsin, a protein-digesting enzyme, is inactivated by alkaline pH. In each case, the structure needed for function has been destroyed by the improper pH.

Except for enzymes, most important types of functional proteins are described with the organ system or functional process to which they are closely related. For instance, protein hormones are discussed in Chapter 9 (Endocrine System), hemoglobin is considered in Chapter 10 (Blood), and antibodies are described in Chapter 12 (The Lymphatic System and Body Defenses). However, enzymes are important in the functioning of all body cells, and so these incredibly complex molecules are considered here.

Enzymes and Enzyme Activity **Enzymes** are functional proteins that act as biological catalysts. A **catalyst** is a substance that increases the rate of a chemical reaction without becoming part of the product or being changed itself. Enzymes accomplish this feat by binding to and "holding" the reacting molecules (the substrates) in the proper position for chemical interaction. While the substrates are bound to the enzyme's active site (see Figure 2.17a), they undergo structural changes that result in a new product. Once the reaction has occurred, the enzyme releases the product. Because enzymes are not changed in doing their job, they are reusable, and only small amounts of each enzyme are needed by the cells.

Enzymes are capable of catalyzing millions of reactions each minute. However, they do more than just increase the speed of chemical reactions; they also determine just which reactions are possible at a particular time. No enzyme, no reaction! Enzymes can be compared to a bellows used to fan a sluggish fire into flaming activity. Without enzymes, biochemical reactions would occur far too slowly to sustain life.

Q *How does an enzyme recognize its substrate?*

Substrate "fits" with active site

Active site

Enzyme (globular functional protein)

(a)

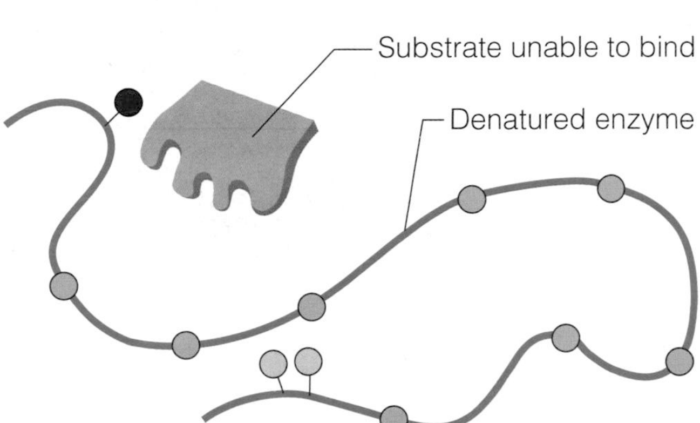

(b)

FIGURE 2.17 Simple diagram illustrating denaturation of a functional protein molecule such as an enzyme. (a) The molecule's three-dimensional globular structure is maintained by intramolecular bonds. Atoms composing the active site of the enzyme are shown as stalked particles. The substrate, or molecule the enzyme acts on, has a corresponding binding site, and the two sites fit together very precisely. **(b)** Breaking the intramolecular bonds that maintain the three-dimensional structure of the enzyme results in a linear molecule, with the atoms of the former active site widely separated. Enzyme-substrate binding can no longer occur.

A *The shape of the substrate and the enzyme's active site are complementary.*

Although there are hundreds of different kinds of enzymes in body cells, they are very specific in their activities, each controlling only one (or a small group of) chemical reaction(s) and acting only on specific molecules. Most enzymes are named according to that specific type of reaction they catalyze. There are "hydrolases," which add water; "oxidases," which cause oxidation; and so on. (In most cases, an enzyme can be recognized by the suffix **-ase** forming part of its name.)

Many enzymes are produced in an inactive form and must be activated in some way before they can function. In other cases, enzymes are inactivated immediately after they have performed their catalytic function. Both events are true of enzymes that promote blood clotting when a blood vessel has been damaged. If this were not so, large numbers of unneeded and potentially lethal blood clots would be formed.

Nucleic Acids

The role of **nucleic** (nu-kle′ik) **acids** is fundamental: they make up the genes, which provide the basic blueprint of life. Not only do they determine

Prove It Yourself

Demonstrate the Presence of an Enzyme in Your Saliva

You can detect the presence of an enzyme (a catalytic protein) in one of your own body fluids, saliva, with the following demonstration.

Place a small cracker in your mouth. Don't swallow it, but move the cracker around in your mouth to moisten it with saliva. After a minute you will experience a sweet taste that was not present initially. This is because your saliva contains an enzyme that breaks the bonds in the starch of the cracker, producing the sweet-tasting disaccharide maltose. What you have just demonstrated is that the hydrolysis (digestion) of starch actually begins in your mouth, even before food reaches your stomach.

To prove that the starch was broken down by an enzyme and not by the watery component of your saliva, wet another cracker with tap water. Moisten it for the same length of time as the first cracker, and then place it in your mouth. It should not taste sweet.

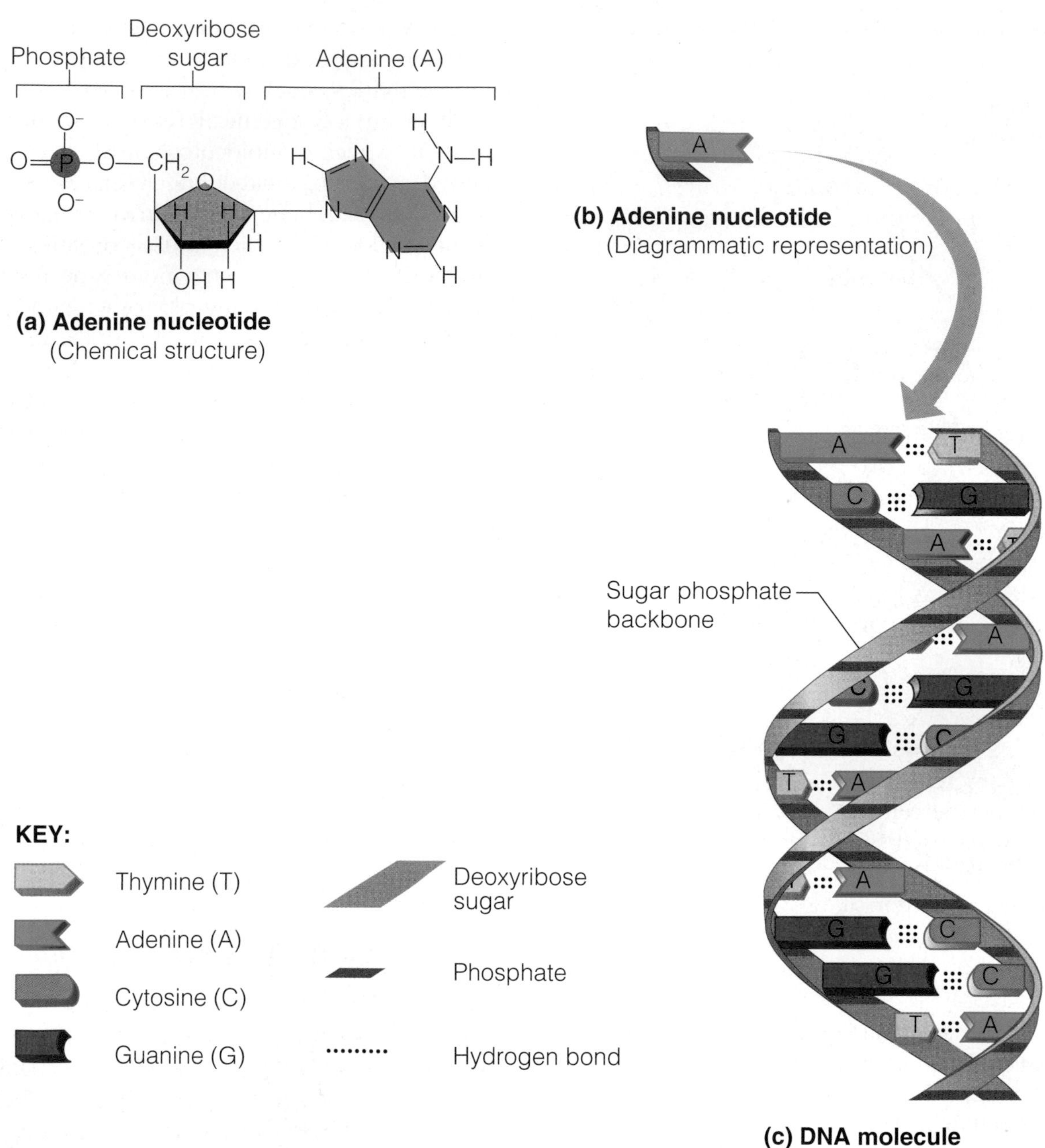

FIGURE 2.18 Structure of DNA. (a) The unit of DNA (deoxyribonucleic acid) is the nucleotide, composed of a linked deoxyribose sugar molecule, a phosphate group, and a nitrogen-containing base (attached to the sugar). The nucleotide illustrated, both in its **(a)** chemical and **(b)** diagrammatic structures, contains the base adenine. **(c)** Structure of a DNA molecule—two nucleotide chains coiled into a double helix. The "backbones" of DNA are formed by alternating sugar and phosphate molecules. The "rungs" are formed by the binding together of complementary bases (A to T, G to C) by hydrogen bonds.

what type of organism you will be, but they also direct your growth and development—and they do this largely by dictating protein structure. (Remember that enzymes, which catalyze all the chemical reactions that occur in the body, are proteins.)

Nucleic acids, composed of carbon, oxygen, hydrogen, nitrogen, and phosphorus atoms, are the largest biological molecules in the body. Their building blocks, the **nucleotides** (nu′kle-o-tīdz), are quite complex. Each consists of three basic parts: (1) a nitrogen-containing base, (2) a pentose (5-carbon) sugar, and (3) a phosphate group (Figure 2.18a and b).

The bases come in five varieties: *adenine* (A), *guanine* (G), *cytosine* (C), *thymine* (T), and *uracil* (U). A and G are large, two-ring bases, whereas the others are smaller, single-ring structures. The nucleotides are named according to the base they contain: A-containing bases are adenine nucleotides, C-containing bases are cytosine nucleotides, and so on.

The two major kinds of nucleic acid are **deoxyribonucleic** (de-ok″sĭ-ri″bo-nu-kle′ik) **acid (DNA)** and **ribonucleic acid (RNA).** DNA and RNA differ in many respects. DNA is the genetic material found within the cell nucleus (the control center of the cell). It has two fundamental roles: (1) It replicates itself exactly before a cell divides, thus ensuring that the genetic information in every body cell is identical. (2) It provides the instructions for building every protein in the body. For the most part, RNA is located outside the nucleus and can be considered the "molecular slave" of DNA; that is, RNA carries out the orders for protein synthesis issued by DNA.

Although both RNA and DNA are formed by the joining together of nucleotides, their final structures are different. As shown in Figure 2.18c, DNA is a long double chain of nucleotides. Its bases are A, G, T, and C, and its sugar is *deoxyribose*. Its two nucleotide chains are held together by hydrogen bonds between the bases, so that a ladderlike molecule is formed. Alternating sugar and phosphate molecules form the "uprights," or backbones, of the ladder, and each "rung" is formed by two joined bases (one base pair). Binding of the bases is very specific: A always binds to T, and G always binds to C. Thus, A and T are said to be *complementary bases,* as are C and G. A base sequence of ATGA on one nucleotide chain would necessarily be bonded to the complementary base sequence TACT on the other nucleotide strand. The whole molecule is then coiled into a spiral-staircaselike structure called a *double helix.*

Whereas DNA is double-stranded, RNA molecules are single nucleotide strands. The RNA bases are A, G, C, and U (U replaces the T found in DNA), and its sugar is *ribose* instead of deoxyribose. Three major varieties of RNA exist—*messenger, ribosomal,* and *transfer RNA*—and each has a specific role to play in carrying out DNA's instructions for building proteins. Messenger RNA carries the information for building the protein from the DNA genes to the ribosomes, the protein-synthesizing sites. Transfer RNA ferries amino acids to the ribosomes. Ribosomal RNA forms part of the ribosomes, where it oversees the translation of the message and the binding together of amino acids to form the proteins. Protein synthesis is described in greater detail in Chapter 3.

Adenosine Triphosphate (ATP)

The synthesis of **adenosine triphosphate** (ah-den′o-sēn tri-fos′fāt), or **ATP,** is all-important because it provides a form of chemical energy that is usable by all body cells. Without ATP, molecules cannot be made or broken down, cells cannot maintain their boundaries, and all life processes grind to a halt.

Although glucose is the most important fuel for body cells, none of the chemical energy contained in its bonds can be used directly to power cellular work. Instead, energy released as glucose is catabolized, captured, and stored in the bonds of ATP molecules as small packets of energy.

Structurally, ATP is a modified nucleotide; it consists of an adenine base, ribose sugar, and three phosphate groups (Figure 2.19a). The phosphate groups are attached by unique chemical bonds called *high-energy phosphate bonds.* When these bonds are ruptured by hydrolysis, energy that can be used immediately by the cell to do work or power a particular activity—such as synthesizing proteins, transporting substances across its membrane, or, in the case of muscle cells, contracting—is liberated (Figure 2.19b). ATP can be compared to a tightly coiled spring that is ready to uncoil with tremendous energy when the "catch" is released. The consequence of cleavage of its terminal phosphate bond can be represented as follows:

ATP	→	ADP	+	Ⓟ	+	E
(adenosine triphosphate)		(adenosine diphosphate)		(inorganic phosphate)		(energy)

As ATP is used to provide cellular energy, **adenosine diphosphate (ADP)** accumulates, and ATP supplies

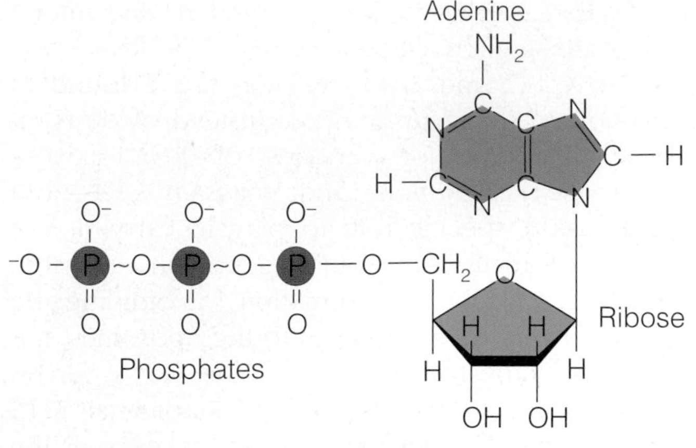

(a) **Adenosine triphosphate (ATP)**

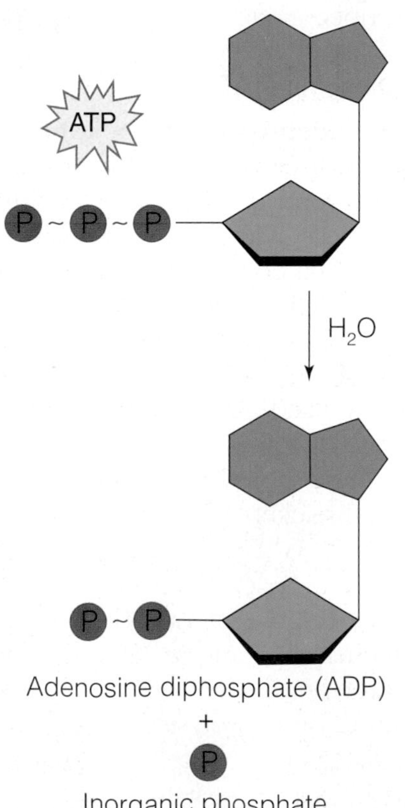

(b) **Hydrolysis of ATP**

FIGURE 2.19 ATP. **(a)** The structure of ATP (adenosine triphosphate). **(b)** Hydrolysis of ATP to yield ADP (adenosine diphosphate) and inorganic phosphate. High-energy bonds are indicated by a red ~.

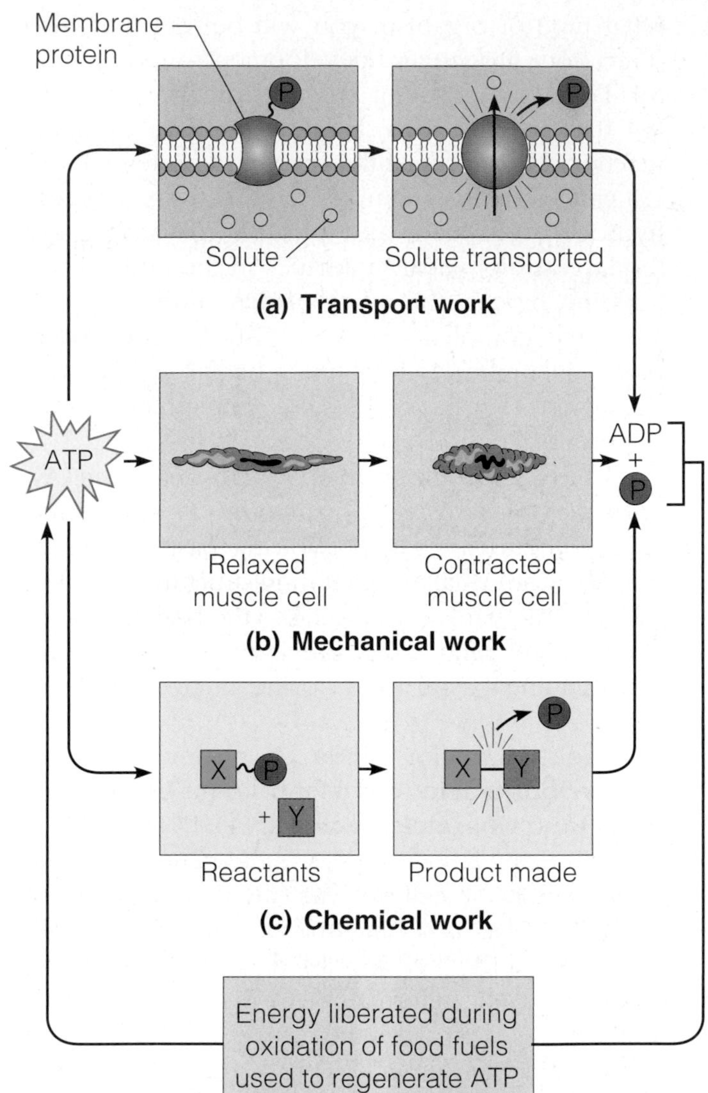

FIGURE 2.20 Three examples of how ATP drives cellular work. The high-energy bonds of ATP release energy for use by the cell when they are broken. **(a)** ATP drives the transport of certain solutes (amino acids, for example) across cell membranes. **(b)** ATP activates contractile proteins in muscle cells so that the cells can shorten and perform mechanical work. **(c)** ATP provides the energy needed to drive energy-absorbing chemical reactions. ATP is regenerated (phosphate is bound to ADP) as energy is released by the oxidation of food fuels and captured in the ADP–P bond.

are replenished by oxidation of food fuels (Figure 2.20). Essentially the same amount of energy must be captured and used to reattach a phosphate group to ADP (that is, to reverse the reaction) as is liberated when the terminal phosphate is cleaved off.

Focus on Careers

Food Technologist

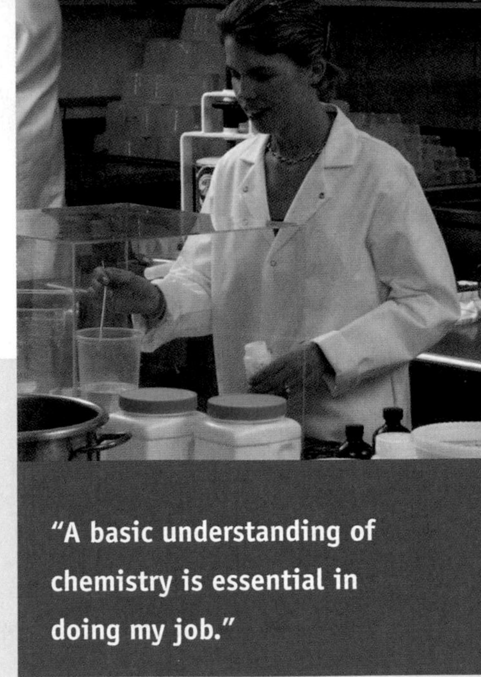

Food technologists regularly use chemistry to determine product stability and shelf life.

HAVE you ever picked a new product off the supermarket shelf or tried the latest coffee and wondered how these items were developed—how they got from an idea in someone's head to the shelf? Anne Cristofano, a food technologist with a San Francisco Bay Area food and beverage product developer, is a central part of this process. Essential to Anne's job is her understanding of basic chemistry and an ability to determine pH, titratable acidity, and viscosity.

Anne's choice of occupation reflects her interest in both science *and* food. "I have a degree in Foods (a combination of culinary arts, food science, and nutrition) from California Polytechnic State University, which could have led me to a number of interesting occupations. I, however, am very interested in food! Not only do I like the culinary aspects—cooking and taking cooking classes—I also enjoy food science and product development. I love the challenge of figuring out how to create a completely new product or how to take an existing product and make it far superior."

"While I enjoy creating fun and exciting products for the general consumer, I think the more significant contribution is in specialty foods," she continues. "For example, at Mattson and Company, we developed a line of food entrées that are low in fat and sodium and fortified with essential vitamins for people with cardiovascular disease. We also have created products that are gluten-free, an area in which gluten-restricted consumers have few choices. I can't think of a more personally satisfying way to use my science and math background."

The terms *food science* and *food technology* often are used interchangeably, but the two disciplines are not identical. Food science is the discipline in which the physical, biological, and engineering sciences are used to study the nature of foods, the causes of deterioration, the principles underlying food processing, and the improvement of foods for the consuming public. Food technology is the application of food science to creating new foods and improving existing products. It applies basic science to preserve, process, package, and distribute safe, nutritious, and wholesome food to the consumer.

A new food or beverage begins as a concept, which is usually created by a marketing team, then given to a food technologist to develop. "The criteria," Anne explains, "range from specific nutritional needs and consumer demands, for example, whether or not it must be kosher or natural, to processing constraints and cost; the list can go on and on. It is very important to not only keep within the range of parameters I have been given and develop a great tasting and visually appealing product, but also one that is designed to go easily from my benchtop to the manufacturing plant."

"A basic understanding of chemistry is essential in doing my job."

Undergraduate programs in Food Science and Technology prepare people for occupations in both the private and the public sector; for example, the Food and Drug Administration and the Department of Agriculture. A search on the Web for "food science careers" reveals a wide range of career opportunities: nutritional science and education, sports nutrition, dietetics, and food management. Are you an animal lover? One university's program will prepare you to become a "flavor chemist," which involves creating flavors "to improve the palatability of dog and cat foods." Other jobs range from assuring the wholesomeness and safety of food to developing new food products and process methods.

The Institute of Food Technologists (IFT), the principal organization for food science and food technology professionals, offers undergraduate and graduate scholarships in these fields. For further information, contact the IFT at:
525 W. Van Buren, Suite 1000
Chicago, IL 60607
(312) 782-8424
http://www.ift.org

For additional information on this career, click the Focus on Careers link at www.anatomyandphysiology.com.

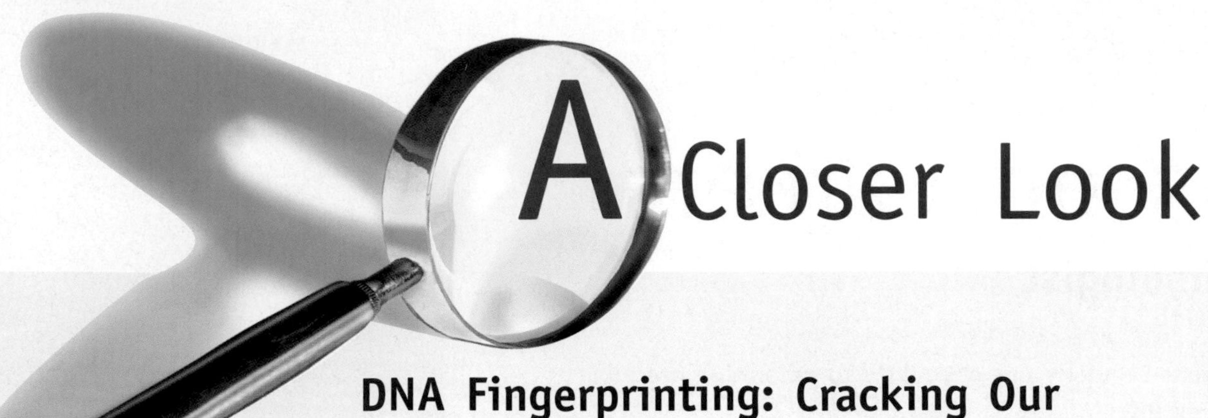

DNA Fingerprinting: Cracking Our Genetic "Barcode"

THE terrorist attacks on New York City's World Trade Center killed more than 3,000 people, their bodies buried in millions of tons of rubble. As weeks passed, it became clear that even if victims could be recovered from the wreckage, their bodies would probably be mangled, crushed, burned, or decomposed to a point where even family members would not recognize them.

In a situation like this, how can we identify individuals with any certainty? The New York Medical Examiner's Office turned to **DNA fingerprinting,** a technique for analyzing tiny samples of DNA taken from semen, skin, blood, or other body tissues. DNA fingerprinting is based on the fact that no two human beings, except for identical twins, possess identical sets of genetic material. In effect, DNA fingerprinting creates a unique genetic "barcode" that distinguishes each of us from all other humans. Let's see how it works.

Creating a DNA Profile

Recall that DNA contains four nucleotides—A, G, C, and T—that form complementary base pairs. In members of the same species, 99.9 percent of DNA is identical. This means that only 0.1 percent of your DNA differs from that of other humans—even close relatives, but this is enough to make you genetically unique. In a DNA string 3 billion units long, that 0.1 percent translates into 3 million variations that differ slightly from everyone else's. Unless you're an identical sibling, your set of DNA is yours alone. DNA fingerprinting involves analyzing an individual's DNA, mapping its unique pattern, and comparing it to other DNA profiles to determine whether there's a match.

A standard technique for creating a DNA profile focuses on 13 specific sites on our chromosomes. At these sites, short segments of nuclear DNA are arranged in a repeating sequence. Individuals vary greatly in the number of repetitions, from 1 to 30 sets or more. While it is theoretically possible that unrelated people could show identical repeats at all 13 sites, the odds are less than 1 in one trillion.

Sometimes it can be difficult to obtain sufficient nuclear DNA for analysis. DNA samples recovered from crime scenes, for example, are frequently contaminated with dirt, fibers, and debris. Many body parts recovered from the World Trade Center were charred, ground into dust, or badly decomposed, limiting the amount of testable tissue. DNA retrieval can become a race against time as microbes, enzymes, insects, and environmental factors such as heat and humidity accelerate the process of decomposition. However, DNA is also contained in mitochondria, some of which can survive in bones, teeth, and hair. Mitochondrial DNA, too, can be profiled to find a match, and sometimes it can be combined with a partial nuclear profile to give a positive identification.

Sorting and Identifying DNA

For DNA to be profiled, it must first be cut into manageable fragments by *restriction enzymes,* enzymes that recognize a specific base sequence and cleave the DNA at this location. This breaks down chromosomes into millions of differently-sized pieces that are then subjected to *gel electrophoresis,* which sorts

DNA electrophoresis. A scientist looking at DNA fragments in an electrophoresis gel.

these pieces by length. The DNA is placed on a gel and positioned in an electric field. The negatively charged fragments of DNA are attracted to the positively charged electrode and migrate toward it. Because the smaller pieces move more quickly through the gel than do the larger pieces, the fragments end up sorted by size.

To locate a specific repeating sequence, researchers make a *DNA probe* with a complementary sequence and tag it with a radioactive compound. Since their sequences are complementary, the probe binds to the site; and when exposed to X-ray film, the image shows dark bands where the probe bound to the DNA.

A victim's DNA profile is then compared to known references to find one that matches. In the case of the World Trade Center attack, DNA references were obtained from victims' personal effects (such as toothbrushes and combs), entered into a computer, and sorted to find a match.

DNA Fingerprinting and Forensics

DNA fingerprinting has become a vital tool in forensic medicine (the application of medical knowledge to questions of law). For example, DNA fingerprinting is used to identify "John and Jane Does," unknown human remains. The U.S. military takes blood and saliva samples from every recruit so it can identify soldiers killed in the line of duty. DNA fingerprinting can also identify victims of mass disasters such as airplane crashes. The World Trade Center tragedy called for genetic analysis on an unprecedented scale: up to one million DNA tests may be necessary to identify all the missing.

DNA fingerprinting can prove that a suspect was actually at the scene of a crime. In the United States, some communities now require certain criminal offenders to provide DNA samples, which are classified and stored. DNA profiles can also establish innocence. At least 10 people in the United States have been released from death row after genetic evidence exonerated them.

DNA fingerprinting can also verify relationships in cases of disputed property, identify long-lost relatives, and establish paternity, even in paternity cases that are centuries old. For example, historians have fiercely debated whether Thomas Jefferson, our third president, fathered any children by his slave Sally Hemings. Modern DNA researchers entered the fray by profiling Jefferson's Y chromosome. A comparison of nineteen genetic markers on the Jefferson Y chromosomes and those of Hemings' descendants found identical matches between the Jefferson line and Hemings' youngest son. Could it be chance? The probability is less than 1 percent.

SUMMARY

Media study tools that could provide you with additional help in reviewing specific key topics of Chapter 2 are referenced below.

IP = Interactive Physiology;

WEB = A&P Place website.

CONCEPTS OF MATTER AND ENERGY (pp. 26–27)

1. Matter
   a. Matter is anything that occupies space and has mass.
   b. Matter exists in three states: gaseous, liquid, and solid.
2. Energy
   a. Energy is the capacity to do work or to move matter. Energy has kinetic (active) and potential (stored) work capacities.
   b. Energy forms that are important in body function include chemical, electrical, mechanical, and radiant.
   c. Energy forms are interconvertible, but some energy is always unusable (lost as heat) in such transformations.

COMPOSITION OF MATTER (pp. 27–32)

1. Elements and atoms
   a. Each element is a unique substance that cannot be decomposed into simpler substances by ordinary chemical methods. A total of 112 (to 116) elements exist; they differ from one another in their chemical and physical properties.
   b. Four elements (carbon, hydrogen, oxygen, and nitrogen) make up 96 percent of living matter. Several other elements are present in small or trace amounts.

   WEB Exercise: Chapter 2, Common Elements in Living Organisms.

   c. The building blocks of elements are atoms. Each atom is described by an atomic symbol consisting of one or two letters.
2. Atomic structure
   a. Atoms are composed of three subatomic particles: protons, electrons, and neutrons. Because all atoms are electrically neutral, the number of protons in any atom is equal to its number of electrons.
   b. The planetary model of the atom portrays all the mass of the atom (protons and neutrons) concentrated in a minute central nucleus. Electrons orbit the nucleus along specific orbits. The orbital model also locates protons and electrons in a central nucleus, but it depicts electrons as occupying areas of space called orbitals and forming an electron cloud of negative charge around the nucleus.

   WEB Exercise: Chapter 2, The Structure of Atoms.

   c. Each atom can be identified by an atomic number, which is equal to the number of protons contained in the atom's nucleus.
   d. The atomic mass number is equal to the sum of the protons and neutrons in the atom's nucleus.
   e. Isotopes are different atomic forms of the same element; they differ only in the number of neutrons in the nucleus. Many of the heavier isotopes are unstable and decompose to a more stable form by ejecting particles or energy from the nucleus, a phenomenon called radioactivity. Such radioisotopes are useful in medical diagnosis and treatment and in biochemical research.
   f. The atomic weight is approximately equal to the mass number of the most abundant isotope of any element.

MOLECULES AND COMPOUNDS (pp. 32–33)

1. A molecule is the smallest unit resulting from the binding of two or more atoms. If the atoms are different, a molecule of a compound is formed.
2. Compounds exhibit properties different from those of the atoms they comprise.

CHEMICAL BONDS AND CHEMICAL REACTIONS (pp. 33–39)

1. Bond formation
   a. Chemical bonds are energy relationships. Electrons in the outermost energy level (valence shell) of the reacting atoms are active in the bonding.
   b. Atoms with a full valence shell (2 electrons in shell 1, or 8 in the subsequent shells) are chemically inactive. Those with an incomplete valence shell interact by losing, gaining, or sharing electrons to achieve stability (that is, to fill the valence shell).

   WEB Exercise: Chapter 2, Electron Arrangement.

c. Ions are formed when valence-shell electrons are completely transferred from one atom to another. The oppositely charged ions thus formed attract each other, forming an ionic bond. Ionic bonds are common in salts.

WEB Exercise: Chapter 2, Ionic Bonds.

d. Covalent bonds involve the sharing of electron pairs between atoms. If the electrons are shared equally, the molecule is a nonpolar covalent molecule. If the electrons are not shared equally, the molecule is a polar covalent molecule. Polar molecules orient themselves toward charged particles.

WEB Exercise: Chapter 2, Covalent Bonds.

e. Hydrogen bonds are fragile bonds that bind together water molecules or different parts of the same molecule (intramolecular bonds). They are common in large, complex organic molecules, such as proteins and nucleic acids.

2. Patterns of chemical reactions
   a. Chemical reactions involve the formation or breaking of chemical bonds. They are indicated by the writing of a chemical equation, which provides information about the atomic composition (formula) of the reactant(s) and product(s).
   b. Chemical reactions that result in larger, more complex molecules are synthesis reactions; they involve bond formation.
   c. In decomposition reactions, larger molecules are broken down into simpler molecules or atoms. Bonds are broken.
   d. Exchange reactions involve both the making and breaking of bonds. Atoms are replaced by other atoms.

BIOCHEMISTRY: THE CHEMICAL COMPOSITION OF LIVING MATTER

(pp. 39–52)

1. Inorganic compounds
   a. Inorganic compounds making up living matter do not contain carbon (exceptions include CO_2 and CO). They include water, salts, acids, and bases.

WEB Exercise: Chapter 2, Characteristics of Acids, Bases, and Salts.

   b. Water is the single most abundant compound in the body. It acts as a universal solvent in which electrolytes (salts, acids, and bases) ionize and in which chemical reactions occur, and it is the basis of transport and lubricating fluids. It slowly absorbs and releases heat, thus helping to maintain homeostatic body temperature, and it protects certain body structures (such as the brain) by forming a watery cushion. Water is also a reactant in hydrolysis reactions.
   c. Salts in ionic form are involved in nerve transmission, muscle contraction, blood clotting, transport of oxygen by hemoglobin, cell permeability, metabolism, and many other reactions. Additionally, calcium salts (as bone salts) contribute to bone hardness.
   d. Acids are proton donors. When dissolved in water, they release hydrogen ions. Strong acids dissociate completely; weak acids dissociate incompletely.
   e. Bases are proton acceptors. The most important inorganic bases are hydroxides. Bicarbonate ions are important bases in the body. When bases and acids interact, neutralization occurs—that is, a salt and water are formed.
   f. The relative concentrations of hydrogen and hydroxyl ions in various body fluids is measured using a pH scale. Each change of one pH unit represents a tenfold change in hydrogen ion concentration. A pH of 7 is neutral (that is, the concentrations of hydrogen and hydroxyl ions are equal). A pH below 7 is acidic; a pH above 7 is alkaline (basic).
   g. Normal blood pH ranges from 7.35 to 7.45. Slight deviations outside this range can be fatal.
2. Organic compounds
   a. Organic compounds are the carbon-containing compounds that comprise living matter. Carbohydrates, lipids, proteins, and nucleic acids are examples. They all contain carbon, oxygen, and hydrogen. Proteins and nucleic acids also contain substantial amounts of nitrogen.
   b. Carbohydrates contain carbon, hydrogen, and oxygen in the general relationship $(CH_2O)_n$; their building blocks are monosaccharides. Monosaccharides include glucose, fructose, galactose, deoxyribose, and ribose. Disaccharides include sucrose, maltose, and lactose; and polysaccharides include starch and glycogen. Carbohydrates are ingested as sugars and starches. Carbohydrates, and in particular glucose, are the major energy source for the formation of ATP.
   c. Lipids include triglycerides (glycerol plus three fatty acid chains), phospholipids, and steroids (most importantly, cholesterol). Triglycerides (neutral fats) are found primarily in adipose tissue, where they provide insulation and reserve

body fuel. Phospholipids and cholesterol are found in all cell membranes. Cholesterol also forms the basis of certain hormones, bile salts, and vitamin D. Like carbohydrates, the lipids are degraded by hydrolysis and synthesized by dehydration synthesis.

d. Proteins are constructed from building blocks called amino acids; 20 common types of amino acids are found in the body. Amino acid sequence determines the proteins constructed. Fibrous, or structural, proteins are the basic structural materials of the body. Globular proteins are also called functional proteins; examples of these include enzymes, some hormones, and hemoglobin. Disruption of the hydrogen bonds of functional proteins leads to their denaturation and inactivation.

e. Enzymes increase the rates of chemical reactions by combining specifically with the reactants and holding them in the proper position to interact. They do not become part of the product. Many enzymes are produced in an inactive form or are inactivated immediately after use.

f. Nucleic acids include deoxyribonucleic acid (DNA) and ribonucleic acid (RNA). The building unit of nucleic acids is the nucleotide; each nucleotide consists of a nitrogenous base, a sugar (ribose or deoxyribose), and a phosphate group. DNA (the "stuff" of the genes) maintains genetic heritage by replicating itself before cell division and contains the code-specifying protein structure. RNA acts in protein synthesis to ensure that instructions of the DNA are executed.

g. ATP (adenosine triphosphate) is the universal energy compound used by all cells of the body. When energy is liberated by the oxidation of glucose, some of that energy is captured in the high-energy phosphate bonds of ATP molecules and is stored for later use.

REVIEW QUESTIONS

MULTIPLE CHOICE

More than one choice may apply.

1. Which of the following is (are) true concerning the atomic nucleus?
   a. Contains the mass of the atom
   b. The negatively charged particles are here
   c. Particles can be ejected
   d. Contains particles that determine atomic number
   e. Contains particles that interact with other atoms

2. Pick out the correct match(es) of element and number of valence-shell electrons.
   a. Oxygen—6
   b. Chlorine—8
   c. Phosphorus—5
   d. Nitrogen—3
   e. Carbon—4

3. Important functions of water include:
   a. cushioning
   b. acts as a transport medium
   c. participates in chemical reactions
   d. solvent for sugars, salts, and other solutes
   e. reduces temperature fluctuations

4. Alkaline substances include:
   a. gastric juice
   b. water
   c. blood
   d. orange juice
   e. ammonia

5. Glucose is to starch as:
   a. a steroid is to a lipid
   b. a nucleotide is to a nucleic acid
   c. an amino acid is to a protein
   d. a polypeptide is to an amino acid

6. What lipid type is the foundation of cell membranes?
   a. Triglyceride
   b. Steroid
   c. Vitamin D
   d. Phospholipid
   e. Prostaglandin

7. Absence of which of the following nitrogen-containing bases would prevent protein synthesis?
   a. Adenine
   b. Cytosine
   c. Guanine
   d. Thymine
   e. Uracil

8. ATP is *not* associated with:
   a. a basic nucleotide structure
   b. high-energy phosphate bonds
   c. deoxyribose
   d. inorganic phosphate
   e. reversible reactions

9. The element essential for normal thyroid function is:
   a. iodine
   b. iron
   c. copper
   d. selenium
   e. zinc

10. What chemical element is present in all proteins but absent from carbohydrates and fats?
    a. carbon
    b. oxygen
    c. hydrogen
    d. nitrogen

SHORT ANSWER ESSAY

1. Why is a study of basic chemistry essential to understanding human physiology?
2. Matter and your body—how are they interrelated?
3. Matter occupies space and has mass. Explain how energy *must be* described in terms of these two factors. Then define energy.
4. Identify the energy form in use in each of the following examples:
   a. Chewing food
   b. Vision (two types, please—think!)
   c. Bending the fingers to make a fist
   d. Breaking the bonds of ATP molecules to energize your muscle cells to make that fist
5. The statement has been made that "some energy is lost in every energy transformation." Explain the meaning of this statement. (Direct your response to answering the questions: Is it really lost? If not, where is it?)
6. According to Greek history, a Greek scientist went running through the streets announcing that he had transformed lead into gold. Both lead and gold are elements. On the basis of what you know about the nature of elements, explain why his rejoicing was short-lived.
7. What four elements make up the bulk of all living matter? (Give both their names and their atomic symbols.) Which of these is found primarily in proteins and nucleic acids?
8. What is the relationship of an atom to an element?
9. All atoms are neutral. Explain the basis of this fact.
10. Fill in the following table to fully describe an atom's subatomic particles.

Particle	Position in the atom	Charge	Mass
Proton			
Neutron			
Electron			

11. Define radioactivity. If an element has three isotopes, which of them (the lightest, the one with an intermediate mass, or the heaviest) is most likely to be a radioisotope, and why?
12. Define molecule. Distinguish between a molecule of an element and a molecule of a compound.
13. Explain the basis of ionic bonding. How do ionic bonds differ from covalent bonds?
14. What are the hydrogen bonds, and how are they important in the body?
15. The two oxygen atoms forming molecules of oxygen gas that you breathe are joined by a polar covalent bond. Explain why this statement is true or false.
16. Identify each of the following reactions as a synthesis, decomposition, or exchange reaction:

$$2Hg + O_2 \rightarrow 2HgO$$

$$Fe^{2+} + CuSO_4 \rightarrow FeSO_4 + Cu^{2+}$$

$$HCl + NaOH \rightarrow NaCl + H_2O$$

$$HNO_3 \rightarrow H^+ + NO_3^-$$

17. Distinguish between inorganic and organic compounds, and list the major categories of each in the body.
18. Salts, acids, and bases are electrolytes. What is an electrolyte?
19. Compare and contrast acids and bases.
20. Define pH. The pH range of blood is from 7.35 to 7.45. Circle the correct answer to complete the sentence: This is slightly (acidic / basic).
21. A pH of 3.3 is (1 / 10 / 100 / 1000) times more acidic than a pH of 4.3.
22. Define monosaccharide, disaccharide, and polysaccharide. Give at least two examples of each. Which of these is the building unit of carbohydrates? What is the primary function of carbohydrates in the body?
23. What are the general structures of neutral fats, phospholipids, and steroids? Give one or two important uses of each of these lipid types in the body.
24. The building block of proteins is the amino acid. Draw a diagram of the structure of a generalized amino acid. What is the importance of the R-group?
25. Name the two protein classes based on structure and function in the body, and give two examples of each.
26. Define enzyme and describe the mechanism of enzyme activity.
27. Virtually no chemical reaction can occur in the body in the absence of enzymes. How might excessively high body temperature or acidosis (acidic blood pH) interfere with enzyme activity?
28. What is the structural unit of nucleic acids? Name the two major classes of nucleic acid found in the body, and then compare and contrast them in terms of **(a)** bases and sugar content, **(b)** general three-dimensional structure, and **(c)** relative functions.

29. What is ATP, and what is its central role in the body?

30. Explain why you can "stack" water slightly above the rim of a glass if you pour the water in very carefully.

31. Water is a precious natural resource in Florida, and it is said that supplies are dwindling. Desalinization (salt removal) of ocean water has been recommended as a solution to the problem. Why shouldn't we drink salt water?

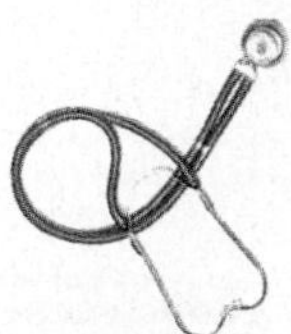

AT THE CLINIC

1. A number of antibiotics act by binding to certain essential enzymes in the target bacteria. How might these antibiotics influence the chemical reaction controlled by the enzyme? What might be the effect on the bacteria? On the person taking the antibiotic prescription?

2. Mrs. Roberts, who is in a diabetic coma, has just been admitted to Noble Hospital. Her blood pH indicates that she is in severe acidosis, and measures are quickly instituted to bring her blood pH back within normal limits. Note the normal pH of blood, and discuss why severe acidosis is a problem.

3. Evelyn is quite proud of her slender, model-like figure and boasts that she doesn't have an "ounce of excess body fat." Barbara, on the other hand, is grossly overweight. She complains of feeling hot most of the time, and on a hot day she is miserable. Evelyn generally feels chilled except on very hot days. Explain the relative sensitivity to environmental temperature of these two women on the basis of information you have been given in the Organic Compounds section of this chapter.

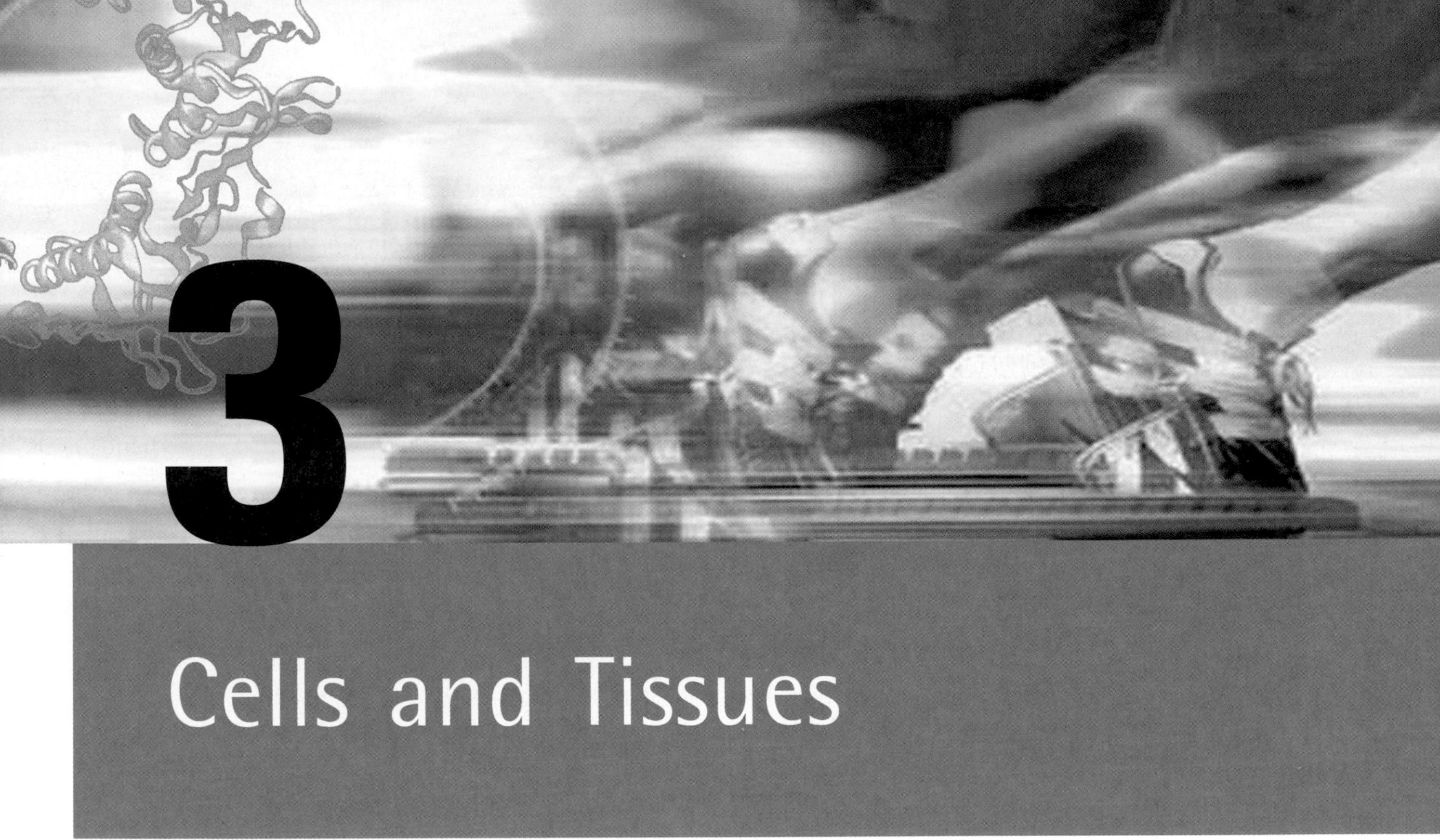

3 Cells and Tissues

YOUR GOALS

After completing this chapter, you will have a working knowledge of the functions of cells and tissues and will have mastered the objectives listed below.

FUNCTION PREVIEW

- Cells carry out all the chemical activities needed to sustain life.
- Tissues provide for a division of labor among body cells.

OBJECTIVE CHECKLIST

PART I: CELLS

OVERVIEW OF THE CELLULAR BASIS OF LIFE (pp. 62–63)

- Name the four elements that make up the bulk of living matter, and list several trace elements.

ANATOMY OF A GENERALIZED CELL (pp. 63–72)

- Define *cell, organelle,* and *inclusion.*
- Identify on a cell model or diagram the three major cell regions (nucleus, cytoplasm, and plasma membrane).
- List the structures of the nucleus, and explain the function of chromatin and nucleoli.
- Identify the organelles on a cell model or describe them, and discuss the major function of each.

CELL PHYSIOLOGY (pp. 72–85)

- Define *selective permeability, diffusion* (including *simple* and *facilitated diffusion* and *osmosis*), *active transport, passive transport, solute pumping, exocytosis, endocytosis, phagocytosis, bulk-phase endocytosis, hypertonic, hypotonic,* and *isotonic.*

>>

(Objective Checklist, continued)

- Describe the structure of the plasma membrane, and explain how the various transport processes account for the directional movements of specific substances across the plasma membrane.
- Describe briefly the process of DNA replication and of mitosis. Explain the importance of mitotic cell division.
- In relation to protein synthesis, describe the roles of DNA and of the three major varieties of RNA.
- Name some cell types, and relate their overall shape and internal structure to their special functions.

PART II: BODY TISSUES (pp. 85–98)

- Name the four major tissue types and their chief subcategories. Explain how the four major tissue types differ structurally and functionally.
- Give the chief locations of the various tissue types in the body.
- Describe the process of tissue repair (wound healing).

PART III: DEVELOPMENTAL ASPECTS OF CELLS AND TISSUES (pp. 98–99)

- Define *neoplasm*, and distinguish between benign and malignant neoplasms.
- Explain the significance of the fact that some tissue types (muscle and nerve) are largely amitotic after the growth stages are over.

PART I: CELLS

In the late 1600s, Robert Hooke was looking through a crude microscope at some plant tissue—cork. He saw some cubelike structures that reminded him of the long rows of monk's rooms (or cells) at the monastery, so he named these structures **cells.** The living cells that had formed the cork were long since dead. However, the name stuck and is still used to describe the smallest unit, or the building block, of all living things, plants and animals alike. Whatever its form, however it behaves, the cell contains all the parts necessary to survive in a changing world. The human body has trillions of these microscopic building blocks.

Overview of the Cellular Basis of Life

Perhaps the most striking thing about a cell is its organization. If we chemically analyze cells, we find that they are made up primarily of four elements—carbon, oxygen, hydrogen, and nitrogen—plus much smaller amounts of several other elements. Although the four major elements build most of the cell's structure (which is largely protein), the trace elements are very important for certain cell functions. For example, calcium is needed for blood clotting (among other things), and iron is necessary to make hemoglobin, which carries oxygen in the blood. Iodine is required to make the thyroid hormone that controls metabolism. In their ionic form, many of the metals (such as calcium, sodium, and potassium) can carry an electrical charge; when they do they are called *electrolytes* (e-lek′tro-līts). Sodium and potassium ions are essential if nerve impulses are to be transmitted and muscles are to contract. (A more detailed account of body chemistry appears in Chapter 2.)

Strange as it may seem, especially when we feel our firm muscles, living cells are about 60 percent water, which is one of the reasons water is essential

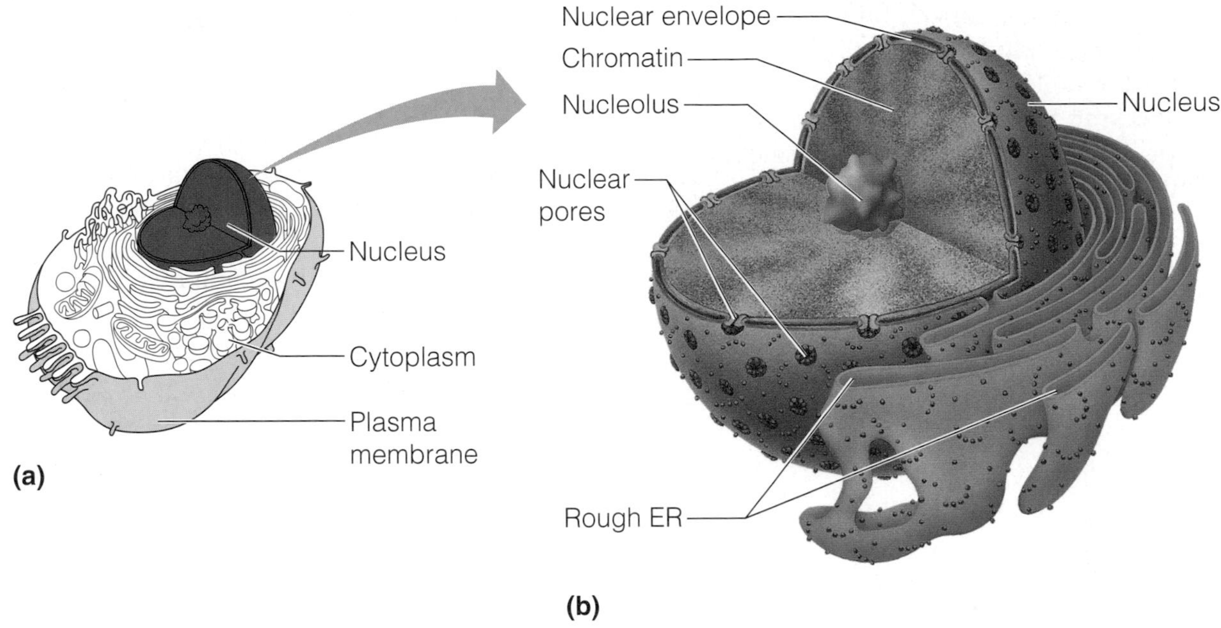

FIGURE 3.1 Anatomy of the generalized animal cell nucleus.
(a) Orientation diagram: The three main regions of the generalized cell.
(b) Structure of the nucleus.

for life. In addition to containing large amounts of water, all the body cells are constantly bathed in a dilute saltwater solution (something like seawater) called *interstitial fluid,* which is derived from the blood. All exchanges between cells and blood are made through this fluid.

Cells vary tremendously in length—ranging from 2 micrometers (1/12,000th of an inch) in the smallest cells to over a meter (3 feet) or more in the nerve cells that cause you to wiggle your toes. Furthermore, a cell's structure often reflects its function; this will become clear later in this chapter. Cells can have amazingly different shapes. Some are disk-shaped (red blood cells), some have threadlike extensions (nerve cells), others are like toothpicks pointed at each end (smooth muscle cells), and still others are cubelike (some types of epithelial cells).

Cells also vary dramatically in the functions, or roles, they play in the body. For example, white blood cells wander freely through the body tissues and protect the body by destroying bacteria and other foreign substances. Some cells make hormones or chemicals that regulate other body cells. Still others take part in gas exchanges in the lungs or cleanse the blood (kidney tubule cells).

Anatomy of a Generalized Cell

Although no one cell type is exactly like all others, cells *do* have the same basic parts, and there are certain functions common to *all* cells. Here we will talk about the **generalized cell,** which demonstrates these many typical features.

In general, all cells have three main regions or parts—a *nucleus* (nu′kle-us), *cytoplasm* (si′to-plazm″), and a *plasma membrane* (Figure 3.1). The nucleus is usually located near the center of the cell. It is surrounded by the semifluid cytoplasm, which in turn is enclosed by the plasma membrane, which forms the outer cell boundary. (Figure 3.4 on p. 67 shows the more detailed structure of the generalized cell as revealed by the electron microscope.)

The Nucleus

Anything that works, works best when it is controlled. For cells, "headquarters," or the control center, is the gene-containing **nucleus** (*nucle* = kernal). The genetic material, or *deoxyribonucleic acid (DNA),* is much like a blueprint that contains all the instructions needed for building the whole body; so, as one might expect, human DNA differs from that of a frog. More specifically, DNA has the instructions for building *proteins*. DNA is also absolutely necessary for cell reproduction. A cell that has lost or ejected its nucleus (for whatever reason) is programmed only to die.

While most often oval or spherical, the shape of the nucleus usually conforms to the shape of the cell. For example, if the cell is elongated, the nucleus is usually extended as well. The nucleus has three recognizable regions or structures: the nuclear envelope, nucleoli, and chromatin.

Nuclear Envelope

The nucleus is bound by a double membrane barrier called the **nuclear envelope,** or **nuclear membrane** (see Figure 3.1). Between the two membranes is a fluid-filled "moat," or space. At various points, the two layers of the nuclear envelope fuse, and **nuclear pores** penetrate through the fused regions. Like other cellular membranes, the nuclear envelope is selectively permeable, but passage of substances through it is much freer than elsewhere because of its relatively large pores. The nuclear membrane encloses a jellylike fluid called *nucleoplasm* (nu′kle-o-plazm″) in which other nuclear elements are suspended.

Nucleoli

The nucleus contains one or more small, dark-staining, essentially round bodies called **nucleoli** (nu-kle′o-li; "little nuclei"). Nucleoli are sites where ribosomes are assembled. The *ribosomes,* most of which eventually migrate into the cytoplasm, serve as the actual sites of protein synthesis, as described shortly.

Chromatin

When a cell is not dividing, its DNA is combined with protein and forms a loose network of bumpy threads called **chromatin** (kro′mah-tin) that is scattered throughout the nucleus. When a cell is dividing to form two daughter cells, the chromatin threads coil and condense to form dense, rodlike bodies called **chromosomes**—much the way a stretched spring becomes shorter and thicker when allowed to relax. The functions of DNA and the mechanism of cell division are discussed in the Cell Physiology section beginning on p. 72.

The Plasma Membrane

The flexible **plasma membrane** is a fragile, transparent barrier that contains the cell contents and separates them from the surrounding environment. (The term *cell membrane* is often used instead, but since nearly all cellular organelles are composed of membranes, we will specifically refer to the cell's surface or outer limiting membrane as the plasma membrane.) Although the plasma membrane is important in defining the limits of the cell, it is much more than a passive envelope, or "baggie." As you will see, its unique structure allows it to play a dynamic role in many cellular activities.

The structure of the plasma membrane consists of two lipid (fat) layers arranged "tail to tail" in which protein molecules float (Figure 3.2). Although most of the lipid portion is *phospholipids* (some with attached sugar groups), a substantial amount of *cholesterol* is also found in plasma membranes. (The characteristics of these specialized lipids are described in Chapter 2.) The olive oil–like lipid bilayer forms the basic "fabric" of the membrane. The polar "heads" of the lollipop-shaped phospholipid molecules are *hydrophilic* ("water loving") and are attracted to water, the main component of both the intercellular and extracellular fluids, and so they lie on both the inner and outer surfaces of the membrane. Their nonpolar "tails" are *hydrophobic* ("water hating") and avoid water, lining up in the center of the membrane. It is the hydrophobic makeup of the membrane interior that makes the plasma membrane relatively impermeable to most water-soluble molecules. The cholesterol has a stabilizing effect and helps keep the membrane fluid.

The proteins scattered in the lipid bilayer are responsible for most of the specialized functions of the membrane. Some proteins are enzymes. Many of the proteins mounted on the cell exterior are receptors for hormones or other chemical messengers or are binding sites for anchoring the cell to fibers or to other extracellular structures. Most proteins that span the membrane are involved in transport functions. For example, some cluster together to form protein channels (tiny *pores*) through which water and small water-soluble molecules or ions can move; others act as *carriers* that bind to a substance and move it through the membrane. Branching sugar groups are attached to most of the proteins abutting the extracellular space. Such

Q *Why do the phospholipids organize into a bilayer, tail to tail, in an aqueous environment?*

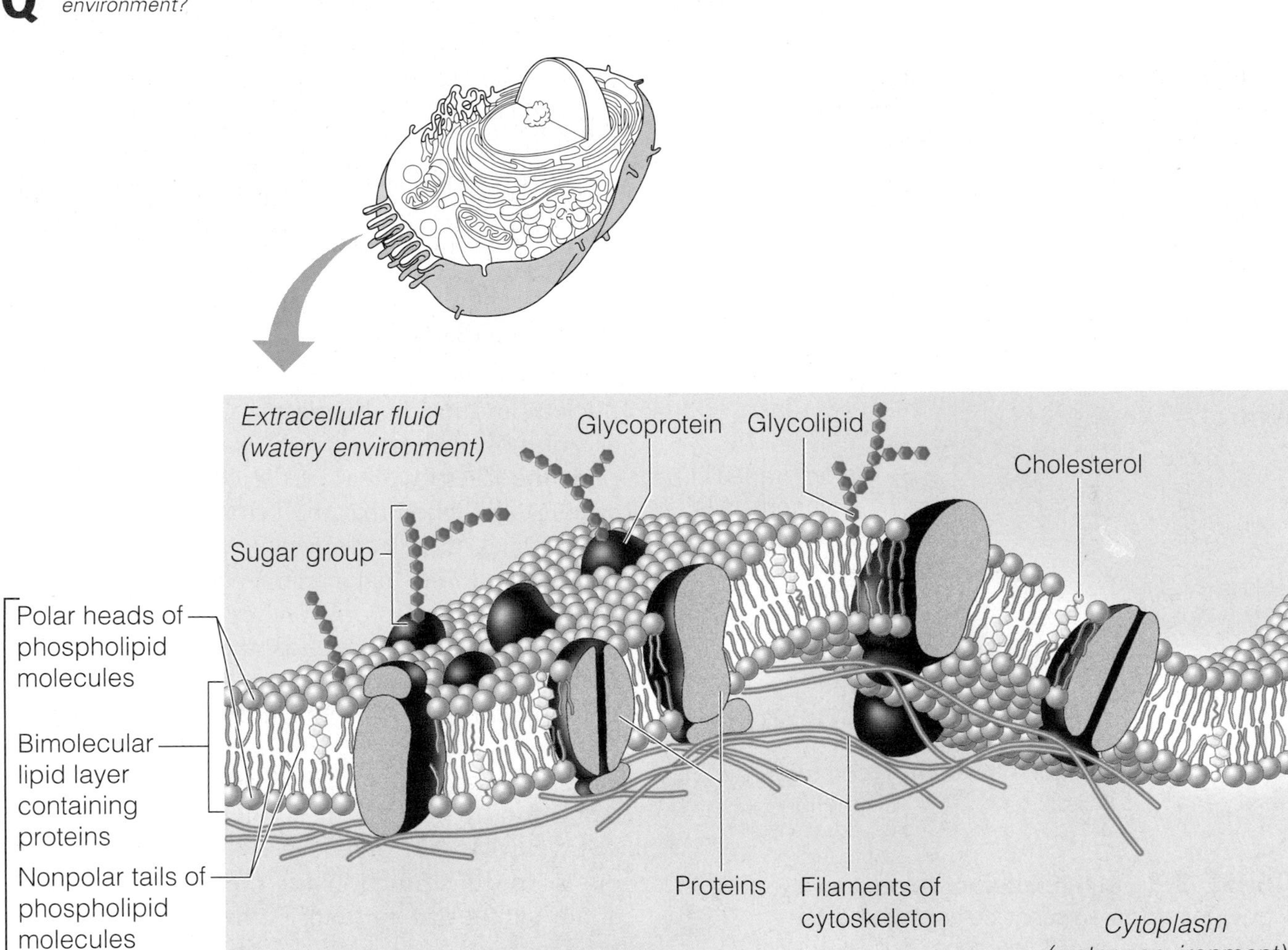

FIGURE 3.2 Structure of the plasma membrane.

"sugar-proteins" are called *glycoproteins,* and because of their presence, the cell surface is a fuzzy, sticky, sugar-rich area. (You can think of your cells as being sugar-coated.) Among other things, these glycoproteins determine your blood type, act as receptors that certain bacteria, viruses, or toxins can bind to, and play a role in cell-to-cell interactions. Definite changes in glycoproteins occur in cells that are being transformed into cancer cells. (Cancer is discussed in "A Closer Look" on pp. 100–101.)

A *The phospholipids have hydrophilic and hydrophobic regions. The hydrophobic (tail) regions shun water and form the inner portion of the membrane in an aqueous environment.*

Specializations of the Plasma Membrane

Specializations of the plasma membrane—such as *microvilli* and *membrane junctions*—are commonly displayed by the (epithelial) cells that form the linings of hollow body organs, such as the small intestine (Figure 3.3). **Microvilli** (mi″kro-vil′i; "little shaggy hairs") are tiny fingerlike projections that greatly increase the cell's surface area for absorption so that the process occurs more quickly.

The **membrane junctions** vary structurally depending on their roles.

- **Tight junctions** are impermeable junctions that bind cells together into leakproof sheets that prevent substances from passing through

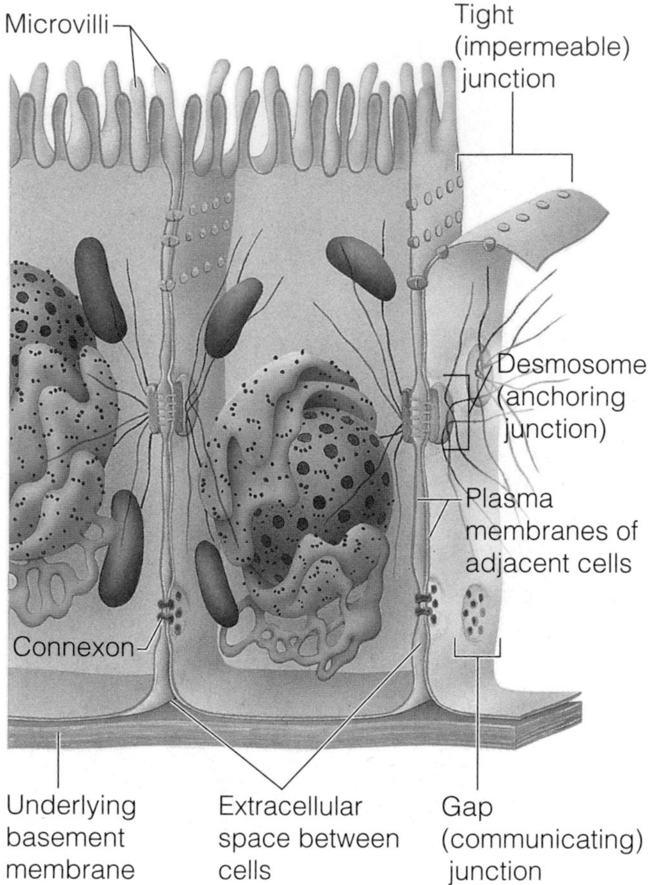

FIGURE 3.3 Cell junctions. An epithelial cell is shown joined to adjacent cells by the three common types of cell junctions: tight junctions, desmosomes, and gap junctions. Also illustrated are microvilli (seen projecting from the free cell surface).

the extracellular space between cells. In tight junctions, adjacent plasma membranes fuse together tightly like a zipper. In the small intestine, for example, these junctions prevent digestive enzymes from seeping into the bloodstream.

- **Desmosomes** (des′mo-somz) are anchoring junctions that prevent cells subjected to mechanical stress (such as skin cells) from being pulled apart. Structurally, these junctions are plaques, buttonlike thickenings of adjacent plasma membranes, which are connected by fine protein filaments. Thicker protein filaments extend from the plaques inside the cells to the plaques on the cells' opposite sides, thus forming an internal system of strong guy wires.
- **Gap junctions,** commonly seen in the heart and between embryonic cells, function mainly to allow communication. Chemical molecules, such as nutrients or ions, can pass directly from one cell to another through them. In gap junctions, the neighboring cells are connected by **connexons,** which are hollow cylinders composed of proteins that span the entire width of the abutting membranes.

The Cytoplasm

The **cytoplasm** is the cellular material outside the nucleus and inside the plasma membrane. It is the site of most cellular activities, so it can be thought of as the "factory area" of the cell. Although early scientists believed that the cytoplasm was a structureless gel, the electron microscope has revealed that it has three major elements: the *cytosol, organelles,* and *inclusions*. The **cytosol** is semitransparent fluid that suspends the other elements. Dissolved in the cytosol, which is largely water, are nutrients and a variety of other solutes (dissolved substances).

The **organelles** (or″gah-nelz′), described in detail shortly, are the metabolic machinery of the cell. Each type of organelle is specialized to carry out a specific function for the cell as a whole. Some synthesize proteins, others package those proteins, and so on.

Inclusions are not functioning units, but instead are chemical substances that may or may not be present, depending on the specific cell type. Most inclusions are stored nutrients or cell products. They include the lipid droplets common in fat cells, glycogen granules abundant in liver and muscle cells, pigments such as melanin seen in skin and hair cells, mucus and other secretory products, and various kinds of crystals.

Cytoplasmic Organelles

The cytoplasmic organelles, literally "little organs," are specialized cellular compartments (Figure 3.4), each performing its own job to maintain the life of the cell. Many organelles are bounded by a membrane similar to the plasma membrane. The membrane boundaries of such organelles allow them to maintain an internal environment quite different from that of the surrounding cytosol. This compartmentalization is crucial to their ability to perform

Q *Which nuclear component contains your genes?*

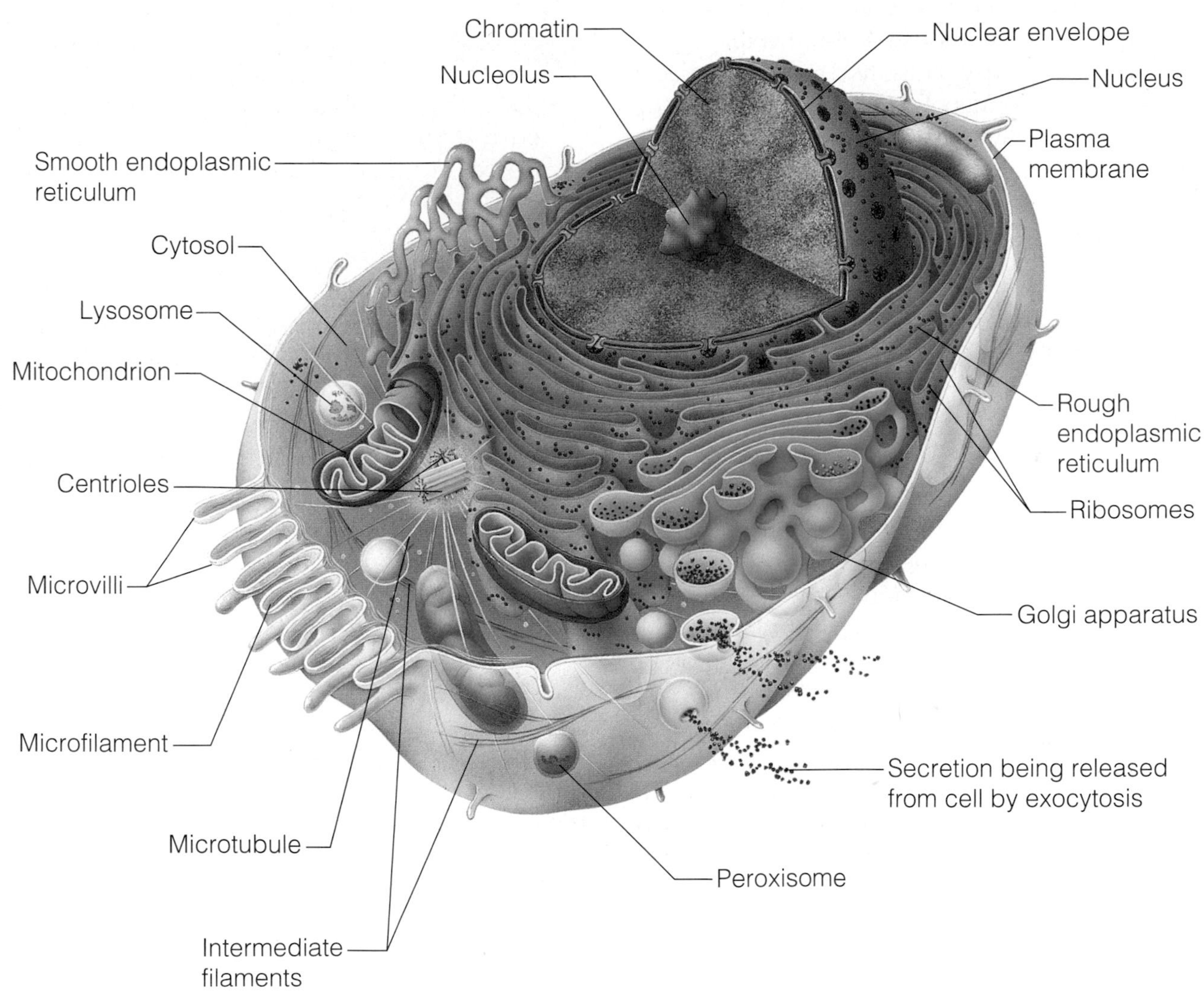

FIGURE 3.4 Structure of the generalized cell. No cell is exactly like this one, but this generalized cell drawing illustrates features common to many human cells.

their specialized functions for the cell. Let us consider what goes on in each of the workshops of our cellular factory.

Mitochondria **Mitochondria** (mi″to-kon′dre-ah) are usually depicted as tiny threadlike (*mitos* = thread) or sausage-shaped organelles (see Figure 3.4), but in living cells they squirm, lengthen, and change shape almost continuously. The mitochondrial wall consists of a double membrane, equal to *two* plasma membranes, placed side by side. The outer membrane is smooth and featureless, but the inner membrane has shelflike protrusions called *cristae* (kris′te; crests). Enzymes dissolved in the fluid within the mitochondria, as well as enzymes that form part of the cristae membranes, carry out the reactions in which oxygen is used to break down foods. As the foods are broken down, energy is released. Much of this energy escapes as heat, but some is captured and used to form *ATP molecules*. ATP provides the energy for all cellular

A *The nucleus.*

work, and every living cell requires a constant supply of ATP for its many activities. Because the mitochondria supply most of this ATP, they are referred to as the "powerhouses" of the cell.

Metabolically "busy" cells, like liver and muscle cells, use huge amounts of ATP and have hundreds of mitochondria. By contrast, cells that are relatively inactive (an unfertilized egg, for instance) have just a few.

Ribosomes **Ribosomes** (ri′bo-sōmz) are tiny, bilobed, dark bodies made of proteins and one variety of RNA called *ribosomal RNA*. Ribosomes are the actual sites of protein synthesis in the cell. Some ribosomes float free in the cytoplasm where they manufacture proteins that function in the cytoplasm. Others attach to membranes and the whole ribosome-membrane combination is called the *rough endoplasmic reticulum*.

Endoplasmic Reticulum The **endoplasmic reticulum** (en″do-plas′mik rĕ-tik′u-lum; "network within the cytoplasm") **(ER)** is a system of fluid-filled cisterns (tubules, or canals) that coil and twist through the cytoplasm. It accounts for about half of a cell's membranes. It serves as a minicirculatory system for the cell because it provides a network of channels for carrying substances (primarily proteins) from one part of the cell to another. There are two forms of ER; a particular cell may have both forms or only one, depending on its specific functions.

The **rough ER** is so called because it is studded with ribosomes. Because essentially all of the building materials of cellular membranes are formed either in it or on it, the rough ER can be thought of as the cell's membrane factory. The proteins made on its ribosomes migrate into the tubules of the rough ER where they fold into their functional three-dimensional shapes and then are dispatched to other areas of the cell in *transport vesicles* (Figure 3.5). In general the amount of rough ER a cell has is a good clue to the amount of protein that cell makes. Rough ER is especially abundant in cells that export protein products—for example, pancreas cells, which produce digestive enzymes to be delivered to the small intestine. The enzymes that catalyze the synthesis of membrane lipids reside on the external face of the rough ER, where the needed building blocks are readily available.

Although the **smooth ER** communicates with the rough variety, it plays no role in protein synthesis. Instead it functions in lipid metabolism (cholesterol and fat synthesis and breakdown), and detoxification of drugs and pesticides. Hence it is not surprising that the liver cells are chock-full of smooth ER. So too are body cells that produce steroid-based hormones—for instance, cells of the male testes that manufacture testosterone.

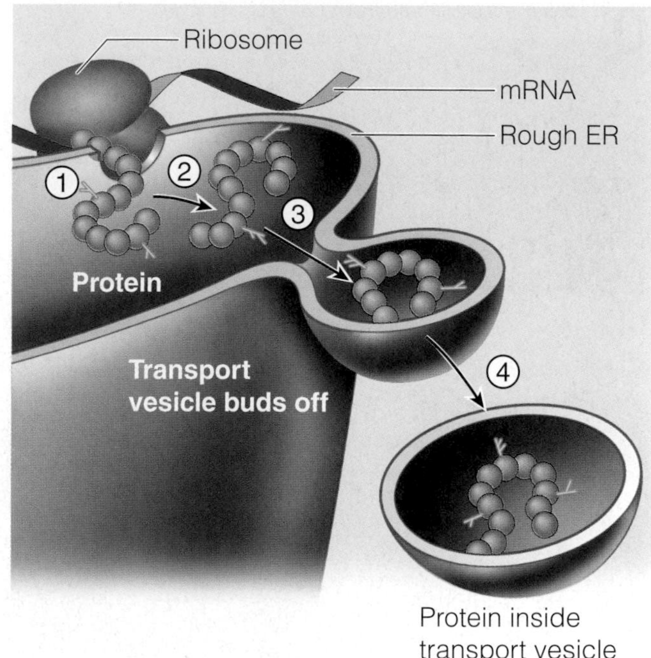

FIGURE 3.5 Synthesis and export of a protein by the rough ER. ① As the protein is synthesized on the ribosome, it migrates into the rough ER cisterna. ② In the cisterna, short sugar chains may be attached to the protein (forming a glycoprotein) and the protein folds into its functional shape. ③ The protein is then packaged in a tiny membranous sac called a transport vesicle. ④ The transport vesicle buds from the rough ER and travels to the Golgi apparatus for further processing or directly to the plasma membrane where its contents are secreted.

Golgi Apparatus The **Golgi** (gol′je) **apparatus** appears as a stack of flattened membranous sacs, associated with swarms of tiny vesicles. It is generally found close to the nucleus and is the principal "traffic director" for cellular proteins. Its major function is to modify and package proteins (sent to it by the rough ER via **transport vesicles**) in specific ways, depending on their final destination (Figure 3.6).

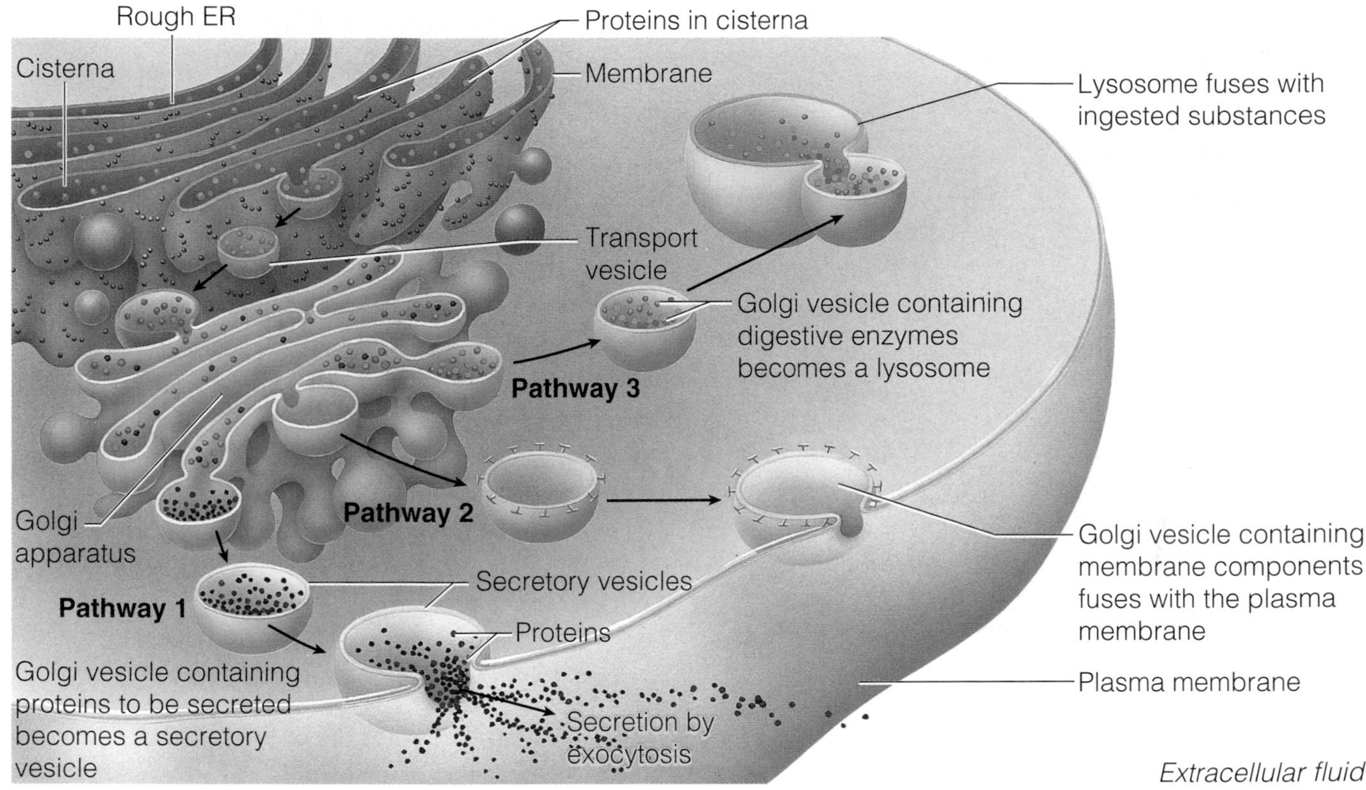

FIGURE 3.6 Role of the Golgi apparatus in packaging the products of the rough ER. Protein-containing transport vesicles pinch off the rough ER and migrate to fuse with the Golgi apparatus. As it passes through the Golgi apparatus, the protein product is sorted (and slightly modified). The product is then packaged within vesicles, which leave the Golgi apparatus and head for various destinations (pathways 1–3), as shown.

As proteins "tagged" for export accumulate in the Golgi apparatus, the sacs swell. Then their swollen ends, filled with protein, pinch off and form **secretory vesicles** (ves′ĭ-kuls), which travel to the plasma membrane. When the vesicles reach the plasma membrane, they fuse with it, the membrane ruptures, and the contents of the sac are ejected to the outside of the cell (pathway 1 in Figure 3.6). Mucus is packaged this way, as are digestive enzymes made by pancreas cells.

In addition to its packaging-for-release functions, the Golgi apparatus pinches off sacs containing proteins and phospholipids destined to become part of the plasma membrane (pathway 2 in Figure 3.6) and packages hydrolytic enzymes into membranous sacs called *lysosomes* that remain in the cell (pathway 3 in Figure 3.6).

Lysosomes **Lysosomes** (li′so-sōmz; "breakdown bodies"), which appear in different sizes, are membranous "bags" containing powerful digestive enzymes. Because lysosomal enzymes are capable of digesting worn-out or nonusable cell structures and most foreign substances that enter the cell, lysosomes function as the cell's demolition sites. Lysosomes are especially abundant in white blood cells that engulf bacteria and other potentially harmful substances because they digest and rid the body of such foreign invaders. As described above, the enzymes they contain are formed by ribosomes and packaged by the Golgi apparatus.

Homeostatic Imbalance

The lysosomal membrane is ordinarily quite stable, but it becomes fragile when the cell is injured or deprived of oxygen and when excessive amounts of vitamin A are present. Lysosomal rupture results in self-digestion of the cell. ▲

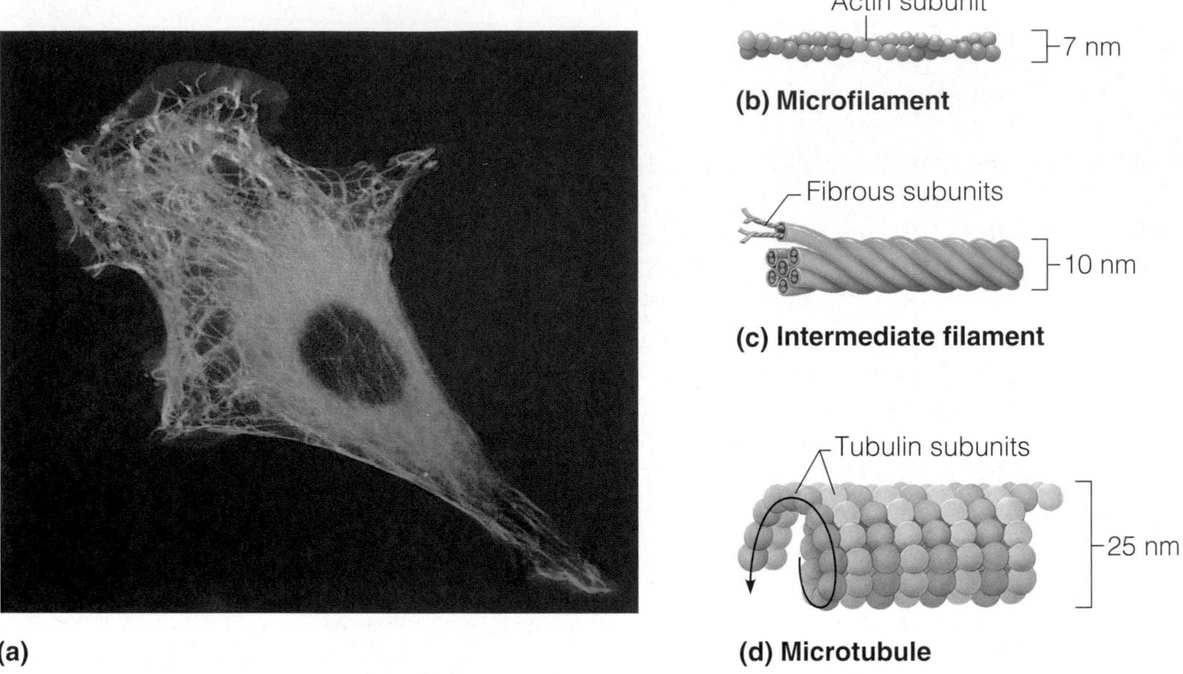

FIGURE 3.7 The cytoskeleton. (a) In this light micrograph of the cytoskeleton of a nerve cell, the microtubules appear green; the microfilaments are blue. Intermediate filaments form most of the rest of the network. **(b–d)** Diagrammatic views of the three types of cytoskeletal elements.

Peroxisomes **Peroxisomes** (per-ok′sih-sōmz) are membranous sacs containing powerful oxidase (ok′sĭ-dāz) enzymes that use molecular oxygen (O_2) to detoxify a number of harmful or poisonous substances, including alcohol and formaldehyde. However, their most important function is to "disarm" dangerous free radicals. **Free radicals** are highly reactive chemicals with unpaired electrons that can scramble the structure of proteins and nucleic acids. Although free radicals are normal by-products of cellular metabolism, if allowed to accumulate, they can have devastating effects on cells. Peroxisomes convert free radicals to hydrogen peroxide (H_2O_2), a function indicated in their naming (*peroxisomes* = "peroxide bodies"). The enzyme *catalase* (kat′ah-lās) then converts excess hydrogen peroxide to water. Peroxisomes are especially numerous in liver and kidney cells, which are very active in detoxification.

Although peroxisomes look like small lysosomes (see Figure 3.4), they do not arise by budding from the Golgi apparatus. Instead, they appear to replicate themselves by simply pinching in half, as do mitochondria.

Cytoskeleton An elaborate network of protein structures extends throughout the cytoplasm (see Figure 3.2). This network, or **cytoskeleton,** acts as a cell's "bones and muscles" by furnishing an internal framework that determines cell shape, supports other organelles, and provides the machinery needed for intracellular transport and various types of cellular movements. From its largest to its smallest elements, the cytoskeleton is made up of microtubules, intermediate filaments, and microfilaments (Figures 3.4 and 3.7). Although there is some overlap in roles, generally speaking the strong, stable ropelike **intermediate filaments** help form desmosomes (see Figure 3.3) and provide internal guy wires to resist pulling forces on the cell. **Microfilaments** (such as *actin* and *myosin*) are most involved in cell motility and in producing changes in cell shape. You could say that cells move when they get their act(in) together. The tubelike **microtubules** determine the overall shape of a cell and the distribution of organelles. They are very important during cell division, as described on pp. 80–81 and 83.

Centrioles The paired **centrioles** (sen′tre-ōlz) lie close to the nucleus (see Figure 3.4). They are rod-shaped bodies that lie at right angles to each other; internally they are made up of fine microtubules. During cell division, the centrioles direct the formation of the *mitotic spindle* (see Figure 3.15, p. 81).

In addition to the cell structures described above, some cells have projections called **cilia** (sil′e-ah; "eyelashes"), whiplike cellular extensions that move substances along the cell surface. For example, the ciliated cells of the respiratory system lining move mucus up and away from the lungs. Where cilia appear, there are usually many of them projecting from the exposed cell surface. When a cell is about to make cilia, its centrioles multiply and then line up beneath the plasma membrane at the free cell surface. Microtubules then begin to "sprout" from the centrioles and put pressure on the membrane, forming the projections. If the projections formed by the centrioles are substantially longer, they are called **flagella** (flah-jel′ah). The only example of a flagellated cell in the human body is the sperm, which has a single propulsive flagellum called its *tail* (Figure 3.8g). *Notice that cilia propel other substances across a cell's surface, whereas a flagellum propels the cell itself.*

Cell Diversity

So far in this chapter, we have focused on an average human cell. However, the trillions of cells in the human body are made up of some 200 different cell types that vary greatly in size, shape, and function. They include sphere-shaped fat cells, disk-shaped red blood cells, branching nerve cells, and cube-shaped cells of kidney tubules. Figure 3.8 illustrates how the shapes of cells and the relative numbers of the various organelles they contain relate to specialized cell functions. Let's take a look at some of these cell specialists.

1. **Cells that connect body parts:**
   - *Fibroblast.* The elongated shape of this cell lies along the cable-like fibers that it secretes. It also has an abundant rough ER and

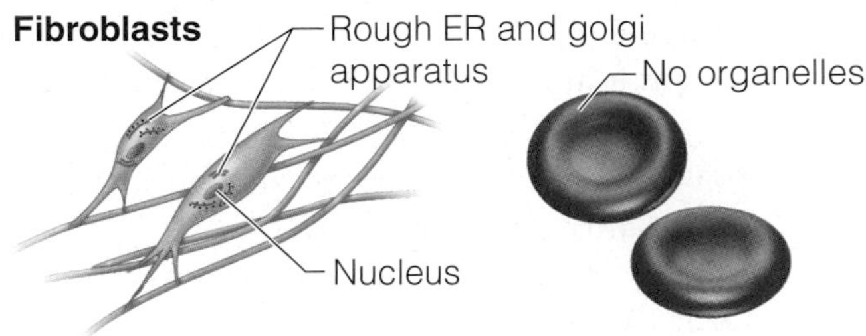

(a) Cells that connect body parts

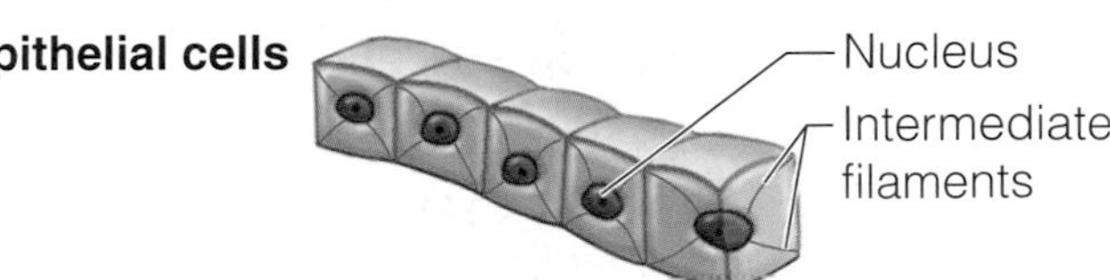

(b) Cells that cover and line body organs

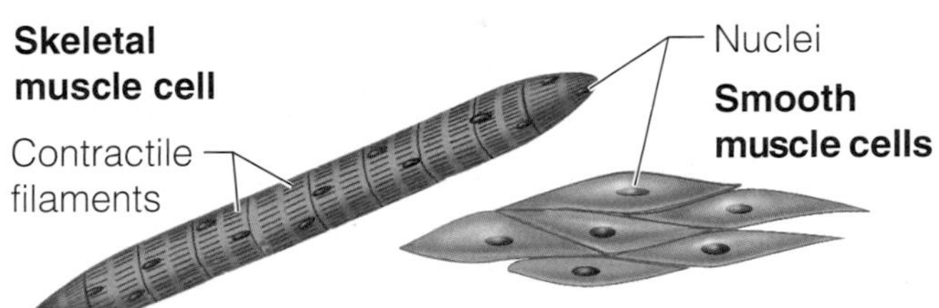

(c) Cells that move organs and body parts

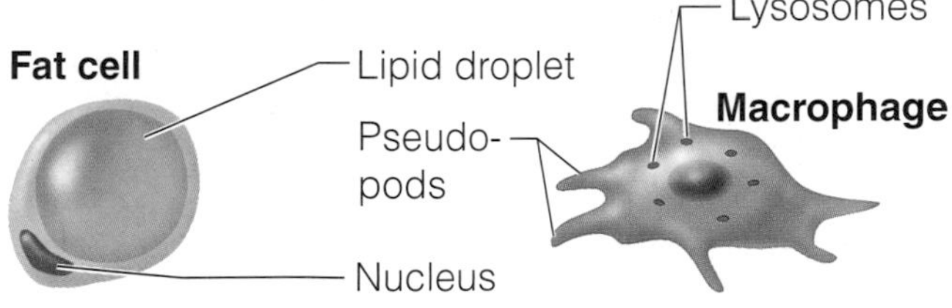

(d) Cell that stores nutrients

(e) Cell that fights disease

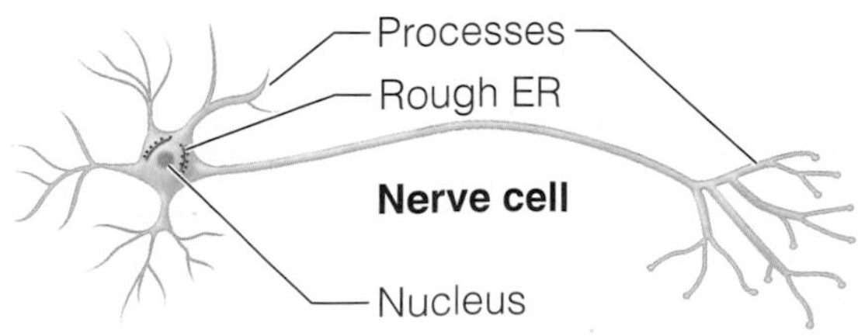

(f) Cell that gathers information and controls body functions

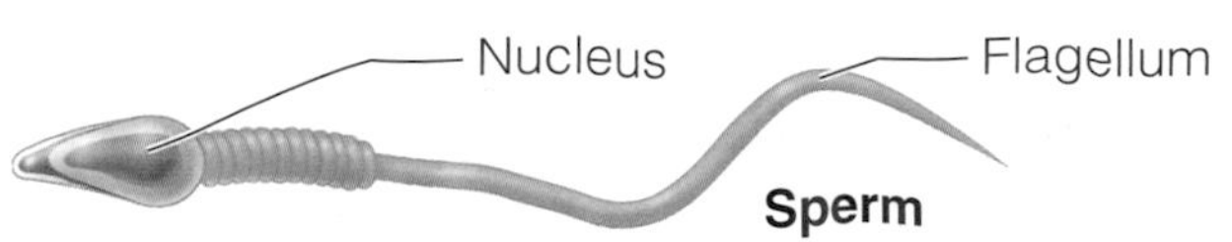

(g) Cell of reproduction

FIGURE 3.8 Cell diversity. The shape of human cells and the relative abundances of their various organelles relate to their function in the body.

a large Golgi apparatus, to make and secrete the protein building blocks of these fibers.

- *Erythrocyte (red blood cell).* This cell carries oxygen in the bloodstream. Its concave disk shape provides extra surface area for the uptake of oxygen and streamlines the cell so it flows easily through the bloodstream. So much oxygen-carrying pigment is packed in erythrocytes that all other organelles have been shed to make room.

2. **Cell that covers and lines body organs:**
   - *Epithelial cell.* The hexagonal shape of this cell is exactly like a "cell" in a honeycomb of a beehive. This shape allows epithelial cells to pack together in sheets. An epithelial cell has abundant intermediate filaments that resist tearing when the epithelium is rubbed or pulled.
3. **Cells that move organs and body parts:**
   - *Skeletal muscle* and *smooth muscle cells.* These cells are elongated and filled with abundant contractile filaments, so they can shorten forcefully and move the bones or change the size of internal organs.
4. **Cell that stores nutrients:**
   - *Fat cell.* The huge spherical shape of a fat cell is produced by a large lipid droplet in its cytoplasm.
5. **Cell that fights disease:**
   - *Macrophage (a phagocytic cell).* This cell extends long pseudopods ("false feet") to crawl through tissue to reach infection sites. The many lysosomes within the cell digest the infectious microorganisms it takes up.
6. **Cell that gathers information and controls body functions:**
   - *Nerve cell (neuron).* This cell has long processes for receiving messages and transmitting them to other structures in the body. The processes are covered with an extensive plasma membrane, and a plentiful rough ER is present to synthesize membrane components.
7. **Cells of reproduction:**
   - *Oocyte (female).* The largest cell in the body, this egg cell contains many copies of all organelles, for distribution to the daughter cells that arise when the fertilized egg divides to become an embryo.
   - *Sperm (male).* This cell is long and streamlined, built for swimming to the egg for fertilization. Its flagellum acts as a motile whip to propel the sperm.

Cell Physiology

As mentioned earlier, each of the cell's internal parts is designed to perform a specific function for the cell. Most cells have the ability to *metabolize* (use nutrients to build new cell material, break down substances, and make ATP), *digest foods, dispose of wastes, reproduce, grow, move,* and *respond to a stimulus* (irritability). Most of these functions are considered in detail in later chapters. For example, metabolism is covered in Chapter 14, and the ability to react to a stimulus is covered in Chapter 7. Here, we will consider only the functions of membrane transport (the means by which substances get through plasma membranes), protein synthesis, and cell reproduction (cell division).

Membrane Transport

The fluid environment on both sides of the plasma membrane is an example of a solution. It is important that you really understand solutions before we dive into an explanation of membrane transport. In the most basic sense, a **solution** is a homogeneous mixture of two or more components. Examples include the air we breathe (a mixture of gases), seawater (a mixture of water and salts), and rubbing alcohol (a mixture of water and alcohol). The substance present in the largest amount in a solution is called the **solvent** (or dissolving medium). Water is the body's chief solvent. Components or substances present in smaller amounts are called **solutes.** The solutes in a solution are so tiny; they do not settle out.

Intracellular fluid (collectively, the nucleoplasm and the cytosol) is a solution containing small amounts of gases (oxygen and carbon dioxide), nutrients, and salts, dissolved in water. So too is **interstitial fluid,** the fluid that continuously bathes the exterior of our cells. Interstitial fluid can be thought of as a rich, nutritious, and rather unusual "soup." It contains thousands of ingredients, including nutrients (amino acids, sugars, fatty acids, vitamins), regulatory substances such as hormones and neurotransmitters, salts, and waste products. To remain healthy, each cell must extract from this soup the exact amounts of the substances it needs at specific times and reject the rest.

The plasma membrane is a selectively permeable barrier. **Selective permeability** means that a barrier allows some substances to pass through it while excluding others. Thus, it allows nutrients to enter the cell but keeps many undesirable substances out. At the same time, valuable cell proteins and other substances are kept within the cell, and wastes are allowed to pass out of it.

Homeostatic Imbalance

The property of selective permeability is typical only of healthy, unharmed cells. When a cell dies or is badly damaged, its plasma membrane can no longer be selective and becomes permeable to nearly everything. This phenomenon is evident when someone has been severely burned. Precious fluids, proteins, and ions "weep" (leak out) from the dead and damaged cells of the burned areas. ▲

Movement of substances through the plasma membrane happens in basically two ways—passively or actively. In **passive transport processes,** substances are transported across the membrane without any energy input from the cell. In **active transport processes,** the cell provides the metabolic energy (ATP) that drives the transport process.

Passive Transport Processes: Diffusion and Filtration

Diffusion (dĭ-fu′zhun) is an important means of passive membrane transport for every cell of the body. The other passive transport process, *filtration,* generally occurs only across capillary walls. Let us examine how these two types of passive transport differ.

Diffusion **Diffusion** is the process by which molecules (and ions) tend to scatter themselves throughout the available space. All molecules possess *kinetic energy* (energy of motion), as described in Chapter 2. As the molecules move about randomly at high speeds, they collide and change direction with each collision. Since the overall effect of this erratic movement is that molecules move away from a region where they are more concentrated (more numerous) to a region where they are less concentrated (fewer of them), we say that molecules move *down* their **concentration gradient.** Because the driving force (source of energy) is the kinetic energy of the molecules themselves, the speed of diffusion is affected by the size of the molecules (the smaller the faster) and temperature (the warmer the faster).

FIGURE 3.9 Diffusion. Particles in solution move continuously and collide constantly with other particles. As a result, particles tend to move away from areas where they are most highly concentrated and to become evenly distributed, as illustrated by the diffusion of dye molecules in a beaker of water.

An example should help you understand diffusion. Picture yourself pouring a cup of coffee, and then adding a cube of sugar (but not stirring the cup). After adding the sugar, the phone rings, and you are called in to work. You never do get to drink the coffee. Upon returning that evening, you find that the coffee tastes sweet even though it was never stirred. This is because the sugar molecules moved around all day and eventually, as a result of their activity, became sufficiently distributed throughout the coffee to sweeten the entire cup. A laboratory example that might be familiar to some students is illustrated in Figure 3.9.

The plasma membrane is a physical barrier to diffusion. Molecules will move *passively* through the plasma membrane by diffusion if (1) they are small enough to pass through its pores or (2) they can dissolve in the fatty portion of the membrane (Figure 3.10). The unassisted diffusion of solutes through the plasma membrane (or any selectively permeable membrane) is called **simple diffusion** (Figure 3.10a). Solutes transported this way are either lipid-soluble (fats, fat-soluble vitamins, oxygen, carbon dioxide) or small enough to pass through the membrane pores (some small ions such as chloride ions, for example).

Q *What "facilitates" facilitated diffusion?*

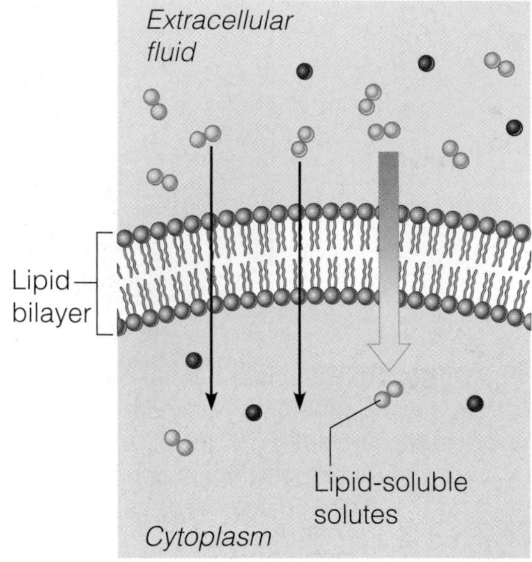

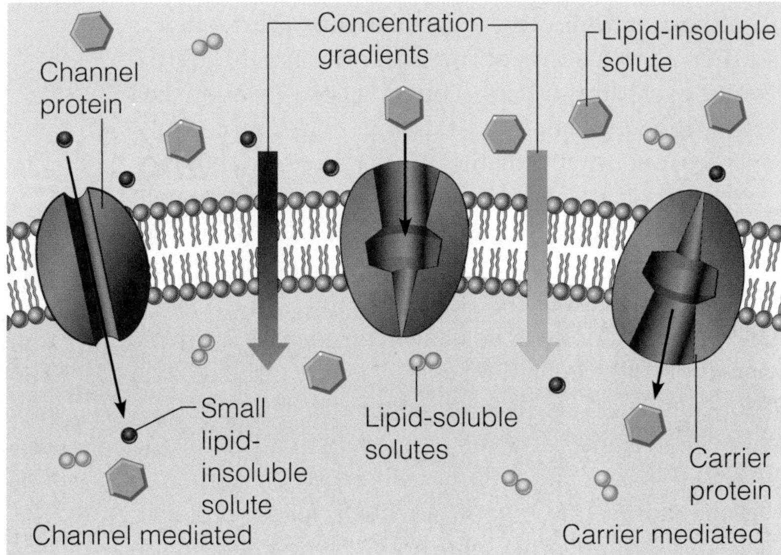

(a) Simple diffusion **(b) Facilitated diffusion**

FIGURE 3.10 Diffusion through the plasma membrane. (a) Simple diffusion. Lipid-soluble molecules diffuse directly through the lipid bilayer of the plasma membrane, in which they can dissolve. **(b)** Facilitated diffusion. On the left, small lipid-insoluble substances (water molecules or small ions) are shown diffusing through channels constructed by membrane proteins. On the right, facilitated diffusion moves large, lipid-insoluble molecules (e.g., glucose) across the membrane. The substance to be transported binds to a transmembrane protein carrier in this example.

Diffusion of water through a selectively permeable membrane such as the plasma membrane is specifically called **osmosis** (oz-mo′sis). Because water is highly polar, it is repelled by the (nonpolar) lipid core of the plasma membrane, but it can and does pass easily through special pores called *aquaporins* ("water pores") created by the proteins in the membrane. Osmosis into and out of cells is occurring all the time as water moves down its concentration gradient.

Still another example of diffusion is **facilitated diffusion** (see Figure 3.10b). Facilitated diffusion provides passage for certain needed substances (notably glucose) that are both lipid-insoluble and too large to pass through the membrane pores. Although facilitated diffusion follows the laws of diffusion—that is, the substances move down their own concentration gradient—a protein membrane channel is used or a protein molecule that acts as a carrier is needed as a transport vehicle. Hence, some of the proteins in the plasma membrane form channels or act as carriers to move glucose and certain other solutes passively across the membrane and make it available for cell use.

Substances that pass into and out of cells by diffusion save the cell a great deal of energy. When you consider how vitally important water, glucose, and oxygen are to cells, it becomes apparent just how necessary these passive transport processes really are. Glucose and oxygen continually move into the cells (where they are in lower concentration because the cells keep using them up), and carbon dioxide (a waste product of cellular activity) continually moves out of the cells into the blood (where it is in lower concentration).

A *Carrier proteins or protein channels.*

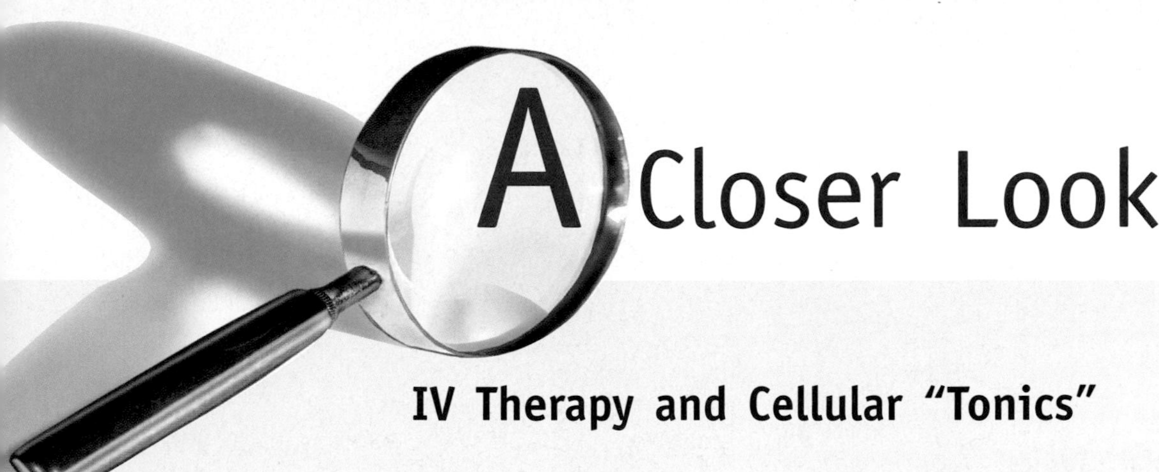

IV Therapy and Cellular "Tonics"

WHY is it essential that medical personnel give only the proper *intravenous (IV)*, or into-the-vein, *solutions* to patients?

Consider that there is a steady traffic of small molecules across the plasma membrane. Although diffusion of solutes across the membrane is rather slow, osmosis, which moves water across the membrane, occurs very quickly. Anyone administering an IV must use the correct solution to protect the patient's cells from life-threatening dehydration or rupture due to excessive water entry.

The tendency of a solution to hold water or "pull" water into it is called osmotic pressure. Osmotic pressure is directly related to the concentration of solutes in the solution. The higher the solute concentration, the greater the osmotic pressure and the greater the tendency of water to move into the solution. Many molecules, particularly proteins and some ions, are prevented from diffusing through the plasma membrane. Consequently, any change in their concentration on one side of the membrane forces water to move from one side of the membrane to the other, causing cells to lose or gain water. The ability of a solution to change the size and shape of cells by altering the amount of water they contain is called *tonicity* (ton-is'i-te; *ton* = strength).

Isotonic (i"so-ton'ik; "same tonicity") solutions (such as Ringer's lactate, 5 percent dextrose, and 0.9 percent saline) have the same solute and water concentrations as cells do. Isotonic solutions cause no visible changes in cells, and when such solutions are infused into the bloodstream, red blood cells retain their normal size and disclike shape

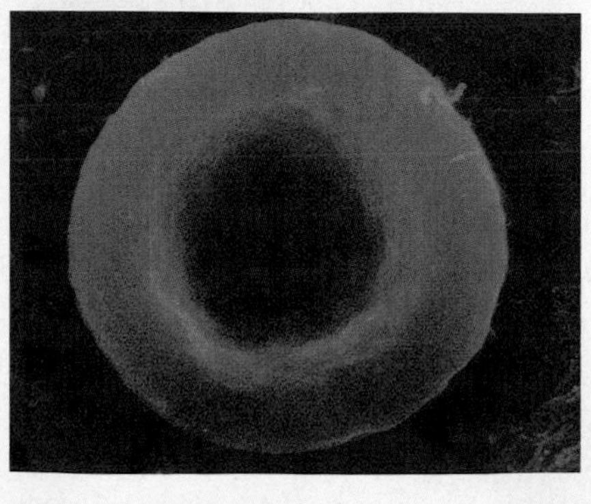

(a)

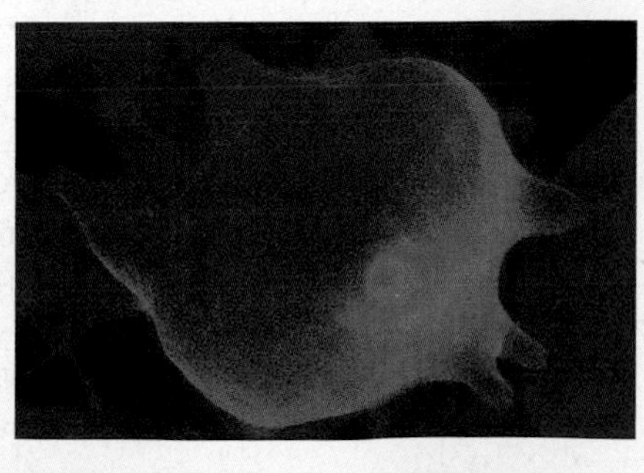

(b)

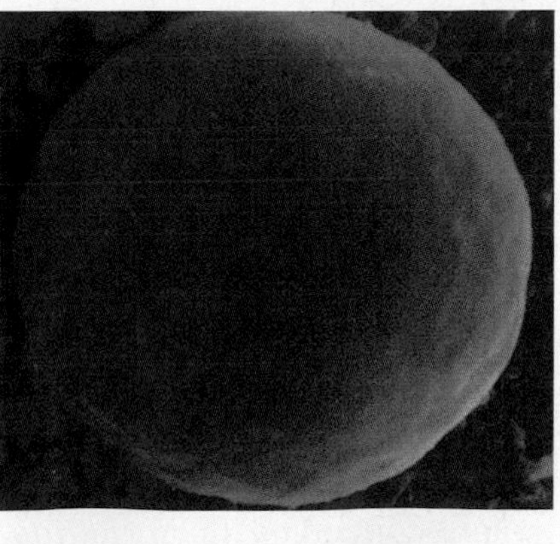

(c)

The effect of IV solutions of varying tonicity on living red blood cells.

IV Therapy and Cellular "Tonics" *(continued)*

(Photo a). As you might guess, interstitial fluid and most intravenous solutions are isotonic solutions.

If red blood cells are exposed to a *hypertonic* (hi″per-ton′ik) solution—a solution that contains more solutes, or dissolved substances, than there are inside the cells—the cells will begin to shrink, or *crenate* (kre′nāt). This is because water is in higher concentration inside the cell than outside, so it follows its concentration gradient and leaves the cell (Photo b). Hypertonic solutions are sometimes given to patients who have *edema* (swollen feet and hands because of fluid retention). Such solutions draw water out of the tissue spaces into the bloodstream so that excess fluid can be eliminated by the kidneys.

When a solution contains fewer solutes (and therefore more water) than the cell does, it is said to be *hypotonic* (hi″po-ton′ik) to the cell. Cells placed in hypotonic solutions plump up rapidly as water rushes into them (Photo c). Distilled water represents the most extreme example of a hypotonic fluid. Since it contains no solutes at all, water will enter cells until they finally burst, or *lyse*. Hypotonic solutions are sometimes infused intravenously (slowly and with care) to rehydrate the tissues of extremely dehydrated patients. In less extreme cases, drinking hypotonic fluids usually does the trick. (Many fluids that humans tend to drink regularly, such as tea, colas, apple juice, and sport drinks, are hypotonic.)

Filtration **Filtration** is the process by which water and solutes are forced through a membrane (or capillary wall) by *fluid,* or *hydrostatic, pressure.* In the body, hydrostatic pressure is usually exerted by the blood. Like diffusion, filtration is a passive process, and a gradient is involved. In filtration, however, the gradient is a **pressure gradient** that actually pushes solute-containing fluid *(filtrate)* from the higher-pressure area to the lower-pressure area. Filtration is necessary for the kidneys to do their job properly. In the kidneys, water and small solutes filter out of the capillaries into the kidney tubules because the blood pressure in the capillaries is greater than the fluid pressure in the tubules. Part of the filtrate formed in this way eventually becomes urine. Filtration is not very selective. For the most part, only blood cells and protein molecules too large to pass through the membrane pores are held back.

Active Transport Processes

Whenever a cell uses some of its ATP supply to move substances across the membrane, the process is referred to as *active.* Substances moved actively are usually unable to pass in the desired direction by diffusion. They may be too large to pass through membrane channels or the membrane lacks special protein carriers for their transport, they may not be able to dissolve in the fat core, or they may have to move "uphill" *against* their concentration gradients. The two most important examples of active transport mechanisms, *solute pumping* and *bulk transport,* are described next.

Solute Pumping *Solute pumping* (more simply called *active transport* by some) is similar to the carrier-mediated facilitated diffusion described earlier in that both processes require protein carriers

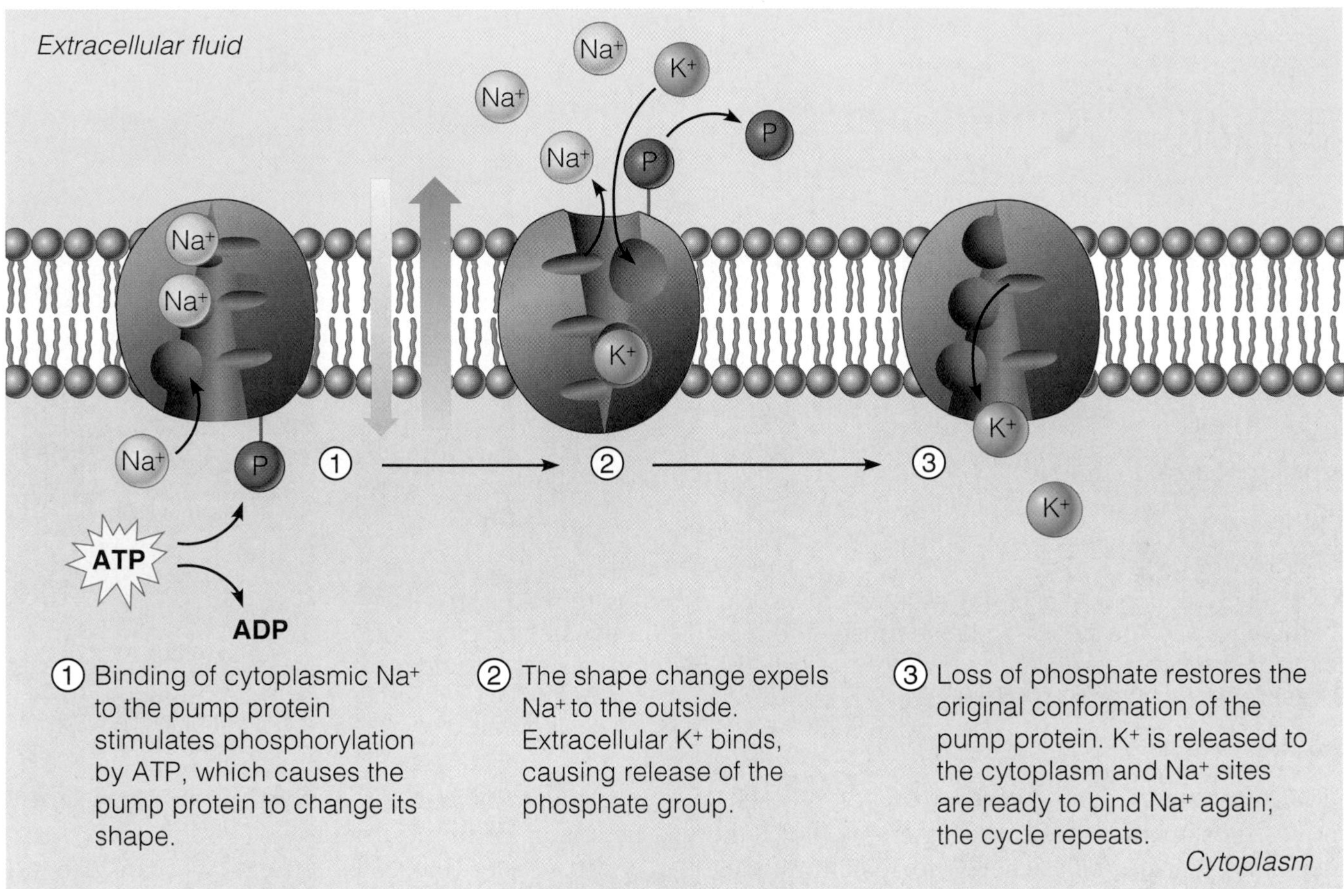

FIGURE 3.11 Operation of the sodium-potassium pump, a solute pump. ATP provides the energy for a "pump" protein to move three sodium ions out of the cell and two potassium ions into the cell. Both ions are moved against their concentration gradients.

that combine reversibly with the substances to be transported across the membrane. However, facilitated diffusion is driven by the kinetic energy of the diffusing molecules, whereas solute pumping uses ATP to energize its protein carriers, which are called **solute pumps.** Amino acids, some sugars, and most ions are transported by solute pumps, and in most cases these substances move *against* concentration (or electrical) gradients. This is opposite to the direction in which substances would naturally flow by diffusion, which explains the need for energy in the form of ATP. Amino acids are needed to build cellular proteins but are too large to pass through the membrane channels and are not lipid-soluble. The **sodium-potassium pump** that simultaneously carries sodium ions out of and potassium ions into the cell is absolutely necessary for normal transmission of impulses by nerve cells. Sodium ions (Na^+) are moved out of cells by solute pumps (Figure 3.11). There are more sodium ions outside the cells than inside, so they tend to remain in the cell unless the cell uses ATP to force, or "pump," them out. Likewise, there are relatively more potassium ions inside cells than in the interstitial (extracellular) fluid, and potassium ions that leak out of cells must be actively pumped back inside. Since each of the pumps in the plasma membrane transports only specific substances, solute pumping provides a way for the cell to be very selective in cases where substances cannot pass by diffusion. (No pump—no transport.)

Bulk Transport Some substances that cannot get through the plasma membrane in any other way are transported with the help of ATP out of or into cells by *bulk transport.* The two types of bulk transport are *exocytosis* and *endocytosis.*

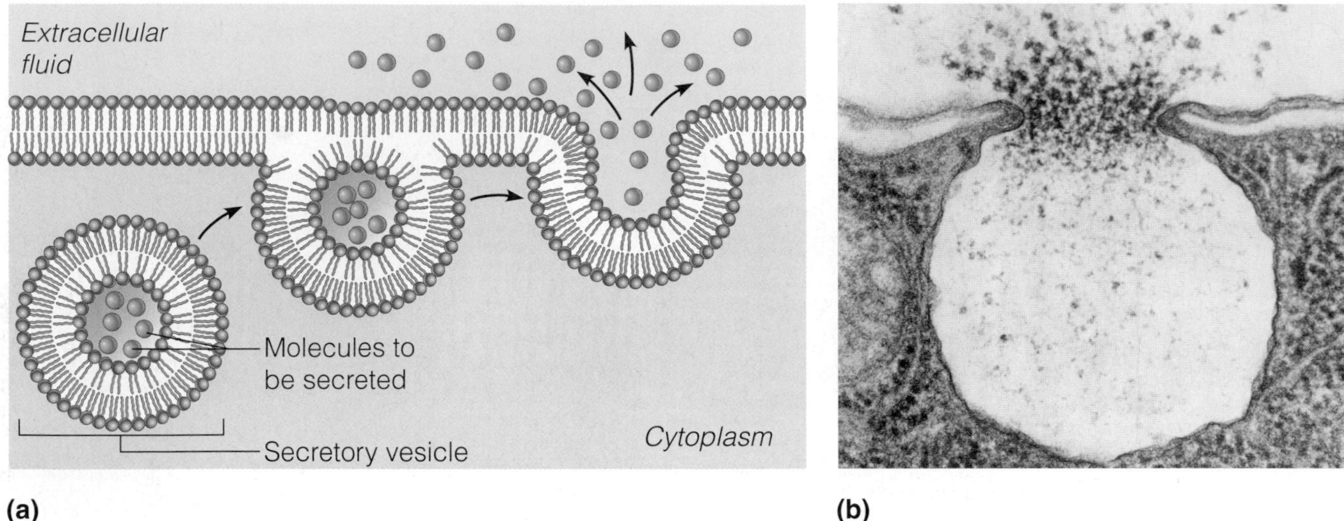

FIGURE 3.12 Exocytosis. (a) A secretory vesicle migrates to the plasma membrane, and the two membranes fuse. The fused site opens and releases the contents to the outside of the cell. **(b)** Electron micrograph of a vesicle in exocytosis (100,000×).

Exocytosis (ek″so-si-to′sis; "out of the cell") moves substances out of cells (Figure 3.12). It is the means by which cells actively secrete hormones, mucus, and other cell products or eject certain cellular wastes. The product to be released is first "packaged" (typically by the efforts of the Golgi apparatus) into a small membranous vesicle or sac. The sac migrates to the plasma membrane and fuses with it. The fused area then ruptures, spilling the sac contents out of the cell (also see Figure 3.6).

Endocytosis (en″do-si-to′sis; "into the cell") includes those ATP-requiring processes that take up, or engulf, extracellular substances by enclosing them in a small membranous vesicle (Figure 3.13). Once the vesicle, or sac, is formed, it detaches from the plasma membrane and moves into the cytoplasm, where it fuses with a lysosome and its contents are digested (by lysosomal enzymes). If the engulfed substances are relatively large particles such as bacteria or dead body cells, which are separated from the external environment by flowing cytoplasmic extensions called pseudopods, the endocytosis process is more specifically called **phagocytosis** (fag″o-si-to′sis), a term that means "cell eating" (Figure 3.13b). Certain white blood cells and other "professional" phagocytes of the body act as scavenger cells that police and protect the body by ingesting bacteria and other foreign debris as well as dead body cells. Hence, phagocytosis is a protective mechanism, not a means of getting nutrients.

If we say that cells can eat, we can also say that they can drink. This is **fluid-phase endocytosis,** also called **pinocytosis** (pi″no-si-to′sis; "cell drinking"). In this process the plasma membrane invaginates to form a tiny pit and then its edges fuse around the droplet of extracellular fluid containing dissolved proteins or fats (Figure 3.13a). Unlike phagocytosis, it is a routine activity of most cells. It is especially important in cells that function in absorption (for example, cells forming the lining of the small intestine and kidney tubule cells).

Receptor-mediated endocytosis is the main cellular mechanism for taking up specific target molecules (Figure 3.13c). In this process, plasma membrane receptor proteins bind only with certain substances, and both the receptors and high concentrations of the attached target molecules are internalized in a vesicle and then the contents of the vesicle are dealt with in one of the ways shown in Figure 3.13a. Although phagocytosis and pinocytosis are important, compared to receptor-mediated endocytosis, they are pretty unselective. Substances endocytosed by receptor-mediated endocytosis include enzymes, some hormones, cholesterol, and iron. Unfortunately, flu viruses also use this route to enter and attack our cells.

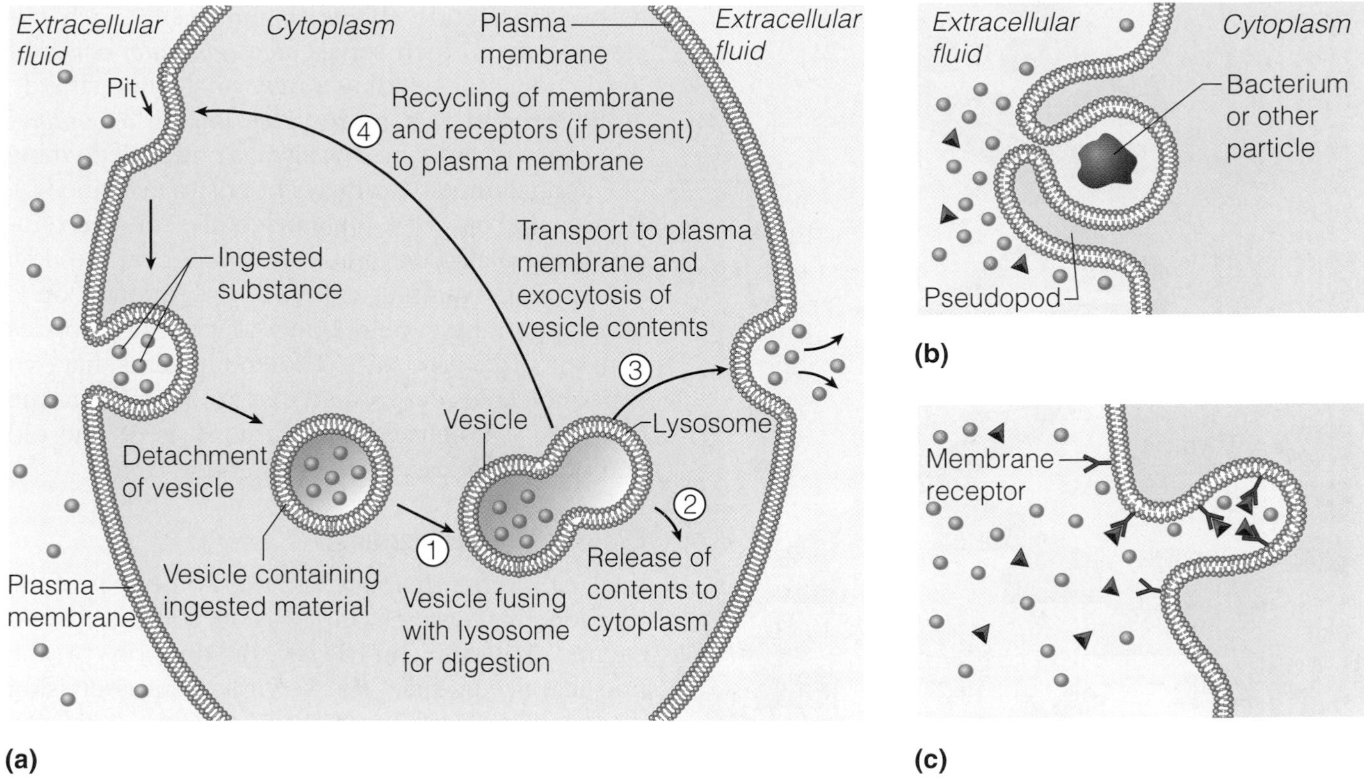

FIGURE 3.13 Events and types of endocytosis. **(a)** Sequence of events in endocytosis. Once the vesicle has detached from the plasma membrane its contents may be digested within a lysosome and then released to the cytoplasm (its membrane components, and receptors if present, are recycled to the plasma membrane); or the vesicle may be transported across the cell intact and then released to the cell exterior by exocytosis. The type illustrated is pinocytosis, also called fluid-phase endocytosis. **(b)** Phagocytosis. **(c)** Receptor-mediated endocytosis.

Cell Division

The **cell life cycle** is the series of changes a cell goes through from the time it is formed until it divides. The cycle has two major periods: **interphase,** in which the cell grows and carries on its usual metabolic activities, and **cell division,** during which it reproduces itself. Although the term *interphase* leads one to believe that it is merely a resting time between the phases of cell division, this is not the case. During interphase, which is by far the longer phase of the cell cycle, the cell is very active and is resting *only* from division. A more accurate name for interphase would be *metabolic phase.*

Preparations: DNA Replication

The function of cell division is to produce more cells for growth and repair processes. Because it is essential that all body cells have the same genetic material, an important event *always precedes* cell division: the genetic material (the DNA molecules that form part of the chromatin) is duplicated exactly. This occurs toward the end of the cell's interphase period.

You will recall from Chapter 2 that DNA is a very complex molecule. It is composed of building blocks called *nucleotides,* each consisting of deoxyribose sugar, a phosphate group, and a nitrogen-containing base. Essentially DNA is a *double helix,* a ladderlike molecule that is coiled into a spiral staircase shape. The upright parts of the DNA "ladder" are alternating phosphate and sugar units, and the rungs of the ladder are made of pairs of nitrogen-containing bases.

The precise trigger for DNA synthesis is unknown, but once it starts, it continues until all the DNA has been replicated. The process begins as the DNA helix uncoils and gradually separates into

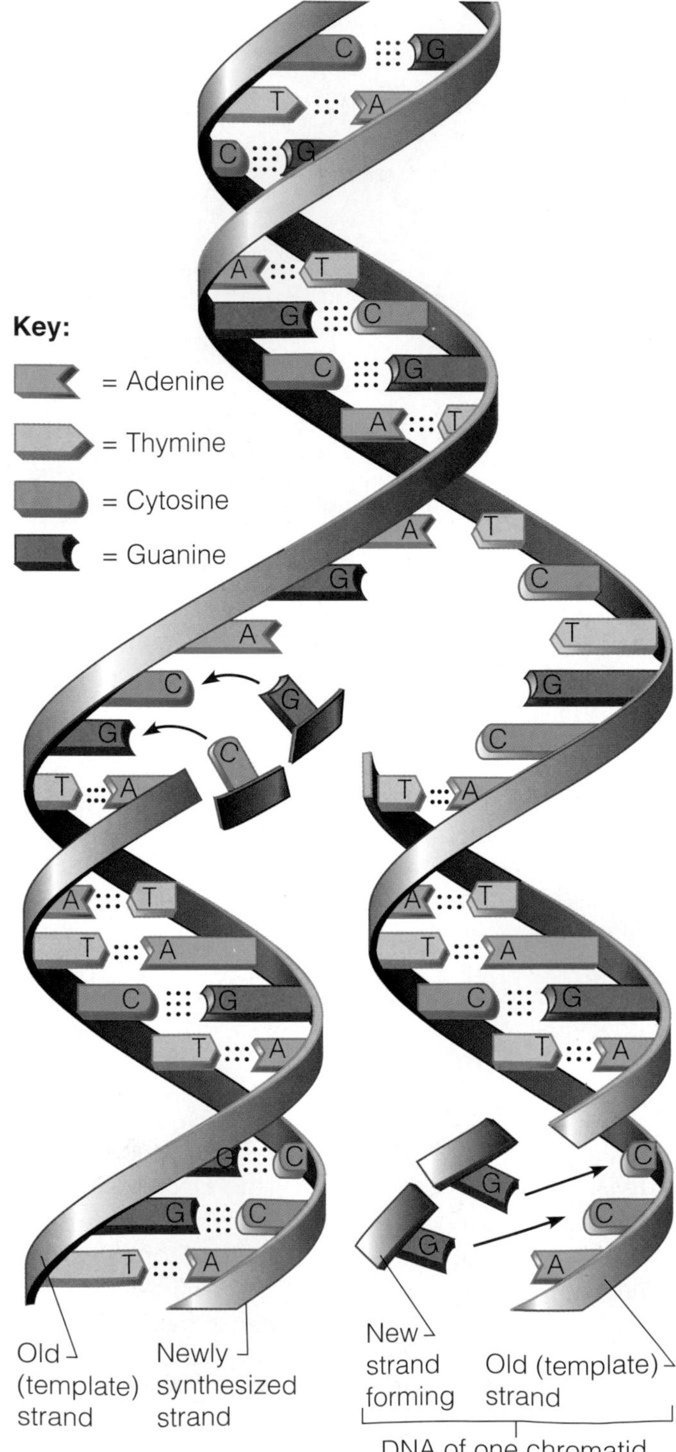

FIGURE 3.14 Replication of the DNA molecule during interphase. The DNA helix unwinds (center), and its nucleotide strands are separated. Each strand then acts as a template for building a new complementary strand. As a result, two helixes, each identical to the original DNA helix, are formed.

its two nucleotide chains (Figure 3.14). Each nucleotide strand then serves as a *template,* or set of instructions, for building a new nucleotide strand.

Remember that nucleotides join in a *complementary* way: adenine (A) always bonds to thymine (T), and guanine (G) always bonds to cytosine (C). Hence, the order of the nucleotides on the template strand also determines the order on the new strand. For example, a TACTGC sequence on a template strand would bond to new nucleotides with the order ATGACG. The end result is that two DNA molecules are formed that are identical to the original DNA helix, and each consists of one old and one newly assembled nucleotide strand.

Events of Cell Division

In all cells other than bacteria and some cells of the reproductive system, cell division consists of two events. **Mitosis** (mi-to″sis), or division of the nucleus, occurs first. The second event is division of the cytoplasm, **cytokinesis** (si′to-kĭ-ne″sis), which begins when mitosis is nearly completed.

Mitosis Mitosis results in the formation of two daughter nuclei with exactly the same genes as the mother nucleus. As explained above, DNA replication precedes mitosis, so that for a short time the cell nucleus contains a double dose of genes. When the nucleus divides, each *daughter cell* ends up with *exactly* the same genetic information as the original mother cell and the original fertilized egg from which it came.

The stages of mitosis, diagrammed in Figure 3.15, include the following events:

- **Prophase** (pro′faz). As cell division begins, the chromatin threads coil and shorten so that visible barlike bodies called **chromosomes** (*chromo* = colored; *soma* = body) appear. Because DNA replication has already occurred, each chromosome is actually made up of two strands, each called a **chromatid** (kro′mah-tid), held together by a small buttonlike body called a **centromere** (sen′tro-mer). The centrioles separate from each other and begin to move toward opposite sides of the cell, directing the assembly of a **mitotic spindle** (composed of thin microtubules) between them as they move. The spindle provides a scaffolding for the attachment and movement of the chromosomes during the later mitotic stages. By the end of prophase, the nuclear envelope and the nucleoli have broken down and disappeared,

Q *When in the cell cycle is each chromosome made up of two chromatids?*

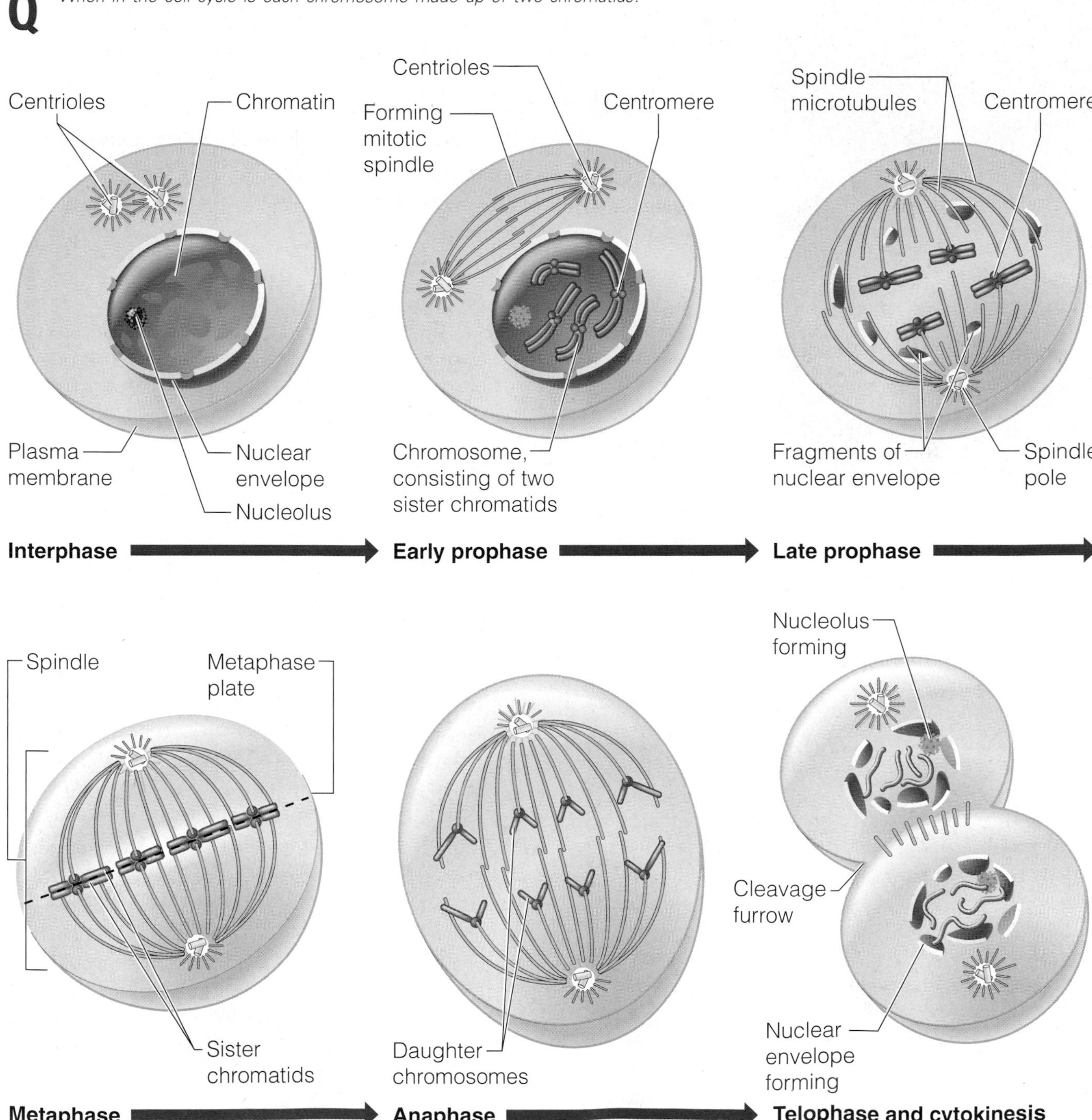

FIGURE 3.15 Stages of mitosis.

and the chromosomes have attached randomly to the spindle fibers by their centromeres.

- **Metaphase** (met′ah-faz). In this short stage, the chromosomes cluster and become aligned at the *metaphase plate* (the center of the spindle midway between the centrioles) so that a straight line of chromosomes is seen.
- **Anaphase** (an′ah-faz). During anaphase, the centromeres that have held the chromatids together split. The chromatids (now called chromosomes again) begin to move slowly

A *From the end of interphase to early anaphase.*

"Anatomy and physiology classes make up a large part of the curriculum in medical transcription training programs."

Focus on Careers

Medical Transcriptionist

"If you have a basic understanding of anatomy and medical terminology, you will be much more accurate at interpreting and transcribing what you hear."

EVERY time you consult a doctor or are hospitalized, records are kept to help healthcare professionals evaluate and monitor your condition. Medical transcriptionists play a key role in creating and maintaining these vital documents.

A medical transcriptionist is a medical language specialist who interprets and transcribes notes dictated by physicians and other healthcare professionals. These reports, which cover all aspects of a patient's assessment, diagnosis, treatment, and outcome, become part of the person's confidential medical record. Medical transcriptionists work in hospitals, clinics, doctors' offices, transcription services, insurance companies, and home healthcare agencies. Some freelance as independent contractors or subcontractors.

What does it take to be a transcriptionist? "Certainly, you need a good English background," says Pamela Shull, an experienced transcriptionist and administrative assistant at Santa Clara Valley Medical Center in San Jose, California. "Strong grammar, spelling, and punctuation skills are crucial. Physicians often dictate these records on the go, perhaps as they're walking down a hospital corridor or driving back to the office, and a good transcriptionist must be able to edit the dictated material for grammar and clarity."

Knowledge of anatomy and physiology, however, is even more important. Notes Shull, "If you have a basic understanding of anatomy and medical terminology, you will be much more accurate at interpreting and transcribing what you hear. A hospital transcriptionist deals with terms from a wide variety of medical specialties—one dictation might be from a gynecologist, the next from an orthopedic surgeon, and the next from a pediatrician. You never know what you'll encounter next."

All health professionals who treat a patient, now and in the future, must rely on these typed documents, so accurate transcription is vital: "I see the transcriptionist as a partner with physicians. We work with them to create excellent medical records, so patients will always be assured of receiving the best and most appropriate care possible."

This is why anatomy and physiology, medical terminology, and disease processes make up most of the curriculum in medical transcription training programs. Classes for medical transcription are offered through community colleges, proprietary schools, and home-study programs. Training programs vary in length from several months to two years and accreditation procedures vary from state to state. The American Association for Medical Transcription (AAMT) recommends a two-year program that includes at least 240 hours of on-the-job training in a healthcare facility. The AAMT also offers a voluntary certification exam to become a certified medical transcriptionist.

Shull enjoys the variety of medical transcription work. "It's fascinating because you get to follow each patient's story, from the initial problem to diagnosis to treatment," she says. "You feel like you get to know these people. It's like watching a gripping television drama—only this is real life!"

For more information, contact the American Association for Medical Transcription:
100 Sycamore Avenue
Modesto, CA 95354
Telephone: (800) 982-2182 or
(209) 527-9620
Fax: (209) 527-9633
http://www.aamt.org

For additional information on this career, click the Focus on Careers link at www.anatomyandphysiology.com.

apart, drawn toward opposite ends of the cell. The chromosomes seem to be pulled by their half-centromeres, with their "arms" dangling behind them. Anaphase is over when chromosome movement ends.

- **Telophase** (tel′o-faz). Telophase is essentially prophase in reverse. The chromosomes at opposite ends of the cell uncoil to become threadlike chromatin again. The spindle breaks down and disappears, a nuclear envelope forms around each chromatin mass, and nucleoli appear in each of the daughter nuclei.

Mitosis is basically the same in all animal cells. Depending on the type of tissue, it takes from 5 minutes to several hours to complete, but typically it lasts about 2 hours. Centriole replication is deferred until late interphase of the next cell cycle, when DNA replication begins before the onset of mitosis.

Cytokinesis Cytokinesis, or the division of the cytoplasm, usually begins during late anaphase and completes during telophase. Due to the activity of a contractile ring made of microfilaments, a **cleavage furrow** appears over the midline of the spindle, and it eventually squeezes or pinches the original cytoplasmic mass into two parts. Thus, at the end of cell division, two daughter cells exist. Each is smaller and has less cytoplasm than the mother cell, but it is genetically identical to it. The daughter cells grow and carry out normal cell activities until it is their turn to divide.

Although mitosis and division of the cytoplasm usually go hand in hand, in some cases the cytoplasm is not divided. This condition leads to the formation of *binucleate* (two nuclei) or *multinucleate* cells. This is fairly common in the liver.

As mentioned earlier, mitosis provides the "new" cells for body growth in youth and is necessary to repair body tissue all through life. Mitosis gone wild is the basis for tumors and cancers.

Protein Synthesis

Genes: The Blueprint for Protein Structure

In addition to replicating itself for cell division, DNA serves as the master blueprint for protein syntheses. Traditionally, a **gene** is defined as a DNA segment that carries the information for building one protein or polypeptide chain.

Proteins are key substances for all aspects of cell life. As described in Chapter 2, *fibrous (structural) proteins* are the major building materials for cells. Other proteins, the *globular (functional) proteins,* do things other than build structures. For example, all **enzymes,** biological catalysts that regulate chemical reactions in the cells, are functional proteins. The importance of enzymes cannot be overstated. Every chemical reaction that goes on in the body requires an enzyme. It follows that DNA regulates cell activities largely by specifying the structure of enzymes, which in turn control or direct the chemical reactions in which carbohydrates, fats, other proteins, and even DNA itself are made and broken down.

How does DNA bring about its miracles? It appears that DNA's information is encoded in the sequence of bases along each side of the ladderlike DNA molecules. Each sequence of *three* bases (a *triplet*) calls for a particular *amino acid* (Figure 3.16). (Amino acids are the building blocks of proteins that are joined during protein synthesis.) For example, a DNA base sequence of AAA specifies an amino acid called phenylalanine, while CCT calls for glycine. Just as different arrangements of notes on sheet music are played as different melodies, variations in the arrangements of A, C, T, and G in each gene allow cells to make all the different kinds of proteins needed. It has been estimated that a single gene has between 300 and 3000 base pairs in sequence.

The Role of RNA

By itself, DNA is rather like a strip of magnetic recording tape; its information is not useful until it is decoded. Furthermore, most ribosomes—the manufacturing sites for proteins—are in the cytoplasm, but in interphase cells DNA never leaves the nucleus. Thus, DNA requires not only a decoder but also a messenger to achieve its task of specifying the structure of proteins to be built at the ribosomes. These messenger and decoder functions are carried out by the second type of nucleic acid, called **ribonucleic** (ri″bo-nu-kle′ik) **acid,** or **RNA.**

As you learned in Chapter 2, RNA differs from DNA in being single-stranded and in having ribose sugar instead of deoxyribose and a uracil (U) base instead of thymine (T). Three varieties of RNA play a special role in protein synthesis. **Transfer RNA (tRNA) molecules** are small cloverleaf-shaped

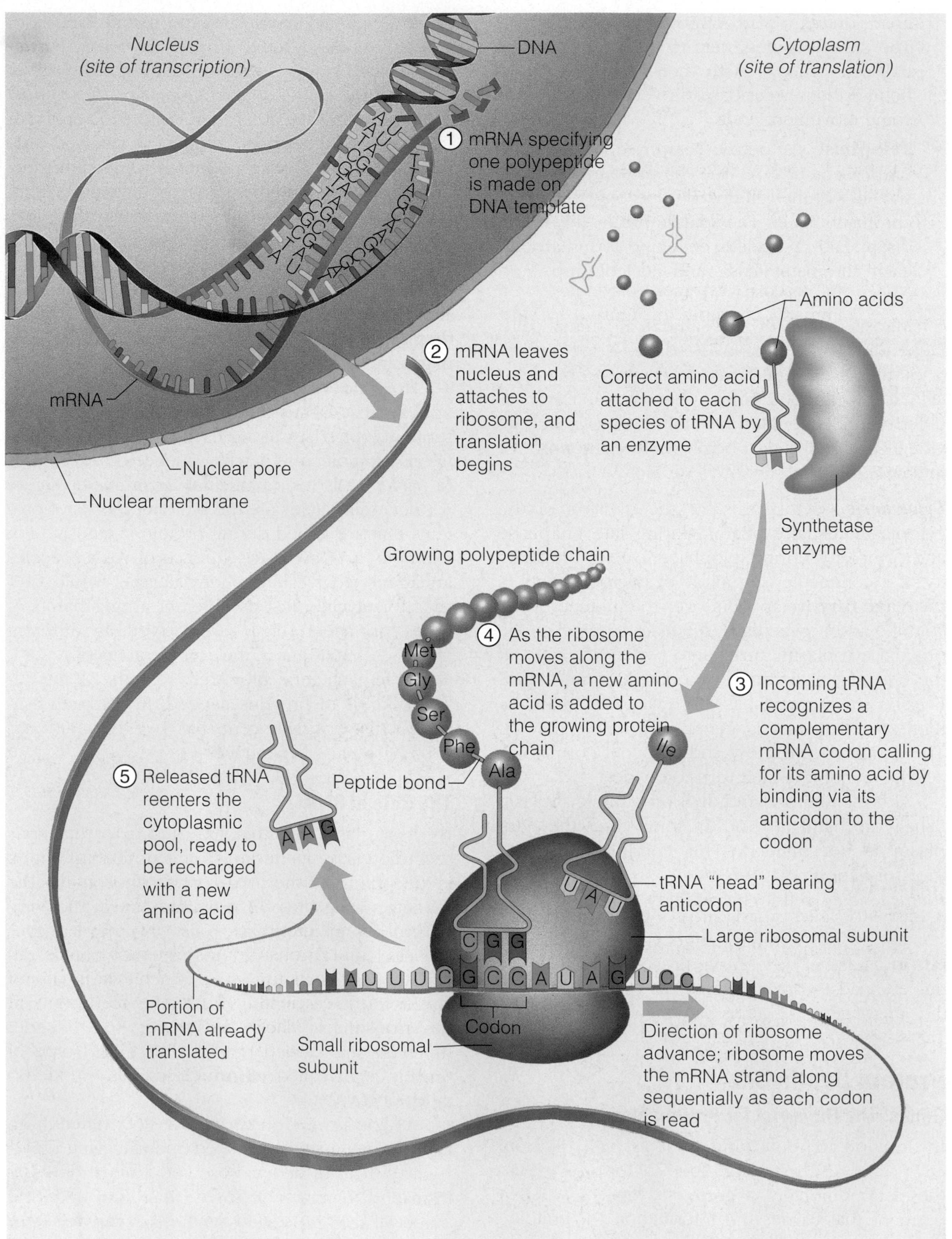

FIGURE 3.16 Protein synthesis. (①) Transcription. (②–⑤) Translation.

molecules. **Ribosomal RNA (rRNA)** helps form the ribosomes, where proteins are built. **Messenger RNA (mRNA) molecules** are long, single nucleotide strands that resemble half of a DNA molecule and carry the "message" containing instructions for protein synthesis from the DNA gene in the nucleus to the ribosomes in the cytoplasm.

Protein synthesis involves two major phases: *transcription,* when complementary mRNA is made at the DNA gene, and *translation,* when the information carried in mRNA molecules is "decoded" and used to assemble proteins. These steps are summarized simply in Figure 3.16, and described in more detail next.

Transcription

The word *transcription* often refers to one of the jobs done by a secretary—converting notes from one form (shorthand notes or an audiotape recording) into another form (a typewritten letter, for example). In other words, the same information is transformed from one form or format to another. In cells, **transcription** involves the transfer of information from DNA's base sequence into the *complementary* base sequence of mRNA (Figure 3.16, step 1). Only DNA and mRNA are involved in transcription. Whereas each three-base sequence specifying a particular amino acid on the DNA gene is called a **triplet,** the corresponding three-base sequences on mRNA are called **codons.** The form is different, but the same information is being conveyed. Thus, if the (partial) sequence of DNA triplets is AAT-CGT-TCG, the related codons on mRNA would be UUA-GCA-AGC.

Translation

A translator takes words in one language and restates them in another language. In the **translation phase** of protein synthesis, the language of nucleic acids (base sequence) is "translated" into the language of proteins (amino acid sequence). Translation occurs in the cytoplasm and involves three major varieties of RNA. As illustrated in Figure 3.16, steps 2–5, translation consists of the following events. Once the mRNA attaches to the ribosome (step 2), tRNA comes into the picture. Its job is to transfer, or ferry, amino acids to the ribosome, where they are bound together by enzymes in the exact sequence specified by the gene (and its mRNA). There are about 45 common types of tRNAs, each capable of carrying one of the 20 or so common types of amino acid to the protein synthesis sites. But that is not the only job of the tiny tRNAs. They also have to recognize the mRNA codons "calling for" the amino acid they are toting. They can do this because they have a special three-base sequence called an **anticodon** on their "head" that can bind to the complementary codons (step 3).

Once the first tRNA has maneuvered itself into the correct position at the beginning of the mRNA message, the ribosome moves the mRNA strand along, bringing the next codon into position to be read by another tRNA. As amino acids are brought to their proper positions along the length of mRNA, they are joined together by enzymes (step 4). As an amino acid is bonded to the chain, its tRNA is released and moves away from the ribosome to pick up another amino acid (step 5). When the last codon (the termination, or "stop," codon) is read, the protein is released.

PART II: BODY TISSUES

The human body, complex as it is, starts out as a single cell, the fertilized egg, which divides almost endlessly. The millions of cells that result become specialized for particular functions. Some become muscle cells, others the transparent lens of the eye, still others skin cells, and so on. Thus, there is a division of labor in the body, with certain groups of highly specialized cells performing functions that benefit the organism as a whole.

Cell specialization carries with it certain hazards. When a small group of cells is indispensable, its loss can disable or even destroy the body. For example, the action of the heart depends on a very specialized cell group in the heart muscle that controls its contractions. If those particular cells are damaged or stop functioning, the heart will no longer work efficiently, and the whole body will suffer or die from lack of oxygen.

Groups of cells that are similar in structure and function are called **tissues.** The four primary tissue types—epithelium, connective tissue, nervous tissue, and muscle—interweave to form the fabric of the body. If we had to assign a single term to each primary tissue type that would best describe its overall role, the terms would most likely be *covering* (epithelium), *support* (connective), *movement* (muscle), and *control* (nervous). However, these terms reflect only a tiny fraction of the functions that each of these tissues performs.

As explained in Chapter 1, tissues are organized into *organs* such as the heart, kidneys, and lungs. Most organs contain several tissue types, and the arrangement of the tissues determines each organ's structure and what it is able to do. Thus, a study of tissues should be helpful in your later study of the body's organs and how they work.

For now, we want to become familiar with the major similarities and differences in the primary tissues. Because epithelium and some types of connective tissue will not be considered again, they are emphasized more in this section than are muscle, nervous tissues, and bone (a connective tissue), which are covered in more depth in later chapters.

Epithelial Tissue

Epithelial tissue, or **epithelium** (ep″ĭ-the′le-um; *epithe* = laid on, covering) is the *lining, covering,* and *glandular tissue* of the body. Glandular epithelium forms various glands in the body. Covering and lining epithelium covers all free body surfaces and contains versatile cells. One type forms the outer layer of the skin. Others dip into the body to line its cavities. Since epithelium forms the boundaries that separate us from the outside world, nearly all substances given off or received by the body must pass through epithelium.

Epithelial functions include *protection, absorption, filtration,* and *secretion.* For example, the epithelium of the skin protects against bacterial and chemical damage and that lining the respiratory tract has cilia, which sweep dust and other debris away from the lungs. Epithelium specialized to absorb substances lines some digestive system organs such as the stomach and small intestine, which absorb food into the body. In the kidneys, epithelium both absorbs and filters. Secretion is a specialty of the glands, which produce such substances as perspiration, oil, digestive enzymes, and mucus.

Special Characteristics of Epithelium

Epithelium generally has the characteristics listed below:

- Except for glandular epithelium (described on p. 90), epithelial cells fit closely together to form continuous sheets. Neighboring cells are bound together at many points by cell junctions, including desmosomes and tight junctions.
- The membranes always have one free (unattached) surface or edge. This so-called **apical surface** is exposed to the body's exterior or to the cavity of an internal organ. The exposed surfaces of some epithelia are slick and smooth, but others exhibit cell surface modifications, such as microvilli or cilia.
- The lower surface of an epithelium rests on a **basement membrane,** a structureless material secreted by the cells.
- Epithelial tissues have no blood supply of their own (that is, they are *avascular*) and depend on diffusion from the capillaries in the underlying connective tissue for food and oxygen.
- If well nourished, epithelial cells regenerate themselves easily.

Classification of Epithelium

Each epithelium is given two names. The first indicates the relative number of cell layers it has (Figure 3.17a). The classifications by cell arrangement (layers) are **simple epithelium** (one layer of cells) and **stratified epithelium** (more than one cell layer). The second describes the shape of its cells (Figure 3.17b). On this basis there are *squamous* (skwa′mus) *cells,* flattened like fish scales (*squam* = scale), *cuboidal* (ku-boi′dal) *cells,* which are cube-shaped like dice, and *columnar cells,* shaped like columns. The terms describing the shape and arrangement are then combined to describe the epithelium fully. Stratified epithelia are named for the cells at the *free surface* of the epithelial membrane, not those resting on the basement membrane.

Simple Epithelia

The simple epithelia are most concerned with absorption, secretion, and filtration. Because simple epithelia are usually very thin, protection is not one of their specialties.

Simple Squamous Epithelium **Simple squamous epithelium** is a single layer of thin squamous cells resting on a basement membrane. The cells fit closely together, much like floor tiles. This type of epithelium usually forms membranes where filtration or exchange of substances by rapid diffusion occurs. It is in the air sacs of the lungs, where

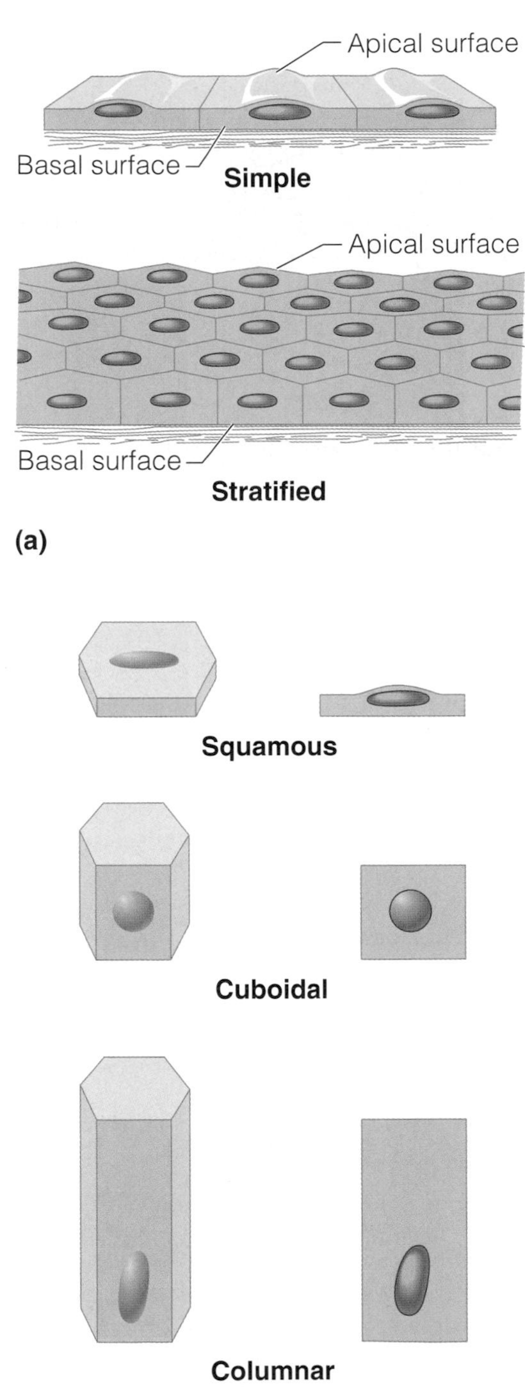

FIGURE 3.17 Classification of epithelia. (a) Classification on the basis of arrangement (layers). **(b)** Classification on the basis of cell shape; for each category, a whole cell is shown on the left and a longitudinal section is shown on the right.

oxygen and carbon dioxide are exchanged (Figure 3.18a), and it forms the walls of capillaries, where nutrients and gases pass between the tissue cells and the blood in the capillaries. Simple squamous epithelium also forms **serous membranes,** or **serosae** (se-ro′se), the slick membranes that line the ventral body cavity and cover the organs in that cavity. The serous membranes are described in more detail in Chapter 4.

Simple Cuboidal Epithelium **Simple cuboidal epithelium,** which is one layer of cuboidal cells resting on a basement membrane, is common in glands and their ducts (for example, the salivary glands and pancreas). It also forms the walls of the kidney tubules and covers the surface of the ovaries (Figure 3.18b).

Simple Columnar Epithelium **Simple columnar epithelium** is made up of a single layer of tall cells that fit closely together. **Goblet cells,** which produce a lubricating mucus, are often seen in this type of epithelium. Simple columnar epithelium lines the entire length of the digestive tract from the stomach to the anus (Figure 3.18c). Epithelial membranes that line body cavities open to the body exterior are called **mucosae** (mu-ko′se) or **mucous membranes.**

Pseudostratified Columnar Epithelium All of the cells of **pseudostratified** (soo″do-stră′tĭ-fīd) **columnar epithelium** rest on a basement membrane. However, some of its cells are shorter than others, and their nuclei appear at different heights above the basement membrane. As a result, this epithelium gives the false (*pseudo*) impression that it is stratified; hence its name. Like simple columnar epithelium, this variety mainly functions in absorption and secretion. A ciliated variety (more precisely called *pseudostratified ciliated columnar epithelium*) lines most of the respiratory tract (Figure 3.18d). The mucus produced by the goblet cells in this epithelium traps dust and other debris, and the cilia propel the mucus upward and away from the lungs.

Stratified Epithelia

Stratified epithelia consist of two or more cell layers. Being considerably more durable than the simple epithelia, these epithelia function primarily to protect.

Stratified Squamous Epithelium **Stratified squamous epithelium** is the most common stratified epithelium in the body. It usually consists of several layers of cells. The cells at the free edge are squamous cells, whereas those close to the basement membrane are cuboidal or columnar. Stratified

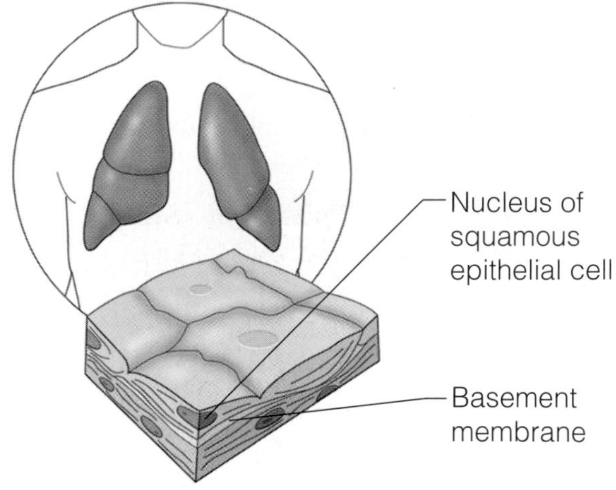

(a) **Diagram:** Simple squamous

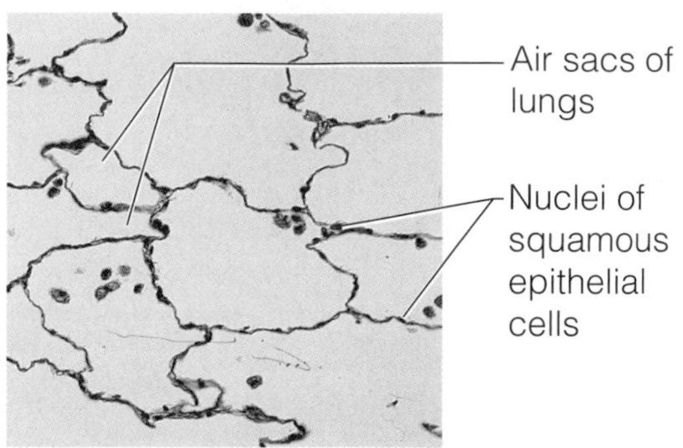

Photomicrograph: Simple squamous epithelium forming part of the alveolar (air sac) walls (400x).

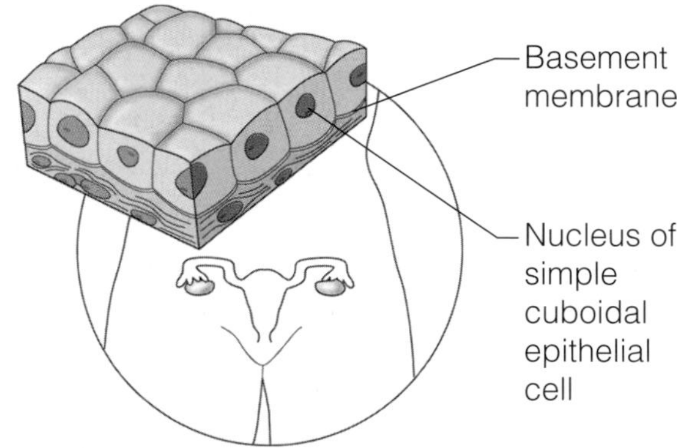

(b) **Diagram:** Simple cuboidal

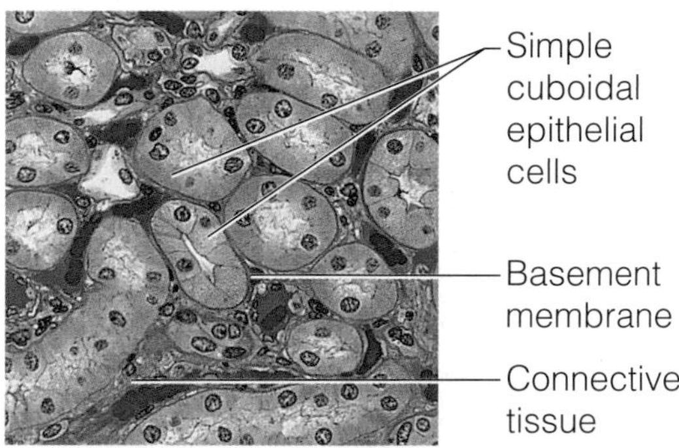

Photomicrograph: Simple cuboidal epithelium in kidney tubules (400x).

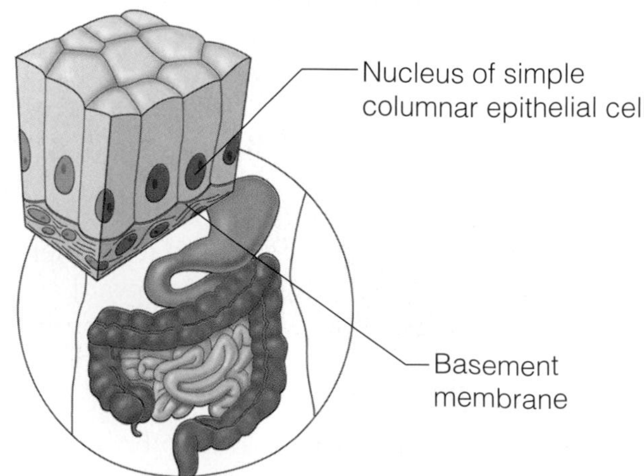

(c) **Diagram:** Simple columnar

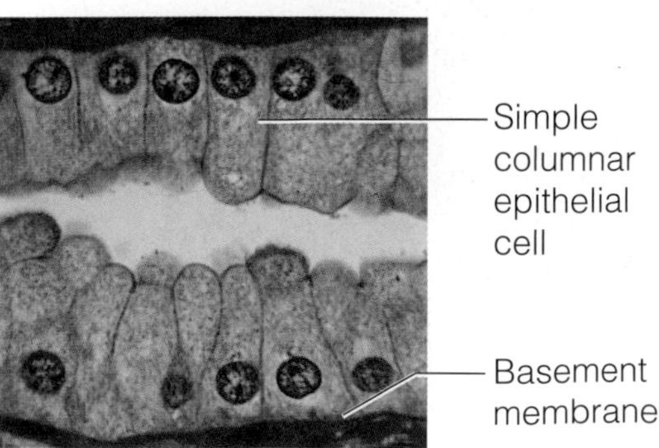

Photomicrograph: Simple columnar epithelium of the stomach lining (1300x).

FIGURE 3.18 Types of epithelia and their common locations in the body.

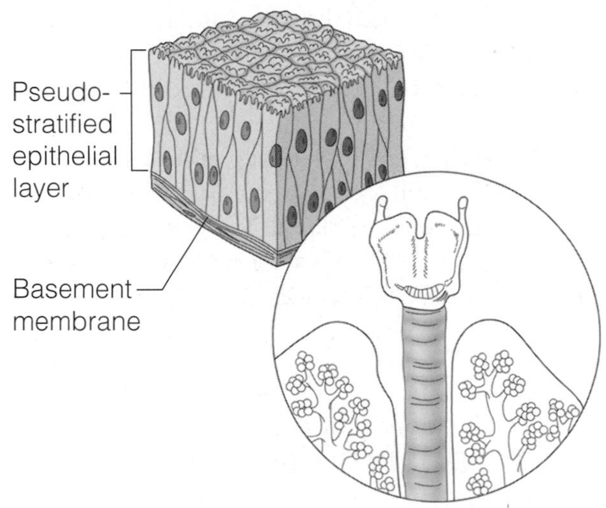

(d) **Diagram:** Pseudostratified (ciliated) columnar

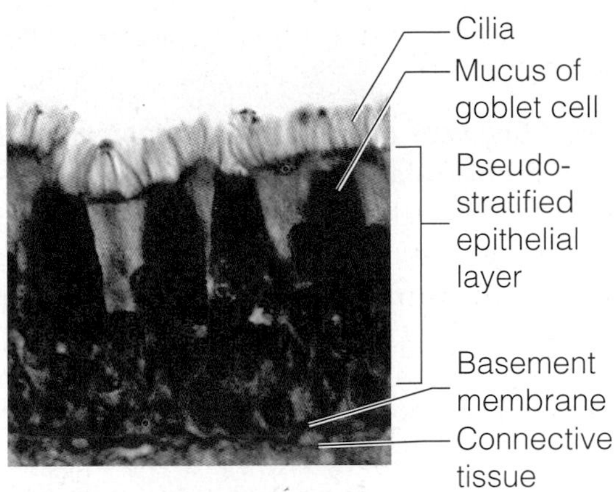

Photomicrograph: Pseudostratified ciliated columnar epithelium lining the human trachea (400x).

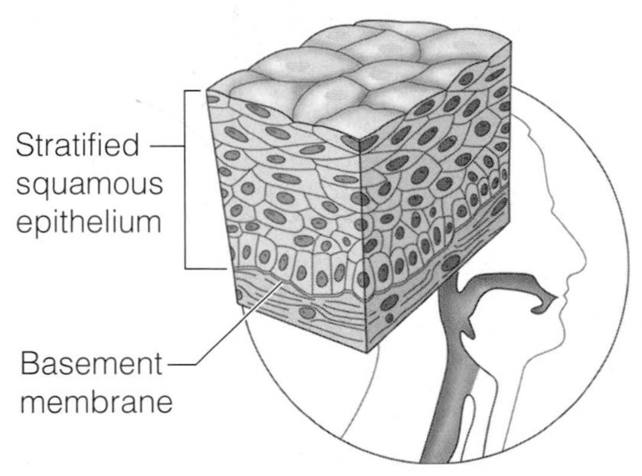

(e) Diagram: Stratified squamous

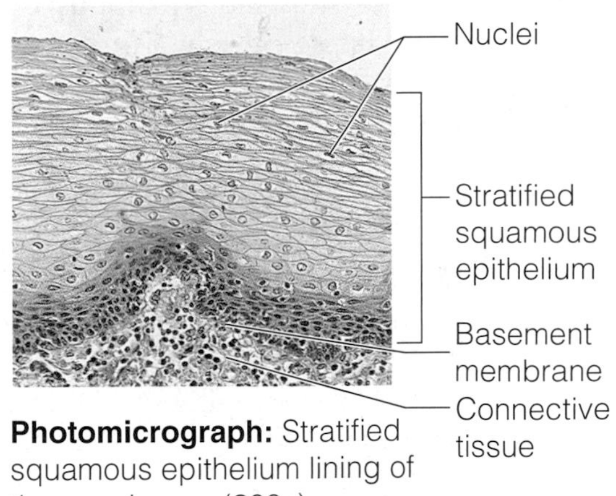

Photomicrograph: Stratified squamous epithelium lining of the esophagus (300x).

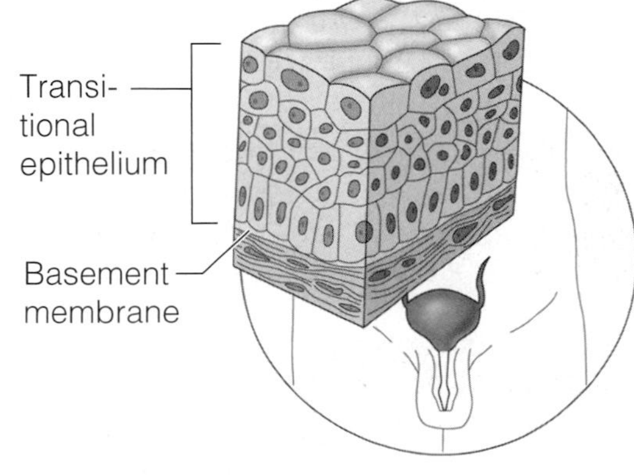

(f) Diagram: Transitional

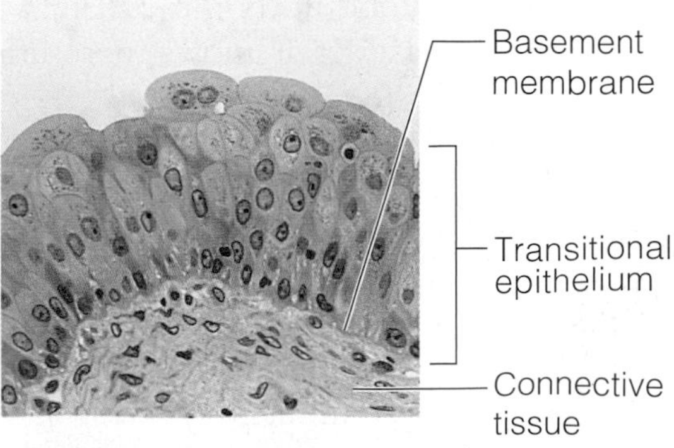

Photomicrograph: Transitional epithelium lining of the bladder, relaxed state (500x); note the bulbous, or rounded, appearance of the cells at the surface; these cells flatten and become elongated when the bladder is filled with urine.

squamous epithelium is found in sites that receive a good deal of abuse or friction, such as the esophagus, the mouth, and the outer portion of the skin (Figure 3.18e).

Stratified Cuboidal and Stratified Columnar Epithelia **Stratified cuboidal epithelium** typically has just two cell layers with (at least) the surface cells being cuboidal in shape. The surface cells of **stratified columnar epithelium** are columnar cells, but its basal cells vary in size and shape. Both of these epithelia are fairly rare in the body, being found mainly in the ducts of large glands. (Because the distribution of these two epithelia is extremely limited, they are not illustrated in Figure 3.18. They are described here only to provide a complete listing of the epithelial tissues.)

Transitional Epithelium **Transitional epithelium** is a highly modified, stratified squamous epithelium that forms the lining of only a few organs—the urinary bladder, the ureters, and part of the urethra. *All* these organs are part of the urinary system and are subject to considerable stretching (Figure 3.18f). Cells of the basal layer are cuboidal or columnar; those at the free surface vary in appearance. When the organ is not stretched, the membrane is many-layered, and the superficial cells are rounded and domelike. When the organ is distended with urine, the epithelium thins, and the surface cells flatten and become squamous-like. This ability of transitional cells to slide past one another and change their shape (undergo "transitions") allows the ureter wall to stretch as a greater volume of urine flows through that tube-like organ. In the bladder, it allows more urine to be stored.

Glandular Epithelium

A **gland** consists of one or more cells that make and secrete a particular product. This product, called a **secretion,** typically contains protein molecules in an aqueous (water-based) fluid. The term *secretion* also indicates an active *process* in which the glandular cells obtain needed materials from the blood and use them to make their secretion, which they then discharge.

Two major types of glands develop from epithelial sheets. **Endocrine** (en′do-krin) **glands** lose their connection to the surface (duct); thus they are often called *ductless* glands. Their secretions (all hormones) diffuse directly into the blood vessels that weave through the glands. Examples of endocrine glands include the thyroid, adrenals, and pituitary.

Exocrine (ek′so-krin) **glands** retain their ducts, and their secretions empty through the ducts to the epithelial surface. Exocrine glands, which include the sweat and oil glands, liver, and pancreas, are both internal and external. They are discussed with the organ systems to which their products are related.

Connective Tissue

Connective tissue, as its name suggests, connects body parts. It is found everywhere in the body. It is the most abundant and widely distributed of the tissue types. Connective tissues perform many functions but they are primarily involved in *protecting, supporting,* and *binding together* other body tissues.

Common Characteristics of Connective Tissue

The characteristics of connective tissue include the following:

- Variations in blood supply. Most connective tissues are well *vascularized* (that is, they have a good blood supply), but there are exceptions. Tendons and ligaments have a poor blood supply, and cartilages are avascular. Consequently, all these structures heal very slowly when injured. (This is why some people say that, given a choice, they would rather have a broken bone than a torn ligament.)
- Extracellular matrix. Connective tissues are made up of many different types of cells plus varying amounts of a nonliving substance found outside the cells, called the extracellular matrix.

Extracellular Matrix

The **extracellular matrix** deserves a bit more explanation because it is what makes connective tissue so different from the other tissue types. The matrix, which is produced by the connective tissue cells and then secreted to their exterior, has two main elements, a structureless ground substance and fibers. The *ground substance* of the matrix is composed largely of water plus some adhesion proteins and large, charged polysaccharide molecules. The cell adhesion proteins serve as a glue that allows the connective tissue cells to attach themselves to the matrix fibers embedded in the ground substance. The charged polysaccharide molecules trap water as they intertwine. As the relative abundance of these polysaccharides increases, they cause the matrix to vary from fluid to gel-like to firm to rock-hard in its consistency. The ability of the ground substance to absorb large amounts of water allow it to serve as a water reservoir for the body. Various types and amounts of fibers are deposited in the matrix and form part of it. These include collagen (white) fibers distinguished by their high tensile strength, elastic (yellow) fibers (the key characteristic of which is an ability to be stretched and then recoil), and reticular fibers (fine collagen fibers that form the internal "skeleton" of soft organs such as the spleen), depending on the connective tissue type. The building blocks, or monomers, of these fibers are made by the connective tissue cells and secreted into the ground substance in the extracellular space, where they spontaneously join together to form the various fiber types.

Because of its extracellular matrix, connective tissue is able to form a soft packing tissue around other organs, to bear weight, and to withstand stretching and other abuses, such as abrasion, that no other tissue could endure. But there is variation. At one extreme, fat tissue is composed mostly of cells, and the matrix is soft. At the opposite extreme, bone and cartilage have very few cells and large amounts of hard matrix, which makes them extremely strong. Find the various types of connective tissues in Figure 3.19 as you read their descriptions that follow.

Types of Connective Tissue

As noted above, all connective tissues consist of living cells surrounded by a matrix. Their major differences reflect fiber type and the number of fibers in the matrix. From most rigid to softest, the major connective tissue classes are *bone, cartilage, dense connective tissue, loose connective tissue,* and *blood.*

Bone

Bone, sometimes called *osseous* (os'e-us) *tissue,* is composed of bone cells sitting in cavities called *lacunae* (lah-ku'ne) and surrounded by layers of a very hard matrix that contains calcium salts in addition to large numbers of collagen fibers (Figure 3.19a). Because of its rocklike hardness, bone has an exceptional ability to protect and support other body organs (for example, the skull protects the brain).

Cartilage

Cartilage is less hard and more flexible than bone. It is found in only a few places in the body. Most widespread is **hyaline** (hi'ah-lin) **cartilage,** which has abundant collagen fibers hidden by a rubbery matrix with a glassy (*hyalin* = glass), blue-white appearance (Figure 3.19b). It forms the supporting structures of the larynx, or voice box, attaches the ribs to the breastbone, and covers the ends of bones where they form joints. The skeleton of a fetus is made largely of hyaline cartilage; but, by the time the baby is born, most of that cartilage has been replaced by bone.

Although hyaline cartilage is the most abundant type of cartilage in the body, there are others. Highly compressible **fibrocartilage** forms the cushionlike disks between the vertebrae of the spinal column (Figure 3.19c). **Elastic cartilage** is found where a structure with elasticity is desired. For example, it supports the external ear. (Elastic cartilage is not illustrated in Figure 3.19.)

Dense Connective Tissue

Dense connective tissue, also called **dense fibrous tissue,** has collagen fibers as its main matrix element (Figure 3.19d). Crowded between the collagen fibers are rows of *fibroblasts* (fiber-forming cells) that manufacture the building blocks of the fibers. Dense connective tissue forms strong, ropelike structures such as tendons and ligaments. **Tendons** attach skeletal muscles to bones; **ligaments** connect bones to bones at joints. Ligaments are more stretchy and contain more elastic fibers than tendons. Dense connective tissue also makes up the lower layers of the skin (dermis), where it is arranged in sheets.

Loose Connective Tissue

Relatively speaking, the **loose connective tissues** are softer and have more cells and fewer fibers than any other connective tissue type except blood.

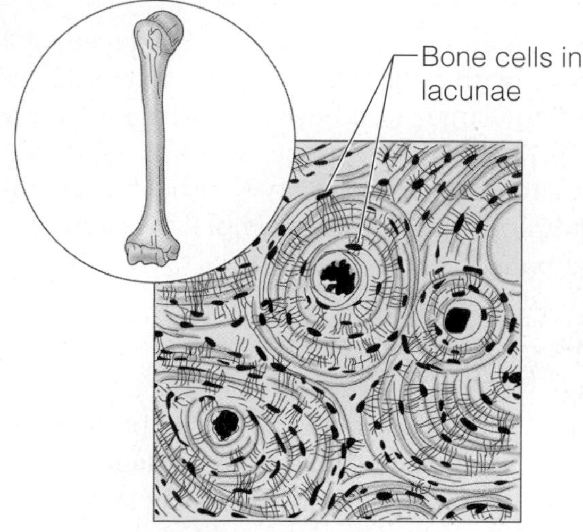

(a) **Diagram:** Bone

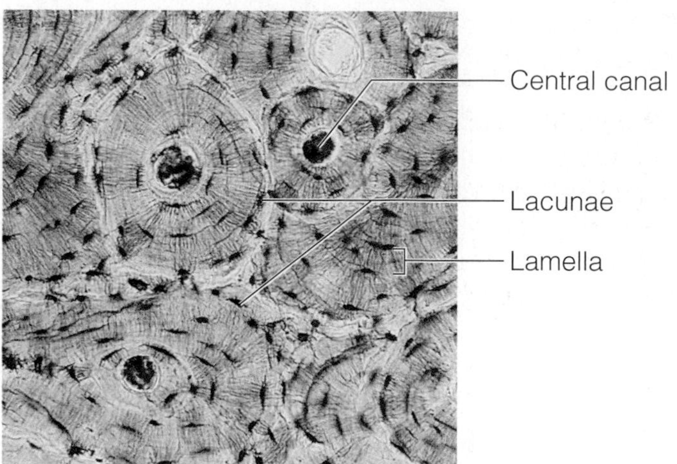

Photomicrograph: Cross-sectional view of ground bone (70x).

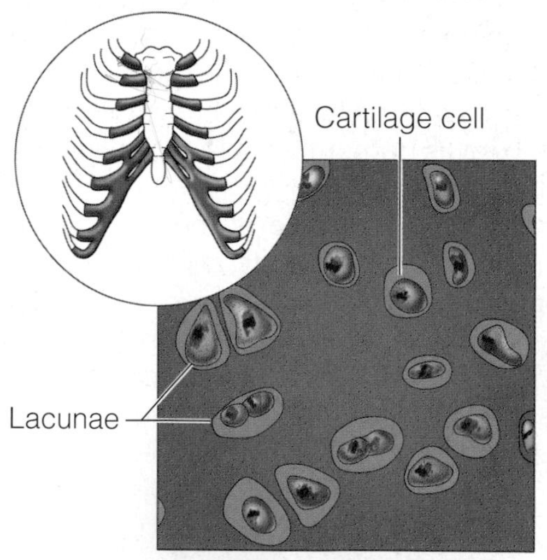

(b) Diagram: Hyaline cartilage

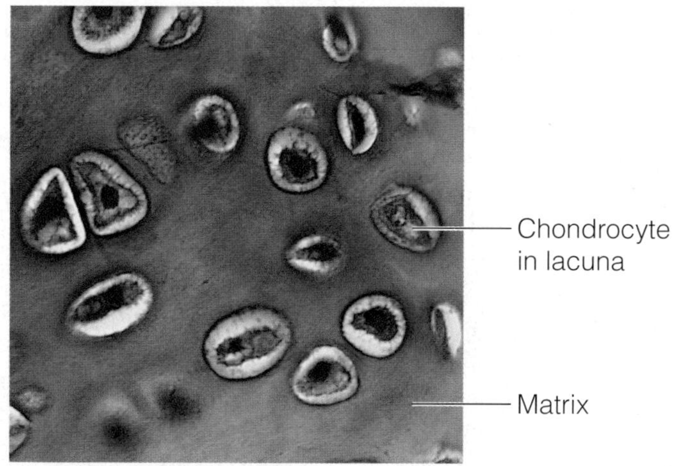

Photomicrograph: Hyaline cartilage from the trachea (300x).

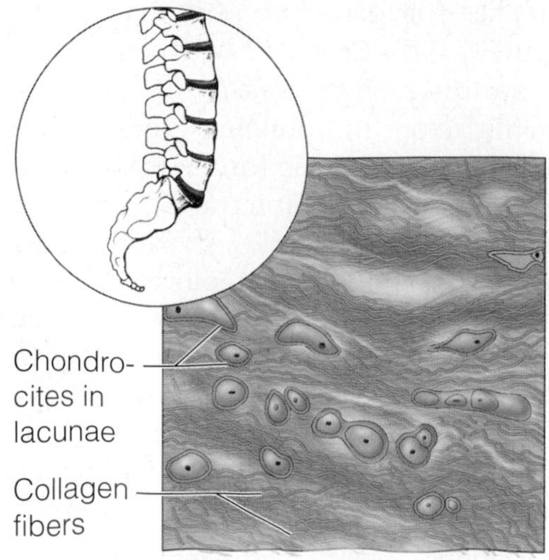

(c) Diagram: Fibrocartilage

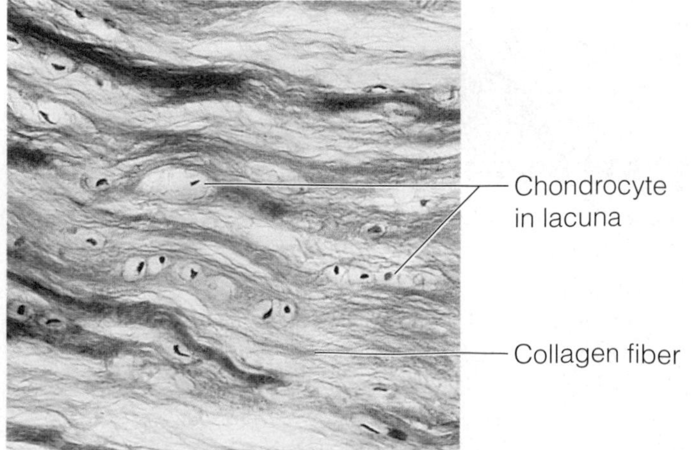

Photomicrograph: Fibrocartilage of an intervertebral disc (200x).

FIGURE 3.19 Connective tissues and their common body locations. (**e, f,** and **g** are subclasses of loose connective tissues.)

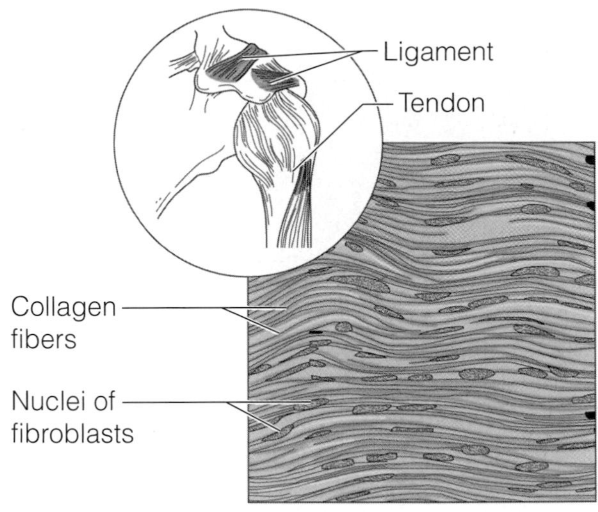

(d) Diagram: Dense fibrous

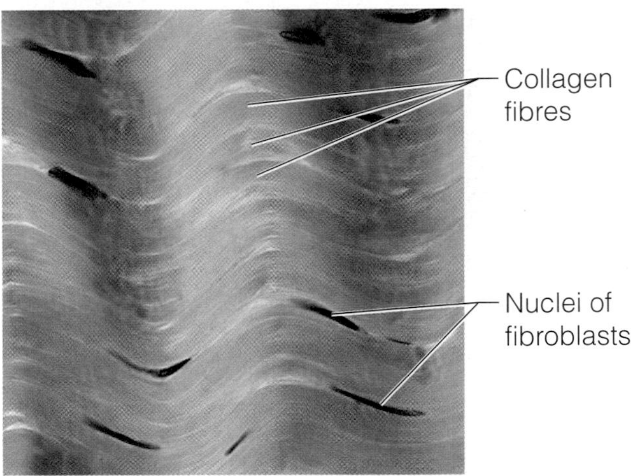

Photomicrograph: Dense fibrous connective tissue from a tendon (1000x).

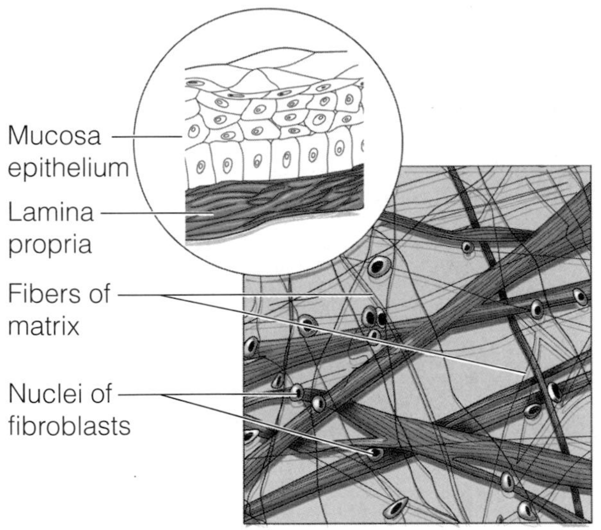

(e) Diagram: Areolar

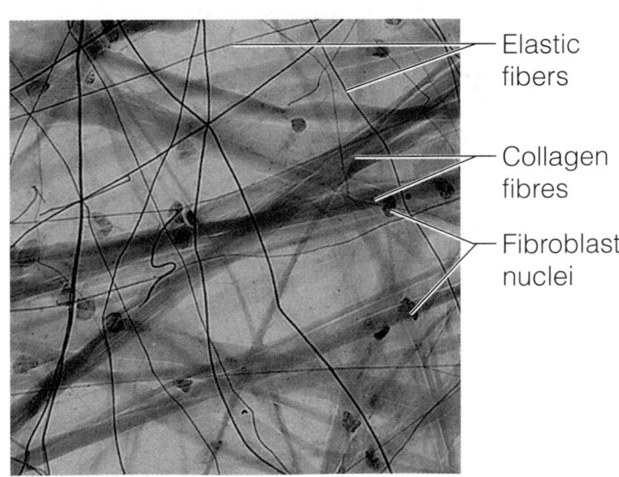

Photomicrograph: Areolar connective tissue, a soft packaging tissue of the body (400x).

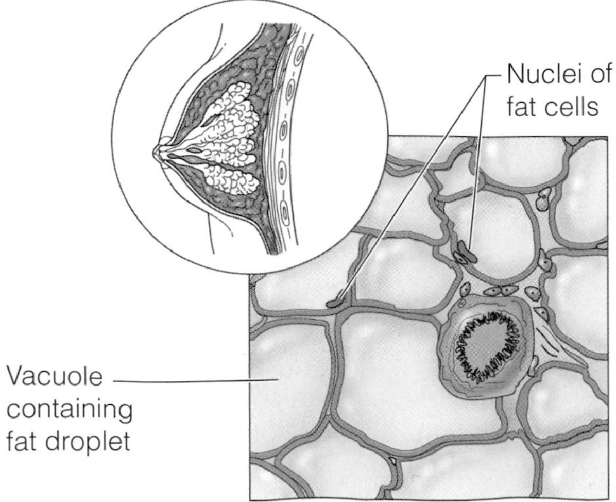

(f) Diagram: Adipose

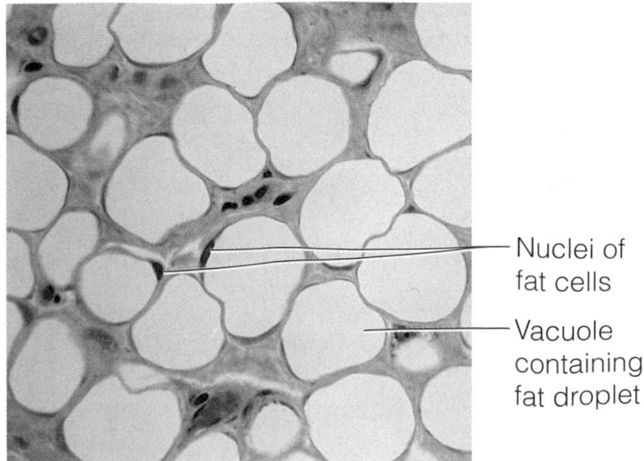

Photomicrograph: Adipose tissue from the subcutaneous layer beneath the skin (600x).

(Continues on page 94)

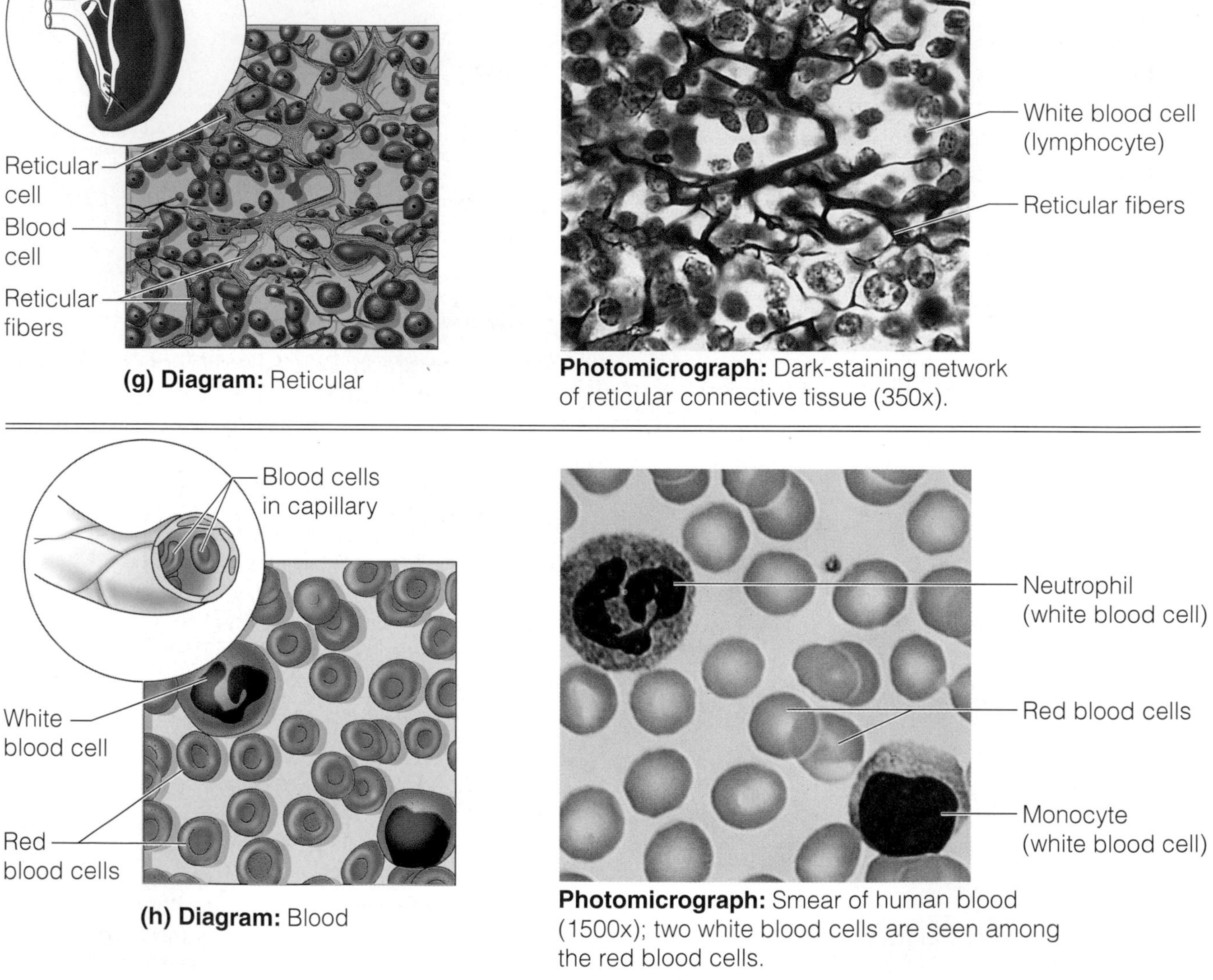

(g) Diagram: Reticular

Photomicrograph: Dark-staining network of reticular connective tissue (350x).

(h) Diagram: Blood

Photomicrograph: Smear of human blood (1500x); two white blood cells are seen among the red blood cells.

FIGURE 3.19 (*continued*) Connective tissues and their common body locations.

Areolar Tissue **Areolar** (ah-re′o-lar) **tissue,** the most widely distributed connective tissue variety in the body, is a soft, pliable, "cobwebby" tissue that cushions and protects the body organs it wraps (Figure 3.19e). It functions as a universal packing tissue and connective tissue "glue" because it helps to hold the internal organs together and in their proper positions. A soft layer of areolar connective tissue called the *lamina propria* (lah′mĭ-nah pro′pre-ah) underlies all mucous membranes. Its fluid matrix contains all types of fibers, which form a loose network. In fact, when viewed through a microscope, most of the matrix appears to be empty space, which explains the name of this tissue type (*areola* = small open space). Because of its loose and fluid nature, areolar connective tissue provides a reservoir of water and salts for the surrounding tissues, and essentially all body cells obtain their nutrients from and release their wastes into this "tissue fluid." When a body region is inflamed, the areolar tissue in the area soaks up the excess fluid like a sponge, and the area swells and becomes puffy, a condition called **edema.** Many types of *phagocytes* wander through this tissue, scavenging for bacteria, dead cells, and other debris, which they destroy.

Adipose Tissue **Adipose** (ad′ĭ-pōs) **tissue** is commonly called *fat.* Basically, it is an areolar tissue in which fat cells predominate (Figure 3.19f). A glistening droplet of stored oil occupies most of a fat cell's volume and compresses the nucleus, displacing it to one side. Since the oil-containing region looks empty and the thin rim of cytoplasm containing the bulging nucleus looks like a ring with a seal, fat cells are sometimes called *signet ring cells.*

Adipose tissue forms the subcutaneous tissue beneath the skin, where it insulates the body and protects it from extremes of both heat and cold. Adipose tissue also protects some organs individually. For example, the kidneys are surrounded by a capsule of fat, and adipose tissue cushions the eyeballs in their sockets. There are also fat "depots" in the body, such as the hips and breasts, where fat is stored and available for fuel if needed.

Reticular Connective Tissue **Reticular connective tissue** consists of a delicate network of interwoven reticular fibers associated with *reticular cells,* which resemble fibroblasts (Figure 3.19g). Reticular tissue is limited to certain sites: it forms the **stroma** (literally, "bed" or "mattress"), or internal framework, which can support many free blood cells (largely lymphocytes) in lymphoid organs such as lymph nodes, the spleen, and bone marrow.

Blood

Blood, or *vascular tissue,* is considered a connective tissue because it consists of *blood cells,* surrounded by a nonliving, fluid matrix called *blood plasma* (Figure 3.19h). The "fibers" of blood are soluble protein molecules that become visible only during blood clotting. Still, we must recognize that blood is quite atypical as connective tissues go. Blood is the transport vehicle for the cardiovascular system, carrying nutrients, wastes, respiratory gases, and many other substances throughout the body. Blood is considered in detail in Chapter 10.

Muscle Tissue

Muscle tissues are highly specialized to *contract,* or *shorten,* to produce movement.

Types of Muscle Tissue

The three types of muscle tissue are illustrated in Figure 3.20. Notice their similarities and differences as you read through the descriptions that follow.

Skeletal Muscle

Skeletal muscle tissue is packaged by connective tissue sheets into organs called *skeletal muscles,* which are attached to the skeleton. These muscles, which can be controlled *voluntarily* (or consciously), form the flesh of the body, the so-called muscular system (see Chapter 6). When the skeletal muscles contract, they pull on bones or skin. The result of their action is gross body movements or changes in our facial expressions. The cells of skeletal muscle are long, cylindrical, multinucleate, and they have obvious *striations* (stripes). Because skeletal muscle cells are elongated to provide a long axis for contraction, they are often called *muscle fibers.*

Cardiac Muscle

Cardiac muscle, covered in more detail in Chapter 11, is found only in the heart. As it contracts, the heart acts as a pump and propels blood through the blood vessels. Like skeletal muscle, cardiac muscle has striations, but cardiac cells are uninucleate, relatively short, branching cells that fit tightly together (like clasped fingers) at junctions called **intercalated disks.** These intercalated disks contain gap junctions that allow ions to pass freely from cell to cell, resulting in rapid conduction of the exciting electrical impulse across the heart. Cardiac muscle is under *involuntary control,* which means that we cannot consciously control

Q *Cell division typically yields two daughter cells, each with one nucleus. How is the multinuclear condition of skeletal muscle explained?*

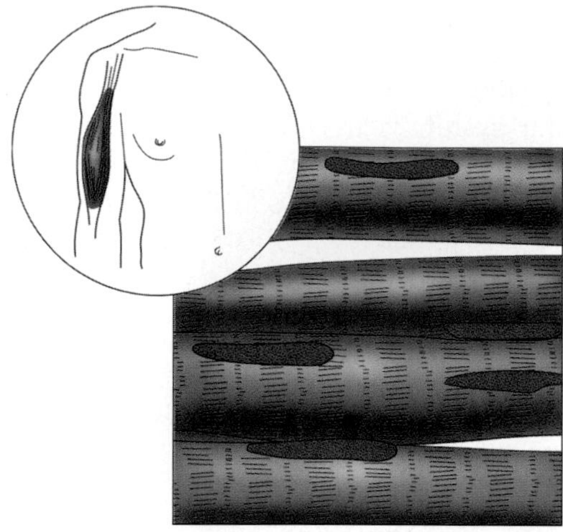

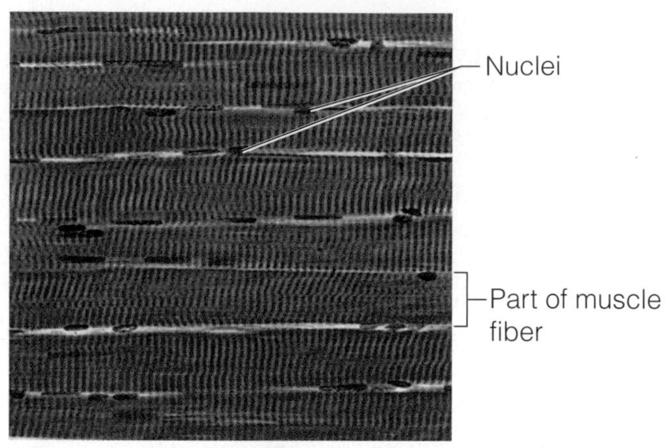

(a) Diagram: Skeletal muscle

Photomicrograph: Skeletal muscle (approx. 300x).

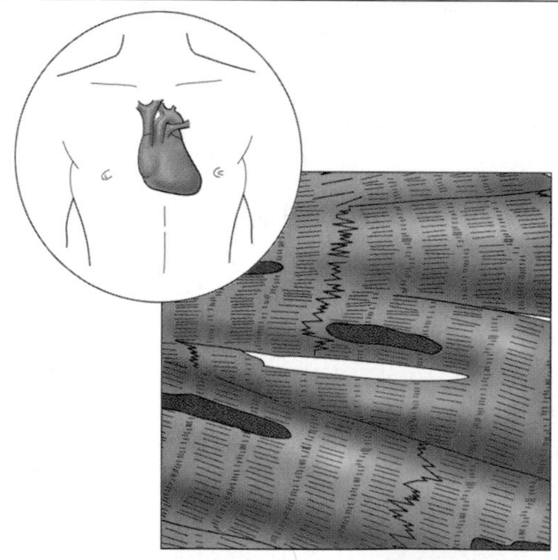

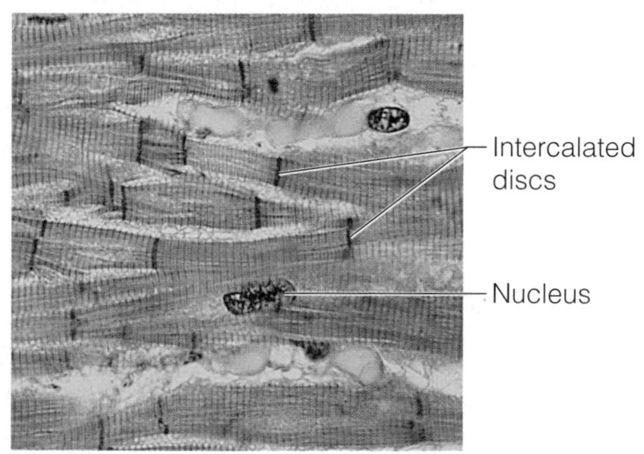

(b) Diagram: Cardiac muscle

Photomicrograph: Cardiac muscle (800x).

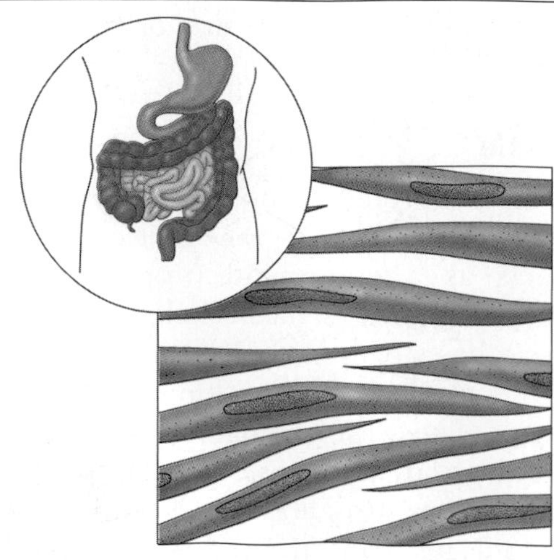

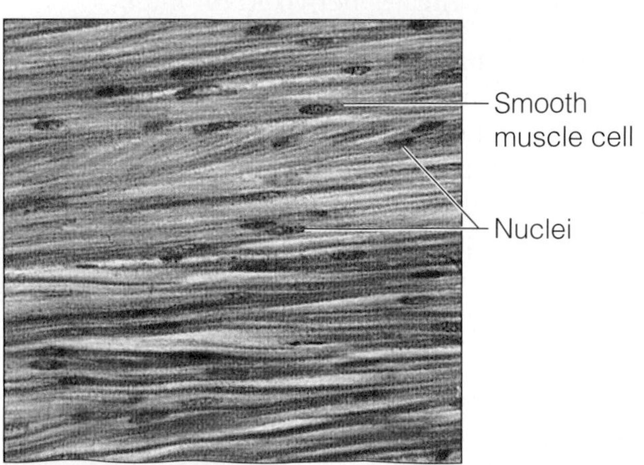

(c) Diagram: Smooth muscle

Photomicrograph: Sheet of smooth muscle (approx. 600x).

FIGURE 3.20 Types of muscle tissue and their common locations in the body.

A *Skeletal muscle cells repeatedly undergo mitosis unaccompanied by cytokinesis.*

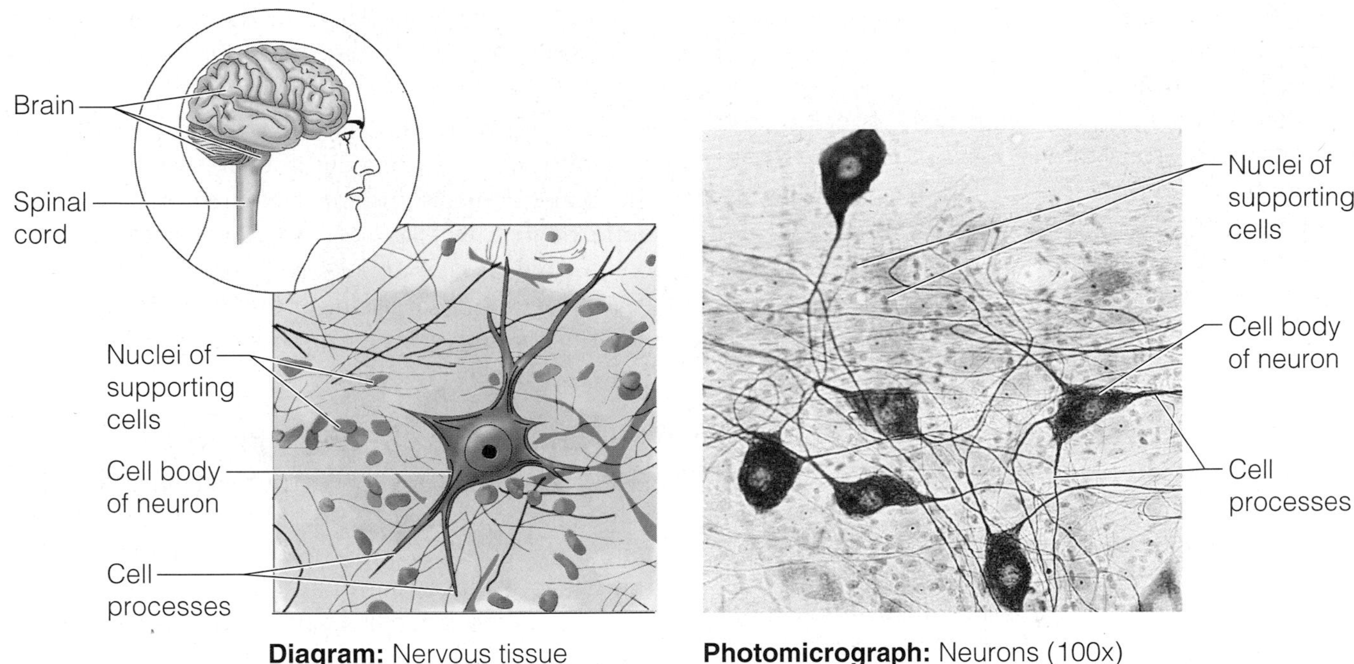

FIGURE 3.21 Nervous tissue. Neurons and supporting cells form the brain, spinal cord, and nerves.

the activity of the heart. (There are, however, rare individuals who claim they have such an ability.)

Smooth Muscle

Smooth, or **visceral, muscle** is so called because no striations are visible. The individual cells have a single nucleus and are spindle-shaped (pointed at each end). Smooth muscle is found in the walls of hollow organs such as the stomach, bladder, uterus, and blood vessels. As smooth muscle contracts, the cavity of an organ alternately becomes smaller (constricts on smooth muscle contraction) or enlarges (dilates on smooth muscle relaxation) so that substances are propelled through the organ along a specific pathway. Smooth muscle contracts much more slowly than the other two muscle types. *Peristalsis* (per″ĭ-stal′sis), a wavelike motion that keeps food moving through the small intestine, is typical of its activity.

Nervous Tissue

When we think of **nervous tissue,** we think of cells called **neurons.** All neurons receive and conduct electrochemical impulses from one part of the body to another; thus *irritability* and *conductivity* are their two major functional characteristics. The structure of neurons is unique (Figure 3.21). Their cytoplasm is drawn out into long processes (extensions), as much as 3 feet or more in the leg, which allows a single neuron to conduct an impulse over long distances in the body. Neurons, along with a special group of **supporting cells** that insulate, support, and protect the delicate neurons, make up the structures of the nervous system—the brain, spinal cord, and nerves.

Tissue Repair (Wound Healing)

The body has many techniques for protecting itself from uninvited guests or injury. Intact physical barriers such as the skin and mucous membranes, cilia, and the strong acid produced by stomach glands are just three examples of body defenses exerted at the local tissue level. When tissue injury does occur, it stimulates the body's inflammatory and immune responses, and the healing process begins almost immediately. Inflammation is a generalized (nonspecific) body response that attempts to prevent further injury. The immune response, on the other hand, is extremely specific and mounts a vigorous attack against recognized invaders (bacteria, viruses, toxins). These protective responses are considered in detail in Chapter 12. Here we will concentrate on the process of tissue repair itself.

Tissue repair, or wound healing, occurs in two major ways: by regeneration and by fibrosis. **Regeneration** is the replacement of destroyed tissue by the same kind of cells, whereas **fibrosis** involves repair by dense (fibrous) connective tissue, that is, by the formation of *scar tissue*. Which occurs depends on (1) the type of tissue damaged and (2) the severity of the injury. Generally speaking, clean cuts (incisions) heal much more successfully than ragged tears of the tissue.

Tissue injury sets a series of events into motion.

- **The capillaries become very permeable.** This allows fluid rich in clotting proteins and other substances to seep into the injured area from the bloodstream. Then leaked clotting proteins construct a clot, which stops the loss of blood, holds the edges of the wound together, and walls off the injured area, preventing bacteria or other harmful substances from spreading to surrounding tissues. Where the clot is exposed to air, it quickly dries and hardens, forming a scab.
- **Granulation tissue forms.** *Granulation tissue* is a delicate pink tissue composed largely of new capillaries that grow into the damaged area from undamaged blood vessels nearby. These capillaries are fragile and bleed freely, as when a scab is picked away from a skin wound. Granulation tissue also contains phagocytes that eventually dispose of the blood clot and connective tissue cells (fibroblasts) that synthesize the building blocks of collagen fibers (scar tissue) to permanently bridge the gap.
- **The surface epithelium regenerates.** As the surface epithelium begins to regenerate, it makes its way across the granulation tissue just beneath the scab. The scab soon detaches and the final result is a fully regenerated surface epithelium that covers an underlying area of fibrosis (the scar). The scar is either invisible or visible as a thin white line, depending on the severity of the wound.

The ability of the different tissue types to regenerate varies widely. Epithelial tissues such as the skin epidermis and mucous membranes regenerate beautifully. So, too, do most of the fibrous connective tissues and bone. Skeletal muscle regenerates poorly, if at all, and cardiac muscle and nervous tissue within the brain and spinal cord are replaced largely by scar tissue.

Homeostatic Imbalance

Scar tissue is strong, but it lacks the flexibility of most normal tissues. Perhaps even more important is its inability to perform the normal functions of the tissue it replaces. Thus, if scar tissue forms in the wall of the bladder, heart, or another muscular organ, it may severely hamper the functioning of that organ. ▲

PART III: DEVELOPMENTAL ASPECTS OF CELLS AND TISSUES

We all begin life as a single cell, which divides thousands of times to form our multicellular embryonic body. Very early in embryonic development, the cells begin to specialize to form the primary tissues, and by birth, most organs are well formed and functioning. The body continues to grow and enlarge by forming new tissue throughout childhood and adolescence.

Cell division is extremely important during the body's growth period. Most cells (except neurons) undergo mitosis until the end of puberty, when adult body size is reached and overall body growth ends. After this time, only certain cells routinely divide—for example, cells exposed to abrasion that continually wear away, such as skin and intestinal cells. Liver cells stop dividing; but they retain this ability should some of them die or become damaged and need to be replaced. Still other cell groups (for example, heart muscle and nervous tissue) almost completely lose their ability to divide when they are fully mature; that is, they become *amitotic* (am″ĭ-tot′ik). Amitotic tissues are severely handicapped by injury because the lost cells cannot be replaced by the same type of cells. This is why the heart of an individual who has had several severe heart attacks becomes weaker and weaker. Damaged cardiac muscle does not regenerate and is replaced by scar tissue that cannot contract, so the heart becomes less and less capable of acting as an efficient blood pump.

The aging process begins once maturity has been reached. (Some believe it begins at birth.) No one has been able to explain just *what* causes aging, but there have been many suggestions. Some believe it is a result of little "chemical insults," which occur continually through life—for example, the presence of toxic chemicals (such as alcohol, certain drugs, or carbon monoxide) in the blood, or the temporary absence of needed substances such as glucose or oxygen. Perhaps the effect of these chemical insults is cumulative and finally succeeds in upsetting the delicate chemical balance of the body cells. Others think that external physical factors such as radiation (X rays or ultraviolet waves) contribute to the aging process. Several believe that the aging "clock" is genetically programmed, or built into our genes. We all know of cases like the radiant woman of 50 who looks about 35 or the barely-out-of-adolescence man of 24 who looks 40. It appears that such traits can run in families.

There is no question that certain events are part of the aging process. For example, with age, epithelial membranes thin and are more easily damaged, and the skin loses its elasticity and begins to sag. The exocrine glands of the body (epithelial tissue) become less active and we begin to "dry out" as less oil, mucus, and sweat are produced. Some endocrine glands produce decreasing amounts of hormones, and the body processes that they control (such as metabolism and reproduction) become less efficient or stop altogether.

Connective tissue structures also show changes with age. Bones become porous and weaken, and the repair of tissue injuries slows. Muscles begin to atrophy. Although a poor diet may contribute to some of these changes, there is little doubt that decreased efficiency of the circulatory system, which reduces nutrient and oxygen delivery to body tissues, is a major factor.

Besides the tissue changes associated with aging, which accelerate in the later years of life, other modifications of cells and tissues may occur at any time. For example, when cells fail to honor normal controls on cell division and multiply wildly, an abnormal mass of proliferating cells, known as a **neoplasm** (ne′o-plazm″; "new growth"), results. Neoplasms may be benign or malignant (cancerous). See "A Closer Look," pp. 100–101, for more information on cancer.

However, not all increases in cell number involve neoplasms. Certain body tissues (or organs) may enlarge because there is some local irritant or condition that stimulates the cells. This is called **hyperplasia** (hi″per-pla′ze-ah). For example, a woman's breasts enlarge during pregnancy in response to increased hormones; this is a normal but temporary situation that doesn't have to be treated. On the other hand, **atrophy** (at′ro-fe), or decrease in size, can occur in an organ or body area that loses its normal stimulation. For example, muscles that are not used or that have lost their nerve supply begin to atrophy and waste away rapidly.

Prove It Yourself

Demonstrate that Oil and Water Don't Mix

Place water and cooking oil in a jar and shake it. Notice that the oil separates from the water, first as small droplets, then as larger drops that join together, and finally as a single layer of oil on top.

Water is a polar molecule, and oils are uncharged, nonpolar molecules. When water and oil mingle, the water molecules are attracted to each other and join together, excluding oil from the regions they occupy. Over time the oil is forced into ever-larger droplets until it is completely separated from the water. It rises to the top because oil is less dense than water.

Although you probably don't have any phospholipids available, you would get a different result if you repeated this experiment with phospholipids because they do not separate from water. On microscopic examination you would see numerous small spheres, each comprised of a bilayer of phospholipids enclosing a small volume of water. In effect, this is a nonliving version of a cell's outer membrane and internal contents.

A Closer Look

Cancer—The Intimate Enemy

THE word *cancer* elicits dread in nearly everyone. Why does cancer strike some and not others? Before attempting to answer that question, let's define some terms. An abnormal cell mass that develops when controls of the cell cycle and cell division malfunction is called a *neoplasm* ("new growth") or *tumor.* However, not all neoplasms are cancerous. *Benign* (be-nīn′: "kindly") neoplasms are strictly local affairs. They tend to be surrounded by a capsule, grow slowly, and seldom kill their hosts if they are removed before they compress vital organs. In contrast, *malignant* ("bad") neoplasms (cancers) are nonencapsulated masses that grow more relentlessly and may become killers. Their cells resemble immature cells, and they invade their surroundings rather than pushing them aside, as reflected in the name *cancer,* from the Latin word for "crab." Malignant cells also tend to break away from the parent mass and spread via the blood to distant parts of the body, where they form new masses. This last capability is called *metastasis* (mĕ-tas′tă-sis).

What causes transformation—the changes that convert a normal cell into a cancerous one? It is well known that radiation, mechanical trauma, certain viral infections, and many chemicals (tobacco tars, saccharine) can act as carcinogens (cancer-causers). What all of these factors have in common is that they all cause *mutations*—changes in DNA that alter the expression of certain genes. However, most carcinogens are eliminated by peroxisomal or lysosomal enzymes or the immune system. Furthermore, one mutation doesn't do it—apparently it takes a sequence of several genetic changes to change a normal cell to a full-fledged cancer cell (Figure a).

Clues to the role of genes were provided by the discovery of *oncogenes* (cancer-causing [*onco* = tumor] genes), and then *proto-oncogenes.* Proto-oncogenes code for proteins that are needed for normal cell division and growth. However, many have fragile sites that break when they are exposed to carcinogens, and this event converts them into oncogenes. An example of a problem that might result from this conversion is the switching on of dormant genes that allow cells to become invasive (an ability of embryonic cells—and cancer cells—but not normal adult cells). However, oncogenes have been discovered in only 15 to 20 percent of human cancers, so the more recent discovery of *tumor suppressor genes,* which work to suppress or prevent cancer, was not too surprising. The tumor suppressor genes (such as *p*53) aid DNA repair, put the "brakes" on cell division, help to inactivate carcinogens, or enhance the ability of the immune system to destroy cancer cells. When the tumor suppressor genes are damaged or changed in some way, the oncogenes are free to "do their thing." One of the best-understood of human cancers, colon cancer, illustrates this principle (Figure b). The first sign of colon cancer is a polyp (benign tumor), due to an unusual increase in the division rate of apparently normal cells of the colon lining. In time, a malignant neoplasm makes its appearance at the site. In most cases, these changes parallel cellular changes at the DNA level and include activation of an oncogene and inactivation of two tumor suppressor genes. Whatever the precise genetic factor at work, the seeds of cancer do appear to be in our own genes. Thus, as you can see, cancer is an intimate enemy indeed.

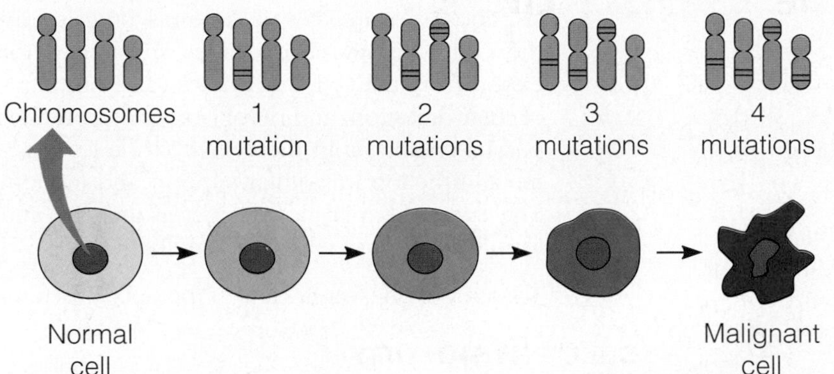

(a) Accumulation of mutations in the development of a cancer cell.

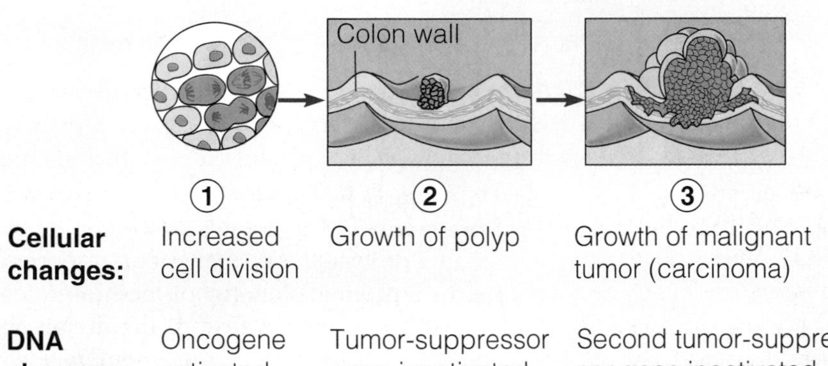

(b) Stepwise development of a typical colon cancer.

Almost half of all Americans develop cancer in their lifetime, and a fifth of us will die of it. Cancer can arise from almost any cell type, but the most common cancers originate in the skin, lung, colon, breast, and male prostate gland.

Screening procedures, such as self-examination of one's breasts or testicles for lumps and checking fecal samples for blood, aid in early detection of cancers. However, most cancers are diagnosed only after they have begun to cause symptoms (pain, bloody discharge, lump, etc.), and the diagnostic method most used is the biopsy. In a biopsy, a sample of the primary tumor is removed surgically (or by scraping) and examined microscopically for structural changes typical of malignant cells.

The treatment of choice for either type of neoplasm is surgical removal. If surgery is not possible—as in cases where the cancer has spread widely or is inoperable—radiation and drugs (chemotherapy) are used. Anticancer drugs have unpleasant side effects because most target *all* rapidly dividing cells, including normal ones. The side effects include nausea, vomiting, and loss of hair. X rays, even if highly localized, also have side effects because, in passing through the body, they kill healthy cells that lie in the path to the cancer cells.

Current cancer treatments—"cut, burn, and poison"—are recognized as crude and painful. Promising new methods focus on delivering anticancer drugs more precisely to the cancer (via monoclonal antibodies that respond to one type of protein on a cancer cell) and on increasing the immune system's ability to fend off cancer. The most recent research focuses on starving tumors by inhibiting their ability to attract a rich blood supply.

SUMMARY

Media study tools that could provide you with additional help in reviewing specific key topics of Chapter 3 are referenced below.

IP = Interactive Physiology;

WEB = A&P Place website.

PART I: CELLS (pp. 62–85)

OVERVIEW OF THE CELLULAR BASIS OF LIFE

1. A cell is composed primarily of four elements—carbon, hydrogen, oxygen, and nitrogen—plus many trace elements. Living matter is over 60 percent water. The major building material of the cell is protein.
2. Cells vary in size from microscopic to over a meter in length. Shape often reflects function. For example, muscle cells have a long axis to allow shortening.

ANATOMY OF A GENERALIZED CELL

1. Cells have three major regions—nucleus, cytoplasm, and plasma membrane.
   a. The nucleus, or control center, directs cell activity and is necessary for reproduction. The nucleus contains genetic material (DNA), which carries instructions for synthesis of proteins.
   b. The plasma membrane limits and encloses the cytoplasm and acts as a selective barrier to the movement of substances into and out of the cell. It is composed of a lipid bilayer containing proteins. The water-impermeable lipid portion forms the basic membrane structure. The proteins (many of which are glycoproteins) act as enzymes or carriers in membrane transport, form membrane channels, provide receptor sites for hormones and other chemicals, or play a role in cellular recognition and interactions during development and immune reactions.

      Specializations of the plasma membrane include microvilli (which increase the absorptive area) and cell junctions (desmosomes, tight junctions, and gap junctions).

   WEB Exercise: Chapter 3, Structure of the Plasma Membrane.

   c. The cytoplasm is where most cellular activities occur. Its fluid substance, the cytosol, contains inclusions, stored or inactive materials in the cytoplasm (fat globules, water vacuoles, crystals, and the like) and specialized bodies called organelles, each with a specific function. For example, mitochondria are sites of ATP synthesis, ribosomes are sites of protein synthesis, and the Golgi apparatus packages substances for export from the cell. Lysosomes carry out intracellular digestion, and peroxisomes disarm dangerous chemicals in the cells. Cytoskeletal elements function in cellular support and motion. The centrioles play a role in cell division and form the bases of cilia and flagella.

   WEB Exercise: Chapter 3, Parts of the Cell: Structure.

CELL PHYSIOLOGY

1. All cells exhibit irritability, digest foods, excrete wastes, and are able to reproduce, grow, move, and metabolize.
2. Transport of substances through the cell membrane:

   WEB Exercise: Chapter 3, Membrane Transport.

   a. Passive transport processes include diffusion and filtration.
      (1) Diffusion is the movement of a substance from an area of its higher concentration to an area of its lower concentration. It occurs because of kinetic energy of the molecules themselves. The diffusion of dissolved solutes through the plasma membrane is simple diffusion. The diffusion of water across the plasma membrane is osmosis. Diffusion that requires a protein channel or carrier is facilitated diffusion.

   WEB Exercise: Chapter 3, Passive Transport.

      (2) Filtration is the movement of substances through a membrane from an area of high hydrostatic pressure to an area of lower fluid pressure. In the body, the driving force of filtration is blood pressure.

   b. Active transport processes use energy (ATP) provided by the cell.
      (1) In solute pumping, substances are moved across the membrane against an electrical or a concentration gradient by proteins called solute pumps. This accounts for the transport of amino acids, some sugars, and most ions.
      (2) The two types of ATP-activated bulk transport are exocytosis and endocytosis. Exocytosis

moves secretions and other substances out of cells; a membrane-bound vesicle fuses with the plasma membrane, ruptures, and ejects its contents to the cell exterior. Endocytosis, in which particles are taken up by enclosure in a plasma membrane sac, includes phagocytosis (uptake of solid particles), fluid-phase endocytosis (uptake of fluids), and the highly-selective receptor-mediated endocytosis. In the latter, membrane receptors bind with and internalize only selected target molecules.

3. Osmotic pressure, which reflects the solute concentration of a solution, determines whether cells gain or lose water. (Discussion in "A Closer Look")
   a. Hypertonic solutions contain more solutes (and less water) than do cells. In these solutions, cells lose water by osmosis and crenate.
   b. Hypotonic solutions contain fewer solutes (and more water) than do the cells. In these solutions, cells swell and may rupture (lyse) as water rushes in by osmosis.
   c. Isotonic solutions, which have the same solute-to-solvent ratio as cells, cause no changes in cell size or shape.
4. Cell division has two phases, mitosis (nuclear division) and cytokinesis (division of the cytoplasm).
   a. Mitosis begins after DNA has been replicated (during interphase); it consists of four stages—prophase, metaphase, anaphase, and telophase. The result is two daughter nuclei, each identical to the mother nucleus.
   b. Cytokinesis usually begins during anaphase and progressively pinches the cytoplasm in half. Cytokinesis does not always occur; in such cases bi- or multinucleate cells result.
   c. Mitotic cell division provides an increased number of cells for growth and repair.
5. Protein synthesis involves both DNA (the genes) and RNA.
   a. A gene is a segment of DNA that carries the instructions for building one protein. The information is in the sequence of bases in the nucleotide strands. Each three-base sequence (triplet) specifies one amino acid in the protein.
   b. Messenger RNA carries the instructions for protein synthesis from the DNA (gene) to the ribosomes. Transfer RNA transports amino acids to the ribosomes. Ribosomal RNA forms part of the ribosomal structure and helps coordinate the protein building process.

PART II: BODY TISSUES (pp. 85–98)

1. Epithelium is the covering, lining, and glandular tissue. Its functions include protection, absorption, and secretion. Epithelia are named according to arrangement (simple, stratified) and cell shape (squamous, cuboidal, columnar).
2. Connective tissue is the supportive, protective, and binding tissue. It is characterized by the presence of a nonliving, extracellular matrix (ground substance plus fibers) produced and secreted by the cells; it varies in amount and consistency. Fat, ligaments and tendons, bones, and cartilage are all connective tissues or connective tissue structures.

   WEB Exercise: Chapter 3, Identifying Connective Tissue.

3. Muscle tissue is specialized to contract, or shorten, which causes movement. There are three types—skeletal (attached to the skeleton), cardiac (forms the heart), and smooth (in the walls of hollow organs).
4. Nervous tissue is composed of supporting cells and irritable cells called neurons, which are highly specialized to receive and transmit nerve impulses and supporting cells. Neurons are important in control of body processes. Nervous tissue is located in nervous system structures—brain, spinal cord, and nerves.
5. Tissue repair (wound healing) may involve regeneration, fibrosis, or both. In regeneration, the injured tissue is replaced by the same type of cells. In fibrosis, the wound is repaired with scar tissue. Epithelia and connective tissues regenerate well. Mature cardiac muscle and nervous tissue are repaired by fibrosis.

PART III: DEVELOPMENTAL ASPECTS OF CELLS AND TISSUES (pp. 98–99)

1. Growth through cell division continues through puberty. Cell populations exposed to friction (such as epithelium) replace lost cells throughout life. Connective tissue remains mitotic and forms repair (scar) tissue. For the most part, muscle tissue becomes amitotic by the end of puberty, and nervous tissue becomes amitotic shortly after birth. Amitotic tissues are severely handicapped by injury.
2. The cause of aging is unknown, but chemical and physical insults, as well as genetic programming, are suggested.
3. Neoplasms, both benign and cancerous, represent abnormal cell masses in which normal controls on cell division are not working. Hyperplasia (increase in size) of tissue or organ may occur when tissue is

strongly stimulated or irritated. Atrophy (decrease in size) of a tissue or organ occurs when the organ is no longer stimulated normally.

REVIEW QUESTIONS

MULTIPLE CHOICE

More than one choice may apply.

1. Which of the following would you expect to find in or on cells whose main function is absorption?
   a. Microvilli
   b. Cilia
   c. Gap junctions
   d. Secretory vesicles
2. Adult cell types you might expect to have gap junctions include:
   a. skeletal muscle
   b. bone
   c. heart muscle
   d. smooth muscle
3. Which of the following are possible functions of the glycoproteins in the plasma membrane?
   a. Determination of blood groups
   b. Binding sites for toxins or bacteria
   c. Aiding the binding of sperm to egg
   d. Increasing the efficiency of absorption
4. A cell with abundant peroxisomes would most likely be involved in:
   a. secretion
   b. storage of glycogen
   c. ATP manufacture
   d. movement
   e. detoxification activities
5. A cell stimulated to increase its steroid production will have abundant:
   a. ribosomes
   b. rough ER
   c. smooth ER
   d. Golgi apparatus
   e. secretory vesicles
6. For diffusion to occur, there must be:
   a. a selectively permeable membrane
   b. equal amounts of solute
   c. a concentration difference
   d. some sort of carrier system
   e. all of the above
7. In which of the following tissue types might you expect to find goblet cells?
   a. Simple cuboidal
   b. Simple columnar
   c. Simple squamous
   d. Stratified squamous
   e. Transitional
8. An epithelium "built" to withstand friction is:
   a. simple squamous
   b. stratified squamous
   c. simple cuboidal
   d. simple columnar
   e. pseudostratified
9. What kind of connective tissue acts as a sponge, soaking up fluid when edema occurs?
   a. Areolar connective
   b. Adipose connective
   c. Dense irregular connective
   d. Reticular connective
   e. Vascular
10. What type of connective tissue prevents muscles from pulling away from bones during contraction?
    a. Dense connective
    b. Areolar
    c. Elastic connective
    d. Hyaline cartilage
11. Which of the following terms describe cardiac muscle?
    a. Striated
    b. Intercalated disks
    c. Multinucleated
    d. Involuntary
    e. Branching

SHORT ANSWER ESSAY

1. Name the four elements making up the bulk of living matter.
2. Define *cell* and *organelle.*
3. Although cells have differences that reflect their special functions in the body, what functional abilities do *all* cells exhibit?
4. Describe the general function of the nucleus. Describe the special function of DNA found in the nucleus. What nuclear structures contain DNA? Help to form ribosomes?
5. Describe the general structure and function of the plasma membrane.
6. Describe the general composition and function of the cytosol and inclusions of the cytoplasm.
7. Name the cellular organelles, and explain the function of each.
8. What is the difference between active and passive transport processes?
9. Define *diffusion, osmosis, simple diffusion, filtration, solute pumping, exocytosis, endocytosis, phagocytosis, fluid-phase endocytosis,* and *receptor-mediated endocytosis.*
10. What two structural characteristics of cell membranes determine if substances can pass through

them passively? What determines whether or not a substance can be actively transported through the membrane?

11. Explain the effect of the following solutions on living cells: hypertonic, hypotonic, and isotonic.
12. Briefly describe the process of DNA replication.
13. Define *mitosis*. Why is mitosis important?
14. What is the role of the spindle in mitosis?
15. Why can an organ be permanently damaged if its cells are amitotic?
16. Describe the relative roles of DNA and RNA in protein synthesis.
17. Define *tissue*. List the four major types of tissue. Which of these types is most widely distributed in the body?
18. Describe the general characteristics of epithelial tissue. List the most important functions of epithelial tissues, and give examples of each.
19. How are epithelial tissues classified?
20. Where is ciliated epithelium found, and what role does it play?
21. How do the endocrine and exocrine glands differ in structure and function?
22. What are the general structural characteristics of connective tissues? What are the functions of connective tissues? How are their functions reflected in their structures?
23. Name a connective tissue with (a) a soft fluid matrix, and (b) a stony hard matrix.
24. What is the function of muscle tissue?
25. Name the three types of muscle tissue, and tell where each would be found in the body. What is meant by "smooth muscles are involuntary in action"? Which muscle type is voluntary in action?
26. What two functional characteristics are highly developed in neurons?
27. Define *neoplasm, hyperplasia,* and *atrophy*.

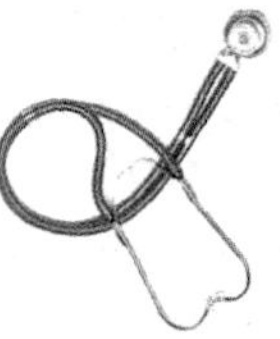

AT THE CLINIC

1. Two examples of chemotherapeutic drugs (used to treat cancer) and their cellular actions are given below. Explain why each drug could be fatal to a cell.
   - Vincristine: damages the mitotic spindle.
   - Adriamycin: binds to DNA and blocks messenger RNA synthesis.
2. Hydrocortisone is an anti-inflammatory drug that stabilizes lysosomal membranes. Explain how this effect reduces cell damage and inflammation.
3. John has severely injured his knee during football practice. He is told that he has a torn knee cartilage and to expect that recovery and repair will take a long time. Why will it take a long time?
4. Three patients in an intensive care unit are examined by the resident doctor. One patient has brain damage from a stroke, another had a heart attack that severely damaged his heart muscle, and the third has a severely damaged liver (a gland) from a crushing injury in a car accident. All three patients have stabilized and will survive, but only one will have full functional recovery through regeneration. Which one and why?

4 Skin and Body Membranes

YOUR GOALS

After completing this chapter, you will have a working knowledge of the functions of the skin and body membranes system and will have mastered the objectives listed below.

FUNCTION PREVIEW

- Body membranes line or cover, protect, and lubricate body surfaces.
- As the outermost boundary of the body, the skin protects against injuries of many types.

OBJECTIVE CHECKLIST

CLASSIFICATION OF BODY MEMBRANES (pp. 108–110)

- List the general functions of each membrane type—cutaneous, mucous, serous, and synovial—and give its location in the body.
- Compare the structure (tissue makeup) of the major membrane types.

INTEGUMENTARY SYSTEM (SKIN) (pp. 110–124)

- List several important functions of the integumentary system, and explain how these functions are accomplished.
- When provided with a model or diagram of the skin, recognize and name the following skin structures: epidermis, dermis (papillary and reticular layers), hair and hair follicle, sebaceous gland, and sweat gland.
- Name the layers of the epidermis, and describe the characteristics of each.
- Describe the distribution and function of the epidermal derivatives—sebaceous glands, sweat glands, and hair.
- Name the factors that determine skin color, and describe the function of melanin.
- Differentiate between first-, second-, and third-degree burns.

>>

(Objective Checklist, continued)

- Explain the importance of the "rule of nines."
- Summarize the characteristics of basal cell carcinoma, squamous cell carcinoma, and malignant melanoma.

DEVELOPMENTAL ASPECTS OF SKIN AND BODY MEMBRANES (p. 124)

- List several examples of integumentary system aging.

Body membranes, which cover surfaces, line body cavities, and form protective (and often lubricating) sheets around organs, fall into two major groups. There are (1) *epithelial membranes,* which include the cutaneous, mucous, and serous membranes, and (2) *connective tissue membranes,* represented by synovial membranes. The cutaneous membrane, generally called the skin or integumentary system, will receive most of our attention in this chapter, but first we will consider the other body membranes.

Classification of Body Membranes

The two major categories of body membranes—epithelial and connective tissue—are classified in part according to their tissue makeup.

Epithelial Membranes

The **epithelial membranes,** also called *covering and lining membranes,* include the cutaneous membrane (skin), the mucous membranes, and the serous membranes (Figure 4.1). However, calling these membranes "epithelial" is not only misleading but also inaccurate. Although they all *do* contain an epithelial sheet, it is always combined with an underlying layer of connective tissue. Hence these membranes are actually simple organs. Since the skin will be discussed in some detail shortly, it will be listed here solely as a subcategory of the epithelial membranes.

Cutaneous Membrane

The **cutaneous** (ku-ta′ne-us) **membrane** is your skin. Its superficial epidermis is composed of a keratinizing stratified squamous epithelium. The underlying dermis is mostly dense (fibrous) connective tissue. Unlike the other epithelial membranes, the cutaneous membrane is exposed to air and is a *dry* membrane.

Mucous Membranes

A **mucous** (myu′kus) **membrane (mucosa)** is composed of epithelium (the type varies with the site) resting on a loose connective tissue membrane called a *lamina propria.* This membrane type lines all body cavities that open to the exterior, such as those of the hollow organs of the respiratory, digestive, urinary, and reproductive tracts (Figure 4.1b). Notice that the term *mucosa* refers only to the location of the epithelial membranes, *not* their cellular makeup, which varies. However, most mucosae contain either stratified squamous epithelium (as in the mouth and esophagus) or simple columnar epithelium (as in the rest of the digestive tract). In all cases, they are "wet," or moist, membranes that are almost continuously bathed in secretions or, in the case of the urinary mucosae, urine.

The epithelium of mucosae is often adapted for absorption or secretion. Although many mucosae secrete mucus, this is not a requirement. The mucosae of the respiratory and digestive tracts secrete large amounts of protective, lubricating mucus; that of the urinary tract does not.

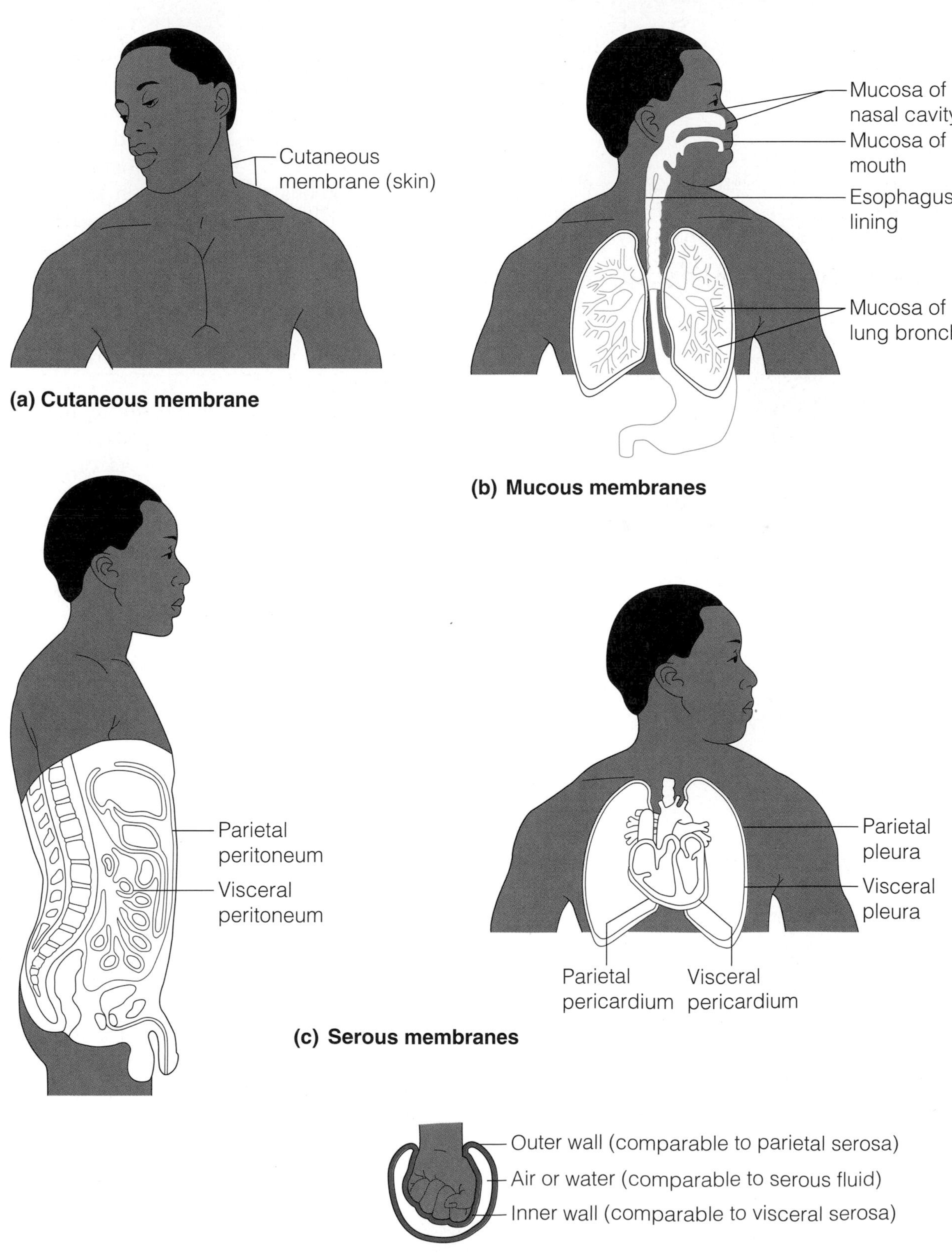

FIGURE 4.1 Classes of epithelial membranes. (a) Cutaneous membrane, or skin. **(b)** Mucous membranes (blue) line body cavities that are open to the exterior. **(c)** Serous membranes (shown in red) line ventral body cavities that are closed to the exterior. **(d)** A fist thrust into a flaccid balloon demonstrates the relationship between the parietal and visceral serous membrane layers.

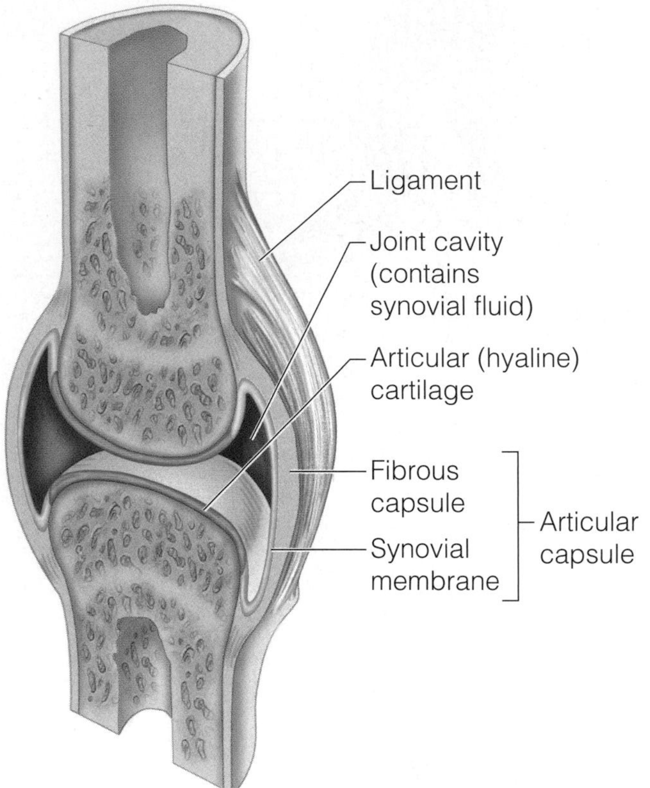

FIGURE 4.2 A typical synovial joint.

Serous Membranes

A **serous membrane (serosa)** is composed of a layer of simple squamous epithelium resting on a thin layer of areolar connective tissue. In contrast to mucous membranes, which line open body cavities, serous membranes line body cavities that are closed to the exterior (except for the dorsal body cavity and joint cavities).

Serous membranes occur in pairs (Figure 4.1c). The *parietal* (pah-ri′e-tal: *parie* = wall) *layer* lines a specific portion of the wall of the ventral body cavity. It folds in on itself to form the *visceral* (vis′er-al) *layer,* which covers the outside of the organs in that cavity.

You can visualize the relationship between the serosal layers by wrapping your fist with a limp balloon only partially filled with air or water (Figure 4.1d). The part of the balloon that clings closely to your fist can be compared to the visceral serosa clinging to the organ's external surface. The outer wall of the balloon represents the parietal serosa that lines the walls of the cavity and that, unlike the balloon, is never exposed but is always fused to the cavity wall. In the body, the serous layers are separated not by air but by a scanty amount of thin, clear fluid, called **serous fluid,** which is secreted by both membranes. Although there is a potential space between the two membranes, they tend to lie very close to each other.

The serous fluid allows the organs to slide easily across the cavity walls and one another without friction as they carry out their routine functions. This is extremely important when mobile organs such as the pumping heart and a churning stomach are involved.

The specific names of the serous membranes depend on their locations. The serosa lining the abdominal cavity and covering its organs is the **peritoneum** (per″ĭ-to-ne′um). In the thorax, serous membranes isolate the lungs and heart from one another. That surrounding the lungs (Figure 4.1c) is the **pleura** (ploo′rah); that around the heart is the **pericardium** (per″ĭ-kar′de-um).

Connective Tissue Membranes

Synovial (sĭ-no′ve-al) **membranes** are composed of soft areolar connective tissue and contain no epithelial cells at all. These membranes line the fibrous capsules surrounding joints (Figure 4.2), where they provide a smooth surface and secrete a lubricating fluid. They also line small sacs of connective tissue called *bursae* (ber′se) and the tubelike *tendon sheaths.* Both of these structures cushion organs moving against each other during muscle activity—such as the movement of a tendon across a bone's surface.

Integumentary System (Skin)

Would you be enticed by an advertisement for a coat that is waterproof, stretchable, washable, and permanent-press, that invisibly repairs small cuts, rips, and burns, and that is guaranteed to last a lifetime with reasonable care? Sounds too good to be true, but you already have such a coat—your *cutaneous membrane,* or **skin.** The skin and its derivatives (sweat and oil glands, hair, and nails) serve a number of functions, mostly protective. Together, these organs are called the **integumentary** (in-teg″u-men′ta-re) **system.**

TABLE 4.1 Functions of the Skin

Functions	How accomplished
Protects deeper tissues from	
• Mechanical damage (bumps)	Physical barrier contains keratin, which toughens cells, and pressure receptors, which alert the nervous system to possible damage.
• Chemical damage (acids and bases)	Has relatively impermeable keratinized cells; contains pain receptors, which alert the nervous system to possible damage.
• Bacterial damage	Has an unbroken surface and "acid mantle" (skin secretions are acidic, and thus inhibit bacteria). Phagocytes ingest foreign substances and pathogens, preventing them from penetrating into deeper body tissues.
• Ultraviolet radiation (damaging effects of sunlight)	Melanin produced by melanocytes offers protection from UV damage.
• Thermal (heat or cold) damage	Contains heat/cold/pain receptors.
• Desiccation (drying out)	Contains a waterproofing glycolipid and keratin.
Aids in body heat loss or heat retention (controlled by the nervous system)	*Heat loss:* By activating sweat glands and allowing blood to flush into skin capillary beds. *Heat retention:* By not allowing blood to flush into skin capillary beds.
Aids in excretion of urea and uric acid	Contained in perspiration produced by sweat glands.
Synthesizes vitamin D	Modified cholesterol molecules in skin converted to vitamin D by sunlight.

Basic Skin Functions

Also called the **integument** (in-teg′u-ment), which simply means "covering," the skin is much more than an external body covering. It is absolutely essential because it keeps water and other precious molecules in the body. It also keeps water (and other things) out. (This is why one can swim for hours without becoming waterlogged.) Structurally, the skin is a marvel. It is pliable yet tough, which allows it to take constant punishment from external agents. Without our skin, we would quickly fall prey to bacteria and perish from water and heat loss.

The skin has many functions; most, but not all, are protective (Table 4.1). It insulates and cushions the deeper body organs and protects the entire body from mechanical damage (bumps and cuts), chemical damage (such as from acids and bases), thermal damage (heat and cold), ultraviolet radiation (in sunlight), and bacteria. The uppermost layer of the skin is full of **keratin** and *cornified,* or hardened, in order to prevent water loss from the body surface.

The skin's rich capillary network and sweat glands (both controlled by the nervous system) play an important role in regulating heat loss from the body surface. The skin acts as a mini-excretory system; urea, salts, and water are lost when we sweat. The skin also manufactures several proteins important to immunity and synthesizes vitamin D. (Modified cholesterol molecules located in the skin are converted to vitamin D by sunlight.) Finally, the *cutaneous sensory receptors,* which are actually part of the nervous system, are located in the skin. These tiny sensors, which include touch, pressure, temperature, and pain receptors, provide us with a great deal of information about our external environment. They alert us to bumps and the presence of tissue-damaging factors as well as to the feel of wind in our hair and a caress.

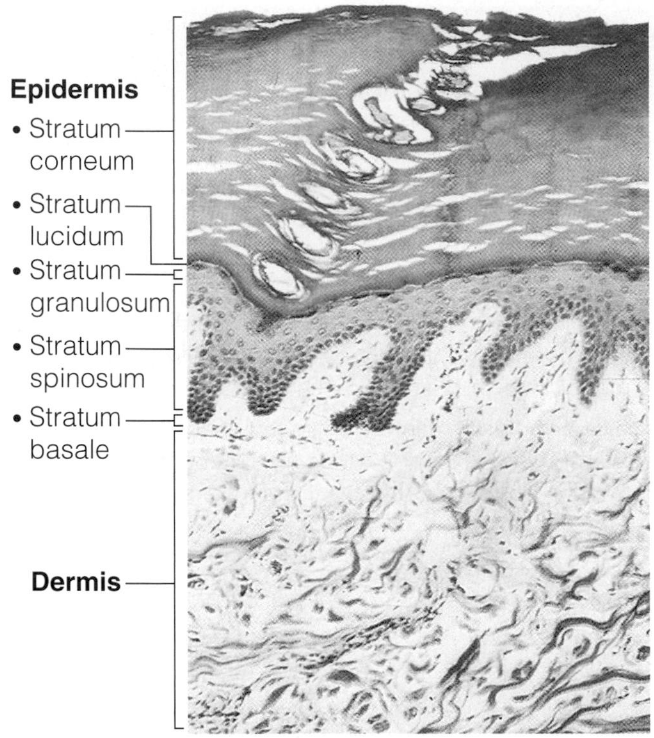

FIGURE 4.3 The epidermis of thick skin (150×). From *Gray's Anatomy,* Henry Gray. Churchill Livingstone, UK.

Structure of the Skin

The skin is composed of two kinds of tissue. The outer **epidermis** (ep″ĭ-der′mis) is made up of stratified squamous epithelium that is capable of *keratinizing* (ker′ah-tin-īz-ing), or becoming hard and tough. The underlying **dermis** is mostly made up of dense connective tissue. The epidermis and dermis are firmly connected. However, a burn or friction (such as the rubbing of a poorly fitting shoe) may cause them to separate, allowing interstitial fluid to accumulate in the cavity between the layers, which results in a *blister.*

Deep to the dermis is the **subcutaneous tissue,** or **hypodermis,** which essentially is adipose tissue. It is not considered part of the skin, but it does anchor the skin to underlying organs. Subcutaneous tissue serves as a shock absorber and insulates the deeper tissues from extreme temperature changes occurring outside the body. It is also responsible for the curves that are more a part of a woman's anatomy than a man's. The main skin areas and structures are described next. As you read, locate the described areas or structures on Figures 4.3 and 4.4.

Epidermis

The epidermis is composed of up to five zones or layers called *strata* (strah′tah). From the inside out these are the stratum basale, spinosum, granulosum, lucidum, and corneum (see Figure 4.3).

Like all epithelial tissues, the epidermis is *avascular;* that is, it has no blood supply of its own. This explains why a man can shave daily and not bleed even though he is cutting off many cell layers each time he shaves.

Most cells of the epidermis are **keratinocytes** (keratin cells), which produce keratin, the fibrous protein that makes the epidermis a tough protective layer. The deepest cell layer of the epidermis, the **stratum basale** (stra′tum bă-sah′le), lies closest to the dermis and contains epidermal cells that receive the most adequate nourishment via diffusion of nutrients from the dermis. These cells are constantly undergoing cell division, and millions of new cells are produced daily; hence its alternate name, *stratum germinativum* (jer″min-ah-tiv′um; "germinating layer"). The daughter cells are pushed upward, away from the source of nutrition, to become part of the epidermal layers closer to the skin surface. They move away from the dermis and become part of the more superficial layers, the **stratum spinosum** and then the **stratum granulosum.** Then they become flatter, increasingly full of keratin (keratinized), and finally die, forming the clear **stratum lucidum** (lu′sid-um). This latter epidermal layer is not seen in all skin regions; it occurs only where the skin is hairless and extra thick, that is, on the palms of the hands and soles of the feet. The combination of accumulating keratin inside them, secreting a water-repellent glycoprotein into the extracellular space, and their increasing distance from the blood supply (in the dermis) effectively dooms the stratum lucidum cells and the more superficial epidermal cells because they are unable to get adequate nutrients and oxygen.

The outermost layer, the **stratum corneum** (kor′ne-um), is 20 to 30 cell layers thick. It accounts for about three-quarters of the epidermal thickness. The shinglelike dead cell remnants, completely filled with keratin, are referred to as *cornified* or *horny cells* (*cornu* = horn). The common saying "Beauty is only skin deep" is especially interesting in light of the fact that nearly everything we see when we look at someone is dead! Keratin is an exceptionally tough protein. Its abundance in

Q *What component of the hypodermis makes it a good insulator and shock absorber?*

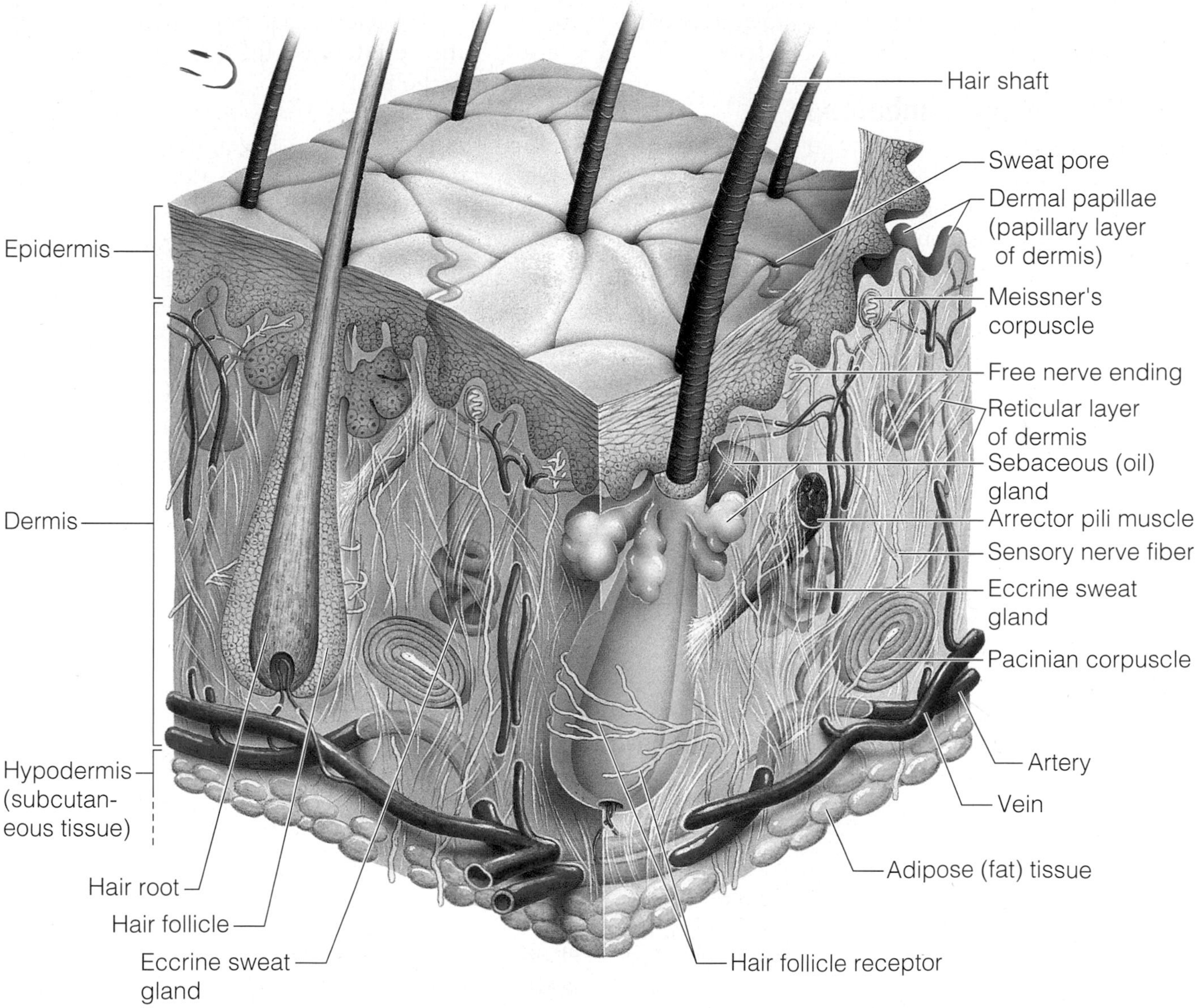

FIGURE 4.4 Skin structure. Three-dimensional view of the skin and underlying subcutaneous tissue.

the stratum corneum allows that layer to provide a durable "overcoat" for the body, which protects deeper cells from the hostile external environment (air) and from water loss and helps the body resist biological, chemical, and physical assaults. The stratum corneum rubs and flakes off slowly and steadily and is replaced by cells produced by the division of the deeper stratum basale cells. Indeed, we have a totally "new" epidermis every 25 to 45 days.

Melanin (mel′ah-nin), a pigment that ranges in color from yellow to brown to black, is produced by special cells called **melanocytes** (mel′ah-no sītz), found chiefly in the stratum basale. When the skin is exposed to sunlight, which stimulates the melanocytes to produce more of the melanin pigment, tanning occurs. The stratum basale cells phagocytize (eat) the pigment, and as it accumulates within them, the melanin forms a protective

A *Its fatty tissue.*

pigment umbrella over the superficial, or "sunny," side of their nuclei that shields their genetic material (DNA) from the damaging effects of ultraviolet radiation in sunlight. *Freckles* and *moles* are seen where melanin is concentrated in one spot.

Homeostatic Imbalance

Despite melanin's protective effects, excessive sun exposure eventually damages the skin. It causes the elastic fibers to clump, leading to leathery skin. It also depresses the immune system. This may help to explain why many people infected with the *herpes simplex*, or *cold sore*, virus are more likely to have an eruption after sunbathing. Overexposure to the sun can also alter the DNA of skin cells and in this way lead to skin cancer. Black people seldom have skin cancer, attesting to melanin's amazing effectiveness as a natural sunscreen. ▲

Dermis

The dermis is your "hide." It is a strong, stretchy envelope that helps to hold the body together. When you purchase leather goods (bags, belts, shoes, and the like), you are buying the treated dermis of animals.

The dense (fibrous) connective tissue making up the dermis consists of two major regions—the *papillary* and the *reticular* areas. Like the epidermis, the dermis varies in thickness. For example, it is particularly thick on the palms of the hands and soles of the feet but is quite thin on the eyelids.

The **papillary layer** is the upper dermal region. It is uneven and has fingerlike projections from its superior surface, called **dermal papillae** (pah-pil′e; *papill* = nipple), which indent the epidermis above. Many of the dermal papillae contain capillary loops, which furnish nutrients to the epidermis. Others house pain receptors *(free nerve endings)* and touch receptors called *Meissner's corpuscles* (mīs′nerz kor′puh-sulz). On the palms of the hands and soles of the feet, the papillae are arranged in definite patterns that form looped and whorled ridges on the epidermal surface that increase friction and enhance the gripping ability of the fingers and feet. Papillary patterns are genetically determined. The ridges of the fingertips are well provided with sweat pores and leave unique, identifying films of sweat called *fingerprints* on almost anything they touch.

The **reticular layer** is the deepest skin layer. It contains blood vessels, sweat and oil glands, and deep pressure receptors called *Pacinian* (pah-sin′e-an) *corpuscles* (see Figure 4.4). Phagocytes found here (and, in fact, throughout the dermis) act to prevent bacteria that have managed to get through the epidermis from penetrating any deeper into the body.

Both *collagen* and *elastic fibers* are found throughout the dermis. Collagen fibers are responsible for the toughness of the dermis; they also attract and bind water and thus help to keep the skin hydrated. Elastic fibers give the skin its elasticity when we are young. As we age, the number of collagen and elastic fibers decreases, and the subcutaneous tissue loses fat. As a result, the skin loses its elasticity and begins to sag and wrinkle.

The dermis is abundantly supplied with blood vessels that play a role in maintaining body temperature homeostasis. When body temperature is high, the capillaries of the dermis become engorged, or swollen, with heated blood, and the skin becomes reddened and warm. This allows body heat to radiate from the skin surface. If the environment is cool and body heat must be conserved, blood bypasses the dermis capillaries temporarily, allowing internal body temperature to stay high.

Homeostatic Imbalance

Any restriction of the normal blood supply to the skin results in cell death and, if severe or prolonged enough, skin ulcers. *Decubitus* (de-ku′bĭ-tus) *ulcers* (bedsores) occur in bedridden patients who are not turned regularly or who are dragged or pulled across the bed repeatedly. The weight of the body puts pressure on the skin, especially over bony projections. Because this restricts the blood supply, the skin becomes pale or blanched at pressure points. At first, the skin reddens when pressure is released, but if the situation is not corrected, the cells begin to die, and typically small cracks or breaks in the skin appear at compressed sites. Permanent damage to the superficial blood vessels and tissue eventually results in degeneration and ulceration of the skin (Figure 4.5). ▲

The dermis also has a rich nerve supply. As mentioned earlier, many of the nerve endings have specialized receptor end-organs that send messages to the central nervous system for interpretation when they are stimulated by environmental factors (pressure, temperature, and the like). These cutaneous receptors are discussed in more detail in Chapter 7.

A Closer Look

Tattoos

TATTOOS are made by using a needle to deposit pigment within the dermis. Tattooing is an ancient practice, believed to have originated around 10,000 years ago. These days, tattoos are symbols of club membership for some males (street gangs, the military, fraternities); other people view them as symbols of individuality. In recent years, females have increasingly acquired tattoos as a means of expression and for cosmetic purposes; permanent eyeliner and tattooed liplines now account for over 125,000 tattoos a year.

But what if a tattoo becomes unfashionable or the pigment migrates? Tattoo removal has been and still is a pain—both physically and financially. Until recently, once you had one, you were essentially stuck with it, because attempts at removal—dermabrasion, cryosurgery (freezing), or applying caustic chemicals—left nasty scars. Using new laser-based technologies, dermatologists have no problem destroying the black or blue pigments in tattoos applied a generation ago, but newer, multicolored tattoos pose a larger problem. The multitude of pigments in tattoos today require several different lasers to be used over seven to nine treatments spaced about a month apart, each costing $75 to $150 and at a cost in pain roughly equal to getting tattooed in the first place. Nonetheless, tattoo removal across the United States is skyrocketing.

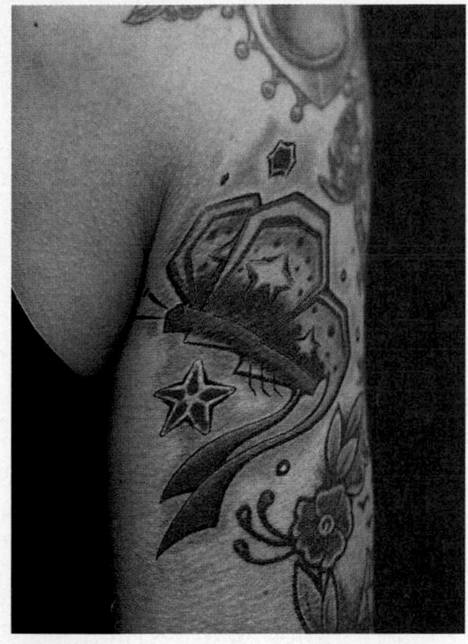

Tattoos present some other risks. Even though the FDA has some regulations concerning the composition of tattoo pigments, their safety is not well established. Indeed, studies of dyes collected from tattooing studios have been found to contain cancer-causing agents that could be activated during tattoo removal. Statutory regulations vary widely (from none to complete prohibition) from state to state. Still, in each case, needles are used and bleeding occurs, and practitioners' competence varies significantly. If strict sterile procedures are not adhered to, tattooing can spread infections. The risk of hepatitis C infection (a chronic liver infection) is 15 times higher in those who have been tattooed than in those who have not.

So if you're thinking about getting a tattoo, look into it carefully. Even with the new laser removal techniques, is it worth the risk?

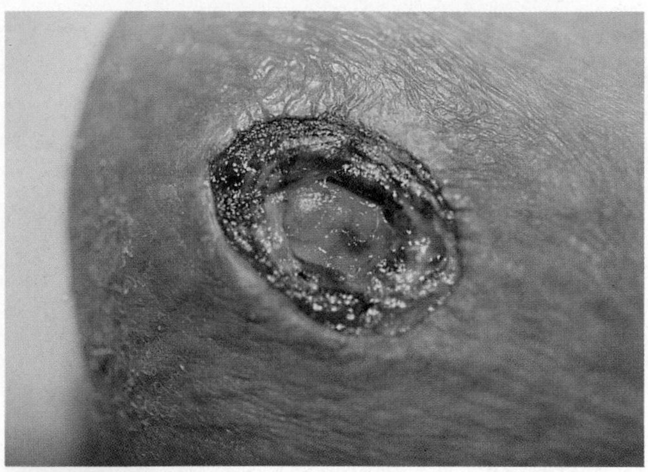

FIGURE 4.5 Photograph of a deep (stage III) decubitus ulcer.

Skin Color

Three pigments contribute to skin color:

1. The amount and kind (yellow, reddish brown, or black) of *melanin* in the epidermis.
2. The amount of *carotene* deposited in the stratum corneum and subcutaneous tissue. (Carotene is an orange-yellow pigment found in abundant amounts in carrots and other orange, deep yellow, or leafy green vegetables.) The skin tends to take on a yellow-orange cast when large amounts of carotene-rich foods are eaten.
3. The amount of *oxygen-rich hemoglobin* (pigment in red blood cells) in the dermal blood vessels.

People who produce a lot of melanin have brown-toned skin. In light-skinned (Caucasian) people, who have less melanin, the crimson color of oxygen-rich hemoglobin in the dermal blood supply flushes through the transparent cell layers above and gives the skin a rosy glow.

Homeostatic Imbalance

When hemoglobin is poorly oxygenated, both the blood and the skin of Caucasians appear blue, a condition called *cyanosis* (si″ah-no′sis). Cyanosis is common during heart failure and severe breathing disorders. In black people, the skin does not appear cyanotic in the same situations because of the masking effects of melanin, but cyanosis is apparent in their mucous membranes and nail beds. ▲

Skin color is also influenced by emotional stimuli, and many alterations in skin color signal certain disease states:

- *Redness,* or *erythema* (er″ĭ-the′mah): Reddened skin may indicate embarrassment (blushing), fever, hypertension, inflammation, or allergy.
- *Pallor,* or *blanching:* Under certain types of emotional stress (fear, anger, and others), some people become pale. Pale skin may also signify anemia, low blood pressure, or impaired blood flow into the area.
- *Jaundice* (jon′dis) or a *yellow cast:* An abnormal yellow skin tone usually signifies a liver disorder in which excess bile pigments are absorbed into the blood, circulated throughout the body, and deposited in body tissues.
- *Bruises* or *black-and-blue marks:* Black-and-blue marks reveal sites where blood has escaped from the circulation and has clotted in the tissue spaces. Such clotted blood masses are called *hematomas.* An unusual tendency to bruising may signify a deficiency of vitamin C in the diet or hemophilia (bleeder's disease).

Appendages of the Skin

The **skin appendages** include cutaneous glands, hair and hair follicles, and nails (see Figure 4.4). Each of these appendages arises from the epidermis and plays a unique role in maintaining body homeostasis.

Cutaneous Glands

The cutaneous glands are all **exocrine glands** that release their secretions to the skin surface via ducts. They fall into two groups: *sebaceous glands* and *sweat glands.* As these glands are formed by the cells of the stratum basale, they push into the deeper skin regions and ultimately reside almost entirely in the dermis.

Sebaceous (Oil) Glands The **sebaceous** (seh-ba′shus) **glands,** or oil glands, are found all over the skin, except on the palms of the hands and the soles of the feet. Their ducts usually empty into a hair follicle (see Figures 4.4 and 4.6), but some open directly onto the skin surface.

The product of the sebaceous glands, **sebum** (se′bum; *seb* = grease), is a mixture of oily substances and fragmented cells. Sebum is a lubricant

Q *Which of these gland types can make your hair lank and oily?*

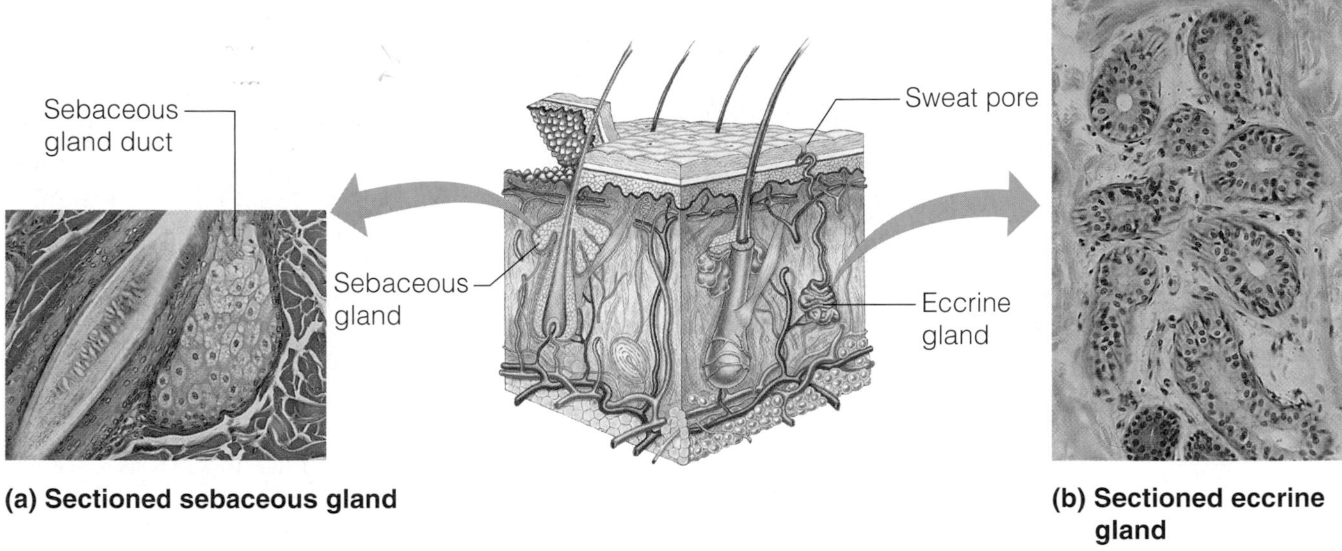

FIGURE 4.6 Cutaneous glands. (a) Photomicrograph of a sebaceous gland (104×). **(b)** Photomicrograph of eccrine sweat glands (148×).

that keeps the skin soft and moist and prevents the hair from becoming brittle. Sebum also contains chemicals that *kill* bacteria, so it is important in preventing the bacteria present on the skin surface from invading the deeper skin regions. The sebaceous glands become very active when male sex hormones are produced in increased amounts (in both sexes) during adolescence. Thus, the skin tends to become oilier during this period of life.

Homeostatic Imbalance

If a sebaceous gland's duct becomes blocked by sebum, a *whitehead* appears on the skin surface. If the accumulated material oxidizes and dries, it darkens, forming a *blackhead. Acne* is an active infection of the sebaceous glands accompanied by pimples on the skin. It can be mild or extremely severe, leading to permanent scarring. *Seborrhea* (seb"o-re'ah), known as "cradle cap" in infants, is caused by overactivity of the sebaceous glands. It begins on the scalp as pink, raised lesions that gradually form a yellow to brown crust that sloughs off as oily dandruff. Careful washing to remove the excessive oil often helps. ▲

A *The sebaceous glands, which produce oily secretions.*

Sweat Glands **Sweat glands,** also called **sudoriferous** (su"do-rif'er-us; *sudor* = sweat) **glands,** are widely distributed in the skin. Their number is staggering—more than 2.5 million per person. There are two types of sweat glands, *eccrine* and *apocrine.*

The **eccrine** (ek'rin) **glands** are far more numerous and are found all over the body. They produce **sweat,** a clear secretion that is primarily water plus some salts (sodium chloride), vitamin C, traces of metabolic wastes (ammonia, urea, uric acid), and lactic acid (a chemical that accumulates during vigorous muscle activity). Sweat is acidic (pH from 4 to 6), a characteristic that inhibits the growth of bacteria, which are always present on the skin surface. Typically, sweat reaches the skin surface via a duct that opens externally as a funnel-shaped *pore* (see Figures 4.4 and 4.6). Notice, however, that the facial "pores" commonly referred to when we talk about our complexion are *not* these sweat pores, but the external outlets of hair follicles.

The eccrine sweat glands are an important and highly efficient part of the body's heat-regulating equipment. They are supplied with nerve endings that cause them to secrete sweat when the external temperature or body temperature is high. When sweat evaporates off the skin surface, it carries

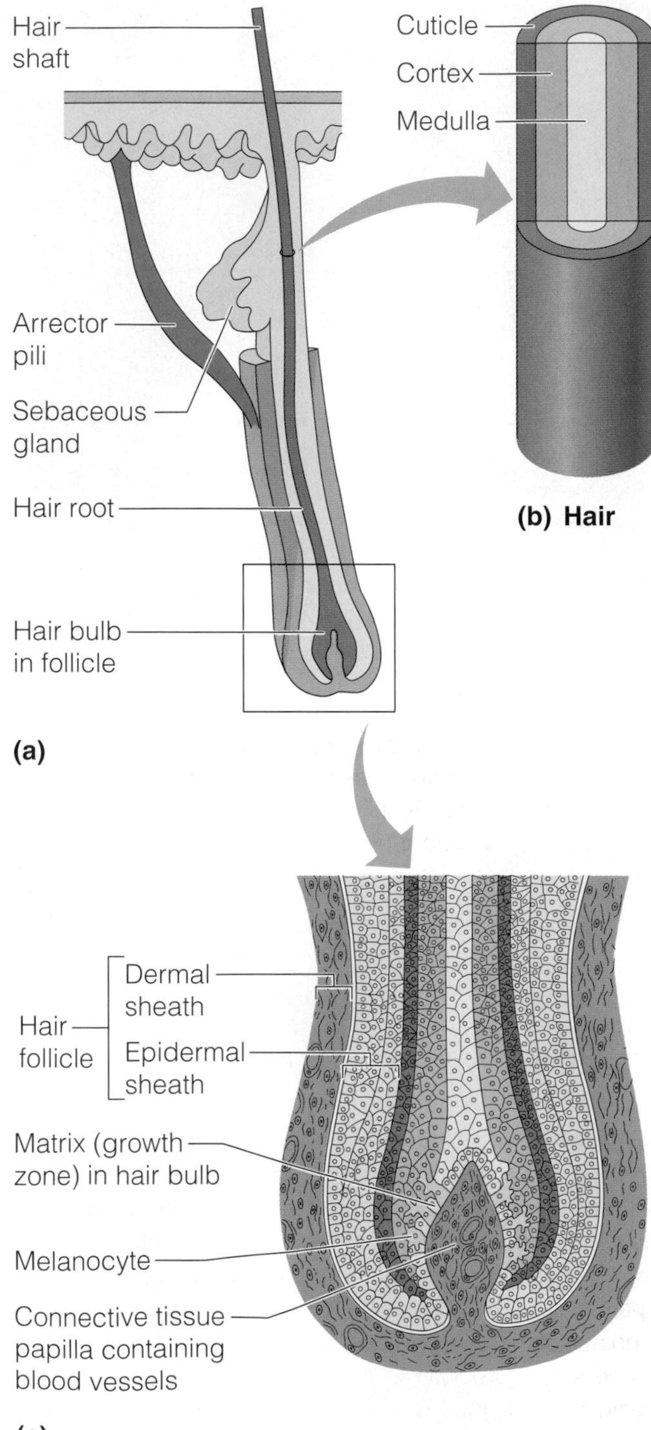

FIGURE 4.7 Structure of a hair and hair follicle. **(a)** Longitudinal section of a hair within its follicle. **(b)** Enlarged longitudinal section of a hair. **(c)** Enlarged longitudinal view of the expanded hair bulb in the follicle showing the matrix, the region of actively dividing epithelial cells that produces the hair.

large amounts of body heat with it. On a hot day, it is possible to lose up to 7 liters of body water in this way. The heat-regulating functions of the body are important—if internal temperature changes more than a few degrees from the normal 37°C (98.2°F), life-threatening changes occur in the body. Body temperature regulation is discussed in more detail in Chapter 14.

Apocrine (ap′o-krin) **glands** are largely confined to the axillary and genital areas of the body. They are usually larger than eccrine glands, and their ducts empty into hair follicles. Their secretion contains fatty acids and proteins, as well as all the substances present in eccrine secretion; consequently, it may have a milky or yellowish color. The secretion is odorless, but when bacteria that live on the skin use its proteins and fats as a source of nutrients for their growth, it takes on a musky, unpleasant odor.

Apocrine glands begin to function during puberty under the influence of *androgens* (male sex hormones). Although their secretion is produced almost continuously, apocrine glands play a minimal role in thermoregulation. Their precise function is not yet known, but they are activated by nerve fibers during pain and stress and during sexual foreplay.

Hair and Hair Follicles

There are millions of **hairs** scattered all over the body. But, other than serving a few minor protective functions—such as guarding the head against bumps, shielding the eyes (via eyelashes), and helping to keep foreign particles out of the respiratory tract (via nose hairs)—our body hair has lost much of its usefulness. Hair served early humans (and still serves hairy animals) by providing insulation in cold weather, but now we have other means of keeping warm.

A hair, produced by a *hair follicle,* is a flexible epithelial structure. That part of the hair enclosed in the follicle is called the *root.* The part projecting from the surface of the scalp or skin is called the *shaft* (Figure 4.7). A hair is formed by division of the well-nourished stratum basale epithelial cells in the **matrix** (growth zone), of the hair bulb at the inferior end of the follicle. As the daughter cells are pushed farther away from the growing region, they become keratinized and die. Thus the bulk of the hair shaft, like the bulk of the epidermis, is dead material and almost entirely protein.

Each hair consists of a central core called the *medulla* (me-dul'ah) surrounded by a bulky *cortex* layer. The cortex is, in turn, enclosed by an outermost *cuticle* formed by a single layer of cells that overlap one another like shingles on a roof. This arrangement of the cuticle cells helps to keep the hairs apart and keeps them from matting (see Figures 4.7b and 4.8). The cuticle is the most heavily keratinized region; it provides strength and helps keep the inner hair layers tightly compacted. Because it is most subject to abrasion, the cuticle tends to wear away at the tip of the shaft, allowing the keratin fibrils in the inner hair regions to frizz out, a phenomenon called "split ends." Hair pigment is made by melanocytes in the hair bulb, and varying amounts of different types of melanin (yellow, rust, brown, and black) combine to produce *all* varieties of hair color from pale blond to pitch black.

Hairs come in a variety of sizes and shapes. They are short and stiff in the eyebrows, long and flexible on the head, and usually nearly invisible almost everywhere else. When the hair shaft is oval, hair is smooth and silky and the person has wavy hair. When the shaft is flat and ribbonlike, the hair is curly or kinky. If it is perfectly round, the hair is straight and tends to be coarse. Hairs are found all over the body surface except the palms of the hands, soles of the feet, nipples, and lips. Humans are born with as many hair follicles as they will ever have, and hairs are among the fastest growing tissues in the body. Hormones account for the development of hairy regions—the scalp and, in the adult, the pubic and axillary (armpit) areas.

Hair follicles are actually compound structures. The inner *epidermal sheath* is composed of epithelial tissue and forms the hair. The outer *dermal sheath* is actually dermal connective tissue. This dermal region supplies blood vessels to the epidermal portion and reinforces it. Its nipplelike *papilla* provides the blood supply to the matrix in the hair bulb.

Look carefully at the structure of the hair follicle at the front corner of Figure 4.4. Notice that it is slanted. Small bands of smooth muscle cells—**arrector pili** (ah-rek'tor pi'li)—connect each side of the hair follicle to the dermal tissue. When these muscles contract (as when we are cold or frightened), the hair is pulled upright, dimpling the skin surface with "goose bumps." This action helps keep animals warm in winter by adding a layer of

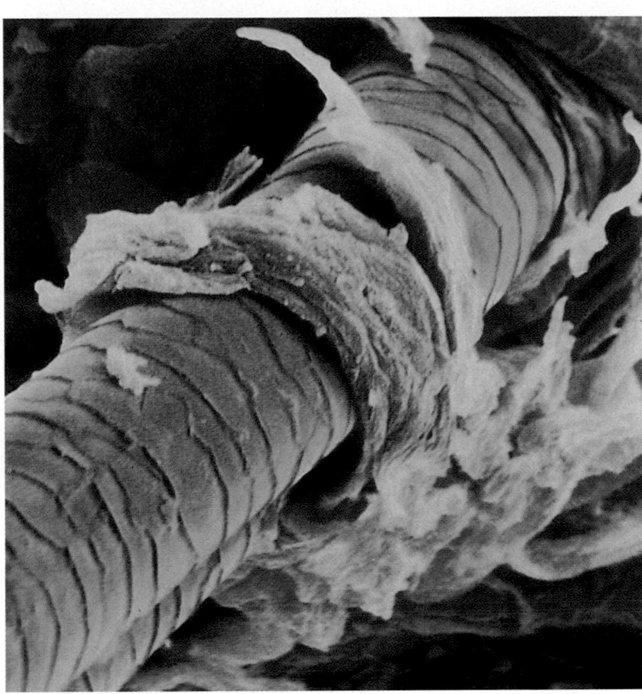

FIGURE 4.8 Scanning electron micrograph showing a hair shaft emerging from a follicle at the skin surface. Notice how the scalelike cells of the cuticle overlap one another (1500×).

Prove It Yourself

Plot the Distribution of Sweat Glands

For this experiment investigating the distribution of sweat glands in different body areas, you will need two squares of bond paper (each 1 cm by 1 cm), adhesive tape, an iodine solution, and a cotton-tipped swab.

First, paint an area of the medial aspect of your left palm (avoid the crease lines) and left forearm with the iodine solution, and let it dry thoroughly. Now have a friend tape a square of bond paper securely over each iodine-painted area, and leave the paper in place for 20 minutes.

After 20 minutes, remove the paper squares and count the number of blue-black dots on each square. Each blue-black dot indicates an active sweat gland. (The iodine in the pore dissolves in your sweat and reacts chemically with the starch in the bond paper to produce the blue-black color). Thus, you have now produced "sweat maps" of two areas of your skin. Which region has the greater density of sweat glands?

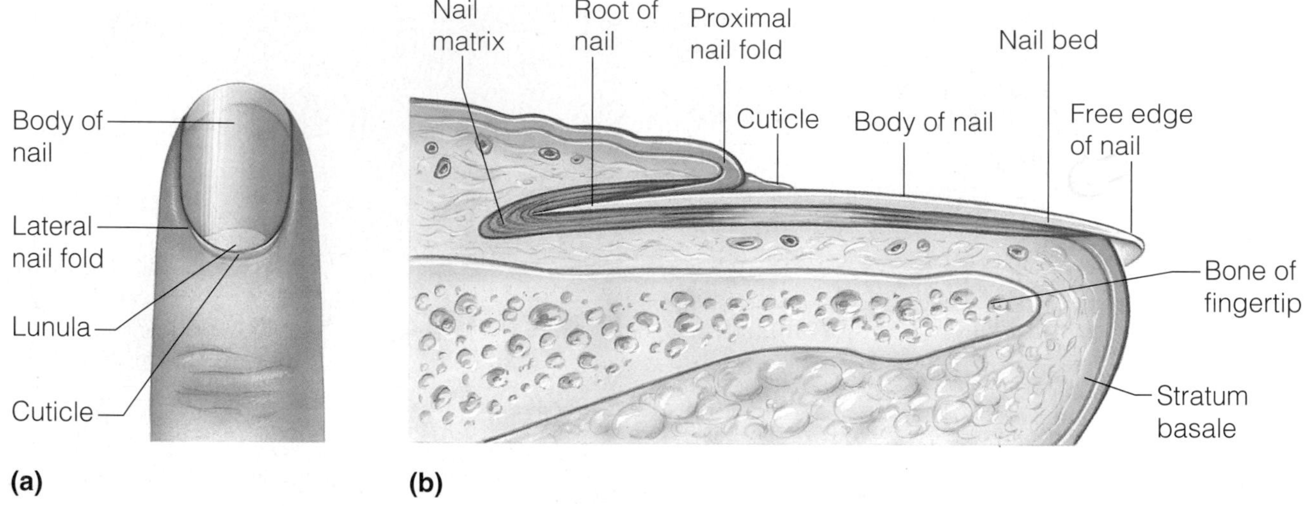

FIGURE 4.9 Structure of a nail. (a) Surface view. **(b)** Longitudinal section of the distal part of a finger, showing nail parts and the nail matrix that forms the nail.

insulating air to the fur. It is especially dramatic in a scared cat, whose fur actually stands on end to make it look larger to scare off its enemy. However, this hair-raising phenomenon is not very useful to human beings.

Nails

A **nail** is a scalelike modification of the epidermis that corresponds to the hoof or claw of other animals. Each nail has a *free edge*, a *body* (visible attached portion), and a *root* (embedded in the skin). The borders of the nail are overlapped by skin folds, called *nail folds*. The thick proximal nail fold is commonly called the *cuticle* (Figure 4.9).

The stratum basale of the epidermis extends beneath the nail as the *nail bed*. Its thickened proximal area, called the *nail matrix*, is responsible for nail growth. As the nail cells are produced by the matrix, they become heavily keratinized and die. Thus, nails, like hairs, are mostly nonliving material.

Nails are transparent and nearly colorless, but they look pink because of the rich blood supply in the underlying dermis. The exception to this is the region over the thickened nail matrix that appears as a white crescent and is called the *lunula* (loo′nyu-luh; *lunul* = crescent). As noted earlier, when the supply of oxygen in the blood is low, the nail beds take on a cyanotic (blue) cast.

Homeostatic Imbalances of Skin

It is difficult to scoff at anything that goes wrong with the skin because, when it rebels, it is quite a visible revolution. Loss of homeostasis in body cells and organs can reveal itself on the skin in ways that are sometimes almost unbelievable. The skin can develop more than 1000 different ailments. The most common skin disorders result from allergies or bacterial, viral, or fungal infections. Less common, but far more damaging, are burns and skin cancers. A number of the homeostatic imbalances of the skin are summarized briefly in the sections just below.

Infections and Allergies

- **Athlete's foot:** An itchy, red, peeling condition of the skin between the toes, resulting from fungus infection. Also called *tinea pedis*.
- **Boils and carbuncles** (kar′bun-kulz): Inflammation of hair follicles and sebaceous glands, common on the dorsal neck. Carbuncles are composite boils typically caused by bacterial infection (often *Staphylococcus aureus*).
- **Cold sores** (fever blisters): Small fluid-filled blisters that itch and sting, caused by a herpes simplex infection. The virus localizes in a cutaneous nerve, where it remains dormant until activated by emotional upset, fever, or UV radiation. Cold sores usually occur around the lips and in the oral mucosa of the mouth (Figure 4.10a).

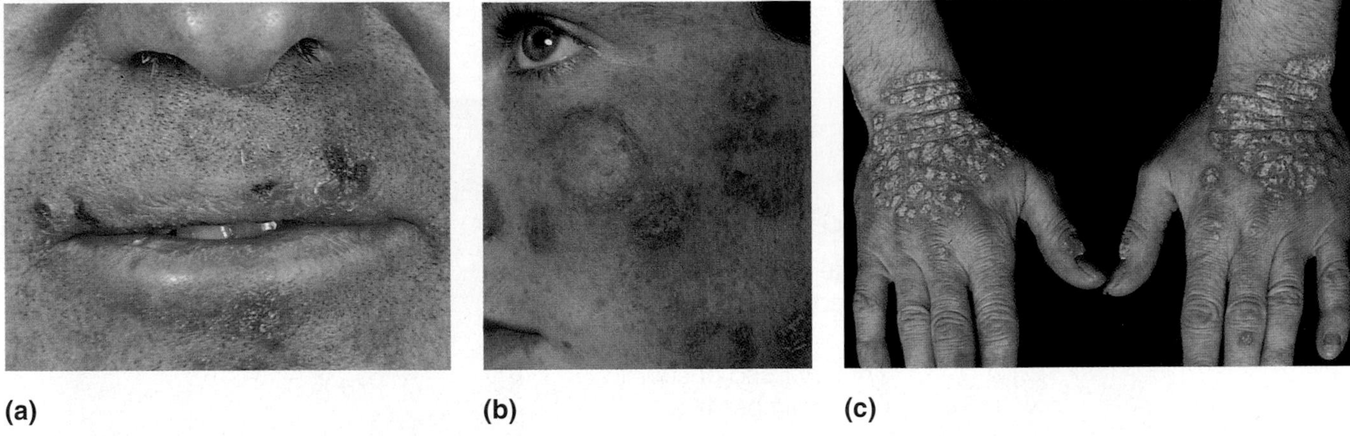

FIGURE 4.10 Cutaneous lesions. (a) Cold sores. **(b)** Impetigo. **(c)** Psoriasis.

- **Contact dermatitis:** Itching, redness, and swelling of the skin, progressing to blistering. Caused by exposure of the skin to chemicals (such as those in poison ivy) that provoke allergic responses in sensitive individuals.
- **Impetigo** (im peh ti'go; *impet* = an attack): Pink, water-filled, raised lesions (commonly around the mouth and nose) that develop a yellow crust and eventually rupture (Figure 4.10b). Caused by a highly contagious staphylococcus infection; common in elementary school–aged children.
- **Psoriasis** (so-ri'ah-sis): A chronic condition, characterized by overproduction of skin cells that results in reddened epidermal lesions covered with dry, silvery scales (Figure 4.10c). When severe, may be disfiguring. It is believed to be an autoimmune disorder in which the immune system attacks a person's own tissues. Attacks often triggered by trauma, infection, hormonal changes, and stress.

Burns

The skin is only about as thick as a paper towel—not too impressive as organ systems go. And yet, when it is severely damaged, nearly every body system suffers. Metabolism accelerates or may be impaired, changes in the immune system occur, and the cardiovascular system may falter. Such severe damage can be caused by burns. A **burn** is tissue damage and cell death caused by intense heat, electricity, UV radiation (sunburn), or certain chemicals (such as acids).

There are few threats to skin more serious than burns. When the skin is burned and its cells are destroyed, two life-threatening problems result. First, the body loses its precious supply of fluids containing proteins and electrolytes as these seep from the burned surfaces. Dehydration and electrolyte imbalance follow and can lead to a shutdown of the kidneys and *circulatory shock* (inadequate circulation of blood caused by low blood volume). To save the patient, the lost fluids must be replaced immediately.

The volume of fluid lost can be estimated indirectly by determining how much of the body surface is burned (extent of burns), using the **rule of nines.** This method divides the body into 11 areas, each accounting for 9 percent of the total body surface area, plus an additional area surrounding the genitals (the perineum) representing 1 percent of body surface area (Figure 4.11a). Later, infection becomes the most important threat and is the leading cause of death in burn victims. Burned skin is sterile for about 24 hours. But after that, *pathogens* (path'o-jenz) such as bacteria and fungi easily invade areas where the skin has been destroyed and multiply rapidly in the nutrient-rich environment of dead tissues. To make matters worse, the patient's immune system becomes depressed within one to two days after severe burn injury.

Burns are classified according to their severity (depth) as first-, second-, or third-degree burns (Figure 4.11b). In **first-degree burns,** only the epidermis is damaged. The area becomes red and swollen. Except for temporary discomfort, first-degree burns are not usually serious and generally heal in two to three days without any special attention. Sunburn is usually a first-degree burn. **Second-degree burns** involve injury to the epidermis and the upper region of the dermis. The skin is red and painful, and *blisters* appear. Because sufficient

Q *Would you expect to see hair regeneration in an area that suffers third-degree burns? Why or why not?*

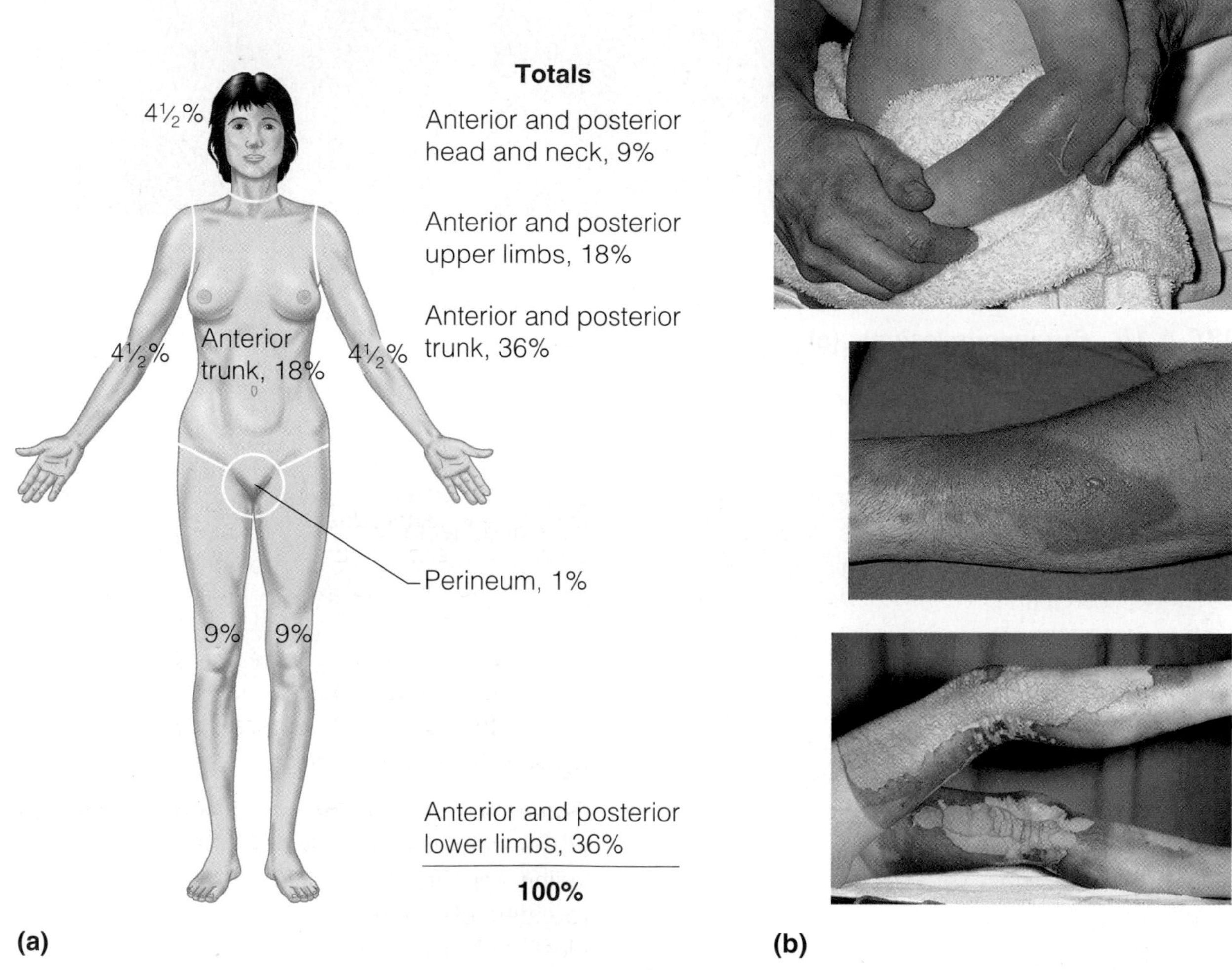

FIGURE 4.11 Burns. (a) Estimating the extent of burns using the rule of nines. The surface areas for the anterior body surface are indicated on the human figure. Total surface area (anterior and posterior body surfaces) is tabulated to the right of the figure. **(b)** Burns of increasing severity, from top to bottom: first-degree, second-degree, third-degree.

numbers of epithelial cells are still present, regrowth (regeneration) of the epithelium can occur. Ordinarily, no permanent scars result if care is taken to prevent infection. First- and second-degree burns are referred to as **partial-thickness burns.**

Third-degree burns destroy the entire thickness of the skin, so these burns are also called **full-thickness burns.** The burned area appears blanched (gray-white) or blackened, and since the nerve endings in the area are destroyed, the burned area is not painful. In third-degree burns, regeneration is not possible, and skin grafting must be done to cover the underlying exposed tissues.

In general, burns are considered *critical* if any of the following conditions exists:

1. Over 25 percent of the body has second-degree burns,
2. Over 10 percent of the body has third-degree burns, or

A *No, because third-degree burns destroy both the epidermis and dermis in that region.*

Q *Which of these cancers arises from the most superficial epidermal cells?*

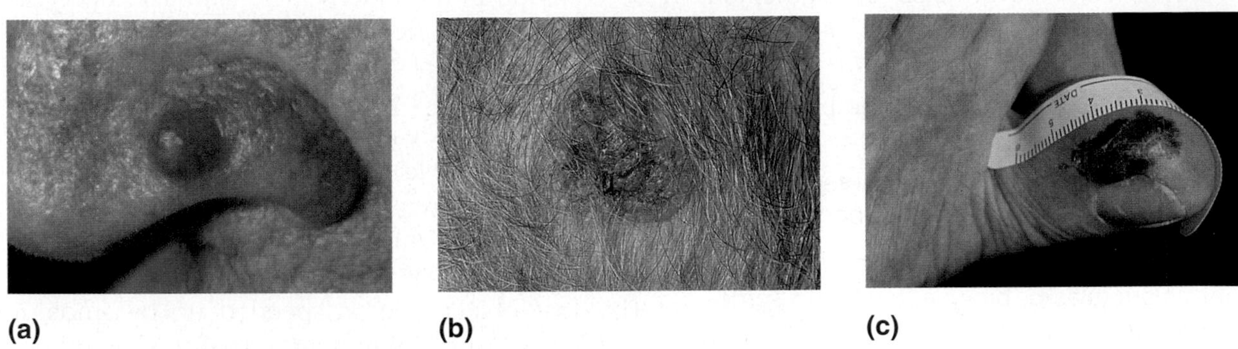

FIGURE 4.12 Photographs of skin cancers. (a) Basal cell carcinoma. **(b)** Squamous cell carcinoma. **(c)** Melanoma.

3. There are third-degree burns of the face, hands, or feet.

Facial burns are dangerous because of the possibility of burned respiratory passageways, which can swell and cause suffocation. Joint injuries are troublesome because the scar tissue that eventually forms can severely limit joint mobility.

Skin Cancer

Numerous types of neoplasms (tumors) arise in the skin. Most skin neoplasms are benign and do not spread (metastasize) to other body areas. (A *wart* caused by a virus is one such example.) However, some skin neoplasms are malignant, or cancerous, and they tend to invade other body areas. Indeed, skin cancer is the single most common type of cancer in humans. One in five Americans now develops skin cancer at some point in his or her life. The cause of most skin cancers is not known, but the most important risk factor is overexposure to ultraviolet radiation in sunlight. Frequent irritation of the skin by infections, chemicals, or physical trauma also seems to be a predisposing factor.

Basal Cell Carcinoma *Basal cell carcinoma* (kar″sĭ no′mah) is the least malignant and most common skin cancer. Cells of the stratum basale, altered so that they cannot form keratin, no longer honor the boundary between epidermis and dermis. They proliferate, invading the dermis and subcutaneous tissue. The cancer lesions occur most often on sun-exposed areas of the face and appear as shiny, dome-shaped nodules that later develop a central ulcer with a "pearly" beaded edge (Figure 4.12a). Basal cell carcinoma is relatively slow-growing, and metastasis seldom occurs before it is noticed. Full cure is the rule in 99 percent of cases in which the lesion is removed surgically.

Squamous Cell Carcinoma *Squamous cell carcinoma* arises from the cells of the stratum spinosum. The lesion appears as a scaly, reddened *papule* (small, rounded elevation) that gradually forms a shallow ulcer with a firm, raised border (Figure 4.12b). This variety of skin cancer appears most often on the scalp, ears, dorsum of the hands, and lower lip. It grows rapidly and metastasizes to adjacent lymph nodes if not removed. This epidermal cancer is also believed to be sun-induced. If it is caught early and removed surgically or by radiation therapy, the chance of complete cure is good.

Malignant Melanoma *Malignant melanoma* (mel″ah-no′mah) is a cancer of melanocytes. It accounts for only about 5 percent of skin cancers, but its incidence is increasing rapidly and it is often deadly. Melanoma can begin wherever there is pigment; most such cancers appear spontaneously, but some develop from pigmented moles. It arises from accumulated DNA damage in a skin cell and usually appears as a spreading brown to black patch (Figure 4.12c) that metastasizes rapidly to surrounding lymph and blood vessels. The chance for survival is about 50 percent,

A *Squamous cell carcinoma.*

and early detection helps. The American Cancer Society suggests that sun worshippers periodically examine their skin for new moles or pigmented spots and apply the **ABCD rule** for recognizing melanoma:

(A) Asymmetry: the two sides of the pigmented spot or mole do not match.

(B) Border irregularity: the borders of the lesion are not smooth but exhibit indentations.

(C) Color: the pigmented spot contains areas of different colors (blacks, browns, tans, and sometimes blues and reds).

(D) Diameter: the spot is larger than 6 millimeters (mm) in diameter (the size of a pencil eraser).

The usual therapy for malignant melanoma is wide surgical excision along with immunotherapy. ▲

Developmental Aspects of Skin and Body Membranes

During the fifth and sixth months of fetal development, the soon-to-be-born infant is covered with a downy type of hair called *lanugo* (lah-noo′go), but this hairy cloak has usually been shed by birth. When a baby is born, its skin is covered with *vernix caseosa* (ver′niks kah-se-o′sah). This white, cheesy-looking substance, produced by the sebaceous glands, protects the baby's skin while it is floating in its water-filled sac inside the mother. The newborn's skin is very thin, and blood vessels can easily be seen through it. Commonly, there are accumulations in the sebaceous glands, which appear as small white spots called *milia* (mil′e-ah), on the baby's nose and forehead. These normally disappear by the third week after birth. As the baby grows, its skin becomes thicker and moist, and more subcutaneous fat is deposited.

During adolescence, the skin and hair become more oily as sebaceous glands are activated, and acne may appear. Acne usually subsides in early adulthood, and the skin reaches its optimal appearance when we are in our twenties and thirties. Then visible changes in the skin begin to appear as it is continually assaulted by abrasion, chemicals, wind, sun, and other irritants and as its pores become clogged with air pollutants and bacteria. As a result, pimples, scales, and various kinds of *dermatitis* (der″mah-ti′tis), or skin inflammation, become more common.

During old age, the amount of subcutaneous tissue decreases, leading to the intolerance to cold so common in the elderly. The skin also becomes drier (because of decreased oil production and declining numbers of collagen fibers), and as a result, it may become itchy and bothersome. Thinning of the skin, another result of the aging process, makes it more susceptible to bruising and other types of injuries. The decreasing elasticity of the skin, along with the loss of subcutaneous fat, allows bags to form under our eyes, and our jowls begin to sag. This loss of elasticity is speeded up by smoking and by sunlight, so two of the best things you can do for your skin are to stop smoking if you have that habit and to shield it from the sun by wearing sunscreens and protective clothing. In doing so, you will also be decreasing the chance of skin cancer. Although there is no way to avoid the aging of the skin, good nutrition, plenty of fluids, and cleanliness help delay the process.

Hair loses its luster as we age, and by age 50 the number of hair follicles has dropped by one-third and continues to decline, resulting in hair thinning and some degree of baldness, or *alopecia* (al″o-pe′she-ah), in most people. Many men become obviously bald as they age, a phenomenon called *male pattern baldness*. A bald man is not really hairless—he does have hairs in the bald area. But, because those hair follicles have begun to degenerate, the hairs are colorless and very tiny (and may not even emerge from the follicle). Such hairs are called *vellus* (*vell* = wool) hairs. Another phenomenon of aging is graying hair. Like balding, this is usually genetically controlled by a "delayed-action" gene. Once the gene takes effect, the amount of melanin deposited in the hair decreases or becomes entirely absent, which results in gray-to-white hair.

Homeostatic Imbalance

Certain events can cause hair to gray or fall out prematurely. For example, many people have claimed that they turned gray nearly overnight because of some emotional crisis in their life. In addition, we know that anxiety, protein-deficient diets, therapy with certain chemicals (chemotherapy), radiation, excessive vitamin A, and certain fungal diseases (ringworm) can cause both graying and hair loss. However, when the cause of these conditions is not genetic, hair loss is usually not permanent. ▲

Systems in Sync

Homeostatic Relationships between the Integumentary System and Other Body Systems

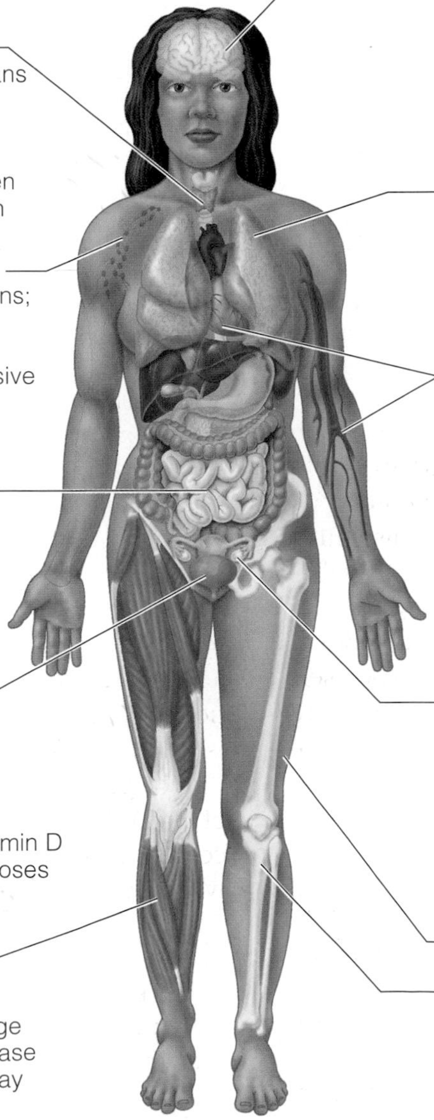

Endocrine System
- Skin protects endocrine organs
- Androgens produced by the endocrine system activate sebaceous glands and help regulate hair growth; estrogen helps maintain skin hydration

Lymphatic System/Immunity
- Skin protects lymphatic organs; prevents pathogen invasion
- Lymphatic system prevents edema by picking up excessive leaked fluid; immune system protects skin cells

Digestive System
- Skin protects digestive organs; provides vitamin D needed for calcium absorption
- Digestive system provides needed nutrients for the skin

Urinary System
- Skin protects urinary organs; excretes salts and some nitrogenous wastes in sweat
- Urinary system activates vitamin D made by keratinocytes; disposes of nitrogenous wastes of skin metabolism

Muscular System
- Skin protects muscles
- Active muscles generate large amounts of heat which increase blood flow to the skin and may promote activation of sweat glands of skin

Nervous System
- Skin protects nervous system organs; cutaneous sensory receptors located in skin
- Nervous system regulates diameter of blood vessels in skin; activates sweat glands, contributing to thermoregulation; interprets cutaneous sensation; activates arrector pili muscles

Respiratory System
- Skin protects respiratory organs
- Respiratory system furnishes oxygen to skin cells and removes carbon dioxide via gas exchange with blood

Cardiovascular System
- Skin protects cardiovascular organs; prevents fluid loss from body surface; serves as blood reservoir
- Cardiovascular system transports oxygen and nutrients to skin and removes wastes from skin; provides substances needed by skin glands to make their secretions

Reproductive System
- Skin protects reproductive organs; highly modified sweat glands (mammary glands) produce milk. During pregnancy, skin stretches to accommodate growing fetus; changes in skin pigmentation may occur

Integumentary System (Skin)

Skeletal System
- Skin protects bones; skin synthesizes vitamin D bones need for normal calcium absorption and deposit of bone (calcium) salts which make bones hard
- Skeletal system provides support for the skin

SUMMARY

Media study tools that could provide you with additional help in reviewing specific key topics of Chapter 4 are referenced below.

IP = Interactive Physiology;

WEB = A&P Place website.

CLASSIFICATION OF BODY MEMBRANES (pp. 108–110)

1. Epithelial: Simple organs, epithelium and connective tissue components.
   a. Cutaneous (the skin): epidermis (stratified squamous epithelium) underlain by the dermis (dense connective tissue); protects body surface.
   b. Mucous: epithelial sheet underlain by a lamina propria (areolar connective tissue); lines body cavities open to the exterior.
   c. Serous: simple squamous epithelium resting on a scant connective tissue layer; lines the ventral body cavity.
2. Connective tissue: Synovial; lines joint cavities.

INTEGUMENTARY SYSTEM (SKIN) (pp. 110–124)

1. Skin functions include protection of the deeper tissue from chemicals, bacteria, bumps, and drying; regulation of body temperature through radiation and sweating; and synthesis of defensive proteins and vitamin D. The cutaneous sensory receptors are located in the skin.
2. The epidermis, the more superficial part of the skin, is formed of stratified squamous keratinizing epithelium and is avascular. Moving from its superficial to deep region, its layers are the stratum corneum, stratum lucidum (in thick skin only), stratum granulosum, stratum spinosum, and stratum basale. Cells at its surface are dead and continually flake off. They are replaced by division of cells in the basal cell layer. As the cells move away from the basal layer, they accumulate keratin and die. Melanin, a pigment produced by melanocytes, protects the nuclei of epithelial cells from the damaging rays of the sun.
3. The dermis is composed of dense connective tissue. It is the site of blood vessels, nerves, and epidermal appendages. It has two regions, the papillary and reticular layers. The papillary layer has ridges, which produce fingerprints.
4. Skin appendages are formed from the epidermis but reside in the dermis.
   a. Sebaceous glands produce an oily product (sebum), usually ducted into a hair follicle. Sebum keeps the skin and hair soft and contains bacteria-killing chemicals.
   b. Sweat (sudoriferous) glands, under the control of the nervous system, produce sweat, which is ducted to the epithelial surface. These glands are part of the body's heat-regulating apparatus. There are two types: eccrine (the most numerous) and apocrine (their product includes fatty acids and proteins, which skin bacteria metabolize).
   c. A hair is primarily dead keratinized cells and is produced by the matrix in the hair bulb. The root is enclosed in a sheath, the hair follicle.
   d. Nails are hornlike derivatives of the epidermis. Like hair, nails are primarily dead keratinized cells.

   WEB Exercise: Chapter 4, Structure of the Skin.
5. Most minor afflictions of the skin result from infections or allergic responses; more serious are burns and skin cancer. Because they interfere with skin's protective functions, burns represent a major threat to the body.

   WEB Exercise: Chapter 4, Integumentary I Case Study.
   a. Burns result in loss of body fluids and invasion of bacteria. The extent of burns is assessed by the rule of nines. The severity (depth) of burns is described as first-degree (epidermal damage only), second-degree (epidermal and some dermal injury), and third-degree (epidermis and dermis totally destroyed). Third-degree burns require skin grafting.
   b. The most common cause of skin cancer is exposure to ultraviolet radiation. Cure of basal cell carcinoma and squamous cell carcinoma is complete if they are removed before metastasis. Malignant melanoma, a cancer of melanocytes, is still fairly rare but is fatal in about half the cases.

   WEB Exercise: Chapter 4, Integumentary II Case Study; At the Clinic: Integumentary.

DEVELOPMENTAL ASPECTS OF SKIN AND BODY MEMBRANES (p. 124)

1. The skin is thick, resilient, and well hydrated in youth but loses its elasticity and thins as aging

occurs. Skin cancer is a major threat to skin exposed to excessive sunlight.

2. Balding and/or graying occurs with aging. Both are genetically determined, but other factors (drugs, emotional stress, and so on) can result in either.

REVIEW QUESTIONS

MULTIPLE CHOICE

More than one choice may apply.

1. Select the one false statement about mucous and serous membranes.
   a. The epithelial type is the same in all serous membranes, but there are different epithelial types in different mucous membranes.
   b. Serous membranes line closed body cavities, while mucous membranes line body cavities open to the outside.
   c. Serous membranes always produce serous fluid, and mucous membranes always secrete mucus.
   d. Both membranes contain an epithelium plus a layer of loose connective tissue.

2. Serous membranes:
   a. line the mouth
   b. have parietal and visceral layers
   c. consist of epidermis and dermis
   d. have a connective tissue layer called the lamina propria
   e. secrete a lubricating fluid

3. Which is *not* a component of sweat?
   a. Water
   b. Sodium chloride
   c. Sebum
   d. Ammonia
   e. Vitamin D

4. Which structure is *not* associated with a hair?
   a. Shaft
   b. Cortex
   c. Lunula
   d. Matrix
   e. Cuticle

5. In investigating the cause of thinning hair, which of the following questions needs to be asked?
   a. Is the diet deficient in proteins?
   b. Is the person taking megadoses of vitamin C?
   c. Has the person been exposed to excessive radiation?
   d. Has the person recently suffered severe emotional trauma?

6. Which structure is *not* associated with a nail?
   a. Nail bed
   b. Lunula
   c. Nail folds
   d. Nail follicle

7. Which one of the following is *not* associated with the production of perspiration?
   a. Sweat glands
   b. Sweat pores
   c. Arrector pili
   d. Eccrine gland
   e. Apocrine gland

8. Which of the following is not a skin structure?
   a. Pacinian corposcle
   b. Dermal papilla
   c. Hair
   d. Nail

SHORT ANSWER ESSAY

1. What is the name of the connective tissue membrane found lining the joint cavities?
2. What primary tissues are destroyed when the skin is damaged?
3. From what types of damage does the skin protect the body?
4. Explain why we become tanned after sitting in the sun.
5. What is a decubitus ulcer? Why does it occur?
6. Name two different categories of skin secretions and the glands that manufacture them.
7. How does the skin help to regulate body temperature?
8. What is a blackhead?
9. What are arrector pili? What do they do?
10. What are the life-threatening consequences of severe burns?
11. Distinguish between first-, second-, and third-degree burns.
12. Why does hair turn gray?
13. Name three changes that occur in the skin as one ages.
14. Is a bald man really hairless? Explain.

AT THE CLINIC

1. A nurse tells a doctor that a patient is cyanotic. What is cyanosis? What does its presence indicate?
2. Both newborn infants and aged individuals have very little subcutaneous tissue. How does this affect their sensitivity to cold environmental temperature?
3. A 40-year-old beachboy is complaining to you that his suntan made him popular when he was

young—but now his face is all wrinkled, and he has several darkly pigmented moles that are growing rapidly and are as big as large coins. He shows you the moles, and immediately you think "ABCD." What does that mean, and why should he be concerned?

4. Martha, the mother of a 13-month-old infant, brings her child to the clinic because his skin has turned orange. Why does the pediatrician inquire about the child's diet?

5. If the water of a swimming pool is hypotonic to our cells and body fluids (and it is), then why do we not swell and pop when we go for a swim?

6. Mr. Bellazono, a fisherman in his late sixties, comes to the clinic to complain of small ulcers on both forearms as well as on his face and ears. Although he has had them for several years, he has not had any other problems. What is the likely diagnosis, and what is the likely cause?

7. Mr. Grayson is receiving a drug treatment transdermally (through the skin). Explain why drugs delivered by this route are fat-soluble rather than water-soluble.

5

The Skeletal System

YOUR GOALS

After completing this chapter, you will have a working knowledge of the functions of the skeletal system and will have mastered the objectives listed below.

FUNCTION PREVIEW

- The skeletal system provides an internal framework for the body, protects organs by enclosure, and anchors skeletal muscles so that muscle contraction can cause movement.

OBJECTIVE CHECKLIST

BONES: AN OVERVIEW (pp. 130–139)

- Identify the subdivisions of the skeleton as axial or appendicular.
- List at least three functions of the skeletal system.
- Name the four main classifications of bones.
- Identify the major anatomical areas of a long bone.
- Explain the role of bone salts and the organic matrix in making bone both hard and flexible.
- Describe briefly the process of bone formation in the fetus, and summarize the events of bone remodeling throughout life.
- Name and describe the various types of fractures.

AXIAL SKELETON (pp. 139–153)

- On a skull or diagram, identify and name the bones of the skull.
- Describe how the skull of a newborn infant (or fetus) differs from that of an adult, and explain the function of fontanels.
- Name the parts of a typical vertebra, and explain in general how the cervical, thoracic, and lumbar vertebrae differ from one another.

>>

(Objective Checklist, continued)

- Discuss the importance of the intervertebral discs and spinal curvatures.
- Explain how the abnormal spinal curvatures (scoliosis, lordosis, and kyphosis) differ from one another.

APPENDICULAR SKELETON (pp. 153–160)

- Identify on a skeleton or diagram the bones of the shoulder and pelvic girdles and their attached limbs.
- Describe important differences between a male and female pelvis.

JOINTS (pp. 163–168)

- Name the three major categories of joints, and compare the amount of movement allowed by each.

DEVELOPMENTAL ASPECTS OF THE SKELETON (pp. 168–170)

- Identify some of the causes of bone and joint problems throughout life.

Although the word *skeleton* comes from the Greek word meaning "dried-up body," our internal framework is so beautifully designed and engineered that it puts any modern skyscraper to shame. Strong, yet light, it is perfectly adapted for its functions of body protection and motion. Shaped by an event that happened more than one million years ago—when a being first stood erect on hind legs—our skeleton is a tower of bones arranged so that we can stand upright and balance ourselves. No other animal has such relatively long legs (compared to the arms or forelimbs) or such a strange foot, and few have such remarkable grasping hands. Even though the infant's backbone is like an arch, it soon changes to the swayback, or S-shaped, structure that is required for the upright posture.

The skeleton is subdivided into two divisions: the **axial skeleton,** the bones that form the longitudinal axis of the body, and the **appendicular skeleton,** the bones of the limbs and girdles. In addition to bones, the **skeletal system** includes *joints, cartilages,* and *ligaments* (fibrous cords that bind the bones together at joints). The joints give the body flexibility and allow movement to occur.

Bones: An Overview

At one time or another, all of us have heard the expressions "bone tired," "dry as a bone," or "bag of bones"—pretty unflattering and inaccurate images of some of our most phenomenal organs. Our brains, not our bones, convey feelings of fatigue, and bones are far from dry. As for "bag of bones," they are indeed more obvious in some of us, but without them to form our internal skeleton, we would creep along the ground like slugs. Let's examine how our bones contribute to overall body homeostasis.

Functions of the Bones

Besides contributing to body shape and form, our bones perform several important body functions:

1. **Support.** Bones, the "steel girders" and "reinforced concrete" of the body, form the internal framework that supports and anchors all soft organs. The bones of the legs act as pillars to support the body trunk when we stand, and the rib cage supports the thoracic wall.
2. **Protection.** Bones protect soft body organs. For example, the fused bones of the skull

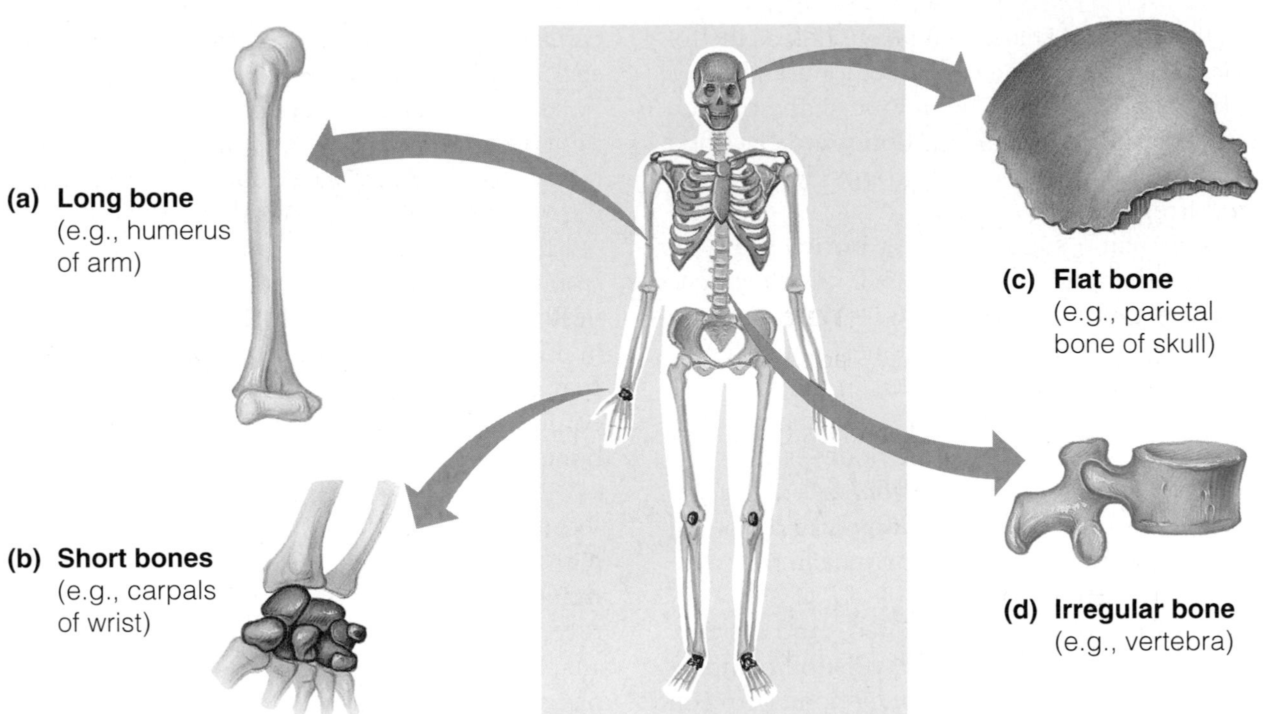

FIGURE 5.1 Classification of bones on the basis of shape.

provide a snug enclosure for the brain, allowing one to head a soccer ball without worrying about injuring the brain. The vertebrae surround the spinal cord, and the rib cage helps protect the vital organs of the thorax.

3. **Movement.** Skeletal muscles, attached to bones by tendons, use the bones as levers to move the body and its parts. As a result, we can walk, swim, throw a ball, and breathe. Before continuing, take a moment to imagine that your bones have turned to putty. What if you were running when this change took place? Now imagine your bones forming a rigid metal framework inside your body, somewhat like a system of plumbing pipes. What problems could you envision with this arrangement? These images should help you understand how well our skeletal system provides support and protection while allowing movement.

4. **Storage.** Fat is stored in the internal cavities of bones. Bone itself serves as a storehouse for minerals, the most important being calcium and phosphorus, although others are also stored. A small amount of calcium in its ion form (Ca^{2+}) must be present in the blood at all times for the nervous system to transmit messages, for muscles to contract, and for blood to clot. Because most of the body's calcium is deposited in the bones as calcium salts, the bones are a convenient place to get more calcium ions for the blood as they are used up. Problems occur not only when there is too little calcium in the blood, but also when there is too much. Hormones control the movement of calcium to and from the bones and blood according to the needs of the body. Indeed, "deposits" and "withdrawals" of calcium (and other minerals) to and from bones go on almost all the time.

5. **Blood cell formation.** Blood cell formation, or hematopoiesis (hem″ah-to-poi-e′sis), occurs within the marrow cavities of certain bones.

Classification of Bones

The adult skeleton is composed of 206 bones. There are two basic types of osseous, or bone, tissue: **Compact bone** is dense and looks smooth and homogeneous. **Spongy bone** is composed of small needlelike pieces of bone and lots of open space.

Bones come in many sizes and shapes (Figure 5.1). For example, the tiny pisiform bone of the wrist is the size and shape of a pea, whereas the femur, or thigh bone, is nearly 2 feet long and has a large, ball-shaped head. The unique shape of each bone fulfills a particular need. Bones are classified according to shape into four groups: long, short, flat, and irregular (Figure 5.1).

As their name suggests, **long bones** are typically longer than they are wide. As a rule they have a shaft with heads at both ends. Long bones are mostly compact bone. All the bones of the limbs, except the wrist and ankle bones, are long bones.

Short bones are generally cube-shaped and contain mostly spongy bone. The bones of the wrist and ankle are short bones. *Sesamoid* (ses′ah-moyd) *bones,* which form within tendons, are a special type of short bone. The best-known example is the patella, or kneecap.

Flat bones are thin, flattened, and usually curved. They have two thin layers of compact bone sandwiching a layer of spongy bone between them. Most bones of the skull, the ribs, and the sternum (breastbone) are flat bones.

Bones that do not fit one of the preceding categories are called **irregular bones.** The vertebrae, which make up the spinal column, and the hip bones fall into this group.

Structure of a Long Bone

Gross Anatomy

The gross structure of a long bone is shown in Figure 5.2. The **diaphysis** (di-af′ĭ-sis), or shaft, makes up most of the bone's length and is composed of compact bone. The diaphysis is covered and protected by a fibrous connective tissue membrane, the **periosteum** (per-e-ŏs′te-um). Hundreds of connective tissue fibers, called **perforating,** or **Sharpey's, fibers,** secure the periosteum to the underlying bone. The **epiphyses** (ĕ-pif′ĭ-sēz) are the ends of the long bone. Each epiphysis consists of a thin layer of compact bone enclosing an area filled with spongy bone. **Articular cartilage,** instead of a periosteum, covers its external surface. Because the articular cartilage is glassy hyaline cartilage, it provides a smooth, slippery surface that decreases friction at joint surfaces.

In adult bones, there is a thin line of bony tissue spanning the epiphysis that looks a bit different from the rest of the bone in that area. This is the **epiphyseal line.** The epiphyseal line is a remnant of the **epiphyseal plate** (a flat plate of hyaline cartilage) seen in a young, growing bone. Epiphyseal plates cause the lengthwise growth of a long bone. By the end of puberty, when hormones inhibit long bone growth, epiphyseal plates have been completely replaced by bone, leaving only the epiphyseal lines to mark their previous location.

In adults the cavity of the shaft is primarily a storage area for adipose (fat) tissue. It is called the **yellow marrow,** or **medullary, cavity.** However, in infants this area forms blood cells, and **red marrow** is found there. In adult bones, red marrow is confined to the cavities of spongy bone of flat bones and the epiphyses of some long bones.

Even when looking casually at bones, one can see that their surfaces are not smooth but scarred with bumps, holes, and ridges. These **bone markings,** described and illustrated in Table 5.1, reveal where muscles, tendons, and ligaments were attached and where blood vessels and nerves passed. There are two categories of bone markings: (a) *projections,* or *processes,* which grow out from the bone surface, and (b) *depressions,* or *cavities,* which are indentations in the bone. These terms do not have to be learned now, but they can help you remember some of the specific markings on bones to which you will be introduced later in this chapter.

There is a little trick for remembering the bone markings listed in the table: All the terms beginning with **T** are projections. The terms beginning with **F** (except *facet*) are depressions.

Microscopic Anatomy

To the naked eye, spongy bone has a spiky, open appearance, whereas compact bone appears to be very dense. Looking at compact bone tissue through a microscope, however, one can see that it has a complex structure (Figure 5.3). It is riddled with passageways carrying nerves, blood vessels, and the like, which provide the living bone cells with nutrients and a route for waste disposal. The mature bone cells, **osteocytes** (os′te-o-sītz″), are found in tiny cavities within the matrix called **lacunae** (lah-ku′ne). The lacunae are arranged in concentric circles called **lamellae** (lah-mel′e) around **central (Haversian) canals.** Each complex consisting of central canal and matrix rings is called an **osteon,** or **Haversian system.** Central canals run lengthwise through the bony matrix, carrying blood vessels and nerves to all areas of the bone. Tiny canals, **canaliculi** (kan″ah-lik′u-li), radiate

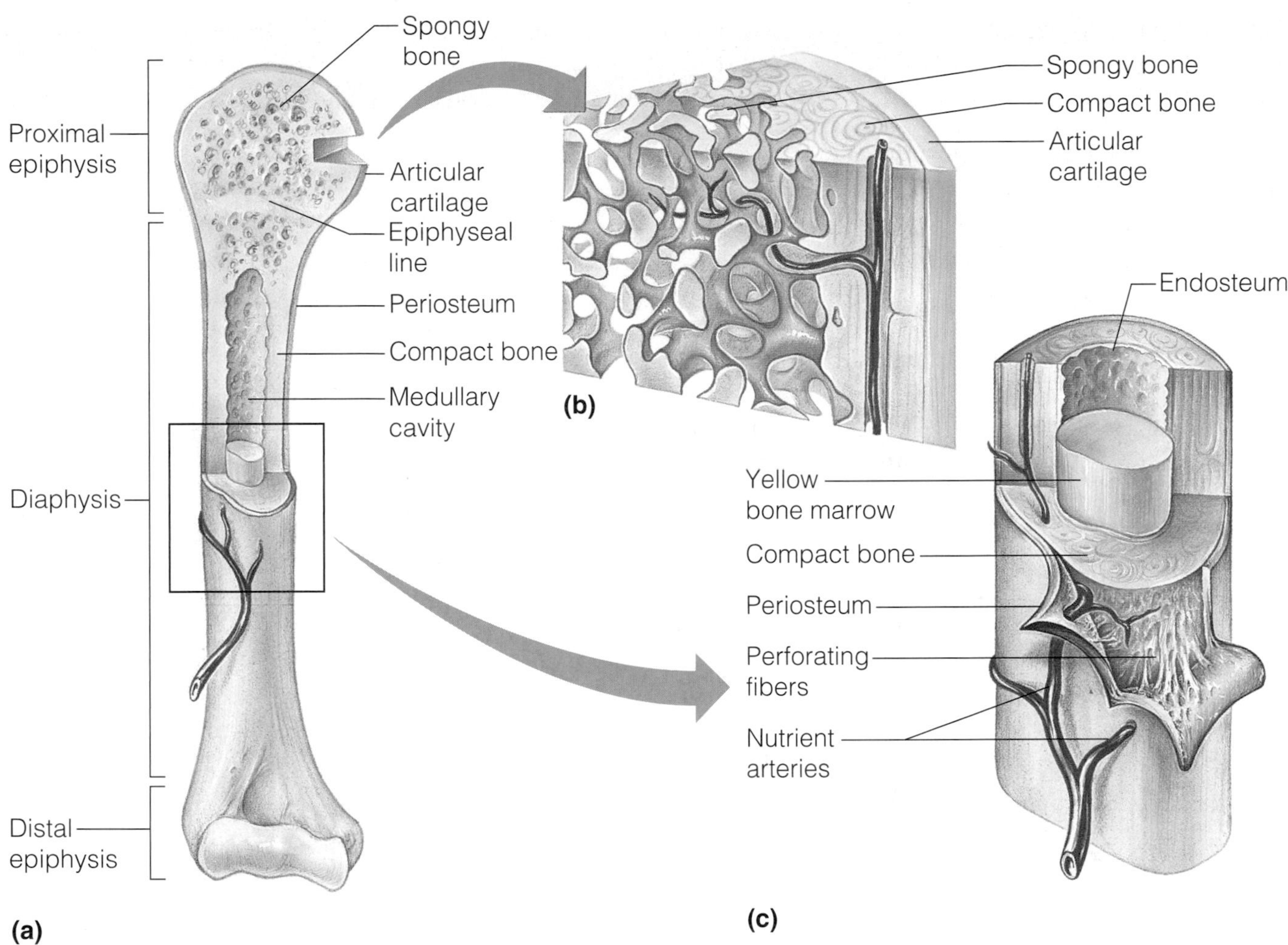

FIGURE 5.2 The structure of a long bone (humerus). (a) Anterior view with longitudinal section cut away at the proximal end. **(b)** Pie-shaped, three-dimensional view of spongy bone and compact bone of the epiphysis. **(c)** Cross section of the shaft (diaphysis). Note that the external surface of the diaphysis is covered by a periosteum, but the articular surface of the epiphysis (see b) is covered with hyaline cartilage.

outward from the central canals to all lacunae. The canaliculi form a transportation system that connects all the bone cells to the nutrient supply through the hard bone matrix. Because of this elaborate network of canals, bone cells are well nourished in spite of the hardness of the matrix, and bone injuries heal quickly and well. The communication pathway from the outside of the bone to its interior (and the central canals) is completed by **perforating (Volkmann's) canals,** which run into the compact bone at right angles to the shaft.

Bone is one of the hardest materials in the body, and although relatively light in weight, it has a remarkable ability to resist tension and other forces acting on it. Nature has given us an extremely strong and exceptionally simple (almost crude) supporting system without giving up mobility. The calcium salts deposited in the matrix give bone its hardness, whereas the organic parts (especially the collagen fibers) provide for bone's flexibility and great tensile strength.

TABLE 5.1 Bone Markings

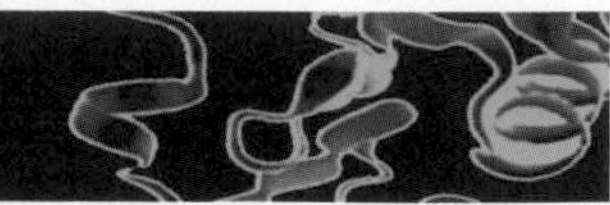

Name of bone marking	Description	Illustration
Projections that are sites of muscle and ligament attachment		Tibia of leg: Tuberosity, Anterior crest. Femur of thigh: Trochanters, Intertrochanteric line, Adductor tubercle, Medial epicondyle
Tuberosity	Large, rounded projection; may be roughened.	
Crest	Narrow ridge of bone; usually prominent.	
Trochanter (tro-kan'ter)	Very large, blunt, irregularly shaped process. (The only examples are on the femur.)	
Line	Narrow ridge of bone; less prominent than a crest.	
Tubercle (too'ber-kl)	Small, rounded projection or process.	
Epicondyle	Raised area on or above a condyle.	
Spine	Sharp, slender, often pointed projection.	Vertebra: Spinous process
Process	Any bony prominence.	
Projections that help to form joints		Rib: Facets, Head. Mandible: Condyle, Ramus
Head	Bony expansion carried on a narrow neck.	
Facet	Smooth, nearly flat articular surface.	
Condyle (kon'dīl)	Rounded articular projection.	
Ramus (ra'mus)	Armlike bar of bone.	
Depressions and openings allowing blood vessels and nerves to pass		Skull: Meatus, Sinus, Fossa, Groove, Inferior orbital fissure, Foramen
Meatus (me-a'tus)	Canal-like passageway.	
Sinus	Cavity within a bone, filled with air and lined with mucous membrane.	
Fossa (fos'ah)	Shallow, basinlike depression in a bone, often serving as an articular surface.	
Groove	Furrow.	
Fissure	Narrow, slitlike opening.	
Foramen (fo-ra'men)	Round or oval opening through a bone.	

Bone Formation, Growth, and Remodeling

The skeleton is formed from two of the strongest and most supportive tissues in the body—cartilage and bone. In embryos, the skeleton is primarily made of hyaline cartilage, but in the young child most of the cartilage has been replaced by bone. Cartilage remains only in isolated areas such as the bridge of the nose, parts of the ribs, and the joints.

Except for flat bones, which form on fibrous membranes, most bones develop using hyaline

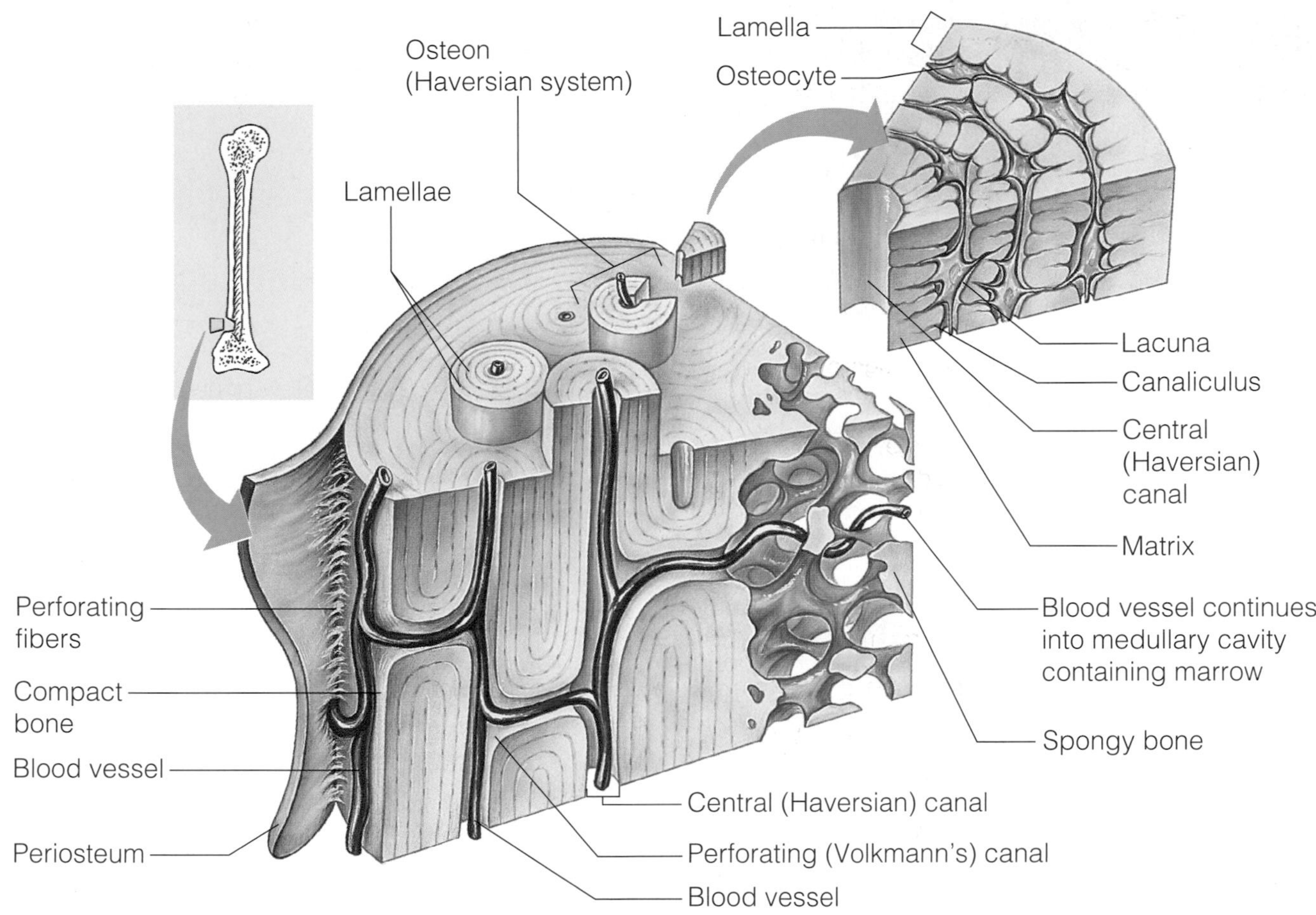

FIGURE 5.3 Microscopic structure of compact bone. Diagram of a pie-shaped segment of compact bone. (The inset shows a more highly magnified view.) Notice the position of osteocytes in lacunae (cavities in the matrix).

cartilage structures as their "models." Most simply, this process of bone formation, or **ossification** (os″ĭ-fĭ-ka′shun), involves two major phases (Figure 5.4a). First, the hyaline cartilage model is completely covered with bone matrix (a bone "collar") by bone-forming cells called **osteoblasts.** So, for a short period, the fetus has cartilage "bones" enclosed by "bony" bones. Then, the enclosed hyaline cartilage model is digested away, opening up a medullary cavity within the newly formed bone.

By birth or shortly after, most hyaline cartilage models have been converted to bone except for two regions—the **articular cartilages** (that cover the bone ends) and the **epiphyseal plates.** The articular cartilages persist for life, reducing friction at the joint surfaces. The epiphyseal plates provide for longitudinal growth of the long bones during childhood. New cartilage is formed continuously on the external face of the articular cartilage and on the epiphyseal plate surface that is farther away from the medullary cavity. At the same time, the old cartilage abutting the internal face of the articular cartilage and the medullary cavity is broken down and replaced by bony matrix (Figure 5.4b). Growing bones also must widen as they lengthen. How do they widen? Simply, osteoblasts in the periosteum add bone tissue to the external face of the diaphysis as osteoclasts in the endosteum remove bone from the inner face of the diaphysis wall (see Figure 5.4b). Since these two processes occur at about the same rate, the circumference of the long bone expands and the bone widens. This process by which bones increase in diameter is called *appositional growth*. This process of long-bone growth is controlled by hormones, most importantly *growth hormone* and, during puberty, the *sex hormones*. It ends during adolescence, when the epiphyseal plates are completely converted to bone.

Q *(a) What specific cell types form the bone collar? (b) What do you think a long bone would look like at the end of adolescence if bone remodeling did not occur?*

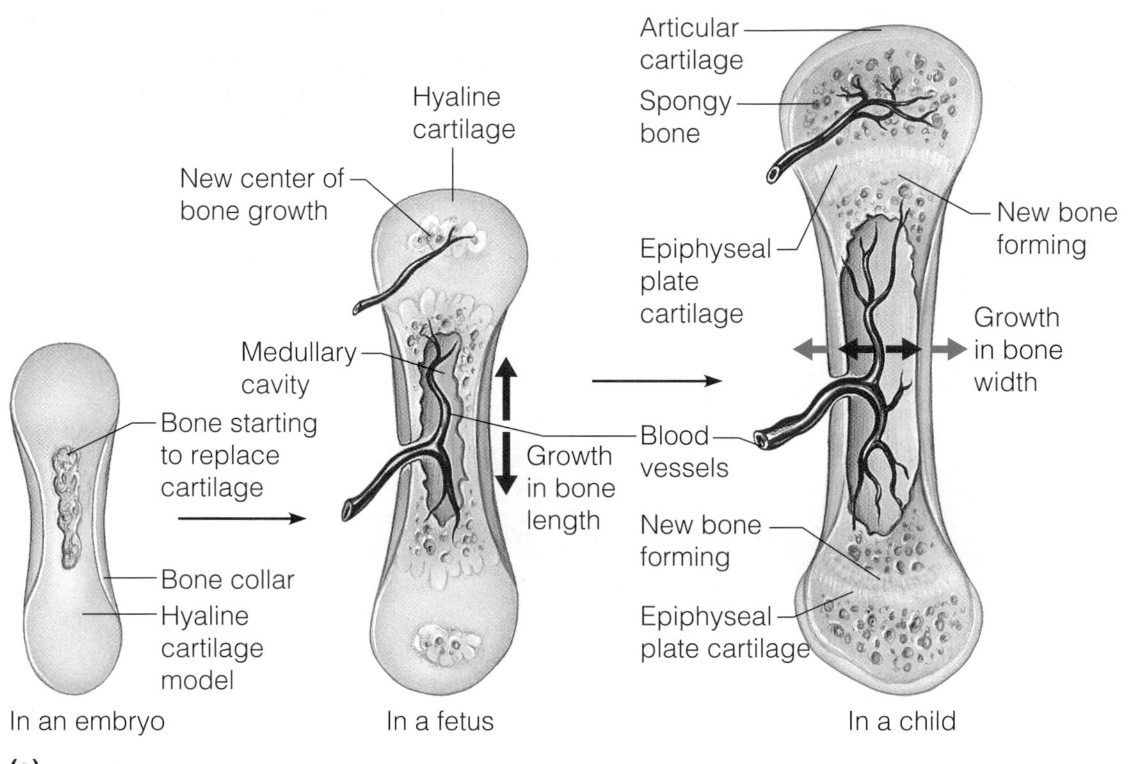

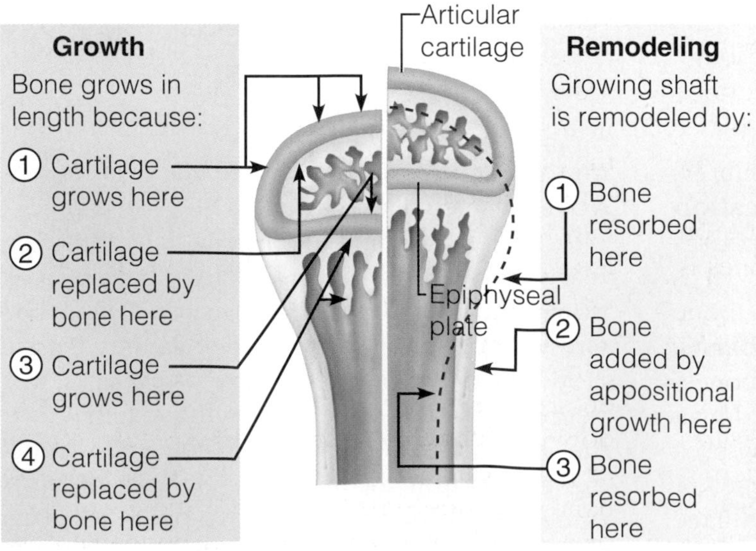

FIGURE 5.4 Long-bone formation and growth. (a) Stages of long-bone formation in an embryo, fetus, and young child. **(b)** The events indicated at the left depict the process of ossification that occurs at the articular cartilages and epiphyseal plates as the bone grows in length. The events indicated at the right reveal the process of appositional growth that occurs during long-bone growth to maintain proper bone proportions by enlarging long-bone diameter.

A *(a) Osteoblasts form the bone collar. (b) It would have a relatively short slender shaft with elongated clublike ends.*

TABLE 5.2 Common Types of Fractures

Fracture type	Illustration	Description	Comment
Comminuted		Bone breaks into many fragments.	Particularly common in the aged, whose bones are more brittle.
Compression		Bone is crushed.	Common in porous bones (i.e., osteoporotic bones).
Depressed		Broken bone portion is pressed inward.	Typical of skull fracture.
Impacted		Broken bone ends are forced into each other.	Commonly occurs when one attempts to break a fall with outstretched arms.
Spiral		Ragged break occurs when excessive twisting forces are applied to a bone.	Common sports fracture.
Greenstick		Bone breaks incompletely, much in the way a green twig breaks.	Common in children, whose bones are more flexible than those of adults.

Many people mistakenly think that bones are lifeless structures that never change once long-bone growth has ended. Nothing could be further from the truth; bone is a dynamic and active tissue. Bones are remodeled continually in response to changes in two factors: (1) calcium levels in the blood, and (2) the pull of gravity and muscles on the skeleton. How these factors influence bones is outlined next.

When blood calcium levels drop below homeostatic levels, the parathyroid glands (located in the throat) are stimulated to release parathyroid hormone (PTH) into the blood. PTH activates **osteoclasts,** giant bone-destroying cells in bones, to break down bone matrix and release calcium ions into the blood. On the other hand, when blood calcium levels are too high (*hypercalcemia* [hi″per-kal-se′me-ah]), calcium is deposited in bone matrix as hard calcium salts.

Bone remodeling is essential if bones are to retain normal proportions and strength during long-bone growth as the body increases in size and weight. It also accounts for the fact that bones become thicker and form large projections to increase their strength in areas where bulky muscles are attached. At such sites, osteoblasts lay down new matrix and become trapped within it. (Once they are trapped, they become osteocytes, or mature bone cells.) On the other hand, the bones of bedridden or physically inactive people tend to lose mass and to atrophy because they are no longer subjected to stress.

To explain the interaction between these two controlling mechanisms as simply as possible, PTH determines *when* (or *if*) bone is to be broken down or formed in response to the need for more or fewer calcium ions in the blood. On the other hand, the stresses of muscle pull and gravity acting on the skeleton determine *where* bone matrix is to be broken down or formed so that the skeleton can remain as strong and vital as possible.

Homeostatic Imbalance

Rickets is a disease of children in which the bones fail to calcify. As a result, the bones soften and a definite bowing of the weight-bearing bones of the legs occurs. Rickets is usually due to a lack of calcium

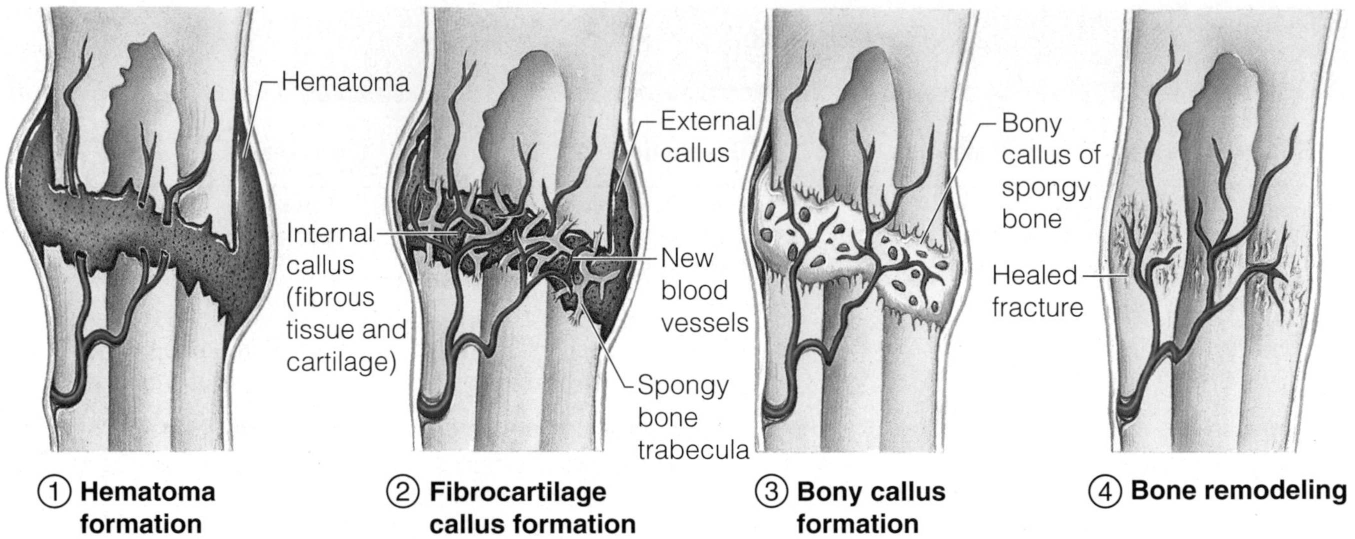

FIGURE 5.5 Stages in the healing of a bone fracture.

in the diet or lack of vitamin D, which is needed to absorb calcium into the bloodstream. Rickets is not seen very often in the United States, where great stress is put on good nutrition. Milk, bread, and other foods are fortified with vitamin D, and most children drink enough calcium-rich milk. However, it can happen in infants nursed by mothers who become vitamin D-deficient over the course of a long gray winter, and it remains a problem in some other parts of the world. ▲

Bone Fractures

Homeostatic Imbalance

For their relatively low mass, bones are amazingly strong. Consider, for example, the forces endured in touch football and professional hockey. Despite their remarkable strength, bones are susceptible to **fractures,** or breaks, all through life. During youth, most fractures result from exceptional trauma that twists or smashes the bones. Sports activities such as football, skating, and skiing jeopardize the bones, and automobile accidents certainly take their toll. In old age, bones thin and weaken, and fractures occur more often.

A fracture in which the bone breaks cleanly but does not penetrate the skin is a *closed* (or *simple*) *fracture*. When the broken bone ends penetrate through the skin, the fracture is *open* (or *compound*). Some of the many common types of fractures are illustrated and described in Table 5.2. ▲

A fracture is treated by **reduction,** which is the realignment of the broken bone ends. In *closed reduction,* the bone ends are coaxed back into their normal position by the physician's hands. In *open reductions,* surgery is performed and the bone ends are secured together with pins or wires. After the broken bone is reduced, it is immobilized by a cast or traction to allow the healing process to begin. The healing time for a simple fracture is 6 to 8 weeks, but it is much longer for large bones and for the bones of elderly people (because of their poorer circulation).

The repair of bone fractures involves four major events (Figure 5.5):

1. **A hematoma is formed.** Blood vessels are ruptured when the bone breaks. As a result, a blood-filled swelling called a **hematoma** (he-mah-to′mah) forms. Bone cells deprived of nutrition die.
2. **The break is splinted by a fibrocartilage callus.** As described in Chapter 3, an early event of tissue repair (and bone is no exception) is the growth of new capillaries (granulation tissue) into the clotted blood at the site of the damage, and the disposal of dead tissue by phagocytes. As this goes on, connective tissue cells of various types form a mass of repair tissue, the **fibrocartilage callus** (kal′us), that contains several elements—some cartilage matrix, some bony matrix, and collagen fibers—and acts to "splint" the broken bone, closing the gap.

3. **The bony callus is formed.** As more osteoblasts and osteoclasts migrate into the area and multiply, the fibrocartilage callus is gradually replaced by one made of spongy bone, the **bony callus.**
4. **Bone remodeling occurs.** Over the next few weeks to months depending on the bone's size and site of the break, the bony callus is remodeled in response to the mechanical stresses placed on it, so that it forms a strong permanent "patch" at the fracture site.

Axial Skeleton

As noted earlier, the skeleton is divided into two parts, the *axial* and *appendicular skeletons.* The axial skeleton, which forms the longitudinal axis of the body, is shown as the green portion of Figure 5.6. It can be divided into three parts—the *skull,* the *vertebral column,* and the *bony thorax.*

Skull

The **skull** is formed by two sets of bones. The **cranium** encloses and protects the fragile brain tissue. The **facial bones** hold the eyes in an anterior position and allow the facial muscles to show our feelings through smiles or frowns. All but one of the bones of the skull are joined together by *sutures,* which are interlocking, immovable joints. Only the mandible (jawbone) is attached to the rest of the skull by a freely movable joint.

Cranium

The boxlike cranium is composed of eight large, flat bones. Except for two paired bones (the parietal and temporal), they are all single bones.

Frontal Bone The frontal bone forms the forehead, the bony projections under the eyebrows, and the superior part of each eye's orbit (Figure 5.7).

Parietal Bones The paired parietal bones form most of the superior and lateral walls of the cranium (see Figure 5.7). They meet in the midline of the skull at the **sagittal suture** and form the **coronal suture** where they meet the frontal bone.

Temporal Bones The temporal bones lie inferior to the parietal bones; they join them at the **squamous sutures.** Several important bone markings appear on the temporal bone (see Figure 5.7):

- The **external acoustic (auditory) meatus** is a canal that leads to the eardrum and the middle ear.
- The **styloid process,** a sharp, needlelike projection, is just inferior to the external auditory meatus. Many neck muscles use the styloid process as an attachment point.
- The **zygomatic** (zi″go-mat′ik) **process** is a thin bridge of bone that joins with the cheekbone (zygomatic bone) anteriorly.
- The **mastoid** (mas′toid) **process** is a rough projection posterior and inferior to the external acoustic meatus, which is full of air cavities (mastoid sinuses). It provides an attachment site for some muscles of the neck. The mastoid sinuses are so close to the middle ear—a high-risk spot for infections—that they may become infected too, a condition called *mastoiditis.* Also, this area is so close to the brain, mastoiditis may spread to the brain itself.
- The **jugular foramen,** at the junction of the occipital and temporal bones (Figures 5.8 and 5.9), allows passage of the jugular vein, the largest vein of the head, which drains the brain. Just anterior to it in the cranial cavity is the **internal acoustic meatus** (Figure 5.8) which transmits cranial nerves VII and VIII (the facial and vestibulocochlear nerves). Anterior to the jugular foramen on the skull's inferior aspect is the **carotid canal** (see Figure 5.9), through which the internal carotid artery runs, supplying blood to most of the brain.

Occipital Bone If you look at Figures 5.7, 5.8, and 5.9, you can see that the occipital (ok-sip′ĭ-tal) bone is the most posterior bone of the cranium. It forms the floor and back wall of the skull. The occipital bone joins the parietal bones anteriorly at the **lambdoid** (lam′doyd) **suture.** In the base of the occipital bone is a large opening, the **foramen magnum** (literally, "large hole"). The foramen magnum surrounds the lower part of the brain and allows the spinal cord to connect with the brain. Lateral to the foramen magnum on each side

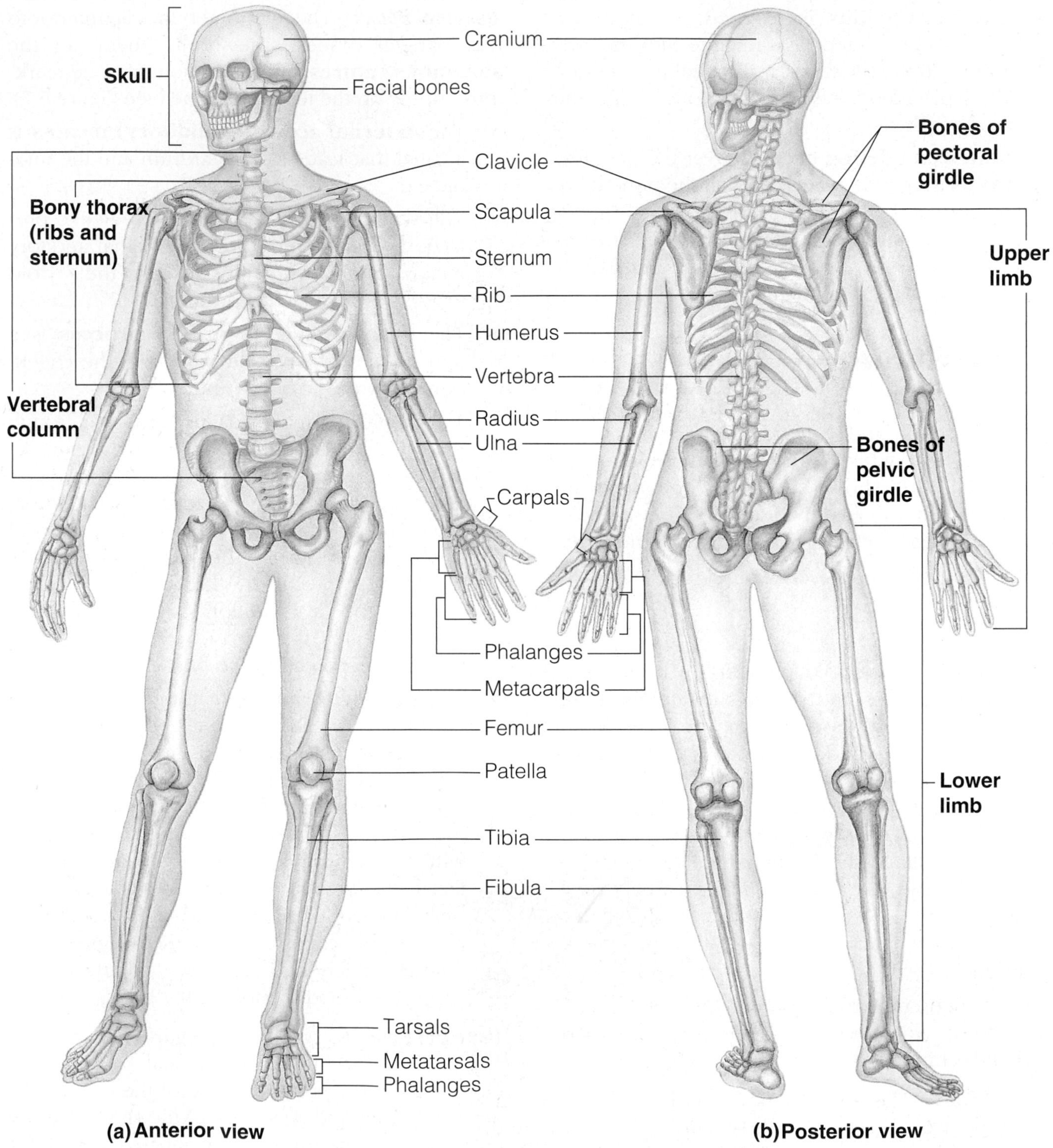

FIGURE 5.6 The human skeleton. The bones of the axial skeleton are colored green to distinguish them from the bones of the appendicular skeleton.

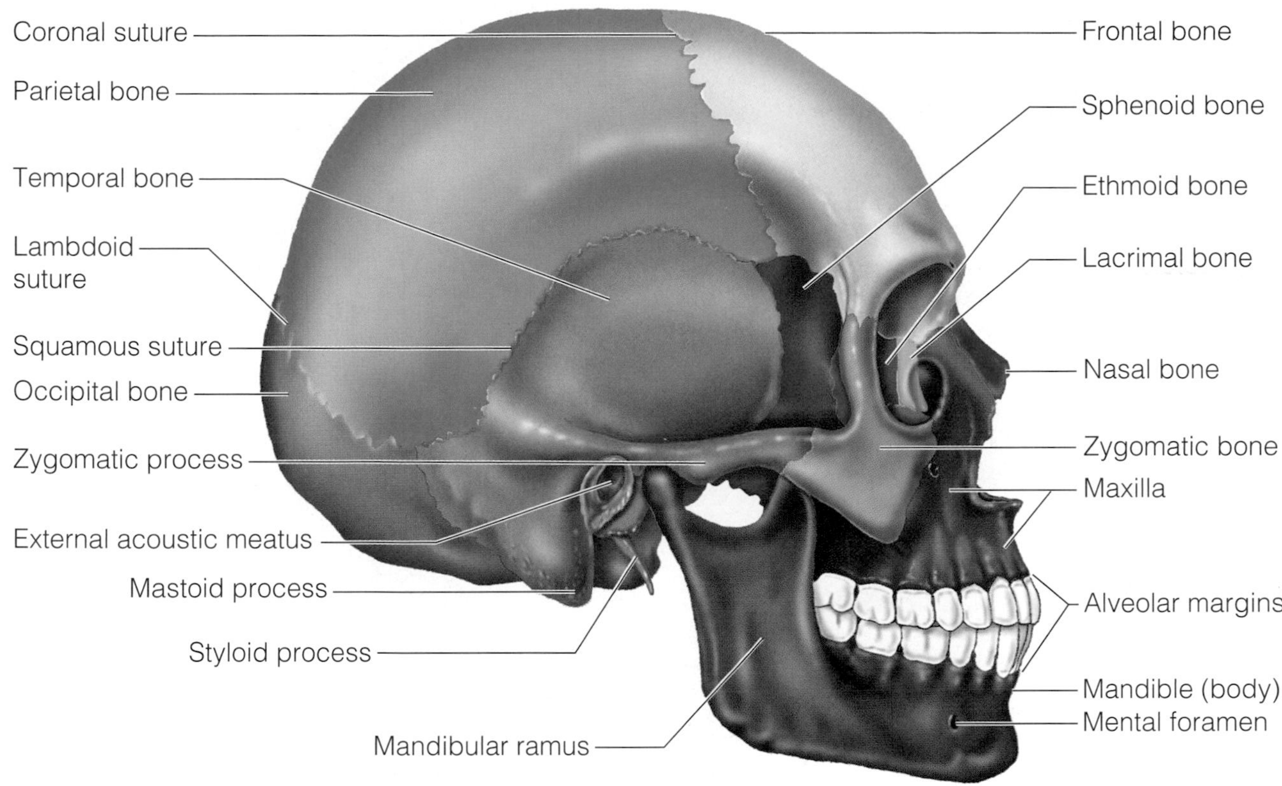

FIGURE 5.7 Human skull, lateral view.

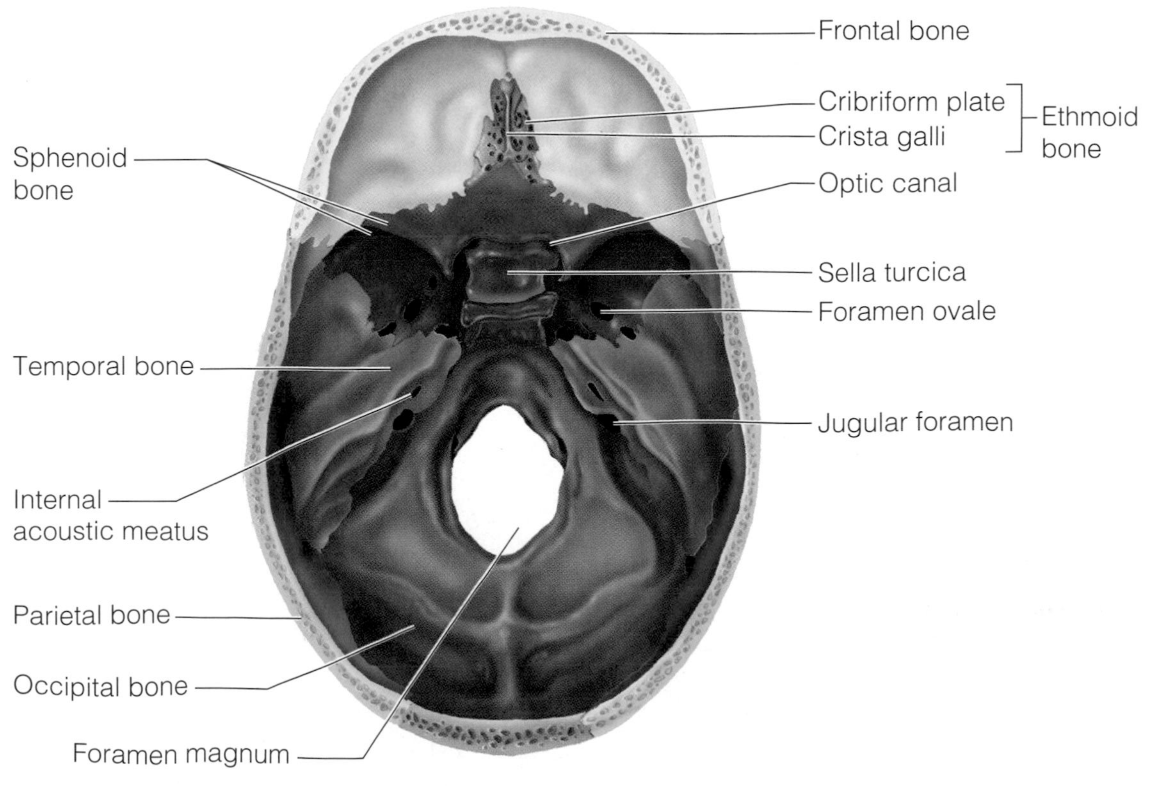

FIGURE 5.8 Human skull, superior view (top of cranium removed).

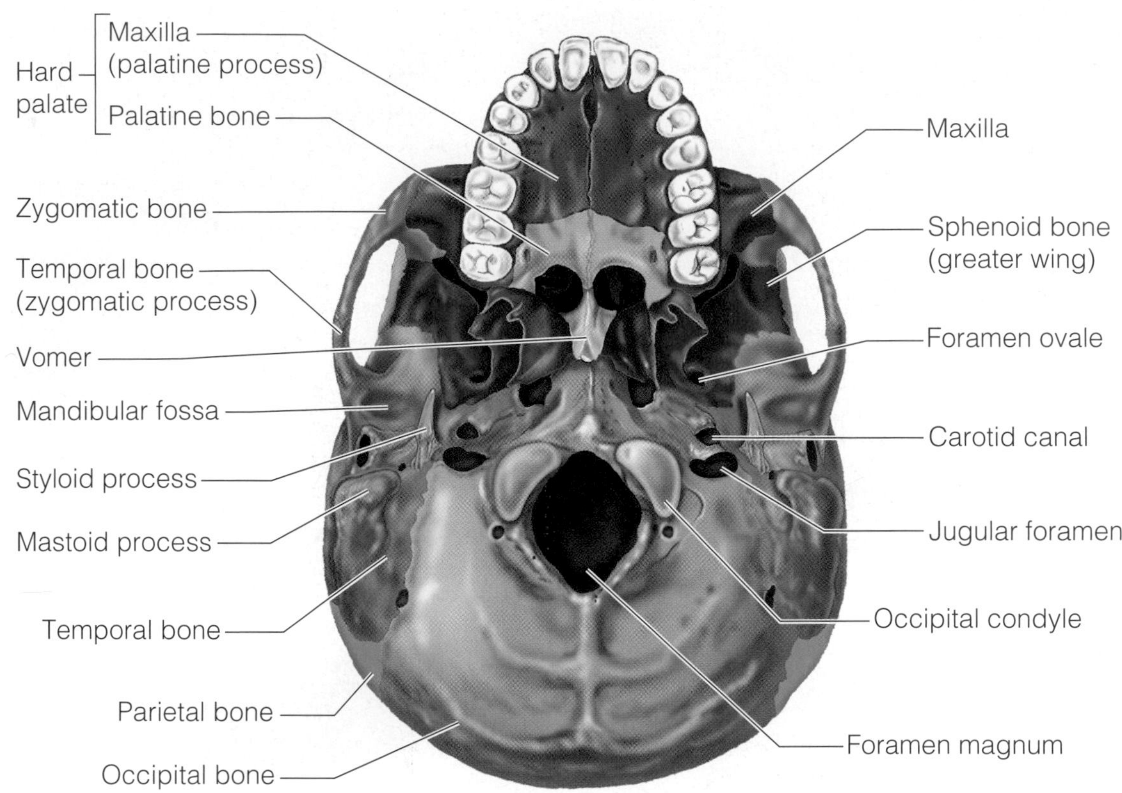

FIGURE 5.9 Human skull, inferior view (mandible removed).

are the rockerlike **occipital condyles** (see Figure 5.9), which rest on the first vertebra of the spinal column.

Sphenoid Bone The butterfly-shaped sphenoid (sfe′noid) bone spans the width of the skull and forms part of the floor of the cranial cavity (see Figure 5.8). In the midline of the sphenoid is a small depression, the **sella turcica** (sel′ah tur′sĭ-kah), or *Turk's saddle,* which holds the pituitary gland in place. The **foramen ovale,** a large oval

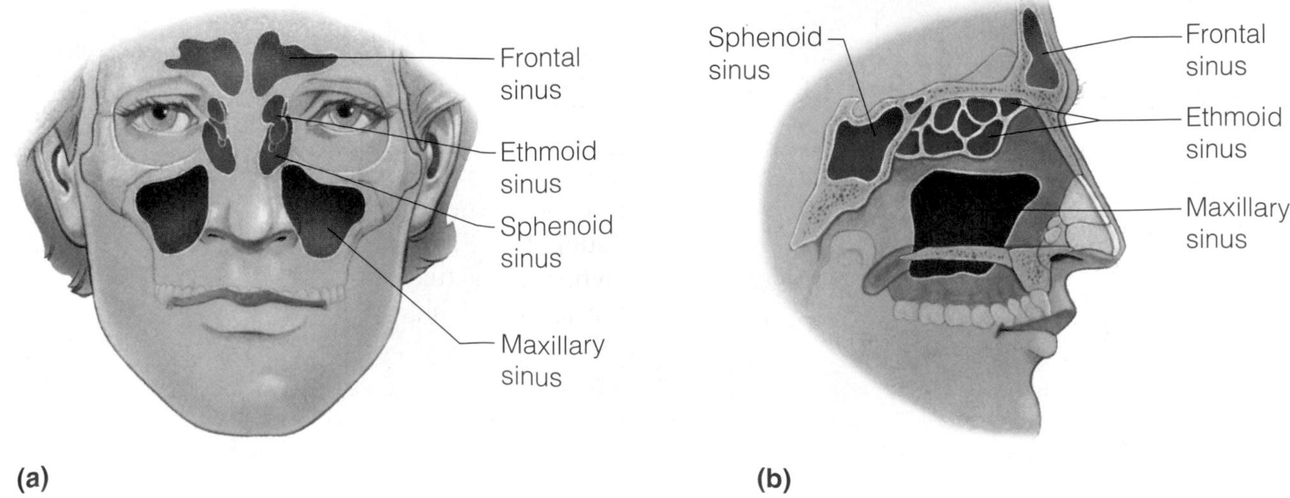

FIGURE 5.10 Paranasal sinuses. (a) Anterior view. **(b)** Medial view.

Q *What bone articulates with every other facial bone?*

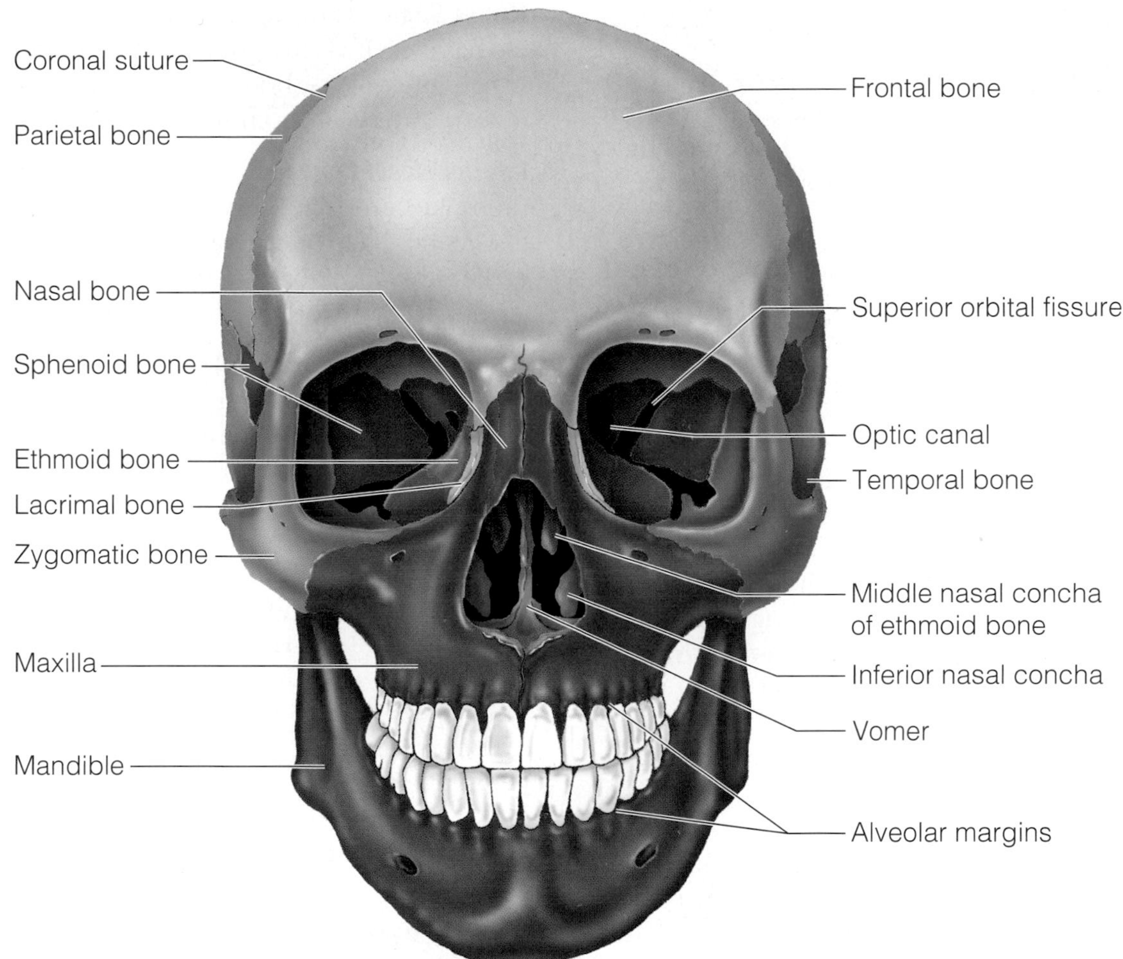

FIGURE 5.11 Human skull, anterior view.

opening in line with the posterior end of the sella turcica (Figure 5.9), allows fibers of cranial nerve V (the trigeminal nerve) to pass to the chewing muscles of the lower jaw (mandible). Parts of the sphenoid, seen exteriorly forming part of the eye orbits, have two important openings, the **optic canal** which allows the optic nerve to pass to the eye, and the slitlike **superior orbital fissure** through which the cranial nerves controlling eye movements (III, IV, and VI) pass (see Figures 5.7 and 5.11). The central part of the sphenoid bone is riddled with air cavities, the **sphenoid sinuses** (Figure 5.10).

A *The maxilla.*

Ethmoid Bone The ethmoid (eth′moid) bone is very irregularly shaped and lies anterior to the sphenoid (see Figures 5.7, 5.8, and 5.11). It forms the roof of the nasal cavity and part of the medial walls of the orbits. Projecting from its superior surface is the **crista galli** (kris′tah gah′le), literally "cock's comb" (see Figure 5.8). The outermost covering of the brain attaches to this projection. On each side of the crista galli are many small holes. These holey areas, the **cribriform** (krib′rĭ-form) **plates,** allow nerve fibers carrying impulses from the olfactory (smell) receptors of the nose to reach the brain. Extensions of the ethmoid bone, the **superior** and **middle nasal conchae** (see Figure 5.11), form part of the lateral walls of the nasal cavity and increase the turbulence of air flowing through the nasal passages.

Facial Bones

Fourteen bones compose the face. Twelve are paired; only the mandible and vomer are single. Figures 5.7 and 5.11 show most of the facial bones.

Maxillae The two maxillae (mak-si′le), or **maxillary bones,** fuse to form the upper jaw. All facial bones except the mandible join the maxillae; thus they are the main, or "keystone," bones of the face. The maxillae carry the upper teeth in the **alveolar margin.**

Extensions of the maxillae called the **palatine** (pal′ah-tīn) **processes** form the anterior part of the hard palate of the mouth (see Figure 5.9). Like many other facial bones, the maxillae contain **sinuses,** which drain into the nasal passages (see Figure 5.10). These **paranasal sinuses,** whose naming reveals their position surrounding the nasal cavity, lighten the skull bones and probably act to amplify the sounds we make as we speak. They also cause many people a great deal of misery. Since the mucosa lining these sinuses is continuous with that in the nasal passages and throat, infections in these areas tend to migrate into the sinuses, causing *sinusitis*. Depending on which sinuses are infected, a headache or upper jaw pain is the usual result.

Palatine Bones The paired palatine bones lie posterior to the palatine processes of the maxillae. They form the posterior part of the hard palate (see Figure 5.9). Failure of these or the palatine processes to fuse medially results in *cleft palate.*

Zygomatic Bones The zygomatic bones are commonly referred to as the cheekbones. They also form a good-sized portion of the lateral walls of the orbits, or eye sockets.

Lacrimal Bones The lacrimal (lak′rĭ-mal) bones are fingernail-size bones forming part of the medial walls of each orbit. Each lacrimal bone has a groove that serves as a passageway for tears (*lacrima* = tear).

Nasal Bones The small rectangular bones forming the bridge of the nose are the nasal bones. (The lower part of the skeleton of the nose is made up of cartilage.)

Vomer Bone The single bone in the median line of the nasal cavity is the vomer. (*Vomer* means "plow," which refers to the bone's shape.) The vomer forms most of the nasal septum.

Inferior Nasal Conchae The inferior nasal conchae (kong′ke) are thin, curved bones projecting from the lateral walls of the nasal cavity. (As mentioned earlier, the superior and middle conchae are similar but are parts of the ethmoid bone.)

Mandible The mandible, or lower jaw, is the largest and strongest bone of the face. It joins the temporal bones on each side of the face, forming the only freely movable joints in the skull. You can find these joints on yourself by placing your fingers over your cheekbones and opening and closing your mouth. The horizontal part of the mandible (the *body*) forms the chin. Two upright bars of bone (the *rami*) extend from the body to connect the mandible with the temporal bone. The lower teeth lie in *alveoli* (sockets) in the **alveolar margin** at the superior edge of the mandibular body.

The Hyoid Bone

Though not really part of the skull, the **hyoid** (hi′oid) **bone** (Figure 5.12) is closely related to the mandible and temporal bones. The hyoid bone is unique in that it is the only bone of the body that does not articulate directly with any other bone. Instead, it is suspended in the midneck region about 2 cm (1 inch) above the larynx, where it is anchored by ligaments to the styloid processes of the temporal bones. Horseshoe-shaped, with a *body* and two pairs of *horns,* or *cornua,* the hyoid bone serves as a movable base for the tongue and an attachment point for neck muscles that raise and lower the larynx when we swallow and speak.

Fetal Skull

The skull of a fetus or newborn infant is different in many ways from an adult skull. As Figure 5.13b illustrates, the infant's face is very small compared to the size of its cranium, but the skull as a whole is large compared to the infant's total body length. The adult skull represents only one-eighth of the total body length, whereas that of a newborn infant

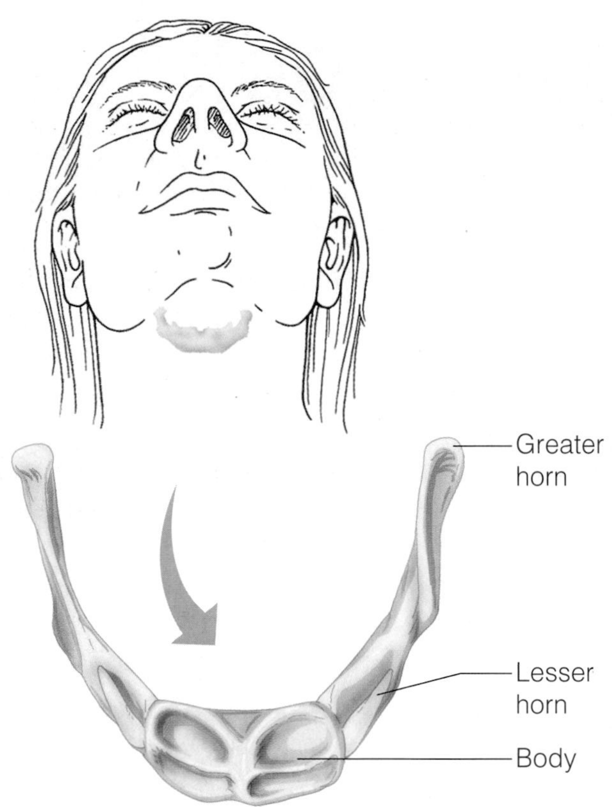

FIGURE 5.12 Anatomical location and structure of the hyoid bone. The hyoid bone is suspended in the midanterior neck by ligaments attached to the lesser horns and the styloid processes of the temporal bones.

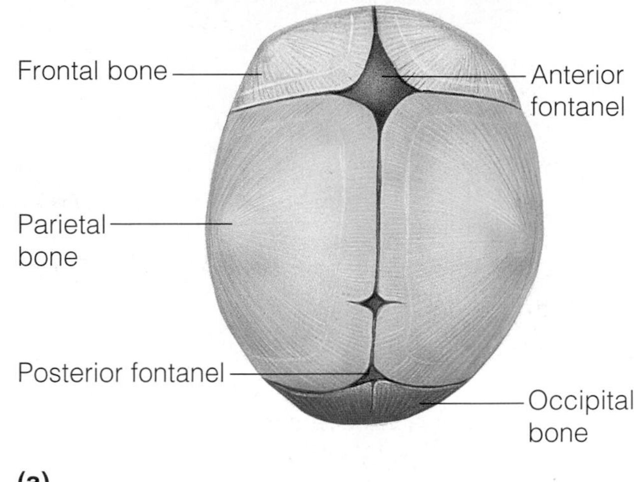

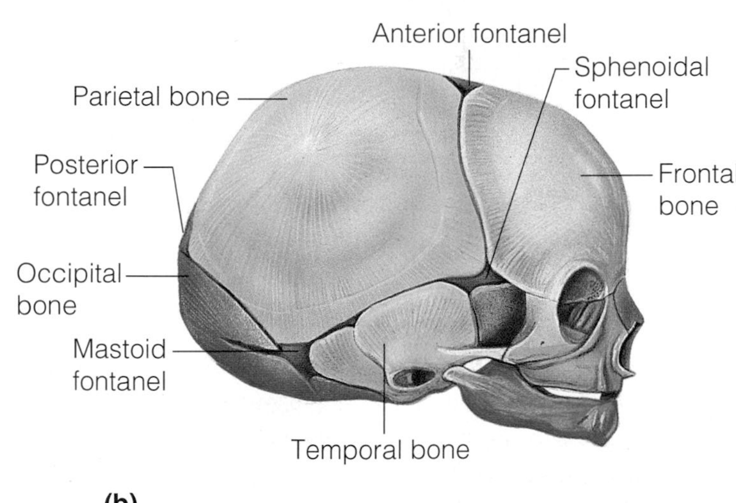

FIGURE 5.13 The fetal skull. (a) Superior view. **(b)** Lateral view.

is one-fourth as long as its entire body. When a baby is born, its skeleton is still unfinished. As noted earlier, some areas of hyaline cartilage still remain to be ossified, or converted to bone. In the newborn, the skull also has fibrous regions that have yet to be converted to bone. These fibrous membranes connecting the cranial bones are called **fontanels** (fon″tah-nelz′). The rhythm of the baby's pulse can be felt in these "soft spots," which explains their name (*fontanel* = little fountain). The largest fontanels are the diamond-shaped *anterior fontanel* and the smaller triangular *posterior fontanel.* The fontanels allow the fetal skull to be compressed slightly during birth. In addition, because they are flexible, they allow the infant's brain to grow during the later part of pregnancy and early infancy. This would not be possible if the cranial bones were fused in sutures as in the adult skull. The fontanels are gradually converted to bone during the early part of infancy and can no longer be felt by 22 to 24 months after birth.

Vertebral Column (Spine)

Serving as the axial support of the body, the **vertebral column,** or **spine,** extends from the skull, which it supports, to the pelvis, where it transmits the weight of the body to the lower limbs. Some people think of the vertebral column as a rigid supporting rod, but that picture is inaccurate. Instead, the spine is formed from 26 irregular bones connected and reinforced by ligaments in such a way that a flexible, curved structure results

A Closer Look

Protect Your Back—It's the Only One You've Got!

REGULAR exercise is vital to keep the spine strong and to protect its beautifully proportioned structure. Unfortunately, most of us neglect our backs until pains and aches demand our attention. Next to sore throats and the common cold, backaches are the most common reason for visits to the physician's office. Millions of U.S. citizens suffer from back pain, and back injuries account for a quarter of all disability insurance payments.

Although backache may be symptomatic of many ailments, from spinal tumor to kidney disease, over 80 percent of back problems result from weak muscles or are anxiety-related. This is not surprising since strong abdominal muscles and flexible back and hip muscles play a major role in reinforcing the spine's delicate architecture, and stress causes muscles to "knot." Tense muscles can aggravate a back problem even if they do not cause it. Obese individuals and those who rarely exercise are prime candidates for lower back pain. Indeed, some orthopedists (bone specialists) look upon back pain as an index of affluence. The more gadgets we accumulate to do our work for us, the more we can sit or lie down—and the more our muscles deteriorate. On the other side of the coin, people who work at jobs that require heavy lifting are also "sitting ducks" for lower back pain if they do not lift properly.

As the saying goes, "Prevention is worth a pound of cure," and preventing back problems is better than attempting to treat agonizing back pain. Perhaps 70 to 80 percent of all cases of lower back pain are preventable with only 10 minutes of daily exercise, if proper body alignment and certain precautions are observed. Here are a few well-documented guidelines for protecting your back.

1. Keep your weight down. Just a few extra pounds of weight, particularly the unbalanced weight of a protruding potbelly, exert a strain on the spine that is much greater than that of the weight itself. The protruding abdomen has the effect of pulling the body forward, forcing the back muscles to contract more strongly to counterbalance the weight. For example, 10 extra pounds located 10 inches anterior to the spine in the abdominal region force the back muscles to exert 50 extra pounds of force to counteract the 10-pound pulling force.
2. Wear high-heel and negative-heel shoes or boots infrequently, and avoid them altogether if you have back problems. Both change the alignment of the spine. High-heel shoes tilt the pelvis forward and increase the stress on the abdominal and back muscles. Negative-heel

Lowering the center of gravity by squatting.

shoes tilt the pelvis backward so that weight transfer during walking is difficult.

3. Maintain good posture. Your head and back should be aligned, with abdomen and buttocks tucked inward.
4. Lift heavy objects using proper body mechanics. Every object, including a human being, has a center of gravity around which its mass is equally distributed. In a standing adult, the center of gravity is inside the pelvis, slightly posterior to the anterior border of the joint between the sacrum and the fifth lumbar vertebra. A basic principle of body mechanics is the broader your base of support and the lower your center of gravity, the more stable you are. Applying this principle to lifting demands that you take a wider than normal stance (spread your feet apart slightly), and then bend your knees (rather than your back) to reach for and pick up the object (see the photograph on previous page). The weight is then transmitted to your stronger legs, sparing your more delicate spine. If you must move large objects, push rather than pull them.
5. Avoid sitting for prolonged periods. Sitting puts much more stress on the spine than standing: truck drivers have five times the normal risk for lower back problems and herniated discs. If you must sit for an extended time, resting your feet on a small stool (or the lowest desk drawer) will reduce the stress on your vertebral column.
6. Take 10 minutes a day to stretch your lower back extensor muscles and hip flexor muscles and to strengthen your abdominal muscles as illustrated and described in instructions **a–c** at the end of this text. If you have any history of back problems, avoid any exercise that strains your lower back, including (1) sit-ups with straight legs, (2) double leg raises, and (3) lying on your abdomen and then raising your head, arms, and legs.

Dare to be different! Be good to your spine and you may well be saved from uttering that nearly universal complaint, "Oh, my aching back!"

(a) *Stretching the extensor muscles of the lower back.* Lie on back with knees bent and feet flat on the floor. Keeping your arms at your sides, raise one knee to the chest. Lower the foot to the floor with knee bent, and then slide the foot along the floor until the leg is fully extended. Rotate the leg gently from side to side.

Resume the starting position and repeat the exercise with the alternate leg. Repeat the entire exercise 5 to 6 times.

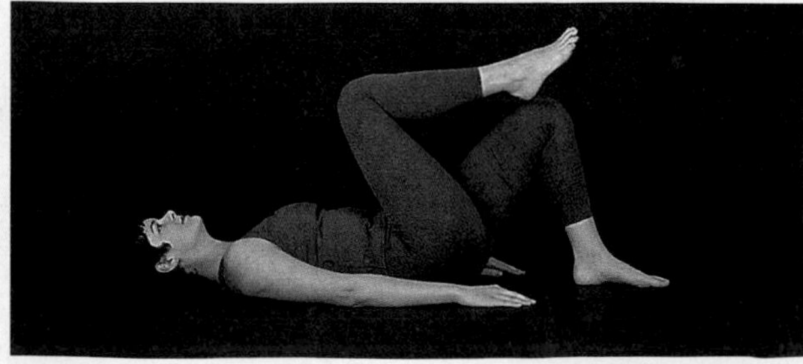

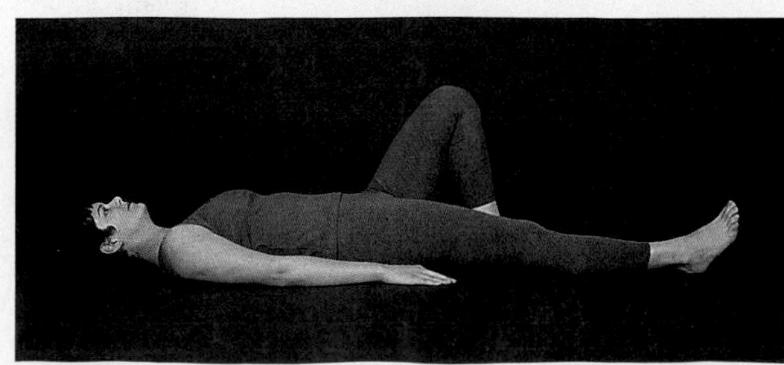

(a)

Protect Your Back—It's the Only One You've Got! *(continued)*

(b) *Stretching the flexor muscles of the hip.* Lie on back. Exhale as you bring both knees toward the chest. Then, while holding one knee to the chest, slide the opposite leg along the floor until it is fully extended. Attempt to touch the back of the knee (popliteal region) of the extended leg to the floor. Hold for a count of 6.

Resume the starting position and repeat the exercise with the alternate leg. Repeat the entire exercise 5 to 6 times.

(c) *Strengthening the abdominal muscles.* Lie on back with knees bent and feet flat on floor. Keeping your arms at your sides, raise one knee toward your chest. Exhale as you lift your head, attempting to touch your raised knee to your forehead. Count to 6. Resume the starting position and roll your head gently from side to side. Inhale, and repeat the exercise with the alternate leg.

Repeat the entire exercise 8 to 10 times at the beginning; work up to 25 repetitions per day.

(b)

(c)

(Figure 5.14). Running through the central cavity of the vertebral column is the delicate spinal cord, which it surrounds and protects.

Before birth, the spine consists of 33 separate bones called **vertebrae,** but 9 of these eventually fuse to form the two composite bones, the *sacrum* and the *coccyx,* that construct the inferior portion of the vertebral column. Of the 24 single bones, the 7 vertebrae of the neck are *cervical vertebrae,* the next 12 are the *thoracic vertebrae,* and the remaining 5 supporting the lower back are *lumbar vertebrae.*

- Remembering common meal times, 7 AM, 12 noon, and 5 PM, may help you to recall the number of bones in these three regions of the vertebral column.

The single vertebrae are separated by pads of flexible fibrocartilage—**intervertebral discs**—which cushion the vertebrae and absorb shocks while allowing the spine flexibility. In a young person, the discs have a high water content (about 90 percent) and are spongy and compressible. But as a person ages, the water content of the discs decreases (as it does in other tissues throughout the body), and the discs become harder and less compressible.

Homeostatic Imbalance

Drying of the discs, along with a weakening of the ligaments of the vertebral column, predisposes older people to *herniated* ("slipped") *discs.* However, herniation also may result when the vertebral column is subjected to exceptional twisting forces. If the protruding disc presses on the spinal cord or the spinal nerves exiting from the cord, numbness and excruciating pain can result. ▲

The discs and the S-shaped structure of the vertebral column work together to prevent shock to the head when we walk or run. They also make the body trunk flexible. The spinal curvatures in the thoracic and sacral regions are referred to as **primary curvatures** because they are present when we are born. Later, the **secondary curvatures** develop. The cervical curvature appears when a baby begins to raise its head, and the lumbar curvature develops when the baby begins to walk.

Homeostatic Imbalance

Why do they do "spine checks" in middle school? The answer is that they are looking for abnormal spinal curvatures. There are several types of abnormal spinal curvatures. Figure 5.15 shows three of these—*scoliosis* (sko″le-o′sis), *kyphosis* (ki-fo′sis), and *lordosis* (lor-do′sis). These abnormalities may be congenital (present at birth) or result from disease, poor posture, or unequal muscle pull on the spine. As you look at these diagrams, try to pinpoint how each of these conditions differs from the normal healthy spine. ▲

Q *What is a slipped disc?*

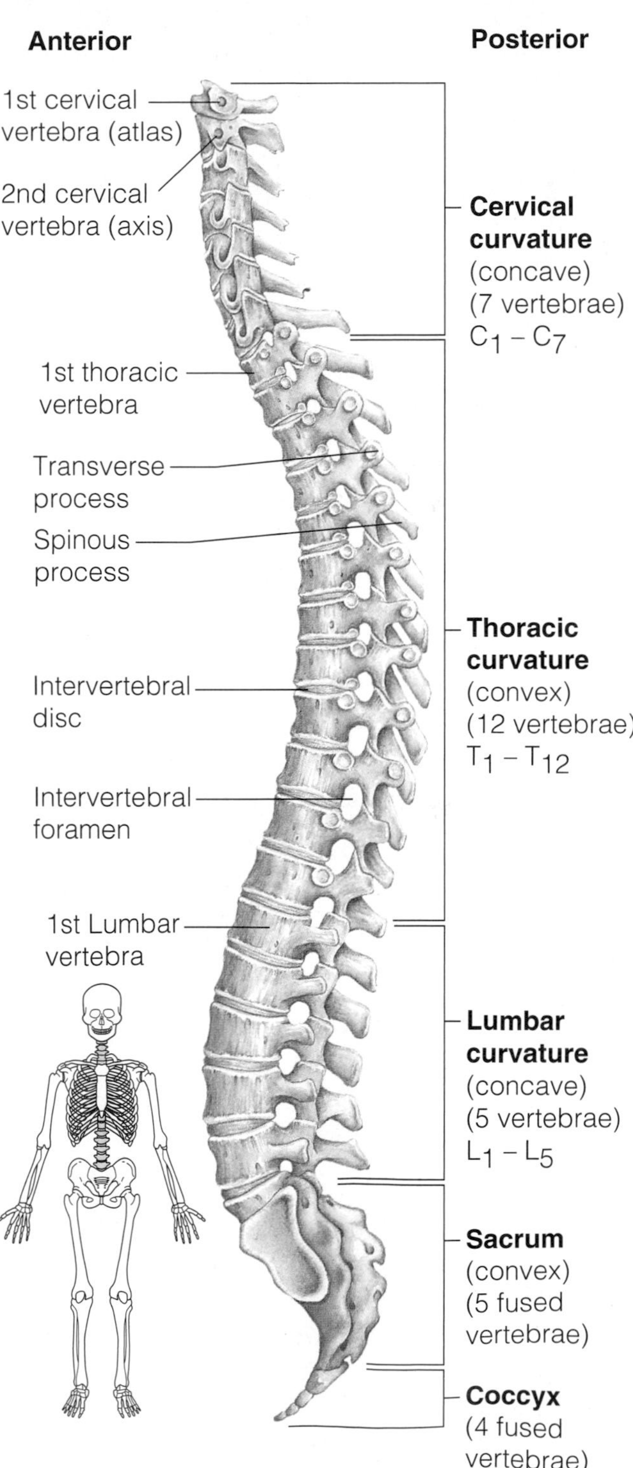

FIGURE 5.14 The vertebral column. Thin discs between the thoracic vertebrae allow great flexibility in the thoracic region; thick discs between the lumbar vertebrae reduce flexibility. Notice that the terms *convex* and *concave* refer to the curvature of the posterior aspect of the vertebral column.

A *A disc that protrudes outward from its normal position in the vertebral column and which may cause pain by pressing on adjacent nerves.*

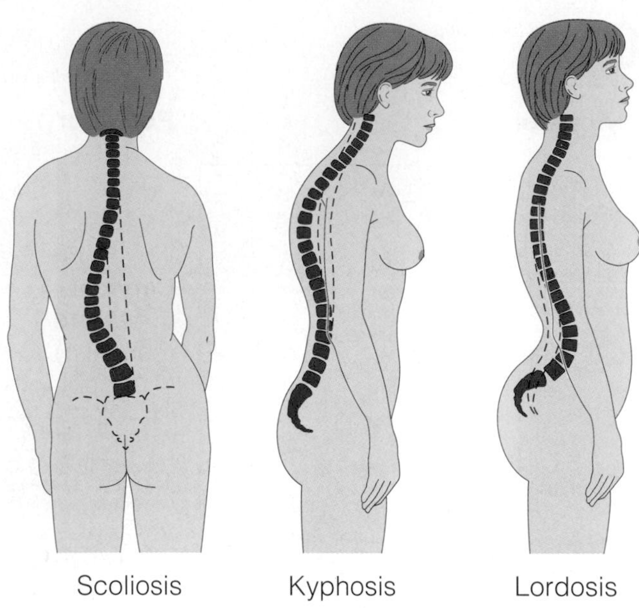

FIGURE 5.15 Abnormal spinal curvatures.

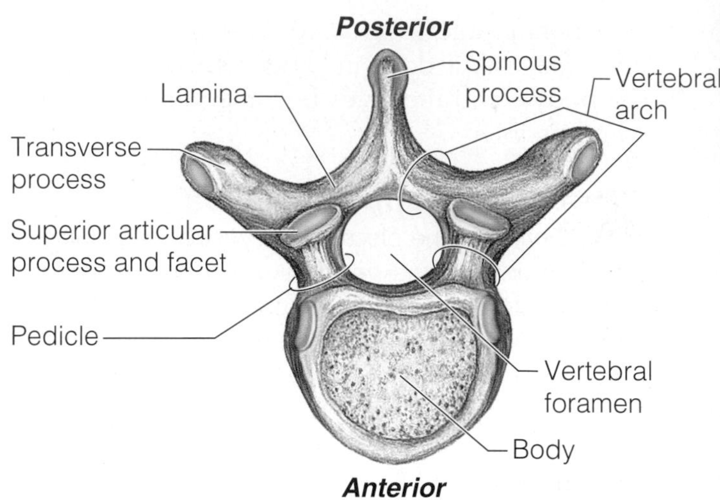

FIGURE 5.16 A typical vertebra, superior view. (Inferior articulating surfaces are not shown.)

All vertebrae have a similar structural pattern (Figure 5.16). The common features are listed below:

- **Body** or **centrum:** disclike, weight-bearing part of the vertebra facing anteriorly in the vertebral column.
- **Vertebral arch:** arch formed from the joining of all posterior extensions, the **laminae** and **pedicles,** from the vertebral body.
- **Vertebral foramen:** canal through which the spinal cord passes.
- **Transverse processes:** two lateral projections from the vertebral arch.
- **Spinous process:** single projection arising from the posterior aspect of the vertebral arch (actually the fused laminae).
- **Superior and inferior articular processes:** paired projections lateral to the vertebral foramen, allowing a vertebra to form joints with adjacent vertebrae (see also Figure 5.17).

In addition to the common features just described, vertebrae in the different regions of the spine have very specific structural characteristics. These unique regional characteristics of the vertebrae are described next.

Cervical Vertebrae

The seven **cervical vertebrae** (identified as C_1 to C_7) form the neck region of the spine. The first two vertebrae (*atlas* and *axis*) are different because they perform functions not shared by the other cervical vertebrae. As you can see in Figure 5.17a, the **atlas** (C_1) has no body. The superior surfaces of its transverse processes contain large depressions that receive the occipital condyles of the skull. This joint allows you to nod "yes." The **axis** (C_2) acts as a pivot for the rotation of the atlas (and skull) above. It has a large upright process, the **dens,** or **odontoid** (o-don′toid) **process,** which acts as the pivot point. The joint between C_1 and C_2 allows you to rotate your head from side to side to indicate "no."

The "typical" cervical vertebrae (C_3 through C_7) are shown in Figure 5.17b. They are the smallest, lightest vertebrae, and most often their spinous processes are short and divided into two branches. The transverse processes of the cervical vertebrae contain foramina (openings) through which the vertebral arteries pass on their way to the brain above. Any time you see these foramina in a vertebra, you should know immediately that it is a cervical vertebra.

Thoracic Vertebrae

The 12 **thoracic vertebrae** (T_1 to T_{12}) are all typical. As seen in Figure 5.17c, they are larger than the cervical vertebrae. The body is somewhat heart-shaped and has two costal facets (articulating surfaces) on each side, which receive the heads of the ribs. The spinous process is long and hooks sharply downward, causing the vertebra to look like a giraffe's head viewed from the side.

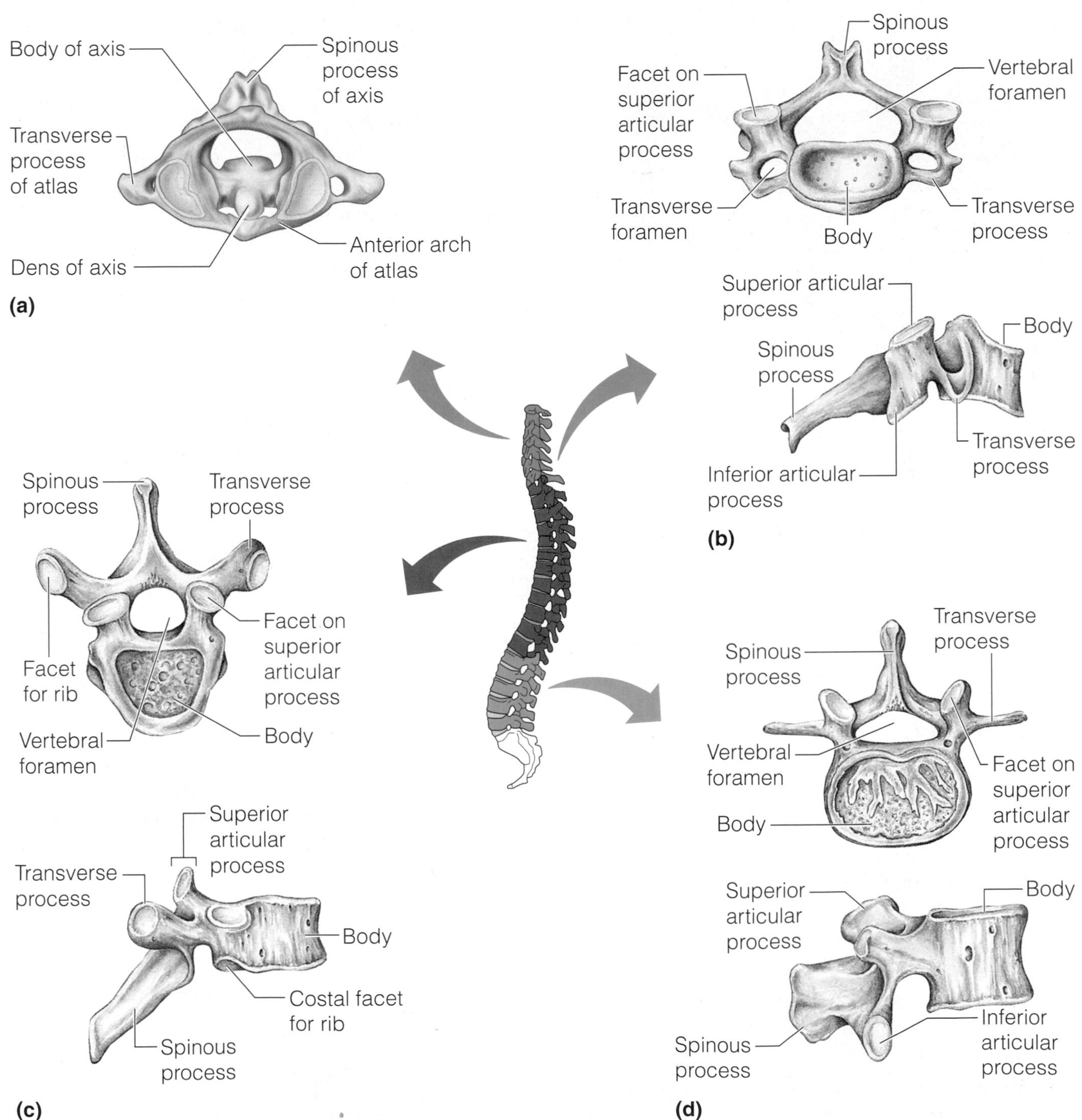

FIGURE 5.17 Regional characteristics of vertebrae. (a) Superior view of the articulated atlas and axis. **(b)** Cervical vertebrae; superior view above, lateral view below. **(c)** Thoracic vertebrae; superior view above, lateral view below. **(d)** Lumbar vertebrae; superior view above, lateral view below.

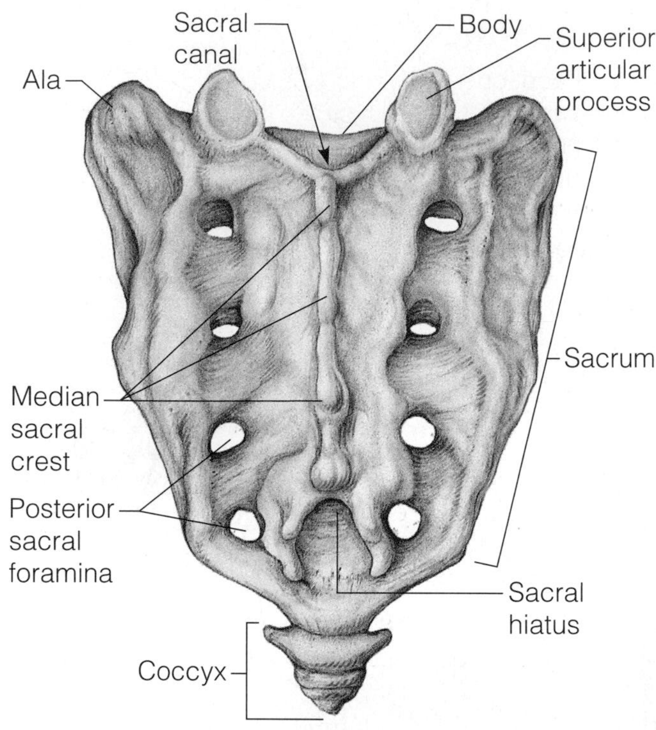

FIGURE 5.18 Sacrum and coccyx, posterior view.

Lumbar Vertebrae

The five **lumbar vertebrae** (L_1 to L_5) have massive, blocklike bodies. Their short, hatchet-shaped spinous processes (Figure 5.17d) make them look like a moose head from the lateral aspect. Since most of the stress on the vertebral column occurs in the lumbar region, these are the sturdiest of the vertebrae.

Sacrum

The **sacrum** (sa′krum) is formed by the fusion of five vertebrae (Figure 5.18). Superiorly it articulates with L_5, and inferiorly it connects with the coccyx. The winglike **alae** articulate laterally with the hip bones, forming the sacroiliac joints. The sacrum forms the posterior wall of the pelvis. Its posterior midline surface is roughened by the **median sacral crest,** the fused spinous processes of the sacral vertebrae. This is flanked laterally by the posterior sacral foramina. The vertebral canal continues inside the sacrum as the **sacral canal** and terminates in a large inferior opening called the **sacral hiatus.**

Coccyx

The **coccyx** is formed from the fusion of three to five tiny, irregularly shaped vertebrae (Figure 5.18). It is the human "tailbone," a remnant of the tail that other vertebrate animals have.

Bony Thorax

The sternum, ribs, and thoracic vertebrae make up the **bony thorax.** The bony thorax is often called the **thoracic cage** because it forms a protective, cone-shaped cage of slender bones around the organs of the thoracic cavity (heart, lungs, and major blood vessels). The bony thorax is shown in Figure 5.19.

Sternum

The **sternum** (breastbone) is a typical flat bone and the result of the fusion of three bones—the **manubrium** (mah-nu′bre-um), **body,** and **xiphoid** (zif′oid) **process.** It is attached to the first seven pairs of ribs.

The sternum has three important bony landmarks—the jugular notch, the sternal angle, and the xiphisternal joint.

- The **jugular notch** (concave upper border of the manubrium) can be palpated easily; generally it is at the level of the third thoracic vertebra.
- The **sternal angle** results where the manubrium and body meet at a slight angle to each other, so that a transverse ridge is formed at the level of the second ribs. It provides a handy reference point for counting ribs to locate the second intercostal space for listening to certain heart valves.
- The **xiphisternal** (zi′fe-ster″nal) **joint,** the point where the sternal body and xiphoid process fuse, lies at the level of the ninth thoracic vertebra.

Palpate your sternal angle and jugular notch.

Because the sternum is so close to the body surface, it is easy to obtain samples of blood-forming (hematopoietic) tissue for the diagnosis of suspected blood diseases from this bone. A needle is inserted into the marrow of the sternum, and the sample is withdrawn; this procedure is called a *sternal puncture.* Because the heart lies immediately posterior to the sternum, the physician must take extreme care not to penetrate through the sternum during this procedure.

Ribs

Twelve pairs of **ribs** form the walls of the bony thorax. (Contrary to popular misconception, males do *not* have one rib less than females!) All the ribs

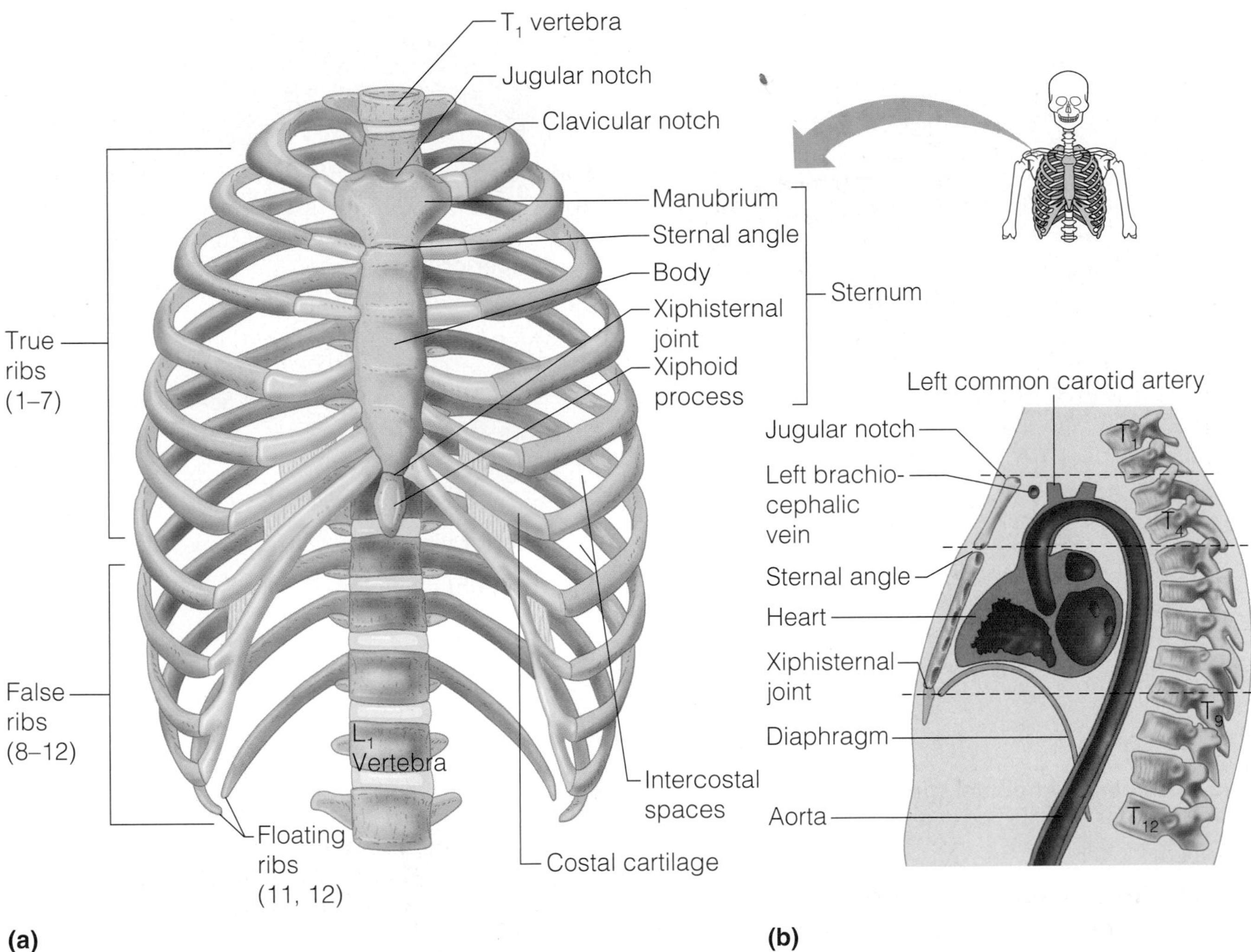

FIGURE 5.19 The bony thorax. (a) Skeleton of the bony thorax, anterior view (costal cartilages are shown in blue). **(b)** Left lateral view of the thorax, showing the relationship of the surface landmarks of the thorax to the vertebral column (thoracic portion).

articulate with the vertebral column posteriorly and then curve downward and toward the anterior body surface. The **true ribs,** the first seven pairs, attach directly to the sternum by costal cartilages. **False ribs,** the next five pairs, either attach indirectly to the sternum or are not attached to the sternum at all. The last two pairs of false ribs lack the sternal attachments, and so they are also called **floating ribs.**

The intercostal spaces (spaces between the ribs) are filled with the intercostal muscles that aid in breathing.

Appendicular Skeleton

The *appendicular skeleton* is shaded gold in Figure 5.6. It is composed of 126 bones of the limbs (appendages) and the pectoral and pelvic girdles, which attach the limbs to the axial skeleton.

Bones of the Shoulder Girdle

Each **shoulder girdle,** or **pectoral girdle,** consists of two bones—a clavicle and a scapula (Figure 5.20).

The **clavicle** (klav′ĭ-kl), or *collarbone,* is a slender, doubly curved bone. It attaches to the manubrium of the sternum medially (at its sternal end) and to the scapula laterally, where it helps to form the shoulder joint (at its acromial end). The clavicle acts as a brace to hold the arm away from the top of the thorax and helps prevent shoulder dislocation. When the clavicle is broken, the whole shoulder region caves in medially, which shows how important its bracing function is.

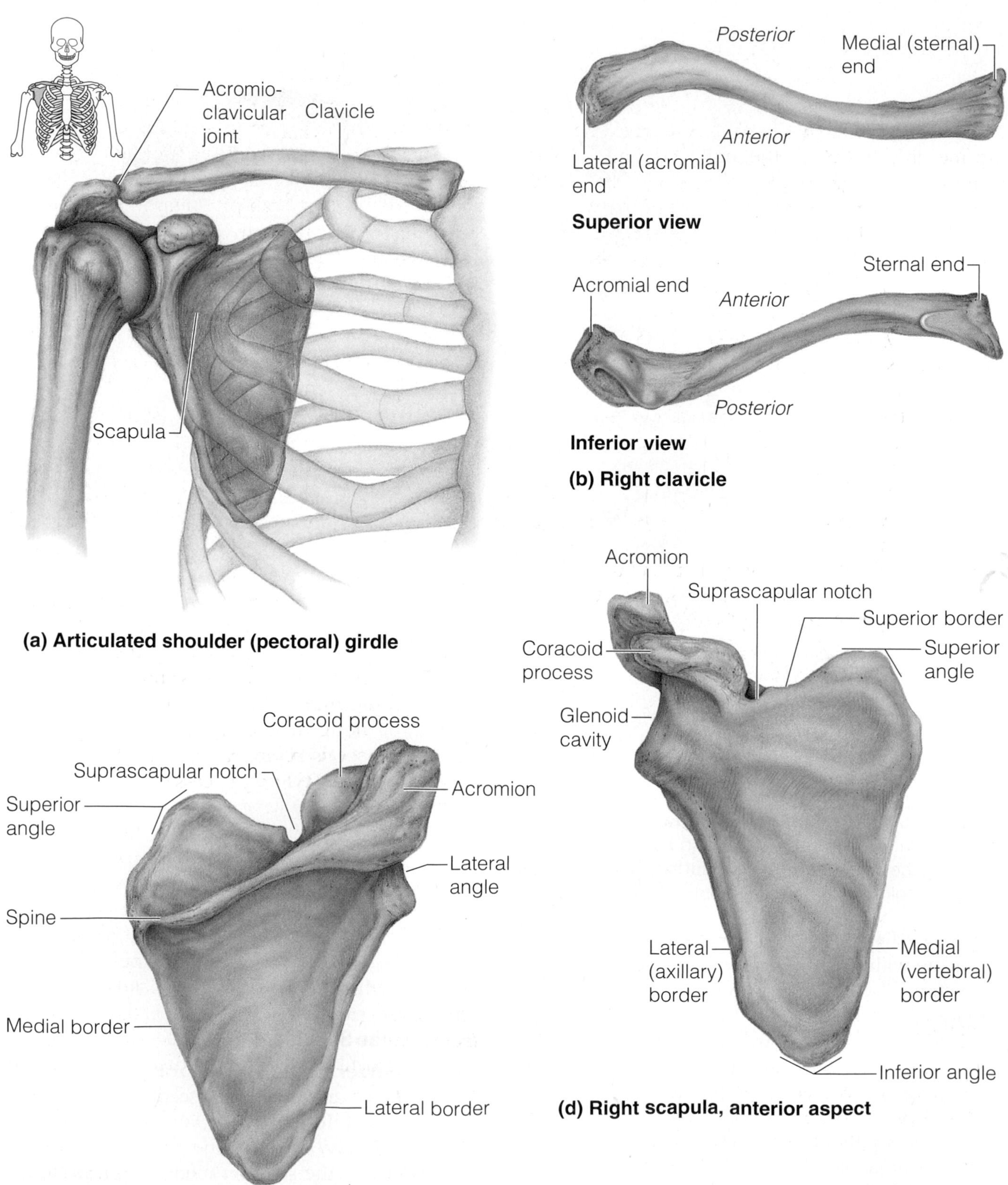

FIGURE 5.20 Bones of the shoulder girdle. **(a)** Relationship of the right shoulder girdle to the bones of the thorax and arm. **(b)** Right clavicle, superior and inferior views. **(c)** Right scapula, posterior view. **(d)** Right scapula, anterior view.

The **scapulae** (skap′u-le), or *shoulder blades,* are triangular and are commonly called "wings" because they flare when we move our arms posteriorly. Each scapula has a flattened body and two important processes—the **acromion** (ah-kro′me-on), which is the enlarged end of the spine of the scapula, and the beaklike **coracoid** (kor′ah-koid) **process.** The acromion connects with the clavicle laterally at the **acromioclavicular joint.** The coracoid process points over the top of the shoulder and anchors some of the muscles of the arm. Just medial to the coracoid process is the large **suprascapular notch,** which serves as a nerve passageway. The scapula is not directly attached to the axial skeleton; it is loosely held in place by trunk muscles. The scapula has three borders—superior, medial (vertebral), and lateral (axillary). It also has three angles—superior, inferior, and lateral. The **glenoid cavity,** a shallow socket that receives the head of the arm bone, is in the lateral angle.

The shoulder girdle is very light and allows the upper limb to have exceptionally free movement. This is due to the following factors:

1. Each shoulder girdle attaches to the axial skeleton at only one point—the *sternoclavicular joint.*
2. The loose attachment of the scapula allows it to slide back and forth against the thorax as muscles act.
3. The glenoid cavity is shallow, and the shoulder joint is poorly reinforced by ligaments.

However, this exceptional flexibility also has a drawback; the shoulder girdle is very easily dislocated.

Bones of the Upper Limbs

Thirty separate bones form the skeletal framework of each upper limb (Figures 5.21 and 5.22). They form the foundations of the arm, forearm, and hand.

Arm

The arm is formed by a single bone, the **humerus** (hu′mer-us), which is a typical long bone (see Figure 5.21a and b). At its proximal end is a rounded head that fits into the shallow glenoid cavity of the scapula. Opposite the head are two bony projections—the **greater** and **lesser tubercles,** which are sites of muscle attachment. In the midpoint of the shaft is a roughened area called the **deltoid tuberosity,** where the large, fleshy deltoid muscle of the shoulder attaches. Nearby, the **radial groove** runs obliquely down the posterior aspect of the shaft. This groove marks the course of the radial nerve, an important nerve of the upper limb. At the distal end of the humerus is the medial **trochlea** (trok′le-ah), which looks somewhat like a spool, and the lateral ball-like **capitulum** (kah-pit′u-lum). Both of these processes articulate with bones of the forearm. Above the trochlea anteriorly is a depression, the **coronoid fossa;** on the posterior surface is the **olecranon** (o-lek′rah-non) **fossa.** These two depressions, which are flanked by **medial** and **lateral epicondyles,** allow the corresponding processes of the ulna to move freely when the elbow is bent and extended.

Forearm

Two bones, the radius and the ulna, form the skeleton of the forearm (see Figure 5.21c). When the body is in the anatomical position, the **radius** is the lateral bone; that is, it is on the thumb side of the forearm. When the hand is rotated so that the palm faces backward, the distal end of the radius crosses over and ends up medial to the ulna. Both proximally and distally the radius and ulna articulate at small **radioulnar joints,** and the two bones are connected along their entire length by the flexible **interosseous membrane.** Both the ulna and the radius have a **styloid process** at their distal end.

The disc-shaped head of the radius also forms a joint with the capitulum of the humerus. Just below the head is the **radial tuberosity,** where the tendon of the biceps muscle attaches.

When the upper limb is in the anatomical position, the **ulna** is the medial bone (on the little-finger side) of the forearm. On its proximal end are the anterior **coronoid process** and the posterior **olecranon process,** which are separated by the **trochlear notch.** Together these two processes grip the trochlea of the humerus in a pliers-like joint.

Hand

The skeleton of the hand consists of the carpals, the metacarpals, and the phalanges (Figure 5.22). The eight **carpal bones,** arranged in two irregular rows of four bones each, form the part of the hand called the **carpus** or, more commonly, the *wrist.* The carpals are bound together by ligaments that restrict movements between them. (In case you

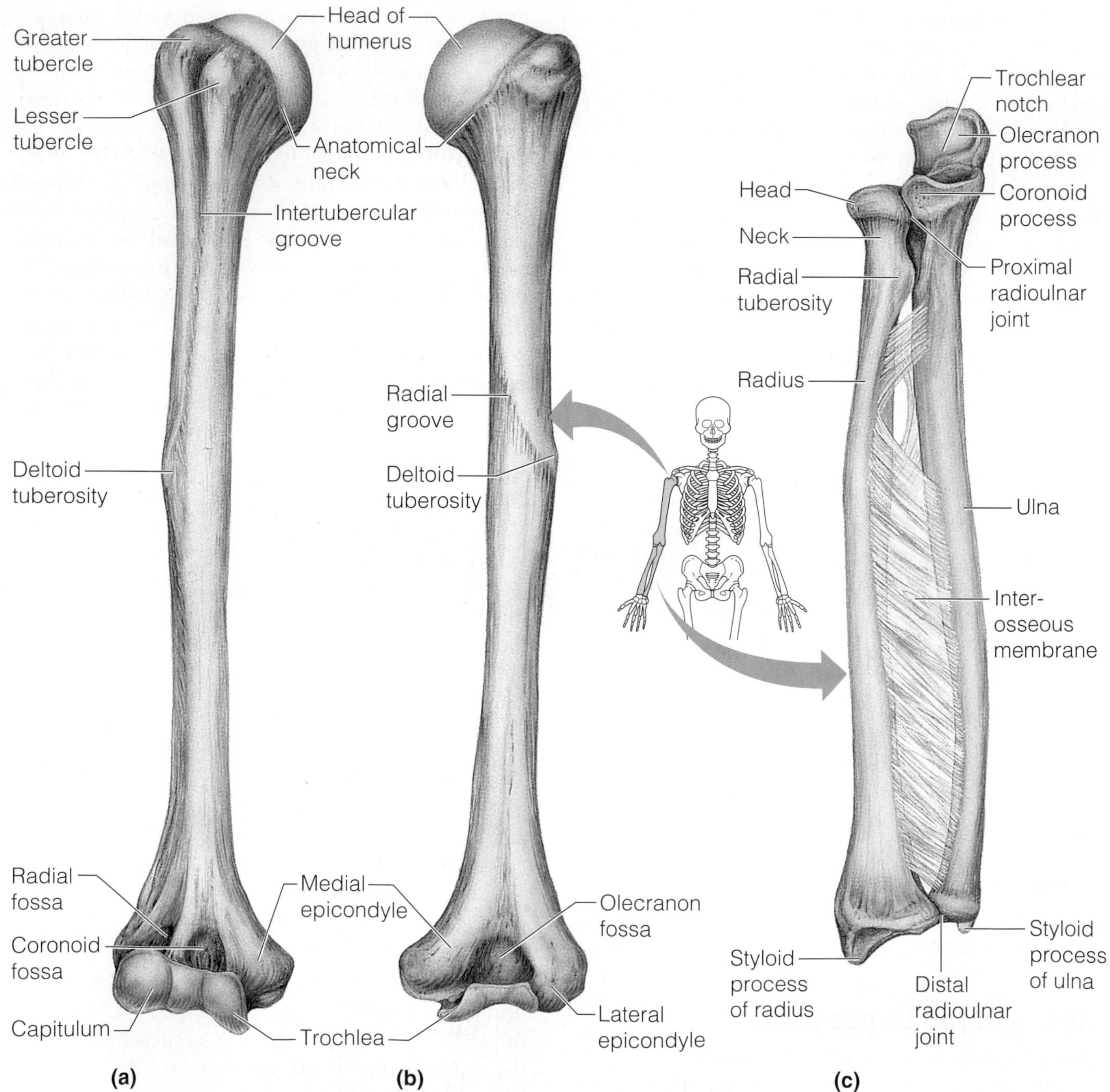

FIGURE 5.21 Bones of the right arm and forearm. **(a)** Humerus, anterior view. **(b)** Humerus, posterior view. **(c)** Anterior view of the bones of the forearm: the radius and the ulna.

need to learn their names, the individual carpal bones are identified in Figure 5.22.)

The palm of the hand consists of the **metacarpals.** The **phalanges** (fah-lan′jēz) are the bones of the fingers. The metacarpals are numbered 1 to 5 from the thumb side of the hand toward the little finger. When the fist is clenched, the heads of the metacarpals become obvious as the "knuckles." Each hand contains 14 phalanges. There are three in each finger (proximal, middle, and distal), except in the thumb, which has only two (proximal and distal).

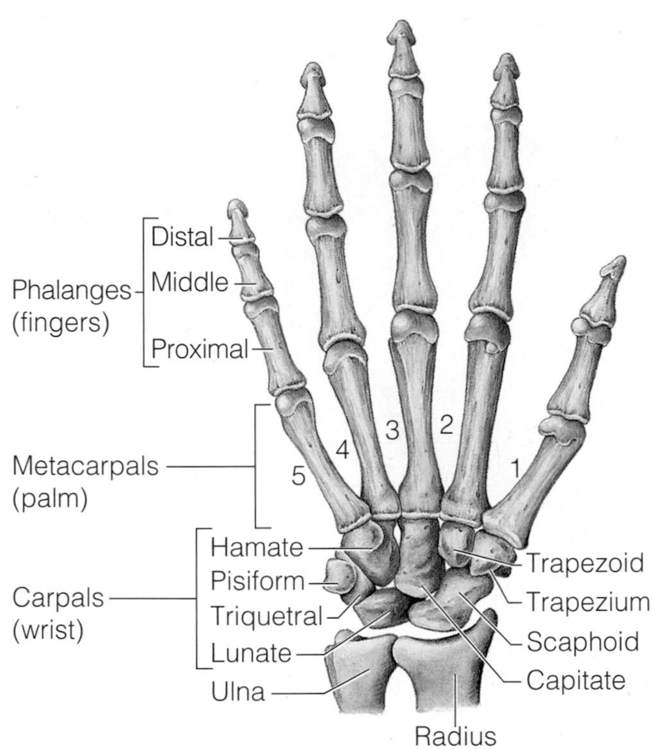

FIGURE 5.22 Bones of the right hand, anterior view.

Bones of the Pelvic Girdle

The **pelvic girdle** is formed by two **coxal** (kok′sal) **bones,** or **ossa coxae,** commonly called **hip bones.** Together with the sacrum and the coccyx, the hip bones form the *bony pelvis* (Figure 5.23). Note that the terms *pelvic girdle* and *pelvis* have slightly different meanings.

The bones of the pelvic girdle are large and heavy, and they are attached securely to the axial skeleton. The sockets, which receive the thigh bones, are deep and heavily reinforced by ligaments that attach the limbs firmly to the girdle. Bearing weight is the most important function of this girdle; the total weight of the upper body rests on the pelvis. The reproductive organs, urinary bladder, and part of the large intestine lie within and are protected by the bony pelvis.

Each hip bone is formed by the fusion of three bones: the *ilium, ischium,* and *pubis.* The **ilium** (il′e-um), which connects posteriorly with the sacrum at the **sacroiliac** (sak″ro-il′e-ac) **joint,** is a large, flaring bone that forms most of the hip bone. When you put your hands on your hips, they are resting over the *alae* or winglike portions of the ilia. The upper edge of the alae, the **iliac crest,** is an important anatomical landmark that is always kept in mind by those who give injections. The iliac crest ends anteriorly in the **anterior superior iliac spine** and posteriorly in the **posterior superior iliac spine.** Small inferior spines are located below these.

The **ischium** (is′ke-um) is the "sit-down bone," since it forms the most inferior part of the coxal bone. The **ischial tuberosity** is a roughened area that receives body weight when you are sitting. The **ischial spine,** superior to the tuberosity, is another important anatomical landmark, particularly in the pregnant woman, because it narrows the outlet of the pelvis through which the baby must pass during the birth process. Another important structural feature of the ischium is the **greater sciatic notch,** which allows blood vessels and the large sciatic nerve to pass from the pelvis posteriorly into the thigh. Injections in the buttock should always be given well away from this area.

The **pubis** (pu′bis), or **pubic bone,** is the most anterior part of a coxal bone. Fusion of the *rami* of the pubis anteriorly and the ischium posteriorly forms a bar of bone enclosing the **obturator** (ob′tu-ra″tor) **foramen,** an opening that allows blood vessels and nerves to pass into the anterior part of the thigh. The pubic bones of each hip bone fuse anteriorly to form a cartilaginous joint, the **pubic symphysis** (pu′bik sim′fĭ-sis).

The ilium, ischium, and pubis fuse at the deep socket called the **acetabulum** (as″ĕ-tab′u-lum), which means "vinegar cup." The acetabulum receives the head of the thigh bone.

The bony pelvis is divided into two regions. The **false pelvis** is superior to the true pelvis; it is the area medial to the flaring portions of the ilia. The **true pelvis** is surrounded by bone and lies inferior to the flaring parts of the ilia and the pelvic brim. The dimensions of the true pelvis of a woman are very important because they must be large enough to allow the infant's head (the largest part of the infant) to pass during childbirth. The dimensions of the cavity, particularly the **outlet** (the inferior opening of the pelvis), and the **inlet** (superior opening) are critical, and thus they are carefully measured by the obstetrician.

Of course, individual pelvic structures vary, but there are fairly consistent differences between a male and a female pelvis. Look at Figure 5.23c and

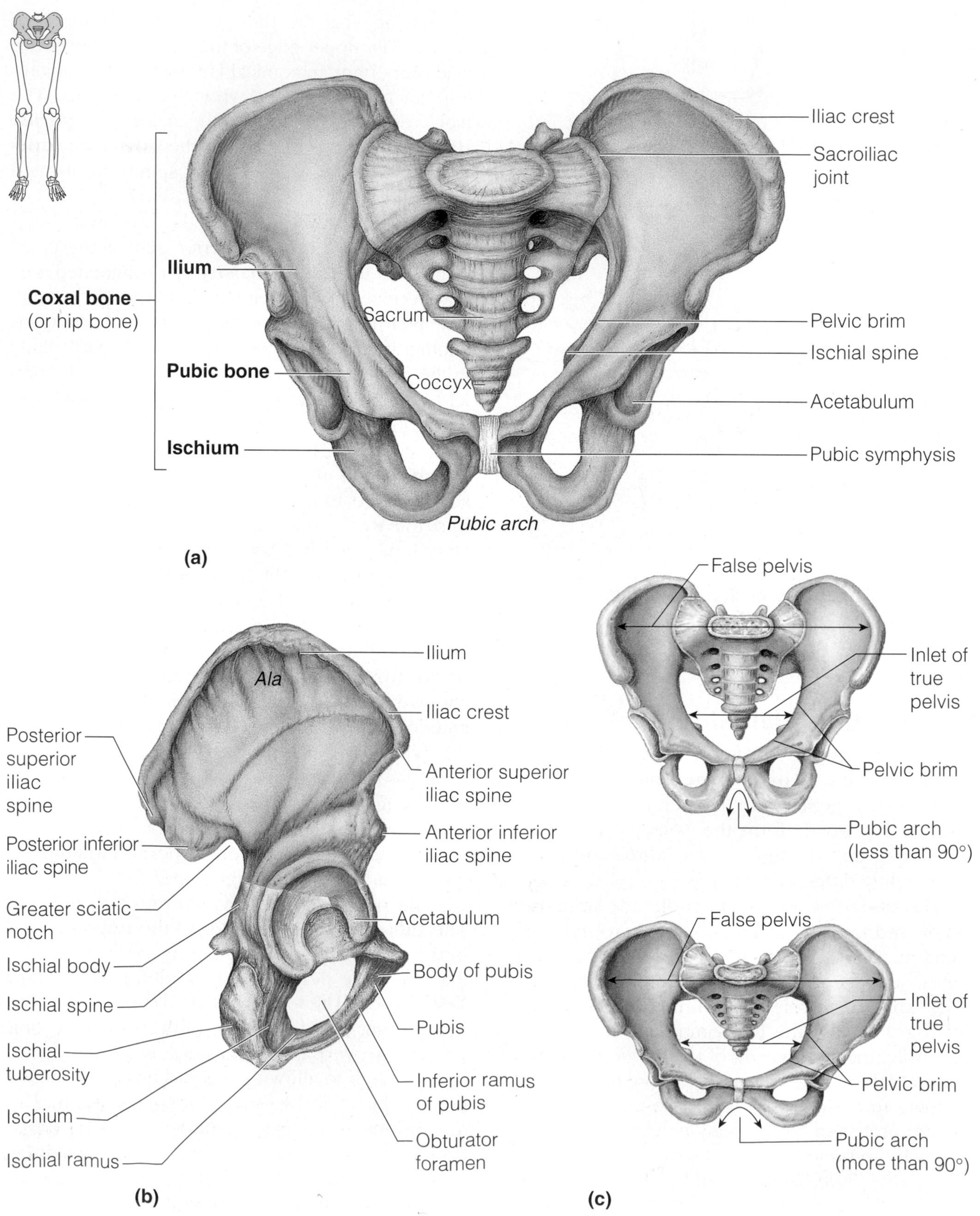

FIGURE 5.23 The pelvis. (a) Articulated male pelvis. **(b)** Right coxal bone, showing the point of fusion of the ilium, ischium, and pubic bones. **(c)** Comparison of the male (above) and female (below) pelves.

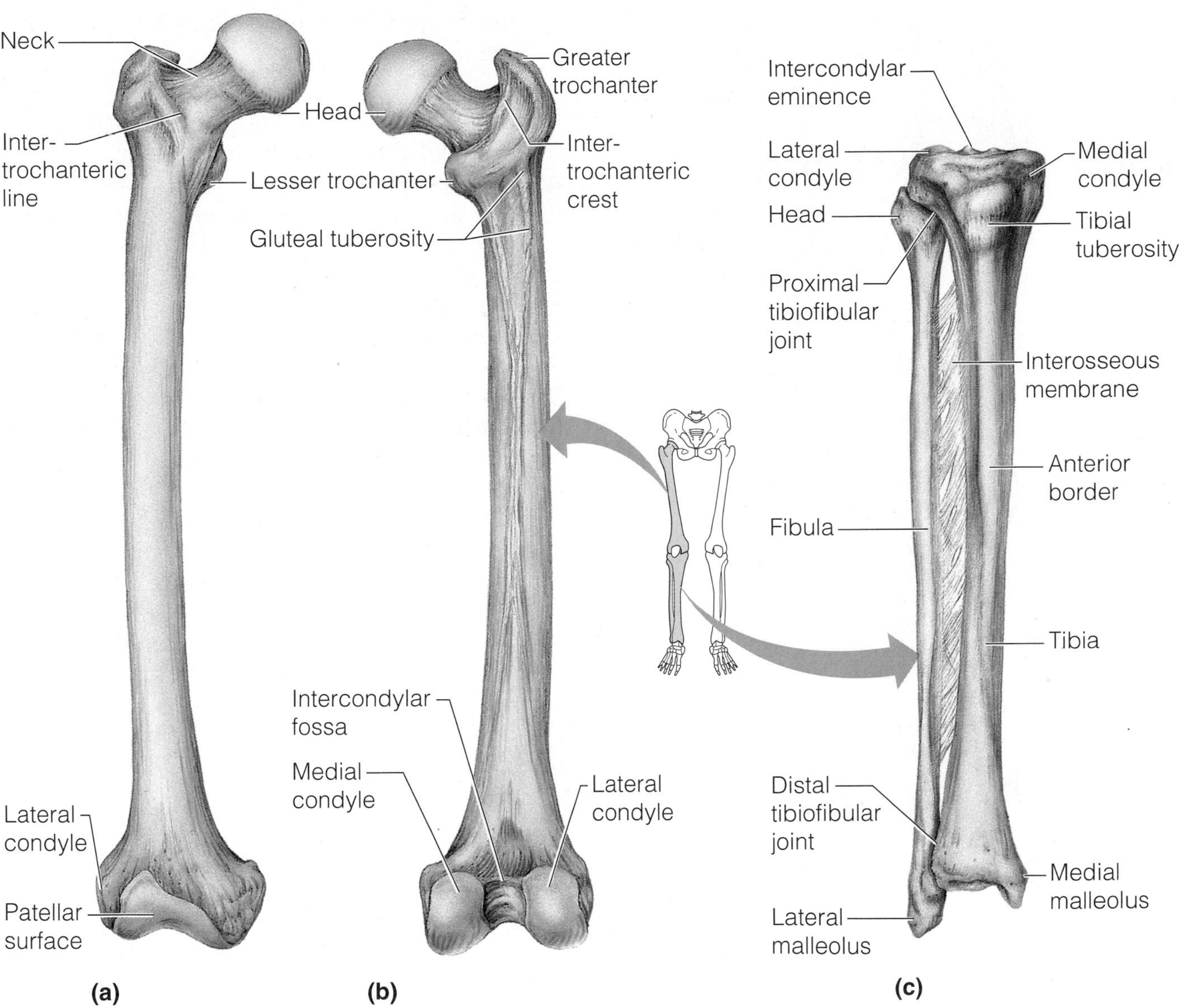

FIGURE 5.24 Bones of the right thigh and leg. (a) Femur (thigh bone), anterior view. **(b)** Femur, posterior view. **(c)** Tibia and fibula of the leg, anterior view.

notice the following characteristics that differ in the pelvis of the male and female.

- The female inlet is larger and more circular.
- The female pelvis as a whole is shallower, and the bones are lighter and thinner.
- The female ilia flare more laterally.
- The female sacrum is shorter and less curved.
- The female ischial spines are shorter and farther apart; thus the outlet is larger.
- The female pubic arch is more rounded because the angle of the pubic arch is greater.

Bones of the Lower Limbs

The lower limbs carry our total body weight when we are erect. Hence, it is not surprising that the bones forming the three segments of the lower limbs (thigh, leg, and foot) are much thicker and stronger than the comparable bones of the upper limb.

Thigh

The **femur** (fe′mur), or *thigh bone,* is the only bone in the thigh (Figure 5.24a and b). It is the

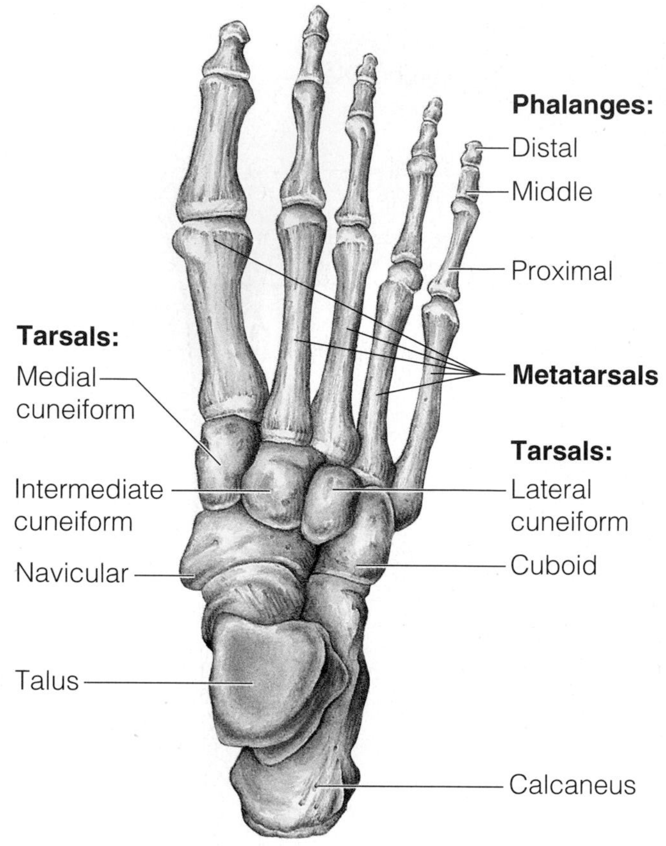

FIGURE 5.25 Bones of the right foot, superior view.

heaviest, strongest bone in the body. Its proximal end has a ball-like head, a neck, and **greater** and **lesser trochanters** (separated anteriorly by the **intertrochanteric line** and posteriorly by the **intertrochanteric crest**). The trochanters, intertrochanteric crest, and the **gluteal tuberosity,** located on the shaft, all serve as sites for muscle attachment. The head of the femur articulates with the acetabulum of the hip bone in a deep, secure socket. However, the neck of the femur is a common fracture site, especially in old age.

The femur slants medially as it runs downward to join with the leg bones; this brings the knees in line with the body's center of gravity. The medial course of the femur is more noticeable in females because of the wider female pelvis. Distally on the femur are the **lateral** and **medial condyles,** which articulate with the tibia below. Posteriorly these condyles are separated by the deep **intercondylar fossa.** Anteriorly on the distal femur is the smooth **patellar surface,** which forms a joint with the patella, or kneecap.

Leg

Connected along their length by an **interosseous membrane,** two bones, the tibia and fibula, form the skeleton of the leg (see Figure 5.24c). The **tibia,** or *shinbone,* is larger and more medial. At the proximal end, the **medial** and **lateral condyles** (separated by the **intercondylar eminence**) articulate with the distal end of the femur to form the knee joint. The patellar (kneecap) ligament attaches to the **tibial tuberosity,** a roughened area on the anterior tibial surface. Distally, a process called the **medial malleolus** (mal-le′o-lus) forms the inner bulge of the ankle. The anterior surface of the tibia is a sharp ridge, the **anterior border,** that is unprotected by muscles; thus, it is easily felt beneath the skin.

The **fibula,** which lies alongside the tibia and forms joints with it both proximally and distally, is thin and sticklike. The fibula has no part in forming the knee joint. Its distal end, the **lateral malleolus,** forms the outer part of the ankle.

Foot

The foot, composed of the tarsals, metatarsals, and phalanges, has two important functions. It supports our body weight and serves as a lever that allows us to propel our bodies forward when we walk and run.

The **tarsus,** forming the posterior half of the foot, is composed of seven **tarsal bones** (Figure 5.25). Body weight is mostly carried by the two largest tarsals, the **calcaneus** (kal-ka′ne-us), or heelbone, and the **talus** (ta′lus; "ankle"), which lies between the tibia and the calcaneus. Five **metatarsals** form the sole, and 14 **phalanges** form the toes. Like the fingers of the hand, each toe has three phalanges, except the great toe, which has two.

The bones in the foot are arranged to form three strong arches: two longitudinal (medial and lateral) and one transverse (Figure 5.26, p. 163). *Ligaments,* which bind the foot bones together, and *tendons* of the foot muscles help to hold the bones firmly in the arched position but still allow a certain amount of give or springiness. Weak arches are referred to as "fallen arches" or "flat feet."

A Closer Look

Them Bones, Them Bones Goin' to Walk Around—Clinical Advances in Bone Repair

ALTHOUGH bones have remarkable self-regenerative powers, some conditions are just too severe for bones to effect repair. Examples include extensive shattering (as in automobile accidents), poor circulation in old bones, and certain birth defects. Here we address healing problems that bones cannot handle themselves. Let's take a look at some techniques currently used to expedite bone repair.

Electrical stimulation of fracture sites dramatically increases the speed and completeness of healing in slowly healing fractures. For years it has been known that bone tissue is deposited in regions of negative electrical charge (its stressed regions) and absorbed in regions of positive charge, but we are still not sure how electricity promotes healing. One theory is that negative fields prevent parathyroid hormone from stimulating the bone-absorbing osteoclast cells at the fracture site. Another theory is that the fields induce production of growth chemicals that stimulate the osteoblasts.

Ultrasound, introduced as the basis of an imaging technique in Chapter 1, can speed the repair of fresh fractures. Daily exposure to pulsed low-power ultrasound waves reduces the healing time of broken arm and shin bones by 35 to 45 percent. It apparently stimulates cartilage cells to make a fibrocartilage callus.

The most troublesome injuries to bones are non-union fractures, in which the two parts of a split bone fail to join. Such fractures traditionally have been treated with grafts, in which sections of bone are taken from the hip and inserted into the gap. However, this requires several painful grafting sessions, and one-third of the grafts fail to heal. A potential improvement is the **free vascular fibular graft technique,** which uses pieces of the fibula to replace missing bone. One reason that traditional grafts often fail is that a blood supply cannot reach their interior. This new technique grafts normal blood vessels along with the bone sections, and subsequent remodeling leads to a good replica of the pre-injury bone.

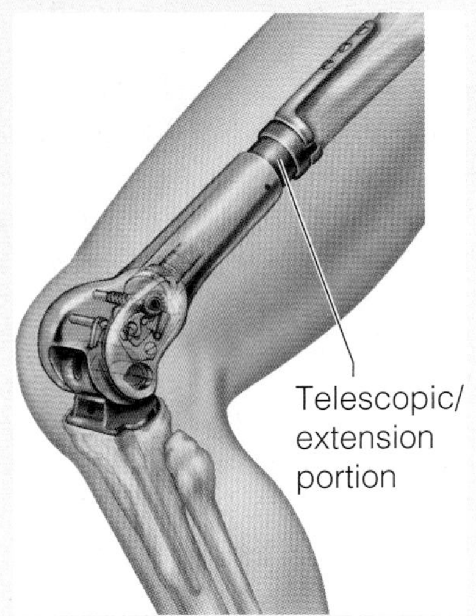

The self-extending leg implant. Lengthening occurs at the telescopic/extension portion.

Although bone implants have proved effective in adults, they have been less satisfactory for children with growing bones. This problem has been partly resolved, ➤

Them Bones, Them Bones Goin' to Walk Around—Clinical Advances in Bone Repair *(continued)*

at least for knee replacement candidates, with **self-extendible endoprostheses.** The telescopic sleeve of these devices (see the figure) undergoes continual automatic elongation of the limb enforced by knee bending. Overlengthening of the prosthesis is prevented by tension in the surrounding tissue which increases after each elongation and then gradually decreases as the soft tissues grow.

Much research has gone into developing **bone substitutes** (crushed cadaver bone or synthetic materials) to fill the gaps in non-union defects. Crushed bone from human cadavers is mixed with water to form a paste that can be molded into the desired shape or packed into small, difficult-to-reach spaces. However, cadaver bone is a foreign tissue that the immune system may reject, and the body sometimes fails to replace it with new bone, as it must for healing to occur. Furthermore, there is a slight but real risk that the cadaver bone contains disease organisms.

Presently several types of artificial bone materials, which can serve as a scaffolding on which new bone can grow, are available. ProOsteon, made from coral, avoids the rejection problems seen with cadaveric bone. The coral is heat-treated to kill its living cells and convert its calcium carbonate to hydroxyapatite, the mineral in bone. The coral graft is then carved to the desired shape, coated with a natural substance that induces bone formation (bone morphogenic protein), and implanted. Osteoblasts and blood vessels migrate from the adjacent natural bone into the coral implant, gradually replacing it with living bone. Research has also produced several types of ceramic bone substitutes. One is TCP, a biodegradable ceramic substance soft enough to be shaped but, like the "coral bone," not very strong. TCP's biggest application has been to replace parts of nonweight-bearing bones, such as skull bones.

Norian SRS, a bone cement made of calcium phosphate, provides immediate structural support to fractured or osteoporotic sites. Mixed at the time of surgery, Norian SRS paste is injected into areas of damaged bone to create an internal "cast." The paste hardens in minutes and cures into a substance with greater compressive strength than spongy bone. Because its crystalline structure is the same as that of natural bone, it continues to provide support as it is gradually remodeled and replaced by host bone. However, Norian SRS can be used only on bone ends because it cannot resist the strong compressive and bending stresses occurring at the central shaft.

Perhaps the most promising product being tested is Osteo-Medica's bioceramic called Megagraft 1000. Made from mildly heated, chemically synthesized hydroxyapatite, bioceramic is stronger than the bone substitutes just described above. Better still, it can be molded and carved into long pieces that are placed in gaps in the shafts of long bones. It essentially works as an artificial bone graft. Thus far in animal tests, it has been shown to bear weight well after several months of initial healing and natural remodeling.

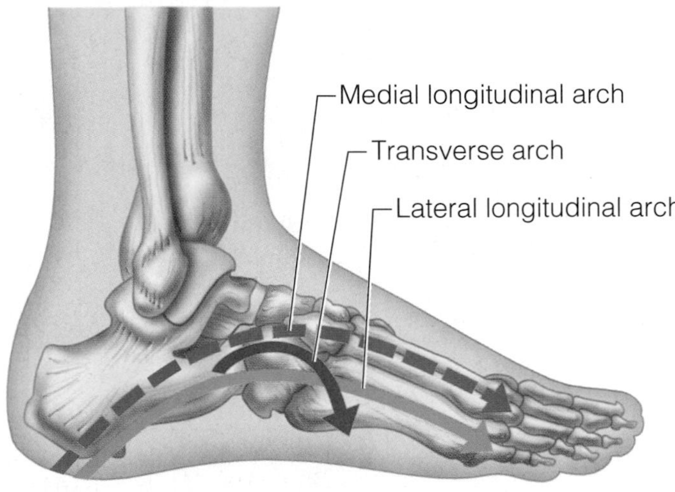

FIGURE 5.26 Arches of the foot.

Joints

With one exception (the hyoid bone of the neck), every bone in the body forms a joint with at least one other bone. **Joints,** also called **articulations,** have two functions: they hold the bones together securely but also give the rigid skeleton mobility.

The graceful movements of a ballet dancer and the rough-and-tumble grapplings of a football player illustrate the great variety of motion allowed by joints, the sites where two or more bones meet. With fewer joints, we would move like robots. Nevertheless, the bone-binding function of joints is just as important as their role in providing mobility. The immovable joints of the skull, for instance, form a snug enclosure for our vital brain.

Joints are classified in two ways—functionally and structurally. The functional classification focuses on the amount of movement allowed by the joint. On this basis, there are **synarthroses** (sin″ar thro′sēz) or immovable joints, **amphiarthroses** (am″fe ar thro′sēz) or slightly movable joints, and **diarthroses** (di″ar thro′sēz) or freely movable joints. Freely movable joints predominate in the limbs, where mobility is important. Immovable and slightly movable joints are restricted mainly to the axial skeleton, where firm attachments and protection of internal organs are priorities.

Structurally, there are *fibrous, cartilaginous,* and *synovial joints* based on whether fibrous tissue, cartilage, or a joint cavity separates the bony regions at the joint. As a general rule, fibrous joints are immovable and synovial joints are freely movable. Although cartilaginous joints have both immovable and slightly movable examples, most are amphiarthrotic. Since the structural classification is more clear-cut, we will focus on that classification scheme here. The joint types are shown in Figure 5.27 and described next.

Fibrous Joints

In **fibrous joints,** the bones are united by fibrous tissue. The best examples of this type of joint are the *sutures* of the skull (Figure 5.27a). In sutures, the irregular edges of the bones interlock and are bound tightly together by connective tissue fibers, allowing essentially no movement to occur. In **syndesmoses** (sin-dez-mo′sēz), the connecting fibers are longer than those of sutures; thus the joint has more "give." The joint connecting the distal ends of the tibia and fibula is a syndesmosis (Figure 5.27b).

Cartilaginous Joints

In **cartilaginous joints,** the bone ends are connected by cartilage. Examples of this joint type that are slightly movable (amphiarthrotic) are the *pubic symphysis* of the pelvis (Figure 5.27e) and *intervertebral joints* of the spinal column (Figure 5.27d), where the articulating bone surfaces are connected by pads (discs) of fibrocartilage. The hyaline-cartilage epiphyseal plates of growing long bones and the cartilaginous joints between the first ribs and the sternum are immovable (synarthrotic) cartilaginous joints (Figure 5.27c).

Synovial Joints

Synovial joints are those in which the articulating bone ends are separated by a joint cavity containing synovial fluid (see Figure 5.27f–h). They account for all joints of the limbs.

All synovial joints have four distinguishing features (Figure 5.28):

- **Articular cartilage.** Articular (hyaline) cartilage covers the ends of the bones forming the joint.
- **Fibrous articular capsule.** The joint surfaces are enclosed by a sleeve or capsule of fibrous connective tissue, and the capsule is lined with a smooth *synovial membrane* (the reason these joints are called synovial joints).
- **Joint cavity.** The articular capsule encloses a cavity, called the joint cavity, which contains lubricating synovial fluid.

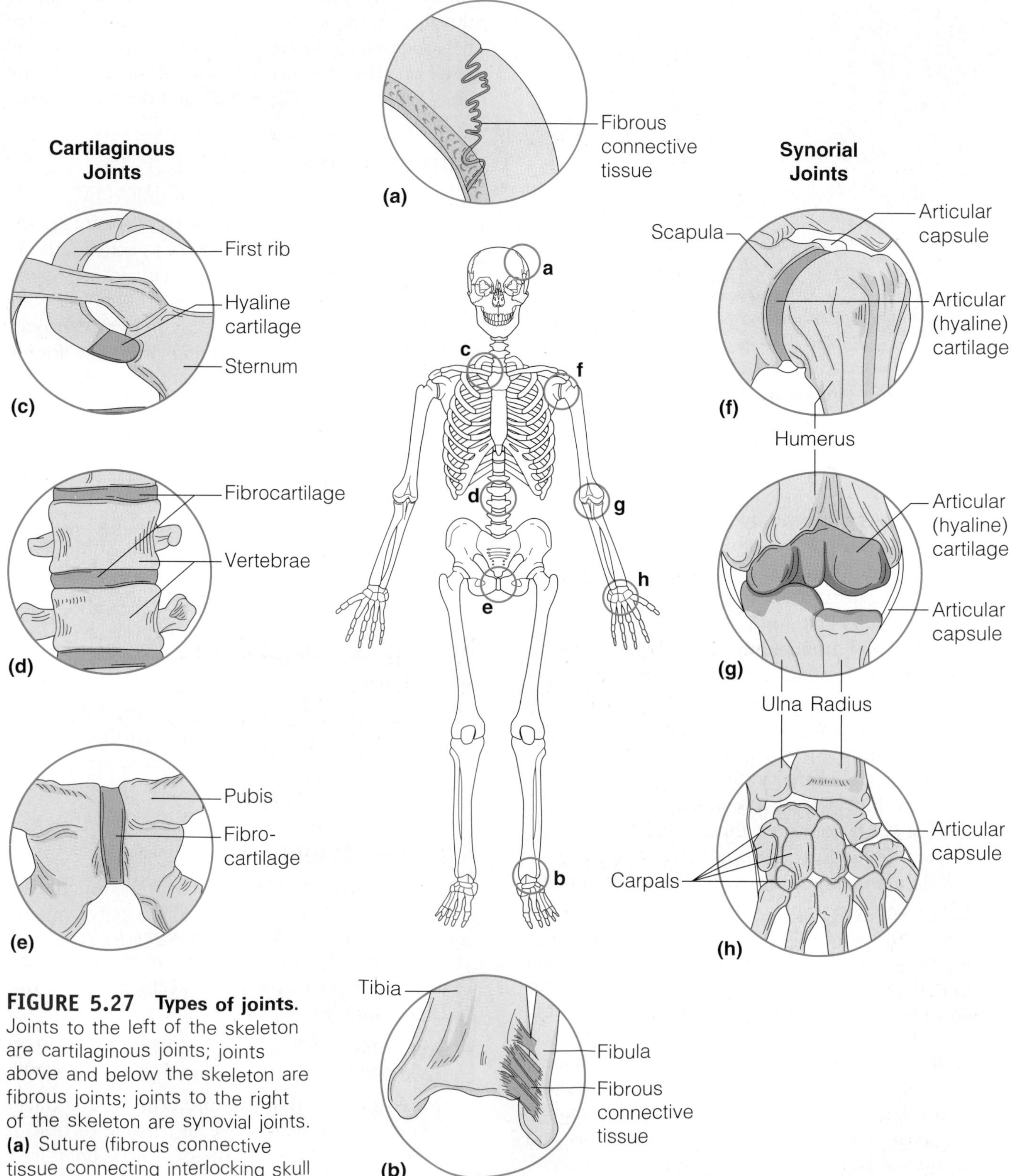

FIGURE 5.27 Types of joints. Joints to the left of the skeleton are cartilaginous joints; joints above and below the skeleton are fibrous joints; joints to the right of the skeleton are synovial joints. **(a)** Suture (fibrous connective tissue connecting interlocking skull bones). **(b)** Syndesmosis (fibrous connective tissue connecting the distal ends of the tibia and fibula). **(c)** Synchondrosis (joint between costal cartilage of rib 1 and the sternum). **(d)** Symphyses (intervertebral discs of fibrocartilage connecting adjacent vertebrae). **(e)** Symphysis (fibrocartilaginous pubic symphysis connecting the pubic bones anteriorly). **(f)** Synovial joint (multiaxial shoulder joint). **(g)** Synovial joint (uniaxial elbow joint). **(h)** Synovial joints (biaxial intercarpal joints of the hand).

Q *How does this joint type differ structurally from cartilaginous and fibrous joints?*

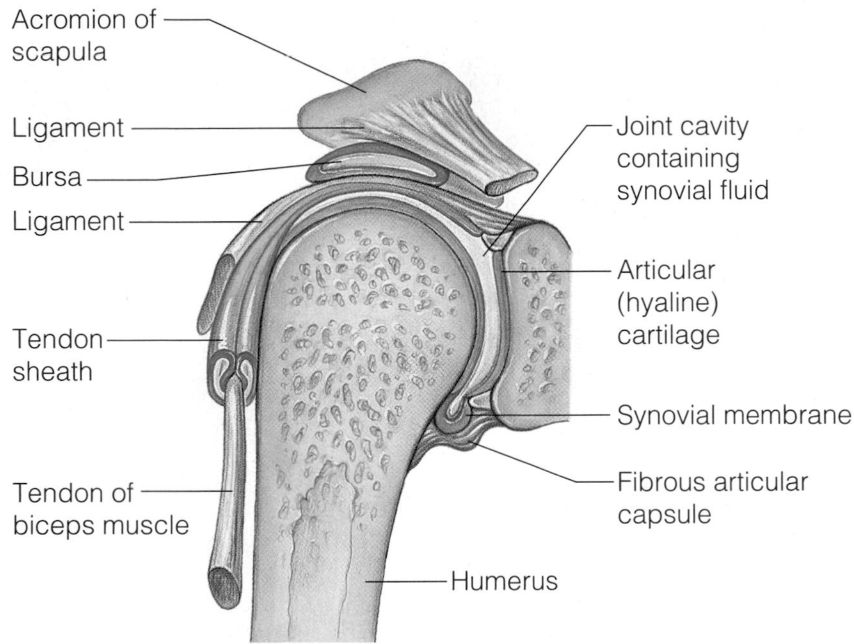

FIGURE 5.28 General structure of a synovial joint.

- **Reinforcing ligaments.** The fibrous capsule is usually reinforced with ligaments.

Bursae and tendon sheaths are not strictly part of synovial joints, but they are often found closely associated with them (Figure 5.28). Essentially bags of lubricant, they act like ball bearings to reduce friction between adjacent structures during joint activity. **Bursae** (ber′se; "purse") are flattened fibrous sacs lined with synovial membrane and containing a thin film of synovial fluid. They are common where ligaments, muscles, skin, tendons, or bones rub together. A **tendon sheath,** also shown in Figure 5.28, is essentially an elongated bursa that wraps completely around a tendon subjected to friction, like a bun around a hot dog.

Homeostatic Imbalance

A *dislocation* happens when a bone is forced out of its normal position in the joint cavity. The process of returning the bone to its proper position, called *reduction,* should be done only by a physician. Attempts by an untrained person to "snap the bone back into its socket" are usually more harmful than helpful. ▲

A *It has a joint cavity instead of cartilage or fibrous tissue separating the articulating bones.*

Types of Synovial Joints Based on Shape

The shapes of the articulating bone surfaces determine what movements are allowed at a joint. Based on such shapes, our synovial joints can be classified as *plane, hinge, pivot, condyloid, saddle,* and *ball-and-socket joints* (Figure 5.29).

- In a **plane joint** (Figure 5.29a), the articular surfaces are essentially flat, and only short slipping or gliding movements are allowed. The movements of plane joints are *nonaxial;* that is, gliding does not involve rotation around any axis. The intercarpal joints of the wrist are the best examples of plane joints.
- In a **hinge joint** (Figure 5.29b), the cylindrical end of one bone fits into a trough-shaped surface on another bone. Angular movement is allowed in just one plane, like a mechanical hinge. Examples are the elbow joint, ankle joint, and the joints between the phalanges of the fingers. Hinge joints are classified as *uniaxial* (u″ne-aks′e-al; "one axis"); they allow movement around one axis only, as indicated by the single magenta arrow in Figure 5.29b.

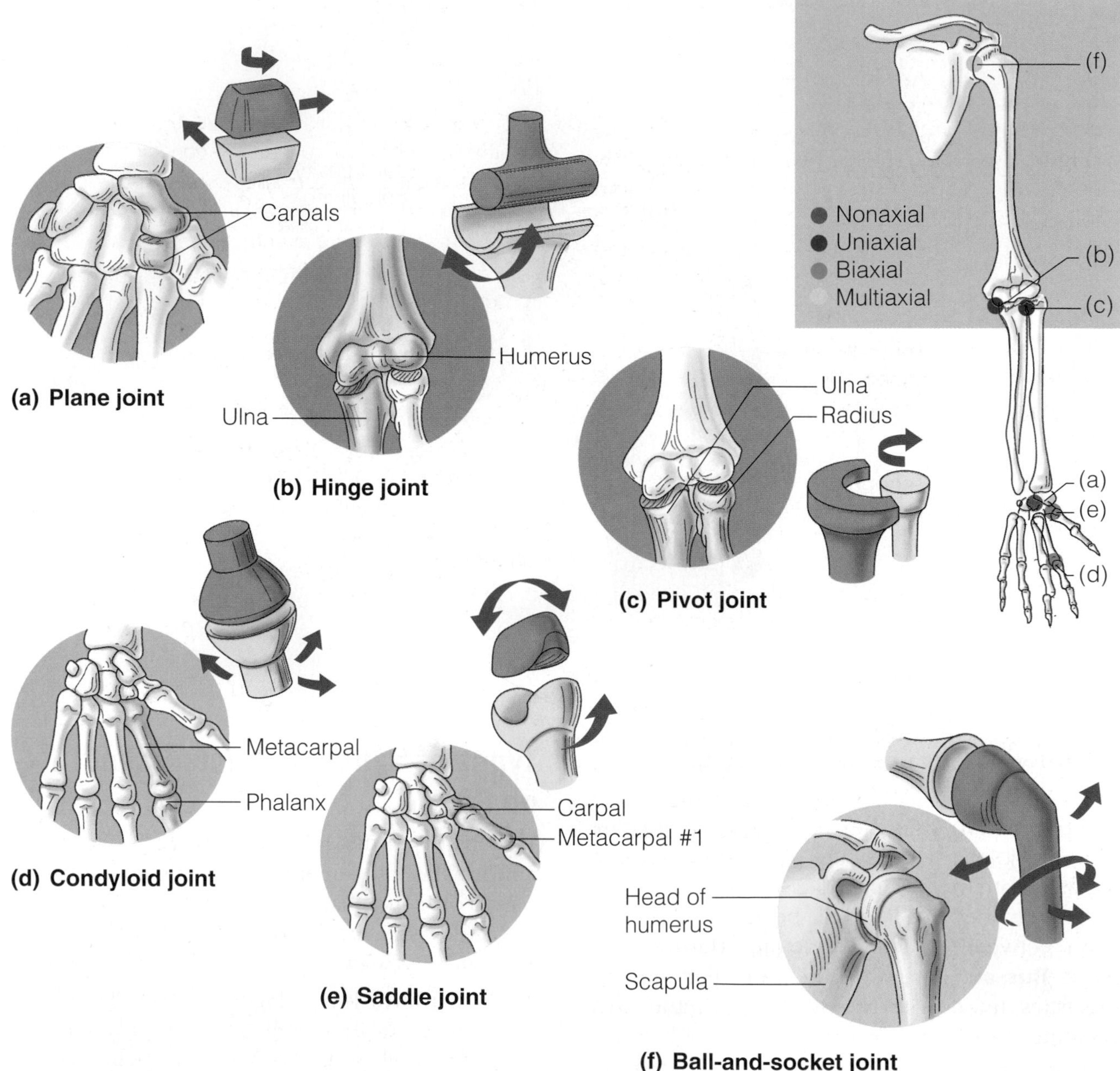

FIGURE 5.29 Types of synovial joints. (a) Plane joint (intercarpal and intertarsal joints). **(b)** Hinge joint (elbow and interphalangeal joints). **(c)** Pivot joint (proximal joint between the radius and the ulna). **(d)** Condyloid joint (knuckles). **(e)** Saddle joint (carpometacarpal joint of the thumb). **(f)** Ball-and-socket joint (shoulder and hip joints).

- In a **pivot joint** (Figure 5.29c), the rounded end of one bone fits into a sleeve or ring of bone (and possibly ligaments). Because the rotating bone can turn only around its long axis, pivot joints are also uniaxial joints (see the single arrow in Figure 5.29c). The proximal radioulnar joint and the joint between the atlas and the dens of the axis are examples.
- In a **condyloid joint** (kon′dĭ-loid; "knuckle-like"), the egg-shaped articular surface of one bone fits into an oval concavity in another (Figure 5.29d). Both of these articular surfaces are oval. Condyloid joints allow the moving bone to travel (1) from side to side and (2) back and forth, but the bone cannot rotate around its

long axis. Movement occurs around two axes, hence these joints are *biaxial* (*bi* = two), as in knuckle (metacarpophalangeal) joints.

- In **saddle joints,** each articular surface has both convex and concave areas, like a saddle (Figure 5.29e). These biaxial joints allow essentially the same movements as condyloid joints. The best examples of saddle joints are the carpometacarpal joints in the thumb, and the movements of these joints are clearly demonstrated by twiddling your thumbs.
- In a **ball-and-socket joint** (Figure 5.29f), the spherical head of one bone fits into a round socket in another. These *multiaxial* joints allow movement in all axes, including rotation (see the three arrows in Figure 5.29f), and are the most freely moving synovial joints. The shoulder and hip are examples.

The various types of movements that occur at synovial joints are discussed in detail in the next chapter, because they relate to muscle activity.

Homeostatic Imbalance of Joints

Few of us pay attention to our joints unless something goes wrong with them. Joint pain and inflammation may be caused by many things. For example, falling on one's knee can cause a painful *bursitis,* called "water on the knee," due to inflammation of bursae or synovial membrane. Sprains and dislocations are other types of joint problems that result in swelling and pain. In a *sprain,* the ligaments or tendons reinforcing a joint are damaged by excessive stretching, or they are torn away from the bone. Since both tendons and ligaments are cords of dense fibrous connective tissue with a poor blood supply, sprains heal slowly and are extremely painful.

Few inflammatory joint disorders cause more pain and suffering than arthritis. The term **arthritis** (*arth* = joint; *itis* = inflammation) describes over 100 different inflammatory or degenerative diseases that damage the joints. In all its forms, arthritis is the most widespread, crippling disease in the United States. One out of seven Americans suffers its ravages. All forms of arthritis have the same initial symptoms: pain, stiffness, and swelling of the joint. Then, depending on the specific form, certain changes in the joint structure occur.

Acute forms of arthritis usually result from bacterial invasion and are treated with antibiotic drugs. The synovial membrane thickens and fluid production decreases, leading to increased friction and pain. Chronic forms of arthritis include osteoarthritis, rheumatoid arthritis, and gouty arthritis, which differ substantially in their later symptoms and consequences. We will focus on these forms here.

Osteoarthritis (OA), the most common form of arthritis, is a chronic degenerative condition that typically affects the aged. OA, also called "wear-and-tear arthritis," affects the articular cartilages. Over the years, there is a softening, fraying, and eventual breakdown of the cartilage. As the disease progresses, the exposed bone thickens and extra bone tissue, called *bone spurs,* grows around the margins of the eroded cartilage and restricts joint movement. Patients complain of stiffness on arising that lessens with activity, and the affected joints may make a crunching noise (*crepitus*) when moved. The joints most commonly affected are those of the fingers, the cervical and lumbar joints of the spine, and the large, weight-bearing joints of the lower limbs (knees and hips).

The course of osteoarthritis is usually slow and irreversible, but it is rarely crippling. In most cases, its symptoms are controllable with a mild analgesic such as aspirin, moderate activity to maintain joint mobility, and rest when the joint becomes very painful. Some people with OA claim that rubbing capsaicin (a hot pepper extract) on the skin over painful joints provides relief. Others swear to the pain-reducing ability of glucosamine sulfate, a nutritional supplement.

Rheumatoid (roo'mah-toid) **arthritis (RA)** is a chronic inflammatory disorder. Its onset is insidious and usually occurs between the ages of 40 and 50, but it may occur at any age. It affects three times as many women as men. Many joints, particularly those of the fingers, wrists, ankles, and feet, are affected at the same time and usually in a symmetrical manner. For example, if the right elbow is affected, most likely the left elbow will be affected also. The course of RA varies and is marked by remissions and flare-ups (*rheumat* = susceptible to change or flux).

RA is an autoimmune disease—a disorder in which the body's immune system attempts to destroy its own tissues. The initial trigger for this reaction is unknown, but some suspect that it results from certain bacterial or viral infections.

RA begins with inflammation of the synovial membranes. The membranes thicken and the joints swell as synovial fluid accumulates. Inflammatory cells (white blood cells and others) enter the joint cavity from the blood and release a deluge of inflammatory chemicals that destroy body tissues when released inappropriately as in RA. In time the inflammed synovial membrane

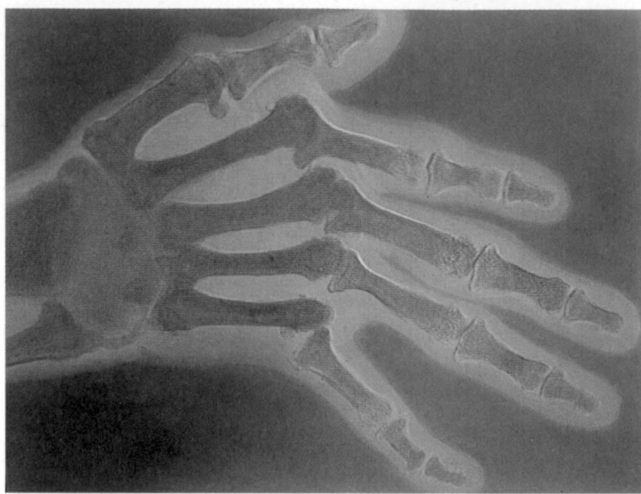

FIGURE 5.30 X ray of a hand deformed by rheumatoid arthritis.

thickens into a *pannus* ("rag"), an abnormal tissue that clings to and erodes articular cartilages. As the cartilage is destroyed, scar tissue forms and connects the bone ends. The scar tissue eventually ossifies, and the bone ends become firmly fused (*ankylosis*) and often deformed (see Figure 5.30). Not all cases of RA progress to the severely crippling ankylosis stage, but all cases involve restricted joint movement and extreme pain.

Current therapy for RA involves many different kinds of drugs. Some, like methotrexate and cyclosporin, are powerful drugs that can neutralize the inflammatory chemicals in the joint space and (hopefully) prevent joint deformity. However, drug therapy is usually begun with aspirin, which in large doses is an effective anti-inflammatory agent. Exercise is recommended to maintain as much joint mobility as possible. Cold packs are used to relieve the swelling and pain, and heat helps to relieve morning stiffness. Replacement joints or bone removal are the last resort for severely crippled RA patients.

Gouty (gow'te) **arthritis,** or **gout,** is a disease in which uric acid (a normal waste product of nucleic acid metabolism) accumulates in the blood and may be deposited as needle-shaped crystals in the soft tissues of joints. This leads to an agonizingly painful attack that typically affects a single joint, often in the great toe. Gout is most common in males and rarely appears before the age of thirty. It tends to run in families, so genetic factors are definitely implicated.

Untreated gout can be very destructive; the bone ends fuse and the joint becomes immobilized. Fortunately, several drugs (colchicine, ibuprofen, and others) are successful in preventing acute gout attacks. Patients are advised to lose weight if obese, to avoid foods such as liver, kidneys, and sardines, which are high in nucleic acids, and to avoid alcohol, which inhibits excretion of uric acid by the kidneys. ▲

Developmental Aspects of the Skeleton

As described earlier, the first "long bones" in the very young fetus are formed of hyaline cartilage, and the earliest "flat bones" of the skull are actually fibrous membranes. As the fetus develops and grows, both the flat and the long bone models are converted to bone (Figure 5.31). At birth, some fontanels still remain in the skull to allow for brain growth, but these areas are usually fully ossified by 2 years of age. By the end of adolescence, the epiphyseal plates of long bones that provide for longitudinal growth in childhood have become fully ossified, and long-bone growth ends.

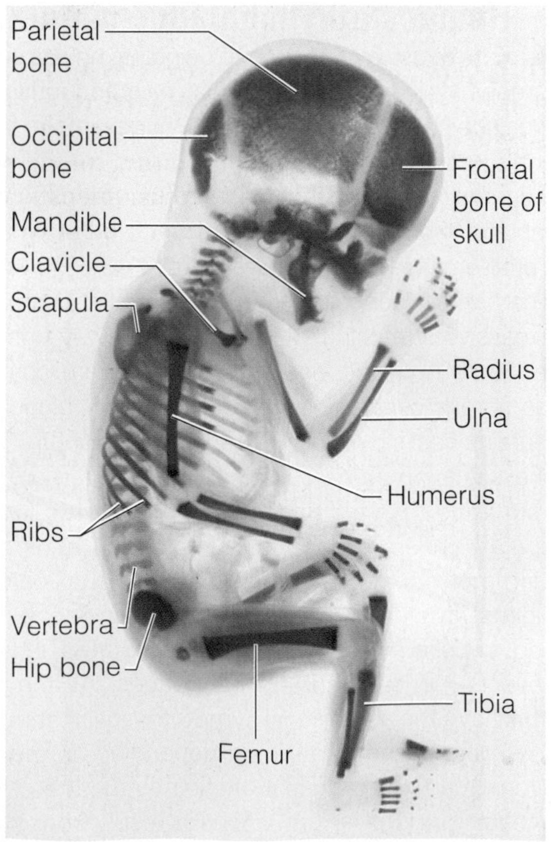

FIGURE 5.31 Ossification centers in the skeleton of a 12-week-old fetus are indicated by the darker areas. Lighter regions are still fibrous or cartilaginous.

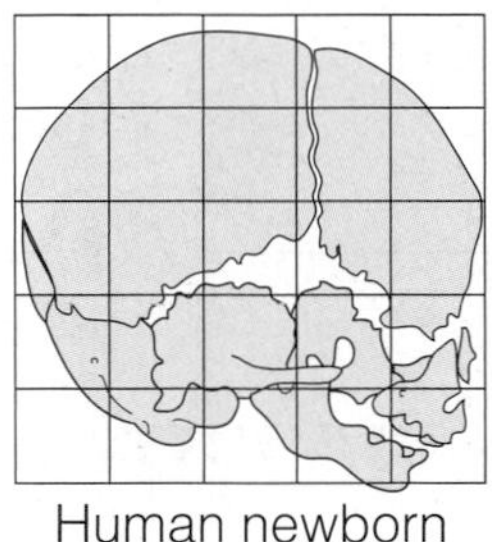

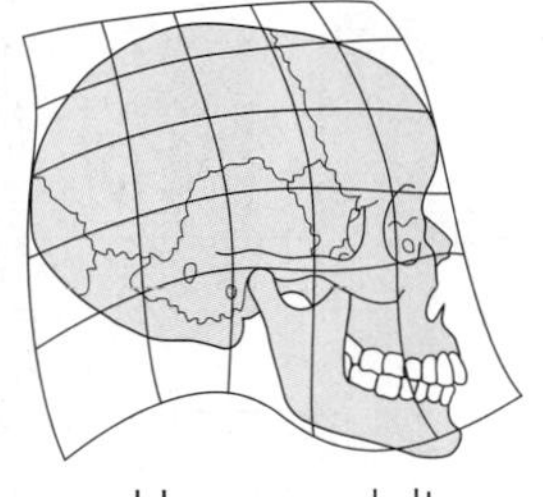

(a)

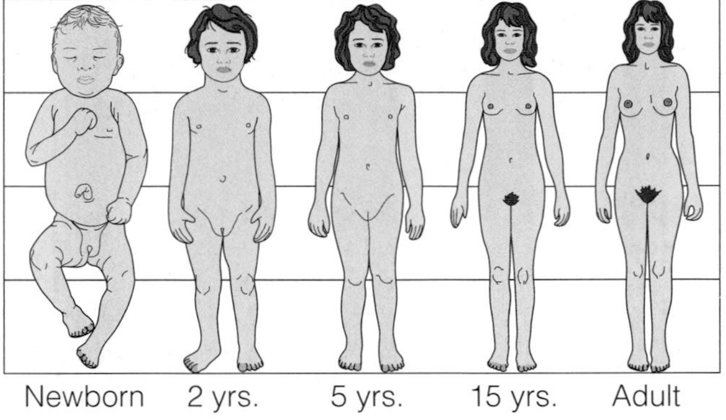

(b)

FIGURE 5.32 Differences in the growth rates for some parts of the body compared to others determine body proportions. (a) Differential growth transforms the rounded, foreshortened skull of a newborn to the sloping skull characteristic of adults. **(b)** During growth of a human, the arms and legs grow faster than the head and trunk, as can be seen in this conceptualization of different-aged individuals all drawn at the same height.

The skeleton changes throughout life, but the changes in childhood are most dramatic. At birth, the baby's cranium is huge relative to its face (Figure 5.32). The rapid growth of the cranium before and after birth is related to the growth of the brain. By 2 years, the skull is three-quarters of its adult size; and, by 8 to 9 years, the skull is almost of adult size and proportions. However, between the ages of 6 and 11, the head appears to enlarge substantially as the face literally grows out from the skull. The jaws increase in size, and the cheekbones and nose become more prominent as respiratory passages expand and the permanent teeth develop.

The so-called primary curvatures of the vertebral column are present at birth and are convex posteriorly, so an infant's spine is arched, like that of a four-legged animal. The secondary curvatures are convex anteriorly and are associated with a child's later development. They result from reshaping of the intervertebral discs rather than from modifications of the bony vertebrae and produce the S-shaped spine typical of the adult.

Most cases of abnormal spinal curvatures, such as scoliosis and lordosis (see Figure 5.15), are congenital, but some can result from injuries. The abnormal curvatures are usually treated by surgery, braces, or casts when diagnosed. Generally speaking, young, healthy people have no skeletal problems, assuming that their diet is nutritious and they stay reasonably active.

During youth, growth of the skeleton not only increases overall body height and size, but changes body proportions as well (Figure 5.32). At birth, the UL (upper/lower) ratio is 1.7:1—that is, the head and trunk are approximately 1½ times as long as the lower limbs. The lower limbs grow more rapidly than the trunk from this time on, and by the age of 10, the UL ratio is approximately 1:1 and changes little thereafter. During puberty, the female pelvis broadens in preparation for childbearing, and the entire male skeleton becomes more robust. Once adult height is reached, a healthy skeleton changes very little until late middle age.

Prove It Yourself

Demonstrate That Joints Are Stabilized by Muscles and Ligaments

To demonstrate the role that ligaments, tendons, and muscles play in stabilizing a joint, try this simple experiment. Sit in a low chair, stretch one leg in front of you with your heel resting on the floor, and relax your muscles. Try to move your kneecap (patella) from side to side gently with your hand. Notice how easily you can shift it out of position. Now, without changing position, tense the muscle of your thigh. Can you still move your patella?

The patella is attached to the tibia by a ligament and to the muscles of the thigh by a tendon. When you contract your thigh muscles, the contraction puts tension on the tendon and the ligament, and the increased tension holds the patella firmly in place. Now move your hand to just below the kneecap. Notice the tightening of the patellar ligament as you alternately contract and relax your thigh muscle.

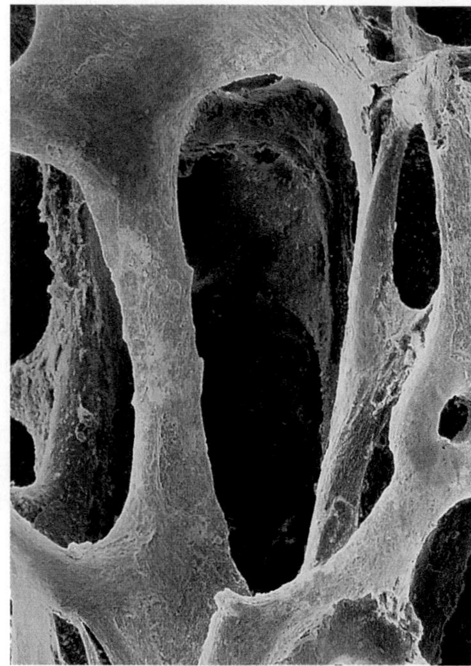

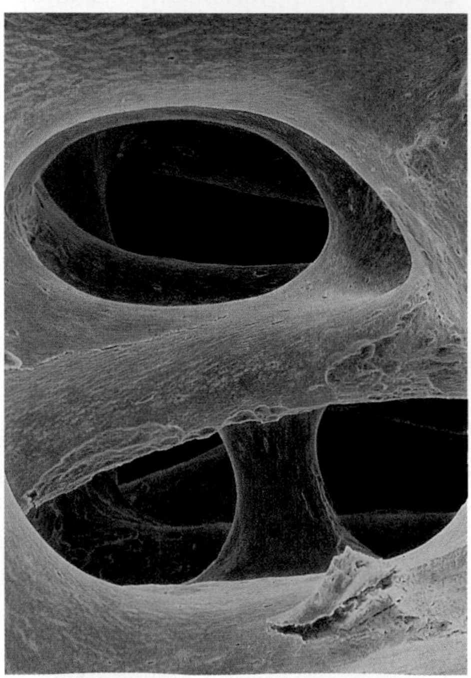

FIGURE 5.33 Osteoporosis. The architecture of osteoporotic bone at top, is contrasted with that of normal bone below.

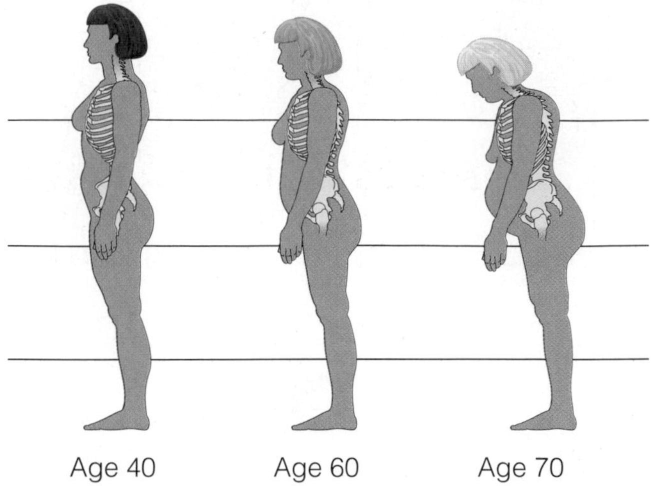

FIGURE 5.34 Vertebral collapse due to osteoporosis. Women with postmenopausal osteoporosis are at risk for fractures in the bones in their spine (vertebrae) as they age. Eventually these vertebrae tend to collapse, causing spinal curvature. Such curvature causes loss of height, a tilted rib cage, a dowager's hump, and a protruding abdomen.

It cannot be emphasized too strongly that bones have to be physically stressed to remain healthy. When we remain active physically and muscles and gravity pull on the skeleton, the bones respond by becoming stronger. On the other hand, if we are totally inactive, they become thin and fragile. **Osteoporosis** is a bone-thinning disease that afflicts half of women over 65 and some 20 percent of men over the age of 70. Osteoporosis makes the bones so fragile that even a hug or a sneeze can cause bones to fracture (Figure 5.33). The bones of the spine and the neck of the femur are particularly susceptible. Vertebral collapse frequently results in a hunched-over posture (kyphosis) familiarly known as dowager's hump (Figure 5.34).

Estrogen helps to maintain the health and normal density of a woman's skeleton, and the estrogen deficiency that occurs after a woman goes through menopause ("change of life" when menstruation ceases) is strongly implicated as a cause of osteoporosis. Other factors that may contribute to osteoporosis are a diet poor in calcium and protein, lack of vitamin D, smoking, and insufficient weight-bearing exercise to stress the bones. Sadly, many elderly people feel that they are helping themselves by "saving their strength" and not doing anything too physical. Their reward for this is *pathologic fractures* (spontaneous breaks without apparent injury), which increase dramatically with age and are the single most common skeletal problem for this age group.

Advancing years also take their toll on joints. Weight-bearing joints in particular begin to degenerate and *osteoarthritis* is common. Such degenerative joint changes lead to the complaint often heard from the aging person: "My joints are getting so stiff. . . ."

Systems in Sync

Homeostatic Relationships between the Skeletal System and Other Body Systems

Endocrine System
- Skeletal system provides some bony protection
- Hormones regulate uptake and release of calcium from bone; hormones promote long-bone growth and maturation

Lymphatic System/Immunity
- Skeletal system provides some protection to lymphatic organs; lymphocytes involved in immune response originate in bone marrow
- Lymphatic system drains leaked tissue fluids; immune cells protect against pathogens

Digestive System
- Skeletal system provides some bony protection to intestines, pelvic organs, and liver
- Digestive system provides nutrients needed for bone health and growth

Urinary System
- Skeletal system protects pelvic organs (bladder, etc.)
- Urinary system activates vitamin D; disposes of nitrogenous wastes

Muscular System
- Skeletal system provides levers plus calcium for muscle activity
- Muscle pull on bones increases bone strength and viability; helps determine bone shape

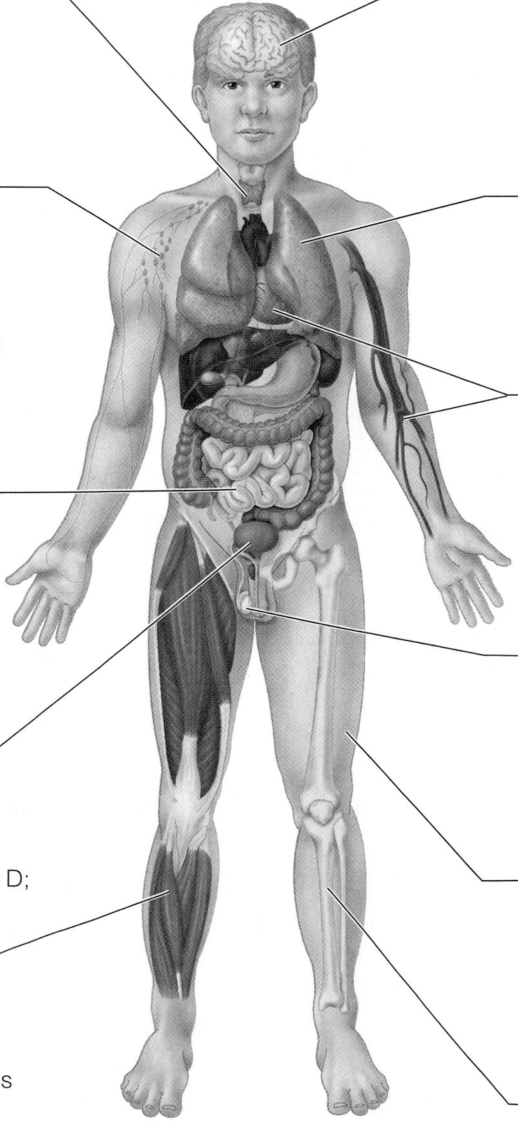

Nervous System
- Skeletal system protects brain and spinal cord; depot for calcium ions needed for neural function
- Nerves innervate bone and joint capsules, providing for pain and joint sense

Respiratory System
- Skeletal system (rib cage) protects lungs by enclosure
- Respiratory system provides oxygen; disposes of carbon dioxide

Cardiovascular System
- Bone marrow cavities provide site for blood cell formation; matrix stores calcium needed for cardiac muscle activity
- Cardiovascular system delivers nutrients and oxygen to bones; carries away wastes

Reproductive System
- Skeletal system protects some reproductive organs by enclosure
- Gonads produce hormones that influence form of skeleton and epiphyseal closure

Integumentary System
- Skeletal system provides support for body organs including the skin
- Skin provides vitamin D needed for proper calcium absorption and use

Skeletal System

SUMMARY

Media study tools that could provide you with additional help in reviewing specific key topics of Chapter 5 are referenced below.

IP = Interactive Physiology;

WEB = A&P Place website.

BONES: AN OVERVIEW (pp. 130–139)

1. Bones support and protect body organs; serve as levers for the muscles to pull on to cause movement at joints; store calcium, fats, and other substances for the body; and contain red marrow, the site of blood cell production.
2. Bones are classified into four groups—long, short, flat, and irregular—on the basis of their shape and the amount of compact or spongy bone they contain. Bone markings are important anatomical landmarks that reveal where muscles attach and where blood vessels and nerves pass.
3. A long bone is composed of a shaft (diaphysis) with two ends (epiphyses). The shaft is compact bone; its cavity contains yellow marrow. The epiphyses are covered with hyaline cartilage; they contain spongy bone (where red marrow is found).

   WEB Exercise: Chapter 5, Microscopic Structure of Compact Bone.
4. The organic parts of the matrix make bone flexible; calcium salts deposited in the matrix make bone hard.
5. Bones form on hyaline cartilage "models" or fibrous membranes. Eventually these initial supporting structures are replaced by bone tissue. Epiphyseal plates persist to provide for longitudinal growth of long bones during childhood and become inactive when adolescence ends.
6. Bones change in shape throughout life. This remodeling occurs in response to hormones (for example, PTH, which regulates blood calcium levels) and mechanical stresses acting on the skeleton.
7. A fracture is a break in a bone. Common types of fractures include simple, compound, compression, comminuted, and greenstick. Bone fractures must be reduced to heal properly.

   WEB Exercise: Chapter 5, Common Types of Fractures.

AXIAL SKELETON (pp. 139–153)

1. The skull is formed by cranial and facial bones. Eight cranial bones protect the brain: frontal, occipital, ethmoid, and sphenoid bones, and the pairs of parietal and temporal bones. The 14 facial bones are all paired (maxillae, zygomatics, palatines, nasals, lacrimals, and inferior nasal conchae), except for the vomer and mandible. The hyoid bone, not really a skull bone, is supported in the neck by ligaments.

   WEB Exercise: Chapter 5, Facial Bones.
2. Skulls of newborns contain fontanels (membranous areas), which allow brain growth. The infant's facial bones are very small compared to the size of the cranium.
3. The vertebral column is formed from 24 vertebrae, the sacrum, and the coccyx. There are 7 cervical vertebrae, 12 thoracic vertebrae, and 5 lumbar vertebrae, which have common as well as unique features. The vertebrae are separated by fibrocartilage discs that allow the vertebral column to be flexible. The vertebral column is S-shaped to allow for upright posture. Primary spinal curvatures present at birth are the thoracic and sacral curvatures; secondary curvatures (cervical and lumbar) develop after birth.

   WEB Exercise: Chapter 5, Typical Vertebra.
4. The bony thorax is formed from the sternum and 12 pairs of ribs. All ribs attach posteriorly to thoracic vertebrae. Anteriorly, the first 7 pairs attach directly to the sternum (true ribs); the last 5 pairs attach indirectly or not at all (false ribs). The bony thorax encloses the lungs, heart, and other organs of the thoracic cavity.

APPENDICULAR SKELETON (pp. 153–160)

1. The shoulder girdle, composed of two bones—the scapula and the clavicle—attaches the upper limb to the axial skeleton. It is a light, poorly reinforced girdle that allows the upper limb a great deal of freedom.

   WEB Exercise: Chapter 5, Articulations Case Study.
2. The bones of the upper limb include the humerus of the arm, the radius and ulna of the forearm, and the carpals, metacarpals, and phalanges of the hand.
3. The pelvic girdle is formed by the two coxal bones, or hip bones. Each hip bone is the result of fusion of the ilium, ischium, and pubis bones. The pelvic girdle is securely attached to the sacrum of the axial

skeleton, and the socket for the thigh bone is deep and heavily reinforced. This girdle receives the weight of the upper body and transfers it to the lower limbs. The female pelvis is lighter and broader than the male's; its inlet and outlet are larger, which reflects the childbearing function of the female.

4. The bones of the lower limb include the femur of the thigh, the tibia and fibula of the leg, and the tarsals, metatarsals, and phalanges of the foot.

JOINTS (pp. 163–168)

1. Joints hold bones together and allow movement of the skeleton.
2. Joints fall into three functional categories: synarthroses (immovable), amphiarthroses (slightly movable), and diarthroses (freely movable).
3. Joints also can be classified structurally as fibrous, cartilaginous, or synovial joints depending on the substance separating the articulating bones.
4. Most fibrous joints are synarthrotic, and most cartilaginous joints are amphiarthrotic. Fibrous and cartilaginous joints occur mainly in the axial skeleton.
5. Most joints of the body are synovial joints, which predominate in the limbs. In synovial joints, the articulating bone surfaces are covered with articular cartilage and enclosed within the joint cavity by a fibrous capsule lined with a synovial membrane. All synovial joints are diarthroses.

   WEB Exercise: Chapter 5, Types of Synovial Joints.

6. The most common joint problem is arthritis, or inflammation of the joints. Osteoarthritis, or degenerative arthritis, is a result of the "wear and tear" on joints over many years and is a common affliction of the aged. Rheumatoid arthritis occurs in both young and older adults; it is believed to be an autoimmune disease. Gouty arthritis, caused by the deposit of uric acid crystals in joints, typically affects a single joint.

DEVELOPMENTAL ASPECTS OF THE SKELETON (pp. 168–170)

1. Fontanels, which allow brain growth and ease birth passage, are present in the skull at birth. Growth of the cranium after birth is related to brain growth; the increase in size of the facial skeleton follows tooth development and enlargement of the respiratory passageways.
2. The vertebral column is C-shaped at birth (thoracic and sacral curvatures are present); the secondary curvatures form when the baby begins to lift its head and walk.
3. Long bones continue to grow in length until late adolescence. The UL ratio changes from 1.7:1 to 1:1 by the age of 10.
4. Fractures are the most common bone problem in elderly persons. Osteoporosis, a condition of bone wasting that results mainly from hormone deficit or inactivity, is also common in elderly individuals.

REVIEW QUESTIONS

MULTIPLE CHOICE

More than one choice may apply.

1. Which of the following are correctly matched?
   a. Short bone—wrist
   b. Long bone—leg
   c. Irregular bone—sternum
   d. Flat bone—cranium
2. A passageway connecting neighboring osteocytes in an osteon is a:
   a. central canal
   b. lamella
   c. lacuna
   d. canaliculus
   e. perforating canal
3. Which of the following would you expect to be prominent in osteoclasts?
   a. Golgi apparatus
   b. Lysosomes
   c. Microfilaments
   d. Exocytosis
4. Bone pain behind the external acoustic meatus probably involves the:
   a. maxilla
   b. ethmoid
   c. sphenoid
   d. temporal
   e. lacrimal
5. Bones that articulate with the sphenoid include:
   a. parietal
   b. vomer
   c. maxilla
   d. zygomatic
   e. ethmoid
6. Which humeral process articulates with the radius?
   a. Trochlea
   b. Greater tubercle
   c. Lesser tubercle
   d. Capitulum
   e. Olecranon fossa
7. Which of the following bones or bone parts articulate with the femur?
   a. Ischial tuberosity
   b. Pubis
   c. Patella
   d. Fibula
   e. Tibia

8. Which bone of the arm corresponds to the femur of the leg?
   a. Ulna
   b. Humerus
   c. Radius
   d. Tibia
   e. Fibula
9. At what stage of life does the UL ratio become 1:1?
   a. at birth
   b. by 10 years of age
   c. at puberty
   d. when the epiphyseal plates fuse
   e. never

SHORT ANSWER ESSAY

1. Name three functions of the skeletal system.
2. What is the anatomical name for the shaft of a long bone? for its ends? What is yellow marrow? How do spongy and compact bone look different?
3. Why do bone injuries heal much more rapidly than injuries to cartilage?
4. Compare and contrast the role of PTH (hormone) and mechanical forces acting on the skeleton in bone remodeling.
5. Define *fracture*. Which fracture types are most common in the elderly? Why are greenstick fractures more common in children?
6. Name the three major parts of the axial skeleton.
7. Name the eight bones of the cranium.
8. What bones are connected by the coronal suture? by the sagittal suture?
9. With one exception, all skull bones are joined by sutures. What is the exception?
10. What facial bone forms the chin? the cheekbone? the upper jaw? the bony eyebrow ridges?
11. Name two ways in which the fetal skull differs from the adult skull.
12. Name the five major regions of the vertebral column.
13. Diagram the normal spinal curvatures and then the curvatures seen in scoliosis and lordosis.
14. What is the function of the intervertebral discs?
15. Name the major components of the thorax.
16. What is a true rib? a false rib? Is a floating rib a true or a false rib? Why are floating ribs easily broken?
17. Name the bones of the shoulder girdle.
18. Name all the bones with which the ulna articulates.
19. The major function of the shoulder girdle is to provide flexibility. What is the major function of the pelvic girdle?
20. What bones make up each hip bone (coxal bone)? Which of these is the largest? Which has tuberosities that we sit on? Which is the most anterior?
21. List three differences between the male and the female pelvis.
22. Name the bones of the lower limb from superior to inferior.
23. What is the function of joints?
24. Compare the amount of movement possible in synarthrotic, amphiarthrotic, and diarthrotic joints. Relate these terms to the structural classification of joints; that is, to fibrous, cartilaginous, and synovial joints.
25. Describe the structure of a synovial joint.
26. Define *arthritis*. What type of arthritis is most common in the elderly? What type is believed to result from the immune system's attack on one's own joint tissues?
27. Professor Rogers pointed to the foramen magnum of the skull and said, "The food passes through this hole when you swallow." Some students believed him, but others said that this was a big mistake. What do you think? Support your answer.
28. Yolanda is asked to review a bone slide that has been set up under a microscope. She sees concentric layers surrounding a central cavity or canal. Is this bone section taken from the diaphysis or the epiphyseal plate of the bone specimen?
29. List two factors that keep bones healthy. List two factors that can cause bones to become soft or to atrophy.

AT THE CLINIC

1. A 75-year-old woman and her 9-year-old granddaughter were in a train crash in which both sustained trauma to the chest while seated next to each other. X rays showed that the grandmother had several fractured ribs, while her granddaughter had none. Explain these surprisingly (?) different findings.
2. The pediatrician at the clinic explains to parents of a newborn that their son suffers from cleft palate. She tells them that the normal palate fuses in an anterior-to-posterior pattern. The child's palatine processes have not fused. Have his palatine bones fused normally?

3. After having a severe cold accompanied by nasal congestion, Helen complained that she had a frontal headache and the right side of her face ached. What bony structures probably became infected by the bacteria or viruses causing the cold?
4. Bernice, a 75-year-old woman, stumbled slightly while walking, then felt a terrible pain in her left hip. At the hospital, X rays revealed that the hip was broken. Also, the compact and spongy bone throughout her spine were very thin. What was her probable condition?
5. At work, a box fell from a shelf onto Bertha's acromial region. In the emergency room, the physician felt that the head of her humerus had moved into the axilla. What had happened to Bertha?
6. An X ray of the arm of an accident victim reveals a faint line curving around and down the shaft. What kind of fracture might this indicate?
7. Bone X rays are sometimes used to determine if a person has reached his or her final height. What are the clinicians checking out?
8. A patient complains of pain starting in the jaw and radiating down the neck. Upon questioning, he states that when he is under stress he grinds his teeth. What joint is causing his pain?

6

The Muscular System

YOUR GOALS

After completing this chapter, you will have a working knowledge of the functions of the muscular system and will have mastered the objectives listed below.

FUNCTION PREVIEW

- The muscular system provides for movement of the body and its parts, maintains posture, generates heat, and stabilizes joints.

OBJECTIVE CHECKLIST

OVERVIEW OF MUSCLE TISSUES (pp. 178–182)

- Describe similarities and differences in the structure and function of the three types of muscle tissue, and indicate where they are found in the body.
- Define *muscular system.*
- Define and explain the role of the following: *endomysium, perimysium, epimysium, tendon,* and *aponeurosis.*

MICROSCOPIC ANATOMY OF SKELETAL MUSCLE (pp. 182–184)

- Describe the microscopic structure of skeletal muscle, and explain the role of actin- and myosin-containing myofilaments.

SKELETAL MUSCLE ACTIVITY (pp. 184–192)

- Describe how an action potential is initiated in a muscle cell.
- Describe the events of muscle cell contraction.
- Define *graded response, tetanus, isotonic* and *isometric contractions,* and *muscle tone* as these terms apply to a skeletal muscle.
- Describe three ways in which ATP is regenerated during muscle activity.

>>

(Objective Checklist, continued)

- Define *oxygen debt* and *muscle fatigue,* and list possible causes of muscle fatigue.
- Describe the effects of aerobic and resistance exercise on skeletal muscles and other body organs.

MUSCLE MOVEMENTS, TYPES, AND NAMES (pp. 192–200)

- Define *origin, insertion, prime mover, antagonist, synergist,* and *fixator* as they relate to muscles.
- Demonstrate or identify the different types of body movements.
- List some criteria used in naming muscles.

GROSS ANATOMY OF SKELETAL MUSCLES (pp. 200–213)

- Name and locate the major muscles of the human body (on a torso model, muscle chart, or diagram), and state the action of each.

DEVELOPMENTAL ASPECTS OF THE MUSCULAR SYSTEM (p. 214)

- Explain the importance of a nerve supply and exercise in keeping muscles healthy.
- Describe the changes that occur in aging muscles.

Because flexing muscles look like mice scurrying beneath the skin, some scientist long ago dubbed them *muscles,* from the Latin word *mus* meaning "little mouse." Indeed, the rippling muscles of professional boxers or weight lifters is often the first thing that comes to mind when one hears the word *muscle.* But muscle is also the dominant tissue in the heart and in the walls of other hollow organs of the body. In all its forms, it makes up nearly half the body's mass.

The essential function of muscle is *contraction,* or *shortening*—a unique characteristic that sets it apart from any other body tissue. As a result of this ability, muscles are responsible for essentially all body movement and can be viewed as the "machines" of the body.

Overview of Muscle Tissues

Muscle Types

There are three types of muscle tissue—skeletal, cardiac, and smooth. As summarized in Table 6.1, these differ in their cell structure, body location, and how they are stimulated to contract. But, before we explore their differences, let's look at some of the ways they are the same.

First, skeletal and smooth muscle cells are elongated. For this reason, these types of muscle cells (but not cardiac muscle cells) are called **muscle fibers.** Second, the ability of muscle to shorten, or contract, depends on two types of *myofilaments,* the muscle cell equivalents of the microfilaments of the cytoskeleton studied in Chapter 3. A third similarity has to do with terminology. Whenever you see the prefixes *myo-* and *mys-* ("muscle") and *sarco-* ("flesh"), you will know that muscle is being referred to. For example, in muscle cells the cytoplasm is called *sarcoplasm* (sar′ko-plaz″um).

Skeletal Muscle

Skeletal muscle fibers are packaged into the organs called *skeletal muscles* that attach to the body's skeleton. As the skeletal muscles cover our bony "underpinnings," they help form the much smoother contours of the body. Skeletal muscle fibers are cigar-shaped, multinucleate cells, and the largest of the muscle fiber types—some ranging up to 30 cm (nearly 1 foot) in length. Indeed, the

TABLE 6.1 Comparison of Skeletal, Cardiac, and Smooth Muscles

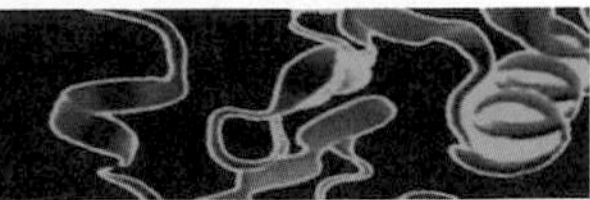

Characteristic	Skeletal	Cardiac	Smooth
Body location	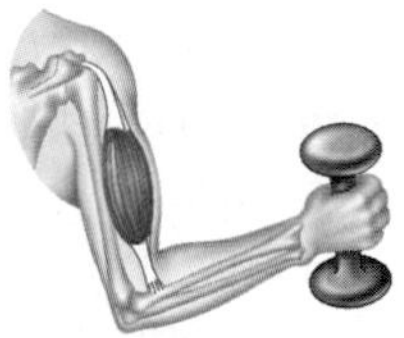 Attached to bones or, for some facial muscles, to skin	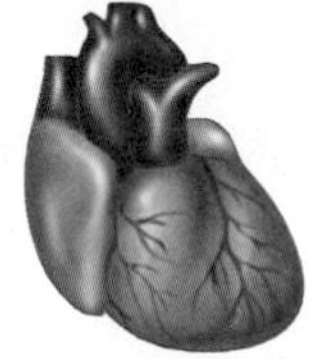 Walls of the heart	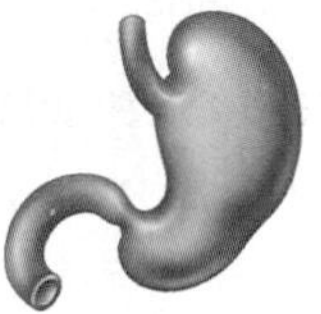 Mostly in walls of hollow visceral organs (other than the heart)
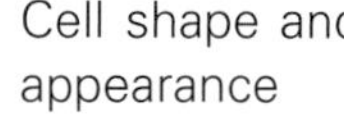 Cell shape and appearance	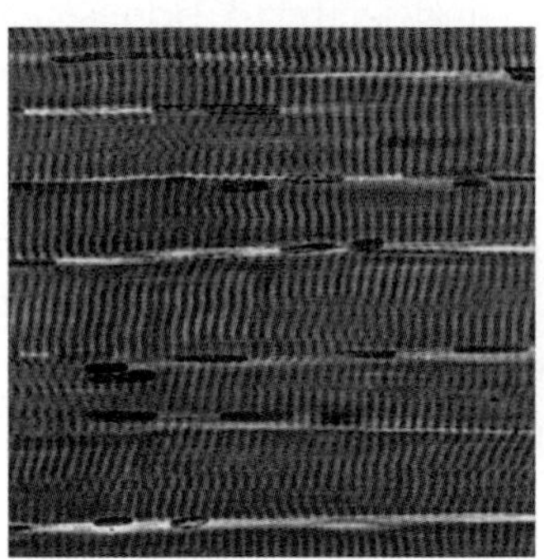  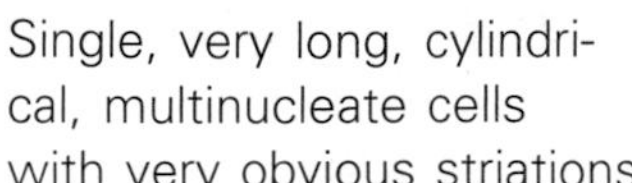 Single, very long, cylindrical, multinucleate cells with very obvious striations	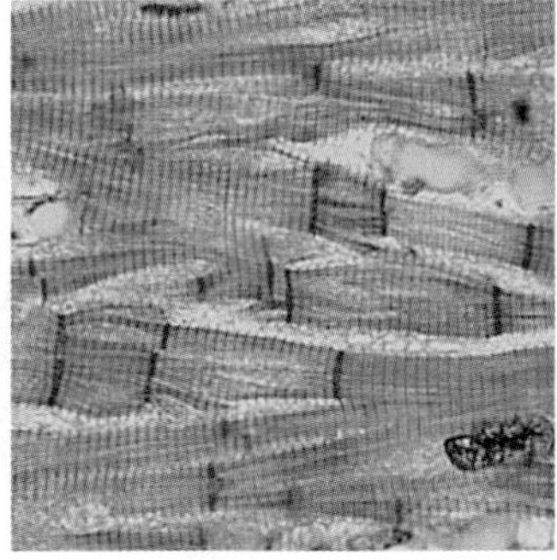 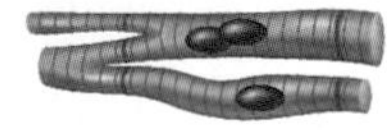 Branching chains of cells; uninucleate, striations; intercalated discs	  Single, fusiform, uninucleate; no striations
Regulation of contraction	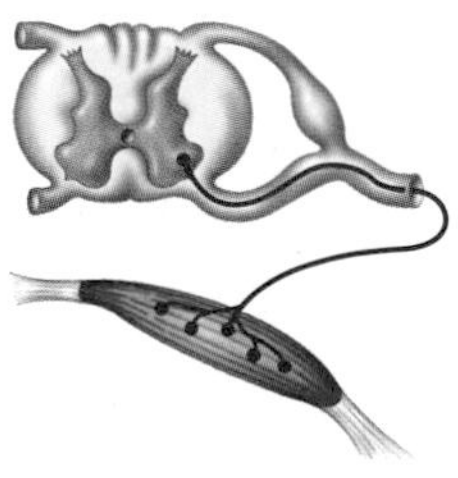 Voluntary; via nervous system controls	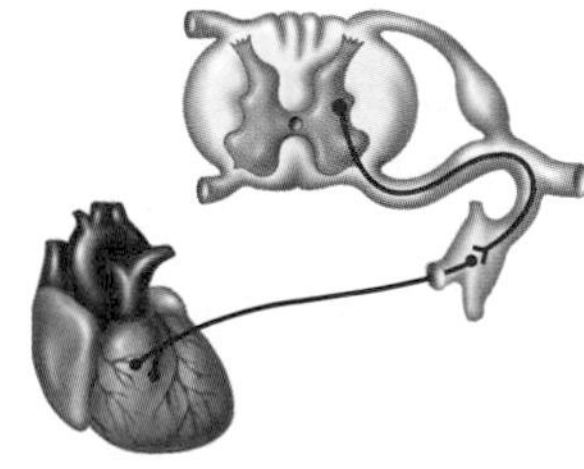 Involuntary; the heart has a pacemaker; also nervous system controls; hormones	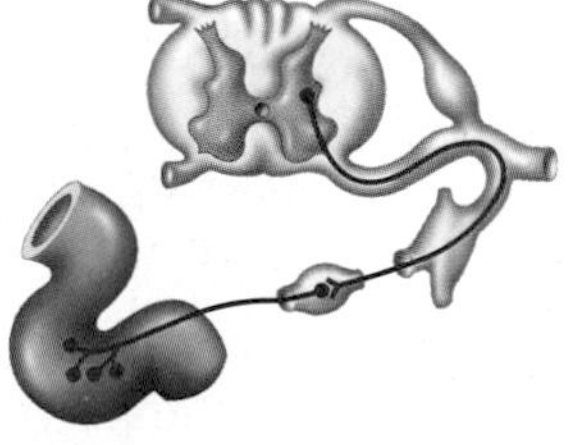 Involuntary; nervous system controls; hormones, chemicals, stretch
Speed of contraction	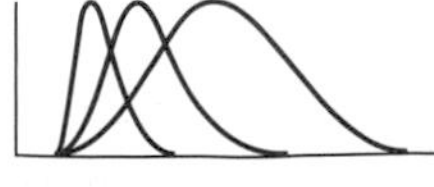 Slow to fast	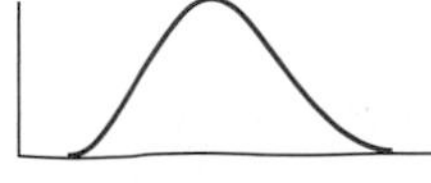 Slow	 Very slow
Rhythmic contraction	No	Yes	Yes, in some

Q *What is the meaning of* epi*? Of* mys*? How do these word roots relate to the role and position of the epimysium?*

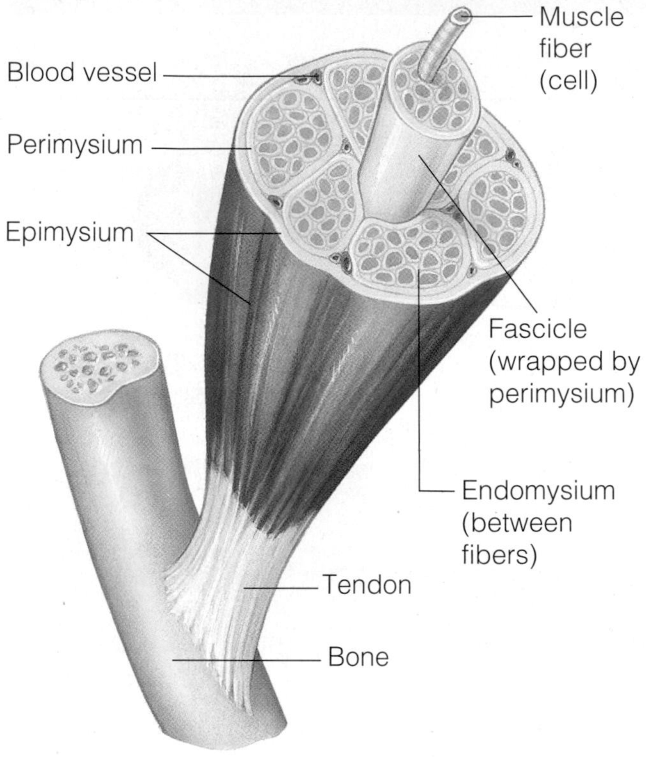

FIGURE 6.1 Connective tissue wrappings of skeletal muscle.

fibers of large, hardworking muscles, such as the antigravity muscles of the hip, are so big and coarse that they can be seen with the naked eye.

Skeletal muscle is also known as **striated muscle** (because its fibers appear to be striped) and as **voluntary muscle** (because it is the only muscle type subject to conscious control). However, it is important to recognize that skeletal muscles are often activated by reflexes (without our "willed command") as well. When you think of skeletal muscle tissue, the key words to keep in mind are *skeletal, striated,* and *voluntary*. Skeletal muscle tissue can contract rapidly and with great force, but it tires easily and must rest after short periods of activity.

Skeletal muscle fibers, like most living cells, are soft and surprisingly fragile. Yet skeletal muscles can exert tremendous power—indeed, the force they generate is often much greater than that required to lift the weight. How so? The reason they are not ripped apart as they exert force is that thousands of their fibers are bundled together by connective tissue, which provides strength and support to the muscle as a whole (Figure 6.1). Each muscle fiber is enclosed in a delicate connective tissue sheath called an **endomysium** (en″do-mis′e-um). Several sheathed muscle fibers are then wrapped by a coarser fibrous membrane called a **perimysium** to form a bundle of fibers called a **fascicle** (fas′ĭ-kul). Many fascicles are bound together by an even tougher "overcoat" of connective tissue called an **epimysium,** which covers the entire muscle. The epimysia blend into the strong, cordlike **tendons,** or into sheetlike **aponeuroses** (ap″o-nu-ro′sēz), which attach muscles indirectly to bones, cartilages, or connective tissue coverings of each other.

Besides simply acting to anchor muscles, tendons perform several functions. The most important are providing durability and conserving space. Tendons are mostly tough collagenic fibers, so they can cross rough bony projections, which would tear the more delicate muscle tissues. Because of their relatively small size, more tendons than fleshy muscles can pass over a joint.

Many people think of muscles as always having an enlarged "belly" that tapers down to a tendon at each end. However, muscles vary considerably in the way their fibers are arranged. Many are spindle-shaped as just described, but in others, the fibers are arranged in a fan shape or a circle, as described on p. 199.

Smooth Muscle

Smooth muscle has no striations and is involuntary, which means that we cannot consciously control it. Found mainly in the walls of hollow visceral organs such as the stomach, urinary bladder, and respiratory passages, smooth muscle propels substances along a definite tract, or pathway, within the body. We can best describe smooth muscle using the terms *visceral, nonstriated,* and *involuntary*.

As described in Chapter 3, smooth muscle cells are spindle-shaped and have a single nucleus (see also Table 6.1 on p. 179). They are arranged in sheets or layers. Most often there are two such layers, one running circularly and the other longitudinally, as shown in Figure 6.2a. As the two layers alternately contract and relax, they change the size and shape of the organ. Movement of food through the digestive tract and emptying the bowels and

A *Epi = upon, over, beside and mys = muscle. The epimysium is a sheath upon or over a muscle.*

bladder are examples of "housekeeping" activities normally handled by smooth muscles. Smooth muscle contraction is slow and sustained. If skeletal muscle is like a speedy wind-up car that quickly runs down, then smooth muscle is like a steady, heavy-duty engine that lumbers along tirelessly.

Cardiac Muscle

Cardiac muscle is found in only one place in the body—the heart. The heart serves as a pump, propelling blood into the blood vessels and to all tissues of the body. Cardiac muscle is like skeletal muscle in that it is striated and like smooth muscle in that it is involuntary and cannot be consciously controlled by most of us. Important key words to jog your memory for this muscle type are *cardiac, striated,* and *involuntary*.

The cardiac fibers are cushioned by small amounts of soft connective tissue and arranged in spiral or figure 8–shaped bundles, as shown in Figure 6.2b. When the heart contracts, its internal chambers become smaller, forcing the blood into the large arteries leaving the heart. Recall that cardiac muscle fibers are branching cells joined by special junctions called *intercalated discs* (see Figure 3.20 on p. 96). These two structural features and the spiral arrangement of the muscle bundles in the heart allow heart activity to be closely coordinated. Cardiac muscle usually contracts at a fairly steady rate set by the heart's "in-house" pacemaker, but the heart can also be stimulated by the nervous system to shift into "high gear" for short periods, as when you race to catch a bus.

Muscle Functions

Muscle plays four important roles in the body: it *produces movement, maintains posture, stabilizes joints,* and *generates heat.*

Producing Movement

Just about all movements of the human body are a result of muscle contraction. Mobility of the body as a whole reflects the activity of skeletal muscles, which are responsible for all locomotion (walking, swimming, and cross-country skiing, for example) and manipulation. They enable us to respond quickly to changes in the external environment. For example, their speed and power enable us to jump out of the way of a runaway car and then follow its flight with our eyes. They also allow us

(a)

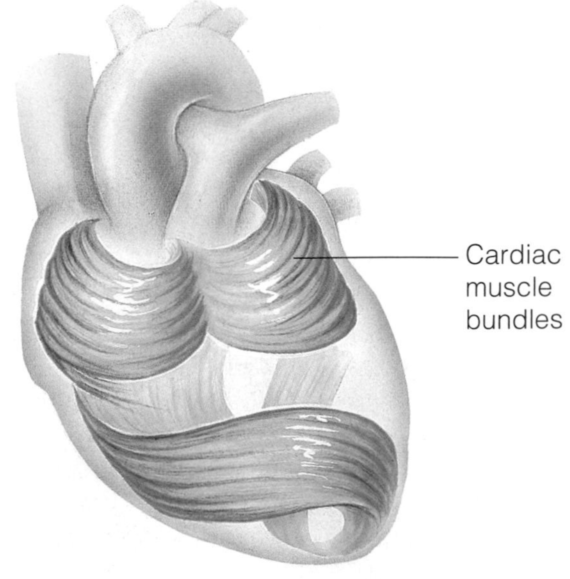

(b)

FIGURE 6.2 Arrangement of smooth and cardiac muscle cells. (a) Diagrammatic view of a cross section of the intestine. The longitudinal and circular layers of muscles yield cross-sectional and longitudinal views, respectively, of the smooth muscle cells. **(b)** Longitudinal view of the heart, showing the spiral arrangement of the cardiac muscle cells in its walls.

to express our emotions with the silent language of smiles and frowns.

These are distinct from the smooth muscle of blood vessel walls and cardiac muscle of the heart, which work together to circulate blood and maintain blood pressure, and the smooth muscle of other hollow organs, which forces fluids (urine, bile) and other substances (food, a baby) through internal body channels.

Maintaining Posture

We are rarely aware of the workings of the skeletal muscles that maintain body posture. Yet, they function almost continuously, making one tiny adjustment after another so that we can maintain an erect or seated posture despite the never-ending downward pull of gravity.

Stabilizing Joints

As the skeletal muscles pull on bones to cause movements, they also stabilize the joints of the skeleton. Indeed, muscle tendons are extremely important in reinforcing and stabilizing joints that have poorly fitting articulating surfaces (the shoulder joint, for example).

Generating Heat

The fourth function of muscle, generation of body heat, is a by-product of muscle activity. As ATP is used to power muscle contraction, nearly three-quarters of its energy escapes as heat. This heat is vital in maintaining normal body temperature. Since skeletal muscle accounts for at least 40 percent of body mass, it is the muscle type most responsible for heat generation.

As you can see, each of the three muscle types has a structure and function well suited for its job in the body. But since the term ***muscular system*** applies specifically to skeletal muscle, we will be concentrating on this muscle type in this chapter. The most important structural and functional aspects of the three muscle types are outlined in Table 6.1 (p. 179).

Microscopic Anatomy of Skeletal Muscle

As mentioned above and illustrated in Figure 6.3a, skeletal muscle cells are multinucleate. Many oval nuclei can be seen just beneath the plasma membrane, which is called the **sarcolemma** (sar″ko-lem′ah; "muscle husk") in muscle cells. The nuclei are pushed aside by long ribbonlike organelles, the **myofibrils** (mi″o-fi′brilz), which nearly fill the cytoplasm. Alternating **light (I)** and **dark (A) bands** along the length of the perfectly aligned myofibrils give the muscle cell as a whole its striped appearance. (Think of the second letter of *light,* I, and the second letter of *dark,* A, to help you remember which band is which.) A closer look at the banding pattern reveals that the light I band has a midline interruption, a darker area called the *Z disc,* and the dark A band has a lighter central area called the *H zone* (Figure 6.3b). The *M line* in the center of the H zone contains tiny protein rods that hold adjacent thick filaments together.

So why are we bothering with all these terms—dark this and light that? Because the banding pattern reveals the working structure of the myofibrils. First, we find that the myofibrils are actually chains of tiny contractile units called **sarcomeres** (sar′ko-merz), which are aligned end-to-end like boxcars in a train along the length of the myofibrils. Second, it is the arrangement of even smaller structures (myofilaments) *within* sarcomeres that actually produces the banding pattern.

Let's examine how the arrangement of the myofilaments leads to the banding pattern. There are two types of threadlike protein **myofilaments** within each of our "boxcar" sarcomeres (Figure 6.3c). The larger **thick filaments,** also called *myosin filaments,* are made mostly of bundled molecules of the protein **myosin,** but they also contain ATPase enzymes, which split ATP to generate the power for muscle contraction. Notice that the thick filaments extend the entire length of the dark A band. Also, notice that the midparts of the thick filaments are smooth, but their ends are studded with small projections (Figure 6.3d). These projections, or myosin *heads,* are called **cross bridges** when they link the thick and thin filaments together during contraction. The **thin filaments** are composed of the contractile protein called **actin,** plus some regulatory proteins that play a role in allowing (or preventing) myosin head–binding to actin. The thin filaments, also called *actin filaments,* are anchored to the Z disc (a disclike membrane). Notice that the light I band includes parts of two adjacent sarcomeres and contains *only* the thin filaments. Although the thin filaments overlap the ends of the thick filaments, they do not extend into the middle of a relaxed

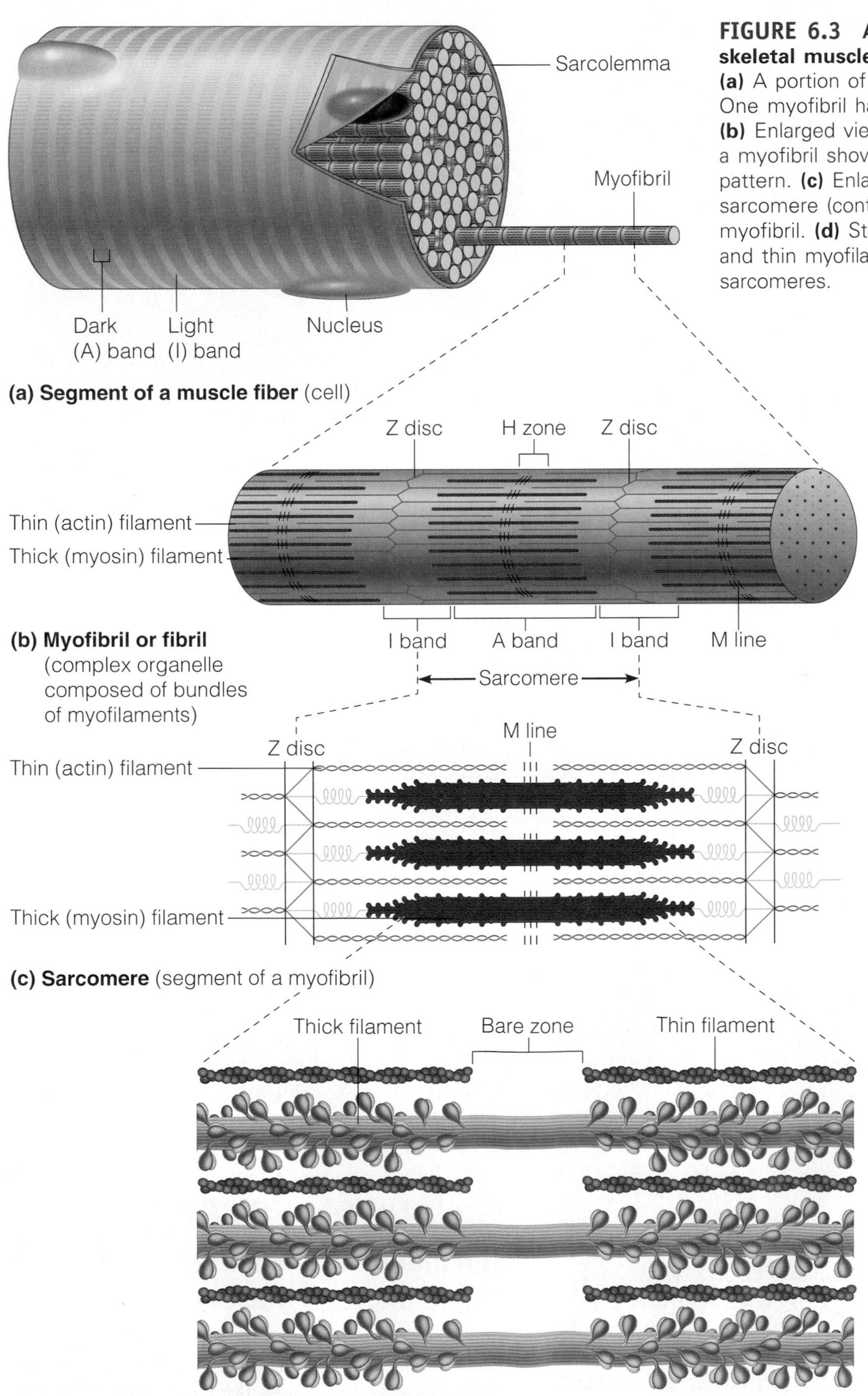

FIGURE 6.3 Anatomy of a skeletal muscle fiber (cell). **(a)** A portion of a muscle fiber. One myofibril has been extended. **(b)** Enlarged view of a section of a myofibril showing its banding pattern. **(c)** Enlarged view of one sarcomere (contractile unit) of a myofibril. **(d)** Structure of the thick and thin myofilaments found in the sarcomeres.

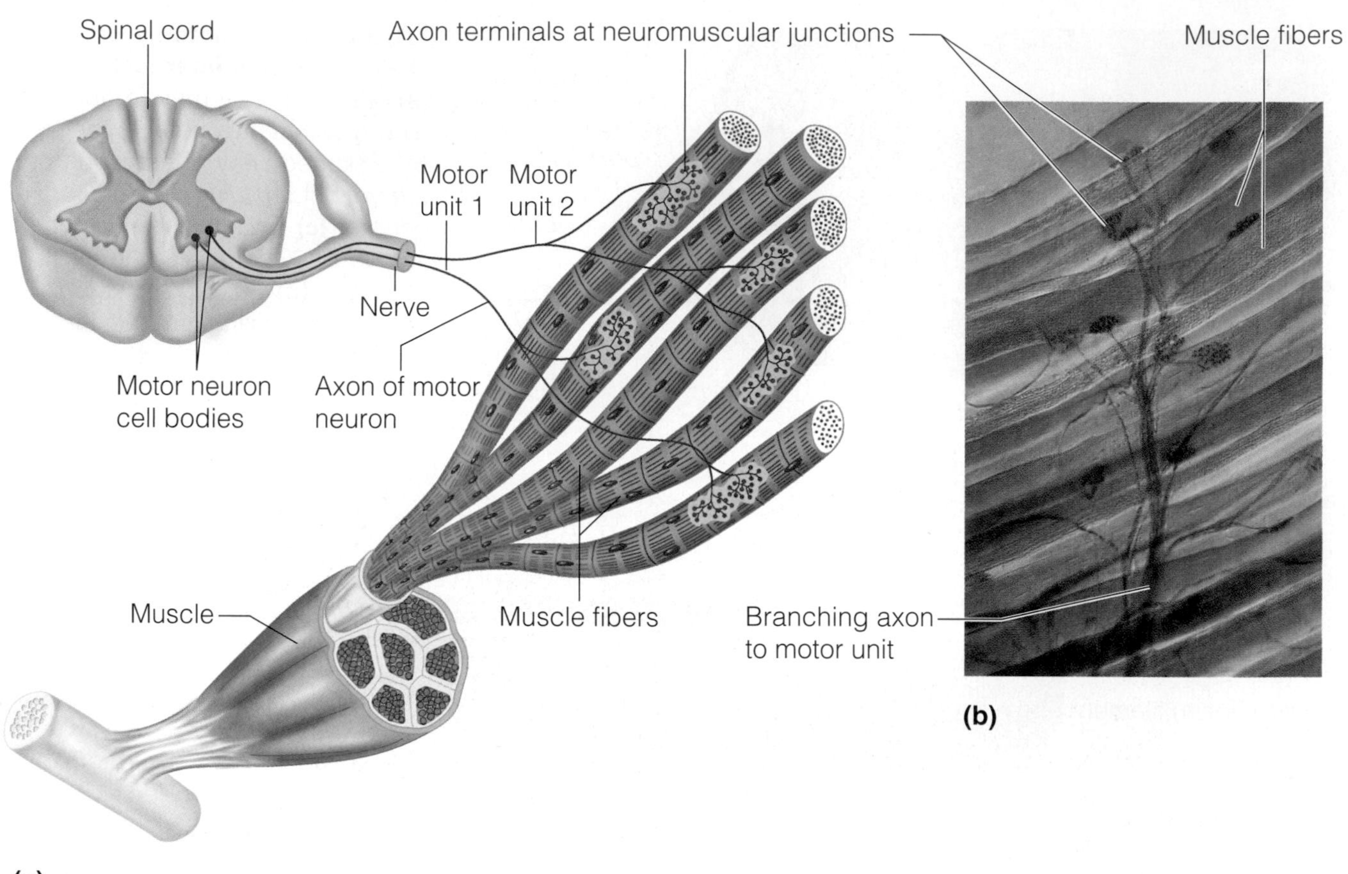

FIGURE 6.4 Motor units. Each motor unit consists of a motor neuron and all the muscle fibers it activates. **(a)** Portions of two motor units are shown. The motor neurons reside in the spinal cord, and their axons extend to the muscle. Within the muscle, each axon divides into a number of axon terminals, distributed to muscle fibers scattered throughout the muscle. **(b)** Photomicrograph of a portion of a motor unit (110×). Notice the diverging axon terminals and the neuromuscular junctions with the muscle fibers.

sarcomere, and thus the central region (the H zone, which lacks actin filaments and looks a bit lighter) is sometimes called the *bare zone*. When contraction occurs, and the actin-containing filaments slide toward each other into the center of the sarcomeres, these light zones disappear because the actin and myosin filaments are completely overlapped. For now, however, just recognize that it is the precise arrangement of the myofilaments in the myofibrils that produces the banding pattern, or striations, in skeletal muscle cells.

Not shown in Figure 6.3 is another very important muscle fiber organelle—the **sarcoplasmic reticulum (SR),** a specialized smooth endoplasmic reticulum. The interconnecting tubules and sacs of the SR surround each and every myofibril just as the sleeve of a loosely crocheted sweater surrounds your arm. The major role of this elaborate system is to store calcium and to release it on demand when the muscle fiber is stimulated to contract. As you will see, calcium provides the final "go" signal for contraction.

Skeletal Muscle Activity

Stimulation and Contraction of Single Skeletal Muscle Cells

Muscle cells have some special functional properties that enable them to perform their duties. The first of these is *irritability,* the ability to receive and

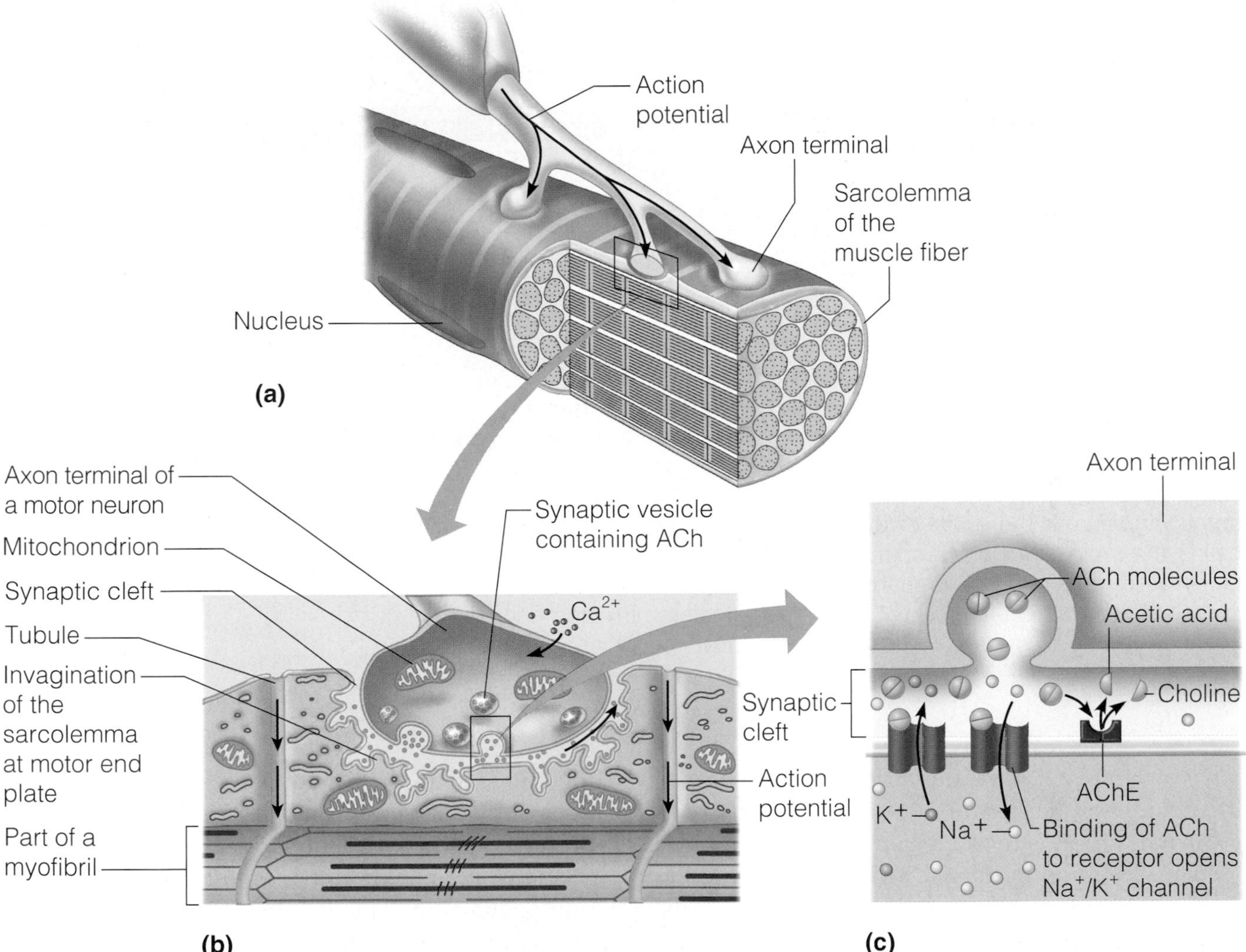

FIGURE 6.5 The neuromuscular junction. (a) Axon terminal of a motor neuron forming a neuromuscular junction with a muscle fiber. **(b)** The axon terminal contains vesicles filled with the neurotransmitter acetylcholine (ACh), which is released when the nerve impulse reaches the axon terminal. The sarcolemma is highly invaginated (folded) adjacent to the synaptic cleft and acetylcholine receptors are present in these folds. **(c)** Acetylcholine diffuses across the synaptic cleft and attaches to ACh receptors on the sarcolemma, initiating changes in the electrical condition of the sarcolemma.

respond to a stimulus. The second, *contractility,* is the ability to shorten (forcibly) when an adequate stimulus is received.

The Nerve Stimulus and the Action Potential

Skeletal muscle cells must be stimulated by nerve impulses to contract. One motor neuron (nerve cell) may stimulate a few muscle cells or hundreds of them, depending on the particular muscle and the work it does. One neuron and all the skeletal muscle cells it stimulates are a **motor unit** (Figure 6.4). When a long threadlike extension of the neuron, called the *nerve fiber* or **axon,** reaches the muscle, it branches into a number of **axon terminals,** each of which forms junctions with the sarcolemma of a different muscle cell (Figure 6.5). These junctions are called **neuromuscular** (literally, "nerve-muscle") **junctions.** Although the nerve endings and the muscle cells' membranes are very close, they never touch. The gap between them, the **synaptic cleft,** is filled with tissue (interstitial) fluid.

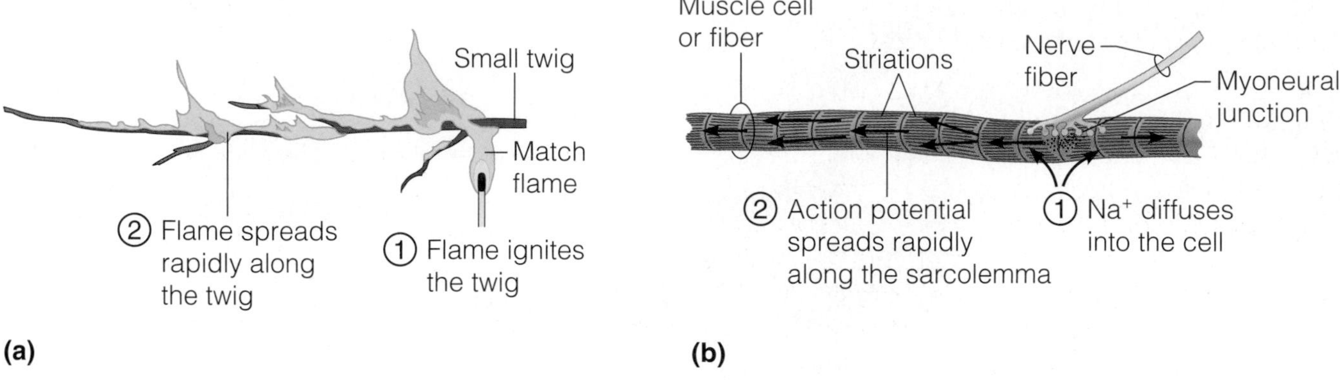

FIGURE 6.6 Comparison of the action potential to a flame consuming a dry twig. (a) The first event in igniting a dry twig is holding the match flame under one area of the twig. The second event is the twig's bursting into flame when it has been heated enough and the flame's spreading to burn the entire twig. **(b)** The first event in exciting a muscle cell is the rapid diffusion of sodium ions (Na^+) into the cell when the permeability of the sarcolemma changes. The second event is the spreading of the action potential along the sarcolemma when enough sodium ions have entered to upset the electrical conditions in the cell.

Now that we have described the structure of the neuromuscular junction, we are ready to examine what happens there. When the nerve impulse reaches the axon terminals, a chemical referred to as a **neurotransmitter** is released. The specific neurotransmitter that stimulates skeletal muscle cells is **acetylcholine** (as″e-til-ko′len), or **ACh.** Acetylcholine diffuses across the synaptic cleft and attaches to receptors (membrane proteins) that are part of the sarcolemma. If enough acetylcholine is released, the sarcolemma at that point becomes *temporarily* more permeable to sodium ions (Na^+), which rush into the muscle cell and to potassium ions (K^+) which diffuse out of the cell. However, more Na^+ enters than K^+ leaves. This gives the cell interior an excess of positive ions, which reverses the electrical conditions of the sarcolemma and opens more channels that allow Na^+ entry only. This "upset" generates an electrical current called an **action potential.** Once begun, the action potential is unstoppable; it travels over the entire surface of the sarcolemma, conducting the electrical impulse from one end of the cell to the other. The result is contraction of the muscle cell.

It should be mentioned that while the action potential is occurring, acetylcholine, which began the process, is broken down to acetic acid and choline by enzymes (acetylcholinesterase, or AchE) present on the sarcolemma (see Figure 6.5c). For this reason, a single nerve impulse produces only one contraction. This prevents continued contraction of the muscle cell in the absence of additional nerve impulses. The muscle cell relaxes until stimulated by the next round of acetylcholine release.

This series of events is explained more fully on pp. 231–233 in the discussion of nerve physiology, but perhaps it would be helpful to compare this to some common event, such as lighting a match under a small dry twig (Figure 6.6). The charring of the twig by the flame can be compared to the change in membrane permeability that allows sodium ions into the cell. When that part of the twig becomes hot enough (when enough sodium ions have entered the cell), the twig will suddenly burst into flame, and the flame will consume the twig (the action potential will be conducted along the entire length of the sarcolemma). The events that return the cell to its resting state include (1) diffusion of potassium ions (K^+) out of the cell, and (2) operation of the sodium-potassium pump, the active transport mechanism that moves the sodium and potassium ions back to their initial positions.

Mechanism of Muscle Contraction: The Sliding Filament Theory

What causes the filaments to slide? This question brings us back to the myosin heads that protrude all around the ends of the thick filaments. When muscle fibers are activated by the nervous system as just described, the myosin heads attach to binding sites on the thin filaments, and the sliding begins. Energized by ATP, each cross bridge attaches and detaches several times during a contraction, acting much like a tiny oar to generate tension and pull the thin filaments toward the center of the sarcomere. As this event occurs simultaneously in sarcomeres throughout the cell, the muscle cell shortens (Figure 6.7). The attachment of the myosin cross bridges to actin requires calcium ions (Ca^{2+}). So where does the calcium come from? As indicated in Figure 6.5b, action potentials (black arrows) pass deep into the muscle cell along membranous tubules that fold inward from the sarcolemma. Inside the cell, the action potentials stimulate the sarcoplasmic reticulum to release calcium ions into the cytoplasm. The calcium ions trigger the binding of myosin to actin initiating filament sliding. This sliding process and the precise role of calcium are depicted in Figure 6.8. When the action potential ends, calcium ions are immediately reabsorbed into the SR storage areas, and the muscle cell relaxes and settles back to its original length. This whole series of events takes just a few thousandths of a second.

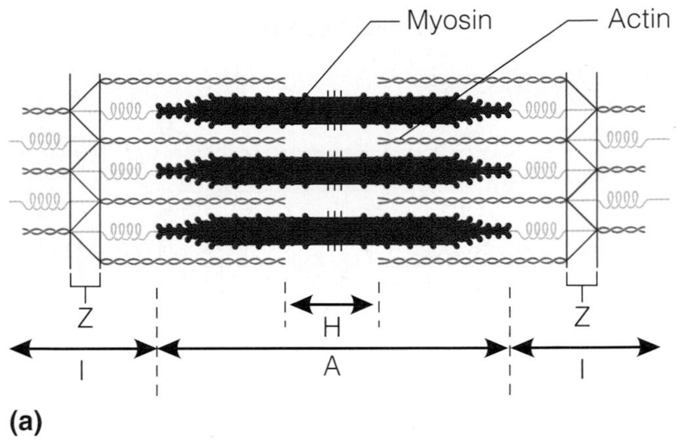

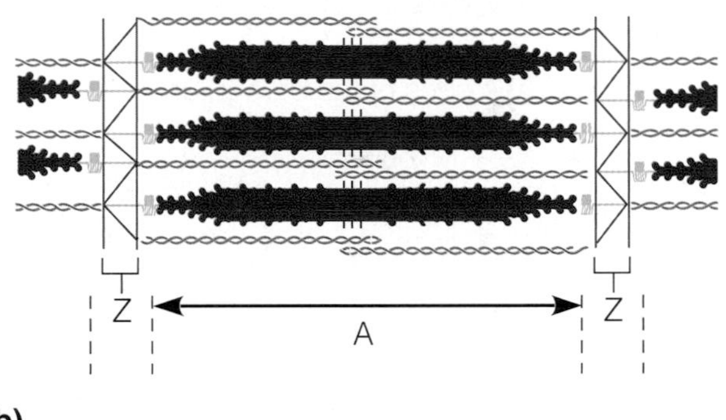

FIGURE 6.7 Diagrammatic views of a sarcomere. (a) Relaxed; **(b)** fully contracted. Notice that in the contracted sarcomere, the light H zone in the center of the A band has disappeared, the Z discs are closer to the thick filaments, and the I bands have nearly disappeared. The A bands move closer together but do not change in length.

Contraction of a Skeletal Muscle as a Whole

Graded Responses

In skeletal muscles, the "all-or-none" law of muscle physiology applies to the *muscle cell,* not to the whole muscle. It states that a muscle cell will contract to its fullest extent when it is stimulated adequately; it never partially contracts. However, skeletal muscles are organs that consist of thousands of muscle cells, and they react to stimuli with **graded responses,** or different degrees of shortening. In general, graded muscle contractions can be produced two ways: (1) by changing the *frequency* of muscle stimulation, and (2) by changing the *number* of muscle cells being stimulated. A muscle's response to each of these is briefly described next.

Muscle Response to Increasingly Rapid Stimulation

Although **muscle twitches** (single, brief, jerky contractions) sometimes occur as a result of certain nervous system problems, this is *not* the way our muscles normally operate. In most types of muscle activity, nerve impulses are delivered to the muscle at a very rapid rate—so rapid that the muscle does not get a chance to relax completely between stimuli. As a result, the effects of the successive contractions are "summed" (added) together, and the contractions of the muscle get stronger and smoother. When the muscle is stimulated so rapidly that no evidence of relaxation is seen and the contractions are completely smooth and sustained, the muscle is said to be in **fused,** or **complete,**

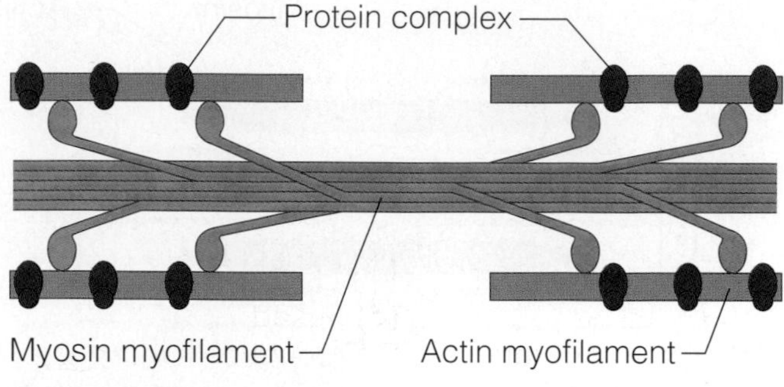

(a)

In a relaxed muscle cell, the regulatory proteins forming part of the actin myofilaments prevent myosin binding (see **a**). When an action potential sweeps along its sarcolemma and a muscle cell is excited, calcium ions (Ca^{2+}) are released from intracellular storage areas (the sacs of the sarcoplasmic reticulum).

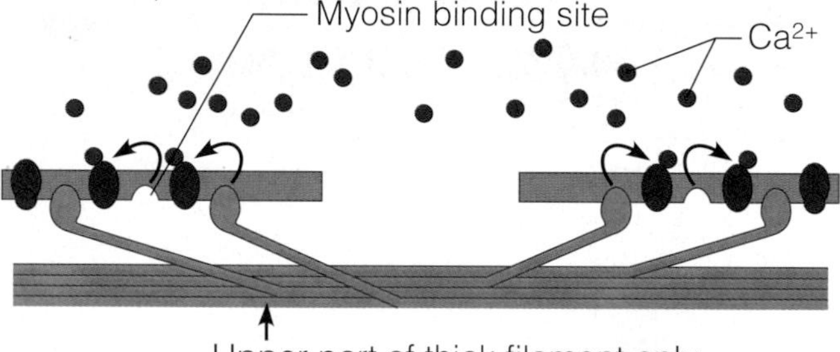

(b)

The flood of calcium acts as the final trigger for contraction, because as calcium binds to the regulatory proteins on the actin filaments, they change both their shape and their position on the thin filaments. This action exposes myosin binding sites on the actin, to which the myosin heads can attach (see **b**), and the myosin heads immediately begin seeking out binding sites.

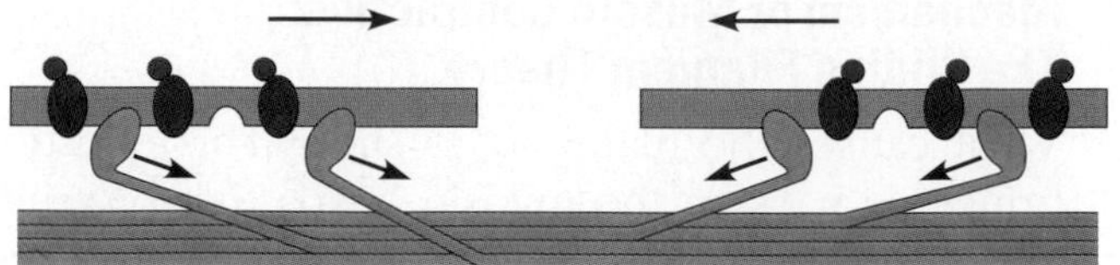

(c)

Free myosin has a unique property: Its heads are "cocked," much like a set mousetrap. The physical attachment of myosin to actin "springs the trap," causing the myosin heads to snap (pivot) toward the center of the sarcomere. Since the actin and myosin are still firmly bound to each other when this happens, the thin filaments are slightly pulled toward the center of the sarcomere (see **c**). ATP provides the energy needed to release and recock each myosin head so that it is ready to take another "step" and attach to a binding site farther along the thin filament. This "walking" of the myosin cross bridges or heads along the thin filaments during muscle shortening is much like a centipede's gait. Some myosin heads ("legs") are always in contact with actin ("the ground"), so that the thin filaments cannot slide backward as this cycle is repeated again and again during contraction. Notice that the myofilaments themselves do not shorten during contraction; they simply slide past each other. When the action potential ends and calcium ions are reabsorbed into the SR storage areas, the regulatory proteins resume their original shape and position, and again block myosin binding to the thin filaments. Since myosin now has nothing to attach to, the muscle cell relaxes and settles back to its original length.

FIGURE 6.8 Schematic representation of contraction mechanism: The sliding filament theory.

tetanus (tet′ah-nus),* or in tetanic contraction. Until this point is reached the muscle is said to be exhibiting **unfused,** or **incomplete, tetanus** (Figure 6.9).

*Tetanic contraction is normal and desirable and is quite different from the pathological condition of tetanus (commonly called *lockjaw*) caused by a toxin made by bacterium. Lockjaw causes muscles to go into uncontrollable spasms, finally causing respiratory arrest.

Muscle Response to Stronger Stimuli Although tetanus also produces stronger muscle contractions, its primary role is to produce smooth and prolonged muscle contractions. How forcefully a muscle contracts depends to a large extent on how many of its cells are stimulated. When only a few cells are stimulated, the contraction of the muscle as a whole will be slight. In the strongest contractions, when all the motor units are active and all the muscle cells are being stimulated, the muscle

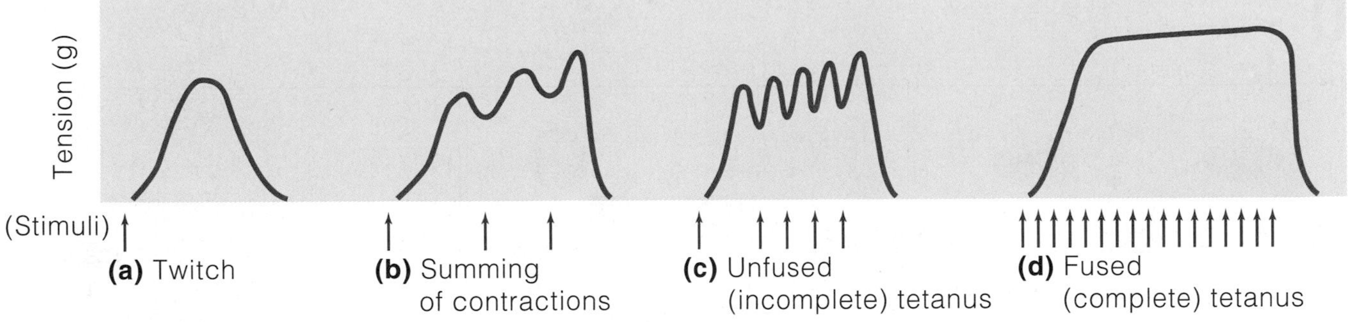

FIGURE 6.9 A whole muscle's response to different rates of stimulation. In **(a),** a single stimulus is delivered, and the muscle contracts and relaxes (a twitch contraction). In **(b),** stimuli are delivered more frequently, so the muscle does not have time to completely relax; contraction force increases because the effects of the individual twitches are summed. In **(c),** more complete fusion of the twitches (unfused tetanus) occurs as stimuli are delivered at a still faster rate. In **(d),** fused tetanus, a smooth continuous contraction without any evidence of relaxation, results from a very rapid rate of stimulation. (Points at which stimuli are delivered are indicated by red arrows. Tension [measured in grams] on the vertical axis refers to the relative force of muscle contraction.)

contraction is as strong as it can get. Thus, muscle contractions can be slight or vigorous depending on what work has to be done. The same hand that gently soothes can also deliver a stinging slap!

Providing Energy for Muscle Contraction

As a muscle contracts, the bonds of ATP molecules are hydrolyzed to release the needed energy. Surprisingly, muscles store very limited supplies of ATP—only 4 to 6 seconds' worth, just enough to get you going. Since ATP is the *only* energy source that can be used directly to power muscle activity, ATP must be regenerated continuously if contraction is to continue.

Essentially, working muscles use three pathways for ATP regeneration:

1. **Direct phosphorylation of ADP by creatine phosphate** (Figure 6.10a). The unique high-energy molecule **creatine phosphate (CP)** is found in muscle fibers but not other cell types. As ATP is being depleted, interactions between CP and ADP result in transfers of a high-energy phosphate group from CP to ADP, thus regenerating more ATP in a fraction of a second. Although muscle cells store perhaps five times as much CP as ATP, the CP supplies are also soon exhausted (in about 20 seconds).
2. **Aerobic respiration** (Figure 6.10b). At rest and during light to moderate exercise, some 95 percent of the ATP used for muscle activity comes from aerobic respiration. **Aerobic respiration** occurs in the mitochondria and involves a series of metabolic pathways that use oxygen. These pathways are collectively referred to as *oxidative phosphorylation.* During aerobic respiration, glucose is broken down completely to carbon dioxide and water, and some of the energy released as the bonds are broken is captured in the bonds of ATP molecules. Although aerobic respiration provides a rich ATP harvest (36 ATP per 1 glucose), it is fairly slow and requires continuous delivery of oxygen and nutrient fuels to the muscle to keep it going.
3. **Anaerobic glycolysis and lactic acid formation** (Figure 6.10c). The initial steps of glucose breakdown occur via a pathway called *glycolysis,* which does not use oxygen and hence is an *anaerobic* (literally "without oxygen") part of the metabolic pathway. During glycolysis, which occurs in the cytosol, glucose is broken down to pyruvic acid, and small amounts of energy are captured in ATP bonds (2 ATP per 1 glucose molecule). As long as enough

Q *Which of these methods of ATP generation is commonly used by the leg muscles of a long-distance cyclist?*

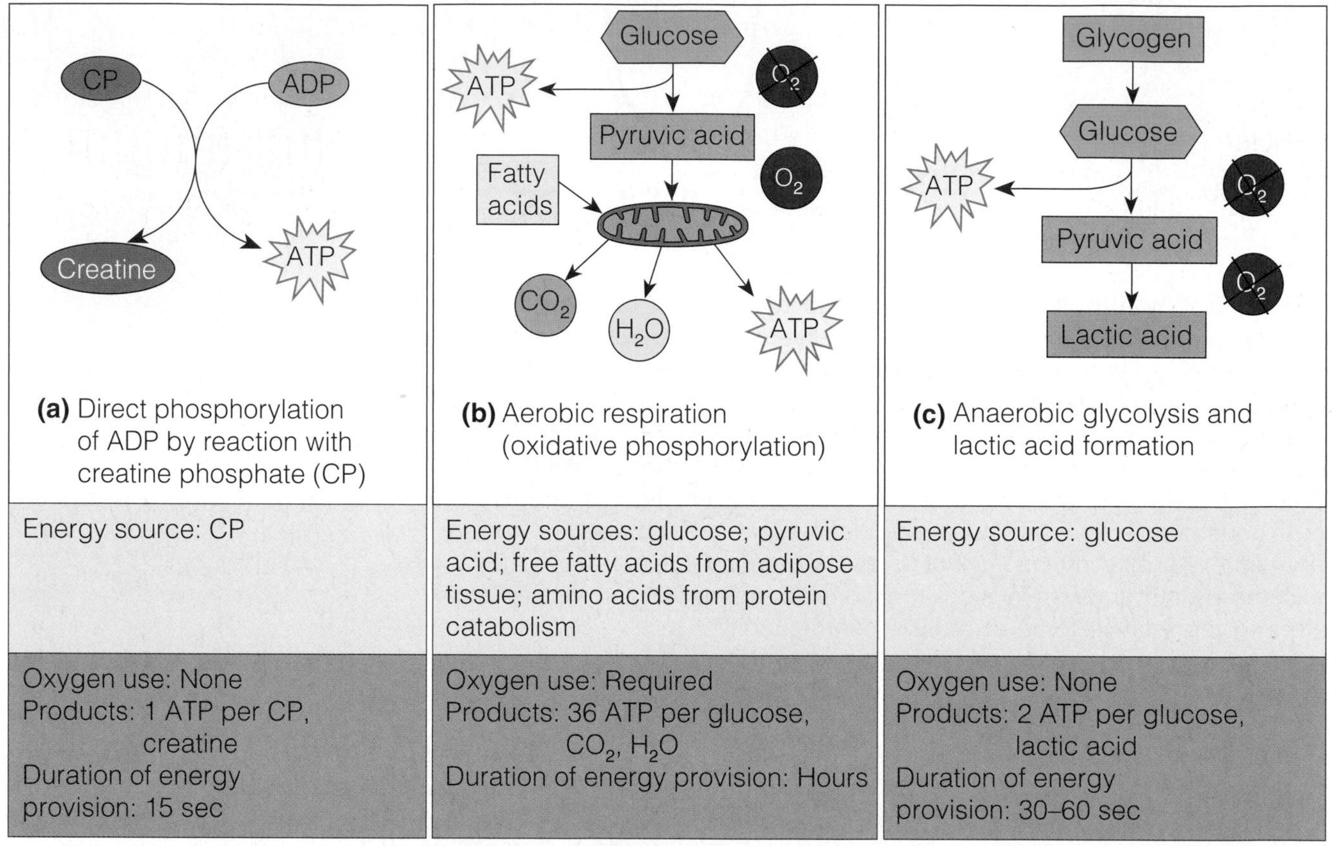

FIGURE 6.10 Methods of regenerating ATP during muscle activity. The fastest mechanism is **(a)** direct phosphorylation; the slowest is **(b)** aerobic respiration.

oxygen is present, the pyruvic acid then enters the oxygen-requiring aerobic pathways that occur within the mitochondria to produce more ATP as described above. However, when muscle activity is intense, or oxygen and glucose delivery is temporarily inadequate to meet the needs of the working muscles, the sluggish aerobic mechanisms cannot keep up with the demands for ATP. Under these conditions, the pyruvic acid generated during glycolysis is converted to **lactic acid,** and the overall process is referred to as **anaerobic glycolysis.**

Anaerobic glycolysis produces only about 5 percent as much ATP from each glucose molecule as aerobic respiration. However, it is some 2½ times faster, and it can provide most of the ATP needed for 30 to 60 seconds of strenuous muscle activity. The main shortcomings of anaerobic glycolysis are that it uses huge amounts of glucose for a small ATP harvest, and accumulating lactic acid promotes muscle fatigue and muscle soreness.

A *The aerobic mechanism (b).*

Muscle Fatigue and Oxygen Debt

If we exercise our muscles strenuously for a long time, **muscle fatigue** occurs. A muscle is fatigued when it is unable to contract even though it is still being stimulated. Without rest, an active or working muscle begins to tire and contracts more weakly until it finally ceases reacting and stops contracting. Muscle fatigue is believed to result from the **oxygen debt** that occurs during prolonged muscle activity: A person is not able to take in oxygen fast enough to keep the muscles supplied with all the oxygen they need when they are working vigorously. Obviously, then, the work that a muscle can do and how long it can work without becoming fatigued depend on how good its blood supply is. When muscles lack oxygen, lactic acid begins to accumulate in the muscle via the anaerobic mechanism described above. In addition, the muscle's ATP supply starts to run low. The increasing acidity in the muscle and the lack of ATP cause the muscle to contract less and less effectively and finally to stop contracting altogether.

True muscle fatigue, in which the muscle quits entirely, rarely occurs in most of us because we feel fatigued long before it happens and we simply slow down or stop our activity. It *does* happen commonly in marathon runners. Many of them have literally collapsed when their muscles became fatigued and could no longer work.

Oxygen debt, which always occurs to some extent during vigorous muscle activity, must be "paid back" whether or not fatigue occurs. During the recovery period after activity, the individual breathes rapidly and deeply. This continues until the muscles have received the amount of oxygen needed to get rid of the accumulated lactic acid and make ATP and creatine phosphate reserves.

Types of Muscle Contractions—Isotonic and Isometric

Until now, we have been discussing contraction in terms of shortening behavior, but muscles do not always shorten when they contract. (I can hear you saying, "What kind of double-talk is that?"—but pay attention.) The event that is common to all muscle contractions is that *tension* develops in the muscle as the actin and myosin myofilaments interact and the myosin cross bridges attempt to slide the actin-containing filaments past them within the muscle fibers.

Isotonic contractions (literally, "same tone" or tension) are more familiar to most of us. In isotonic contractions, the myofilaments are successful in their sliding movements, the muscle shortens, and movement occurs. Bending the knee, rotating the arms, and smiling are all examples of isotonic contractions.

Contractions in which the muscles do not shorten are called **isometric contractions** (literally, "same measurement" or length). In isometric contractions, the myosin myofilaments are "skidding their wheels," and the tension in the muscle keeps increasing. They are trying to slide, but the muscle is pitted against some more or less immovable object. For example, muscles are contracting isometrically when you try to lift a 400-pound dresser alone. When you straighten a bent elbow, the triceps muscle is contracting isotonically. But when you push against a wall with bent elbows, the wall doesn't move, and the triceps muscles, which cannot shorten to straighten the elbows, are contracting isometrically.

Muscle Tone

One aspect of skeletal muscle activity cannot be consciously controlled. Even when a muscle is voluntarily relaxed, some of its fibers are contracting—first one group and then another. Their contraction is not visible, but, as a result of it, the muscle remains firm, healthy, and constantly ready for action. This state of continuous partial contractions is called **muscle tone.** Muscle tone is the result of different motor units, which are scattered through the muscle, being stimulated by the nervous system in a systematic way.

Homeostatic Imbalance

If the nerve supply to a muscle is destroyed (as in an accident), the muscle is no longer stimulated in this manner, and it loses tone and becomes paralyzed. Soon after, it becomes *flaccid* (flak'sid), or soft and flabby, and begins to *atrophy* (waste away). ▲

Effect of Exercise on Muscles

The amount of work done by a muscle is reflected in changes in the muscle itself. Muscle inactivity (due to a loss of nerve supply, immobilization, or whatever the cause) always leads to muscle weakness and wasting. Muscles are no exception to the saying "Use it or lose it!"

Conversely, regular exercise increases muscle size, strength, and endurance. However, not all types of exercise produce these effects—in fact, there are important differences in the benefits of exercise.

Aerobic, or **endurance,** types of exercise such as participating in an aerobics class, jogging, or biking (Figure 6.11a), result in stronger, more flexible muscles with greater resistance to fatigue. These changes come about, at least partly, because the blood supply to the muscles increases, and the individual muscle cells form more mitochondria

(a)

(b)

FIGURE 6.11 The effects of aerobic training versus strength training. (a) A marathon runner. **(b)** A weight lifter.

and store more oxygen. However, aerobic exercise benefits much more than the skeletal muscles. It makes overall body metabolism more efficient, improves digestion (and elimination), enhances neuromuscular coordination, and makes the skeleton stronger. The heart enlarges (*hypertrophies*) so that more blood is pumped out with each beat, fat deposits are cleared from the blood vessel walls, and the lungs become more efficient in gas exchange. These benefits may be permanent or temporary, depending on how often and how vigorously one exercises.

Aerobic exercise does *not* cause the muscles to increase much in size, even though the exercise may go on for hours. The bulging muscles of a bodybuilder or professional weight lifter result mainly from **resistance,** or **isometric, exercises** (Figure 6.11b) in which the muscles are pitted against some immovable object (or nearly so). Resistance exercises require very little time and little or no special equipment. A few minutes every other day is usually sufficient. A wall can be pushed against, and buttock muscles can be strongly contracted even while standing in line at the grocery store. The key is forcing the muscles to contract with as much force as possible. The increased muscle size and strength that results is due mainly to enlargement of individual muscle cells (they make more contractile filaments), rather than an increase in their number. The amount of connective tissue that reinforces the muscle also increases.

Because endurance and resistance exercises produce different patterns of muscle response, it is important to know what your exercise goals are. Lifting weights will not improve your endurance for a marathon. By the same token, jogging will do little to improve your muscle definition for competing in the Mr. or Ms. Muscle contest, nor will it make you stronger for moving furniture. Obviously, the best exercise program for most people is one that includes both types of exercise.

Muscle Movements, Types, and Names

This section is a bit of a hodge-podge. It includes some topics that don't really fit together, but they don't fit anywhere else any better. For example, there are five very basic understandings about gross muscle activity. I call these the *Five Golden Rules* of skeletal muscle activity because until you understand them, comprehending muscle movements and appreciating muscle interactions is nearly impossible. These golden rules are summarized for your quick review in Table 6.2.

Types of Body Movements

Every one of our 600-odd skeletal muscles is attached to bone, or to other connective tissue structures, at no less than two points. One of these

TABLE 6.2 The Five Golden Rules of Skeletal Muscle Activity

1. With a few exceptions, all muscles cross at least one joint.
2. Typically, the bulk of the muscle lies proximal to the joint crossed.
3. All muscles have at least two attachments: the origin and the insertion.
4. Muscles can only pull; they never push.
5. During contraction, the muscle insertion moves toward the origin.

points, the **origin,** is attached to the immovable or less movable bone (Figure 6.12). The **insertion** is attached to the movable bone, and when the muscle contracts, the insertion moves toward the origin. Some muscles have interchangeable origins and insertions. For example, the rectus femoris muscle of the anterior thigh crosses both the hip and knee joints. Its most common action is to extend the knee, in which case the proximal pelvic attachment is the origin. However, when the knee is bent (by other muscles), the rectus femoris can flex the hip, and then its distal attachment on the leg is considered the origin.

Generally speaking, body movement occurs when muscles contract across joints. The type of movement depends on the mobility of the joint and on where the muscle is located in relation to the joint. The most obvious examples of the action of muscles on bones are the movements that occur at the joints of the limbs. However, less freely movable bones are also tugged into motion by the muscles, such as the vertebrae's movements when the torso is bent to the side.

The most common types of body movements are described next and shown in Figure 6.13. Try to demonstrate each movement as you read the following descriptions:

- **Flexion.** A movement, generally in the sagittal plane, that decreases the angle of the joint and brings two bones closer together (Figure 6.13a and b). Flexion is typical of hinge joints (bending the knee or elbow), but it is also common at ball-and-socket joints (bending forward at the hip).
- **Extension.** Extension is the opposite of flexion, so it is a movement that increases the angle, or the distance, between two bones or parts of the body (straightening the knee or elbow). If extension is greater than 180° (as when you tip your head or your torso posteriorly so that your chin points toward the ceiling), it is hyperextension (Figure 6.13a and b).
- **Rotation.** Rotation is movement of a bone around its longitudinal axis (Figure 6.13c). Rotation is a common movement of ball-and-socket joints and describes the movement of the atlas around the dens of the axis (as in shaking your head "no").
- **Abduction.** Abduction is moving a limb away (generally on the frontal plane) from the midline, or median plane, of the body (Figure 6.13d). The terminology also applies to the fanning movement of the fingers or toes when they are spread apart.
- **Adduction.** Adduction is the opposite of abduction, so it is the movement of a limb toward the body midline (Figure 6.13d).
- **Circumduction.** Circumduction is a combination of flexion, extension, abduction, and adduction commonly seen in ball-and-socket joints such as the shoulder. The proximal end of the limb is stationary, and its distal end moves in a circle. The limb as a whole outlines a cone (Figure 6.13d).

Q *The other movement that the biceps brachii muscle (shown in this illustration) can bring about is to move the torso toward the bar when you chin yourself. Would the forearm still be the insertion for that movement?*

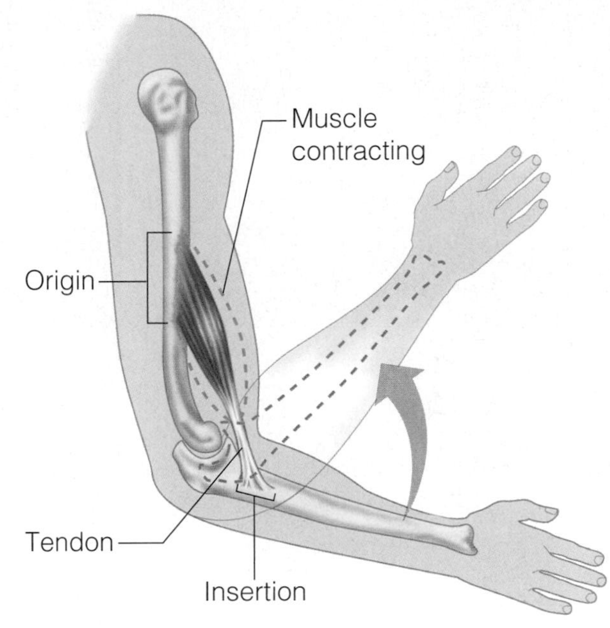

FIGURE 6.12 Muscle attachments (origin and insertion). When a skeletal muscle contracts, its insertion moves toward its origin.

Special Movements

Certain movements do not fit into any of the previous categories and occur at only a few joints.

A *No, the insertion in this case would be its attachment to the humerus, and the attachment on the forearm (which is held steady during this movement) is the insertion.*

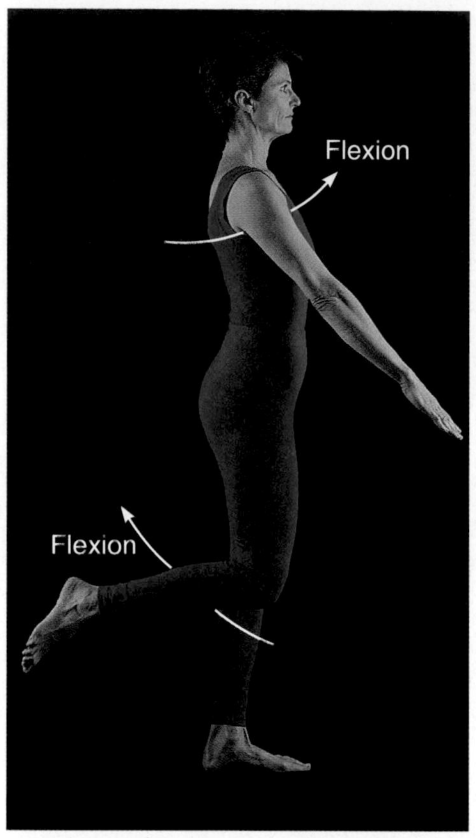

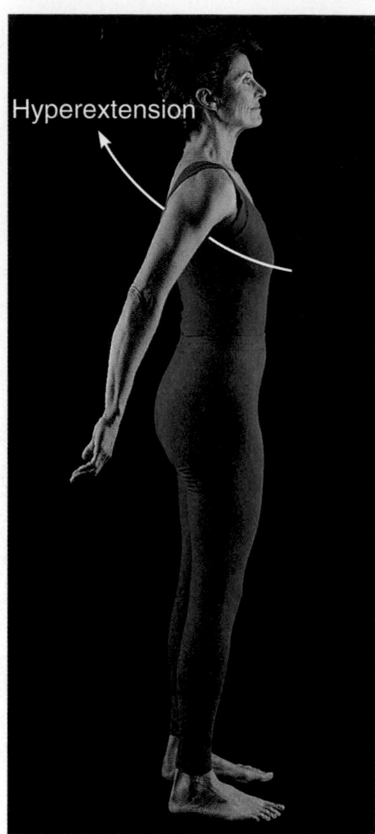

(a) Flexion and extension of the shoulder and knee

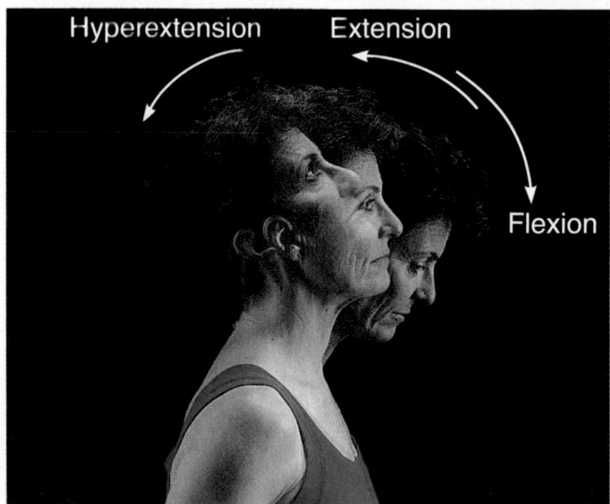

(b) Flexion, extension, and hyperextension

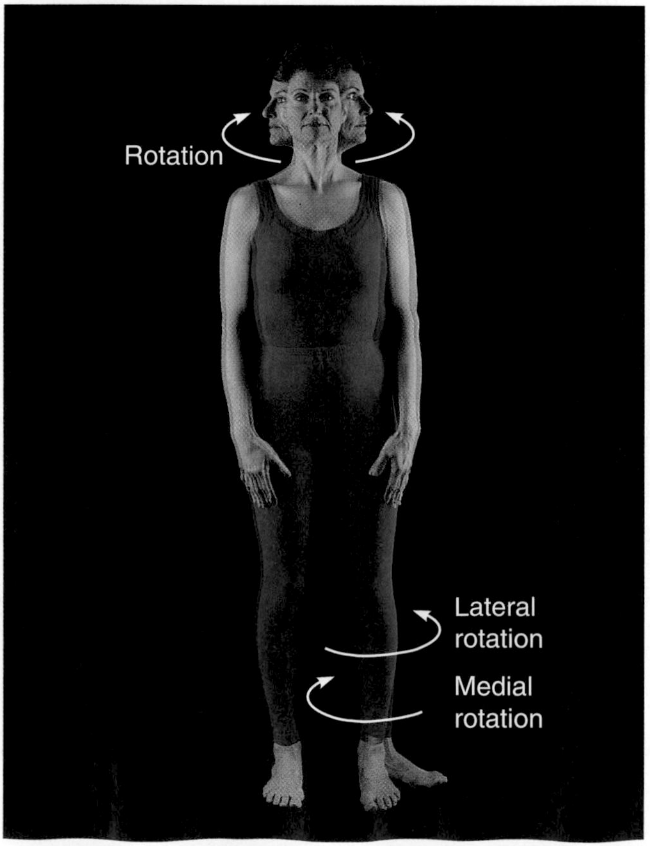

(c) Rotation

FIGURE 6.13 Body movements.

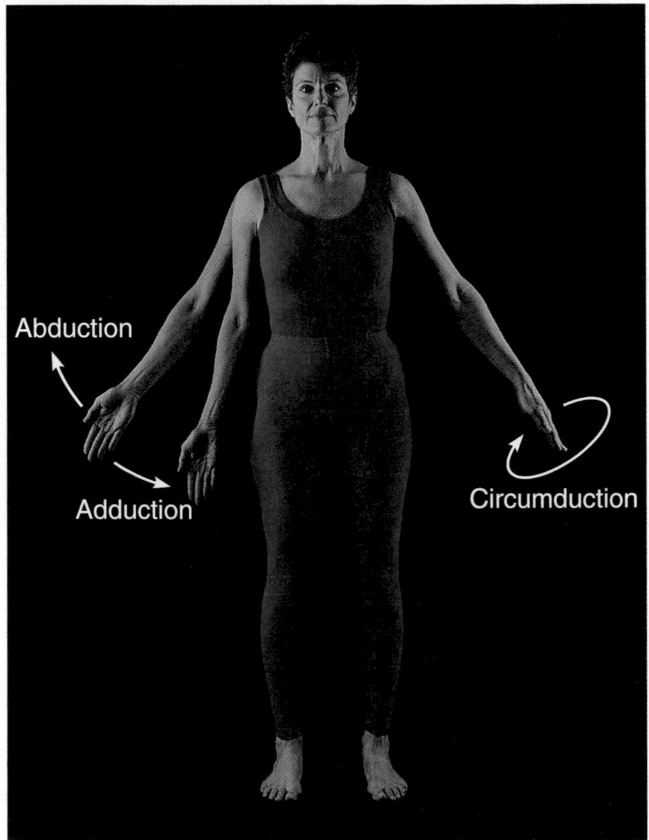

(d) Abduction, adduction, and circumduction

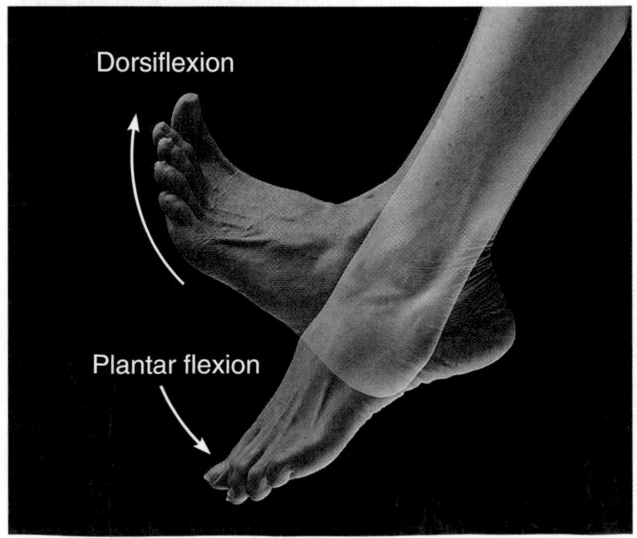

(e) Dorsiflexion and plantar flexion

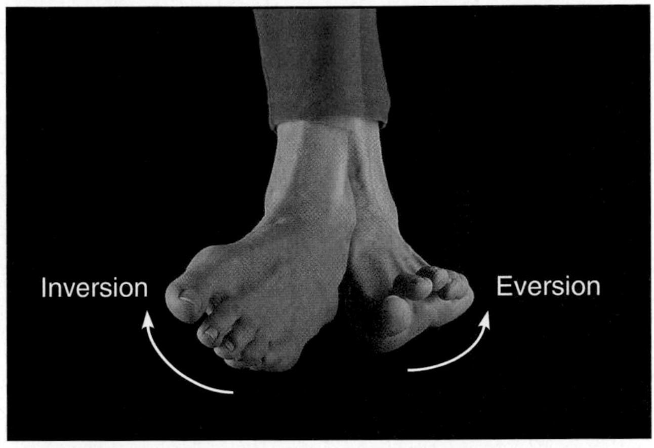

(f) Inversion and eversion

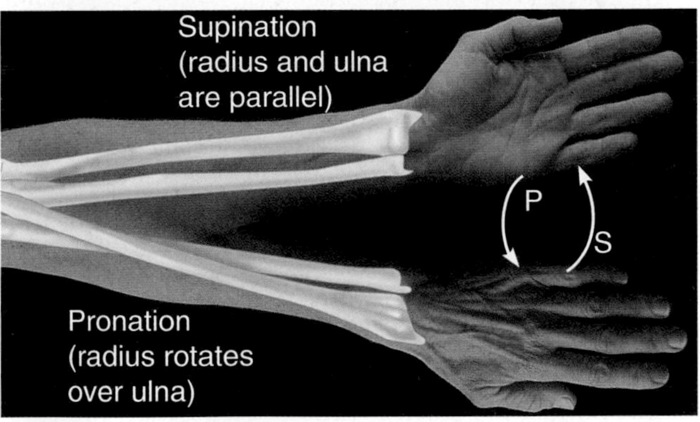

(g) Supination (S) and pronation (P)

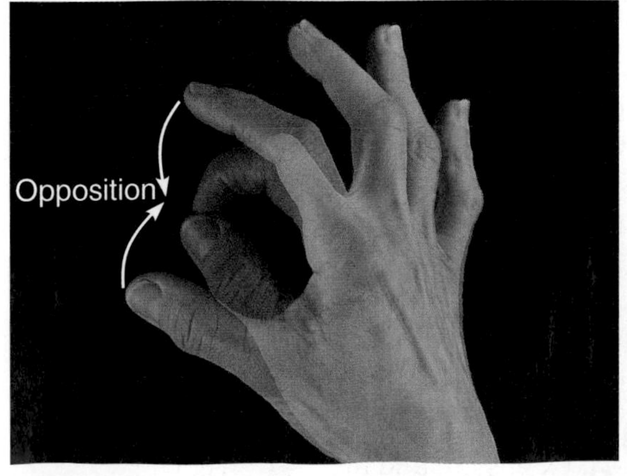

(h) Opposition

FIGURE 6.13 ***(continued)***

Some of these special movements are shown in Figure 6.13.

- **Dorsiflexion and plantar flexion.** Up and down movements of the foot at the ankle are given special names. Lifting the foot so that its superior surface approaches the shin (standing on your heels) is called dorsiflexion, whereas depressing the foot (pointing the toes) is called plantar flexion (Figure 6.13e). Dorsiflexion of the foot corresponds to extension of the hand at the wrist, whereas plantar flexion of the foot corresponds to flexion of the hand.
- **Inversion and eversion.** Inversion and eversion are also special movements of the foot (Figure 6.13f). To invert the foot, turn the sole medially. To evert the foot, turn the sole laterally.
- **Supination and pronation.** The terms supination (soo″pĭ-na′shun; "turning backward") and pronation (pro-na′shun; "turning forward") refer to movements of the radius around the ulna (Figure 6.13g). Supination occurs when the forearm rotates laterally so that the palm faces anteriorly, and the radius and ulna are parallel. Pronation occurs when the forearm rotates medially so that the palm faces posteriorly. Pronation brings the radius across the ulna so that the two bones form an X. A helpful memory trick: If you lift a cup of soup up to your mouth *on your palm,* you are supinating ("soup"-inating).
- **Opposition.** In the palm of the hand, the saddle joint between metacarpal 1 and the carpals allows opposition of the thumb (Figure 6.13h). This is the action by which you move your thumb to touch the tips of the other fingers on the same hand. It is this unique action that makes the human hand such a fine tool for grasping and manipulating things.

Interactions of Skeletal Muscles in the Body

Muscles can't push—they can only pull as they contract—so most often body movements are the result of the activity of two or more muscles acting together or against each other. Muscles are arranged in such a way that whatever one muscle (or group of muscles) can do, other muscles can reverse. Because of this, muscles are able to bring about an immense variety of movements.

The muscle that has the major responsibility for causing a particular movement is called the **prime mover.** (This physiological term has been borrowed by the business world to label a person who gets things done.) Muscles that oppose or reverse a movement are **antagonists** (an-tag′o-nists). When a prime mover is active, its antagonist is stretched and relaxed. Antagonists can be prime movers in their own right. For example, the biceps of the arm (prime mover of elbow flexion) is antagonized by the triceps (a prime mover of elbow extension).

Synergists (sin′er-jists; *syn* = together, *erg* = work) help prime movers by producing the same movement or by reducing undesirable movements. When a muscle crosses two or more joints, its contraction will cause movement in all the joints crossed unless synergists are there to stabilize them. For example, the finger-flexor muscles cross both the wrist and the finger joints. You can make a fist without bending your wrist because synergist muscles stabilize the wrist joints and allow the prime mover to act on the finger joints.

Fixators are specialized synergists. They hold a bone still or stabilize the origin of a prime mover so all the tension can be used to move the insertion bone. The postural muscles that stabilize the vertebral column are fixators, as are the muscles that anchor the scapulae to the thorax.

In summary, although prime movers seem to get all the credit for causing certain movements, the actions of antagonistic and synergistic muscles are also important in effecting smooth, coordinated, and precise movements.

Naming Skeletal Muscles

Like bones, muscles come in many shapes and sizes to suit their particular tasks in the body. Muscles are named on the basis of several criteria, each of which focuses on a particular structural or functional characteristic. Paying close attention to these cues can greatly simplify your task of learning muscle names and actions:

- **Direction of the muscle fibers.** Some muscles are named in reference to some imaginary line, usually the midline of the body or the long axis of a limb bone. When a muscle's name includes the term *rectus* (straight), its fibers run parallel to that imaginary line. For example, the rectus femoris is the straight

A Closer Look

Are Athletes Looking Good and Doing Better with Anabolic Steroids?

EVERYONE loves a winner, and top athletes are popular and make lots of money. It is not surprising that some will grasp at anything to increase their performance—including anabolic steroids. These hormones, engineered by pharmaceutical companies, were introduced in the 1950s to treat victims of certain muscle-wasting diseases and anemia and to prevent muscle atrophy in patients immobilized after surgery. Testosterone, a natural anabolic steroid hormone made by the body, triggers the increase in muscle and bone mass and other physical changes that occur during puberty and convert boys into men. Convinced that huge doses of the anabolic steroids could enhance masculinizing effects in grown men, many athletes were using the steroids by the early 1960s, and the practice is still going strong today. Indeed, it is estimated that one out of every ten young men has tried steroids, so use is no longer confined to athletes looking for the edge.

The use of these drugs has been banned by most international athletic competitions, and users (and prescribing physicians or drug dealers) are naturally reluctant to talk about it. Nonetheless, there is little question that many professional bodybuilders and athletes competing in events that require great muscle strength (such as discus throwing and weight lifting) are heavy users. Sports figures such as football players have also admitted to using steroids to help them prepare for games. Advantages of anabolic steroids cited by athletes include increased muscle mass and strength, increased oxygen-carrying capacity of the blood (because of greater red blood cell volume), and an increase in aggressive behavior (the urge to "steamroller the other guy").

But do the drugs do all that is claimed for them? Research studies have reported increases in isometric strength and body weight in steroid users. Although these are results weight lifters dream about, there is a hot dispute over whether the drugs also enhance the fine muscle coordination and endurance needed by runners and others.

Do the claimed slight advantages conferred by steroid use outweigh the risks? Absolutely not! Physicians say they cause bloated faces (a sign of steroid excess), shriveled testes, and infertility; damage the liver and promote liver cancer; and cause changes in blood-cholesterol levels (which may place long-term users at risk for coronary heart disease). Additionally, about one-third of anabolic steroid users develop serious psychiatric problems. Manic behavior in which the users undergo Jekyll–Hyde personality swings and become extremely violent (the so-called 'roid rage) is common; so, too, are depression and delusions.

A recent arrival on the scene, sold over the counter as a "nutritional performance enhancer," is ➤

Are Athletes Looking Good and Doing Better with Anabolic Steroids? *(continued)*

androstenedione, which is converted to testosterone in the body. It is taken orally and much of it is destroyed by the liver soon after ingestion, but the few milligrams that survive temporarily boost testosterone levels. "Wannabe" athletes from the fifth grade up are said to be sweeping the supplement off the drugstore shelves. This is troubling; androstenedione is not regulated by the U.S. Food and Drug Administration (FDA) and its long-term effects are unpredictable. Ongoing studies have found that males who took the supplement developed elevated levels of the female hormone estrogen as well as testosterone (raising their risk of feminizing effects such as enlarged breasts), early puberty, and stunted bone growth.

The question of why athletes use these drugs is easy to answer. Some say they are willing to do almost anything to win, short of killing themselves. Are they unwittingly doing this as well?

muscle of the thigh, or femur. Similarly, the term *oblique* as part of a muscle's name tells you that the muscle fibers run obliquely (at a slant) to the imaginary line.

- **Relative size of the muscle.** Such terms as *maximus* (largest), *minimus* (smallest), and *longus* (long) are often used in the names of muscles—for example, the gluteus maximus is the largest muscle of the gluteus muscle group.
- **Location of the muscle.** Some muscles are named for the bone with which they are associated. For example, the temporalis and frontalis muscles overlie the temporal and frontal bones of the skull, respectively.
- **Number of origins.** When the term *biceps, triceps,* or *quadriceps* forms part of a muscle name, one can assume that the muscle has two, three, or four origins, respectively. For example, the biceps muscle of the arm has two heads, or origins, and the triceps muscle has three.
- **Location of the muscle's origin and insertion.** Occasionally, muscles are named for their attachment sites. For example, the sternocleidomastoid muscle has its origin on the sternum (*sterno*) and clavicle (*cleido*) and inserts on the *mastoid* process of the temporal bone.
- **Shape of the muscle.** Some muscles have a distinctive shape that helps to identify them. For example, the deltoid muscle is roughly triangular (*deltoid* means "triangular").
- **Action of the muscle.** When muscles are named for their actions, terms such as *flexor, extensor,* and *adductor* appear in their names. For example, the adductor muscles of the thigh all bring about its adduction, and the extensor muscles of the wrist all extend the wrist.

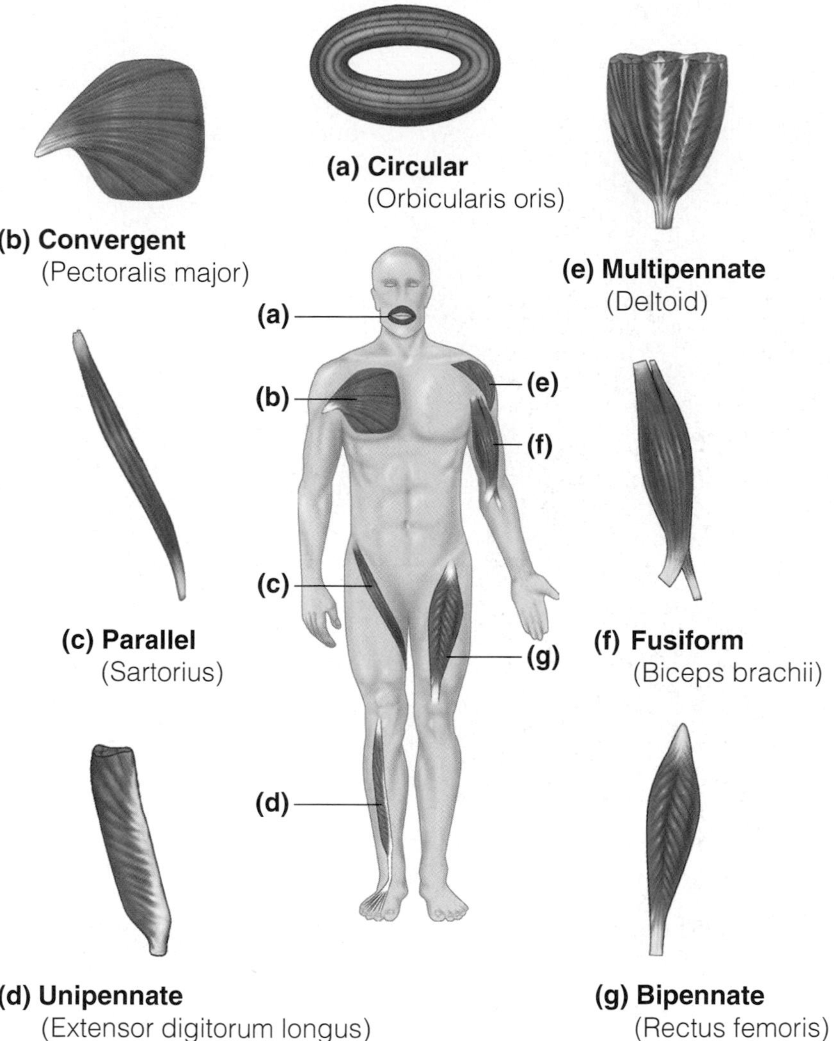

FIGURE 6.14 Relationship of fascicle arrangement to muscle structure.

Arrangement of Fascicles

Skeletal muscles consist of fascicles, but fascicle arrangements vary, producing muscles with different structures and functional properties. The most common patterns of fascicle arrangement are described next.

The pattern is **circular** when the fascicles are arranged in concentric rings (Figure 6.14a). Circular muscles are typically found surrounding external body openings which they close by contracting. A general term for such muscles is *sphincters* ("squeezers"). Examples are the orbicularis muscles surrounding the eyes and mouth.

In a **convergent** muscle, the fascicles converge toward a single insertion tendon. Such a muscle is triangular or fan-shaped like the pectoralis major muscle of the anterior thorax (Figure 6.14b).

In a **parallel** arrangement, the length of the fascicles run parallel to the long axis of the muscle. These muscles are straplike (Figure 6.14c). A modification of the parallel arrangement, called **fusiform,** results in a spindle-shaped muscle with an expanded belly (midsection) like the biceps brachii muscle of the arm (Figure 6.14f).

In a **pennate** (pen′āt; "feather") pattern, short fascicles attach obliquely to a central tendon. In the extensor digitorum muscle of the leg, the fascicles insert into only one side of the tendon and the muscle is *unipennate* (Figure 6.14d). If the fascicles insert into opposite sides of the tendon or from several different sides, the muscle is *bipennate* (Figure 6.14g) or *multipennate* (Figure 6.14e), respectively.

A muscle's fascicle arrangement determines its range of motion and power. The longer and the more nearly parallel the fascicles are to a muscle's long axis, the more the muscle can shorten, but such muscles are not usually very powerful. Muscle power depends more on the total number of muscle

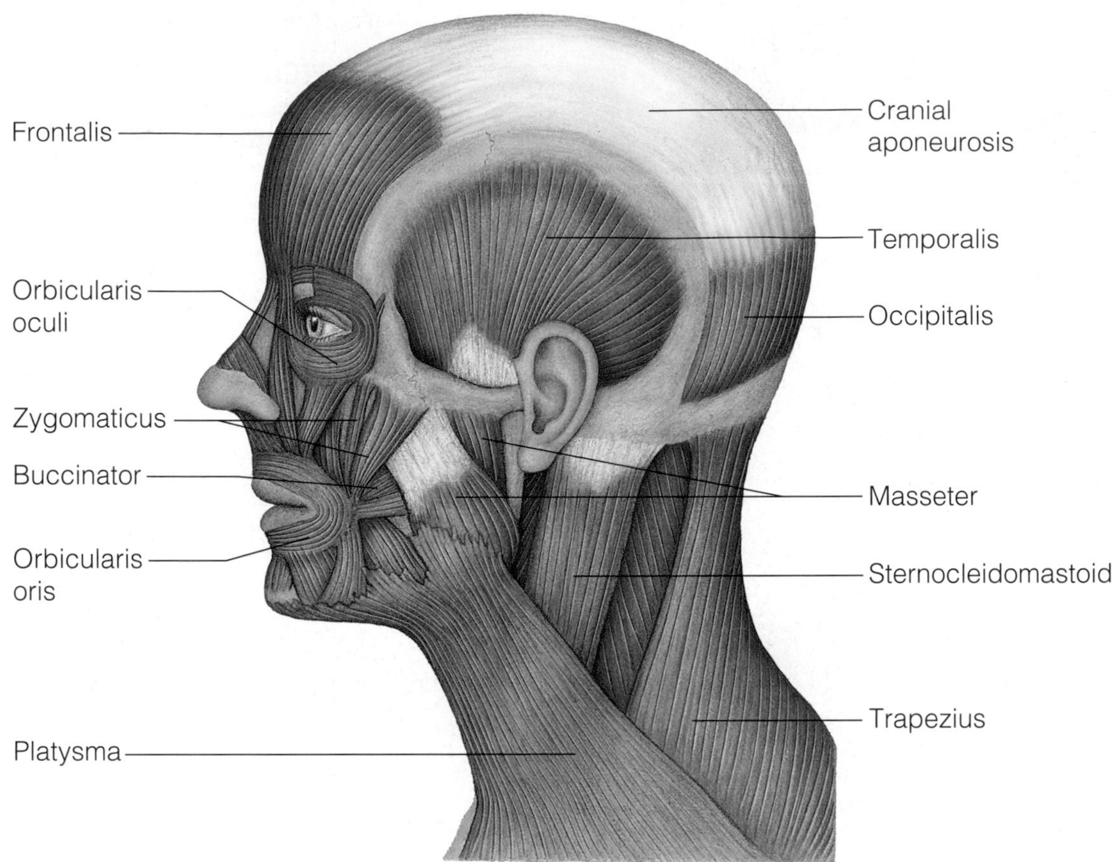

FIGURE 6.15 Superficial muscles of the face and neck.

cells in the muscle. The stocky bipennate and multipennate muscles, which pack in the most fibers, shorten very little but are very powerful.

Gross Anatomy of Skeletal Muscles

It is beyond the scope of this book to describe the hundreds of skeletal muscles of the human body. Only the most important muscles are described here. In addition, all the superficial muscles considered are summarized in Tables 6.3 and 6.4 and are illustrated in overall body views in Figures 6.21 and 6.22, which accompany the tables (pp. 210–213).

Head and Neck Muscles

The head muscles (Figure 6.15) are an interesting group. They have many specific functions but are usually grouped into two large categories—facial muscles and chewing muscles. Facial muscles are unique because they are inserted into soft tissues such as other muscles or skin. When they pull on the skin of the face, they permit us to smile faintly, grin widely, frown, pout, deliver a kiss, and so forth. The chewing muscles begin the breakdown of food for the body.

Facial Muscles

Frontalis The frontalis covers the frontal bone as it runs from the cranial aponeurosis to the skin of the eyebrows, where it inserts. This muscle allows you to raise your eyebrows, as in surprise, and to wrinkle your forehead. At the posterior end of the cranial aponeurosis is the small **occipitalis** muscle, which covers the posterior aspect of the skull and pulls the scalp posteriorly.*

Orbicularis Oculi The orbicularis oculi (or-bik″u-la′ris ok′u-li) has fibers that run in circles around

*Although the current references on anatomic terminology refer to the frontalis and occipitalis as the *frontal* and *occipital bellies* of the *epicranius* ("over the cranium") muscle, we will continue to use the terms frontalis and occipitalis here.

the eyes. It allows you to close your eyes, squint, blink, and wink.

Orbicularis Oris The orbicularis oris is the circular muscle of the lips. Because it closes the mouth and protrudes the lips, it is often called the "kissing" muscle.

Buccinator The fleshy buccinator (bu′sĭ-na″tor) muscle runs horizontally across the cheek and inserts into the orbicularis oris. It flattens the cheek (as in whistling or blowing a trumpet). It is also listed as a chewing muscle because it compresses the cheek to hold the food between the teeth during chewing.

Zygomaticus The zygomaticus (zi″go-mat′i-kus) extends from the corner of the mouth to the cheekbone. It is often referred to as the "smiling" muscle because it raises the corners of the mouth upward.

Chewing Muscles

The buccinator muscle, which is a member of this group, is described with the facial muscles.

Masseter The masseter (mă-se′ter) covers the angle of the lower jaw as it runs from the zygomatic process of the temporal bone to the mandible. This muscle closes the jaw by elevating the mandible.

Temporalis The temporalis is a fan-shaped muscle overlying the temporal bone. It inserts into the mandible and acts as a synergist of the masseter in closing the jaw.

Neck Muscles

For the most part, the neck muscles, which move the head and shoulder girdle, are small and straplike. Only two neck muscles are considered here.

Platysma The platysma is a single sheetlike muscle that covers the anterolateral neck (see Figure 6.15). It originates from the connective tissue covering of the chest muscles and inserts into the area around the mouth. Its action is to pull the corners of the mouth inferiorly, producing a downward sag of the mouth.

Sternocleidomastoid The paired sternocleidomastoid (ster″no-kli″do-mas′toid) muscles are two-headed muscles, one found on each side of the neck. Of the two heads of each muscle, one arises from the sternum and the other arises from the clavicle. The heads fuse before inserting into the mastoid process of the temporal bone. When both sternocleidomastoid muscles contract together, they flex your neck. (It is this action of bowing the head that has led some people to call these muscles the "prayer" muscles.) If just one muscle contracts, the head is rotated toward the opposite side.

Homeostatic Imbalance

In some difficult births, one of these muscles may be injured and develop spasms. A baby injured in this way has *torticollis* (tor″ti-kol′is), or wryneck. ▲

Trunk Muscles

The trunk muscles include (1) those that move the vertebral column (most of which are posterior antigravity muscles); (2) anterior thorax muscles, which move the ribs, head, and arms; and (3) muscles of the abdominal wall, which help to move the vertebral column and, most importantly, form the muscular "natural girdle" of the abdominal body wall.

Anterior Muscles (Figure 6.16)

Pectoralis Major The pectoralis (pek″to-ra′lis) major is a large fan-shaped muscle covering the upper part of the chest. Its origin is from the sternum, shoulder girdle, and the first six ribs. It inserts on the proximal end of the humerus. This muscle forms the anterior wall of the axilla and acts to adduct and flex the arm.

Intercostal Muscles The intercostal muscles are deep muscles found between the ribs. (Although they are not shown in Figure 6.16, which only shows superficial muscles, they are illustrated in Figure 6.21.) The external intercostals are important in breathing because they help to raise the rib cage for breathing air in. The internal intercostals, which lie deep to the external intercostals, depress the rib cage, which helps to move air out of the lungs when you exhale forcibly.

Muscles of the Abdominal Girdle The anterior abdominal muscles (rectus abdominis, external and internal obliques, and transversus abdominis) form a natural "girdle" that reinforces the body trunk. Taken together, they resemble the structure of plywood because the fibers of each muscle or muscle pair run in a different direction. Just as plywood is exceptionally strong for its thickness, the abdominal muscles form a muscular wall that is well suited for its job of containing and protecting the abdominal contents.

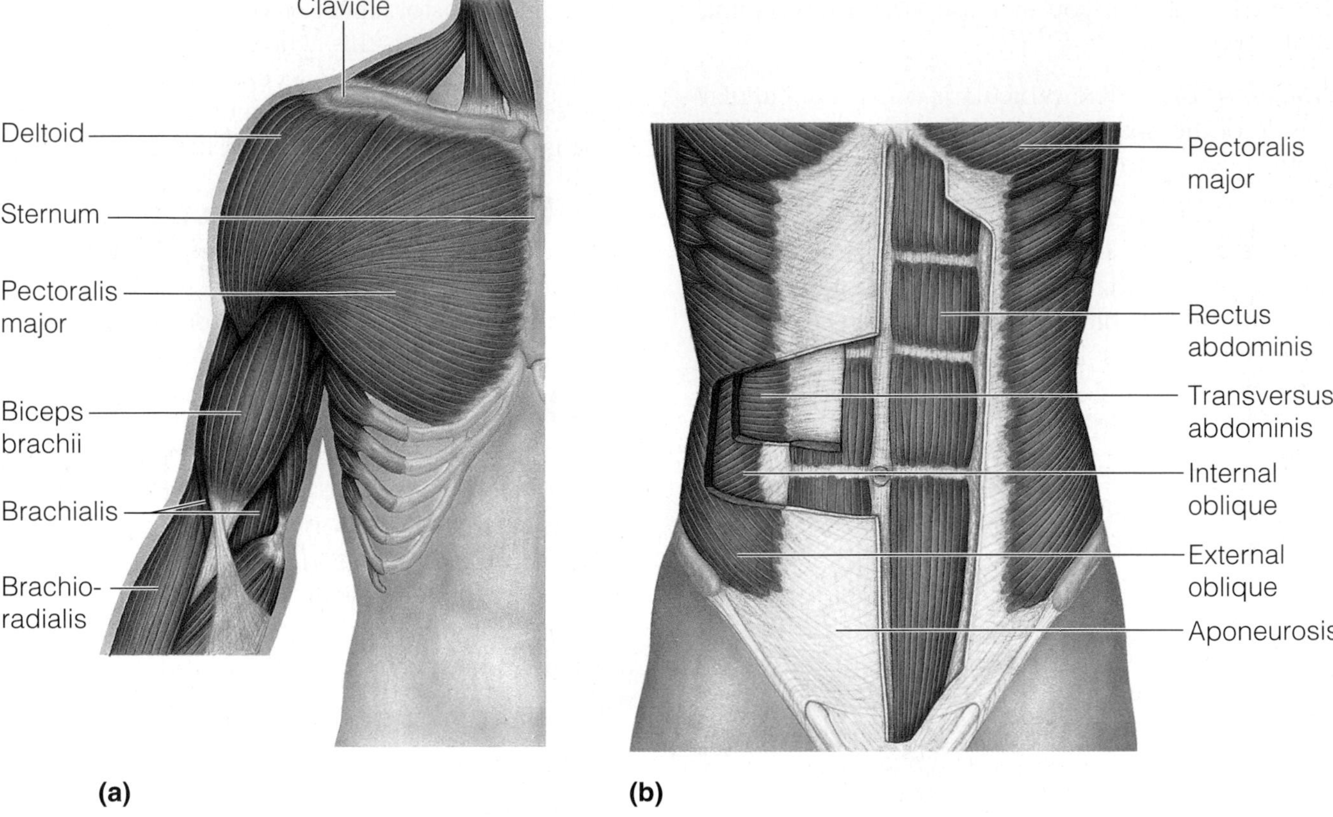

FIGURE 6.16 Muscles of the anterior trunk, shoulder, and arm. **(a)** Muscles crossing the shoulder joint, causing movements of the arm. The platysma of the neck is removed. **(b)** Muscles of the abdominal wall. Portions of the superficial muscles of the right side of the abdomen are cut away to reveal the deeper muscles.

- **Rectus abdominis.** The paired straplike rectus abdominis muscles are the most superficial muscles of the abdomen. They run from the pubis to the rib cage, enclosed in an aponeurosis. Their main function is to flex the vertebral column. They also compress the abdominal contents during defecation and childbirth and are involved in forced breathing.
- **External oblique.** The external oblique muscles are paired superficial muscles that make up the lateral walls of the abdomen. Their fibers run downward and medially from the last eight ribs and insert into the ilium. Like the rectus abdominis, they flex the vertebral column, but they also rotate the trunk and bend it laterally.
- **Internal oblique.** The internal oblique muscles are paired muscles deep to the external obliques. Their fibers run at right angles to those of the external obliques. They arise from the iliac crest and insert into the last three ribs. Their functions are the same as those of the external obliques.
- **Transversus abdominis.** The transversus abdominis is the deepest muscle of the abdominal wall and has fibers that run horizontally across the abdomen. It arises from the lower ribs and iliac crest and inserts into the pubis. This muscle compresses the abdominal contents.

Posterior Muscles (Figure 6.17)

Trapezius The trapezius (trah-pe′ze-us) muscles are the most superficial muscles of the posterior neck and upper trunk. When seen together, they form a diamond- or kite-shaped muscle mass. Their origin is very broad. Each muscle runs from the occipital bone of the skull down the vertebral column to the end of the thoracic vertebrae. They then flare laterally to insert on the scapular spine

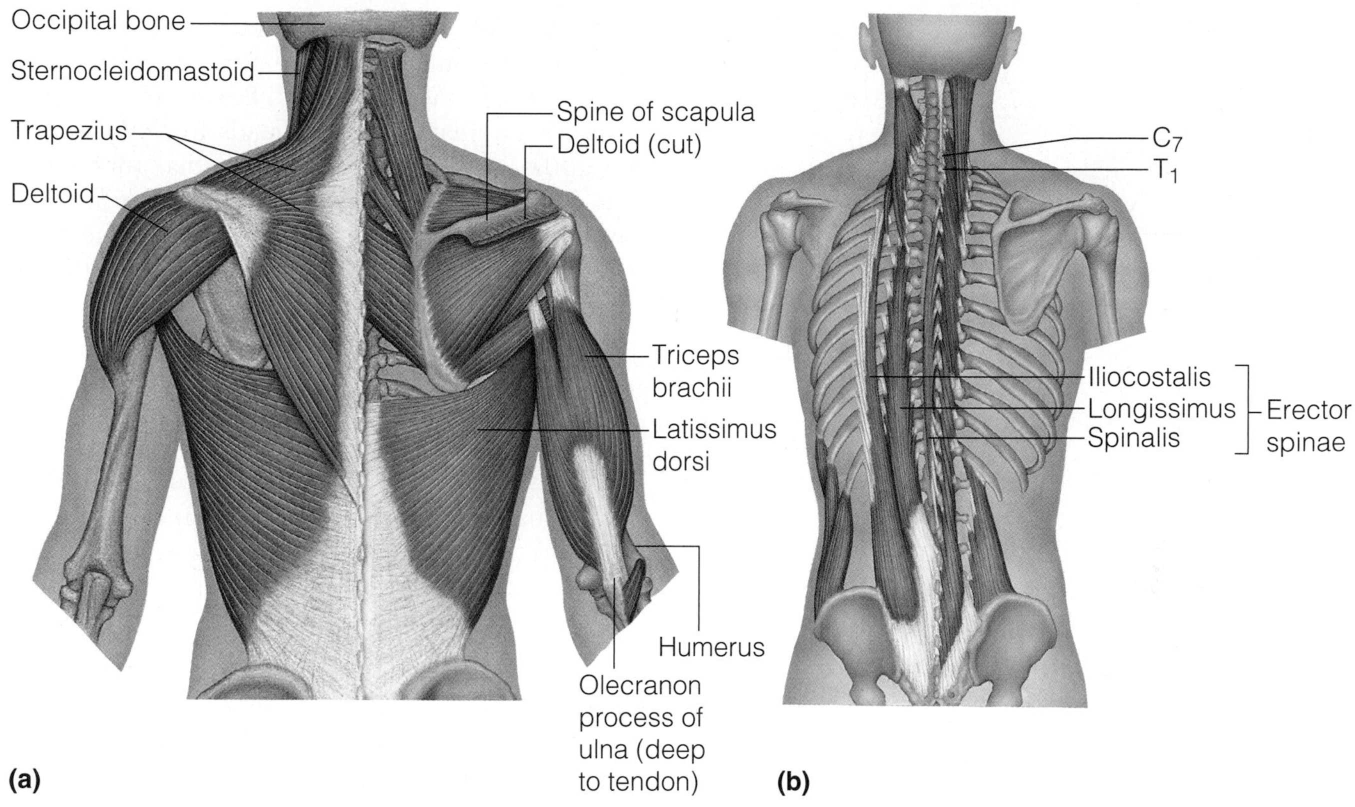

FIGURE 6.17 Muscles of the posterior neck, trunk, and arm. **(a)** Superficial muscles. **(b)** The erector spinae muscles (longissimus, iliocostalis, and spinalis), deep muscles of the back.

and clavicle. The trapezius muscles extend the head (thus they are antagonists of the sternocleidomastoids). They also can elevate, depress, adduct, and stabilize the scapula.

Latissimus Dorsi The latissimus (lah-tis'ĭ-mus) dorsi is the large, flat muscle pair that covers the lower back. It originates on the lower spine and ilium and then sweeps superiorly to insert into the proximal end of the humerus. The latissimus dorsi extends and adducts the humerus. These are very important muscles when the arm must be brought down in a power stroke, as when swimming or striking a blow.

Erector Spinae The erector spinae (e-rek'tor spi'ne) group is a prime mover of back extension. These paired muscles are deep muscles of the back; they are shown in Figure 6.17b. Each erector spinae is a composite muscle consisting of three muscle columns (longissimus, iliocostalis, and spinalis) that collectively span the entire length of the vertebral column. These muscles not only act as powerful back extensors ("erectors"), but also provide resistance that helps control the action of bending over at the waist. Following injury to back structures, these muscles go into spasms, a common source of lower back pain.

Deltoid The deltoids are fleshy, triangle-shaped muscles that form the rounded shape of your shoulders (see Figure 6.17a). Because they are so bulky, they are a favorite injection site (Figure 6.18) when relatively small amounts of medication (less than 5 ml) must be given intramuscularly (into muscle). The origin of each deltoid winds across the shoulder girdle from the spine of the scapula to the clavicle. It inserts into the proximal humerus. The deltoids are the prime movers of arm abduction.

Muscles of the Upper Limb

The upper limb muscles fall into three groups. The first group includes muscles that arise from the shoulder girdle and cross the shoulder joint to insert into the humerus (see Figures 6.16 and 6.17a). These muscles, which move the arm, have

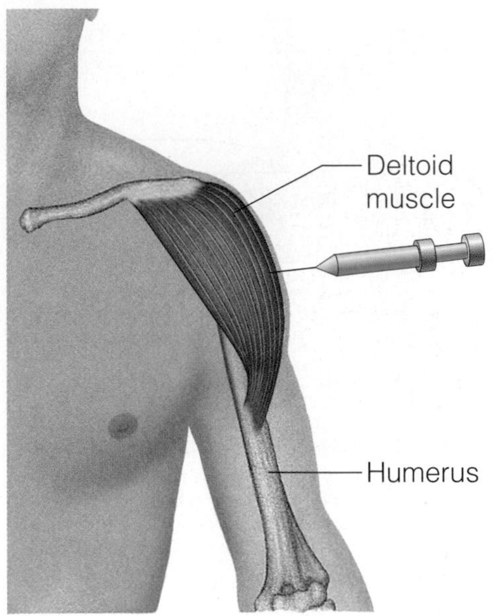

FIGURE 6.18 The fleshy deltoid muscle is a favored site for administering intramuscular injections.

already been considered—the pectoralis major, latissimus dorsi, and deltoid.

The second group causes movement at the elbow joint. These muscles enclose the humerus and insert on the forearm bones. Only the muscles of this second group will be described in this section. The third group includes the muscles of the forearm, which insert on the hand bones and cause their movement. The muscles of this last group are thin and spindle-shaped, and there are many of them. They will not be considered here except to mention their general naming and function. As a rule, the forearm muscles have names that reflect their activities. For example, the flexor carpi and flexor digitorum muscles, found on the anterior aspect of the forearm, cause flexion of the wrist and fingers, respectively. The extensor carpi and extensor digitorum muscles, found on the lateral and posterior aspect of the forearm, extend the same structures. (Some of these muscles are described briefly in Table 6.4 and illustrated in Figure 6.22.)

Muscles of the Humerus That Act on the Forearm

All *anterior* arm muscles cause elbow flexion. In order of decreasing strength these are the brachialis, biceps brachii, and brachioradialis (Figures 6.16a and 6.21).

Biceps Brachii The biceps brachii (bra′ke-i) is the most familiar muscle of the forearm because it bulges when the elbow is flexed (see Figure 6.16a). It originates by two heads from the shoulder girdle and inserts into the radial tuberosity. This muscle is the powerful prime mover for flexion of the forearm and acts to supinate the forearm. The best way to remember its action is that "it turns the corkscrew *and* pulls the cork."

Brachialis The brachialis lies deep to the biceps muscle and is as important as the biceps in elbow flexion.

Brachioradialis The brachioradialis is a fairly weak muscle that arises on the humerus and inserts into the distal forearm (see Figures 6.16a and 6.21). Hence, it resides mainly in the forearm.

Triceps Brachii The triceps muscle is the only muscle fleshing out the posterior humerus (see Figure 6.17a). Its three heads arise from the shoulder girdle and proximal humerus, and it inserts into the olecranon process of the ulna. Being the powerful prime mover of elbow extension, it is the antagonist of the biceps brachii. This muscle is often called the "boxer's" muscle because it can deliver a straight-arm knockout punch.

Muscles of the Lower Limb

Muscles that act on the lower limb cause movement at the hip, knee, and foot joints. They are among the largest, strongest muscles in the body and are specialized for walking and balancing the body. Because the pelvic girdle is composed of heavy, fused bones that allow little movement, no special group of muscles is necessary to stabilize it. This is very different from the shoulder girdle, which requires several fixator muscles.

Many muscles of the lower limb span two joints and can cause movement at both of them. Therefore, the terms *origin* and *insertion* are often interchangeable in referring to these muscles.

Muscles acting on the thigh are massive muscles that help hold the body upright against the pull of gravity and cause various movements at the hip joint. Muscles acting on the leg form the flesh of the thigh. (Recall that in common usage the term *leg* refers to the whole lower limb, but anatomically the term refers only to that part between the knee and the ankle.) The thigh muscles cross the knee and cause its flexion or extension. Because many

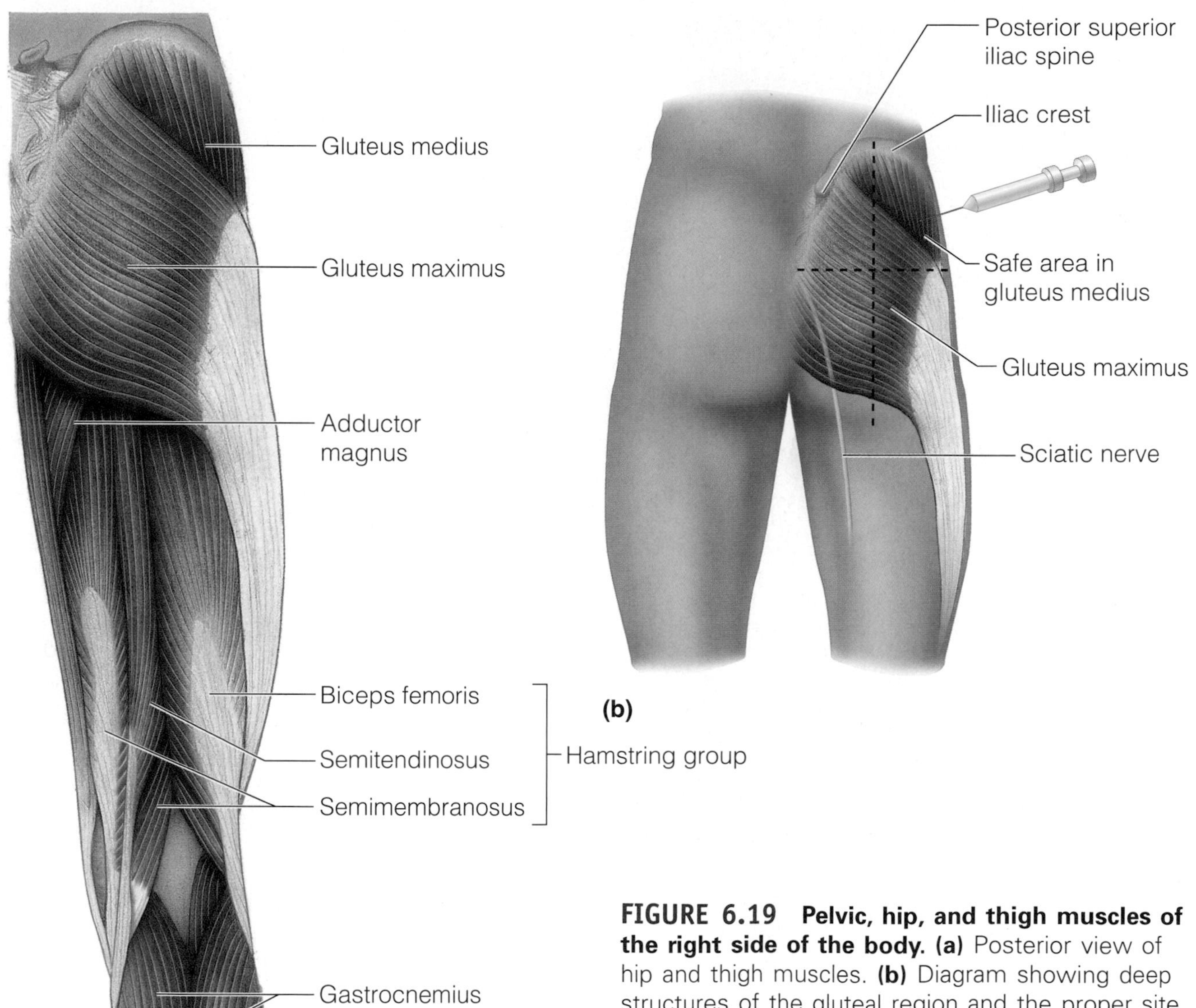

FIGURE 6.19 Pelvic, hip, and thigh muscles of the right side of the body. (a) Posterior view of hip and thigh muscles. **(b)** Diagram showing deep structures of the gluteal region and the proper site for administering an injection into the gluteus medius muscle. (Continued on page 206.)

of the thigh muscles also have attachments on the pelvic girdle, they can cause movement at the hip joint as well.

Muscles originating on the leg cause assorted movements of the ankle and foot. Only three muscles of this group will be considered, but there are many others that act to extend and flex the ankle and toe joints.

Muscles Causing Movement at the Hip Joint (Figure 6.19)

Gluteus Maximus The gluteus maximus (gloo′te-us max′ĭ-mus) is a superficial muscle of the hip that forms most of the flesh of the buttock (Figure 6.19a). It is a powerful hip extensor that acts to bring the thigh in a straight line with the pelvis. Although it is not very important in walking, it is probably the most important muscle for extending the hip when power is needed, as when climbing stairs and when jumping. It originates from the sacrum and iliac bones and inserts on the gluteal tuberosity of the femur.

Gluteus Medius The gluteus medius runs from the ilium to the femur, beneath the gluteus maximus for most of its length. The gluteus medius is a hip abductor and is important in steadying the pelvis during walking. The gluteus medius is an important site for giving intramuscular injections, particularly when more than 5 ml is administered (see Figure 6.19b). Although it might appear that the

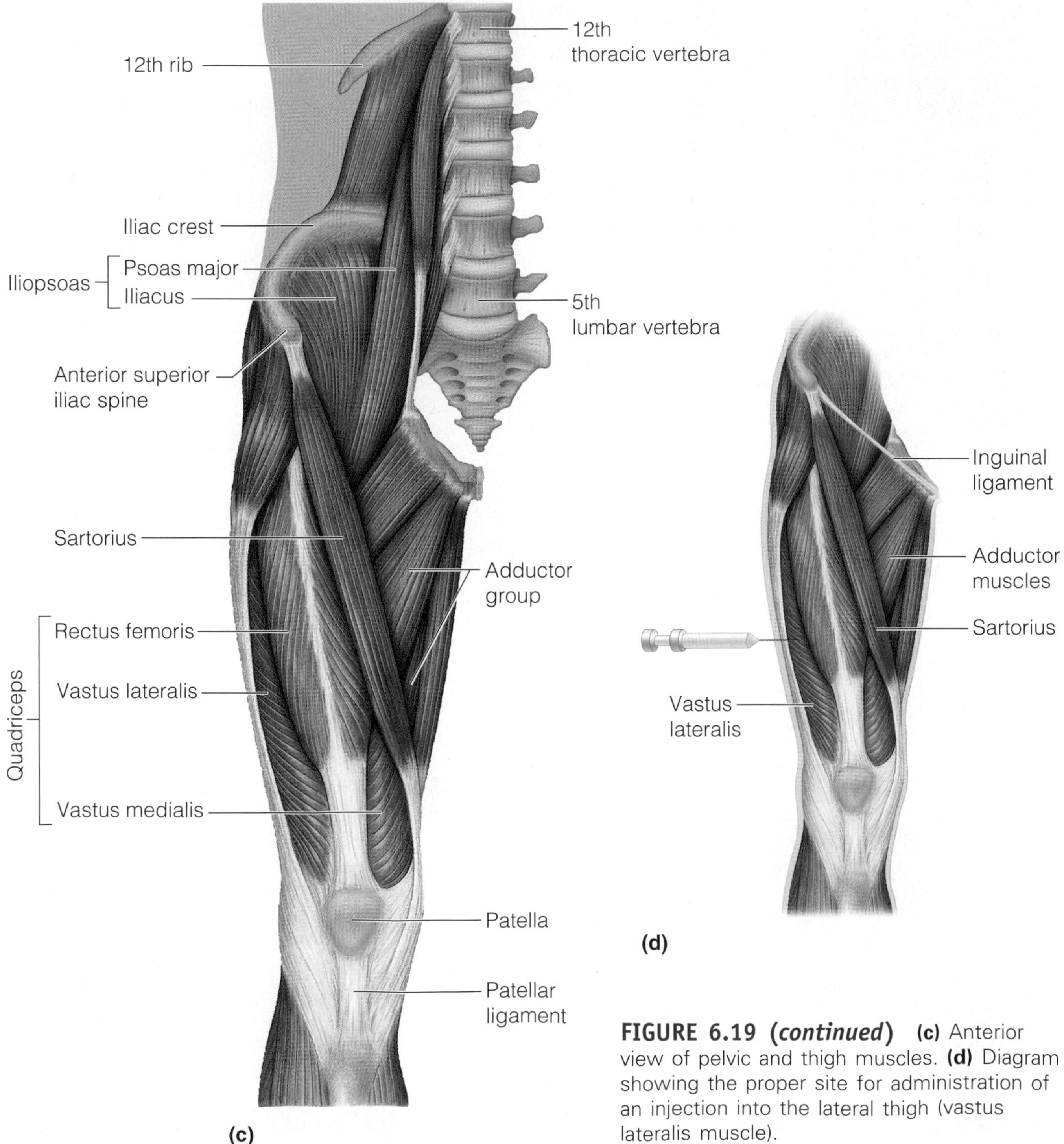

FIGURE 6.19 (*continued*) **(c)** Anterior view of pelvic and thigh muscles. **(d)** Diagram showing the proper site for administration of an injection into the lateral thigh (vastus lateralis muscle).

large, fleshy gluteus maximus that forms the bulk of the buttock mass would be a better choice, notice that the medial part of each buttock overlies the large *sciatic nerve;* hence this area must be carefully avoided. This can be accomplished by *mentally* dividing the buttock into four equal quadrants (shown by the division lines on Figure 6.19b). The upper outer quadrant then overlies the gluteus medius muscle, which is usually a very safe site for an intramuscular injection.

Iliopsoas The iliopsoas (il″e-o-so′as; the *p* is silent) is a fused muscle composed of two muscles, the *iliacus* and the *psoas major* (Figure 6.19c). It runs from the iliac bone and lower vertebrae deep inside the pelvis to insert on the lesser trochanter of the femur. It is a prime mover of hip flexion. It also acts to keep the upper body from falling backward when we are standing erect.

Adductor Muscles The muscles of the adductor group form the muscle mass at the medial side of each thigh (Figure 6.19c). As their name indicates, they adduct or press the thighs together. However, since gravity does most of the work for them, they tend to become flabby very easily. Special exercises are usually needed to keep them toned. The adductors have their origin on the pelvis and insert on the proximal aspect of the femur.

Muscles Causing Movement at the Knee Joint (Figure 6.19)

Hamstring Group The muscles forming the muscle mass of the posterior thigh are the hamstrings (Figure 6.19a). The group consists of three muscles, the **biceps femoris, semimembranosus,** and **semitendinosus,** which originate on the ischial tuberosity and run down the thigh to insert on both sides of the proximal tibia. Their name comes from the fact that butchers use their tendons to hang hams (consisting of thigh and hip muscles) for smoking. These tendons can be felt at the back of the knee.

Sartorius Compared to other thigh muscles described here, the thin, straplike sartorius (sar-to′re-us) muscle is not too important. However, it is the most superficial muscle of the thigh so it is rather hard to miss (Figure 6.19c). It runs obliquely across the thigh from the anterior iliac crest to the medial side of the tibia. It is a weak thigh flexor. The sartorius is commonly referred to as the "tailor's" muscle because it acts as a synergist to bring about the cross-legged position in which old-time tailors are often shown.

Quadriceps Group The quadriceps (kwod′rĭ-seps) group consists of four muscles—the **rectus femoris** and three **vastus muscles**—that flesh out the anterior thigh. The vastus muscles originate from the femur; the rectus femoris originates on the pelvis. All four muscles insert into the tibial tuberosity via the patellar ligament. The group as a whole acts to extend the knee powerfully, as when kicking a football. Because the rectus femoris crosses two joints, the hip and the knee, it can also help to flex the hip. The vastus lateralis and rectus femoris are sometimes used as intramuscular injection sites (Figure 6.19d), particularly in infants, who have poorly developed gluteus muscles.

Muscles Causing Movement at the Ankle and Foot (Figure 6.20)

Tibialis Anterior The tibialis anterior is a superficial muscle on the anterior leg. It arises from the upper tibia and then parallels the anterior crest as it runs to the tarsal bones, where it inserts by a long tendon. It acts to dorsiflex and invert the foot.

Extensor Digitorum Longus Lateral to the tibialis anterior, this muscle arises from the lateral tibial condyle and proximal radius and inserts into the phalanges of toes 2 to 5. It is a prime mover of toe extension and a dorsiflexor of the foot.

Fibularis Muscles The three fibularis muscles—**longus, brevis,** and **tertius**—are found on the lateral part of the leg. They arise from the fibula and insert into the metatarsal bones of the foot. The group as a whole plantar flexes and everts the foot.

Prove It Yourself

Palpate Muscles as They Contract

The following demonstrations will help you locate and identify specific muscles discussed in this chapter:

- Go into a deep bend. Now palpate your *gluteus maximus* muscle as you extend your hip to stand up again.
- Sit down and have a friend hold on to your leg. Demonstrate the contraction of the anterior *rectus femoris* by trying to extend your knee against resistance. Note how the patellar tendon reacts. The *biceps femoris* of the posterior thigh comes into play when you flex your knee against your friend's applied resistance.
- Now stand on your toes. Have a friend palpate the lateral and medical heads of your *gastrocnemius* and follow it to its insertion in the calcaneal tendon.
- Dorsiflex and invert your foot while palpating your *tibialis anterior* muscle, which parallels the sharp anterior crest of the tibia laterally.

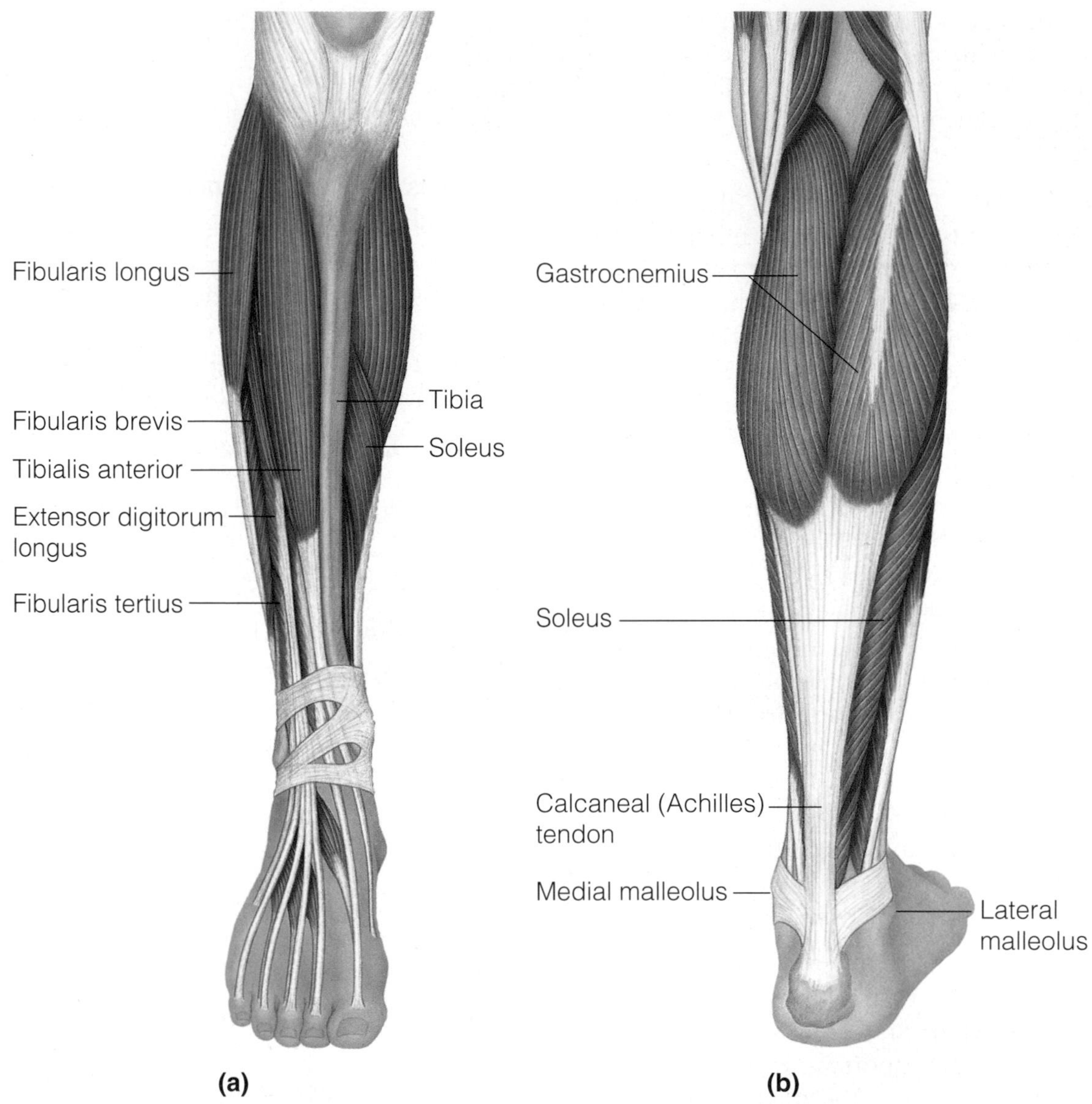

FIGURE 6.20 Superficial muscles of the right leg. (a) Anterior view; **(b)** posterior view.

Gastrocnemius The gastrocnemius (gas″trok-ne′me-us) muscle is a two-bellied muscle that forms the curved calf of the posterior leg. It arises by two heads, one from each side of the distal femur, and inserts through the large *calcaneal (Achilles) tendon* into the heel of the foot. It is a prime mover for plantar flexion of the foot; for this reason it is often called the "toe dancer's" muscle. If its insertion tendon is cut, walking is very difficult. The foot drags because the heel cannot be lifted.

Soleus Deep to the gastrocnemius is the fleshy soleus muscle. Because it arises on the tibia and fibula (rather than the femur), it does not affect knee movement, but like the gastrocnemius, it is a strong plantar flexor of the foot.

Remember that most of the superficial muscles previously described are shown in anterior and posterior views of the body as a whole in Figures 6.21 and 6.22 and are summarized in Tables 6.3 and 6.4. Take the time to review these muscles again before continuing with this chapter.

Focus on Careers

Massage Therapist

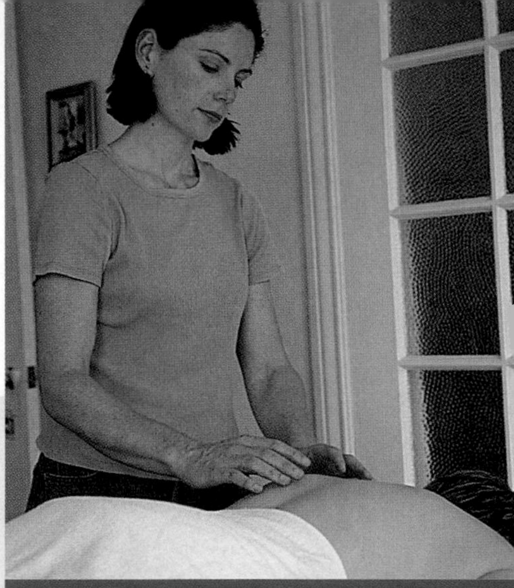

To be effective, a massage therapist needs a thorough understanding of anatomy and physiology.

MANY of us think of massage simply as something that feels good. While this is certainly true, there's a lot of skill involved in massaging the body correctly.

Diana Syverud, part-owner of the Berkeley Massage and Self-Healing Center in Berkeley, California, doesn't just make her clients feel good; she helps them recover from injuries and prevent further problems. Many of them suffer from pain due to overusing certain muscles. "Most often, I see people who spend hours at the computer," she explains. "This can lead to trouble with the flexor carpi ulnaris, trapezius, and other muscles in the neck and back. But you don't need to have a desk job to experience muscle and tendon problems. I treat musicians who are sore from spending hours playing the violin and waitresses who have painful wrists from carrying heavy trays."

Training for massage therapists varies widely. State requirements for hours of training vary from 200 to 500 hours or more. Some states also require therapists to be licensed, while others do not. Syverud studied in Washington, where she completed an 850-hour program plus an optional internship. Regardless of their length, all effective massage programs require an understanding of anatomy and physiology; Syverud estimates it made up more than one-third of her course work.

"When I was studying anatomy and physiology, sometimes I wondered why I had to learn all that," she admits. "But today, I'm glad I did. It's important for me to know where muscles are located in the body and where they attach. For instance, sometimes a muscle may contract and pinch the sciatic nerve, causing excruciating pain. I check to determine which muscle is contracted and use various massage techniques to release it. I also help clients who have no specific muscular problems stretch and strengthen their muscles, increasing their range of motion." She notes another reason she's glad she took those classes: some parts of the body are better left untouched. "It's important to know the areas of the body where massage is contraindicated, such as in the area of the esophagus or hyoid bone."

"Anatomy and physiology classes are fascinating because you're really learning how your own body works," she comments. "This knowledge is valuable, for your clients' sake and for your own welfare."

Syverud's favorite aspect of her work is the one-on-one contact. "Typically I meet with a client for at least an hour at a time, often over a period of several months," she says. "Many initially come for a specific reason, then stay with me for years, long after we've resolved their original problem. We might meet weekly while they're in pain, then switch to monthly appointments to help prevent future problems." Some customers are referred to her by chiropractors, but most come on their own. If she feels their problems require more than massage, she refers them to appropriate health-care practitioners such as chiropractors, psychiatrists, or internists.

"Anatomy and physiology classes are fascinating because you're really learning how your own body works."

While Syverud is in private practice, many massage therapists work in spas, health clubs, or chiropractic offices. Often employers prefer to hire graduates of programs accredited by the American Massage Therapy Association (AMTA). Please note that accreditation procedures vary from state to state. For more information about the field, contact

AMTA
820 Davis Street, Suite 100
Evanston, IL 60201
(847) 864-0123
http://www.amtamassage.org

For additional information on this career, click the Focus on Careers link at www.anatomyandphysiology.com.

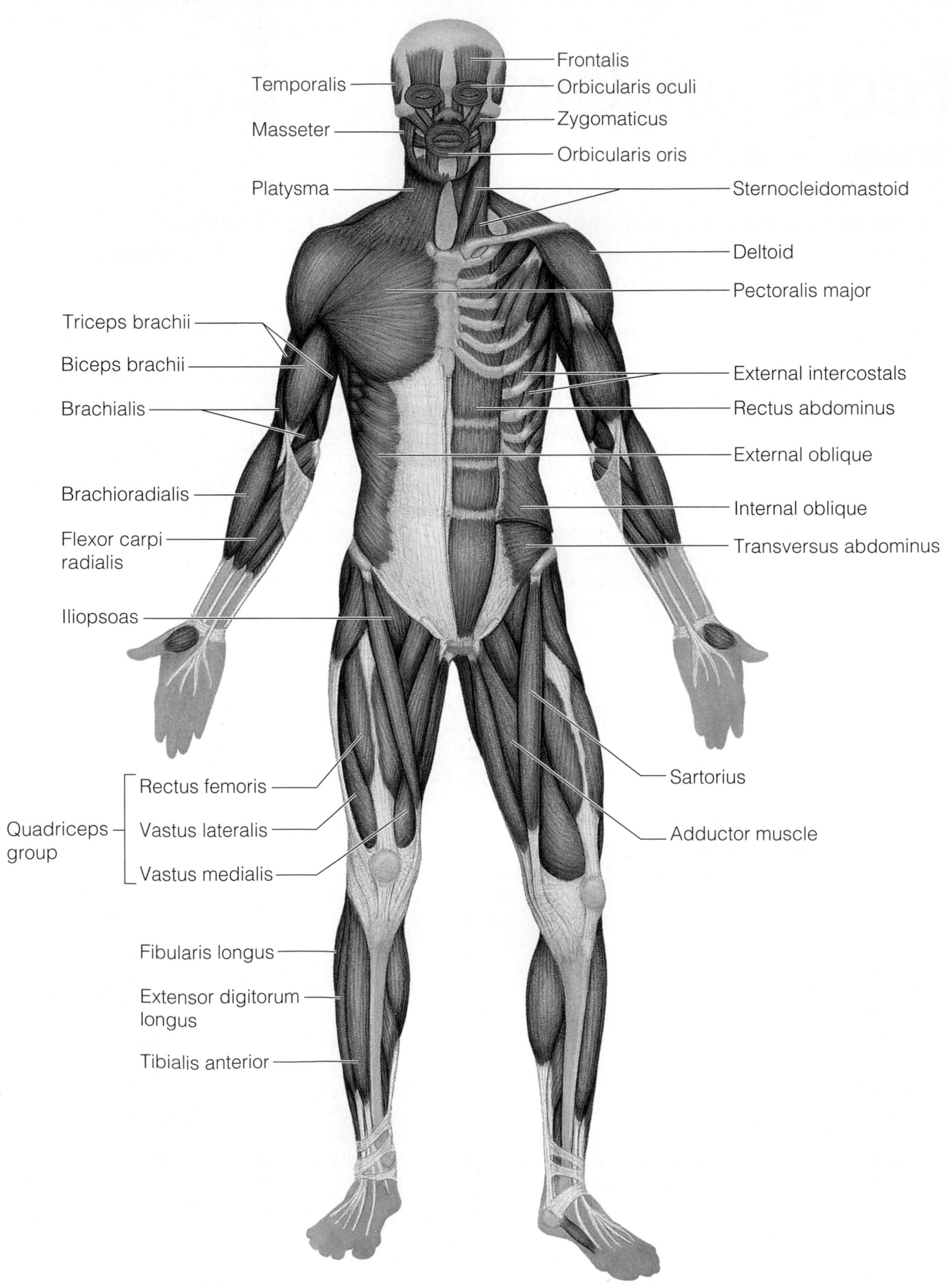

FIGURE 6.21 Major superficial muscles of the anterior surface of the body.

TABLE 6.3 Superficial Anterior Muscles of the Body (See Figure 6.21)

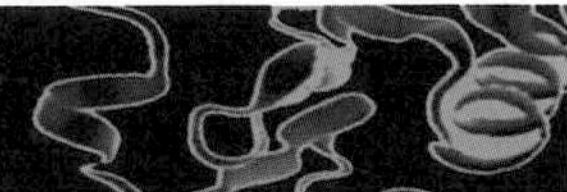

Name	Origin	Insertion	Primary Action(s)
Head/Neck Muscles			
Frontalis	Cranial aponeurosis	Skin of eyebrows	Raises eyebrows
Orbicularis oculi	Frontal bone and maxilla	Tissue around eyes	Blinks and closes eyes
Orbicularis oris	Mandible and maxilla	Skin and muscle around mouth	Closes and protrudes lips
Temporalis	Temporal bone	Mandible	Closes jaw
Zygomaticus	Zygomatic bone	Skin and muscle at corner of lips	Raises corner of mouth
Masseter	Temporal bone	Mandible	Closes jaw
Buccinator	Maxilla and mandible near molars	Orbicularis oris	Compresses cheek as in whistling and sucking; holds food between teeth during chewing
Sternocleidomastoid	Sternum and clavicle	Temporal bone (mastoid process)	Flexes neck; rotates head
Platysma	Connective tissue covering of superior chest muscles	Tissue around mouth	Pulls corners of mouth inferiorly
Trunk Muscles			
Pectoralis major	Sternum, clavicle, and first to sixth ribs	Proximal humerus	Adducts and flexes humerus
Rectus abdominis	Pubis	Sternum and fifth to seventh ribs	Flexes vertebral column
External oblique	Lower eight ribs	Iliac crest	Flexes and rotates vertebral column
Arm/Shoulder Muscles			
Biceps brachii	Scapula of shoulder girdle	Proximal radius	Flexes elbow and supinates forearm
Brachialis	Distal humerus	Proximal ulna	Flexes elbow
Deltoid	See Table 6.4		Abducts arm
Hip/Thigh/Leg Muscles			
Iliopsoas	Ilium and lumbar vertebrae	Femur (lesser trochanter)	Flexes hip
Adductor muscles	Pelvis	Proximal femur	Adduct thigh
Sartorius	Ilium	Proximal tibia	Flexes thigh on hip
Quadriceps group (vastus medialis, intermedius, and lateralis;	Vasti: Femur	Tibial tuberosity via patellar ligament	All extend knee; rectus femoris also flexes hip on thigh
and the rectus femoris)	Rectus femoris: Pelvis	Tibial tuberosity via patellar ligament	
Tibialis anterior	Proximal tibia	First cuneiform (tarsal) and first metatarsal of foot	Dorsiflexes and inverts foot
Extensor digitorum longus	Proximal tibia and radius	Distal toes 2–5	Extends toes and dorsiflexes foot
Fibularis muscles	Fibula	Metatarsals of foot	Plantar flex and evert foot

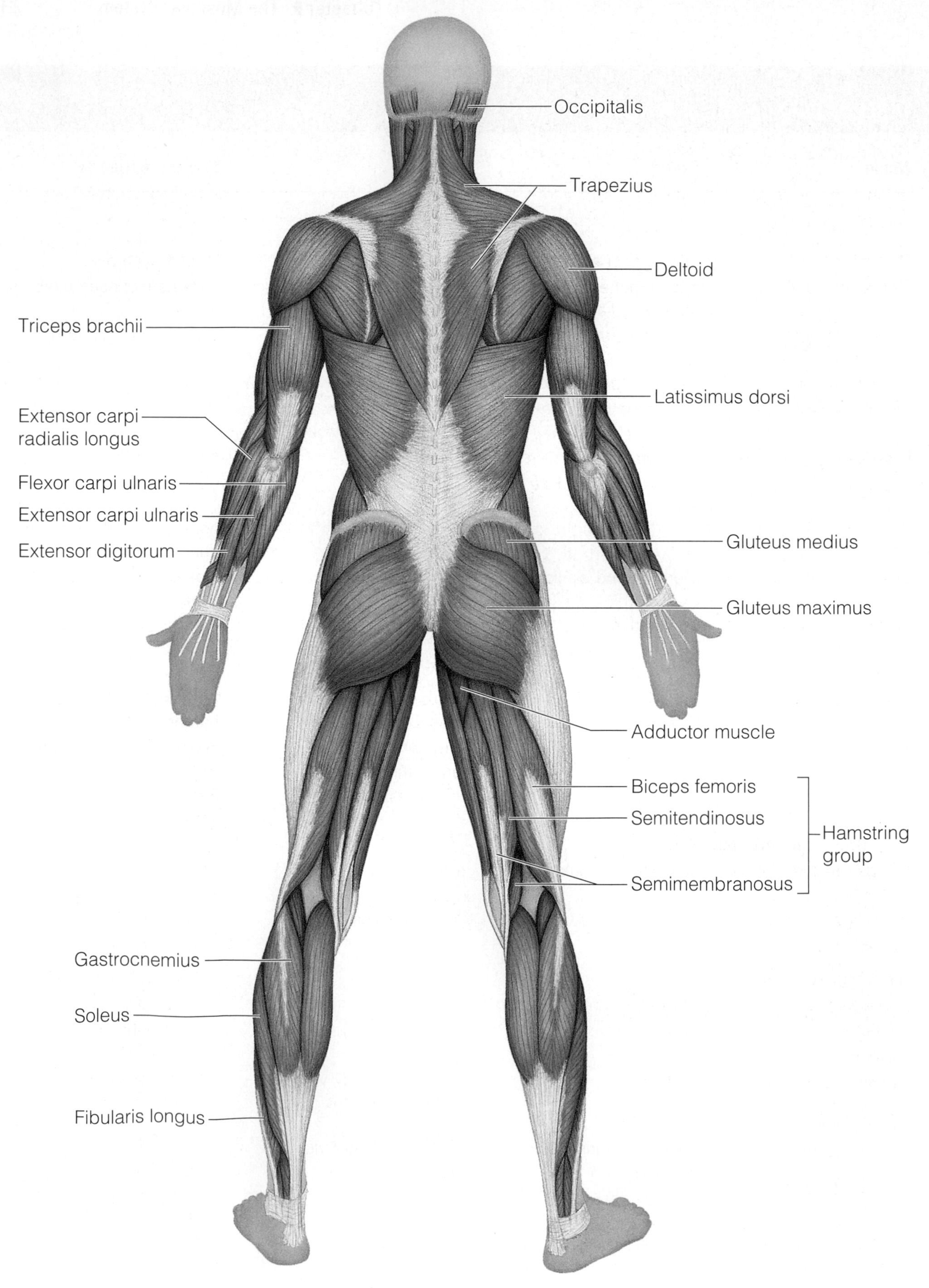

FIGURE 6.22 Major superficial muscles of the posterior surface of the body.

TABLE 6.4 Superficial Posterior Muscles of the Body (Some Forearm Muscles Also Shown) (See Figure 6.22)

Name	Origin	Insertion	Primary Action(s)
Neck/Trunk/Shoulder Muscles			
Trapezius	Occipital bone and all cervical and thoracic vertebrae	Scapular spine and clavicle	Extends neck and adducts scapula
Latissimus dorsi	Lower spine and iliac crest	Proximal humerus	Extends and adducts humerus
Erector spinae*	Iliac crests, ribs 3–12, and vertebrae	Ribs, thoracic and cervical vertebrae	Extends back
Deltoid	Scapular spine and clavicle	Humerus (deltoid tuberosity)	Abducts humerus
Arm/Forearm Muscles			
Triceps brachii	Shoulder girdle and proximal humerus	Olecranon process of ulna	Extends elbow
Flexor carpi radialis	Distal humerus	Second and third metacarpals	Flexes wrist and abducts hand (see Figure 6.21)
Flexor carpi ulnaris	Distal humerus and posterior ulna	Carpals of wrist and fifth metacarpal	Flexes wrist and adducts hand
Flexor digitorum superficialis†	Distal humerus, ulna and radius	Middle phalanges of second to fifth fingers	Flexes wrist and fingers
Extensor carpi radialis	Humerus	Base of second and third metacarpals	Extends wrist and abducts hand
Extensor digitorum	Distal humerus	Distal phalanges of second to fifth fingers	Extends fingers and wrist
Hip/Thigh/Leg Muscles			
Gluteus maximus	Sacrum and ilium	Proximal femur (gluteal tuberosity)	Extends hip (when forceful extension is required)
Gluteus medius	Ilium	Proximal femur	Abducts thigh; steadies pelvis during walking
Hamstring muscles (semitendinosus, semimembranosus, biceps femoris)	Ischial tuberosity	Proximal tibia (head of fibula in the case of biceps femoris)	Flex knee and extend hip
Gastrocnemius	Distal femur	Calcaneus (heel via calcaneal tendon)	Plantar flexes foot and flexes knee
Soleus	Proximal tibia and fibula	Calcaneus	Plantar flexes foot

*Erector spinae is a deep muscle group and not shown in Figure 6.22.

†Although its name indicates that it is a superficial muscle, the flexor digitorum superficialis lies deep to the flexor carpi radialis and is not visible in a superficial view.

Developmental Aspects of the Muscular System

In the developing embryo, the muscular system is laid down in segments (much like the structural plan of an earthworm), and then each segment is invaded by nerves. The muscles of the thoracic and lumbar regions become very extensive since they must cover and move the bones of the limbs. Development of the muscles and their control by the nervous system occur rather early in pregnancy. The expectant mother is often astonished by the first movements (called the *quickening*) of the fetus, which usually occur by the sixteenth week of pregnancy.

Homeostatic Imbalance

Very few congenital muscular problems occur. The exception to this is *muscular dystrophy*—a group of inherited muscle-destroying diseases that affect specific muscle groups. The muscles enlarge due to fat and connective tissue deposit, but the muscle fibers degenerate and atrophy.

The most common and serious form is *Duchenne's muscular dystrophy,* which is expressed almost exclusively in males. This tragic disease is usually diagnosed between the ages of 2 and 6 years. Active, normal-appearing children become clumsy and begin to fall frequently as their muscles weaken. The disease progresses relentlessly from the extremities upward, finally affecting the head and chest muscles. Most victims must use wheelchairs by the age of 12 and generally do not live beyond young adulthood. Although the cause of muscular dystrophy has been pinned down (the diseased muscle fibers lack a protein [called dystrophin] that helps maintain the sarcolemma), a cure is still elusive. ▲

Initially after birth, a baby's movements are all gross reflex types of movements. Because the nervous system must mature before the baby can control muscles, we can trace the increasing efficiency of the nervous system by observing a baby's development of muscle control. This development proceeds in a cephalic/caudal direction, and gross muscular movements precede fine ones. Babies can raise their heads before they can sit up and can sit up before they can walk. Muscular control also proceeds in a proximal/distal direction; that is, babies can wave "bye-bye" and pull objects to themselves before using the pincher grasp to pick up a pin. All through childhood, the control of the skeletal muscles by the nervous system becomes more and more precise. By midadolescence, we have reached the peak level of development of this natural control and can simply accept it or bring it to a fine edge by athletic training.

Because of its rich blood supply, skeletal muscle is amazingly resistant to infection throughout life, and given good nutrition, relatively few problems afflict skeletal muscles. It should be repeated, however, that muscles, like bones, *will* atrophy, even with normal tone, if they are not used continually. On the other hand, a lifelong program of regular exercise keeps the whole body operating at its best possible level.

Homeostatic Imbalance

One rare disease that can affect muscles during adulthood is *myasthenia gravis* (mi"as-the'ne-ah gra'vis; *asthen* = weakness; *gravi* = heavy), a disease characterized by drooping of the upper eyelids, difficulty in swallowing and talking, and generalized muscle weakness and fatigability. The disease involves a shortage of acetylcholine receptors at the neuromuscular junction. The blood of many of these patients contains antibodies to acetylcholine receptors, which suggests that myasthenia gravis is an autoimmune disease. Although the receptors may initially be present in normal numbers, they appear to be destroyed as the disease progresses. Whatever the case, the muscle cells are not stimulated properly and get progressively weaker. Death usually occurs as a result of the inability of the respiratory muscles to function. This is called *respiratory failure.* ▲

As we age, the amount of connective tissue in the muscles increases and the amount of muscle tissue decreases; thus the muscles become stringier, or more sinewy. Since the skeletal muscles represent so much of the body weight, body weight begins to decline in the elderly person as this natural loss in muscle mass occurs. Another result of the loss in muscle mass is a decrease in muscle strength; strength decreases by about 50 percent by the age of 80. Regular exercise can help offset the effects of aging on the muscular system, and frail elders who begin to "pump iron" (use leg and hand weights) can rebuild muscle mass and dramatically increase their strength.

Systems in Sync

Homeostatic Relationships between the Muscular System and Other Body Systems

Nervous System
- Facial muscle activity allows emotions to be expressed
- Nervous system stimulates and regulates muscle activity

Endocrine System
- Growth hormone and androgens influence skeletal muscle strength and mass

Respiratory System
- Muscular exercise increases respiratory capacity
- Respiratory system provides oxygen and disposes of carbon dioxide

Lymphatic System/Immunity
- Physical exercise may enhance or depress immunity depending on its intensity
- Lymphatic vessels drain leaked tissue fluids; immune system protects muscles from disease

Cardiovascular System
- Skeletal muscle activity increases efficiency of cardiovascular functioning; helps prevent atherosclerosis and causes cardiac hypertrophy
- Cardiovascular system delivers oxygen and nutrients to muscles; carries away wastes

Digestive System
- Physical activity increases gastrointestinal mobility when at rest
- Digestive system provides nutrients needed for muscle health; liver metabolizes lactic acid

Reproductive System
- Skeletal muscle helps support pelvic organs (e.g., uterus in females); assists erection of penis and clitoris
- Testicular androgen promotes increased skeletal muscle size

Urinary System
- Physical activity promotes normal voiding behavior; skeletal muscle forms the voluntary sphincter of the urethra
- Urinary system disposes of nitrogenous wastes

Integumentary System
- Muscular exercise enhances circulation to skin and improves skin health; exercise also increases body heat, which the skin helps dissipate
- Skin protects the muscles by external enclosure

Muscular System

Skeletal System
- Skeletal muscle activity maintains bone health and strength
- Bones provide levers for muscle activity

SUMMARY

Media study tools that could provide you with additional help in reviewing specific key topics of Chapter 6 are referenced below.

IP = Interactive Physiology;

WEB = A&P Place website.

OVERVIEW OF MUSCLE TISSUES (pp. 178–182)

1. Skeletal muscle forms the muscles attached to the skeleton, which move the limbs and other body parts. Its cells are long, striated, and multinucleate. Skeletal muscle fibers are subject to voluntary control. Connective tissue coverings (endomysium, perimysium, and epimysium) enclose and protect the muscle fibers and increase the strength of skeletal muscles. Skeletal muscles make up the muscular system.

   WEB Exercise: Chapter 6, Connective Tissue Wrappings of Skeletal Muscle.

2. Smooth muscle cells are uninucleate, spindle-shaped, and arranged in opposing layers in the walls of hollow organs. When they contract, substances (food, urine, a baby) are moved along internal pathways. Smooth muscle control is involuntary.
3. Cardiac muscle cells are striated, branching cells that fit closely together and are arranged in spiral bundles in the heart. Their contraction pumps blood through the blood vessels. Control is involuntary.
4. The sole function of muscle tissue is to contract or shorten. As it contracts, it causes movement, maintains posture, stabilizes joints, and generates heat.

MICROSCOPIC ANATOMY OF SKELETAL MUSCLE (pp. 182–184)

1. The multinucleate cylindrical skeletal muscle fibers are packed with unique organelles called myofibrils. The banding pattern (striations) of the myofibrils and the cell as a whole reflects the regular arrangement of thin (actin-containing) and thick (myosin) filaments within the sarcomeres, the contractile units composing the myofibrils.

   WEB Exercise: Chapter 6, Microscopic Anatomy of a Skeletal Fiber.
   IP Muscular System Topic: Anatomy Review: Skeletal Muscle Tissue, pages 7–9.

2. Each myofibril is loosely enclosed by a specialized ER, called the sarcoplasmic reticulum (SR), which plays an important role in storing and releasing calcium ions. Calcium ions are the final trigger for muscle fiber contraction.

SKELETAL MUSCLE ACTIVITY (pp. 184–192)

1. All skeletal muscle cells are stimulated by motor neurons. When the neuron releases a neurotransmitter (acetylcholine), the permeability of the sarcolemma changes, allowing sodium ions to enter the muscle cell. This produces an electrical current (action potential), which flows across the entire sarcolemma, resulting in release of calcium ions from the SR.

   IP Muscular System Topic: The Neuromuscular Junction, pages 3–6.

2. Calcium binds to regulatory proteins on the thin filaments and exposes myosin binding sites, allowing the myosin heads on the thick filaments to attach. The attached heads pivot, sliding the thin filaments toward the center of the sarcomere, and contraction occurs. ATP provides the energy for the sliding process, which continues as long as ionic calcium is present.

   IP Muscular System Topic: Sliding Filament Theory, pages 3–28.

3. Although individual muscle cells contract completely when adequately stimulated, a muscle (an organ) responds to stimuli to different degrees, that is, it exhibits graded responses.

   WEB Exercise: Chapter 6, Graded Muscle Responses.

4. Most skeletal muscle contractions are tetanic (smooth and sustained) because rapid nerve impulses are reaching the muscle, and the muscle cannot relax completely between contractions. The strength of muscle contraction reflects the relative number of muscle cells contracting (more = stronger).
5. ATP, the immediate source of energy for muscle contraction, is stored in muscle fibers in small amounts that are quickly used up. ATP is regenerated via three routes. From the fastest to the slowest, these are via a coupled reaction of creatine phosphate with ADP, via anaerobic glycolysis and lactic acid formation, and via aerobic respiration. Only aerobic respiration requires oxygen.

   IP Muscular System Topic: Muscle Metabolism, pages 6–13.

6. If muscle activity is strenuous and prolonged, muscle fatigue occurs due to an accumulation of lactic acid in the muscle and a decrease in its energy (ATP) supply. After exercise, the oxygen debt is repaid by rapid deep breathing.
7. Muscle contractions are isotonic (the muscle shortens and movement occurs) or isometric (the muscle does not shorten, but its tension increases).
8. Muscle tone keeps muscles healthy and ready to react. It is a result of a staggered series of nerve impulses delivered to different cells within the muscle. If the nerve supply is destroyed, the muscle loses tone, becomes paralyzed, and atrophies.
9. Inactive muscles atrophy. Muscles challenged by resistance exercise to respond (almost) beyond their ability increase in size and strength. Muscles subjected to regular aerobic exercise become more efficient and stronger and can work longer without tiring. Aerobic exercise also benefits other body organ systems.

MUSCLE MOVEMENTS, TYPES, AND NAMES (pp. 192–200)

1. All muscles are attached to bones at two points. The origin is the immovable attachment; the insertion is the movable bony attachment. When contraction occurs, the insertion moves toward the origin.
2. Body movements include flexion, extension, abduction, adduction, circumduction, rotation, pronation, supination, inversion, eversion, dorsiflexion, plantar flexion, and opposition.
3. On the basis of their general functions in the body, muscles are classified as prime movers, antagonists, synergists, and fixators.
4. Muscles are named according to several criteria, including muscle size, shape, number and location of origins, associated bones, and action of the muscle.
5. Muscles have several fascicle arrangements that influence their force and degree of shortening.

   WEB Exercise: Chapter 6, Muscles of the Body; Posterior Surface Musculature.

GROSS ANATOMY OF SKELETAL MUSCLES (pp. 200–213)

1. Muscles of the head fall into two groups. The muscles of facial expression include the frontalis, orbicularis oris and oculi, and zygomaticus. The chewing muscles are the masseter, temporalis, and buccinator (which is also a muscle of facial expression).
2. Muscles of the trunk and neck move the head, shoulder girdle, and trunk and form the abdominal girdle. Anterior neck and trunk muscles include the sternocleidomastoid, pectoralis major, intercostals, rectus abdominis, external and internal obliques, and transversus abdominis. Posterior trunk and neck muscles include the trapezius, latissimus dorsi, and deltoid. Deep muscles of the back are the erector spinae muscles.
3. Muscles of the upper limb include muscles that cause movement at the shoulder joint, elbow, and hand. Muscles causing movement at the elbow include the brachialis, biceps brachii, brachioradialis, and triceps brachii.
4. Muscles of the lower extremity cause movement at the hip, knee, and foot. They include the iliopsoas, gluteus maximus and medius, adductors, quadriceps and hamstring groups, gastrocnemius, tibialis anterior, fibularis muscles, soleus, and extensor digitorum longus.

DEVELOPMENTAL ASPECTS OF THE MUSCULAR SYSTEM (p. 214)

1. Increasing muscular control reflects the maturation of the nervous system. Muscle control is achieved in a cephalic/caudal and proximal/distal direction.
2. To remain healthy, muscles must be regularly exercised. Without exercise, they atrophy; with extremely vigorous exercise, they hypertrophy.
3. As we age, muscle mass decreases, and the muscles become more sinewy. Exercise helps to retain muscle mass and strength.

REVIEW QUESTIONS

MULTIPLE CHOICE

More than one choice may apply.

1. In comparing electron micrographs of a relaxed skeletal muscle fiber and a fully contracted muscle fiber, which would be seen only in the *relaxed* fiber?
   a. Z discs
   b. Triads
   c. I bands
   d. A bands
   e. H zones
2. After ACh attaches to its receptors at the neuromuscular junction, the next step is:
   a. sodium channels open
   b. calcium binds to regulatory proteins on the thin filaments
   c. cross bridges attach
   d. ATP is hydrolyzed

3. Your ability to lift that heavy couch would be increased by which type of exercise?
   a. Aerobic
   b. Endurance
   c. Resistance
   d. Swimming
4. Someone who sticks out a thumb to hitch a ride is ______ the thumb.
   a. extending
   b. abducting
   c. adducting
   d. opposing
5. Which are ways in which muscle names have been derived?
   a. Attachments
   b. Size
   c. Function
   d. Location
6. Which of the following muscles attach to the hip bones?
   a. Rectus abdominis
   b. Rectus femoris
   c. Vastus medialis
   d. Longissimus group of erector spinae
7. Which of these thigh muscles causes movement at the hip joint?
   a. Rectus femoris
   b. Biceps femoris
   c. Vastus lateralis
   d. Semitendinosus
8. Which of the following insert on the arm?
   a. Biceps brachii
   b. Triceps brachii
   c. Trapezius
   d. Latissimus dorsi

SHORT ANSWER ESSAY

1. What is the major function of muscle?
2. Compare skeletal, smooth, and cardiac muscles as to their microscopic anatomy, location and arrangement in body organs, and function in the body.
3. Specifically, what is responsible for the banding pattern seen in skeletal muscle cells?
4. Why are the connective tissue wrappings of skeletal muscles important? Name these connective tissue coverings beginning with the finest and ending with the most coarse.
5. What is the function of tendons? How is a tendon different from an aponeurosis? How is it similar?
6. Define *neuromuscular junction, motor unit, tetanus, graded response, aerobic respiration, anaerobic glycolysis, muscle fatigue,* and *neurotransmitter.*
7. Describe the events that occur from the time a motor neuron releases acetylcholine at the neuromuscular junction until muscle cell contraction occurs.
8. How do isotonic and isometric contractions differ?
9. Muscle tone keeps muscles healthy. What is muscle tone, and what causes it? What happens to a muscle that loses its tone?
10. A skeletal muscle is attached to bones at two points. Name each of these attachment points, and indicate which is movable and which is nonmovable.
11. List the 12 body movements studied in this chapter and demonstrate each.
12. How is a prime mover different from a synergist muscle? How can a prime mover also be considered an antagonist?
13. If you were alternately contracting and relaxing your masseter muscle, what would you be doing? Name three other muscles of the face and give the location and function of each.
14. The sternocleidomastoid muscles help to flex the neck. What are their antagonists?
15. Name two muscles that reverse the movement of the deltoid muscle.
16. Name the prime mover of elbow flexion. Name its antagonist.
17. Other than acting to flex the spine and compress the abdominal contents, the abdominal muscles are extremely important in protecting and containing the abdominal viscera. What is it about the arrangement of these muscles that makes them so well suited for their job?
18. The hamstring and quadriceps muscle groups are antagonists of each other, and each group is a prime mover in its own right. What action does each muscle group perform?
19. What two-bellied muscle makes up the calf region of the leg? What is its function?
20. What happens to muscles when they are exercised regularly? Exercised vigorously as in weight lifting? Not used?
21. What is the effect of aging on skeletal muscles?
22. Should a triathlete engage in aerobic or resistance training? Explain.

AT THE CLINIC

1. Name three muscles or muscle groups used as sites for intramuscular injections. Which is most often used in babies?

2. Mr. Ahmadi was advised by his physician to lose weight and start jogging. He began to jog daily. On the sixth day, he was forced to jump out of the way of a speeding car. He heard a snapping sound that was immediately followed by pain in his right lower calf. A gap was visible between his swollen calf and his heel, and he was unable to plantar flex that foot. What do you think happened?

3. While painting her house, Sally fell off the ladder and fractured her right clavicle. Treatment prescribed by the emergency room physician included use of a sling for her right arm to immobilize the clavicle and speed its healing. What muscles are temporarily "put out of business" by the sling?

4. When Eric returned from jogging, he was breathing heavily and sweating profusely, and he complained that his legs ached and felt weak. His wife poured him a sport drink and urged him to take it easy until he could "catch his breath." On the basis of what you have learned about muscle energy metabolism, respond to the following questions:
   a. Why is Eric breathing heavily?
   b. What ATP harvesting pathway have his working muscles been using that leads to such a breathing pattern?
   c. What metabolic product(s) might account for his sore muscles and his feeling of muscle weakness?

5. Drugs are used to relax muscles during major surgery. Which of the two chemicals described next would make the best muscle relaxant and why? Chemical A binds and blocks acetylcholine receptors of muscle cells. Chemical B floods the cytoplasm of muscle cells with calcium ions.

6. Mr. Posibo has had colon surgery. Now he is experiencing weakness of the muscles on his right side only, the side in which the incision was made through the abdominal musculature. Consequently, the abdominal muscles on his left side contract more strongly, throwing his torso into a lateral flexion. Mr. Posibo needs physical therapy. What abnormal spinal curvature will result if he doesn't get it, and why?

7. When a person dies, rigor mortis sets in as ATP synthesis ceases. Explain why the lack of ATP in muscle cells would cause the muscles to become rigid rather than limp soon after death.

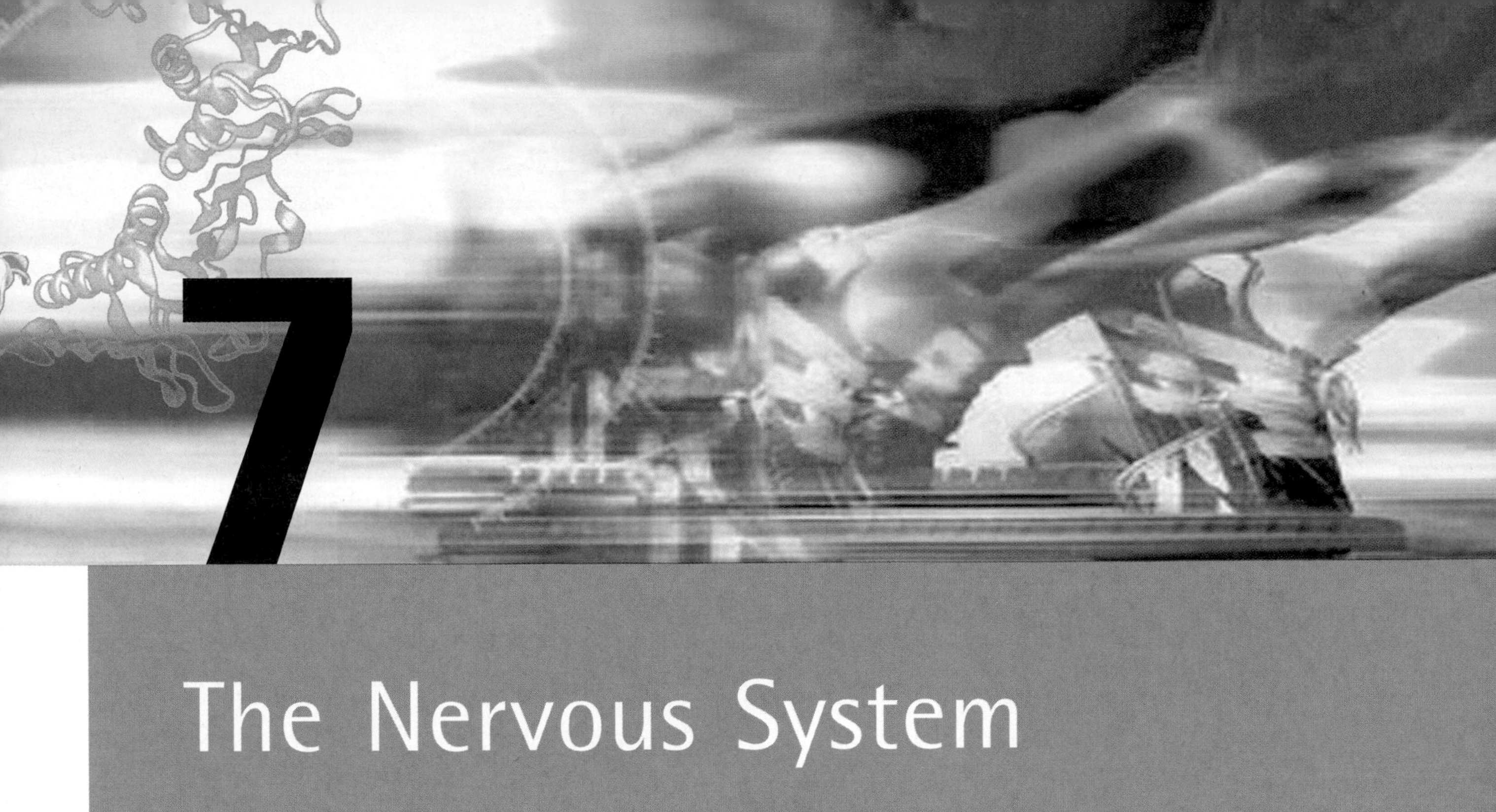

7 The Nervous System

YOUR GOALS

After completing this chapter, you will have working knowledge of the functions of the nervous system and will have mastered the objectives listed below.

FUNCTION PREVIEW

- The nervous system maintains body homeostasis with electrical signals; provides for sensation, higher mental functioning, and emotional response; and activates muscles and glands.

OBJECTIVE CHECKLIST

ORGANIZATION OF THE NERVOUS SYSTEM (pp. 223–224)

- List the general functions of the nervous system.
- Explain the structural and functional classifications of the nervous system.
- Define *central nervous system* and *peripheral nervous system,* and list the major parts of each.

NERVOUS TISSUE: STRUCTURE AND FUNCTION (pp. 224–235)

- State the function of neurons and neuroglia.
- Describe the general structure of a neuron, and name its important anatomical regions.
- Describe the composition of gray matter and white matter.
- List the two major functional properties of neurons.
- Classify neurons according to structure and function.
- List the types of general sensory receptors, and describe their functions.
- Describe the events that lead to the generation of a nerve impulse and its conduction from one neuron to another.
- Define *reflex arc,* and list its elements.

>>

(Objective Checklist, continued)

CENTRAL NERVOUS SYSTEM (pp. 235–249)

- Identify and indicate the functions of the major regions of the cerebral hemispheres, diencephalon, brain stem, and cerebellum on a human brain model or diagram.
- Name the three meningeal layers, and state their functions.
- Discuss the formation and function of cerebrospinal fluid and the blood-brain barrier.
- Compare the signs of a CVA with those of Alzheimer's disease; of a contusion with those of a concussion.
- Define *EEG,* and explain how it evaluates neural functioning.
- List two important functions of the spinal cord.
- Describe spinal cord structure.

PERIPHERAL NERVOUS SYSTEM (pp. 249–263)

- Describe the general structure of a nerve.
- Identify the cranial nerves by number and by name, and list the major functions of each.
- Describe the origin and fiber composition of (a) ventral and dorsal roots, (b) the spinal nerve proper, and (c) ventral and dorsal rami.
- Discuss the distribution of the dorsal and ventral rami of spinal nerves.
- Name the four major nerve plexuses, give the major nerves of each, and describe their distribution.
- Identify the site of origin, and explain the function of the sympathetic and parasympathetic divisions of the autonomic nervous system.
- Contrast the effect of the parasympathetic and sympathetic divisions on the following organs: heart, lungs, digestive system, blood vessels.

DEVELOPMENTAL ASPECTS OF THE NERVOUS SYSTEM (pp. 263–266)

- List several factors that may have harmful effects on brain development.
- Briefly describe the cause, signs, and consequences of the following congenital disorders: spina bifida, anencephaly, and cerebral palsy.
- Explain the decline in brain size and weight that occurs with age.
- Define *senility,* and list some possible causes.

You are driving down the freeway, and a horn blares on your right. You swerve to your left. Charlie leaves a note on the kitchen table: "See you later—have the stuff ready at 6." You know that the "stuff" is chili with taco chips. You are dozing, and your infant son makes a soft cry. Instantly, you awaken. What do all these events have in common? They are all everyday examples of the functioning of your nervous system, which has your body cells humming with activity nearly all the time.

The **nervous system** is the master controlling and communicating system of the body. Every thought, action, and emotion reflects its activity. Its signaling device, or means of communicating with body cells, is electrical impulses, which are rapid and specific and cause almost immediate responses.

To carry out its normal role, the nervous system has three overlapping functions (Figure 7.1): (1) Much like a sentry, it uses its millions of sensory receptors to *monitor changes* occurring both inside and outside the body. These changes are called *stimuli,* and the gathered information is called **sensory input.** (2) It *processes and interprets* the sensory input and makes decisions about what should be done at each moment—a process called **integration.** (3) It then *effects a response* by activating muscles or glands (effectors) via **motor output.** An example will illustrate how these functions work together. When you are driving and see a red light just ahead (sensory input), your nervous system integrates this information (red light means "stop") and sends motor output to the muscles of your right leg and foot, and your foot goes for the brake pedal (the response).

The nervous system does not work alone to regulate and maintain body homeostasis; the endocrine system is a second important regulating system. While the nervous system controls with rapid electrical nerve impulses, the endocrine system organs produce hormones that are released into the blood. Thus, the endocrine system typically brings about its effects in a more leisurely way.

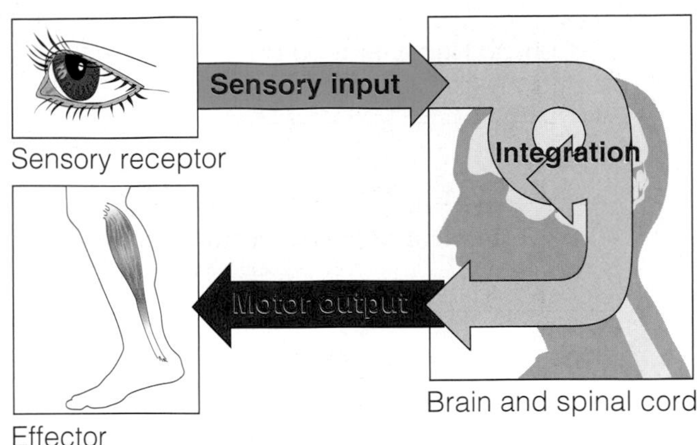

FIGURE 7.1 The nervous system's functions.

Organization of the Nervous System

We have only one nervous system, but, because of its complexity, it is difficult to consider all its parts at the same time. So, to simplify its study, we divide it in terms of its structures (structural classification) or in terms of its activities (functional classification). Each of these classification schemes is described briefly below, and their relationships are illustrated in Figure 7.2. It is not necessary to memorize this whole scheme now, but as you are reading the descriptions, try to get a "feel" for the major parts and how they fit together. This will make your learning task easier as you make your way through this chapter. Later you will meet all these terms and concepts again and in more detail.

Structural Classification

The structural classification, which includes all nervous system organs, has two subdivisions—the central nervous system and the peripheral nervous system (see Figure 7.2).

The **central nervous system (CNS)** consists of the brain and spinal cord, which occupy the dorsal body cavity and act as the integrating and command centers of the nervous system. They interpret incoming sensory information and issue instructions based on past experience and current conditions.

The **peripheral** (pĕ-rif′er-al) **nervous system (PNS),** the part of the nervous system outside the CNS, consists mainly of the nerves that extend from the brain and spinal cord. *Spinal nerves* carry impulses to and from the spinal cord. *Cranial* (kra′ne-al) *nerves* carry impulses to and from the brain. These nerves serve as communication lines. They link all parts of the body by carrying impulses from the sensory receptors to the CNS and from the CNS to the appropriate glands or muscles.

The organs making up the CNS and PNS are discussed at length later in this chapter.

Functional Classification

The functional classification scheme is concerned only with PNS structures. It divides them into two principal subdivisions (see Figure 7.2).

The **sensory,** or **afferent** (af′er-ent), **division** consists of nerve fibers that convey impulses *to* the central nervous system from sensory receptors located in various parts of the body. Sensory fibers delivering impulses from the skin, skeletal muscles, and joints are called *somatic* (*soma* = body) *sensory* (afferent) *fibers,* whereas those transmitting impulses from the visceral organs are called *visceral sensory fibers,* or *visceral afferents.* The sensory division keeps the CNS constantly informed of events going on both inside and outside the body.

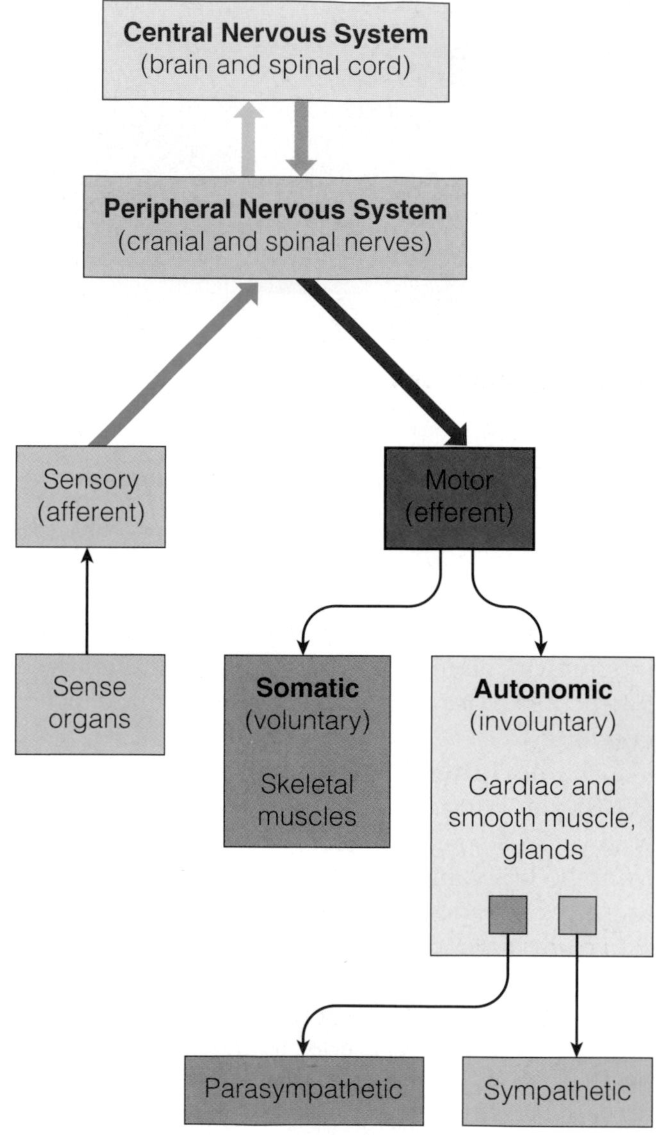

FIGURE 7.2 Organization of the nervous system. Organizational flowchart showing that the central nervous system receives input via sensory fibers and issues commands via motor fibers. The sensory and motor fibers together form the nerves that constitute the peripheral nervous system.

The **motor,** or **efferent** (ef′er-rent), **division** carries impulses *from* the CNS to effector organs, the muscles and glands. These impulses activate muscles and glands; that is, they *effect* (bring about) a motor response.

The motor division in turn has two subdivisions (see Figure 7.2):

1. The **somatic** (so-mat′ik) **nervous system** allows us to consciously, or voluntarily, control our skeletal muscles. Hence, this subdivision is often referred to as the **voluntary nervous system.** However, not all skeletal muscle activity controlled by this motor division is voluntary. Skeletal muscle reflexes, like the stretch reflex for example, are initiated involuntarily by these same fibers.
2. The **autonomic** (aw″to-nom′ik) **nervous system (ANS)** regulates events that are automatic, or involuntary, such as the activity of smooth and cardiac muscles and glands. This subdivision, commonly called the **involuntary nervous system,** itself has two parts, the *sympathetic* and *parasympathetic,* which typically bring about opposite effects. What one stimulates, the other inhibits. These will be described later.

Although it is simpler to study the nervous system in terms of its subdivisions, you should recognize that these subdivisions are made for the sake of convenience only. Remember that the nervous system acts as a coordinated unit, both structurally and functionally.

Nervous Tissue: Structure and Function

Even though it is complex, nervous tissue is made up of just two principal types of cells—*supporting cells* and *neurons.*

Supporting Cells

Supporting cells in the CNS are "lumped together" as **neuroglia** (nu-rog′le-ah), literally, "nerve glue." Neuroglia includes many types of cells that generally support, insulate, and protect the delicate neurons (Figure 7.3). In addition, each of the different types of neuroglia, also simply called **glia** (gle′ah) or glial cells, has special functions. The CNS glia include:

- **Astrocytes:** abundant star-shaped cells that account for nearly half of the neural tissue. Their numerous projections have swollen ends that cling to neurons, bracing them and anchoring them to their nutrient supply lines, the blood capillaries (Figure 7.3a). Astrocytes form a living barrier between capillaries and neurons and play a role in making exchanges between the two. In this way, they help protect the neurons from harmful substances that might be in the

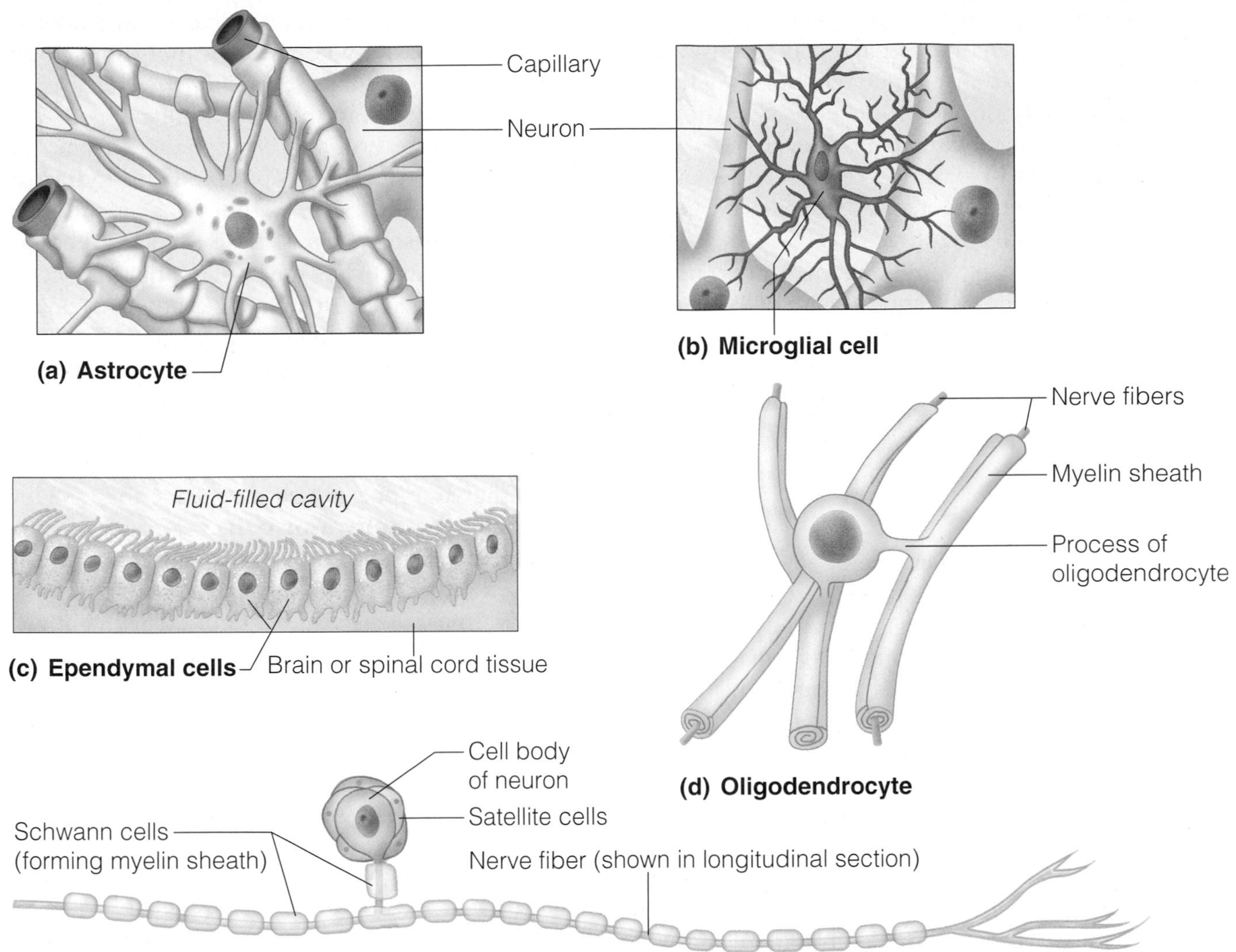

FIGURE 7.3 Supporting (glial) cells of nervous tissues. Astrocytes **(a)** form a living barrier between neurons and capillaries in the CNS. Microglia **(b)** are phagocytes, whereas ependymal cells **(c)** line the fluid-filled cavities of the CNS. The oligodendrocytes **(d)** form myelin sheaths around nerve fibers in the CNS. **(e)** The relationship of Schwann cells (myelinating cells) and satellite cells to a neuron in the peripheral nervous system.

blood. Astrocytes also help control the chemical environment in the brain by picking up excess ions and recapturing released neurotransmitters.

- **Microglia** (mi-krog′le-ah): spiderlike phagocytes that dispose of debris, including dead brain cells and bacteria (Figure 7.3b).
- **Ependymal** (ĕ-pen′dĭ-mal) **cells:** these glial cells line the cavities of the brain and the spinal cord (Figure 7.3c). The beating of their cilia helps to circulate the cerebrospinal fluid that fills those cavities and forms a protective cushion around the CNS.
- **Oligodendrocytes** (ol″ĭ-go-den′dro-sītz): glia that wrap their flat extensions tightly around the nerve fibers, producing fatty insulating coverings called *myelin sheaths* (Figure 7.3d).

Although they somewhat resemble neurons structurally (both cell types have cell extensions), glia are not able to transmit nerve impulses, a function that is highly developed in neurons. Another important difference is that glia never lose their ability to divide, whereas most neurons do. Consequently, most brain tumors are *gliomas,* or tumors formed by glial cells (neuroglia).

Supporting cells in the PNS come in two major varieties—Schwann cells and satellite cells (Figure 7.3e). **Schwann cells** form the myelin sheaths around nerve fibers that are found in the PNS. **Satellite cells** act as protective, cushioning cells.

Neurons

Anatomy

Neurons, also called **nerve cells,** are highly specialized to transmit messages (nerve impulses) from one part of the body to another. Although neurons differ structurally, they have many common features (Figure 7.4). All have a cell body, which contains the nucleus and is the metabolic center of the cell, and one or more slender processes extending from the cell body.

The **cell body** is the metabolic center of the neuron. It contains the usual organelles except for centrioles (which confirms the amitotic nature of most neurons). The rough ER, called **Nissl** (nis′l) **substance,** and **neurofibrils,** intermediate filaments that are important in maintaining cell shape, are particularly abundant in the cell body.

The armlike **processes,** or **fibers,** vary in length from microscopic to 3 to 4 feet. The longest ones in humans reach from the lumbar region of the spine to the great toe. Neuron processes that convey incoming messages (electrical signals) *toward* the cell body are **dendrites** (den′drītz), whereas those that generate nerve impulses and typically conduct them *away* from the cell body are **axons** (ak′sonz). Neurons may have hundreds of the branching dendrites (*dendr* = tree), depending on the neuron type, but each neuron has only one axon, which arises from a conelike region of the cell body called the **axon hillock.**

An occasional axon gives off a *collateral branch* along its length, but all axons branch profusely at their terminal end, forming hundreds to thousands of **axon terminals.** These terminals contain hundreds of tiny vesicles, or membranous sacs, that contain chemicals called **neurotransmitters.** As we said, axons transmit nerve impulses away from the cell body. When these impulses reach the axon terminals, they stimulate the release of neurotransmitters into the extracellular space.

Each axon terminal is separated from the next neuron by a tiny gap called the **synaptic** (sĭ-nap′tik) **cleft.** Such a functional junction is called a **synapse** (*syn* = to clasp or join). Although they are close, neurons never actually touch other neurons. We will learn more about synapses and the events that occur there a bit later.

Most long nerve fibers are covered with a whitish, fatty material, called **myelin** (mi′ĕ-lin), which has a waxy appearance. Myelin protects and insulates the fibers and increases the transmission rate of nerve impulses. Axons outside the CNS are myelinated by **Schwann cells,** specialized supporting cells that wrap themselves tightly around the axon jelly-roll fashion (Figure 7.5). When the wrapping process is done, a tight coil of wrapped membranes, the **myelin sheath,** encloses the axon. Most of the Schwann cell cytoplasm ends up just beneath the outermost part of its plasma membrane. This part of the Schwann cell, external to the myelin sheath, is called the **neurilemma** (nu″rĭ-lem′mah, "neuron husk"). Since the myelin sheath is formed by many individual Schwann cells, it has gaps or indentations, called **nodes of Ranvier** (rahn-vēr), at regular intervals (see Figure 7.4).

Myelinated fibers are also found in the central nervous system. However, there it is oligodendrocytes that form CNS myelin sheaths (see Figure 7.3d). In contrast to Schwann cells, each of which deposits myelin around a small segment of one nerve fiber, the oligodendrocytes with their many flat extensions can coil around as many as 60 different fibers at the same time. Although the myelin sheaths formed by oligodendrocytes and those formed by Schwann cells are quite similar, the CNS sheaths lack a neurilemma. Because the neurilemma remains intact (for the most part) when a peripheral nerve fiber is damaged, it plays an important role in fiber regeneration, an ability that is largely lacking in the central nervous system.

Homeostatic Imbalance

The importance of the myelin insulation to nerve transmission is best illustrated by observing what happens when it is not there. In people with *multiple sclerosis (MS),* the myelin sheaths around the fibers are gradually destroyed, converted to hardened

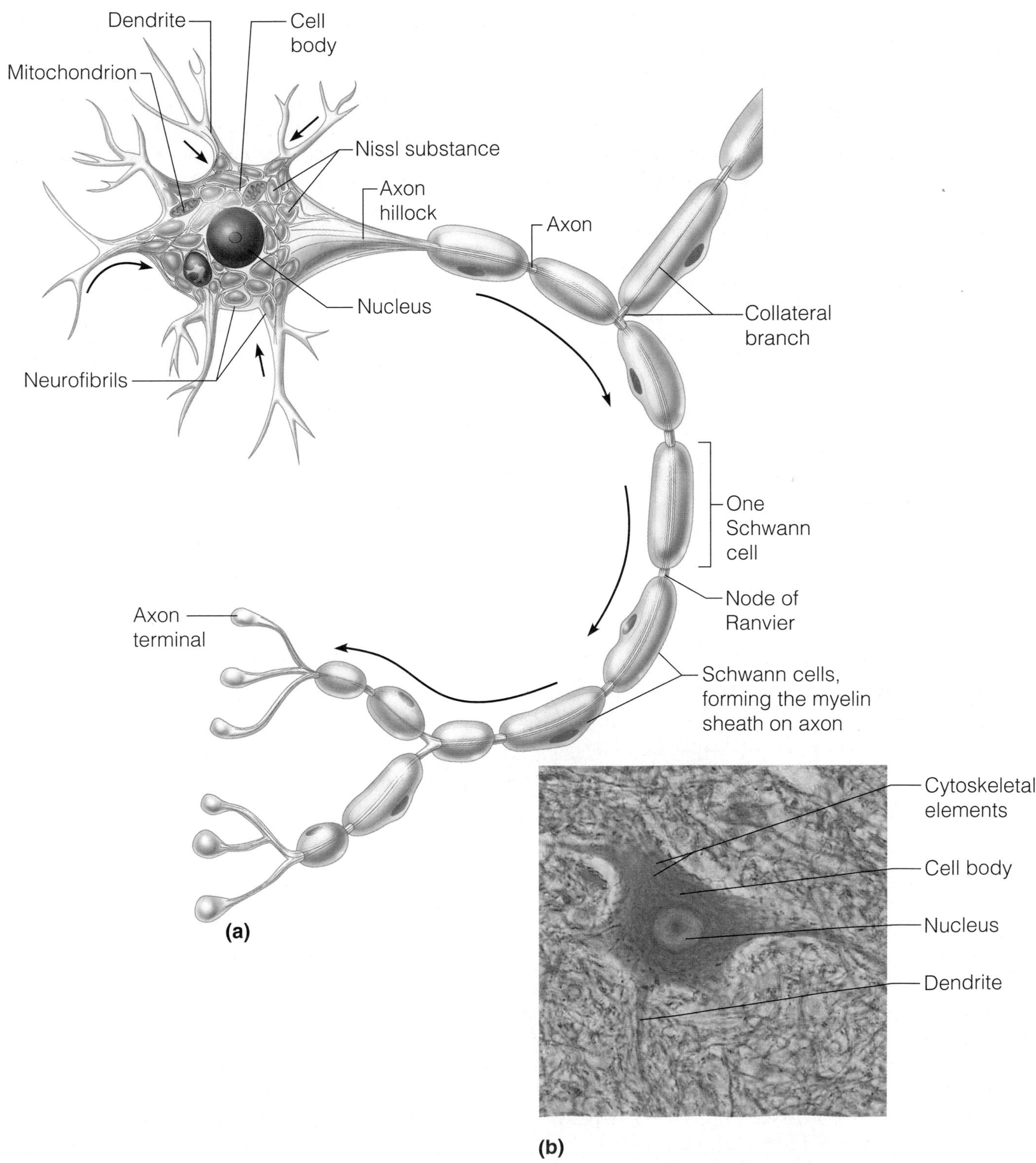

FIGURE 7.4 Structure of a typical motor neuron. (a) Diagrammatic view. **(b)** Photomicrograph (265×).

Q *Why does the myelin sheath that is produced by Schwann cells have gaps in it?*

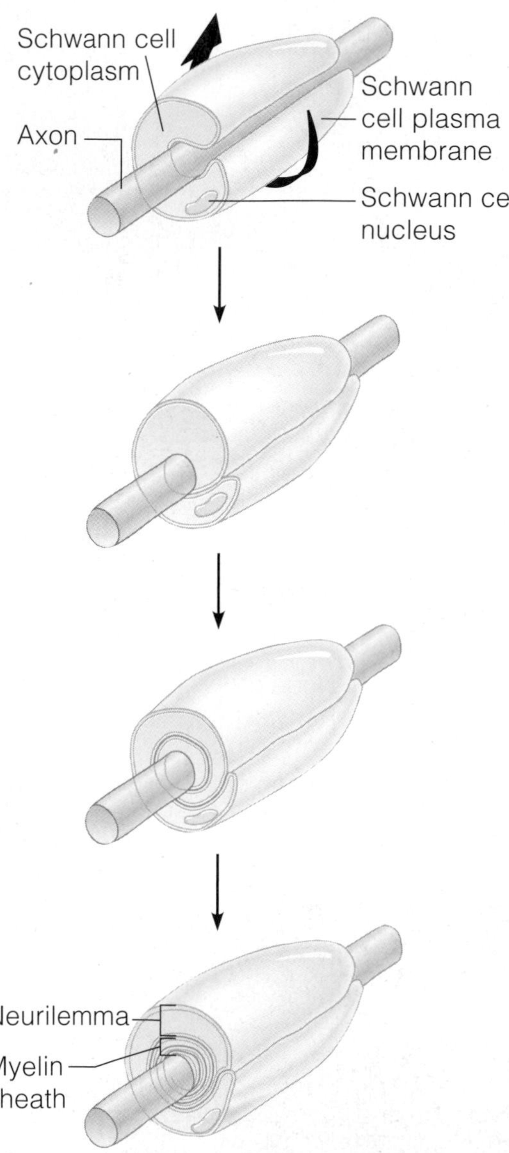

FIGURE 7.5 Relationship of Schwann cells to axons in the peripheral nervous system. As illustrated (top to bottom), a Schwann cell envelops part of an axon in a trough and then rotates around the axon. Most of the Schwann cell cytoplasm comes to lie just beneath the exposed part of its plasma membrane. The tight coil of plasma membrane material surrounding the axon is the myelin sheath. The Schwann cell cytoplasm and exposed membrane are referred to as the *neurilemma*.

A *Because the sheath is produced by several Schwann cells that arrange themselves end to end along the nerve fiber, each Schwann cell forming only one tiny segment of the sheath.*

sheaths called *scleroses*. As this happens, the current is short-circuited, and the affected person loses the ability to control his or her muscles and becomes increasingly disabled. Multiple sclerosis is an autoimmune disease in which a protein component of the sheath is attacked. As yet there is no cure, but injections of interferon (a hormonelike substance released by some immune cells) appears to provide some relief. ▲

Clusters of neuron cell bodies and collections of nerve fibers are named differently when they are in the CNS than when they are part of the PNS. For the most part, cell bodies are found in the CNS in clusters called **nuclei.** This well-protected location within the bony skull or vertebral column is essential to the well-being of the nervous system—remember that neurons do not routinely undergo cell division after birth. The cell body carries out most of the metabolic functions of a neuron, so if it is damaged the cell dies and is not replaced. Small collections of cell bodies called **ganglia** (gang′le-ah; **ganglion,** singular) are found in a few sites outside the CNS in the PNS. Bundles of nerve fibers (neuron processes) running through the CNS are called **tracts,** whereas in the PNS they are called **nerves.** The terms *white matter* and *gray matter* refer respectively to myelinated versus unmyelinated regions of the CNS. As a general rule, the **white matter** consists of dense collections of myelinated fibers (tracts), and **gray matter** contains mostly unmyelinated fibers and cell bodies.

Classification

Neurons may be classified either according to how they function or according to their structure.

Functional Classification Functional classification groups neurons according to the direction the nerve impulse is traveling relative to the CNS. On this basis, there are sensory, motor, and association neurons (Figure 7.6). Neurons carrying impulses from sensory receptors (in the internal organs or the skin) to the CNS are **sensory,** or **afferent, neurons.** (*Afferent* literally means "to go toward.") The *cell bodies* of sensory neurons are always found in a *ganglion* outside the CNS. Sensory neurons keep us informed about what is happening both inside and outside the body.

The *dendrite* endings of the sensory neurons are usually associated with specialized **receptors** that are activated by specific changes occurring

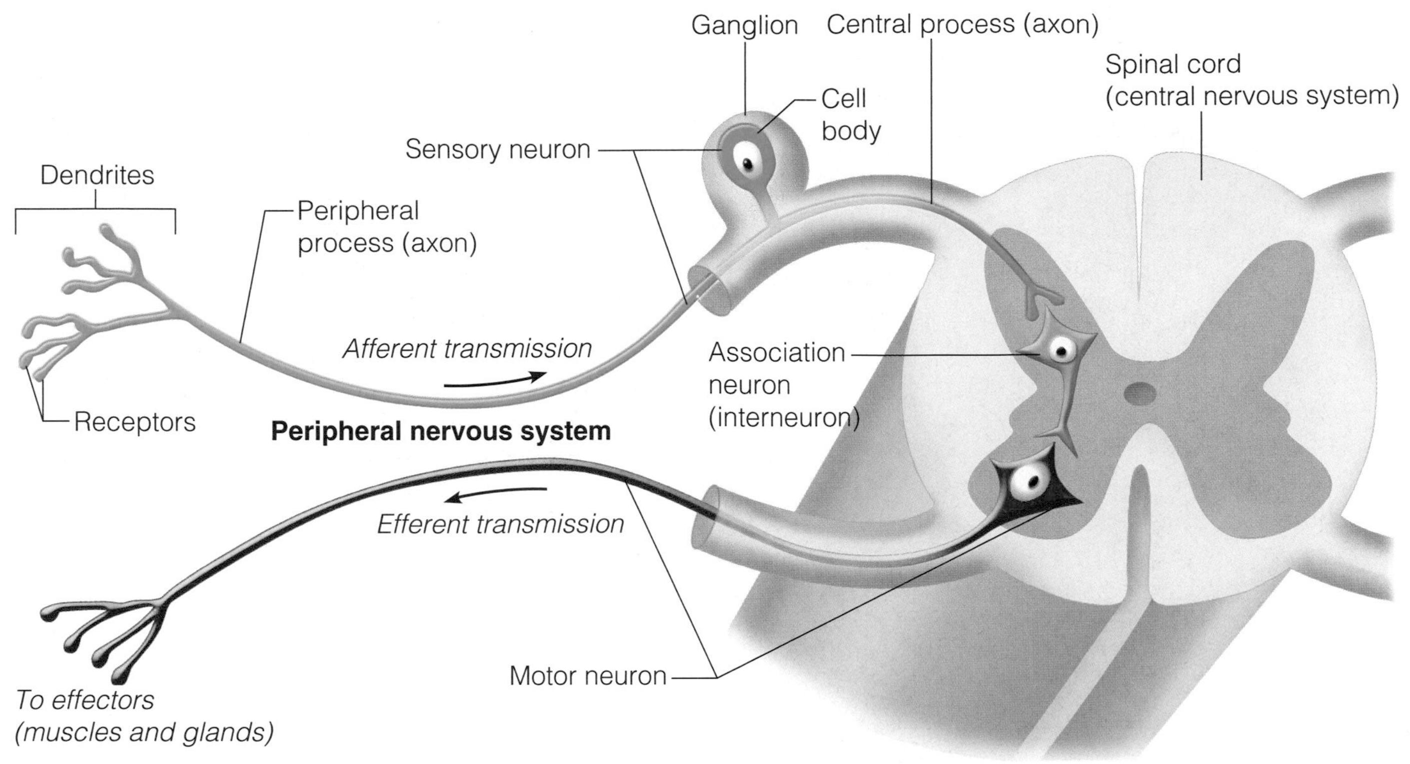

FIGURE 7.6 Neurons classified by function. Sensory (afferent) neurons conduct impulses from sensory receptors (in the skin, viscera, muscles) to the central nervous system; most cell bodies are in ganglia in the PNS. Motor (efferent) neurons transmit impulses from the CNS (brain or spinal cord) to effectors in the body periphery. Association neurons (interneurons) complete the communication pathway between sensory and motor neurons; their cell bodies reside in the CNS.

nearby. The very complex receptors of the special sense organs (vision, hearing, equilibrium, taste, and smell) are covered separately in Chapter 8. The simpler types of sensory receptors seen in the skin (**cutaneous sense organs**) and in the muscles and tendons (**proprioceptors** [pro″pre-o-sep′torz]) are shown in Figure 7.7. The pain receptors (actually bare dendrite endings) are the least specialized of the cutaneous receptors. They are also the most numerous, because pain warns us that some type of body damage is occurring or is about to occur. However, strong stimulation of any of the cutaneous receptors (for example, by searing heat, extreme cold, or excessive pressure) is also interpreted as pain.

The proprioceptors detect the amount of stretch, or tension, in skeletal muscles, their tendons, and joints. They send this information to the brain so that the proper adjustments can be made to maintain balance and normal posture. *Propria* comes from the Latin word meaning "one's own," and the proprioceptors constantly advise our brain of our own movements.

Neurons carrying impulses from the CNS to the viscera and/or muscles and glands are **motor,** or **efferent, neurons** (see Figure 7.6). The cell bodies of motor neurons are always located in the CNS.

The third category of neurons is the **association neurons,** or **interneurons.** They connect the motor and sensory neurons in neural pathways. Like the motor neurons, their cell bodies are always located in the CNS.

Structural Classification Structural classification is based on the number of processes extending from

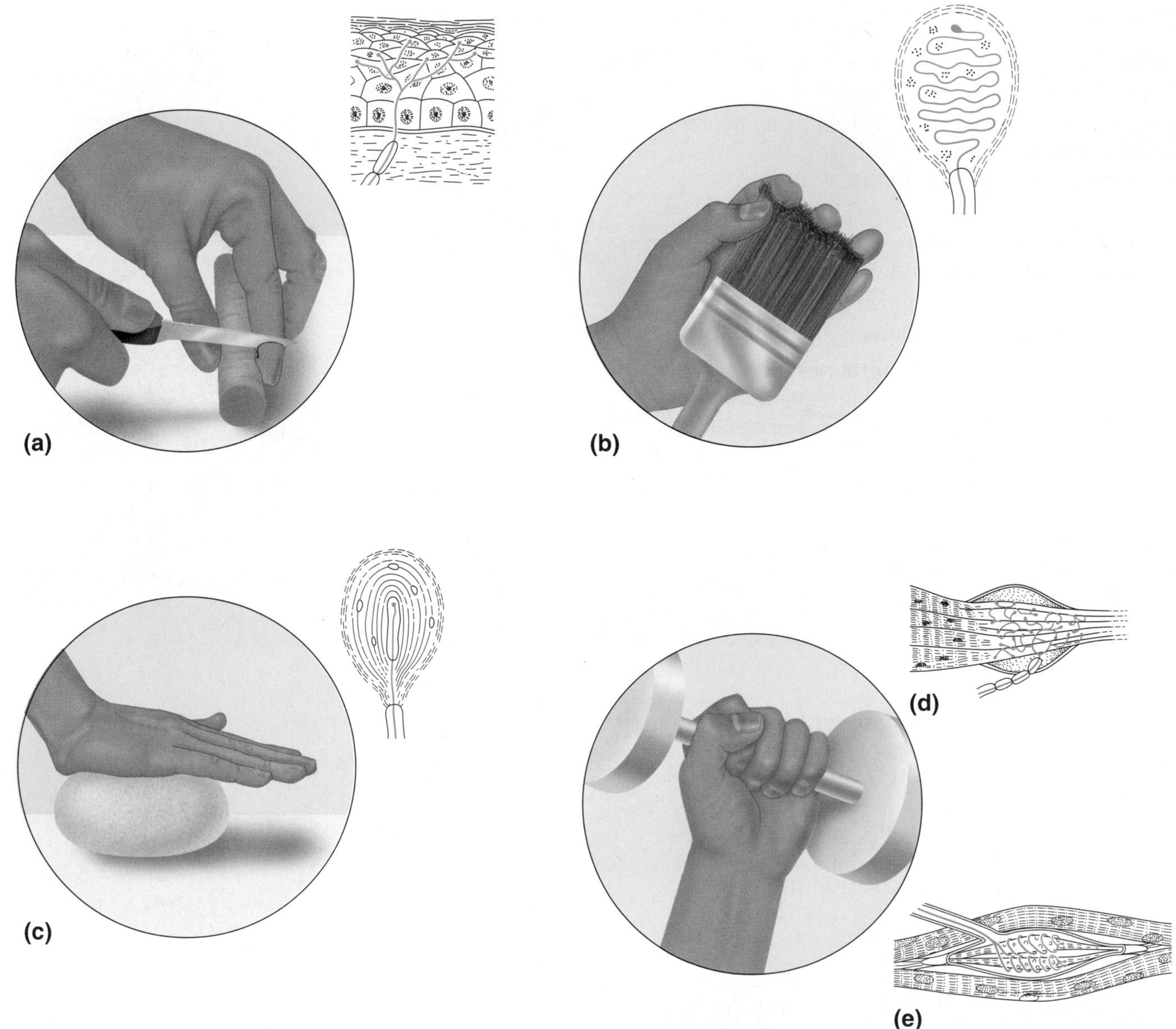

FIGURE 7.7 Types of sensory receptors. (a) Naked nerve endings (pain and temperature receptors). **(b)** Meissner's corpuscle (touch receptor). **(c)** Pacinian corpuscle (deep pressure receptor). **(d)** Golgi tendon organ (proprioceptor). **(e)** Muscle spindle (proprioceptor).

the cell body (Figure 7.8). If there are several, the neuron is a **multipolar neuron.** Since all motor and association neurons are multipolar, this is the most common structural type. Neurons with two processes—an axon and a dendrite—are called **bipolar neurons.** Bipolar neurons are rare in adults, found only in some special sense organs (eye, nose), where they act in sensory processing as receptor cells. **Unipolar neurons** have a single process emerging from the cell body. However, it is very short and divides almost immediately into proximal (central) and distal (peripheral) processes. Unipolar neurons are unique in that only the small branches at the end of the peripheral process are dendrites. The remainder of the peripheral process and the central process function as axons; thus, in this case, the axon conducts nerve impulses both toward *and* away from the cell body. Sensory neurons found in PNS ganglia are unipolar.

Physiology

Nerve Impulses Neurons have two major functional properties: *irritability,* the ability to respond to a stimulus and convert it into a nerve impulse, and *conductivity,* the ability to transmit the impulse to other neurons, muscles, or glands. We will consider these functional abilities next.

The plasma membrane of a resting, or inactive, neuron is **polarized,** which means that there are fewer positive ions sitting on the inner face of the neuron's plasma membrane than there are on its outer face in the tissue fluid that surrounds it (Figure 7.9). The major positive ions inside the cell are potassium (K^+), whereas the major positive ions outside the cell are sodium (Na^+). As long as the inside remains more negative as compared to the outside, the neuron will stay inactive.

Many different types of stimuli excite neurons to become active and generate an impulse. For example, light excites the eye receptors, sound excites some of the ear receptors, and pressure excites some cutaneous receptors of the skin. However, *most* neurons in the body are excited by neurotransmitters released by other neurons, as will be described shortly. Regardless of what the stimulus is, the result is always the same—the permeability properties of the cell's plasma membrane change for a very brief period. *Normally,* sodium ions cannot diffuse through the plasma membrane to any great extent; but when the neuron is adequately stimulated, the "gates" of sodium channels in the membrane open. Because sodium is in much higher concentration outside the cell, it will then diffuse quickly into the neuron. (Remember the laws of diffusion?) This inward rush of sodium ions changes the polarity of the neuron's membrane at that site, an event called **depolarization.** Locally, the inside is now more positive, and the outside is less positive, a situation called a **graded potential.** However, if the stimulus is strong enough and the sodium influx is great enough, the local depolarization (graded potential) activates the neuron to initiate and transmit a long distance signal called an **action potential,** also called a **nerve impulse** in neurons. The nerve impulse is an *all-or-none response* like firing a gun. It is either propagated (conducted) over the entire axon, or it doesn't happen at all. The nerve impulse never goes partway along an axon's length, nor does it die out with distance as do graded potentials.

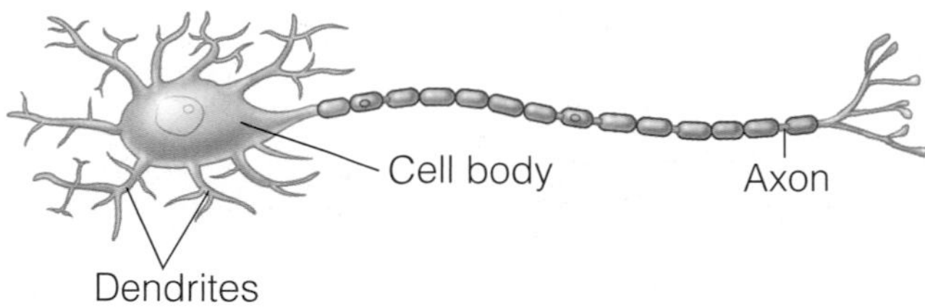

(a) Multipolar neuron

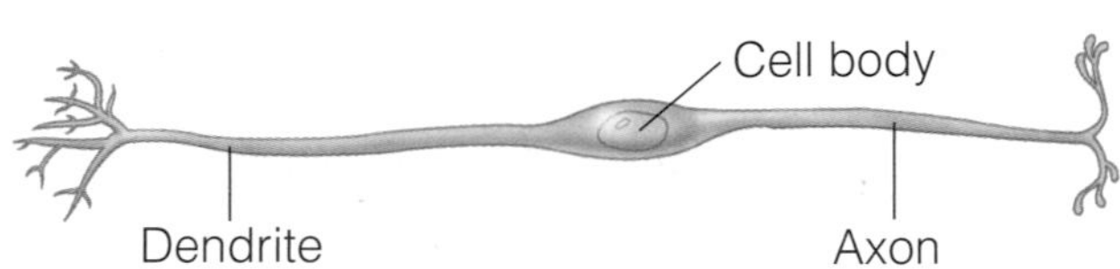

(b) Bipolar neuron

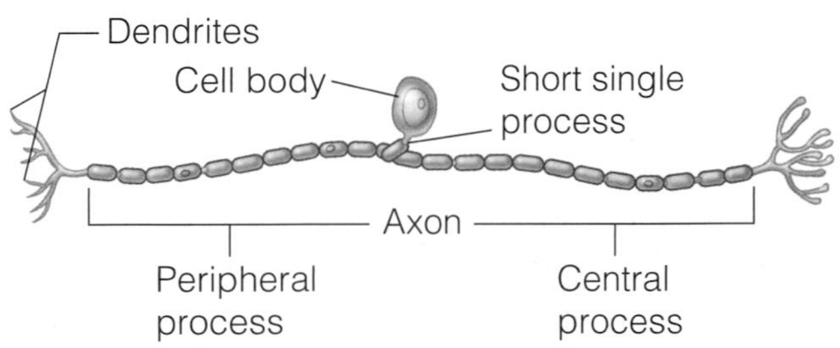

(c) Unipolar neuron

FIGURE 7.8 Classification of neurons on the basis of structure. (a) Multipolar. **(b)** Bipolar. **(c)** Unipolar.

Almost immediately after the sodium ions rush into the neuron, the membrane permeability changes again, becoming impermeable to sodium ions but permeable to potassium ions. So potassium ions are allowed to diffuse out of the neuron into the tissue fluid, and they do so very rapidly. This outflow of positive ions from the cell restores the electrical conditions at the membrane to the polarized, or resting, state, an event called **repolarization.** *Until repolarization occurs, a neuron cannot conduct another impulse.* After repolarization occurs, the initial concentrations of the sodium and potassium ions inside and outside the neuron are restored by activation of the sodium-potassium pump. This pump uses ATP (cellular energy) to pump excess sodium ions out of the cell and to bring potassium ions back into it. Once begun, these sequential events spread along the entire neuronal membrane.

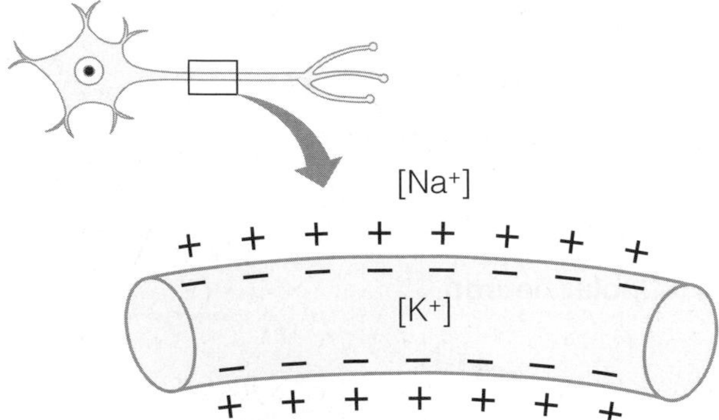

(a) Resting membrane

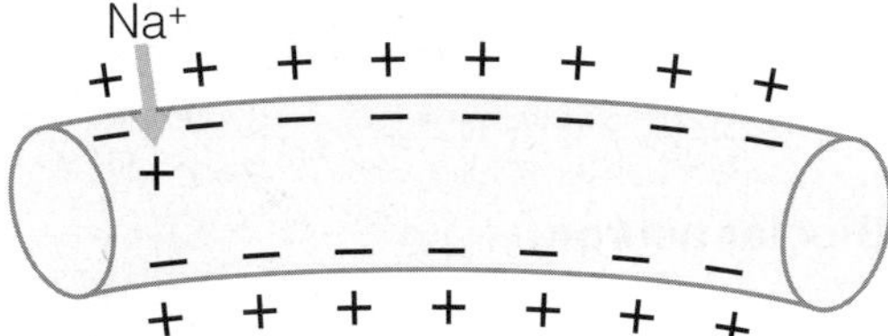

(b) Stimulus initiates local depolarization

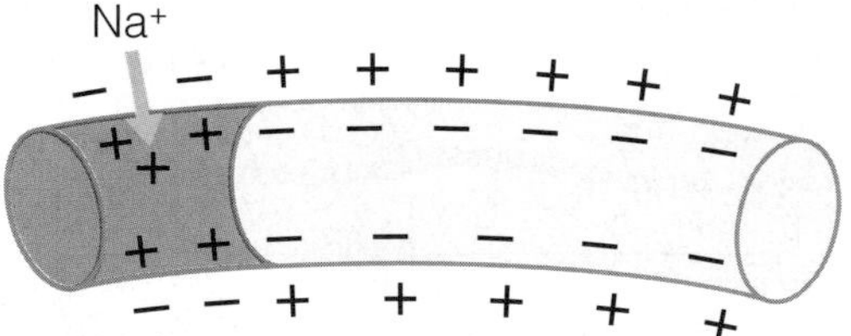

(c) Depolarization and generation of the action potential

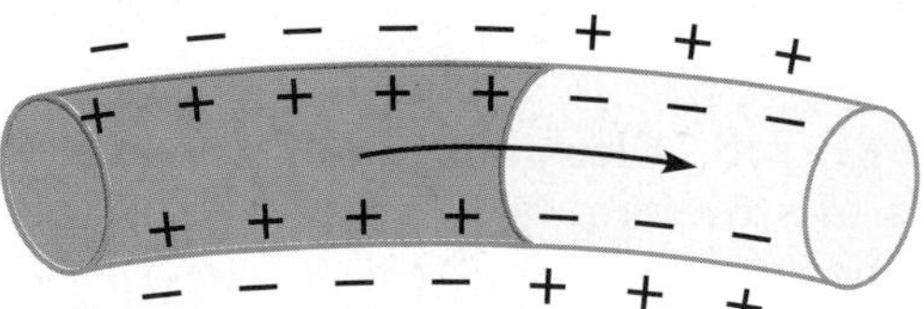

(d) Propagation of the action potential

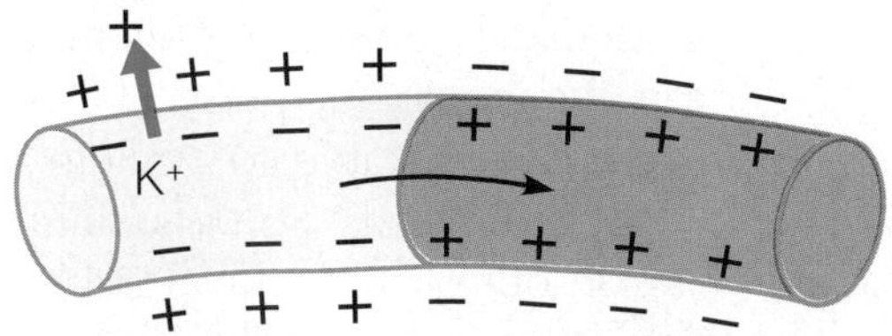

(e) Repolarization

(f) Restoration of ionic concentrations via sodium-potassium pump

FIGURE 7.9 The nerve impulse. (a) Resting membrane electrical conditions. The external face of the membrane is slightly positive; its internal face is slightly negative. The chief extracellular ion is sodium (Na^+), whereas the chief intracellular ion is potassium (K^+). The membrane is relatively impermeable to both ions. **(b)** Stimulus initiates local depolarization. A stimulus changes the permeability of a "patch" of the membrane, and sodium ions diffuse rapidly into the cell. This changes the polarity of the membrane (the inside becomes more positive; the outside becomes more negative). **(c)** Depolarization and generation of an action potential. If the stimulus is strong enough, depolarization causes membrane polarity to be completely reversed and an action potential is initiated. **(d)** Propagation of the action potential. Depolarization of the first membrane patch causes permeability changes in the adjacent membrane, and the events described in (b) are repeated. Thus, the action potential propagates rapidly along the entire length of the membrane. **(e)** Repolarization. Potassium ions diffuse out of the cell as membrane permeability changes again, restoring the negative charge on the inside of the membrane and the positive charge on the outside surface. Repolarization occurs in the same direction as depolarization. **(f)** The ionic conditions of the resting state are restored later by the activity of the sodium-potassium pump.

The events just described explain propagation of a nerve impulse along unmyelinated fibers. Fibers that have myelin sheaths conduct impulses much faster because the nerve impulse literally jumps, or leaps, from node to node along the length of the fiber. This occurs because no current can flow across the axon membrane where there is fatty myelin insulation. This faster type of impulse propagation is called *saltatory* (sal'tah-to"re) *conduction* (*saltare* = to dance or leap).

Homeostatic Imbalance

A number of factors can impair the conduction of impulses. For example, sedatives and anesthetics block nerve impulses by altering membrane permeability to ions, mainly sodium ions. As we have seen, no sodium entry = no action potential.

Cold and continuous pressure hinder impulse conduction because they interrupt blood circulation (and hence the delivery of oxygen and nutrients) to the neurons. For example, your fingers get numb when you hold an ice cube for more than a few seconds. Likewise, when you sit on your foot, it "goes to sleep." When you warm the fingers or remove the pressure from your foot, the impulses begin to be transmitted once again, leading to an unpleasant prickly feeling. ▲

So far we have explained only the irritability aspect of neuronal functioning. What about conductivity—how does the electrical impulse traveling along one neuron get across the synapse to the next neuron (or effector cell) to influence its activity? The answer is that *it* doesn't! When the action potential reaches the axon terminal, the tiny vesicles containing the neurotransmitter chemical fuse with the axonal membrane, causing a pore-like opening to form and releasing the neurotransmitter. The neurotransmitter molecules diffuse across the synapse* and bind to receptors on the membrane of the next neuron (Figure 7.10). If enough neurotransmitter is released, the whole series of events described above (sodium entry, depolarization, etc.) will occur, leading to generation of a nerve impulse in the neuron beyond the synapse. The electrical changes prompted by neurotransmitter binding are very brief because the neurotransmitter is quickly removed from the synapse, either by re-uptake into the axonal terminal or by enzymatic breakdown. This limits the effect of each nerve impulse to a period shorter than the blink of an eye.

Notice that the transmission of an impulse is an *electrochemical event.* Transmission down the length of the neuron's membrane is basically *electrical,* but the next neuron is stimulated by a neurotransmitter, which is a *chemical.* Since each neuron both receives signals from and sends signals to scores of other neurons, it carries on "conversations" with many different neurons at the same time.

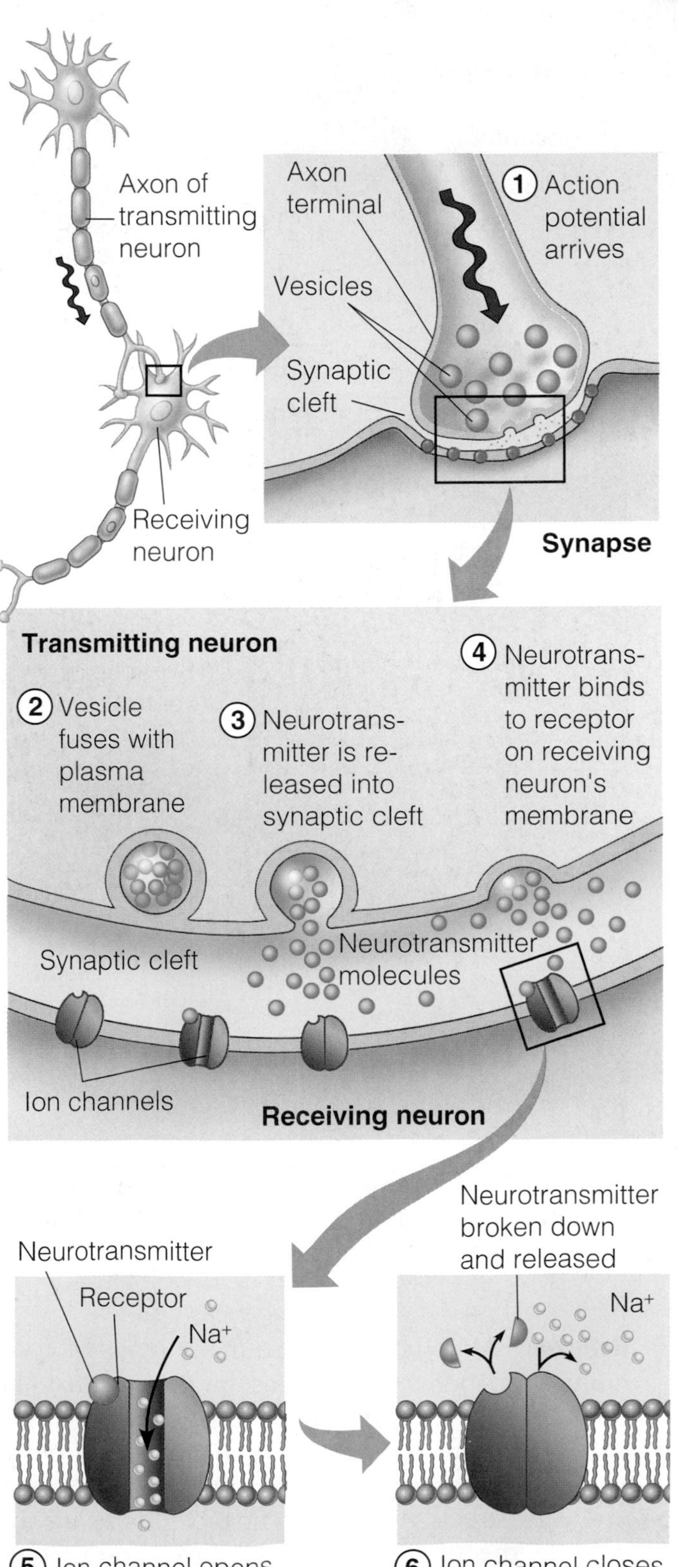

FIGURE 7.10 How neurons communicate at chemical synapses. The events occurring at the synapse are numbered in order.

---

*Although most neurons communicate via the *chemical* type of synapse described above, there are some examples of *electrical* synapses in which the neurons are physically joined by gap junctions and electrical currents actually flow from one neuron to the next.

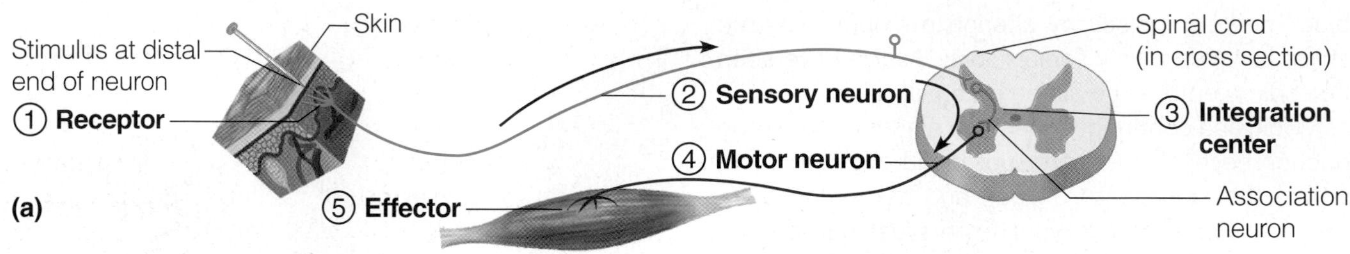

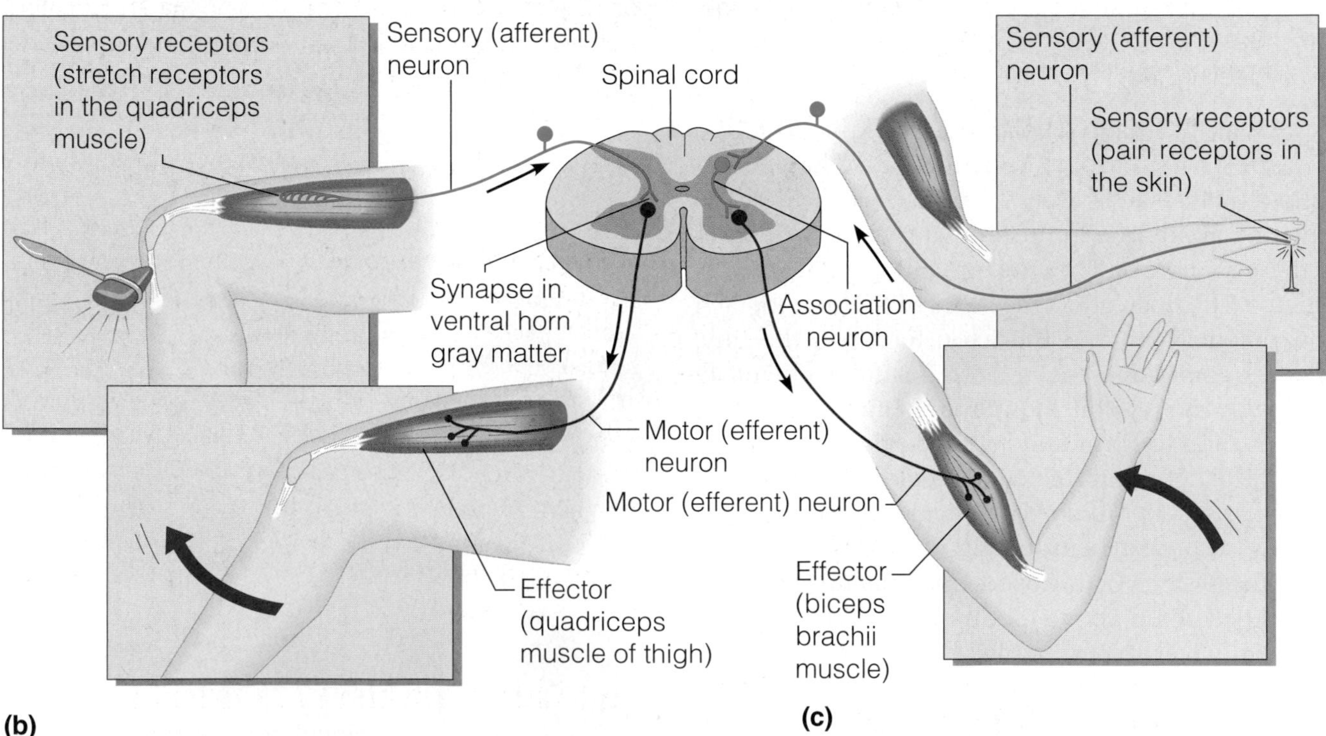

FIGURE 7.11 Simple reflex arcs. (a) The five basic elements of all reflex arcs. **(b)** A two-neuron reflex arc (example, patellar reflex). **(c)** A three-neuron reflex arc (example, the flexor reflex).

Reflexes Although there are many types of communication between neurons, much of what the body *must* do every day is programmed as reflexes. **Reflexes** are *rapid, predictable, and involuntary responses* to stimuli. They are much like one-way streets—once a reflex begins, it always goes in the same direction. Reflexes occur over neural pathways called **reflex arcs,** and involve both CNS and PNS structures.

The types of reflexes that occur in the body are classed as either somatic or autonomic reflexes. **Somatic reflexes** include all reflexes that stimulate the skeletal muscles. When you quickly pull your hand away from a hot object, a somatic reflex is working. **Autonomic reflexes** regulate the activity of smooth muscles, the heart, and glands. Secretion of saliva (salivary reflex) and changes in the size of the eye pupils (pupillary reflex) are two such reflexes. Autonomic reflexes regulate such body functions as digestion, elimination, blood pressure, and sweating.

All reflex arcs have a minimum of five elements (Figure 7.11a): a *sensory receptor* (which reacts to a stimulus), an *effector organ* (the muscle or gland eventually stimulated), and *sensory* and *motor neurons* to connect the two. The synapse between the sensory and motor neurons represents the central element—the CNS *integration center.*

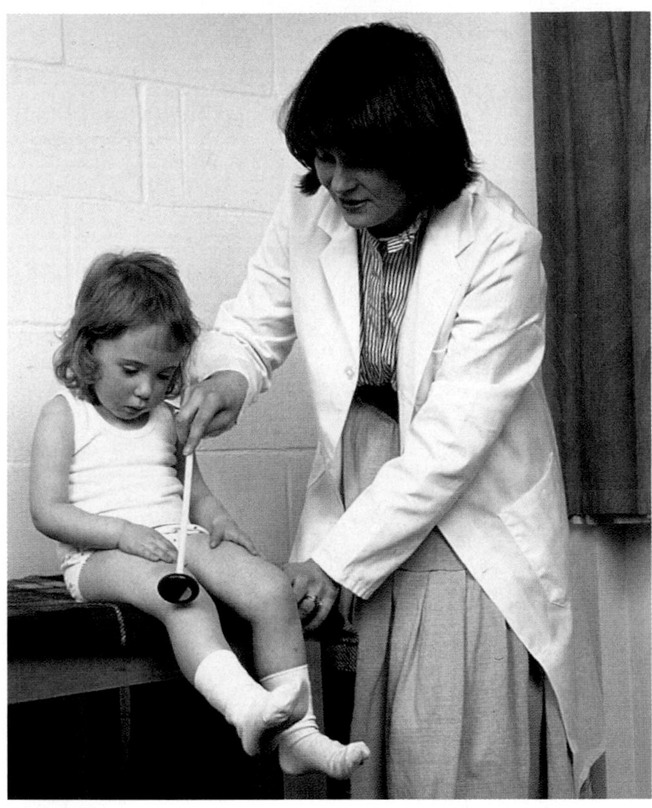

(d)

FIGURE 7.11 (*continued*) **(d)** Photo of a physician testing the patellar (knee jerk) reflex.

The simple *patellar* (pah-tel′ar), or *knee-jerk, reflex,* shown in Figure 7.11b and d, is an example of a two-neuron reflex arc, the most simple type in humans. The patellar reflex (in which the quadriceps muscle attached to the hit tendon is stretched) is familiar to most of us. It is usually tested during a physical exam to determine the general health of the motor portion of the nervous system. Most reflexes are much more complex than the two-neuron reflex, involving synapses between one or more association neurons in the CNS (integration center). A three-neuron reflex arc, the *flexor,* or *withdrawal, reflex,* in which the limb is withdrawn from a painful stimulus, is diagrammed in Figure 7.11c. The three-neuron reflex arc consists of five elements—receptor, sensory neuron, association neuron, motor neuron, and effector. Since there is always a delay at synapses (it takes time for the neurotransmitter to diffuse through the synaptic cleft), the more synapses there are in a reflex pathway, the longer the reflex takes to happen.

Many spinal reflexes involve only spinal cord neurons and occur without brain involvement. As long as the spinal cord is functional, spinal reflexes such as the flexor reflex will work. On the other hand, some reflexes require that the brain become involved because many different types of information have to be evaluated to arrive at the "right" response. The response of the pupils of the eyes to light is a reflex of this type.

As noted earlier, reflex testing is an important tool in evaluating the condition of the nervous system. Whenever reflexes are exaggerated, distorted, or absent, nervous system disorders are indicated. Reflex changes often occur before the pathological condition has become obvious in other ways.

Central Nervous System

During embryonic development, the CNS first appears as a simple tube, the **neural tube,** which extends down the dorsal median plane of the developing embryo's body. By the fourth week, the anterior end of the neural tube begins to expand, and brain formation begins. The rest of the neural tube posterior to the forming brain becomes the spinal cord. The central canal of the neural tube, which is continuous between the brain and spinal cord, becomes enlarged in four regions of the brain to form chambers called **ventricles** (see Figure 7.17a and b, p. 243).

Functional Anatomy of the Brain

The adult brain's unimpressive appearance gives few hints of its remarkable abilities. It is about two good fistfuls of pinkish gray tissue, wrinkled like a walnut, and with the texture of cold oatmeal. It weighs a little over three pounds. Because the brain is the largest and most complex mass of nervous tissue in the body, it is commonly discussed in terms of its four major regions—*cerebral hemispheres, diencephalon* (di″en-sef′ah-lon), *brain stem,* and *cerebellum* (Figure 7.12).

Cerebral Hemispheres

The paired **cerebral** (ser′e-bral) **hemispheres,** collectively called the **cerebrum,** are the most superior part of the brain and together are a good deal larger than the other three brain regions combined. In fact, as the cerebral hemispheres develop and grow, they enclose and obscure most of the brain stem, so many brain stem structures cannot normally be seen unless a sagittal section is

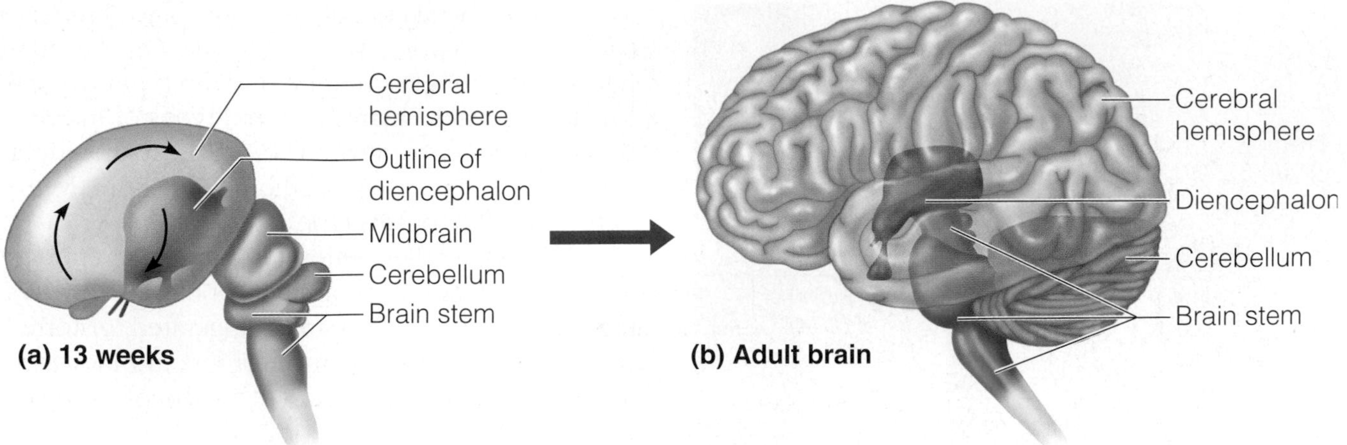

FIGURE 7.12 Development and regions of the human brain. The brain can be considered in terms of four main parts: cerebral hemispheres, diencephalon, brain stem, and cerebellum. In the developing brain **(a),** the cerebral hemispheres are forced to grow posteriorly and laterally over the other brain regions by the bones of the skull. In the adult brain **(b),** the cerebral hemispheres hide the diencephalon and the upper part of the brain stem. The left cerebral hemisphere is drawn so that it looks transparent, to reveal the location of the deeply situated diencephalon and superior part of the brain stem.

made. Picture how a mushroom cap covers the top of its stalk, and you have a fairly good idea of how the cerebral hemispheres cover the diencephalon and the superior part of the brain stem (see Figure 7.12).

The entire surface of the cerebral hemispheres exhibits elevated ridges of tissue called **gyri** (ji′re; **gyrus,** singular; "twisters"), separated by shallow grooves called **sulci** (sul′ki; **sulcus,** singular; "furrows"). Less numerous are the deeper grooves called **fissures** (Figure 7.13a), which separate large regions of the brain. Many of the fissures and gyri are important anatomical landmarks. The cerebral hemispheres are separated by a single deep fissure, the *longitudinal fissure.* Other fissures or sulci divide each cerebral hemisphere into a number of **lobes,** named for the cranial bones that lie over them (see Figure 7.13a and b).

Speech, memory, logical and emotional response, as well as consciousness, interpretation of sensation, and voluntary movement, are all functions of neurons of the cerebral cortex, and many of the functional areas of the cerebral hemispheres have been identified (Figure 7.13c). The **somatic sensory area** is located in the **parietal lobe** posterior to the **central sulcus.** Impulses traveling from the body's sensory receptors (except for the special senses) are localized and interpreted in this area of the brain. The somatic sensory area allows you to recognize pain, coldness, or a light touch. As illustrated in Figure 7.14, the body is represented in an upside-down manner in the sensory area. This spatial map is called the **sensory** homunculus (ho-mung′ku-lus; "little man"). Body regions with the most sensory receptors—the lips and fingertips—send impulses to neurons that make up a large part of the sensory cortex. Furthermore, the sensory pathways are crossed pathways—meaning that the left side of the sensory cortex receives impulses from the right side of the body, and vice versa.

Impulses from the special sense organs are interpreted in other cortical areas (see Figure 7.13b and c). For example, the visual area is located in the posterior part of the **occipital lobe,** the auditory area is in the **temporal lobe** bordering the *lateral sulcus,* and the olfactory area is found deep inside the temporal lobe.

The **primary motor area** that allows us to consciously move our skeletal muscles is anterior to the central sulcus in the **frontal lobe.** The axons of these motor neurons form the major voluntary motor tract—the **corticospinal** (kor″tĭ-ko-spi′nal), or **pyramidal tract,** which descends to the cord. As in the somatic sensory cortex, the body is represented upside-down and the pathways are

Q *What sense(s) would be affected by temporal lobe damage? By occipital lobe damage?*

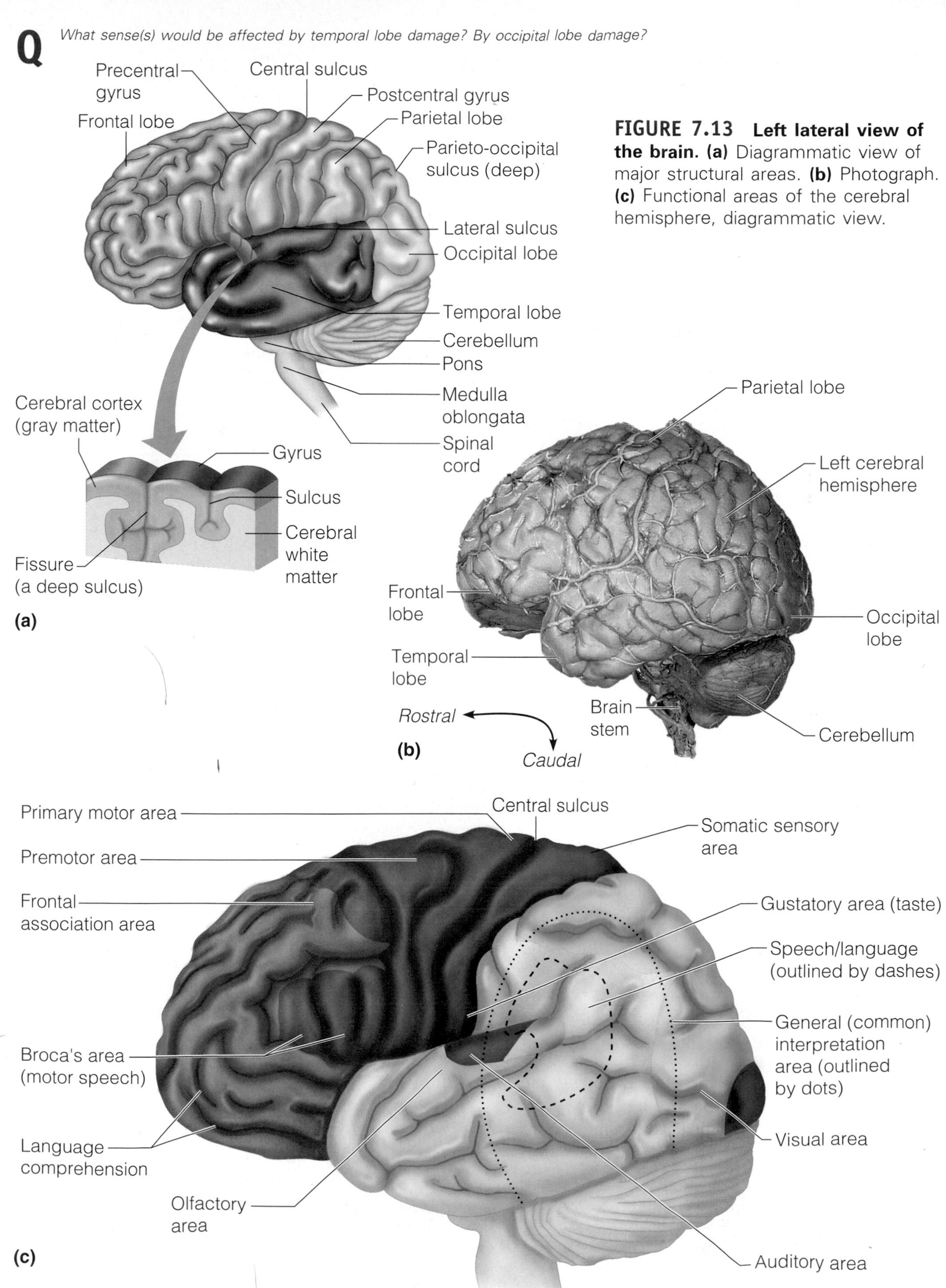

FIGURE 7.13 Left lateral view of the brain. (a) Diagrammatic view of major structural areas. **(b)** Photograph. **(c)** Functional areas of the cerebral hemisphere, diagrammatic view.

A *Damage to the temporal lobe (depending on its site) might affect hearing and/or the sense of smell. Occipital lobe damage would cause visual problems.*

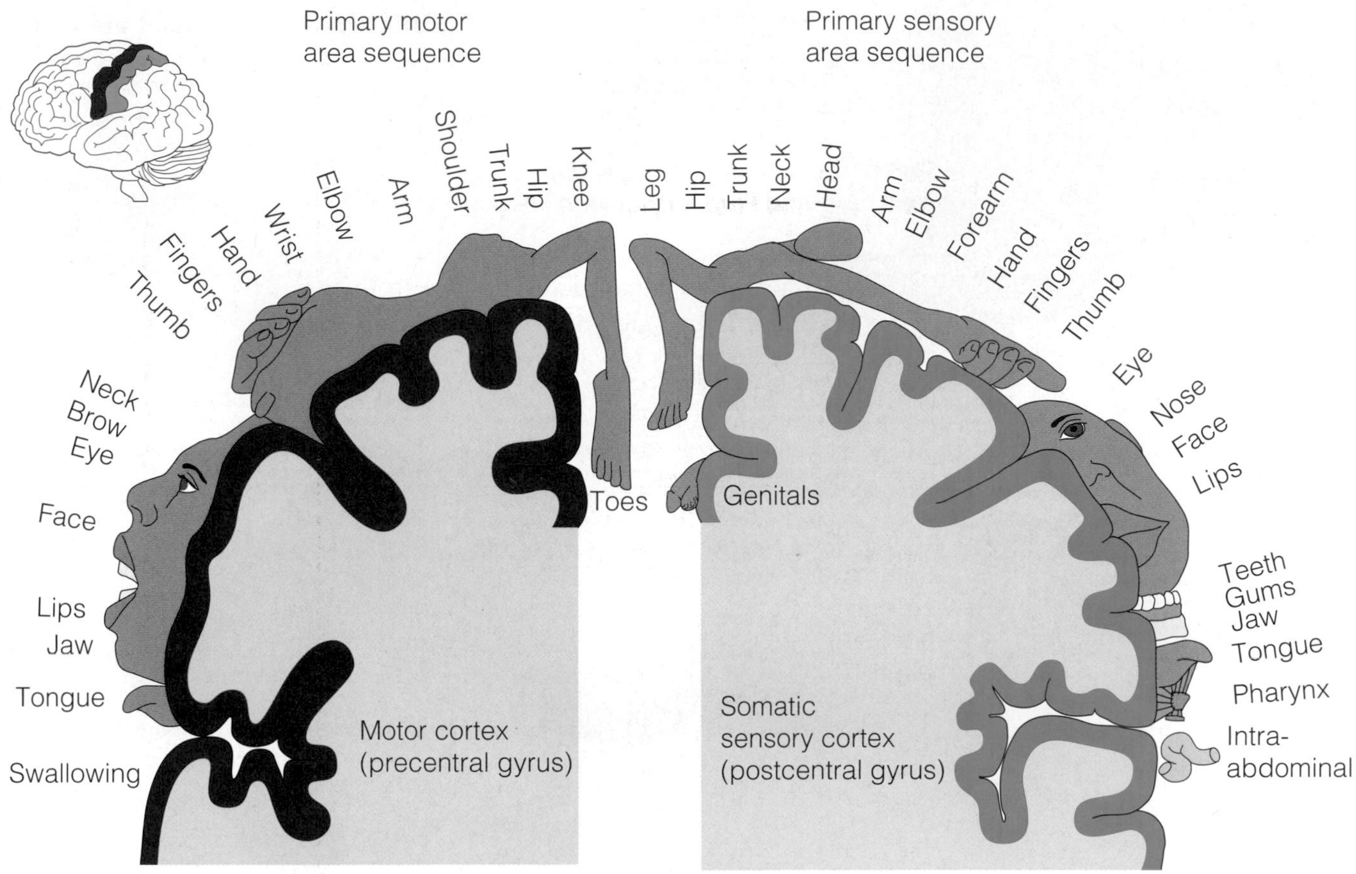

FIGURE 7.14 Sensory and motor areas of the cerebral cortex. The relative amount of cortical tissue devoted to each function is indicated by the amount of the gyrus occupied by the body area diagrams (homunculi). The primary motor cortex is shown on the right, the somatic sensory cortex is on the left.

crossed. Most of the neurons in this primary motor area control body areas having the finest motor control; that is, the face, mouth, and hands (see Figure 7.14). The body map on the motor cortex, as you might guess, is called the **motor** homunculus.

A specialized area that is very involved in our ability to speak, **Broca's** (bro'kahz) **area** (see Figure 7.13c), is found at the base of the precentral gyrus (the gyrus anterior to the central sulcus). Damage to this area, which is located in only one cerebral hemisphere (usually the left), causes inability to say words properly. You know what you want to say, but you can't vocalize the words.

Areas involved in higher intellectual reasoning and socially acceptable behavior are believed to be in the anterior part of the frontal lobes. Complex memories appear to be stored in the temporal and frontal lobes. The **speech area** is located at the junction of the temporal, parietal, and occipital lobes. The speech area allows one to sound out words. This area (like Broca's area) is usually in only one cerebral hemisphere. The frontal lobes house areas involved with language comprehension (word meanings).

The cell bodies of neurons involved in the cerebral hemisphere functions named above are found only in the outermost **gray matter** of the cerebrum, the **cerebral cortex** (see Figure 7.13a). As noted earlier, the cortical region is highly ridged and convoluted, providing more room for the thousands of neurons found there.

Most of the remaining cerebral hemisphere tissue—the deeper **cerebral white matter** (see Figure 7.13a)—is composed of fiber *tracts* (bundles of nerve fibers) carrying impulses to or from the cortex. One very large fiber tract, the **corpus**

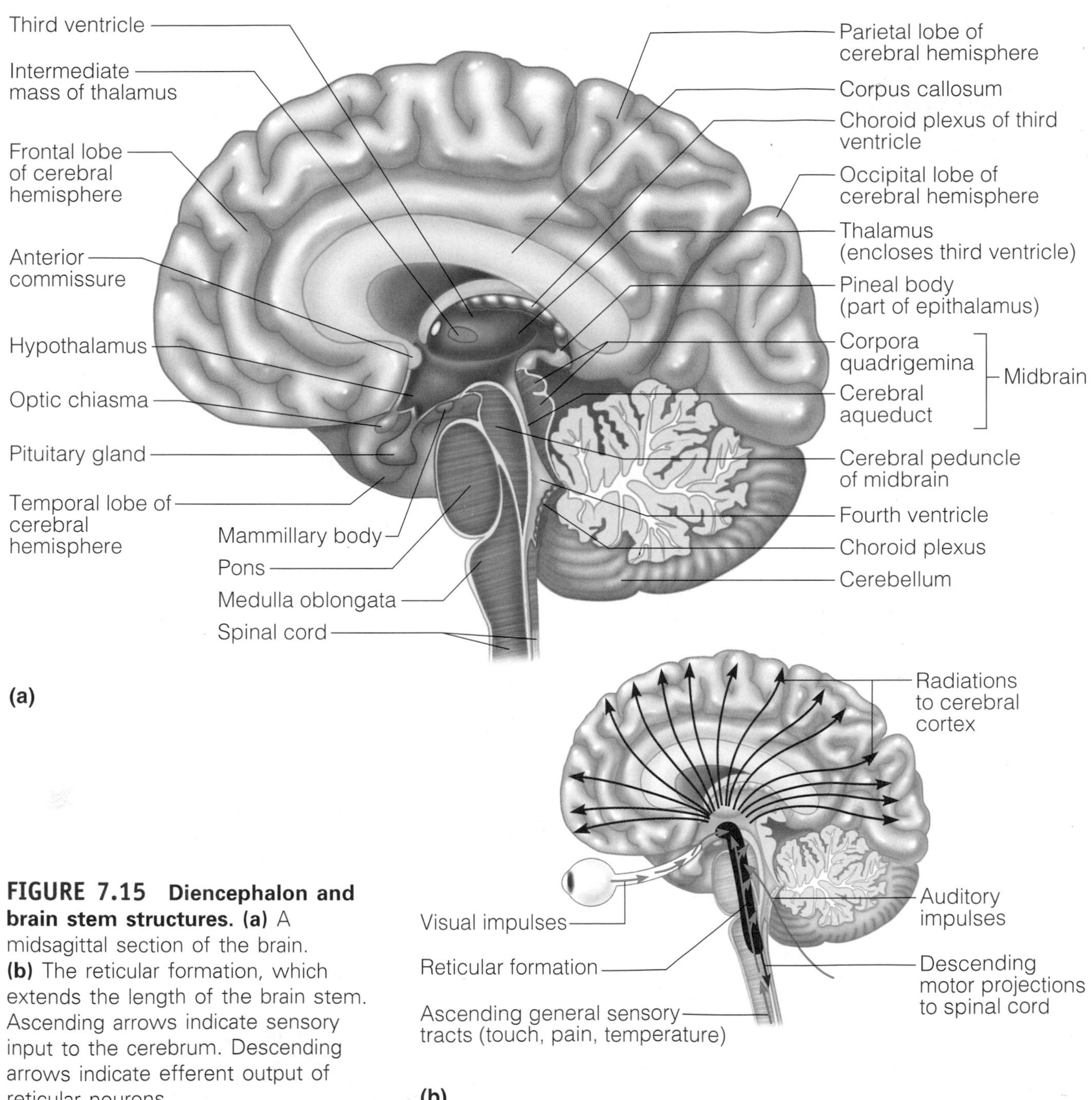

FIGURE 7.15 Diencephalon and brain stem structures. (a) A midsagittal section of the brain. **(b)** The reticular formation, which extends the length of the brain stem. Ascending arrows indicate sensory input to the cerebrum. Descending arrows indicate efferent output of reticular neurons.

callosum (kah-lo′sum), connects the cerebral hemispheres (Figure 7.15). The corpus callosum arches above the structures of the brain stem and allows the cerebral hemispheres to communicate with one another. This is important because, as already noted, some of the cortical functional areas are in only one hemisphere.

Although most of the gray matter is in the cerebral cortex, there are several "islands" of gray matter, called the **basal nuclei,** or **basal ganglia,*** buried deep within the white matter of the cerebral hemispheres. The basal nuclei help regulate voluntary

*The term *basal ganglia* is a historical misnomer. Ganglia are peripheral nervous system structures but the basal ganglia are collections of nerve cell bodies *in the CNS.* Hence, *basal nuclei* is a more accurate term.

motor activities by modifying instructions sent to the skeletal muscles by the primary motor cortex.

Homeostatic Imbalance

Individuals who have problems with their basal nuclei are often unable to walk normally or carry out other voluntary movements in the usual normal way. *Huntington's disease* (or *Huntington's chorea*) and *Parkinson's disease,* two examples of such syndromes, are discussed in the "A Closer Look" box on pages 245–246. ▲

Diencephalon

The **diencephalon,** or **interbrain,** sits atop the brain stem and is enclosed by the cerebral hemispheres (see Figure 7.12). The major structures of the diencephalon are the *thalamus, hypothalamus,* and *epithalamus* (see Figure 7.15). The **thalamus,** which encloses the shallow *third ventricle* of the brain, is a relay station for sensory impulses passing upward to the sensory cortex. As impulses surge through the thalamus, we have a crude recognition of whether the sensation we are about to have is pleasant or unpleasant. The actual localization and interpretation of the sensation is done by the neurons of the sensory cortex.

The **hypothalamus** (literally, "under the thalamus") makes up the floor of the diencephalon. It is an important autonomic nervous system center because it plays a role in the regulation of body temperature, water balance, and metabolism. The hypothalamus is also the center for many drives and emotions, and as such it is an important part of the so-called **limbic system,** or "emotional-visceral brain." For example, thirst, appetite, sex, pain, and pleasure centers are in the hypothalamus. Additionally, the hypothalamus regulates the pituitary gland (an endocrine organ) and produces two hormones of its own. The **pituitary gland** hangs from the anterior floor of the hypothalamus by a slender stalk. (Its function is discussed in Chapter 9.) The **mammillary bodies,** reflex centers involved in olfaction (the sense of smell), bulge from the floor of the hypothalamus posterior to the pituitary gland.

The **epithalamus** (ep″ĭ-thal′ah-mus) forms the roof of the third ventricle. Important parts of the epithalamus are the **pineal body** (part of the endocrine system) and the **choroid** (ko′roid) **plexus** of the third ventricle. The choroid plexuses, knots of capillaries within each ventricle, form the cerebrospinal fluid.

Brain Stem

The **brain stem** is about the size of a thumb in diameter and approximately 3 inches (approximately 7.5 cm) long. Its structures are the *midbrain, pons,* and *medulla oblongata.* In addition to providing a pathway for ascending and descending tracts, the brain stem has many small gray matter areas. These nuclei are part of the cranial nerves and control vital activities such as breathing and blood pressure. Identify the brain stem areas in Figure 7.15 as you read their descriptions that follow.

Midbrain The **midbrain** is a relatively small part of the brain stem. It extends from the mammillary bodies to the pons inferiorly. The **cerebral aqueduct** is a tiny canal that travels through the midbrain and connects the third ventricle of the diencephalon to the fourth ventricle below. Anteriorly, the midbrain is composed primarily of two bulging fiber tracts, the **cerebral peduncles** (literally, "little feet of the cerebrum"), which convey ascending and descending impulses. Dorsally located are four rounded protrusions called the **corpora quadrigemina** (kor′por-ah kwah″drĭ-jem′ĭ-nah) because they reminded some anatomist of two pairs of twins (*gemini*). These bulging nuclei are reflex centers involved with vision and hearing.

Pons The **pons** (ponz) is the rounded structure that protrudes just below the midbrain. *Pons* means "bridge," and this area of the brain stem is mostly fiber tracts. However, it does have important nuclei involved in the control of breathing.

Medulla Oblongata The **medulla oblongata** (mĕ-dul′ah ob″long-gă′tah) is the most inferior part of the brain stem. It merges into the spinal cord below without any obvious change in structure. Like the pons, the medulla is an important fiber tract area. The medulla also contains many nuclei that regulate vital visceral activities. It contains centers that control heart rate, blood pressure, breathing, swallowing, and vomiting, among others. The **fourth ventricle** lies posterior to the pons and medulla and anterior to the cerebellum.

Reticular Formation Extending the entire length of the brain stem is a diffuse mass of gray matter, the **reticular formation.** The neurons of the reticular formation are involved in motor control of the visceral organs. A special group of reticular formation neurons, the **reticular activating system (RAS),** plays a role in consciousness and the awake/sleep cycles (Figure 7.15b). Damage to this area can result in permanent unconsciousness (coma).

Cerebellum

The large, cauliflowerlike **cerebellum** (ser″e-bel′um) projects dorsally from under the occipital lobe of the cerebrum. Like the cerebrum, it has two hemispheres and a convoluted surface. The cerebellum also has an outer cortex made up of gray matter and an inner region of white matter.

The cerebellum provides the precise timing for skeletal muscle activity and controls our balance and equilibrium. Because of its activity, body movements are smooth and coordinated. Fibers reach the cerebellum from the equilibrium apparatus of the inner ear, the eye, the proprioceptors of the skeletal muscles and tendons, and many other areas. The cerebellum can be compared to an automatic pilot, continuously comparing the brain's "intentions" with actual body performance by monitoring body position and amount of tension in various body parts. When needed, it sends messages to initiate the appropriate corrective measures.

Homeostatic Imbalance

If the cerebellum is damaged (for example, by a blow to the head, a tumor, or a stroke), movements become clumsy and disorganized—a condition called *ataxia.* Victims cannot keep their balance and may appear to be drunk because of the loss of muscle coordination. They are no longer able to touch their finger to their nose with eyes closed—a feat that normal individuals accomplish easily. ▲

Protection of the Central Nervous System

Nervous tissue is very soft and delicate, and the irreplaceable neurons are injured by even the slightest pressure. Nature has tried to protect the brain and spinal cord by enclosing them within bone (the skull and vertebral column), membranes (the meninges), and a watery cushion (cerebrospinal fluid). Protection from harmful substances in the blood is provided by the so-called blood-brain barrier. Since we have already considered the bony enclosures (Chapter 5), we will focus on the other protective devices here.

Meninges

The three connective tissue membranes covering and protecting the CNS structures are **meninges** (mĕ-nin′jēz) (Figure 7.16). The outermost layer, the leathery dura mater (du′rah ma′ter), meaning "tough or hard mother," is a double-layered membrane where it surrounds the brain. One of its layers is attached to the inner surface of the skull, forming the periosteum (*periosteal layer*). The other, called the *meningeal layer,* forms the outermost covering of the brain and continues as the dura mater of the spinal cord. The dural layers are fused together except in three areas where they separate to enclose *dural sinuses* that collect venous blood.

In several places, the inner dural membrane extends inward to form a fold that attaches the brain to the cranial cavity. One of these folds, the **falx** (falks) **cerebri,** is shown in Figure 7.16a. Another such fold, the **tentorium cerebelli** separating the cerebellum from the cerebrum, is shown in Figures 7.16b and 7.17c.

The middle meningeal layer is the weblike **arachnoid** (ah-rak′noid) **mater** (see Fig. 7.16). *Arachnida* means "spider," and some think the arachnoid membrane looks like a cobweb. Its threadlike extensions span the **subarachnoid space** to attach it to the innermost membrane, the **pia** (pi′ah) **mater** ("gentle mother"). The delicate pia mater clings tightly to the surface of the brain and spinal cord, following every fold.

The subarachnoid space is filled with cerebrospinal fluid. Specialized projections of the arachnoid membrane, **arachnoid villi** (vih′li), protrude through the dura mater. The cerebrospinal fluid is absorbed into the venous blood in the dural sinuses through the arachnoid villi.

Homeostatic Imbalance

Meningitis, an inflammation of the meninges, is a serious threat to the brain because bacterial or viral meningitis may spread into the nervous tissue of the CNS. This condition of brain inflammation is called *encephalitis* (en-sef-ah-li′tis). Meningitis is usually diagnosed by taking a sample of cerebrospinal fluid from the subarachnoid space. ▲

Cerebrospinal Fluid

Cerebrospinal (ser″e-bro-spi′nal) **fluid (CSF)** is a watery "broth" similar in its makeup to blood plasma, from which it forms. However, it contains less protein, more vitamin C, and its ion composition is different.

CSF is continually formed from blood by the choroid plexuses. Choroid plexuses are clusters of

Q *What would be the consequence of blocked arachnoid villi?*

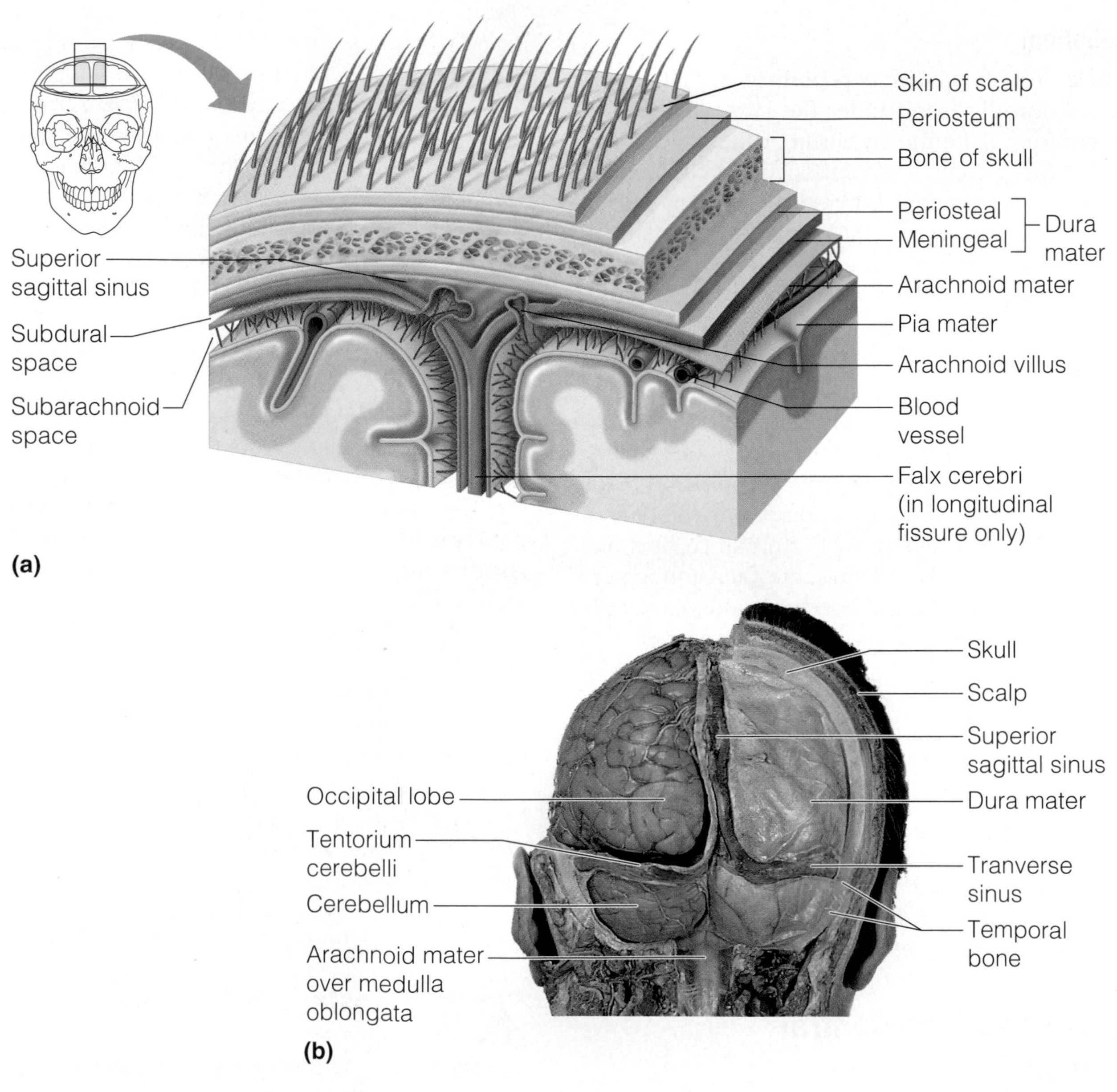

FIGURE 7.16 Meninges of the brain. (a) Three-dimensional frontal section showing the meninges—the dura mater, arachnoid mater, and pia mater—that surround and protect the brain. The relationship of the dura mater to the falx cerebri and the superior sagittal (dural) sinus is also shown. **(b)** Posterior view of the brain in place surrounded by the dura mater.

capillaries hanging from the "roof" in each of the brain's ventricles. The CSF in and around the brain and cord forms a watery cushion that protects the fragile nervous tissue from blows and other trauma.

A *Hydrocephalus ("water on the brain"). The ventricles would expand as cerebrospinal fluid (unable to drain into the dural sinus) accumulated.*

Inside the brain, CSF is continually moving (see Figure 7.17c). It circulates from the two lateral ventricles (in the cerebral hemispheres) into the third ventricle (in the diencephalon), and then through the cerebral aqueduct of the midbrain into the fourth ventricle dorsal to the pons and medulla oblongata. Some of the fluid reaching the fourth ventricle continues down the **central canal** of the

Q *Why are the lateral ventricles horn-shaped rather than oriented vertically like the third and fourth ventricles?*

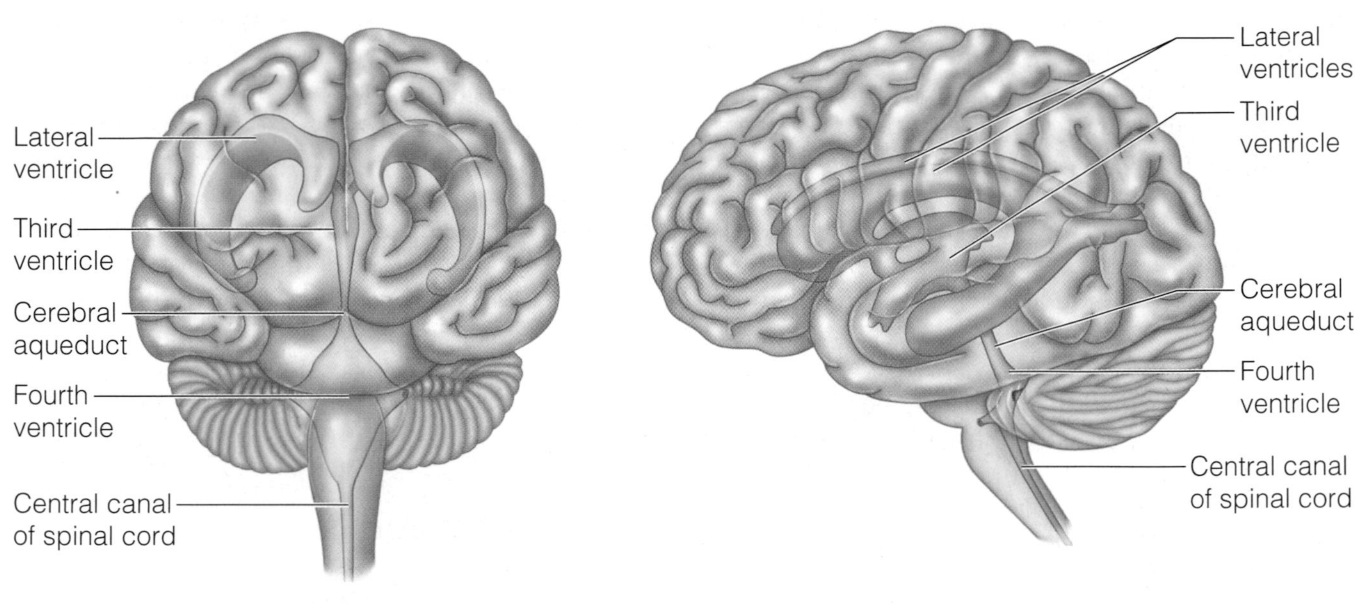

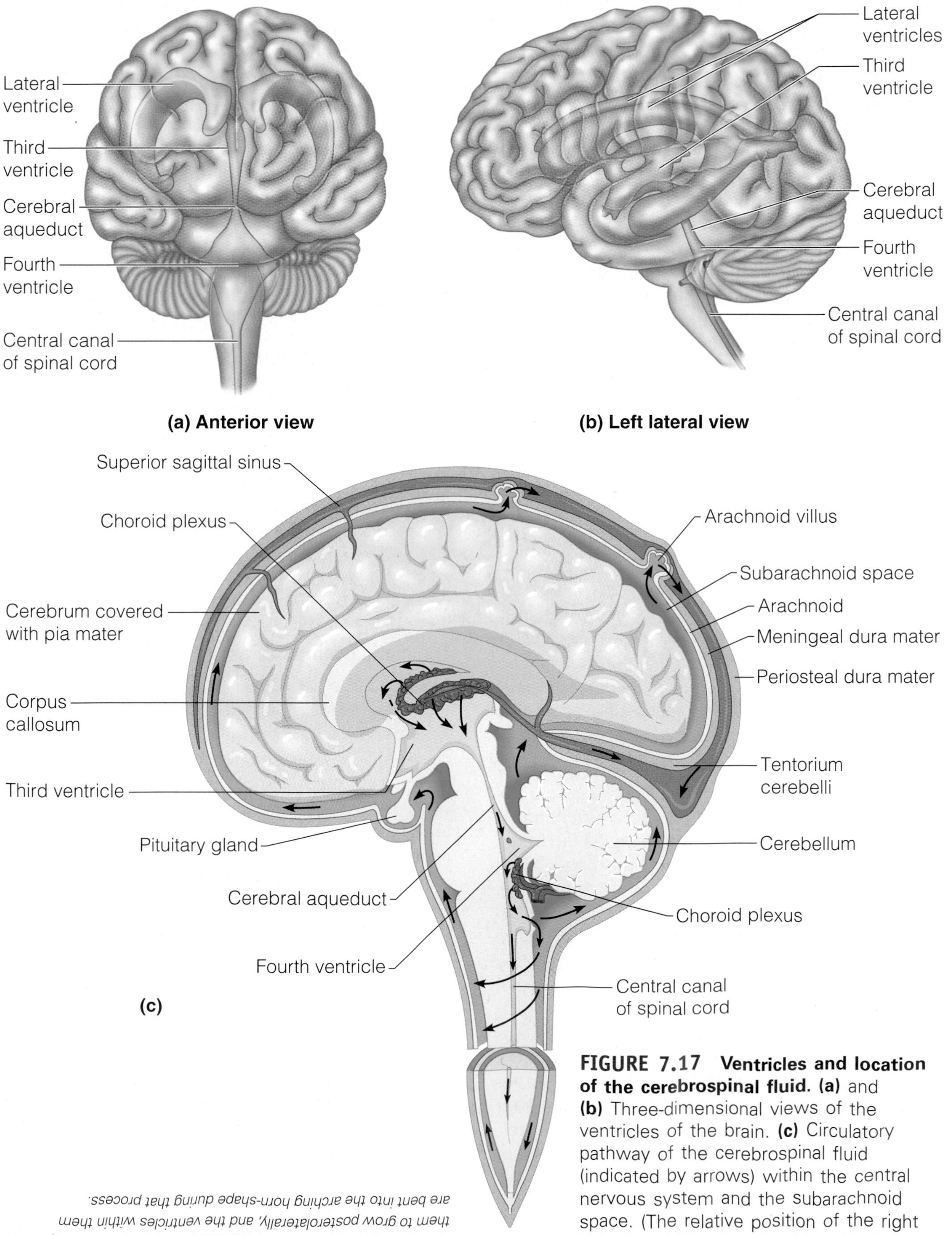

FIGURE 7.17 Ventricles and location of the cerebrospinal fluid. (a) and **(b)** Three-dimensional views of the ventricles of the brain. **(c)** Circulatory pathway of the cerebrospinal fluid (indicated by arrows) within the central nervous system and the subarachnoid space. (The relative position of the right lateral ventricle is indicated by the pale blue area deep to the corpus callosum.)

A *The bones of the skull restrict superior growth of the cerebral hemispheres (and their ventricles) during development, forcing them to grow posterolaterally, and the ventricles within them are bent into the arching horn-shape during that process.*

spinal cord, but most of it circulates into the **subarachnoid space** through three openings in the walls of the fourth ventricle. The fluid returns to the blood in the dural sinuses through the arachnoid villi. Ordinarily, CSF forms and drains at a constant rate so that its normal pressure and volume (150 ml—about half a cup) are maintained. Any significant changes in CSF composition (or the appearance of blood cells in it) may be a sign of meningitis or certain other brain pathologies (such as tumors and multiple sclerosis). The CSF sample for testing is obtained by a procedure called a *lumbar (spinal) tap.* Since the withdrawal of fluid for testing decreases CSF fluid pressure, the patient must remain in a horizontal position (lying down) for 6 to 12 hours after the procedure to prevent an agonizingly painful "spinal headache."

Homeostatic Imbalance

If something obstructs its drainage (for example, a tumor), CSF begins to accumulate and exert pressure on the brain. This condition is *hydrocephalus* (hi-dro-sef'ah-lus), literally, "water on the brain." Hydrocephalus in a newborn baby causes the head to enlarge as the brain increases in size. This is possible in an infant because the skull bones have not yet fused. However, in an adult this condition is likely to result in brain damage because the skull is hard, and the accumulating fluid crushes soft nervous tissue. Today hydrocephalus is treated surgically by inserting a shunt (a plastic tube) to direct the excess fluid into a vein in the neck. ▲

The Blood-Brain Barrier

No other body organ is so absolutely dependent on a constant internal environment as is the brain. Other body tissues can withstand the rather small fluctuations in the concentrations of hormones, ions, and nutrients that continually occur, particularly after eating or exercising. If the brain were exposed to such chemical changes, uncontrolled neural activity might result—remember that certain ions (sodium and potassium) are involved in initiating nerve impulses, and some amino acids serve as neurotransmitters. Consequently, neurons are kept separated from bloodborne substances by a so-called **blood-brain barrier,** composed of the *least* permeable capillaries in the whole body. Of water-soluble substances, only water, glucose, and essential amino acids pass easily through the walls of these capillaries. Metabolic wastes, such as urea, toxins, proteins, and most drugs are prevented from entering the brain tissue. Nonessential amino acids and potassium ions are not only prevented from entering the brain, but also are actively pumped from the brain into the blood across capillary walls. Although the bulbous "feet" of the astrocytes that cling to the capillaries may contribute to the barrier, the relative impermeability of the brain capillaries is most responsible for providing this protection.

The blood-brain barrier is virtually useless against fats, respiratory gases, and other fat-soluble molecules that diffuse easily through all plasma membranes. This explains why bloodborne alcohol, nicotine, and anesthetics can affect the brain.

Brain Dysfunctions

Homeostatic Imbalance

Brain dysfunctions are unbelievably varied. We mention some of them (the "terrible three" in the "A Closer Look" box on pages 245–246), and we will discuss developmental problems in the final section of this chapter. Here, we will focus on traumatic brain injuries and cerebrovascular accidents. Techniques used to diagnose many brain disorders are described in the "A Closer Look" box on pages 262–263.

Traumatic Brain Injuries

Head injuries are a leading cause of accidental death in the United States. Consider, for example, what happens when you forget to fasten your seat belt and then crash into the rear end of another car. Your head is moving and then is suddenly stopped as it hits the windshield. Brain damage is caused not only by injury at the site of the blow, but also by the effect of the ricocheting brain hitting the opposite end of the skull.

A *concussion* occurs when brain injury is slight. The victim may be dizzy, "see stars," or lose consciousness briefly, but no permanent brain damage occurs. A brain *contusion* is the result of marked tissue destruction. If the cerebral cortex is injured, the individual may remain conscious, but severe brain stem contusions always result in a coma lasting from hours to a lifetime because of injury to the reticular activating system.

After head blows, death may result from *intracranial hemorrhage* (bleeding from ruptured vessels) or *cerebral edema* (swelling of the brain due to inflammatory response to injury). Individuals who are initially alert and lucid following head trauma and then begin to deteriorate neurologically later are most

A Closer Look

Alzheimer's, Parkinson's, and Huntington's—The Terrible Three

WHAT do former President Reagan and actor Michael J. Fox have in common? Not much, other than the fact that each was or is a victim of one of the terrible three CNS degenerative diseases.

Alzheimer's (altz'hi-merz) **disease (AD)** is a progressive degenerative disease of the brain that ultimately results in dementia (mental deterioration). Alzheimer's patients represent nearly half of all people in nursing homes. Between 5 and 15 percent of people over 65 develop this condition, and up to half of those over 85 die of it.

Its victims, which include the recently deceased President Reagan, exhibit memory loss (particularly for recent events), a short attention span and disorientation, and eventual language loss. Over a period of several years, formerly good-natured people become irritable, moody, confused, and sometimes violent. Ultimately, hallucinations occur.

AD is associated with a shortage of acetylcholine (ACh) and with structural changes in the brain, particularly in areas involved with cognition and memory. The gyri shrink, and the brain atrophies. Its precise cause is unknown, but some cases of AD appear to run in families.

Microscopic examinations of brain tissue reveal abnormal protein deposits (masses of cells and degenerated fibers around a *beta amyloid peptide* core) littering the brain like shrapnel between the neurons. They also show *neurofibrillary tangles* (twisted fibers within neuron cell bodies). It has been frustratingly difficult for researchers to uncover how beta amyloid peptide acts as a neurotoxin, particularly since it is also present in healthy brain cells (but in lower amounts). Just what tips things off balance to favor production of more beta amyloid is not understood, but it is known that this tiny peptide does its damage by enhancing calcium entry into certain brain neurons.

Another line of research has implicated a protein called *tau,* which appears to function like railroad ties to bind microtubule "tracks" together. In the brains of AD victims, tau abandons its microtubule stabilizing role and grabs onto other tau molecules, forming the spaghetti-like neurofibrillary tangles. These degenerative changes develop over a period of several years, during which time the family members watch the person they love "disappear." It is a long and painful process. It is hoped that the lines of investigation, particularly stem cell research, will eventually merge and point to a treatment, but at present only drugs that ease symptoms by inhibiting ACh breakdown are useful.

Parkinson's disease, an example of basal nuclei problems, typically strikes people in their 50s and 60s (Michael J. Fox is an exception). It results from a degeneration of the dopamine-releasing neurons of the substantia nigra, and as those neurons deteriorate, the dopamine-deprived basal nuclei they target become overactive, causing the well-known symptoms of the disease. Afflicted individuals, including Michael J. Fox, have a persistent tremor at rest (exhibited by head nodding and a "pill-rolling" movement of the fingers), a forward-bent walking posture and shuffling gait, a stiff facial expression, and they have trouble initiating movement or getting their muscles going.

The cause of Parkinson's disease is still unknown. The drug *L-dopa* helps to alleviate some of the symptoms; however, L-dopa is not a cure, and as more and more neurons die off, it becomes ineffective. It also has undesirable side effects: severe nausea, dizziness, and in some, liver damage. A newer treatment drug is deprenyl. When given early on in the disease, deprenyl slows the neurological deterioration to some extent and delays the need to administer L-dopa for up to 18 months.

Deep-brain (thalamic) stimulation via implanted electrodes has proved helpful in alleviating tremors ➤

Alzheimer's, Parkinson's, and Huntington's—The Terrible Three *(continued)*

but does little else. More promising for long-term results are the intrabrain transplants of embryonic substantia nigra tissue, genetically engineered adult nigral cells, or dopamine-producing cells from fetal pigs; all these have produced some regression of disease symptoms. However, the use of fetal tissue is controversial and riddled with ethical and legal roadblocks.

Huntington's disease is a genetic disease that strikes during middle age and leads to massive degeneration of the basal nuclei and later of the cerebral cortex. Its initial symptoms in many are wild, jerky, and almost continuous flapping movements called *chorea* (Greek for "dance"). Although the movements appear to be voluntary, they are not. Late in the disease, marked mental deterioration occurs. Huntington's disease is progressive and usually fatal within 15 years of onset of symptoms.

The signs and symptoms of Huntington's disease are essentially the opposite of those of Parkinson's disease (overstimulation rather than inhibition of the motor drive), and Huntington's is usually treated with drugs that block, rather than enhance, dopamine's effects. As you can see, neurotransmitters, which are the "vocabulary" of neurons, can cause garbled neural language when things go wrong. As with Parkinson's disease, fetal tissue implants may provide promise for its treatment in the future.

likely hemorrhaging or suffering the consequences of edema, both of which compress vital brain tissue.

Cerebrovascular Accident Commonly called *strokes,* **cerebrovascular** (ser″e-bro-vas′ku-lar) **accidents (CVAs)** are the third leading cause of death in the United States. CVAs occur when blood circulation to a brain area is blocked, as by a blood clot or a ruptured blood vessel, and vital brain tissue dies. After a CVA, it is often possible to determine the area of brain damage by observing the patient's symptoms. For example, if the patient has left-sided paralysis, the right motor cortex of the frontal lobe is most likely involved. *Aphasias* (ah-fa′ze-ahz) are a common result of damage to the left cerebral hemisphere, where the language areas are located. There are many types of aphasias, but the most common are *motor aphasia,* which involves damage to Broca's area and a loss of ability to speak, and *sensory aphasia,* in which a person loses the ability to understand written or spoken language. Aphasias are maddening to the victims because, as a rule, their intellect is unimpaired. Brain lesions can also cause marked changes in a person's disposition (for example, a change from a sunny to a foul personality). In such cases, a tumor as well as a CVA might be suspected.

Fewer than a third of those surviving a CVA are alive three years later. Even so, the picture is not hopeless. Some patients recover at least part of their lost faculties because undamaged neurons spread into areas where neurons have died and take over some lost functions. Indeed, most of the recovery seen after brain injury is due to this phenomenon.

Not all strokes are "completed." Temporary brain ischemia, or restriction of blood flow, is called a *transient ischemic attack (TIA).* TIAs last from 5 to 50 minutes and are characterized by symptoms such as numbness, temporary paralysis, and impaired speech. Although these defects are not permanent, they do constitute "red flags" that warn of impending, more serious CVAs. ▲

Spinal Cord

The cylindrical **spinal cord,** which is approximately 17 inches (42 cm) long, is a glistening white continuation of the brain stem. The spinal cord provides a two-way conduction pathway to and from the brain, and it is a major reflex center (the spinal reflexes are completed at this level). Enclosed within the vertebral column, the spinal cord extends from the foramen magnum of the skull to the first or second lumbar vertebra, where it ends just below the ribs (Figure 7.18). Like the brain, the spinal cord is cushioned and protected by meninges. Meningeal coverings do not end at the second lumbar vertebra (L_2), but instead extend well beyond the end of the spinal cord in the vertebral canal. Because there is no possibility of damaging the cord beyond L_3, the meningeal sac inferior to that point provides a nearly ideal spot for removing CSF for testing.

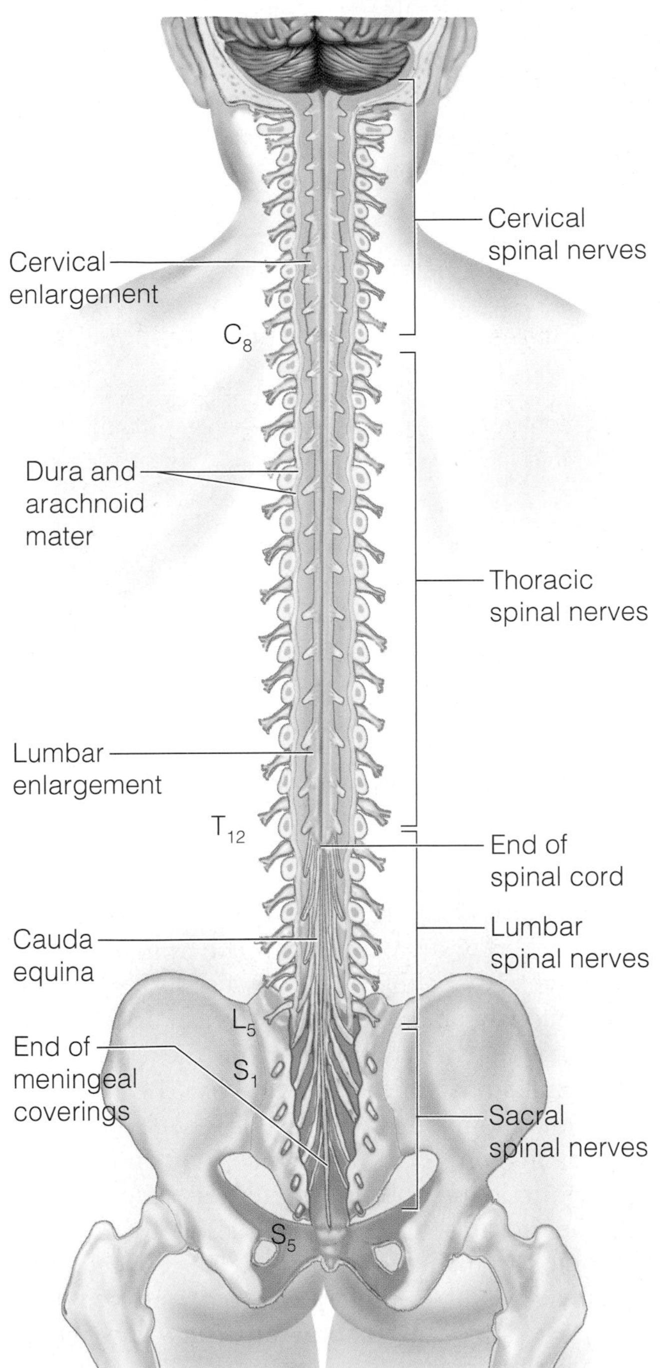

FIGURE 7.18 Anatomy of the spinal cord, posterior view.

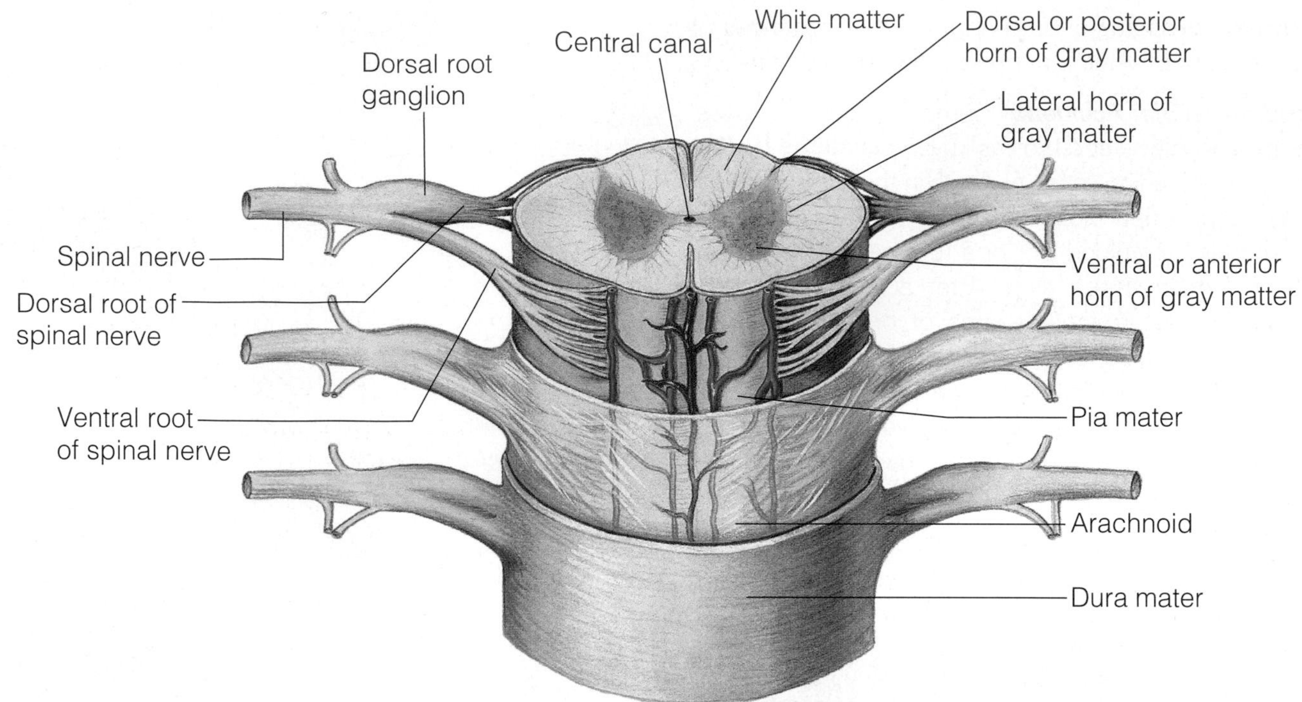

FIGURE 7.19 Spinal cord with meninges (three-dimensional view).

In humans, 31 pairs of spinal nerves arise from the cord and exit from the vertebral column to serve the body area close by. The spinal cord is about the size of a thumb for most of its length, but it is obviously enlarged in the cervical and lumbar regions where the nerves serving the upper and lower limbs arise and leave the cord. Because the vertebral column grows faster than the spinal cord, the spinal cord does not reach the end of the vertebral column and the spinal nerves leaving its inferior end must travel through the vertebral canal for some distance before exiting. This collection of spinal nerves at the inferior end of the vertebral canal is called the **cauda equina** (kaw′da e-kwi′nah) because it looks so much like a horse's tail (the literal translation of *cauda equina*).

Gray Matter of the Spinal Cord and Spinal Roots

The gray matter of the spinal cord looks like a butterfly or the letter H in cross section (Figure 7.19). The two posterior projections are the **dorsal,** or **posterior, horns;** the two anterior projections are the **ventral,** or **anterior, horns.** The gray matter surrounds the **central canal** of the cord, which contains CSF.

Neurons with specific functions can be located in the gray matter. The dorsal horns contain association neurons, or interneurons. The cell bodies of the sensory neurons, whose fibers enter the cord by the **dorsal root,** are found in an enlarged area called the **dorsal root ganglion.** If the dorsal root or its ganglion is damaged, sensation from the body area served will be lost. The ventral horns of the gray matter contain cell bodies of motor neurons of the somatic (voluntary) nervous system, which send their axons out the **ventral root** of the cord. The dorsal and ventral roots fuse to form the **spinal nerves.**

Homeostatic Imbalance

Damage to the ventral root results in a *flaccid paralysis* of the muscles served. In flaccid paralysis, nerve impulses do not reach the muscles affected; thus, no voluntary movement of those muscles is possible. The muscles begin to atrophy because they are no longer stimulated. ▲

White Matter of the Spinal Cord

White matter of the spinal cord is composed of myelinated fiber tracts—some running to higher centers, some traveling from the brain to the cord, and some conducting impulses from one side of the spinal cord to the other.

Because of the irregular shape of gray matter, the white matter on each side of the cord is divided into three regions—the **posterior, lateral,** and

anterior columns. Each of the columns contains a number of fiber tracts made up of axons with the same destination and function. Tracts conducting sensory impulses to the brain are *sensory,* or *afferent, tracts.* Those carrying impulses from the brain to skeletal muscles are *motor,* or *efferent, tracts.* All tracts in the posterior columns are ascending tracts that carry sensory input to the brain. The lateral and anterior tracts contain both ascending and descending (motor) tracts.

Homeostatic Imbalance

If the spinal cord is transected (cut crosswise) or crushed, *spastic paralysis* results. The affected muscles stay healthy because they are still stimulated by spinal reflex arcs, and movement of those muscles does occur. However, movements are involuntary and not controllable, and this can be as much of a problem as complete lack of mobility. In addition, because the spinal cord carries both sensory and motor impulses, a loss of feeling or sensory input occurs in the body areas below the point of cord destruction. Physicians often use a pin to see if a person can feel pain after spinal cord injury—to find out if regeneration is occurring. Pain is a hopeful sign in such cases. If the spinal cord injury occurs high in the cord, so that all four limbs are affected, the individual is a *quadriplegic* (kwod″rĭ-ple′jik). If only the legs are paralyzed, the individual is a *paraplegic* (par″ă-ple′jik). ▲

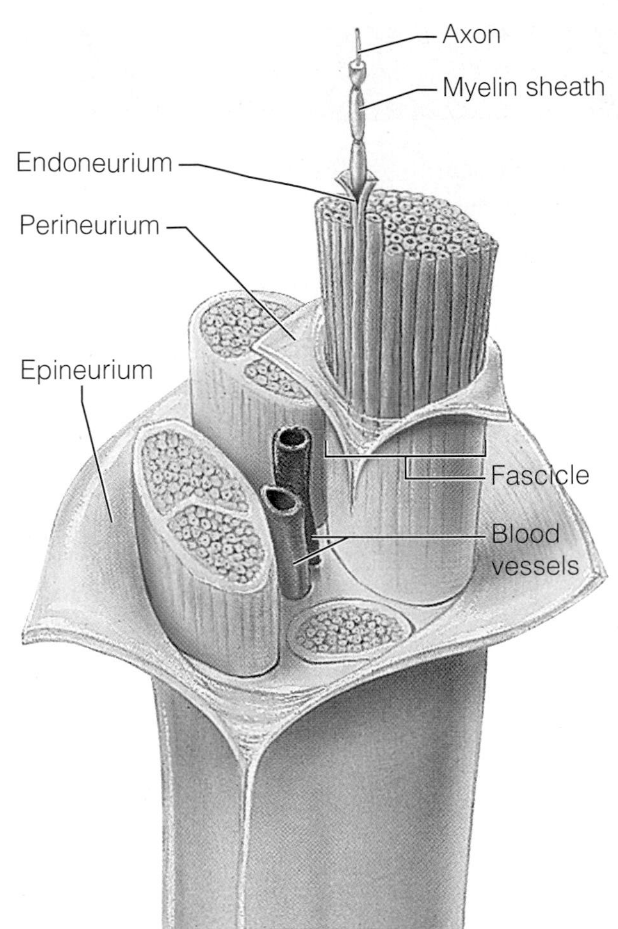

FIGURE 7.20 Structure of a nerve. Three-dimensional view of a portion of a nerve, showing its connective tissue wrappings.

Peripheral Nervous System

The **peripheral nervous system (PNS)** consists of nerves and scattered groups of neuronal cell bodies (ganglia) found outside the CNS. One type of ganglion has already been considered—the dorsal root ganglion of the spinal cord. Others will be covered in the discussion of the autonomic nervous system. Here, we will concern ourselves only with nerves.

Structure of a Nerve

A **nerve** is a bundle of neuron fibers found outside the CNS. Within a nerve, neuron fibers, or processes, are wrapped in protective connective tissue coverings. Each fiber is surrounded by a delicate connective tissue sheath, an **endoneurium** (en″do-nu′re-um). Groups of fibers are bound by a coarser connective tissue wrapping, the **perineurium** (per″ĭ-nu′re-um), to form fiber bundles, or **fascicles.** Finally, all the fascicles are bound together by a tough fibrous sheath, the **epineurium,** to form the cordlike nerve (Figure 7.20).

Like neurons, nerves are classified according to the direction in which they transmit impulses. Nerves carrying both sensory and motor fibers are called **mixed nerves;** all spinal nerves are mixed nerves. Nerves that carry impulses toward the CNS only are called **sensory,** or **afferent, nerves,** whereas those that carry only motor fibers are **motor,** or **efferent, nerves.**

Cranial Nerves

The 12 pairs of **cranial nerves** primarily serve the head and neck. Only one pair (the vagus nerves) extends to the thoracic and abdominal cavities.

The cranial nerves are numbered in order, and in most cases their names reveal the most important structures they control. The cranial nerves are described by name, number, course, and major function in Table 7.1. The last column

TABLE 7.1 The Cranial Nerves

Name/Number	Origin/Course	Function	Test
I. Olfactory	Fibers arise from olfactory receptors in the nasal mucosa and synapse with the olfactory bulbs (which, in turn, send fibers to the olfactory cortex)	Purely sensory; carries impulses for the sense of smell	Subject is asked to sniff and identify aromatic substances, such as oil of cloves or vanilla
II. Optic	Fibers arise from the retina of the eye and form the optic nerve. The two optic nerves form the optic chiasma by partial crossover of fibers; the fibers continue to the optic cortex as the optic tracts	Purely sensory; carries impulses for vision	Vision and visual field are tested with an eye chart and by testing the point at which the subject first sees an object (finger) moving into the visual field; eye interior is viewed with an ophthalmoscope
III. Oculomotor	Fibers run from the midbrain to the eye	Supplies motor fibers to four of the six muscles (superior, inferior, and medial rectus, and inferior oblique) that direct the eyeball; to the eyelid; and to the internal eye muscles controlling lens shape and pupil size	Pupils are examined for size, shape, and size equality; pupillary reflex is tested with a penlight (pupils should constrict when illuminated); eye convergence is tested, as is the ability to follow moving objects
IV. Trochlear	Fibers run from the midbrain to the eye	Supplies motor fibers for one external eye muscle (superior oblique)	Tested in common with cranial nerve III for the ability to follow moving objects
V. Trigeminal	Fibers emerge from the pons and form three divisions that run to the face	Conducts sensory impulses from the skin of the face and mucosa of the nose and mouth; also contains motor fibers that activate the chewing muscles	Sensations of pain, touch, and temperature are tested with a safety pin and hot and cold objects; corneal reflex tested with a wisp of cotton; motor branch tested by asking the subject to open mouth against resistance and move jaw from side to side
VI. Abducens	Fibers leave the pons and run to the eye	Supplies motor fibers to the lateral rectus muscle, which rolls the eye laterally	Tested in common with cranial nerve III for the ability to move each eye laterally

TABLE 7.1 *(continued)*

Name/Number	Origin/Course	Function	Test
VII. Facial	Fibers leave the pons and run to the face	Activates the muscles of facial expression and the lacrimal and salivary glands; carries sensory impulses from the taste buds of anterior tongue	Anterior two-thirds of tongue is tested for ability to taste sweet, salty, sour, and bitter substances; subject is asked to close eyes, smile, whistle, etc.; tearing is tested with ammonia fumes
VIII. Vestibulocochlear	Fibers run from the equilibrium and hearing receptors of the inner ear to the brain stem	Purely sensory; vestibular branch transmits impulses for the sense of balance, and cochlear branch transmits impulses for the sense of hearing	Hearing is checked by air and bone conduction, using a tuning fork
IX. Glossopharyngeal	Fibers emerge from the medulla and run to the throat	Supplies motor fibers to the pharynx (throat) that promote swallowing and saliva production; carries sensory impulses from taste buds of the posterior tongue and from pressure receptors of the carotid artery	Gag and swallowing reflexes are checked; subject is asked to speak and cough; posterior tongue may be tested for taste
X. Vagus	Fibers emerge from the medulla and descend into the thorax and abdominal cavity	Fibers carry sensory impulses from and motor impulses to the pharynx, larynx, and the abdominal and thoracic viscera; most motor fibers are parasympathetic fibers that promote digestive activity and help regulate heart activity	Tested in common with cranial nerve IX, since they both serve muscles of the throat
XI. Accessory	Fibers arise from the medulla and superior spinal cord and travel to muscles of the neck and back	Mostly motor fibers that activate the sternocleidomastoid and trapezius muscles	Sternocleidomastoid and trapezius muscles are checked for strength by asking the subject to rotate head and shrug shoulders against resistance
XII. Hypoglossal	Fibers run from the medulla to the tongue	Motor fibers control tongue movements; sensory fibers carry impulses from the tongue	Subject is asked to stick out tongue, and any position abnormalities are noted

Q *If you can't shrug your right shoulder, what cranial nerve is involved?*

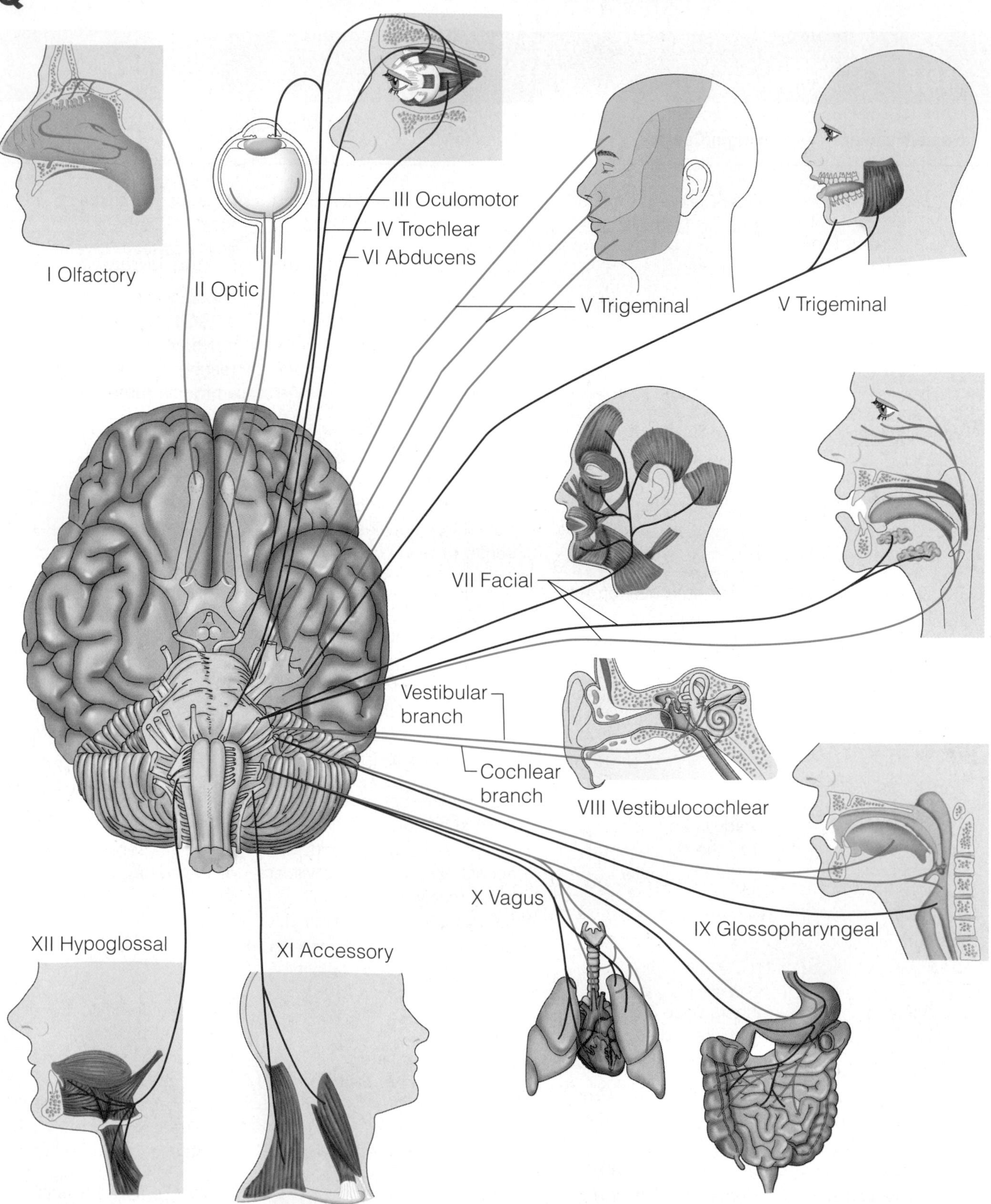

FIGURE 7.21 Distribution of cranial nerves. Sensory nerves are shown in blue, motor nerves in red. Although cranial nerves III, IV, and VI have sensory fibers, these are not shown because the sensory fibers account for only minor parts of these nerves.

A *Left member of cranial nerve XI.*

of the table describes how cranial nerves are tested, which is an important part of any neurologic examination. You do not need to memorize these tests, but this information may help you understand cranial nerve function. As you read through the table, also look at Figure 7.21, which shows the location of the cranial nerves on the brain's anterior surface.

Most cranial nerves are mixed nerves; however, three pairs, the optic, olfactory, and vestibulocochlear (ves-tib″u-lo-kok′le-ar) nerves, are purely sensory in function. (The older name for the vestibulocochlear nerve is *acoustic nerve,* a name that reveals its role in hearing but not in equilibrium.) I give my students the following little saying as a memory jog to help them remember the cranial nerves in order; perhaps it will help you, too. The first letter of each word in the saying (and both letters of "ah") is the first letter of the cranial nerve to be remembered: "**O**h, **o**h, **o**h, **t**o **t**ouch **a**nd **f**eel **v**ery **g**ood **v**elvet, **ah.**"

Spinal Nerves and Nerve Plexuses

The 31 pairs of human **spinal nerves** are formed by the combination of the ventral and dorsal roots of the spinal cord. Although each of the cranial nerves issuing from the brain is named specifically, the spinal nerves are named for the region of the cord from which they arise. Figure 7.22 shows how the nerves are named in this scheme.

Almost immediately after being formed, each spinal nerve divides into **dorsal** and **ventral rami** (ra′mi), making each spinal nerve only about ½ inch long. The rami, like the spinal nerves, contain both motor and sensory fibers. Thus, damage to a spinal nerve or either of its rami results both in loss of sensation and flaccid paralysis of the area of the body served. The smaller dorsal rami serve the skin and muscles of the posterior body trunk. The ventral rami of spinal nerves T_1 through T_{12} form the *intercostal nerves,* which supply the muscles between the ribs and the skin and muscles of the anterior and lateral trunk. The ventral rami of all other spinal nerves form complex networks of nerves called **plexuses,** which serve the motor and sensory needs of the limbs. The four nerve plexuses are described in Table 7.2; three of the four plexuses are shown in Figure 7.23.

Autonomic Nervous System

The **autonomic nervous system (ANS)** is the motor subdivision of the PNS that controls body activities automatically. It is composed of a special group of neurons that regulate cardiac muscle (the heart), smooth muscles (found in the walls of the visceral organs and blood vessels), and glands. Although all body systems contribute to homeostasis, the relative stability of our internal environment depends largely on the workings of the ANS. At every moment, signals flood from the visceral organs into the CNS, and the autonomic nerves make adjustments as necessary to best support body activities. For example, blood flow may be shunted to more "needy" areas, heart and breathing rate may be speeded up or slowed down, blood pressure may be adjusted, and stomach secretions may be increased or decreased. Most of this fine-tuning occurs without our awareness or attention—few of us realize when our pupils dilate or our arteries constrict—hence the ANS is also called the **involuntary nervous system.**

Somatic and Autonomic Nervous Systems Compared

Our previous discussions of motor nerves have focused on the activity of the somatic nervous system, the motor subdivision that controls our skeletal muscles. So, before plunging into a description of autonomic nervous system anatomy, we will take the time to point out some important differences between the somatic and autonomic divisions.

Besides differences in their effector organs and in the neurotransmitters released, the patterns of their motor pathways differ. In the somatic division, the cell bodies of the motor neurons are inside the CNS, and their axons (in spinal nerves) extend all the way to the skeletal muscles they serve. The autonomic nervous system, however, has a chain of *two* motor neurons. The first motor neuron of each pair is in the brain or spinal cord. Its axon, the **preganglionic axon** (literally, the "axon before the ganglion"), leaves the CNS to synapse with the second motor neuron in a ganglion outside the CNS. The axon of this neuron, the

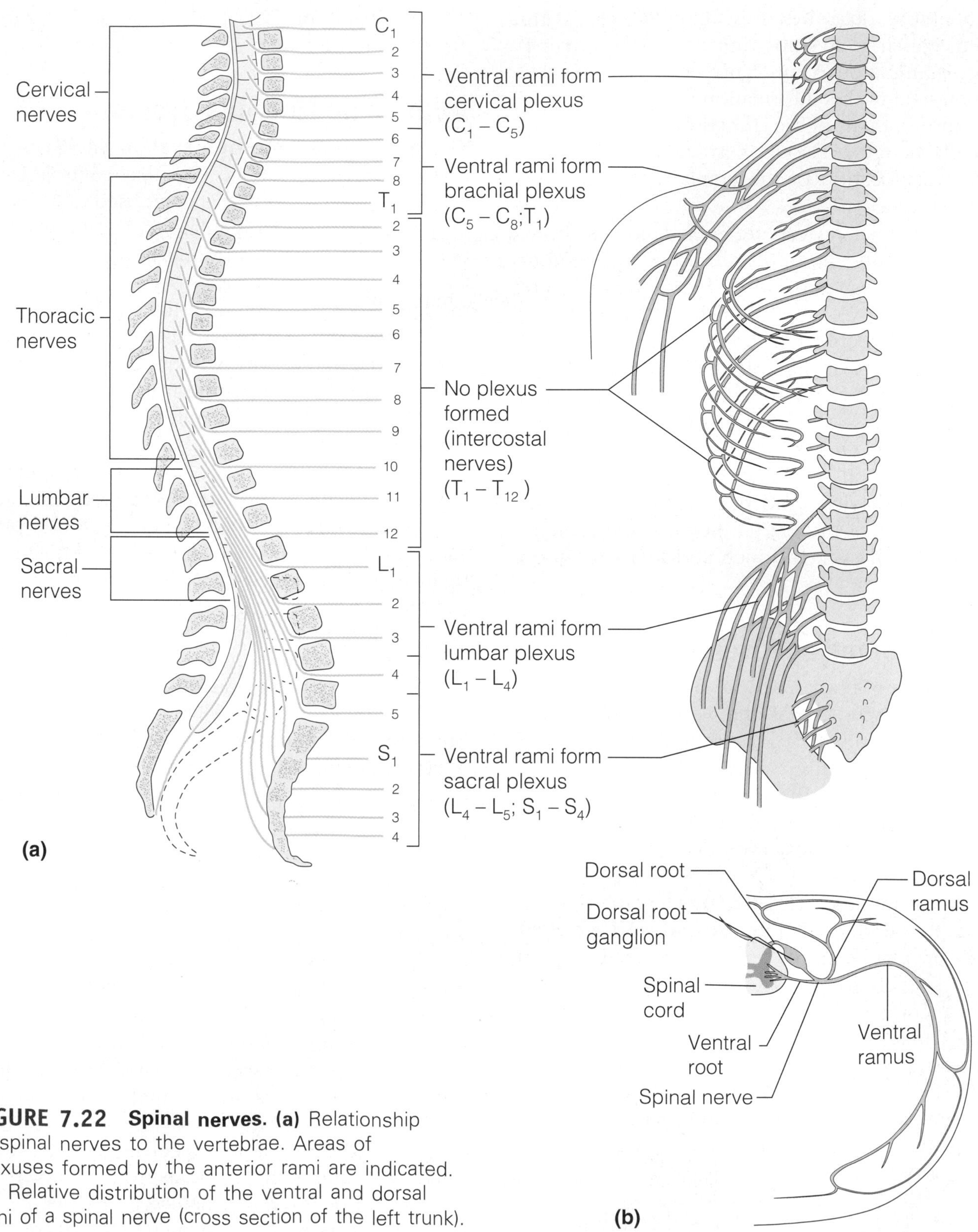

FIGURE 7.22 Spinal nerves. (a) Relationship of spinal nerves to the vertebrae. Areas of plexuses formed by the anterior rami are indicated. **(b)** Relative distribution of the ventral and dorsal rami of a spinal nerve (cross section of the left trunk).

TABLE 7.2 Spinal Nerve Plexuses

Plexus	Origin (from ventral rami)	Important nerves	Body areas served	Result of damage to plexus or its nerves
Cervical	C_1–C_5	Phrenic	Diaphragm and muscles of shoulder and neck	Respiratory paralysis (and death if not treated promptly)
Brachial	C_5–C_8 and T_1	Axillary	Deltoid muscle of shoulder	Paralysis and atrophy of deltoid muscle
		Radial	Triceps and extensor muscles of the forearm	Wristdrop—inability to extend hand at wrist
		Median	Flexor muscles of forearm and some muscles of hand	Decreased ability to flex and abduct hand and flex and abduct thumb and index finger—therefore, inability to pick up small objects
		Musculocutaneous	Flexor muscles of arm	Decreased ability to flex forearm on arm
		Ulnar	Wrist and many hand muscles	Clawhand—inability to spread fingers apart
Lumbar	L_1–L_4	Femoral (including lateral and anterior cutaneous branches)	Lower abdomen, buttocks, anterior thighs, and skin of anteromedial leg and thigh	Inability to extend leg and flex hip; loss of cutaneous sensation
		Obturator	Adductor muscles of medial thigh and small hip muscles; skin of medial thigh and hip joint	Inability to adduct thigh
Sacral	L_4–L_5 and S_1–S_4	Sciatic (largest nerve in body; splits to common fibular and tibial nerves)	Lower trunk and posterior surface of thigh (and leg)	Inability to extend hip and flex knee; sciatica
		• Common fibular (superficial and deep branches)	Lateral aspect of leg and foot	Footdrop—inability to dorsiflex foot
		• Tibial (including sural and plantar branches)	Posterior aspect of leg and foot	Inability to plantar flex and invert foot; shuffling gait
		Superior and inferior gluteal	Gluteus muscles of hip	Inability to extend hip (maximus) or abduct and medially rotate thigh (medius)

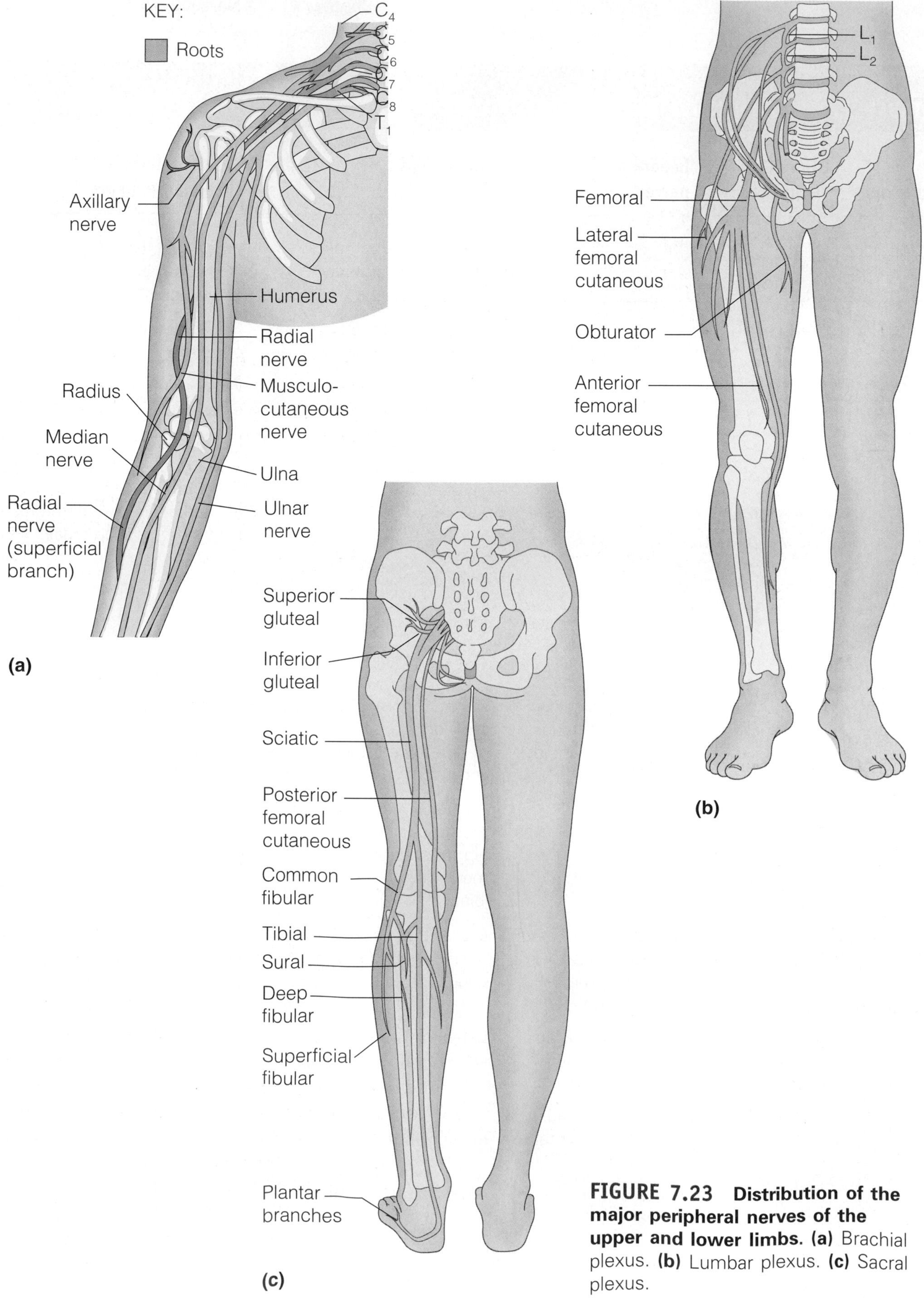

FIGURE 7.23 Distribution of the major peripheral nerves of the upper and lower limbs. (a) Brachial plexus. **(b)** Lumbar plexus. **(c)** Sacral plexus.

Q *Transmission of nerve impulses along ANS pathways is generally much slower than along somatic fibers. Why?*

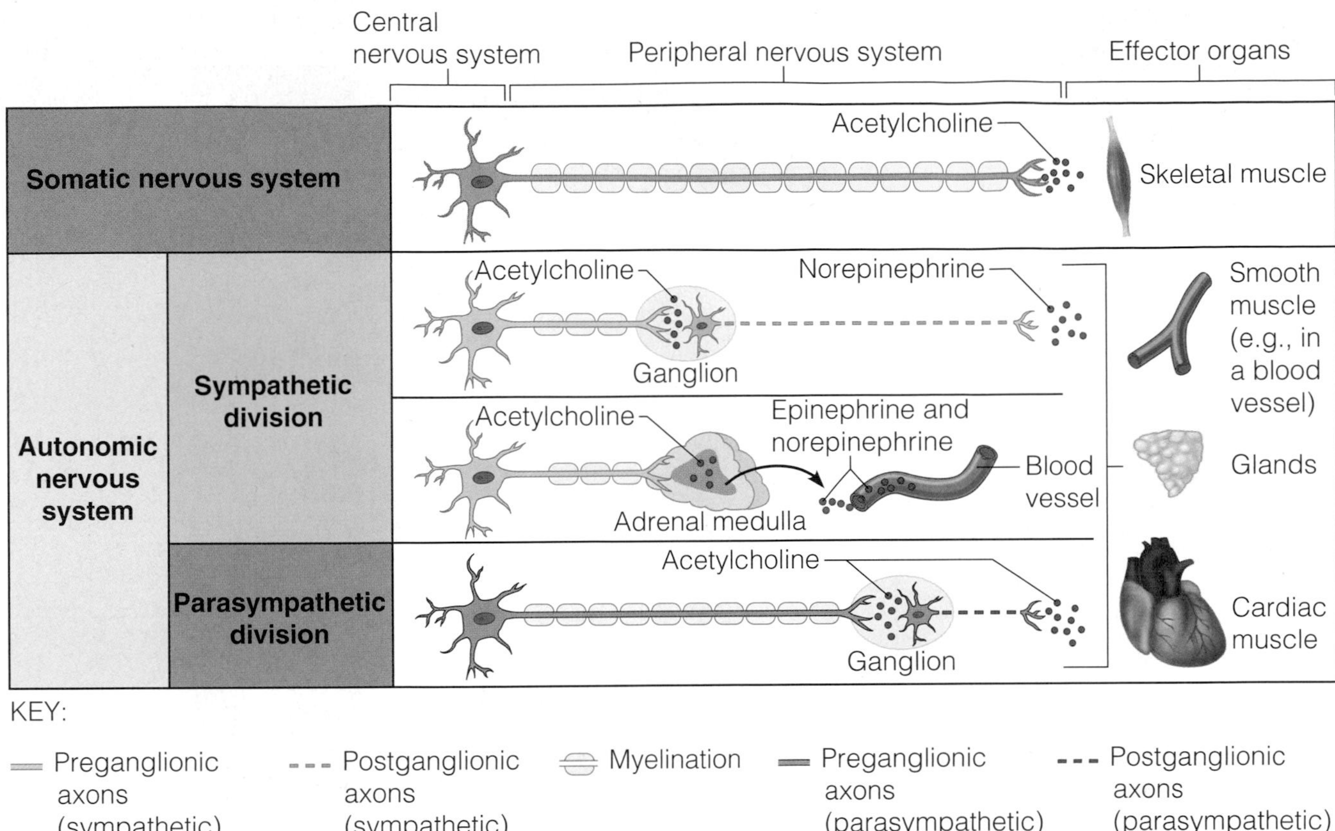

FIGURE 7.24 Comparison of the somatic and autonomic nervous systems.

postganglionic axon, then extends to the organ it serves. These differences are summarized in Figure 7.24.

The autonomic nervous system has two arms, the sympathetic and the parasympathetic (Figure 7.25). Both serve the same organs but cause essentially opposite effects, counterbalancing each other's activities to keep body systems running smoothly. The **sympathetic division** mobilizes the body during extreme situations (such as fear, exercise, or rage), whereas the **parasympathetic division** allows us to "unwind" and conserve energy. These differences are examined in more detail shortly, but first we will consider the structural characteristics of the two arms of the ANS.

A *Postganglionic fibers of the ANS are unmyelinated fibers which conduct much more slowly than the myelinated fibers that are typical of somatic nerve fibers.*

Anatomy of the Parasympathetic Division

The first neurons of the parasympathetic division are located in brain nuclei of several cranial nerves—III, VII, IX, and X (the vagus being the most important of these) and in the S_2 through S_4 levels of the spinal cord (see Figure 7.25). The neurons of the cranial region send their axons out in cranial nerves to serve the head and neck organs. There they synapse with the second motor neuron in a **terminal ganglion.** From the terminal ganglion, the postganglionic axon extends a short distance to the organ it serves. In the sacral region, the preganglionic axons leave the spinal cord and form the *pelvic splanchnic* (splank'nik) *nerves,* also called the *pelvic nerves,* which travel to the pelvic cavity. In the pelvic cavity, the preganglionic axons synapse with the second motor neurons in terminal ganglia on, or close to, the organs they serve.

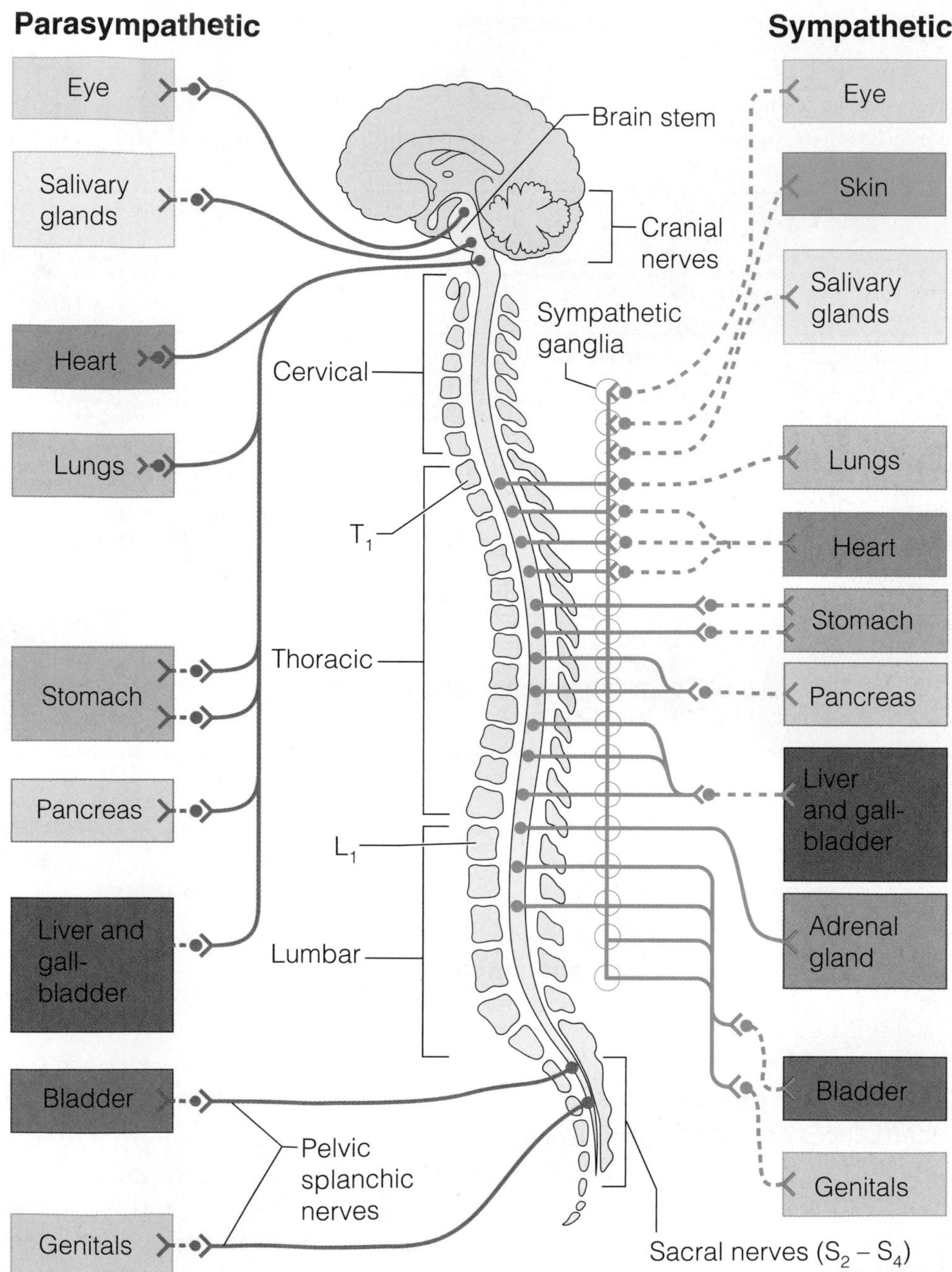

FIGURE 7.25 Anatomy of the autonomic nervous system. Parasympathetic fibers are shown in purple, sympathetic fibers in green. Solid lines represent preganglionic fibers; dashed lines indicate postganglionic fibers.

Anatomy of the Sympathetic Division

The sympathetic division is also called the *thoracolumbar* (tho″rah-ko-lum′bar) *division* because its first neurons are in the gray matter of the spinal cord from T_1 through L_2 (see Figure 7.25). The preganglionic axons leave the cord in the ventral root, enter the spinal nerve, and then pass through a **ramus communicans,** or small communicating branch, to enter a **sympathetic chain ganglion** (Figure 7.26). The **sympathetic chain,** or **trunk,** lies alongside the vertebral column on each side. After it reaches the ganglion, the axon may synapse with the second neuron in the sympathetic chain at the same or a different level (the

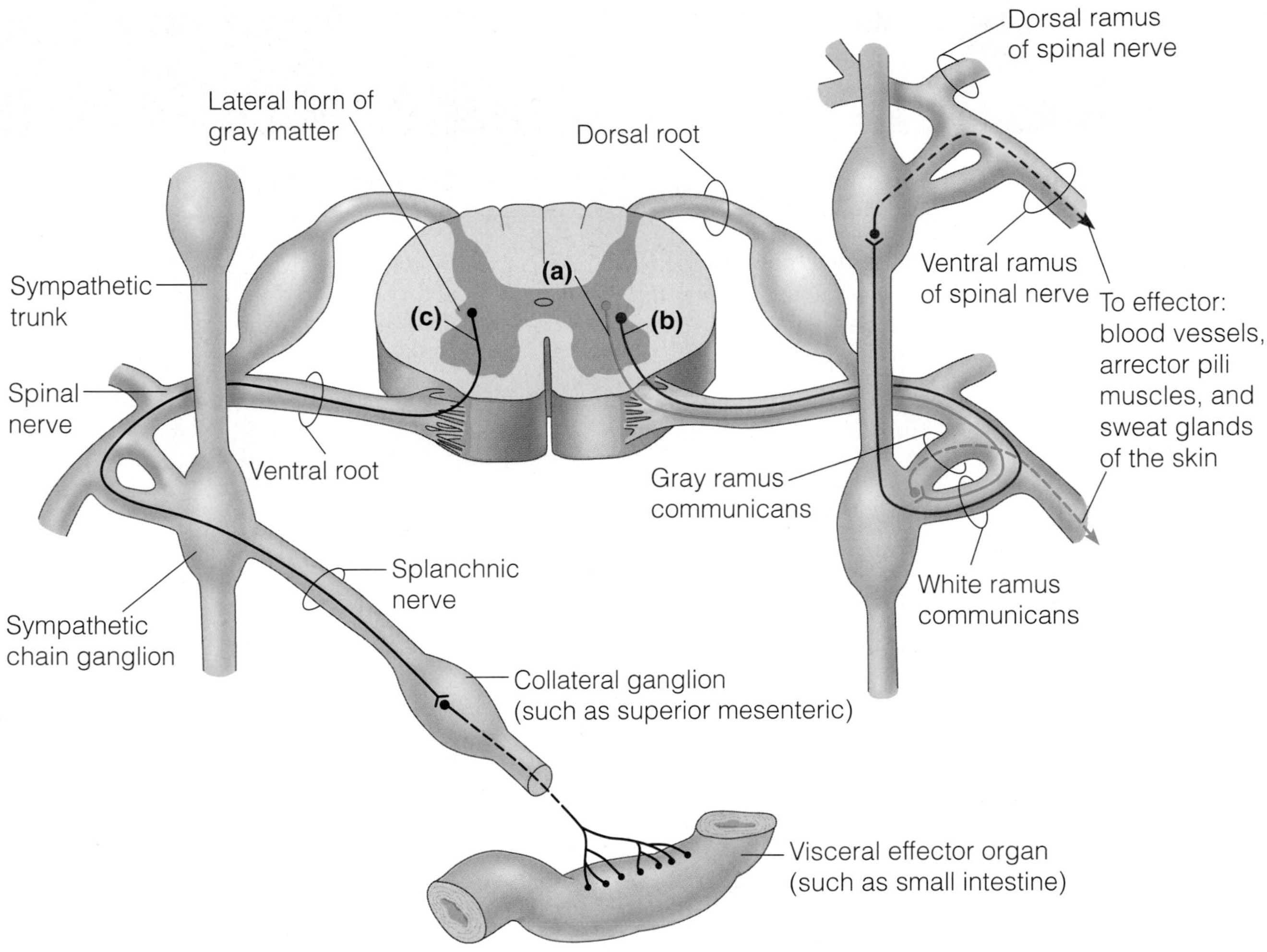

FIGURE 7.26 Sympathetic pathways. (a) Synapse in a sympathetic chain ganglion at the same level. **(b)** Synapse in a sympathetic chain ganglion at a different level. **(c)** Synapse in a collateral ganglion anterior to the vertebral column.

postganglionic axon then reenters the spinal nerve to travel to the skin), or the axon may pass through the ganglion without synapsing and form part of the **splanchnic nerves.** The splanchnic nerves travel to the viscera to synapse with the second neuron, found in a **collateral ganglion** anterior to the vertebral column. The major collateral ganglia—the celiac and the superior and inferior mesenteric ganglia—supply the abdominal and pelvic organs. The postganglionic axon then leaves the collateral ganglion and travels to serve a nearby visceral organ.

Now that the anatomical details have been described, we are ready to examine ANS functions in a little more detail.

Autonomic Functioning

Body organs served by the autonomic nervous system receive fibers from both divisions. Exceptions are most blood vessels and most structures of the skin, some glands, and the adrenal medulla, all of which receive only sympathetic fibers (Table 7.3). When both divisions serve the same organ they cause antagonistic effects, mainly because their postganglionic axons release different neurotransmitters (see Figure 7.24). The parasympathetic fibers, called *cholinergic* (ko″lin-er′jik) *fibers,* release acetylcholine. The sympathetic postganglionic fibers, called *adrenergic* (ad″ren-er′jik) *fibers,* release norepinephrine (nor″ep-ĭ-nef′rin). The preganglionic

TABLE 7.3 Effects of the Sympathetic and Parasympathetic Divisions of the Autonomic Nervous System

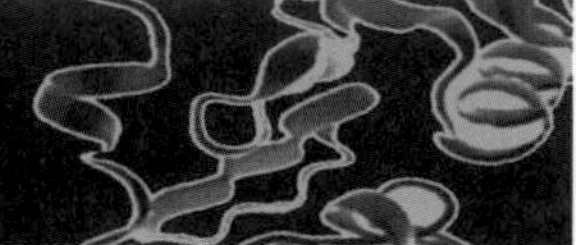

Target organ/system	Parasympathetic effects	Sympathetic effects
Digestive system	Increases smooth muscle mobility (peristalsis) and amount of secretion by digestive system glands; relaxes sphincters	Decreases activity of digestive system and constricts digestive system sphincters (for example, anal sphincter)
Liver	No effect	Causes glucose to be released to blood
Lungs	Constricts bronchioles	Dilates bronchioles
Urinary bladder/ urethra	Relaxes sphincters (allows voiding)	Constricts sphincters (prevents voiding)
Kidneys	No effect	Decreases urine output
Heart	Decreases rate; slows and steadies	Increases rate and force of heartbeat
Blood vessels	No effect on most blood vessels	Constricts blood vessels in viscera and skin (dilates those in skeletal muscle and heart); increases blood pressure
Glands—salivary, lacrimal	Stimulates; increases production of saliva and tears	Inhibits; result is dry mouth and dry eyes
Eye (iris)	Stimulates constrictor muscles; constricts pupils	Stimulates dilator muscles; dilates pupils
Eye (ciliary muscle)	Stimulates to increase bulging of lens for close vision	Inhibits; decreases bulging of lens; prepares for distant vision
Adrenal medulla	No effect	Stimulates medulla cells to secrete epinephrine and norepinephrine
Sweat glands of skin	No effect	Stimulates to produce perspiration
Arrector pili muscles attached to hair follicles	No effect	Stimulates; produces "goose bumps"
Penis	Causes erection due to vasodilation	Causes ejaculation (emission of semen)
Cellular metabolism	No effect	Increases metabolic rate; increases blood sugar levels; stimulates fat breakdown

axons of *both* divisions release acetylcholine. To emphasize the *relative* roles of the two arms of the ANS, we will focus briefly on situations in which each division is "in control."

Sympathetic Division The **sympathetic division** is often referred to as the "fight-or-flight" system. Its activity is evident when we are excited or find ourselves in emergency or threatening situations, such as being frightened by street toughs late at night. A pounding heart; rapid, deep breathing; cold, sweaty skin; a prickly scalp; and dilated eye pupils are sure signs of sympathetic nervous system activity. Under such conditions, the sympathetic nervous system increases heart rate, blood pressure, and blood glucose levels; dilates the bronchioles of the lungs; and brings about many other effects that help the individual cope with the stressor. Dilation of blood vessels in skeletal muscles (so that one can run faster or fight better) and withdrawal of blood from the digestive organs (so that the bulk of the blood can be used to serve the heart, brain, and skeletal muscles) are other examples.

The sympathetic nervous system is working at full speed not only when you are emotionally upset, but also when you are physically stressed. For example, if you have just had surgery or run a marathon, your adrenal glands (activated by the sympathetic nervous system) would be pumping out epinephrine and norepinephrine (see Figure 7.24). The effects of sympathetic nervous system activation continue for several minutes until its hormones are destroyed by the liver. Thus, although sympathetic nerve impulses themselves may act only briefly, the hormonal effects they provoke linger. The widespread and prolonged effects of sympathetic activation help explain why we need time to "come down" after an extremely stressful situation.

The sympathetic division generates a head of steam that enables the body to cope rapidly and vigorously with situations that threaten homeostasis. Its function is to provide the best conditions for responding to some threat, whether the best response is to run, to see better, or to think more clearly.

Homeostatic Imbalance

Some illnesses or diseases are at least aggravated, if not caused, by excessive sympathetic nervous system stimulation. Certain individuals, called Type A people, always work at breakneck speed and push themselves continually. These are people who are likely to have heart disease, high blood pressure, and ulcers, all of which may result from prolonged sympathetic nervous system activity or the rebound from it. ▲

Parasympathetic Division The **parasympathetic division** is most active when the body is at rest and not threatened in any way. This division, sometimes called the "resting-and-digesting" system, is chiefly concerned with promoting normal digestion and elimination of feces and urine and with conserving body energy, particularly by decreasing demands on the cardiovascular system. (This explains why it is a good idea to relax after a heavy meal so that digestion is not inhibited or disturbed by sympathetic activity.) Its activity is best illustrated by a person who relaxes after a meal and reads the newspaper. Blood pressure and heart and respiratory rates are being regulated at low normal levels, the digestive tract is actively digesting food, and the skin is warm (indicating that there is no need to divert blood to skeletal muscles or vital organs). The eye pupils are constricted to protect the retinas from excessive damaging light, and the lenses of the eyes are "set" for close vision. We might also consider the parasympathetic division as the "housekeeping" system of the body.

An easy way to remember the most important roles of the two ANS divisions is to think of the parasympathetic division as the **D** (digestion, defecation, and diuresis [urination]) division and the sympathetic division as the **E** (exercise, excitement,

Prove It Yourself

Improve Your Memory

Can you improve your ability to learn and remember new information? Yes! The following techniques take advantage of the brain's storage and retrieval mechanisms:

- **Concentrate.** This may seem obvious, but paying attention increases brain activity and epinephrine levels, thereby promoting consolidation of information into long-term memory.
- **Minimize interference.** Go where it is quiet. A noisy environment will impair your ability to concentrate.
- **Break down large amounts of information into smaller topics.** Give yourself time to review each topic, and take a break in between.
- **Rephrase material in your own words.** Restate the information in a way that makes sense to you personally.
- **Test yourself.** Create outlines or diagrams. Try to define key terms before looking up their definitions. Use practice and review questions when they are available.

Short-term memory involves quick bursts of action potentials. Every time you read, think about, or test yourself on a concept, more neurons fire. By studying new material actively and repeatedly, you trigger additional action potentials and improve long-term retention because the neural synapses are reinforced by use.

A Closer Look

Tracking Down CNS Problems

ANYONE who has had a routine physical examination is familiar with the reflex tests done to assess neural function. The doctor taps your patellar (Achilles) tendon with a reflex hammer and your leg muscles contract, resulting in the knee- or ankle-jerk response. These responses show that the spinal cord and brain centers are functioning normally. When reflex tests are abnormal or when brain cancer, intracranial hemorrhage, multiple sclerosis, or hydrocephalus are suspected, more sophisticated neurological tests may be ordered to try to localize and identify the problem.

An "oldie-but-goodie" procedure used to diagnose and localize many different types of brain lesions (such as epileptic lesions, tumors, and abscesses) is *electroencephalography* (e-lek"tro-en-sef-ah-lah'grah-fe). Normal brain function involves the continuous transmission of electrical impulses by neurons. A recording of their activity, called an **electroencephalogram,** or **EEG,** can be made by placing electrodes at various points on the scalp and connecting these to a recording device (see the figure). The patterns of electrical activity of the neurons are called *brain waves.* Because people differ genetically, and because everything we have ever experienced has left its imprint in our brain, each of us has a brain wave pattern that is as unique as our fingerprints. The four most commonly seen brain waves are illustrated and described in Figure (b).

As might be expected, brain-wave patterns typical of the alert wide-awake state differ from those that occur during relaxation or deep sleep. Interference with the function of the cerebral cortex is suggested by brain waves that are too fast or too slow, and unconsciousness occurs at both extremes. Sleep and coma result in brain-wave patterns that are slower than normal, whereas fright, epileptic seizures, and some kinds of drug overdose cause abnormally fast brain waves. Since brain waves are seen even during coma, absence of brain waves (a flat EEG) is taken as evidence of clinical death.

Pneumoencephalography (nu"mo-en-sef"ah-lah'grah-fe) provides a fairly clear X-ray picture of the brain ventricles and has long been the procedure of choice for diagnosing hydrocephalus. A small amount of cerebrospinal fluid is withdrawn by lumbar puncture, and air injected into the subarachnoid space floats upward and into the ventricles, allowing them to be visualized. Although the procedure is remarkably simple, it can cause a blinding headache.

A **cerebral angiogram** (an'je-o-gram) is used to assess the condition of the cerebral arteries serving the brain (or the carotid arteries of the neck, which feed most of those vessels). A dye is injected into an artery, and time is allowed for the dye to become dispersed to the brain. Then, an X ray is taken of the arteries of interest. The dye allows arteries narrowed by arteriosclerosis to be localized. This procedure is commonly ordered for individuals who have suffered a stroke or who have a history of transient ischemic attacks.

The new imaging techniques described in Chapter 1 (pp. 20–22) have revolutionized the diagnosis of brain lesions. **CT** and **MRI scans** allow most tumors, intracranial lesions, multiple sclerosis plaques, and areas of dead brain tissue *(infarcts)* to be identified quickly. The CT scanner is also becoming an important tool to enhance the precision and safety of brain surgery. **PET scans,** which use high-energy gamma rays to monitor the brain's biochemical activity, can localize lesions that generate epileptic seizures. PET scans are also being used to diagnose Alzheimer's disease.

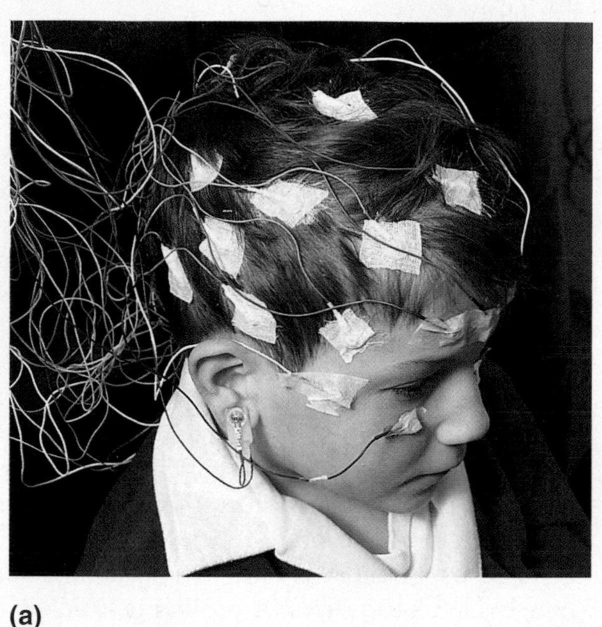

(a)

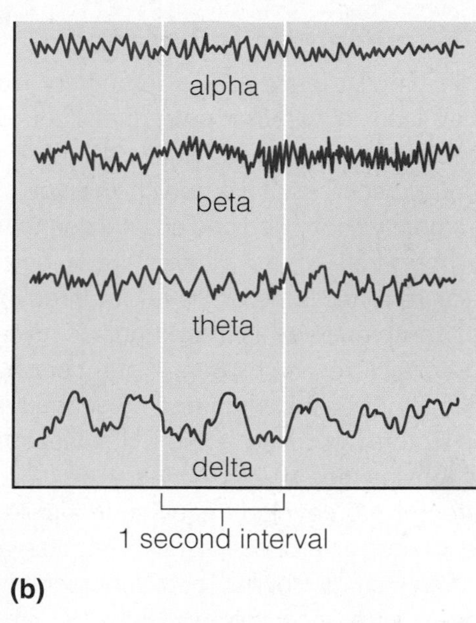

(b)

Electroencephalography and brain waves. (a) To obtain a recording of brain-wave activity (an EEG), electrodes are positioned on the patient's scalp and attached to a recording device called an electroencephalograph. **(b)** Typical EEGs. Alpha waves are typical of the awake, relaxed state; beta waves occur in the awake, alert state; theta waves are common in children but not in normal adults; and delta waves occur during deep sleep.

emergency, and embarrassment) division. Remember, however, while it is easiest to think of the sympathetic and parasympathetic divisions as working in an all-or-none fashion, this is rarely the case. A dynamic balance exists between the two divisions, and fine adjustments are continuously made by both. Also, although we have described the parasympathetic division as the "at-rest" system, most blood vessels are controlled only by the sympathetic fibers regardless of whether the body is "on alert" or relaxing. A summary of the major effects of each division appears in Table 7.3.

Developmental Aspects of the Nervous System

Because the nervous system is formed during the first month of embryonic development, any maternal infection early in pregnancy can have extremely harmful effects on the fetal nervous system. For example, maternal measles *(rubella)* often causes deafness and other types of CNS damage. Also, since nervous tissue has the highest metabolic rate in the body, lack of oxygen for even a few minutes leads to death of neurons. (Because smoking decreases the amount of oxygen that can be carried in the blood, a smoking mother may be sentencing her infant to possible brain damage.) Radiation and various drugs (alcohol, opiates, cocaine, and others) can also be very damaging if administered during early fetal development.

Homeostatic Imbalance

In difficult deliveries, a temporary lack of oxygen may lead to *cerebral palsy* (pawl'ze), but this is only one of the suspected causes. Cerebral palsy is a neuromuscular disability in which the voluntary muscles are poorly controlled and spastic due to brain damage. About half of its victims have seizures, are mentally retarded, and/or have impaired hearing or vision. Cerebral palsy is the largest single cause of physical disabilities in children. A number of other congenital malformations, triggered by genetic or environmental factors, also plague the CNS. Most serious are hydrocephalus (see p. 244), *anencephaly*—a failure of the cerebrum to develop, resulting in a child who cannot hear, see, or process sensory inputs—and spina bifida. *Spina bifida* (spi'nah bi'fĭ-dah) results when the vertebrae form incompletely (typically in the lumbosacral region). There are several varieties. In the least serious, a dimple, and perhaps a tuft of hair, appears over the site of malformation, but no neurological problems occur. In the most serious, meninges, nerve roots, and even parts of the spinal cord protrude from the spine, rendering the lower part of the spinal cord functionless. The child is unable to control the bowels or bladder, and the lower limbs are paralyzed. ▲

One of the last areas of the CNS to mature is the hypothalamus, which contains centers for regulating body temperature. For this reason, premature babies usually have problems in controlling their loss of body heat and must be carefully monitored. Few neurons are formed after birth (because neurons are amitotic), but growth and maturation of the nervous system continues all through childhood, largely as a result of myelination that goes on during this period. A good indication of the degree of myelination of particular neural pathways is the level of neuromuscular control in that body area. As described in Chapter 6, neuromuscular coordination progresses in a superior to inferior (craniocaudal) direction and in a proximal to distal direction, and we know that myelination occurs in the same sequence.

The brain reaches its maximum weight in the young adult. Over the next 60 years or so, neurons are damaged and die; and our store of neurons continually decreases. However, an unlimited number of neural pathways are always available and ready to be developed. We never run out of "recording tape" and can continue to learn throughout life.

As we grow older, the sympathetic nervous system gradually becomes less and less efficient, particularly in its ability to constrict blood vessels. When elderly people stand up quickly after sitting or lying down, they often become lightheaded or faint. This is because the sympathetic nervous system is not able to react quickly enough to counteract the pull of gravity by activating the vasoconstrictor fibers, and blood pools in the feet. This condition, *orthostatic hypotension,* is a type of low blood pressure resulting from changes in body position as described. Orthostatic hypotension can be prevented to some degree if changes in position are made slowly. This gives the sympathetic nervous system a little more time to adjust and react.

The usual cause of nervous system deterioration is circulatory system problems. For example, *arteriosclerosis* (ar-ter″e-o-skle-ro′sis) and high blood pressure result in a decreasing supply of oxygen to the brain neurons. A gradual lack of oxygen due to the aging process finally leads to *senility,* characterized by forgetfulness, irritability, difficulty in concentrating and thinking clearly, and confusion. A sudden loss of blood and oxygen delivery to the brain results in a CVA, as described earlier (p. 247). However, many people continue to enjoy intellectual lives and to work at mentally demanding tasks through their entire life. In fact, fewer than 5 percent of people over the age of 65 demonstrate true senility.

Sadly, many cases of "reversible senility," caused by certain drugs, low blood pressure, constipation, poor nutrition, depression, dehydration, and hormone imbalances, go undiagnosed. The best way to maintain one's mental abilities in old age may be to seek regular medical checkups throughout life.

Although eventual shrinking of the brain is normal, it seems that some individuals (professional boxers and chronic alcoholics, for example) accelerate the process long before aging plays its part. Whether a boxer wins the match or not, the likelihood of brain damage and atrophy increases

Systems in Sync

Homeostatic Relationships between the Nervous System and Other Body Systems

Endocrine System
- Sympathetic division of the ANS activates the adrenal medulla; hypothalamus helps regulate the activity of the anterior pituitary gland and produces two hormones
- Hormones influence metabolism of neurons

Lymphatic System/Immunity
- Nerves innervate lymphoid organs; the brain plays a role in regulating immune function
- Lymphatic vessels carry away leaked tissue fluids from tissues surrounding nervous system structures; immune elements protect all body organs from pathogens (CNS has additional mechanisms)

Digestive System
- ANS (particularly the parasympathetic division) regulates digestive system activity
- Digestive system provides nutrients needed for health of neurons

Urinary System
- ANS regulates bladder emptying and renal blood pressure
- Kidneys help to dispose of metabolic wastes and maintain proper electrolyte composition and pH of blood for neural functioning

Muscular System
- Somatic division of nervous system activates skeletal muscles; maintains muscle health
- Skeletal muscles are the effectors of the somatic division

Nervous System

Respiratory System
- Nervous system regulates respiratory rhythm and depth
- Respiratory system provides life-sustaining oxygen; disposes of carbon dioxide

Cardiovascular System
- ANS helps regulate heart rate and blood pressure
- Cardiovascular system provides blood containing oxygen and nutrients to the nervous system; carries away wastes

Reproductive System
- ANS regulates sexual erection and ejaculation in males; erection of the clitoris in females
- Testosterone masculinizes the brain and underlies sex drive and aggressive behavior

Integumentary System
- Sympathetic division of the ANS regulates sweat glands and blood vessels of skin (therefore heat loss/retention)
- Skin serves as heat loss surface

Skeletal System
- Nerves innervate bones
- Bones serve as depot for calcium needed for neural function; protect CNS structures

with each fight as the brain bounces and rebounds within the skull with every blow. The expression "punch drunk" reflects the symptoms of slurred speech, tremors, abnormal gait, and dementia (mental illness) seen in many retired boxers.

Everyone recognizes that alcohol has a profound effect on the mind as well as the body. However, these effects may not be temporary. CT scans of chronic alcoholics reveal reduced brain size at a fairly early age. Like boxers, chronic alcoholics tend to exhibit signs of *senile* (age-related) *dementia* unrelated to the aging process.

The human cerebral hemispheres—our "thinking caps"—are awesome in their complexity. No less amazing are the brain regions that oversee all our subconscious, autonomic body functions—the diencephalon and brain stem—particularly when you consider their relatively insignificant size. The spinal cord, which acts as a reflex center, and the peripheral nerves, which provide communication links between the CNS and body periphery, are equally important to body homeostasis.

A good deal of new terminology has been introduced in this chapter and, as you will see, much of it will come up again in later chapters as we study the other organ systems of the body and examine how the nervous system helps to regulate their activity. The terms *are* essential, so try to learn them as you go along. Use the Glossary at the back of the book as often as you find it helpful.

SUMMARY

Media study tools that could provide you with additional help in reviewing specific key topics of Chapter 7 are referenced below.

IP = Interactive Physiology;

WEB = A&P Place website.

ORGANIZATION OF THE NERVOUS SYSTEM (pp. 223–224)

1. Structural: All nervous system structures are classified as part of the CNS (brain and spinal cord) or PNS (nerves and ganglia).
2. Functional: Motor nerves of the PNS are classified on the basis of whether they stimulate skeletal muscle (somatic division) or smooth/cardiac muscle and glands (autonomic division).

NERVOUS TISSUE: STRUCTURE AND FUNCTION (pp. 224–235)

1. Supportive connective tissue cells
   a. Neuroglia support and protect neurons in the CNS. Specific glial cells are phagocytes; others myelinate neuron processes in the CNS or line cavities.
   b. Schwann cells myelinate neuron processes in the PNS.

   WEB Exercise: Chapter 7, Glial Cells and Their Functions.
2. Neurons
   a. Anatomy: All neurons have a cell body containing the nucleus and processes (fibers) of two types: (1) axons (one per cell) typically generate and conduct impulses away from the cell body and release a neurotransmitter, and (2) dendrites (one to many per cell) typically carry electrical currents toward the cell body. Most large fibers are myelinated; myelin increases the rate of nerve impulse transmission.

   IP Nervous System Topic: Anatomy Review, pages 3–11.

   b. Classification
      (1) On the basis of function (direction of impulse transmission) there are sensory (afferent) and motor (efferent) neurons and association neurons (interneurons). Dendritic endings of sensory neurons are bare (pain receptors) or are associated with sensory receptors.
      (2) On the basis of structure, there are unipolar, bipolar, and multipolar neurons; the terminology reveals the number of processes extending from the cell body. Motor and association neurons are multipolar; most sensory neurons are unipolar. The exceptions are sensory neurons in certain special sense organs (ear, eye), which are bipolar.

   WEB Exercise: Chapter 7, Classification of Neurons.

   c. Physiology
      (1) A nerve impulse is an electrochemical event (initiated by various stimuli) that causes a change in neuron plasma membrane permeability, allowing sodium ions (Na^+) to enter the cell (depolarization). Once begun, the action potential, or nerve impulse, continues over the entire surface of the cell. Electrical conditions of the rest-

ing state are restored by the diffusion of potassium ions (K^+) out of the cell (repolarization). Ion concentrations of the resting state are restored by the sodium-potassium pump.

IP Nervous System Topic: The Membrane Potential, pages 3–13.

(2) A neuron influences other neurons or effector cells by releasing neurotransmitters, chemicals that diffuse across the synaptic cleft and attach to membrane receptors on the postsynaptic cell. The result is opening of specific ion channels and activation or inhibition, depending on the neurotransmitter released and the target cell.

(3) A reflex is a rapid, predictable response to a stimulus. There are two types—autonomic and somatic. The minimum number of components of a reflex arc is four: receptor, effector, and sensory and motor neurons (most, however, have one or more association neurons). Normal reflexes indicate normal nervous system function.

CENTRAL NERVOUS SYSTEM (pp. 235–249)

1. The brain is located within the cranial cavity of the skull and consists of the cerebral hemispheres, diencephalon, brain stem structures, and cerebellum.

   WEB Exercise: Chapter 7, Parts of the Brain; The Human Brain: Sagittal Section.

   a. The two cerebral hemispheres form the largest part of the brain. Their surface, or cortex, is gray matter, and their interior is white matter. The cortex is convoluted and has gyri, sulci, and fissures. The cerebral hemispheres are involved in logical reasoning, moral conduct, emotional responses, sensory interpretation, and the initiation of voluntary muscle activity. Several functional areas of the cerebral lobes have been identified (see p. 237). The basal nuclei, regions of gray matter deep within the white matter of the cerebral hemispheres, modify voluntary motor activity. Parkinson's disease and Huntington's disease are disorders of the basal nuclei.

   b. The diencephalon is superior to the brain stem and is enclosed by the cerebral hemispheres. The major structures include the following:

      (1) The thalamus encloses the third ventricle and is the relay station for sensory impulses passing to the sensory cortex for interpretation.

      (2) The hypothalamus makes up the "floor" of the third ventricle and is the most important regulatory center of the autonomic nervous system (regulates water balance, metabolism, thirst, temperature, and the like).

      (3) The epithalamus includes the pineal body (an endocrine gland) and the choroid plexus of the third ventricle.

   c. The brain stem is the short region inferior to the hypothalamus that merges with the spinal cord.

      (1) The midbrain is most superior and is primarily fiber tracts.

      (2) The pons is inferior to the midbrain and has fiber tracts and nuclei involved in respiration.

      (3) The medulla oblongata is the most inferior part of the brain stem. In addition to fiber tracts, it contains autonomic nuclei involved in the regulation of vital life activities (breathing, heart rate, blood pressure, etc.).

   d. The cerebellum is a large, cauliflower-like part of the brain posterior to the fourth ventricle. It coordinates muscle activity and body balance.

2. Protection of the CNS

   a. Bones of the skull and vertebral column are the most external protective structures.

   b. Meninges are three connective tissue membranes—dura mater (tough outermost), arachnoid mater (middle weblike), and pia mater (innermost delicate). The meninges extend beyond the end of the spinal cord.

   WEB Exercise: Chapter 7, Meninges of the Brain.

   c. Cerebrospinal fluid (CSF) provides a watery cushion around the brain and cord. CSF is formed by the choroid plexuses of the brain. It is found in the subarachnoid space, ventricles, and central canal. CSF is continually formed and drained.

   d. The blood-brain barrier is composed of relatively impermeable capillaries.

3. Brain dysfunctions

   a. Head trauma may cause concussions (reversible damage) or contusions (nonreversible damage). When the brain stem is affected, unconsciousness (temporary or permanent) occurs. Trauma-induced brain injuries may be aggravated by intracranial hemorrhage or cerebral edema, both of which compress brain tissue.

   b. Cerebrovascular accidents (CVAs, or strokes) result when blood circulation to brain neurons is blocked and brain tissue dies. The result may be visual impairment, paralysis, and aphasias.

   c. Alzheimer's disease is a degenerative brain disease in which abnormal protein deposits and other structural changes appear. It results in slow, progressive loss of memory and motor control plus increasing dementia.

d. Techniques used to diagnose brain dysfunctions include the EEG, simple reflex tests, pneumoencephalography, angiography, and CT, PET, and MRI scans.

4. The spinal cord is a reflex center and conduction pathway. Found within the vertebral canal, the cord extends from the foramen magnum to L_1 or L_2. The cord has a central bat-shaped area of gray matter surrounded by columns of white matter, which carry motor and sensory tracts from and to the brain.

WEB Exercise: Chapter 7, Anatomy of the Spinal Cord.

PERIPHERAL NERVOUS SYSTEM (pp. 249–263)

1. A nerve is a bundle of neuron processes wrapped in connective tissue coverings (endoneurium, perineurium, and epineurium).

WEB Exercise: Chapter 7, Structure of a Nerve.

2. Cranial nerves: 12 pairs of nerves that extend from the brain to serve the head and neck region. The exception is the vagus nerves, which extend into the thorax and abdomen.

WEB Exercise: Chapter 7, Cranial Nerves; Descriptions of Cranial Nerves.

3. Spinal nerves: 31 pairs of nerves are formed by the union of the dorsal and ventral roots of the spinal cord on each side. The spinal nerve proper is very short and splits into dorsal and ventral rami. Dorsal rami serve the posterior body trunk; ventral rami (except T_1 through T_{12}) form plexuses (cervical, brachial, lumbar, sacral) that serve the limbs.

WEB Exercise: Chapter 7, Distributions of Spinal Nerves.

4. Autonomic nervous system: Part of the PNS, composed of neurons that regulate the activity of smooth and cardiac muscle and glands. This system differs from the somatic nervous system in that there is a chain of two motor neurons from the CNS to the effector. Two subdivisions serve the same organs with different effects.

   a. The parasympathetic division is the "housekeeping" system and is in control most of the time. This division maintains homeostasis by seeing that normal digestion and elimination occur and that energy is conserved. The first motor neurons are in the brain or the sacral region of the cord. The second motor neurons are in the terminal ganglia close to the organ served. Postganglionic axons secrete acetylcholine.

   b. The sympathetic division is the "fight-or-flight" subdivision, which prepares the body to cope with some threat. Its activation results in increased heart rate and blood pressure. The preganglionic neurons are in the gray matter of the cord. The postganglionic neurons are in sympathetic chains or in collateral ganglia. Postganglionic axons secrete norepinephrine.

DEVELOPMENTAL ASPECTS OF THE NERVOUS SYSTEM (pp. 263–266)

1. Maternal and environmental factors may impair embryonic brain development. Oxygen deprivation destroys brain cells. Severe congenital brain diseases include cerebral palsy, anencephaly, hydrocephalus, and spina bifida.

2. Premature babies have trouble regulating body temperature because the hypothalamus is one of the last brain areas to mature prenatally.

3. Development of motor control indicates the progressive myelination and maturation of a child's nervous system. Brain growth ends in young adulthood. Neurons die throughout life and are not replaced; thus, brain mass declines with age.

4. Healthy, aged people maintain nearly optimal intellectual function. Disease—particularly cardiovascular disease—is the major cause of declining mental function with age.

REVIEW QUESTIONS

MULTIPLE CHOICE

More than one choice may apply.

1. An example of integration by the nervous system is:
   a. the feel of a cold breeze
   b. the shivering and goose bumps in response to cold
   c. perceiving the sound of rain
   d. the decision to go back for an umbrella

2. Where might a gray matter nucleus be located?
   a. Alongside the vertebral column
   b. Within the brain
   c. Within the spinal cord
   d. In the sensory receptors

3. Histological examination of a slice of neural tissue reveals a bundle of nerve fibers held together by cells whose multiple processes wrap around several fibers and form a myelin sheath. The specimen is likely to be:
   a. a nucleus c. a nerve
   b. a ganglion d. a tract
4. The pineal gland is located in the:
   a. hypothalamus c. epithalamus
   b. mesencephalon d. corpus callosum
5. A type of brain trauma likely to result in a long-term coma is:
   a. concussion
   b. cortical contusion
   c. brain stem contusion
   d. uncontrolled subdural hemorrhage
6. The spinal cord feature associated with the leash of nerves supplying the upper limbs is the:
   a. brachial plexus
   b. brachial enlargement
   c. cervical enlargement
   d. lateral gray horns
7. Which contains only motor fibers?
   a. Dorsal root c. Ventral root
   b. Dorsal ramus d. Ventral ramus
8. Cranial nerves that have some function in vision include the:
   a. trochlear c. abducens
   b. trigeminal d. facial
9. Motor functions of the extensor muscles of the arm, forearm, and fingers would be affected by damage to which one of these nerves?
   a. Radial c. Ulnar
   b. Axillary d. Median
10. Glial cells present in large numbers in areas of bacterial infection in the brain would most likely be:
    a. oligodendrocytes
    b. astrocytes
    c. ependymal cells
    d. microglia
11. Which of the following is true of the autonomic, but not the somatic, nervous system?
    a. Neurotransmitter is acetylcholine
    b. Axons are myelinated
    c. Effectors are muscle cells
    d. Has motor neurons located in ganglia

SHORT ANSWER ESSAY

1. What are the two great controlling systems of the body?
2. Explain both the structural and functional classifications of the nervous system. In your explanation, include the subdivisions of each.
3. List the structures of the CNS and PNS.
4. Two major cell groups make up the nervous system—neurons and connective tissue cells such as astrocytes and Schwann cells. Which are "nervous" cells? Why? What are the major functions of the other cell group?
5. Give the basis for the functional classification of neurons.
6. One neuron transmits a nerve impulse at 40 meters per second; another conducts at the rate of 1 meter per second. Which neuron has a myelinated axon?
7. Briefly explain how nerve impulses are initiated and transmitted, and why one-way conduction at synapses always happens.
8. Name four types of cutaneous sensory receptors. Which of the cutaneous receptor types is most numerous? Why?
9. What is a reflex arc? Name its minimum components.
10. Make a rough drawing of the left cerebral hemisphere. On your drawing, locate at least five different functional areas, and then indicate their specific functions.
11. Other than serving as a conduction pathway, what is a major function of the pons? Why is the medulla the most vital part of the brain?
12. What is the function of the thalamus? the hypothalamus? the cerebellum?
13. Discuss how the brain is protected by bone, membranes, fluid, and capillaries.
14. What is gray matter? white matter? How does the arrangement of gray and white matter differ in the cerebral hemispheres and the spinal cord?
15. What are two functions of the spinal cord?
16. How many pairs of cranial nerves are there? Which are purely sensory? Which activates the chewing muscles? Which helps regulate heart rate and activity of the digestive tract?
17. Except for the vagus nerves, what general area of the body do the cranial nerves serve?
18. How many pairs of spinal nerves are there? How do they arise?

19. What region of the body is served by the dorsal rami of the spinal nerves? the ventral rami?

20. Name the four major nerve plexuses formed by the ventral rami and the body region served by each.

21. How does the autonomic nervous system differ from the somatic nervous system?

22. What is the difference in function of the sympathetic and parasympathetic divisions of the autonomic nervous system **(a)** generally speaking, and **(b)** as specifically relates to the operation of the cardiovascular and digestive systems?

23. Since the sympathetic and parasympathetic fibers serve the same organs, how can their opposing effects be explained?

24. How would a drug that inhibits the parasympathetic division affect a person's pulse?

25. Differentiate between a *concussion* and a *contusion*.

26. Compare CVAs and TIAs in terms of causes, symptoms, and consequences.

27. Define *senility*. Name possible causes of permanent and reversible senility.

AT THE CLINIC

1. Mrs. Jones has had a progressive decline in her mental capabilities in the past five or six years. At first, her family attributed her occasional memory lapses, confusion, and agitation to grief over her husband's death six years earlier. When examined, Mrs. Jones was aware of her cognitive problems and was shown to have an IQ score approximately 30 points lower than would be predicted by her work history. A CT scan showed diffuse cerebral atrophy. The physician prescribed a mild tranquilizer for Mrs. Jones and told her family that there was little else he could recommend. What is Mrs. Jones's problem?

2. Joseph, a man in his early 70s, was having problems chewing his food. He was asked to stick out his tongue. It deviated to the right, and its right side was quite wasted. What cranial nerve was impaired?

3. Jed, a couch potato, likes to eat a very large meal in the evening. After the meal, his wife asks him to help clean the dishes, but Jed explains that he is "too tired" and promptly goes to sleep. What seems to be his problem?

4. A semiconscious young woman is brought to the hospital by friends after falling from a roof. She did not lose consciousness immediately, and she was initially lucid. After a while, though, she became confused and then unresponsive. What is a likely explanation of her condition?

5. In John's checkup, one year after an accident severed his right accessory nerve, severe muscle atrophy was noted. What two prominent muscles have been affected?

6. Mrs. Tonegawa, a new mother, brings her infant to the clinic because he has suffered repeated seizures. Upon questioning, she states that her labor was unusually long and difficult. What condition do you suspect? Will the infant's condition worsen?

7. Three-year-old Samantha is sobbing that her right arm is "gone," and the examination shows her to have little muscle strength in that limb. Questioning the parents reveals that her father had been swinging her by the arms. What part of the PNS has been damaged?

8. Mr. Dingledine is an 82-year-old bedridden gentleman who has discovered a new interest in learning about his body. While being tended by the visiting nurse, he remarks that the supporting cells in nervous tissue (such as Schwann cells and oligodendrocytes) act like the rubber coating around household wiring. What does he mean by this analogy?

9. Jemal, an elderly man with a history of TIAs, complained to his daughter that he had a severe headache. Shortly thereafter, he lapsed into a coma. At the hospital, he was diagnosed as having a brain hemorrhage. Which part of the brain was damaged by the hemorrhage?

10. Why does exposure to toxins have more devastating neural effects during early pregnancy than in late pregnancy?

11. Clancy is the star of his hometown ice hockey team. During a game, he is walloped on his backside with a hockey stick so hard that he hits the ice. When he tries to get up he is unable to flex his left hip or extend his left knee, but he has no pain. Which nerve has been damaged?

8 Special Senses

YOUR GOALS

After completing this chapter, you will have a working knowledge of the functions of special senses and will have mastered the objectives listed below.

FUNCTION PREVIEW

- The special senses respond to different types of energetic stimuli involved in vision, hearing, balance, smell, and taste.

OBJECTIVE CHECKLIST

THE EYE AND VISION (pp. 272–284)

- When provided with a model or diagram, identify the accessory eye structures, and list the functions of each.
- Name the eye tunics, and indicate the major function of each.
- Explain how rod and cone function differ.
- Describe image formation on the retina.
- Trace the pathway of light through the eye to the retina.
- Discuss the importance of an ophthalmoscopic examination.
- Define the following terms: *accommodation, astigmatism, blind spot, cataract, emmetropia, glaucoma, hyperopia, myopia,* and *refraction.*
- Trace the visual pathway to the optic cortex.
- Discuss the importance of the pupillary and convergence reflexes.

THE EAR: HEARING AND BALANCE (pp. 284–291)

- Identify the structures of the external, middle, and internal ear, and list the functions of each.
- Describe how the equilibrium organs help maintain balance.
- Explain the function of the organ of Corti in hearing.

>>

(Objective Checklist, continued)

- Define *sensorineural* and *conductive deafness,* and list possible causes of each.
- Explain how one is able to localize the source of a sound.

CHEMICAL SENSES: TASTE AND SMELL (pp. 291–293)

- Describe the location, structure, and function of the olfactory and taste receptors.
- Name the four basic taste sensations and list factors that modify the sense of taste.

DEVELOPMENTAL ASPECTS OF THE SPECIAL SENSES (pp. 293–295)

- Describe changes that occur with age in the special sense organs.

People are responsive creatures. Hold freshly baked bread before us, and our mouths water. A sudden clap of thunder makes us jump. These "irritants" (the bread and the thunderclap) and many others are the stimuli that continually greet us and are interpreted by our nervous system.

We are usually told that we have five senses that keep us in touch with what is going on in the external world: touch, taste, smell, sight, and hearing. Actually touch is a mixture of the general senses that we considered in Chapter 7—the temperature, pressure, and pain receptors of the skin and the proprioceptors of muscles and joints. The other four "traditional" senses—*smell, taste, sight,* and *hearing*—are called **special senses.** Receptors for a fifth special sense, *equilibrium,* are housed in the ear, along with the organ of hearing. In contrast to the small and widely distributed general receptors, the **special sense receptors** are either large, complex sensory organs (eyes and ears) or localized clusters of receptors (taste buds and olfactory epithelium).

This chapter focuses on the functional anatomy of each of the special sense organs individually, but keep in mind that sensory inputs are overlapping. What we finally experience—our "feel" of the world—is a blending of stimulus effects.

The Eye and Vision

How we see has captured the curiosity of many researchers. Vision is the sense that has been studied most. Of all the sensory receptors in the body, 70 percent are in the eyes. The optic tracts that carry information from the eyes to the brain are massive bundles, containing over a million nerve fibers. Vision is the sense that requires the most "learning," and the eye appears to delight in being fooled. The old expression "You see what you expect to see" is often very true.

Anatomy of the Eye

External and Accessory Structures

The adult eye is a sphere that measures about 1 inch (2.5 cm) in diameter. Only the anterior one-sixth of the eye's surface can normally be seen. The rest of it is enclosed and protected by a cushion of fat and the walls of the bony orbit. The accessory structures of the eye include the extrinsic eye muscles, eyelids, conjunctiva, and lacrimal apparatus.

Anteriorly the eyes are protected by the **eyelids,** which meet at the medial and lateral corners of the eye, the **medial** and **lateral canthus** respectively (see Figure 8.1b). Projecting from the border of each eyelid are the **eyelashes.** Modified sebaceous glands associated with the eyelid edges are the **tarsal glands.** These glands produce an oily secretion that lubricates the eye (Figure 8.1a). *Ciliary glands,* modified sweat glands, lie between the eyelashes (*cilium* = eyelash).

A delicate membrane, the **conjunctiva** (kon-junk″ti′vah), lines the eyelids and covers part of the outer surface of the eyeball (Figures 8.1a and 8.2).

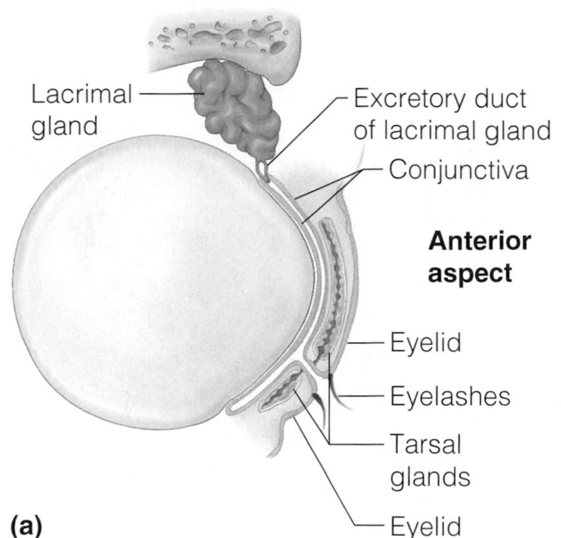

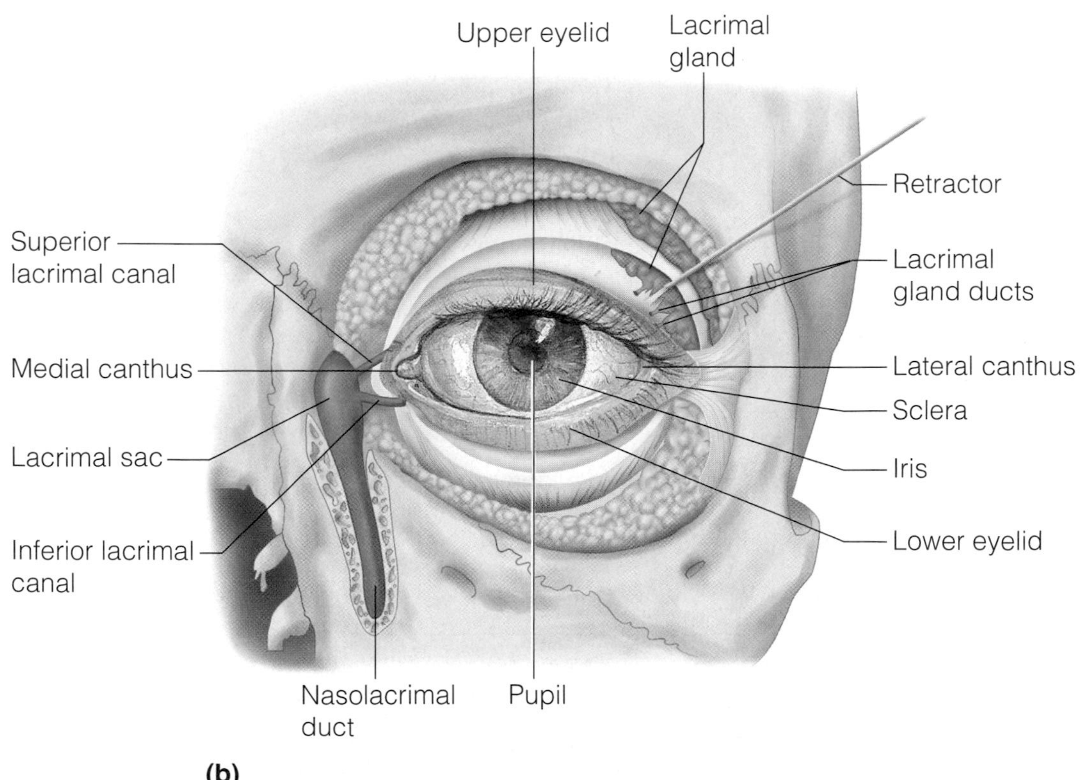

FIGURE 8.1 External anatomy of the eye and accessory structures. (a) Sagittal section of the accessory structures associated with the anterior part of the eye. **(b)** Anterior view of the lacrimal apparatus. The lacrimal gland is shown pulled superiorly by a retractor to expose its ducts.

It ends at the edge of the cornea by fusing with the corneal epithelium. The conjunctiva secretes mucus, which helps to lubricate the eyeball and keep it moist.

Homeostatic Imbalance

Inflammation of the conjunctiva, called *conjunctivitis*, results in reddened, irritated eyes. *Pinkeye*, its infectious form caused by bacteria or viruses, is highly contagious. ▲

The **lacrimal apparatus** consists of the lacrimal gland and a number of ducts that drain the lacrimal secretions into the nasal cavity. The **lacrimal glands** are located above the lateral end of each eye. They continually release a dilute salt solution *(tears)* onto the anterior surface of the eyeball through several small ducts (see Figure 8.1). The tears flush across the eyeball into the **lacrimal canals** medially, then into the **lacrimal sac**, and finally into the **nasolacrimal duct**, which

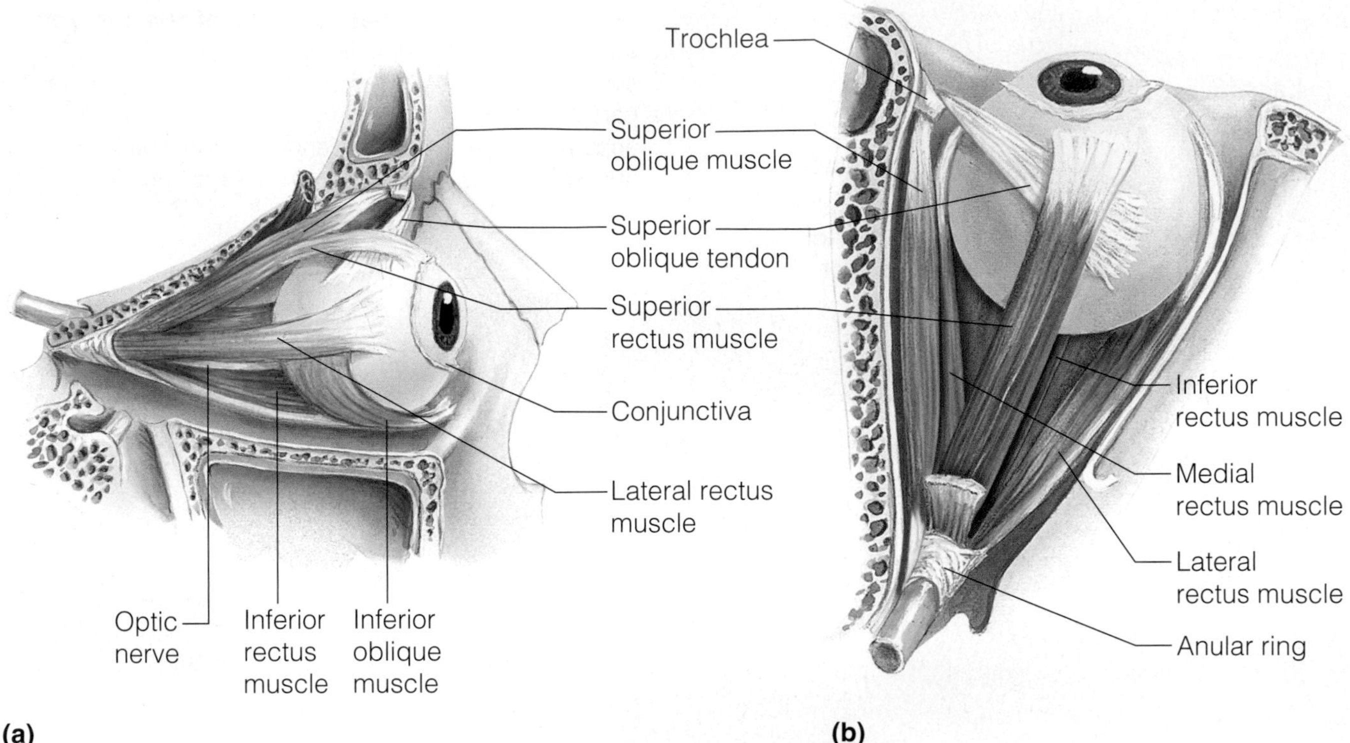

Name	Action	Controlling cranial nerve
Lateral rectus	Moves eye laterally	VI (abducens)
Medial rectus	Moves eye medially	III (oculomotor)
Superior rectus	Elevates eye	III (oculomotor)
Inferior rectus	Depresses eye	III (oculomotor)
Inferior oblique	Elevates eye and turns it laterally	III (oculomotor)
Superior oblique	Depresses eye and turns it laterally	IV (trochlear)

(c)

FIGURE 8.2 Extrinsic muscles of the eye. **(a)** Lateral view of the right eye. **(b)** Superior view of the right eye. The four rectus muscles originate from the anular ring, a ringlike tendon at the back of the eye socket. **(c)** Summary of cranial nerve supply and actions of the extrinsic eye muscles.

empties into the nasal cavity (see Figure 8.1b). Lacrimal secretion also contains antibodies and **lysozyme** (li′so-zīm), an enzyme that destroys bacteria. Thus, it cleanses and protects the eye surface as it moistens and lubricates it. When lacrimal secretion increases substantially, tears spill over the eyelids and fill the nasal cavities, causing congestion and the "sniffles." This happens when the eyes are irritated by foreign objects or chemicals and when we are emotionally upset. In the case of irritation, the enhanced tearing acts to wash away or dilute the irritating substance. The importance of "emotional tears" is poorly understood, but some suspect that crying is important in reducing stress. Anyone who has had a good cry would probably agree, but this has been difficult to prove scientifically.

Homeostatic Imbalance

Because the nasal cavity mucosa is continuous with that of the lacrimal duct system, a cold or nasal inflammation often causes the lacrimal mucosa to become inflamed and swell. This impairs the drainage of tears from the eye surface, causing "watery" eyes. ▲

Six **extrinsic,** or **external, eye muscles** are attached to the outer surface of each eye. These muscles produce gross eye movements and make it

Q *Which layer of the eye would be the first to be affected by deficient tear production?*

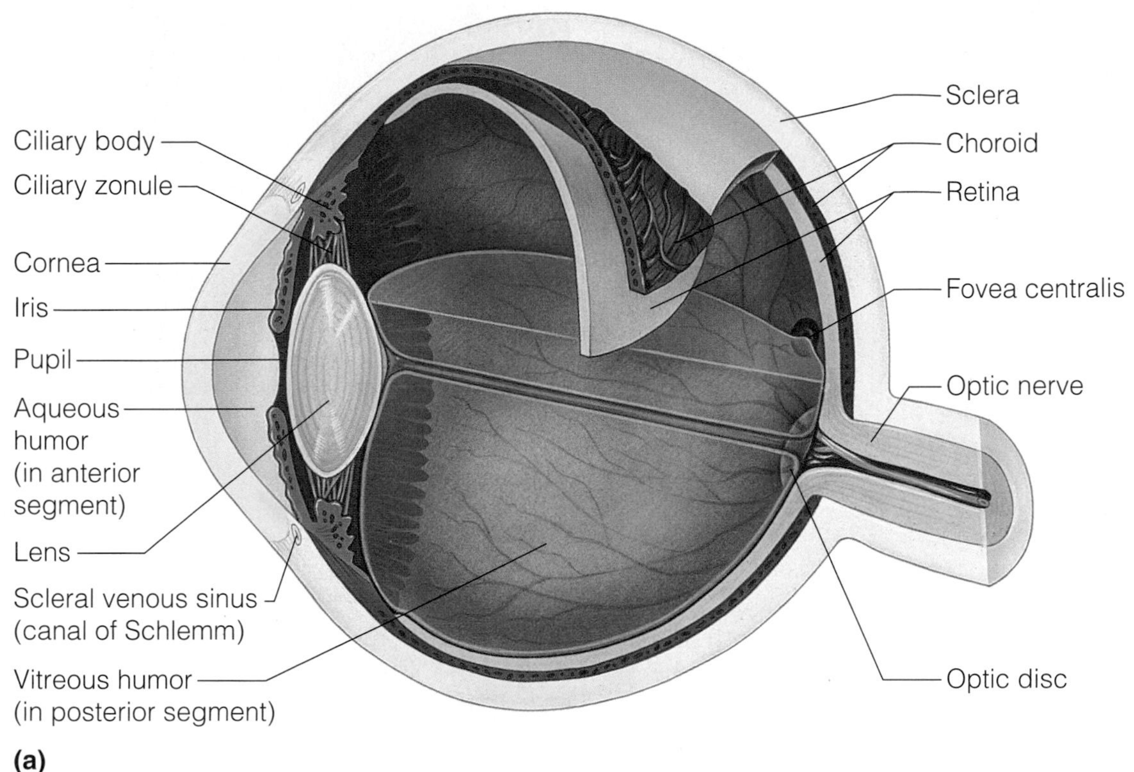

FIGURE 8.3 Internal anatomy of the eye (sagittal section).
(a) Diagrammatic view.

possible for the eyes to follow a moving object. The names, locations, actions, and cranial nerve serving each of the extrinsic muscles are given in Figure 8.2.

Internal Structures: The Eyeball

The eye itself, commonly called the **eyeball,** is a hollow sphere (Figure 8.3). Its wall is composed of three *tunics,* or coats, and its interior is filled with fluids called *humors* that help to maintain its shape. The lens, the main focusing apparatus of the eye, is supported upright within the eye cavity, dividing it into two chambers. Now that we have covered the general anatomy of the eyeball, we are ready to get specific.

Tunics of the Eyeball The outermost tunic, the protective **sclera** (skle′rah), is thick, white connective tissue. Also called the *fibrous tunic,* the sclera is seen anteriorly as the "white of the eye." Its central anterior portion is modified so that it is crystal clear. This transparent "window" is the **cornea** (kor′ne-ah) through which light enters the eye. The cornea is well supplied with nerve endings. Most are pain fibers, and when the cornea is touched, blinking and increased tearing occur. Even so, the cornea is the most exposed part of the eye, and it is very vulnerable to damage. Luckily, its ability to repair itself is extraordinary. Furthermore, the cornea is the only tissue in the body that can be transplanted from one person to another without the worry of rejection. Since it has no blood vessels, it is beyond the reach of the immune system.

The middle coat of the eyeball is the *vascular tunic* which has three distinguishable regions. Most posteriorly is the **choroid** (ko′roid), a blood-rich nutritive tunic that contains a dark pigment. The pigment prevents light from scattering inside the eye. Moving anteriorly, the choroid is modified to form two smooth muscle structures, the **ciliary** (sil′e-er-e) **body,** to which the **lens** is attached by

A *The outermost sclera (mostly its cornea) which is normally continuously washed by tears.*

FIGURE 8.3 (*continued*) Internal anatomy of the eye (sagittal section). (b) Photograph.

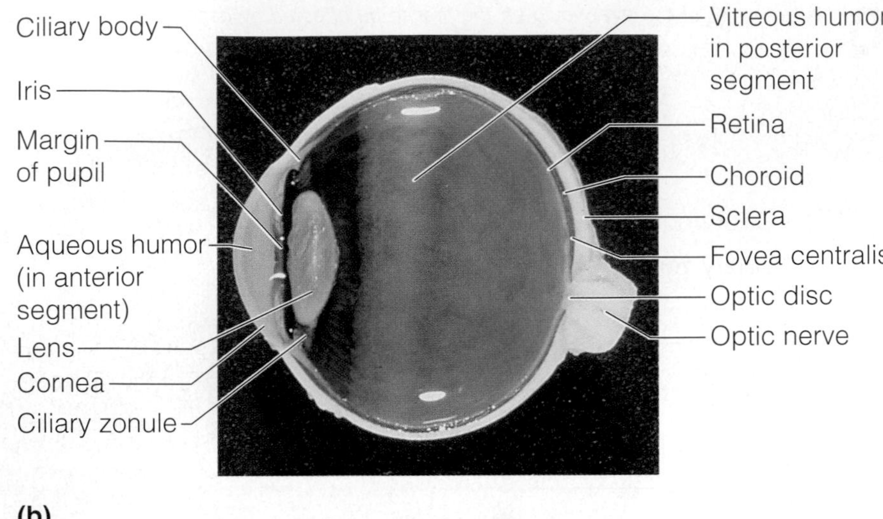

a suspensory ligament called the **ciliary zonule,** and then the **iris.** The pigmented iris has a rounded opening, the **pupil,** through which light passes. Circularly and radially arranged smooth muscle fibers form the iris, which acts like the diaphragm of a camera. That is, it regulates the amount of light entering the eye so that one can see as clearly as possible in the available light. In close vision and bright light, the circular muscles contract, and the pupil constricts. In distant vision and dim light, the radial fibers contract to enlarge (dilate) the pupil, which allows more light to enter the eye.

The innermost *sensory tunic* of the eye is the delicate white **retina** (ret′ĭ-nah), which extends anteriorly only to the ciliary body. The retina contains millions of receptor cells, the **rods** and **cones.** Rods and cones are called **photoreceptors** because they respond to light (Figure 8.4). Electrical signals pass from the photoreceptors via a two-neuron chain—**bipolar cells** and then **ganglion cells**—before leaving the retina via the **optic nerve** as nerve impulses that are transmitted to the optic cortex. The result is vision.

The photoreceptor cells are distributed over the entire retina, except where the **optic nerve** (composed of ganglion cell axons) leaves the eyeball; this site is called the **optic disc,** or **blind spot.** When light from an object is focused on the optic disc, it disappears from our view and we cannot see it. To illustrate this, perform the blind spot demonstration by holding Figure 8.5 about 18 inches (45 cm) from your eyes. Close your left eye and stare at the X with your right eye. Move the figure slowly toward your face, keeping your right eye focused on the X. When the dot focuses on your blind spot, which lacks photoreceptors, the dot will disappear. Repeat the test for your left eye. This time close your right eye and stare at the dot with your left eye. Move the page as before until the X disappears.

The rods and cones are not evenly distributed in the retina. The rods are most dense at the periphery, or edge, of the retina and decrease in number as the center of the retina is approached. The rods allow us to see in gray tones in dim light, and they provide for our peripheral vision.

Homeostatic Imbalance

Anything that interferes with rod function hinders our ability to see at night, a condition called *night blindness.* Night blindness dangerously impairs the ability to drive safely at night. Its most common cause is prolonged vitamin A deficiency, which eventually results in deterioration of much of the neural retina. As described in the "A Closer Look" box on page 279, vitamin A is one of the building blocks of the pigments the photoreceptor cells need to respond to light. Vitamin A supplements will restore function if taken before degenerative changes occur. ▲

Cones are discriminatory receptors that allow us to see the details of our world in color under bright light conditions. They are densest in the center of the retina and decrease in number toward the retinal edge. Lateral to each blind spot is the **fovea centralis** (fo′ve-ah sen-tra′lis), a tiny pit that contains only cones (see Figure 8.3). Consequently, this is the area of greatest **visual acuity,** or point of sharpest vision, and anything we wish to view critically is focused on the fovea centralis.

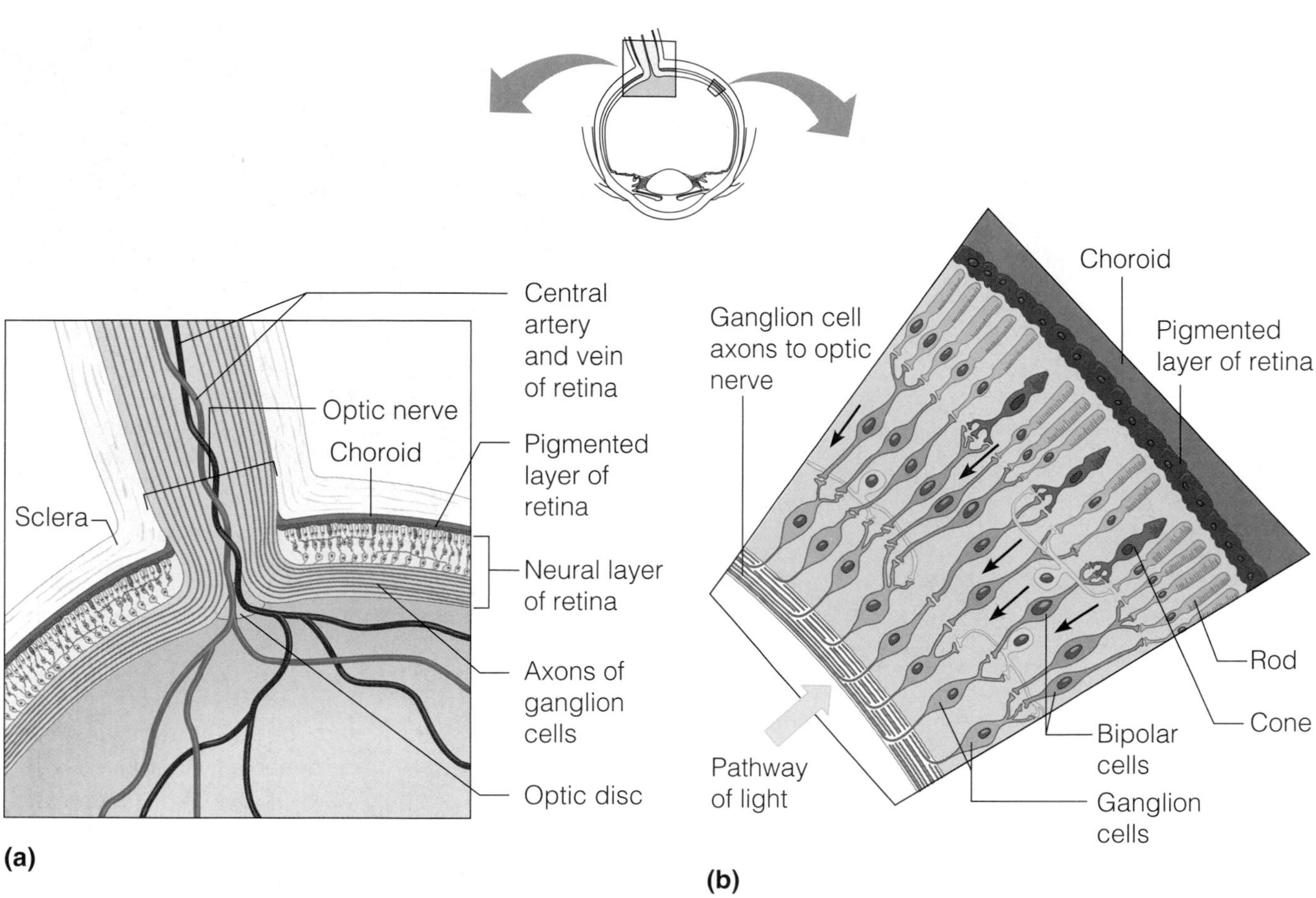

FIGURE 8.4 The three major types of neurons composing the retina. **(a)** Schematic view of the posterior part of the eyeball illustrating how the axons of the ganglion cells form the optic nerve, which leaves the back of the eyeball at the optic disc. **(b)** Notice in this schematic view of the neural layer of the retina that light passes through the thickness of the retina to excite the rods and cones. The flow of electrical signals (black arrows) occurs in the opposite direction: from the rods and cones to the bipolar cells and finally to the ganglion cells. The ganglion cells generate the nerve impulses that leave the eye via the optic nerve.

There are three varieties of cones. Each type is most sensitive to particular wavelengths of visible light (Figure 8.6). One type responds most vigorously to blue light, another to green light. The third cone variety responds to a range including both green and red wavelengths of light. However, this is the only cone population to respond to red light at all, so these are called the "red cones." Impulses received at the same time from more than one type of cone by the visual cortex are interpreted as *intermediate* colors. For example, simultaneous impulses from blue and red color receptors are seen as purple or violet tones. When all three cone types are being stimulated, we see white. If someone

FIGURE 8.5 Figure for blind spot test.

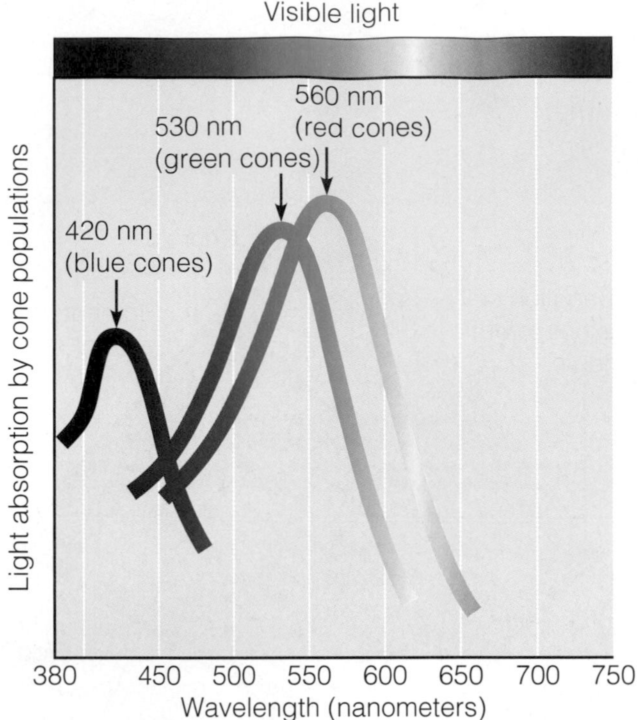

FIGURE 8.6 Sensitivities of the three cone types to the different wavelengths of visible light.

shines red light into one of your eyes and green into the other, you will see yellow, indicating that the "mixing" and interpretation of colors occurs in the brain, not in the retina.

Homeostatic Imbalance

Lack of all three cone types results in total *color blindness,* whereas lack of one cone type leads to partial color blindness. Most common is the lack of red or green receptors, which leads to two varieties of red-green color blindness. Red and green are seen as the same color—either red or green, depending on the cone type *present.* Many color-blind people are unaware of their condition because they have learned to rely on other cues—such as differences in intensities of the same color—to distinguish something green from something red, for example on traffic signals. Since the genes regulating color vision are on the X (female) sex chromosome, color blindness is a sex-linked condition. It occurs almost exclusively in males. ▲

Lens Light entering the eye is focused on the retina by the lens, a flexible biconvex crystal-like structure. The **lens** is held upright in the eye by a suspensory ligament, the ciliary zonule, attached to the ciliary body (see Figure 8.3).

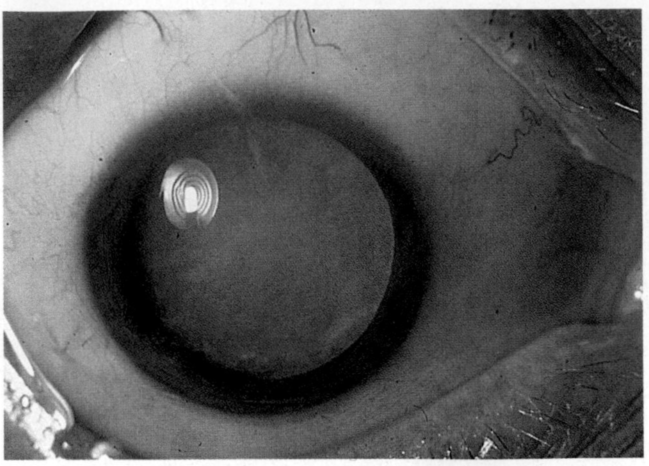

FIGURE 8.7 Photograph of a cataract. The cataract appears as a milky structure that seems to fill the pupil.

Homeostatic Imbalance

In youth, the lens is perfectly transparent and has the consistency of hardened jelly, but as we age it becomes increasingly hard and opaque. **Cataracts,** which result from this process, cause vision to become hazy and eventually cause blindness in the affected eye (Figure 8.7). Current treatment of cataracts is either surgical removal of the lens and replacement with a lens implant or special cataract glasses. ▲

The lens divides the eye into two segments or chambers. The *anterior (aqueous) segment,* anterior to the lens, contains a clear watery fluid called **aqueous humor.** The *posterior (vitreous) segment,* posterior to the lens, is filled with a gel-like substance called **vitreous** (vit′re-us) **humor,** or the **vitreous body** (see Figure 8.3). Vitreous humor helps prevent the eyeball from collapsing inward by reinforcing it internally. Aqueous humor is similar to blood plasma and is continually secreted by a special area of the choroid. Like the vitreous humor, it helps maintain *intraocular* (in″trah-ok′u-lar) *pressure,* or the pressure inside the eye. It also provides nutrients for the lens and cornea, which lack a blood supply. Aqueous humor is reabsorbed into the venous blood through the **scleral venous sinus (canal of Schlemm** [shlĕm]**),** which is located at the junction of the sclera and cornea.

Homeostatic Imbalance

If drainage of aqueous humor is blocked, pressure within the eye increases dramatically and begins

A Closer Look

Visual Pigments—The Actual Photoreceptors

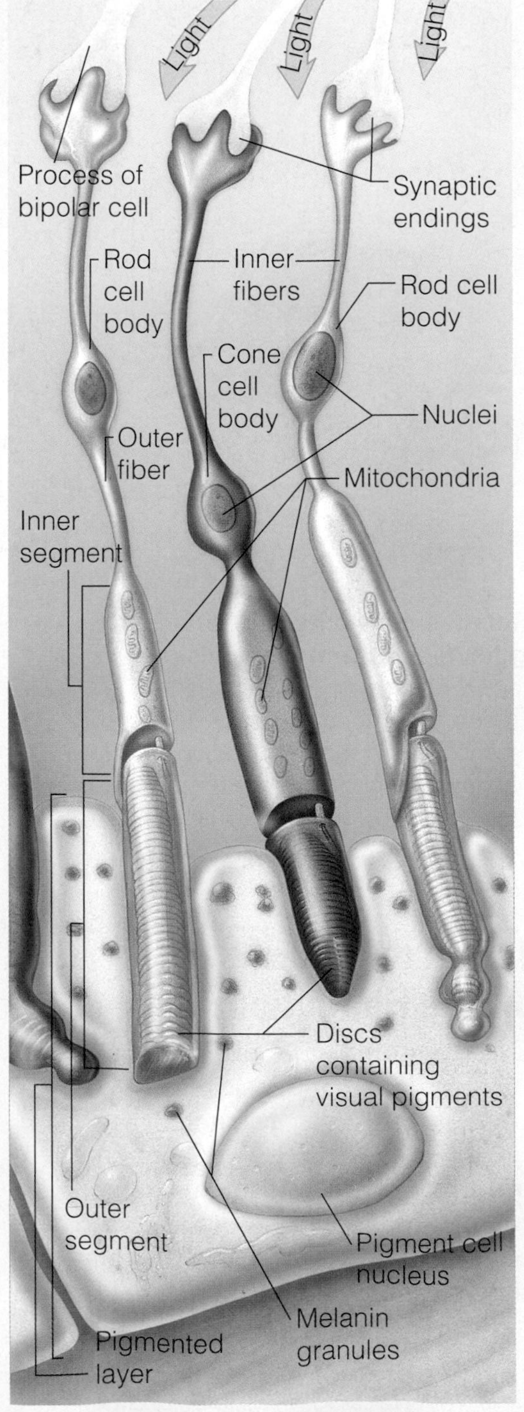

THE tiny photoreceptor cells of the retina have names that reflect their general shapes. As shown to the left, rods are slender, elongated neurons, whereas the fatter cones taper to more pointed tips. In each type of photoreceptor, there is a region called an *outer segment,* attached to the cell body. The outer segment corresponds to a light-trapping dendrite, in which the discs containing the visual pigments are stacked like a row of pennies.

The behavior of the visual pigments is dramatic. When light strikes them, they lose their color, or are "bleached"; shortly afterward, they regenerate their pigment. Absorption of light and pigment bleaching cause electrical changes in the photoreceptor cells that ultimately cause nerve impulses to be transmitted to the brain for visual interpretation. Pigment regeneration ensures that one is not blinded and unable to see in bright sunlight.

A good deal is known about the structure and function of **rhodopsin,** the purple pigment found in rods (see figure below). It is formed from the union of a protein **(opsin)** and a modified vitamin A product **(retinal)**. When combined in rhodopsin, retinal has a kinked shape that allows it to bind to opsin. But when light strikes rhodopsin, retinal straightens out and releases the protein. Once straightened out, the retinal continues its conversion until it is once again vitamin A. As these changes occur, the purple color of rhodopsin changes to the yellow of retinal and finally becomes colorless as the change to vitamin A occurs. Thus the term "bleaching of the pigment" accurately describes the color changes that occur when light hits the pigment. Regeneration of rhodopsin occurs as vitamin A is again converted to the kinked form of retinal and recombined with opsin in an ATP-requiring process. The cone pigments, while similar to rhodopsin, differ in the specific kinds of proteins they contain.

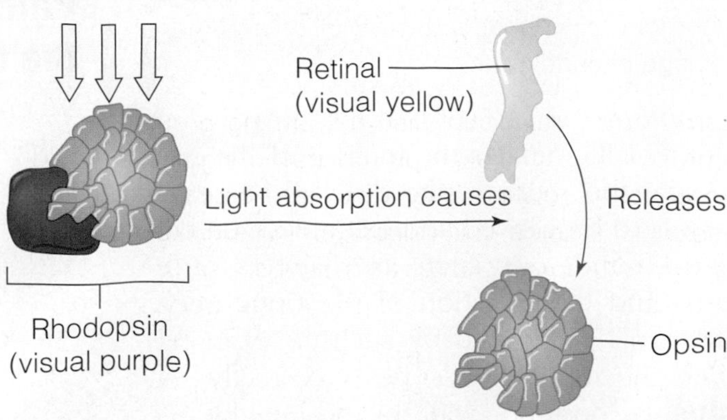

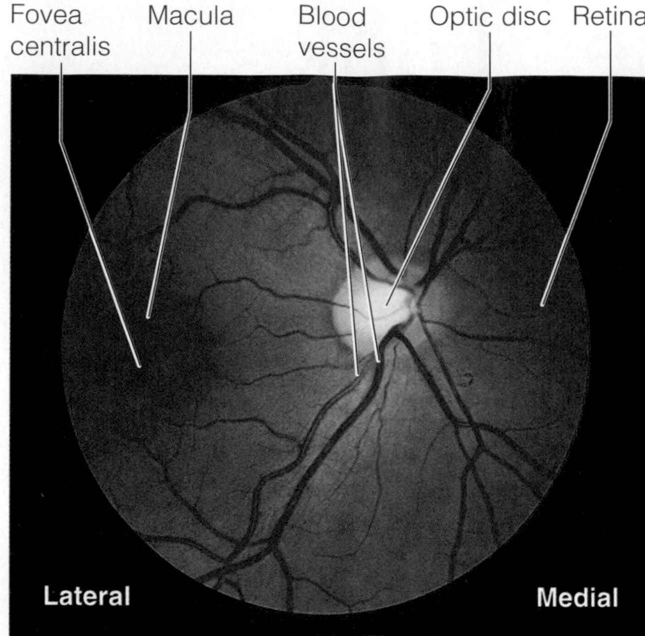

FIGURE 8.8 The posterior wall (fundus) of the retina as seen with an ophthalmoscope. Notice the optic disc, from which the blood vessels radiate.

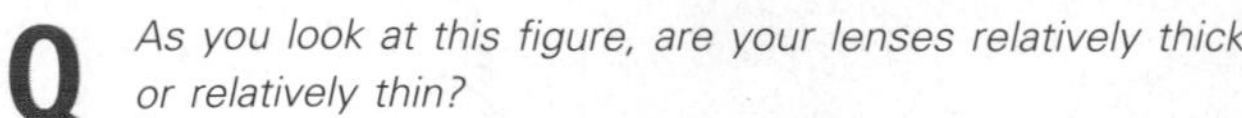

Q *As you look at this figure, are your lenses relatively thick or relatively thin?*

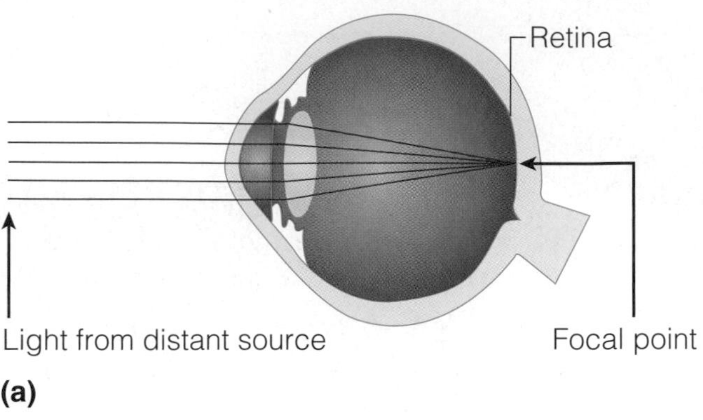

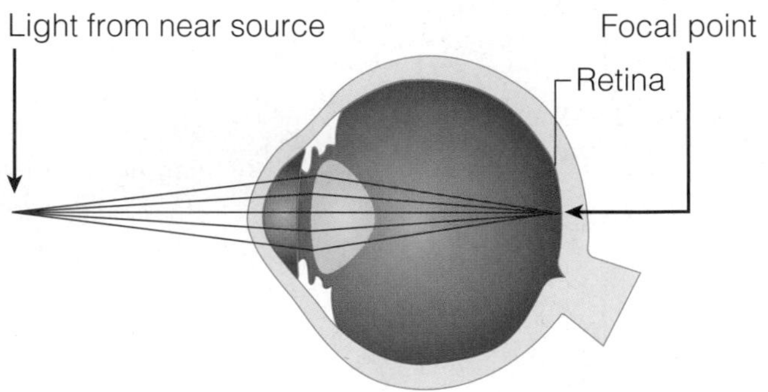

FIGURE 8.9 Relative convexity of the lens during focusing for distant and close vision. **(a)** Light rays from a distant object are nearly parallel as they reach the eye and can be focused without requiring changes in lens convexity. **(b)** Diverging light rays from close objects require that the lens bulge more to focus the image sharply on the retina.

to compress the delicate retina and optic nerve. The resulting condition, **glaucoma** (glaw-ko′mah), eventually causes pain and possibly blindness. Glaucoma is a common cause of blindness in elderly persons. Since it progresses slowly and has almost no symptoms at first, it tends to occur without obvious signs. Later signs include seeing halos around lights, headaches, and blurred vision. A simple instrument called a *tonometer* (to-nom′e-ter) is used to measure the intraocular pressure. This examination should be performed yearly in people over 40. Glaucoma is treated with eyedrops (miotics), which increase the rate of aqueous humor drainage, or with surgical enlargement of the drainage channel. ▲

The *ophthalmoscope* (of-thal′mo-skōp) is an instrument that illuminates the interior of the eyeball, allowing the retina, optic disc, and internal blood vessels to be viewed and examined. Certain pathological conditions such as diabetes, arteriosclerosis, and degeneration of the optic nerve and retina can be detected by such an examination. When the ophthalmoscope is correctly set, the **fundus,** or posterior wall, of the healthy eye should appear as shown in Figure 8.8.

Pathway of Light through the Eye and Light Refraction

When light passes from one substance to another substance that has a different density, its speed changes and its rays are bent, or **refracted.** Light rays are bent in the eye as they encounter the cornea, aqueous humor, lens, and vitreous humor.

The refractive, or bending, power of the cornea and humors is constant. However, that of the lens can be changed by changing its shape—that is, by

A *You would be using your close vision so your lenses would be bulged and thus relatively thick.*

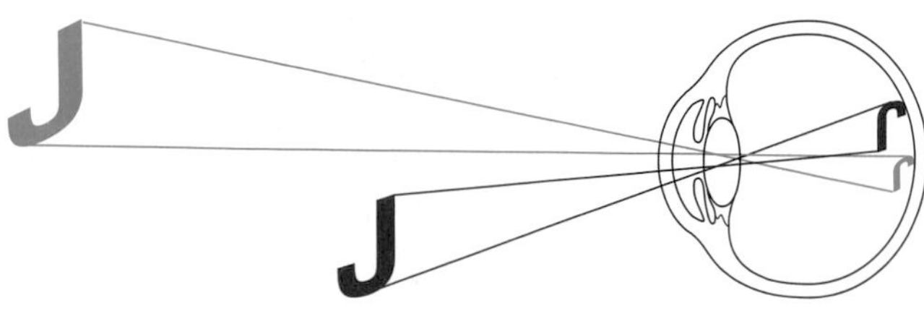

FIGURE 8.10 Real image (reversed left to right, and upside down) formed on the retina. Notice that the farther away the object, the smaller its image on the retina.

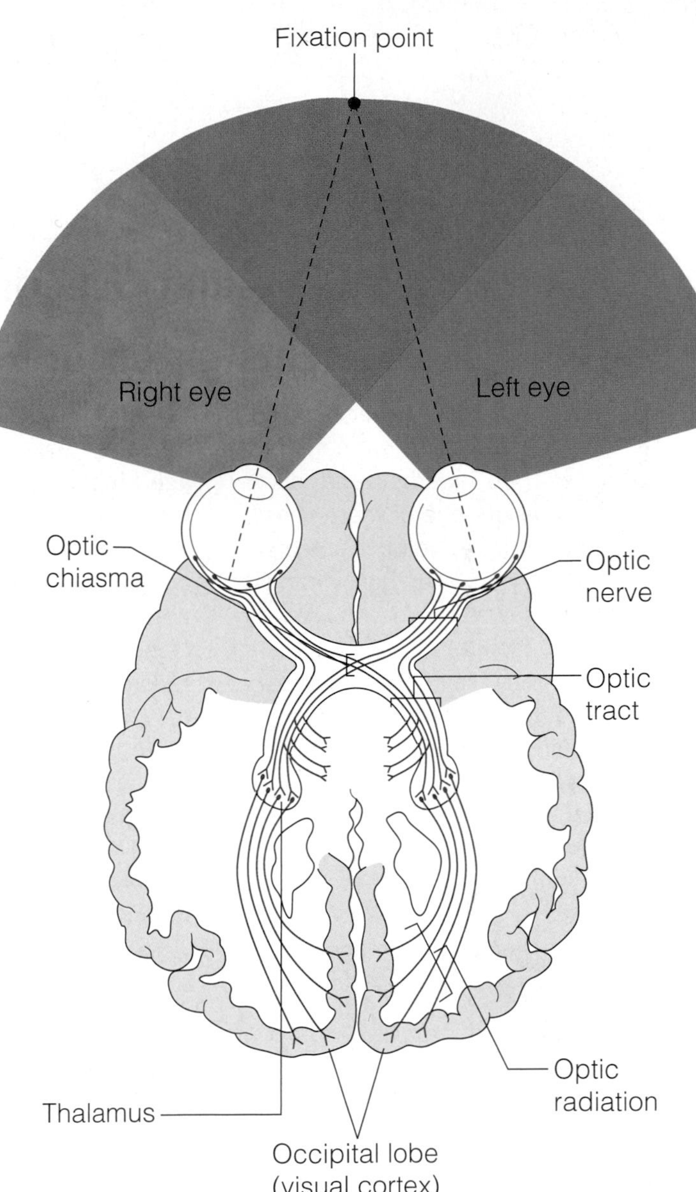

FIGURE 8.11 Visual fields of the eyes and visual pathway to the brain. Notice that the visual fields overlap considerably (area of binocular vision). Notice also the retinal sites at which a real image would be focused when both eyes are fixed on a close, pointlike object.

making it more or less convex, so that light can be properly focused on the retina. The greater the lens convexity, or bulge, the more it bends the light. On the other hand, the flatter the lens, the less it bends the light.

The resting eye is "set" for distant vision. In general, light from a distant source (over 20 feet away) approaches the eye as parallel rays (Figure 8.9a), and no change in lens shape is necessary for it to be focused properly on the retina. However, light from a close object tends to scatter and to *diverge,* or spread out, and the lens must bulge more to make close vision possible (Figure 8.9b). To achieve this, the ciliary body contracts, allowing the lens to become more convex. This ability of the eye to focus specifically for close objects (those less than 20 feet away) is called **accommodation.** The image formed on the retina as a result of the light-bending activity of the lens is a *real image*—that is, it is reversed from left to right, upside down (inverted), and smaller than the object (Figure 8.10). The normal eye is able to accommodate properly. However, vision problems occur when a lens is too strong or too weak (overconverging and underconverging, respectively) or from structural problems of the eyeball (as described in the "A Closer Look" box on near- and farsightedness on page 282).

Visual Fields and Visual Pathways to the Brain

Axons carrying impulses from the retina are bundled together at the posterior aspect of the eyeball and issue from the back of the eye as the optic nerve. At the **optic chiasma** (ki-as′mah; *chiasm* = cross) the fibers from the medial side of each eye cross over to the opposite side. The fiber tracts that result are the **optic tracts.** Each optic tract contains fibers from the lateral side of the eye on the same side and the medial side of the opposite eye. The optic tract fibers synapse with neurons in the thalamus, whose axons form the **optic radiation,** which runs to the occipital lobe of the brain. There they synapse with the cortical cells, and visual interpretation, or seeing, occurs. The visual pathway from the eye to the brain is shown in Figure 8.11.

A Closer Look

If I Can't See Things Far Away, Am I Nearsighted or Farsighted?

IT seems that whenever people who wear glasses or contact lenses get together and discuss their vision, one of them says something like, "Nearby objects appear blurry to me, but I can't remember if that means I'm nearsighted or farsighted." Or, someone else may say, "My glasses allow me to see faraway objects more clearly, so does that mean I am farsighted?" Here we will explain the meaning of nearsightedness and farsightedness as we explore the basis of eye-focusing disorders.

The eye that focuses images correctly on the retina is said to have **emmetropia** (em″ĕ-tro′pe-ah), literally, "harmonious vision." Such an eye is shown in part (a) of the figure.

Nearsightedness is formally called **myopia** (mi″o′pe-ah; "short vision"). It occurs when the parallel light rays from distant objects fail to reach the retina and instead are focused in front of it (see part (b) in the figure). Therefore, *distant* objects appear blurry to myopic people. Nearby objects are in focus, however, because the lens "accommodates" (bulges) to focus the image properly on the retina. Myopia results from an eyeball that is too long, a lens that is too strong, or a cornea that is too curved. Correction requires *concave* corrective lenses that diverge the light rays before they enter the eye, so that they converge farther back. To answer the first question posed above, *near*sighted people see *near* objects clearly and need corrective lenses to focus distant objects.

Farsightedness is formally called **hyperopia** (hi″per-o′pe-ah; "far vision"). It occurs when the parallel light rays from distant objects are focused *behind* the retina—at least in the resting eye in which the lens is flat and the ciliary muscle is relaxed (see part (c) in the figure). Hyperopia usually results from an eyeball that is too short or a "lazy" lens. People with hyperopia can see distant objects clearly because their ciliary muscles contract continuously to increase the light-bending power of the lens, which moves the focal point forward onto the retina. However, the diverging rays from *nearby* objects are focused so far behind the retina that even at full "bulge," the lens cannot focus the image on the retina. Therefore, nearby objects appear blurry. Furthermore, hyperopic individuals are subject to eyestrain as their endlessly contracting ciliary muscles tire from overwork. Correction of hyperopia requires *convex* corrective lenses that converge the light rays before they enter the eye. To answer the second question posed at the beginning of this essay, *far*sighted people can see *far*away objects clearly and require corrective lenses to focus on nearby objects. Unequal curvatures in different parts of the cornea or lens cause **astigmatism** (ah-stig′mah-tizm). In this condition, blurry images occur because points of light are focused not as points on the retina but as lines (*astigma* = not a point). Special cylindrically ground lenses or contacts are used to correct this problem. Eyes that are myopic or hyperopic *and* astigmatic require a more complex correction.

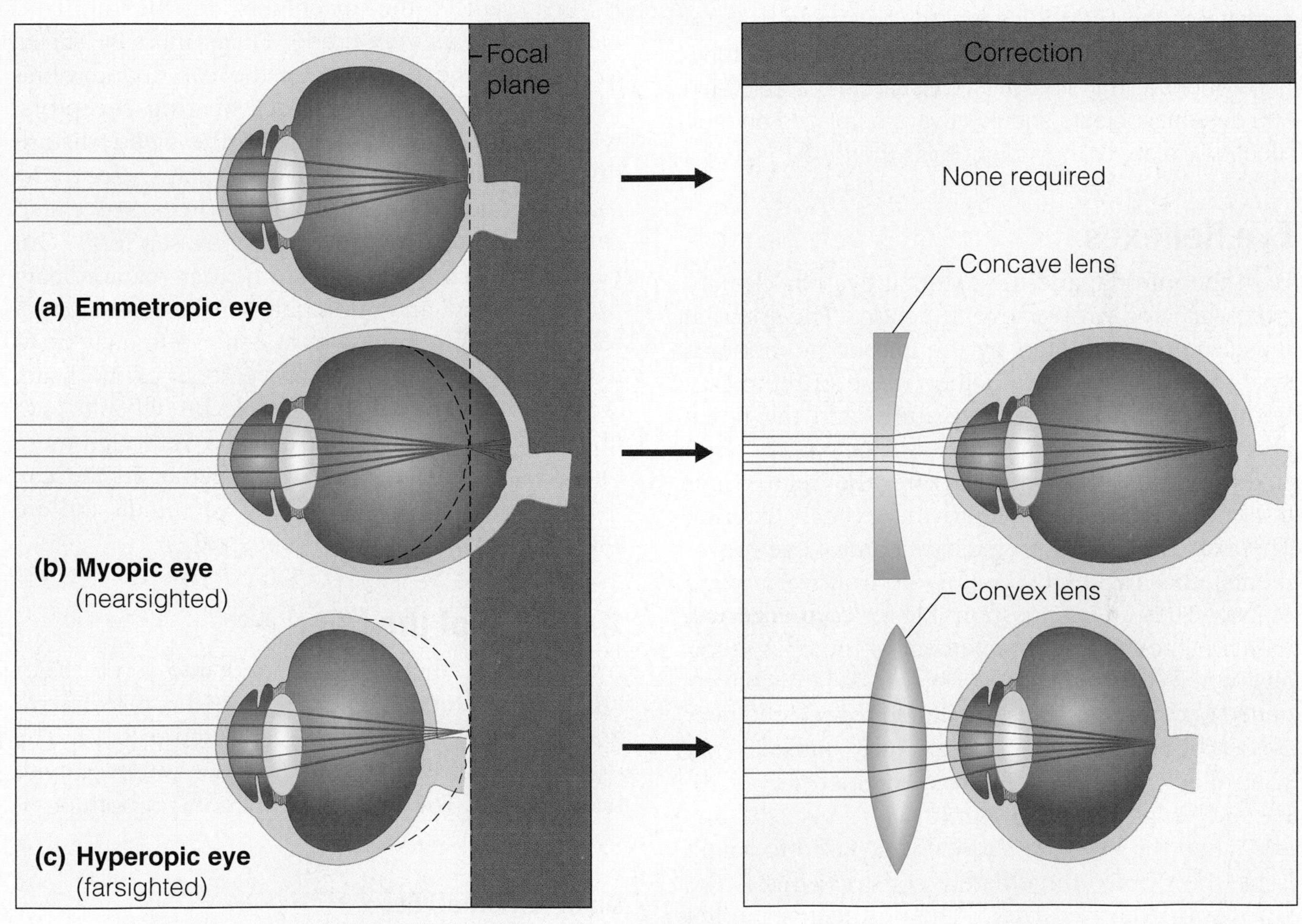

As you can see, each side of the brain receives visual input from both eyes—from the lateral field of vision of the eye on its own side and the medial field of the other eye. Also notice that each eye "sees" a slightly different view, but their *visual fields* overlap quite a bit. As a result of these two facts, humans have *binocular vision*. Binocular vision, literally "two-eyed vision," provides for depth perception, also called "three-dimensional" vision, as our visual cortex fuses the two slightly different images delivered by the two eyes.

Homeostatic Imbalance

Hemianopia (hem″e-ah-no′pe-ah) is the loss of the same side of the visual field of both eyes, which results from damage to the visual cortex on one side only (as occurs in some CVAs). Thus, the person would not be able to see things past the middle of his or her visual field on either the right or left side, depending on the site of the CVA. Such individuals should be carefully attended and warned of objects in the nonfunctional (nonseeing) side of the visual field. Their food and personal objects should always be placed on their functional side, or they might miss them. ▲

Eye Reflexes

Both the internal and the external eye muscles are necessary for proper eye function. The internal muscles are controlled by the autonomic nervous system. These muscles include those of the ciliary body, which alters lens curvature, and the radial and circular muscles of the iris, which control pupil size. The external muscles are the rectus and oblique muscles attached to the eyeball exterior. The external (extrinsic) muscles control eye movements and make it possible to follow moving objects. They are also responsible for **convergence,** which is the reflexive movement of the eyes medially when we view close objects. When convergence occurs, both eyes are aimed toward the near object being viewed. The extrinsic muscles are controlled by somatic fibers of cranial nerves III, IV, and VI, as shown in Figure 8.2.

When the eyes are suddenly exposed to bright light, the pupils immediately constrict; this is the **photopupillary reflex.** This protective reflex prevents excessively bright light from damaging the delicate photoreceptors. The pupils also constrict reflexively when we view close objects; this **accommodation pupillary reflex** provides for more acute vision.

Reading requires almost continuous work by both sets of muscles. The muscles of the ciliary body bring about the lens bulge, and the circular (or *constrictor*) muscles of the iris produce the accommodation pupillary reflex. In addition, the extrinsic muscles must converge the eyes as well as move them to follow the printed lines. This is why long periods of reading tire the eyes and often result in what is commonly called *eyestrain.* When you read for an extended time, it is helpful to look up from time to time and stare into the distance. This temporarily relaxes all the eye muscles.

The Ear: Hearing and Balance

At first glance, the machinery for hearing and balance appears very crude. Fluids must be stirred to stimulate the receptors of the ear: sound vibrations move fluid to stimulate hearing receptors, whereas gross movements of the head disturb fluids surrounding the balance organs. Receptors that respond to such physical forces are called **mechanoreceptors** (mek″ah-no-re-sep′terz). Our hearing apparatus allows us to hear an extraordinary range of sound, and our highly sensitive equilibrium receptors keep our nervous system continually up to date on the position and movements of the head. Without this information, it would be difficult if not impossible to maintain our balance. Although these two sense organs are housed together in the ear, their receptors respond to different stimuli and are activated independently of one another.

Anatomy of the Ear

Anatomically, the ear is divided into three major areas: the outer, or external, ear; the middle ear; and the inner, or internal, ear (Figure 8.12). The outer and middle ear structures are involved with hearing *only.* The inner ear functions in both equilibrium and hearing.

Outer (External) Ear

The **outer ear** is composed of the pinna and the external acoustic meatus. The **pinna** (pin′nah), or **auricle** (aw′ri-kul), is what most people call the "ear"—the shell-shaped structure surrounding the auditory canal opening. In many animals, it collects and directs sound waves into the auditory canal, but in humans this function is largely lost.

The **external acoustic meatus** (or **external auditory canal**) is a short, narrow chamber (about 1 inch long by ¼ inch wide) carved into the temporal bone of the skull. In its skin-lined walls are the **ceruminous** (sĕ-roo′mĭ-nus) **glands,** which secrete a waxy yellow substance called **earwax,** or **cerumen.** Sound waves entering the

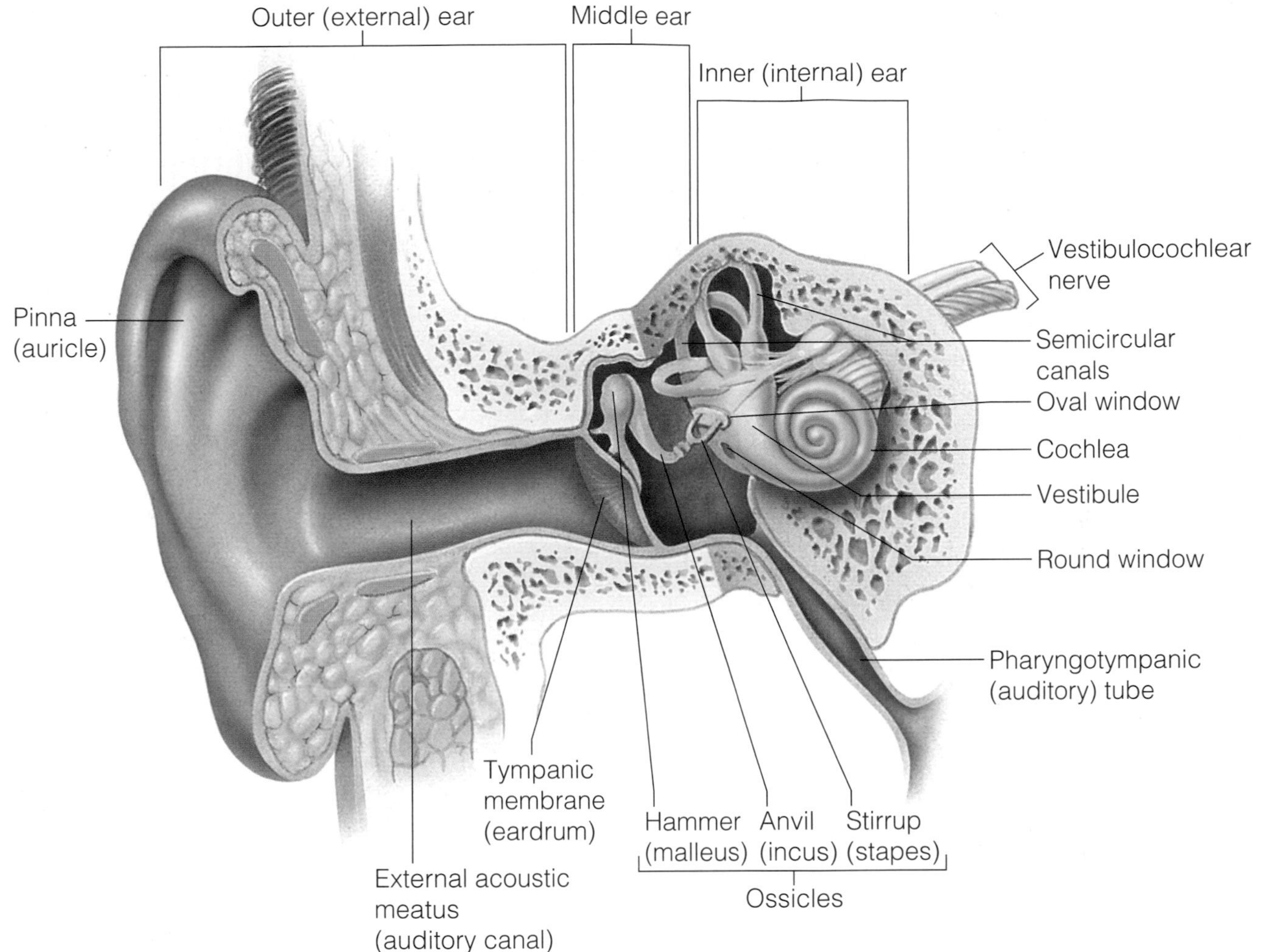

FIGURE 8.12 Anatomy of the ear.

external auditory canal eventually hit the **tympanic** (tim-pan′ik; *tympanum* = drum) **membrane,** or **eardrum,** and cause it to vibrate. The canal ends at the eardrum, which separates the outer from the middle ear.

Middle Ear

The **middle ear,** or **tympanic cavity,** is a small, air-filled cavity within the temporal bone. It is flanked laterally by the eardrum and medially by a bony wall with two openings, the **oval window** and the inferior, membrane-covered **round window.** The **pharyngotympanic** (or **auditory**) **tube** runs obliquely downward to link the middle ear cavity with the throat, and the mucosae lining the two regions are continuous. Normally, the pharyngotympanic tube is flattened and closed, but swallowing or yawning can open it briefly to equalize the pressure in the middle ear cavity with the external, or atmospheric, pressure. This is an important function because the eardrum does not vibrate freely unless the pressure on both of its surfaces is the same. When the pressures are unequal, the eardrum bulges inward or outward, causing hearing difficulty (voices may sound far away) and sometimes earaches. The ear-popping sensation of the pressures equalizing is familiar to anyone who has flown in an airplane.

Homeostatic Imbalance

Inflammation of the middle ear, *otitis media* (o-ti′tis me′de-ah), is a fairly common result of a sore throat, especially in children, whose pharyngotympanic tubes run more horizontally. In otitis media, the eardrum bulges and often becomes inflamed. When large amounts of fluid or pus accumulate in the cavity, an emergency *myringotomy* (lancing of the eardrum) may be required to relieve the pressure. During myringotomy, a tiny tube is implanted in the eardrum that allows pus formed in the middle ear to continue to drain into the external ear canal. The tube usually falls out by itself within the year. ▲

The more horizontal course of the pharyngotympanic tube in infants also explains why it is never a good idea to "prop" a bottle or feed them when they are lying flat (a condition that favors the entry of the food into that tube).

The tympanic cavity is spanned by the three smallest bones in the body, the **ossicles** (os′sĭ-kulz), which transmit the vibratory motion of the eardrum to the fluids of the inner ear (see Figure 8.12). These bones, named for their shape, are the **hammer,** or **malleus** (mă′le-us), the **anvil,** or **incus** (in′kus), and the **stirrup,** or **stapes** (sta′pēz). When the eardrum moves, the hammer moves with it and transfers the vibration to the anvil. The anvil, in turn, passes it on to the stirrup, which presses on the oval window of the inner ear. The movement at the oval window sets the fluids of the inner ear into motion, eventually exciting the hearing receptors.

Inner (Internal) Ear

The **inner ear** is a maze of bony chambers called the **osseous,** or **bony, labyrinth** (lab′ĭ-rinth), located deep within the temporal bone, and just behind the eye socket. The three subdivisions of the bony labyrinth are the **cochlea** (kok′le-ah), the **vestibule** (ves′ti-būl), and the **semicircular canals.** The vestibule is situated between the semicircular canals and the cochlea. The views of the bony labyrinth typically seen in textbooks, including this one, are somewhat misleading because we are really talking about a cavity. The view seen in Figure 8.12 can be compared to a *cast* of the bony labyrinth; that is, a labyrinth that was filled with plaster of paris and then had the bony walls removed after the plaster hardened. The shape of the plaster then reveals the shape of the *cavity* that worms through the temporal bone.

The bony labyrinth is filled with a plasmalike fluid called **perilymph** (per′ĭ-limf). Suspended in the perilymph is a **membranous labyrinth,** a system of membrane sacs that more or less follows the shape of the bony labyrinth. The membranous labyrinth itself contains a thicker fluid called **endolymph** (en′do-limf).

Mechanisms of Equilibrium

If a cat is released upside down from a certain height, it will land on its feet. If an infant is tilted backward, its eyes will roll downward so that its gaze remains fixed (the *doll's-eye reflex*). Both of these reactions, and countless others, are compensations for a disturbance in balance, reflexes that depend on the sensory receptors within the vestibule and semicircular canals.

The equilibrium sense is not easy to describe because it does not see, hear, or feel. What it *does* is respond (frequently without our awareness) to the various movements of our head. The equilibrium receptors of the inner ear, sometimes called the **vestibular apparatus,** can be divided into two functional arms—one arm responsible for monitoring *static equilibrium,* and the other involved with *dynamic equilibrium.*

Static Equilibrium

Within the membrane sacs of the vestibule are receptors called **maculae** (mak′u-le; "spots") that are essential to our sense of **static equilibrium** (Figure 8.13). The maculae report on the position of the head with respect to the pull of gravity when the body is not moving (*static* = at rest). Since they provide information on which way is up or down, they help us keep our head erect. They are extremely important to divers swimming in the dark depths (where most other orienting cues are absent), enabling them to tell which way is up (to the surface). Each macula is a patch of receptor cells with their "hairs" embedded in the **otolithic membrane,** a gel or jellylike material containing **otoliths** (o′to-lithz), tiny stones made of calcium salts. As the head moves, the otoliths roll in response to changes in the pull of gravity. This movement creates a pull on the gel, which in turn slides like a greased plate over the hair cells, bending their hairs. This event activates the hair cells, which send impulses along the **vestibular nerve** (a division of cranial nerve VIII) to the cerebellum of the brain, informing it of the position of the head in space.

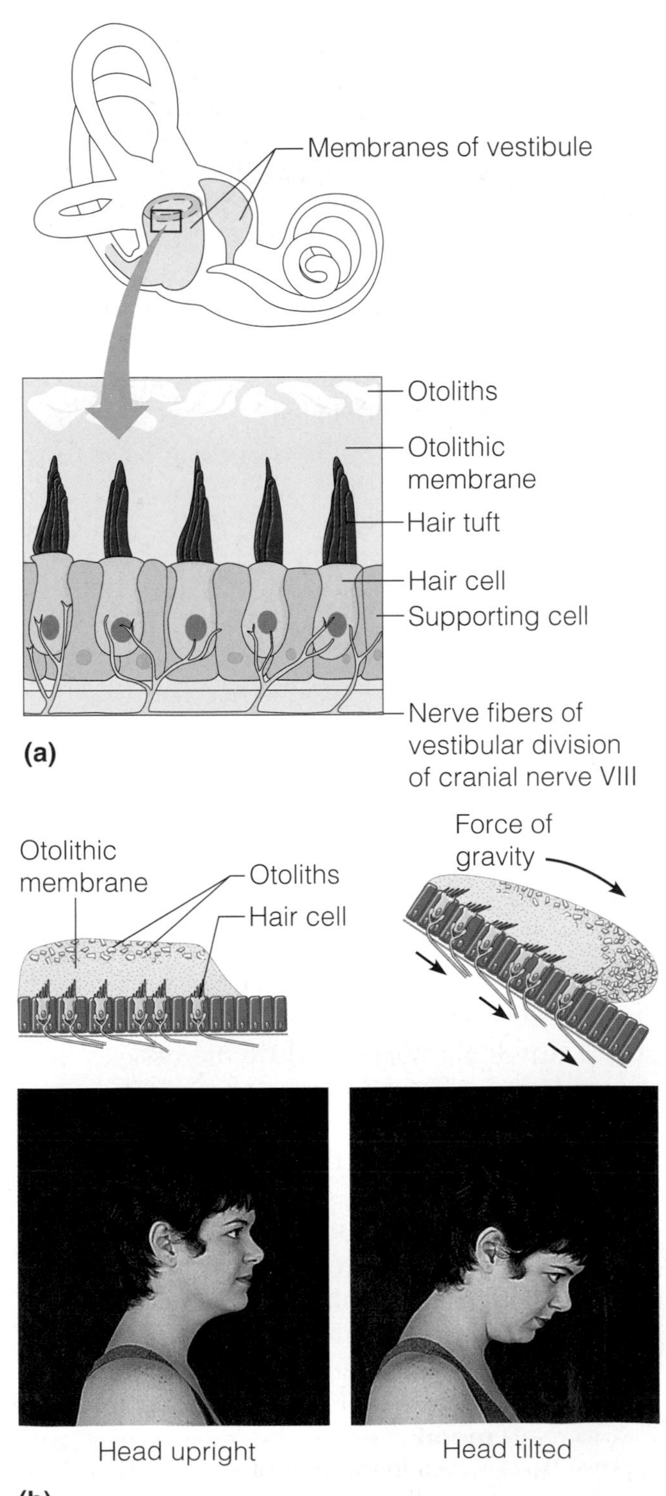

FIGURE 8.13 Structure and function of maculae (static equilibrium receptors). (a) Diagrammatic view of part of a macula. **(b)** When the head is tipped, the maculae are stimulated by movement of the otoliths in the gelatinous otolithic membrane in the direction of gravitational pull, which creates a pull on the hair cells.

Dynamic Equilibrium

The **dynamic equilibrium** receptors, found in the semicircular canals, respond to angular or rotatory movements of the head, rather than to straight-line movements. When you twirl on the dance floor or suffer through a rough boat ride, these receptors are working overtime. The semicircular canals (each about ½ inch, or 1.3 cm, around) are oriented in the three planes of space. Thus, regardless of which plane one moves in, there will be receptors to detect the movement.

Within each membranous semicircular canal is a receptor region called a **crista ampullaris** (kris'tah am″pu-lar'is), which consists of a tuft of hair cells covered with a gelatinous cap called the **cupula** (ku'pu-lah) (Figure 8.14). When your head moves in an arclike or angular direction, the endolymph in the canal lags behind. Then, as the cupula drags against the stationary endolymph, the cupula bends—like a swinging door—with the body's motion. This stimulates the hair cells, and impulses are transmitted up the **vestibular nerve** to the cerebellum. When you are moving at a constant rate, the receptors gradually stop sending impulses,

Prove It Yourself

Fluids Move in the Semicircular Canals in Response to Head Movement

You can easily demonstrate that fluids move in the semicircular canals as you move your head. First, fill a glass half-full with water and place a few objects in it, such as peppercorns or peas. Now twist the glass a half-turn or so. Notice that the objects (and the water) move more slowly than the glass, especially if the movement of the glass is rapid. The water and the objects appear to be flowing in the opposite direction from the glass, but in fact they are just slow to get started.

When you stop rotating the glass, the opposite happens. The water and the objects continue to spin for a short time.

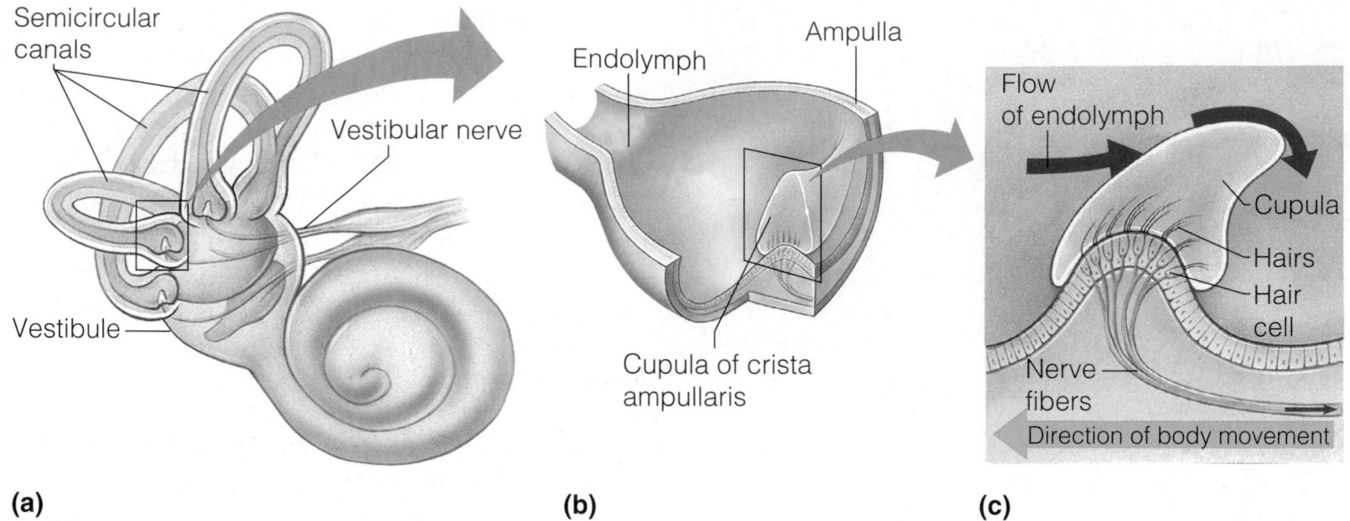

FIGURE 8.14 Structure and function of the crista ampullaris (dynamic equilibrium receptor region). **(a)** Arranged in the three spatial planes, the semicircular ducts in the semicircular canals each have a swelling called an ampulla at their base. **(b)** Each ampulla contains a crista ampullaris, a receptor that is essentially a cluster of hair cells with hairs projecting into a gelatinous cap called the cupula. **(c)** When head position changes in an angular direction, inertia causes the endolymph in the semicircular ducts to lag behind, and as the cupula moves it drags across the endolymph bending the hair cells in the opposite direction. The bending results in increased impulse transmission in the sensory neurons. This mechanism adjusts quickly if the angular motion (or rotation) continues at a constant speed.

and you no longer have the sensation of motion until your speed or direction of movement changes.

Although the receptors of the semicircular canals and vestibule are responsible for dynamic and static equilibrium, respectively, they usually act together. Besides these equilibrium senses, sight and the proprioceptors of the muscles and tendons are also important in providing information used to control balance to the cerebellum.

Mechanism of Hearing

Within the **cochlear duct,** the endolymph-containing membranous labyrinth of the snail-like cochlea is the **organ of Corti** (kor′te), which contains the hearing receptors, or **hair cells** (Figure 8.15a). The chambers (scalae) above and below the cochlear duct contain perilymph. Sound waves that reach the cochlea through vibrations of the eardrum, ossicles, and oval window set the cochlear fluids into motion (Figure 8.16). As the sound waves are transmitted by the ossicles from the eardrum to the oval window, their force (amplitude) is increased by the lever activity of the ossicles. In this way, nearly the total force exerted on the much larger eardrum reaches the tiny oval window, which in turn sets the fluids of the inner ear into motion, and these pressure waves set up vibrations in the **basilar membrane.** The receptor cells, positioned on the basilar membrane in the organ of Corti, are stimulated when their "hairs" are bent or tweaked by the movement of the gel-like **tectorial** (tek-to′re-al) **membrane** that lies over them (Figure 8.15b). The length of the fibers spanning the basilar membrane "tunes" specific regions to vibrate at specific frequencies. In general, high-pitch sounds disturb the shorter fibers of the basilar membrane and stimulate receptor cells close to the oval window, whereas low-pitch sounds affect longer fibers and activate specific hair cells further along the cochlea. Once stimulated, the hair cells transmit impulses along the **cochlear nerve** (a division of cranial nerve VIII—the vestibulocochlear nerve) to

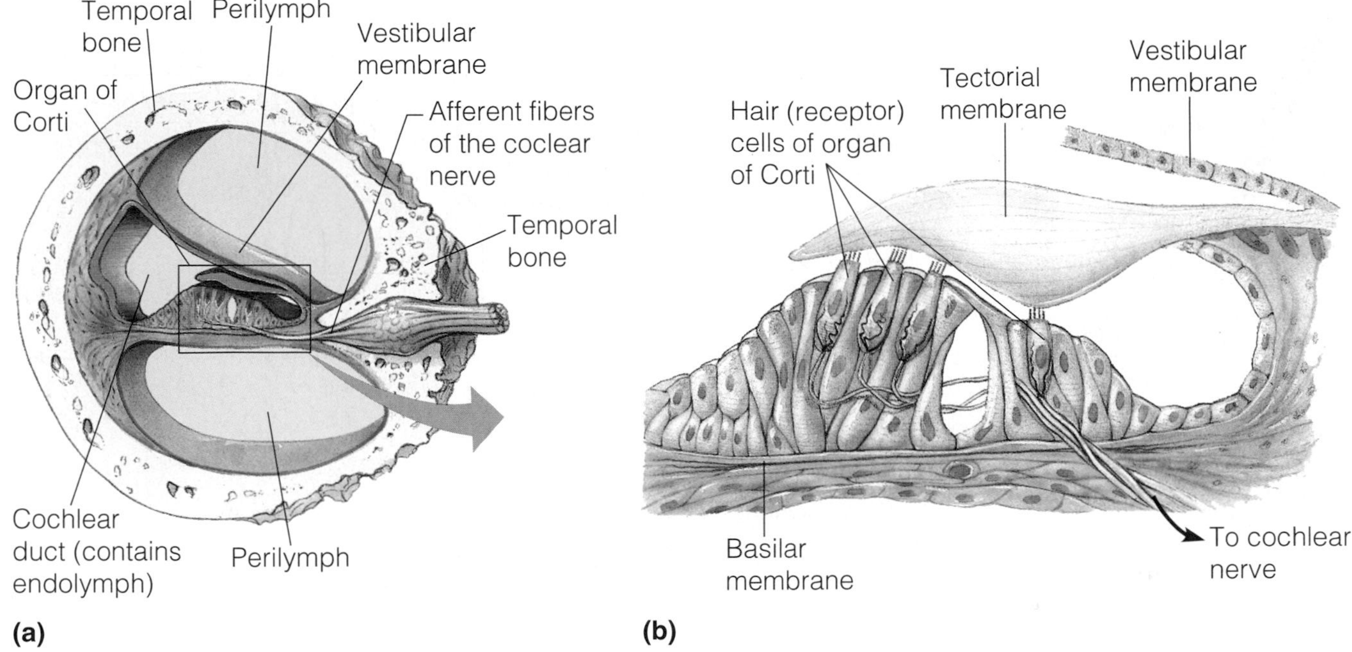

FIGURE 8.15 Anatomy of the cochlea. (a) A cross-sectional view of one turn of the cochlea, showing the position of the organ of Corti in the cochlear duct. The cavities of the bony labyrinth contain perilymph. The cochlear duct contains endolymph. **(b)** Detailed structure of the organ of Corti. The receptor cells (hair cells) rest on the basilar membrane.

the auditory cortex in the temporal lobe, where interpretation of the sound, or hearing, occurs. Since sound usually reaches the two ears at different times, we could say that we hear "in stereo." Functionally, this helps us to determine where sounds are coming from in our environment.

When the same sounds, or tones, keep reaching the ears, the auditory receptors tend to *adapt* or stop responding to those sounds, and we are no longer aware of them. This is why the drone of a continually running motor does not demand our attention after the first few seconds. However, hearing is the last sense to leave our awareness when we fall asleep or receive anesthesia (or die) and is the first to return as we awaken.

Hearing and Equilibrium Deficits

Homeostatic Imbalance

Children with ear problems or hearing deficits often pull on their ears or fail to respond when spoken to. Under such conditions, tuning fork or audiometry testing is done to try to diagnose the problem. *Deafness* is defined as *hearing loss of any degree*—from a slight loss to a total inability to hear sound. Generally speaking, there are two kinds of deafness, conduction and sensorineural. Temporary or permanent *conduction deafness* results when something interferes with the conduction of sound vibrations to the fluids of the inner ear. Something as simple as a buildup of earwax may be the cause. Other causes of conduction deafness include fusion of the ossicles (a problem called *otosclerosis* [o″to-sklĕ-ro′sis]), a ruptured eardrum, and *otitis media.*

Sensorineural deafness occurs when there is degeneration or damage to the receptor cells in the organ of Corti, to the cochlear nerve, or to neurons of the auditory cortex. This often results from extended listening to excessively loud sounds. Thus, whereas conduction deafness results from mechanical factors, sensorineural deafness is a problem of nervous system structures.

A person who has a hearing loss due to conduction deafness will still be able to hear by bone conduction, even though his or her ability to hear air-conducted sounds (the normal conduction route) is decreased or lost. On the other hand, individuals with sensorineural deafness cannot hear better by *either* conduction

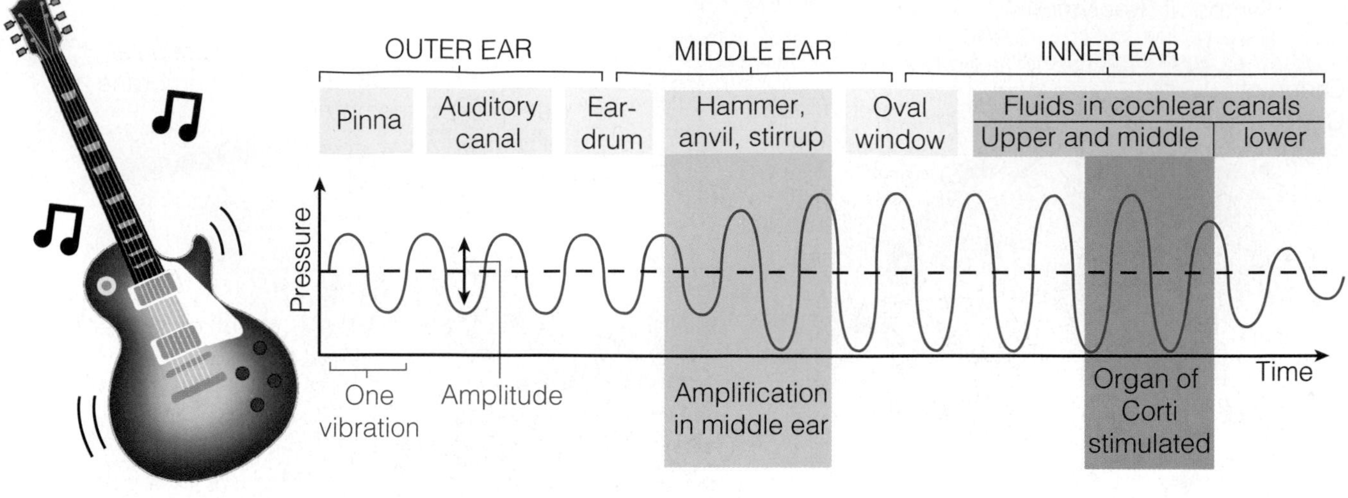

(a)

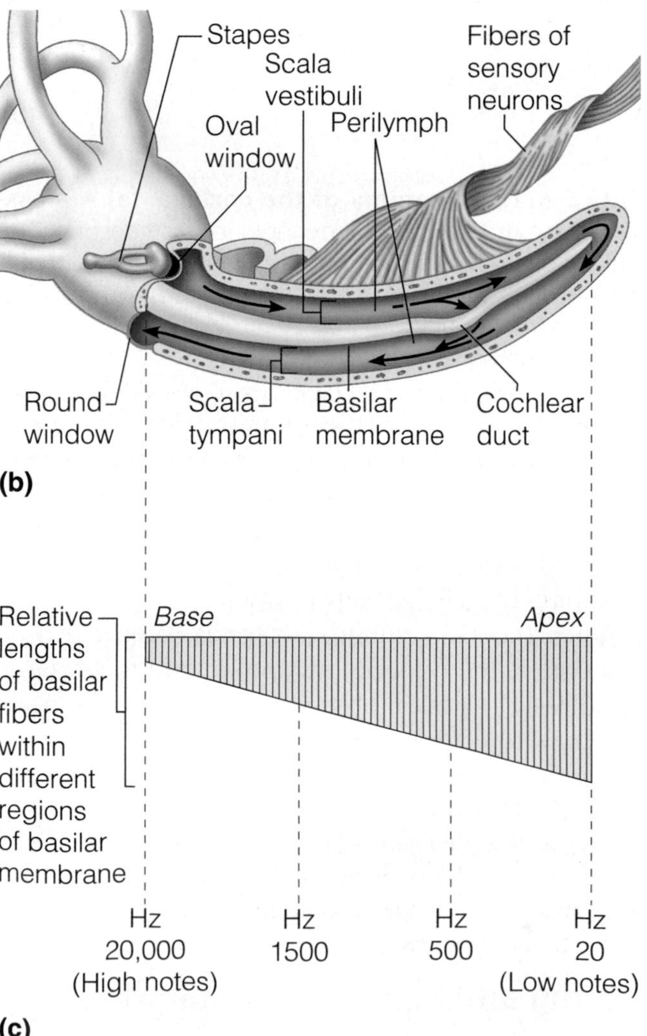

FIGURE 8.16 Route of sound waves through the ear and activation of the cochlear hair cells. **(a)** To excite the hair cells in the organ of Corti in the inner ear, sound wave vibrations must pass through air, membranes, bone and fluid. **(b)** The cochlea is drawn as if it is uncoiled to make the events of sound transmission occurring there easier to follow. Sound waves of low frequency that are below the level of hearing travel entirely around the cochlear duct without exciting hair cells. But sounds of higher frequency result in pressure waves that penetrate through the cochlear duct and basilar membrane to reach the scala tympani. This causes the basilar membrane to vibrate maximally in certain areas in response to certain frequencies of sound, stimulating particular hair cells and sensory neurons. The differential stimulation of hair cells is perceived in the brain as sound of a certain pitch. **(c)** The length of the fibers spanning the basilar membrane tune specific regions to vibrate at specific frequencies. The higher notes—20,000 Hertz (Hz)—are detected by shorter hair cells along the base of the basilar membrane.

route. Hearing aids, which use skull bones to conduct sound vibrations to the inner ear, are generally very successful in helping those with conduction deafness to hear. They are less helpful for sensorineural deafness.

Equilibrium problems are usually obvious. Nausea, dizziness, and problems in maintaining balance are common symptoms, particularly when impulses from the vestibular apparatus "disagree" with what we see

Q *How does sniffing help to identify scents?*

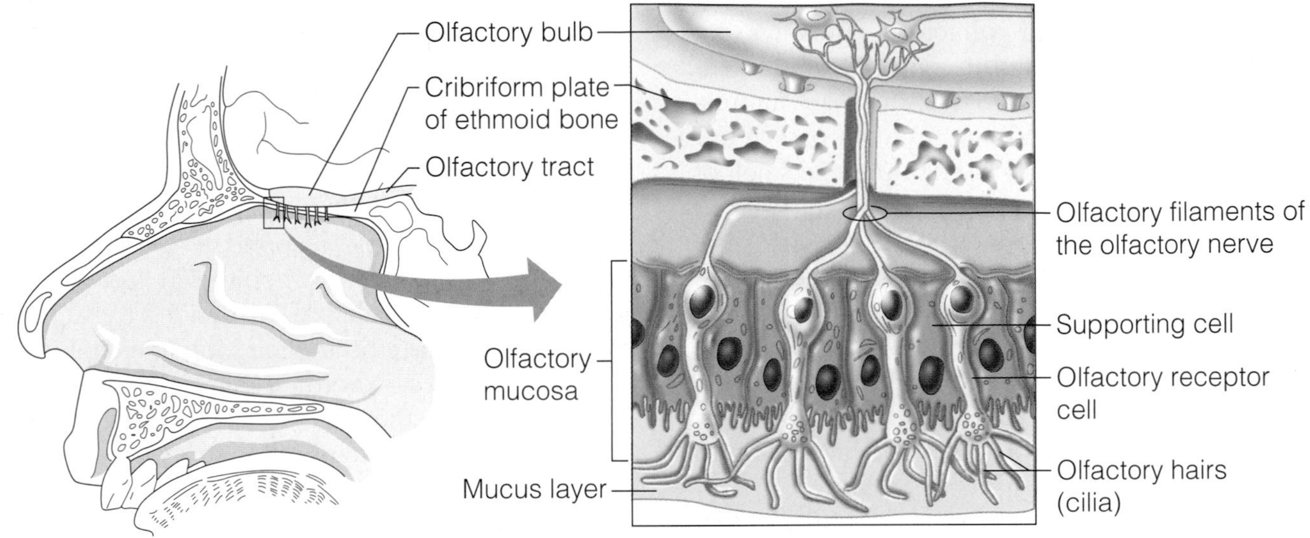

FIGURE 8.17 Location and cellular makeup of the olfactory epithelium.

(visual input). There also may be strange (jerky or rolling) eye movements.

A serious pathology of the inner ear is *Ménière's* (mān″e-airz′) *disease.* The exact cause of this condition is not fully known, but suspected causes are arteriosclerosis, degeneration of cranial nerve VIII, and increased pressure of the inner ear fluids. In Ménière's disease, progressive deafness occurs. Affected individuals become nauseated and often have *vertigo* (a sensation of spinning) that is so severe that they cannot stand up without extreme discomfort. Anti–motion sickness drugs are often prescribed to decrease the discomfort. ▲

Chemical Senses: Taste and Smell

The receptors for taste and olfaction are classified as **chemoreceptors** (ke″mo-re-sep′terz) because they respond to chemicals in solution. Five types of taste receptors have been identified, but the olfactory receptors (for smell) are believed to be sensitive to a much wider range of chemicals. The receptors for smell and taste complement each other and respond to many of the same stimuli.

Olfactory Receptors and the Sense of Smell

Even though our sense of smell is far less acute than that of many other animals, the human nose is still much keener than any machine in picking up small differences in odors. Some people capitalize on this ability by becoming tea and coffee blenders, perfumers, or wine tasters.

The thousands of **olfactory receptors,** receptors for the sense of smell, occupy a postage stamp–sized area in the roof of each nasal cavity (Figure 8.17). Air entering the nasal cavities must make a hairpin turn to enter the respiratory passageway below, so sniffing, which causes more air to flow superiorly across the olfactory receptors, intensifies the sense of smell.

The **olfactory receptor cells** are neurons equipped with **olfactory hairs,** long cilia that protrude from the nasal epithelium and are continually bathed by a layer of mucus secreted by underlying glands. When the receptors are stimulated by chemicals dissolved in the mucus, they transmit impulses along the **olfactory filaments,** which

A *It brings more odor-containing air into contact with the olfactory receptors in the superior part of the nasal cavity.*

collectively make up the **olfactory nerve** (cranial nerve I), to the olfactory cortex of the brain. There interpretation of the odor occurs, and an "odor snapshot" is made. The olfactory pathways are closely tied into the limbic system (emotional-visceral part of the brain). Thus, olfactory impressions are long-lasting and very much a part of our memories and emotions. For example, the smell of chocolate chip cookies may remind you of your grandmother, and the smell of a special pipe tobacco may make you think of your father. There are hospital smells, school smells, baby smells, travel smells. The list can be continued almost without end. Our reactions to odors are rarely neutral. We tend to either like or dislike certain odors, and we change, avoid, or add odors according to our preferences.

The olfactory receptors are exquisitely sensitive—just a few molecules can activate them. Like the auditory receptors, the olfactory neurons tend to adapt rather quickly when they are exposed to an unchanging stimulus, in this case, an odor. This is why a woman stops smelling her own perfume after a while but will quickly pick up the scent of another perfume on someone else.

Homeostatic Imbalance

While it is possible to have either taste or smell deficits, most people seeking medical help for loss of chemical senses have olfactory disorders, or *anosmias* (ah-noz′me-uz). Most anosmias result from head injuries, the aftereffects of nasal cavity inflammation (due to a cold, an allergy, or smoking), or aging. Some brain disorders can destroy the sense of smell or mimic it. For example, *olfactory auras* (olfactory hallucinations) are experienced by some epileptics just before they go into seizures. ▲

Taste Buds and the Sense of Taste

The word *taste* comes from the Latin word *taxare,* which means "to touch, estimate, or judge." When we taste things, we are, in fact, testing or judging our environment in an intimate way, and the sense of taste is considered by many to be the most pleasurable of our special senses. There is no question that what does not taste good to us will usually not be allowed to enter the body.

The **taste buds,** or specific receptors for the sense of taste, are widely scattered in the oral cavity. Of the 10,000 or so taste buds that we have, most are located on the tongue. A few are found on the soft palate and inner surface of the cheeks.

The dorsal tongue surface is covered with small peglike projections, or **papillae** (pah-pil′e), of three types: sharp *filiform* (fil′ĭ-form) papillae and the rounded *fungiform* (fun′jĭ-form) and *circumvallate* (ser″kum-val′at) papillae. The taste buds are found on the sides of the circumvallate papillae and on the more numerous fungiform papillae (Figure 8.18). The specific cells that respond to chemicals dissolved in the saliva are epithelial cells called **gustatory cells,** which are surrounded by supporting cells in the taste bud. Their long microvilli—the **gustatory hairs**—protrude through the **taste pore,** and when they are stimulated, they depolarize and impulses are transmitted to the brain. Three cranial nerves—VII, IX, and X—carry taste impulses from the various taste buds to the gustatory cortex. The **facial nerve** (VII) serves the anterior part of the tongue. The other two cranial nerves—the **glossopharyngeal** and **vagus,** respectively—serve the other taste bud–containing areas.

There are five basic taste sensations, each corresponding to stimulation of one of the five major types of taste buds. The *sweet receptors* respond to substances such as sugars, saccharine, and some amino acids. Some believe that the common factor is the hydroxyl (OH^-) group. *Sour receptors* respond to hydrogen ions (H^+), or the acidity of the solution; *bitter receptors* to alkaloids; and *salty receptors* to metal ions in solution. *Umami* (u-mah′me; "delicious"), a taste discovered by the Japanese, is elicited by the amino acid glutamate which appears to be responsible for the "beef taste" of steak and the flavor of monosodium glutamate, a food additive. Historically, the tip of the tongue was believed to be most sensitive to sweet and salty substances, its sides to sour, the back of the tongue to bitter and the pharynx to umami. Actually there are only slight differences in the locations of the taste receptors in different regions of the tongue, and most taste buds respond to two, three, four, or even all five taste modalities.

Taste likes and dislikes have homeostatic value. A liking for sugar and salt will satisfy the body's need for carbohydrates and minerals (as well as some amino acids). Many sour, naturally acidic foods (such as oranges, lemons, and tomatoes) are rich sources of vitamin C, an essential vitamin. Umami guides the intake of proteins; and, since many natural poisons and spoiled foods are bitter,

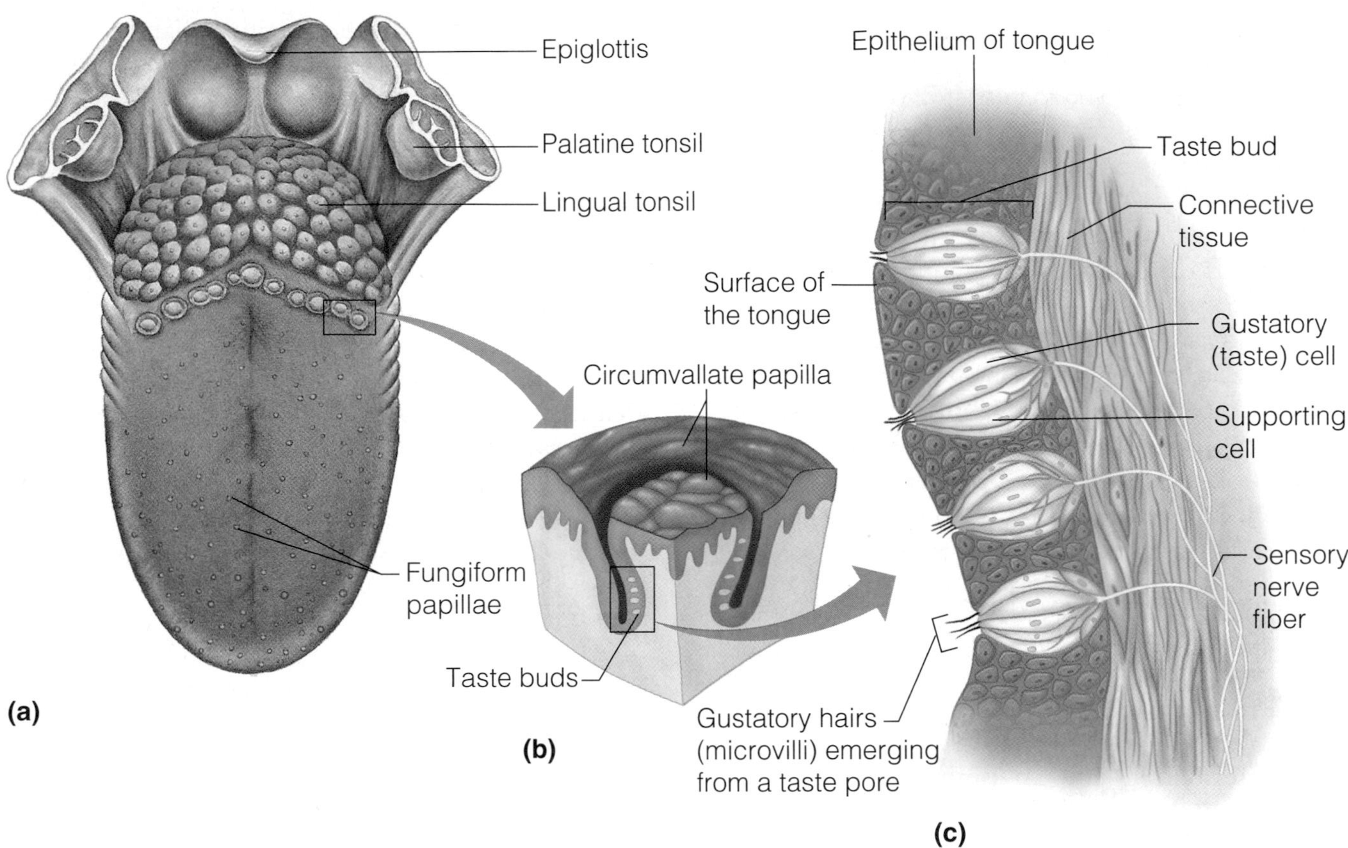

FIGURE 8.18 Location and structure of taste buds. (a) Taste buds on the tongue are associated with papillae, projections of the tongue mucosa. **(b)** A sectioned circumvallate papilla shows the position of the taste buds in its lateral walls. **(c)** An enlarged view of four taste buds.

our dislike for bitterness is protective. (The position of large numbers of bitter receptors on the most posterior part of the tongue seems strange, however, because usually by the time we actually taste a bitter substance, we have swallowed some of it.)

Taste is affected by many factors, and what is commonly referred to as our sense of taste depends heavily on stimulation of our olfactory receptors by aromas. Think of how bland food is when your nasal passages are congested by a cold. Without the sense of smell, our morning coffee would simply taste bitter. In addition the temperature and texture of food can enhance or spoil its taste for us. For example, some people will not eat foods that have a pasty texture (avocados) or that are gritty (pears), and almost everyone considers a cold greasy hamburger unfit to eat. "Hot" foods like chili peppers actually excite pain receptors in the mouth.

Developmental Aspects of the Special Senses

The special sense organs, essentially part of the nervous system, are formed very early in embryonic development. For example, the eyes, which are literally outgrowths of the brain, are developing by the fourth week. All of the special senses are functional, to a greater or lesser degree, at birth.

Homeostatic Imbalance

Congenital eye problems are relatively uncommon, but some examples can be given. *Strabismus* (strah-biz′mus), commonly called "crossed eyes," results from unequal pulls by the external eye muscles that prevent the baby from coordinating movement of the two eyes. First, exercises are used to strengthen the weaker eye muscles and/or the stronger eye may be covered with an eye patch to force the weaker muscles to become stronger. If these measures are not successful, surgery is always used to correct the condition because, if it is allowed to persist, the brain may stop recognizing signals from the deviating eye, causing that eye to become functionally blind.

Maternal infections, particularly *rubella* (measles), that occur during early pregnancy may lead to congenital blindness or cataracts. If the mother has a type of sexually transmitted disease called *gonorrhea* (gon″o-re′ah), the baby's eyes will be infected by the bacteria during delivery. In the resulting *conjunctivitis,* specifically called *ophthalmia neonatorum* (of-thal′me-ah ne″o-na-to′rum), the baby's eyelids become red and swollen, and pus is produced. All states legally require that all newborn babies' eyes be routinely treated with silver nitrate or antibiotics shortly after birth. ▲

Generally speaking, vision is the only special sense that is not fully functional when the baby is born, and many years of "learning" are needed before the eyes are fully mature. The eyeballs continue to enlarge until the age of 8 or 9, but the lens grows throughout life. At birth, the eyeballs are foreshortened, and all babies are hyperopic (farsighted). As the eyes grow, this condition usually corrects itself. The newborn infant sees only in gray tones, eye movements are uncoordinated, and often only one eye at a time is used. Because the lacrimal glands are not fully developed until about two weeks after birth, the baby is tearless for this period, even though he or she may cry lustily.

By 5 months, the infant is able to focus on articles within easy reach and to follow moving objects, but visual acuity is still poor (the infant sees an object that is 20 feet away clearly that someone with mature vision would see clearly when it is 200 feet away—such vision is said to be 20/200). By the time the child is 5 years old, color vision is well developed, visual acuity has improved to about 20/30, and depth perception is present, providing a readiness to begin reading. By school age, the earlier hyperopia has usually been replaced by emmetropia. This condition continues until about age 40, when **presbyopia** (pres″be-o′pe-ah) begins to set in. Presbyopia (literally, "old vision") results from decreasing lens elasticity that accompanies aging. This condition makes it difficult to focus for close vision; it is basically farsightedness. The person who holds the newspaper at arm's length to read it provides the most familiar example of this developmental change in vision.

As aging occurs, the lacrimal glands become less active and the eyes tend to become dry and more vulnerable to bacterial infection and irritation. The lens loses its crystal clarity and becomes discolored. The dilator muscles of the iris become less efficient; thus, the pupils are always somewhat constricted. These last two conditions work together to decrease by half the amount of light reaching the retina, so that visual acuity is dramatically lower by one's seventies. In addition to these changes, elderly people are susceptible to certain conditions that may result in blindness, such as glaucoma and cataracts. Other common aging-related problems, such as arteriosclerosis and diabetes, may lead to the death of the delicate photoreceptors because of an increasing lack of oxygen and nutrients.

Homeostatic Imbalance

Congenital abnormalities of the ears are fairly common. Examples include partly or completely missing pinnas and closed or absent external acoustic meatuses. Maternal infections can have a devastating effect on ear development, and maternal rubella during the early weeks of pregnancy results in sensorineural deafness. ▲

A newborn infant can hear after his or her first cry, but early responses to sound are mostly reflexive—for example, crying and clenching the eyelids in response to a loud noise. By the age of 3 or 4 months, the infant is able to localize sounds and will turn to the voices of family members. Critical listening occurs in the toddler as he or she begins to imitate sounds, and good language skills are very closely tied to an ability to hear well.

Except for ear inflammations *(otitis)* resulting from bacterial infections or allergies, few problems affect the ears during childhood and adult life. By the sixties, however, a gradual deterioration and atrophy of the organ of Corti begins and leads to a loss in the ability to hear high tones and speech sounds. This condition, **presbycusis** (pres″bĭ-ku′sis), is a type of sensorineural deafness. In some cases, the ear ossicles fuse *(otosclerosis),* which compounds the hearing problem by interfering with sound conduction to the inner ear. Because many elderly people refuse to accept their hearing loss and resist using hearing aids, they begin to rely more and more on their vision for clues as to what is going on around them and may be accused of ignoring people. Although presbycusis was once considered a disability of old age, it is becoming much more common in younger people as our world grows noisier day by day. Noise pollution has become a major health problem, and the damage caused by excessively loud sounds is progressive and cumulative. Each insult causes a bit more damage. Music played and heard at deafening levels is definitely a contributing factor to the deterioration of the hearing receptors.

The chemical senses, taste and smell, are sharp at birth, and infants relish some food that adults consider bland or tasteless. Some researchers claim the sense of smell is just as important as the sense of touch in guiding a newborn baby to its mother's breast. However, very young children seem indifferent to odors and can play happily with their own feces. As they get older, their emotional responses to specific odors increase.

There appear to be few problems with the chemical senses throughout childhood and young adulthood. Beginning in the midforties, our ability to taste and smell diminishes, which reflects the gradual decrease in the number of these receptor cells. Almost half of people over the age of 80 cannot smell at all, and their sense of taste is poor. This may explain their inattention to formerly disagreeable odors, and why older adults often prefer highly seasoned (although not necessarily spicy) foods or lose their appetite entirely.

SUMMARY

Media study tools that could provide you with additional help in reviewing specific key topics of Chapter 8 are referenced below.

IP = Interactive Physiology;

WEB = A&P Place website.

THE EYE AND VISION (pp. 272–284)

1. External/accessory structures of the eye:
   a. Extrinsic eye muscles aim the eyes for following moving objects and for convergence.
   b. The lacrimal apparatus includes a series of ducts and the lacrimal glands that produce a saline solution, which washes and lubricates the eyeball.
   c. Eyelids protect the eyes. Associated with the eyelashes are the ciliary glands (modified sweat glands) and the tarsal glands (which produce an oily secretion that helps keep the eye lubricated).
   d. The conjunctiva is a mucous membrane that covers the anterior eyeball and lines the eyelids. It produces a lubricating mucus.
2. Three tunics form the eyeball.
   a. The sclera is the outer, tough, protective tunic. Its anterior portion is the cornea, which is transparent to allow light to enter the eye.
   b. The choroid is the middle coat, which provides nutrition to the internal eye structures and prevents light's scattering in the eye. Anterior modifications include two smooth muscle structures: the ciliary body and the iris (which controls the size of the pupil).
   c. The retina is the innermost (sensory) coat, which contains the photoreceptors. Rods are dim light receptors. Cones are receptors that provide for color vision and high visual acuity. The fovea centralis, on which acute focusing occurs, contains only cones.

   WEB Exercise: Chapter 8, Internal Structures of the Eye.
3. The blind spot (optic disc) is the point where the optic nerve leaves the back of the eyeball.
4. The lens is the major light-bending (refractory) structure of the eye. Its convexity is increased by the ciliary body for close focus. Anterior to the lens is the aqueous humor; posterior to the lens is the vitreous humor. Both humors reinforce the eye internally. The aqueous humor also provides nutrients to the avascular lens and cornea.
5. Errors of refraction include myopia, hyperopia, and astigmatism. All are correctable with specially ground lenses.

   WEB Exercise: Chapter 8, Optics of the Eye; Special Senses Case Study.
6. The pathway of light through the eye is cornea → aqueous humor → (through pupil) → aqueous humor → lens → vitreous humor → retina.
7. Overlap of the visual fields and inputs from both eyes to each optic cortex provide for depth perception.

8. The pathway of nerve impulses from the retina of the eye is optic nerve → optic chiasma → optic tract → thalamus → optic radiation → visual cortex in occipital lobe of brain.

9. Eye reflexes include the photopupillary, accommodation pupillary, and convergence.

THE EAR: HEARING AND BALANCE (pp. 284–291)

1. The ear is divided into three major areas.
   a. Outer ear structures are the pinna, external acoustic meatus, and tympanic membrane. Sound entering the external acoustic meatus sets the eardrum into vibration. These structures are involved with sound transmission only.
   b. Middle ear structures are the ossicles and pharyngotympanic tube within the tympanic cavity. Ossicles transmit the vibratory motion from the eardrum to the oval window. The pharyngotympanic tube allows pressure to be equalized on both sides of the eardrum. These structures are involved with sound transmission only.
   c. The inner ear, or bony labyrinth, consists of bony chambers (cochlea, vestibule, and semicircular canals) in the temporal bone. The bony labyrinth contains perilymph and membranous sacs filled with endolymph. Within the membranous sacs of the vestibule and semicircular canals are equilibrium receptors. Hearing receptors are found within the membranes of the cochlea.

   WEB Exercise: Chapter 8, Internal Structures of the Ear.

2. Receptors of the semicircular canals (cristae ampullaris) are dynamic equilibrium receptors, which respond to angular or rotational body movements. Receptors of the vestibule (maculae) are static equilibrium receptors, which respond to the pull of gravity and report on head position. Visual and proprioceptor input are also necessary for normal balance.

3. Symptoms of equilibrium apparatus problems include involuntary rolling of the eyes, nausea, vertigo, and an inability to stand erect.

4. Hair cells of the organ of Corti (the receptor for hearing within the cochlea) are stimulated by sound vibrations transmitted through air, membranes, bone, and fluids.

5. Deafness is any degree of hearing loss. Conduction deafness results when the transmission of sound vibrations through the external and middle ears is hindered. Sensorineural deafness occurs when there is damage to the nervous system structures involved in hearing.

CHEMICAL SENSES: SMELL AND TASTE (pp. 291–293)

1. Chemical substances must be dissolved in water to excite the receptors for smell and taste.

2. The olfactory (smell) receptors are located in the superior aspect of each nasal cavity. Sniffing helps to bring more air (containing odors) over the olfactory mucosa.

3. Olfactory pathways are closely linked to the limbic system; odors stimulate the recall of memories and arouse emotional responses.

4. Gustatory (taste) cells are located in the taste buds, primarily on the tongue. The five major taste sensations are sweet, salt, sour, bitter, and umami.

5. Taste and appreciation of foods is influenced by the sense of smell and the temperature and texture of foods.

DEVELOPMENTAL ASPECTS OF THE SPECIAL SENSES (pp. 293–295)

1. Special sense organs are formed early in embryonic development. Maternal infections during the first five or six weeks of pregnancy may cause visual abnormalities as well as sensorineural deafness in the developing child. An important congenital eye problem is strabismus. The most important congenital ear problem is lack of the external acoustic meatus.

2. Vision requires the most learning. The infant has poor visual acuity (is farsighted) and lacks color vision and depth perception at birth. The eye continues to grow and mature until the eighth or ninth year of life.

3. Problems of aging associated with vision include presbyopia, glaucoma, cataracts, and arteriosclerosis of the eye's blood vessels.

4. The newborn infant can hear sounds, but initial responses are reflexive. By the toddler stage, the child is listening critically and beginning to imitate sounds as language development begins.

5. Sensorineural deafness (presbycusis) is a normal consequence of aging.

6. Taste and smell are most acute at birth and decrease in sensitivity after the age of 40 as the number of olfactory and gustatory receptors decreases.

REVIEW QUESTIONS

MULTIPLE CHOICE

More than one choice may apply.

1. Gustatory cells are:
   a. bipolar neurons
   b. multipolar neurons
   c. unipolar neurons
   d. epithelial cells
2. Alkaloids excite gustatory hairs at the:
   a. tip of the tongue
   b. back of the tongue
   c. circumvallate papillae
   d. fungiform papillae
3. Cranial nerves that are part of the gustatory pathway include:
   a. trigeminal
   b. facial
   c. hypoglossal
   d. glossopharyngeal
4. Which cranial nerve controls contraction of the circular smooth muscle of the iris?
   a. Trigeminal
   b. Facial
   c. Oculomotor
   d. Abducens
5. The cornea is nourished by:
   a. corneal blood vessels
   b. aqueous humor
   c. vitreous humor
   d. scleral blood vessels
6. In focusing for far vision:
   a. the lens is at its thinnest
   b. the ciliary muscles contract
   c. the light rays are nearly parallel
   d. suspensory fibers of the ciliary zonule are slack
7. Convergence:
   a. requires contraction of the medial rectus muscles of both eyes
   b. is needed for near vision
   c. involves transmission of impulses along the abducens nerves
   d. can promote eyestrain
8. Which of the following are paired *incorrectly*?
   a. Cochlear duct—cupula
   b. Vestibule—macula
   c. Ampulla—otoliths
   d. Semicircular duct—ampulla
9. Movement of the ___________ membrane triggers bending of hairs of the hair cells in the organ of Corti.
   a. tympanic
   b. tectorial
   c. basilar
   d. vestibular
10. Sounds entering the external acoustic meatus are eventually converted to nerve impulses via a chain of events including:
    a. vibration of the eardrum
    b. vibratory motion of the ossicles against the round window
    c. stimulation of hair cells in the organ of Corti
    d. resonance of the cupula

SHORT ANSWER ESSAY

1. Name three accessory eye structures that help to lubricate the eyeball, and name the secretion of each.
2. Why do you often have to blow your nose after crying?
3. Diagram and label the internal structures of the eye, and give the major function of each structure.
4. Name the extrinsic eye muscles that allow you to direct your eyes.
5. Locate and describe the functions of the two humors of the eye.
6. What is the blind spot, and why is it called this?
7. How do the functions of the rods and cones differ?
8. What is the fovea centralis, and why is it important?
9. Trace the pathway of light from the time it hits the cornea until it excites the rods and cones.
10. Trace the pathway of nerve impulses from the photoreceptors in the retina to the visual cortex of the brain.
11. How is the right optic *tract* anatomically different from the right optic *nerve*?

12. Define *refraction,* and name the refractory structures or substances of the eye.
13. Define *hyperopia, myopia,* and *emmetropia.*
14. Why do most people develop presbyopia as they age? Which of the conditions in Question 13 does it most resemble?
15. Since there are only three types of cones, how can you explain the fact that we see many more colors?
16. Why are ophthalmoscopic examinations important?
17. Many students struggling through mountains of reading assignments are told that they need glasses for eyestrain. Why is it more of a strain on the extrinsic and intrinsic muscles to look at close objects than at far objects?
18. When a light is shone into one eye, both pupils constrict simultaneously. Why is this an important protective reflex?
19. Name the structures of the outer, middle, and inner ears and give the general function of each structure and each group of structures.
20. Sound waves hitting the eardrum set it into motion. Trace the pathway of vibrations from the eardrum to the organ of Corti, where the hair cells are stimulated.
21. Explain the difference between *sensorineural* deafness and *conductive* deafness, and then give two causes of each type of deafness.
22. Distinguish between *static* and *dynamic* types of equilibrium.
23. Normal balance depends on information transmitted from a number of sensory receptor types. Name at least three of these receptors.
24. What name is given to the taste receptors, and where are they found?
25. Name the five primary taste sensations.
26. Where are the olfactory receptors located, and why is that site poorly suited for their job?
27. What common name is used to describe both the taste and the olfactory receptors, and why?
28. Describe the effects or results of aging on the special sense organs.
29. Which special sense requires the most learning?

AT THE CLINIC

1. An engineering student has been working in a disco to earn money to pay for his education. After about eight months, he notices that he is having problems hearing high-pitched tones. What is the cause-and-effect relationship here?
2. Nine children attending the same day-care center developed red, inflamed eyes and eyelids. What is the most likely cause and name of this condition?
3. Dr. Nakvarati used an instrument to press on Mr. Cruz's eye during his annual physical examination on his sixtieth birthday. The eye deformed very little, indicating the intraocular pressure was too high. What was Mr. Cruz's probable condition?
4. Lionel suffered a ruptured artery in his middle cranial fossa, and a pool of blood compressed his left optic tract, destroying its axons. What part of the visual field was blinded?
5. Sylvia Marcus, age 70, recently underwent surgery for otosclerosis. The operation was a failure and did not improve her condition. What was the purpose of the surgery, and exactly what was it trying to accomplish?
6. Janie is referred to the eye clinic by her teacher, who suspects a need for glasses. Examination demonstrates that Janie is myopic. Will she need concave or convex lenses? Explain.
7. Julie and her father loved to find the constellations in the sky on starry nights. One evening, Julie came running into the house and whispered excitedly to her mother. "Mom, I've got power! When I look hard at a star it disappears!" What was happening?
8. While visiting at her father's office on the 25th floor of the Harris Building, five-year-old Katie wandered away into the hall. Fascinated by the buttons in the elevator, she entered, pressed 1, and the high-speed elevator plummeted to the first floor. Later she told her father that she felt like she "kept on going" when the elevator stopped. Explain her sensation.
9. Mrs. Miniver has an immune disorder that causes dry mouth, and she complains to her doctor that she's lost her sense of taste. How might her symptoms be explained?

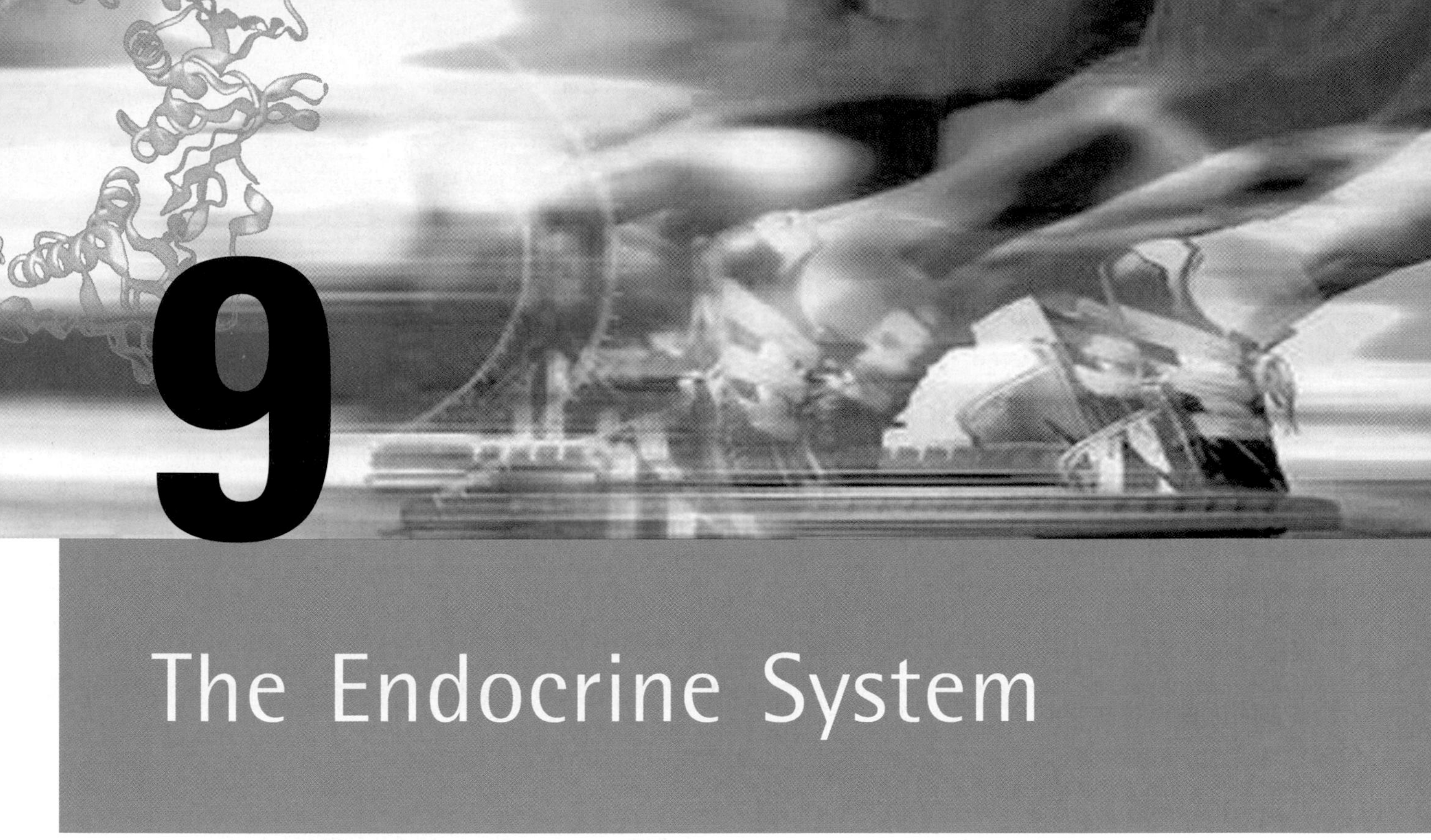

The Endocrine System

YOUR GOALS

After completing this chapter, you will have a working knowledge of the functions of the endocrine system and will have mastered the objectives listed below.

FUNCTION PREVIEW

- The endocrine system maintains homeostasis by releasing chemicals called *hormones,* and controls prolonged or continuous processes such as growth and development, reproduction, and metabolism.

OBJECTIVE CHECKLIST

THE ENDOCRINE SYSTEM AND HORMONE FUNCTION—AN OVERVIEW (pp. 300–303)

- Define *hormone* and *target organ.*
- Describe how hormones bring about their effects in the body.
- Explain how various endocrine glands are stimulated to release their hormonal products.
- Define *negative feedback,* and describe its role in regulating blood levels of the various hormones.

THE MAJOR ENDOCRINE ORGANS (pp. 303–320)

- Describe the difference between endocrine and exocrine glands.
- On an appropriate diagram, identify the major endocrine glands and tissues.
- List hormones produced by the endocrine glands, and discuss their general functions.
- Discuss ways in which hormones promote body homeostasis by giving examples of hormonal actions.
- Describe the functional relationship between the hypothalamus and the pituitary gland.

>>

(Objective Checklist, continued)

- Describe major pathological consequences of hypersecretion and hyposecretion of the hormones considered in this chapter.

OTHER HORMONE-PRODUCING TISSUES AND ORGANS (pp. 320–321)

- Indicate the endocrine role of the kidneys, the stomach and intestine, the heart, and the placenta.

DEVELOPMENTAL ASPECTS OF THE ENDOCRINE SYSTEM (p. 323)

- Describe the effect of aging on the endocrine system and body homeostasis.

When insulin molecules, carried passively along in the blood, leave the blood and bind tightly to protein receptors of nearby cells, the response is dramatic: bloodborne glucose molecules begin to disappear into the cells, and cellular activity accelerates. Such is the power of the second great controlling system of the body, the **endocrine system.** Along with the nervous system, it coordinates and directs the activity of the body's cells. However, the speed of control in these two great regulating systems is very different. The nervous system is "built for speed." It uses nerve impulses to prod the muscles and glands into immediate action so that rapid adjustments can be made in response to changes occurring both inside and outside the body. On the other hand, the more slowly acting endocrine system uses chemical messengers called **hormones,** which are released into the blood to be transported leisurely throughout the body.

Although hormones have widespread and varied effects, the major processes controlled by hormones are reproduction; growth and development; mobilizing body defenses against stressors; maintaining electrolyte, water, and nutrient balance of the blood; and regulating cellular metabolism and energy balance. As you can see, the endocrine system regulates processes that go on for relatively long periods and, in some cases, continuously.

The Endocrine System and Hormone Function—An Overview

Compared to other organs of the body, the organs of the endocrine system are small and unimpressive. Indeed, to collect 1 kg (about 2.2 pounds) of hormone-producing tissue, you would need to collect *all* the endocrine tissue from eight or nine adults! The endocrine organs also lack the structural or anatomical continuity typical of most organ systems. Instead, bits and pieces of endocrine tissue are tucked away in widely separated regions of the body (see Figure 9.3, p. 303). However, functionally the endocrine organs are very impressive, and when their role in maintaining body homeostasis is considered, they are true giants.

The Chemistry of Hormones

The key to the incredible power of the endocrine glands is the hormones they produce and secrete. *Hormones* may be defined as chemical substances that are secreted by cells into the extracellular fluids and regulate the metabolic activity of other cells in the body. Although many different hormones are produced, nearly all of them can be classified chemically as either **amino acid–based molecules** (including proteins, peptides, and amines)

Q *What determines if a hormone will influence a given body cell?*

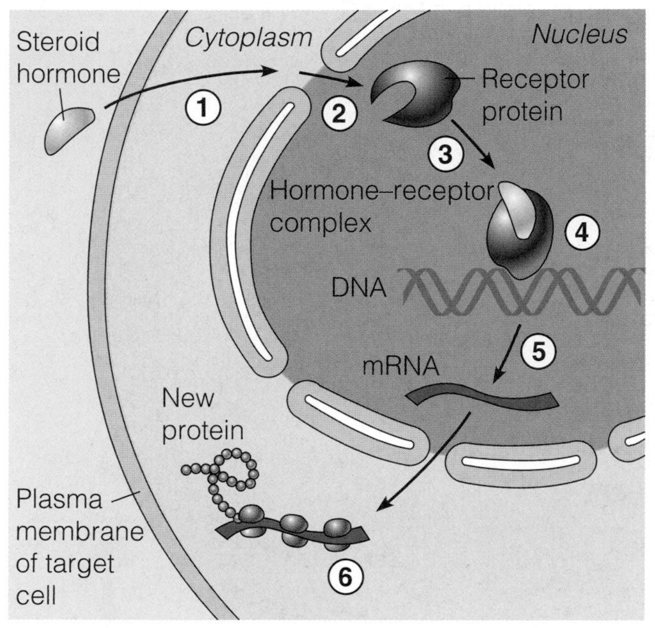

(a) Steroid hormone action

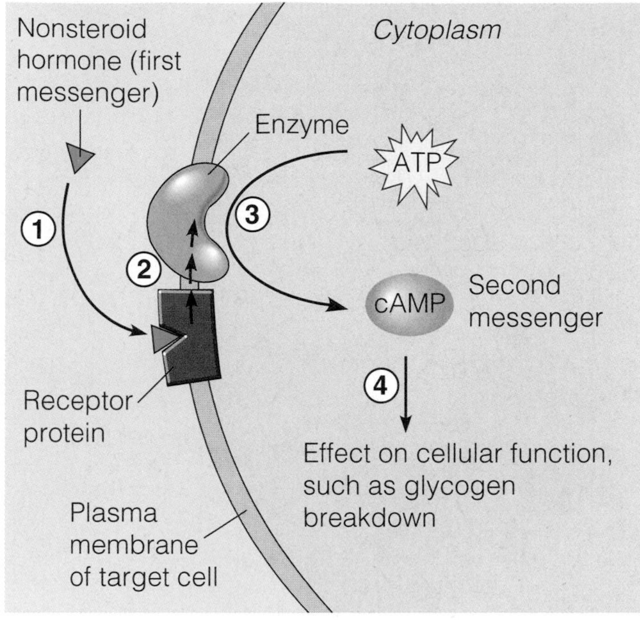

(b) Nonsteroid hormone action

FIGURE 9.1 Mechanisms of hormone action. (a) Direct gene activation: the steroid hormone mechanism. **(b)** A second messenger system: the nonsteroid (amino acid–based) hormone mechanism.

or **steroids.** Steroid hormones (made from cholesterol) include the sex hormones made by the gonads (ovaries and testes) and the hormones produced by the adrenal cortex. All others are nonsteroidal amino acid derivatives. If we also consider the local hormones called **prostaglandins** (pros″tah-glan′dinz), described later in the chapter (see Table 9.2, p. 321), we must add a third chemical class, because the prostaglandins are made from highly active lipids found in the cells' plasma membranes.

Mechanisms of Hormone Action

Although the bloodborne hormones circulate to all the organs of the body, a given hormone affects only certain tissue cells or organs, referred to as its **target cells** or **target organs.** In order for a target cell to respond to a hormone, specific protein receptors must be present on its plasma membrane or in its interior to which *that* hormone can attach.

A *A hormone can only influence a body cell if that cell has receptors for that hormone on its plasma membrane or internally.*

Only when this binding occurs can the hormone influence the workings of a cell.

The term *hormone* comes from a Greek word meaning "to arouse." In fact, the body's hormones do just that. They "arouse" or bring about their effects on the body's cells primarily by *altering* cellular activity—that is, by increasing or decreasing the rate of a normal, or usual, metabolic process rather than by stimulating a new one. The precise changes that follow hormone binding depend on the specific hormone and the target cell type, but typically one or more of the following occurs:

1. Changes in plasma membrane permeability or electrical state.
2. Synthesis of proteins or certain regulatory molecules (such as enzymes) in the cell.
3. Activation or inactivation of enzymes.
4. Stimulation of mitosis.

Despite the huge variety of hormones, there are really only two mechanisms by which hormones trigger changes in cells. Steroidal hormones (and, strangely, thyroid hormone) use the mechanism shown in Figure 9.1a. Being lipid-soluble

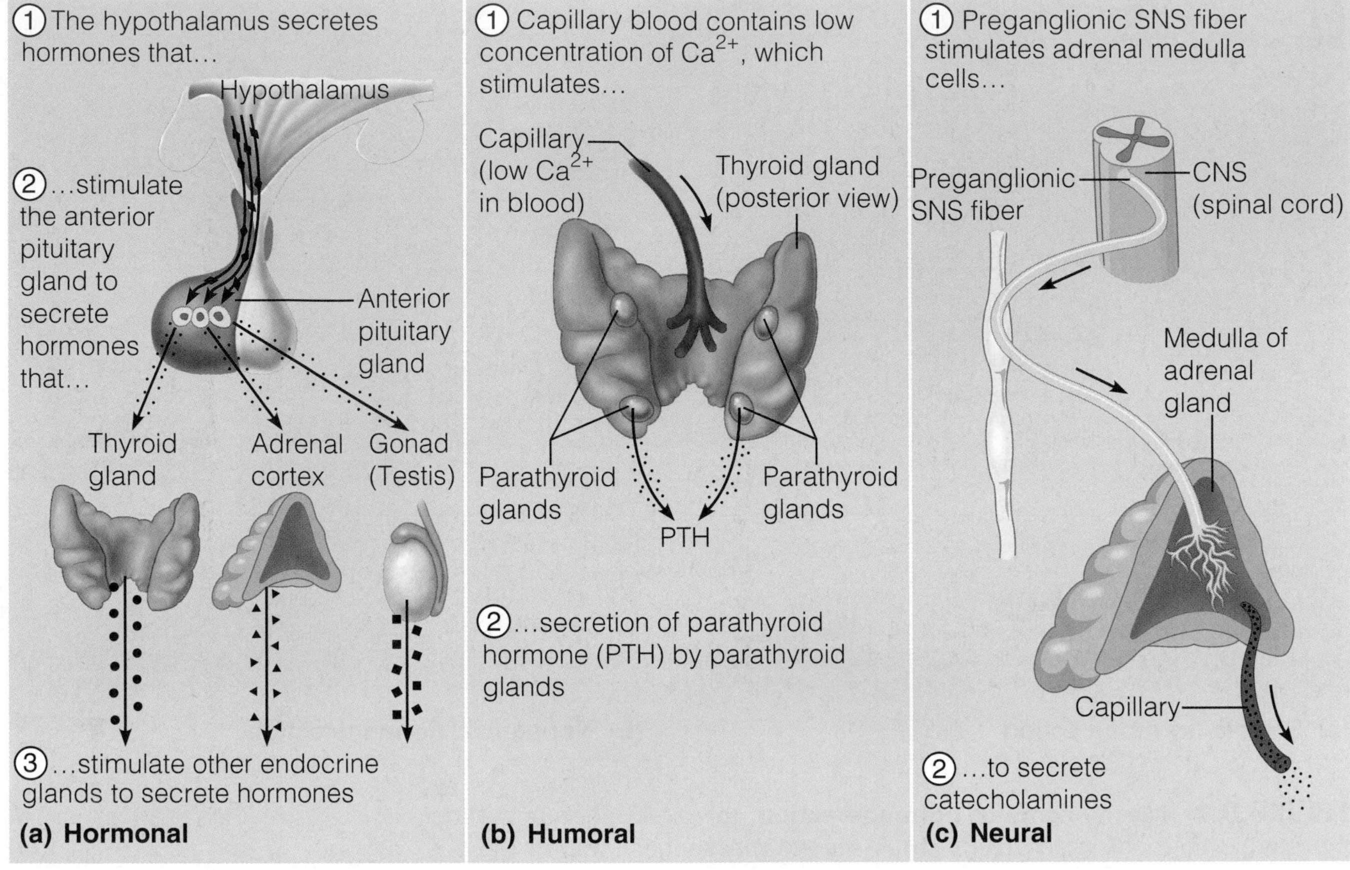

FIGURE 9.2 Endocrine gland stimuli. (a) Hormonal stimulus. In this example, hormones released by the hypothalamus stimulate the anterior pituitary to release hormones that stimulate other endocrine organs to secrete hormones. **(b)** Humoral stimulus. Low blood calcium levels trigger parathyroid hormone (PTH) release from the parathyroid glands. PTH causes blood calcium levels to rise by stimulating release of Ca^{2+} from bone. Consequently, the stimulus for PTH secretion ends. **(c)** Neural stimulus. The stimulation of adrenal medullary cells by sympathetic nervous system (SNS) fibers triggers the release of catecholamines (epinephrine and norepinephrine) to the blood.

molecules, the steroid hormones can (1) diffuse through the plasma membranes of their target cells. Once inside, the steroid hormone (2) enters the nucleus, and (3) binds to a specific receptor protein there. The hormone-receptor complex then (4) binds to specific sites on the cell's DNA, (5) activating certain genes to transcribe messenger RNA (mRNA). The mRNA then (6) is translated in the cytoplasm, resulting in the synthesis of new proteins.

Nonsteroidal hormones—protein and peptide hormones—are unable to enter the target cells. Instead they bind to receptors situated on the target cell's plasma membrane and utilize a **second-messenger system.** In these cases (Figure 9.1b), (1) the hormone binds to the membrane receptor, (2) setting off a series of reactions that activates an enzyme. The enzyme, in turn, (3) catalyzes a reaction that produces a second messenger molecule (in this case, *cyclic AMP,* also known as *cAMP* or cyclic adenine monophosphate) that (4) oversees additional intracellular changes that promote the typical response of the target cell to the hormone. As you might guess, there are a variety of possible second messengers (including *G proteins* and *calcium ions*) and many possible target cell responses to the same hormone, depending on the tissue type stimulated.

Control of Hormone Release

Now that we've discussed *how* hormones work, the next question is, "What prompts the endocrine glands to release or not release their hormones?" Let's take a look.

Negative feedback mechanisms are the chief means of regulating blood levels of nearly all hormones (see Chapter 1, p. 11). In such systems, hormone secretion is triggered by some internal or external stimulus; then rising hormone levels inhibit further hormone release (even while promoting responses in their target organs). As a result, blood levels of many hormones vary only within a very narrow range.

The stimuli that activate the endocrine organs fall into three major categories—hormonal, humoral, and neural (Figure 9.2). The most common stimulus is a *hormonal stimulus,* in which endocrine organs are prodded into action by other hormones. For example, hypothalamic hormones stimulate the anterior pituitary gland to secrete its hormones, and many anterior pituitary hormones stimulate other endocrine organs to release their hormones into the blood (Figure 9.2a). As the hormones produced by the final target glands increase in the blood, they "feed back" to inhibit the release of anterior pituitary hormones and thus their own release. Hormone release promoted by this mechanism tends to be rhythmic, with hormone blood levels rising and falling again and again.

Changing blood levels of certain ions and nutrients may also stimulate hormone release. Such stimuli are referred to as *humoral* (hyoo-mor′al) *stimuli* to distinguish them from hormonal stimuli, which are also bloodborne chemicals. The term *humoral* refers to the ancient use of the word *humor* to indicate the various body fluids (blood, bile, and others). For example, the release of parathyroid hormone (PTH) by cells of the parathyroid glands is prompted by decreasing blood calcium levels. Because PTH acts by several routes to reverse that decline, blood Ca^{2+} levels soon rise, ending the stimulus for PTH release (Figure 9.2b). Other hormones released in response to humoral stimuli include calcitonin, released by the thyroid gland, and insulin, produced by the pancreas.

In isolated cases, nerve fibers stimulate hormone release and the target cells are said to respond to *neural stimuli.* The classic example is sympathetic nervous system stimulation of the adrenal medulla to release norepinephrine and epinephrine during periods of stress (Figure 9.2c). Although these three mechanisms typify most systems that control hormone release, they by no means explain all of them, and some endocrine organs respond to many different stimuli.

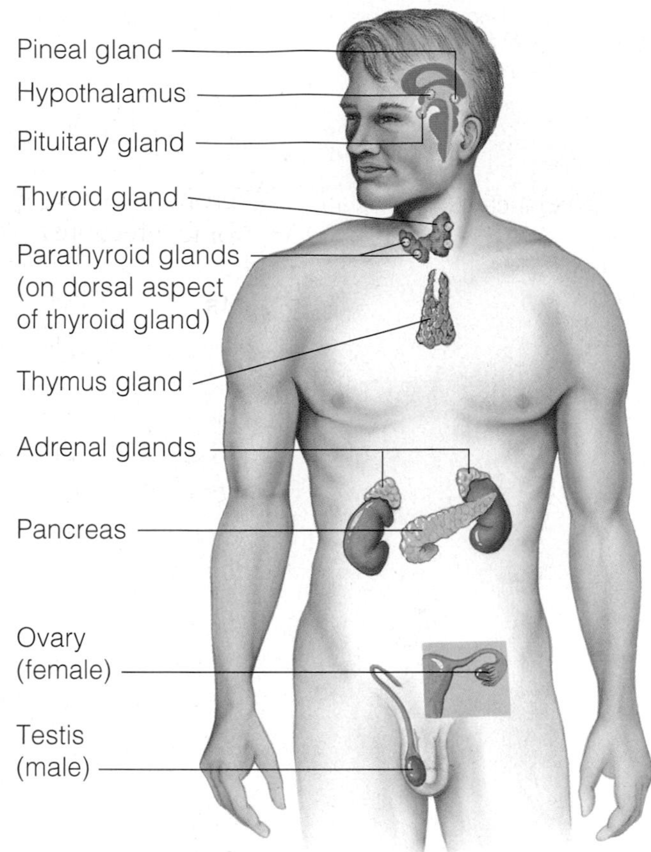

FIGURE 9.3 Location of the major endocrine organs of the body. (The parathyroid glands, which appear to be on the anterior surface of the thyroid gland in this illustration, are actually located on its posterior aspect in most cases.)

The Major Endocrine Organs

The major endocrine organs of the body include the **pituitary, thyroid, parathyroid, adrenal, pineal,** and **thymus glands,** the **pancreas,** and the gonads (**ovaries** and **testes**) (Figure 9.3). The **hypothalamus,** which is part of the nervous system, is also recognized as a major endocrine organ because it produces several hormones. Although the function

Q *What effect would elevated levels of thyroid hormone in the blood have on TSH secretion?*

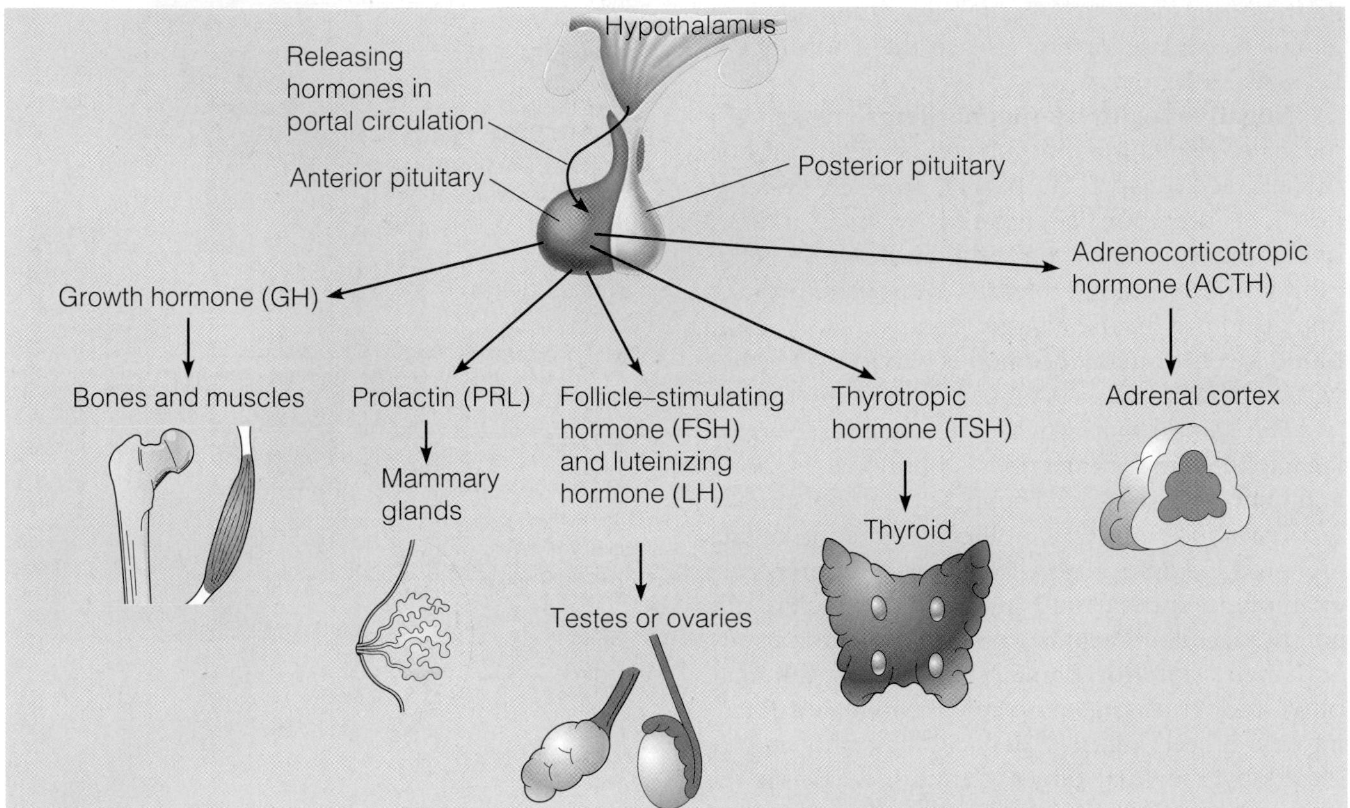

FIGURE 9.4 Hormones of the anterior pituitary and their major target organs. Secretion of anterior pituitary hormones is stimulated by releasing hormones secreted by hypothalamic neurons. The releasing hormones are secreted into a capillary network that connects via portal veins to a second capillary bed in the anterior lobe of the pituitary gland.

of some hormone-producing glands (the anterior pituitary, thyroid, adrenals, and parathyroids) is purely endocrine, the function of others (pancreas and gonads) is mixed—both endocrine and exocrine. Both types of glands are formed from epithelial tissue, but the endocrine glands are **ductless glands** that produce hormones that they release into the blood or lymph. (As you might expect, the endocrine glands have a very rich blood supply.) Conversely, the exocrine glands release their products at the body's surface or into body cavities through ducts. The formation of and differences and similarities between these two types of glands have already been discussed in Chapter 3. Here we will direct our attention to the endocrine glands only.

Besides the more detailed descriptions of the endocrine organs provided next, a summary of their hormones' main actions and regulatory factors appears in Table 9.1 (pp. 318–319).

A *It would inhibit TSH secretion by the anterior pituitary.*

Pituitary Gland

The **pituitary** (pĭ-tu′ĭ-tār″e) **gland** is approximately the size of a grape. It hangs by a stalk from the inferior surface of the hypothalamus of the brain, where it is snugly surrounded by the "Turk's saddle" of the sphenoid bone. It has two functional lobes—the anterior pituitary (glandular tissue) and the posterior pituitary (nervous tissue).

Hormones of the Anterior Pituitary

As shown in Figure 9.4, there are several anterior pituitary hormones that affect many body organs. Two of the six anterior pituitary lobe hormones—growth hormone and prolactin—exert their major effects on nonendocrine targets. The remaining four—thyrotropic hormone, adrenocorticotropic hormone, and the two gonadotropic hormones—are all **tropic** (tro′pik) **hormones.** Tropic hormones stimulate their target organs, which are also endocrine glands, to secrete their hormones, which in turn exert their effects on other body organs and tissues. All anterior pituitary hormones (1) are proteins (or peptides), (2) act through second-messenger systems, and (3) are regulated by hormonal stimuli and, in most cases, negative feedback.

Growth hormone (GH) is a general metabolic hormone. However, its major effects are directed to the growth of skeletal muscles and long bones of the body, and thus it plays an important role in determining final body size. GH is a protein-sparing and anabolic hormone that causes amino acids to be built into proteins and stimulates most target cells to grow in size and divide. At the same time, it causes fats to be broken down and used for energy while it spares glucose, helping to maintain blood sugar homeostasis.

Homeostatic Imbalance

If untreated, both deficits and excesses of GH may result in structural abnormalities. Hyposecretion of GH during childhood leads to *pituitary dwarfism.* Body proportions are fairly normal, but the person as a whole is a living miniature (with a maximum adult height of 4 feet). Hypersecretion during childhood results in *gigantism.* The individual becomes extremely tall; 8 to 9 feet is common. Again, body proportions are fairly normal. If hypersecretion occurs after long-bone growth has ended, **acromegaly** (ak″ro-meg′ah-le) results. The facial bones, particularly the lower jaw and the bony ridges underlying the eyebrows, enlarge tremendously, as do the feet and hands. Thickening of soft tissues leads to coarse or malformed facial features. Most cases of hypersecretion by endocrine organs (the pituitary and the other endocrine organs) result from tumors of the affected gland. The tumor cells act in much the same way as the normal glandular cells do; that is, they produce the hormones normally made by that gland. The use of pharmacological doses of GH to reverse some of the effects of aging is highlighted in the "A Closer Look" box on p. 306. ▲

Prolactin (PRL) is a protein hormone structurally similar to growth hormone. Its only known target in humans is the breast (*pro* = for; *lact* = milk). After childbirth, it stimulates and maintains milk production by the mother's breasts. Its function in males is not known.

Adrenocorticotropic (ad-re″no-kor″tĭ-ko-tro′pik) **hormone (ACTH)** regulates the endocrine activity of the cortex portion of the adrenal gland. **Thyroid-stimulating hormone (TSH),** also called **thyrotropic hormone (TH),** influences the growth and activity of the thyroid gland.

The **gonadotropic** (gon″ă-do-trop′ik) **hormones** regulate the hormonal activity of the **gonads** (ovaries and testes). In females, **follicle-stimulating hormone (FSH)** stimulates follicle development in the ovaries. As the follicles mature, they produce estrogen, and eggs are readied for ovulation. In males, FSH stimulates sperm development by the testes. **Luteinizing** (lu′te-in-īz″ing) **hormone (LH)** triggers ovulation of an egg from the female ovary and causes the ruptured follicle to produce progesterone and some estrogen. In men, LH is also referred to as **interstitial cell–stimulating hormone (ICSH)** because it stimulates testosterone production by the interstitial cells of the testes.

Homeostatic Imbalance

Hyposecretion of FSH or LH leads to sterility in both males and females. In general, hypersecretion does not appear to cause any problems. However, some drugs used to promote fertility stimulate the release of the gonadotropic hormones, and multiple births (indicating multiple ovulations at the same time rather than the usual single ovulation each month) are fairly common after their use. ▲

Pituitary-Hypothalamus Relationship

Despite its insignificant size, the anterior pituitary gland controls the activity of so many other endocrine glands that it has often been called the "master endocrine gland." Its removal or destruction has a dramatic effect on the body. The adrenal and thyroid glands and gonads atrophy, and results of hyposecretion by those glands quickly become obvious. However, the anterior pituitary is not as all-powerful in its control as it might appear because the release of each of its hormones is controlled by **releasing** and **inhibiting hormones** produced by the hypothalamus. The hypothalamus liberates these regulatory hormones into the blood of the

A Closer Look

Potential Uses for Growth Hormone

GROWTH hormone (GH) has been used for pharmaceutical purposes (that is, as a drug) since its discovery in the 1950s. Originally obtained from the pituitary glands of cadavers, it is now biosynthesized and administered by injection. Although GH is widely used in clinical trials, its use as a prescription drug is restricted until its helpful and harmful effects—many of which are very intriguing—can be fully documented.

GH is administered legally to children who do not produce it naturally or who have chronic kidney failure, to allow these children to grow to near-normal heights. Unfortunately, some physicians succumb to parental pressures to prescribe GH to children who *do* produce it but are extremely short.

When GH is administered to *adults* with a growth-hormone deficiency, body fat decreases and lean body mass, bone density, and muscle mass increase. It also appears to increase the performance and muscle mass of the heart, and it decreases blood cholesterol, boosts the immune system, and perhaps improves one's psychological outlook. Such effects (particularly those involving increased muscle mass and decreased body fat) have led to abuse of GH by bodybuilders and athletes, which is one reason why this substance remains restricted.

Because GH may also reverse some effects of aging, anti-aging clinics using GH injections to delay aging are springing up. Many people naturally stop producing GH after age 60, and this may explain why their ratio of lean-to-fat mass declines and their skin thins. GH already is the drug treatment of choice for many aging Hollywood stars who dread the loss of their youth and vitality. Administration of GH to elderly patients reverses these declines. However, clinical studies reveal that the administered GH does not increase strength or exercise tolerance in elderly patients, and a careful study of very sick patients in intensive care units (where GH is routinely given to restore nitrogen balance) found that high doses of GH are associated with increased mortality. For these reasons, earlier media claims that GH is a "youth potion" have proven to be dangerously misleading, and GH should not be administered to the very old or the critically ill.

GH may help AIDS patients. Because of improved antibiotics, fewer AIDS patients are dying from opportunistic infections. The other side of this picture is that more die from the weight loss called "wasting." It has been shown that injections of GH can actually reverse wasting during AIDS, leading to weight gain—a gain of lean muscle. In 1996, the U.S. Food and Drug Administration approved the use of GH to treat such wasting.

GH is not a wonder drug, even in cases where it is clearly beneficial. GH treatment is expensive and has undesirable side effects. It can lead to fluid retention and edema, joint and muscle pain, high blood sugar, glucose intolerance, and gynecomastia (breast enlargement in males). Hypertension, heart enlargement, diabetes, and cancer of the colon are other possible results of high doses of GH, and edema and headaches accompany even the lowest doses. Carefully tailored dosages can avoid most of these side effects, however.

Intensive research into the potential benefits of GH is ongoing and should keep this hormone in the public eye for years to come. Let's hope its unbridled use does not become a public health problem.

Can growth hormone help elderly patients?

portal circulation, which connects the blood supply of the hypothalamus with that of the anterior pituitary. (In a *portal circulation,* two capillary beds are connected by vein(s); in this case, the capillaries of the hypothalamus are drained by veins which empty into the capillaries of the anterior pituitary.)

The hypothalamus also makes two additional hormones, oxytocin and antidiuretic hormone, which are transported along the axons of the hypothalamic **neurosecretory cells** to the posterior pituitary for storage (Figure 9.5). They are later released into the blood in response to nerve impulses from the hypothalamus.

Hormones of the Posterior Pituitary

The posterior pituitary is not an endocrine gland in the strict sense because it *does not make* the peptide hormones it releases. Instead, as mentioned above, it simply acts as a storage area for hormones made by hypothalamic neurons.

Oxytocin (ok″se-to′sin) is released in significant amounts only during childbirth and in nursing women. It stimulates powerful contractions of the uterine muscle during labor, during sexual relations, and when a woman breast-feeds her baby. It also causes milk ejection (the *let-down reflex*) in a nursing woman. Both natural and synthetic oxytocic drugs (Pitocin and others) are used to induce labor or to hasten labor that is progressing normally but at a slow pace. Less frequently, oxytocics are used to stop postpartum bleeding (by causing constriction of the ruptured blood vessels at the placental site) and to stimulate the milk ejection reflex.

The second hormone released by the posterior pituitary is **antidiuretic** (an″ti-di″u-ret′ik) **hormone (ADH).** *Diuresis* is urine production. Thus, an antidiuretic is a chemical that inhibits or prevents urine production. ADH causes the kidneys to reabsorb more water from the forming urine; as a result, urine volume decreases and blood volume increases. In larger amounts, ADH also increases blood pressure by causing constriction of the arterioles (small arteries). For this reason, it is sometimes referred to as **vasopressin** (vas″o-pres′in).

Drinking alcoholic beverages inhibits ADH secretion and results in output of large amounts of urine. The dry mouth and intense thirst experienced "the morning after" reflect this dehydrating effect of alcohol. Certain drugs, classed together as *diuretics,* antagonize the effects of ADH, causing water to be flushed from the body. These drugs are used to manage the edema (water retention in tissues) typical of congestive heart failure.

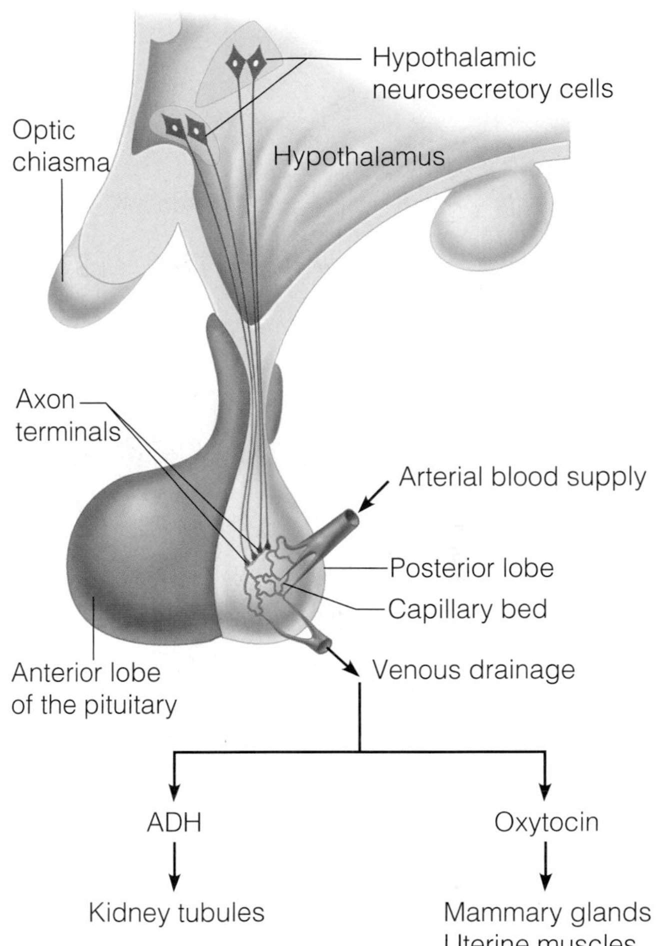

FIGURE 9.5 Hormones released by the posterior lobe of the pituitary and the target organs of such hormones. Neurosecretory cells in the hypothalamus synthesize oxytocin and antidiuretic hormone (ADH) and transport them down their axons to the posterior pituitary. There, the hormones are stored until their release is triggered by nerve impulses from the hypothalamus.

Homeostatic Imbalance

Hyposecretion of ADH leads to a condition of excessive urine output called *diabetes insipidus* (di″ah-be′tez in-sip′ĭ-dus). People with this problem are continually thirsty and drink huge amounts of water. ▲

Thyroid Gland

The **thyroid gland** is a hormone-producing gland that is familiar to most people primarily because many obese individuals blame their overweight

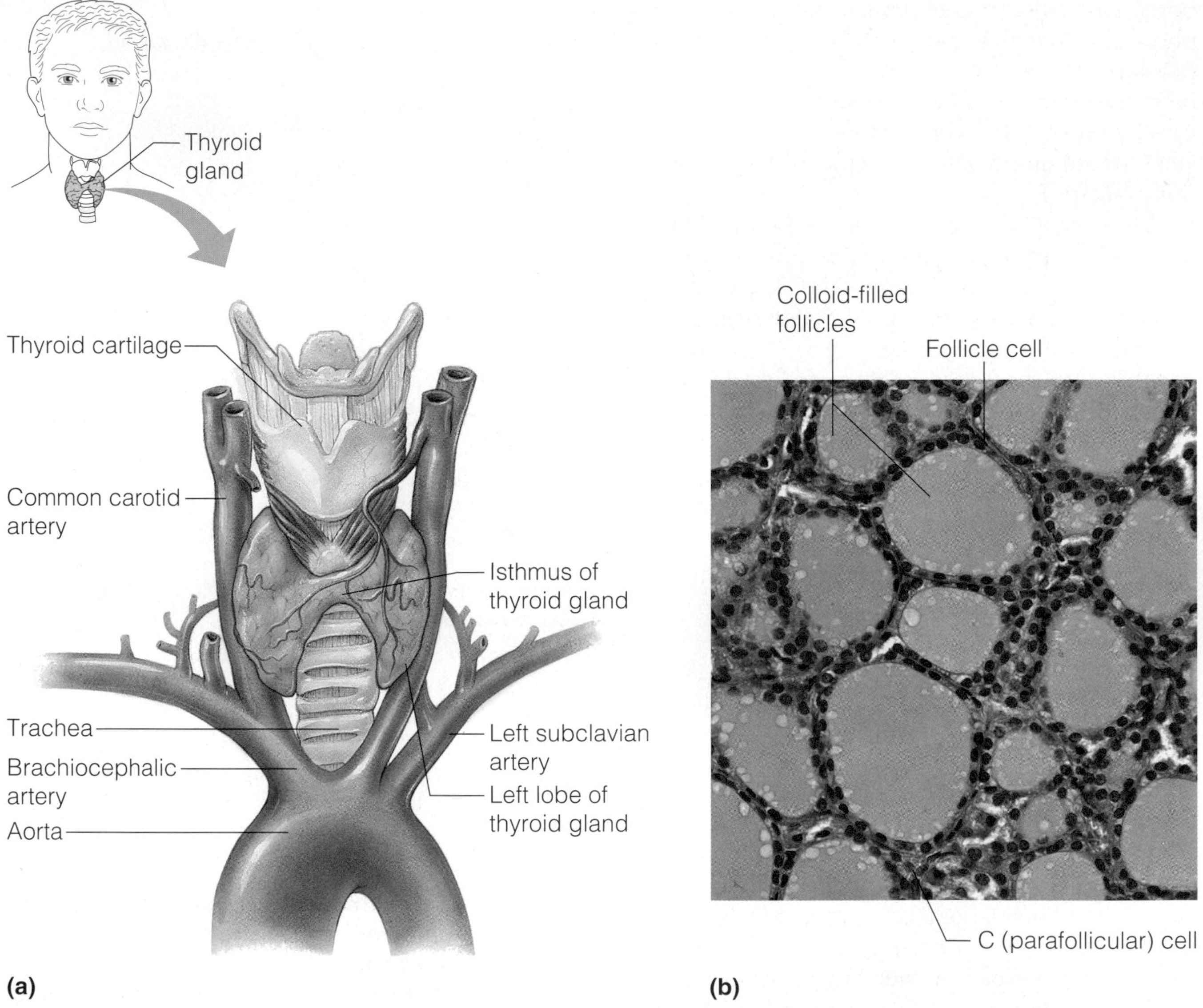

FIGURE 9.6 Anatomy of the thyroid gland. (a) Location of the thyroid gland, anterior view. **(b)** Photomicrograph of the thyroid gland (250×).

condition on their "glands" (meaning the thyroid). Actually, the effect of thyroid hormones on body weight is not as great as many believe it to be.

The thyroid gland is located at the base of the throat, just inferior to the Adam's apple, where it is easily palpated during a physical examination. It is a fairly large gland consisting of two lobes joined by a central mass, or *isthmus* (Figure 9.6). The thyroid gland makes two hormones, one called *thyroid hormone,* the other called *calcitonin.* Internally, the thyroid gland is composed of hollow structures called **follicles,** which store a sticky colloidal material (Figure 9.6b). Thyroid hormone is derived from this colloid.

Thyroid hormone, often referred to as the body's major metabolic hormone, is actually two active iodine-containing hormones, **thyroxine** (thi-rok′sin), or $\mathbf{T_4}$, and **triiodothyronine** (tri″i″o-do-thi′ro-nēn), or $\mathbf{T_3}$. Thyroxine is the major hormone secreted by the thyroid follicles. Most triiodothyronine is formed at the target tissues by conversion of thyroxine to triiodothyronine. These two hormones are very much alike. Each is constructed from two tyrosine amino acids linked together, but thyroxine has four bound iodine atoms, whereas triiodothyronine has three (thus, T_4 and T_3, respectively).

Thyroid hormone controls the rate at which glucose is "burned," or oxidized, and converted to

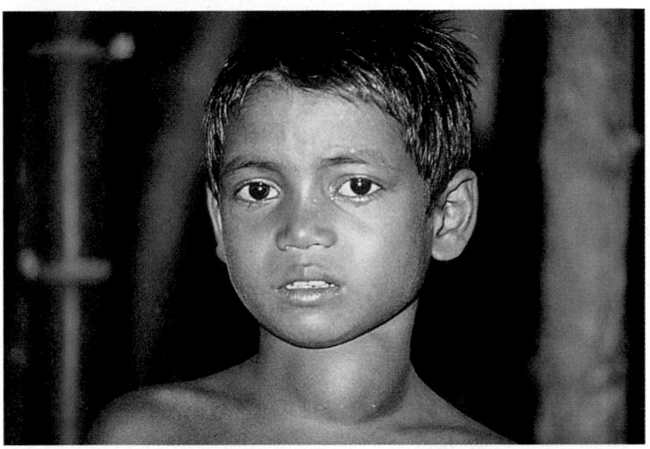

FIGURE 9.7 Goiter. An enlarged thyroid (goiter) of a boy from Bangladesh.

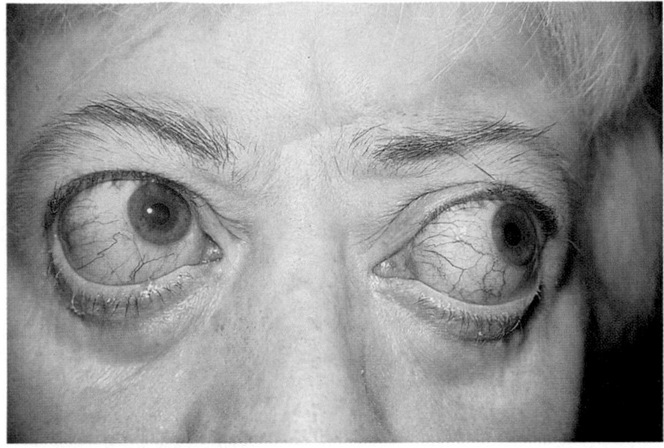

FIGURE 9.8 The exophthalmos of Graves' disease.

body heat and chemical energy. Since all body cells depend on a continuous supply of chemical energy to power their activities, every cell in the body is a target. Thyroid hormone is also important for normal tissue growth and development, especially in the reproductive and nervous systems.

Homeostatic Imbalance

Without iodine, functional hormones cannot be made. The source of iodine is our diet, and the foods richest in iodine are seafoods. Years ago many people who lived in the Midwest, in areas with iodine-deficient soil that were far from the seashore (and a supply of fresh seafood), developed **goiters** (goy'terz). That region of the country came to be known as the "goiter belt." A goiter is an enlargement of the thyroid gland (Figure 9.7) that results when the diet is deficient in iodine. TSH keeps "calling" for thyroxine, and the thyroid gland continues to enlarge so that it can put out more thyroxine. Without iodine, however, the thyroid makes only the peptide part of the molecule, which is nonfunctional and thus fails to provide negative feedback to inhibit TSH release. Simple goiter is uncommon in the United States today because most of our salt is iodized, but it is still a problem in some other areas of the world.

Hyposecretion of thyroxine may indicate problems other than iodine deficiency, such as lack of stimulation by TSH. If it occurs in early childhood, the result is *cretinism* (kre'tin-izm). Cretinism results in dwarfism in which adult body proportions remain childlike. Together the head and trunk are about 1½ times the length of the legs rather than approximately the same length, as in normal adults. Untreated cretins are mentally retarded. Their hair is scanty, and their skin is dry. If the hyposecretion problem is discovered early, hormone replacement will prevent mental retardation and other signs and symptoms of the deficiency. Hypothyroidism occurring in adults results in *myxedema* (mik"se-de'mah), in which there is both physical and mental sluggishness (however, mental retardation does not occur). Other signs are a puffiness of the face, fatigue, poor muscle tone, low body temperature (the person is always cold), obesity, and dry skin. Oral thyroxine is prescribed to treat this condition.

Hyperthyroidism generally results from a tumor of the thyroid gland. Extreme overproduction of thyroxine results in a high basal metabolic rate, intolerance of heat, rapid heartbeat, weight loss, nervous and agitated behavior, and a general inability to relax. *Graves' disease* is one form of hyperthyroidism. In addition to the symptoms of hyperthyroidism given earlier, the thyroid gland enlarges and the eyes may bulge, or protrude anteriorly (a condition called *exophthalmos* (ek"sof-thal'mos) (Figure 9.8). Hyperthyroidism may be treated surgically by removal of part of the thyroid (and/or a tumor if present) or chemically by administering thyroid-blocking drugs or radioactive iodine, which destroy some of the thyroid cells. ▲

The second important hormone product of the thyroid gland, **calcitonin,** or **thyrocalcitonin,** decreases blood calcium levels by causing calcium to be deposited in the bones. It acts antagonistically to parathyroid hormone, the hormone produced by the parathyroid glands. Whereas thyroxine is made and stored in follicles before it is released to the blood, calcitonin is made by the so-called **C (parafollicular) cells** found in the connective tissue *between* the

Q *What effect would removal of the parathyroid glands have on blood calcium levels?*

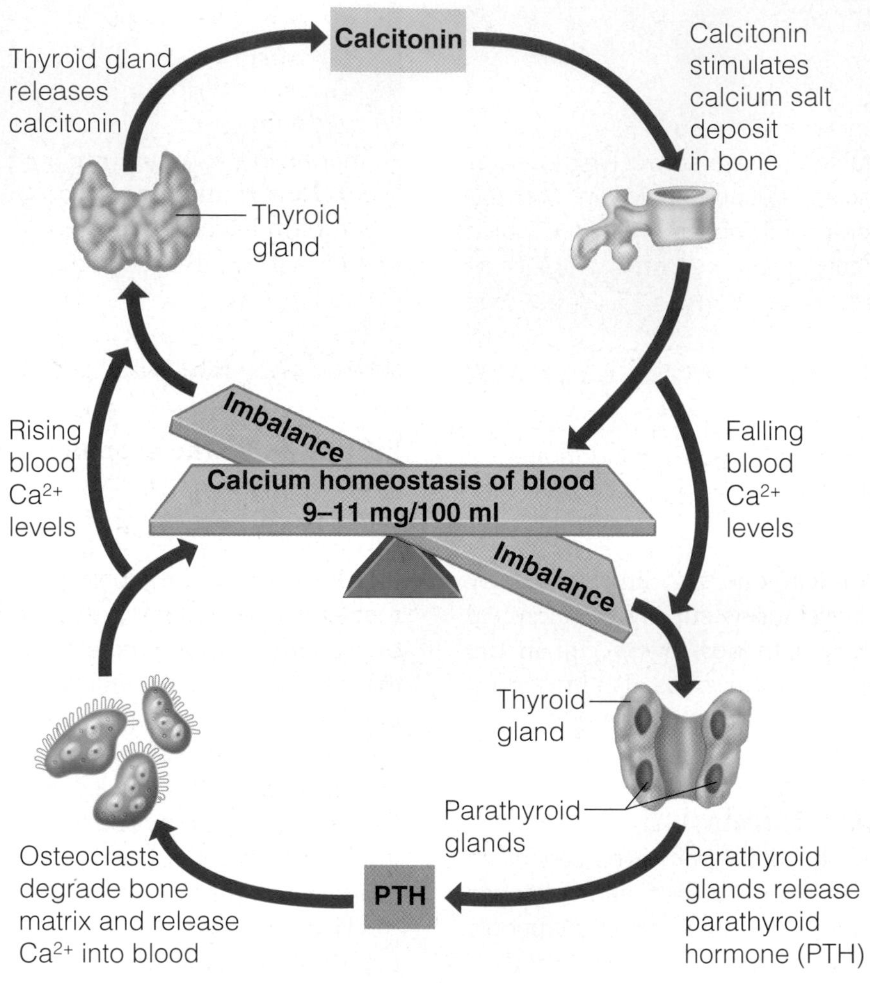

FIGURE 9.9 Hormonal controls of ionic calcium levels in the blood. PTH and calcitonin operate in negative feedback control systems that influence each other.

follicles (Figure 9.6b). It is released directly to the blood in response to increasing levels of blood calcium. Few effects of hypo- or hypersecretion of calcitonin are known, but it is believed that calcitonin production is meager or ceases in elderly adults. This may help to explain (at least in part) the progressive decalcification of bones that accompanies aging.

Parathyroid Glands

The **parathyroid glands** are tiny masses of glandular tissue most often found on the posterior surface of the thyroid gland (see Figure 9.3). Typically, there are two glands on each thyroid lobe; that is, a total of four, but as many as eight have been reported, and some may be in other regions of the neck. The parathyroids secrete **parathyroid hormone (PTH),** or **parathormone** (par″ah-thor′mōn), which is the most important regulator of calcium ion (Ca^{2+}) homeostasis of the blood. When blood calcium levels drop below a certain level, the parathyroids release PTH, which stimulates bone destruction cells (osteoclasts) to break down bone matrix and release calcium into the blood. Thus, PTH is a *hypercalcemic* hormone (that is, it acts to increase blood levels of calcium), whereas calcitonin is a *hypocalcemic* hormone. The negative feedback interaction between these two hormones as they control blood calcium level is illustrated in Figure 9.9. Although the skeleton

A *Blood calcium levels would drop because the bones would no longer be exposed to the bone-digesting stimulus of PTH.*

is the major PTH target, PTH also stimulates the kidneys and intestine to absorb more calcium (from urinary filtrate and foodstuffs, respectively).

Homeostatic Imbalance

If blood calcium levels fall too low, neurons become extremely irritable and overactive. They deliver impulses to the muscles at such a rapid rate that the muscles go into uncontrollable spasms *(tetany)*, which may be fatal. Before surgeons knew the importance of these tiny glands on the backside of the thyroid, they would remove a hyperthyroid patient's gland entirely. Many times this resulted in death. Once it was revealed that the parathyroids are functionally very different from the thyroid gland, surgeons began to leave at least some parathyroid-containing tissue (if at all possible) to take care of blood calcium homeostasis.

Severe hyperparathyroidism causes massive bone destruction—an X-ray examination of the bones shows large punched-out holes in the bony matrix. The bones become very fragile, and spontaneous fractures begin to occur. ▲

Adrenal Glands

As illustrated in Figure 9.3, the two bean-shaped **adrenal glands** curve over the top of the kidneys. Although the adrenal gland looks like a single organ, it is structurally and functionally two endocrine organs in one. Much like the pituitary gland, it has glandular (cortex) and neural tissue (medulla) parts. The central medulla region is enclosed by the adrenal cortex, which contains three separate layers of cells.

Hormones of the Adrenal Cortex

The **adrenal cortex** produces three major groups of steroid hormones collectively called **corticosteroids** (kor″ti-ko-ster′oidz)—mineralocorticoids, glucocorticoids, and sex hormones.

The **mineralocorticoids**, mainly **aldosterone** (al″dos-ter′ōn), are produced by the outermost adrenal cortex cell layer. As their name suggests, the mineralocorticoids are important in regulating the mineral (or salt) content of the blood, particularly the concentrations of sodium and potassium ions. Their target is the kidney tubules that selectively reabsorb the minerals or allow them to be flushed out of the body in urine. When blood levels of aldosterone rise, the kidney tubule cells reclaim increasing amounts of sodium ions and secrete more potassium ions into the urine. When sodium is reabsorbed, water follows. Thus, the mineralocorticoids help regulate both water and electrolyte balance in body fluids. As shown in Figure 9.10, the release of aldosterone is stimulated by humoral factors such as fewer sodium ions or more potassium ions in the blood (and by ACTH to a lesser degree). **Renin,** an enzyme produced by the kidneys when blood pressure drops, also causes the release of aldosterone by triggering a series of reactions that form **angiostensin II,** a potent stimulator of aldosterone release. A hormone released by the heart, **atrial natriuretic** (na″tre-u-ret′ik) **peptide (ANP),** prevents aldosterone release, its goal being to *reduce* blood volume and blood pressure.

The middle cortical layer produces **glucocorticoids,** which include **cortisone** and **cortisol.** Glucocorticoids promote normal cell metabolism and help the body to resist *long-term stressors,* primarily by increasing blood glucose levels. When blood levels of glucocorticoids are high, fats and even proteins are broken down by body cells and converted to glucose, which is released to the blood. For this reason, glucocorticoids are said to be *hyperglycemic hormones.* Glucocorticoids also seem to control the more unpleasant effects of inflammation by decreasing edema, and they reduce pain by inhibiting some pain-causing molecules called *prostaglandins* (see Table 9.2, p. 321). Because of their anti-inflammatory properties, glucocorticoids are often prescribed as drugs to suppress inflammation for patients with rheumatoid arthritis. Glucocorticoids are released from the adrenal cortex in response to rising blood levels of ACTH.

Regardless of one's gender, both male and female **sex hormones** are produced by the adrenal cortex throughout life in relatively small amounts. Although the bulk of the sex hormones produced by the innermost cortex layer are **androgens** (male sex hormones), some **estrogens** (female sex hormones) are also formed.

Homeostatic Imbalance

A generalized hyposecretion of all the adrenal cortex hormones leads to *Addison's disease.* A major sign of Addison's disease is a peculiar bronze tone of the skin. Because aldosterone levels are low, sodium and water are lost from the body, which leads to problems with electrolyte and water balance. This, in turn, causes the muscles to become weak, and shock is a possibility. Other signs and symptoms of Addison's

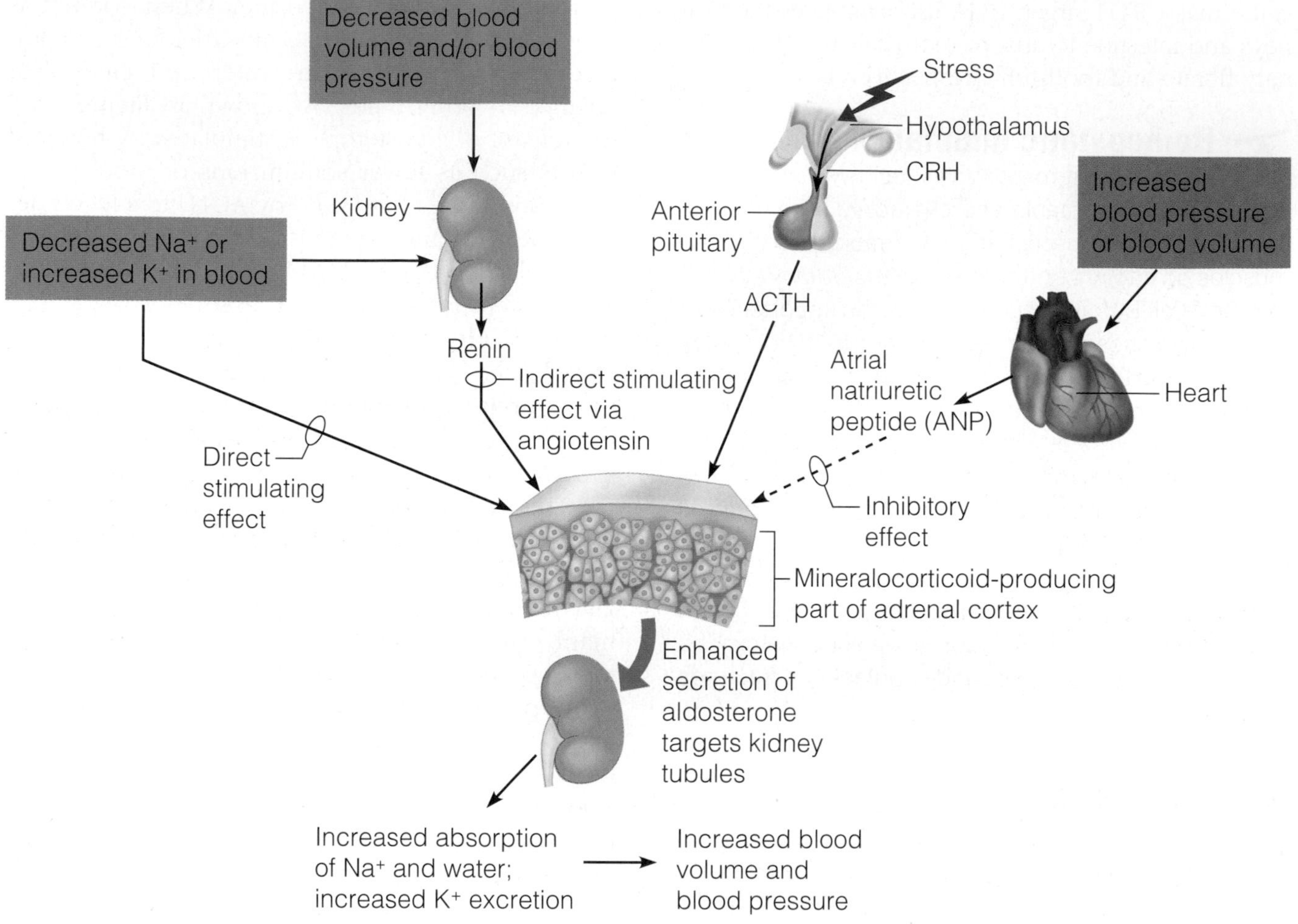

FIGURE 9.10 Major mechanisms controlling aldosterone release from the adrenal cortex. Solid arrows indicate factors that stimulate aldosterone release; dashed arrow indicates an inhibitory factor.

disease include those resulting from deficient levels of glucocorticoids, such as hypoglycemia, a lessened ability to cope with stress (burnout), and suppression of the immune system (and thereby increased susceptibility to infection). A complete lack of glucocorticoids is incompatible with life.

Hypersecretion problems are generally the result of tumors, and the resulting condition depends on the cortical area involved. Hyperactivity of the outermost cortical area results in *hyperaldosteronism* (hi"per-al"dos-ter'on-izm). Excessive water and sodium are retained, leading to high blood pressure and edema, and potassium is lost to such an extent that the activity of the heart and nervous system may be disrupted. When the tumor is in the middle cortical area, *Cushing's syndrome* occurs. Excessive output of glucocorticoids results in a "moon face" and the appearance of a "buffalo hump" of fat on the upper back (Figure 9.11). Other common and undesirable effects include high blood pressure, hyperglycemia and possible diabetes, weakening of the bones (as protein is withdrawn to be converted to glucose), and severe depression of the immune system.

Hypersecretion of the sex hormones leads to *masculinization,* regardless of sex. In adult males these effects may be masked, but in females the results are often dramatic. A beard develops, and a masculine pattern of body hair distribution occurs, among other things. ▲

Hormones of the Adrenal Medulla

The **adrenal medulla,** like the posterior pituitary, develops from a knot of nervous tissue. When the

(a)

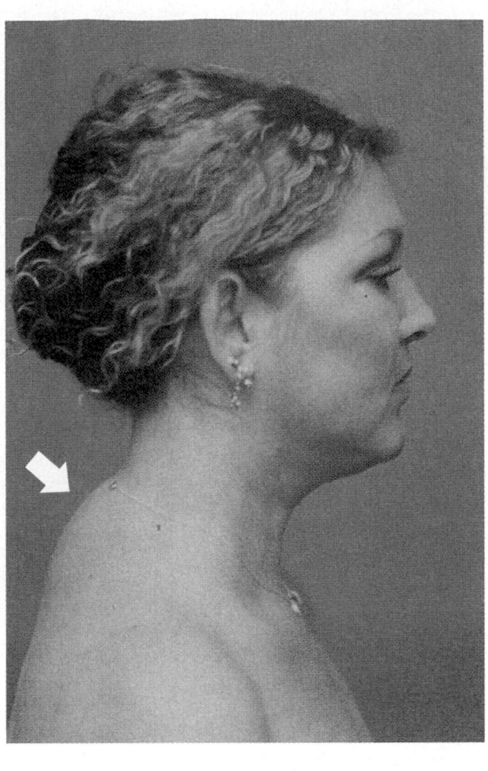

(b)

FIGURE 9.11 Appearance of a woman (a) before and (b) during Cushing's disease. The characteristic "buffalo hump" of fat on the woman's upper back is indicated by a white arrow.

medulla is stimulated by sympathetic nervous system neurons, its cells release two similar hormones, **epinephrine** (ep″ĭ-nef′rin), also called **adrenaline,** and **norepinephrine (noradrenaline),** into the bloodstream. Collectively, these hormones are referred to as **catecholamines** (kat″ĕ-kol-ah′menz). Since some sympathetic neurons also release norepinephrine as a neurotransmitter, the adrenal medulla is often thought of as a "misplaced sympathetic nervous system ganglion."

When you are (or feel) threatened physically or emotionally, your sympathetic nervous system brings about the "fight-or-flight" response to help you cope with the stressful situation. One of the organs it stimulates is the adrenal medulla, which literally pumps its hormones into the bloodstream to enhance and prolong the effects of the neurotransmitters of the sympathetic nervous system. Basically, the catecholamines increase heart rate, blood pressure, and blood glucose levels and dilate the small passageways of the lungs. These events result in more oxygen and glucose in the blood and a faster circulation of blood to the body organs (most importantly, to the brain, muscles, and heart). Thus, the body is better able to deal with a short-term stressor, whether the job at hand is to fight, begin the inflammatory process, or make you more alert so you think more clearly.

The catecholamines of the adrenal medulla prepare the body to cope with a brief or short-term stressful situation and cause the so-called *alarm stage* of the stress response. Glucocorticoids, by contrast, are produced by the adrenal cortex and are more important in helping the body to cope with prolonged or continuing stressors, such as dealing with the death of a family member or having a major operation. Glucocorticoids operate primarily during the *resistance stage* of the stress response. If they are successful in protecting the body, the problem will eventually be resolved without lasting damage to the body. When the stress continues on and on, the adrenal cortex may simply "burn out," which is

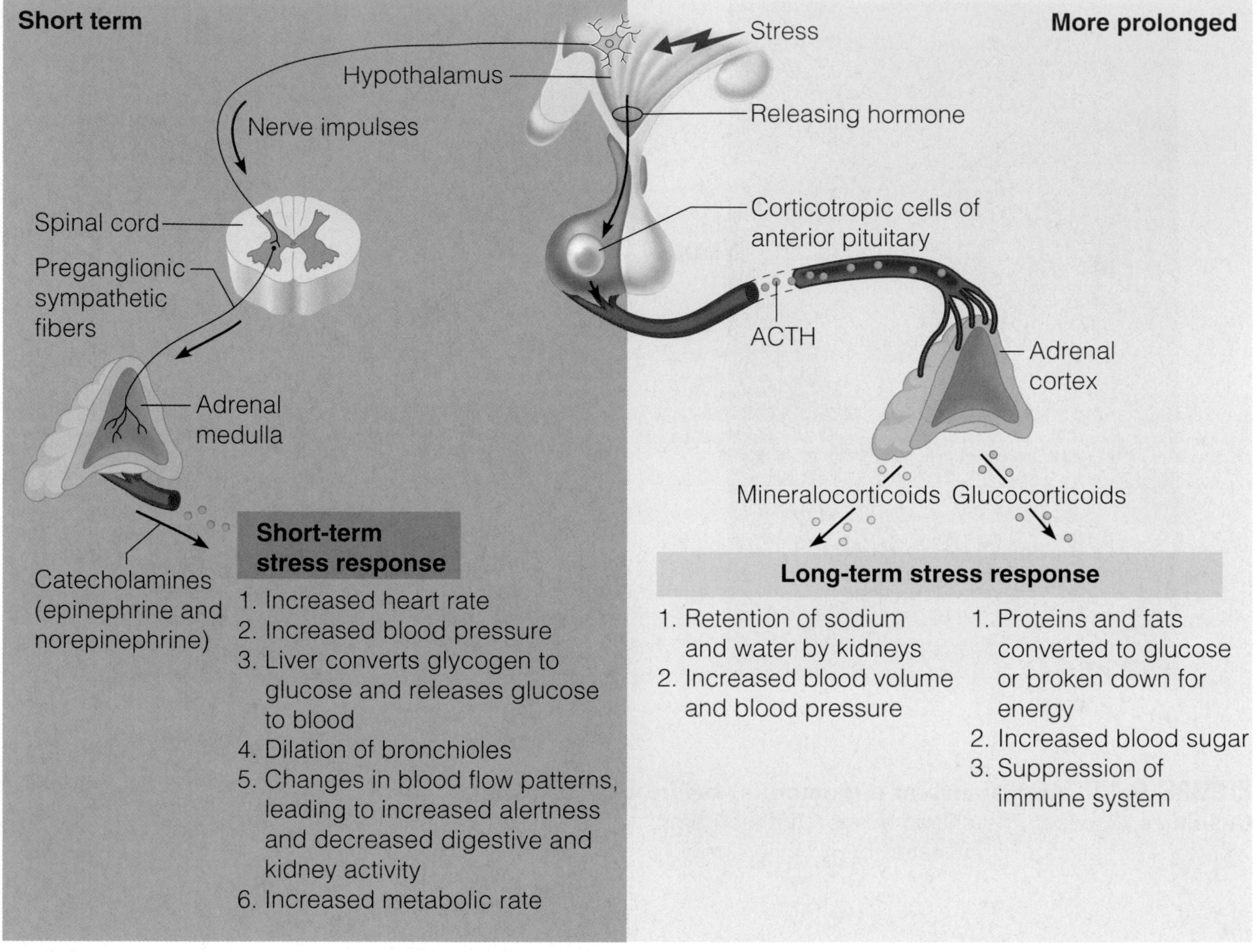

FIGURE 9.12 Roles of the hypothalamus, adrenal medulla, and adrenal cortex in the stress response. (Note that ACTH is only a weak stimulator of mineralocorticoid release under normal conditions.)

usually fatal. The relationship of catecholamines and glucocorticoids in the stress response is shown in Figure 9.12.

Homeostatic Imbalance

Damage or destruction of the adrenal medulla has no major effects as long as the sympathetic nervous system neurons continue to function normally. However, hypersecretion of catecholamines leads to symptoms typical of excessive sympathetic nervous system activity—a rapidly beating heart, high blood pressure, and a tendency to perspire and be very irritable. Surgical removal of the catecholamine-secreting cells corrects this condition. ▲

Pancreatic Islets

The **pancreas,** located close to the stomach in the abdominal cavity (see Figure 9.3), is a mixed gland. Probably the best-hidden endocrine glands in the body are the **pancreatic islets,** formerly called the *islets of Langerhans* (lahng′er-hanz). These little masses of hormone-producing tissue are scattered among the enzyme-producing acinar tissue of the pancreas. The **exocrine** (enzyme-producing) part of the pancreas, which acts as part of the digestive system, will be discussed later; only the pancreatic islets will be considered here.

Although there are more than a million islets, separated by exocrine cells, each of these tiny

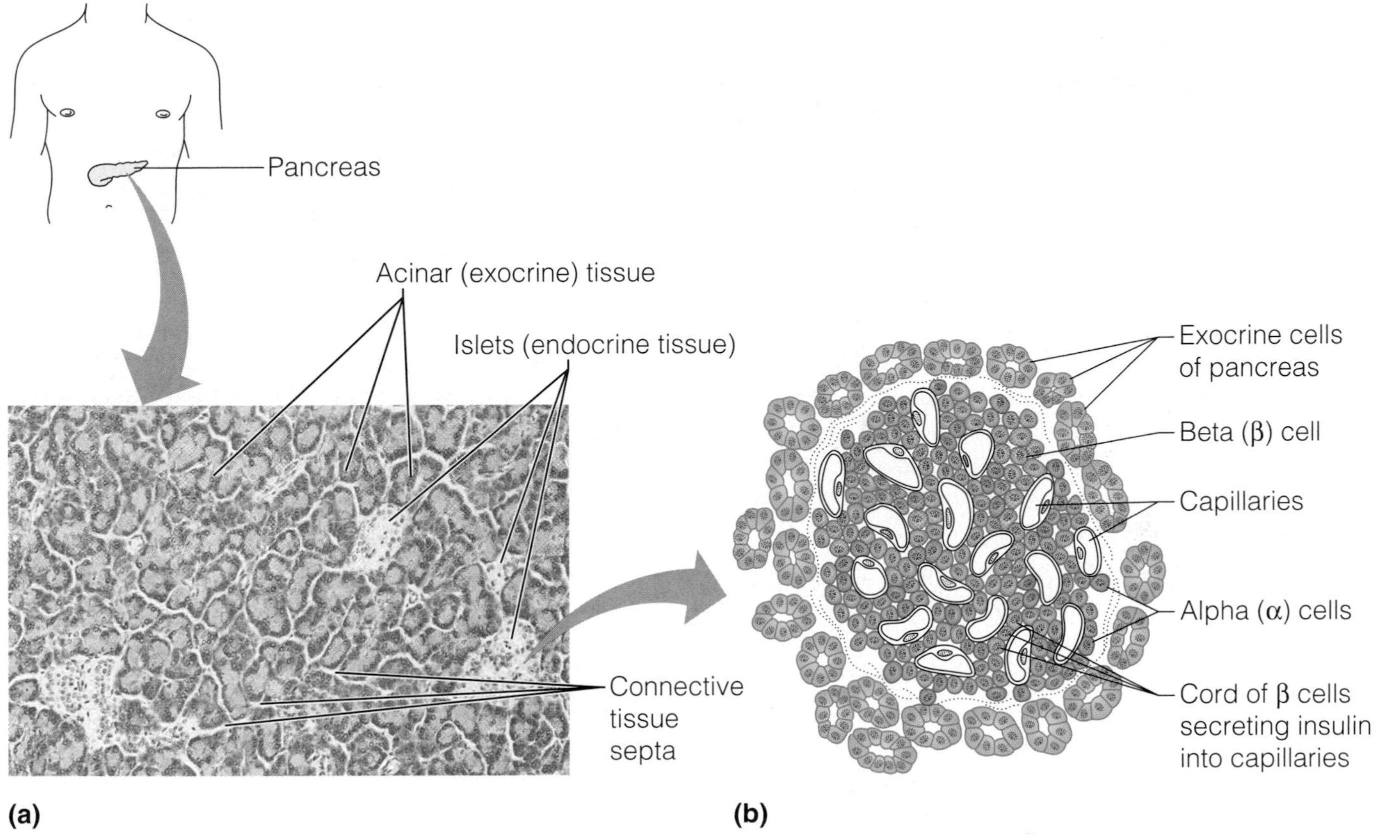

FIGURE 9.13 **Pancreatic tissue. (a)** Photomicrograph of pancreas with exocrine and endocrine (islets) areas clearly visible (110×). **(b)** Diagrammatic view of a pancreatic islet. Beta cells produce insulin, which aids cells in taking up glucose from the blood. Alpha cells produce glucagon, which stimulates liver cells to release glucose to the blood.

clumps of cells busily manufactures its hormones and works like an organ within an organ. Two important hormones produced by the islet cells are **insulin** and **glucagon** (gloo′kah-gon). The islets also produce small amounts of other hormones, but those will not be discussed here.

High levels of glucose in the blood stimulate the release of insulin from the **beta** (ba′tah) **cells** (Figure 9.13) of the islets. Insulin acts on just about all body cells and increases their ability to transport glucose across their plasma membranes. Once inside the cells, glucose is oxidized for energy or converted to glycogen or fat for storage. These activities are also speeded up by insulin. Since insulin sweeps the glucose out of the blood, its effect is said to be *hypoglycemic*. As blood glucose levels fall, the stimulus for insulin release ends—another classic case of negative feedback control. Many hormones have hyperglycemic effects (glucagon, glucocorticoids, and epinephrine, to name a few), but insulin is the only hormone that decreases blood glucose levels. Insulin is absolutely necessary for the use of glucose by the body cells. Without it, essentially no glucose can get into the cells to be used.

Homeostatic Imbalance

Without insulin, blood levels of glucose (which normally range from 80 to 120 mg/100 ml of blood) rise to dramatically high levels (for example, 600 mg/100 ml of blood). In such instances, glucose begins to spill into the urine because the kidney tubule cells cannot reabsorb it fast enough. As glucose flushes from the body, water follows, leading to dehydration. The clinical name for this condition is **diabetes mellitus** (me-li′tus), which literally means that something sweet (*mel* = honey) is passing through or siphoning (*diabetes* = Greek "siphon") from the body. Because cells cannot use glucose, fats and even proteins are broken down and used to meet the energy requirements of the body. As a result, body weight begins to decline. Loss of body proteins leads to a decreased ability to fight infections, so diabetics must be careful with their hygiene and in caring for even small cuts and bruises. When large amounts of fats (instead of sugars) are used for energy, the blood becomes very acidic (**acidosis** [as″ĭ-do′sis]) as ketones (intermediate products of fat breakdown) appear in the blood. On the basis of cause, this condition of acidosis (as″i-do′sis) is referred to as **ketosis.** Unless corrected, coma and death result. The three cardinal signs of diabetes mellitus are (1) *polyuria* (pol″e-u′re-ah)—excessive urination to flush out the glucose and ketones; (2) *polydipsia* (pol″e-dip′se-ah)—excessive thirst resulting from water loss; and (3) *polyphagia* (pol″e-fa′je-ah)—hunger due to inability to use sugars and the loss of fat and proteins from the body.

Those with mild cases of diabetes mellitus (most cases of type II, or adult-onset, diabetes) produce insulin, but for some reason their insulin receptors are unable to respond to it, a situation called **insulin resistance.** Type II diabetics are treated with special diets or oral hypoglycemic medications that prod the sluggish islets into action and increase the sensitivity of the target tissues to insulin and of beta cells to the stimulating effects of glucose. To regulate blood glucose levels in the more severe type I (juvenile, or brittle) diabetic, insulin is infused continuously by an insulin pump worn externally, or a regimen of carefully planned insulin injections is administered throughout the day. ▲

Glucagon acts as an antagonist of insulin; that is, it helps to regulate blood glucose levels but in a way opposite to that of insulin (Figure 9.14). Its release by the **alpha** (al′fah) **cells** (see Figure 9.13b) of the islets is stimulated by low blood levels of glucose. Its action is basically hyperglycemic. Its primary target organ is the liver, which it stimulates to break down stored glycogen to glucose and to release the glucose into the blood. No important disorders resulting from hypo- or hypersecretion of glucagon are known.

Pineal Gland

The **pineal** (pin′e-al) **body,** also called the **pineal gland,** is a small, cone-shaped gland found in the roof of the third ventricle of the brain (see Figure 9.3). The endocrine function of this tiny gland is still somewhat of a mystery. Although many chemical substances have been identified in the pineal gland, only the hormone **melatonin** (mel″ah-to′nin) appears to be secreted in substantial amounts. The levels of melatonin rise and fall during the course of the day and night. Peak levels occur at night and make us drowsy; the lowest levels occur during daylight around noon. Melatonin is believed to be a "sleep trigger" that plays an important role in establishing the body's day-night cycle. In some animals, melatonin also helps regulate mating behavior and rhythms. In humans, it is believed to coordinate the hormones of fertility and to inhibit the reproductive system (especially the ovaries of females) so that sexual maturation is prevented from occurring before adult body size has been reached.

Thymus Gland

The **thymus gland** is located in the upper thorax, posterior to the sternum. Large in infants and children, it decreases in size throughout adulthood. By old age, it is composed mostly of fibrous connective tissue and fat. The thymus produces a hormone called **thymosin** (thi′mo-sin), and during childhood the thymus acts as an incubator for the maturation of a special group of white blood cells (*T lymphocytes,* or *T cells*) that are important in the immune response. The role of the thymus (and its hormones) in immunity is described in Chapter 12.

Gonads

The female and male gonads (see Figure 9.3) produce sex hormones that are identical to those produced by adrenal cortex cells. The major differences are the source and relative amounts produced.

Hormones of the Ovaries

The female *gonads* (go′nadz), or **ovaries,** are paired, almond-sized organs located in the pelvic cavity. Besides producing female sex cells (ova, or eggs), ovaries produce two groups of steroid hormones, *estrogens* and *progesterone.* The ovaries do not really begin to function until puberty, when the anterior pituitary gonadotropic hormones stimulate them into activity. This results in the rhythmic ovarian cycles in which ova develop and blood levels of ovarian hormones rise and fall.

Q *What happens to the liver's ability to synthesize and store glycogen when glucagon blood levels rise?*

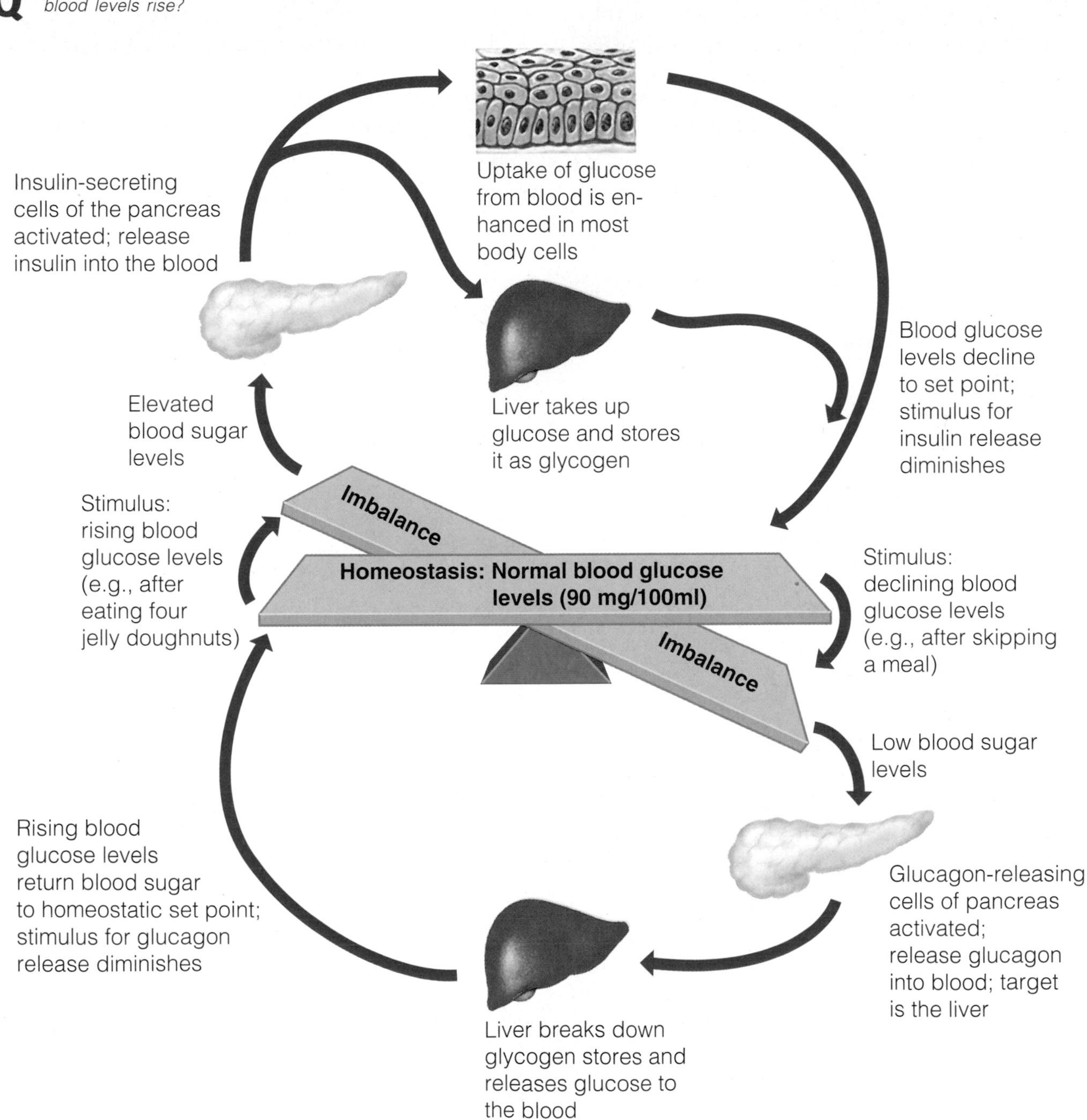

FIGURE 9.14 Regulation of blood glucose levels by a negative feedback mechanism involving pancreatic hormones.

Estrogens, primarily **estrone** (es′trōn) and **estradiol** (es″trah-di′ol), produced by the **Graafian follicles** of the ovaries, stimulate the development of the secondary sex characteristics in females (primarily growth and maturation of the reproductive organs and the appearance of hair in the pubic and axillary regions). In addition, the estrogens work with progesterone to prepare the uterus to receive a fertilized egg. This results in cyclic changes in the uterine lining, which is called the **menstrual cycle.** Estrogens also help

A Glucagon inhibits those activities in the liver, so the liver's ability to perform them would decrease as glucagon levels rise.

TABLE 9.1 Major Endocrine Glands and Some of Their Hormones

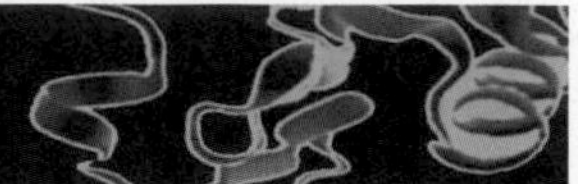

Gland	Hormone	Chemical class	Representative actions	Regulated by
Pineal body	Melatonin	Amine	Involved in biological rhythms (daily and seasonal)	Light/dark cycles
Hypothalamus	Hormones released by the posterior pituitary; releasing and inhibiting hormones that regulate the anterior pituitary (see below)			
Pituitary gland				
• Posterior lobe (releases hormones made by the hypothalamus)	Oxytocin	Peptide	Stimulates contraction of uterus and the milk "let-down" reflex	Nervous system (hypothalamus) in response to uterine stretching and/or suckling of a baby
	Antidiuretic hormone (ADH)	Peptide	Promotes retention of water by kidneys	Hypothalamus in response to water/salt imbalance
• Anterior lobe	Growth hormone (GH)	Protein	Stimulates growth (especially of bones and muscles) and metabolism	Hypothalamic releasing and inhibiting hormones
	Prolactin (PRL)	Protein	Stimulates milk production	Hypothalamic hormones
	Follicle-stimulating hormone (FSH)	Protein	Stimulates production of ova and sperm	Hypothalamic hormones
	Luteinizing hormone (LH), also called interstitial cell stimulating hormone (ICSH)	Protein	Stimulates ovaries and testes	Hypothalamic hormones
	Thyroid-stimulating hormone (TSH)	Protein	Stimulates thyroid gland	Thyroxine in blood; hypothalamic hormones
	Adrenocorticotropic hormone (ACTH)	Protein	Stimulates adrenal cortex to secrete glucocorticoids	Glucocorticoids; hypothalamic hormones
Thyroid gland	Thyroxine (T_4) and triiodothyronine (T_3)	Amine	Stimulates metabolism	TSH

TABLE 9.1 *(continued)*

Gland	Hormone	Chemical class	Representative actions	Regulated by
Thyroid gland	Calcitonin	Peptide	Reduces blood calcium level	Calcium level in blood
Parathyroid glands	Parathyroid hormone (PTH)	Peptide	Raises blood calcium level	Calcium level in blood
Thymus gland	Thymosin	Peptide	"Programs" T lymphocytes	Not known
Adrenal glands				
• Adrenal medulla	Epinephrine and norepinephrine	Amines	Raise blood glucose level; increase rate of metabolism; constrict certain blood vessels	Nervous system (sympathetic division)
• Adrenal cortex	Glucocorticoids	Steroids	Increase blood glucose	ACTH
	Mineralocorticoids	Steroids	Promote reabsorption of Na^+ and excretion of K^+ in kidneys	Changes in blood volume or blood pressure; K^+ (potassium) or Na^+ levels in blood
Pancreas	Insulin	Protein	Reduces blood glucose	Glucose level in blood
	Glucagon	Protein	Raises blood glucose	Glucose level in blood
Gonads				
• Testes	Androgens	Steroids	Support sperm formation; development and maintenance of male secondary sex characteristics	FSH and LH
• Ovaries	Estrogens	Steroids	Stimulate uterine lining growth; development and maintenance of female secondary sex characteristics	FSH and LH
	Progesterone	Steroids	Promotes growth of uterine lining	FSH and LH

maintain pregnancy and prepare the breasts to produce milk (lactation). However, the placenta and not the ovaries is the source of the estrogens at this time.

Progesterone (pro-jes′tĕ-rōn), as already noted, acts with estrogen to bring about the menstrual cycle. During pregnancy, it quiets the muscles of the uterus so that an implanted embryo will not be aborted and helps prepare breast tissue for lactation. Progesterone is produced by another glandular structure of the ovaries, the **corpus luteum** (lu′te-um). The corpus luteum produces both estrogen and progesterone, but progesterone is secreted in larger amounts.

Ovaries are stimulated to release their estrogens and progesterone in a cyclic way by the anterior pituitary gonadotropic hormones. More detail on this feedback cycle and on the structure and function of the ovaries is given in Chapter 16, but it should be obvious that hyposecretion of the ovarian hormones severely hampers the ability of a woman to conceive and bear children.

Hormones of the Testes

The paired oval **testes** of the male are suspended in a sac, the *scrotum,* outside the pelvic cavity. In addition to male sex cells, or *sperm,* the testes also produce male sex hormones, or **androgens,** of which **testosterone** (tes-tos′tĕ-rōn) is the most important. Testosterone, made by the **interstitial cells** of the testes, causes development of the adult male sex characteristics. It promotes the growth and maturation of the reproductive system organs to prepare the young man for reproduction. It also causes the male's secondary sex characteristics (growth of facial hair, development of heavy bones and muscles, and lowering of the voice) to appear and stimulates the male sex drive.

In adults, testosterone is necessary for continuous production of sperm. In cases of hyposecretion, the man becomes sterile; such cases are usually treated by testosterone injections. Both the endocrine and exocrine functions of the testes begin at puberty under the influence of the anterior pituitary gonadotropic hormones. Testosterone production is specifically stimulated by LH. Chapter 16, which deals with the reproductive system, contains more information on the structure and exocrine function of the testes.

Other Hormone-Producing Tissues and Organs

Besides the major endocrine organs, pockets of hormone-producing cells are found in fatty tissue and in the walls of the small intestine, stomach, kidneys, and heart—organs whose chief functions have little to do with hormone production. The placenta, a temporary organ formed during pregnancy, produces hormones generally thought of as ovarian hormones (estrogen and progesterone). Additionally, certain tumor cells, such as those of some lung and pancreatic cancers, make hormones identical to those made in normal endocrine glands but in an excessive and uncontrolled fashion.

Because most of these hormones are described in later chapters, their chief characteristics are only summarized in Table 9.2. Only the placenta is considered further here.

Placenta

The **placenta** (plah-sen′tah) is a remarkable organ formed temporarily in the uterus of pregnant women. In addition to its roles as the respiratory, excretory, and nutrition-delivery systems for the fetus, it also produces hormones that help to maintain the pregnancy and pave the way for delivery of the baby.

During very early pregnancy, a hormone called **human chorionic** (ko″re-on′ik) **gonadotropin (hCG)** is produced by the developing embryo and then by the fetal part of the placenta. Similar to LH (luteinizing hormone), hCG stimulates the corpus luteum of the ovary to *continue* producing estrogen and progesterone so that the lining of the uterus is not sloughed off in menses. (The home pregnancy tests sold over the counter test for the presence of hCG in the woman's urine.) In the third month, the placenta assumes the job of producing *estrogen* and *progesterone,* and the ovaries become inactive for the rest of the pregnancy. The high estrogen and progesterone blood levels maintain the lining of the uterus (thus, the pregnancy) and prepare the breasts for producing milk. *Human placental lactogen (hPL)* works cooperatively with estrogen and progesterone in preparing the breasts for lactation, and *relaxin,* another placental hormone, causes the mother's pelvic ligaments and the pubic symphysis to relax and become more flexible, which eases birth passage.

TABLE 9.2 Hormones Produced by Organs Other Than the Major Endocrine Organs

Hormone	Chemical composition	Source	Stimulus for secretion	Target organ/Effects
Prostaglandins (PGs); several groups indicated by letters A–I (PGA–PGI)	Derived from fatty acid molecules	Plasma membranes of virtually all body cells	Various (local irritation, hormones, etc.)	Have many targets, but act locally at site of release. Examples of effects include: increase blood pressure by acting as vasoconstrictors; cause constriction of respiratory passageways; stimulate muscle of the uterus promoting labor; enhance blood clotting; promote inflammation and pain; increase output of digestive secretions by stomach; cause fever
Gastrin	Peptide	Stomach	Food	*Stomach:* stimulates glands to release hydrochloric acid (HCl)
Intestinal gastrin	Peptide	Duodenum of small intestine	Food, especially fats	*Stomach:* inhibits HCl secretion and gastrointestinal tract mobility
Secretin	Peptide	Duodenum	Food	*Pancreas:* stimulates release of bicarbonate-rich juice *Liver:* increases release of bile *Stomach:* inhibits secretory activity
Cholecystokinin (CCK)	Peptide	Duodenum	Food	*Pancreas:* stimulates release of enzyme-rich juice *Gallbladder:* stimulates expulsion of stored bile *Duodenal papilla:* causes sphincter to relax, allowing bile and pancreatic juice to enter duodenum
Erythropoietin	Glycoprotein	Kidney	Hypoxia	*Bone marrow:* stimulates production of red blood cells
Active vitamin D_3	Steroid	Kidney (activates vitamin D made by epidermal cells of skin)	PTH	*Intestine:* stimulates active transport of dietary calcium across intestinal cell membranes
Atrial natriuretic peptide (ANP)	Peptide	Heart	Stretching of heart	*Kidney:* inhibits sodium ion reabsorption and renin release *Adrenal cortex:* inhibits secretion of aldosterone
Leptin	Peptide	Adipose tissue	Fatty foods	*Brain:* suppresses appetite and increases energy expenditure

Systems in Sync

Homeostatic Relationships between the Endocrine System and Other Body Systems

Endocrine System

Lymphatic System/Immunity
- Lymphocytes "programmed" by thymic hormones seed the lymph nodes; glucocorticoids depress the immune response and inflammation
- Lymph provides a route for transport of hormones

Digestive System
- Local gastro-intestinal (GI) hormones influence GI function; activated vitamin D necessary to absorb calcium from diet; catecholamines influence digestive system activity
- Digestive system provides nutrients to endocrine organs

Urinary System
- Aldosterone and ADH influence renal function; erythropoietin released by kidneys promotes red blood cell formation
- Kidneys activate vitamin D (considered a hormone)

Muscular System
- Growth hormone is essential for normal muscular development; other hormones (thyroxine and catecholamines) influence muscle metabolism
- Muscular system mechanically protects some endocrine glands; muscular activity promotes catecholamine release

Nervous System
- Many hormones (growth hormone, thyroxine, sex hormones) influence normal maturation and function of the nervous system
- Hypothalamus controls anterior pituitary function

Respiratory System
- Epinephrine influences ventilation (dilates bronchioles)
- Respiratory system provides oxygen; disposes of carbon dioxide; converting enzyme in lungs converts angiotensin I to angiotensin II

Cardiovascular System
- Several hormones influence blood volume, blood pressure, and heart contractility; erythropoietin stimulates red blood cell production
- Blood is the main transport medium of hormones; heart produces atrial natriuretic peptide

Reproductive System
- Hypothalamic, anterior pituitary, and gonadal hormones direct reproductive system development and function; oxytocin and prolactin involved in birth and breastfeeding
- Gonadal hormones feed back to influence endocrine system function

Integumentary System
- Androgens activate sebaceous glands; estrogen increases skin hydration
- Skin produces a precursor of vitamin D (cholecalciferol or provitamin D)

Skeletal System
- PTH and calcitonin regulate calcium blood levels; growth hormone, T_3, T_4, and sex hormones are necessary for normal skeletal development
- The skeleton protects some endocrine organs, especially those in brain, chest, and pelvis

Developmental Aspects of the Endocrine System

The embryonic development of the endocrine glands varies. The pituitary gland is derived from epithelium of the oral cavity and a neural tissue projection of the hypothalamus. The pineal body is entirely neural tissue. Most strictly epithelial glands develop as little saclike outpocketings of the mucosa of the digestive tract. These would the thyroid, thymus, and pancreas. Formation of the gonads and the adrenal and parathyroid glands is much more complex and is not considered here.

Barring outright malfunctions of the endocrine glands, most endocrine organs seem to operate smoothly until old age. In late middle age, the efficiency of the ovaries begins to decline, causing **menopause** (commonly called "the change of life"). During this period, a woman's reproductive organs begin to atrophy, and the ability to bear children ends. Problems associated with estrogen deficiency begin to occur, such as arteriosclerosis, osteoporosis, decreased skin elasticity, and changes in the operation of the sympathetic nervous system that result in "hot flashes." In addition, fatigue, nervousness, and mood changes such as depression are common. No such dramatic changes seem to happen in men. In fact, many men remain fertile throughout their lifespan, indicating that testosterone is still being produced in adequate amounts.

The efficiency of the endocrine system as a whole gradually declines in old age. Striking changes occur in aging women due to decreasing levels of female hormones, and there is no question that growth hormone output by the anterior pituitary declines, which partially explains muscle atrophy in old age. Elderly persons are less able to resist stress and infection. This may result from overproduction or underproduction of the defensive hormones, since both "derail" the stress defense equilibrium and alter general body metabolism. Older people are often mildly hypothyroid, and all older people have some decline in insulin production, and type II diabetes is most common in the elderly.

SUMMARY

Media study tools that could provide you with additional help in reviewing specific key topics of Chapter 9 are referenced below.

IP = Interactive Physiology;

WEB = A&P Place website.

THE ENDOCRINE SYSTEM AND HORMONE FUNCTION—AN OVERVIEW (pp. 300–303)

1. The endocrine system is a major controlling system of the body. Through hormones, it stimulates such long-term processes as growth and development, metabolism, reproduction, and body defense.

2. Endocrine organs are small and widely separated in the body. Some are part of mixed glands (both endocrine and exocrine in function). Others are purely hormone producing.

3. All hormones are amino acid–based or steroids.

4. Endocrine organs are activated to release their hormones into the blood by hormonal, humoral, or neural stimuli. Negative feedback is important in regulating hormone levels in the blood.

5. Bloodborne hormones alter the metabolic activities of their target organs. The ability of a target organ to respond to a hormone depends on the presence of receptors in or on its cells to which the hormone binds or attaches.

6. Amino acid–based hormones act through second messengers. Steroid hormones directly influence the target cell's DNA.

   IP Endocrine System; Topic: Endocrine System Review, pp. 1–5.

   WEB Exercise: Chapter 9, Hormones and Their Target Cells.

THE MAJOR ENDOCRINE ORGANS (pp. 303–320)

1. Pituitary gland
   a. The pituitary gland hangs from the base of the brain by a stalk and is enclosed by bone. It consists of a glandular (anterior) portion and a neural (posterior) portion.
   b. Except for growth hormone and prolactin, hormones of the anterior pituitary are all tropic hormones.
      (1) Growth hormone (GH): An anabolic and protein-conserving hormone that promotes total body growth. Its most important effect

is on skeletal muscles and bones. Untreated hyposecretion during childhood results in pituitary dwarfism; hypersecretion produces gigantism (in childhood) and acromegaly (in adulthood).

(2) Prolactin (PRL): Stimulates production of breast milk.

(3) Adrenocorticotropic hormone (ACTH): Stimulates the adrenal cortex to release its hormones.

(4) Thyroid-stimulating hormone (TSH): Stimulates the thyroid gland to release thyroid hormone.

(5) Gonadotropic hormones

(a) Follicle-stimulating hormone (FSH): Beginning at puberty, stimulates follicle development and estrogen production by the female ovaries; promotes sperm production in the male.

(b) Luteinizing hormone (LH): Beginning at puberty, stimulates ovulation, converts the ruptured ovarian follicle to a corpus luteum, and causes the corpus luteum to produce progesterone; stimulates the male's testes to produce testosterone.

c. Releasing and inhibiting hormones made by the hypothalamus regulate release of hormones made by the anterior pituitary. The hypothalamus also makes two hormones that are transported to the posterior pituitary for storage and later release.

IP Endocrine System; Topic: Hypothalamic-Pituitary Axis, pp. 1–3.

d. The posterior pituitary stores and releases hypothalamic hormones on command.

(1) Oxytocin: Stimulates powerful uterine contractions and causes milk ejection in the nursing woman.

(2) Antidiuretic hormone (ADH): Causes kidney tubule cells to reabsorb and conserve body water and increases blood pressure by constricting blood vessels. Hyposecretion leads to diabetes insipidus.

WEB Exercise: Chapter 9, Endocrine Case Study.

2. Thyroid gland

a. The thyroid gland is located in the anterior throat.

b. Thyroid hormone (thyroxine [T_4] and triiodothyronine [T_3]) is released from the thyroid follicles when blood levels of TSH rise. Thyroid hormone is the body's metabolic hormone. It increases the rate at which cells oxidize glucose and is necessary for normal growth and development. Lack of iodine leads to goiter. Hyposecretion of thyroxine results in cretinism in children and myxedema in adults. Hypersecretion results from Graves' disease or other forms of hyperthyroidism.

IP Endocrine System; Topic: Hormone Imbalances, p. 3.

WEB Exercise: Chapter 9, At the Clinic: Endocrine System.

c. Calcitonin is released by C cells surrounding the thyroid follicles in response to high blood levels of calcium. It causes calcium to be deposited in bones.

3. Parathyroid glands

a. The parathyroid glands are four small glands located on the posterior aspect of the thyroid gland.

b. Low blood levels of calcium stimulate the parathyroid glands to release parathyroid hormone (PTH). It causes bone calcium to be liberated into the blood. Hyposecretion of PTH results in tetany; hypersecretion leads to extreme bone wasting and fractures.

WEB Exercise: Chapter 9, Ionic Calcium Levels in the Blood.

4. Adrenal glands

a. The adrenal glands are paired glands perched on the kidneys. Each gland has two functional endocrine portions, the cortex and the medulla.

b. Adrenal cortex hormones include:

(1) Mineralocorticoids, primarily aldosterone, regulate sodium ion (Na^+) and potassium ion (K^+) reabsorption by the kidneys. Their release is stimulated primarily by low Na^+ and/or high K^+ levels in blood.

(2) Glucocorticoids enable the body to resist long-term stress by increasing blood glucose levels and depressing the inflammatory response.

(3) Sex hormones (mainly androgens) are produced in small amounts throughout life.

c. Generalized hypoactivity of the adrenal cortex results in Addison's disease. Hypersecretion can result in hyperaldosteronism, Cushing's disease, and/or masculinization.

d. The adrenal medulla produces catecholamines (epinephrine and norepinephrine) in response to sympathetic nervous system stimulation. Its catecholamines enhance and prolong the effects of the "fight-or-flight" (sympathetic nervous system) response to short-term stress. Hypersecretion leads to symptoms typical of sympathetic nervous system overactivity.

IP Endocrine System; Topic: Hormone Imbalances, pp. 4–5.

5. Pancreatic islets
   a. Located in the abdomen close to the stomach, the pancreas is both an exocrine and endocrine gland. The endocrine portion (islets) releases insulin and glucagon to blood.
   b. Insulin is released when blood levels of glucose are high. It increases the rate of glucose uptake and metabolism by body cells. Hyposecretion of insulin results in diabetes mellitus, which severely disturbs body metabolism. Cardinal signs are polyuria, polydipsia, and polyphagia.
   c. Glucagon, released when blood levels of glucose are low, stimulates the liver to release glucose to blood, thus increasing blood glucose levels.

   IP Endocrine System; Topic: Hormone Imbalances, p. 2.

   WEB Exercise: Chapter 9, Regulation of Blood Sugar Levels by Insulin and Glycogen.

6. The pineal gland, located in the third ventricle of the brain, releases melatonin, which affects biological rhythms and reproductive behavior.
7. The thymus gland, located in the upper thorax, functions during youth but atrophies in old age. Its hormone, thymosin, promotes maturation of T lymphocytes, important in body defense.
8. Gonads
   a. The ovaries of the female release:
      (1) Estrogens: Release of estrogens by ovarian follicles begins at puberty under the influence of FSH. Estrogens stimulate maturation of the female reproductive organs and female secondary sex characteristics. With progesterone, they cause the menstrual cycle.
      (2) Progesterone: Progesterone is released from the corpus luteum of the ovary in response to high blood levels of LH. It works with estrogens in establishing the menstrual cycle.
   b. The testes of the male begin to produce testosterone at puberty in response to LH stimulation. Testosterone promotes maturation of the male reproductive organs, male secondary sex characteristics, and production of sperm by the testes.
   c. Hyposecretion of gonadal hormones results in sterility in both females and males.

OTHER HORMONE-PRODUCING TISSUES AND ORGANS (pp. 320–321)

1. The placenta is a temporary organ formed in the uterus of pregnant women. Its primary endocrine role is to produce estrogen and progesterone, which maintain pregnancy and ready breasts for lactation.
2. Several organs that are generally nonendocrine in overall function, such as the stomach, small intestine, kidneys, and heart, have cells that secrete hormones.
3. Certain cancer cells secrete hormones.

DEVELOPMENTAL ASPECTS OF THE ENDOCRINE SYSTEM (p. 323)

1. Excluding pathology, efficiency of the endocrine system remains high until old age.
2. Decreasing function of female ovaries at menopause leads to such symptoms as osteoporosis, increased chance of heart disease, and possible mood changes.
3. Efficiency of all endocrine glands gradually decreases with aging, which leads to a generalized increase in incidence of diabetes mellitus, immune system depression, and lower metabolic rate.

REVIEW QUESTIONS

MULTIPLE CHOICE

More than one choice may apply.

1. The major endocrine organs of the body:
   a. tend to be very large organs
   b. are closely connected with each other
   c. all contribute to the same function (digestion)
   d. tend to lie near the midline of the body
2. Which is generally true of hormones?
   a. Exocrine glands produce them.
   b. They travel throughout the body in the blood.
   c. They affect only nonhormone-producing organs.
   d. All steroid hormones produce very similar physiological effects in the body.
3. Which of the following hormones is (are) secreted by neurons?
   a. Oxytocin c. ADH
   b. Insulin d. Cortisol
4. ANP, the hormone secreted by the heart, has exactly the opposite function to this hormone secreted by the outermost zone of the adrenal cortex.
   a. Epinephrine c. Aldosterone
   b. Cortisol d. Testosterone
5. Hormones that act directly or indirectly to elevate blood glucose include:
   a. GH c. insulin
   b. cortisol d. ACTH

6. Hypertension may result from hypersecretion of:
   a. thyroxine c. aldosterone
   b. cortisol d. ADH
7. Hormones that regulate mineral (salt) levels include:
   a. calcitonin c. atrial natriuretic peptide
   b. aldosterone d. glucagon
8. Which of the following is given as a drug to reduce inflammation?
   a. Epinephrine c. Aldosterone
   b. Cortisol d. ADH
9. The element needed for thyroid gland function is:
   a. potassium c. calcium
   b. iodine d. manganese

SHORT ANSWER ESSAY

1. Explain how the nervous and endocrine systems differ in **(a)** the rate of their control, **(b)** the way in which they communicate with body cells, and **(c)** the types of body processes they control.
2. Which endocrine organs are actually mixed (endocrine and exocrine) glands? Which are purely endocrine?
3. Define *hormone,* and describe the chemical nature of hormones.
4. Name three ways in which endocrine glands are stimulated to release their hormones, and give one example for each way.
5. Define *negative feedback,* and explain how it regulates blood levels of the various hormones.
6. Define *target organ,* and explain why all organs are not target organs for all hormones.
7. Describe the body location of each of the following endocrine organs: anterior pituitary, pineal gland, thymus, pancreas, ovaries, testes. Then, for each organ, name its hormones and their effect(s) on body processes. Finally, for each hormone, list the important results of its hypersecretion or hyposecretion.
8. Name two endocrine-producing glands (or regions) that are important in the stress response, and explain *why* they are important.
9. The anterior pituitary is often referred to as the master endocrine gland, but it too has a "master." What controls the release of hormones by the anterior pituitary?
10. What are tropic hormones?
11. The posterior pituitary is not really an endocrine gland. Why not? What is it?
12. What is the most common cause of hypersecretion by endocrine organs?
13. Name three hormone antagonists of insulin and one of PTH.
14. Two hormones are closely involved in the regulation of the fluid and electrolyte balance of the body. Name them, and explain their effects on their common target organ.
15. What causes a simple goiter?
16. In general, the endocrine system becomes less efficient as we age. List some examples of problems that elderly individuals have as a result of decreasing hormone production.

AT THE CLINIC

1. A woman with excessive body hair and a deep voice shows the outward symptoms of which hormonal dysfunction?
2. The parents of 14-year-old Megan are concerned about her height because she is only 4 feet tall and they are both close to 6 feet tall. After tests by their doctor, certain hormones are prescribed for the girl. What is the probable diagnosis? What hormones are prescribed, and explain why the girl might expect to reach normal height?
3. Paula, a 28-year-old, has been in the first stage of labor for 15 hours. Her uterine contractions are weak, and her labor is not progressing normally. Since Paula and her doctor desire a vaginal delivery, the physician orders that Pitocin (a synthetic oxytocin) be infused. What is the effect of this hormone?
4. Mr. Holdt brings his wife to the clinic, concerned about her nervousness, heart palpitations, and excessive sweating. Tests show hyperglycemia and hypertension. What hormones are probably being hypersecreted? What is the cause? What physical factors allow you to rule out thyroid problems?
5. What are the possible harmful effects of using anabolic steroids to increase muscle mass and strength?
6. Bertha Wise, age 40, comes to the clinic, troubled by swelling in her face and unusual fat deposition on her back and abdomen. She reports that she bruises easily. Blood tests show elevated glucose levels. What is your diagnosis, and what glands might be causing the problem?
7. Maryanne, a street person, is pregnant. She has had no prenatal care and her diet consists of what she is able to salvage from trash cans. What could you surmise about the PTH levels in her blood?

10 Blood

YOUR GOALS

After completing this chapter, you will have a working knowledge of the functions of blood and will have mastered the objectives listed below.

FUNCTION PREVIEW

- Blood serves as a vehicle for distributing body heat and for transporting nutrients, respiratory gases, and other substances throughout the body.

OBJECTIVE CHECKLIST

COMPOSITION AND FUNCTIONS OF BLOOD (pp. 328–335)

- Indicate the composition and volume of whole blood.
- Describe the composition of plasma, and discuss its importance in the body.
- List the cell types making up the formed elements, and describe the major functions of each type.
- Define *anemia, polycythemia, leukopenia,* and *leukocytosis,* and list possible causes for each condition.
- Explain the role of the hemocytoblast.

HEMOSTASIS (pp. 335–338)

- Describe the blood-clotting process.
- Name some factors that may inhibit or enhance the blood-clotting process.

BLOOD GROUPS AND TRANSFUSIONS (pp. 338–341)

- Describe the ABO and Rh blood groups.
- Explain the basis for a transfusion reaction.

>>

(Objective Checklist, continued)

DEVELOPMENTAL ASPECTS OF BLOOD (pp. 341–344)

- Explain the basis of physiologic jaundice seen in some newborn babies.
- Indicate blood disorders that increase in frequency in the aged.

Blood is the "river of life" that surges within us. It transports everything that must be carried from one place to another within the body—nutrients, wastes (headed for elimination from the body), and body heat—through blood vessels. For centuries, long before modern medicine, people recognized that blood was vital (some believed "magical"), and its loss was always considered to be a possible cause of death. In this chapter, we consider the composition and function of this life-sustaining fluid. The means by which it is propelled throughout the body is discussed in Chapter 11.

Composition and Functions of Blood

Components

Among all of the body's tissues, blood is unique: It is the only *fluid* tissue. Although blood appears to be a thick, homogeneous liquid, the microscope reveals it has both solid and liquid components. Essentially, blood is a complex connective tissue in which living blood cells, the **formed elements,** are suspended in a nonliving fluid matrix called **plasma** (plaz′muh).

If a sample of blood is spun in a centrifuge, the heavier formed elements are packed down by centrifugal force and the plasma rises to the top (Figure 10.1). Most of the reddish mass at the bottom of the tube consists of *erythrocytes* (eh-rith′ro-sīts), the red blood cells that function in oxygen transport. Although it is barely visible in Figure 10.1, there is a thin, whitish layer called the **buffy coat** at the junction between the formed elements and the plasma. This layer contains *leukocytes* (lu′ko-sīts; *leuko* = white), the white blood cells that act in various ways to protect the body, and *platelets,* cell fragments that function in the blood-clotting process. Erythrocytes normally account for about 45 percent of the total volume of a blood sample, a percentage known as the **hematocrit.** White blood cells and platelets contribute less than 1 percent, and plasma makes up most of the remaining 55 percent of whole blood.

Physical Characteristics and Volume

Blood is a sticky opaque fluid with a characteristic metallic taste. As children, we discover its saltiness the first time we stick a cut finger into our mouth. Depending on the amount of oxygen it is carrying, the color of blood varies from scarlet (oxygen-rich) to a dull red (oxygen-poor). Blood is heavier than water and about five times thicker, or more viscous, largely because of its formed elements. Blood is slightly alkaline, with a pH between 7.35 and 7.45. Its temperature (38°C, or 100.4°F) is always slightly higher than body temperature.

Blood accounts for approximately 8 percent of body weight, and its volume in healthy males is 5 to 6 liters, or approximately 6 quarts.

Plasma

Plasma, which is approximately 90 percent water, is the liquid part of the blood. Over 100 different substances are dissolved in this straw-colored fluid. Examples of dissolved substances include nutrients, salts (electrolytes), respiratory gases, hormones, plasma proteins, and various wastes and products of cell metabolism.

Plasma proteins are the most abundant solutes in plasma. Except for antibodies and protein-based hormones, most plasma proteins are made by the

Q *How would a decrease in the amount of plasma proteins affect plasma volume?*

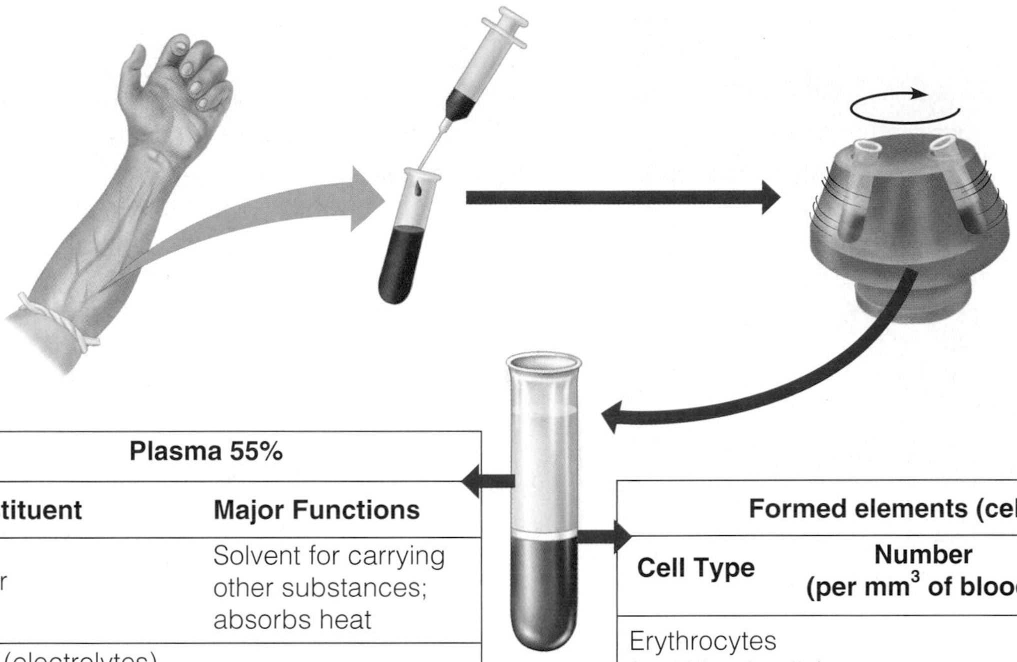

Plasma 55%	
Constituent	**Major Functions**
Water	Solvent for carrying other substances; absorbs heat
Salts (electrolytes) Sodium Potassium Calcium Magnesium Chloride Bicarbonate	Osmotic balance, pH buffering, regulation of membrane permeability
Plasma proteins Albumin	Osmotic balance, pH buffering
Fibrinogen	Clotting of blood
Globulins	Defense (antibodies) and lipid transport
Substances transported by blood Nutrients (glucose, fatty acids, amino acids, vitamins) Waste products of metabolism (urea, uric acid) Respiratory gases (O_2 and CO_2) Hormones	

Formed elements (cells) 45%		
Cell Type	**Number (per mm^3 of blood)**	**Functions**
Erythrocytes (red blood cells)	4 – 6 million	Transport oxygen and help transport carbon dioxide
Leukocytes (white blood cells) Basophil Eosinophil Neutrophil Lymphocyte Monocyte	4000 – 11,000	Defense and immunity
Platelets	250,000 – 500,000	Blood clotting

FIGURE 10.1 The composition of blood.

Plasma proteins create the osmotic pressure that helps to maintain plasma volume and draws leaked fluid back into the circulation. Hence, a decrease in the amount of plasma proteins would result in a reduced plasma volume.

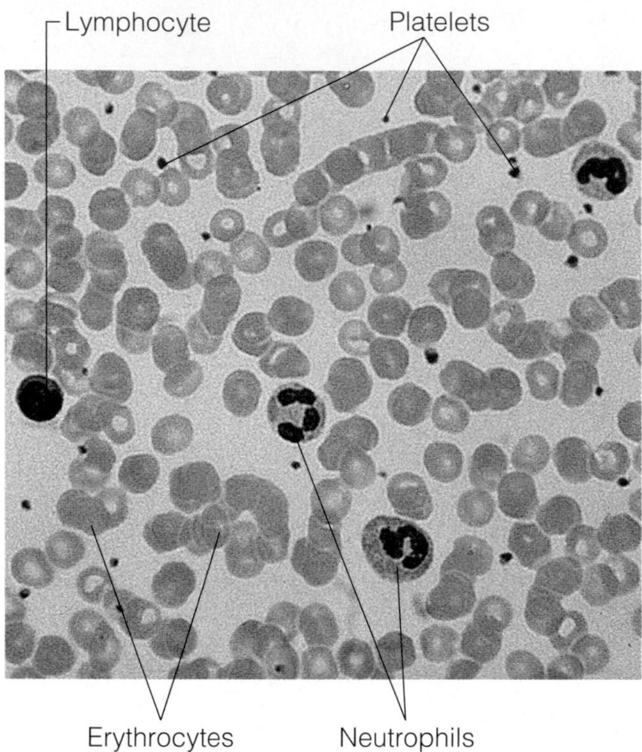

FIGURE 10.2 Photomicrograph of a blood smear. Most of the cells in this view are erythrocytes (red blood cells). Two kinds of leukocytes (white blood cells) are also present: lymphocytes and neutrophils. Also note the platelets.

liver. The plasma proteins serve a variety of functions. For instance, **albumin** (al-bu'min) contributes to the osmotic pressure of blood, which acts to keep water in the bloodstream; clotting proteins help stem blood loss when a blood vessel is injured; and antibodies help protect the body from pathogens. Plasma proteins are *not* taken up by cells to be used as food fuels or metabolic nutrients, as are other solutes such as glucose, fatty acids, and oxygen.

The composition of plasma varies continuously as cells remove or add substances to the blood. Assuming a healthy diet, however, the composition of plasma is kept relatively constant by various homeostatic mechanisms of the body. For example, when blood proteins drop to undesirable levels, the liver is stimulated to make more proteins; when the blood starts to become too acid *(acidosis)* or too basic *(alkalosis)*, both the respiratory system and the kidneys are called into action to restore it to its normal, slightly alkaline pH range of 7.35 to 7.45. Various body organs make literally dozens of adjustments day in and day out to maintain the many plasma solutes at life-sustaining levels. Besides transporting various substances around the body, plasma helps to distribute body heat evenly throughout the body.

Formed Elements

If you observe a stained smear of human blood under a light microscope, you will see smooth, disc-shaped red blood cells, a variety of gaudily stained white blood cells, and, most likely, some scattered platelets that look like debris (Figure 10.2). However, erythrocytes vastly outnumber the other types of formed elements. Table 10.2 on p. 333 provides a summary of the important characteristics of the various formed elements that make up about 45 percent of whole blood.

Erythrocytes

Erythrocytes, or **red blood cells (RBCs),** function primarily to ferry oxygen in blood to all cells of the body. They are superb examples of the "fit" between cell structure and function. RBCs differ from other blood cells because they are *anucleate* (a-nu'kle-at); that is, they lack a nucleus. They also contain very few organelles. In fact, mature RBCs circulating in the blood are literally sacs of hemoglobin molecules. **Hemoglobin** (he"mo-glo'bin) **(Hb),** an iron-bearing protein, transports the bulk of the oxygen that is carried in the blood. (It also binds with a small amount of carbon dioxide.) Moreover, because erythrocytes lack mitochondria and make ATP by anaerobic mechanisms, they do not use up any of the oxygen they are transporting, making them very efficient oxygen transporters indeed.

Erythrocytes are small cells shaped like biconcave discs—flattened discs with depressed centers on both sides (see Figures 10.2 and 10.3a). Because of their thinner centers, they look like miniature doughnuts when viewed with a microscope. Their small size and peculiar shape provide a large surface area relative to their volume, making them ideally suited for gas exchange.

RBCs outnumber white blood cells by about 1000 to 1 and are the major factor contributing to blood viscosity. Although the numbers of RBCs in the circulation do vary, there are normally about 5 million cells per cubic millimeter of blood. (A cubic millimeter [mm^3] is a very tiny drop of blood, almost not enough to be seen.) When the number of RBC/mm^3 increases, blood viscosity increases. Similarly, as the number of RBCs decreases, blood thins and flows more rapidly. However, let's not get carried away talking about RBC *numbers*. Although their numbers are important, it is the amount of hemoglobin in the bloodstream at any time that really determines how well the erythrocytes are performing their role of oxygen transport.

The more hemoglobin molecules the RBCs contain, the more oxygen they will be able to carry. Perhaps the most accurate way of measuring the oxygen-carrying capacity of the blood is to determine how much hemoglobin it contains. Since a single red blood cell contains about 250 million hemoglobin molecules, each capable of binding 4 molecules of oxygen, each of these tiny cells can carry about 1 billion molecules of oxygen! This information is astounding but not very practical. Much more important clinically is the fact that normal blood contains 12–18 grams (g) hemoglobin per 100 milliliters (ml) blood. The hemoglobin content is slightly higher in men (13–18 g) than in women (12–16 g).

Homeostatic Imbalance

A decrease in the oxygen-carrying ability of the blood, whatever the reason, is **anemia** (ah-ne′me-ah). Anemia may be the result of (1) a lower-than-normal *number* of RBCs, or (2) abnormal or deficient *hemoglobin content* in the RBCs. Several types of anemia are classified and described briefly in Table 10.1, but one of these, *sickle cell anemia,* deserves a little more attention because people with this genetic disorder are frequently seen in hospital emergency rooms.

In sickle cell anemia (SCA), the abnormal hemoglobin formed becomes spiky and sharp (Figure 10.3b) when the RBCs unload oxygen molecules or when the oxygen content of the blood is lower than normal, as during vigorous exercise, anxiety, or other stressful situations. The deformed (crescent-shaped) erythrocytes rupture easily and dam up in small blood vessels. These events interfere with oxygen delivery and cause extreme pain. It is amazing that this havoc results from a change in just *one* of the amino acids in each of the beta chains of the globin molecule!

Sickle cell anemia occurs chiefly in black people who live in the malaria belt of Africa and among their descendants. Apparently, the same gene that causes sickling makes red blood cells infected by the malaria-causing parasite stick to the capillary walls and then lose potassium, an essential nutrient for survival of the parasite. Hence, the malaria-causing parasite is prevented from multiplying within the red blood cells, and individuals with the sickle cell gene have a better chance of surviving where malaria is prevalent. Only

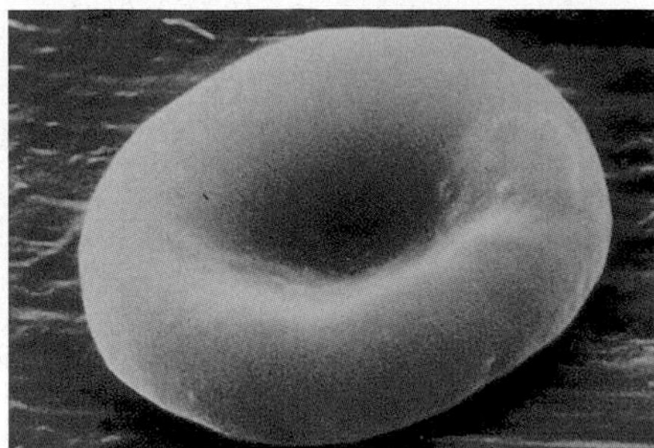

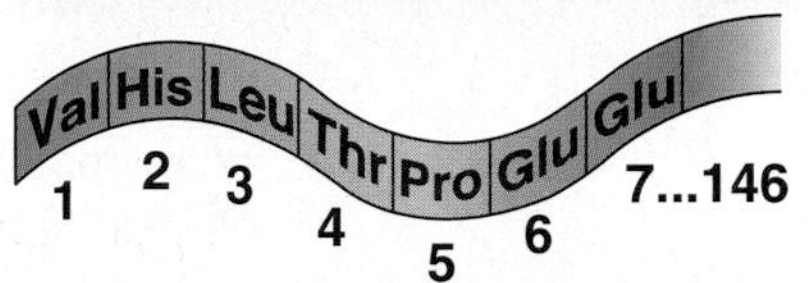

(a) Normal hemoglobin

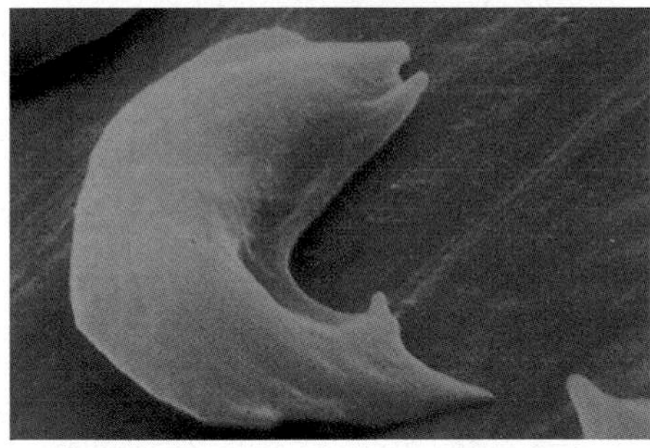

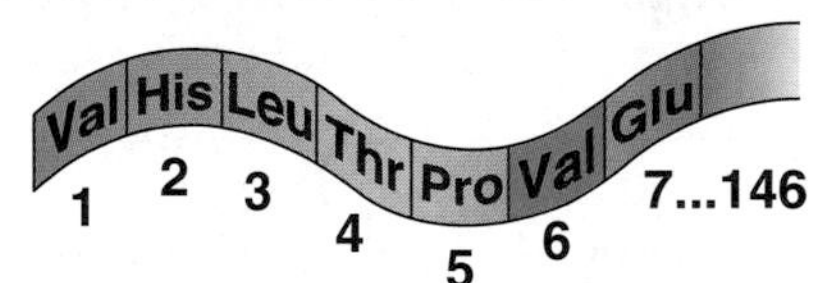

(b) Sickle cell hemoglobin

FIGURE 10.3 Comparison of (a) normal erythrocyte to a (b) sickled erythrocyte (31,500×).

those carrying two copies of the defective gene have sickle cell anemia. Those carrying just one sickling gene have *sickle cell trait (SCT);* they generally do not display the symptoms but can pass on the sickling gene to their offspring.

An excessive or abnormal increase in the number of erythrocytes is **polycythemia** (pol″e-si-the′me-ah). Polycythemia may result from bone marrow cancer *(polycythemia vera).* It may also be a normal physiologic (homeostatic) response to living at high altitudes where the air is thinner and less oxygen is available *(secondary polycythemia).* The major problem that results from excessive numbers of RBCs is increased blood viscosity, which causes it to flow sluggishly in the body and impairs circulation. ▲

TABLE 10.1 Types of Anemia

Direct cause	Resulting from	Leading to
Decrease in RBC number	Sudden hemorrhage	Hemorrhagic anemia
	Lysis of RBCs as a result of bacterial infections	Hemolytic (he″mo-lit′ik) anemia
	Lack of vitamin B_{12} (usually due to lack of intrinsic factor required for absorption of the vitamin; intrinsic factor is formed by stomach mucosa cells)	Pernicious (per-nish′us) anemia
	Depression/destruction of bone marrow by cancer, radiation, or certain medications	Aplastic anemia
Inadequate hemoglobin content in RBCs	Lack of iron in diet or slow/prolonged bleeding (such as heavy menstrual flow or bleeding ulcer), which depletes iron reserves needed to make hemoglobin; RBCs are small and pale because they lack hemoglobin	Iron deficiency anemia
Abnormal hemoglobin in RBCs	Genetic defect leads to abnormal hemoglobin, which becomes sharp and sickle-shaped under conditions of increased oxygen use by body; occurs mainly in people of African descent	Sickle cell anemia

Leukocytes

Although **leukocytes,** or **white blood cells (WBCs),** are far less numerous than red blood cells, they are crucial to body defense against disease. On average, there are 4000 to 11,000 WBCs/mm^3, and they account for less than 1 percent of total blood volume. White blood cells are the only complete cells in blood; that is, they contain nuclei and the usual organelles.

Leukocytes form a protective, movable army that helps defend the body against damage by bacteria, viruses, parasites, and tumor cells. As such, they have some very special characteristics. Red blood cells are confined to the bloodstream and carry out their functions in the blood. White blood cells, by contrast, are able to slip into and out of the blood vessels—a process called *diapedesis* (di″ah-peh-de′sis; "leaping across"). The circulatory system is simply their means of transportation to areas of the body where their services are needed for inflammatory or immune responses (as described in Chapter 12).

In addition, WBCs can locate areas of tissue damage and infection in the body by responding to certain chemicals that diffuse from the damaged cells. This capability is called *positive chemotaxis* (ke″mo-tax′is). Once they have "caught the scent," the WBCs move through the tissue spaces by *ameboid* (ah-me′boid) *motion* (they form flowing cytoplasmic extensions that help move them along). By following the diffusion gradient, they pinpoint areas of tissue damage and rally round in large numbers to destroy microorganisms or dead cells.

Whenever WBCs mobilize for action, the body speeds up their production, and as many as twice the normal number of WBCs may appear in the blood within a few hours. A total WBC count above 11,000 cells/mm^3 is referred to as **leukocytosis** (lu″ko-si-to′sis). Leukocytosis generally indicates that a bacterial or viral infection is stewing in the body. The opposite condition, **leukopenia** (lu″ko-pe′ne-ah), is an abnormally low WBC count. It is commonly caused by certain drugs, such as corticosteroids and anticancer agents.

Homeostatic Imbalance

Leukocytosis is a normal and desirable response to infectious threats to the body. By contrast, the excessive production of abnormal WBCs that

TABLE 10.2 Characteristics of Formed Elements of the Blood

Cell type	Occurrence in blood (per mm³)	Cell anatomy*	Function
Erythrocytes (red blood cells, or RBCs)	4–6 million	Salmon-colored biconcave disks; anucleate; literally, sacs of hemoglobin; most organelles have been ejected	Transport oxygen bound to hemoglobin molecules; also transport small amount of carbon dioxide
Leukocytes (white blood cells, or WBCs)	4000–11,000		
Granulocytes			
• Neutrophils	3000–7000 (40–70% of WBCs)	Cytoplasm stains pale pink and contains fine granules, which are difficult to see; deep purple nucleus consists of three to seven lobes connected by thin strands of nucleoplasm	Active phagocytes; number increases rapidly during short-term or acute infections
• Eosinophils	100–400 (1–4% of WBCs)	Red coarse cytoplasmic granules; figure-8 or bilobed nucleus stains blue-red	Kill parasitic worms; increase during allergy attacks; might phagocytize antigen-antibody complexes and inactivate some inflammatory chemicals
• Basophils	20–50 (0–1% of WBCs)	Cytoplasm has a few large blue-purple granules; U- or S-shaped nucleus with constrictions, stains dark blue	Granules contain histamine (vasodilator chemical), which is discharged at sites of inflammation
Agranulocytes			
• Lymphocytes	1500–3000 (20–45% of WBCs)	Cytoplasm pale blue and appears as thin rim around nucleus; spherical (or slightly indented) dark purple-blue nucleus	Part of immune system; one group (B lymphocytes) produces antibodies; other group (T lymphocytes) involved in graft rejection, fighting tumors and viruses, and activating B lymphocytes
• Monocytes	100–700 (4–8% of WBCs)	Abundant gray-blue cytoplasm; dark blue-purple nucleus often kidney-shaped	Active phagocytes that become macrophages in the tissues; long-term "clean-up team"; increase in number during chronic infections such as tuberculosis
Platelets	250,000–500,000	Essentially irregularly shaped cell fragments; stain deep purple	Needed for normal blood clotting; initiate clotting cascade by clinging to broken area; help to control blood loss from broken blood vessels

*Appearance when stained with Wright's stain.

occurs in infectious mononucleosis and leukemia is distinctly pathological. In **leukemia** (lu-ke′me-ah), literally "white blood," the bone marrow becomes cancerous, and huge numbers of WBCs are turned out rapidly. Although this might not appear to present a problem, the "newborn" WBCs are immature and incapable of carrying out their normal protective functions. Consequently, the body becomes the easy prey of disease-causing bacteria and viruses. ▲

WBCs are classified into two major groups, depending on whether or not they contain visible granules in their cytoplasm. Specific characteristics of the leukocytes are listed in Table 10.2. Microscopic views can be seen in Figure 10.1.

Granulocytes (gran′u-lo-sītz″) are granule-containing WBCs. They have lobed nuclei, which typically consist of several rounded nuclear areas connected by thin strands of nuclear material. The granules in their cytoplasm stain specifically with Wright's stain. The granulocytes include the neutrophils (nu′tro-filz), eosinophils (e″o-sin′o-filz), and basophils (ba′so-filz).

1. **Neutrophils** have a multilobed nucleus and very fine granules that respond to both acid and basic stains. Consequently, the cytoplasm as a whole stains pink. Neutrophils are avid phagocytes at sites of acute infection.
2. **Eosinophils** have a blue-red nucleus that resembles an old-fashioned telephone receiver and sport large brick-red cytoplasmic granules. Their number increases rapidly during allergies and infections by parasitic worms (flatworms, tapeworms, etc.).
3. **Basophils,** the rarest of the WBCs, contain large histamine-containing granules that stain dark blue. **Histamine** is an inflammatory chemical that makes blood vessels leaky and attracts other WBCs to the inflammatory site.

The second group, **agranulocytes,** lack visible cytoplasmic granules. Their nuclei are closer to the norm—that is, they are spherical, oval, or kidney-shaped. The agranulocytes include lymphocytes (lim′fo-sīts) and monocytes (mon′o-sīts).

1. **Lymphocytes** have a large dark purple nucleus that occupies most of the cell volume. Only slightly larger than RBCs, lymphocytes tend to take up residence in lymphatic tissues, where they play an important role in the immune response.
2. **Monocytes** are the largest of the WBCs. Except for their more abundant cytoplasm and indented nucleus, they resemble large lymphocytes. When they migrate into the tissues, they change into macrophages with huge appetites. Macrophages are very important in fighting chronic infections, such as tuberculosis.

Platelets

Platelets are not cells in the strict sense. They are fragments of bizarre multinucleate cells called **megakaryocytes** (meg″ah-kar′e-o-sītz), which pinch off thousands of anucleate platelet "pieces" that quickly seal themselves off from the surrounding fluids. The platelets appear as darkly staining, irregularly shaped bodies scattered among the other blood cells. The normal platelet count in blood is about 300,000/mm^3. As indicated in Table 10.2, platelets are needed for the clotting process that occurs in plasma when blood vessels are ruptured or broken. (This process is explained on pp. 335–337.)

Hematopoiesis (Blood Cell Formation)

Blood cell formation, or **hematopoiesis** (hem″ah-to-poi-e′sis), occurs in red bone marrow, or *myeloid* tissue. In adults, this tissue is found chiefly in the flat bones of the skull and pelvis, the ribs, sternum, and proximal epiphyses of the humerus and femur. Each type of blood cell is produced in different numbers in response to changing body needs and different stimuli. After they mature, they are discharged into the blood vessels surrounding the area.

All the formed elements arise from a common type of *stem cell,* the **hemocytoblast** (he″mo-si′to-blast), which resides in the red bone marrow. Their development differs, however, and once a cell is committed to a specific blood pathway it cannot change. As indicated in the flowchart in Figure 10.4, the hemocytoblast forms two types of descendants—the *lymphoid stem cell,* which produces lymphocytes, and the *myeloid stem cell,* which can produce all other classes of formed elements.

Because they are anucleate, RBCs are unable to synthesize proteins, grow, or divide. As they age, RBCs become more rigid and begin to fragment, or fall apart, in 100 to 120 days. Their remains are eliminated by phagocytes in the spleen, liver, and other body tissues. Lost cells are replaced more or less continuously by the division of hemocytoblasts in the red bone marrow. The developing RBCs divide many times and then begin synthesizing huge amounts of hemoglobin. Suddenly, when enough hemoglobin has been accumulated, the nucleus and most organelles are ejected and the cell collapses inward. The result is the young RBC, called a *reticulocyte* (rĕ-tik′u-lo-sīt) because it still contains some rough endoplasmic reticulum (ER). The reticulocytes enter the bloodstream to begin their task of transporting oxygen. Within 2 days of release, they have ejected the remaining ER and have become fully functioning erythrocytes. The entire developmental process from hemocytoblast to mature RBC takes 3 to 5 days.

The rate of erythrocyte production is controlled by a hormone called **erythropoietin** (ĕ-rith″ro-poi-e′tin). Normally a small amount of erythropoietin circulates in the blood at all times, and red blood cells are formed at a fairly constant rate. Although the liver produces some, the kidneys play the major role in producing this hormone. When blood levels of oxygen begin to decline for any reason, the kidneys step up their release of erythropoietin (Figure 10.5). Erythropoietin targets the bone marrow, prodding it into "high gear" to turn out more RBCs. As you might expect, an overabundance of erythrocytes, or an excessive amount of oxygen in the bloodstream, depresses erythropoietin release and red blood cell production. An important point to remember is that it is *not* the relative number of RBCs in the blood that controls RBC production. Control is based on their ability to transport enough oxygen to meet the body's demands.

Like erythrocyte production, the formation of leukocytes and platelets is stimulated by hormones. These *colony stimulating factors (CSFs)* and *interleukins* not only prompt red bone marrow to turn out leukocytes, but also marshal up an army of WBCs to ward off attacks by enhancing the ability of mature leukocytes to protect the body. Apparently, they are released in response to specific chemical signals in the environment such as inflammatory chemicals and certain bacteria or their toxins. The hormone *thrombopoietin* accelerates the production of platelets, but little is known about how that process is regulated.

When bone marrow problems or disease conditions such as aplastic anemia or leukemia are suspected, a special needle is used to withdraw a small sample of red marrow from one of the flat bones (ilium or sternum) close to the body surface. This procedure provides cells for a microscopic examination called a *bone marrow biopsy.*

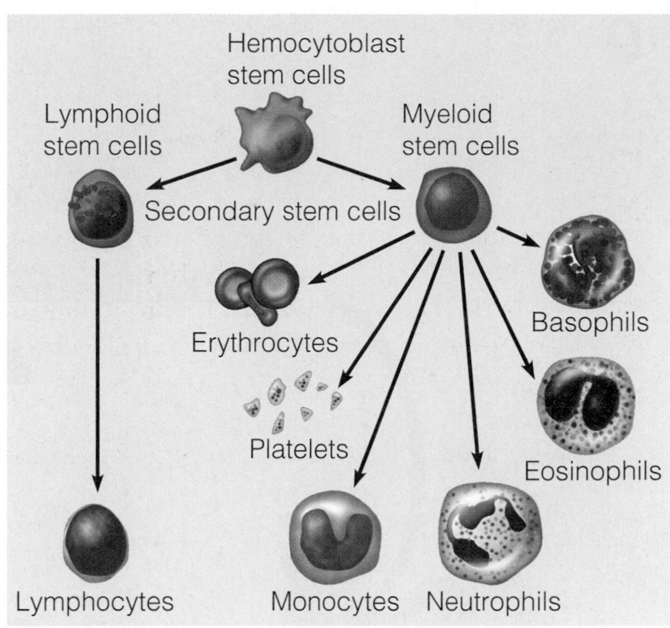

FIGURE 10.4 The development of blood cells. All blood cells differentiate from a common source, hemocytoblast stem cells in red bone marrow. The population of stem cells renews itself by mitosis. Some of these cells become lymphoid stem cells, which then develop into two classes of lymphocytes that function in the immune response. All other blood cells differentiate from myeloid stem cells, which are also derived from the stem cells.

Hemostasis

Normally, blood flows smoothly past the intact lining (endothelium) of blood vessel walls. But if a blood vessel wall breaks, a series of reactions is set in motion to accomplish **hemostasis** (*hem* = blood; *stasis* = standing still), or stoppage of blood flow. This response, which is fast and localized, involves many substances normally present in plasma, as well as some that are released by platelets and injured tissue cells.

Hemostasis involves three major phases, which occur in rapid sequence: **platelet plug formation, vascular spasms,** and **coagulation,** or **blood clotting.** Blood loss at the site is permanently prevented when fibrous tissue grows into the clot and seals the hole in the blood vessel.

Q *Why do many people with advanced kidney disease become anemic?*

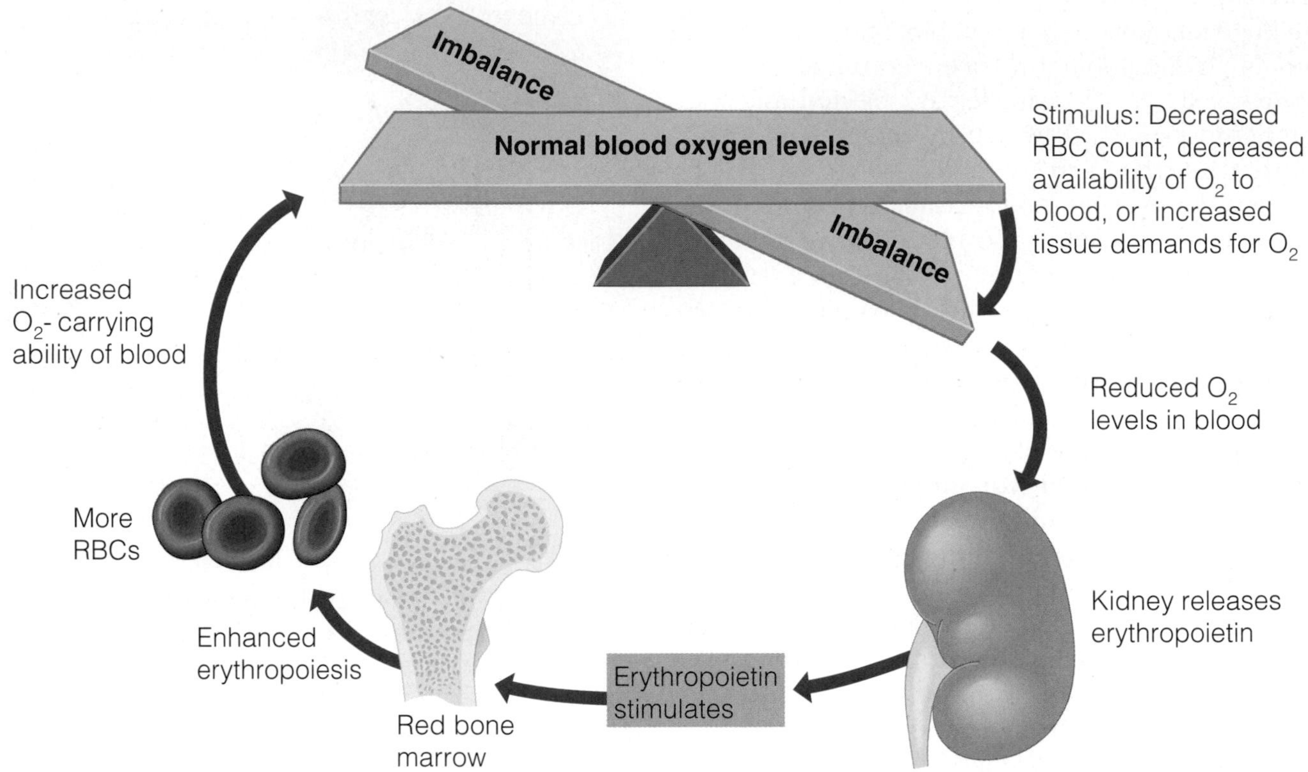

FIGURE 10.5 Mechanism for regulating the rate of RBC production. Increased erythropoietin release, which stimulates RBC production in bone marrow, occurs when oxygen levels in the blood become inadequate to support normal cellular activity, whatever the cause.

Basically, hemostasis occurs as follows (Figure 10.6):

1. **Platelet plug forms.** Platelets are repelled by an intact endothelium, but when it is broken so that the underlying collagen fibers are exposed, the platelets become "sticky" and cling to the damaged site. Anchored platelets release chemicals that attract more platelets to the site, and as more and more platelets pile up, a small mass called a *platelet plug,* or *white thrombus,* is formed.
2. **Vascular spasms occur.** The anchored platelets also release **serotonin** (ser″o-to′nin), which causes that blood vessel to go into spasms. The spasms narrow the blood vessel at that point, decreasing blood loss until clotting can occur. (Other factors causing vessel spasms include direct injury to the smooth muscle cells and stimulation of local pain receptors.)
3. **Coagulation events occur.**

   a. At the same time, the injured tissues are releasing **tissue factor (TF),** a substance that plays an important role in clotting.

   b. **PF_3,** a phospholipid that coats the surfaces of the platelets, interacts with TF, vitamin K and other blood protein clotting factors, and calcium ions (Ca^{2+}) to form an activator that triggers the *clotting cascade.*

   c. This **prothrombin activator** converts **prothrombin** (pro-throm′bin), present in the plasma, to **thrombin,** an enzyme.

   d. Thrombin then joins soluble **fibrinogen** (fi-brin′o-jen) proteins into long hairlike molecules of insoluble **fibrin,** which forms a meshwork that traps the RBCs and forms the basis of

A *The kidneys produce most of the erythropoietin that stimulates red blood cell production by the bone marrow.*

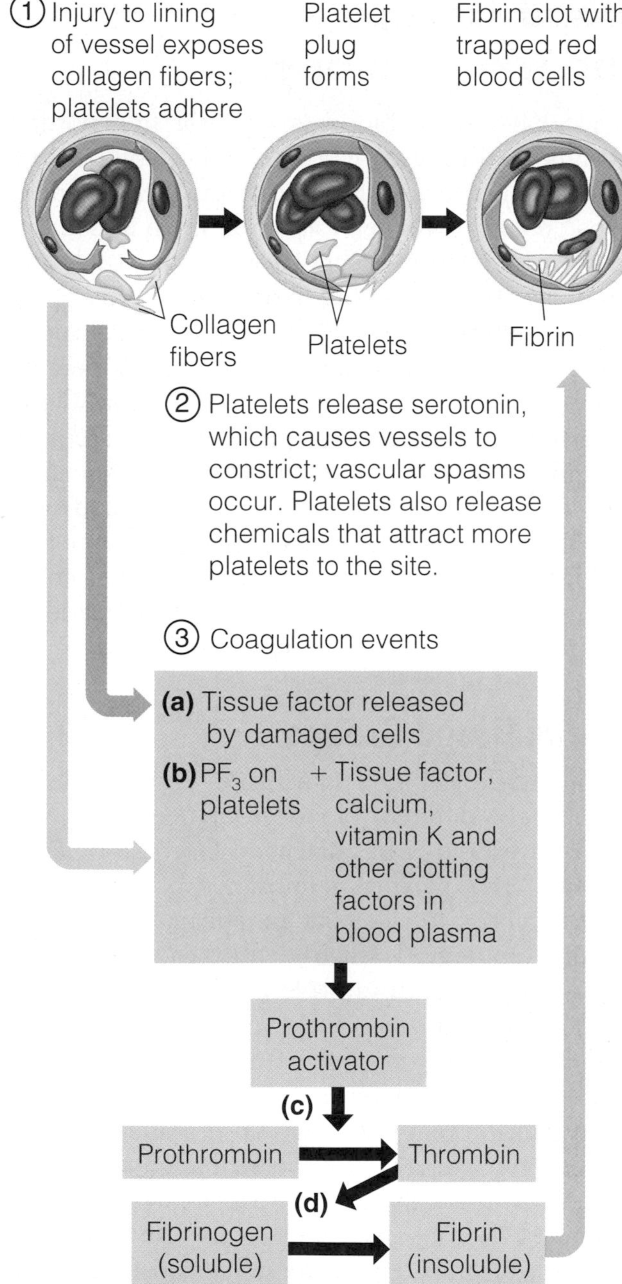

FIGURE 10.6 Hemostasis. The multistep process, detailed in the text, begins when a blood vessel is damaged and connective tissue in the vessel wall is exposed to blood.

the clot (Figure 10.7). Within the hour, the clot begins to retract, squeezing **serum** (plasma minus the clotting proteins) from the mass and pulling the ruptured edges of the blood vessel closer together.

Normally, blood clots within 3 to 6 minutes. As a rule, once the clotting cascade has started, the triggering factors are rapidly inactivated to prevent widespread clotting ("solid blood"). Eventually, the endothelium regenerates, and the clot is broken down. Once these events of the clotting cascade were understood, it became clear that placing a sterile gauze over a cut or applying pressure to a wound would speed up the clotting process. The gauze provides a rough surface to which the platelets can adhere, and the pressure fractures cells, increasing the release of tissue factor locally.

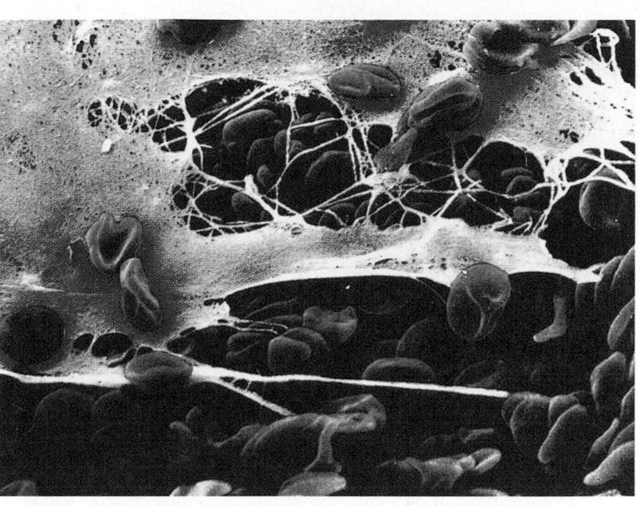

FIGURE 10.7 Fibrin clot. Scanning electron micrograph (artificially colored) of red blood cells trapped in a mesh of fibrin threads.

Disorders of Hemostasis

Homeostatic Imbalance

The two major disorders of hemostasis—undesirable clot formation and bleeding disorders—are at opposite poles.

Undesirable Clotting

Despite the body's safeguards against abnormal clotting, undesirable clots sometimes form in intact blood vessels, particularly in the legs. A clot that develops and persists in an unbroken blood vessel is called a **thrombus** (throm'bus). If large enough, it may prevent blood flow to the cells beyond the blockage. For example, if a thrombus forms in the blood vessels serving the heart *(coronary thrombosis),* the consequences may be death of heart muscle and a fatal heart attack. If a thrombus breaks away from the vessel wall and floats freely in the bloodstream, it becomes an **embolus** (em'bo-lus; plural, *emboli*). An embolus is usually no problem unless or until it lodges in a blood vessel too narrow for it to pass through. For example, a *cerebral embolus* may cause a stroke.

Undesirable clotting may be caused by anything that roughens the endothelium of a blood vessel and encourages clinging of platelets, such as severe burns, physical blows, or an accumulation of fatty material. Slowly flowing blood, or blood pooling, is another risk factor, especially in immobilized patients. In this case, clotting factors are not washed away as usual and accumulate so that clot formation becomes possible. A number of anticoagulants, most importantly aspirin, heparin, and dicumarol, are used clinically for thrombus-prone patients.

Bleeding Disorders

The most common causes of abnormal bleeding are platelet deficiency (thrombocytopenia) and deficits of some of the clotting factors, such as might result from impaired liver function or certain genetic conditions.

Thrombocytopenia results from an insufficient number of circulating platelets. Even normal movements cause spontaneous bleeding from small blood vessels. This is evidenced by many small purplish blotches, called *petechiae* (pe-te′ke-e), on the skin. It can arise from any condition that suppresses myeloid tissue, such as bone marrow cancer, radiation, or certain drugs.

When the liver is unable to synthesize its usual supply of clotting factors, abnormal and often severe bleeding episodes occur. If vitamin K (needed by the liver cells to produce the clotting factors) is deficient, the problem is easily corrected with vitamin K supplements. However, when liver function is severely impaired (as in hepatitis and cirrhosis), only whole blood transfusions are helpful.

The term **hemophilia** (he″mo-fil′e-ah) applies to several different hereditary bleeding disorders that result from a lack of any of the factors needed for clotting. Commonly called "bleeder's disease," the hemophilias have similar signs and symptoms that begin early in life. Even minor tissue trauma results in prolonged bleeding and can be life-threatening. Repeated bleeding into joints causes them to become disabled and painful. When a bleeding episode occurs, hemophiliacs are given a transfusion of fresh plasma or injections of the purified clotting factor they lack. Because hemophiliacs are absolutely dependent on one or the other of these therapies, some have become the victims of blood-transmitted viral diseases such as hepatitis and AIDS. (AIDS, acquired immunodeficiency syndrome, is a condition of depressed immunity and is described in Chapter 12.) These problems have been largely resolved because of the availability of genetically engineered clotting factors. ▲

Blood Groups and Transfusions

As we have seen, blood is vital for transporting substances through the body. When blood loss occurs, the blood vessels constrict and the bone marrow steps up blood cell formation in an attempt to keep the circulation going. However, the body can compensate for a loss of blood volume only up to a certain limit. Losses of 15 to 30 percent lead to pallor and weakness. Losses of over 30 percent cause severe shock, which can be fatal. Whole blood transfusions are routinely given to replace substantial blood loss and to treat severe anemia or thrombocytopenia. The usual blood bank procedure involves collecting blood from a donor and mixing it with an anticoagulant to prevent clotting. The treated blood can be stored (refrigerated at 4°C, or 39.2°F) for about 35 days until needed.

Human Blood Groups

Although whole blood transfusion can save lives, people have different blood groups, and transfusing incompatible or mismatched blood can be fatal. How so? The plasma membranes of RBCs, like those of all body cells, bear genetically determined proteins (antigens), which identify each person as unique. An **antigen** (an′tĭ-jen) is a substance that the body recognizes as foreign; it stimulates the immune system to release antibodies or use other means to mount a defense against it. Most antigens are foreign proteins, such as those that are part of viruses or bacteria that have managed to invade the body. Although each of us tolerates our own cellular (self) antigens, one person's RBC proteins will be recognized as foreign if transfused into another person with different RBC antigens. The "recognizers" are **antibodies** present in the plasma that attach to RBCs bearing surface antigens different from those on the patient's (blood recipient's) RBCs. Binding of the antibodies causes the RBCs to clump, a phenomenon called **agglutination*** (ah-gloo″tĭ-na′shun), which leads to the clogging of small blood vessels throughout the body. During the next few hours, the foreign RBCs are lysed

*The RBC antigens that promote this clumping are sometimes called **agglutinogens** (ag″loo-tin′o-jenz), and the antibodies that bind them together are called **agglutinins** (ag″loo′tĭ-ninz).

Focus on Careers

Phlebotomy Technician

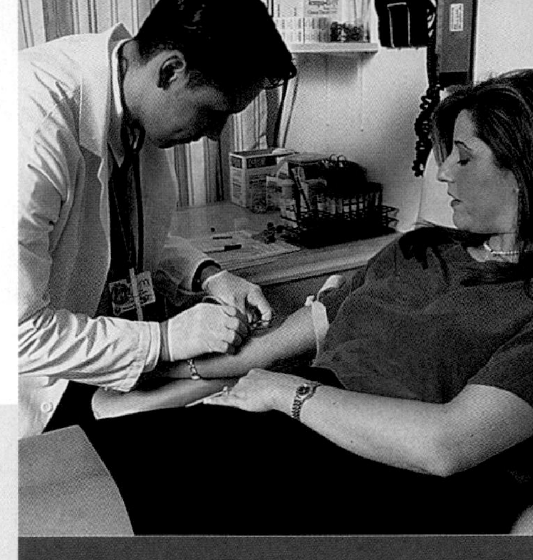

Phlebotomists must know where all the arteries and veins are located in the body.

"PHLEBOTOMY is the most important procedure done for a medical laboratory," says Michael Coté, who supervises the phlebotomy staff at Palo Alto Veterans Administration Hospital in California. "To make accurate diagnosis and effective treatment possible, it's vital to draw a good blood sample, place it correctly in a clean container, and process it accurately in the lab. Without a high-quality specimen, none of this can happen."

Phlebotomy is not exactly a household word. It derives from the Greek terms for "vein" and "to cut." A phlebotomy technician, or phlebotomist, is trained to collect and process blood samples that will be subjected to laboratory analysis.

Coté appreciates how important it is for phlebotomists to understand anatomy and physiology. "Anatomy is a key requirement in phlebotomy training," he says, "because you have to learn where all the arteries and veins are located in the body. Some patients' veins are easy to find, but others have veins that are practically invisible. You need to know the right place to insert that needle. Although 90 percent of the blood we draw comes from the antecubital region inside the elbow, we may also draw blood from the cephalic vein in the forearm, or from veins in the hands."

Coté notes that knowledge of physiology is also important. "I have to be able to assess patients' overall health and physical condition, because this affects their ability to give an adequate blood sample and may demand that a different needle size be used to draw the blood sample. There are several types and sizes of needles I can use, and I need to select the best one for each patient. People who are dehydrated can be difficult because their blood pressure is lower and venous return is impaired. Patients with poor circulation are also harder to work with. The blood tends to stay in the trunk of the body rather than flowing freely into the extremities because they're cold, and it's difficult to get a good return from veins in the arms. Cancer patients often show increased sensitivity to pain, so we have to be very gentle and use the smallest needle possible." Patients with a history of drug abuse pose other challenges. "Frequent 'sticking' with needles causes scar tissue to form and hardens vein walls. You can tell people who have used intravenous drugs—their veins feel rock-hard and are much more difficult to penetrate with a needle."

Coté says that a good phlebotomist must also possess effective interpersonal skills: "People are apprehensive about being stuck with needles, so you have to be patient and be able to put them at ease. Above all, you have to be confident. If the phlebotomist is nervous, the patient will sense it and get nervous too."

"Anatomy is a key requirement in phlebotomy training."

To become certified, a phlebotomy technician must be a high school graduate, complete a phlebotomy training program or acquire equivalent experience, and pass a certification exam offered by the American Society of Clinical Pathologists. Anatomy and physiology topics make up a significant portion of the certifying exam, particularly concerning the structure and function of the circulatory system, composition and function of blood and other body fluids, and how various health conditions affect the sample collection process.

Accreditation procedures for phlebotomists vary from state to state. For more information, contact the American Society of Clinical Pathologists at
2100 W. Harrison Street
Chicago, IL 60612-3798
(312) 738-1336
http://www.ascp.org

For additional information on this career, click the Focus on Careers link at www.anatomyandphysiology.com.

TABLE 10.3 ABO Blood Groups

Blood group	Frequency (% U.S. population) White	Black	Asian	RBC antigens (agglutinogens)	Illustration	Plasma antibodies (agglutinins)	Blood that can be received
AB	4	4	5	A B	A, B	None	A, B, AB, O Universal recipient
B	11	20	27	B	Anti-A, B	Anti-A	B, O
A	40	27	28	A	Anti-B, A	Anti-B	A, O
O	45	49	40	None	Anti-B, Anti-A	Anti-A Anti-B	O Universal donor

(ruptured) and their hemoglobin is released into the bloodstream. Although the transfused blood is unable to deliver the increased oxygen-carrying capacity hoped for and some tissue areas may be deprived of blood, the most devastating consequence of severe transfusion reactions is that the freed hemoglobin molecules may block the kidney tubules and cause kidney failure. Transfusion reactions can also cause fever, chills, nausea, and vomiting, but in the absence of kidney shutdown these reactions are rarely fatal. Treatment is aimed at preventing kidney damage by infusing alkaline fluids to dilute and dissolve the hemoglobin and diuretics to flush it out of the body in urine.

There are over 30 common RBC antigens in humans, allowing each person's blood cells to be classified into different blood groups. However, it is the antigens of the ABO and Rh blood groups that cause the most vigorous transfusion reactions. These two blood groups are described here.

As shown in Table 10.3, the **ABO blood groups** are based on which of two antigens, type A or type B, a person inherits. Absence of both antigens results in type O blood, presence of both antigens leads to type AB, and the possession of either A or B antigen yields type A or B blood, respectively. In the ABO blood group, antibodies are formed during infancy against the ABO antigens *not* present on your own RBCs. As shown in the table, a baby with neither the A nor the B antigen (group O) forms both anti-A and anti-B antibodies, while those with type A antigens (group A) form anti-B antibodies, and so on.

The **Rh blood groups** are so named because one of the eight Rh antigens (agglutinogen D) was originally identified in **Rh**esus monkeys. Later the same antigen was discovered in human beings. Most Americans are Rh^+ (Rh positive), meaning that their RBCs carry the Rh antigen. Unlike the antibodies of the ABO system, anti-Rh antibodies are *not* automatically formed and present in the blood of Rh^- (Rh negative) individuals. However, if an Rh^- person receives mismatched blood (that is, Rh^+), shortly after the transfusion his or her immune system becomes sensitized and begins producing antibodies (anti-Rh^+ antibodies) against the foreign blood type. **Hemolysis** (rupture of RBCs) does not occur with the first transfusion because it takes time for the body to react and start making antibodies. However, the second time and every time thereafter, a typical transfusion reaction occurs in which the patient's antibodies attack and rupture the donor's Rh^+ RBCs.

An important Rh-related problem occurs in pregnant Rh^- women who are carrying Rh^+ babies. The *first* such pregnancy usually results in the delivery of a healthy baby. But because the mother is sensitized by Rh^+ antigens that have passed through the placenta into her bloodstream, she will form anti-Rh^+ antibodies unless treated with RhoGAM shortly after giving birth. RhoGAM is an immune serum that prevents this sensitization and her subsequent immune response. If she is not treated and becomes pregnant again with an Rh^+ baby, her antibodies will cross through the placenta and destroy the baby's RBCs, producing a condition known as *hemolytic disease of the newborn.* The baby is anemic and becomes hypoxic and cyanotic (the skin takes on a blue cast). Brain damage and even death may result unless fetal transfusions are done before birth to provide more RBCs for oxygen transport.

Blood Typing

The importance of determining the blood group of both the donor and the recipient *before* blood is transfused is glaringly obvious. The general procedure for determining ABO blood type is briefly outlined in Figure 10.8. Essentially, it involves testing the blood by mixing it with two different types of immune serum—anti-A and anti-B. Agglutination occurs when RBCs of a group A person are mixed with the anti-A serum but not when they are mixed with the anti-B serum. Likewise, RBCs of type B blood are clumped by anti-B serum but not by anti-A serum. Because it is critical that blood groups be compatible, cross matching is also done. *Cross matching* involves testing for agglutination of donor RBCs by the recipient's serum and of the recipient's RBCs by the donor serum. Typing for the Rh factors is done in the same manner as ABO blood typing.

Developmental Aspects of Blood

In the young embryo, development of the entire circulatory system occurs early. Before birth, there are many sites of blood cell formation—the fetal liver and spleen, among others—but by the seventh month of development, the red marrow has become the chief site of hematopoiesis, and it remains so throughout life. Generally, embryonic blood cells are circulating in the newly formed blood vessels by day 28 of development. Fetal hemoglobin (HbF) differs from the hemoglobin formed after birth. It has a greater ability to pick up oxygen, a characteristic that is highly desirable since fetal blood is less oxygen rich than that of the mother. After birth, fetal blood cells are gradually replaced by RBCs that contain the more typical hemoglobin A (HbA). In situations in which the fetal RBCs are destroyed at such a rapid rate that

Q *Which blood types can be transfused into a person with type B blood?*

FIGURE 10.8 Blood typing of ABO blood groups. When serum containing anti-A or anti-B antibodies is added to a blood sample diluted with saline, agglutination will occur between the antibody and the corresponding antigen (if present).

A Types B and O.

A Closer Look

Concocting Blood: Artificial Blood Substitutes

THE term "blood substitute" is somewhat misleading. Blood has many components that play such a wide variety of roles—from fighting infection to transporting oxygen—that no single artificial substitute yet engineered can fulfill all those functions. However, substitute liquids are available that can transport oxygen from the lungs through the body and can "stretch" a limited blood supply while sidestepping transfusion reactions. An important benefit of these blood substitutes is that the recipient is not at risk for transmission of bloodborne disease factors. Four of these oxygen-transporting products—Fluosol, chemically altered hemoglobin, artificial red blood cells, and Hemopure—are described briefly.

Fluosol

The main ingredient of *Fluosol,* a milky artificial blood substitute, is perfluorocarbons (PFCs), chemicals related to Teflon, the nonstick coating used in cookware. Developed in Japan, the product was first tested in the United States in 1982. Many of the early recipients were people who needed surgery but refused blood transfusions on religious grounds.

Fluosol serves as a dissolving medium for oxygen. In order to "load" sufficient amounts of oxygen into it, patients must breathe pure oxygen by mask or must be in a hyperbaric (high-pressure) chamber. The developers claim that oxygen transported by Fluosol is used more easily by the body tissues because the slippery particles are much smaller than erythrocytes and glide through the capillaries at a faster rate. PFCs are cleared from the circulation and stored in the spleen and liver until they are exhaled as vapor by the lungs 4 to 12 hours after injection. Although Fluosol was initially promising for treatment of heart attack, carbon monoxide poisoning, and sickle cell anemia, its use has been clouded by research indicating that it depresses the immune system.

Currently under development is Oxyfluor, a new generation of PFC-based compounds designed to resolve the problems of Fluosol. It has a long shelf life (two years if refrigerated) and delivers up to four times as much oxygen as the earlier versions did. However, increasing the oxygen-carrying capacity of blood can have diminishing returns because when oxygen accumulates in body tissues, it causes tissue damage.

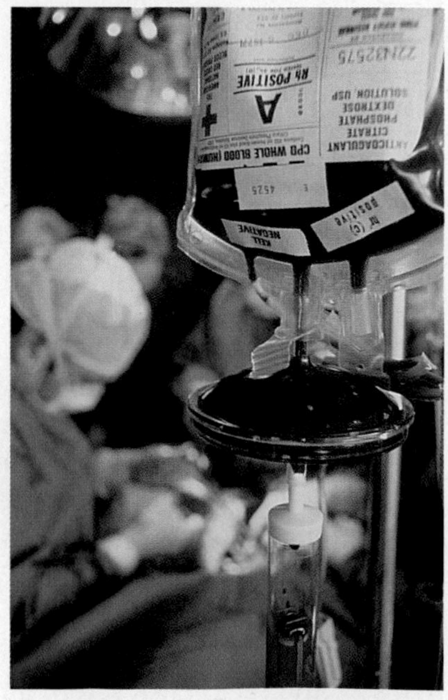

Chemically Altered Hemoglobin

In their search to find a substitute to boost oxygen delivery, scientists have harvested hemoglobin from erythrocytes, and altered its chemistry by creating a chemical bridge between two of its four peptide chains (as in HemAssist) and by linking several hemoglobin molecules together (the polymer PolyHeme). The altered hemoglobin gives up more oxygen to the tissues than normal hemoglobin, even at low temperatures (50°F, or 10°C). Since body temperature is routinely lowered in patients undergoing heart surgery, using the modified hemoglobin during such procedures might boost oxygen delivery to the patient's tissues. An additional plus is that because the cross-linked form of hemoglobin does not fragment in the blood (as natural hemoglobin does), the necessity for a container—the RBC plasma membrane—is eliminated.

Although test results are promising, there are still important problems to be solved. For example, bacterial endotoxins, potent poisons produced by some bacteria, tend to cling to the modified hemoglobin, and there is some evidence that the free hemoglobin provokes a generalized constriction of blood vessels, making oxygen delivery more difficult.

Artificial RBCs (Neohemocytes)

Researchers at the University of California at San Francisco have created artificial RBCs, which they call *neohemocytes,* by packaging natural hemoglobin molecules in fat bubbles made from phospholipids and cholesterol. The resulting "red cells" are about 1⁄12 the size of human erythrocytes. Although the neohemocytes are destroyed and cleared more rapidly from the bloodstream than are real RBCs, they have a shelf life of six months (versus about 35 days for whole blood). This would make neohemocyte infusion a viable choice for trauma patients in immediate need of blood, but clinical trials on humans are still in the future.

Hemopure

Hemopure, a natural but nonhuman blood substitute recently approved by the FDA, contains purified hemoglobin extracted from cattle blood. Because there are no plasma membranes to deal with, cross-reactions should not be a problem. However, this bovine-based substitute must address the danger of transmission of mad cow disease (bovine spongiform encephalopathy) and perhaps other diseases not yet identified.

Although the U.S. Food and Drug Administration has encouraged the development of artificial blood for over 20 years, no marketable product, including those described above, has been approved for anything but experimental use in humans. This may change. Early in 1997, researchers at Albany Medical Center reported that they had developed a process to coat RBCs with the harmless polymer PEG to make all blood compatible with all patients. Researchers are also proceeding down another path—in 2001, it was documented that stem cells had been converted to red blood cells. As of yet, however, blood is still a priceless commodity, and its beautiful complexity has yet to be replaced by modern medical technology.

the immature infant liver cannot rid the body of hemoglobin breakdown products in the bile fast enough, the infant becomes *jaundiced* (jawn′dist). This type of jaundice generally causes no major problems and is referred to as *physiologic jaundice,* to distinguish it from more serious disease conditions that result in jaundiced, or yellowed, tissues.

Homeostatic Imbalance

Various congenital diseases result from genetic factors (such as hemophilia and sickle cell anemia) and from interactions with maternal blood factors (such as hemolytic disease of the newborn). Dietary factors can lead to abnormalities in blood cell formation as well as hemoglobin production. Iron deficiency anemia is especially common in women because of their monthly blood loss during menses. The young and the old are particularly at risk for leukemia. With increasing age, chronic types of leukemias, anemias, and diseases involving undesirable clot formation are more prevalent. However, these are usually secondary to disorders of the heart, blood vessels, or immune system. The elderly are particularly at risk for pernicious anemia because the stomach mucosa (which produces intrinsic factor) atrophies with age. ▲

SUMMARY

Media study tools that could provide you with additional help in reviewing specific key topics of Chapter 10 are referenced below.

IP = Interactive Physiology;

WEB = A&P Place website.

COMPOSITION AND FUNCTIONS OF BLOOD (pp. 328–335)

1. Blood is composed of a nonliving fluid matrix (plasma) and formed elements. It is scarlet to dull red, depending on the amount of oxygen carried. Normal adult blood volume is 5 to 6 liters.
2. Dissolved in plasma (primarily water) are nutrients, gases, hormones, wastes, proteins, salts, and so on. Plasma composition changes as body cells remove or add substances to it, but homeostatic mechanisms act to keep it relatively constant. Plasma makes up 55 percent of whole blood.
3. Formed elements, the living blood cells that make up about 45 percent of whole blood, include:
   a. Erythrocytes, or RBCs—disc-shaped, anucleate cells that transport oxygen bound to their hemoglobin molecules. Their lifespan is 100 to 120 days.
   b. Leukocytes, or WBCs—ameboid cells involved in protection of the body.
   c. Platelets—cell fragments that act in blood clotting.

   WEB Exercise: Chapter 10, Formed Elements.
4. A decrease in oxygen-carrying ability of blood is anemia. Possible causes are decrease in number of functional RBCs or decrease in amount of hemoglobin they contain. Polycythemia is an excessive number of RBCs that may result from bone marrow cancer or a move to a location where less oxygen is available in the air (at high altitude, for example).
5. Leukocytes are nucleated cells, classed into two groups:
   a. Granulocytes include neutrophils, eosinophils, and basophils.
   b. Agranulocytes include monocytes and lymphocytes.
6. When bacteria, viruses, or other foreign substances invade the body, WBCs increase in number (leukocytosis) and fight them in various ways.
7. An abnormal decrease in number of WBCs is leukopenia. An abnormal increase in WBCs is seen in infectious mononucleosis and leukemia (cancer of leukocytes).
8. All formed elements arise in red bone marrow from a common stem cell, the hemocytoblast. However, their developmental pathways differ. The stimulus for hematopoiesis is hormonal (erythropoietin in the case of RBCs).

HEMOSTASIS (pp. 335–338)

1. Stoppage of blood loss from an injured blood vessel, or hemostasis, involves three steps: platelet plug formation, vascular spasms, and blood clot formation.
2. Hemostasis is started by a tear or interruption in the blood vessel lining. Platelets adhere to the damaged

site and release serotonin, which causes vasoconstriction. Platelet PF_3 and tissue factor initiate the clotting cascade, leading to formation of fibrin threads. Fibrin traps RBCs as they flow past, forming the clot.

3. Normally, clots are digested when a vessel has been permanently repaired. An attached clot that forms or persists in an unbroken blood vessel is a thrombus; a clot traveling in the bloodstream is an embolus.
4. Abnormal bleeding may reflect a deficit of platelets (thrombocytopenia), genetic factors (hemophilia), or inability of the liver to make clotting factors.

BLOOD GROUPS AND TRANSFUSIONS (pp. 338–341)

1. Blood groups are classified on the basis of proteins (antigens) on RBC membranes. Complementary antibodies may (or may not) be present in blood. Antibodies act to agglutinate (clump) and lyse foreign RBCs.
2. The blood group most commonly typed for is ABO. Type O is most common; least common is AB. ABO antigens are accompanied by preformed antibodies in plasma, which act against RBCs that have "foreign" antigens.
3. Rh factor is found in most Americans. Rh^- people do not have preformed antibodies to Rh^+ RBCs but form them once exposed to Rh^+ blood.

DEVELOPMENTAL ASPECTS OF BLOOD (pp. 341–344)

1. Congenital blood defects include various types of hemolytic anemias and hemophilia. Incompatibility between maternal and fetal blood can result in fetal cyanosis, resulting from destruction of fetal blood cells.
2. Fetal hemoglobin (HbF) binds more readily with oxygen than does HbA.
3. Physiologic jaundice in a newborn reflects immaturity of the infant's liver.
4. Excessive leukocytosis may be indicative of malignancy of blood-forming organs or leukemia. Leukemias are most common in the very young and very old.
5. The elderly are at risk for anemia and clotting disorders.

REVIEW QUESTIONS

MULTIPLE CHOICE

More than one choice may apply.

1. Which would lead to increased erythropoiesis?
   a. Chronic bleeding ulcer
   b. Reduction in respiratory ventilation
   c. Decreased level of physical activity
   d. Reduced blood flow to the kidneys
2. Sickling of RBCs can be induced in those with sickle cell anemia by:
   a. blood loss
   b. vigorous exercise
   c. stress
   d. fever
3. A child is diagnosed with sickle cell anemia. This means that:
   a. one parent had sickle cell anemia
   b. one parent carried the sickle cell gene
   c. both parents had sickle cell anemia
   d. both parents carried the sickle cell gene
4. Polycythemia vera will result in:
   a. overproduction of WBCs
   b. exceptionally high blood volume
   c. abnormally high blood viscosity
   d. abnormally low hematocrit
5. Which of the following is not typical of leukocytes?
   a. Ameboid movement
   b. Phagocytic (some)
   c. Nucleated
   d. The most numerous cells in the bloodstream
6. The leukocyte that releases histamine and other inflammatory chemicals is the:
   a. basophil
   b. eosinophil
   c. monocyte
   d. neutrophil
7. Which of the following formed elements are phagocytic?
   a. Erythrocytes
   b. Neutrophils
   c. Monocytes
   d. Lymphocytes
8. A condition resulting from thrombocytopenia is:
   a. thrombus formation
   b. embolus formation
   c. petechiae
   d. hemophilia

9. Which of the following can cause problems in a transfusion reaction?
   a. Donor antibodies attacking recipient RBCs
   b. Clogging of small vessels by agglutinated clumps of RBCs
   c. Lysis of donated RBCs
   d. Blockage of kidney tubules

10. If an Rh^- mother becomes pregnant, when can hemolytic disease of the newborn *not possibly* occur in the child?
    a. If the child is Rh^-
    b. If the child is Rh^+
    c. If the father is Rh^+
    d. If the father is Rh^-

11. Plasma without the clotting proteins is called:
    a. serum
    b. whole blood
    c. fibrin
    d. tissue factor

SHORT ANSWER ESSAY

1. What is the blood volume of an average-sized adult?
2. What determines whether blood is bright red (scarlet) or dull red in color?
3. Name as many different categories of substances carried in plasma as you can.
4. Define *formed elements,* and list their three major categories. Which category is most numerous? Which makes up the buffy coat?
5. What is the average lifespan of an RBC? How does the fact that it has no nucleus affect its lifespan?
6. Define *anemia,* and give three possible causes.
7. Name the granular and agranular WBCs. Give the major function of each type in the body.
8. If you had a severe infection, would you expect your total WBC count to be closest to 5000, 10,000, or 15,000/mm^3? Why? What is this condition called?
9. Name the stem cell that produces virtually all formed elements. Name the formed elements that arise from myeloid stem cells. Name those arising from lymphoid stem cells.
10. Describe the process of hemostasis. Indicate what starts the process.
11. How can liver dysfunction cause bleeding disorders?
12. What is the basis of blood groups? What are agglutinins?
13. Name the four ABO blood groups. Which is most common? Which is least common?
14. What is a transfusion reaction? Why does it happen?
15. Explain why an Rh^- person does not have a transfusion reaction on the first exposure to Rh^+ blood. Why is there a transfusion reaction the second time he or she receives the Rh^+ blood?
16. If you had a high hematocrit, would you expect your hemoglobin determination to be high or low? Why?
17. What blood-related problems are most common in the aged?

AT THE CLINIC

1. A patient on renal dialysis has a low RBC count. What hormone, secreted by the kidney, can be assumed to be deficient?
2. A bone marrow biopsy of Mr. Bongalonga, a man on a long-term drug therapy, shows an abnormally high percentage of nonhematopoietic connective tissue. What condition does this indicate? If the symptoms are critical, what short-term and long-term treatments are indicated? Will infusion of whole blood or packed red cells be more likely?
3. A woman comes to the clinic complaining of fatigue, shortness of breath, and chills. Blood tests show anemia, and a bleeding ulcer is diagnosed. What type of anemia is this?
4. A patient is diagnosed with bone marrow cancer and has a hematocrit of 70 percent. What is this condition called?
5. A middle-aged college professor from Boston is in the Swiss Alps studying astronomy. He arrived two days ago and plans to stay the entire year. However, he notices that he is short of breath when he walks up steps and that he tires easily with any physical activity. His symptoms gradually disappear; after two months, he feels fine. Upon returning to the United States, he has a complete physical exam and is told that his erythrocyte count is higher than normal. (a) Attempt to explain this finding. (b) Will his RBC count remain at this higher-than-normal level? Why or why not?
6. Why is someone more likely to bleed to death when an artery is cleanly severed than when it is crushed and torn?
7. Explain how fetal hemoglobin, HbF, enhances oxygen transfer across the placenta from the mother to the fetus.

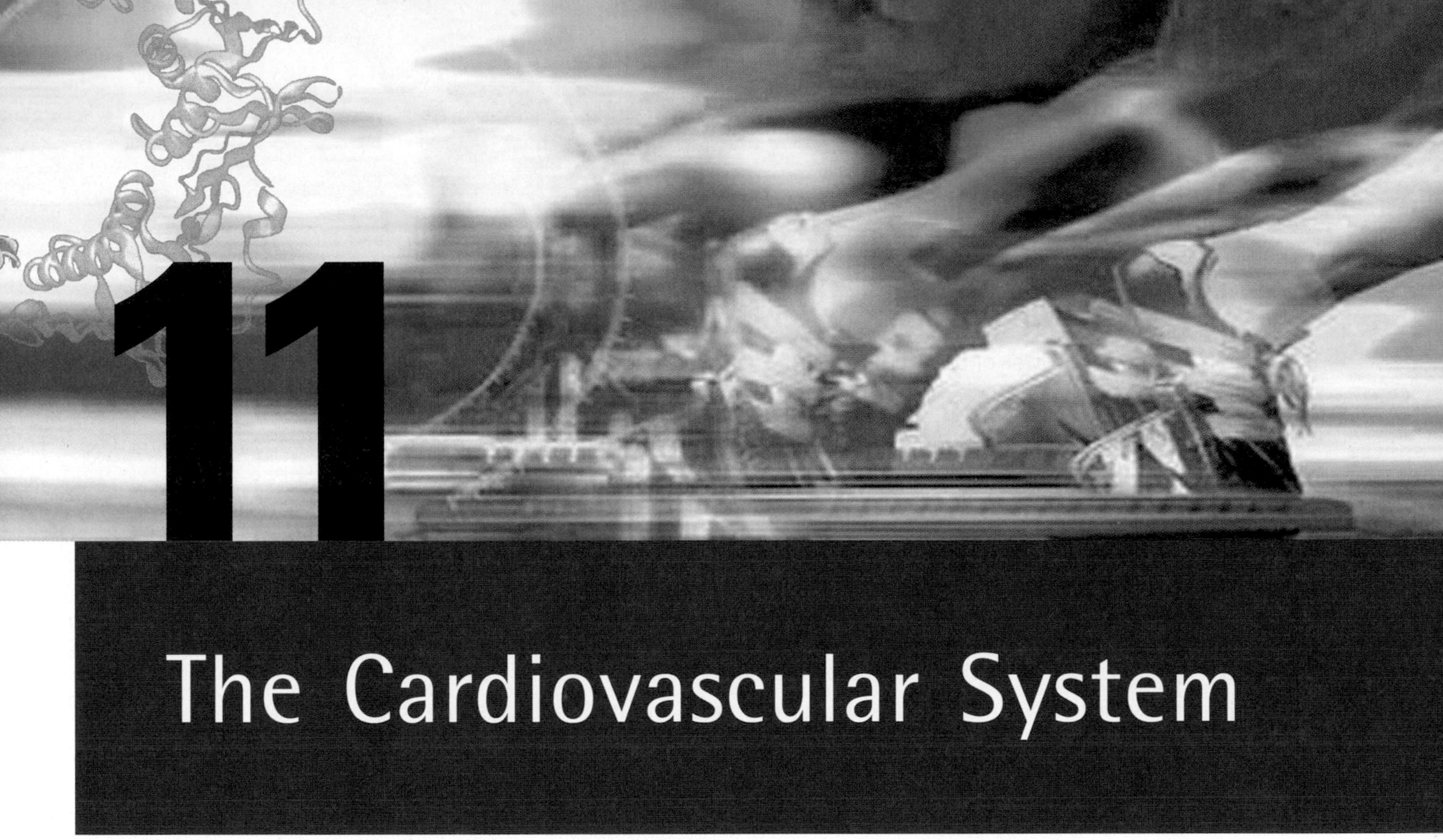

11 The Cardiovascular System

YOUR GOALS

After completing this chapter, you will have a working knowledge of the functions of the cardiovascular system and will have mastered the objectives listed below.

FUNCTION PREVIEW

- The heart pumps blood.
- Blood vessels provide the conduits within which blood circulates to all body tissues.

OBJECTIVE CHECKLIST

THE HEART (pp. 348–360)

- Describe the location of the heart in the body, and identify its major anatomical areas on an appropriate model or diagram.
- Trace the pathway of blood through the heart.
- Compare the pulmonary and systemic circuits.
- Explain the operation of the heart valves.
- Name the functional blood supply of the heart.
- Name the elements of the intrinsic conduction system of the heart, and describe the pathway of impulses through this system.
- Define *systole, diastole, stroke volume,* and *cardiac cycle.*
- Define *heart sounds* and *murmur.*
- Explain what information can be gained from an electrocardiogram.
- Describe the effect of each of the following on heart rate: stimulation by the vagus nerve, exercise, epinephrine, and various ions.

BLOOD VESSELS (pp. 360–381)

- Compare and contrast the structure and function of arteries, veins, and capillaries.
- Identify the body's major arteries and veins, and name the body region supplied by each.

>>

(Objective Checklist, continued)

- Discuss the unique features of special circulations of the body: arterial circulation of the brain, hepatic portal circulation, and fetal circulation.
- Define *blood pressure* and *pulse,* and name several pulse points.
- List factors affecting and/or determining blood pressure.
- Define *hypertension* and *atherosclerosis,* and describe possible health consequences of these conditions.
- Describe the exchanges that occur across capillary walls.

DEVELOPMENTAL ASPECTS OF THE CARDIOVASCULAR SYSTEM
(pp. 381–383)

- Briefly describe the development of the cardiovascular system.
- Name the fetal vascular modifications, or "fetal shunts," and describe their function before birth.
- Explain how regular exercise and a diet low in fats and cholesterol may help maintain cardiovascular health.

When most people hear the term *cardiovascular system,* they immediately think of the heart. We have all felt our own heart "pound" from time to time, and we tend to get a bit nervous when this happens. The crucial importance of the heart has been recognized for a long time. However, the **cardiovascular system** is much more than just the heart, and from a scientific and medical standpoint, it is important to understand *why* this system is so vital to life.

The almost continuous traffic into and out of a busy factory at rush hour occurs at a snail's pace compared to the endless activity going on within our bodies. Night and day, minute after minute, our trillions of cells take up nutrients and excrete wastes. Although the pace of these exchanges slows during sleep, they must go on continuously, because when they stop we die. Cells can make such exchanges only with the tissue fluid in their immediate vicinity. Thus, some means of changing and "refreshing" these fluids is necessary to renew the nutrients and prevent pollution caused by the buildup of wastes. Like the bustling factory, the body must have a transportation system to carry its various "cargos" back and forth. Instead of roads, railway tracks, and airways, the body's delivery routes are its hollow blood vessels.

Most simply stated, the major function of the cardiovascular system is transportation. Using blood as the transport vehicle, the system carries oxygen, nutrients, cell wastes, hormones, and many other substances vital for body homeostasis to and from the cells. The force to move the blood around the body is provided by the beating heart.

The cardiovascular system can be compared to a muscular pump equipped with one-way valves and a system of large and small plumbing tubes within which the blood travels. Blood (the substance transported) is discussed in Chapter 10. Here we will consider the heart (the pump) and the blood vessels (the network of tubes).

The Heart

Anatomy of the Heart

Location and Size

The relative size and weight of the heart give few hints of its incredible strength. Approximately the size of a person's fist, the hollow, cone-shaped heart

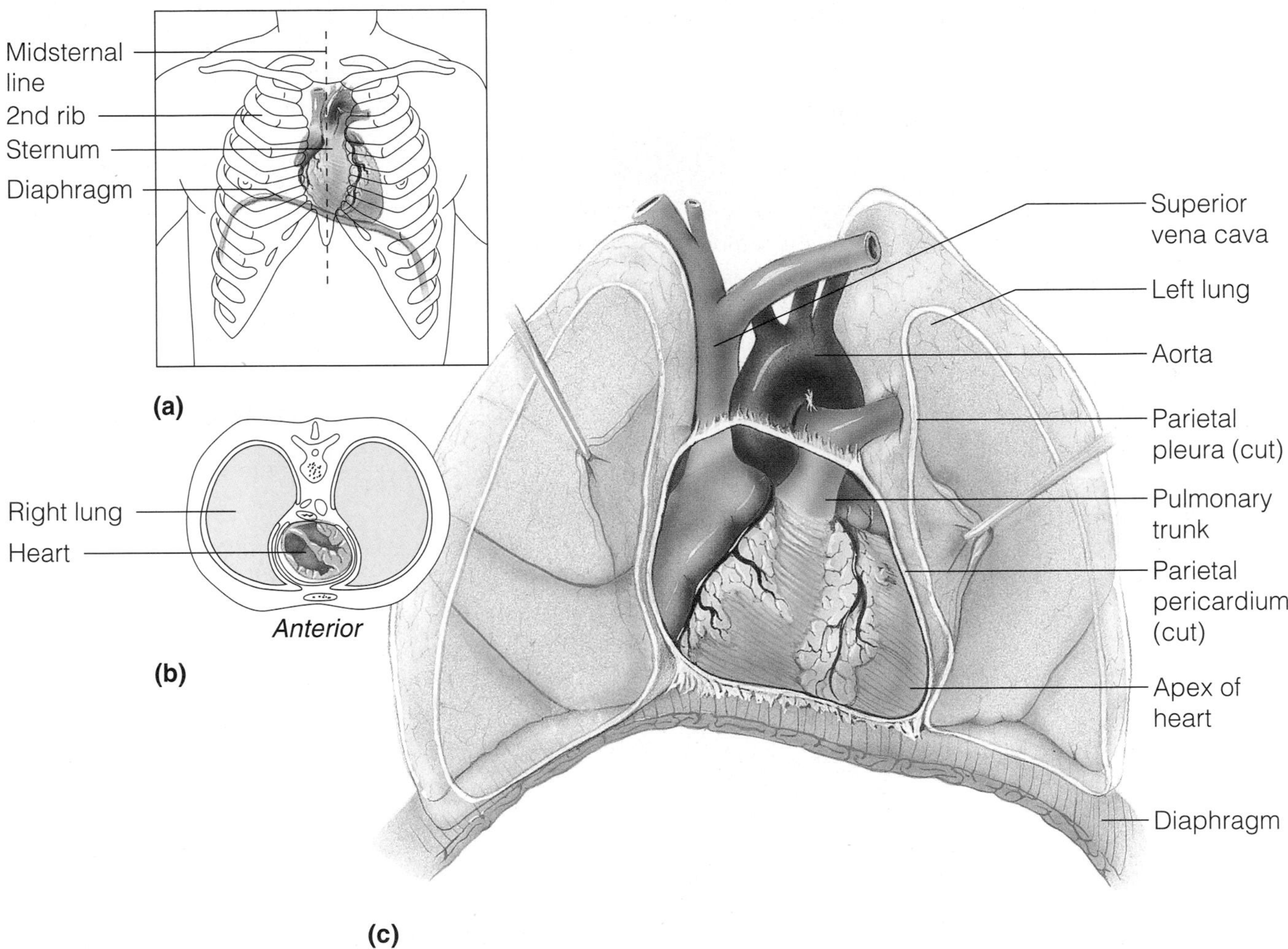

FIGURE 11.1 Location of the heart within the thorax. (a) Relationship of the heart to the sternum and ribs. **(b)** Cross-sectional view showing relative position of the heart in the thorax. **(c)** Relationship of the heart and great vessels to the lungs.

weighs less than a pound. Snugly enclosed within the inferior **mediastinum** (me″de-ah-sti′num), the middle cavity of the thorax, the heart is flanked on each side by the lungs (Figure 11.1). Its more pointed **apex** is directed toward the left hip and rests on the diaphragm, approximately at the level of the fifth intercostal space. (This is exactly where one would place a stethoscope to count the heart rate for an apical pulse.) Its broader posterosuperior aspect, or **base,** from which the great vessels of the body emerge, points toward the right shoulder and lies beneath the second rib.

Coverings and Wall

The heart is enclosed by a double sac of serous membrane, the **pericardium** (per″i-kar′de-um). The thin **epicardium,** or **visceral pericardium,** tightly hugs the external surface of the heart and is actually part of the heart wall (Figure 11.2b). It is continuous at the heart base with the loosely applied **parietal pericardium,** which is reinforced on its superficial face by dense connective tissue, referred to as the **fibrous pericardium.** This fibrous layer helps protect the heart and anchors it to surrounding structures, such as the diaphragm and sternum. A slippery lubricating fluid (serous fluid) is produced by the serous pericardial

Q *Which heart chamber has the thickest walls? What is the functional significance of this structural difference?*

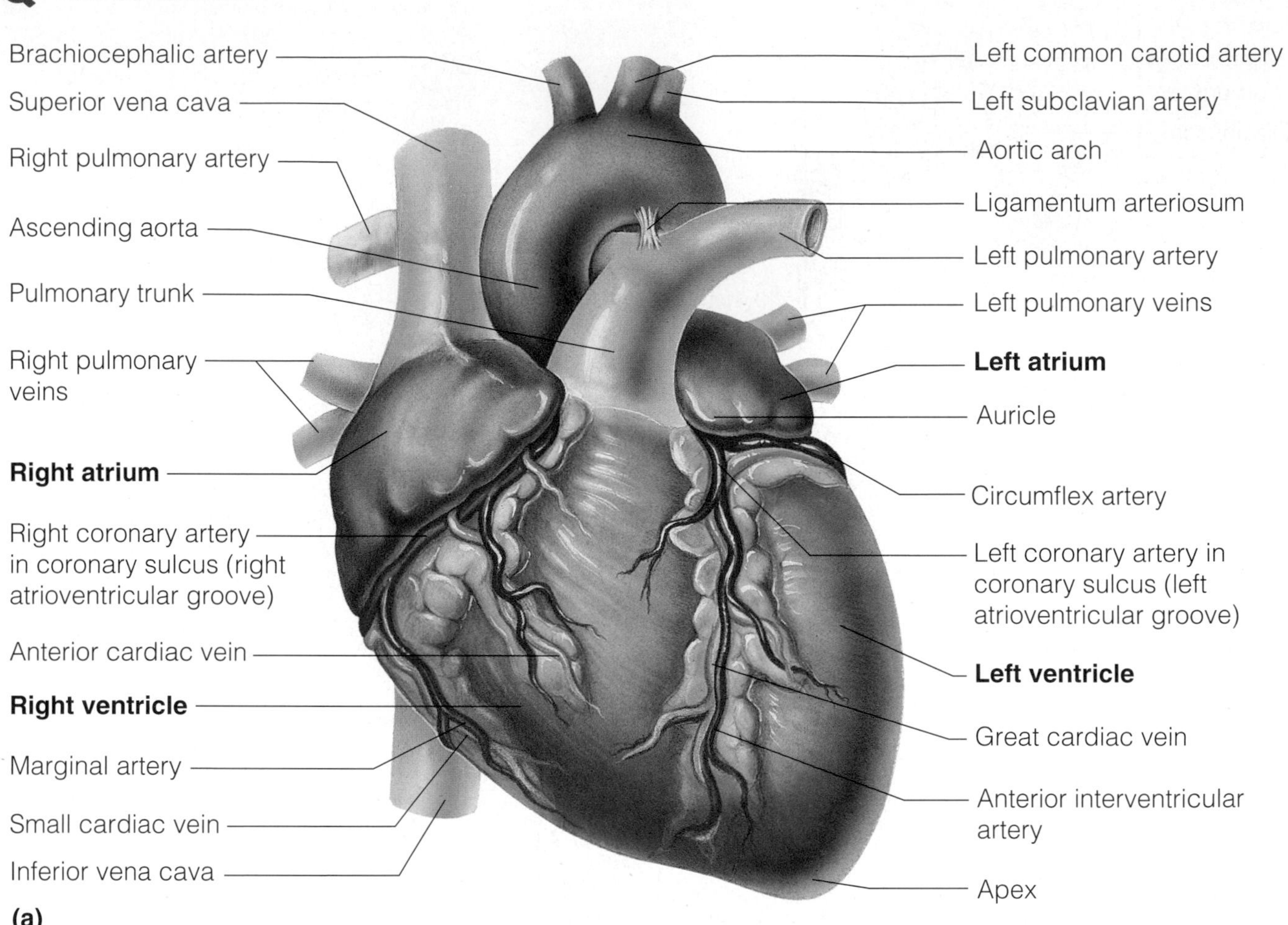

FIGURE 11.2 Gross anatomy of the heart. (a) Anterior surface view.

membranes. This fluid allows the heart to beat easily in a relatively frictionless environment as the pericardial layers slide smoothly across each other.

Homeostatic Imbalance

Inflammation of the pericardium, *pericarditis* (per″ĭ-kar-di′tis), often results in a decrease in the amount of serous fluid. This causes the pericardial layers to bind and stick to each other, forming painful *adhesions* that interfere with heart movements. ▲

The heart walls are composed of three layers: the outer *epicardium* (the visceral pericardium described above), the *myocardium,* and the innermost *endocardium* (Figure 11.2b). The **myocardium** (mi″o-kar′de-um) consists of thick bundles of cardiac muscle twisted and whorled into ringlike arrangements (see Figure 6.2b, p. 181). It is the layer that actually contracts. The myocardium is reinforced internally by a dense, fibrous connective tissue network called the "skeleton of the heart." The **endocardium** (en″do-kar′de-um) is a thin, glistening sheet of endothelium that lines the heart chambers. It is continuous with the linings of the blood vessels leaving and entering the heart. Figure 11.2 shows two views of the heart—an external anterior view and a frontal section. As the anatomical areas of the heart are described in the next section, keep referring to Figure 11.2 to locate each of the heart structures or regions.

A *The left ventricle's walls are the thickest; that chamber pumps blood throughout the entire body and back to the heart; the right ventricle serves a short circuit through the lungs and back to the heart so does not require as much muscle tissue.*

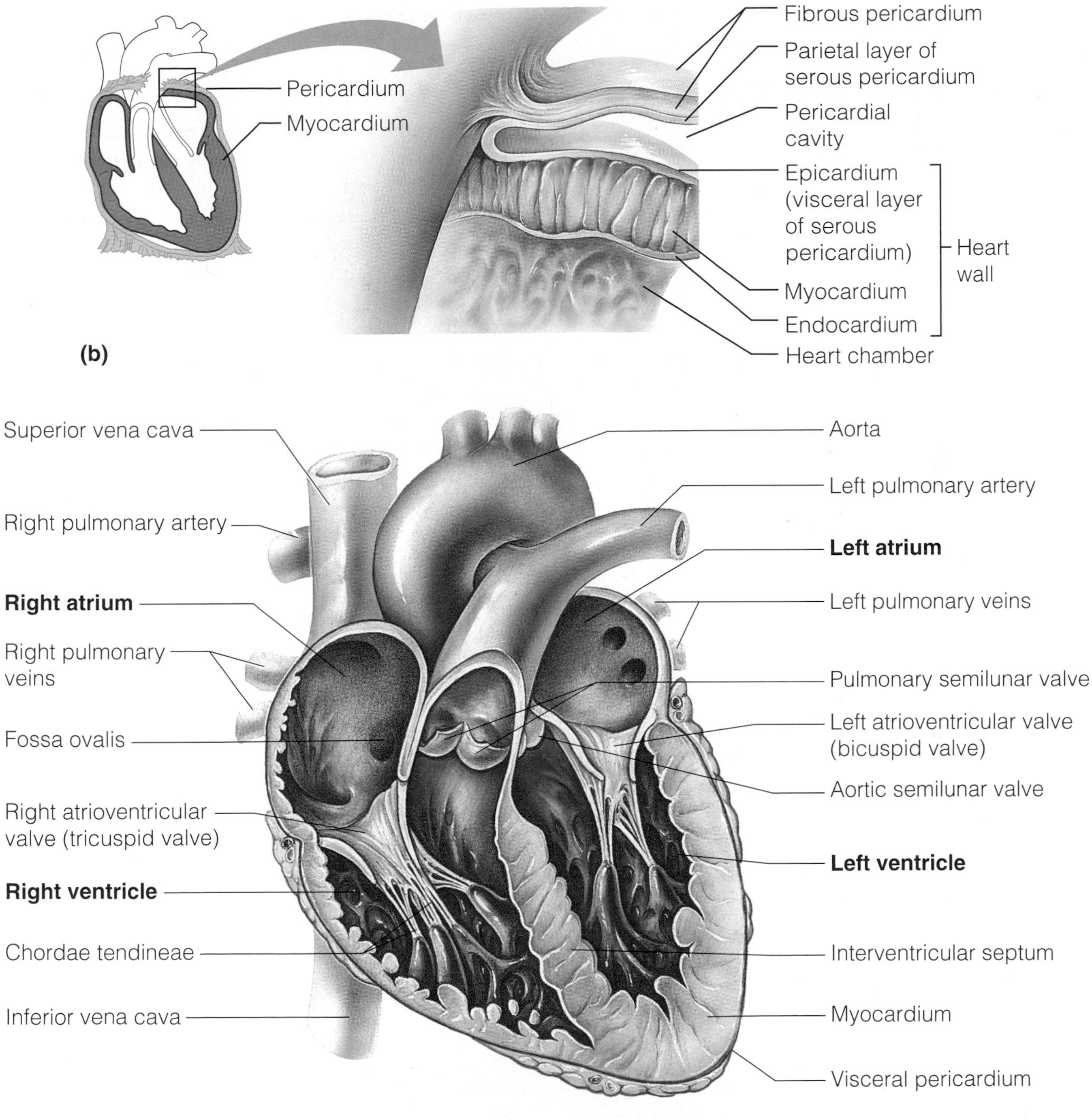

FIGURE 11.2 (*continued*) **(b)** Heart wall and coverings. **(c)** Frontal section showing interior chambers and valves.

Chambers and Associated Great Vessels

The heart has four hollow chambers or cavities—two **atria** (a'tre-ah, singular *atrium*) and two **ventricles** (ven'trĭ-kulz). Each of these chambers is lined with endocardium, which helps blood flow smoothly through the heart. The superior atria are primarily *receiving chambers*. As a rule, they are not important in the pumping activity of the heart. Blood flows into the atria under low pressure from the veins of the body and then continues on to fill the ventricles. The inferior, thick-walled ventricles are the *discharging chambers*, or actual pumps of the heart. When they contract, blood is propelled out of the heart and into the circulation. As illustrated in

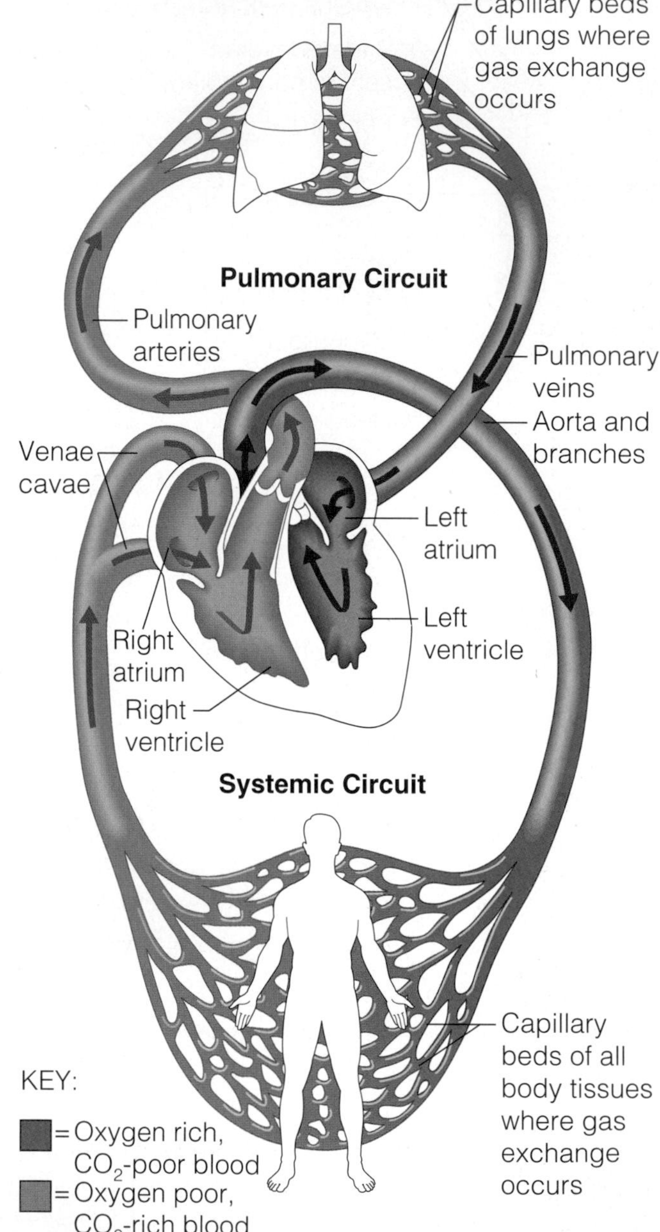

FIGURE 11.3 The systemic and pulmonary circulations. The left side of the heart is the systemic pump; the right side is the pulmonary circuit pump. (Although there are two pulmonary arteries, one each to the right and left lung, for simplicity only one is shown.)

Figure 11.2a, the right ventricle forms most of the heart's anterior surface; the left ventricle forms its apex. The septum that divides the heart longitudinally is referred to as the **interventricular** or **interatrial septum,** depending on which chamber it divides and separates.

Although it is a single organ, the heart functions as a double pump. The right side works as the pulmonary circuit pump. It receives relatively oxygen-poor blood from the veins of the body through the large **superior** and **inferior venae cavae** (ka′ve) and pumps it out through the **pulmonary trunk.** The pulmonary trunk splits into the right and left **pulmonary arteries,** which carry blood to the lungs, where oxygen is picked up and carbon dioxide is unloaded. Oxygen-rich blood drains from the lungs and is returned to the left side of the heart through the four **pulmonary veins.** The circulation just described, from the right side of the heart to the lungs and back to the left side of the heart, is called the **pulmonary circulation** (Figure 11.3). Its only function is to carry blood to the lungs for gas exchange and then return it to the heart.

Blood returned to the left side of the heart is pumped out of the heart into the **aorta** (a-or′tah), from which the systemic arteries branch to supply essentially all body tissues. Oxygen-poor blood circulates from the tissues back to the right atrium via the systemic veins, which finally empty their cargo into either the superior or inferior vena cava. This second circuit, from the left side of the heart through the body tissues and back to the right side of the heart, is called the **systemic circulation** (see Figure 11.3). It supplies oxygen- and nutrient-rich blood to all body organs. Because the left ventricle is the systemic pump that pumps blood over a much longer pathway through the body, its walls are substantially thicker than those of the right ventricle, and it is a much more powerful pump.

Valves

The heart is equipped with four valves, which allow blood to flow in only one direction through the heart chambers—from the atria through the ventricles and out the great arteries leaving the heart (see Figure 11.2a). The **atrioventricular** (a″tre-o-ven-trik′u-lar), or **AV, valves** are located between the atrial and ventricular chambers on each side. The AV valves prevent backflow into the atria when the ventricles contract. The left AV valve—the **bicuspid,** or **mitral** (mi′tral), **valve**—consists of two flaps, or cusps, of endocardium. The right AV valve, the **tricuspid valve,** has three flaps. Tiny white cords, the **chordae tendineae** (kor′de ten-din′e)—literally, "tendinous cords" (but I like to think of them as the "heart strings" of song)—anchor the flaps to the walls of the ventri-

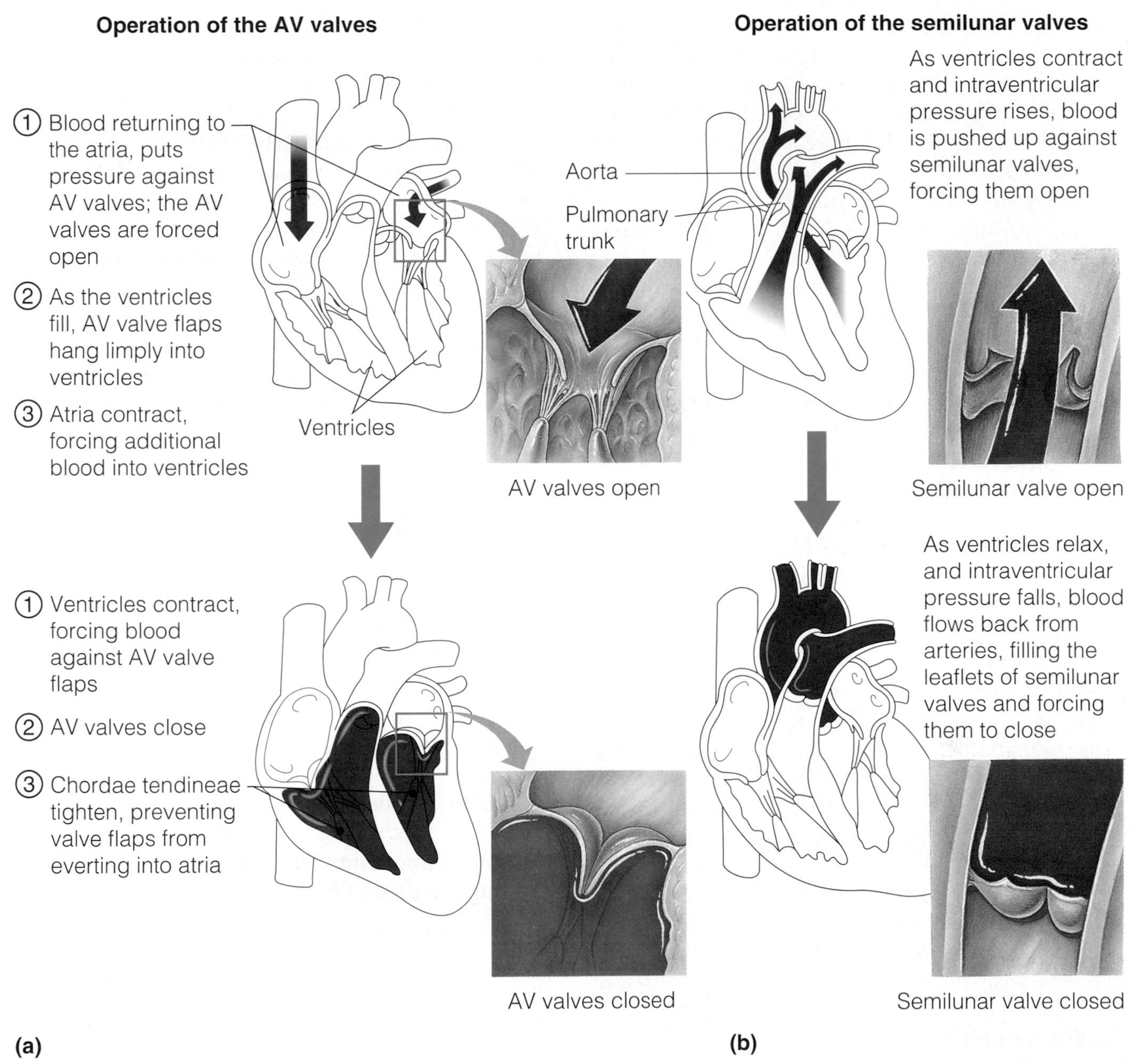

FIGURE 11.4 Operation of the heart valves. (a) Atrioventricular (AV) valves. **(b)** Semilunar valves.

cles. When the heart is relaxed and blood is passively filling its chambers, the AV-valve flaps hang limply into the ventricles (Figure 11.4a). As the ventricles contract, they press on the blood in their chambers, and the intraventricular pressure (pressure inside the ventricles) begins to rise. This causes the AV-valve flaps to be forced upward, closing the valves. At this point the chordae tendineae are working to anchor the flaps in a closed position. If the flaps were unanchored, they would blow upward into the atria like an umbrella being turned inside out by a gusty wind. In this manner, the AV valves prevent backflow into the atria when the ventricles are contracting.

The second set of valves, the **semilunar** (sem″e-lu′nar) **valves,** guards the bases of the two large arteries leaving the ventricular chambers. Thus, they are known as the **pulmonary** and **aortic semilunar valves** (see Figure 11.2b). Each semilunar valve has three leaflets that fit tightly together when the valves are closed. When the ventricles are contracting and forcing blood out of the heart, the leaflets are forced open and flattened against the walls of the arteries by the tremendous force of rushing blood (Figure 11.4b). Then, when the ventricles relax, the blood begins to flow backward toward the heart, and the leaflets fill with blood, closing the valves. This prevents arterial blood from reentering the heart.

Each set of valves operates at a different time. The AV valves are open during heart relaxation and closed when the ventricles are contracting. The semilunar valves are closed during heart relaxation and are forced open when the ventricles contract. As they open and close in response to pressure changes in the heart, the valves force blood to continually move forward in its journey through the heart.

Homeostatic Imbalance

Heart valves are basically simple devices, and the heart—like any mechanical pump—can function with "leaky" valves as long as the damage is not too great. However, severely deformed valves can seriously hamper cardiac function. For example, an *incompetent valve* forces the heart to pump and repump the same blood because the valve does not close properly and blood backflows. In *valvular stenosis,* the valve flaps become stiff, often because of repeated bacterial infection of the endocardium **(endocarditis).** This forces the heart to contract more vigorously than normal. In each case, the heart's workload increases, and ultimately the heart weakens and may fail. Under such conditions, the faulty valve is replaced with a synthetic valve or a valve taken from a pig heart. ▲

Cardiac Circulation

Although the heart chambers are bathed with blood almost continuously, the blood contained in the heart does not nourish the myocardium. The blood supply that oxygenates and nourishes the heart is provided by the right and left coronary arteries. The **coronary arteries** branch from the base of the aorta and encircle the heart in the **coronary sulcus (atrioventricular groove)** at the junction of the atria and ventricles (see Figure 11.2a). The coronary arteries and their major branches (the **anterior interventricular** and **circumflex arteries** on the left, and the **posterior interventricular** and **marginal arteries** on the right) are compressed when the ventricles are contracting and fill when the heart is relaxed. The myocardium is drained by several **cardiac veins,** which empty into an enlarged vessel on the backside of the heart called the **coronary sinus.** The coronary sinus, in turn, empties into the right atrium.

Homeostatic Imbalance

When the heart beats at a very rapid rate, the myocardium may receive an inadequate blood supply because the relaxation periods (when the blood is able to flow to the heart tissue) are shortened. Situations in which the myocardium is deprived of oxygen often result in crushing chest pain called **angina pectoris** (an-ji′nah pek′tor-is). This pain is a warning that should *never* be ignored because, if angina is prolonged, the ischemic heart cells may die, forming an **infarct.** The resulting **myocardial infarction** (in-fark′shun) is commonly called a "heart attack" or "coronary." ▲

Physiology of the Heart

As the heart beats or contracts, the blood makes continuous round trips—into and out of the heart, through the rest of the body, and then back to the heart—only to be sent out again. The amount of work that a heart does is almost too incredible to believe. In one day it pushes the body's supply of 6 quarts or so of blood (6 liters [L]) through the blood vessels over 1000 times, meaning that it actually pumps about 6000 quarts of blood in a single day!

Intrinsic Conduction System of the Heart: Setting the Basic Rhythm

Unlike skeletal muscle cells that must be stimulated by nerve impulses before they will contract, cardiac muscle cells can and do contract spontaneously and independently, even if all nervous connections are severed. Moreover, these spontaneous contractions occur in a regular and continuous way. Although cardiac muscle *can* beat independently, the muscle cells in different areas of the heart have different rhythms. The atrial cells beat about 60 times per minute, but the ventricular cells contract much more slowly (20–40/min). Therefore, without some type of unifying control system, the heart would be an uncoordinated and inefficient pump.

Two systems act to regulate heart activity. One of these involves the nerves of the autonomic nervous system that act like brakes and accelerators to decrease or increase the heart rate depending on which division is activated. This topic is considered later (see pp. 359–360). The second system is the **intrinsic conduction system,** or **nodal system,** that is built into the heart tissue (Figure 11.5) and sets

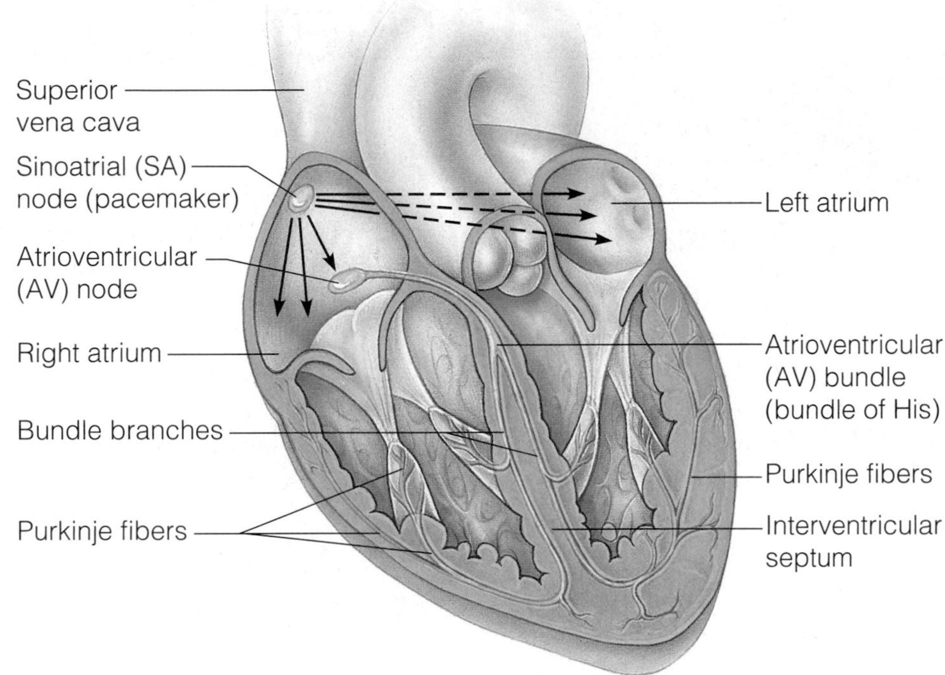

FIGURE 11.5 The intrinsic conduction system of the heart. The depolarization wave is initiated by the sinoatrial (SA) node and then passes successively through the atrial myocardium to the atrioventricular (AV) node, the AV bundle, the right and left bundle branches, and the Purkinje fibers in the ventricular walls.

its basic rhythm. The intrinsic conduction system is composed of a special tissue found nowhere else in the body; it is much like a cross between muscle and nervous tissue. This system causes heart muscle depolarization in only one direction—from the atria to the ventricles. In addition, it enforces a contraction rate of approximately 75 beats per minute on the heart; thus, the heart beats as a coordinated unit.

One of the most important parts of the intrinsic conduction system is a crescent-shaped node of tissue called the **sinoatrial** (si″no-a′tre-al) **(SA) node,** located in the right atrium. Other components include the **atrioventricular (AV) node** at the junction of the atria and ventricles, the **atrioventricular (AV) bundle (bundle of His)** and the right and left **bundle branches** located in the interventricular septum, and finally the **Purkinje** (pur-kin′je) **fibers,** which spread within the muscle of the ventricle walls.

The SA node is a tiny cell mass with a mammoth job. Because it has the highest rate of depolarization in the whole system, it starts each heartbeat and sets the pace for the whole heart. Consequently, the SA node is often called the **pacemaker.** From the SA node, the impulse spreads through the atria to the AV node, and then the atria contract. At the AV node, the impulse is delayed briefly to give the atria time to finish contracting. It then passes rapidly through the AV bundle, the bundle branches, and the Purkinje fibers, resulting in a "wringing" contraction of the ventricles that begins at the heart apex and moves toward the atria. This contraction effectively ejects blood superiorly into the large arteries leaving the heart. The "A Closer Look" box on p. 356 describes *electrocardiography,* the clinical procedure for mapping the electrical activity of the heart.

Homeostatic Imbalance

Because the atria and ventricles are separated from one another by "insulating" connective tissue, which is part of the fibrous skeleton of the heart, depolarization waves can reach the ventricles only by traveling through the AV node. Thus, any damage to the AV node can partially or totally release the ventricles from the control of the SA node. When this occurs, the ventricles (thus the heart) begin to beat at their own rate, which is much slower, some or all of the time. This condition is called *heart block.*

A Closer Look

Electrocardiography: (Don't) Be Still My Heart

When impulses pass through the heart, electrical currents are generated that spread throughout the body. These impulses can be detected on the body surface and recorded with an *electrocardiograph*. The recording that is made, the **electrocardiogram (ECG),** traces the flow of current through the heart. The illustration shows a normal ECG tracing.

The typical ECG has three recognizable waves. The first wave, the **P wave,** is small and signals the depolarization of the atria immediately before they contract. The large **QRS complex,** which results from the depolarization of the ventricles, has a complicated shape. It precedes the contraction of the ventricles. The **T wave** results from currents flowing during the repolarization of the ventricles. (Atrial repolarization is generally hidden by the large QRS complex, which is being recorded at the same time.)

Abnormalities in the shape of the waves and changes in their timing send signals that something may be wrong with the intrinsic conduction system or may indicate a *myocardial infarct* (present or past). A myocardial infarct is an area of heart tissue in which the cardiac cells have died; it is generally a result of *ischemia.* During *fibrillation,* the normal pattern of the ECG is totally lost, and the heart ceases to act as a functioning pump.

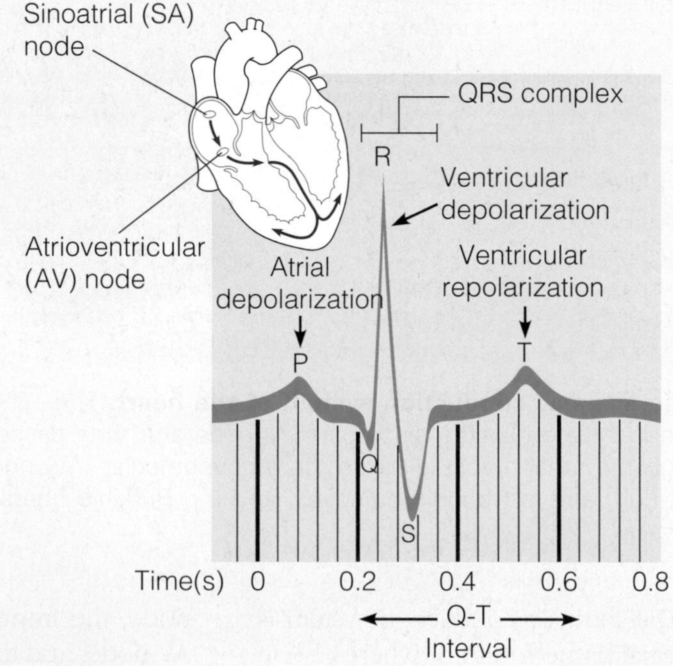

An electrocardiogram tracing showing the three normally recognizable deflection waves—P, QRS, and T.

Q *Are the ventricular cardiac cells contracting isometrically or isotonically during the first part of phase 2?*

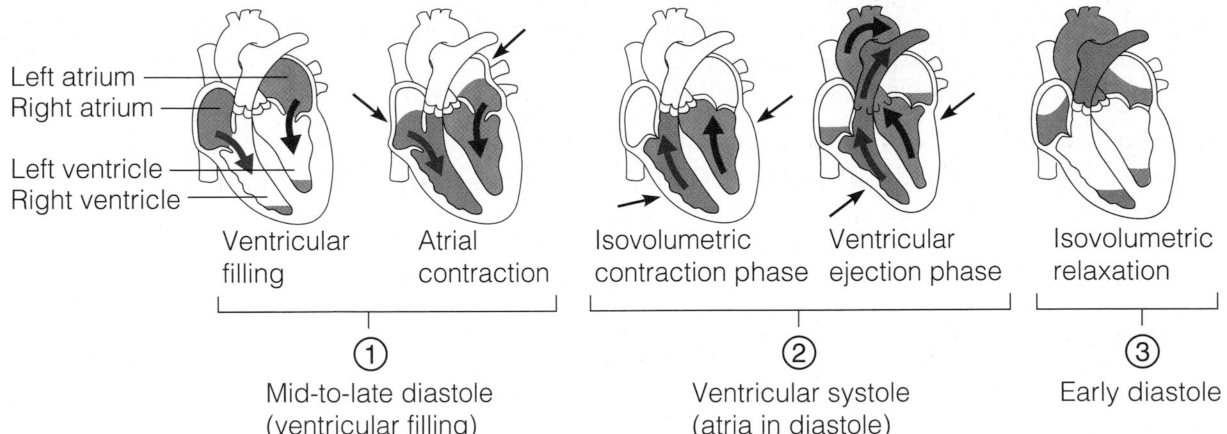

FIGURE 11.6 Summary of events occurring during the cardiac cycle. (Small black arrows indicate the regions of the heart that are contracting; thick red and blue arrows indicate direction of blood flow. During the *isovolumetric* (literally "same volume measurement") phases in periods 2 and 3, the ventricles are closed chambers and the volume of blood they contain is unchanging.)

There are other conditions that can interfere with the regular conduction of impulses across the heart—for example, damage to the SA node results in a slower heart rate. When this is a problem, artificial pacemakers are usually installed surgically. **Ischemia** (is-ke′me-ah), or lack of an adequate blood supply to the heart muscle, may lead to **fibrillation**—a rapid uncoordinated shuddering of the heart muscle (it looks like a bag of worms). Fibrillation makes the heart totally useless as a pump and is a major cause of death from heart attacks in adults. ▲

Tachycardia (tak″e-kar′de-ah) is a rapid heart rate (over 100 beats per minute). **Bradycardia** (brad″e-kar′de-ah) is a heart rate that is substantially slower than normal (less than 60 beats per minute). Neither condition is pathological, but prolonged tachycardia may progress to fibrillation.

Cardiac Cycle and Heart Sounds

In a healthy heart, the atria contract simultaneously. Then, as they start to relax, contraction of the ventricles begins. **Systole** (sis′to-le) and **diastole** (di-as′to-le) mean heart *contraction* and *relaxation,* respectively. Since most of the pumping work is done by the ventricles, these terms always refer to the contraction and relaxation of the *ventricles* unless otherwise stated.

The term **cardiac cycle** refers to the events of one complete heartbeat, during which both atria and ventricles contract and then relax. Since the average heart beats approximately 75 times per minute, the length of the cardiac cycle is normally about 0.8 second. We will consider the cardiac cycle in terms of events occurring during three periods—*mid-to-late diastole, ventricular systole,* and *early diastole* (Figure 11.6).

1. **Mid-to-late diastole.** Our discussion begins with the heart in complete relaxation. At this point, the pressure in the heart is low, and blood is flowing passively into and through the atria into the ventricles from the pulmonary and systemic circulations. The semilunar valves are closed, and the AV valves are open. Then the atria contract and force the blood remaining in their chambers into the ventricles.
2. **Ventricular systole.** Shortly after, ventricular contraction (systole) begins and the pressure within the ventricles increases rapidly, closing the AV valves. When the intraventricular pressure

Isometrically until they have enough force to overcome the back pressure of the blood against the semilunar valves at which point their contraction becomes isotonic.

(pressure in the ventricles) is higher than the pressure in the large arteries leaving the heart, the semilunar valves are forced open, and blood rushes through them out of the ventricles. During ventricular systole, the atria are relaxed, and their chambers are again filling with blood.

3. **Early diastole.** At the end of systole, the ventricles relax, the semilunar valves snap shut (preventing backflow), and for a moment, the ventricles are completely closed chambers. During early diastole, the intraventricular pressure drops. When it drops below the pressure in the atria (which has been increasing as blood has been filling their chambers), the AV valves are forced open, and the ventricles again begin to refill rapidly with blood, completing the cycle.

When using a stethoscope, you can hear two distinct sounds during each cardiac cycle. These **heart sounds** are often described by the two syllables "lub" and "dup," and the sequence is lub-dup, pause, lub-dup, pause, and so on. The first heart sound (lub) is caused by the closing of the AV valves. The second heart sound (dup) occurs when the semilunar valves close at the end of systole. The first heart sound is longer and louder than the second heart sound, which tends to be short and sharp.

Homeostatic Imbalance

Abnormal or unusual heart sounds are called *murmurs.* Blood flows silently as long as the flow is smooth and uninterrupted. If it strikes obstructions, its flow becomes turbulent and generates sounds, such as heart murmurs, that can be heard with a stethoscope. Heart murmurs are fairly common in young children (and some elderly people) with perfectly healthy hearts, probably because their heart walls are relatively thin and vibrate with rushing blood. However, most often, murmurs indicate valve problems. For example, if a valve does not close tightly (is *incompetent*), a swishing sound will be heard *after* that valve has (supposedly) closed, as the blood flows back through the partially open valve. Distinct sounds also can be heard when blood flows turbulently through *stenosed* (narrowed) valves. ▲

Cardiac Output

Cardiac output (CO) is the amount of blood pumped out by *each* side of the heart (actually each ventricle) in 1 minute. It is the product of the **heart rate (HR)** and the **stroke volume (SV).** Stroke volume is the volume of blood pumped out by a ventricle with each heartbeat. In general, stroke volume increases as the force of ventricular contraction increases. If we use the normal resting values for heart rate (75 beats per minute) and stroke volume (70 ml per beat), the average adult cardiac output can be easily figured:

$$CO = HR\ (75\ \text{beats/min}) \times SV\ (70\ \text{ml/beat})$$

$$CO = 5250\ \text{ml/min}$$

Since the normal adult blood volume is about 5000 ml, the entire blood supply passes through the body once each minute. Cardiac output varies with the demands of the body. It rises when the stroke volume is increased or the heart beats faster or both; it drops when either or both of these factors decrease. Since this is so, let's take a look at how stroke volume and heart rate are regulated.

Regulation of Stroke Volume A healthy heart pumps out about 60 percent of the blood present in its ventricles. As noted above, this is approximately 70 ml (about 2 ounces) with each heartbeat. According to *Starling's law of the heart,* the critical factor controlling stroke volume is how much the cardiac muscle cells are stretched just before they contract. The more they are stretched, the stronger the contraction will be. The important factor stretching the heart muscle is *venous return,* the amount of blood entering the heart and distending its ventricles. If one side of the heart suddenly begins to pump more blood than the other, the increased venous return to the opposite ventricle will force it to pump out an equal amount, thus preventing backup of blood in the circulation.

Anything that increases the volume or speed of venous return also increases stroke volume and force of contraction (Figure 11.7). For example, a slow heartbeat allows more time for the ventricles to fill. Exercise speeds venous return because it results in increased heart rate and force. The enhanced squeezing action of active skeletal muscles on the veins returning blood to the heart, the so-called muscular pump, also plays a major role in increasing the venous return. On the other hand, low venous return, such as might result from severe blood loss or an extremely rapid heart rate, decreases stroke volume, causing the heart to beat less forcefully.

Factors Modifying Basic Heart Rate In healthy people, stroke volume tends to be relatively constant. However, when blood volume drops suddenly or

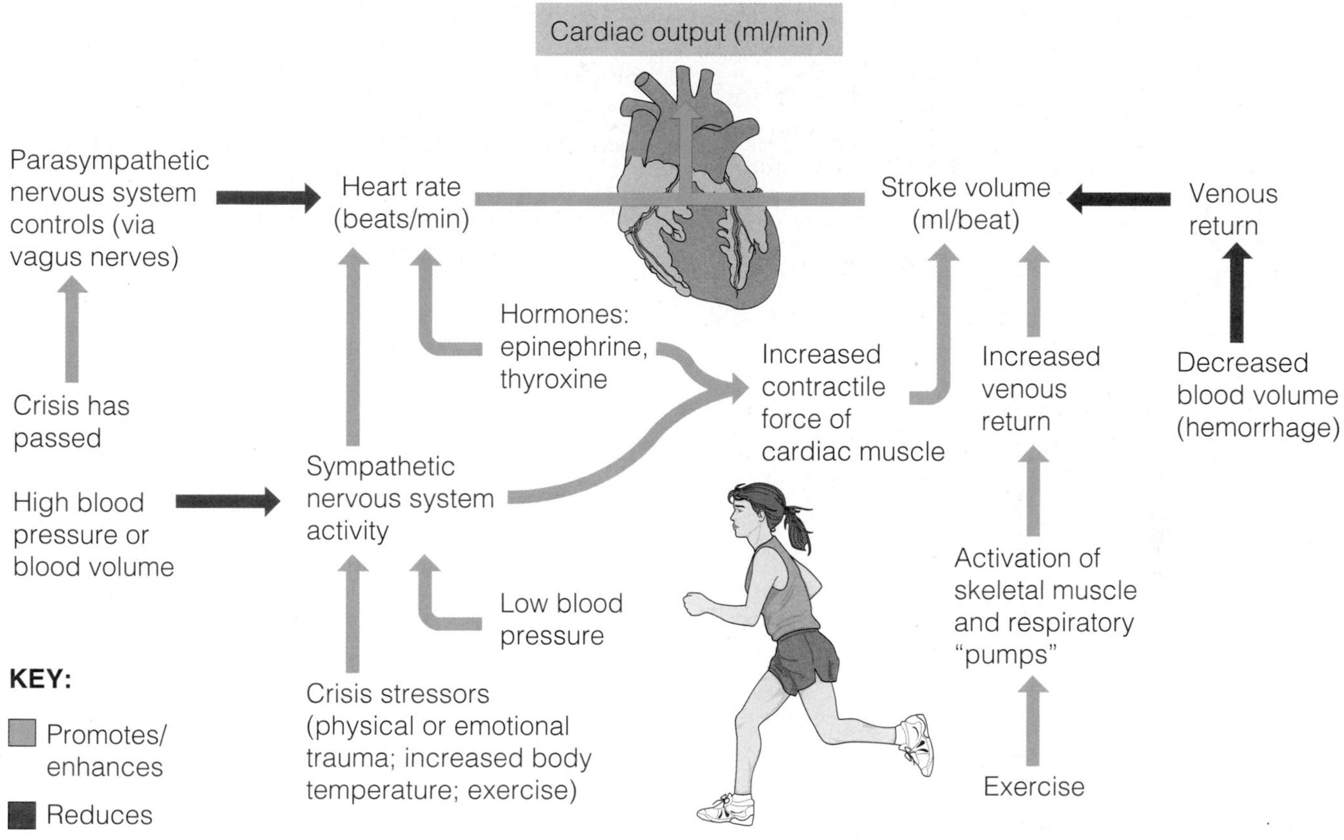

FIGURE 11.7 Influence of selected factors on cardiac output. Note that the direction of the flowchart is from the bottom up.

when the heart has been seriously weakened, stroke volume declines, and cardiac output is maintained by a faster heartbeat. Although heart contraction does not depend on the nervous system, its rate *can* be changed temporarily by the autonomic nerves. Indeed, the most important external influence on heart rate is the activity of the autonomic nervous system. Heart rate is also modified by various chemicals, hormones, and ions. Some of these factors are summarized in Figure 11.7.

Neural (ANS) Controls During times of physical or emotional stress, the nerves of the *sympathetic division* of the autonomic nervous system more strongly stimulate the SA and AV nodes and the cardiac muscle itself. As a result, the heart beats more rapidly. This is a familiar phenomenon to anyone who has ever been frightened or has had to run to catch a bus. As fast as the heart pumps under ordinary conditions, it really speeds up when special demands are placed on it. Since a faster blood flow increases the rate at which fresh blood reaches body cells, more oxygen and glucose are made available to them during periods of stress. When demand declines, the heart adjusts. *Parasympathetic nerves,* primarily the vagus nerves, slow and steady the heart, giving it more time to rest during noncrisis times. In patients with *congestive heart failure,* a condition in which the heart is nearly "worn out" due to age, hypertensive heart disease, or another pathological process, the heart pumps weakly. For those patients, the drug digitalis is routinely prescribed. It enhances contractile force and stroke volume of the heart resulting in greater cardiac output.

Various hormones and ions can have a dramatic effect on heart activity. *Epinephrine,* which mimics the effect of the sympathetic nerves, and *thyroxine* both increase heart rate. Electrolyte imbalances pose a real threat to the heart. For example, reduced levels of ionic calcium in the blood depress the heart, whereas excessive blood calcium causes such prolonged contractions that the heart may stop entirely. Excesses or a lack of needed ions such as sodium and potassium also modify heart activity. A deficit of potassium ions in the blood, for example, causes the heart to beat feebly, and abnormal heart rhythms appear.

Physical Factors A number of physical factors, including age, gender, exercise, and body temperature, influence heart rate. Resting heart rate is fastest in the fetus (140–160 beats per minute) and then gradually decreases throughout life. The average heart rate is faster in females (72–80 beats per minute) than in males (64–72 beats per minute). Heat increases heart rate by boosting the metabolic rate of heart cells. This explains the rapid, pounding heartbeat you feel when you have a high fever and accounts in part for the effect of exercise on heart rate (remember, working muscles generate heat). Cold has the opposite effect; it directly decreases heart rate. As noted above, exercise acts through nervous system controls (sympathetic division) to increase heart rate (and also, through the action of the muscular pump, to increase stroke volume).

Homeostatic Imbalance

The pumping action of the healthy heart maintains a balance between cardiac output and venous return. But when the pumping efficiency of the heart is depressed so that circulation is inadequate to meet tissue needs, **congestive heart failure (CHF)** occurs. Congestive heart failure is usually a progressive condition that reflects weakening of the heart by *coronary atherosclerosis* (clogging of the coronary vessels with fatty buildup), persistent high blood pressure, or multiple myocardial infarctions (leading to repair with noncontracting scar tissue).

Because the heart is a double pump, each side can fail independently of the other. If the left heart fails, *pulmonary congestion* occurs. The right side of the heart continues to propel blood to the lungs, but the left side is unable to eject the returning blood into the systemic circulation. As blood vessels within the lungs become swollen with blood, the pressure within them increases, and fluid leaks from the circulation into the lung tissue, causing **pulmonary edema.** If untreated, the person suffocates.

If the right side of the heart fails, *peripheral congestion* occurs as blood backs up in the systemic circulation. Edema is most noticeable in the distal parts of the body: The feet, ankles, and fingers become swollen and puffy. Failure of one side of the heart puts a greater strain on the opposite side, and eventually the whole heart fails. ▲

Blood Vessels

Blood circulates inside the blood vessels, which form a closed transport system, the so-called **vascular system.** The idea that blood circulates, or "makes rounds," through the body is only about 300 years old. The ancient Greeks believed that blood moved through the body like an ocean tide, first moving out from the heart and then ebbing back to it in the same vessels to get rid of its impurities in the lungs. It was not until the seventeenth century that William Harvey, an English physician, proved that blood did, in fact, move in circles.

Like a system of roads, the vascular system has its freeways, secondary roads, and alleys. As the heart beats, blood is propelled into the large **arteries** leaving the heart. It then moves into successively smaller and smaller arteries and then into the **arterioles** (ar-ter′e-ōlz), which feed the **capillary** (kap′ĭ-lar″e) **beds** in the tissues. Capillary beds are drained by **venules** (ven′ulz), which in turn empty into **veins** that finally empty into the great veins (venae cavae) entering the heart. Thus arteries, which carry blood away from the heart, and veins, which drain the tissues and return the blood to the heart, are simply conducting vessels—the freeways and secondary roads. Only the tiny hairlike capillaries, which extend and branch through the tissues and connect the smallest arteries (arterioles) to the smallest veins (venules), directly serve the needs of the body cells. The capillaries are the side streets or alleys that intimately intertwine among the body cells and provide access to individual "homes." It is only through their walls that exchanges between the tissue cells and the blood can occur.

Notice that this book routinely depicts arteries red and veins blue because, by convention, red indicates oxygen-rich blood and blue indicates relatively oxygen-depleted, carbon dioxide-rich blood, the normal status of blood in most of the body's arteries and veins. However, there are exceptions to this convention which will be pointed out as they are encountered.

Microscopic Anatomy of Blood Vessels

Tunics

Except for the microscopic capillaries, the walls of blood vessels have three coats, or tunics (Figure 11.8). The **tunica intima** (tu′nĭ-kah in-tim′ah),

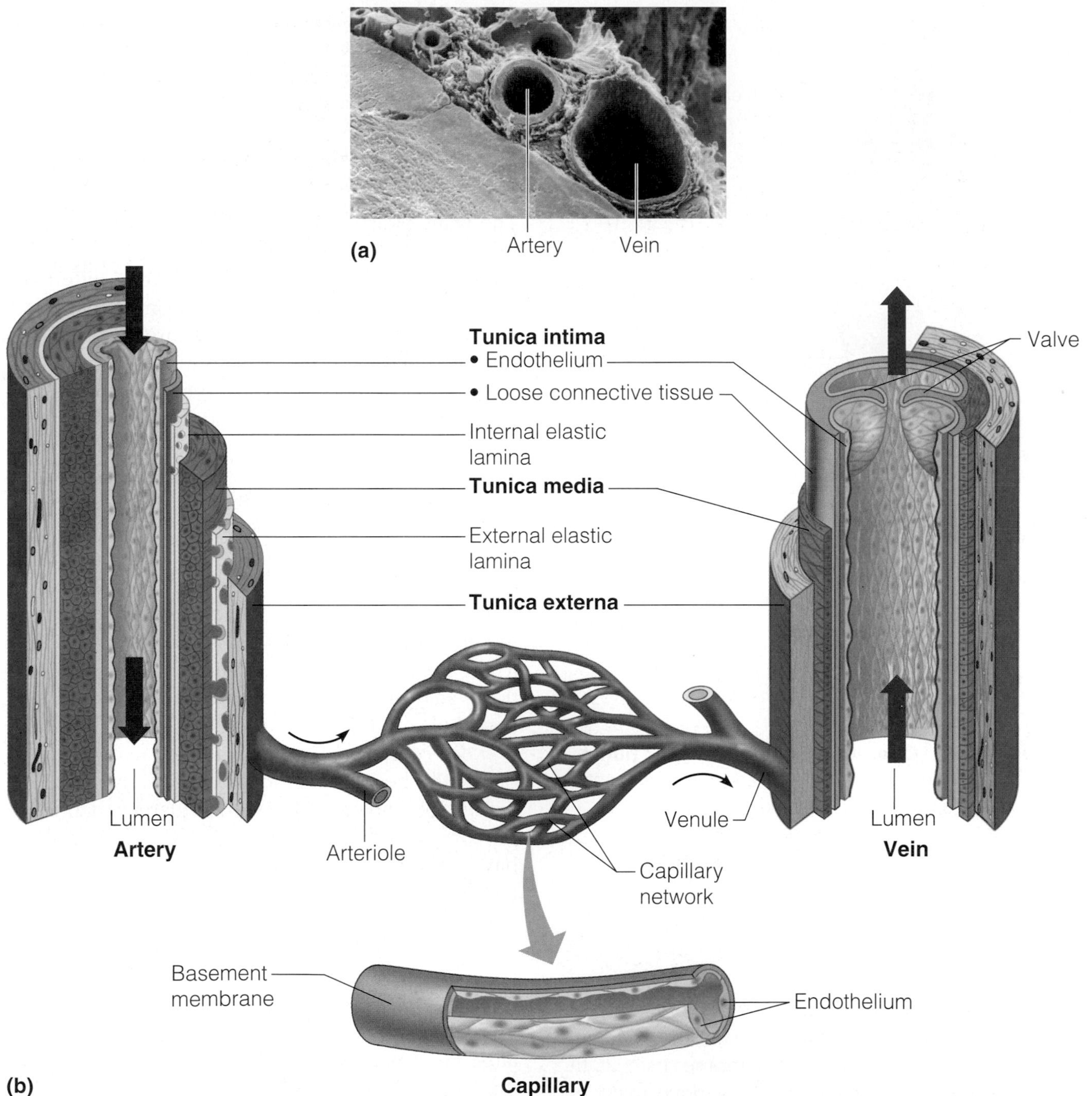

FIGURE 11.8 **Structure of blood vessels. (a)** Scanning electron micrograph of an artery and vein in cross section (120×). **(b)** The walls of arteries and veins are composed of three tunics: the tunica intima (endothelium underlain by a basement membrane), tunica media (smooth muscle and elastic fibers), and tunica externa (largely collagen fibers). Capillaries—between arteries and veins in the circulatory pathway—are composed only of the tunica intima. Notice that the tunica media is thick in arteries and relatively thin in veins. Part **(a):** Copyright by R.G. Kessel and R.H. Kardon, *Tissues and Organs: A Text-Atlas of Scanning Electron Microscopy,* W.H. Freeman and Company, 1979, all rights reserved.

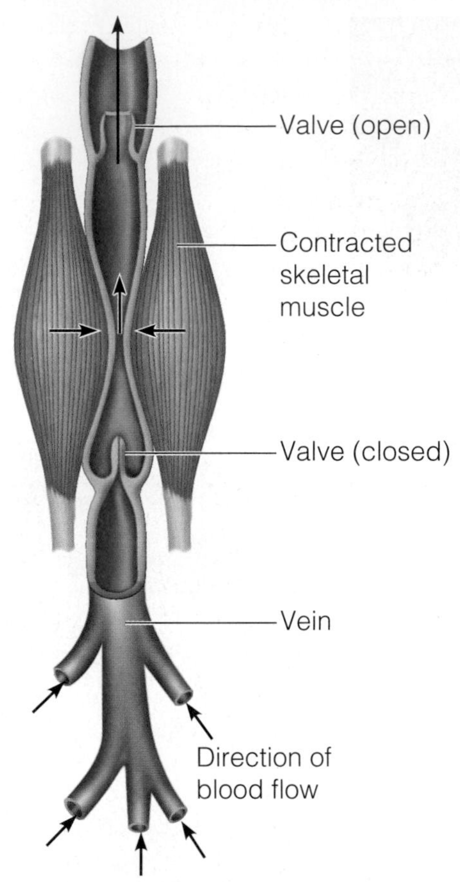

FIGURE 11.9 Operation of the muscular pump. When skeletal muscles contract and press against the flexible veins, the valves proximal to the area of contraction are forced open, and blood is squeezed toward the heart. The valves distal to the point of contraction are closed by the backflowing blood.

which lines the lumen or interior of the vessels, is a thin layer of **endothelium** (squamous epithelial cells) resting on a basement membrane. Its cells fit closely together and form a slick surface that decreases friction as blood flows through the vessel lumen.

The **tunica media** (me′de-ah) is the bulky middle coat. It is mostly smooth muscle and elastic tissue. Some of the larger arteries have *elastic laminae,* complete sheets of elastic tissue in addition to the scattered elastic fibers. The smooth muscle, which is controlled by the sympathetic nervous system, is active in changing the diameter of the vessels. As the vessels constrict or dilate, blood pressure increases or decreases, respectively.

The **tunica externa** (eks′tern-ah) is the outermost tunic; it is composed largely of fibrous connective tissue. Its function is basically to support and protect the vessels.

Structural Differences between Arteries, Veins, and Capillaries

The walls of arteries are usually much thicker than the walls of veins. Their tunica media, in particular, tends to be much heavier. This structural difference is related to a difference in function of these two types of vessels. Arteries, which are closer to the pumping action of the heart, must be able to expand as blood is forced into them and then recoil passively as the blood flows off into the circulation during diastole. Their walls must be strong and stretchy enough to take these continuous changes in pressure (see Figure 11.8).

On the other hand, veins are far from the heart in the circulatory pathway, and the pressure in them tends to be low all the time. Thus veins have thinner walls. However, since the blood pressure in veins is usually too low to force the blood back to the heart and blood returning to the heart often flows against gravity, veins are modified to ensure that the amount of blood returning to the heart *(venous return)* equals the amount being pumped out of the heart *(cardiac output)* at any time. The lumens of veins tend to be much larger than those of corresponding arteries, and the larger veins have **valves** that prevent backflow of blood (see Figure 11.8). Skeletal muscle activity also enhances venous return. As the muscles surrounding the veins contract and relax, the blood is forced or squeezed through the veins toward the heart (Figure 11.9). Finally, when we inhale, the drop in pressure that occurs in the thorax causes the large veins near the heart to expand and fill. Thus, the "respiratory pump" also helps return blood to the heart (see Figure 11.7).

The transparent walls of the capillaries are only one cell layer thick—just the tunica intima. Because of this exceptional thinness, exchanges are easily made between the blood and the tissue cells. The tiny capillaries tend to form interweaving networks called **capillary beds.** The flow of blood from an arteriole to a venule—that is, through a capillary bed—is called **microcirculation.** In most body regions, a capillary bed consists of two types of vessels: (1) a **vascular shunt,** a vessel that directly connects the arteriole and venule at opposite ends

Q *Assume the capillary bed depicted here is in the biceps brachii muscle of your arm. What condition would the capillary bed be in (**a** or **b**) if you were doing push-ups at the gym?*

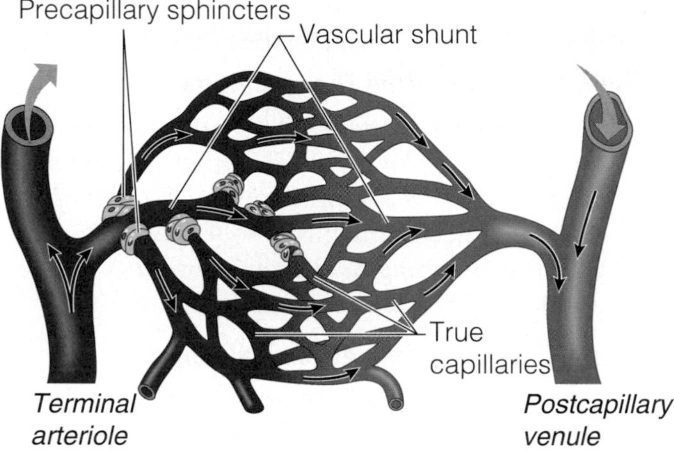

(a) Sphincters open

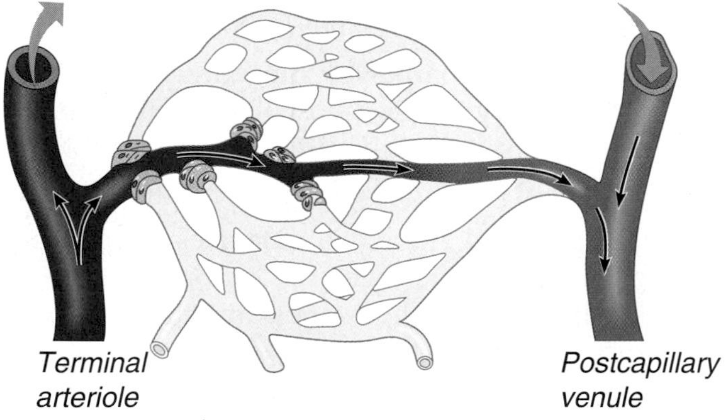

(b) Sphincters closed

FIGURE 11.10 Anatomy of a capillary bed. The vascular shunt bypasses the true capillaries when precapillary sphincters controlling blood entry into the true capillaries are constricted.

of the bed, and (2) **true capillaries,** the actual *exchange vessels* (Figure 11.10).

The *true capillaries* number 10 to 100 per capillary bed, depending on the organ or tissues served. They usually branch off the proximal end of the shunt and return to the distal end, but occasionally they spring from the **terminal arteriole** and empty directly into the **postcapillary venule.** A cuff of smooth muscle fibers, called a **precapillary sphincter,** surrounds the root of each true capillary and acts as a valve to regulate the flow of blood into the capillary. Blood flowing through a terminal arteriole may take one of two routes: through the true capillaries or through the shunt. When the precapillary sphincters are relaxed (open), blood flows through the true capillaries and takes part in exchanges with tissue cells. When the sphincters are contracted (closed), blood flows through the shunts and bypasses the tissue cells.

Homeostatic Imbalance

Varicose veins are common in people who stand for long periods of time (for example, dentists and hairdressers) and in obese (or pregnant) individuals. The common factor is the pooling of blood in the feet and legs and inefficient venous return resulting from inactivity or pressure on the veins. In any case, the overworked valves give way, and the veins become twisted and dilated. A serious complication of varicose veins is *thrombophlebitis* (throm″bo-fle-bi′tis), inflammation of a vein that results when a clot forms in a vessel with poor circulation. Since all venous blood must pass through the pulmonary circulation before traveling

Prove It Yourself

Venous Valves Prevent Backflow

To demonstrate the efficiency of the venous valves in preventing backflow of blood, perform the following simple experiment.

Allow one hand to hang by your side for a minute or two, until the blood vessels on its dorsal aspect become distended (swollen) with blood. Place two fingertips side by side against one of the distended veins. Then, pressing firmly, move your proximal finger along the vein toward your heart. Now release that finger. As you can see, the vein remains collapsed in spite of gravity. Now remove your distal finger, and watch the vein fill rapidly with blood.

A ***a****, The true capillaries would be flushed with blood to serve the working muscle cells.*

through the body tissues again, a common consequence of thrombophlebitis is clot detachment and *pulmonary embolism,* which is a life-threatening condition. ▲

Gross Anatomy of Blood Vessels

Major Arteries of the Systemic Circulation

The **aorta** is the largest artery of the body, and it is a truly splendid vessel. In adults, the aorta is about the size of a garden hose (with an internal diameter the size of your thumb) where it issues from the left ventricle of the heart. It decreases only slightly in size as it runs to its terminus. Different parts of the aorta are named for their location or shape. The aorta springs upward from the left ventricle of the heart as the **ascending aorta,** arches to the left as the **aortic arch,** and then plunges downward through the thorax following the spine **(thoracic aorta)** to finally pass through the diaphragm into the abdominopelvic cavity, where it becomes the **abdominal aorta** (Figure 11.11).

The major branches of the aorta and the organs they serve are listed next in sequence from the heart. Figure 11.11 shows the course of the aorta and its major branches. As you locate the arteries on the figure, be aware of ways to make your learning easier. In many cases the name of the artery tells you the body region or organs served (renal artery, brachial artery, and coronary artery) or the bone followed (femoral artery and ulnar artery).

Arterial Branches of the Ascending Aorta

- The only branches of the ascending aorta are the **right [R.]** and **left [L.] coronary arteries,** which serve the heart.

Arterial Branches of the Aortic Arch

- The **brachiocephalic** (bra″ke-o-se-fal′ik) **trunk** (the first branch off the aortic arch) splits into the **R. common carotid** (kah-ro′tid) **artery** and **R. subclavian** (sub-kla′ve-an) **artery.** (See same-named vessels on left side of body for organs served.)
- The **L. common carotid artery** is the second branch off the aortic arch. It divides, forming the **L. internal carotid,** which serves the brain, and the **L. external carotid,** which serves the skin and muscles of the head and neck.
- The third branch of the aortic arch, the **L. subclavian artery,** gives off an important branch—the **vertebral artery,** which serves part of the brain. In the axilla, the subclavian artery becomes the **axillary artery** and then continues into the arm as the **brachial artery,** which supplies the arm. At the elbow, the brachial artery splits to form the **radial** and **ulnar arteries,** which serve the forearm.

Arterial Branches of the Thoracic Aorta

- The *intercostal arteries* (ten pairs) supply the muscles of the thorax wall. Other branches of the thoracic aorta supply the lungs *(bronchial arteries),* the esophagus *(esophageal arteries),* and the diaphragm *(phrenic arteries).* These arteries are not illustrated in Figure 11.11.

Arterial Branches of the Abdominal Aorta

- The **celiac trunk** is the first branch of the abdominal aorta. It is a single vessel that has three branches: (1) the **L. gastric artery** supplies the stomach, (2) the **splenic artery** supplies the spleen, and (3) the **common hepatic artery** supplies the liver.
- The unpaired **superior mesenteric** (mes″en-ter′ik) **artery** supplies most of the small intestine and the first half of the large intestine, or colon.
- The **renal** (R. and L.) **arteries** serve the kidneys.
- The **gonadal** (R. and L.) **arteries** supply the gonads. They are called the *ovarian arteries* in females (serving the ovaries) and the *testicular arteries* in males (serving the testes).
- The *lumbar arteries* (not illustrated in Figure 11.11) are several pairs of arteries serving the heavy muscles of the abdomen and trunk walls.
- The **inferior mesenteric artery** is a small, unpaired artery supplying the second half of the large intestine.
- The **common iliac** (R. and L.) **arteries** are the final branches of the abdominal aorta. Each divides into an **internal iliac artery,** which supplies the pelvic organs (bladder, rectum, and so on), and an **external iliac artery,** which enters the thigh, where it becomes the **femoral artery.** The femoral artery and its branch, the **deep femoral artery,** serve the thigh. At the knee, the femoral artery becomes the **popliteal artery,** which then splits into the **anterior** and **posterior tibial arteries,** which supply the leg and foot. The anterior tibial artery terminates in the **dorsalis pedis artery,** which supplies the dorsum of the foot. (The dorsalis pedis is often

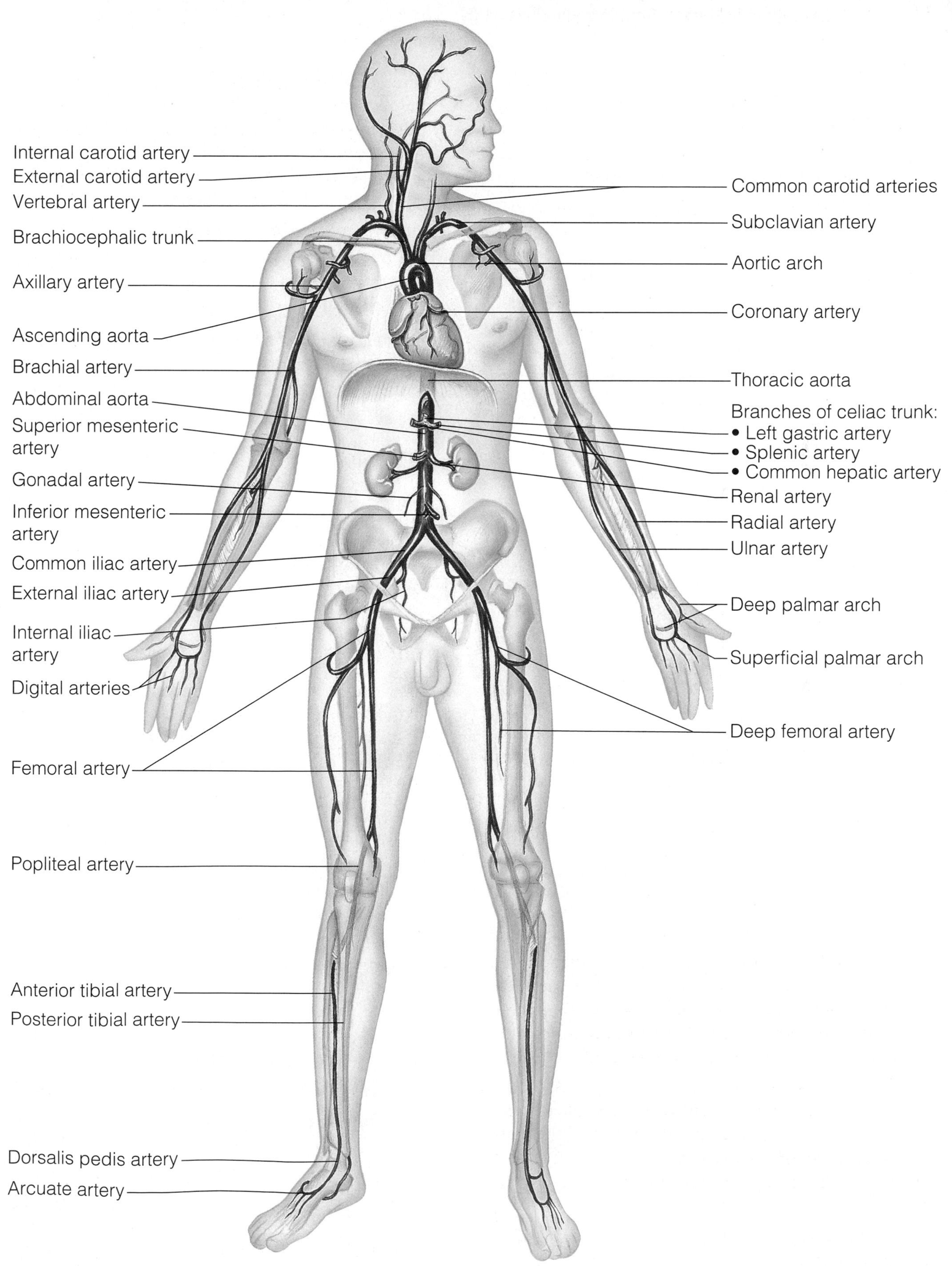

FIGURE 11.11 **Major arteries of the systemic circulation, anterior view.**

palpated in patients with circulatory problems of the legs to determine if the distal part of the leg has adequate circulation.)

Major Veins of the Systemic Circulation

Although arteries are generally located in deep, well-protected body areas, many veins are more superficial and some are easily seen and palpated on the body surface. Most deep veins follow the course of the major arteries; and, with a few exceptions, the naming of these veins is identical to that of their companion arteries. Major systemic arteries branch off the aorta, whereas the veins converge on the venae cavae, which enter the right atrium of the heart. Veins draining the head and arms empty into the **superior vena cava** and those draining the lower body empty into the **inferior vena cava.** These veins are described next and shown in Figure 11.12. As before, locate the veins on the figure as you read their descriptions.

Veins Draining into the Superior Vena Cava Veins draining into the superior vena cava are named in a distal to proximal direction; that is, in the same direction the blood flows into the superior vena cava.

- The **radial** and **ulnar veins** are deep veins draining the forearm. They unite to form the deep **brachial vein,** which drains the arm and empties into the **axillary vein** in the axillary region.
- The **cephalic** (se-fal′ik) **vein** provides for the superficial drainage of the lateral aspect of the arm and empties into the axillary vein.
- The **basilic** (bah-sil′ik) **vein** is a superficial vein that drains the medial aspect of the arm and empties into the brachial vein proximally. The basilic and cephalic veins are joined at the anterior aspect of the elbow by the **median cubital vein.** (The median cubital vein is often chosen as the site for blood removal for the purpose of blood testing.)
- The **subclavian vein** receives venous blood from the arm through the axillary vein and from the skin and muscles of the head through the **external jugular vein.**
- The **vertebral vein** drains the posterior part of the head.
- The **internal jugular vein** drains the dural sinuses of the brain.
- The **brachiocephalic** (R. and L.) **veins** are large veins that receive venous drainage from the subclavian, vertebral, and internal jugular veins on their respective sides. The brachiocephalic veins join to form the superior vena cava, which enters the heart.
- The *azygos* (az′ĭ-gos) *vein* is a single vein that drains the thorax and enters the superior vena cava just before it joins the heart. (This vein is not illustrated in Figure 11.12.)

Veins Draining into the Inferior Vena Cava The inferior vena cava, which is much longer than the superior vena cava, returns blood to the heart from all body regions below the diaphragm. As before, we will trace the venous drainage in a distal to proximal direction.

- The **anterior** and **posterior tibial veins** and the **fibular vein** drain the leg (calf and foot). The posterior tibial vein becomes the **popliteal vein** at the knee and then the **femoral vein** in the thigh. The femoral vein becomes the **external iliac vein** as it enters the pelvis.
- The **great saphenous** (sah-fe′nus) **veins** are the longest veins in the body. They receive the superficial drainage of the leg. They begin at the **dorsal venous arch** in the foot and travel up the medial aspect of the leg to empty into the femoral vein in the thigh.
- Each **common iliac** (R. and L.) **vein** is formed by the union of the **external iliac vein** and the **internal iliac vein** (which drains the pelvis) on its own side. The common iliac veins join to form the inferior vena cava, which then ascends superiorly in the abdominal cavity.
- The *R. gonadal vein* drains the right ovary in females and the right testicle in males. (The *L. gonadal vein* empties into the left renal vein superiorly.) (The gonadal veins are not illustrated in Figure 11.12.)
- The **renal** (R. and L.) **veins** drain the kidneys.
- The **hepatic portal vein** is a single vein that drains the digestive tract organs and carries this blood through the liver before it enters the systemic circulation. (The hepatic portal circulation is discussed in the next section.)
- The **hepatic** (R. and L.) **veins** drain the liver.

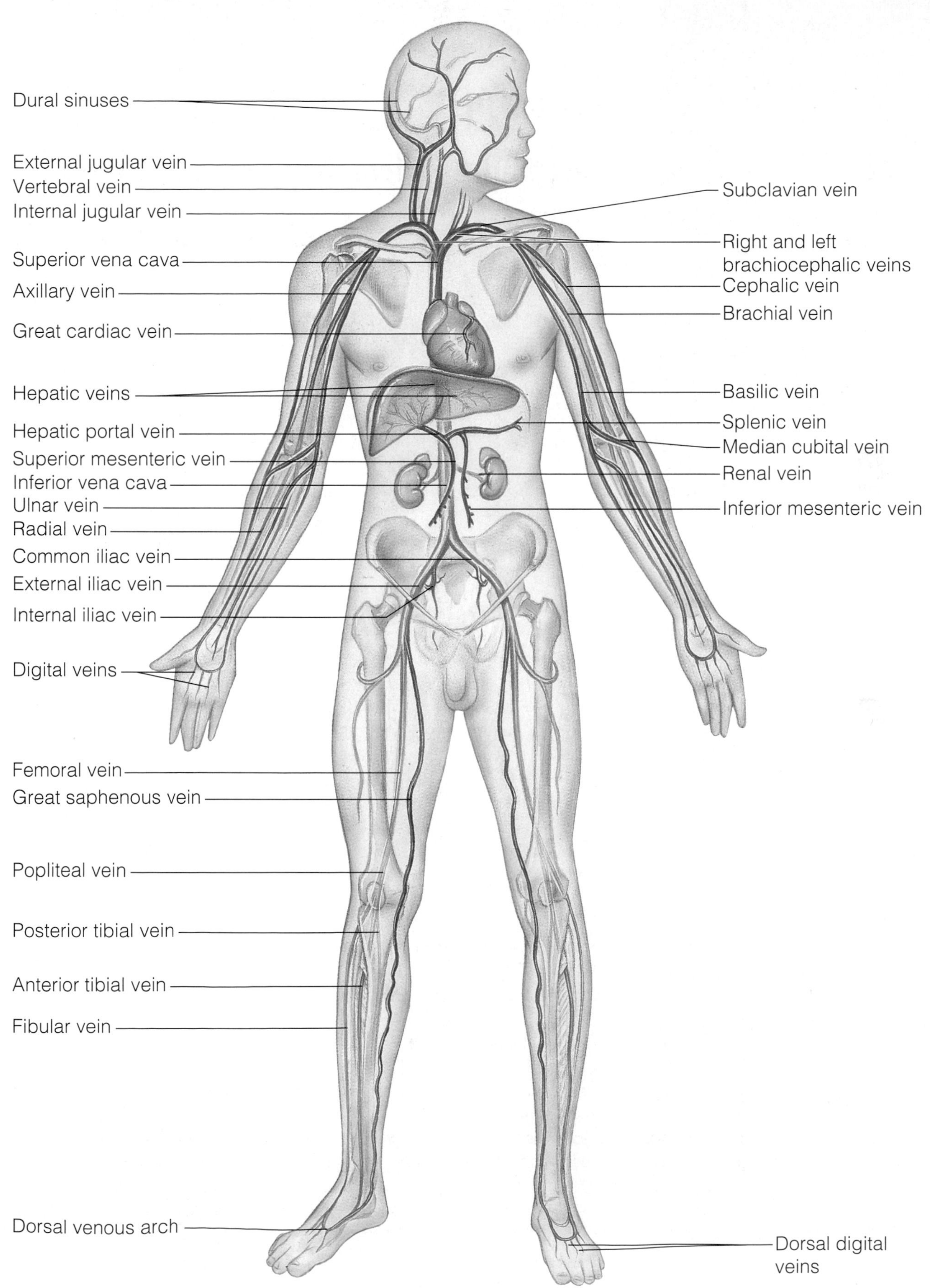

FIGURE 11.12 Major veins of the systemic circulation, anterior view. The vessels of the pulmonary circulation are not illustrated, accounting for the incomplete appearance of the circulation from the heart.

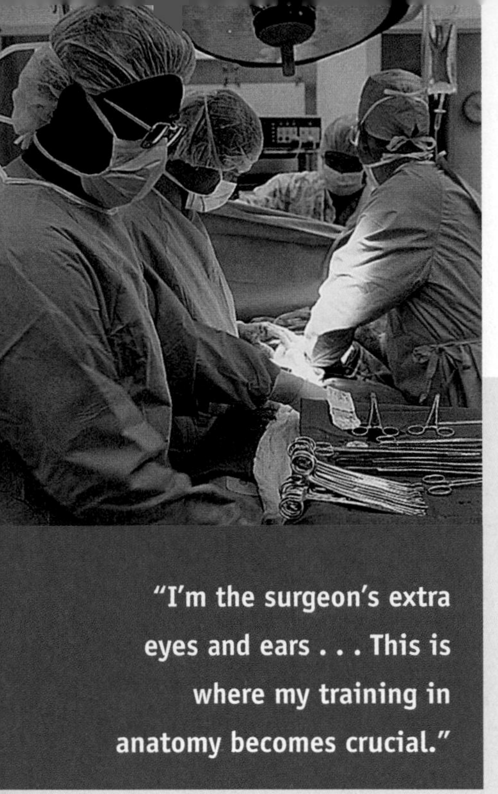

"I'm the surgeon's extra eyes and ears . . . This is where my training in anatomy becomes crucial."

Focus on Careers

Certified Surgical Technologist (CST)

Surgical technologists use their knowledge of anatomy to assist surgeons in the operating room.

ASK Nathan Williams what it takes to be a good surgical technologist, and he will list several key qualities. "Attention to detail is crucial," he says. "You have to be methodical, orderly, and cool. There is a lot of stress in the operating room, and surgeons are under tremendous pressure. You have to remain calm and do your job well, so the surgeons can do theirs."

Surgical technologists (sometimes called surgical technicians) are allied health professionals who work under the supervision of a surgeon. They make sure that the operating room environment is safe, the equipment functions properly, and procedures are conducted safely. Surgical technologists combine their knowledge of anatomy, surgical procedures, and surgical tools and technologies to facilitate and enhance a physician's performance in the operating room.

Says Williams, "My primary job is to maintain the sterility of the operating room environment. I start the day by setting up a sterile field—that is, I carefully open sealed surgical instruments and lay them out in an orderly fashion to prepare for each procedure. I try to anticipate everything the surgeon will need during that operation. Everything remains sterile until the end of the procedure, when I assist the patient onto a stretcher, sort out instruments for decontamination, and take off my surgical gown and gloves. Then I change into a new gown and gloves for the next patient. If this sterile environment is compromised, it can have devastating effects because the patient is contaminated. The most common result is a staph infection—an infection by *Staphylococcus* bacteria that can be life threatening."

Another big part of Williams's job is anticipating the surgeon's next move. "I'm the surgeon's extra eyes and ears, and I have to be alert and stay a step ahead at all times. This is where my training in anatomy becomes crucial because I have to be aware of which body tissues surgeons will encounter next. For instance, when they work in areas where I know there are lots of blood vessels, I always keep a clamp handy to stop bleeding. As another example, there are many types of forceps, some with smooth edges and some with teeth. During the initial incision, surgeons need forceps with teeth because they handle skin most effectively. But after surgeons pierce through the skin and enter the body, I have to be sure to hand them only the smooth type, because forceps with teeth could puncture internal organs."

Surgical technologists must complete an accredited training program, usually 9 to 12 months long, that includes both classroom and clinical education of at least 900 hours. Then they must pass a National Certifying Exam to be certified for a six-year period. Their certification is renewable by earning continuing education credits or by retesting. Accreditation procedures vary from state to state.

For more information, contact the Association of Surgical Technologists:
7108-C South Alton Way
Englewood, CO 80112-2106
Telephone: (303) 694-9130
FAX: (303) 694-9169
http://www.ast.org

For additional information on this career, click the Focus on Careers link at www.anatomyandphysiology.com.

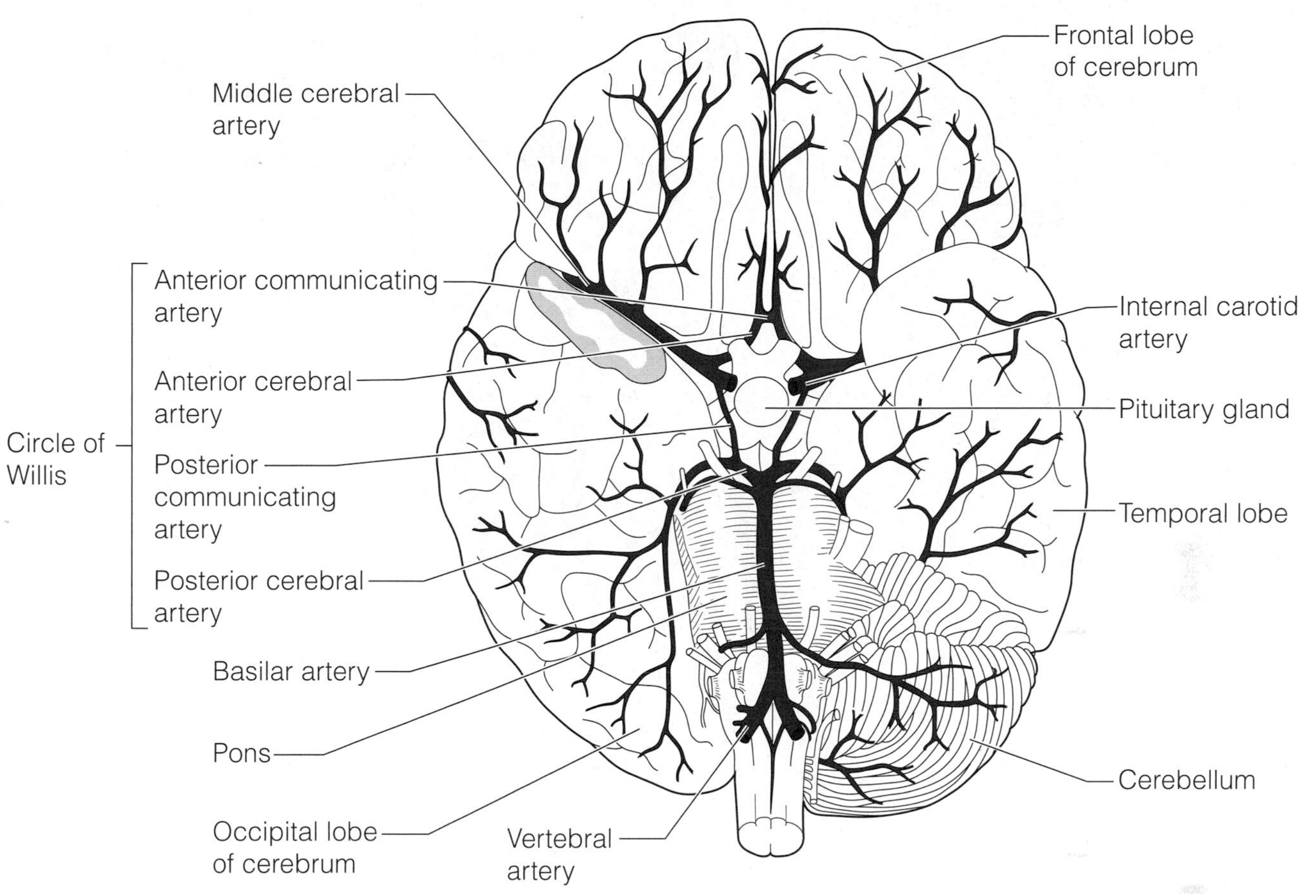

FIGURE 11.13 **Arterial supply of the brain.** (Cerebellum is shown only on the left side of the brain.)

Special Circulations

Arterial Supply of the Brain and the Circle of Willis

A continuous blood supply to the brain is crucial, since a lack of blood for even a few minutes causes the delicate brain cells to die. The brain is supplied by two pairs of arteries, the internal carotid arteries and the vertebral arteries (Figure 11.13).

The **internal carotid arteries,** branches of the common carotid arteries, run through the neck and enter the skull through the temporal bone. Once inside the cranium, each divides into the **anterior** and **middle cerebral arteries,** which supply most of the cerebrum.

The paired **vertebral arteries** pass upward from the subclavian arteries at the base of the neck. Within the skull, the vertebral arteries join to form the single **basilar artery,** which serves the brain stem and cerebellum as it travels upward. At the base of the cerebrum, the basilar artery divides to form the **posterior cerebral arteries,** which supply the posterior part of the cerebrum.

The anterior and posterior blood supplies of the brain are united by small *communicating arterial branches.* The result is a complete circle of connecting blood vessels called the **circle of Willis,** which surrounds the base of the brain. The circle of Willis protects the brain by providing more than one route for blood to reach brain tissue in case of a clot or impaired blood flow anywhere in the system.

Hepatic Portal Circulation The veins of the **hepatic portal circulation** drain the digestive organs, spleen, and pancreas and deliver this blood to the liver through the **hepatic portal vein** (Figure 11.14). When you have just eaten, the hepatic portal blood contains large amounts of nutrients. Since the liver is a key body organ involved in maintaining the proper glucose, fat, and protein concentrations in the blood, this system "takes a detour" to ensure that the liver processes these substances before they enter the systemic circulation. As blood flows slowly through the liver, some of

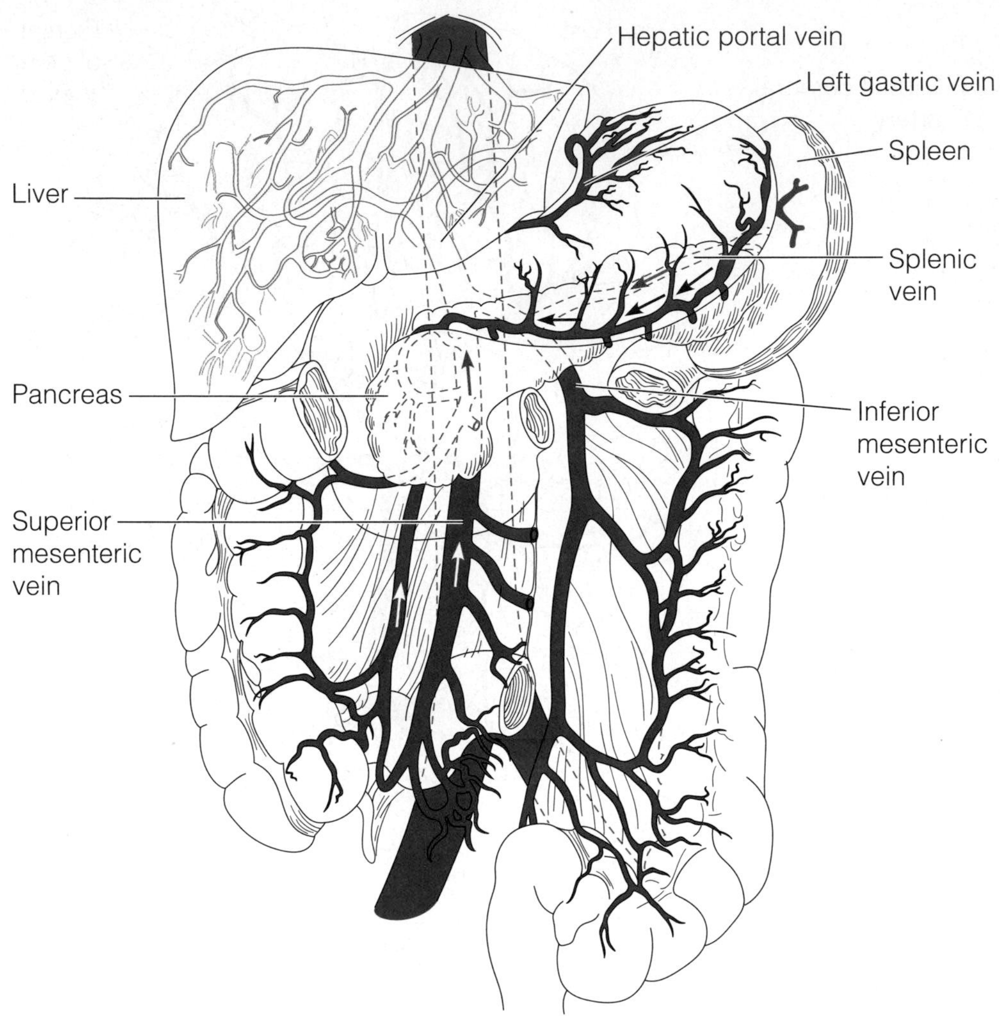

FIGURE 11.14 The hepatic portal circulation.

the nutrients are removed to be stored or processed in various ways for later release to the blood. The liver is drained by the hepatic veins that enter the inferior vena cava. Like the portal circulation that links the hypothalamus of the brain and the anterior pituitary gland (see Chapter 9), the hepatic portal circulation is a unique and unusual circulation. Normally, arteries feed capillary beds, which in turn drain into veins. Here we see *veins* feeding the liver circulation.

The **inferior mesenteric vein,** draining the terminal part of the large intestine, drains into the **splenic vein,** which itself drains the spleen, pancreas, and the left side of the stomach. The splenic vein and **superior mesenteric vein** (which drains the small intestine and the first part of the colon) join to form the hepatic portal vein. The **L. gastric vein,** which drains the right side of the stomach, drains directly into the hepatic portal vein.

Fetal Circulation Since the lungs and digestive system are not yet functioning in a fetus, all nutrient, excretory, and gas exchanges occur through the placenta. Nutrients and oxygen move from the mother's blood into the fetal blood, and fetal wastes move in the opposite direction. As shown in Figure 11.15, the *umbilical cord* contains three blood vessels: one large **umbilical vein** and two smaller **umbilical arteries.** The umbilical vein carries blood rich in nutrients and oxygen to the fetus. The umbilical arteries carry carbon dioxide and debris-laden blood from the fetus to the placenta. As blood flows superiorly toward the heart of the fetus, most of it bypasses the immature liver through the **ductus venosus** (duk′tus ve-no′sus)

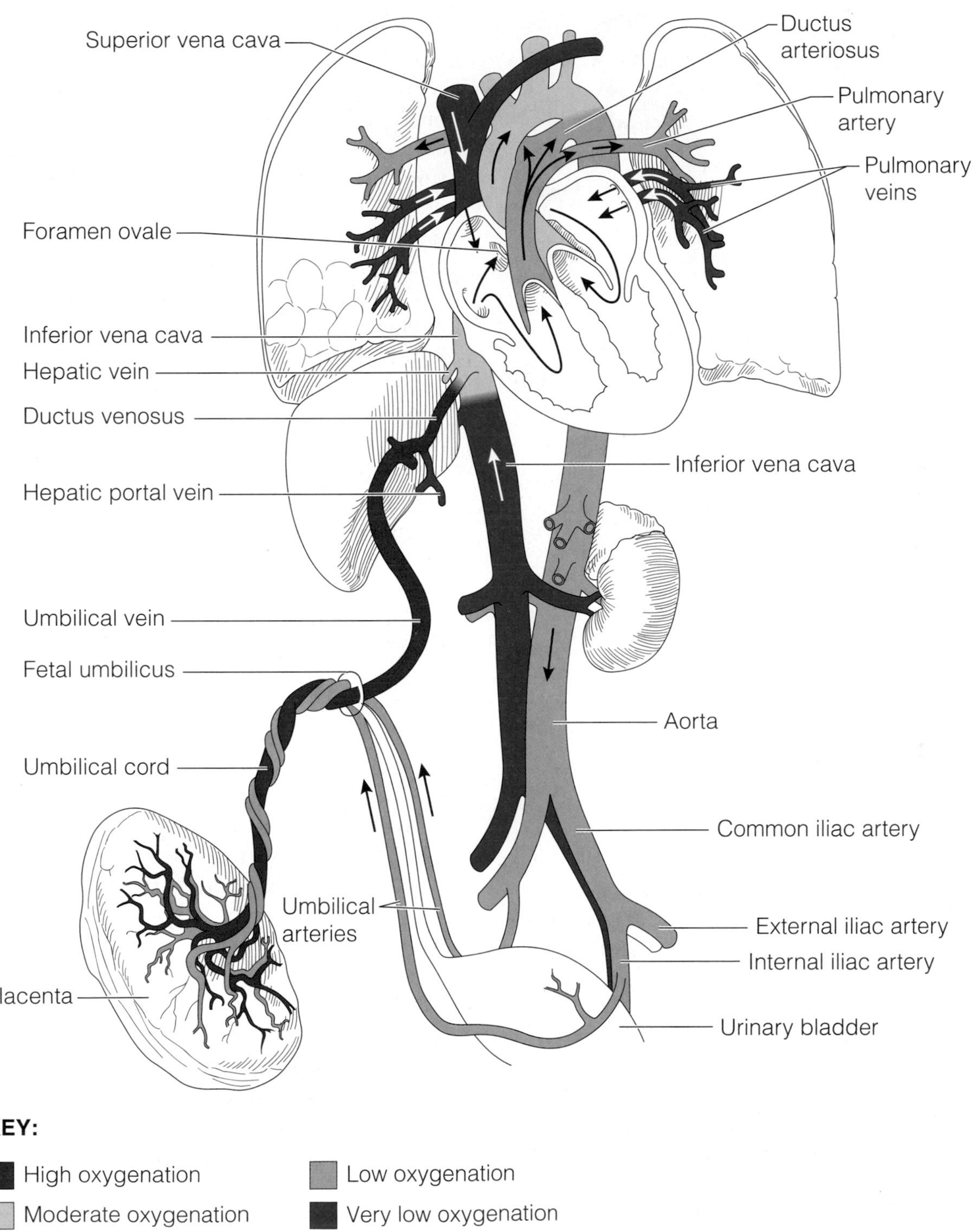

FIGURE 11.15 The fetal circulation.

and enters the inferior vena cava, which carries the blood to the right atrium of the heart.

Since fetal lungs are nonfunctional and collapsed, two shunts see to it that they are almost entirely bypassed. Some of the blood entering the right atrium is shunted directly into the left atrium through the **foramen ovale** (fo-ra′men o-val′e), a flaplike opening in the interatrial septum. Blood that does manage to enter the right ventricle is pumped out the pulmonary trunk, where it meets a second shunt, the **ductus arteriosus** (ar-ter″e-o′sus), a short vessel that connects the aorta and

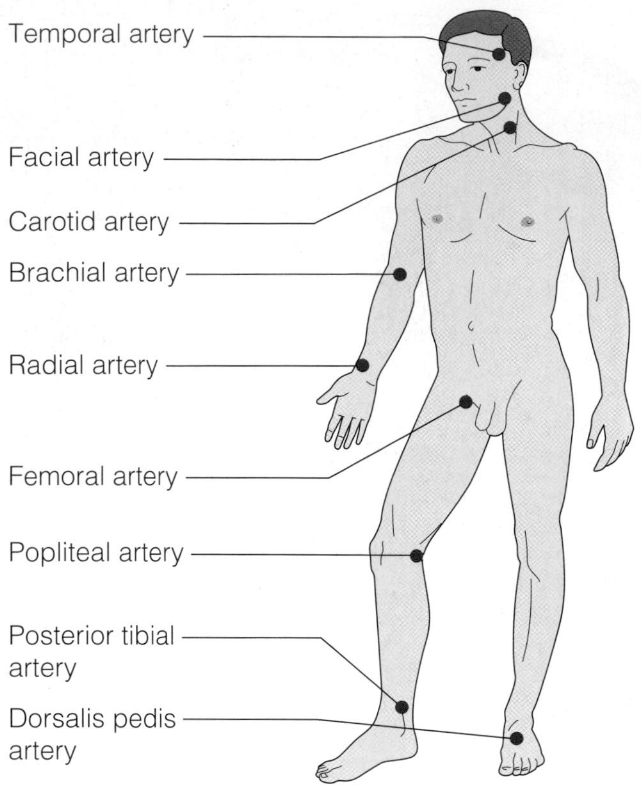

FIGURE 11.16 Body sites where the pulse is most easily palpated. (The specific arteries indicated are discussed on pp. 364.)

the pulmonary trunk. Because the collapsed lungs are a high-pressure area, blood tends to enter the systemic circulation through the ductus arteriosus. The aorta carries blood to the tissues of the fetal body and ultimately back to the placenta through the umbilical arteries.

At birth, or shortly after, the foramen ovale closes, and the ductus arteriosus collapses and is converted to the fibrous **ligamentum arteriosum** (lig″ah-men′tum ar-ter″e-o′sum) (see Figure 11.2a). As blood stops flowing through the umbilical vessels, they become obliterated, and the circulatory pattern converts to that of an adult.

Physiology of Circulation

A fairly good indication of the efficiency of a person's circulatory system can be obtained by taking arterial pulse and blood pressure measurements. These measurements, along with those of respiratory rate and body temperature, are referred to collectively as **vital signs** in clinical settings.

Arterial Pulse

The alternating expansion and recoil of an artery that occurs with each beat of the left ventricle creates a pressure wave—a **pulse**—that travels through the entire arterial system. Normally the pulse rate (pressure surges per minute) equals the heart rate (beats per minute). The pulse averages 70 to 76 beats per minute in a normal resting person. It is influenced by activity, postural changes, and emotions.

You can feel a pulse in any artery lying close to the body surface by compressing the artery against firm tissue; this provides an easy way of counting heart rate. Because it is so accessible, the point where the radial artery surfaces at the wrist (the radial pulse) is routinely used to take a pulse measurement, but there are several other clinically important arterial pulse points (Figure 11.16). Because these same points are compressed to stop blood flow into distal tissues during hemorrhage, they are also called **pressure points.** For example, if you seriously cut your hand, you can stop the bleeding somewhat by compressing the radial artery or the brachial artery.

Try to palpate each of the pulse points shown in Figure 11.16 by placing the tips of your first two or three fingers of one hand over the artery at the site indicated. It helps to compress the artery firmly as you begin and then immediately ease up on your pressure slightly. In each case, notice the regularity of the pulse and its relative strength.

Blood Pressure

Any system equipped with a pump that forces fluid through a network of closed tubes operates under pressure. The closer you get to the pump, the higher the pressure. **Blood pressure** is the pressure the blood exerts against the inner walls of the blood vessels, and it is the force that keeps blood circulating continuously even between heartbeats. Unless stated otherwise, the term *blood pressure* is understood to mean the pressure within the large systemic arteries near the heart.

Blood Pressure Gradient When the ventricles contract, they force blood into large, thick-walled elastic arteries that expand as the blood is pushed into them. The high pressure in these arteries forces the blood to continually move into areas where the pressure is lower. The pressure is highest in the large arteries and continues to drop throughout the pathway, reaching zero or negative pressure at the venae cavae (Figure 11.17). Recall that the blood flows into the smaller arteries, then arterioles, capillaries, venules, veins, and finally back to the large venae cavae entering the right atrium of the heart. It flows continually along a pressure gradient (from high to low pressure) as it makes its circuit day in and day out. Notice that if venous return depended entirely on a high blood pressure throughout the system, blood would probably never be able to complete its circuit back to the heart. This is why the valves in the larger veins, the milking activity of the skeletal muscles, and pressure changes in the thorax are so important.

The pressure differences between arteries and veins become very clear when these vessels are cut. If a vein is cut, the blood flows evenly from the wound; a lacerated artery produces rapid spurts of blood.

Continual blood flow absolutely depends on the stretchiness of the larger arteries and their ability to recoil and keep the pressure on the blood as it flows off into the circulation. To illustrate this, think of a garden hose with relatively hard walls. When the water is turned on, the water spurts out under high pressure because the hose walls don't expand. However, when the water faucet is suddenly turned off, the flow of water stops just as abruptly. This is because the walls of the hose cannot recoil to keep pressure on the water; therefore, the pressure drops and the flow of water stops. The importance of the elasticity of the arteries is best appreciated when it is lost, as happens in *arteriosclerosis*. Arteriosclerosis, also called "hardening of the arteries," is discussed in the "A Closer Look" box on pp. 377–379.

Measuring Blood Pressure Because the heart alternately contracts and relaxes, the off-and-on flow of blood into the arteries causes the blood pressure to rise and fall during each beat. Thus, two arterial blood pressure measurements are usually made: **systolic** (sis-to'lik) **pressure,** the pressure in the arteries at the peak of ventricular contraction, and **diastolic** (di"us-to'lik) **pressure,** the pressure when the ventricles are relaxing. Blood pressures are reported in millimeters of mercury (mm Hg), with the systolic pressure written first—120/80 (120 over 80) translates to a systolic pressure of 120 mm Hg and a diastolic pressure of 80 mm Hg. Most often, systemic arterial blood pressure is measured indirectly by the **auscultatory** (os-kul'tuh-tor-e) **method.** This procedure, as used to measure

Q *How does the pulsating change in blood pressure on the left side of the graph relate to the structure of the largest arteries?*

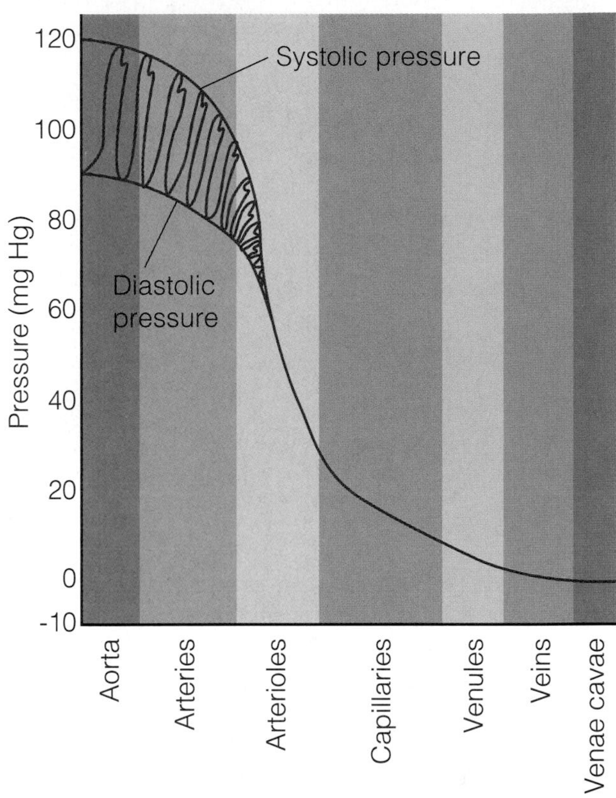

FIGURE 11.17 Blood pressure in various areas of the cardiovascular system.

blood pressure in the brachial artery of the arm, is illustrated and described in Figure 11.18.

Effects of Various Factors on Blood Pressure Arterial blood pressure (BP) is directly related to cardiac output (CO; the amount of blood pumped out of the left ventricle per minute) and peripheral resistance (PR). This relationship is expressed by the equation: BP = CO × PR. Since regulation of cardiac output has already been considered, we will concentrate on peripheral resistance here.

Peripheral resistance is the amount of friction encountered by the blood as it flows through

It reveals their elasticity. When the heart contracts and forces blood into the large arteries near the heart, they stretch to accommodate the greater volume of blood (systolic pressure), and then as the blood continues on in the circuit, their walls recoil, keeping pressure on the blood which keeps it moving (diastolic pressure).

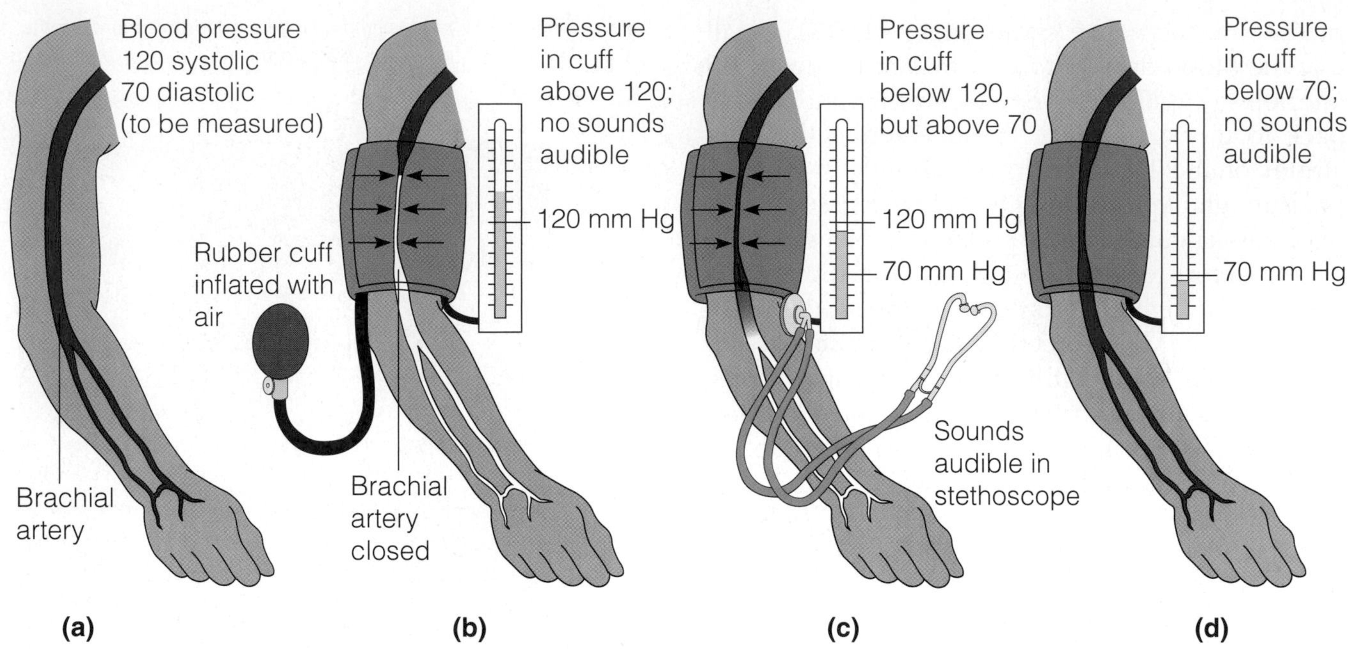

FIGURE 11.18 Measuring blood pressure. (a) The course of the brachial artery of the arm. Assume a blood pressure of 120/70 in a young, healthy person. **(b)** The blood pressure cuff is wrapped snugly around the arm just above the elbow and inflated until the cuff pressure exceeds the systolic blood pressure. At this point, blood flow into the arm is stopped, and a brachial pulse cannot be felt or heard. **(c)** The pressure in the cuff is gradually reduced while the examiner listens (auscultates) carefully for sounds in the brachial artery with a stethoscope. The pressure read as the first soft tapping sounds are heard (the first point at which a small amount of blood is spurting through the constricted artery) is recorded as the systolic pressure. **(d)** As the pressure is reduced still further, the sounds become louder and more distinct; when the artery is no longer constricted and blood flows freely, the sounds can no longer be heard. The pressure at which the sounds disappear is recorded as the diastolic pressure.

the blood vessels. It is increased by many factors, but probably the most important is the constriction, or narrowing, of blood vessels, especially arterioles, as a result of sympathetic nervous system activity or atherosclerosis. Increased blood volume or blood viscosity (thickness) also raises peripheral resistance. Any factor that increases either the cardiac output or peripheral resistance causes an almost immediate reflex rise in blood pressure. Many factors can alter blood pressure—age, weight, time of day, exercise, body position, emotional state, and various drugs, to name a few. The influence of a few of these factors is discussed next.

1. **Neural factors: the autonomic nervous system.** The parasympathetic division of the autonomic nervous system has little or no effect on blood pressure, but the sympathetic division is important and is responsive to many different factors. The major action of the sympathetic nerves on the vascular system is to cause **vasoconstriction** (vas″o-kon-strik′shun), or narrowing of the blood vessels, which increases the blood pressure. The sympathetic center in the medulla of the brain is activated to cause vasoconstriction in many different circumstances (Figure 11.19). For example, when we stand up suddenly after lying down, the effect of gravity causes blood to pool in the vessels of the legs and feet, and the blood pressure drops. This activates *pressoreceptors* (also called *baroreceptors*) in the large arteries of the

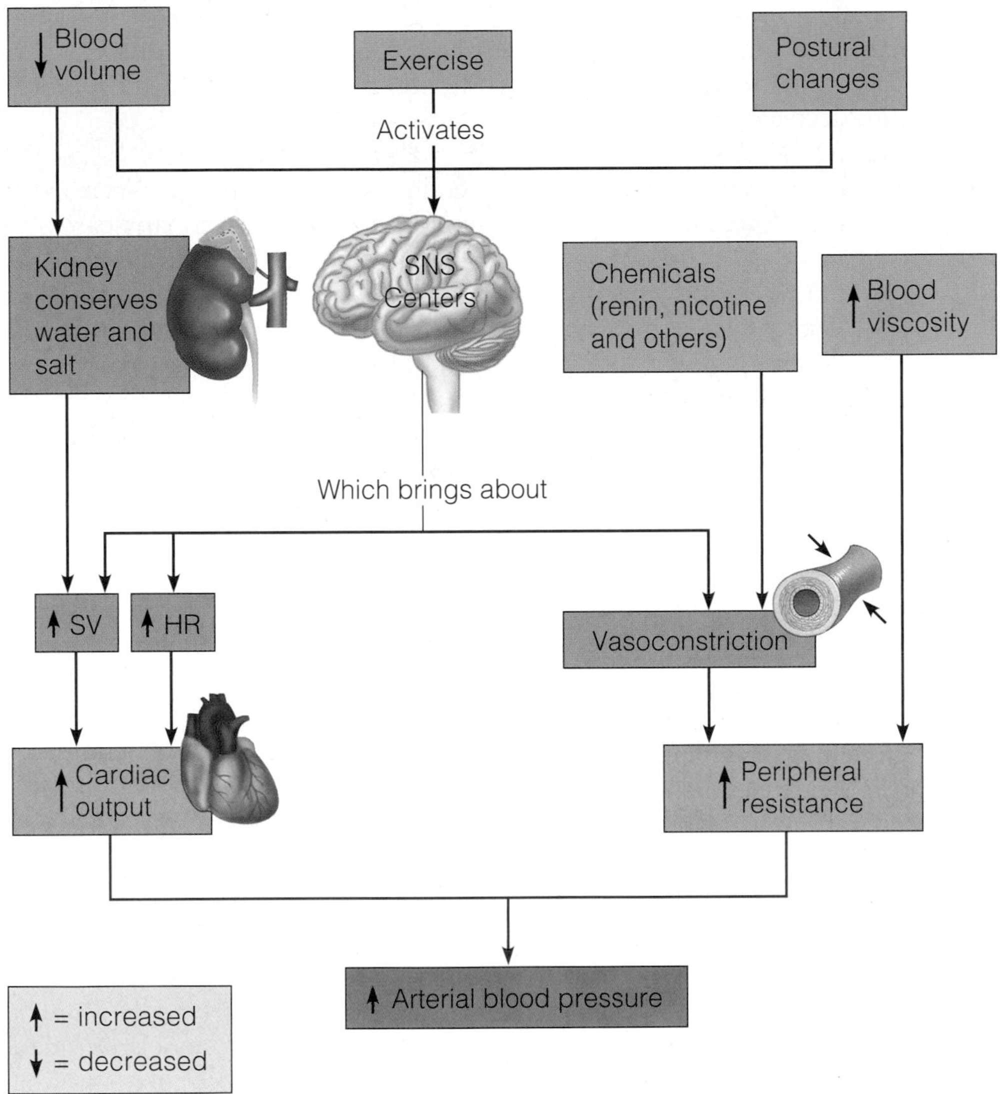

FIGURE 11.19 Summary of factors causing an increase in arterial blood pressure. (SV = stroke volume; HR = heart rate)

neck and chest. They send off warning signals that result in reflexive vasoconstriction, which increases blood pressure back to homeostatic levels.

When blood volume suddenly decreases, as in hemorrhage, blood pressure drops, and the heart begins to beat more rapidly (as it tries to compensate). However, because venous return is reduced by blood loss, the heart also beats weakly and inefficiently. In such cases, the sympathetic nervous system causes vasoconstriction to increase the blood pressure so that (hopefully) venous return increases and circulation can continue.

The final example concerns sympathetic nervous system activity when we exercise vigorously or are frightened and have to make a hasty escape. Under these conditions, there is a generalized vasoconstriction *except* in the skeletal muscles. The vessels of the skeletal muscles dilate to increase the blood flow to the working muscles. (It should be noted that the sympathetic nerves *never* cause vasoconstriction of blood vessels of the heart or brain.)

2. **Renal factors: the kidneys.** The kidneys play a major role in regulating arterial blood pressure by altering blood volume. As blood pressure (and/or blood volume) increases beyond

normal, the kidneys allow more water to leave the body in the urine. Since the source of this water is the bloodstream, blood volume decreases, which in turn decreases blood pressure. However, when arterial blood pressure falls, the kidneys retain body water, increasing blood volume, and blood pressure rises (see Figure 11.19).

In addition, when arterial blood pressure is low, certain kidney cells release the enzyme *renin* into the blood. Renin triggers a series of chemical reactions that result in the formation of *angiotensin II,* a potent vasoconstrictor chemical. Angiotensin also stimulates the adrenal cortex to release aldosterone, a hormone that enhances sodium ion reabsorption by the kidneys. As sodium moves into the blood, water follows. Thus, blood volume and blood pressure both rise.

3. **Temperature.** In general, cold has a vasoconstricting effect. This is why your exposed skin feels cold to the touch on a winter day and why cold compresses are recommended to prevent swelling of a bruised area. On the other hand, heat has a *vasodilating* effect, and warm compresses are used to speed the circulation into an inflamed area.
4. **Chemicals.** The effects of chemical substances, many of which are drugs, on blood pressure are widespread and well known in many cases. Just a few examples will be given here. **Epinephrine** increases both heart rate and blood pressure. *Nicotine* increases blood pressure by causing vasoconstriction. Both *alcohol* and *histamine* cause vasodilation and decrease the blood pressure. The reason a person who has "one too many" becomes flushed is that the skin vessels have been dilated by alcohol.
5. **Diet.** Although medical opinions tend to change and are at odds from time to time, it is generally believed that a diet low in salt, saturated fats, and cholesterol helps to prevent *hypertension,* or high blood pressure.

Variations in Blood Pressure In normal adults at rest, systolic blood pressure varies between 110 and 140 mm Hg, and diastolic pressure between 75 and 80 mm Hg—but blood pressure varies considerably from one person to another. What is normal for you may not be normal for your grandfather or your neighbor. Blood pressure varies with age, weight, race, mood, physical activity, and posture. Nearly all these variations can be explained in terms of the factors affecting blood pressure that have already been discussed.

Hypotension, or low blood pressure, is generally considered to be a systolic blood pressure below 100 mm Hg. In many cases, it simply reflects individual differences and is no cause for concern. In fact, low blood pressure is an expected result of physical conditioning and is often associated with long life and an old age free of illness.

Homeostatic Imbalance

Elderly people may experience temporary low blood pressure and dizziness when they rise suddenly from a reclining or sitting position—a condition called *orthostatic hypotension.* Because an aging sympathetic nervous system reacts more slowly to postural changes, blood pools briefly in the lower limbs, reducing blood pressure and, consequently, blood delivery to the brain. Making postural changes more slowly to give the nervous system time to make the necessary adjustments usually prevents this problem. ▲

Chronic hypotension (not explained by physical conditioning) may hint at poor nutrition and inadequate levels of blood proteins. Because blood viscosity is low, blood pressure is also lower than normal. Acute hypotension is one of the most important warnings of *circulatory shock,* a condition in which the blood vessels are inadequately filled and blood cannot circulate normally. The most common cause is blood loss.

A brief elevation in blood pressure is a normal response to fever, physical exertion, and emotional upset, such as anger and fear. Persistent **hypertension,** or **high blood pressure,** is pathological and is defined as a condition of sustained elevated arterial pressure of 140/90 or higher.

Homeostatic Imbalance

Chronic hypertension is a common and dangerous disease that warns of increased peripheral resistance. Although it progresses without symptoms for the first 10 to 20 years, it slowly and surely strains the heart and damages the arteries. For this reason, hypertension is often called the "silent killer." Because the heart is forced to pump against increased resistance, it must work harder, and in time, the myocardium enlarges. When finally strained beyond its capacity to respond, the heart weakens and its walls

A Closer Look

Atherosclerosis? Get Out the Cardiovascular Drāno!

WHEN pipes get clogged, it is usually because something gets stuck in them—a greasy mass of bone bits or a hair ball. But when arteries are narrowed by **atherosclerosis,** the damming-up process occurs from the inside out: the walls of the vessels thicken and then protrude into the vessel lumen. Once this happens, it does not take much to close the vessel completely. A roaming blood clot or arterial spasms can do it.

Although all blood vessels are susceptible to this serious degenerative condition of blood vessel walls, for some unknown reason the aorta and the coronary arteries serving the heart are most often affected. The disease progresses through many stages before the arterial walls actually become hard and approach the stage of the rigid tube system described in the text, but some of the earlier stages are just as lethal or more so.

Onset and Stages of Atherosclerosis

What triggers this scourge of blood vessels? According to the *response to injury hypothesis,* the initial event is damage to the tunica interna caused by bloodborne chemicals such as carbon monoxide (present in cigarette smoke or auto exhaust); by bacteria or viruses; or by physical factors such as a blow or persistent hypertension. Once a break has occurred, blood platelets cling to the injured site and initiate clotting to prevent blood loss. The injured endothelium sets off the alarm, summoning the immune system and the inflammatory process to repair the damage. If it is a one-time injury, when it's over, it's over. But most plaques grow slowly, through a series of injuries that heal, only to be ruptured again and again. As the plaque grows, the injured endothelial cells release chemotactic agents and chemicals that increase the permeability of the endothelium to fats and cholesterol, allowing them to take up residence just deep to the tunica interna. Monocytes attracted to the area migrate beneath the endothelium, where they become macrophages that gorge themselves on the fat. Soon they are joined by smooth muscle cells migrating from the media of the blood vessel wall. The result is the so-called *fatty streak stage* characterized by thickening of the tunica interna by greasy gray to yellow lesions called *atherosclerotic plaques.* When these small, fatty

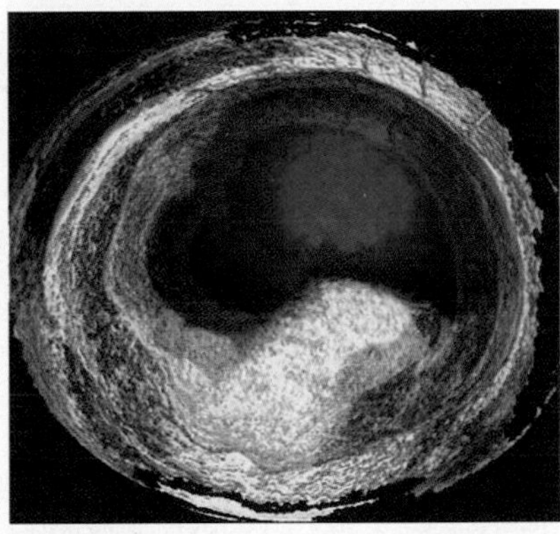

(a) Cross section of an artery that is partially occluded by atherosclerotic plaque.

mounds of muscle begin to protrude into the vessel wall (and ultimately the vessel lumen), the condition is called *atherosclerosis* (see photo a).

Arteriosclerosis is the end stage of the disease. As enlarging plaques hinder diffusion of nutrients from the blood to the deeper tissues of the artery wall, smooth muscle cells in the tunica media die and the elastic fibers deteriorate and are gradually replaced by nonelastic scar tissue. Then, calcium salts are deposited in the lesions. Collectively, these events cause the arterial wall to fray and ulcerate, conditions that encourage thrombus formation. The increased rigidity of the vessels leads to hypertension. Together, these events increase the risk of myocardial infarctions, strokes, and kidney failure.

However, the popular view that most heart attacks are the consequence of severe vessel narrowing and hardening is now being challenged, particularly since some 70 percent of heart attacks are caused by much smaller obstructions, too small to be seen on an arteriogram or to cause any symptoms in most cases. It now appears that the body's defense system betrays it. The inflammatory process that occurs in the still soft, unstable, cholesterol-rich plaques changes the biology of the vessel wall and makes the plaques susceptible to rupture, exploding off fragments that trigger massive clots that can cause lethal heart attacks. The victim appears perfectly healthy until he drops dead!

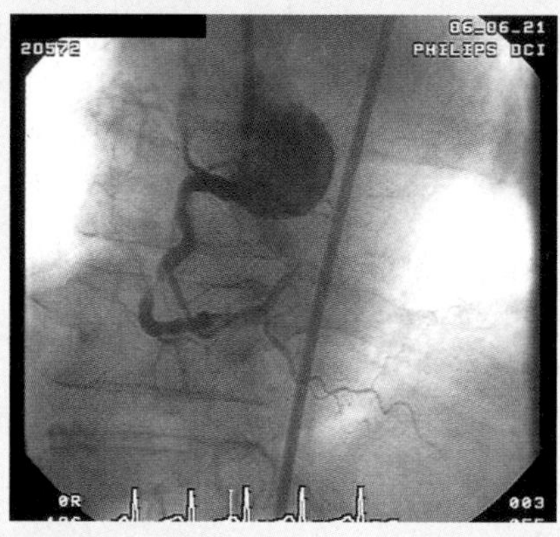

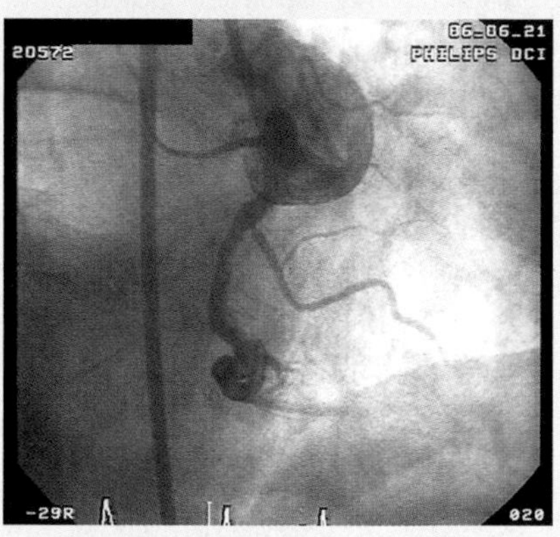

(b) Angiographs of an occluded artery (left) and the same artery after being cleared by balloon angioplasty (right). These views were prepared by a computer imaging technique, digital subtraction angiography (DSA).

Treatment and Prevention

The *vulnerable plaque hypothesis* mentioned above has attracted many medical converts, but the question of what to do about it remains. Some medical centers test heart patients for elevated levels of cholesterol and C-reactive protein, a marker of inflammation. Electron beam CT scans may be able to identify those at risk by detecting calcium deposits in their coronary arteries. Antibiotics and anti-inflammatory drugs are being tested as preventative measures. Even the humble aspirin is gaining new respect, and more cardiologists recommend that people at high risk take one baby aspirin daily (81 mg).

So what can help when the damage is done and the heart is at risk because of atherosclerotic coronary vessels? In the past, the only choice has been coronary artery bypass surgery, in which vessels removed from the legs or thoracic cavity are implanted in the heart to restore circulation. More recently, devices threaded through blood vessels to obstructed sites have become part of the ammunition of cardiovascular medicine. *Balloon angioplasty* uses a catheter with a balloon packed into its tip (see the two photos in b).

When the catheter reaches the blockage, the balloon is inflated, and the fatty mass is compressed against the vessel wall. However, this procedure is useful to clear only a few very localized obstructions. A newer catheter device uses a laser beam to vaporize the arterial clogs. Although these intravascular devices are faster, cheaper, and much less risky than bypass surgery, they carry with them the same major shortcoming: they do nothing to stop the underlying disease, and in time new blockages occur in 30 to 50 percent of cases.

When a blood clot is trapped by the diseased vessel walls, the answer may be a *clot-dissolving agent,* for example, *tissue plasminogen activator (t-PA),* a naturally occurring substance now being produced by genetic engineering techniques. Injecting t-PA directly into the heart restores blood flow quickly and puts an early end to many heart attacks in progress.

There is little doubt that lifestyle factors—emotional stress, smoking, obesity, high-fat and high-cholesterol diets, and lack of exercise—contribute to both atherosclerosis and hypertension. If these are the risk factors (and indeed they are), then why not just have patients at risk change their lifestyle? This is not as easy as it seems. Although taking antioxidants (E and C vitamins and beta carotene) and exercising more may "undo" some of the damage, old habits die hard, and North Americans like their burgers and butter. Can atherosclerosis be reversed to give the heart a longer and healthier life? If so, many more people with diseased arteries may be more willing to trade lifelong habits for a healthy old age!

become flabby. Hypertension also ravages blood vessels, causing small tears in the endothelium that accelerate the progress of **atherosclerosis.**

Although hypertension and atherosclerosis are often linked, it is difficult to blame hypertension on any distinct anatomical pathology. In fact, about 90 percent of hypertensive people have *primary,* or *essential, hypertension,* which cannot be accounted for by any specific organic cause. However, factors such as diet, obesity, heredity, race, and stress appear to be involved. For instance, more women than men and more blacks than whites are hypertensive. Hypertension runs in families. The child of a hypertensive parent is twice as likely to develop high blood pressure as is a child of parents with normal blood pressure. Dietary factors presumed to contribute to hypertension include high cholesterol, saturated fat, and sodium intakes. High blood pressure is common in obese people because the total length of their blood vessels is relatively greater than that in thinner individuals. For each pound of fat, miles of additional blood vessels are required, making the heart work harder to pump blood over longer distances. ▲

Capillary Exchange of Gases and Nutrients

Capillaries form an intricate network among the body's cells such that no substance has to diffuse very far to enter or leave a cell. The substances exchanged first diffuse through an intervening space (Figure 11.20a) filled with **interstitial fluid (tissue fluid).**

Substances tend to move to and from body cells according to their concentration gradients. Thus, oxygen and nutrients leave the blood and move into the tissue cells, and carbon dioxide and other wastes exit the tissue cells and enter the blood. Basically, substances entering or leaving the bloodstream may take one of four routes across the

Q *Assume there is a bacterial infection in the interstitial fluid of* ***(c).*** *How would this affect fluid flows across the capillary walls in the area?*

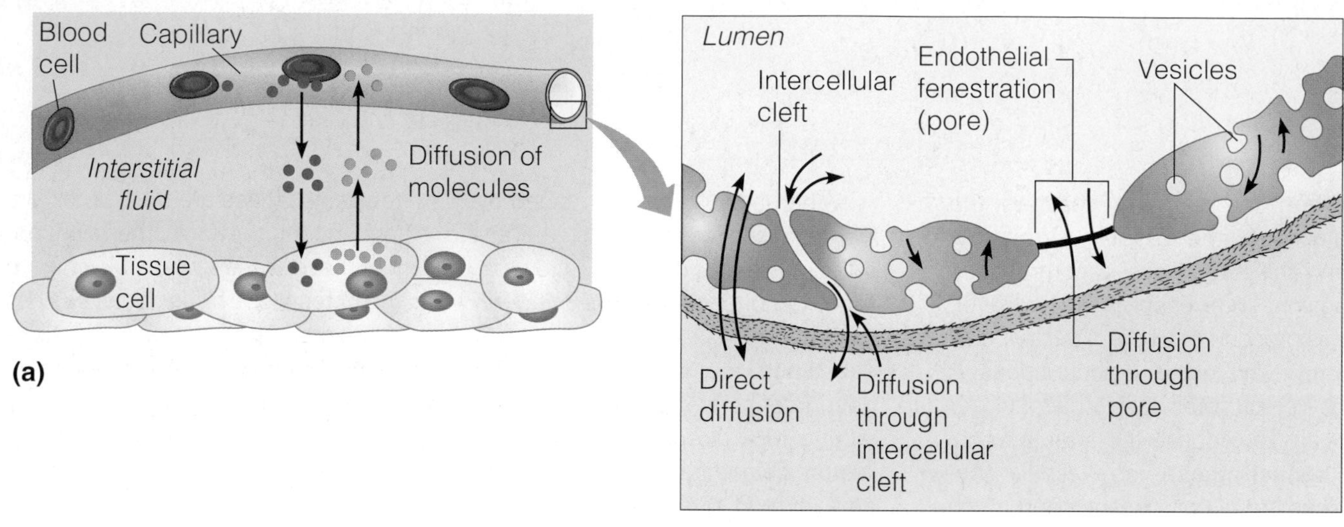

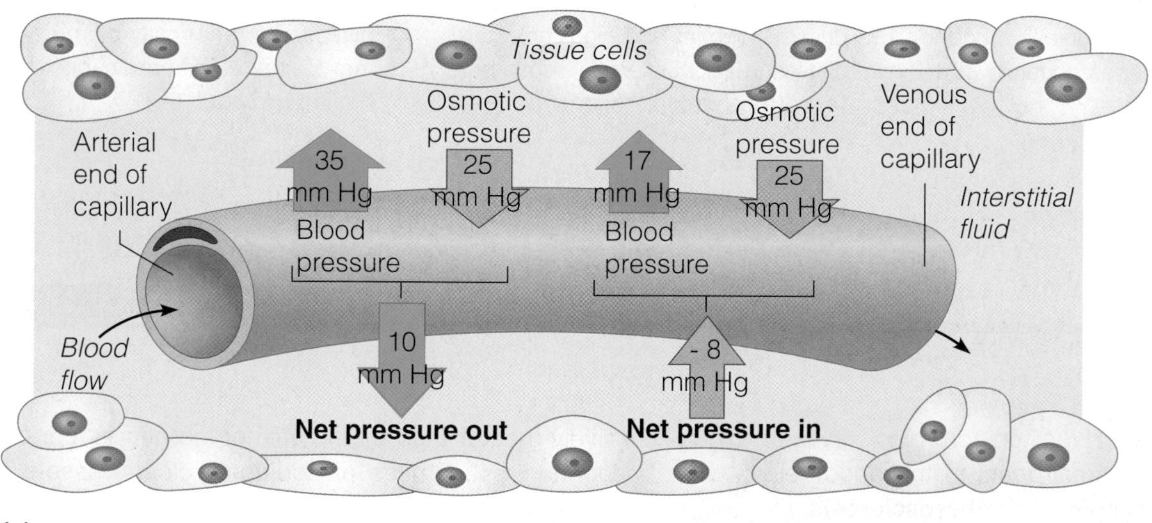

FIGURE 11.20 Capillary transport mechanisms. (a) Substances exchanged between the blood and the tissue cells must diffuse through intervening interstitial fluid. **(b)** The four possible pathways, or routes of transport, across the wall of an endothelial cell. (The endothelial cell is illustrated as if cut in cross section.) **(c)** Bulk fluid flows across capillary walls depend largely on the difference between the hydrostatic (blood) pressure and the osmotic pressure at different regions of the capillary bed. Blood pressure, which predominates at the arterial end of the bed (35 mm Hg), provides an outward force. Osmotic pressure (25 mm Hg), which predominates at the venous end of the bed where blood pressure is lower (17 mm Hg), produces a negative pressure (−8 mm Hg), which draws the fluid back. Hence, fluid is forced out of a capillary through clefts at the arterial end and returns to the blood at the venous end.

A *It would increase it because the osmotic pressure of the interstitial fluid would rise as inflammatory molecules and debris accumulated in the area.*

plasma membranes of the single layer of endothelial cells forming the capillary wall (Figure 11.20b).

1. As with all cells, substances can diffuse directly through (cross) their plasma membranes if they are lipid-soluble (like the respiratory gases).
2. Certain lipid-insoluble substances may enter or leave the blood and/or endothelial cells within vesicles, that is, by endocytosis or exocytosis.

Diffusion of substances via the other two routes depends on the specific structural (and permeability) characteristics of the capillary.

3. Limited passage of fluid and small solutes is allowed by **intercellular clefts** (gaps or areas of plasma membrane not joined by tight junctions). It is safe to say that, with the exception of brain capillaries, which are entirely secured together by tight junctions (the basis of the blood-brain barrier described in Chapter 7), most of our capillaries have intercellular clefts.
4. Very free passage of small solutes and fluids is allowed by *fenestrated capillaries*. These unique capillaries are found where absorption is a priority (intestinal capillaries or those serving endocrine glands) or where filtration occurs (the kidney). A fenestra is an oval pore (*fenestra* = window) or opening and is usually covered by a delicate membrane (see Figure 11.20b). Even so, a fenestra is much more permeable than other regions of the plasma membrane.

Only substances unable to pass by one of these routes are prevented from leaving (or entering) the capillaries. These include protein molecules (in plasma or interstitial fluid) and blood cells.

Fluid Movements at Capillary Beds

Besides the exchanges made via vesicles, and by passive diffusion through endothelial cell plasma membranes, clefts, or fenestrations, there are active forces operating at capillary beds. Because of their intercellular clefts and fenestrations, some capillaries are leaky and bulk fluid flows occur across their plasma membranes. Hence, blood pressure tends to force fluid (and solutes) outward, and osmotic pressure tends to draw fluid back into the bloodstream because blood has a higher solute concentration (due to its plasma proteins) than does interstitial fluid. Whether fluid moves out of or into the capillary depends on the difference between the two pressures. As a rule, blood pressure is higher at the arterial end of the capillary bed, and osmotic pressure is higher at the venous end. Consequently, fluid moves out of the capillaries at the beginning of the bed and is reclaimed at the opposite (venule) end (Figure 11.20c). However, not quite all of the fluid forced out of the bloodstream is reclaimed at the venule end. Returning that lost fluid to the blood is the chore of the lymphatic system, discussed in Chapter 12.

Developmental Aspects of the Cardiovascular System

The heart begins as a simple tube in the embryo. It is beating and busily pumping blood by the fourth week of pregnancy. During the next three weeks, the heart continues to change and mature, finally becoming a four-chambered structure capable of acting as a double pump—all without missing a beat! During fetal life, the collapsed lungs and nonfunctional liver are mostly bypassed by the blood, through special vascular shunts. After the seventh week of development, few changes other than growth occur in the fetal circulation until birth. Shortly after birth, the bypass structures become blocked, and the special umbilical vessels stop functioning.

Homeostatic Imbalance

Congenital heart defects account for about half of infant deaths resulting from all congenital defects. Environmental interferences, such as maternal infection and ingested drugs during the first three months of pregnancy (when the embryonic heart is forming), seem to be the major causes of such problems. Congenital heart defects may include a ductus arteriosus that does not close, septal openings, and other structural abnormalities of the heart. Such problems can usually be corrected surgically. ▲

In the absence of congenital heart problems, the heart usually functions smoothly throughout a long lifetime for most people. Homeostatic mechanisms are so effective that we rarely are aware of when the heart is working harder. The heart will hypertrophy and its cardiac output will increase substantially if we exercise regularly and aerobically (that is, vigorously enough to force it to beat at a higher-than-normal rate for extended periods of time). Not only does the heart become a more powerful pump; it also becomes more efficient: pulse rate and blood pressure decrease. An added

Systems in Sync

Homeostatic Relationships between the Cardiovascular System and Other Body Systems

Endocrine System
- The cardiovascular system delivers oxygen and nutrients; carries away wastes; blood serves as a transport vehicle for hormones
- Several hormones influence blood pressure (epinephrine, ANP, thyroxine, ADH); estrogen maintains vascular health in women

Lymphatic System/Immunity
- The cardiovascular system delivers oxygen and nutrients to lymphatic organs, which house immune cells; transports lymphocytes and antibodies; carries away wastes
- The lymphatic system picks up leaked fluid and plasma proteins and returns them to the cardiovascular system; its immune cells protect cardiovascular organs from specific pathogens

Digestive System
- The cardiovascular system delivers oxygen and nutrients; carries away wastes
- The digestive system provides nutrients to the blood including iron and B vitamins essential for RBC (and hemoglobin) formation

Urinary System
- The cardiovascular system delivers oxygen and nutrients; carries away wastes; blood pressure maintains kidney function
- The urinary system helps regulate blood volume and pressure by altering urine volume and releasing renin

Muscular System
- The cardiovascular system delivers oxygen and nutrients; carries away wastes
- Aerobic exercise enhances cardiovascular efficiency and helps prevent arteriosclerosis; the muscle "pump" aids venous return

Nervous System
- The cardiovascular system delivers oxygen and nutrients; removes wastes
- ANS regulates cardiac rate and force; sympathetic division maintains blood pressure and controls blood distribution according to need

Respiratory System
- The cardiovascular system delivers oxygen and nutrients; carries away wastes
- The respiratory system carries out gas exchange: loads oxygen and unloads carbon dioxide from the blood; respiratory "pump" aids venous return

Cardiovascular System

Reproductive System
- The cardiovascular system delivers oxygen and nutrients; carries away wastes
- Estrogen maintains vascular health in women

Integumentary System
- The cardiovascular system delivers oxygen and nutrients; carries away wastes
- The skin's blood vessels provide an important blood reservoir and a site for heat loss from body

Skeletal System
- The cardiovascular system delivers oxygen and nutrients and carries away wastes
- Bones are the site of hematopoiesis; protect cardiovascular organs by enclosure; provide a calcium depot

benefit of aerobic exercise is that it clears fatty deposits from the blood vessel walls, helping to slow the progress of atherosclerosis. However, let's raise a caution flag here: The once-a-month or once-a-year tennis player or snow shoveler has not built up this type of heart endurance and strength. When such an individual pushes his or her heart too much, it may not be able to cope with the sudden demand. This is why many weekend athletes are myocardial infarction victims.

As we get older, more and more signs of cardiovascular system disturbances start to appear. In some, the venous valves weaken, and purple, snakelike varicose veins appear. Not everyone has varicose veins, but we all have progressive atherosclerosis. Some say the process begins at birth, and there's an old saying that goes, "You are only as old as your arteries," referring to this degenerative process. The gradual loss in elasticity in the blood vessels leads to hypertension and hypertensive heart disease. The insidious filling of the blood vessels with fatty, calcified deposits leads most commonly to **coronary artery disease.** Also, as described in Chapter 10, the roughening of the vessel walls encourages thrombus formation. At least 30 percent of the population in the United States has hypertension by the age of 50, and cardiovascular disease causes more than one-half of the deaths in those over age 65. Although the aging process itself contributes to changes in the walls of the blood vessels that can lead to strokes or myocardial infarctions, most researchers feel that diet, not aging, is the single most important contributing factor to cardiovascular diseases. There is some agreement that the risk is lowered if people eat less animal fat, cholesterol, and salt. Other recommendations include avoiding stress, eliminating cigarette smoking, and taking part in a regular, moderate exercise program.

SUMMARY

Media study tools that could provide you with additional help in reviewing specific key topics of Chapter 11 are referenced below.

IP = Interactive Physiology;

WEB = A&P Place website.

THE HEART (pp. 348–360)

1. The heart, located in the thorax, is flanked laterally by the lungs and enclosed in a pericardium.

2. The bulk of the heart (myocardium) is composed of cardiac muscle. The heart has four hollow chambers—two atria (receiving chambers) and two ventricles (discharging chambers), each lined with endocardium. The heart is divided longitudinally by a septum.

   WEB Exercise: Chapter 11, External Anatomy of the Heart.

   IP Cardiovascular System Topic: Anatomy Review: The Heart, pages 3–5.

3. The heart functions as a double pump. The right heart is the pulmonary pump (right heart to lungs to left heart). The left heart is the systemic pump (left heart to body tissues to right heart).

4. Four valves prevent backflow of blood in the heart. The AV valves (mitral and tricuspid) prevent backflow into the atria when the ventricles are contracting. The semilunar valves prevent backflow into the ventricles when the heart is relaxing. The valves open and close in response to pressure changes in the heart.

   WEB Exercise: Chapter 11, Frontal Section of the Heart.

   IP Cardiovascular System Topic: Cardiac Cycle, page 3.

5. The myocardium is nourished by the coronary circulation, which consists of the right and left coronary arteries and their branches, and is drained by the cardiac veins and the coronary sinus.

6. Cardiac muscle is able to initiate its own contraction in a regular way, but its rate is influenced by both intrinsic and extrinsic factors. The intrinsic conduction system increases the rate of heart contraction and ensures that the heart beats as a unit. The SA node is the heart's pacemaker.

   WEB Exercise: Chapter 11, Intrinsic Conduction System of the Heart.

   IP Cardiovascular System Topic: Intrinsic Conduction System, pages 3–6.

7. The time and events occurring from one heartbeat to the next are the cardiac cycle.

8. As the heart beats, sounds resulting from the closing of the valves ("lub-dup") can be heard. Faulty valves reduce the efficiency of the heart as a pump and result in abnormal heart sounds (murmurs).

IP Cardiovascular System Topic: Cardiac Cycle, pages 4–10.

9. Cardiac output, the amount of blood pumped out by each ventricle in one minute, is the product of heart rate (HR) × stroke volume (SV). SV is the amount of blood ejected by a ventricle with each beat.

10. SV rises or falls with the volume of venous return. HR is influenced by the nerves of the autonomic nervous system, drugs (and other chemicals), and ion levels in the blood.

BLOOD VESSELS (pp. 360–381)

1. Arteries, which transport blood from the heart, and veins, which carry blood back to the heart, are conducting vessels. Only capillaries play a role in actual exchanges with tissue cells.

2. Except for capillaries, blood vessels are composed of three tunics: The tunica intima forms a friction-reducing lining for the vessel. The tunica media is the bulky middle layer of muscle and elastic tissue. The tunica externa is the protective, outermost connective tissue layer. Capillary walls are formed of the tunica interna only.

IP Cardiovascular System Topic: Anatomy Review: Blood Vessel Structure and Function, pages 3–5.

3. Artery walls are thick and strong to withstand pressure fluctuations. They expand and recoil as the heart beats. Vein walls are thinner, their lumens are larger, and they are equipped with valves. These modifications reflect the low-pressure nature of veins.

IP Cardiovascular System Topic: Anatomy Review: Blood Vessel Structure and Function, pages 25–27.

4. Capillary beds have two types of vessels—a vascular shunt and true capillaries the entrance to which are guarded by precapillary sphincters. Exchanges with tissue cells occur across the walls of the true capillaries. When the precapillary sphincters are closed, blood bypasses the local area via the vascular shunt.

WEB Exercise: Chapter 11, Arterial Circulation.

5. Varicose veins, a structural defect due to incompetent valves, is a common vascular problem, especially in the obese and people who stand for long hours. It is a predisposing factor for thrombophlebitis.

WEB Exercise: Chapter 11, Veins of the Systemic Circulation.

6. All the major arteries of the systemic circulation are branches of the aorta, which leaves the left ventricle. They branch into smaller arteries and then into the arterioles, which feed the capillary beds of the body tissues. For the names and locations of the systemic arteries, see pp. 364–366.

7. The major veins of the systemic circulation ultimately converge on one of the venae cavae. All veins above the diaphragm drain into the superior vena cava, and those below the diaphragm drain into the inferior vena cava. Both venae cavae enter the right atrium of the heart. See pp. 366–367 for the names and locations of the systemic veins.

8. The arterial circulation of the brain is formed by branches of paired vertebral and internal carotid arteries. The circle of Willis provides alternate routes for blood flow in case of a blockage in the brain's arterial supply.

9. The hepatic portal circulation is formed by veins draining the digestive organs, which empty into the hepatic portal vein. The hepatic portal vein carries the nutrient-rich blood to the liver, where it is processed before the blood is allowed to enter the systemic circulation.

10. The fetal circulation is a temporary circulation seen only in the fetus. It consists primarily of three special vessels: the single umbilical vein that carries nutrient- and oxygen-laden blood to the fetus from the placenta, and the two umbilical arteries that carry carbon dioxide and waste-laden blood from the fetus to the placenta. Shunts bypassing the lungs and liver are also present.

11. The pulse is the alternate expansion and recoil of a blood vessel wall (the pressure wave) that occurs as the heart beats. It may be felt easily over any superficial artery; such sites are called pressure points.

12. Blood pressure is the pressure that blood exerts on the walls of the blood vessels. It is the force that causes blood to continue to flow in the blood vessels. It is high in the arteries, lower in the capillaries, and lowest in the veins. Blood is forced along a descending pressure gradient. Both systolic and diastolic pressures are recorded.

IP Cardiovascular System Topic: Measuring Blood Pressure, pages 3–12.

13. Arterial blood pressure is directly influenced by heart activity (increased heart rate leads to increased blood pressure) and by resistance to blood flow. The most important factors increasing the peripheral resistance are a decrease in the diameter or stretchiness of the arteries and arterioles and an increase in blood viscosity.

14. Many factors influence blood pressure. Some of these factors are the activity of the sympathetic nerves and kidneys, drugs, and diet.

15. Hypertension, which reflects an increase in peripheral resistance, strains the heart and damages blood vessels. In most cases, the precise cause is unknown.

16. Substances move to and from the blood and tissue cells through capillary walls. Some substances are transported in vesicles, but most move by diffusion—directly through the endothelial cell plasma membranes, through intercellular clefts, or through fenestrations. Fluid is forced from the bloodstream by blood pressure and drawn back into the blood by osmotic pressure.

DEVELOPMENTAL ASPECTS OF THE CARDIOVASCULAR SYSTEM

(pp. 381–383)

1. The heart begins as a tubelike structure that is beating and pumping blood by the fourth week of embryonic development.
2. Congenital heart defects account for half of all infant deaths resulting from congenital problems.
3. Arteriosclerosis is an expected consequence of aging. Gradual loss of elasticity in the arteries leads to hypertension and hypertensive heart disease, and clogging of the vessels with fatty substances leads to coronary artery disease and stroke. Cardiovascular disease is an important cause of death in individuals over age 65.
4. Modifications in diet (decreased fats, cholesterol, and salt), stopping smoking, and regular aerobic exercise may help to reverse the atherosclerotic process and prolong life.

REVIEW QUESTIONS

MULTIPLE CHOICE

More than one choice may apply.

1. Freshly oxygenated blood is first received by the:
   a. right ventricle c. right atrium
   b. left ventricle d. left atrium

2. Given an end-diastolic volume of 150 ml, an end-systolic volume of 50 ml, and a heart rate of 60 bpm, the cardiac output is:
   a. 600 ml/min c. 1200 ml/min
   b. 6 liters/min d. 3 liters/min

3. Which of the following depolarizes next after the AV node?
   a. Atrial myocardium
   b. Ventricular myocardium
   c. Bundle branches
   d. AV bundle

4. During atrial systole:
   a. the atrial pressure exceeds ventricular pressure
   b. 70 percent of ventricular filling occurs
   c. the AV valves are open
   d. valves prevent backflow into the great veins

5. Atrial repolarization coincides in time with the:
   a. P wave c. QRS wave
   b. T wave d. P-Q interval

6. Soon after the onset of ventricular systole the:
   a. AV valves close
   b. semilunar valves open
   c. first heart sound is heard
   d. aortic pressure increases

7. The base of the heart is its _______ surface.
   a. diaphragmatic c. anterior
   b. posterior d. inferior

8. In comparing a parallel artery and vein, you would find that:
   a. the artery wall is thicker
   b. the artery diameter is greater
   c. the artery lumen is smaller
   d. the artery endothelium is thicker

9. Considering the vessels named pulmonary trunk and celiac trunk, the term *trunk* must refer to:
   a. a vessel in the heart wall
   b. a vein
   c. a capillary
   d. a large artery from which other arteries branch

10. Which of these vessels is bilaterally symmetrical (i.e., one vessel of the pair occurs on each side of the body)?
    a. Internal carotid artery
    b. Brachiocephalic trunk
    c. Azygos vein
    d. Superior mesenteric vein

11. A stroke that occludes a posterior cerebral artery will most likely affect:
    a. hearing
    b. vision
    c. smell
    d. higher thought processes
12. Vessels involved in the circulatory pathway to and from the brain are the:
    a. brachiocephalic artery
    b. subclavian artery
    c. internal jugular vein
    d. internal carotid artery
13. Which layer of the artery wall thickens most in atherosclerosis?
    a. Tunica media
    b. Tunica interna
    c. Tunica adventitia
    d. Tunica externa
14. Which of the following are associated with aging?
    a. Increasing blood pressure
    b. Weakening of venous valves
    c. Arteriosclerosis
    d. Stenosis of the ductus arteriosus
15. An increase in BP would be caused by all of the following except:
    a. increase in SV
    b. increase in heart rate
    c. increase in the length of ventricular diastole
    d. constriction of the arterioles

SHORT ANSWER ESSAY

1. Describe the location and position of the heart in the thorax.
2. Draw a diagram of the heart showing the three layers composing its wall and its four chambers. Label each. Show where the AV and semilunar valves are. Show and label all blood vessels entering and leaving the heart chambers.
3. Trace one drop of blood from the time it enters the right atrium of the heart until it enters the left atrium. What is this circuit called?
4. Explain the difference in function of the systemic and pulmonary circulations.
5. Why are the heart valves important? Can the heart function with leaky valves?
6. Why might a thrombus in a coronary artery cause sudden death?
7. What is the function of the fluid that fills the pericardial sac?
8. Define *systole, diastole, stroke volume,* and *cardiac cycle.*
9. To which heart chamber(s) do the terms *systole* and *diastole* most often apply?
10. How does the heart's ability to contract differ from that of other muscles of the body?
11. What is the function of the intrinsic conduction system of the heart? Name its elements, *in order,* beginning with the pacemaker.
12. What causes the lub-dup heart sounds that are heard with a stethoscope?
13. Name three different factors that increase heart rate.
14. Name and describe from the inside out the three tunics making up the walls of arteries and veins, and give the most important function of each layer.
15. Describe the structure of capillary walls. How is their structure related to their function in the body?
16. Why are artery walls so much thicker than those of corresponding veins?
17. Name three factors that are important in promoting venous return.
18. Arteries are often described as vessels that carry oxygen-rich blood, and veins are said to carry oxygen-poor (carbon dioxide–rich) blood. Name two sets of exceptions to this rule that were discussed in this chapter.
19. Trace a drop of blood from the left ventricle of the heart to the wrist of the right hand and back to the heart. Now trace it to the dorsum of the right foot and back to the right heart.
20. What is the circle of Willis, and why is it important?
21. What is the function of the hepatic portal circulation? In what way is a portal circulation a "strange" circulation?
22. The liver and lungs are almost entirely bypassed by blood in a fetus. Why is this? Name the vessel that bypasses the liver. Name two lung bypasses. Three vessels travel in the umbilical cord; which of these carries oxygen- and nutrient-rich blood?
23. Define *pulse.*
24. Which artery is palpated at the following pressure points: wrist? front of the ear? side of the neck? the groin? back of the knee?

25. Define *blood* pressure, *systolic* pressure, and *diastolic* pressure.

26. What vital role does blood pressure play?

27. Two elements determine blood pressure—the cardiac output of the heart and the peripheral resistance, or friction, in the blood vessels. Name two factors that increase cardiac output. Name two factors that increase peripheral resistance.

28. What is the effect of hemorrhage on blood pressure? Why? In which position—sitting, lying down, or standing—is the blood pressure normally highest? lowest?

29. What is different about the capillary exchanges seen in a capillary with fenestrations and intercellular clefts and the exchanges seen in a capillary lacking those modifications?

30. What are varicose veins? What factors seem to promote their formation?

AT THE CLINIC

1. Define *hypertension* and *arteriosclerosis*. How are they often related? Why is hypertension called the "silent killer"? Name three changes in your lifestyle that might help prevent cardiovascular disease in your old age.

2. A middle-aged woman is admitted to the coronary care unit with a diagnosis of left ventricular failure resulting from a myocardial infarction. Her chart indicates that she was awakened in the middle of the night by severe chest pain. Her skin is pale and cold, and moist sounds of pulmonary edema are heard over the lower regions of both lungs. Explain how failure of the left ventricle might cause these signs and symptoms.

3. Linda, a 14-year-old girl undergoing a physical examination before being admitted to summer camp, was found to have a loud heart murmur at the second intercostal space on the left side of the sternum. The murmur takes the form of a swishing sound with no high-pitched whistle. What, exactly, is producing the murmur?

4. Mrs. Johnson is brought to the emergency room after being involved in an auto accident. She is hemorrhaging and has a rapid, thready pulse, but her blood pressure is still within normal limits. Describe the compensatory mechanisms that are maintaining her blood pressure in the face of blood loss.

5. During a lethal heart attack, a blood clot lodges in the first part of the circumflex branch of the left coronary artery, blocking blood flow through this vessel. What parts of the heart will become ischemic and die?

6. Mr. Grimaldi was previously diagnosed as having a posterior pituitary tumor that causes hypersecretion of ADH. He comes to the clinic regularly to have his blood pressure checked. Would you expect his blood pressure to be chronically elevated or depressed? Why?

7. Explain why taking a vigorous swim right after eating is more likely to cause indigestion than cramping of your muscles.

8. The guards at the royal palace in London stand at attention while on duty. On a very hot day, it is not unusual for one (or more) to become lightheaded and faint. Explain this phenomenon.

12 The Lymphatic System and Body Defenses

YOUR GOALS

After completing this chapter, you will have a working knowledge of the functions of the lymphatic system and of body defenses and will have mastered the objectives listed below.

FUNCTION PREVIEW

- The lymphatic system returns leaked plasma to the blood vessels after cleansing it of bacteria and other foreign matter. It also provides sites for surveillance by immune system cells.
- The nonspecific defenses hinder pathogen entry, prevent the spread of disease-causing microorganisms, and strengthen the immune response.
- The immune system protects against disease by destroying "foreign" cells and by inactivating toxins and other foreign chemicals with its antibodies.

OBJECTIVE CHECKLIST

PART I: THE LYMPHATIC SYSTEM (pp. 390–395)

- Name the two major types of structures composing the lymphatic system, and explain how the lymphatic system is functionally related to the cardiovascular and immune systems.
- Describe the composition of lymph, and explain its formation and transport.
- Describe the function(s) of lymph nodes, tonsils, the thymus, Peyer's patches, and the spleen.

PART II: BODY DEFENSES (pp. 395–417)

- Describe the protective functions of skin and mucous membranes.
- Explain the importance of phagocytes and natural killer cells.
- Describe the inflammatory process.
- Name several antimicrobial substances produced by the body that act in nonspecific body defense.

>>

(Objective Checklist, continued)

- Explain how fever helps protect the body against invading bacteria.
- Define *antigen* and *hapten,* and name substances that act as complete antigens.
- Name the two arms of the immune response, and relate each to a specific lymphocyte type (B or T cell).
- Compare and contrast the development of B and T cells.
- State the roles of B cells, T cells, and plasma cells.
- Explain the importance of macrophages in immunity.
- List the five antibody classes, and describe their specific roles in immunity.
- Describe several ways in which antibodies act against antigens.
- Distinguish between active and passive immunity.
- Describe immunodeficiencies, allergies, and autoimmune diseases.

PART III: DEVELOPMENTAL ASPECTS OF THE LYMPHATIC SYSTEM AND BODY DEFENSES (pp. 417–421)

- Describe the origin of the lymphatic vessels.
- Describe the effects of aging on immunity.

PART I: THE LYMPHATIC SYSTEM

They can't all be superstars! When we mentally tick off the names of the body's organ systems, the lymphatic (lim-fat′ik) system is probably not the first to come to mind. Yet without this quietly working system, our cardiovascular system would stop working and our immune system would be hopelessly impaired. The **lymphatic system** actually consists of two semi-independent parts: (1) a meandering network of *lymphatic vessels* and (2) various *lymphoid tissues* and *organs* scattered throughout the body. The lymphatic vessels transport fluids that have escaped from the blood vascular system back to the blood. The lymphoid organs house phagocytic cells and lymphocytes, which play essential roles in body defense and resistance to disease.

Lymphatic Vessels

As blood circulates through the body, exchanges of nutrients, wastes, and gases occur between the blood and the interstitial fluid. As explained in Chapter 11, the hydrostatic and osmotic pressures operating at capillary beds force fluid out of the blood at the arterial ends of capillary beds ("upstream") and cause most of it to be reabsorbed at the venous ends ("downstream"). The fluid that remains behind in the tissue spaces, as much as 3 L daily, becomes part of the interstitial fluid. These leaked fluids, as well as any plasma proteins that escape from the bloodstream, must be carried back to the blood if the vascular system is to have sufficient blood volume to operate properly. If it does not, fluid accumulates in the tissues, producing **edema.** Excessive edema impairs the ability of tissue cells to make exchanges with the interstitial fluid and ultimately the blood. The function of the **lymphatic vessels** is to pick up this excess tissue fluid, now called **lymph** (*lymph* = clear water) and return it to the bloodstream (Figure 12.1).

The lymphatic vessels, also called *lymphatics,* form a one-way system, and lymph flows only toward the heart. The microscopic, blind-ended **lymph capillaries** spiderweb between the tissue cells and blood capillaries in the loose connective tissues of the body (Figures 12.1 and 12.2a) and absorb the leaked fluid (primarily water and a small amount of dissolved proteins). Although similar to blood capillaries, lymphatic capillaries are so remarkably permeable that they were once thought to be open at one end like a straw. Not so—instead, what we find is that the edges of the endothelial cells forming their walls loosely overlap one another, forming flaplike *minivalves* (Figure 12.2b) that act as one-way swinging doors. The flaps, anchored by fine collagen fibers to surrounding structures, gape open when the fluid pressure is higher in the interstitial space, allowing fluid to enter the lymphatic capillary. However, when the pressure is higher inside the lymphatic vessels, the endothelial cell flaps are forced together, preventing the lymph from leaking back out and forcing it along the vessel. Proteins, and even larger particles such as cell debris, bacteria, and viruses, are normally prevented from entering blood capillaries, but they enter the lymphatic capillaries easily, particularly in inflamed areas. But there is a problem here—bacteria and viruses (and cancer cells) that enter the lymphatics can then use them to travel throughout the body. This dilemma is partly resolved by the fact that lymph takes "detours" through the lymph nodes, where it is cleansed of debris and "examined" by cells of the immune system, as explained in more detail shortly.

Lymph is transported from the lymph capillaries through successively larger lymphatic vessels, referred to as *lymphatic collecting vessels,* until it is finally returned to the venous system through one of the two large ducts in the thoracic region. The **right lymphatic duct** drains the lymph from the right arm and the right side of the head and thorax. The large **thoracic duct** receives lymph from the rest of the body, as shown in Figure 12.3. Both ducts empty the lymph into the subclavian vein on their own side of the body.

Like the veins of the cardiovascular system, the lymphatic vessels are thin-walled, and the larger ones have valves. The lymphatic system is a low-pressure, pumpless system. Lymph is transported by the same mechanisms that aid return of venous blood—the milking action of the skeletal muscles and the pressure changes in the thorax during breathing (that is, the muscular and respiratory "pumps"). In addition, smooth muscle in the walls of the larger lymphatics contracts rhythmically, actually helping to "pump" the lymph along.

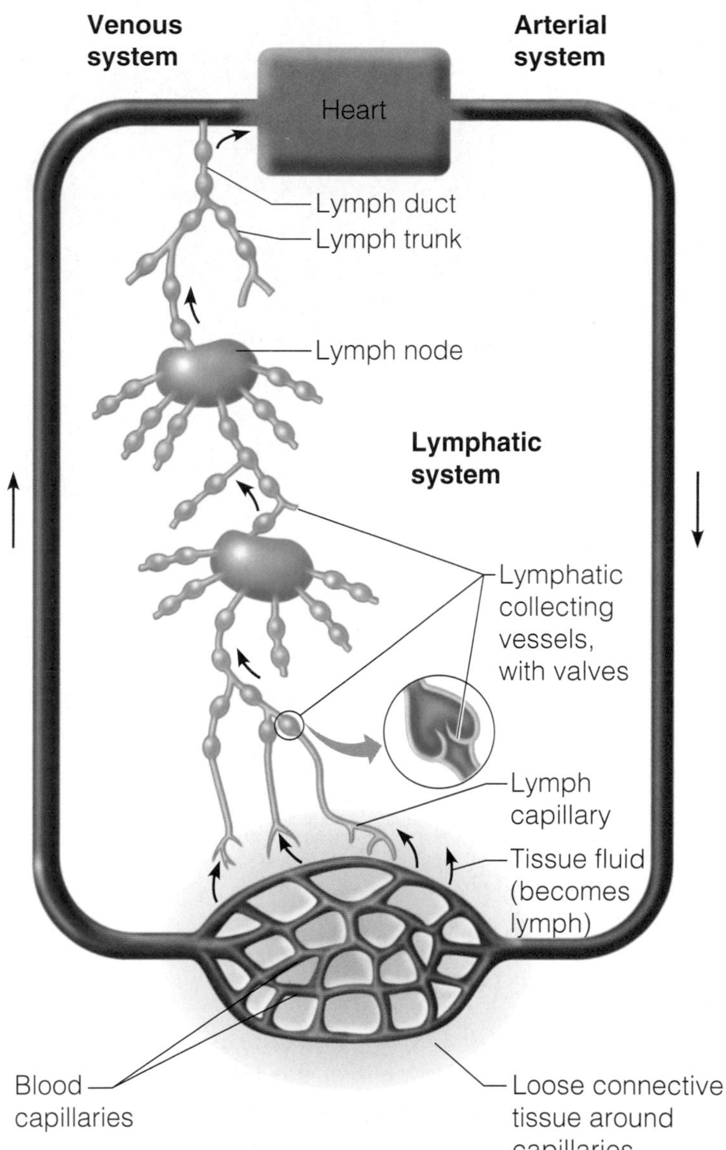

FIGURE 12.1 Relationship of the lymphatic vessels to the blood vessels of the cardiovascular circuit. Beginning at the bottom of this figure, we see that lymph, which begins as tissue fluid derived from blood capillaries, enters the lymph capillaries, travels through the lymphatic vessels and lymph nodes, and enters the bloodstream via the great veins at the root of the neck.

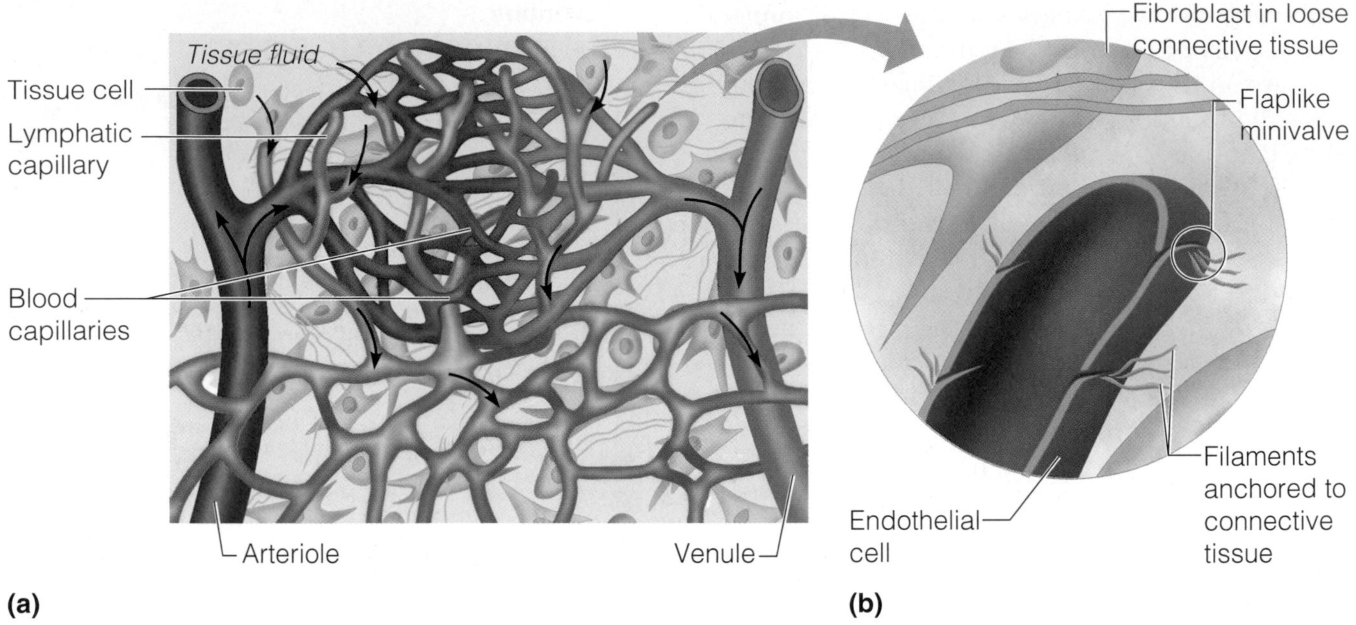

FIGURE 12.2 Distribution and special structural features of lymphatic capillaries. (a) Structural relationship between blood capillaries and lymph capillaries. Black arrows indicate direction of fluid movement. **(b)** Lymph capillaries begin as blind-ended tubes. The endothelial cells forming their walls overlap one another, forming flaplike minivalves.

Lymph Nodes

More closely related to the immune system, the **lymph nodes** help protect the body by removing foreign material such as bacteria and tumor cells from the lymphatic stream and by producing lymphocytes that function in the immune response.

As lymph is transported toward the heart, it is filtered through the thousands of lymph nodes that cluster along the lymphatic vessels (see Figure 12.1). Particularly large clusters are found in the inguinal, axillary, and cervical regions of the body (see Figure 12.3). Within the lymph nodes are **macrophages** (mak′ro-fāj-ez), which engulf and destroy bacteria, viruses, and other foreign substances in the lymph before it is returned to the blood. Collections of **lymphocytes** (a type of white blood cell) are also strategically located in the lymph nodes and respond to foreign substances in the lymphatic stream. Although we are not usually aware of the protective nature of the lymph nodes, most of us have had swollen glands during an active infection. This swelling is a result of the trapping function of the nodes.

Lymph nodes vary in shape and size, but most are kidney-shaped, less than 1 inch (approximately 2.5 centimeters [cm]) long, and "buried" in the connective tissue that surrounds them. Each node is surrounded by a fibrous *capsule* from which strands called *trabeculae* (trah-bek′yu-le) extend inward to divide the node into a number of compartments (Figure 12.4). The internal framework is a network of soft reticular connective tissue that supports a continually changing population of lymphocytes. (As described in Chapter 10, lymphocytes arise from the red bone marrow but then migrate to the lymphatic organs, where they proliferate further.)

The outer part of the node, its **cortex,** contains collections of lymphocytes called **follicles.** Many of the follicles have dark-staining centers called **germinal centers.** These centers enlarge when specific lymphocytes (the *B cells*) are generating daughter cells called **plasma cells,** which release antibodies. The rest of the cortical cells are lym-

Q *What would be the consequence of thoracic duct blockage?*

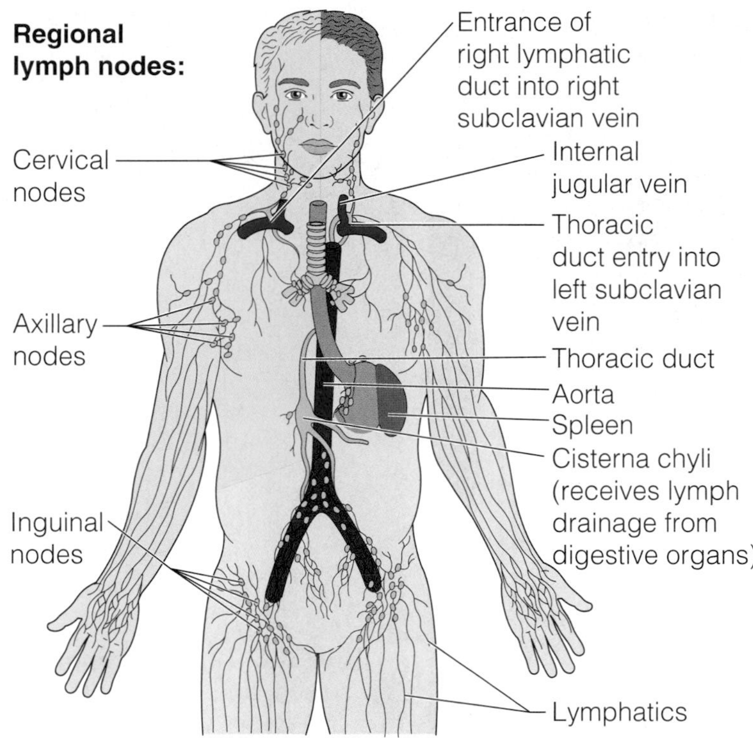

FIGURE 12.3 Distribution of lymphatic vessels and lymph nodes. Shading shows the body area drained by the right lymphatic duct; the rest of the body is drained by the thoracic duct.

phocytes "in transit," the so-called *T cells* that circulate continuously between the blood, lymph nodes, and lymphatic stream, performing their surveillance role. Phagocytic macrophages are located in the central **medulla** of the lymph node. (The precise roles of the cells of the lymph nodes in immunity are discussed shortly.)

Lymph enters the convex side of a lymph node through **afferent lymphatic vessels.** It then flows through a number of **sinuses** that cut through the lymph node and finally exits from the node at its indented region, the **hilus** (hi'lus), via **efferent lymphatic vessels.** Because there are fewer efferent vessels draining the node than afferent vessels feeding it, the flow of lymph through the node is very slow. This allows time for the lymphocytes and macrophages to perform their protective functions. Generally speaking, lymph passes through several nodes before its cleaning process is complete.

A *Edema in the areas that drain into the thoracic duct.*

Homeostatic Imbalance

Lymph nodes help rid the body of infectious agents and cancer cells, but sometimes they are overwhelmed by the very agents they are trying to destroy. For example, when large numbers of bacteria or viruses are trapped in the nodes, the nodes become inflamed and tender to the touch. The lymph nodes can also become secondary cancer sites, particularly in cancers that use lymphatic vessels to spread throughout the body. The fact that cancer-infiltrated lymph nodes are swollen but not painful helps to distinguish cancerous lymph nodes from those infected by microorganisms. ▲

Q *What is the benefit of having fewer efferent than afferent lymphatic vessels?*

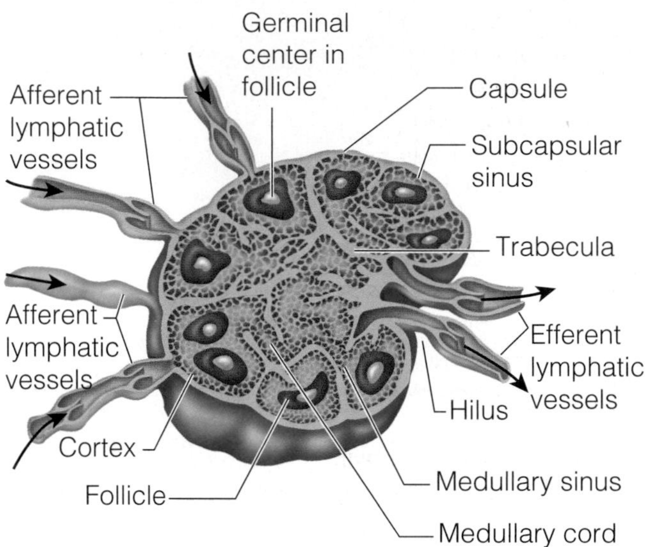

FIGURE 12.4 Structure of a lymph node. Longitudinal view of the internal structure of a lymph node and associated lymphatics. Notice that several afferent lymphatics enter the node, whereas fewer efferent lymphatics exit at its hilus.

FIGURE 12.5 Lymphoid organs. Tonsils, thymus, spleen, and Peyer's patches.

Other Lymphoid Organs

Lymph nodes are just one example of the many types of **lymphoid organs** in the body. Others are the spleen, thymus gland, tonsils, and Peyer's patches of the intestine (Figure 12.5), as well as bits of lymphatic tissue scattered in the epithelial and connective tissues. The common feature of all these organs is a predominance of reticular connective tissue and lymphocytes. Although all lymphoid organs have roles in protecting the body, only the lymph nodes filter lymph.

The **spleen** is a blood-rich organ that filters blood. It is located in the left side of the abdominal cavity and extends to curl around the anterior aspect of the stomach. Instead of filtering lymph, the spleen filters and cleanses blood of bacteria, viruses, and other debris. Its most important function is to destroy worn-out red blood cells and return some of their breakdown products to the liver. For example, iron is used again for making hemoglobin, and the rest of the hemoglobin molecule is secreted in bile. Other functions of the spleen include storing platelets and acting as a blood reservoir (as does the liver). During hemorrhage, both the spleen and liver contract and empty their contained blood into the circulation to help bring the blood volume back to normal levels. In the fetus, the spleen is an important hematopoietic (blood cell–forming) site, but only lymphocytes are produced by the adult spleen.

The **thymus gland,** which functions at peak levels only during youth, is a lymphatic mass found low in the throat overlying the heart. As described in Chapter 9, the thymus produces hormones, *thymosin* and others, that function in the programming of certain lymphocytes so they can carry out their protective roles in the body.

The **tonsils** are small masses of lymphatic tissue that ring the pharynx (the throat), where they are found in the mucosa. Their job is to trap and remove any bacteria or other foreign pathogens entering the throat. They carry out this function so efficiently that sometimes they become congested

A *Since the outlet is smaller than the inlet to the lymph node, the lymph fluid stagnates (stops flowing) briefly in the lymph node giving macrophages and lymphocytes time to monitor and process the lymph for pathogens.*

Nonspecific defense mechanisms		Specific defense mechanisms (immune system)
First line of defense	Second line of defense	Third line of defense
• Skin • Mucous membranes • Secretions of skin and mucous membranes	• Phagocytic cells • Antimicrobial proteins • The inflammatory response	• Lymphocytes • Antibodies • Macrophages

FIGURE 12.6 An overview of the body's defenses.

with bacteria and become red, swollen, and sore, a condition called *tonsillitis.*

Peyer's patches, which resemble tonsils, are found in the wall of the small intestine. The macrophages of Peyer's patches are in an ideal position to capture and destroy bacteria (always present in tremendous numbers in the intestine), thereby preventing them from penetrating the intestinal wall. Peyer's patches and the tonsils are part of the collection of small lymphoid tissues referred to as **mucosa-associated lymphatic tissue (MALT).** Collectively, MALT acts as a sentinel to protect the upper respiratory and digestive tracts from the never-ending attacks of foreign matter entering those cavities.

Now that we have set the stage on which many of the body's defense mechanisms play their roles, we are ready to consider that topic in more detail.

PART II: BODY DEFENSES

Every second of every day, an army of hostile bacteria, viruses, and fungi swarms on our skin and invades our inner passageways—yet we stay amazingly healthy most of the time. The body seems to have developed a single-minded approach toward such foes—if you're not with us, then you're against us!

The body's defenders against these tiny but mighty enemies are two systems, simply called the *nonspecific* and the *specific defense systems* (Figure 12.6). The **nonspecific defense system** responds immediately to protect the body from all foreign substances, whatever they are. The nonspecific defenses are provided by intact skin and mucous membranes, the inflammatory response, and a number of proteins produced by body cells. This system reduces the workload of the second protective arm, the specific defense system, by preventing entry and spread of microorganisms throughout the body.

The **specific defense system,** more commonly called the **immune system,** mounts the attack against *particular* foreign substances. Although certain body organs (lymphatic organs and blood vessels) are intimately involved with the immune response, the immune system is a *functional system* rather than an organ system in an anatomical sense. Its "structures" are a variety of molecules and trillions of immune cells, which inhabit lymphatic tissues and circulate in body fluids. The most important of the immune cells are *lymphocytes* and *macrophages.*

When our immune system is operating effectively, it protects us from most bacteria, viruses, transplanted organs or grafts, and even our own cells that have turned against us. The immune system does this both directly, by cell attack, and indirectly, by releasing mobilizing chemicals and protective antibody molecules. The resulting highly specific resistance to disease is called **immunity** (*immun* = free).

Unlike the nonspecific defenses which are always prepared to defend the body, the immune system must first "meet" or be primed by an initial exposure to a foreign substance (antigen) before it can protect the body against it. Nonetheless, what it lacks in speed it makes up for in the precision of its counterattacks. Although we will consider them separately, keep in mind that specific and nonspecific defenses always work hand-in-hand to protect the body.

Nonspecific Body Defenses

Some nonspecific resistance to disease is inherited. For instance, there are certain diseases that humans never get, such as some forms of tuberculosis that affect birds. Most often, however, the term *nonspecific body defense* refers to the mechanical barriers that cover body surfaces and to cells and chemicals that act on the initial battlefronts to protect the body from invading **pathogens** (harmful or disease-causing microorganisms). Table 12.1 (p. 401) provides a more detailed summary of the most important nonspecific defenses.

Surface Membrane Barriers

The body's *first line of defense* against the invasion of disease-causing microorganisms is the *skin* and *mucous membranes.* As long as the skin is unbroken, its keratinized epidermis is a strong physical barrier to most microorganisms that swarm on the skin. Intact mucous membranes provide similar mechanical barriers within the body. Recall that mucous membranes line all body cavities open to the exterior: the digestive, respiratory, urinary, and reproductive tracts. Besides serving as physical barriers, these membranes produce a variety of protective chemicals:

1. The acid pH of skin secretions inhibits bacterial growth, and sebum contains chemicals that are toxic to bacteria. Vaginal secretions of adult females are also very acidic.
2. The stomach mucosa secretes hydrochloric acid and protein-digesting enzymes. Both kill pathogens.
3. Saliva and lacrimal fluid contain *lysozyme,* an enzyme that destroys bacteria.
4. Sticky mucus traps many microorganisms that enter digestive and respiratory passageways.

Some mucosae also have structural modifications that fend off potential invaders. Mucus-coated hairs inside the nasal cavity trap inhaled particles, and the respiratory tract mucosa is ciliated. The cilia sweep dust- and bacteria-laden mucus superiorly toward the mouth, preventing it from entering the lungs, where the warm, moist environment provides an ideal site for bacterial growth.

Although the surface barriers are very effective, they *are* broken from time to time by small nicks and cuts resulting, for example, from brushing your teeth or shaving. When this happens and microorganisms do invade deeper tissues, other nonspecific mechanisms come into play to defend the body.

Cells and Chemicals

For its *second line of defense,* the body uses an enormous number of cells and chemicals to protect itself. These defenses rely on the destructive powers of *phagocytes* and *natural killer cells,* the inflammatory response, and a variety of chemical substances that kill pathogens and help repair tissue. Fever is also considered to be a nonspecific protective response.

Phagocytes

Pathogens that make it through the mechanical barriers are confronted by **phagocytes** (fa′go-sītz″; *phago* = eat) in nearly every body organ. A phagocyte, such as a *macrophage* or *neutrophil,* engulfs a foreign particle much the way an amoeba ingests a food particle (Figure 12.7). Flowing cytoplasmic extensions bind to the particle and then pull it inside, enclosing it in a vacuole. The vacuole is then fused with the enzymatic contents of a *lysosome,* and its contents are broken down or digested.

Natural Killer Cells

Natural killer (NK) cells, which "police" the body in blood and lymph, are a unique group of lymphocytes that can lyse and kill cancer cells and virus-infected body cells well before the immune system is enlisted in the fight. Unlike the lymphocytes of the immune system, which can recognize and react only against *specific* virus-infected or tumor cells, natural killer cells are far less "picky." They can act spontaneously against *any* such target by recognizing certain sugars on the "intruder's" surface. NK cells are not phagocytic. They attack the target cell's membrane and release a lytic chemical called *perforins.* Shortly thereafter, the target cell's membrane and nucleus disintegrate.

Inflammatory Response

The **inflammatory response** is a nonspecific response that is triggered whenever body tissues

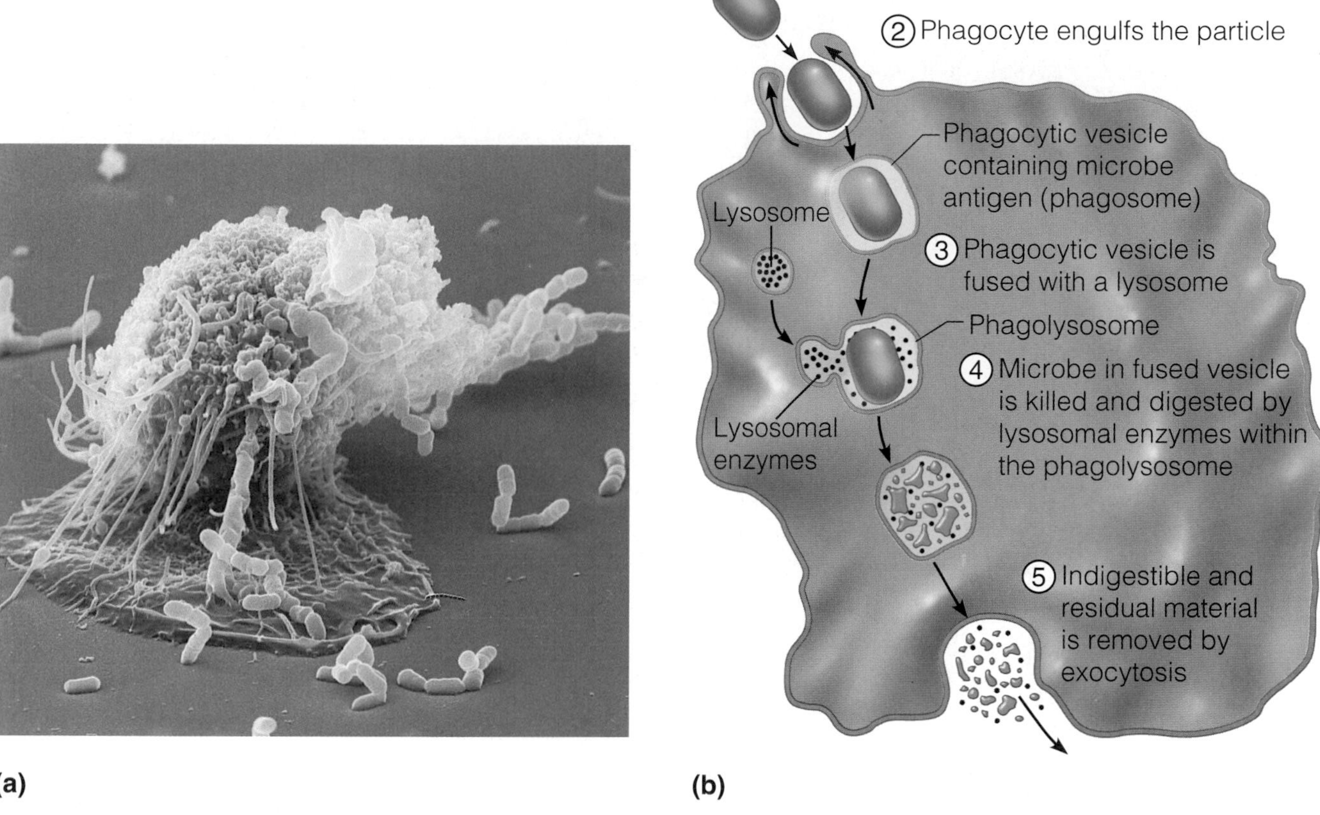

FIGURE 12.7 Phagocytosis by a macrophage. (a) This scanning electron micrograph (4300×) in computer-generated color shows a macrophage pulling sausage-shaped *Escherichia coli* bacteria toward it with its long cytoplasmic extensions. Several bacteria on the macrophage's surface are being engulfed. **(b)** Events of phagocytosis.

are injured (Figure 12.8). For example, it occurs in response to physical trauma, intense heat, and irritating chemicals, as well as to infection by viruses and bacteria. The four *cardinal signs* and major symptoms of an acute inflammation are *redness, heat* (*inflamm* = set on fire), *swelling,* and *pain.* It is easy to understand why these symptoms occur, once the events of the inflammatory response are understood.

The inflammatory process begins with a chemical "alarm." When cells are injured, they release inflammatory chemicals, including **histamine** and **kinins** (ki′ninz), that (1) cause blood vessels in the involved area to dilate and capillaries to become leaky, (2) activate pain receptors, and (3) attract phagocytes and white blood cells to the area. (This latter phenomenon is called **chemotaxis** because the cells are following a chemical gradient.) Dilation of the blood vessels increases the blood flow to the area, accounting for the redness and heat observed. Increased permeability of the capillaries allows plasma to leak from the bloodstream into the tissue spaces, causing local edema (swelling) that also activates pain receptors in the area. If the swollen, painful area is a joint, its function (movement) may be impaired temporarily. This forces the injured part to rest, which aids healing. Some authorities consider limitation of joint movement to be an additional (fifth) cardinal sign of inflammation.

The inflammatory response (1) prevents the spread of damaging agents to nearby tissues, (2) disposes of cell debris and pathogens, and (3) sets the stage for repair. Let's look at how it accomplishes these tasks. Within an hour or so after the inflammatory process has begun, neutrophils are squeezing through the capillary walls, a process called

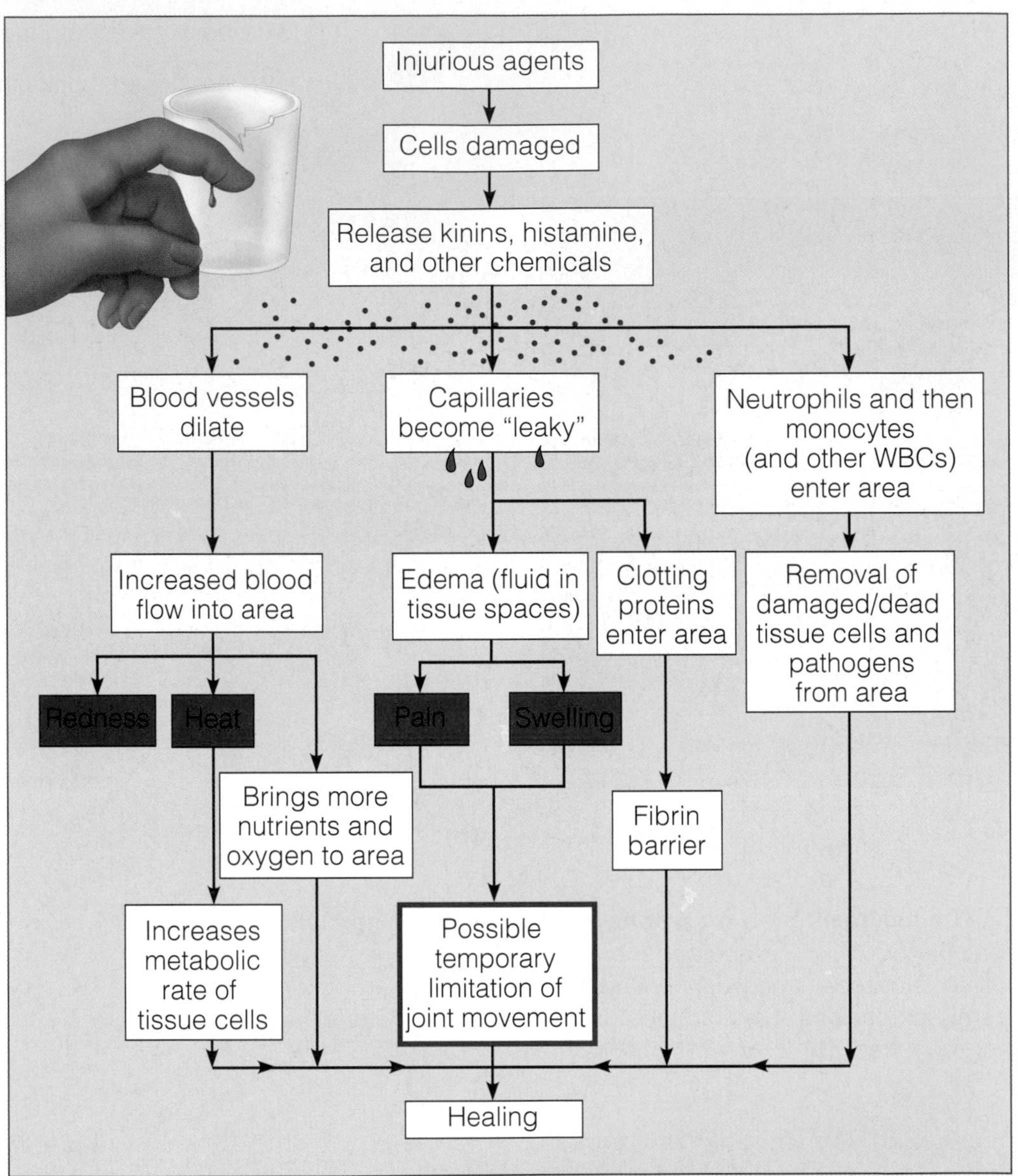

FIGURE 12.8 Flowchart of inflammatory events. The four cardinal signs of acute inflammation are shown in red boxes. Limitation of joint movement (red outlined box) occurs in some cases and is considered to be the fifth cardinal sign of acute inflammation.

diapedesis (Figure 12.9). Drawn to the area by inflammatory chemicals, the neutrophils begin the cleanup detail by engulfing damaged or dead tissue cells and/or pathogens. As the counterattack continues, monocytes begin to leave the bloodstream and follow the neutrophils into the inflamed area. Monocytes are fairly poor phagocytes, but within 8 to 12 hours after entering the tissues they become macrophages with insatiable appetites. The macrophages continue to wage the battle, replacing the short-lived neutrophils on the battlefield. Macrophages are the central actors in the final disposal of cell debris as the inflammation subsides.

Besides phagocytosis, other protective events are also occurring at the inflamed site. Clotting proteins, leaked into the area from the blood, are activated and begin to wall off the damaged area with fibrin to prevent the spread of pathogens or harmful agents to neighboring tissues. The fibrin mesh also forms a scaffolding for permanent repair. The local

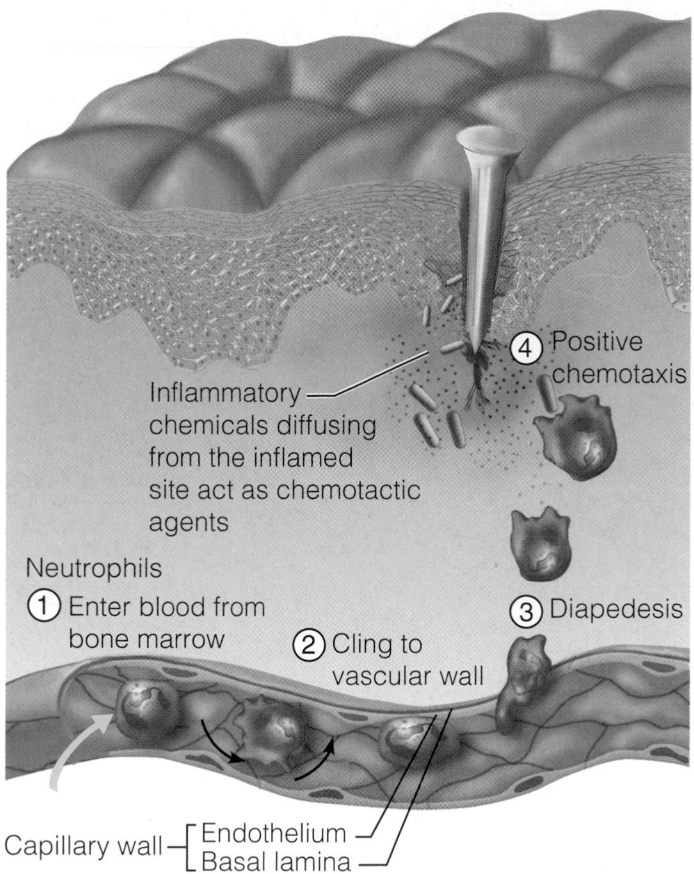

FIGURE 12.9 Phagocyte mobilization.

heat increases the metabolic rate of the tissue cells, speeding up their defensive actions and repair processes.

If the area contains pathogens that have previously invaded the body, the *third line of defense* also comes into play—the immune response mediated by lymphocytes. Both protective antibodies and T cells (T lymphocytes) invade the area to act specifically and directly against the damaging agents. (We will speak more of the immune response shortly.)

Homeostatic Imbalance

In severely infected areas, the battle takes a considerable toll on both sides, and creamy, yellow pus may be formed in the wound. **Pus** is a mixture of dead or dying neutrophils, broken-down tissue cells, and living and dead pathogens. If the inflammatory mechanism fails to fully clear the area of debris, the sac of pus may become walled off, forming an *abscess.* Surgical drainage of abscesses is often necessary before healing can occur. ▲

Antimicrobial Chemicals

The body's most important **antimicrobial chemicals,** apart from those produced in the inflammatory reaction, are *complement proteins* and *interferon*.

Complement The term ***complement*** refers to a group of at least 20 plasma proteins that circulate in the blood in an inactive state. However, when complement becomes attached, or *fixed,* to foreign cells such as bacteria, fungi, or mismatched red blood cells, it is activated and becomes a major factor in the fight against the foreign cells. This **complement fixation** occurs when complement proteins bind to certain sugars or proteins (such as antibodies) on the foreign cell's surface. One result of complement fixation is the formation of *membrane attack complexes (MAC)* that produce lesions, complete with holes, in the foreign cell's surface (Figure 12.10). These allow water to rush into the cell, causing it to burst. Activated complement also amplifies the inflammatory response. Some of the molecules released during the activation process are *vasodilators,* and some are *chemotaxis chemicals*

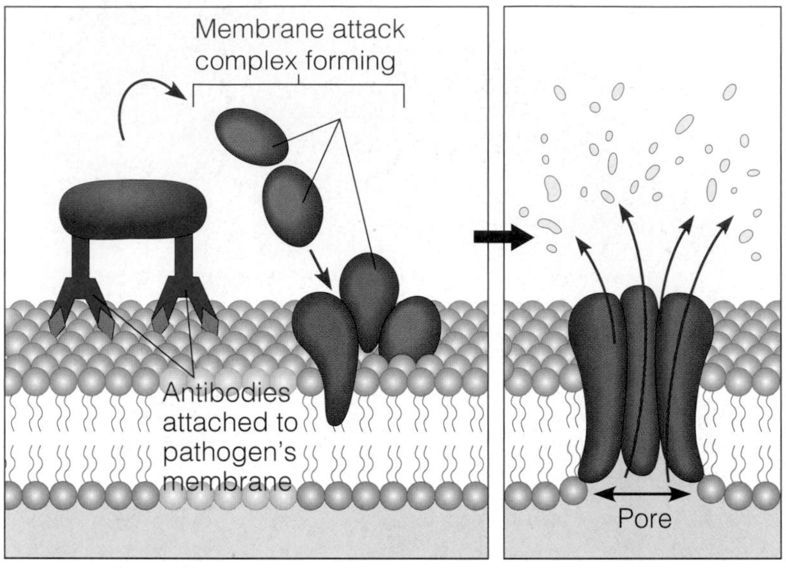

FIGURE 12.10 Activation of complement, resulting in lysis of a target cell.

that attract neutrophils and macrophages into the region. Others cause the cell membranes of the foreign cells to become sticky so they are easier to phagocytize; this effect is called *opsonization*. Although the complement attack is often directed against specific microorganisms that have been "identified" by antibody binding, complement itself is a nonspecific defensive mechanism that "complements," or enhances, the effectiveness of *both* nonspecific and specific defenses.

Interferon Viruses lack the cellular machinery required to generate ATP or make proteins. They do their "dirty work" or damage in the body by entering tissue cells and taking over the cellular machinery needed to reproduce themselves. Although the virus-infected cells can do little to save themselves, they help defend cells that have not yet been infected by secreting small proteins called **interferons** (in-ter-fer′onz). The interferon molecules diffuse to nearby cells and bind to their membrane receptors. Somehow this binding hinders the ability of viruses to multiply within these cells.

Fever

Fever, or abnormally high body temperature, is a systemic response to invading microorganisms. As described further in Chapter 14, body temperature is regulated by a part of the hypothalamus, commonly referred to as the body's "thermostat." Normally the thermostat is set at approximately 36.2°C (98.2°F), but it can be reset upward in response to **pyrogens** (*pyro* = fire), chemicals secreted by white blood cells and macrophages exposed to foreign cells or substances in the body.

Although high fevers are dangerous because excess heat "scrambles" enzymes and other body proteins, mild or moderate fever seems to benefit the body. Bacteria require large amounts of iron and zinc to multiply; but, during a fever the liver and spleen gather up these nutrients, making them less available. Fever also increases the metabolic rate of tissue cells in general, speeding up repair processes.

Specific Body Defenses: The Immune System

Most of us would find it wonderfully convenient if we could walk into a single clothing store and buy a complete wardrobe—hat to shoes—that fit us "to a T" regardless of any special figure problems. We know that such a service would be next to

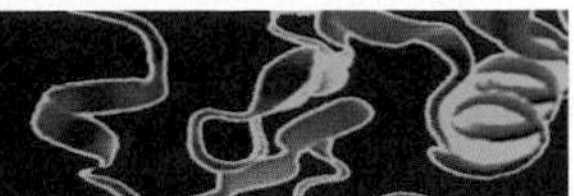

TABLE 12.1 Summary of Nonspecific Body Defenses

Category and associated elements	Protective mechanism
Surface membrane barriers—First line of defense	
Intact skin (epidermis)	Forms mechanical barrier that prevents entry of pathogens and other harmful substances into body.
• Acid mantle	Skin secretions make epidermal surface acidic, which inhibits bacterial growth; sebum also contains bacteria-killing chemicals.
• Keratin	Provides resistance against acids, alkalis, and bacterial enzymes.
Intact mucous membranes	Form mechanical barrier that prevents entry of pathogens.
• Mucus	Traps microorganisms in respiratory and digestive tracts.
• Nasal hairs	Filter and trap microorganisms in nasal passages.
• Cilia	Propel debris-laden mucus away from lower respiratory passages.
• Gastric juice	Contains concentrated hydrochloric acid and protein-digesting enzymes that destroy pathogens in stomach.
• Acid mantle of vagina	Inhibits growth of bacteria and fungi in female reproductive tract.
• Lacrimal secretion (tears); saliva	Continuously lubricate and cleanse eyes (tears) and oral cavity (saliva); contain lysozyme, an enzyme that destroys microorganisms.
Cellular and chemical defenses—Second line of defense	
Phagocytes	Engulf and destroy pathogens that breach surface membrane barriers; macrophages also contribute to immune response.
Natural killer cells	Promote cell lysis by direct cell attack against virus-infected or cancerous body cells; do not depend on specific antigen recognition.
Inflammatory response	Prevents spread of injurious agents to adjacent tissues, disposes of pathogens and dead tissue cells, and promotes tissue repair; releases chemical mediators that attract phagocytes (and immunocompetent cells) to the area.
Antimicrobial chemicals	
• Complement	Group of plasma proteins that lyses microorganisms, enhances phagocytosis by opsonization, and intensifies inflammatory response.
• Interferons	Proteins released by virus-infected cells that protect uninfected tissue cells from viral takeover; mobilize immune system.
• Urine	Normally acid pH inhibits bacterial growth; cleanses the lower urinary tract as it flushes from the body.
Fever	Systemic response triggered by pyrogens; high body temperature inhibits multiplication of bacteria and enhances body repair processes.

impossible to find—yet we take for granted our *immune system,* our built-in *specific defense system,* which stalks and eliminates with nearly equal precision almost any type of pathogen that intrudes into the body.

The immune system's response to a threat, called the **immune response,** tremendously increases the inflammatory response, and it provides protection that is carefully targeted against *specific* antigens. Furthermore, the initial exposure to an antigen "primes" the body to react more vigorously to later meetings with the same antigen.

Sometimes referred to as the body's *third line of defense,* the immune system is a functional system that recognizes foreign molecules (antigens) and acts to inactivate or destroy them. Normally it protects us from a wide variety of pathogens, as well as from abnormal body cells. When it fails, malfunctions, or is disabled, some of the most devastating diseases—such as cancer, rheumatoid arthritis, and AIDS—may result.

Although *immunology,* the study of immunity, is a fairly new science, the ancient Greeks knew that once someone had suffered through a certain infectious disease, that person was unlikely to have the same disease again. The basis of this immunity was revealed in the late 1800s, when it was shown that animals surviving a serious bacterial infection have "factors" in their blood that protect them from future attacks by the same pathogen. (These factors are now known to be unique proteins, called *antibodies.*) Furthermore, it was demonstrated that if antibody-containing serum from the surviving animals (immune serum) was injected into animals that had not been exposed to the pathogen, those animals would also be protected. These landmark experiments revealed three important aspects of the immune response:

1. **It is antigen specific**—It recognizes and acts against *particular* pathogens or foreign substances.
2. **It is systemic**—Immunity is not restricted to the initial infection site.
3. **It has "memory"**—It recognizes and mounts even stronger attacks on previously encountered pathogens.

This was exciting news. But then, in the mid-1900s, it was discovered that injection of serum containing antibodies did not always protect a recipient from diseases the donor had survived. In such cases, however, injection of the donor's lymphocytes *did* provide immunity.

As the pieces began to fall into place, two separate but overlapping arms of immunity were recognized. **Humoral** (hu′mor-al) **immunity,** also called **antibody-mediated immunity,** is provided by antibodies present in the body's "humors," or fluids. When lymphocytes themselves defend the body, the immunity is called **cellular** or **cell-mediated immunity** because the protective factor is living cells. The cellular arm also has cellular targets—virus-infected tissue cells, cancer cells, and cells of foreign grafts. The lymphocytes act against such targets either *directly,* by lysing the foreign cells, or *indirectly,* by releasing chemicals that enhance the inflammatory response or activate other immune cells. However, before we describe the humoral and cellular responses individually, we will consider the antigens that trigger the activity of the remarkable cells involved in these immune responses.

Antigens

An **antigen** (an′tĭ-jen) **(Ag)** is any substance capable of exciting our immune system and provoking an immune response. Most antigens are large, complex molecules that are not normally present in our bodies. Consequently, as far as our immune system is concerned, they are foreign intruders, or **nonself.** An almost limitless variety of substances can act as antigens, including virtually all foreign proteins, nucleic acids, many large carbohydrates, and some lipids. Of these, proteins are the strongest antigens. Pollen grains and microorganisms such as bacteria, fungi, and virus particles are *antigenic* because their surfaces bear such foreign molecules.

It is also important to remember that our own cells are richly studded with a variety of protein molecules (self-antigens). Somehow, as our immune system develops, it takes an inventory of all these proteins so that, thereafter, they are recognized as self. Although these **self-antigens** do not trigger an immune response in us, they *are* strongly antigenic to other people. This helps explain why our bodies reject cells of transplanted organs or foreign grafts unless special measures (drugs and others) are taken to cripple or stifle the immune response.

As a rule, small molecules are not antigenic. But, when they link up with our own proteins, the immune system may recognize the combination as foreign and mount an attack that is harmful rather than protective. (Such reactions, called allergies, are described on pp. 414–416.) In such cases, the troublesome small molecule is called a *hapten* (*haptein* = to grasp), or *incomplete antigen.* Besides certain drugs, chemicals that act as haptens are found in poison ivy, animal dander, and even in some detergents, hair dyes, cosmetics, and other commonly used household and industrial products.

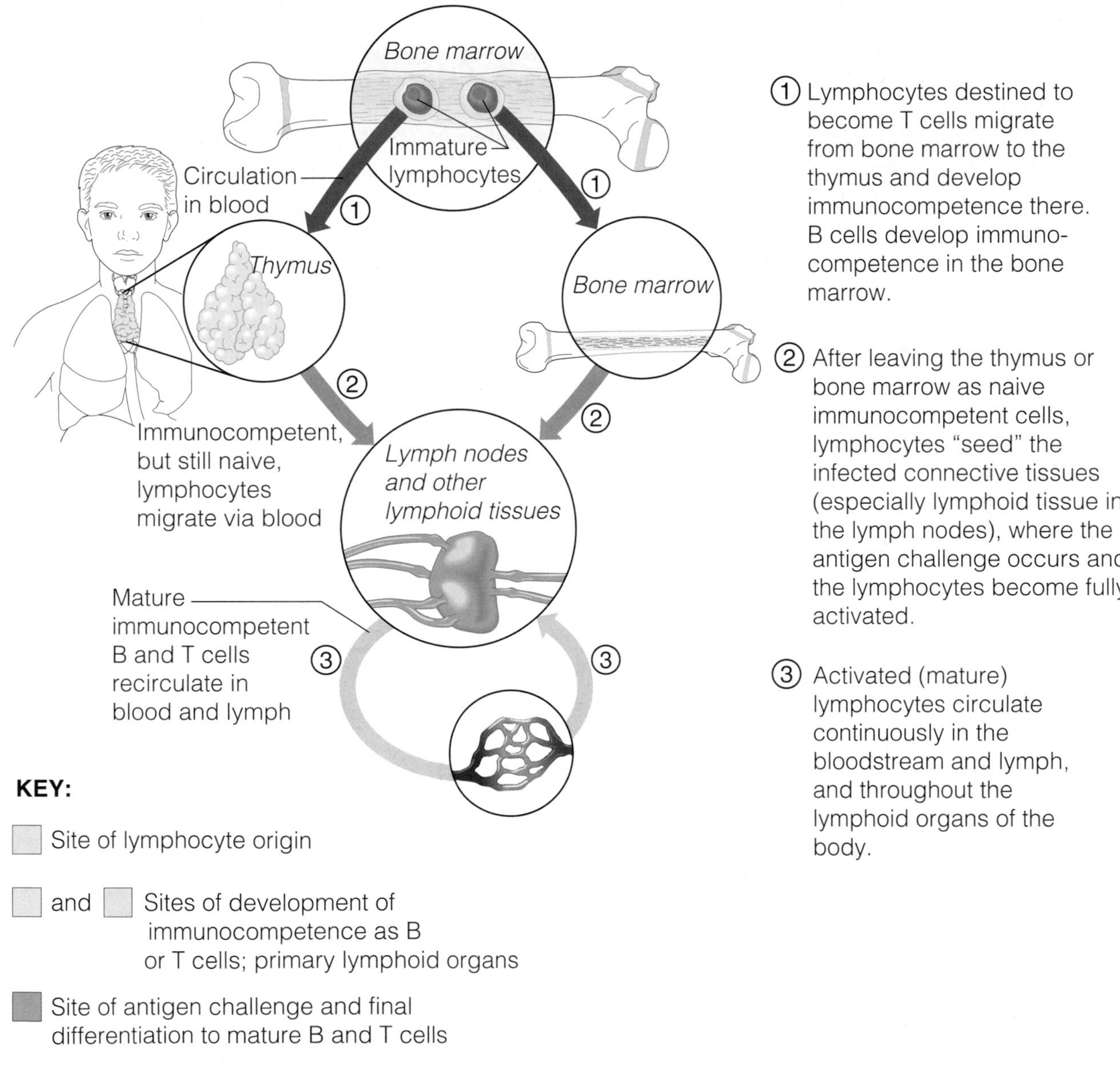

FIGURE 12.11 Lymphocyte differentiation and activation.

Homeostatic Imbalance

Perhaps the most dramatic and familiar example of a drug hapten's provoking an immune response involves the binding of penicillin to blood proteins, which causes a *penicillin reaction* in some people. In such cases, the immune system mounts such a vicious attack that the person's life is endangered. ▲

Cells of the Immune System: An Overview

The crucial cells of the immune system are lymphocytes and macrophages. Lymphocytes exist in two major "flavors." The **B lymphocytes,** or **B cells,** produce antibodies and oversee humoral immunity, whereas the **T lymphocytes,** or **T cells,** are non–antibody-producing lymphocytes that constitute the cell-mediated arm of immunity. Unlike the two types of lymphocytes, macrophages do not respond to specific antigens but instead play an essential role in helping the lymphocytes that do.

Lymphocytes

Like all blood cells, lymphocytes originate from hemocytoblasts in red bone marrow (Figure 12.11). The immature lymphocytes released from the marrow are essentially identical. Whether a given lymphocyte matures into a B cell or a T cell depends on where in the body it becomes **immunocompetent,** that is, capable of responding to a specific antigen by binding to it. **T** cells arise from lymphocytes that migrate to the **t**hymus (see Figure 12.11), where they undergo a maturation process of 2 to 3 days, directed by thymic hormones (*thymosin* and others). Within the thymus, the immature lymphocytes divide rapidly and their numbers increase enormously, but only those maturing T cells with the sharpest ability to identify *foreign* antigens survive. Lymphocytes capable of binding strongly with *self-antigens* (and of acting against body cells) are vigorously weeded out and destroyed. Thus, the development of tolerance for self-antigens is an essential part of a lymphocyte's "education." This is true not only for T cells but also for B cells. **B** cells develop immunocompetence in **b**one marrow, but little is known about the factors that regulate B cell maturation.

Once a lymphocyte is immunocompetent, it will be able to react to one distinct antigen and one only, because *all* the antigen receptors on its external surface are the same. For example, the receptors of one lymphocyte can recognize only a part of the hepatitis A virus, those of another lymphocyte can recognize or bind only to pneumococcus bacteria, and so forth.

Although all the details of the maturation process are still beyond our grasp, we know that lymphocytes become immunocompetent *before* meeting the antigens they may later attack. Thus, *it is our genes, not antigens, that determine what specific foreign substances our immune system will be able to recognize and resist.* Only some of the possible antigens our lymphocytes are programmed to resist will ever invade our bodies. Consequently, only some members of our army of immunocompetent cells will be mobilized during our lifetime. The others will be forever idle. As usual, our bodies have done their best to protect us.

After becoming immunocompetent, both T cells and B cells migrate to the lymph nodes and spleen (and loose connective tissues), where their encounters with antigens will occur (see Figure 12.9). Then, when the lymphoctyes bind with recognized antigens, they complete their differentiation into fully mature T cells and B cells.

Macrophages

Macrophages, which also become widely distributed throughout the lymphoid organs and connective tissues, arise from monocytes formed in the bone marrow. As described earlier, a major role of macrophages (literally, "big eaters") in the nonspecific defense system is to engulf foreign particles and rid them from the area. However, their job doesn't stop there. They also present fragments of those antigens, like signal flags, on their own surfaces, where they can be recognized by immunocompetent T cells. Thus, they act as *antigen presenters* in the specific defense system. Macrophages also secrete cytokine proteins, called *monokines,* that are important in the immune response (see Table 12.4, pp. 412–413). Activated T cells, in turn, release chemicals that cause macrophages to become insatiable phagocytes, or *killer macrophages.* As you will see, interactions between lymphocytes, and between lymphocytes and macrophages, underlie virtually all phases of the immune response.

Macrophages tend to remain fixed in the lymphoid organs (as if waiting for antigens to come to them). But lymphocytes, especially T cells, circulate continuously through the body (see Figure 12.11). This makes sense because it greatly increases a lymphocyte's chance of coming into contact with antigens collected by the lymph capillaries from the tissue spaces, as well as with huge numbers of macrophages and other lymphocytes.

To summarize, the immune system is a two-armed defensive system that uses lymphocytes, macrophages, and specific molecules to identify and destroy all substances—both living and nonliving—that are in the body but are not recognized as being part of the body or as being self. The immune system's ability to respond to such threats depends on the ability of its cells (1) to recognize foreign substances (antigens) in the body by binding to them, and (2) to communicate with one another so that the system as a whole mounts a response specific to those antigens.

Humoral (Antibody-Mediated) Immune Response

An immunocompetent but as yet immature B lymphocyte is stimulated to complete its development into a fully mature B cell when an antigen binds to

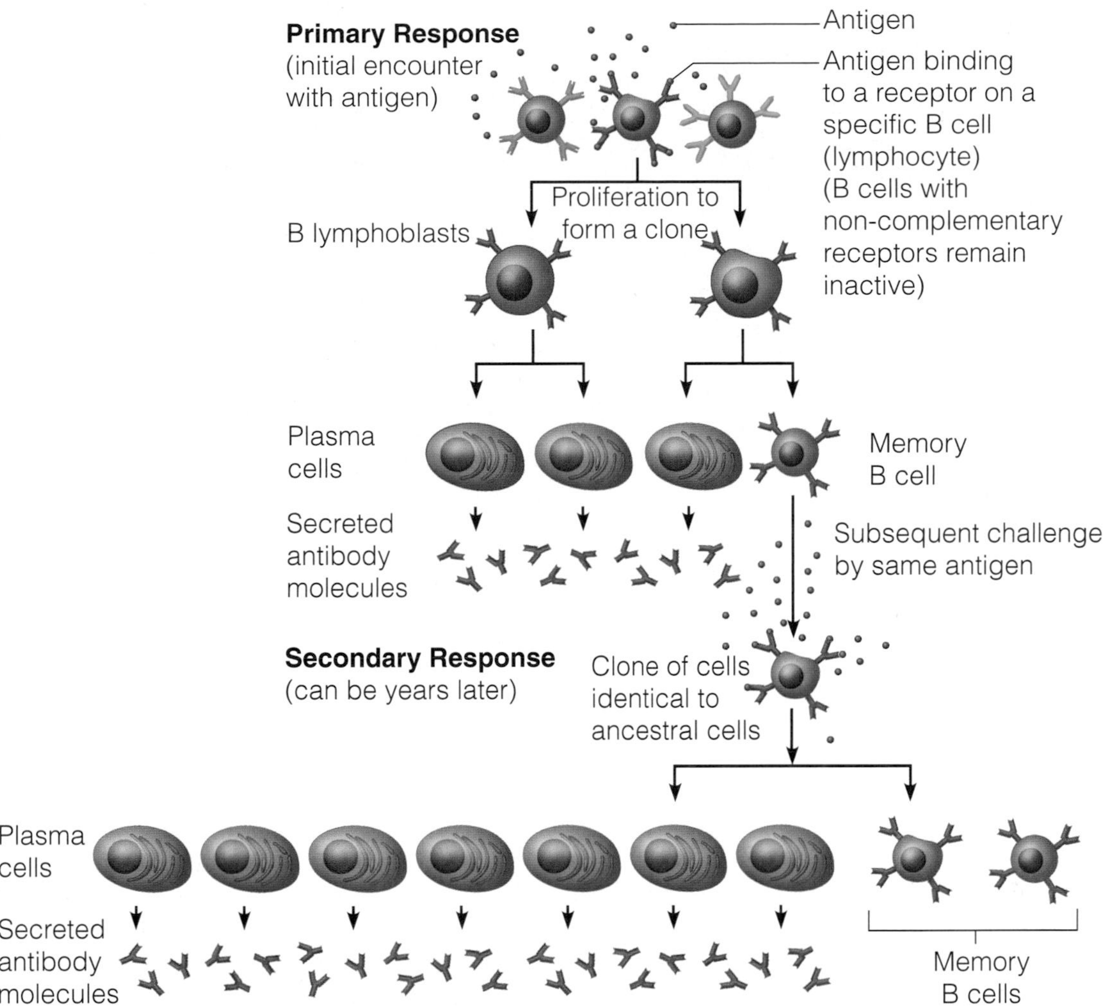

FIGURE 12.12 Clonal selection of a B cell stimulated by antigen binding. The initial meeting stimulates the primary response in which the B cell proliferates rapidly, forming a clone of like cells (clonal expansion), most of which become antibody-producing plasma cells. Cells that do not differentiate into plasma cells become memory cells, which respond to subsequent exposures to the same antigen. Should such a meeting occur, the memory cells quickly produce more memory cells and larger numbers of effector plasma cells with the same antigen specificity. Responses generated by memory cells are called secondary responses.

its surface receptors. This binding event *sensitizes,* or *activates,* the lymphocyte to "switch on" and undergo **clonal selection.** The lymphocyte begins to grow and then multiplies rapidly to form an army of cells all exactly like itself and bearing the same antigen-specific receptors (Figure 12.12). The resulting family of identical cells descended from the *same* ancestor cell is called a **clone,** and clone formation is the **primary humoral response** to that antigen. (As described later, T cells also influence B cell activation.)

Most of the B cell clone members, or descendants, become **plasma cells.** After an initial lag period, these antibody-producing "factories" swing into action, producing the same highly specific antibodies at an unbelievable rate of about 2000 antibody molecules per second. (The B cells themselves produce only very small amounts of antibodies.) However, this flurry of activity lasts only 4 or 5 days; then the plasma cells begin to die. Antibody levels in the blood during this primary response peak in about 10 days and then slowly decline (Figure 12.13).

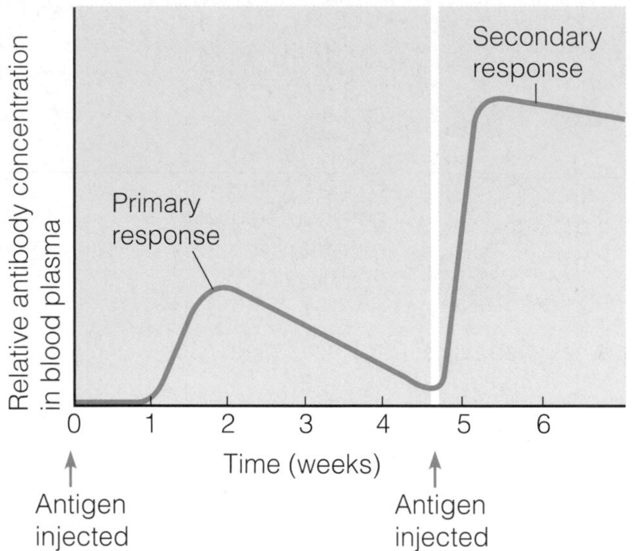

FIGURE 12.13 Primary and secondary humoral responses to an antigen. In the primary response, there is a gradual rise and then a rapid decline in the level of antibodies in the blood. The secondary response is both more rapid and more intense. Additionally, antibody levels remain high for a much longer time.

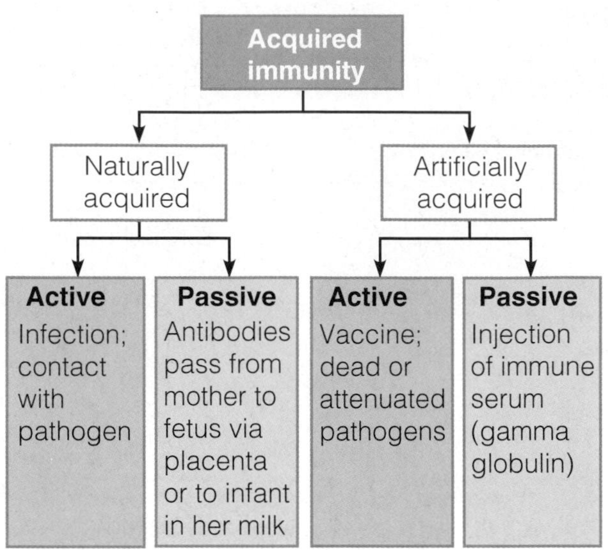

FIGURE 12.14 Types of acquired immunity. Orange boxes signify active types of immunity in which immunological memory is established. Gold boxes signify the short-lived passive types of immunity; no immunological memory is established.

B cell clone members that do not become plasma cells become long-lived **memory cells** capable of responding to the same antigen at later meetings with it. Memory cells are responsible for the immunological "memory" mentioned earlier. These later immune responses, called **secondary humoral responses,** are much faster, more prolonged, and more effective because all the preparations for this attack have already been made. Within hours after recognition of the "old-enemy" antigen, a new army of plasma cells is being generated, and antibodies begin to flood into the bloodstream. Within 2 to 3 days, blood antibody levels peak (at much higher levels than seen in the primary response), and their levels remain high for weeks to months. How antibodies protect the body is described shortly.

Active and Passive Humoral Immunity

When your B cells encounter antigens and produce antibodies against them, you are exhibiting **active immunity** (Figure 12.14). Active immunity is (1) *naturally acquired* during bacterial and viral infections, during which we may develop the symptoms of the disease and suffer a little (or a lot), and (2) *artificially acquired* when we receive **vaccines.** It makes little difference whether the antigen invades the body under its own power or is introduced in the form of a vaccine. The response of the immune system is pretty much the same. Indeed, once it was recognized that secondary responses are so much more vigorous, the race was on to develop vaccines to "prime" the immune response by providing a first meeting with the antigen. Most vaccines contain dead or *attenuated* (living, but extremely weakened) pathogens.

We receive two benefits from vaccines: (1) we are spared most of the signs and symptoms (and discomfort) of the disease that would otherwise occur during the primary response, and (2) the weakened antigens are still able to stimulate antibody production and promote immunological memory. So-called booster shots, which may intensify the immune response at later meetings with the same antigen, are also available. Vaccines are currently available against microorganisms that cause pneumonia, smallpox, polio, tetanus, diphtheria, measles, and many other diseases. In the United States, many potentially serious childhood diseases have been virtually wiped out by active immunization programs. A summary of the currently recommended schedule for administering vaccines to U.S. children is provided in Table 12.2.

TABLE 12.2 Recommended* Immunization Schedule for Children in the United States

Recommended age	Immunizing agent in vaccine
2 months	DPT vaccine (diphtheria toxoid, pertussis [whooping cough] vaccine, and tetanus toxoid); OPV (oral poliomyelitis vaccine)
4 months	DPT vaccine; OPV
6 months	DPT vaccine (OPV optional for areas with high risk of polio exposure)
12–15 months	DPT vaccine; OPV; MMR vaccine (combined mumps vaccine, measles vaccine, and rubella virus vaccine) or individual mumps, measles, and rubella virus vaccines. Completes primary series of DPT and OPV. MMR dose also given one month later.
18 months	HbCV (*Haemophilus* influenza type b); conjugate vaccine preferred over the polysaccharide vaccine (HbPV)
4–6 years	DPT vaccine; OPV (preferably at or before school entry)
11–13 years	MMR (at entrance to middle school or high school); second dose of two.
14–16 years	Td booster (tetanus and diphtheria toxoid); repeat every 10 years throughout life

*Recommended by the American Academy of Pediatrics

Passive immunity is quite different from active immunity, both in the antibody source and in the degree of protection it provides (see Figure 12.14). Instead of being made by your plasma cells, the antibodies are obtained from the serum of an immune human or animal donor. As a result, your B cells are *not* challenged by the antigen, immunological memory does *not* occur, and the temporary protection provided by the "borrowed antibodies" ends when they naturally degrade in the body.

Passive immunity is conferred *naturally* on a fetus when the mother's antibodies cross the placenta and enter the fetal circulation, and after birth during breastfeeding. For several months after birth, the baby is protected from all the antigens to which the mother has been exposed.

Passive immunity is *artificially* conferred when one receives immune serum or gamma globulin. Gamma globulin is commonly administered after exposure to hepatitis. Other immune sera are used to treat poisonous snake bites (an antivenom), botulism, rabies, and tetanus (an antitoxin) because these diseases will kill a person before active immunity can be established. The donated antibodies provide immediate protection, but their effect is short-lived (two to three weeks). Meanwhile, however, the body's own defenses take over.

In addition to their use to provide passive immunity, antibodies are prepared commercially for use in research, clinical testing for diagnostic purposes, and treating certain cancers. **Monoclonal antibodies** used for such purposes are produced by descendants of a single cell and are pure antibody preparations that exhibit specificity for one, and only one, antigen. Besides their use in delivering cancer-fighting drugs to cancerous tissue, monoclonal antibodies are being used for diagnosis of pregnancy, hepatitis, and rabies. They are also used for early diagnosis and to track the extent of cancers hidden deep within the body.

Antibodies

Antibodies, also referred to as **immunoglobulins** (im″mu-no-glob′u-linz), or **Igs,** constitute the *gamma globulin* part of blood proteins. Antibodies are soluble proteins secreted by activated B cells or by their plasma-cell offspring in response to an antigen, and they are capable of binding specifically with that antigen.

Antibodies are formed in response to a huge number of different antigens. Despite their variety,

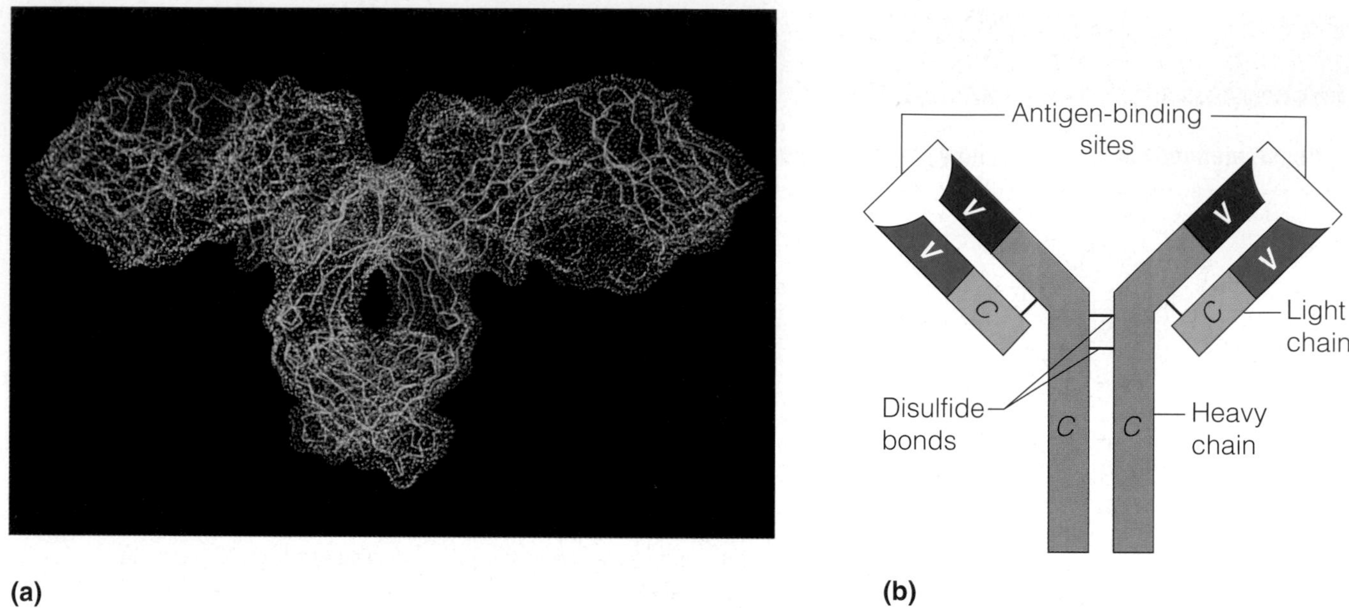

FIGURE 12.15 Basic antibody structure. (a) Computer-generated image. **(b)** Diagrammed structure. The basic structure (monomer) of each type of antibody is formed by four polypeptide chains that are joined by disulfide bonds. Two of the chains are short, light chains; the other two are long, heavy chains. Each chain has a variable (V) region (different in different antibodies) and a constant (C) region (essentially identical in different antibodies of the same class). The variable regions are the antigen-binding sites of the antibody. Hence, each antibody monomer has two antigen-binding sites.

they all have a similar basic anatomy that allows them to be grouped into five Ig classes, each slightly different in structure and function.

Basic Antibody Structure Regardless of its class, every antibody has a basic structure consisting of four amino acid (polypeptide) chains linked together by *disulfide* (sulfur-to-sulfur) *bonds* (Figure 12.15). Two of the four chains are identical and contain approximately 400 amino acids each; these are the *heavy chains*. The other two chains, the *light chains,* are also identical to each other but are only about half as long. When the four chains are combined, the antibody molecule formed has two identical halves, each consisting of a heavy and a light chain, and the molecule as a whole is T- or Y-shaped.

When scientists began investigating antibody structure, they discovered something very peculiar. Each of the four chains forming an antibody had a **variable (V) region** at one end and a much larger **constant (C) region** at the other end. Antibodies responding to different antigens had very different variable regions, but their constant regions were the same or nearly so. This made sense when it was discovered that the variable regions of the heavy and light chains in each arm combine their efforts to form an **antigen-binding site** (Figure 12.15) uniquely shaped to "fit" its specific antigen. Hence, each antibody has two such antigen-binding regions.

The constant regions that form the "stem" of an antibody can be compared to the handle of a key. A key handle has a common function for all keys: it allows you to hold the key and place its tumbler-moving portion into the lock. Similarly, the constant regions of the antibody chains serve common functions in all antibodies: they determine the type of antibody formed (antibody class), as well as how the antibody class will carry out its immune roles in the body, and the cell types or chemicals with which the antibody can bind.

Antibody Classes There are five major immunoglobulin classes—IgM, IgA, IgD, IgG, and IgE. (Remember the woman's name MADGE to recall the five Ig types.) As illustrated in Table 12.3,

TABLE 12.3 Immunoglobin Classes

Class	Generalized structure	Where found	Biological function
IgD		Virtually always attached to B cell.	Believed to be cell surface receptor of immunocompetent B cell; important in activation of B cell.
IgM		Attached to B cell; free in plasma.	When bound to B cell membrane, serves as antigen receptor; first Ig class *released* to plasma by plasma cells during primary response; potent agglutinating agent; fixes complement.
IgG		Most abundant antibody in plasma; represents 75–85% of circulating antibodies.	Main antibody of both primary and secondary responses; crosses placenta and provides passive immunity to fetus; fixes complement.
IgA		Some (monomer) in plasma; dimer in secretions such as saliva, tears, intestinal juice, and milk.	Bathes and protects mucosal surfaces from attachment of pathogens.
IgE		Secreted by plasma cells in skin, mucosae of gastrointestinal and respiratory tracts, and tonsils.	Binds to mast cells and basophils, and triggers release of histamine and other chemicals that mediate inflammation and certain allergic responses.

IgD, IgG, and IgE antibodies have the same basic Y-shaped structure described above and are referred to as *monomers.* IgA antibodies occur in both monomer and *dimer* (two linked monomers) forms. (Only the dimer form is shown in the table.) Compared to the other antibodies, IgM antibodies are huge. Because they are constructed of five linked monomers, IgM antibodies are called *pentamers* (*penta* = five).

The antibodies of each class have slightly different biological roles and locations in the body. For example, IgG is the most abundant antibody in blood plasma and is the only type that can cross the placental barrier. Hence, the passive immunity that a mother transfers to her fetus is "with the compliments" of her IgG antibodies. Only IgM and IgG can fix complement. The IgA dimer, sometimes called *secretory IgA,* is found mainly in mucus and other secretions that bathe body surfaces. It plays a major role in preventing pathogens from gaining entry into the body. IgE antibodies are the "troublemaker" antibodies involved in allergies. These and other characteristics unique to each of the immunoglobulin classes are summarized in Table 12.3.

Antibody Function Antibodies inactivate antigens in a number of ways—by complement fixation, neutralization, agglutination, and precipitation (Figure 12.16). Of these, complement fixation and neutralization are most important to body protection.

Complement is the chief antibody ammunition used against cellular antigens, such as bacteria or mismatched red blood cells. As noted earlier, complement is fixed (activated) during nonspecific body defenses. It is also activated very efficiently when it binds to antibodies attached to cellular targets. This

Q *Complement and agglutination aid phagocytosis, but they assist it in different ways. What is this difference?*

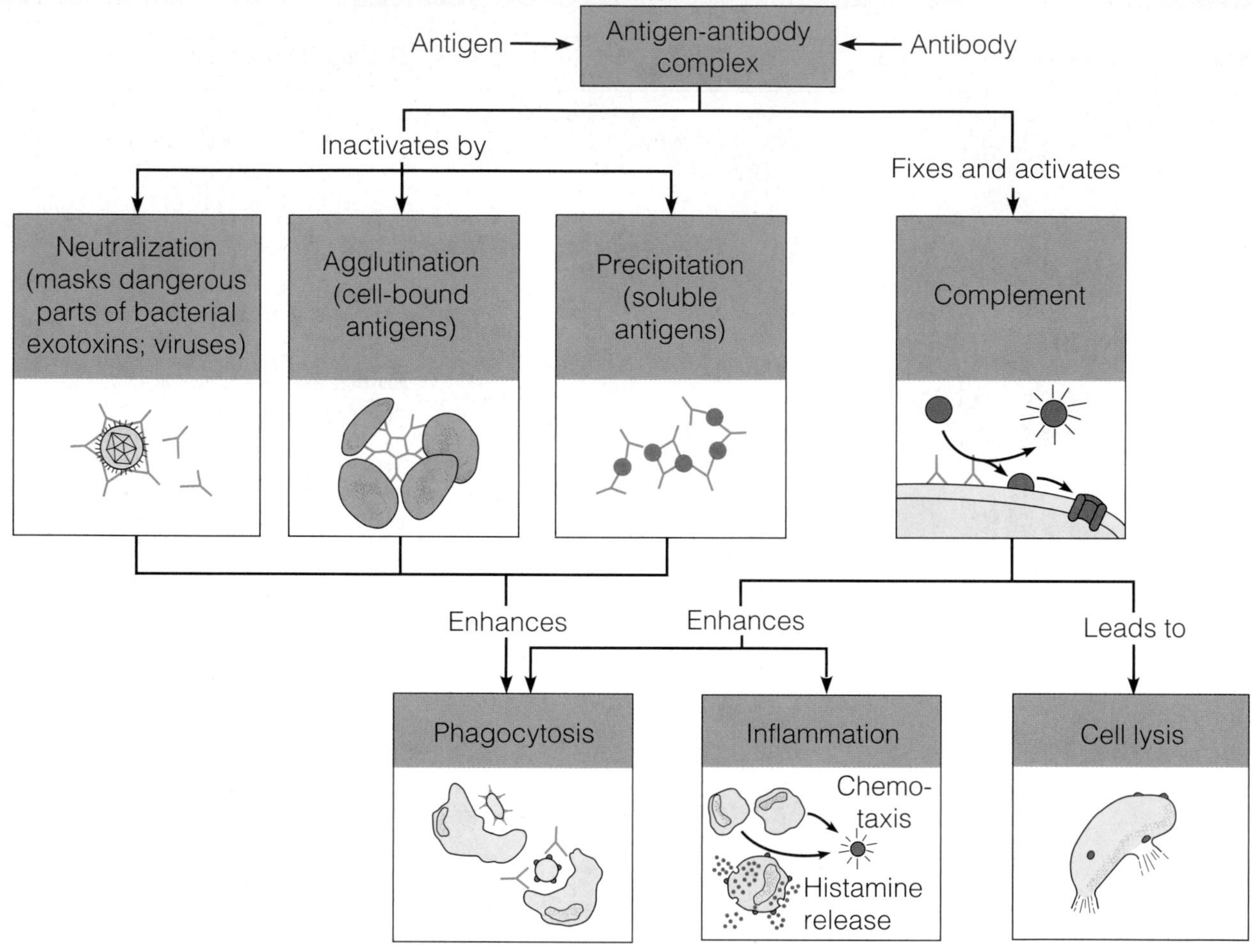

FIGURE 12.16 Mechanisms of antibody action.

triggers events (described earlier) that result in lysis of the foreign cell and release of molecules that tremendously enhance the inflammatory process.

Neutralization occurs when antibodies bind to specific sites on bacterial *exotoxins* (toxic chemicals secreted by bacteria) or on viruses that can cause cell injury. In this way, they block the harmful effects of the exotoxin or virus.

Because antibodies have more than one antigen-binding site, they can bind to more than one antigen at a time; consequently, *antigen-antibody complexes* can be cross-linked into large lattices. When the cross-linking involves cell-bound antigens, the process causes clumping of the foreign cells, a process called **agglutination.** This type of antigen-antibody reaction occurs when mismatched blood is transfused (the foreign red blood cells are clumped) and is the basis of tests used for blood typing. When the cross-linking process involves soluble antigenic molecules, the resulting antigen-antibody complexes are so large that they become insoluble and settle out of solution. This cross-linking reaction is more precisely called **precipitation.** There is little question that agglutinated bacteria and immobilized (precipitated) antigen molecules are much more easily captured and engulfed by the body's phagocytes than are freely moving antigens.

A *Complement fixation provides "knobs" for the phagocyte to attach to whereas agglutination binds antigens (typically microorganisms) together into large masses that become nonmobile and easier to destroy.*

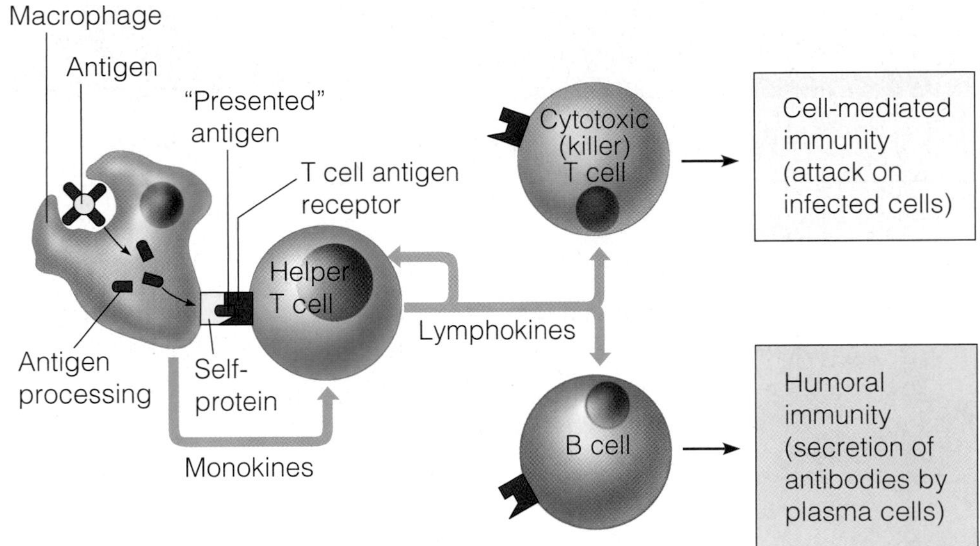

FIGURE 12.17 T cell activation and interactions with other cells of the immune response. Macrophages are important both as phagocytes and as antigen presenters. After they have ingested an antigen, they display parts of it on their surface membranes, where it can be recognized by a helper T cell that bears receptors for the same antigen. During the binding process, the T cell binds simultaneously to the antigen and to the macrophage (self) receptor, which leads to T cell activation and cloning (not illustrated). In addition, the macrophage releases monokines, which enhance T cell activation. Activated helper T cells release lymphokines, which stimulate proliferation and activity of other helper T cells and help activate cytotoxic (killer) T cells and B cells.

Cellular (Cell-Mediated) Immune Response

Like B cells, immunocompetent T cells are activated to form a clone by binding with a "recognized" antigen (see Figure 12.19, p. 415). However, unlike B cells, the T cells are not able to bind with *free* antigens. Instead, the antigens must be "presented" by macrophages, and a *double recognition* must occur. The macrophages engulf the antigens, process them internally, and then finally display parts of the processed antigens on their external surface in combination with one of their own (self) proteins (Figure 12.17).

Apparently, a T cell must recognize "nonself," the antigen fragment presented by the macrophage, and also "self" by coupling with a specific glycoprotein on the macrophage's surface at the same time. Antigen binding alone is not enough to sensitize T cells. They must be "spoon-fed" the antigens by macrophages, and something like a "double handshake" must occur. Although this idea seemed preposterous when it was first suggested, there is no longer any question that **antigen presentation** is a major role of macrophages and is essential for activation and clonal selection of the T cells. Without macrophage "presenters," the immune response is severely impaired. Cytokine chemicals (monokines, particularly interleukin 1) released by macrophages also play important roles in the immune response, as shown in Table 12.4.

The different classes of T cell clones, which provide for cell-mediated immunity, are a diverse lot and produce their deadly effects in a variety of ways (see Table 12.4). Some are **cytotoxic (killer) T cells,** cells that specialize in killing virus-infected, cancer, or foreign graft cells (Figure 12.18). One way they accomplish this is by binding to them and inserting a toxic chemical (*perforin* or others) into the foreign cell's plasma membrane (delivering

TABLE 12.4 Functions of Cells and Molecules Involved in Immunity

Element	Function in the immune response
Cells	
B cell	Lymphocyte that resides in the lymph nodes, spleen, or other lymphoid tissues, where it is induced to replicate by antigen-binding and helper T cell interactions; its progeny (clone members) form plasma cells and memory cells.
Plasma cell	Antibody-producing "machine"; produces huge numbers of the same antibody (immunoglobulin); represents further specialization of B cell clone descendants.
Helper T cell	A *regulatory* T cell that binds with a specific antigen presented by a macrophage; it stimulates the production of other immune cells (cytotoxic T cells and B cells) to help fight the invader; acts both directly and indirectly by releasing lymphokines.
Cytotoxic T cell	Also called a killer T cell; activity enhanced by helper T cells; its specialty is killing virus-invaded body cells, as well as body cells that have become cancerous; involved in graft rejection.
Suppressor T cell	Slows or stops the activity of B and T cells once the infection (or attack by foreign cells) has been conquered.
Memory cell	Descendant of an activated B cell or T cell; generated during the initial immune response (primary response); may exist in the body for years thereafter, enabling it to respond quickly and efficiently to subsequent infections or meetings with the same antigen.
Macrophage	Engulfs and digests antigens that it encounters and presents parts of them on its plasma membrane for recognition by T cells bearing receptors for the same antigen; this function, *antigen presentation,* is essential for normal cell-mediated responses. Also releases chemicals that activate T cells.
Molecules	
Antibody (immunoglobulin)	Protein produced by a B cell or its plasma cell offspring and released into the body fluids (blood, lymph, saliva, mucus, etc.), where it attaches to antigens, causing neutralization, precipitation, or agglutination, which "marks" the antigens for destruction by phagocytes or complement.
Lymphokines	Cytokine chemicals released by sensitized T cells: • Macrophage migration inhibiting factor (MIF)—"inhibits" macrophage migration and keeps them in the local area. • Interleukin 2—stimulates T cells and B cells to proliferate. • Helper factors—enhance antibody formation by plasma cells. • Suppressor factors—suppress antibody formation or T cell–mediated immune responses. • Chemotactic factors—attract leukocytes (neutrophils, eosinophils, and basophils) into inflamed area. • Perforin—a cell toxin; causes cell lysis. • Gamma interferon—helps make tissue cells resistant to viral infection; activates macrophages; activates NK cells; enhances maturation of cytotoxic T cells.

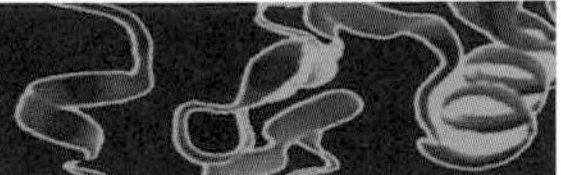

TABLE 12.4 (continued)

Element	Function in the immune response
Monokines	Cytokine chemicals released by activated macrophages: • Interleukin 1—stimulates T cells to proliferate and causes fever. • Tumor necrosis factor (TNF)—like perforin, causes cell killing; attracts granulocytes; activates T cells and macrophages.
Complement	Group of bloodborne proteins activated after binding to antibody-covered antigens; when activated, complement causes lysis of the microorganism and enhances inflammatory response.
Antigen	Substance capable of provoking an immune response; typically a large complex molecule not normally present in the body.

the so-called kiss of death). Shortly thereafter, the target cell ruptures. By that time, the cytotoxic T cell is long gone and is seeking other foreign prey to attack.

Helper T cells are the T cells that act as the "directors" or "managers" of the immune system. Once activated, they circulate through the body, recruiting other cells to fight the invaders. For example, helper T cells interact directly with B cells (that have already attached to antigens), prodding them into more rapid division (clone production) and then, like the "boss" of an assembly line, signaling for antibody formation to begin. They also release a variety of cytokine chemicals called **lymphokines** (see Table 12.4) that act indirectly to rid the body of antigens by (1) stimulating cytotoxic T cells and B cells to grow and divide; (2) attracting other types of protective white blood cells, such as neutrophils, into the area; and (3) enhancing the ability of macrophages to engulf and destroy microorganisms. (Actually, the macrophages are pretty good phagocytes even in the absence of lymphokines, but in their presence they develop an insatiable appetite.) As the released lymphokines summon more and more cells into the battle, the immune response gains momentum, and the antigens are overwhelmed by the sheer numbers of immune elements acting against them.

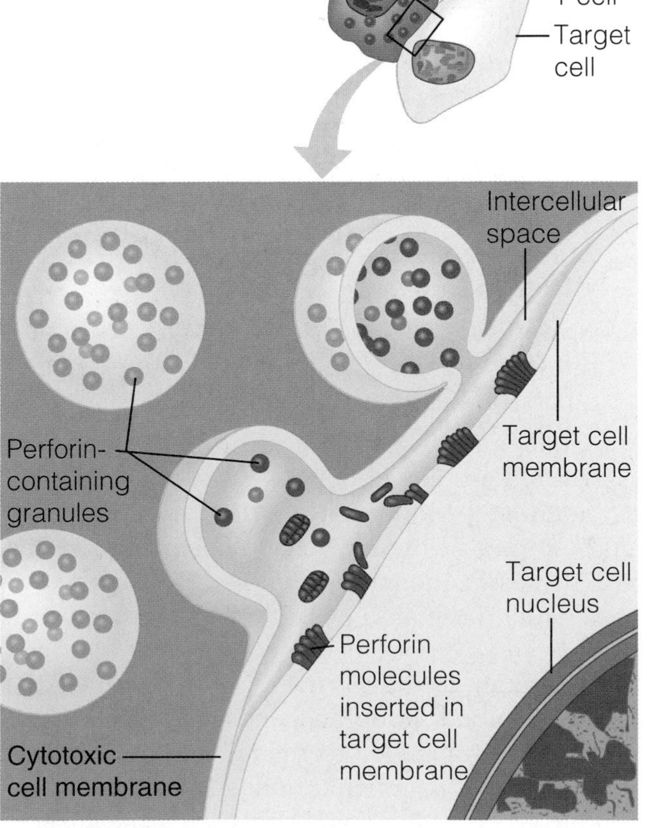

FIGURE 12.18 A proposed mechanism of target cell lysis by cytotoxic T cells.

Another T cell population, the **suppressor T cells,** releases chemicals that suppress the activity of both T and B cells. Suppressor T cells are vital for winding down and finally stopping the immune response after an antigen has been successfully inactivated or destroyed. This helps prevent uncontrolled or unnecessary immune system activity.

Most of the T cells enlisted to fight in a particular immune response are dead within a few days. However, a few members of each clone are long-lived **memory cells** that remain behind to provide the immunological memory for each antigen encountered and enable the body to respond quickly to its subsequent invasions.

A summary of the major elements of the immune response appears in Figure 12.19.

Organ Transplants and Rejection

For those suffering with end-stage heart or kidney disease, organ transplants are a desirable treatment option. However, organ transplants have had mixed success because the immune system is ever vigilant, and rejection is a real problem.

Essentially there are four major types of grafts:

1. **Autografts** are tissue grafts transplanted from one site to another in the same person.
2. **Isografts** are tissue grafts donated by a genetically identical person, the only example being an identical twin.
3. **Allografts** are tissue grafts taken from an unrelated person.
4. **Xenografts** are tissue grafts harvested from a different animal species, such as transplanting a baboon heart into a human being.

Autografts and isografts are ideal donor organs or tissues and are just about always successful given an adequate blood supply and no infection. Although pig heart valves have been transplanted with success, xenografts of whole organs are never successful. The graft type most used is an allograft taken from a recently deceased person.

Before an allograft is even attempted, the ABO and other blood group antigens of both the donor and recipient must be determined and must match. Then, the cell membrane antigens of their tissue cells are typed to determine how closely they match. At least a 75 percent match is needed to attempt a graft; as you might guess, good tissue matches between unrelated people are hard to find.

After surgery, to prevent rejection, the patient receives **immunosuppressive therapy,** including one or more of the following: corticosteroids to suppress inflammation, cytotoxic drugs, radiation (X-ray) therapy, and immunosuppressor drugs. Many of these drugs kill rapidly dividing cells (such as activated lymphocytes), and all of them have severe side effects. However, the major problem with immunosuppressive therapy is that while the immune system is suppressed, it cannot protect the body against other foreign agents. Explosive bacterial and viral infection is the most frequent cause of death in these patients.

Disorders of Immunity

Homeostatic Imbalance

The most important disorders of the immune system are allergies, immunodeficiencies, and autoimmune diseases.

Allergies

At first, the immune response was thought to be purely protective. However, it was not long before its dangerous potentials were discovered. **Allergies** or **hypersensitivities** are abnormally vigorous immune responses in which the immune system causes tissue damage as it fights off a perceived "threat" that would otherwise be harmless to the body. The term *allergen* (*allo* = altered; *erg* = reaction) is used to distinguish this type of antigen from those producing essentially normal responses. People rarely die of allergies; they are just miserable with them.

Although there are several different types of allergies, the most common type is **immediate hypersensitivity** (Figure 12.20). This type of response, also called **acute hypersensitivity,** is triggered by the release of a flood of histamine when IgE antibodies bind to *mast cells.* Histamine causes small blood vessels in the area to become dilated and leaky and is largely to blame for the best-recognized symptoms of allergy: a runny nose, watery eyes, and itching, reddened skin (hives). When the allergen is inhaled, symptoms of asthma appear because smooth muscle in the walls of the bronchioles contracts, constricting the passages and restricting air flow. Over-the-counter (OTC) anti-allergy drugs contain *antihistamines* that counteract these effects. Most of these reactions begin within seconds after contact with the allergen and last about half an hour.

Fortunately, the bodywide or systemic acute allergic response, known as **anaphylactic** (an″ah-fĭ-lak-tik) **shock,** is fairly rare. Anaphylactic shock occurs when the allergen directly enters the blood and circulates rapidly through the body, as might happen with certain

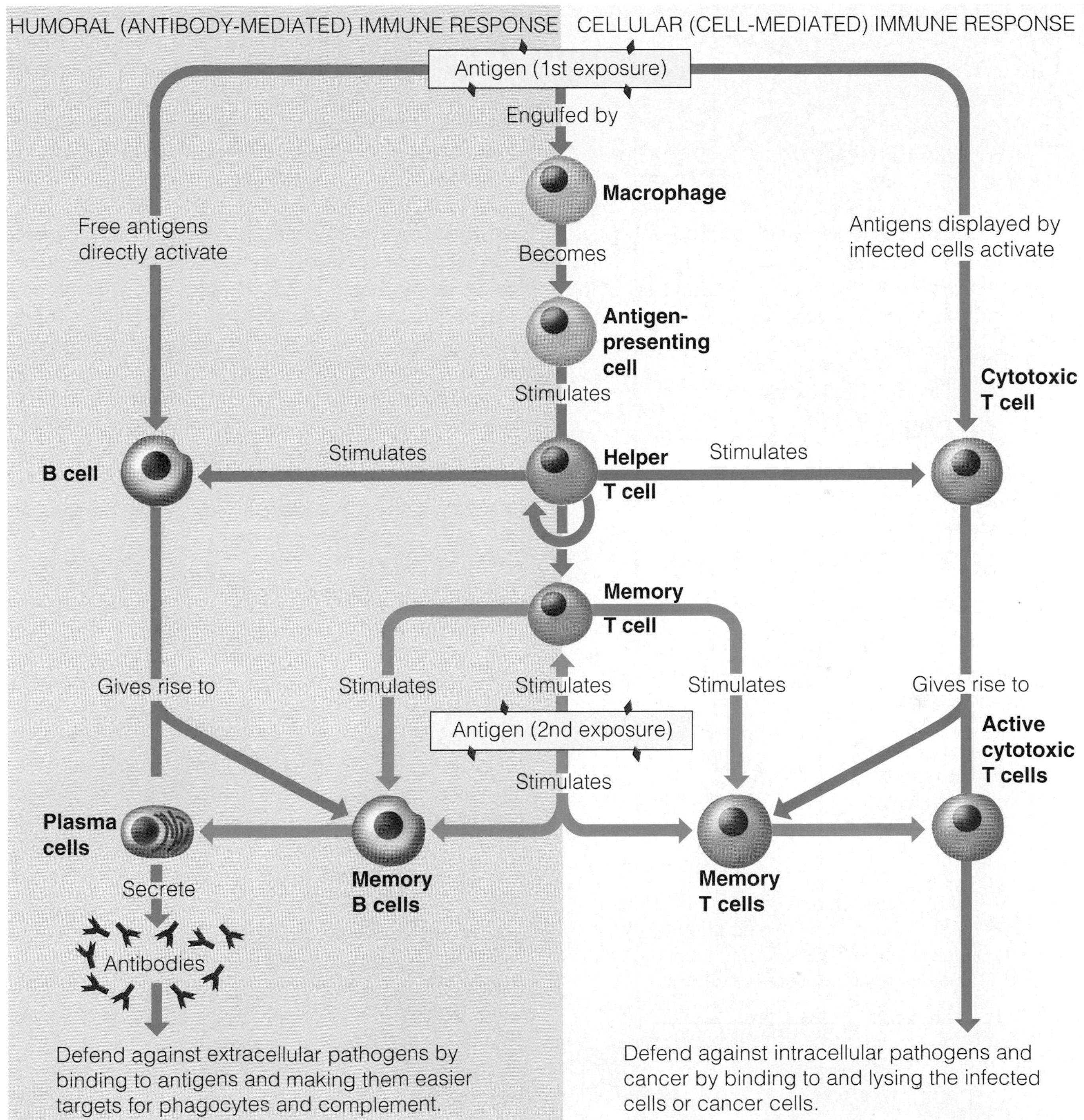

FIGURE 12.19 A summary of the immune responses. In this simple flowchart, green arrows track the primary response, and blue arrows track the secondary response.

bee stings or spider bites. It may also follow an injection of a foreign substance (such as horse serum, penicillin, or other drugs that act as haptens) into susceptible individuals. The mechanism of anaphylactic shock is essentially the same as that of local responses; but when the entire body is involved, the outcome is life-threatening. For example, it is difficult to breathe when the smooth muscles of lung passages contract, and the sudden vasodilation (and fluid loss) that occurs may cause circulatory collapse and death within minutes. Epinephrine is the drug of choice to reverse these histamine-mediated effects.

Delayed hypersensitivities, mediated mainly by a special subgroup of helper T cells, cytotoxic T cells,

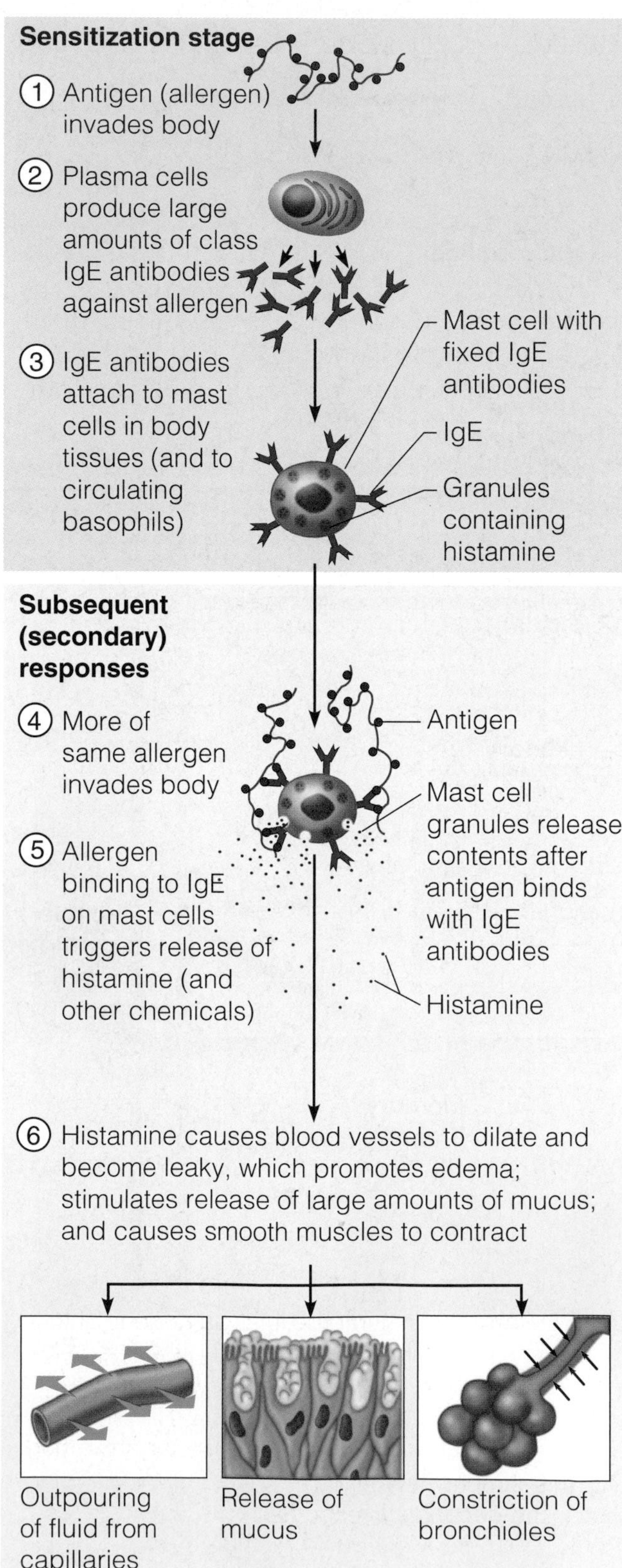

FIGURE 12.20 Mechanism of an immediate (acute) hypersensitivity response.

and macrophages take much longer to appear (1 to 3 days) than any of the acute reactions produced by antibodies. Instead of histamine, the chemicals mediating these reactions are lymphokines released by the activated T cells. Hence, antihistamine drugs are *not* helpful against the delayed types of allergies. Corticosteroid drugs are used to provide relief.

The most familiar examples of delayed hypersensitivity reactions are those classed as *allergic contact dermatitis,* which follow skin contact with poison ivy, some heavy metals (lead, mercury, and others), and certain cosmetic and deodorant chemicals. These agents act as haptens; after diffusing through the skin and attaching to body proteins, they are perceived as foreign by the immune system. The *Mantoux* and *tine tests,* skin tests for detection of tuberculosis, depend on delayed hypersensitivity reactions. When the tubercle antigens are injected just under (or scratched into) the skin, a small, hard lesion forms if the person has been sensitized to the antigen.

Immunodeficiencies

The **immunodeficiencies** include both congenital and acquired conditions in which the production or function of immune cells or complement is abnormal. The most devastating congenital condition is *severe combined immunodeficiency disease (SCID),* in which there is a marked deficit of both B and T cells. Because T cells are absolutely required for normal operation of *both* arms of the immune response, afflicted children have essentially no protection against pathogens of any type. Minor infections easily shrugged off by most children are lethal to those with SCID. Bone marrow transplants and umbilical cord blood, which provide normal lymphocyte stem cells, have helped some SCID victims. Without such treatment, the only hope for survival is living behind protective barriers (in a plastic "bubble") that keep out all infectious agents.

Currently, the most important and most devastating of the acquired immunodeficiencies is **acquired immune deficiency syndrome (AIDS).** AIDS, which cripples the immune system by interfering with the activity of helper T cells, is highlighted in the "A Closer Look" box on pp. 418–419.

Autoimmune Diseases

Occasionally, the immune system loses its ability to distinguish friend from foe, that is, to tolerate self-antigens, while still recognizing and attacking foreign antigens. When this happens, the body produces antibodies (*autoantibodies*) and sensitized T cells that attack and damage its own tissues. This puzzling phenomenon is called **autoimmune disease,** because it is one's own immune system that produces the disorder.

Some 5 percent of adults in North America—two-thirds of them women—are afflicted with autoimmune disease. Most common are:

- *Multiple sclerosis (MS),* which destroys the white matter (myelin sheaths) of the brain and spinal cord (see pp. 226–228)
- *Myasthenia gravis,* which impairs communication between nerves and skeletal muscles (see p. 214)
- *Graves' disease,* in which the thyroid gland produces excessive amounts of thyroxine
- *Type I diabetes mellitus,* which destroys pancreatic beta cells, resulting in deficient production of insulin (see pp. 315–316)
- *Systemic lupus erythematosus (SLE),* a systemic disease that occurs mainly in young females and particularly affects the kidneys, heart, lungs, and skin
- *Glomerulonephritis,* a severe impairment of kidney function
- *Rheumatoid arthritis (RA),* which systematically destroys joints (see p. 167)

Current therapies include treatments that depress certain aspects of the immune response.

How does the normal state of self-tolerance break down? It appears that one or more of the following may be triggers:

1. **Inefficient lymphocyte programming.** Instead of being silenced or eliminated, self-reactive B or T lymphocytes escape to the rest of the body. This is believed to occur in MS.
2. **Appearance of self-proteins in the circulation that were not previously exposed to the immune system.** Such "hidden" antigens are found in sperm cells, the eye lens, and certain proteins in the thyroid gland. In addition, "new self-antigens" may appear as a result of gene mutations or alterations in self-proteins by hapten attachment or by bacterial or viral damage.
3. **Cross-reaction of antibodies produced against foreign antigens with self-antigens.** For instance, antibodies produced during an infection caused by streptococcus bacteria are known to cross-react with heart antigens, causing damage to both the heart muscle and its valves, as well as to joints and kidneys. This age-old disease is called *rheumatic fever.* ▲

PART III: DEVELOPMENTAL ASPECTS OF THE LYMPHATIC SYSTEM AND BODY DEFENSES

The lymphatic vessels, which bud from the veins of the blood vascular system, and the main clusters of lymph nodes are obvious by the fifth week of development. Except for the thymus gland and the spleen, the lymphoid organs are poorly developed before birth. However, shortly after birth, they become heavily populated with lymphocytes as the immune system begins to function.

Lymphatic system problems are relatively uncommon, but when they do occur, they are painfully obvious. For example, when the lymphatic vessels are blocked (as in *elephantiasis,* a tropical disease in which the lymphatics become clogged with parasitic worms) or when lymphatics are removed (as in radical breast surgery), severe edema is the result. However, lymphatic vessels removed surgically do grow back in time.

Stem cells of the immune system originate in the spleen and liver during the first month of embryonic development. Later, the bone marrow becomes the predominant source of stem cells (hemocytoblasts), and it persists in this role into

Prove It Yourself

Swollen Lymph Nodes Usually Indicate Infection

The next time you get a bad cold or the flu, you may notice that your lymph nodes become swollen and tender (try feeling the ones in your neck and under your lower jaw). This is a healthy response to the pathogen—it indicates that your lymphatic system is taking the appropriate action to combat an infection. The entry of the pathogen into your lymphatic system has triggered an explosion in the lymphocyte population, and your lymph nodes are swollen with dividing lymphocytes, clumps of virus particles, and pathogens under attack by lymphocytes.

After the infection subsides, check your lymph nodes again. They should have returned to their normal size.

A Closer Look

AIDS: The Modern Day Plague

IN October 1347, several ships made port in Sicily, and within days all the sailors they carried were dead of bubonic plague. By the end of the fourteenth century, approximately 25 percent of the population of Europe had been wiped out by this "black death." In January 1987, the U.S. Secretary of Health and Human Services warned that acquired immune deficiency syndrome, or AIDS, might be the plague of our time. These are strong words. Are they true?

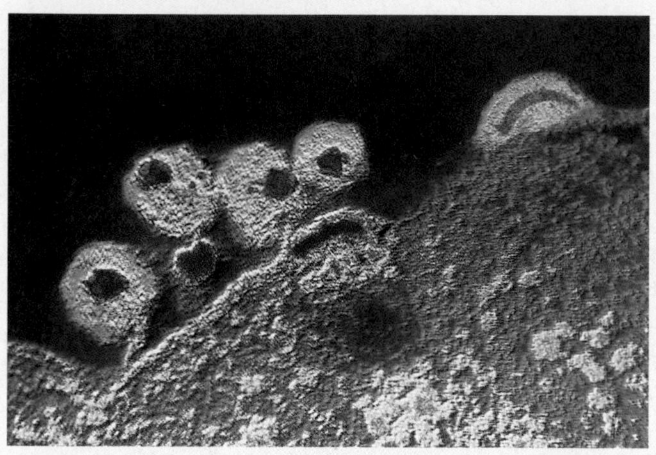

New HIV viruses emerge (yellow and red dots) from an infected human cell.

Although AIDS was first identified in this country in 1981 among homosexual men and intravenous drug users of both sexes, it had begun afflicting the heterosexual population of Africa several years earlier. AIDS is characterized by severe weight loss, night sweats, swollen lymph nodes, and increasingly frequent infections, including a rare type of pneumonia called *pneumocystis pneumonia* and the bizarre malignancy *Kaposi's sarcoma,* a cancerlike condition of blood vessels evidenced by purple lesions of the skin. Some AIDS victims develop slurred speech and severe dementia. The course of AIDS is grim, and thus far inescapable, finally ending in complete debilitation and death from cancer or overwhelming infection.

AIDS is caused by a virus transmitted in blood, semen, vaginal secretions, and saliva. Most commonly, AIDS enters the body via blood transfusions or blood-contaminated needles and during intimate sexual contact in which the mucosa is torn or where open lesions caused by sexually transmitted diseases allow the virus access to the blood.

The virus, named HIV (human immunodeficiency virus), specifically targets and destroys helper T cells, resulting in depression of cell-mediated immunity. Although antibody levels rise and cytotoxic T cells initially mount a vigorous response to viral exposure, in time a profound deficit of normal antibodies develops. Cytotoxic T cells become unresponsive to viral cues, and helper T cells become the prey of the virus. The whole immune system is turned topsy-turvy. It is now clear that the virus multiplies steadily in the lymph nodes throughout most of the chronic asymptomatic period. Symptomatic AIDS appears when the lymph nodes can no longer contain the virus and the immune system collapses. The virus also invades the brain, which accounts for the dementia of some AIDS patients. Although there are exceptions, most AIDS victims have died within a few months to eight years after diagnosis.

The years since 1981 have witnessed a global AIDS epidemic—worldwide, 16,000 people are infected with the HIV virus daily. As of 2001, more than 40 million people were infected worldwide and almost 90 percent of them live in the developing countries of Asia and southern Africa. In the United States, estimates of Americans infected with HIV topped 700,000 by 2001. Furthermore, because there is a six-month "window" during which antibodies may develop after exposure to HIV, there are probably 100 asymptomatic carriers of the virus for every newly diagnosed case. Moreover, the disease has a long incubation period (from a few months to 10 years) between exposure and the appearance of clinical symptoms.

Not only has the number of identified cases jumped explosively in the at-risk populations, but the "face of AIDS" is changing. Victims have begun to include people who do not belong to the original high-risk group. Before reliable testing of donated blood was available, some people contracted the virus from blood transfusions. Hemophiliacs have been especially vulnerable because the blood factors they need are isolated from pooled blood donations. Although manufacturers began taking measures to kill the virus in 1984, by then an estimated 60 percent of the hemophiliacs in this country were already infected. The virus can also be transmitted from an infected mother to her fetus. Though homosexual men still account for the bulk of cases transmitted by sexual contact, more and more heterosexuals are contracting this disease. Particularly disturbing is the near-epidemic increase in diagnosed cases among teenagers and young adults. AIDS is now the fifth leading killer of all Americans ages 25 to 44.

Large-city hospitals are seeing more AIDS patients daily, and statistics reporting the number of cases in inner-city ghettos, where intravenous drug use is the chief means of AIDS transmission, are alarming. Currently the drug-abusing community accounts for 25 percent of all AIDS cases. Furthermore, 75 percent of AIDS cases in newborns occur where drug abuse abounds.

Diagnostic tests to identify carriers of the AIDS virus are increasingly sophisticated. For example, besides the simple test that scrapes tissue from the oral mucosa, an even easier urine test provides a painless alternative to the standard HIV tests. However, no sure cure has yet been found. Over 100 drugs are now in the U.S. Food and Drug Administration pipeline and more than 20 vaccines are undergoing clinical trials, but it is unlikely that an approved vaccine will be available soon.

Several antiviral drugs (named like alphabet soups) that act by inhibiting the enzymes that the HIV virus needs in order to multiply in the body are now clinically available. The *reverse transcriptase inhibitors,* such as AZT, were early on the scene. AZT has been followed by others, including ddl, ddC, d4T, and 3TC. In late 1995 and early 1996, *protease inhibitors* (saquinavir, ritonavir, and others) were approved. Thus far, it appears that combination therapy using drugs from each class delivers a one-two punch to the HIV virus, at least for a while. Combination therapy postpones drug resistance (a problem with mono-drug therapy using AZT alone) and substantially reduces the amount of HIV virus in the blood while boosting the number of helper T cells. Sadly, the treatments are beginning to fail in about half of those treated. New hope comes from the integrase group of new drugs that block HIV's entry into DNA of the target helper cells.

Given AIDS' poor prognosis, prevention of infection is the way to go. Perhaps, as urged in the media, the best defense is to practice "safer sex" by using condoms and knowing one's sexual partner's history, but the only fool-proof alternative is sexual abstinence.

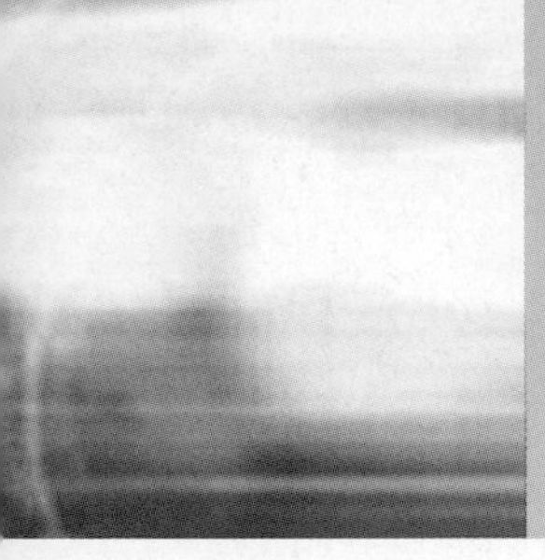

Systems in Sync

Homeostatic Relationships between the Lymphatic System and Other Body Systems

Nervous System
- The lymphatic vessels pick up leaked plasma fluid and proteins in the peripheral nervous system structures; immune cells protect peripheral nervous system structures from specific pathogens
- The nervous system innervates larger lymphatic vessels; the brain helps regulate immune response

Endocrine System
- Lymphatic vessels pick up leaked fluids and proteins; lymph distributes hormones; immune cells protect endocrine organs from pathogens
- The thymus produces hormones that promote development of lymphatic organs and "program" T lymphocytes

Lymphatic System/Immunity

Respiratory System
- Lymphatic vessels pick up leaked fluid and proteins from respiratory organs; immune cells protect respiratory organs from specific pathogens; plasma cells in the respiratory mucosa secrete IgA to prevent pathogen invasion of deeper tissues
- The lungs provide oxygen needed by lymphoid/immune cells and eliminate carbon dioxide; the pharynx houses some lymphoid organs (tonsils); the respiratory "pump" aids lymph flow

Digestive System
- Lymphatic vessels pick up leaked fluids and proteins from digestive organs; lymph transports some products of fat digestion to the blood; lymphoid nodules in the wall of the intestine prevent invasion of pathogens
- The digestive system digests and absorbs nutrients needed by cells of lymphatic organs; gastric acidity inhibits pathogens' entry into blood

Muscular System
- Lymphatic vessels pick up leaked fluid and proteins from skeletal muscle; immune cells protect muscles from specific pathogens
- The skeletal muscle "pump" aids the flow of lymph; muscles protect superficial lymph nodes

Cardiovascular System
- Lymphatic vessels pick up leaked plasma and proteins; spleen destroys aged RBCs, stores iron, and removes debris from blood; immune cells protect cardiovascular organs from specific pathogens
- Blood is the source of lymph; lymphatics develop from veins; blood provides the route for circulation of immune elements

Urinary System
- Lymphatic vessels pick up leaked fluid and proteins from urinary organs; immune cells protect urinary organs from specific pathogens
- Urinary system eliminates wastes and maintains homeostatic water/acid-base/electrolyte balance of the blood for immune cell functioning; urine flushes some pathogens out of the body

Reproductive System
- Lymphatic vessels pick up leaked fluid and proteins from reproductive organs; immune cells protect the organs from specific pathogens
- Acidity of vaginal secretions prevents bacterial multiplication

Integumentary System
- Lymphatic vessels pick up leaked plasma fluid and proteins from the dermis; lymphocytes in lymph enhance the skin's protective role by defending against specific pathogens
- The skin's keratinized epithelium provides a mechanical barrier to pathogens; acid pH of skin secretions inhibits growth of bacteria on skin

Skeletal System
- Lymphatic vessels pick up leaked plasma fluid and proteins from the.periostea; immune cells protect bones from specific pathogens
- The bones house hematopoietic tissue (red marrow) which produces the lymphocytes (and macrophages) that populate the lymphoid organs and provide body immunity

adult life. In late fetal life and shortly after birth, the young lymphocytes develop self-tolerance and immunocompetence in their "programming organs" (thymus and bone marrow) and then populate the other lymphoid tissues. Upon meeting "their" antigens, the T and B cells complete their development to fully mature immune cells.

Although the ability of our immune system to recognize foreign substances is determined by our genes, the nervous system may help to control the activity of the immune response. The immune response is definitely impaired in individuals who are under severe stress—for example, in those mourning the death of a beloved family member or friend. Our immune system normally serves us well throughout our lifetime, until old age. But, during the later years, its efficiency begins to wane, as does that of the nonspecific defenses. As a result, the body becomes less able to fight infections and destroy cells that have become cancerous. Additionally, we become more susceptible to both autoimmune and immunodeficiency diseases.

SUMMARY

Media study tools that could provide you with additional help in reviewing specific key topics of Chapter 12 are referenced below.

IP = Interactive Physiology;

WEB = A&P Place website.

PART I: THE LYMPHATIC SYSTEM (pp. 390–395)

1. The lymphatic system consists of the lymphatic vessels, lymph nodes, and certain other lymphoid organs in the body.

   WEB Exercise: Chapter 12, Lymphatic Collecting Vessels and Regional Lymph Nodes.

2. Extremely porous blind-ended lymphatic capillaries pick up excess tissue fluid leaked from the blood capillaries. The fluid (lymph) flows into the larger lymphatics and finally into the blood vascular system through the right lymphatic duct and the left thoracic duct.
3. Lymph transport is aided by the muscular and respiratory pumps and by contraction of smooth muscle in the walls of the lymphatic vessels.
4. Lymph nodes are clustered along lymphatic vessels, and the lymphatic stream flows through them. Lymph nodes form agranular WBCs (lymphocytes), and phagocytic cells within them remove bacteria, viruses, and the like from the lymph stream before it is returned to the blood.
5. Other lymphoid organs include the tonsils (in the throat), which remove bacteria trying to enter the digestive or respiratory tracts; the thymus, a programming region for some lymphocytes of the body; Peyer's patches, which prevent bacteria in the intestine from penetrating deeper into the body; and the spleen, a RBC graveyard and blood reservoir.

PART II: BODY DEFENSES (pp. 395–417)

NONSPECIFIC BODY DEFENSES (pp. 396–400)

1. Surface membranes (skin and mucous membranes) provide mechanical barriers to pathogens. Some produce secretions and/or have structural modifications that enhance their defensive effects; the skin's acidity, lysozyme, mucus, keratin, and ciliated cells are examples.
2. Phagocytes (macrophages and neutrophils) engulf and destroy pathogens that penetrate epithelial barriers. This process is enhanced when the pathogen's surface is altered by attachment of antibodies and/or complement.
3. Natural killer cells are nonimmune cells that act nonspecifically to lyse virus-infected and malignant cells.
4. The inflammatory response prevents spread of harmful agents, disposes of pathogens and dead tissue cells, and promotes healing. Protective leukocytes enter the area; the area is walled off by fibrin; and tissue repair occurs.
5. When complement (a group of plasma proteins) becomes fixed on the membrane of a foreign cell, lysis of the target cell occurs. Complement also enhances phagocytosis and the inflammatory and immune responses.
6. Interferon is a group of proteins synthesized by virus-infected cells and certain immune cells. It prevents viruses from multiplying in other body cells.
7. Fever enhances the fight against infectious microorganisms by increasing metabolism (which speeds up repair processes) and by causing the liver and spleen to store iron and zinc (which are needed for bacterial multiplication).

SPECIFIC BODY DEFENSES: THE IMMUNE SYSTEM (pp. 400–417)

1. The immune system recognizes something as foreign and acts to inactivate or remove it. Immune response is antigen-specific, is systemic, and has memory. The two arms of immune response are humoral immunity (mediated by antibodies) and cellular immunity (mediated by living cells, the lymphocytes).
2. Antigens
   a. Antigens are large, complex molecules (or parts of them) recognized as foreign by the body. Foreign proteins are the strongest antigens.
   b. Complete antigens provoke an immune response and bind with products of that response (antibodies or sensitized lymphocytes).
   c. Incomplete antigens, or haptens, are small molecules that are unable to cause an immune response by themselves but do so when they bind to body proteins and the complex is recognized as foreign.
3. Cells of the immune system: An overview
   a. Two main cell populations, lymphocytes and macrophages, provide for immunity.
   b. Lymphocytes arise from hemocytoblasts of bone marrow. T cells develop immunocompetence in the thymus and oversee cell-mediated immunity. B cells develop immunocompetence in bone marrow and provide humoral immunity. Immunocompetent lymphocytes seed lymphoid organs, where antigen challenge occurs, and circulate through blood, lymph, and lymphoid organs.
   c. Immunocompetence is signaled by the appearance of antigen-specific receptors on surfaces of lymphocytes.
   d. Macrophages arise from monocytes produced in bone marrow. They phagocytize pathogens and present parts of the antigens on their surfaces, for recognition by T cells.
4. Humoral (antibody-mediated) immune response
   a. Clonal selection of B cells occurs when antigens bind to their receptors, causing them to proliferate. Most clone members become plasma cells, which secrete antibodies. This is called the *primary immune response.*
   b. Other clone members become memory B cells, capable of mounting a rapid attack against the same antigen in subsequent meetings *(secondary immune responses).* These memory cells provide immunological "memory."
   c. Active humoral immunity is acquired during an infection or via vaccination and provides immunological memory. Passive immunity is conferred when a donor's antibodies are injected into the bloodstream or when the mother's antibodies cross the placenta. It does not provide immunological memory.
   d. Basic antibody structure
      (1) Antibodies are proteins produced by sensitized B cells or plasma cells in response to an antigen, and they are capable of binding with that antigen.
      (2) An antibody is composed of four polypeptide chains (two heavy and two light) that form a T- or Y-shaped molecule.
      (3) Each polypeptide chain has a variable and a constant region. Variable regions form antigen-binding sites, one on each arm of the T or Y. Constant regions determine antibody function and class.
      (4) Five classes of antibodies exist: IgA, IgG, IgM, IgD, IgE. They differ structurally and functionally.
      (5) Antibody functions include complement fixation, neutralization, precipitation, and agglutination.
      (6) Monoclonal antibodies are pure preparations of a single antibody type useful in diagnosis of various infectious disorders and cancer and in treatment of certain cancers.
5. Cellular (cell-mediated) immune response
   a. T cells are sensitized by binding simultaneously to an antigen and a self-protein displayed on the surface of a macrophage. Clonal selection occurs, and clone members differentiate into effector T cells or memory T cells.
   b. There are several different classes of T cells. Cytotoxic (killer) T cells directly attack and lyse infected and cancerous cells. Helper T cells interact directly with B cells bound to antigens. They also liberate lymphokines, chemicals that enhance the killing activity of macrophages, attract other leukocytes, or act as helper factors that stimulate activity of B cells and cytotoxic T cells. A special subgroup of helper T cells, cytotoxic T cells, and macrophages promote a delayed allergic reaction. Suppressor T cells terminate the normal immune response by releasing suppressor chemicals.
6. Organ transplants include autografts, isografts, allografts, and xenografts. The most usual graft is an allograft. Blood group and tissue matching are done to assure the best match possible, and organ transplant is followed by immunosuppressive therapy.

7. Disorders of immunity
   a. In allergy or hypersensitivity the immune system overreacts to an otherwise harmless antigen, and tissue destruction occurs. Immediate (acute) hypersensitivity, as seen in hay fever, hives, and anaphylaxis, is due to IgE antibodies. Delayed hypersensitivity (for example, contact dermatitis) reflects activity of T cells, macrophages, and lymphokines.

   WEB Exercise: Chapter 12, Events in Allergic Reaction.

   b. Immunodeficiencies result from abnormalities in any immune element. Most serious is severe combined immunodeficiency disease (a congenital disease) and AIDS, an acquired immunodeficiency disease caused by a virus that attacks and cripples the helper T cells.
   c. Autoimmune disease occurs when the body's self-tolerance breaks down, and antibodies and/or T cells attack the body's own tissues. Most forms of autoimmune disease result from inefficient lymphocyte programming in the fetus, changes in structure of self-antigens or appearance of formerly hidden self-antigens in blood, and cross-reactions with self-antigens and antibodies formed against foreign antigens.

PART III: DEVELOPMENTAL ASPECTS OF THE LYMPHATIC SYSTEM AND BODY DEFENSES (pp. 417–421)

1. Lymphatic vessels form by budding off veins. The thymus gland is the first lymphoid organ to appear in the embryo. Other lymphoid organs remain relatively undeveloped until after birth.
2. Development of immune response occurs around the time of birth.
3. The ability of immunocompetent cells to recognize foreign antigens is genetically determined. Stress appears to interfere with normal immune response.
4. Efficiency of immune response wanes in old age, and infections, cancer, immunodeficiencies, and autoimmune diseases become more prevalent.

REVIEW QUESTIONS

MULTIPLE CHOICE

More than one choice may apply.

1. Lymph capillaries:
   a. are open-ended like drinking straws
   b. have continuous tight junctions like the capillaries of the brain
   c. contain endothelial cells separated by flaplike valves that can open wide
   d. have special barriers that stop cancer cells from entering
2. Which parts of the lymph node show increased activity when antibody production is high?
   a. Germinal centers
   b. Outer follicle
   c. Medullary cords
   d. Medullary sinuses
3. Which of the following connect to the lymph node at the hilus?
   a. Afferent lymphatic vessels
   b. Efferent lymphatic vessels
   c. Trabeculae
   d. Anchoring filaments
4. Which of the following are part of MALT?
   a. Tonsils
   b. Thymus
   c. Peyer's patches
   d. Any lymphoid tissue along the digestive tract
5. Developmentally, embryonic lymphatic vessels are most closely associated with the:
   a. veins
   b. arteries
   c. nerves
   d. thymus gland
6. Which of the following are cardinal signs of inflammation?
   a. Phagocytosis
   b. Edema
   c. Leukocytosis
   d. Pain
7. Chemical mediators of inflammation include:
   a. interferon
   b. complement
   c. histamine
   d. antibodies
8. Against which of the following will interferon do some good?
   a. Infection of body cells by a virus
   b. Circulating free viruses
   c. Some types of cancer
   d. Bacterial infection
9. Which of these antibody classes is usually arranged as a pentamer?
   a. IgG
   b. IgM
   c. IgA
   d. IgD
10. Which parts of an antibody molecule are different for an IgG antibody than for an IgM antibody that attacks the same antigen?
    a. Heavy chain constant region
    b. Heavy chain variable region

c. Light chain constant region
d. Light chain variable region

11. Which of the following antibody capabilities causes a transfusion reaction with A or B erythrocyte antigens?
    a. Neutralization
    b. Precipitation
    c. Complement fixation
    d. Agglutination

12. Which of the following is/are examples of autoimmune disease?
    a. Type I diabetes
    b. Multiple sclerosis
    c. Graves' disease
    d. Rheumatoid arthritis

13. The main cellular target of the HIV virus that causes AIDS is:
    a. helper T cells
    b. cytotoxic T cells
    c. macrophages
    d. B cells

SHORT ANSWER ESSAY

1. What is the most important function of the lymphatic vessels? of the lymph nodes?
2. Where are the lymph nodes most dense?
3. What is the special role of the tonsils? the spleen?
4. Besides acting as mechanical barriers, the skin and mucosae of the body contribute to body protection in other ways. Cite the common body locations and the importance of mucus, lysozyme, keratin, acid pH, and cilia.
5. What is complement? How does it cause bacterial lysis? What are some of the other roles of complement?
6. Interferons are referred to as antiviral proteins. What stimulates their production, and how do they protect uninfected cells?
7. Define *immune response.*
8. Define *antigen.* What is the difference between a complete antigen and an incomplete antigen (hapten)?
9. Differentiate clearly between humoral and cellular immunity and between the roles of B lymphocytes and T lymphocytes.
10. Although the immune system has two arms, it has been said, "No T cells, no immunity." How is this so?
11. Define *immunocompetence.* What indicates that a B cell or T cell has developed immunocompetence? Where does the "programming phase" occur in the case of T cells? B cells?
12. Binding of antigens to receptors of immunocompetent lymphocytes leads to clonal selection. Describe the process of clonal selection. What nonlymphocyte cell is a central actor in this process, and what is its function?
13. Name the cell types that would be present in a B cell clone, and give the function of each type.
14. Describe the specific roles of helper, cytotoxic, and suppressor T cells in cell-mediated immunity. Which is thought to be disabled in AIDS?
15. Compare and contrast a primary and a secondary immune response. Which is more rapid, and why?
16. Describe the structure of an antibody, and explain the importance of its variable and constant regions.
17. Name the five classes of immunoglobulins. Which is most likely to be found attached to a B cell membrane? Which is most abundant in plasma? Which is important in allergic responses? Which is the first Ig to be released during the primary response? Which can cross the placental barrier?
18. How do antibodies help to defend the body?
19. Define *allergy,* and distinguish between immediate types of allergy and delayed allergic reactions relative to cause and consequences.
20. What events can result in the loss of self-tolerance and autoimmune disease?
21. Would lack of memory B cells for a particular antigen impact the primary or secondary humoral response?

AT THE CLINIC

1. As an infant receives her first dose of oral polio vaccine, the nurse explains to her parents that the vaccine is a preparation of weakened virus. What type of immunity will the infant develop?
2. Some people with a deficit of IgA exhibit recurrent paranasal sinus and respiratory tract infections. Explain these symptoms.
3. Mr. James, an 80-year-old man, is grumbling about having to receive a flu shot every year. Flu viruses have a high mutation rate (undergo rapid genetic changes), which results in the appearance of new proteins on the flu virus's "coat." How does this help explain the need to get a flu shot each year?
4. A 59-year-old woman has undergone a left radical mastectomy (removal of the left breast and left axillary lymph nodes and vessels). Her left arm is severely swollen and painful, and she is unable to raise it more than shoulder height. (a) Explain her signs and symptoms. (b) Can she expect relief from these symptoms in time? How so?
5. Lymphocytes continuously circulate through the body using blood and lymph as their transport vehicles. What is the importance of this recirculation behavior?

13

The Respiratory System

YOUR GOALS

After completing this chapter, you will have a working knowledge of the functions of the respiratory system and will have mastered the objectives listed below.

FUNCTION PREVIEW

- The respiratory system supplies oxygen to the blood while removing carbon dioxide.

OBJECTIVE CHECKLIST

FUNCTIONAL ANATOMY OF THE RESPIRATORY SYSTEM (pp. 426–434)

- Name the organs forming the respiratory passageway from the nasal cavity to the alveoli of the lungs (or identify them on a diagram or model), and describe the function of each.
- Describe several protective mechanisms of the respiratory system.
- Describe the structure and function of the lungs and the pleural coverings.

RESPIRATORY PHYSIOLOGY (pp. 434–445)

- Define *cellular respiration, external respiration, internal respiration, pulmonary ventilation, expiration,* and *inspiration.*
- Explain how the respiratory muscles cause volume changes that lead to air flow into and out of the lungs (breathing).
- Define the following respiratory volumes: *tidal volume, vital capacity, expiratory reserve volume, inspiratory reserve volume,* and *residual air.*
- Name several nonrespiratory air movements, and explain how they modify or differ from normal respiratory air movements.
- Describe the process of gas exchanges in the lungs and tissues.
- Describe how oxygen and carbon dioxide are transported in the blood.
- Name the brain areas involved in control of respiration.

>>

(Objective Checklist, continued)

- Name several physical factors that influence respiratory rate.
- Explain the relative importance of oxygen and carbon dioxide in modifying the rate and depth of breathing.
- Explain why it is not possible to stop breathing voluntarily.
- Define *apnea, dyspnea, hyperventilation, hypoventilation,* and *chronic obstructive pulmonary disease (COPD).*

RESPIRATORY DISORDERS **(pp. 445–446)**

- Describe the symptoms and probable causes of COPD and lung cancer.

DEVELOPMENTAL ASPECTS OF THE RESPIRATORY SYSTEM **(pp. 446–449)**

- Describe normal changes that occur in respiratory system functioning from infancy to old age.

The trillions of cells in the body require an abundant and continuous supply of oxygen to carry out their vital functions. We cannot "do without oxygen" for even a little while, as we can without food or water. Furthermore, as cells use oxygen, they give off carbon dioxide, a waste product the body must get rid of.

The *cardiovascular* and *respiratory systems* share responsibility for supplying the body with oxygen and disposing of carbon dioxide. The respiratory system organs oversee the gas exchanges that occur between the blood and the external environment. The transportation of respiratory gases between the lungs and the tissue cells is accomplished by the cardiovascular system organs, using blood as the transporting fluid. If either system fails, body cells begin to die from oxygen starvation and accumulation of carbon dioxide.

Functional Anatomy of the Respiratory System

The organs of the **respiratory system** include the nose, pharynx, larynx, trachea, bronchi and their smaller branches, and the lungs, which contain the *alveoli* (al-ve′o-li), or terminal air sacs. Since gas exchanges with the blood happen only in the alveoli, the other respiratory system structures are really just conducting passageways that allow air to reach the lungs. However, these passageways have another, very important job. They purify, humidify, and warm incoming air. Thus, the air finally reaching the lungs has many fewer irritants (such as dust or bacteria) than when it entered the system, and it is warm and damp. As the respiratory system organs are described in detail next, locate each on Figure 13.1.

The Nose

The **nose,** whether "pug" or "ski-jump" in shape, is the only externally visible part of the respiratory system. During breathing, air enters the nose by passing through the **nostrils,** or **external nares.** The interior of the nose consists of the **nasal cavity,** divided by a midline *nasal septum.* The *olfactory receptors* for the sense of smell are located in the mucosa in the slitlike superior part of the nasal cavity, just beneath the ethmoid bone. The rest of the mucosa lining the nasal cavity, called the *respiratory mucosa,* rests on a rich network of thin-walled veins that warms the air as it flows past. (Because of the superficial location of these blood vessels, nosebleeds are common and

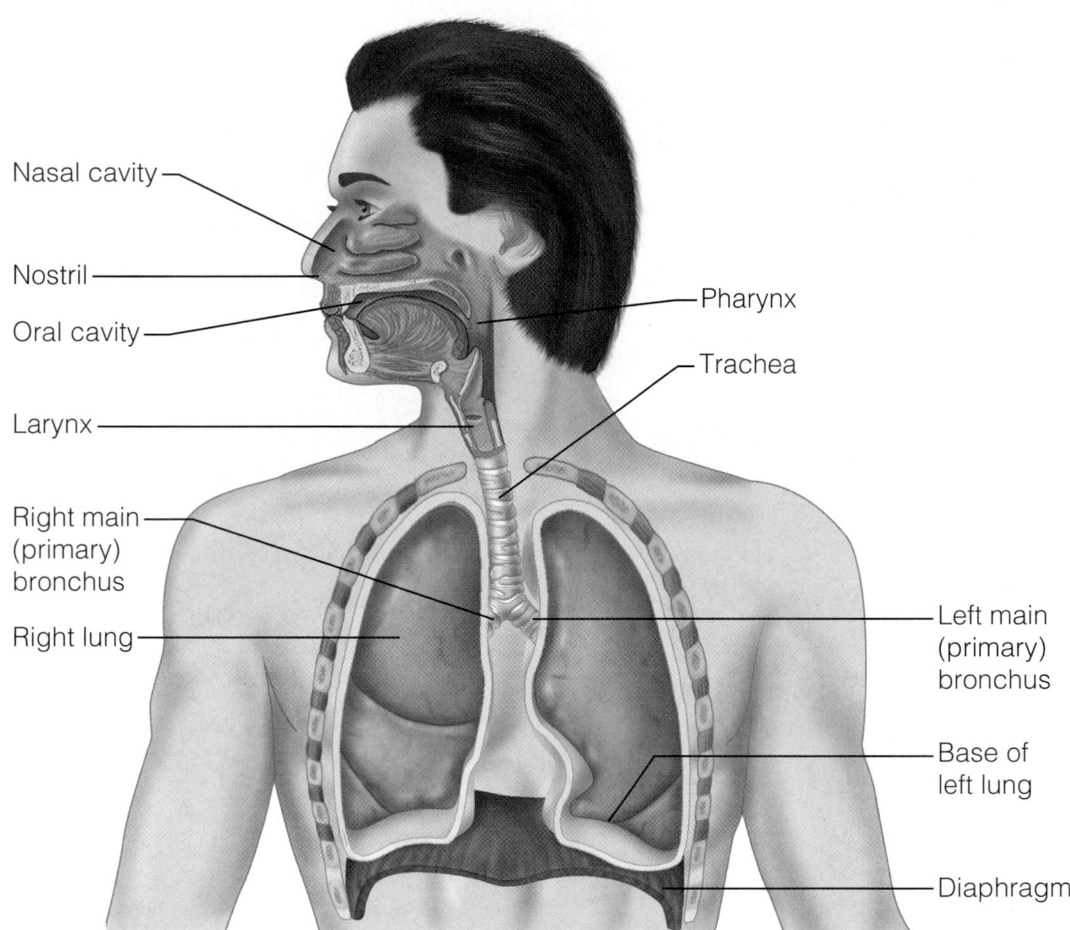

FIGURE 13.1 The major respiratory organs shown in relation to surrounding structures.

often profuse.) In addition, the sticky mucus produced by the mucosa's glands moistens the air and traps incoming bacteria and other foreign debris. The ciliated cells of the nasal mucosa create a gentle current that moves contaminated mucus posteriorly toward the throat (pharynx), where it is swallowed and digested by stomach juices. We are usually unaware of this important ciliary action, but when the external temperature is extremely cold, these cilia become sluggish, allowing mucus to accumulate in the nasal cavity and to dribble outward through the nostrils. This helps explain why you might have a "runny" nose on a crisp, wintry day.

As shown in Figures 13.1 and 13.2, the lateral walls of the nasal cavity are uneven owing to three mucosa-covered projections or lobes, called **conchae** (kong′ke), which greatly increase the surface area of the mucosa exposed to the air. The conchae also increase the air turbulence in the nasal cavity. As the air swirls through the twists and turns, inhaled particles are deflected onto the mucus-coated surfaces, where they are trapped and prevented from reaching the lungs.

The nasal cavity is separated from the oral cavity below by a partition, the **palate** (pal′et). Anteriorly, where the palate is supported by bone, is the **hard palate;** the unsupported posterior part is the **soft palate.**

Homeostatic Imbalance

The genetic defect *cleft palate* (failure of the bones forming the palate to fuse medially) results in breathing difficulty as well as problems with oral cavity functions such as chewing and speaking. ▲

The nasal cavity is surrounded by a ring of **paranasal sinuses** located in the frontal, sphenoid,

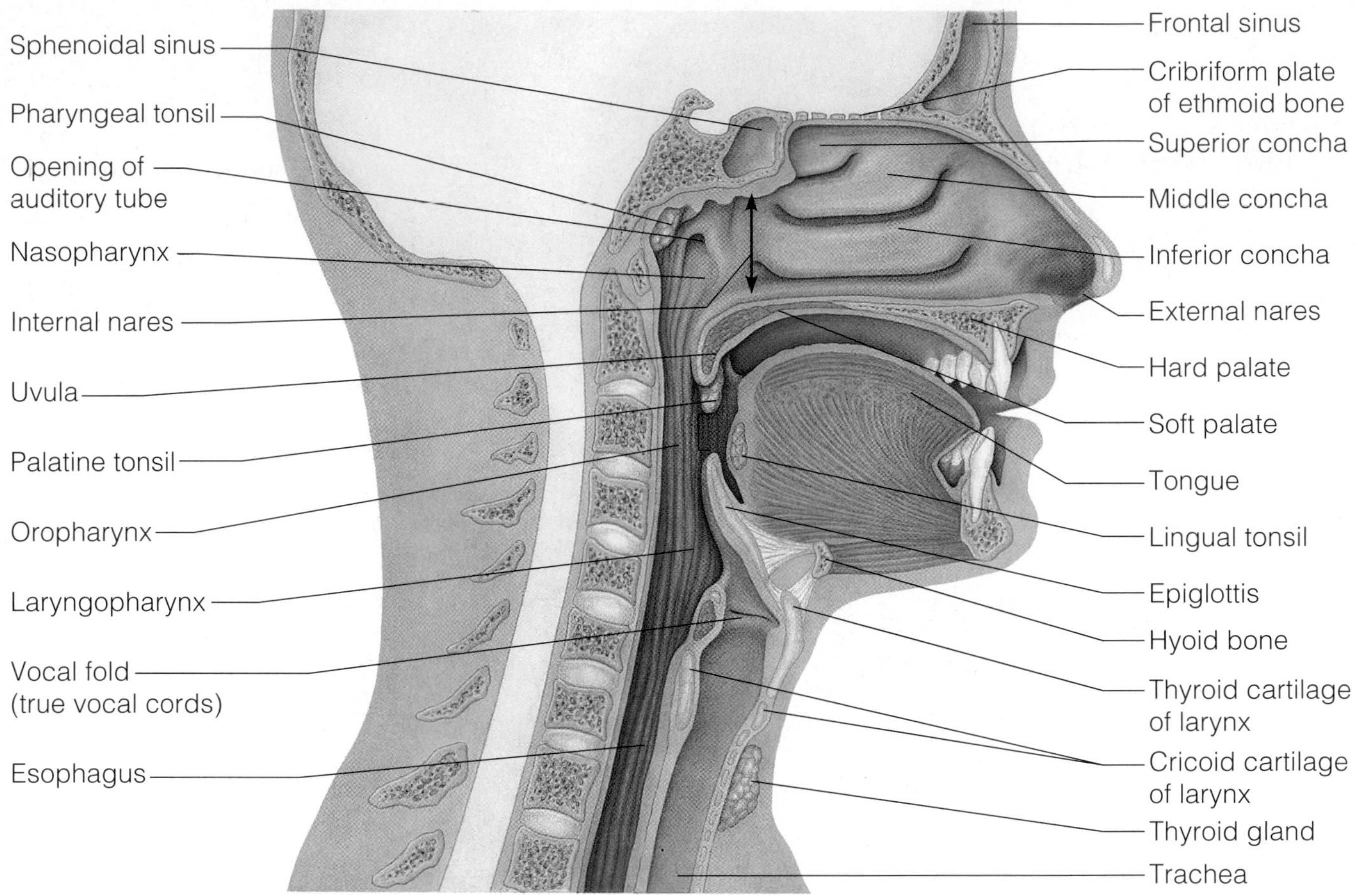

FIGURE 13.2 Basic anatomy of the upper respiratory tract, sagittal section.

ethmoid, and maxillary bones. (See Figure 5.10, p. 142.) The sinuses lighten the skull, and they act as resonance chambers for speech. They also produce mucus, which drains into the nasal cavities. The suctioning effect created by nose blowing helps to drain the sinuses. The *nasolacrimal ducts,* which drain tears from the eyes, also empty into the nasal cavities.

Homeostatic Imbalance

Cold viruses and various allergens can cause *rhinitis* (ri-ni'tis), inflammation of the nasal mucosa. The excessive mucus produced results in nasal congestion and postnasal drip. Because the nasal mucosa is continuous throughout the respiratory tract and extends tentacle-like into the nasolacrimal (tear) ducts and paranasal sinuses, nasal cavity infections often spread to those regions as well. *Sinusitis,* or sinus inflammation, is difficult to treat and can cause marked changes in voice quality. When the passageways connecting the sinuses to the nasal cavity are blocked with mucus or infectious matter, the air in the sinus cavities is absorbed. The result is a partial vacuum and a *sinus headache* localized over the inflamed area. ▲

Pharynx

The **pharynx** (far'inks) is a muscular passageway about 13 cm (5 inches) long that vaguely resembles a short length of red garden hose. Commonly called the *throat,* the pharynx serves as a common passageway for food and air (Figures 13.1 and 13.2). It is continuous with the nasal cavity anteriorly via the **internal nares.**

Air enters the superior portion, the **nasopharynx** (na″zo-far'inks), from the nasal cavity and then descends through the **oropharynx** (o″ro-far'inks) and **laryngopharynx** (lah-ring″go-far'inks) to enter the larynx below. Food enters the mouth and then travels along with air through the oropharynx and laryngopharynx. But instead of entering the larynx, food is directed into the *esophagus* (ĕ-sof'ah-gus) posteriorly.

The pharyngotympanic tubes, which drain the middle ear, open into the nasopharynx. Since the mucosae of these two regions are continuous, ear infections such as *otitis media* (o-ti′tis me′de-ah) may follow a sore throat or other types of pharyngeal infections.

Clusters of lymphatic tissue called *tonsils* are also found in the pharynx. The **pharyngeal** (far-rin′je-al) **tonsil,** often called *adenoid,* is located high in the nasopharynx. The **palatine tonsils** are in the oropharynx at the end of the soft palate; the **lingual tonsils** are at the base of the tongue. The role of the tonsils in body protection is described in Chapter 12 (p. 394).

Homeostatic Imbalance

If the pharyngeal tonsil becomes inflamed and swollen (as during a bacterial infection), it obstructs the nasopharynx and forces the person to breathe through the mouth. In mouth breathing, air is not properly moistened, warmed, or filtered before reaching the lungs. Many children seem to have almost continuous *tonsillitis.* Years ago the belief was that the tonsils, though protective, were more trouble than they were worth in such cases, and they were routinely removed. Presently, because of the widespread use of antibiotics, this is no longer necessary (or true). ▲

Larynx

The **larynx** (lar′inks), or *voice box,* routes air and food into the proper channels and plays a role in speech. Located inferior to the pharynx (see Figures 13.1 and 13.2), it is formed by eight rigid hyaline cartilages and a spoon-shaped flap of elastic cartilage, the epiglottis (ep″ĭ-glot′tis). The largest of the hyaline cartilages is the shield-shaped **thyroid cartilage,** which protrudes anteriorly and is commonly called the *Adam's apple.* Sometimes referred to as the "guardian of the airways," the **epiglottis** protects the superior opening of the larynx. When we are not swallowing, the epiglottis does not restrict the passage of air into the lower respiratory passages. When we swallow food or fluids, the situation changes dramatically; the larynx is pulled upward and the epiglottis tips, forming a lid over the opening of the larynx. This routes food into the esophagus, or food tube, posteriorly. If anything other than air enters the larynx, a *cough reflex* is triggered to expel the substance and prevent it from continuing into the lungs. Because this protective reflex does *not* work when we are unconscious, it is never a good idea to try to give fluids to an unconscious person when attempting to revive him or her.

- Palpate your larynx by placing your hand midway on the anterior surface of your neck. Swallow. Can you feel the larynx rising as you swallow?

Part of the mucous membrane of the larynx forms a pair of folds, called the **vocal folds,** or **true vocal cords,** which vibrate with expelled air. This ability of the vocal folds to vibrate allows us to speak. The slitlike passageway between the vocal folds is the **glottis.**

Trachea

Air entering the **trachea** (tra′ke-ah), or *windpipe,* from the larynx travels down its length (10–12 cm, or about 4 inches) to the level of the fifth thoracic vertebra, which is approximately midchest (Figure 13.1).

The trachea is fairly rigid because its walls are reinforced with C-shaped rings of **hyaline cartilage.** These rings serve a double purpose. The open parts of the rings abut the **esophagus** and allow it to expand anteriorly when we swallow a large piece of food. The solid portions support the trachea walls and keep it *patent,* or open, in spite of the pressure changes that occur during breathing.

Homeostatic Imbalance

Because the trachea is the only way air can enter the lungs, tracheal obstruction is life-threatening. Many people have suffocated after choking on a piece of food that suddenly closed off the trachea (or the glottis of the larynx). The **Heimlich maneuver,** a procedure in which the air in a person's own lungs is used to "pop out," or expel, an obstructing piece of food, has saved many other people from becoming victims of such "café coronaries." The Heimlich maneuver is simple to learn and easy to do. However, it is best learned by demonstration because cracked ribs are a distinct possibility when it is done incorrectly. In some cases of obstructed breathing, an emergency *tracheostomy* (tra′ke-ost′o-me; surgical opening of the trachea) is done to provide an alternate route for air to reach the lungs. Individuals with tracheostomy tubes in place form huge amounts of mucus the first few days because of irritation to the trachea. Thus, they must be suctioned frequently during this time to prevent the mucus from pooling in their lungs. ▲

Q *In what direction is the power stroke of these cilia directed—superiorly toward the mouth or inferiorly toward the lungs?*

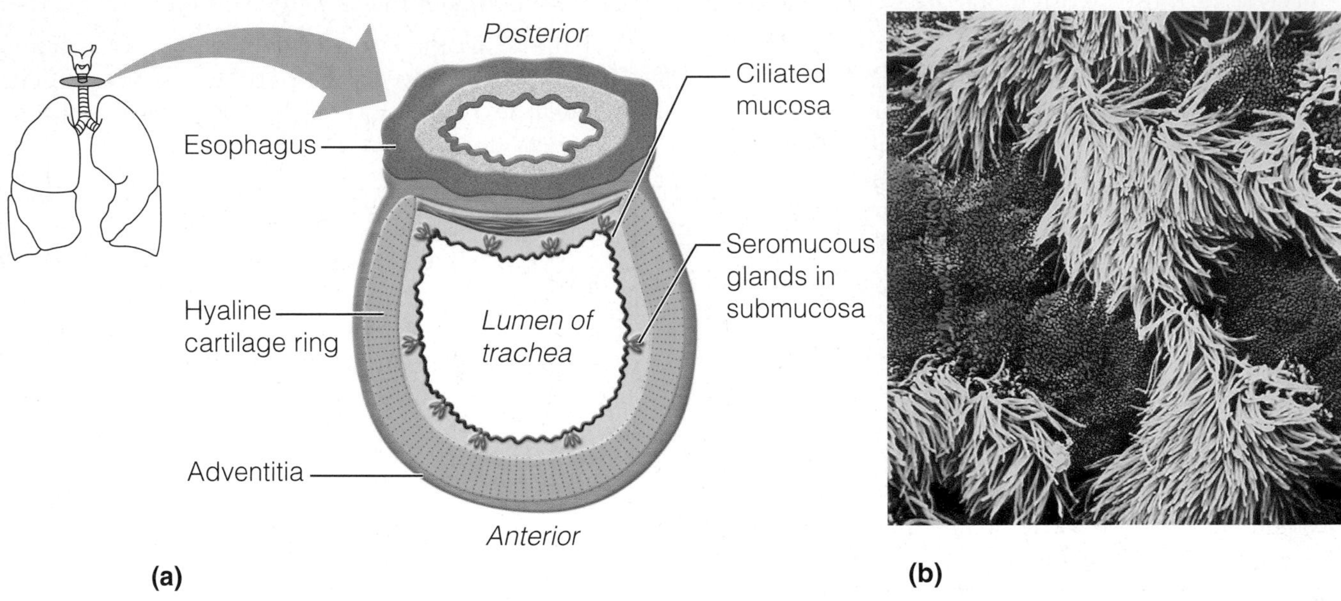

FIGURE 13.3 Structural relationship of the trachea and esophagus. **(a)** Cross-sectional view. **(b)** Cilia in the trachea. The cilia are the yellow, grasslike projections surrounded by the mucus-secreting goblet cells, which exhibit short microvilli (orange). (Scanning electron micrograph, 221,000×.)

The trachea is lined with a ciliated mucosa (Figure 13.3). The cilia beat continuously and in a direction opposite to that of the incoming air. They propel mucus, loaded with dust particles and other debris, away from the lungs to the throat, where it can be swallowed or spat out.

Homeostatic Imbalance

Smoking inhibits ciliary activity and ultimately destroys the cilia. Without these cilia, coughing is the only means of preventing mucus from accumulating in the lungs. Smokers with respiratory congestion should avoid medications that inhibit the cough reflex. ▲

Main Bronchi

The right and left **main (primary) bronchi** (brong′ki) are formed by the division of the trachea. Each main bronchus runs obliquely before it plunges into the medial depression *(hilus)* of the lung on its own side (see Figures 13.1 and 13.4). The right main bronchus is wider, shorter, and straighter than the left. Consequently, it is the more common site for an inhaled foreign object to become lodged. By the time incoming air reaches the bronchi, it is warm, cleansed of most impurities, and well humidified. The smaller subdivisions of the main bronchi within the lungs are direct routes to the air sacs.

A *Superiorly toward the mouth to prevent unwanted substances from entering the lungs.*

Lungs

The paired **lungs** are fairly large organs. They occupy the entire thoracic cavity except for the most central area, the **mediastinum** (me″de-as-ti′num), which houses the heart (in its inferior pericardial cavity region), the great blood vessels, bronchi, esophagus, and other organs (Figure 13.4). The narrow superior portion of each lung, the **apex,** is located just deep to the clavicle. The broad lung area resting on the diaphragm is the **base.** Each lung is divided into lobes by fissures; the left lung has two lobes, and the right lung has three.

The surface of each lung is covered with a visceral serosa called the **pulmonary,** or **visceral, pleura** (ploor′ah), and the walls of the thoracic cavity are lined by the **parietal pleura.** The pleural membranes produce *pleural fluid,* a slippery serous secretion which allows the lungs to glide easily over the thorax wall during breathing movements and

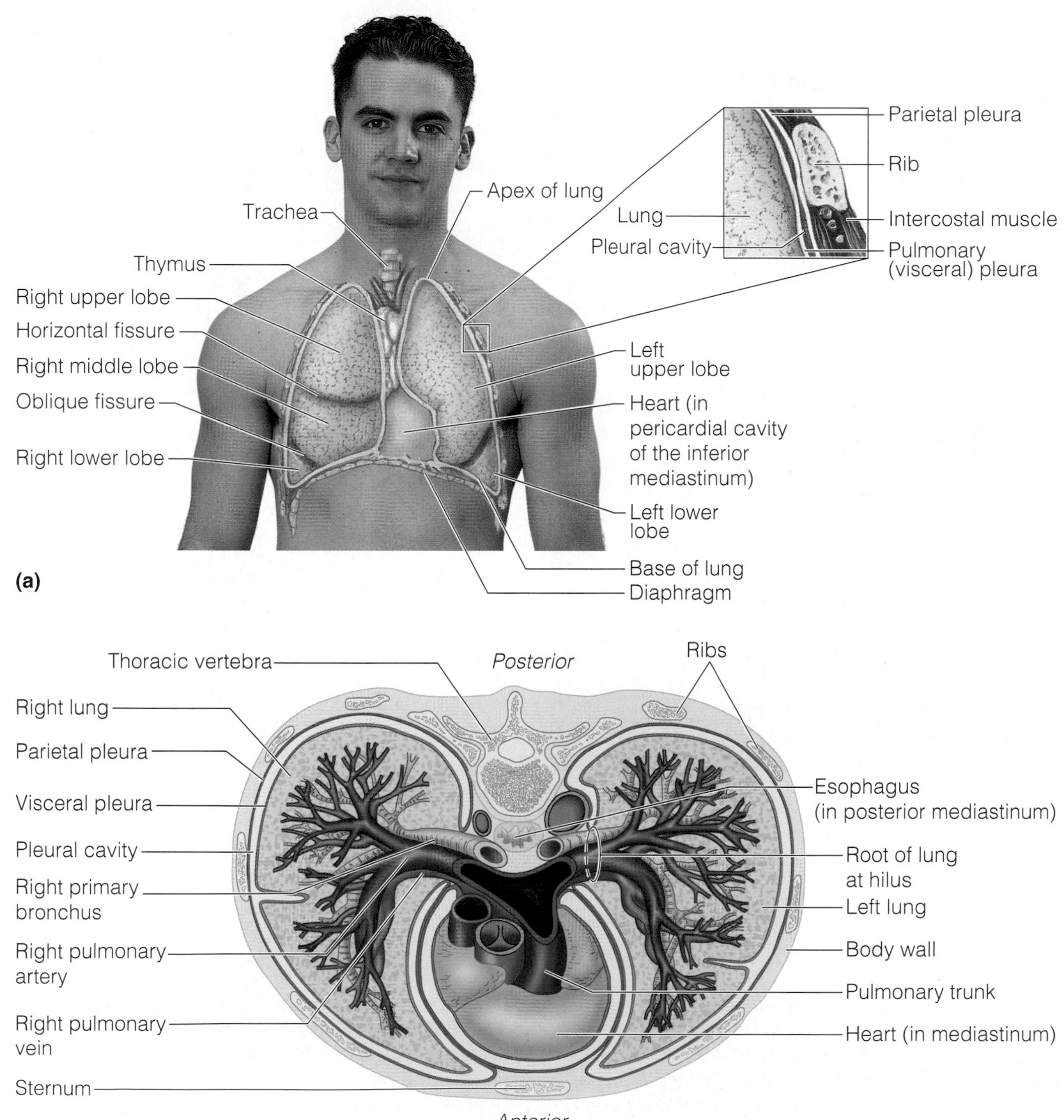

FIGURE 13.4 Anatomical relationships of organs in the thoracic cavity. (a) Anterior view of the thoracic cavity organs, showing the position of the lungs, which flank the heart laterally. **(b)** Transverse section through the thorax, showing the relationship of the lungs, the pleural membranes, the major organs present in the mediastinum, and the thorax.

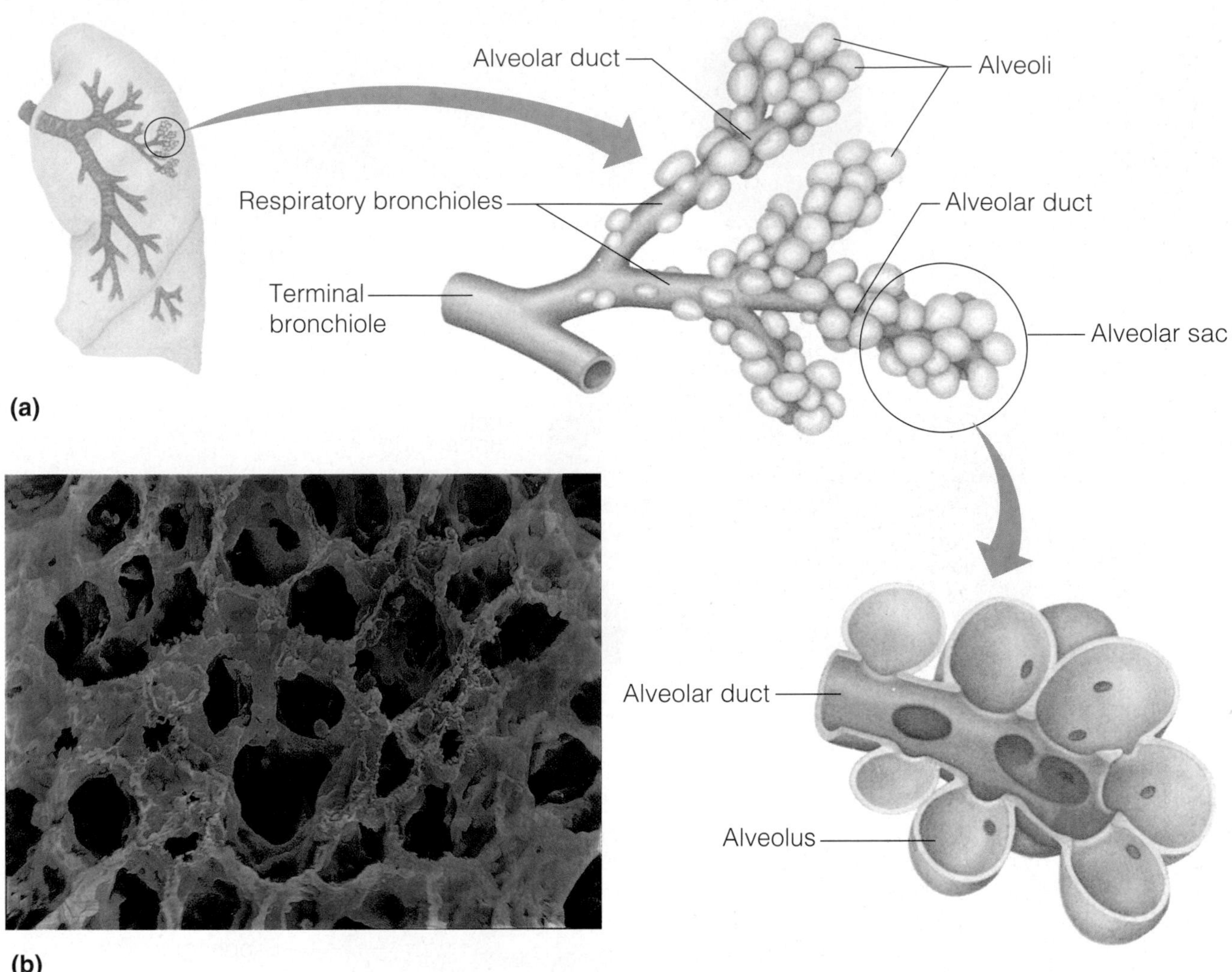

FIGURE 13.5 Respiratory zone structures. (a) Diagrammatic view of respiratory bronchioles, alveolar ducts, and alveoli. **(b)** Scanning electron micrograph (SEM) of human lung tissue, showing the final divisions of the respiratory tree (475×).

causes the two pleural layers to cling together. The pleurae can slide easily from side to side across one another, but they strongly resist being pulled apart. Consequently, the lungs are held tightly to the thorax wall, and the **pleural space** is more of a potential space than an actual one. As described shortly, this condition of tightly adhering pleural membranes is absolutely essential for normal breathing. Figure 13.4 shows the position of the pleura on the lungs and the thorax wall.

Homeostatic Imbalance

Pleurisy (ploo′rĭ-se), inflammation of the pleura, can be caused by decreased secretion of pleural fluid. The pleural surfaces become dry and rough, which results in friction and stabbing pain with each breath. Conversely, the pleurae may produce excessive amounts of fluid, which exerts pressure on the lungs. This type of pleurisy hinders breathing movements, but it is much less painful than the dry rubbing type. ▲

After the primary bronchi enter the lungs, they subdivide into smaller and smaller branches (secondary and tertiary bronchi, and so on), finally ending in the smallest of the conducting passageways, the **bronchioles** (brong′ke-ōlz) (Figure 13.5). Because of this branching and rebranching of the respiratory passageways within the lungs, the network

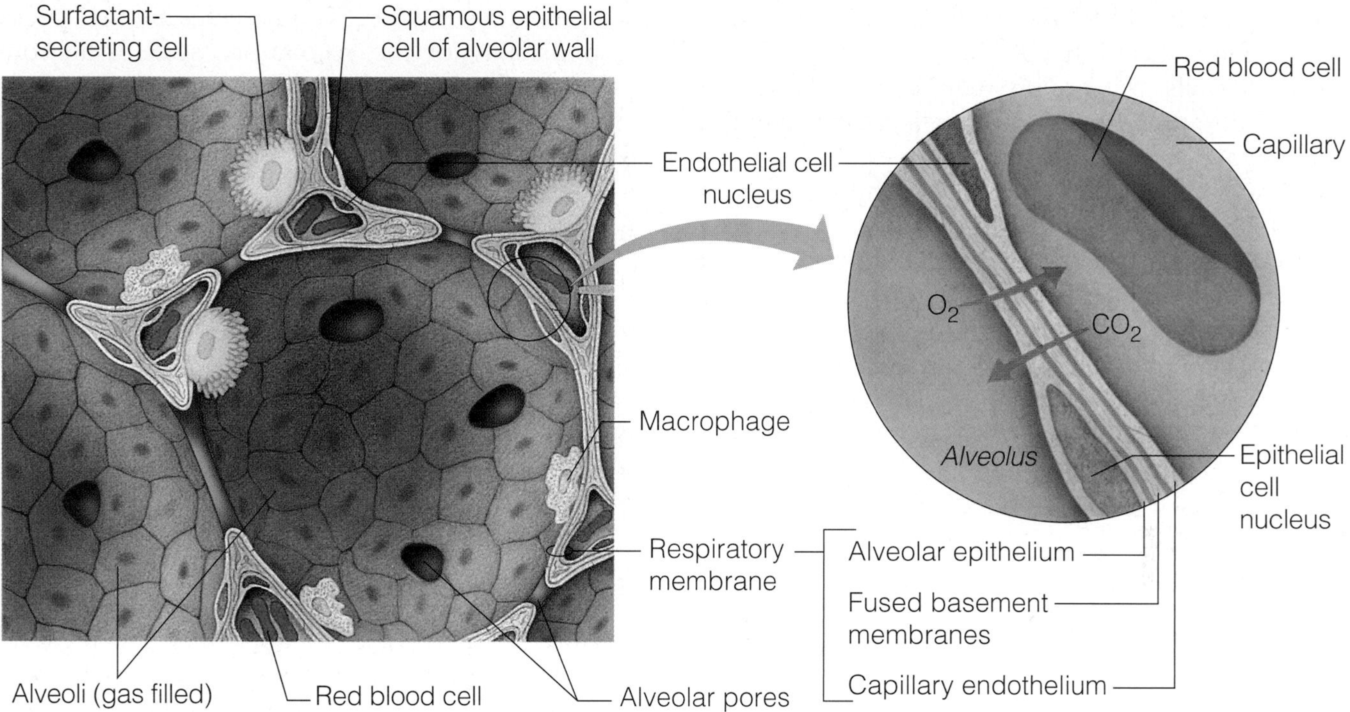

FIGURE 13.6 Anatomy of the respiratory membrane (air-blood barrier). The respiratory membrane is composed of squamous epithelial cells of the alveoli, the capillary endothelium, and the scant basement membranes between. Surfactant-secreting cells are also shown. Oxygen diffuses from the alveolar air into the pulmonary capillary blood; carbon dioxide diffuses from the pulmonary blood into the alveolus. Neighboring alveoli are connected by small pores.

formed is often referred to as the *bronchial* or *respiratory tree.* All but the smallest branches have reinforcing cartilage in their walls.

The *terminal bronchioles* lead into *respiratory zone structures,* even smaller conduits that eventually terminate in **alveoli** (al-ve′o-li; *alveol* = small cavity), or air sacs. The **respiratory zone,** which includes the *respiratory bronchioles, alveolar ducts, alveolar sacs,* and alveoli, is the only site of gas exchange. All other respiratory passages are **conducting zone structures** that serve as conduits to and from the respiratory zone. There are millions of the clustered alveoli, which resemble bunches of grapes, and they make up the bulk of the lungs. Consequently, the lungs are mostly air spaces. The balance of the lung tissue, its *stroma,* is elastic connective tissue. Thus, in spite of their relatively large size, the lungs weigh only about 2½ pounds, and they are soft and spongy.

The Respiratory Membrane

The walls of the alveoli are composed largely of a single, thin layer of squamous epithelial cells. The thinness of their walls is hard to imagine, but a sheet of tissue paper is much thicker. *Alveolar pores* connect neighboring air sacs and provide alternate routes for air to reach alveoli whose feeder bronchioles have been clogged by mucus or otherwise blocked. The external surfaces of the alveoli are covered with a "cobweb" of pulmonary capillaries. Together, the alveolar and capillary walls, their fused basement membranes, and occasional elastic fibers construct the **respiratory membrane (air-blood barrier),** which has gas (air) flowing past on one side and blood flowing past on the other (Figure 13.6). The gas exchanges occur by simple diffusion through the respiratory membrane—oxygen passing from the alveolar air into the capillary blood and carbon dioxide leaving the blood to enter the gas-filled alveoli. It has been estimated that the total gas exchange surface provided by the alveolar walls of a healthy man is 50 to 70 square meters, or approximately 40 times greater than the surface area of his skin.

The final line of defense for the respiratory system is in the alveoli. Macrophages, sometimes called "dust cells," wander in and out of the alveoli picking up bacteria, carbon particles, and other debris. Also scattered amid the epithelial cells that form most of the alveolar walls are chunky cuboidal cells, which look very different. The cuboidal cells produce a lipid (fat) molecule called *surfactant,* which coats the gas-exposed alveolar surfaces and is very important in lung function (as described on p. 446).

Respiratory Physiology

The major function of the respiratory system is to supply the body with oxygen and to dispose of carbon dioxide. To do this, at least four distinct events, collectively called **respiration,** must occur:

1. **Pulmonary ventilation.** Air must move into and out of the lungs so that the gases in the air sacs (alveoli) of the lungs are continuously changed and refreshed. This process of pulmonary ventilation is commonly called **breathing.**
2. **External respiration.** Gas exchange (oxygen loading and carbon dioxide unloading) between the pulmonary blood and alveoli must take place. Remember that in **ex**ternal respiration, gas exchanges are being made between the blood and the body *exterior.*
3. **Respiratory gas transport.** Oxygen and carbon dioxide must be transported to and from the lungs and tissue cells of the body via the bloodstream.
4. **Internal respiration.** At systemic capillaries, gas exchanges must be made between the blood and tissue cells.* In **in**ternal respiration, gas exchanges are occurring between the blood and cells *inside* the body.

Although only the first two processes are the special responsibility of the respiratory system, all four processes are necessary for it to accomplish its goal of gas exchange. Thus, each process is described in turn next.

Mechanics of Breathing

Breathing, or pulmonary ventilation, is a completely mechanical process that depends on volume changes occurring in the thoracic cavity. Here is a rule to keep in mind about the mechanics of breathing: *Volume changes lead to pressure changes, which lead to the flow of gases to equalize the pressure.*

A gas, like a liquid, always conforms to the shape of its container. However, unlike liquid, a gas *fills* its container. Therefore, in a large volume, the gas molecules will be far apart and the pressure (created by the gas molecules hitting each other and the walls of the container) will be low. If the volume is reduced, the gas molecules will be closer together and the pressure will rise. Let us see how this relates to the two phases of breathing—**inspiration,** when air is flowing into the lungs, and **expiration,** when air is leaving the lungs.

Inspiration

When the inspiratory muscles, the **diaphragm** and **external intercostals,** contract, the size of the thoracic cavity increases. As the dome-shaped diaphragm contracts, it moves inferiorly and flattens out (is depressed). As a result, the superior-inferior dimension (height) of the thoracic cavity increases. Contraction of the external intercostals lifts the rib cage and thrusts the sternum forward, which increases the anteroposterior and lateral dimensions of the thorax (Figure 13.7a). Since the lungs adhere tightly to the thorax walls (due to the surface tension of the fluid between the pleural membranes), they are stretched to the new, larger size of the thorax. As *intrapulmonary volume* (the volume within the lungs) increases, the gases within the lungs spread out to fill the larger space. The resulting decrease in the gas pressure in the lungs produces a partial vacuum (pressure less than atmospheric pressure), which sucks air into the lungs (Figure 13.8). Air continues to move into the lungs until the intrapulmonary pressure equals atmospheric pressure. This series of events is called *inspiration* (inhalation).

---

*The actual *use* of oxygen and production of carbon dioxide by tissue cells, that is, **cellular respiration,** is the cornerstone of all energy-producing chemical reactions in the body. Cellular respiration, which occurs in all body cells, is discussed in Chapter 14.

Q *Given that the lungs do not have any skeletal muscle, explain what causes them to inflate.*

Changes in anterior-posterior and superior-inferior dimensions

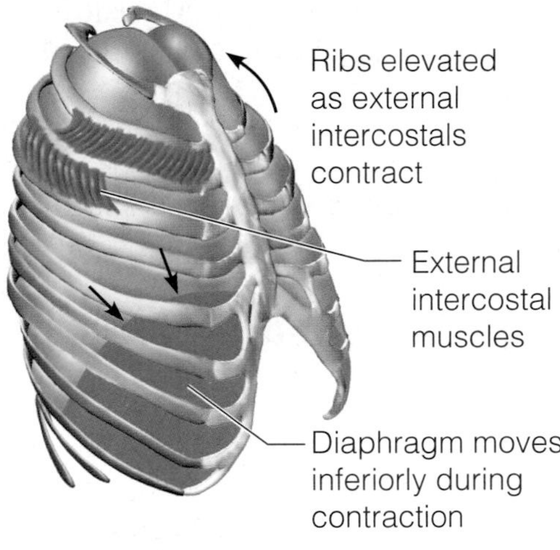

(a) Inspiration: Air (gases) flows into the lungs

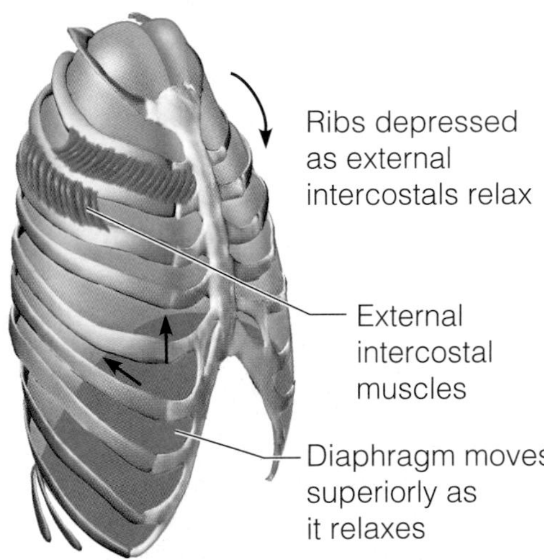

(b) Expiration: Air (gases) flows out of the lungs

Changes in lateral dimensions

FIGURE 13.7 Rib cage and diaphragm positions during breathing. **(a)** At the end of a normal inspiration: Chest is expanded laterally, rib cage is elevated, and diaphragm is depressed and flattened. Lungs are stretched to the larger thoracic volume causing the intrapulmonary pressure to fall and air to flow into the lungs. **(b)** At the end of a normal expiration: Chest is depressed and the lateral dimension is reduced, rib cage is descended, and diaphragm is elevated and dome-shaped. Lungs recoil to smaller volume, intrapulmonary pressure rises and air flows out of lung.

A *The enlargement of the thoracic cavity due to contraction of the diaphragm and intercostal muscles results in lower pressure in the lungs than outside the body. This difference in pressure leads to air flow into the lungs and their expansion.*

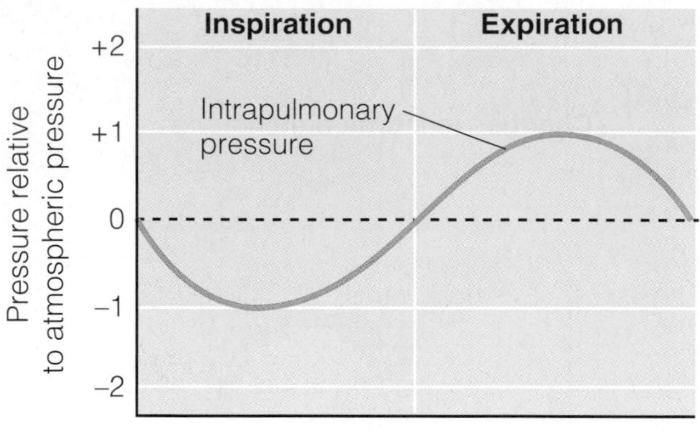

(a)

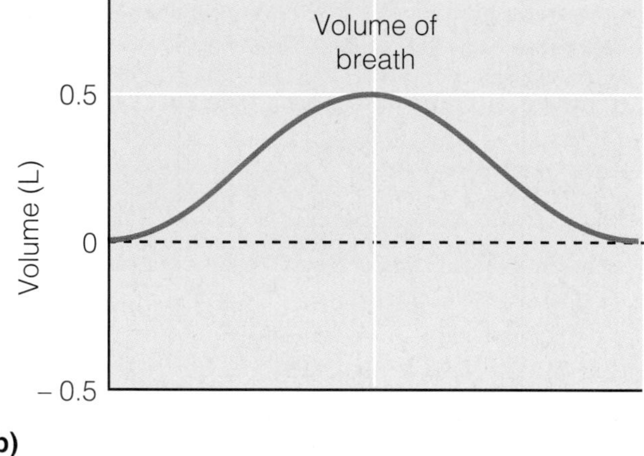

(b)

FIGURE 13.8 Changes in intrapulmonary pressure and air flow during inspiration and expiration.

Expiration

Expiration (exhalation) in healthy people is largely a passive process that depends more on the natural elasticity of the lungs than on muscle contraction. As the inspiratory muscles relax and resume their initial resting length, the rib cage descends and the lungs recoil. Thus, both the thoracic and intrapulmonary volumes decrease (Figure 13.7b). As the intrapulmonary volume decreases, the gases inside the lungs are forced more closely together, and the intrapulmonary pressure rises to a point higher than atmospheric pressure (see Figure 13.8). This causes the gases to flow out to equalize the pressure inside and outside the lungs. Under normal circumstances, expiration is effortless, but if the respiratory passageways are narrowed by spasms of the bronchioles (as in *asthma*) or clogged with mucus or fluid (as in *chronic bronchitis* or *pneumonia*), expiration becomes an active process. In such cases of *forced expiration*, the internal intercostal muscles are activated to help depress the rib cage, and the abdominal muscles contract and help to force air from the lungs by squeezing the abdominal organs upward against the diaphragm.

The normal pressure within the pleural space *(intrapleural pressure)* is *always* negative, and this is the major factor preventing collapse of the lungs. If for any reason the intrapleural pressure becomes equal to the atmospheric pressure, the lungs immediately recoil completely and collapse.

Homeostatic Imbalance

During *atelectasis* (a″teh-lek′tuh-sis), or lung collapse, the lung is useless for ventilation. This phenomenon is seen when air enters the pleural space through a chest wound, but it may also result from a rupture of the visceral pleura, which allows air to enter the pleural space from the respiratory tract. The presence of air in the intrapleural space, which disrupts the fluid bond between the pleurae, is referred to as *pneumothorax* (nu″mo-tho′rakz). Pneumothorax is reversed by drawing air out of the intrapleural space with chest tubes, which allows the lung to reinflate and resume its normal function. ▲

Nonrespiratory Air Movements

Many situations other than breathing move air into or out of the lungs and may modify the normal respiratory rhythm. Coughs and sneezes clear the air passages of debris or collected mucus. Laughing and crying reflect our emotions. For the most part, these **nonrespiratory air movements** are a result of reflex activity, but some may be produced voluntarily. Examples of the most common of these movements are given in Table 13.1.

Respiratory Volumes and Capacities

Many factors affect respiratory capacity—for example, a person's size, sex, age, and physical condition. Normal quiet breathing moves approximately 500 ml of air (about a pint) into and out of the lungs with each breath (see Figure 13.8b). This respiratory volume is referred to as the **tidal volume (TV).**

As a rule, a person *can* inhale much more air than is taken in during a normal, or tidal, breath. The amount of air that can be taken in forcibly

TABLE 13.1 Nonrespiratory Air (Gas) Movements

Movement	Mechanism and result
Cough	Taking a deep breath, closing glottis, and forcing air superiorly from lungs against glottis. Then, the glottis opens suddenly and a blast of air rushes upward. Coughs act to clear the lower respiratory passageways.
Sneeze	Similar to a cough, except that expelled air is directed through nasal cavities instead of through oral cavity. The uvula (u'vu-lah), a tag of tissue hanging from the soft palate, becomes depressed and closes oral cavity off from pharynx, routing the air through nasal cavities. Sneezes clear upper respiratory passages.
Crying	Inspiration followed by release of air in a number of short breaths. Primarily an emotionally induced mechanism.
Laughing	Essentially same as crying in terms of the air movements produced. Also an emotionally induced response.
Hiccups	Sudden inspirations resulting from spasms of diaphragm; initiated by irritation of diaphragm or phrenic nerves, which serve diaphragm. The sound occurs when inspired air hits vocal folds of closed glottis.
Yawn	Very deep inspiration, taken with jaws wide open. Formerly believed to be triggered by need to increase amount of oxygen in blood but this theory is now being questioned; ventilates all alveoli (this is not the case in normal quiet breathing).

over the tidal volume is the **inspiratory reserve volume (IRV).** Normally, the inspiratory reserve volume is between 2100 and 3200 ml.

Similarly, after a normal expiration, more air can be exhaled. The amount of air that can be forcibly exhaled after a tidal expiration, the **expiratory reserve volume (ERV),** is approximately 1200 ml.

Even after the most strenuous expiration, about 1200 ml of air still remains in the lungs, and it cannot be voluntarily expelled. This is the **residual volume.** Residual volume air is important because it allows gas exchange to go on continuously even between breaths and helps to keep the alveoli open (inflated).

The total amount of exchangeable air is typically around 4800 ml in healthy young males, and this respiratory capacity is the **vital capacity (VC).** The vital capacity is the sum of the TV + IRV + ERV. The respiratory volumes are summarized in Figure 13.9.

Obviously, much of the air that enters the respiratory tract remains in the conducting zone passageways and never reaches the alveoli. This is called the **dead space volume,** and during a normal tidal breath, it amounts to about 150 ml. The functional volume—air that actually reaches the respiratory zone and contributes to gas exchange—is about 350 ml.

Respiratory capacities are measured with a *spirometer* (spi-rom'ĕ-ter). As a person breathes, the volumes of air exhaled can be read on an indicator, which shows the changes in air volume inside the apparatus. Spirometer testing is useful for evaluating losses in respiratory functioning and in following the course of some respiratory diseases. In pneumonia, for example, inspiration is obstructed and the IRV and VC decrease. In emphysema, where expiration is hampered, the ERV is much lower than normal and the residual volume is higher.

Respiratory Sounds

As air flows in and out of the respiratory tree, it produces two recognizable sounds that can be picked up with a stethoscope. **Bronchial sounds**

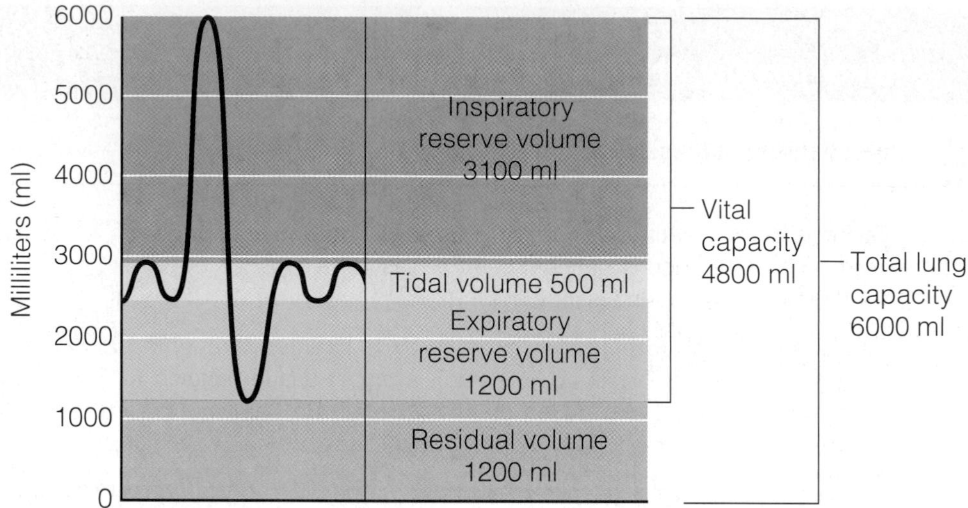

FIGURE 13.9 Idealized tracing of the various respiratory volumes of a healthy young adult male.

are produced by air rushing through the large respiratory passageways (trachea and bronchi). **Vesicular** (vě-sik′u-lar) **breathing sounds** occur as air fills the alveoli. The vesicular sounds are soft and resemble a muffled breeze.

Homeostatic Imbalance

Diseased respiratory tissue, mucus, or pus can produce abnormal sounds such as *crackle* (a bubbling sound) and *wheezing* (a whistling sound). ▲

External Respiration, Gas Transport, and Internal Respiration

As explained earlier, *external respiration* is the actual exchange of gases between the alveoli and the blood (pulmonary gas exchange), and *internal respiration* is the gas exchange process that occurs between the systemic capillaries and the tissue cells. It is important to remember that all gas exchanges are made according to the laws of diffusion; that is, movement occurs *toward* the area of lower concentration of the diffusing substance. The relative amounts of O_2 and CO_2 in the alveolar tissues, and in the arterial and venous blood, are illustrated in Figure 13.10.

External Respiration

During external respiration, dark red blood flowing through the pulmonary circuit is transformed into the scarlet river that is returned to the heart for distribution to the systemic circuit. Although this color change is due to oxygen pickup by hemoglobin in the lungs, carbon dioxide is being unloaded from the blood equally fast. Because body cells continually remove oxygen from blood, there is always more oxygen in the alveoli than in the blood. Thus, oxygen tends to move from the air of the alveoli through the respiratory membrane into the more oxygen-poor blood of the pulmonary capillaries. On the other hand, as tissue cells remove oxygen from the blood in the systemic circulation, they release carbon dioxide into the blood. Because the concentration of carbon dioxide is much higher in the pulmonary capillaries than it is in the alveolar air, it will move from the blood into the alveoli and be flushed out of the lungs during expiration. Relatively speaking, blood draining from the lungs into the pulmonary veins is oxygen-rich and carbon dioxide-poor and is ready to be pumped to the systemic circulation.

Gas Transport in the Blood

Oxygen is transported in the blood in two ways. Most attaches to hemoglobin molecules inside the RBCs to form **oxyhemoglobin** (ok″se-he″mo-glo′bin)—HbO_2 in Figure 13.11a. A very small amount of oxygen is carried dissolved in the plasma.

Most carbon dioxide is transported in plasma as the **bicarbonate ion** (HCO_3^-), which plays a very important role in the blood buffer system. (The enzymatic conversion of carbon dioxide to bicarbonate ion actually occurs within the red blood cells, and then the newly formed bicarbonate ions diffuse into the plasma.) A smaller amount (between 20 and 30 percent of the transported CO_2) is carried inside the RBCs bound to hemoglobin. Carbon dioxide carried inside the RBCs binds to hemoglobin at a different site than oxygen does, and so it does not interfere in any way with oxygen transport. Before carbon dioxide can diffuse out of the blood into the alveoli, it must first be released from its bicarbonate ion form. For this to occur, bicarbonate ions must enter the red blood cells where they combine with hydrogen ions (H^+) to form carbonic acid (H_2CO_3). Carbonic acid quickly splits to form water and carbon dioxide, and carbon dioxide then diffuses from the blood and enters the alveoli.

Homeostatic Imbalance

Impaired oxygen transport: Whatever the cause, inadequate oxygen delivery to body tissues is called **hypoxia** (hi-pok'se-ah). This condition is easy to recognize in fair-skinned people because their skin and mucosae take on a bluish cast (become *cyanotic*). In dark-skinned individuals, this color change can be observed only in the mucosae and nailbeds. Hypoxia may be the result of anemia, pulmonary disease, or impaired or blocked blood circulation.

Carbon monoxide poisoning represents a unique type of hypoxia. Carbon monoxide (CO) is an odorless, colorless gas that competes vigorously with oxygen for the same binding sites on hemoglobin. Moreover, since hemoglobin binds to CO more readily than to oxygen, carbon monoxide is a very successful competitor—so much so that it crowds out or displaces oxygen.

Carbon monoxide poisoning is the leading cause of death from fire. It is particularly dangerous because it kills its victims softly and quietly. It does not produce the characteristic signs of hypoxia—cyanosis and respiratory distress. Instead, the victim becomes confused and has a throbbing headache. In rare cases, the skin becomes cherry red (the color of the hemoglobin-CO complex), which is often interpreted as a healthy "blush." Those with CO poisoning are given 100 percent oxygen until the carbon monoxide has been cleared from the body. ▲

Internal Respiration

Internal respiration, the exchange of gases that takes place between the blood and the tissue cells, is opposite to what occurs in the lungs. This process, in which oxygen is unloaded and carbon dioxide is loaded into the blood, is shown in Figure 13.11b. Carbon dioxide diffusing out of tissue cells enters the blood. In the blood, it combines with water to form carbonic acid, which quickly releases the bicarbonate ions. As previously mentioned, most conversion of carbon dioxide to bicarbonate ions actually occurs *inside* the

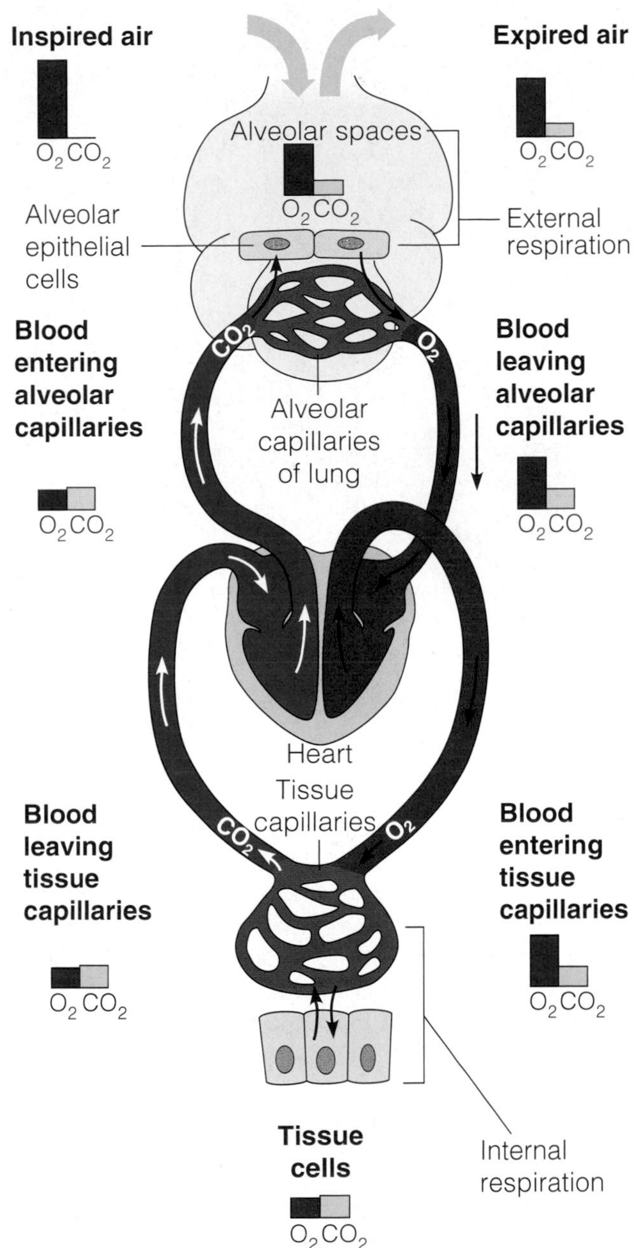

FIGURE 13.10 Gas exchanges in the body occur according to the laws of diffusion.

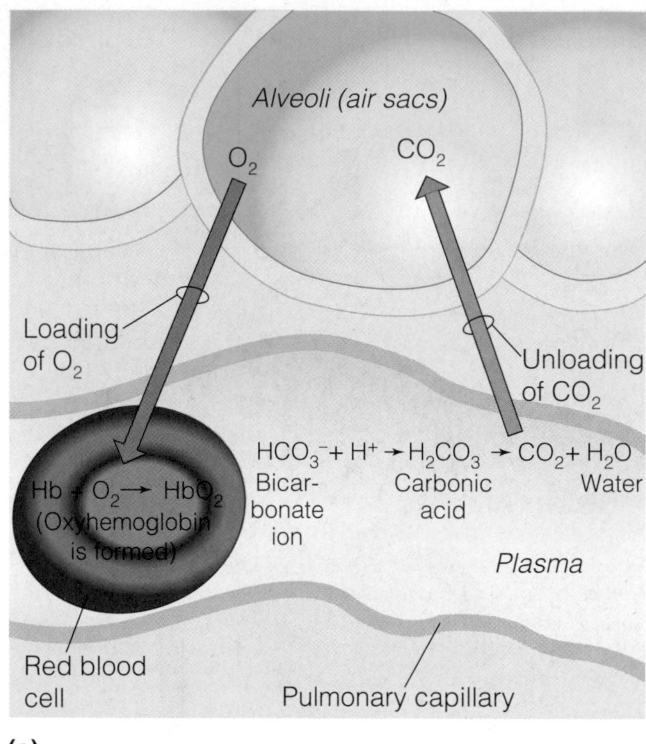

(a)

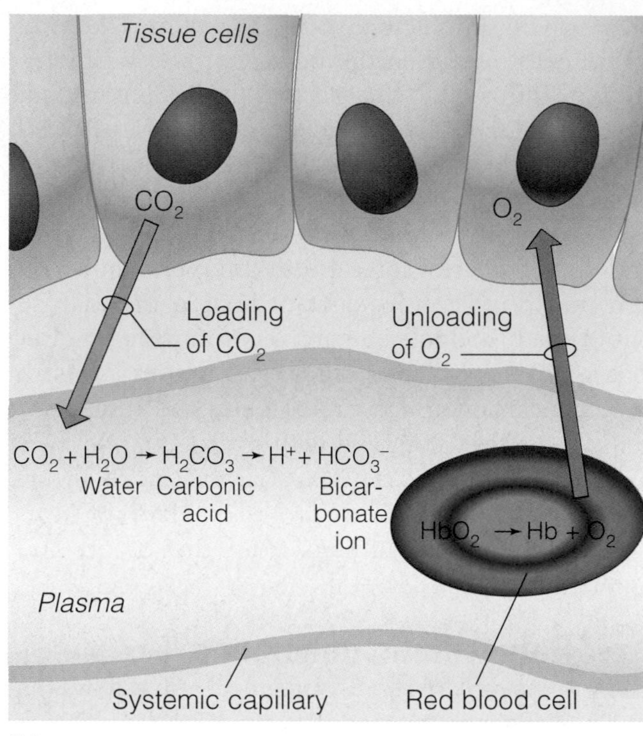

(b)

FIGURE 13.11 Diagrammatic representation of the major means of oxygen (O_2) and carbon dioxide (CO_2) loading and unloading in the body. (a) External respiration in the lungs (pulmonary gas exchange): Oxygen is loaded and carbon dioxide is unloaded. **(b)** Internal respiration in the body tissues (systemic capillary gas exchange): Oxygen is unloaded and carbon dioxide is loaded into the blood. (Note that although the conversion of CO_2 to bicarbonate ion and the reverse reaction are shown occurring in the plasma, most such conversions occur within the red blood cells.) Additionally, though not illustrated, some CO_2 is carried within red blood cells bound to hemoglobin.

RBCs, where a special enzyme (carbonic anhydrase) is available to speed up this reaction. Then the bicarbonate ions diffuse out into plasma, where they are transported. At the same time, oxygen is released from hemoglobin, and the oxygen diffuses quickly out of the blood to enter the tissue cells. As a result of these exchanges, venous blood in the systemic circulation is much poorer in oxygen and richer in carbon dioxide than that leaving the lungs.

Control of Respiration

Neural Regulation: Setting the Basic Rhythm

Although our tidelike breathing seems so beautifully simple, its control is fairly complex. We will cover only the most basic aspects of the respiratory controls. The activity of the respiratory muscles, the diaphragm and external intercostals, is regulated by nerve impulses transmitted to them from the brain by the **phrenic** and **intercostal nerves.**

The neural centers that control respiratory rhythm and depth are located in the *medulla* and *pons* (Figure 13.12). The medulla, which sets the basic rhythm of breathing, contains a **self-exciting inspiratory center,** as well as other respiratory centers. The pons centers appear to smooth out the basic rhythm of inspiration and expiration set by the medulla. Impulses going back and forth between the pons and medulla centers maintain a rate of 12–15 respirations/minute. This normal respiratory rate is referred to as **eupnea** (ūp-ne′ah).

In addition, the bronchioles and alveoli have stretch receptors that respond to extreme overinflation (which might damage the lungs) by initiating protective reflexes. In the case of overinflation, impulses are sent from the stretch receptors to the medulla by the vagus nerves; soon therafter inspiration ends and expiration occurs.

During exercise, we breathe more vigorously and deeply because the brain centers send more impulses to the respiratory muscles. This respiratory pattern is called **hyperpnea** (hy-perp′ne-ah). However, the *rate* of breathing may not be significantly increased with exercising. After strenuous exercise, expiration becomes active, and the abdominal muscles and any other muscles capable of lifting the ribs are used to aid expiration.

Homeostatic Imbalance

If the medullary centers are completely suppressed (as with an overdose of sleeping pills, morphine, or alcohol), respiration stops completely, and death occurs. ▲

Factors Influencing Respiratory Rate and Depth

Physical Factors Although brain centers set the basic rhythm of breathing, there is no question that physical factors such as talking, coughing, and exercising can modify both the rate and depth of breathing. Some of these factors have already been examined in the earlier discussion of nonrespiratory air movements. Increased body temperature causes an increase in the rate of breathing.

Volition (Conscious Control) We all have consciously controlled our breathing pattern at one time or another. During singing and swallowing, breath control is extremely important, and many of us have held our breath for short periods to swim underwater. However, voluntary control of breathing is limited, and the respiratory centers will simply ignore messages from the cortex (our wishes) when the oxygen supply in the blood is getting low or blood pH is falling. All you need do to prove this is to try to talk normally or to hold your breath after running at breakneck speed for a few minutes. It simply cannot be done. Many toddlers try to manipulate their parents by holding their breath "to death." Even though this threat causes many parents to become anxious, they need not worry because the involuntary controls take over and normal respiration begins again.

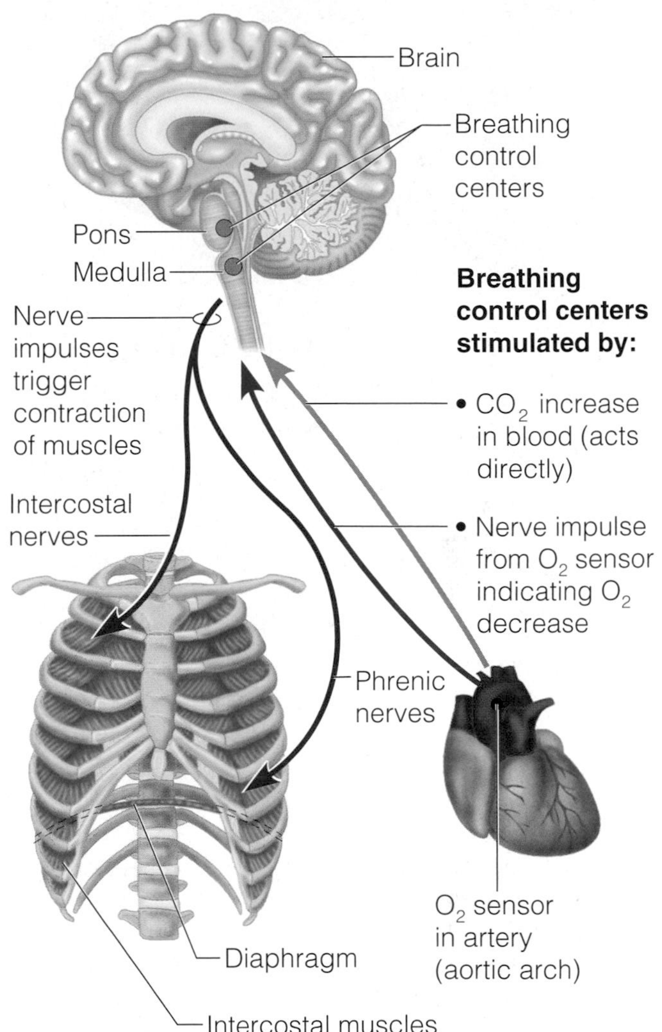

FIGURE 13.12 Breathing control centers, sensory inputs, and effector nerves.

Emotional Factors Emotional factors also modify the rate and depth of breathing. Have you ever watched a horror movie with bated (held) breath or been so scared by what you saw that you were nearly panting? Have you ever touched something cold and clammy and gasped? All of these result from reflexes initiated by emotional stimuli acting through centers in the hypothalamus.

Chemical Factors Although many factors can modify respiratory rate and depth, the most important factors are chemical—the levels of carbon dioxide and oxygen in the blood. Increased levels of carbon dioxide and decreased blood pH are the most important stimuli leading to an increase in

A Closer Look

Lung Cancer: The Facts behind the Smoke Screen

LUNG cancer accounts for fully one-third of all cancer deaths in the United States, and its incidence is increasing daily. It is the most prevalent type of malignancy in both sexes. Most types of lung cancer are tremendously aggressive and metastasize rapidly and widely. Because most cases cannot be diagnosed until they are advanced, it has a notoriously low survival rate: The overall five-year survival rate is just 7 percent, and the average person survives only about nine months after diagnosis. (Only recently have new, whole-lung CT scans been shown to detect early-stage tumors well enough to make screening for this cancer worthwhile.)

For years, Americans were remarkably unaware of the link between lung cancer and cigarette smoking, despite the fact that over 90 percent of lung cancer patients are smokers. Professional athletes still promoted cigarettes in the 1950s, and cigarettes were advertised as a harmless way to keep one's weight down ("Reach for a cigarette rather than a sweet"). As late as the 1960s, smoking was considered socially desirable, even romantic, even though it gives smokers bad breath and yellow teeth, wrinkles their skin, and leaves an unpleasant residue on the clothes and hair of everyone exposed to the smoke. Smoking even a single cigarette increases one's heart rate, constricts peripheral blood vessels throughout the body, disrupts the flow of air in the lungs, and affects one's brain and mood. Long-term smoking contributes to atherosclerosis and heart disease, strokes, cataracts, and early onset of osteoporosis. Furthermore, each year secondhand tobacco smoke causes 3,000 lung cancer deaths among nonsmokers in the United States, and it appears to be linked to heart disease. People who work in bars and restaurants are especially vulnerable.

Ordinarily, sticky mucus and the action of cilia do a fine job of protecting the lungs from chemical and biological irritants, but smoking overwhelms these cleansing devices and they eventually become nonfunctional. Continuous irritation prompts the production of more mucus, but smoking slows the movements of cilia that clear this mucus and depresses the activity of lung macrophages. One result is a pooling of mucus in the lower respiratory tree and an increased frequency of pulmonary infections, including pneumonia and COPD. However, it is the irritating effects of the "cocktail" of free radicals and other carcinogens in tobacco smoke that ultimately lead to lung cancer. The worst of these is a chemical called nitrosamine, but the tars in tobacco also contain carcinogens that eventually cause the epithelial cells lining the bronchial tree to proliferate wildly and lose their normal structure.

The three most common types of lung cancer are (1) *squamous cell carcinoma* (20–40 percent of cases), which arises in the epithelium of the larger bronchi and tends to form masses that cavitate (hollow out) and bleed; (2) *adenocarcinoma* (25–35 percent), which originates in the peripheral areas of the lung as solitary nodules that develop from bronchial glands and alveolar cells; and (3) *small cell carcinoma* (20–25 percent of cases, but rapidly increasing), which contains lymphocyte-like cells that originate in the primary bronchi and grow aggressively in cords or small grapelike clusters within the mediastinum.

The most effective treatment for lung cancer is complete removal of

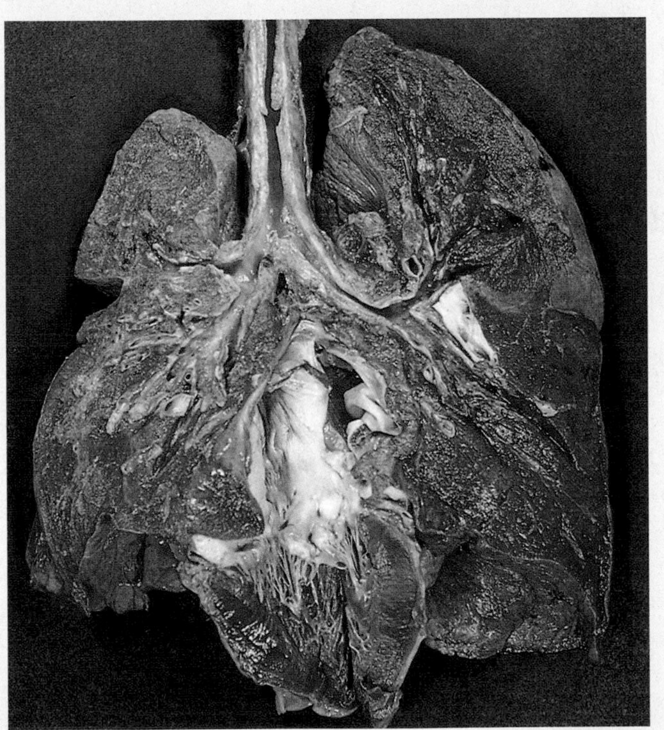

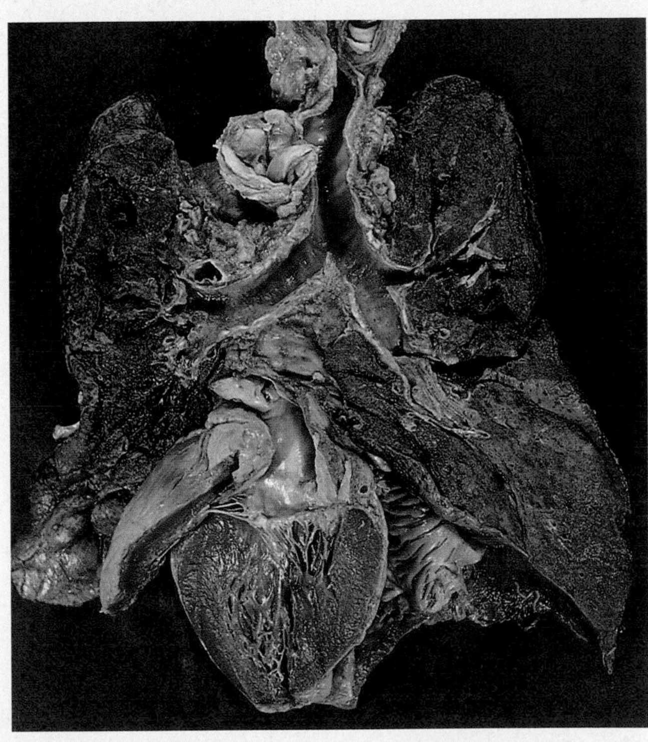

Photographs of healthy lungs of a nonsmoker (left) and cancerous lungs of a smoker (right).

the diseased lung lobes in an attempt to halt metastasis. However, removal is an option open to very few lung cancer patients because metastasis has usually occurred by the time of diagnosis, and most patients' chances of survival are too poor to justify the surgery. In most cases, radiation therapy and chemotherapy are the only options, but most lung cancers are resistant to these treatments. Only small cell carcinoma responds to chemotherapy, but frequently it returns quickly and spawns brain tumors. However, this may change soon. Most lung cancers other than small cell carcinoma result from absence or mutation of the tumor suppressor gene *p53* or from the activity of a specific oncogene. By infusing tumor cells with viruses carrying working *p53* genes or the oncogene inhibitors, an 80 percent cure rate has been achieved in mice.

The extreme addictiveness of nicotine in tobacco may explain why over 50 million Americans smoke cigarettes, even though smoking is estimated to contribute to one-fifth of all deaths in the United States. Surprisingly, a recent survey showed

Lung Cancer: The Facts behind the Smoke Screen *(continued)*

that most smokers do not think they are at increased risk for lung cancer or heart disease. To the contrary, a 40-year lung cancer survey of British men indicates that death rates are three times higher during middle age (35–69 years) for long-term smokers than for nonsmokers. Half of all regular smokers were killed by their habit. Only 43 percent of those who smoked over 25 cigarettes per day lived to age 70, as compared to 79 percent of nonsmokers. These statistics are alarming in light of the fact that more young people are taking up smoking in Great Britain and the United States.

Fortunately, quitting helps: Whereas the incidence of lung cancer is 20:1 for smokers versus nonsmokers, this ratio drops to 2:1 for exsmokers who have not smoked in 15 years. Using such treatments as nicotine patches, nasal sprays, and inhalants has been shown to double the success rate of quitting smoking (from 10–20 percent). Quitting is a valuable goal; consider the alternative!

the rate and depth of breathing. (Actually, an increase in carbon dioxide levels and decreased blood pH are the same thing in this case because carbon dioxide retention leads to increased levels of carbonic acid, which decrease the blood pH.) Changes in carbon dioxide concentrations in the blood seem to act directly on the medulla centers (see Figure 13.12).

Conversely, changes in oxygen concentration in the blood are detected by chemoreceptor regions in the aorta (aortic arch) and carotid artery (carotid body). These, in turn, send impulses to the medulla when blood oxygen levels are dropping. Although every cell in the body must have oxygen to live, it is the body's need to rid itself of carbon dioxide (not to take in oxygen) that is the *most* important stimulus for breathing in a healthy person. Decreases in oxygen levels only become important stimuli when they are dangerously low.

Homeostatic Imbalance

In people who retain carbon dioxide, as in those with chronic lung diseases like emphysema and chronic bronchitis, increased levels of carbon dioxide are no longer recognized as important by the brain, and dropping oxygen levels become the respiratory stimulus. This is an interesting fact because it explains why such patients are always given low levels of oxygen. If they were given high levels, they would stop breathing because their respiratory stimulus (low oxygen levels) would be gone. ▲

Homeostatic mechanisms of the respiratory system, in healthy individuals, are obvious. As carbon dioxide or other sources of acids begin to accumulate in blood and blood pH starts to drop, you begin to breathe more deeply and more rapidly. Notice that this breathing pattern, called **hyperventilation,** is distinct from the hyperpnea of exercise.

This blows off more carbon dioxide and decreases the amount of carbonic acid, which returns blood pH to the normal range. On the other hand, when blood starts to become slightly alkaline, or basic (for whatever reason), breathing slows and becomes more shallow. Slower breathing allows carbon dioxide to accumulate in the blood and brings blood pH back into normal range. Indeed, control of breathing during rest is aimed primarily at regulating the hydrogen ion concentration in the brain. *Hypoventilation* (extremely slow or shallow breathing) or *hyperventilation* can dramatically change the amount of carbonic acid in the blood. Carbonic acid increases dramatically during hypoventilation and decreases substantially during hyperventilation. In both situations, the buffering ability of the blood is likely to be overwhelmed, so that *acidosis* or *alkalosis* occurs.

Homeostatic Imbalance

Hyperventilation, often brought on by anxiety attacks, frequently leads to brief periods of **apnea** (ap′ne-ah), cessation of breathing, until the carbon dioxide builds up in the blood again. If breathing stops for an extended time, **cyanosis** (si″ah-no′sis) may occur due to insufficient oxygen in the blood. In addition, the hyperventilating person may get dizzy and faint because the resulting alkalosis causes cerebral blood vessels to constrict. Such attacks can be prevented by having the hyperventilating individual breathe into a paper bag. Because exhaled air contains more carbon dioxide than atmospheric air, it upsets the normal diffusion gradient that causes carbon dioxide to be unloaded from the blood and leave the body. As a result, carbon dioxide (and thus carbonic acid) levels begin to rise in the blood, ending alkalosis. ▲

Respiratory Disorders

Homeostatic Imbalance

The respiratory system is particularly vulnerable to infections because it is open to airborne pathogens. Because many of these inflammatory conditions, such as rhinitis and tonsillitis, have already been considered, we will turn our attention to the most disabling respiratory disorders, the group of diseases collectively referred to as *chronic obstructive pulmonary disease (COPD)* and *lung cancer.* These disorders are "living proof" of cigarette smoking's devastating effects on the body. Long known to promote cardiovascular disease, cigarettes are perhaps even more effective at destroying the lungs. The photos in the "A Closer Look" box on pp. 442–444, which considers lung cancer, contrast healthy lungs with the blackened, cancer-ridden lungs of a smoker.

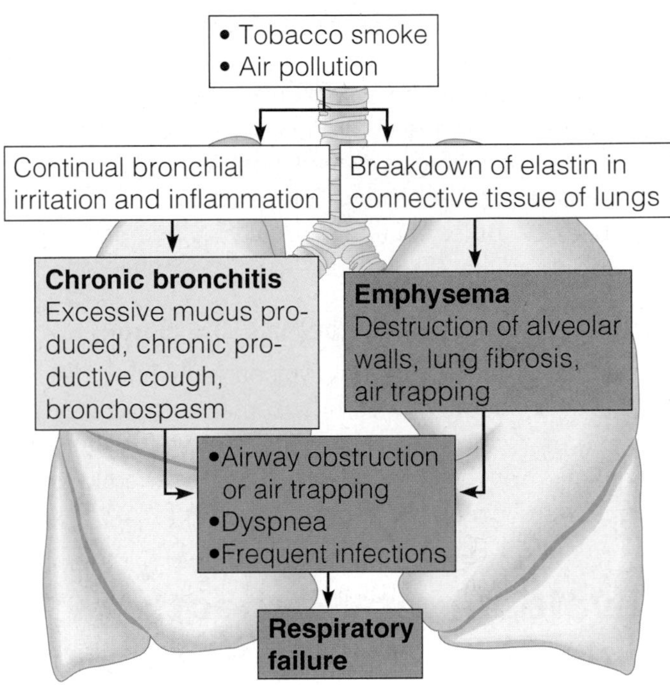

FIGURE 13.13 The pathogenesis of COPD.

Chronic Obstructive Pulmonary Disease (COPD)

The **chronic obstructive pulmonary diseases,** exemplified by *chronic bronchitis* and *emphysema* (em″fĭ-se′mah), are a major cause of death and disability in the United States. These diseases have certain features in common: (1) Patients almost always have a history of smoking; (2) *dyspnea* (disp′ne-ah), difficult or labored breathing, often referred to as "air hunger," occurs and becomes progressively more severe; (3) coughing and frequent pulmonary infections are common; and (4) most COPD victims are hypoxic, retain carbon dioxide and have respiratory acidosis, and ultimately develop respiratory failure (Figure 13.13).

In **chronic bronchitis,** the mucosa of the lower respiratory passages becomes severely inflamed and produces excessive amounts of mucus. The pooled mucus impairs ventilation and gas exchange and dramatically increases the risk of lung infections, including pneumonias. Chronic bronchitis patients are sometimes called "blue bloaters" because hypoxia and carbon dioxide retention occur early in the disease and cyanosis is common.

In **emphysema,** the alveoli enlarge as the walls of adjacent chambers break through, and chronic inflammation promotes fibrosis of the lungs. As the lungs become less elastic, the airways collapse during expiration and obstruct outflow of air. As a result, these patients use an incredible amount of energy to exhale, and they are always exhausted. Because air is retained in the lungs, oxygen exchange is surprisingly efficient, and cyanosis does not usually appear until late in the disease. Consequently, emphysema sufferers are sometimes referred to as "pink puffers." However, overinflation of the lungs leads to a permanently expanded barrel chest. ▲

Developmental Aspects of the Respiratory System

In the fetus, the lungs are filled with fluid, and all respiratory exchanges are made by the placenta. At birth, the fluid-filled pathway is drained, and the respiratory passageways fill with air. The alveoli inflate and begin to function in gas exchange, but the lungs are not fully inflated for two weeks. The success of this change—that is, from nonfunctional to functional respiration—depends on the presence of **surfactant** (sur-fak′tant), a fatty molecule made by the cuboidal alveolar cells (see Figure 13.6). Surfactant lowers the surface tension of the film of water lining each alveolar sac so that the alveoli do not collapse between each breath. Surfactant is not usually present in large enough amounts to accomplish this function until late in pregnancy, that is, between 28 and 30 weeks.

Homeostatic Imbalance

Infants born prematurely (before week 28) or those in which surfactant production is inadequate for other reasons (as in many infants born to diabetic mothers) have *infant respiratory distress syndrome (IRDS).* These infants have dyspnea within a few hours after birth and use tremendous amounts of energy just to keep reinflating their alveoli, which collapse after each breath. Although IRDS still accounts for over 20,000 newborn deaths a year, many of these babies survive now because of the current use of equipment that supplies a positive pressure continuously and keeps the alveoli open and working in gas exchange until adequate amounts of surfactant are produced by the maturing lungs.

Important birth defects of the respiratory system include cleft palate and cystic fibrosis. **Cystic fibrosis (CF),** the most common lethal genetic disease in the United States, strikes 1 out of every 2400 children, and every day two children die of it. CF causes oversecretion of a thick mucus that clogs the respiratory passages and puts the child at risk for fatal respiratory infections. It affects other secretory processes as well. Most importantly, it impairs food digestion by clogging ducts that deliver pancreatic enzymes and bile to the small intestine. Also, sweat glands produce an extremely salty perspiration. At the heart of CF is a faulty gene that codes for the CFTR protein, which works as a chloride (Cl^-) channel to control the flow of Cl^- in and out of cells. In those with the mutated gene, CFTR gets "stuck" in the endoplasmic reticulum and is unable to reach the plasma membrane to perform its normal role. Consequently, less Cl^- is secreted and less water follows, resulting in the thick mucus typical of CF. Conventional therapy for CF is mucus-dissolving drugs, "clapping" the chest to loosen the thick mucus, and antibiotics to prevent infection. ▲

The respiratory rate is highest in newborn infants, about 40 to 80 respirations per minute. It continues to drop through life: in the infant it is around 30/minute, at 5 years it is around 25/minute, and in adults it is 12 to 18/minute. However, the rate often increases again in old age. The lungs continue to mature throughout childhood, and more alveoli are formed until young adulthood. But when smoking is begun during the early teens, the lungs never completely mature, and those additional alveoli are lost forever.

Homeostatic Imbalance

Sudden infant death syndrome (SIDS) claims many newborn infants. Apparently healthy infants stop breathing and die in their sleep, leaving their anguished parents to face charges of child abuse in some cases. Although some cases are believed to be a problem of neural control of respiration or viral infection, more than one-third of SIDS cases appear to result from a heart rhythm abnormality. ▲

Except for sneezes or coughs (responses to irritants), and the occasional common cold that blocks the upper respiratory passageways with mucus, the respiratory system works so efficiently and smoothly that we are not even aware of it. Most

Focus on Careers

Licensed Practical Nurse (LPN)

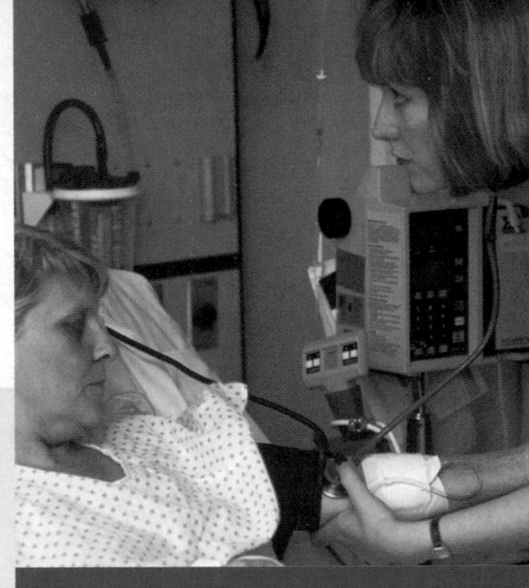

Knowledge of anatomy and physiology helps licensed practical nurses care for their patients.

ORIGINALLY, the term "practical nurse" was used to describe health workers who went into private homes to provide bedside nursing to the ill and injured. Today some licensed practical nurses (LPNs) still provide care in patients' homes, while others play important roles in hospitals, doctors' offices, clinics, and health care administration.

Linda Davila, an LPN employed by Stein Hospice in Sandusky, Ohio, specializes in the challenging task of caring for the terminally ill. "Most of our patients, perhaps 60 percent, have cancer," she says. "Others have life-threatening kidney, lung, or cardiac conditions, or AIDS. Most hospice patients have been diagnosed as having six months or less to live, and they have chosen to forgo aggressive medical treatments and heroic lifesaving measures. My job is to assist them in staying as comfortable and pain-free as possible for the remaining days of their lives and to help maintain the quality of life for them and their families."

Hospice care provides a multidisciplinary approach involving a team of professionals who help patients deal with a variety of physical and emotional issues. Davila's team includes social workers, spiritual advisers, registered nurses (RNs), and other LPNs. Explains Davila, "I provide personal care for our patients by helping them with bathing and grooming and monitoring their medications. I also work closely with their families and teach them how to care for the patient: how to empty catheters, bathe someone in bed, and change the sheets for someone who's bedridden."

Davila's clinical skills and anatomy/physiology training enable her to monitor a patient's condition closely and alert team members to any changes. "Hospice patients take many medications, including narcotics, that can have serious side effects," she notes. "I always monitor patients for side effects and report any symptoms immediately to the RN on my team so we can adjust their medications. The narcotics also slow down the digestive process and can make patients constipated. Based on their symptoms, I assess the degree of constipation and the best way to relieve it."

In cancer patients, Davila can often gauge the extent of the disease and recognize when it spreads. "In class we learned that many types of cancer spread to the bone. Sometimes patients will complain of sudden pain in their ribs. They think it's just a bruise, but it could indicate their cancer has progressed. Another symptom I watch for is a persistent headache, which could signal cancer entering the brain." With lung cancer patients, Davila monitors for shortness of breath, a bluish tinge in skin and fingernails, and mental confusion—signs of reduced oxygen, indicating that the cancer is impairing lung function. She promptly alerts team members so treatment can begin.

Davila's clinical skills and anatomy/physiology training enable her to monitor a patient's condition closely.

LPN training follows a structured curriculum in state-approved schools. Applicants must hold a high school diploma or the equivalent and pass the school's entrance examination. Davila completed an intensive one-year program that included anatomy and physiology as well as mathematics and other general educational requirements. (She notes that the math comes in handy for calculating intravenous medication formulas and drip rates.)

After graduating from training, an LPN must also pass a state licensing examination and earn continuing education credits to maintain and renew licensure. Please note that accreditation procedures vary from state to state. For more information about licensed practical nursing and testing requirements, contact:

National League for Nursing
61 Broadway, 33rd Floor
New York, New York 10006
(800) 669-1656

For additional information on this career, click the Focus on Careers link at www.anatomyandphysiology.com.

Systems in Sync

Homeostatic Relationships between the Respiratory System and Other Body Systems

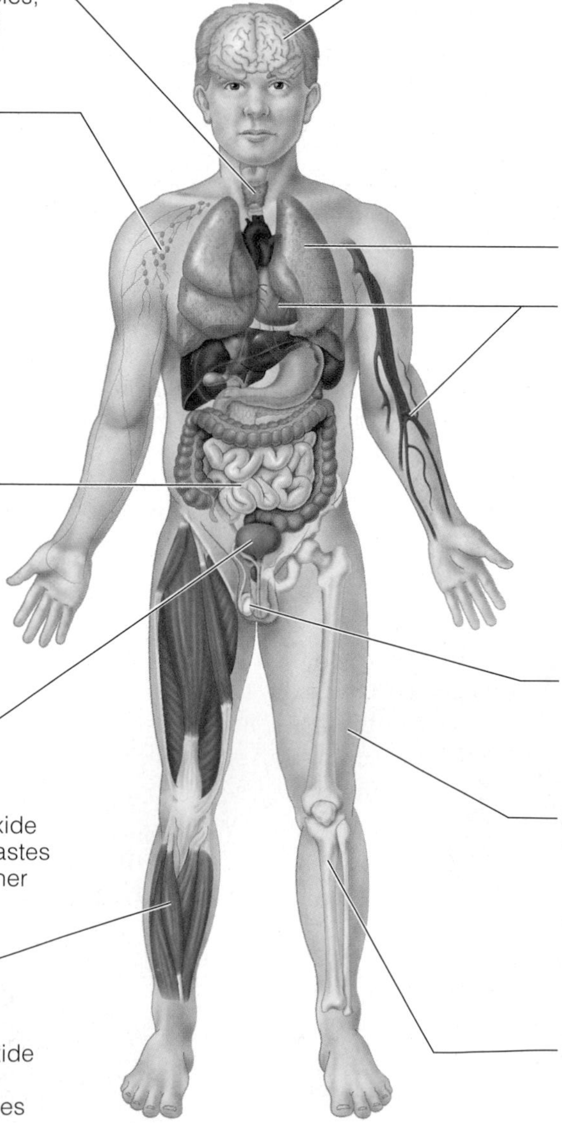

Endocrine System
- Respiratory system provides oxygen; disposes of carbon dioxide
- Epinephrine dilates the bronchioles; testosterone promotes laryngeal enlargement in pubertal males

Lymphatic System/Immunity
- Respiratory system provides oxygen; disposes of carbon dioxide; tonsils in pharynx house immune cells
- Lymphatic system helps to maintain blood volume required for respiratory gas transport; immune system protects respiratory organs from pathogens and cancer

Digestive System
- Respiratory system provides oxygen; disposes of carbon dioxide
- Digestive system provides nutrients needed by respiratory system

Urinary System
- Respiratory system provides oxygen; disposes of carbon dioxide
- Kidneys dispose of metabolic wastes of respiratory system organs (other than carbon dioxide)

Muscular System
- Respiratory system provides oxygen needed for muscle activity; disposes of carbon dioxide
- The diaphragm and intercostal muscles produce volume changes necessary for breathing; regular exercise increases respiratory efficiency

Nervous System
- Respiratory system provides oxygen needed for normal neural activity; disposes of carbon dioxide
- Medullary and pons centers regulate respiratory rate/depth; stretch receptors in lungs and chemoreceptors in large arteries provide feedback

Respiratory System

Cardiovascular System
- Respiratory system provides oxygen; disposes of carbon dioxide; carbon dioxide present in blood as HCO_3^- and H_2CO_3 contributes to blood buffering
- Blood transports respiratory gases

Reproductive System
- Respiratory system provides oxygen; disposes of carbon dioxide

Integumentary System
- Respiratory system provides oxygen; disposes of carbon dioxide
- Skin protects respiratory system organs by forming surface barriers

Skeletal System
- Respiratory system provides oxygen; disposes of carbon dioxide
- Bones enclose and protect lungs and bronchi

problems that occur are a result of external factors—for example, obstruction of the trachea by a piece of food, or aspiration of food particles or vomitus (which leads to aspiration pneumonia). Some unfortunate individuals are plagued by **asthma,** caused by chronically inflamed, hypersensitive bronchial passages that respond to many irritants (such as dust mite and cockroach droppings, dog dander, and fungi) with dyspnea, coughing, and wheezing.

For many years, tuberculosis and pneumonia were the worst killers in the United States. Antibiotics have decreased their lethal threat to a large extent, but they are still dangerous diseases. Newly diagnosed, and frequently drug-resistant, tuberculosis cases in AIDS patients are increasing by leaps and bounds the world over, but *at present* the most damaging and disabling respiratory diseases are still those described above, COPD and lung cancer.

As we age, the chest wall becomes more rigid and the lungs begin to lose their elasticity, resulting in a slowly decreasing ability to ventilate the lungs. Vital capacity decreases by about one-third by the age of 70. In addition, blood oxygen levels decrease and sensitivity to the stimulating effects of carbon dioxide decreases, particularly in a reclining or supine position. As a result, many old people tend to become hypoxic during sleep and exhibit *sleep apnea*.

Additionally, many of the respiratory system's protective mechanisms also become less efficient with age. Ciliary activity of the mucosa decreases, and the phagocytes in the lungs become sluggish. The net result is that the elderly population is more at risk for respiratory tract infections, particularly pneumonia and influenza.

SUMMARY

Media study tools that could provide you with additional help in reviewing specific key topics of Chapter 13 are referenced below.

IP = Interactive Physiology;

WEB = A&P Place website.

FUNCTIONAL ANATOMY OF THE RESPIRATORY SYSTEM (pp. 426–434)

1. The nasal cavity, the chamber within the nose, is divided medially by a nasal septum and separated from the oral cavity by the palate. The nasal cavity is lined with a mucosa, which warms, filters, and moistens incoming air. The mucosa also contains receptors for sense of smell. Paranasal sinuses and nasolacrimal ducts drain into the nasal cavity.
2. The pharynx (throat) is a mucosa-lined, muscular tube with three regions—nasopharynx, oropharynx, and laryngopharynx. The nasopharynx functions in respiration only; the others serve both respiratory and digestive functions. The pharynx contains tonsils, which act as part of the body's defense system.
3. The larynx (voice box) is a cartilaginous structure; most prominent is the thyroid cartilage (Adam's apple). The larynx connects the pharynx with the trachea below. The laryngeal opening (glottis) is hooded by the epiglottis, which prevents entry of food or drink into respiratory passages when swallowing. The larynx contains the vocal folds (true vocal cords), which produce sounds used in speech.
4. The trachea (windpipe) extends from larynx to the main bronchi. The trachea is a smooth-muscle tube lined with a ciliated mucosa and reinforced with C-shaped cartilaginous rings, which keep the trachea patent.
5. Right and left main (primary) bronchi result from subdivision of the trachea. Each plunges into the hilus of the lung on its side.
6. The lungs are paired organs flanking the mediastinum in the thoracic cavity. The lungs are covered with pulmonary (visceral) pleura; the thorax wall is lined with parietal pleura. Pleural secretions decrease friction during breathing. The lungs are primarily elastic tissue and passageways of the respiratory tree. The smallest passageways end in clusters of alveoli.
7. The conducting zone includes all respiratory passages from the nasal cavity to the terminal bronchioles; they conduct air to and from the lungs. Respiratory bronchioles, alveolar ducts and sacs, and alveoli—which have thin walls through which all gas exchanges are made with pulmonary capillary blood—are respiratory zone structures.

WEB Exercise: Chapter 13, Anatomy of the Upper Respiratory Tract.

IP Respiratory System Topic: Anatomy Review: Respiratory Structures, pages 3–12.

RESPIRATORY PHYSIOLOGY (pp. 434–445)

1. Mechanics of breathing: Gas travels from high-pressure to low-pressure areas. Pressure outside the body is atmospheric pressure; pressure inside the lungs is intrapulmonary pressure; pressure in the intrapleural space is intrapleural pressure (which is always negative). Movement of air into and out of the lungs is called pulmonary ventilation, or breathing. When inspiratory muscles contract, intrapulmonary volume increases, its pressure decreases, and air rushes in (inspiration). When inspiratory muscles relax, the lungs recoil and air rushes out (expiration). Expansion of the lungs is helped by cohesion between pleurae and by the presence of surfactant in alveoli.

   IP Respiratory System Topic: Pulmonary Ventilation, pages 3–10.

2. Nonrespiratory air movements: Nonrespiratory air movements are voluntary or reflex activities that move air into or out of the lungs. These include coughing, sneezing, laughing, crying, hiccuping, yawning.

3. Respiratory volumes and capacities: Air volumes exchanged during breathing are TV, IRV, ERV, and VC (see p. 436–437 for values). Residual volume is nonexchangeable respiratory volume and allows gas exchange to go on continually.

4. Respiratory sounds: Bronchial sounds are sounds of air passing through large respiratory passageways. Vesicular breathing sounds occur as air fills alveoli.

5. External respiration, gas transport, and internal respiration: Gases move according to the laws of diffusion. Oxygen moves from alveolar air into pulmonary blood. Most oxygen is transported bound to hemoglobin inside RBCs. Carbon dioxide moves from pulmonary blood into alveolar air. Most carbon dioxide is transported as bicarbonate ion in plasma. At body tissues, oxygen moves from blood to the tissues, whereas carbon dioxide moves from the tissues to blood.

   WEB Exercise: Chapter 13, Gas Transport.

   IP Respiratory System Topic: Gas Transport, pages 3–12.

6. Control of respiration
   a. Nervous control: Neural centers for control of respiratory rhythm are in the medulla and pons. Reflex arcs initiated by stretch receptors in the lungs also play a role in respiration by notifying neural centers of excessive overinflation.
   b. Physical factors: Increased body temperature, exercise, speech, singing, and nonrespiratory air movements modify both rate and depth of breathing.
   c. Volition: To a degree, breathing may be consciously controlled if it does not interfere with homeostasis.
   d. Emotional factors: Some emotional stimuli can modify breathing. Examples are fear, anger, and excitement.
   e. Chemical factors: Changes in blood levels of carbon dioxide are the most important stimuli affecting respiratory rhythm and depth. Carbon dioxide acts directly on the medulla via its effect on reducing blood pH. Rising levels of carbon dioxide in blood result in faster, deeper breathing; falling levels lead to shallow, slow breathing. Hyperventilation may result in apnea and dizziness, due to alkalosis. Oxygen is less important as a respiratory stimulus in normal, healthy people, but it *is* the stimulus for those whose systems have become accustomed to high levels of carbon dioxide as a result of disease.

RESPIRATORY DISORDERS (pp. 445–446)

1. The major respiratory disorders are COPD (emphysema and chronic bronchitis) and lung cancer. A significant cause is cigarette smoking.

2. Emphysema is characterized by permanent enlargement and destruction of alveoli. The lungs lose their elasticity, and expiration becomes an active process.

3. Chronic bronchitis is characterized by excessive mucus production and its pooling in lower respiratory passageways, which severely impairs ventilation and gas exchange. Patients may become cyanotic as a result of chronic hypoxia.

4. Lung cancer is extremely aggressive and metastasizes rapidly. The three most common lung cancers are squamous cell carcinoma, adenocarcinoma, and small cell carcinoma.

DEVELOPMENTAL ASPECTS OF THE RESPIRATORY SYSTEM (pp. 446–449)

1. Premature infants have problems keeping their lungs inflated due to lack of surfactant in their alveoli. (Surfactant is formed late in pregnancy.)

2. The most important birth defects of the respiratory system are cleft palate and cystic fibrosis.

3. The lungs continue to mature until young adulthood.
4. During youth and middle age, most respiratory system problems are a result of external factors, such as infections and substances that physically block respiratory passageways.
5. In old age, the thorax becomes more rigid and lungs become less elastic, leading to decreased vital capacity. Protective mechanisms of the respiratory system decrease in effectiveness in elderly persons, predisposing them to more respiratory tract infections.

REVIEW QUESTIONS

MULTIPLE CHOICE

More than one choice may apply.

1. When you exhale, air flows through respiratory structures in which sequence?
   a. Alveolus, bronchiole, bronchus, larynx, trachea, pharynx, nasal cavity
   b. Alveolus, trachea, bronchus, bronchiole, larynx, pharynx, nasal cavity
   c. Alveolus, bronchus, bronchiole, trachea, larynx, pharynx, nasal cavity
   d. Alveolus, bronchiole, bronchus, trachea, larynx, pharynx, nasal cavity
2. When you inhale, the diaphragm:
   a. relaxes and moves inferiorly
   b. relaxes and moves superiorly
   c. contracts and moves superiorly
   d. contracts and moves inferiorly
3. During inspiration, intrapulmonary pressure is:
   a. greater than atmospheric pressure
   b. less than atmospheric pressure
   c. greater than intrapleural pressure
   d. less than intrapleural pressure
4. Lung collapse is prevented by:
   a. high surface tension of alveolar fluid
   b. adhesion of the pleural membranes
   c. high pressure in the pleural cavities
   d. high elasticity of lung tissue
5. Disorders classified as COPDs include:
   a. pneumonia
   b. emphysema
   c. bronchitis
   d. sleep apnea
6. Which of the following changes will accompany the loss of lung elasticity associated with aging?
   a. Increase in tidal volume
   b. Increase in inspiratory reserve volume
   c. Increase in residual volume
   d. Increase in vital capacity
7. Which of the following is not part of the conducting zone of the respiratory system?
   a. Pharynx
   b. Alveolar sac
   c. Trachea
   d. Secondary bronchioles
   e. Larynx

SHORT ANSWER ESSAY

1. What is the most basic function of respiration?
2. Clearly explain the difference between external and internal respiration.
3. Trace the route of air from the external nares to an alveolus.
4. Why is it important that the trachea is reinforced with cartilaginous rings? What is the advantage of the fact that the rings are incomplete posteriorly?
5. Where in the respiratory tract is the air filtered, warmed, and moistened?
6. The trachea has cilia and goblet cells that produce mucus. What is the specific protective function of each of these?
7. Which primary bronchus is the more likely site for an inspired object to become lodged? Why?
8. In terms of general health, what is the importance of the fact that the pharyngotympanic tubes and the sinuses drain into the nasal cavities and nasopharynx?
9. The lungs are mostly passageways and elastic tissue. What is the role of the passageways? of the elastic tissue?
10. What is it about the structure of the alveoli that makes them an ideal site for gas exchange?
11. What do TV, ERV, and VC mean? Which of these values is the largest? Why?
12. Name several nonrespiratory air movements, and explain how each differs from normal breathing.
13. The contraction of the diaphragm and the external intercostal muscles begins inspiration. Explain exactly what happens, in terms of volume and pressure changes in the lungs, when these muscles contract.
14. What causes air to flow out of the lungs during expiration?

15. What is the major way that oxygen is transported in the blood? in carbon dioxide?

16. What determines in which direction carbon dioxide and oxygen will diffuse in the lungs? in the tissues?

17. Name the two major brain areas involved in the nervous control of breathing.

18. Name three physical factors that can modify respiratory rate or depth.

19. Name two chemical factors that modify respiratory rate and depth. Which is usually more important?

20. Define *hyperventilation*. If you hyperventilate, do you retain or expel more carbon dioxide? What effect does hyperventilation have on blood pH?

21. Compare and contrast the signs and symptoms of emphysema and chronic bronchitis.

22. What would happen to the alveoli if surfactant was not produced?

AT THE CLINIC

1. After putting her 1-year-old boy (who puts virtually everything in his mouth) down for a nap, a mother failed to find one of the larger beads she used to make the custom jewelry she produces for sale. Two days later, the boy developed a cough and became feverish. What is likely to have happened to the bead, and where (anatomically) would you expect it to be found?

2. Why doesn't Mom have to worry when 3-year-old Johnny threatens to "hold his breath till he dies"?

3. Alvin, a smoker, sees his doctor because he has a persistent cough and becomes short of breath after very little exertion. He has a barrel chest and a red face and explains that it is difficult for him to exhale but not to inhale. What diagnosis will the doctor make?

4. Mr. Rasputin bumped a bee's nest while making repairs on his roof. Not surprisingly, he was promptly stung several times. Since he knew he was allergic to bee stings, he rushed to the hospital. While waiting, he went into a state of shock and had extreme difficulty breathing. Examination showed his larynx to be edematous, and a tracheostomy was performed. Why is edema of the larynx likely to obstruct the airway? What is a tracheostomy, and what purpose does it serve?

5. As a result of a stroke, Mrs. Minnick's swallowing is uncoordinated. What detrimental effect might this have on her ability to breathe?

6. Aspirin is an acidic substance. Describe the pathway by which an aspirin overdose causes an increase in respiratory rate, and explain how this increase would be helpful to the victim.

7. Nine-year-old Jerry stumbled into the drug store gasping for breath. Blood was oozing from a small hole in his chest wall. When the paramedics arrived they said that Jerry had suffered a pneumothorax and atelectasis. Just what do both of these terms mean, and how do you explain his respiratory distress. How will it be treated?

14

The Digestive System and Body Metabolism

YOUR GOALS

After completing this chapter, you will have a working knowledge of the functions of the digestive system and body metabolism and will have mastered the objectives listed below.

FUNCTION PREVIEW

- The digestive system breaks down ingested food into particles small enough to be absorbed into the blood.
- Metabolism produces cellular energy (ATP) and accounts for all constructive and degradative cellular activities.

OBJECTIVE CHECKLIST

PART I: ANATOMY AND PHYSIOLOGY OF THE DIGESTIVE SYSTEM

ANATOMY OF THE DIGESTIVE SYSTEM (pp. 454–467)

- Name the organs of the alimentary canal and accessory digestive organs, and identify each on an appropriate diagram or model.
- Identify the overall function of the digestive system as digestion and absorption of foodstuffs, and describe the general activities of each digestive system organ.
- Describe the composition and function(s) of saliva.
- Name the deciduous and permanent teeth, and describe the basic anatomy of a tooth.
- Explain how villi aid digestive processes in the small intestine.

FUNCTIONS OF THE DIGESTIVE SYSTEM (pp. 467–478)

- Describe the mechanisms of swallowing, vomiting, and defecation.
- Describe how foodstuffs in the digestive tract are mixed and moved along the tract.
- Describe the function of local hormones in the digestive process.

>>

(Objective Checklist, continued)

- List the major enzymes or enzyme groups produced by the digestive organs or accessory glands, and name the foodstuffs on which they act.
- Name the end products of protein, fat, and carbohydrate digestion.
- State the function of bile in the digestive process.

PART II: NUTRITION AND METABOLISM

NUTRITION (pp. 478–480)

- Define *nutrient* and *calorie.*
- List the six major nutrient categories. Note important dietary sources and the principal cellular uses of each.

METABOLISM (pp. 480–490)

- Define *enzyme, metabolism, anabolism,* and *catabolism.*
- Describe the metabolic roles of the liver.
- Recognize the sources of carbohydrates, fats, and proteins and their uses in cell metabolism.
- Explain the importance of energy balance in the body, and indicate consequences of energy imbalance.
- List several factors that influence metabolic rate, and indicate the effect of each.
- Describe how body temperature is regulated.

PART III: DEVELOPMENTAL ASPECTS OF THE DIGESTIVE SYSTEM AND METABOLISM (pp. 490–491)

- Name important congenital disorders of the digestive system and significant inborn errors of metabolism.
- Describe the effect of aging on the digestive system.

Children have a special fascination with the workings of the digestive system: They relish crunching a potato chip, delight in making "mustaches" with milk, and giggle when their stomach "growls." As adults, we know that a healthy digestive system is essential for good health because it converts food into the raw materials that build and fuel our body's cells. Specifically, the digestive system takes in food (*ingests* it), breaks it down physically and chemically into nutrient molecules (*digests* it), and *absorbs* the nutrients into the bloodstream. Then it rids the body of the indigestible remains *(defecates).*

PART I: ANATOMY AND PHYSIOLOGY OF THE DIGESTIVE SYSTEM

Anatomy of the Digestive System

The organs of the digestive system can be separated into two main groups: those forming the *alimentary* (al″ĕ-men′tar-e; *aliment* = nourish) *canal,*

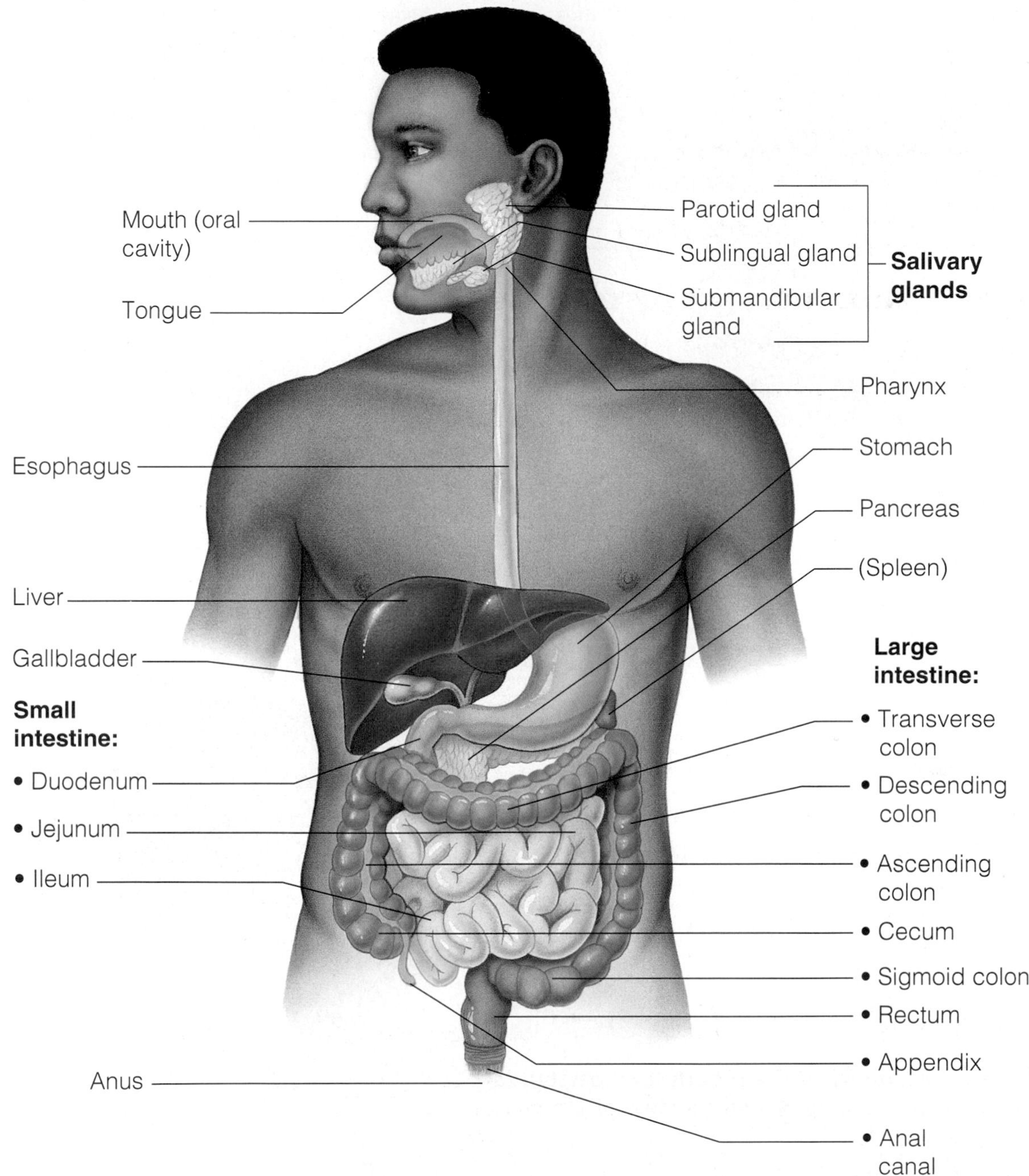

FIGURE 14.1 The human digestive system: Alimentary canal and accessory organs. (Liver and gallbladder are reflected superiorly and to the right side of the body.)

and the *accessory digestive organs* (see Figure 14.1). The alimentary canal performs the whole menu of digestive functions (ingests, digests, absorbs, and defecates). The accessory organs (teeth, tongue, and several large digestive glands) assist the process of digestive breakdown in various ways.

Organs of the Alimentary Canal

The **alimentary canal,** also called the **gastrointestinal (GI) tract,** is a continuous, coiled, hollow, muscular tube that winds through the ventral body cavity and is open at both ends. Its organs are the

Q *What is the protective value of having several sets of tonsils at the oral entrance to the pharynx?*

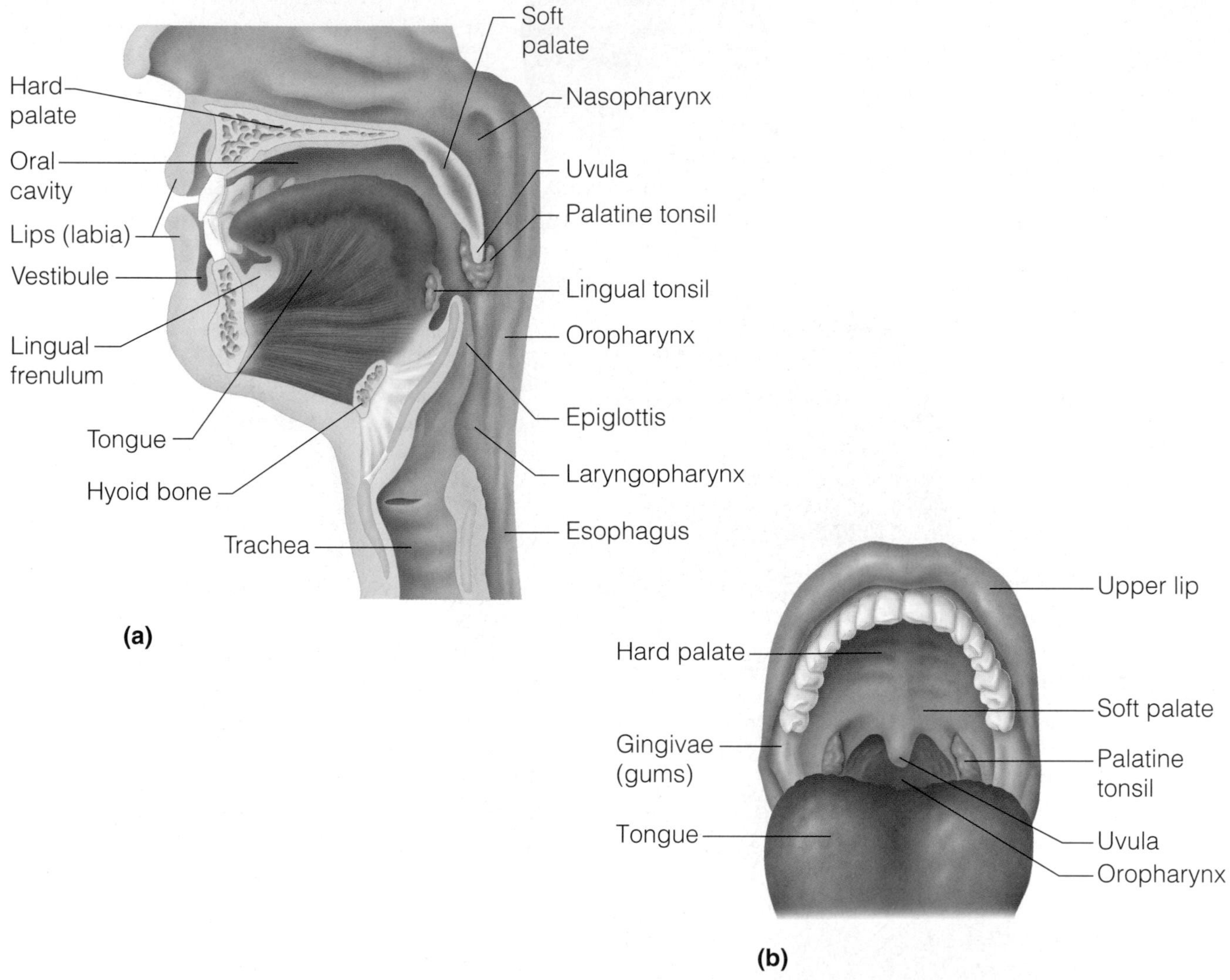

FIGURE 14.2 Anatomy of the mouth (oral cavity). (a) Sagittal view of the oral cavity and pharynx. **(b)** Anterior view of the oral cavity.

mouth, pharynx, esophagus, stomach, small intestine, and *large intestine.* The large intestine leads to the terminal opening, or *anus.* In a cadaver, the alimentary canal is approximately 9 m (about 30 feet) long, but in a living person, it is considerably shorter because of its relatively constant muscle tone. Food material within this tube is technically outside the body, because it has contact only with cells lining the tract and the tube is open to the external environment at both ends. As each organ of the alimentary canal is described next, find it in Figure 14.1.

A *The mouth is a favored site of body entry by bacteria, and the presence of the tonsils (lymphocyte- and macrophage-filled organs) is very effective in preventing many pathogens from getting further into the digestive tract.*

Mouth

Food enters the digestive tract through the **mouth,** or **oral cavity,** a mucous membrane–lined cavity (Figure 14.2). The **lips (labia)** protect its anterior opening, the **cheeks** form its lateral walls, the **hard palate** forms its anterior roof, and the **soft**

palate forms its posterior roof. The **uvula** (u′vu-lah) is a fleshy fingerlike projection of the soft palate, which extends downward from its posterior edge. The space between the lips and cheeks externally and the teeth and gums internally is the **vestibule.** The area contained by the teeth is the **oral cavity proper.** The muscular **tongue** occupies the floor of the mouth. The tongue has several bony attachments—two of these are to the hyoid bone and the styloid processes of the skull. The **lingual frenulum** (ling′gwal fren′u-lum), a fold of mucous membrane, secures the tongue to the floor of the mouth and limits its posterior movements (see Figure 14.2a).

Homeostatic Imbalance

Children born with an extremely short frenulum are often referred to as "tongue-tied" because distorted speech results when tongue movement is restricted. This congenital condition can be corrected surgically by cutting the frenulum. ▲

At the posterior end of the oral cavity are paired masses of lymphatic tissue, the **palatine tonsils.** The **lingual tonsil** covers the base of the tongue just beyond. The tonsils, along with other lymphatic tissues, are part of the body's defense system. When the tonsils become inflamed and enlarge, they partially block the entrance into the throat (pharynx), making swallowing difficult and painful.

As food enters the mouth, it is mixed with saliva and **masticated** (chewed). The cheeks and closed lips hold the food between the teeth during chewing. The nimble tongue continually mixes food with saliva during chewing and initiates swallowing. Thus, the breakdown of food begins before the food has even left the mouth. As noted in Chapter 8, *papillae* containing taste buds, or taste receptors, are found on the tongue surface. And so, besides its food-manipulating function, the tongue allows us to enjoy and appreciate the food as it is eaten.

Pharynx

From the mouth, food passes posteriorly into the *oropharynx* and *laryngopharynx,* both of which are common passageways for food, fluids, and air. As explained in Chapter 13, the pharynx is subdivided into the *nasopharynx,* part of the respiratory passageway; the **oropharynx,** posterior to the oral cavity; and the **laryngopharynx,** which is continuous with the esophagus below.

The walls of the pharynx contain two skeletal muscle layers. The cells of the inner layer run longitudinally; those of the outer layer (the constrictor muscles) run around the wall in a circular fashion. Alternating contractions of these two muscle layers propel food through the pharynx into the esophagus below. This propelling mechanism, called *peristalsis* (per″i-stal′sis), is described later.

Esophagus

The **esophagus** (ĕ-sof′ah-gus), or *gullet,* runs from the pharynx through the diaphragm to the stomach. About 25 cm (10 inches) long, it is essentially a passageway that conducts food (by peristalsis) to the stomach.

The walls of the alimentary canal organs from the esophagus to the large intestine are made up of the same four basic tissue layers, or tunics (Figure 14.3):

1. The **mucosa** is the innermost layer, a moist membrane that lines the cavity, or **lumen,** of the organ. It consists primarily of a *surface epithelium,* plus a small amount of connective tissue *(lamina propria)* and a scanty *smooth muscle layer.* Beyond the esophagus, which has a friction-resisting stratified squamous epithelium, the epithelium is mostly simple columnar.
2. The **submucosa** is found just beneath the mucosa. It is a soft connective tissue layer containing blood vessels, nerve endings, lymph nodules, and lymphatic vessels.
3. The **muscularis externa** is a muscle layer typically made up of an inner *circular layer* and an outer *longitudinal layer* of smooth muscle cells.
4. The **serosa** is the outermost layer of the wall. It consists of a single layer of flat serous fluid-producing cells, the **visceral peritoneum** (per″i-to-ne′um). The visceral peritoneum is continuous with the slick, slippery **parietal peritoneum,** which lines the abdominopelvic cavity by way of a membrane extension, the **mesentery** (mes′en-ter″e). These relationships are illustrated in Figure 14.5.

The alimentary canal wall contains two important *intrinsic nerve plexuses*—the **submucosal nerve**

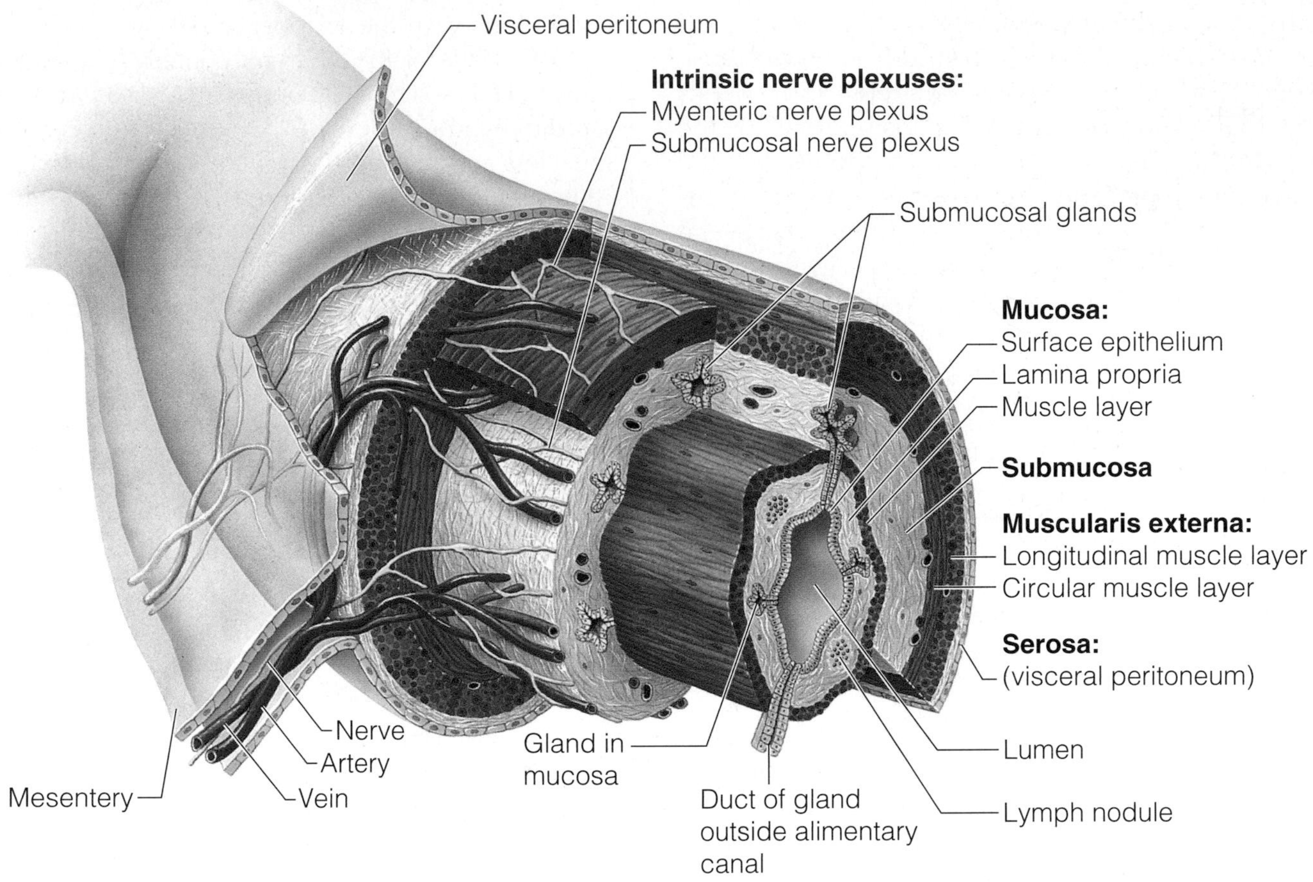

FIGURE 14.3 Basic structure of the alimentary canal wall.

plexus and the **myenteric** (mi-en′ter-ik; "intestinal muscle") **nerve plexus.** An additional small *subserous plexus* is associated with the serosa. These networks of nerve fibers are actually part of the autonomic nervous system. They help regulate the mobility and secretory activity of GI tract organs.

Stomach

The C-shaped **stomach** (Figure 14.4) is on the left side of the abdominal cavity, nearly hidden by the liver and diaphragm. Different regions of the stomach have been named. The *cardiac region* (named for its position near the heart) surrounds the **cardioesophageal** (kar″de-o-ĕ-sof″ah-je′al) **sphincter,** through which food enters the stomach from the esophagus. The *fundus* is the expanded part of the stomach lateral to the cardiac region. The *body* is the midportion, and as it narrows inferiorly, it becomes the *pyloric antrum,* and then the funnel-shaped *pylorus* (pi-lo′rus), the terminal part of the stomach. The pylorus is continuous with the small intestine through the **pyloric sphincter,** or **valve.** The stomach is approximately 25 cm (10 inches) long, but its diameter depends on how much food it contains. When it is full, it can hold about 4 liters (1 gallon) of food. When it is empty, it collapses inward on itself, and its mucosa is thrown into large folds called **rugae** (roo′ge; *ruga* = wrinkle, fold). The convex lateral surface of the stomach is the **greater curvature;** its concave medial surface is the **lesser curvature.**

The **lesser omentum** (o-men′tum), a double layer of peritoneum, extends from the liver to the lesser curvature. The **greater omentum,** another extension of the peritoneum, drapes downward and covers the abdominal organs like a lacy apron before attaching to the posterior body wall (Figure 14.5). The greater omentum is riddled with

What modification of the stomach muscularis externa allows it to mechanically digest food?

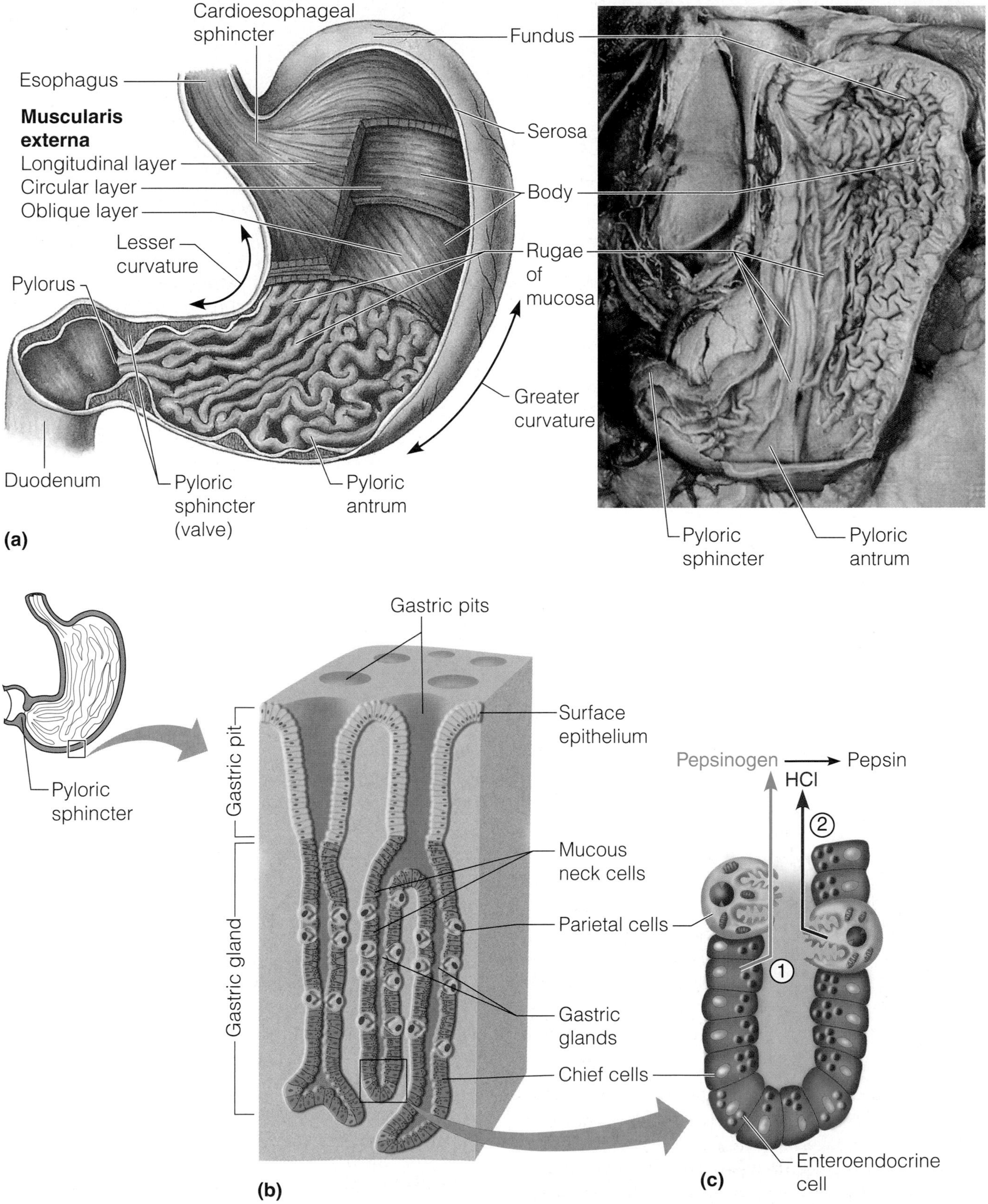

FIGURE 14.4 Anatomy of the stomach. (a) Gross internal anatomy (frontal section). **(b)** Enlarged view of gastric pits and glands (longitudinal section). **(c)** Schematic showing the sequence of events from ① the production of pepsinogen by the chief cells to ② its activation (to pepsin) by HCl secreted by the parietal cells.

A *Its third layer of smooth muscle, the oblique layer which allows it to knead or pummel the food.*

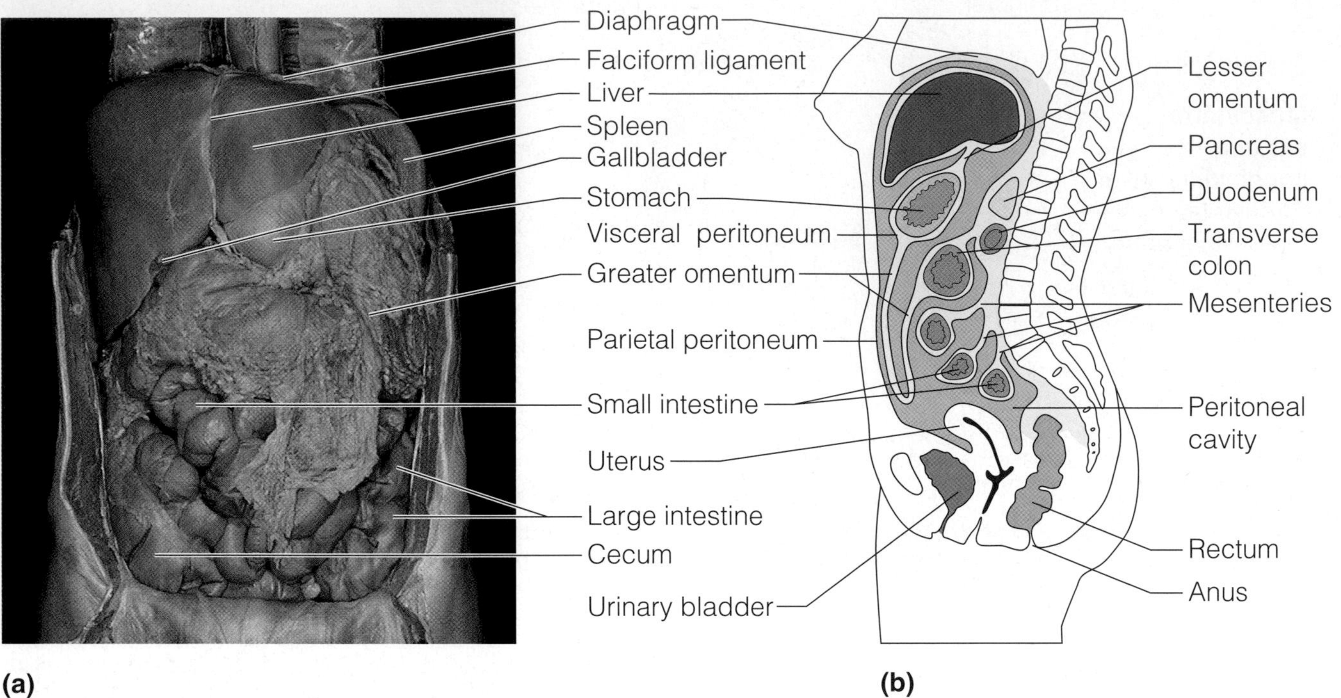

FIGURE 14.5 Peritoneal attachments of the abdominal organs. **(a)** Anterior view; the greater omentum is shown in its normal position, covering the abdominal viscera. **(b)** Sagittal view of the abdominopelvic cavity of a female.

fat, which helps to insulate, cushion, and protect the abdominal organs, and has large collections of lymph nodules containing macrophages and defensive cells of the immune system.

Homeostatic Imbalance

When the peritoneum is infected, a condition called *peritonitis* (per″i-to-ni′tis), the peritoneal membranes tend to stick together around the infection site. This helps to seal off and localize many intraperitoneal infections (at least initially), providing time for macrophages in the lymphatic tissue to mount an attack. ▲

The stomach acts as a temporary "storage tank" for food as well as a site for food breakdown. Besides the usual longitudinal and circular muscle layers, its wall contains a third obliquely arranged layer in the *muscularis externa* (see Figure 14.4a). This arrangement allows the stomach not only to move food along the tract, but also to churn, mix, and pummel the food, physically breaking it down to smaller fragments. In addition, chemical breakdown of proteins begins in the stomach. The mucosa of the stomach is a simple columnar epithelium that produces large amounts of mucus. This otherwise smooth lining is dotted with millions of deep *gastric pits,* which lead into *gastric glands* (Figure 14.4b) that secrete the solution called **gastric juice.** For example, some stomach cells produce *intrinsic factor,* a substance needed for the absorption of vitamin B_{12} from the small intestine. The **chief cells** produce protein-digesting enzymes, mostly **pepsinogens,** and the **parietal cells** produce corrosive hydrochloric acid, which makes the stomach contents acidic and activates the enzymes (Figure 14.4c). The *mucous neck cells* produce a sticky alkaline mucus, which clings to the stomach mucosa and protects the stomach wall itself from being damaged by the acid and digested by the enzymes. Still other cells, the **enteroendocrine cells** (*entero* = gut), produce local hormones, such as *gastrin,* that are important to the digestive activities of the stomach (see Table 14.1 on p. 475).

Most digestive activity occurs in the pyloric region of the stomach. After food has been processed

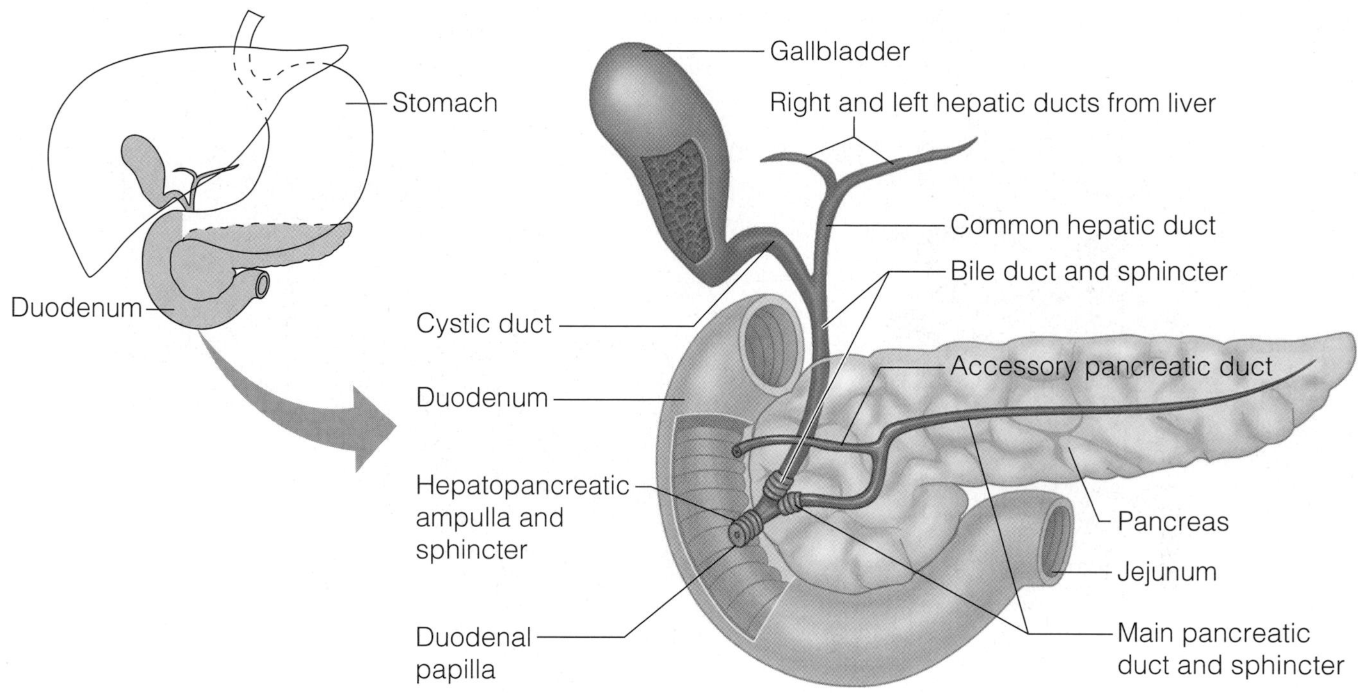

FIGURE 14.6 The duodenum of the small intestine and related organs.

in the stomach, it resembles heavy cream and is called **chyme** (kīm). The chyme enters the small intestine through the pyloric sphincter.

Small Intestine

The **small intestine** is the body's major digestive organ. Within its twisted passageways, usable food is finally prepared for its journey into the cells of the body. The small intestine is a muscular tube extending from the pyloric sphincter to the **ileocecal** (il″e-o-se′kal) **valve** (see Figure 14.8, p. 463). It is the longest section of the alimentary tube, with an average length of 2.5 to 7 m (8–18 feet) in a living person. Except for the initial part of the small intestine (the duodenum), which mostly lies in a retroperitoneal position, the small intestine hangs in sausagelike coils in the abdominal cavity, suspended from the posterior abdominal wall by the fan-shaped mesentery (see Figure 14.5). The large intestine encircles and frames it in the abdominal cavity.

The small intestine has three subdivisions: the **duodenum** (du″o-de′num; "twelve finger widths long"), the **jejunum** (je-joo′num: "empty"), and the **ileum** (il′e-um; "twisted intestine"), which contribute 5 percent, nearly 40 percent, and almost 60 percent of the length of the small intestine, respectively (see Figure 14.1). The ileum joins the large intestine at the ileocecal valve.

Chemical digestion of foods begins in earnest in the small intestine. The small intestine is able to process only a small amount of food at one time. The *pyloric sphincter* (literally, "gatekeeper") controls food movement into the small intestine from the stomach and prevents the small intestine from being overwhelmed. Though the C-shaped duodenum is the shortest subdivision of the small intestine, it has the most interesting features. Some enzymes are produced by the intestinal cells. More important are enzymes produced by the pancreas which are ducted into the duodenum through the **pancreatic ducts,** where they complete the chemical breakdown of foods in the small intestine. *Bile* (formed by the liver) also enters the duodenum through the **bile duct** in the same area (Figure 14.6). The main pancreatic and bile ducts join at the duodenum to form the flasklike *hepatopancreatic ampulla* (he-pah″to-pan-kre-a′tik am-pu′lah), literally, the "liver-pancreatic enlargement." From there, the bile and pancreatic juice travel through the *duodenal papilla* and enter the duodenum together.

Q *What is the functional value of the microvilli in the absorptive cells of the small intestine?*

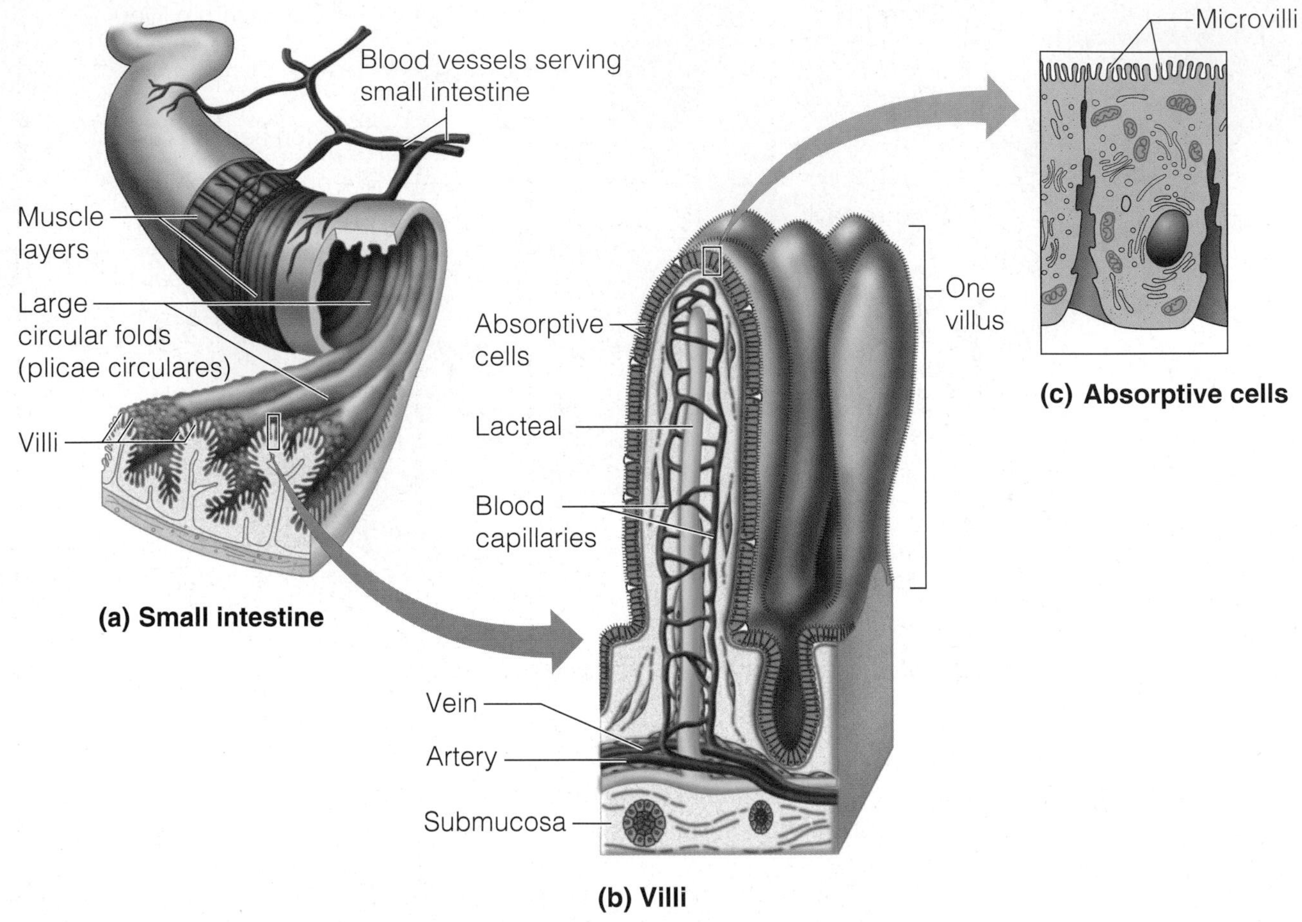

FIGURE 14.7 Structural modifications of the small intestine. **(a)** Several circular folds (plicae circulares), seen on the inner surface of the small intestine. **(b)** Enlargement of one villus extension of the circular fold. **(c)** Enlargement of an absorptive cell to show microvilli (brush border).

Nearly all food absorption occurs in the small intestine. The small intestine is well suited for its function. Its wall has three structures that increase the absorptive surface tremendously—microvilli, villi, and circular folds (Figure 14.7). **Microvilli** (mi″kro-vih′lī) are tiny projections of the plasma membrane of the mucosa cells that give the cell surface a fuzzy appearance, sometimes referred to as the **brush border. Villi** are fingerlike projections of the mucosa that give it a velvety appearance and feel, much like the soft nap of a Turkish towel. Within each villus is a rich capillary bed and a modified lymphatic capillary called a **lacteal.** The digested foodstuffs are absorbed through the mucosal cells into both the capillaries and the lacteal, as discussed on pp. 476–477. **Circular folds,** also called **plicae circulares** (pli′se ser-ku-la′res), are deep folds of both mucosa and submucosa layers. Unlike the rugae of the stomach, the circular folds do not disappear when food fills the small intestine. All these structural modifications, which increase the surface area, decrease in number toward the end of the small intestine. On the other hand, local collections of lymphatic tissue (called **Peyer's patches**) found in the submucosa increase in number toward the end of the small intestine. This reflects the fact that the remaining

A *They tremendously increase the surface area available for absorption of digested foodstuffs.*

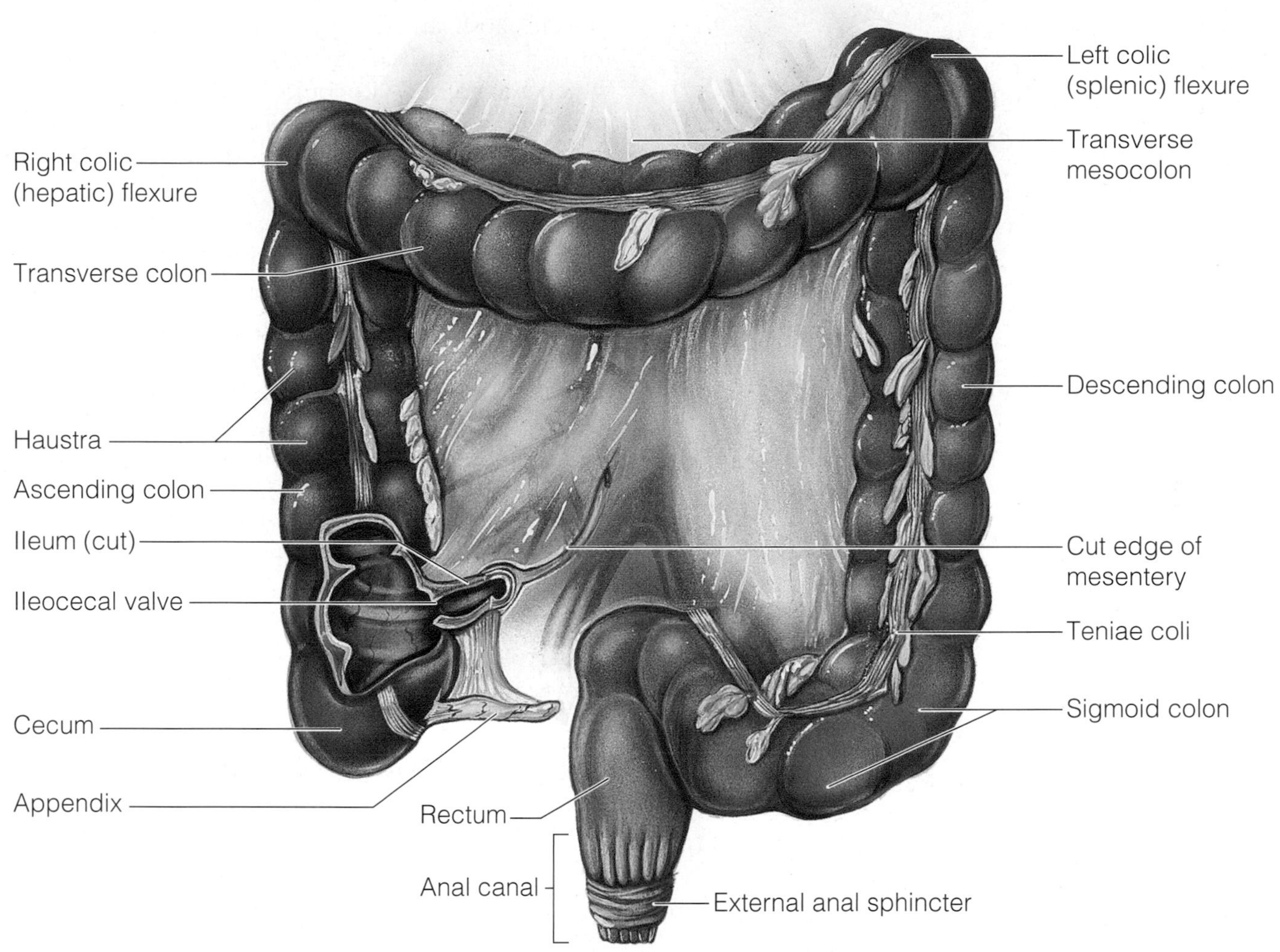

FIGURE 14.8 The large intestine. A section of the cecum is removed to show the ileocecal valve.

(undigested) food residue in the intestine contains huge numbers of bacteria, which must be prevented from entering the bloodstream if at all possible.

Large Intestine

The **large intestine** is much larger in diameter than the small intestine (thus its name, the *large* intestine) but shorter in length. About 1.5 m (5 feet) long, it extends from the ileocecal valve to the anus (Figure 14.8). Its major functions are to dry out the indigestible food residue by absorbing water and to eliminate these residues from the body as feces. It frames the small intestine on three sides and has the following subdivisions: **cecum** (se′kum), **appendix, colon, rectum,** and **anal canal.** The saclike cecum is the first part of the large intestine. Hanging from the cecum is the wormlike ("vermiform") appendix, a potential trouble spot. Since it is usually twisted, it is an ideal location for bacteria to accumulate and multiply. Inflammation of the appendix, *appendicitis,* is the usual result. The colon is divided into several distinct regions. The **ascending colon** travels up the right side of the abdominal cavity and makes a turn, the *right colic* (or *hepatic*) *flexure,* to travel across the abdominal cavity as the **transverse colon.** It then turns again at the *left colic* (or *splenic*) *flexure,* and continues down the left side as the **descending colon,** to enter the pelvis,

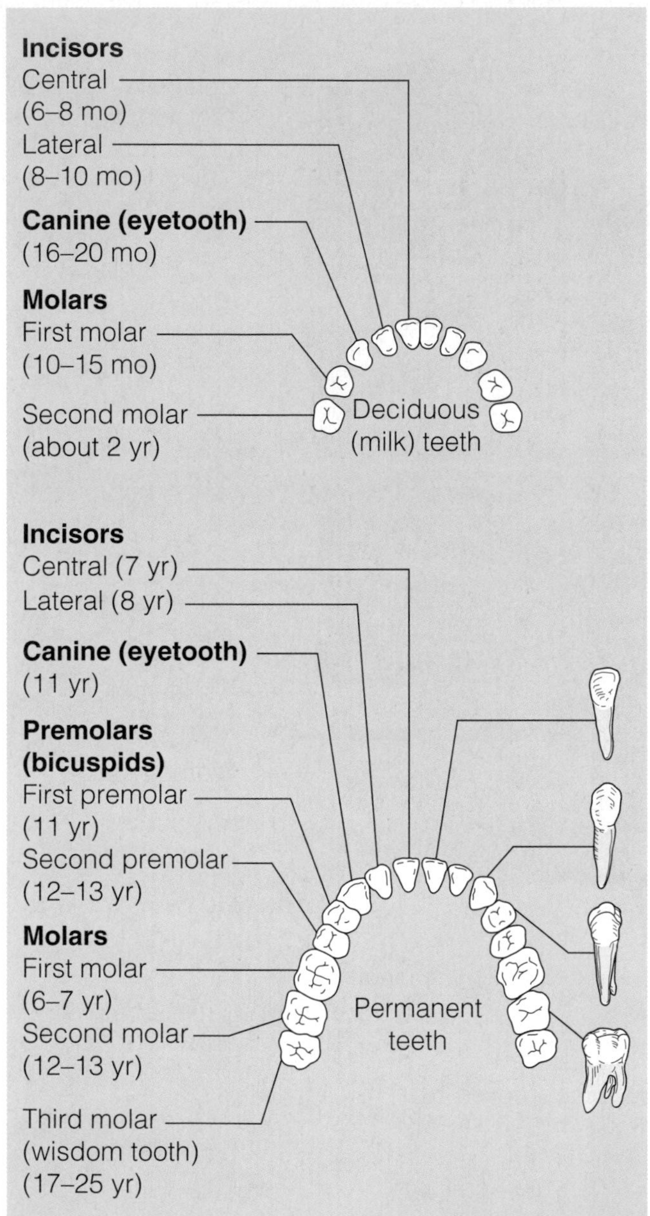

FIGURE 14.9 Human deciduous and permanent teeth. Approximate time of tooth eruption is shown in parentheses. Since the same number and arrangement of teeth exist in both upper and lower jaws, only the lower jaw is shown in each case. The shapes of individual teeth are shown on the right.

where it becomes the S-shaped **sigmoid** (sig′moid) **colon.** The sigmoid colon, rectum, and anal canal lie in the pelvis. The anal canal ends at the **anus** (a′nus), which opens to the exterior. The anal canal has an external *voluntary sphincter* (the **external anal sphincter**) composed of skeletal muscle and an internal *involuntary sphincter* formed by smooth muscle. These sphincters, which act rather like purse strings to open and close the anus, are ordinarily closed except during defecation, when feces are eliminated from the body.

Because most nutrient absorption has occurred before the large intestine is reached, no villi are seen in the large intestine, but there are tremendous numbers of *goblet cells* in its mucosa that produce an alkaline (HCO_3^--rich) mucus. The mucus acts as a lubricant to ease the passage of feces to the end of the digestive tract.

In the large intestine, the longitudinal muscle layer of the muscularis externa is reduced to three bands of muscle called *teniae coli* (ten′ne-e ko′li; "ribbons of the colon"). Since these muscle bands usually display some degree of tone (are partially contracted), they cause the wall to pucker into small pocketlike sacs called **haustra** (haws′trah).

Accessory Digestive Organs

Salivary Glands

Three pairs of **salivary glands** empty their secretions into the mouth. The large **parotid** (pah-rot′id) **glands** lie anterior to the ears. *Mumps,* a common childhood disease, is an inflammation of the parotid glands. If you look at the location of the parotid glands in Figure 14.1, you can readily understand why people with mumps complain that it hurts to open their mouth or chew.

The **submandibular glands** and the small **sublingual** (sub-ling′gwal) **glands** empty their secretions into the floor of the mouth through tiny ducts. The product of the salivary glands, **saliva,** is a mixture of mucus and serous fluids. The mucus moistens and helps to bind food together into a mass called a **bolus** (bo′lus), which makes chewing and swallowing easier. The clear serous portion contains an enzyme, **salivary amylase** (am′ĭ-lās), in a bicarbonate-rich (alkaline) juice that begins the process of starch digestion in the mouth. Saliva also contains substances such as lysozyme and antibodies (IgA) that inhibit bacteria; therefore, it has a protective function as well. Last but not least, saliva dissolves food chemicals so they can be tasted.

Teeth

The role of the teeth in food processing needs little introduction. We **masticate,** or *chew,* by opening and closing our jaws and moving them from side to side while continually using our tongue to move the food between our teeth. In the process, the teeth tear and grind the food, breaking it down into smaller fragments.

Ordinarily, by the age of 21, two sets of teeth have been formed (Figure 14.9). The first set is the **deciduous** (de-sid′u-us) **teeth,** also called **baby teeth** or **milk teeth.** The deciduous teeth begin to erupt around six months, and a baby has a full set (20 teeth) by the age of 2 years. The first teeth to appear are the lower central incisors, an event that is usually anxiously awaited by the child's parents.

As the second set of teeth, the deeper **permanent teeth,** enlarge and develop, the roots of the milk teeth are reabsorbed, and between the ages of 6 and 12 years they loosen and fall out. All of the permanent teeth but the third molars have erupted by the end of adolescence. The third molars, also called *wisdom teeth,* emerge later, between the ages of 17 and 25. Although there are 32 permanent teeth in a full set, the wisdom teeth often fail to erupt; sometimes they are completely absent.

Homeostatic Imbalance

When teeth remain embedded in the jawbone, they are said to be *impacted.* Impacted teeth exert pressure and cause a good deal of pain and must be removed surgically. Wisdom teeth are the most commonly impacted. ▲

The teeth are classified according to shape and function as incisors, canines, premolars, and molars (see Figure 14.9). The chisel-shaped **incisors** are adapted for cutting; the fanglike **canines** (eyeteeth) are for tearing or piercing. The **premolars** (bicuspids) and **molars** have broad crowns with rounded cusps (tips) and are best suited for grinding.

A tooth consists of two major regions, the **crown** and the **root,** as shown in Figure 14.10. The enamel-covered crown is the exposed part of the tooth above the **gingiva** (jin-ji′vah), or **gum. Enamel** is the hardest substance in the body and is fairly brittle because it is heavily mineralized with calcium salts. The portion of the tooth embedded in the jawbone is the root; the root and crown are connected by the tooth region called the **neck.** The outer surface of the root is covered by a substance called **cementum,** which attaches the tooth to the **periodontal** (per″e-o-don′tal) **membrane (ligament).** This ligament holds the tooth in place in the bony jaw. **Dentin,** a bonelike material, underlies the enamel and forms the bulk of the tooth. It surrounds a central **pulp cavity,** which contains a number of structures (connective tissue, blood vessels, and nerve fibers) collectively called **pulp.** Pulp supplies nutrients to the tooth tissues and provides for tooth sensations. Where the pulp cavity extends into the root, it becomes the **root canal,** which provides a route for blood vessels, nerves, and other pulp structures to enter the pulp cavity of the tooth.

Q *What substance forms the bulk of the tooth?*

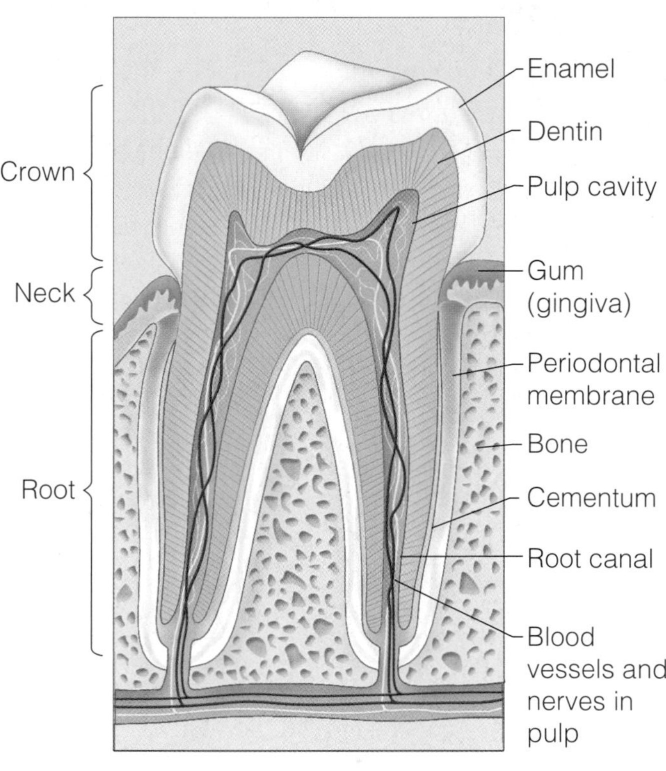

FIGURE 14.10 Longitudinal section of a molar.

A Dentin.

Pancreas

The **pancreas** is a soft, pink, triangular gland that extends across the abdomen from the spleen to the duodenum (see Figures 14.1 and 14.6).

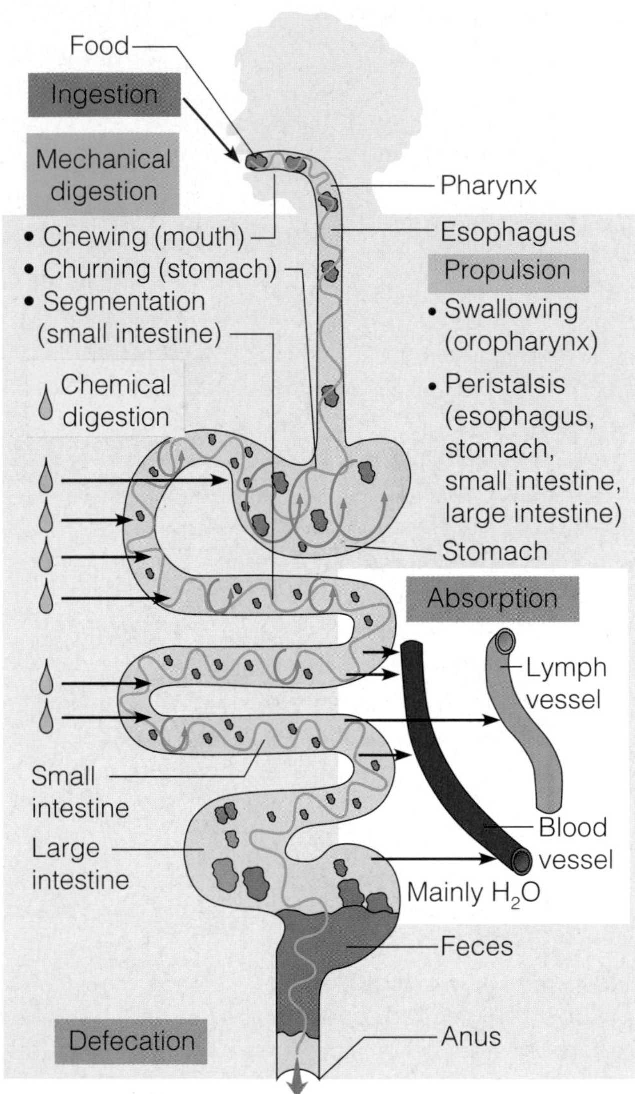

FIGURE 14.11 Schematic summary of gastrointestinal tract activities. Gastrointestinal tract activities include ingestion, mechanical digestion, chemical (enzymatic) digestion, propulsion, absorption, and defecation. Sites of chemical digestion are also sites that produce enzymes or that receive enzymes or other secretions made by accessory organs outside the alimentary canal. The mucosa of virtually the entire GI tract secretes mucus, which protects and lubricates.

Most of the pancreas lies posterior to the parietal peritoneum; hence its location is referred to as *retroperitoneal.*

The pancreas produces enzymes (described later) that break down all categories of digestible foods. The pancreatic enzymes are secreted into the duodenum in an alkaline fluid, which neutralizes the acidic chyme coming in from the stomach. The pancreas also has an endocrine function; it produces the hormones insulin and glucagon, as explained in Chapter 9.

Liver and Gallbladder

The **liver** is the largest gland in the body. It is located under the diaphragm, more to the right side of the body (see Figures 14.1 and 14.5). As described earlier, the liver overlies and almost completely covers the stomach. The liver has four lobes and is suspended from the diaphragm and abdominal wall by a delicate mesentery cord, the **falciform** (fal′si-form) **ligament.**

There is no question that the liver is one of the body's most important organs. It has many metabolic and regulatory roles; however, its digestive function is to produce **bile.** Bile leaves the liver through the **common hepatic duct** and enters the duodenum through the *bile duct* (see Figure 14.6).

Bile is a yellow-to-green, watery solution containing bile salts, bile pigments (chiefly bilirubin, a breakdown product of hemoglobin), cholesterol, phospholipids, and a variety of electrolytes. Of these components, only the bile salts (derived from cholesterol) and phospholipids aid the digestive process. Bile does not contain enzymes, but its bile salts *emulsify* fats by physically breaking large fat globules into smaller ones, thus providing more surface area for the fat-digesting enzymes to work on.

The **gallbladder** is a small, thin-walled green sac that snuggles in a shallow fossa in the inferior surface of the liver (see Figures 14.1 and 14.6). When food digestion is not occurring, bile backs up the **cystic duct** and enters the gallbladder to be stored. While being stored in the gallbladder, bile is concentrated by the removal of water. Later, when fatty food enters the duodenum, a hormonal stimulus prompts the gallbladder to contract and spurt out stored bile, making it available to the duodenum.

Homeostatic Imbalance

If bile is stored in the gallbladder for too long or too much water is removed, the cholesterol it contains may crystallize, forming *gallstones.* Since gallstones tend to be quite sharp, agonizing pain may occur when the gallbladder contracts (the typical *gallbladder attack*).

Blockage of the common hepatic or bile ducts (for example, by wedged gallstones) prevents bile from entering the small intestine, and it begins to accumulate and eventually backs up into the liver. This exerts pressure on the liver cells, and bile salts and bile pigments begin to enter the bloodstream. As the bile pigments circulate through the body, the tissues become yellow, or *jaundiced.* Blockage of the ducts is just one cause of jaundice. More often it results from actual liver problems such as *hepatitis* (an inflammation of the liver) or *cirrhosis* (sir-ro′sis), a chronic inflammatory condition in which the liver is severely damaged and becomes hard and fibrous. Hepatitis is most often due to viral infection resulting from drinking contaminated water or transmitted in blood via transfusion or contaminated needles. Cirrhosis is almost guaranteed when one drinks alcoholic beverages in excess for many years, and it is a common consequence of severe hepatitis. ▲

Functions of the Digestive System

Overview of Gastrointestinal Processes and Controls

The major functions of the digestive tract are usually summarized in two words—*digestion* and *absorption*. However, many of its specific activities (such as smooth muscle activity) and certain regulatory events are not really covered by either term. To describe digestive system processes a little more accurately, we really have to consider a few more functional terms. The essential activities of the GI tract include the following six processes, summarized in Figure 14.11.

1. **Ingestion**—Food must be placed into the mouth before it can be acted on. This is an active, voluntary process called ingestion.
2. **Propulsion**—If foods are to be processed by more than one digestive organ (and indeed they are), they must be propelled from one organ to the next. Swallowing is one example of food movement that depends largely on the propulsive process called **peristalsis.** Peristalsis is involuntary and involves alternating waves of contraction and relaxation of the muscles in the organ wall (Figure 14.12a). The net effect is to squeeze the food along the tract. Although **segmentation** (Figure 14.12b) may help to propel foodstuffs through the small intestine, it normally only moves food back and forth across the internal wall of the organ, serving to mix it with the digestive juices. Thus, segmentation is more an example of mechanical digestion than of propulsion.

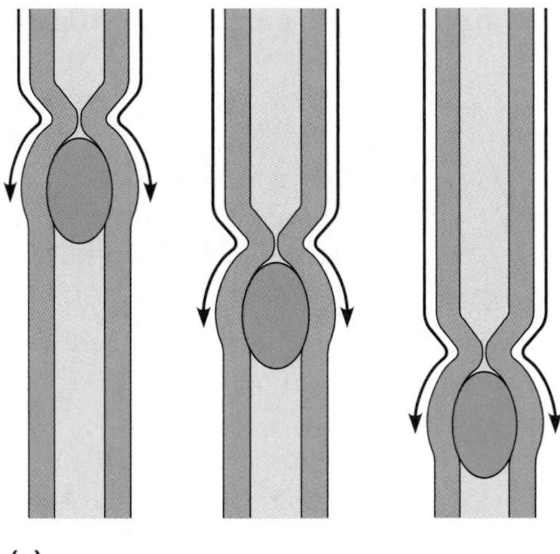

(a)

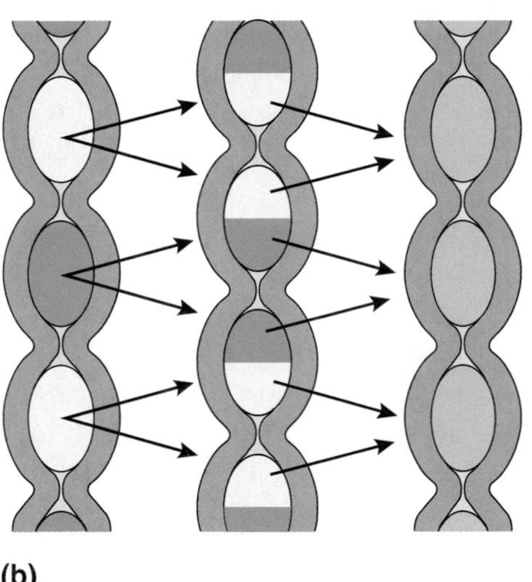

(b)

FIGURE 14.12 Peristaltic and segmental movements of the digestive tract. (a) In peristalsis, adjacent or neighboring segments of the intestine (or other alimentary canal organs) alternately contract and relax, which moves food distally along the tract. **(b)** In segmentation, single segments of the intestine alternately contract and relax. Because active segments are separated by inactive ones, the food is moved forward and then backward. Thus the food is mixed rather than simply propelled along the tract.

3. **Food breakdown: mechanical digestion**—Mixing of food in the mouth by the tongue, churning of food in the stomach, and segmentation in the small intestine are all examples of processes contributing to mechanical digestion. Mechanical digestion prepares food for further degradation by enzymes by physically fragmenting the foods into smaller particles.
4. **Food breakdown: chemical digestion**—The sequence of steps in which large food molecules are broken down to their building blocks by enzymes (protein molecules that act as catalysts) is called chemical digestion. Recall from Chapter 2 that these reactions are called *hydrolysis* reactions, because a water molecule is added to each bond to be broken. Water is also necessary as a dissolving medium and a softening agent for food digestion.

Since each of the major food groups has very different building blocks, it is worth taking a little time to review these chemical units, which were first introduced in Chapter 2. The building blocks, or units, of *carbohydrate* foods are *monosaccharides* (mon″o-sak′ah-rīdz), or simple sugars. We need to remember only three of these that are common in our diet—*glucose, fructose,* and *galactose*. Glucose is by far the most important, and when we talk about blood sugar levels, glucose is the "sugar" being referred to. Fructose is the most abundant sugar in fruits, and galactose is found in milk. Essentially, the only carbohydrates that our digestive system digests, or breaks down to simple sugars, are *sucrose* (table sugar), *lactose* (milk sugar), *maltose* (malt sugar), and *starch*. Sucrose, maltose, and lactose are referred to as *disaccharides,* or double sugars, because each consists of two simple sugars linked together. Starch is a *polysaccharide* (literally, "many sugars") formed of hundreds of glucose units. Although we eat foods containing other polysaccharides, such as *cellulose,* we do not have enzymes capable of breaking them down. The indigestible polysaccharides do not provide us with any nutrients, but they help move the foodstuffs along the gastrointestinal tract by providing bulk, or *fiber,* in our diet.

Proteins are digested to their building blocks, which are amino (ah-me′no) acids. Intermediate products of protein digestion are polypeptides and peptides. When lipids (fats) are digested, they yield two different types of building blocks—fatty acids and an alcohol called glycerol (glis′er-ol). The chemical breakdown of carbohydrates, proteins, and fats is summarized in Figure 14.13 and is described in more detail shortly.

5. **Absorption**—Transport of digested end products from the lumen of the GI tract to the blood or lymph is absorption. For absorption to occur, the digested foods must first enter the mucosal cells by active or passive transport processes. The small intestine is the major absorptive site.
6. **Defecation**—Defecation is the elimination of indigestible residues from the GI tract via the anus in the form of feces (fe′sēz).

Some of these processes are the job of a single organ. For example, only the mouth ingests, and only the large intestine defecates. But most digestive system activities occur bit by bit as food is moved along the tract. Thus, in one sense, the digestive tract can be viewed as a "disassembly line" in which food is carried from one stage of its processing to the next, and its nutrients are made available to the cells in the body en route.

A point that has been stressed throughout this book has been the drive of the body to maintain a constant internal environment, particularly in terms of homeostasis of the blood, which comes into intimate contact with all body cells. The digestive system, however, creates an optimal environment for itself to function in the lumen (cavity) of the alimentary canal, an area that is actually *outside* the body. Conditions in that lumen are controlled so that digestive processes occur efficiently. Digestive activity is mostly controlled by reflexes via the parasympathetic division of the autonomic nervous system. (Recall from Chapter 7 that this is the "resting-and-digesting" arm.) The sensors (mechanoreceptors, chemoreceptors) involved in these reflexes are located in the walls of the alimentary canal organs and respond to a number of stimuli, the most important being stretch of the organ by food in its lumen, pH of the contents, and presence of certain breakdown products of digestion. When these receptors are activated, they start reflexes that activate or inhibit (1) the glands that secrete digestive juices into the lumen or hormones into the blood, and (2) the smooth muscles of the muscularis that mix and propel the foods along the tract.

Now that we have summarized some points that apply to the function of the digestive organs as a group, we are ready to look at their special capabilities.

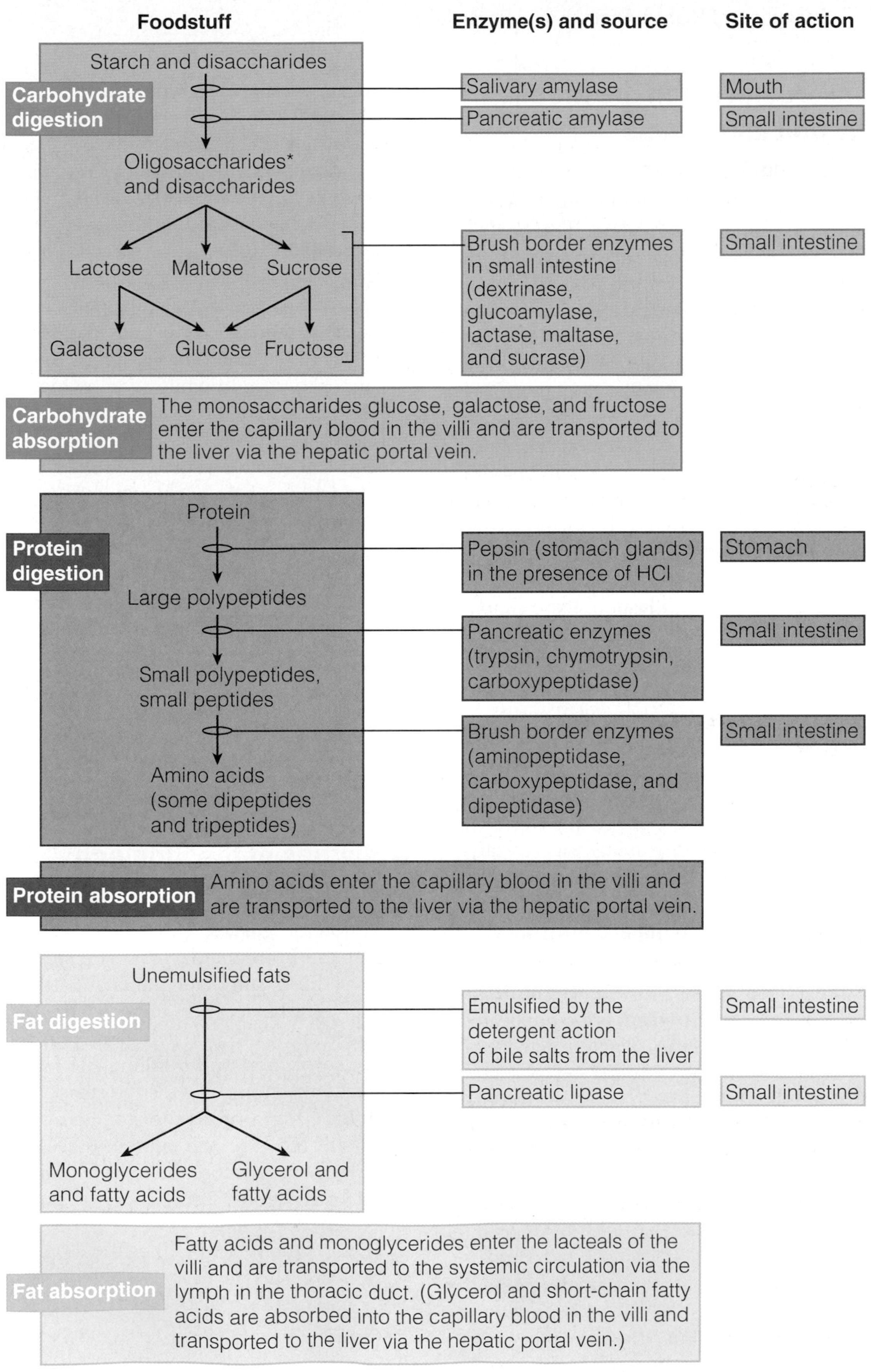

FIGURE 14.13 Flow chart of chemical digestion and absorption of foodstuffs.

Activities Occurring in the Mouth, Pharynx, and Esophagus

Food Ingestion and Breakdown

Once food has been placed in the mouth, both mechanical and chemical digestion begin. First the food is *physically* broken down into smaller particles by chewing. Then, as the food is mixed with saliva, salivary amylase begins the *chemical* digestion of starch, breaking it down into maltose (Figure 14.13). The next time you eat a piece of bread, chew it for a few minutes before swallowing it. You will notice that it begins to taste sweet as the sugars are released.

Saliva is normally secreted continuously to keep the mouth moist; but, when food enters the mouth, much larger amounts of saliva pour out. However, the simple pressure of anything put in the mouth and chewed, such as rubber bands or sugarless gum, will also stimulate the release of saliva. Some emotional stimuli can also cause salivation. For example, the mere thought of a hot fudge sundae will make many a mouth water. All these reflexes, though initiated by different stimuli, are brought about by parasympathetic fibers in cranial nerves V and IX.

Essentially no food absorption occurs in the mouth. (However, some drugs such as nitroglycerine are absorbed easily through the oral mucosa.) The pharynx and esophagus have no digestive function; they simply provide passageways to carry food to the next processing site, the stomach.

Food Propulsion—Swallowing and Peristalsis

In order for food to be sent on its way from the mouth, it must first be swallowed. **Deglutition** (deg″loo-tish′un), or **swallowing,** is a complicated process that involves the coordinated activity of several structures (tongue, soft palate, pharynx, and esophagus). It has two major phases. The first phase, the voluntary **buccal phase,** occurs in the mouth. Once the food has been chewed and well mixed with saliva, the bolus (food mass) is forced into the pharynx by the tongue. As food enters the pharynx, it passes out of our control and into the realm of reflex activity.

The second phase, the involuntary **pharyngeal-esophageal phase,** transports food through the pharynx and esophagus. The parasympathetic division of the autonomic nervous system (primarily the vagus nerves) controls this phase and promotes the mobility of the digestive organs from this point on. All routes that the food might take except the desired route distal into the digestive tract are blocked off. The tongue blocks off the mouth, and the soft palate closes off the nasal passages. The larynx rises so that its opening (into the respiratory passageways) is covered by the flaplike epiglottis. Food is moved through the pharynx and then into the esophagus below by wavelike peristaltic contractions of their muscular walls—first the longitudinal muscles contract, and then the circular muscles contract. The events of the swallowing process are illustrated in Figure 14.14.

If we try to talk while swallowing, our routing mechanisms may be "short-circuited," and food may manage to enter the respiratory passages. This triggers still another protective reflex—coughing—during which air rushes upward from the lungs in an attempt to expel the food.

Once food reaches the distal end of the esophagus, it presses against the cardioesophageal sphincter, causing it to open, and the food enters the stomach. The movement of food through the pharynx and esophagus is so automatic that a person can swallow and food will reach the stomach even if he is standing on his head. Gravity plays no part in the transport of food once it has left the mouth, which explains why astronauts (in the zero gravity of outer space) can still swallow and get nourishment.

Activities of the Stomach

Food Breakdown

Secretion of **gastric juice** is regulated by both neural and hormonal factors. The sight, smell, and taste of food stimulate parasympathetic nervous system reflexes, which increase the secretion of gastric juice by the stomach glands. In addition, the presence of food and a falling pH in the stomach stimulate the stomach cells to release the hormone **gastrin.** Gastrin prods the stomach glands to produce still more of the protein-digesting enzymes (pepsinogens), mucus, and hydrochloric acid. Under normal conditions, 2 to 3 liters of gastric juice are produced every day.

Hydrochloric acid makes the stomach contents very acid. This is (somewhat) dangerous since both hydrochloric acid and the protein-digesting enzymes have the ability to digest the stomach itself, causing *ulcers* (see the "A Closer Look" box on pp. 472–473). However, as long as enough mucus is made, the stomach is "safe" and will remain unharmed.

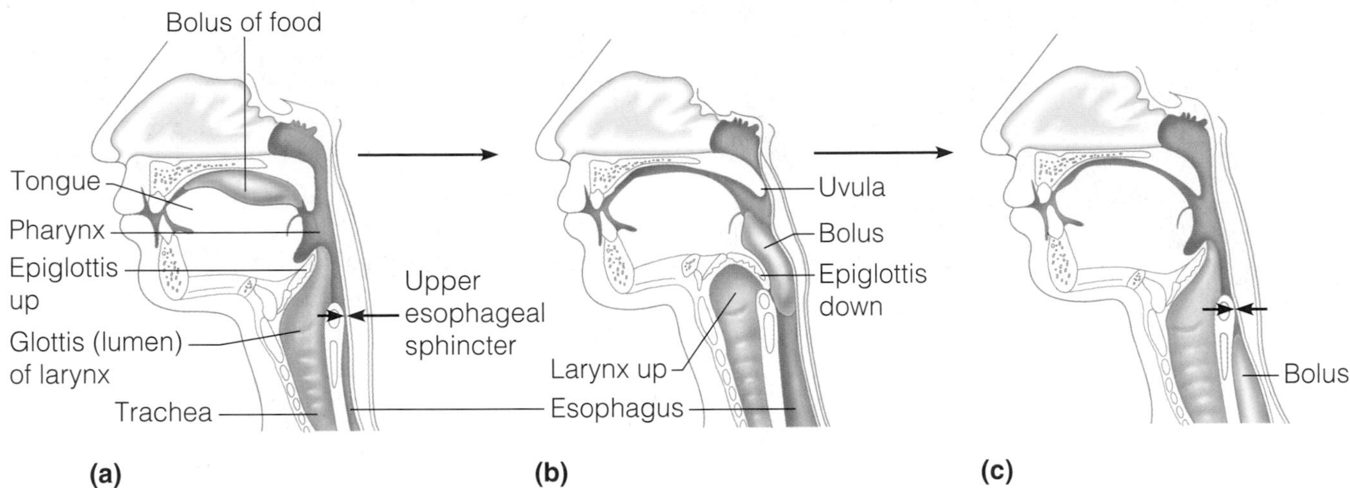

FIGURE 14.14 Swallowing. (a) The tongue pushes the food bolus posteriorly and against the soft palate. **(b)** The soft palate rises to close off the nasal passages as the bolus enters the pharynx. The larynx rises so that the epiglottis covers its opening as peristalsis carries the food through the pharynx and into the esophagus. The upper esophageal sphincter relaxes to allow food entry. **(c)** The upper esophageal sphincter contracts again as the larynx and epiglottis return to their former positions and the food bolus moves inferiorly to the stomach.

Homeostatic Imbalance

Occasionally, the cardioesophageal sphincter fails to close tightly and gastric juice backs up into the esophagus, which has little mucus protection. This results in a characteristic pain known as *heartburn,* which, if uncorrected, leads to inflammation of the esophagus (*esophagitis* [ĕ-sof′ah-ji′tis]) and perhaps even to ulceration of the esophagus. A common cause is a *hiatal hernia,* a structural abnormality in which the superior part of the stomach protrudes slightly above the diaphragm. Since the diaphragm no longer reinforces the cardioesophageal sphincter, which is a weak sphincter to begin with, gastric juice flows into the unprotected esophagus. Conservative treatment involves restricting food intake after the evening meal, taking antacids, and sleeping with the head elevated. ▲

The extremely acidic environment that hydrochloric acid provides *is* necessary, because it activates *pepsinogen* to *pepsin,* the active protein-digesting enzyme. *Rennin,* the second protein-digesting enzyme produced by the stomach, works primarily on milk protein and converts it to a substance that looks like sour milk. Many mothers mistakenly think that when their infants spit up a curdy substance after having their bottle that the milk has soured in their stomach. Rennin is produced in large amounts in infants, but it is not believed to be produced in adults.

Other than the beginning of protein digestion, little chemical digestion occurs in the stomach. With the exception of aspirin and alcohol (which seem somehow to have a "special pass"), virtually no absorption occurs through the stomach walls.

As food enters and fills the stomach, its wall begins to stretch (at the same time the gastric juices are being secreted, as just described). Then the three muscle layers of the stomach wall become active. They compress and pummel the food, breaking it apart physically, all the while continually mixing the food with the enzyme-containing gastric juice so that the semifluid chyme is formed. The process looks something like the preparation of a cake mix, in which the floury mixture is continually folded on itself and mixed with the liquid until it reaches uniform texture.

Food Propulsion

Once the food has been well mixed, a rippling peristalsis begins in the lower half of the stomach, and the contractions increase in force as the pyloric valve is approached. The pylorus of the stomach, which holds about 30 ml of chyme, acts

A Closer Look

Peptic Ulcers: "Something Is Eating at Me"

ARCHIE, a 53-year-old factory worker, began to experience a gnawing pain in his upper abdomen an hour or two after each meal. At first, he blamed the quality of his home cooking, but he experienced the same symptoms after eating at the factory cafeteria or at restaurants. Archie always responded to stress by drinking and smoking heavily, and his abdominal pain became markedly worse during a hectic week when he worked 15 hours of overtime on the assembly line. Finally, after two months of increasingly severe pain, Archie consulted a physician and was told that he had a peptic ulcer.

Peptic ulcers affect one of every eight Americans. A peptic ulcer is a craterlike erosion in the mucosa of any part of the GI tract that is exposed to the secretions of the stomach. Hydrochloric acid and pepsin cause this damage; people whose stomachs fail to secrete these substances never develop peptic ulcers. A few peptic ulcers occur in the lower esophagus, following the regurgitation of stomach contents, but most (98 percent) occur in the pyloric part of the stomach (gastric ulcers) or the first part of the duodenum (duodenal ulcers). Duodenal ulcers are about three times more common than gastric ulcers. Peptic ulcers may appear at any age, but they develop most frequently between the ages of 50 and 70 years. After developing, they tend to recur—healing, then flaring up periodically—for the rest of one's life if not treated.

Gastric and duodenal ulcers may produce a gnawing or burning pain in the epigastric region of the abdomen. This pain often appears one to three hours after a meal (or causes one to awaken at night) and is relieved by eating. Other potential symptoms include loss of appetite, burping, nausea, and vomiting. Not all people with ulcers experience the above symptoms, however, and many exhibit no symptoms at all.

Despite years of intensive study, the cause of peptic ulcers remains incompletely understood. For over a century, it has been "common knowledge" that stress causes ulcers, and the stereotypical ulcer patient has been the overworked business executive. Recent studies have not been able to demonstrate such a link between stress and ulcers, however, and many researchers now doubt the validity of this claim. Nonetheless, a stressful lifestyle does seem to *aggravate* existing ulcers. Recent studies indicate that many ulcers are actually caused by a strain of acid-resistant bacteria *(Helicobacter pylori)* that inhabit the stomachs of 40 percent of healthy people and 70 to 90 percent of those with ulcers.

The anatomy of a peptic ulcer is shown in the figure in this box. It is a round, sharply defined crater in the mucosa. Typical ulcers are 1 to 4 cm in diameter. The base of the ulcer contains dead tissue cells, granulation tissue, and scar tissue. Eroded blood vessels may sometimes be seen there as well.

Peptic ulcers can produce serious complications. In about 20 percent of cases, eroded blood vessels bleed into the GI tract, causing vomiting of blood and blood in the feces. In such cases, anemia may result from a severe loss of blood. In 5 to 10 percent of ulcer patients, scarring within the stomach obstructs the pyloric opening, blocking digestion. About 5 percent of peptic ulcers *perforate,* allowing the contents of

A peptic ulcer.

the stomach or duodenum to leak into the peritoneal cavity. This can cause either peritonitis or digestion and destruction of the nearby pancreas. A perforated ulcer is a life-threatening condition.

In spite of these potential complications, most peptic ulcers heal readily and respond well to treatment. The first steps in treatment are to avoid smoking, alcohol, ibuprofen, and aspirin, all of which aggravate ulcers. Antacid drugs are often suggested to neutralize the stomach acids. In ulcers found to be colonized by *H. pylori,* the goal is to kill the embedded bacteria. Triple drug therapy (a combination of the antibiotics tetracycline and metronidazole, and bismuth subsalicylate, the active ingredient in Pepto-Bismol) effectively promotes healing and prevents recurrence, even in resistant strains. In other cases *H. pylori* is not the cause, such as in esophageal ulcers, which are caused by gastric reflux. Then the therapy of choice is drugs that inhibit HCl secretion (H_2 blockers) such as ranitidine (Zantac). These drugs cure the ulcer by decreasing the production of acid and pepsin.

Recent animal trials using a newly developed vaccine against *H. pylori* have been successful. It is hoped that preventative vaccination, coupled with antibiotic cures, will eradicate peptic ulcers within the next 25 years.

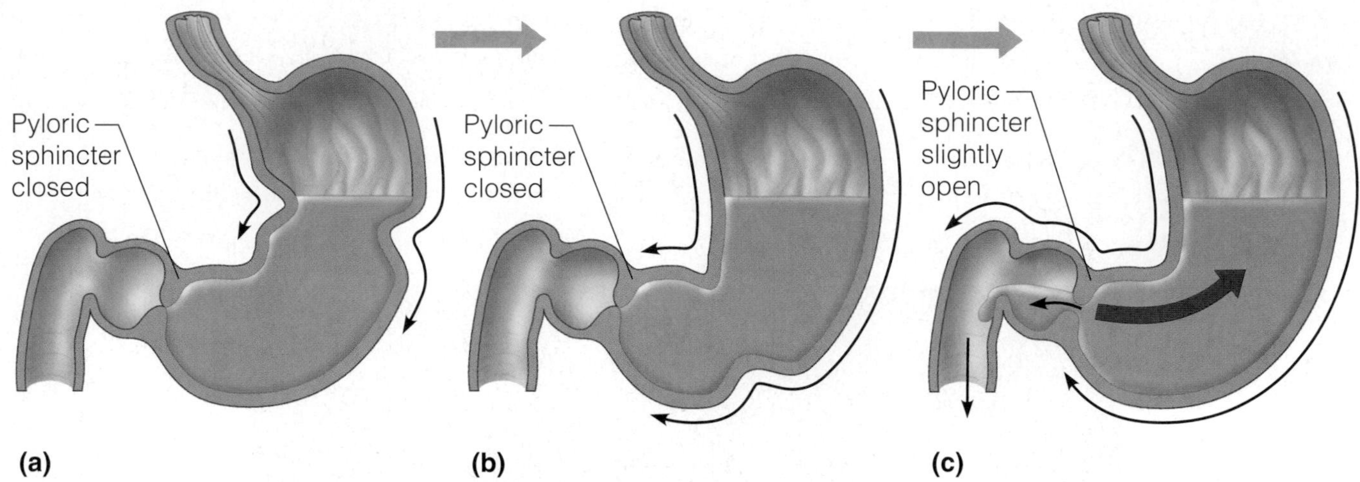

FIGURE 14.15 Peristaltic waves act primarily in the inferior portion of the stomach to mix and move chyme through the pyloric valve. **(a)** Peristaltic waves move toward the pylorus. **(b)** The most vigorous peristalsis and mixing action occurs close to the pylorus. **(c)** The pyloric end of the stomach acts as a pump that delivers small amounts of chyme into the duodenum while forcing most of the contained material backward into the stomach, where it undergoes further mixing.

like a meter that allows only liquids and very small particles to pass through the pyloric sphincter (Figure 14.15). Because the pyloric sphincter barely opens, each contraction of the stomach muscle squirts 3 ml or less of chyme into the small intestine. Since the contraction also *closes* the valve, the rest (about 27 ml) is propelled backward into the stomach for more mixing. When the duodenum is filled with chyme and its wall is stretched, a nervous reflex, the *enterogastric* (en″ter-o-gas′trik) *reflex,* occurs. This reflex "puts the brakes on" gastric activity and slows the emptying of the stomach by inhibiting the vagus nerves and tightening the pyloric sphincter, thus allowing time for intestinal processing to catch up. Generally, it takes about 4 hours for the stomach to empty completely after eating a well-balanced meal and 6 hours or more if the meal has a high fat content.

Homeostatic Imbalance

Local irritation of the stomach, such as occurs with bacterial food poisoning, may activate the *emetic* (ĕ-met′ik) *center* in the brain (medulla). The emetic center, in turn, causes **vomiting (emesis).** Vomiting is essentially a reverse peristalsis occurring in the stomach (and perhaps the small intestine), accompanied by contraction of the abdominal muscles and the diaphragm, which increases the pressure on the abdominal organs. The emetic center may also be activated through other pathways; disturbance of the equilibrium apparatus of the inner ear during a boat ride on rough water is one example. ▲

Activities of the Small Intestine

Food Breakdown and Absorption

Food reaching the small intestine is only partially digested. Carbohydrate and protein digestion have been started, but virtually no fats have been digested up to this point. Here the process of chemical food digestion is accelerated as the food now takes a rather wild 3- to 6-hour journey through the looping coils and twists of the small intestine. By the time the food reaches the end of the small intestine, digestion is complete and nearly all food absorption has occurred.

The microvilli of small intestine cells bear a few important enzymes, the so-called **brush border enzymes** that break down double sugars into simple sugars and complete protein digestion (see Figure 14.13). *Intestinal juice* itself is relatively enzyme-poor, and protective mucus is probably the most important intestinal gland secretion. However, foods entering the small intestine are literally deluged with enzyme-rich **pancreatic**

TABLE 14.1 Hormones and Hormonelike Products That Act in Digestion

Hormone	Source	Stimulus for secretion	Action
Gastrin	Stomach	Food in stomach (chemical stimulus)	• Stimulates release of gastric juice • Stimulates mobility of small intestine • Relaxes ileocecal valve.
Histamine	Stomach	Food in stomach	• Activates parietal cells to secrete hydrochloric acid.
Somatostatin	Stomach	Food in stomach	• Inhibits secretion of gastric juice and pancreatic juice • Inhibits emptying of stomach and gallbladder.
Secretin	Duodenum	Acidic chyme and partially digested foods in duodenum	• Increases output of pancreatic juice rich in bicarbonate ions • Increases bile output by liver • Inhibits gastric mobility and gastric gland secretion.
Cholecystokinin (CCK)	Duodenum	Fatty chyme and partially digested proteins in duodenum	• Increases output of enzyme-rich pancreatic juice • Stimulates gallbladder to expel stored bile • Relaxes sphincter of duodenal papilla to allow bile and pancreatic juice to enter the duodenum.
Gastric inhibitory peptide (GIP)	Duodenum	Fatty chyme in duodenum	• Inhibits gastric mobility and secretion of gastric juice.

juice ducted in from the pancreas, as well as bile from the liver.

Pancreatic juice contains enzymes that (1) along with brush border enzymes, complete the digestion of starch *(pancreatic amylase)*; (2) carry out about half of protein digestion (via the action of *trypsin, chymotrypsin, carboxypeptidase,* and others); (3) are totally responsible for fat digestion since the pancreas is essentially the only source of *lipases;* and (4) digest nucleic acids *(nucleases)*.

In addition to enzymes, pancreatic juice contains a rich supply of bicarbonate, which makes it very basic (about pH 8). When pancreatic juice reaches the small intestine, it neutralizes the acidic chyme coming in from the stomach and provides the proper environment for activation and activity of intestinal and pancreatic digestive enzymes.

Homeostatic Imbalance

Pancreatitis (pan″kre-ah-ti′tis) is a rare but extremely serious inflammation of the pancreas that results from activation of pancreatic enzymes in the pancreatic duct. Since pancreatic enzymes break down all categories of biological molecules, the pancreatic tissue and duct are digested. This painful condition can lead to nutritional deficiencies, because pancreatic enzymes are essential to digestion in the small intestine. ▲

The release of pancreatic juice into the duodenum is stimulated by both the vagus nerves and local hormones. When chyme enters the small intestine, it stimulates the mucosa cells to produce several hormones (Table 14.1). Two of these

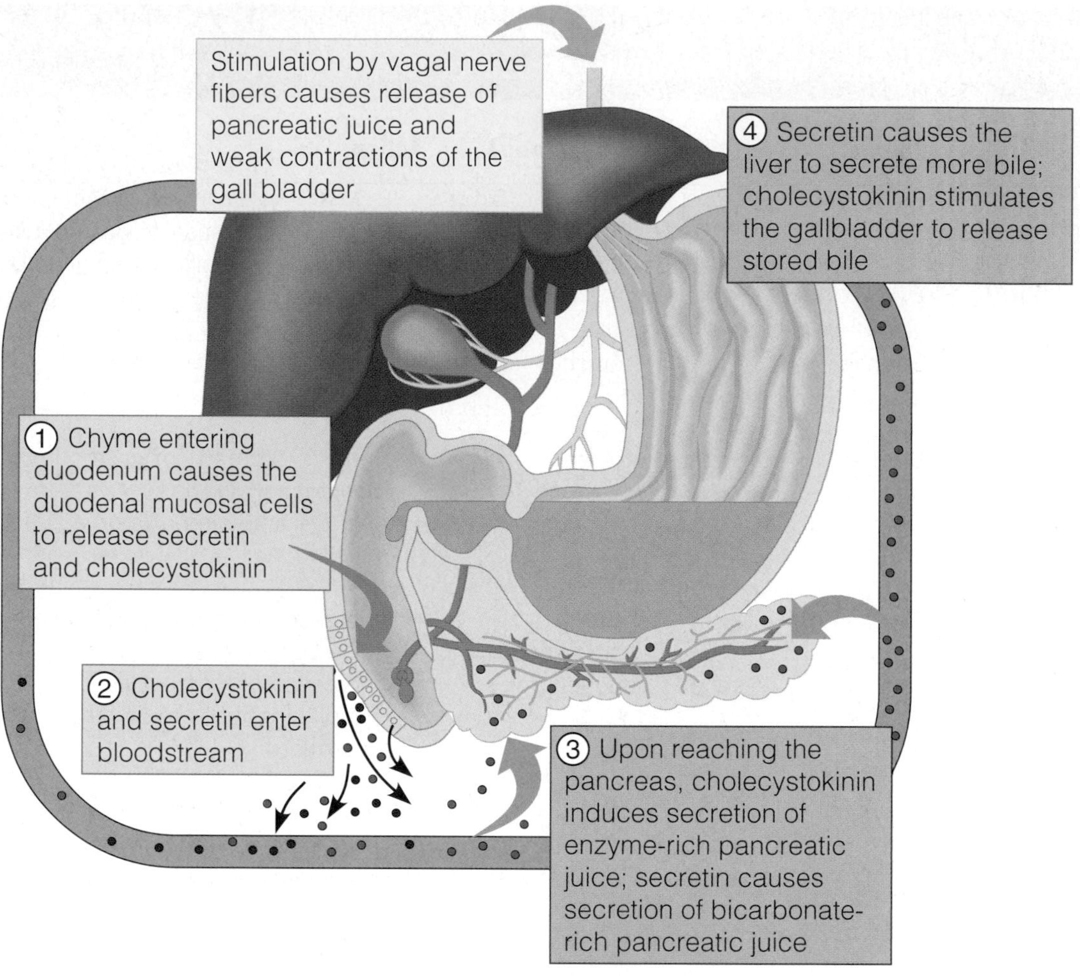

FIGURE 14.16 Regulation of pancreatic juice secretion. Hormonal controls, exerted by secretin and cholecystokinin (steps 1–3), are the main regulatory factors. Neural control is mediated by the vagus nerves.

hormones, **secretin** (se-kre′tin) and **cholecystokinin** (ko″le-sis″to-kin′in) **(CCK),** influence the release of pancreatic juice and bile.

The hormones enter the blood and circulate to their target organs, the pancreas, liver, and gallbladder. Both hormones work together to stimulate the pancreas to release its enzyme- and bicarbonate-rich product (Figure 14.16). In addition, secretin causes the liver to increase its output of bile, and cholecystokinin causes the gallbladder to contract and release stored bile into the bile duct so that bile and pancreatic juice enter the small intestine together. As mentioned before, bile is not an enzyme. Instead, it acts like a detergent to emulsify, or mechanically break down, large fat globules into thousands of tiny ones, providing a much greater surface area for the pancreatic lipases to work on. Bile is also necessary for absorption of fats (and other fat-soluble vitamins [K, D, and A] that are absorbed along with them) from the intestinal tract.

Homeostatic Imbalance

If *either* bile or pancreatic juice is absent, essentially no fat digestion or absorption goes on, and fatty, bulky stools are the result. In such cases, blood-clotting problems also occur because the liver needs vitamin K to make prothrombin, one of the clotting factors. ▲

Absorption of water and of the end products of digestion occurs all along the length of the small intestine. Most substances are absorbed through the intestinal cell plasma membranes by the process of *active transport*. Then they enter the capillary beds in the villi to be transported in the blood to the liver via the hepatic portal vein. The exception seems to be lipids, or fats, which are absorbed passively by the process of *diffusion*. Lipid breakdown products enter both the capillary beds and the lacteals in the villi and are carried to the liver by both blood and lymphatic fluids.

At the end of the ileum, all that remains is some water, indigestible food materials (plant fibers such as cellulose), and large amounts of bacteria. This debris enters the large intestine through the ileocecal valve. The complete process of food digestion and absorption is summarized in Figure 14.13 (p. 469).

Food Propulsion

As mentioned previously, *peristalsis* is the major means of propelling food through the digestive tract. It involves waves of contraction that move along the length of the intestine, followed by waves of relaxation. The net effect is that the food is moved through the small intestine in much the same way that toothpaste is squeezed from a tube. Rhythmic segmental movements produce local constrictions of the intestine (see Figure 14.12b) that mix the chyme with the digestive juices, and help to propel food through the intestine.

Activities of the Large Intestine

Food Breakdown and Absorption

What is finally delivered to the large intestine contains few nutrients, but that residue still has 12 to 24 hours more to spend there. The colon itself produces no digestive enzymes. However, the "resident" bacteria that live within its lumen metabolize some of the remaining nutrients, releasing gases (methane and hydrogen sulfide) that contribute to the odor of feces. About 500 ml of gas (flatus) is produced each day, much more when certain carbohydrate-rich foods (such as beans) are eaten.

Bacteria residing in the large intestine also make some vitamins (vitamin K and some B vitamins). Absorption by the large intestine is limited to the absorption of these vitamins, some ions, and most of the remaining water. **Feces,** the more or less solid product delivered to the rectum, contain undigested food residues, mucus, millions of bacteria, and just enough water to allow their smooth passage.

Propulsion of the Residue and Defecation

Peristalsis and mass movements are the two major types of propulsive movements occurring in the large intestine. Colon peristalsis is sluggish and, compared to the mass movements, probably contributes very little to propulsion. **Mass movements** are long, slow-moving but powerful contractile waves that move over large areas of the colon three or four times daily and force the contents toward the rectum. Typically, they occur during or just after eating, when food begins to fill the stomach and small intestine. Bulk, or fiber, in the diet increases the strength of colon contractions and softens the stool, allowing the colon to act as a well-oiled machine.

Homeostatic Imbalance

When the diet lacks bulk, the colon narrows and its circular muscles contract more powerfully, which increases the pressure on its walls. This encourages formation of *diverticula* (di″ver-tik′u-lah), in which the mucosa protrudes through the colon walls, a condition called *diverticulosis. Diverticulitis,* a condition in which the diverticula become inflamed, can be life-threatening if ruptures occur. ▲

The rectum is generally empty, but when feces are forced into it by mass movements and its wall is stretched, the **defecation reflex** is initiated. The defecation reflex is a spinal (sacral region) reflex that causes the walls of the sigmoid colon and the rectum to contract and the anal sphincters to relax. As the feces are forced through the anal canal, messages reach the brain giving us time to make a decision as to whether the external voluntary sphincter should remain open or be constricted to stop passage of feces. If it is not convenient, defecation (or "moving the bowels") can be delayed temporarily. Within a few seconds, the reflex contractions end and the rectal walls relax. With the next mass movement, the defecation reflex is initiated again.

Homeostatic Imbalance

Watery stools, or *diarrhea* (di″ah-re′ah), result from any condition that rushes food residue through the large intestine before that organ has had sufficient time to absorb the water (as in irritation of the colon by bacteria). Because fluids and ions are lost from the body, prolonged diarrhea may result in dehydration and electrolyte imbalance which, if severe, can be fatal.

If food residue remains in the large intestine for extended periods, too much water is absorbed, and the stool becomes hard and difficult to pass. This condition, called *constipation,* may result from lack of fiber in the diet, poor bowel habits ("failing to heed the call"), and laxative abuse. ▲

PART II: NUTRITION AND METABOLISM

Although it seems at times that people can be divided into two camps—those who live to eat and those who eat to live—we all recognize the vital importance of food for life. It has been said that "you are what you eat," and this is true in that part of the food we eat is converted to our living flesh. In other words, a certain fraction of nutrients is used to build cellular molecules and structures and to replace worn-out parts. However, most foods are used as metabolic fuels. That is, they are oxidized and transformed into **ATP,** the chemical energy form needed by body cells to drive their many activities. The energy value of foods is measured in units called **kilocalories (kcal),** or Calories (with a capital C), the units conscientiously counted by dieters.

We have just considered how foods are digested and absorbed. But what happens to these foods once they enter the blood? Why do we need bread, meat, and fresh vegetables? Why does everything we eat seem to turn to fat? We will try to answer these questions in this section.

Nutrition

A **nutrient** is a substance in food that is used by the body to promote normal growth, maintenance, and repair. The nutrients divide neatly into six categories. The *major nutrients*—carbohydrates, lipids, and proteins—make up the bulk of what we eat. Vitamins and minerals, while equally crucial for health, are required in minute amounts. Water, which accounts for about 60 percent of the volume of the food we eat, is also considered to be a major nutrient. However, because its importance as a solvent and in many other aspects of body functioning is described in Chapter 2, only the other five classes of nutrients will be considered here.

Most foods offer a combination of nutrients. For example, a bowl of cream of mushroom soup contains all of the major nutrients plus some vitamins and minerals. A diet consisting of foods selected from each of the five food groups (Table 14.2), that is, grains, fruits, vegetables, meats and fish, and milk products, normally guarantees adequate amounts of all of the needed nutrients.

Dietary Sources of the Major Nutrients

Carbohydrates

Except for milk sugar (lactose) and small amounts of glycogen in meats, all the **carbohydrates**—sugars and starches—we ingest are derived from plants. Sugars come mainly from fruits, sugar cane, and milk. The polysaccharide starch is found in grains, legumes, and root vegetables. The polysaccharide cellulose, which is plentiful in most vegetables, is not digested by humans, but it provides roughage, or fiber, which increases the bulk of the stool and aids defecation.

Lipids

Although we also ingest cholesterol and phospholipids, most dietary **lipids** are triglycerides. We eat saturated fats in animal products such as meat and dairy foods and in a few plant products, such as coconut. Unsaturated fats are present in seeds, nuts, and most vegetable oils. Major sources of cholesterol are egg yolk, meats, and milk products.

Proteins

Animal products contain the highest-quality **proteins,** molecules that are basically amino acid polymers. Eggs, milk, and most meat proteins are *complete proteins* that meet all of the body's amino acid requirements for tissue maintenance and growth. Legumes (beans and peas), nuts, and cereals are also protein-rich, but their proteins are nutritionally incomplete because they are low in one or more of the essential amino acids. As you can see, strict vegetarians must carefully plan their diets to obtain all the essential amino acids and prevent protein malnutrition. Cereal grains and legumes when ingested together provide all the essential amino acids, and some variety of this combination is found in the diets of all cultures

TABLE 14.2 Five Basic Food Groups and Some of Their Major Nutrients

Group	Example foods	Major nutrients supplied in significant amounts: By all in group	By only some in group
Fruits	Apples, bananas, dates, oranges, tomatoes	Carbohydrate Water	Vitamins: A, C, folic acid Minerals: iron, potassium Fiber
Vegetables	Broccoli, cabbage, green beans, lettuce, potatoes	Carbohydrate Water	Vitamins: A, C, E, K, and B vitamins except B_{12} Minerals: calcium, magnesium, iodine, manganese, phosphorus Fiber
Grain products (preferably whole grain; otherwise, enriched or fortified)	Breads, rolls, bagels; cereals, dry and cooked; pasta; rice, other grains; tortillas, pancakes, waffles; crackers; popcorn	Carbohydrate Protein Vitamins: thiamin (B_1), niacin	Water Fiber Minerals: iron, magnesium, selenium
Milk products	Milk, yogurt; cheese; ice cream, ice milk, frozen yogurt	Protein Fat Vitamins: riboflavin, B_{12} Minerals: calcium, phosphorus Water	Carbohydrate Vitamins: A, D
Meats and meat alternates	Meat, fish, poultry; eggs; seeds; nuts, nut butters; soybeans, tofu; other legumes (peas and beans)	Protein Vitamins: niacin, B_6 Minerals: iron, zinc	Carbohydrate Fat Vitamins: B_{12}, thiamin (B_1) Water Fiber

Source: Christian, Janet, and Janet Greger. *Nutrition for Living,* 3rd ed. San Francisco, CA: Benjamin Cummings, 1991.

(most obviously in the rice and beans seen on nearly every plate in a Mexican restaurant).

Vitamins

Vitamins are organic nutrients of various forms that the body requires in small amounts. Although vitamins are found in all major food groups, no one food contains all the required vitamins. Thus, a balanced diet is the best way to ensure a full vitamin complement, particularly since certain vitamins (A, C, and E) appear to have anticancer effects. Diets rich in broccoli, cabbage, and brussels sprouts (all good sources of vitamins A and C) appear to reduce cancer risk. However, controversy abounds concerning the ability of vitamins to work wonders.

Most vitamins function as **coenzymes** (or parts of coenzymes); that is, they act with an enzyme to accomplish a particular type of catalysis.

Minerals

The body also requires adequate supplies of seven **minerals** (that is, inorganic substances including calcium, phosphorus, potassium, sulfur, sodium, chloride, and magnesium) and trace amounts of about a dozen others.

Fats and sugars have practically no minerals, and cereals and grains are poor sources. The most

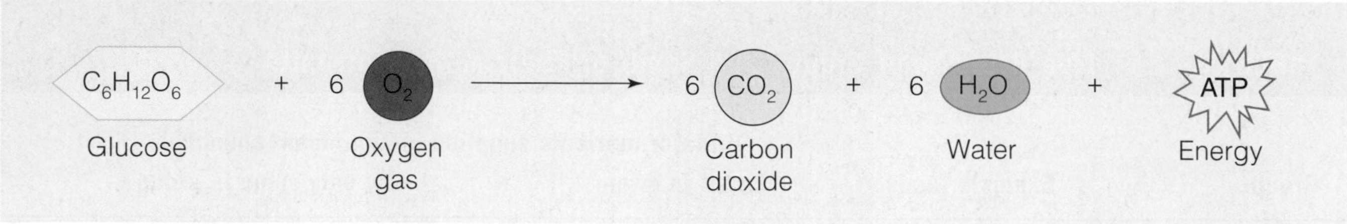

$$C_6H_{12}O_6 + 6\,O_2 \longrightarrow 6\,CO_2 + 6\,H_2O + ATP$$

Glucose, Oxygen gas, Carbon dioxide, Water, Energy

FIGURE 14.17 Summary equation for cellular respiration.

mineral-rich foods are vegetables, legumes, milk, and some meats.

The main uses of the major nutrients in the body are discussed in the section on metabolism. Appendix C details some important roles of vitamins and minerals in the body.

Metabolism

Metabolism (mĕ-tab′o-lizm; *metabol* = change) is a broad term referring to all chemical reactions that are necessary to maintain life. It involves **catabolism** (kah-tab′o-lizm), in which substances are broken down to simpler substances, and **anabolism** (ah-nab′o-lizm), in which larger molecules or structures are built from smaller ones. During catabolism, energy is released and captured to make ATP, the energy-rich molecule used to energize all cellular activities, including catabolic reactions (Figure 14.17).

Not all foodstuffs are treated in the same way by body cells. For example, carbohydrates, particularly glucose, are usually broken down to make ATP. Fats are used to build cell membranes, make myelin sheaths, and insulate the body with a fatty cushion. They are also used as the body's main energy fuel for making ATP when there are inadequate carbohydrates in the diet. Proteins tend to be carefully conserved (even hoarded) by the body cells. This is easy to understand when you recognize that proteins are the major structural materials used for building cell structures.

Carbohydrate, Fat, and Protein Metabolism in Body Cells

Carbohydrate Metabolism

Just as an oil furnace uses oil (its fuel) to produce heat, the cells of the body use carbohydrates as their preferred fuel to produce cellular energy (ATP). **Glucose,** also known as **blood sugar,** is the major breakdown product of carbohydrate digestion. Glucose is also the major fuel used for making ATP in most body cells. The liver is an exception; it routinely uses fats as well, thus saving glucose for other body cells. Essentially, glucose is broken apart piece by piece, and some of the chemical energy released when its bonds are broken is captured and used to bind phosphate to ADP molecules to make ATP.

The overall reaction is summed up simply in Figure 14.17. Basically, the carbon atoms released leave the cells as carbon dioxide, and the hydrogen atoms removed (which contain energy-rich electrons) are eventually combined with oxygen to form water. These oxygen-using events are referred to collectively as **cellular respiration.** The events of the three main metabolic pathways that are involved in cellular respiration—*glycolysis,* the *Krebs cycle,* and the *electron transport chain*—are shown schematically in Figure 14.18.

Oxidation via the removal of hydrogen atoms (which are temporarily passed to vitamin-containing coenzymes) is a major role of glycolysis and the Krebs cycle. **Glycolysis** also energizes each glucose molecule so that it can be split into two pyruvic acid molecules and yield a small amount of ATP in the process (Figure 14.18). The **Krebs cycle** produces virtually all the carbon dioxide and water that results during cell respiration. Like glycolysis, it yields a small amount of ATP by transferring high energy phosphate groups directly from phosphorylated substances to ADP. Free oxygen is not involved.

The **electron transport chain** is where the action is for ATP production. The hydrogen atoms removed during the first two metabolic phases are loaded with energy. These hydrogens are delivered by the coenzymes to the protein carriers of the electron transport chain, which form part of the

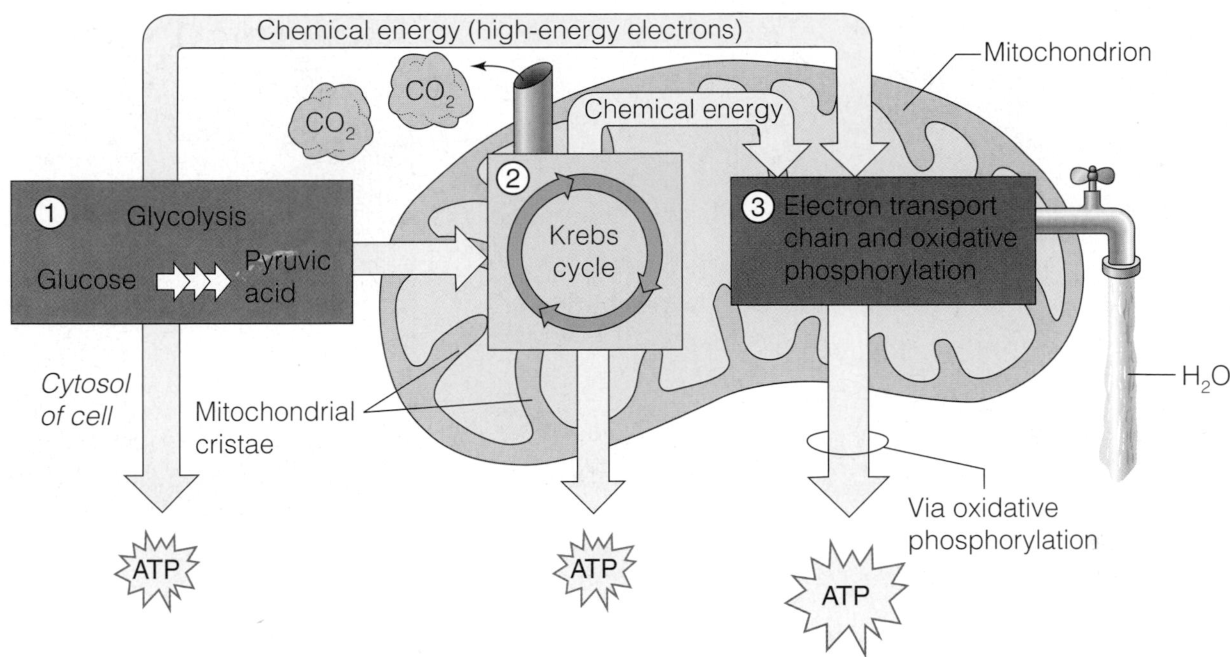

FIGURE 14.18 An overview of sites of ATP formation during cellular respiration. Glycolysis occurs outside the mitochondria in the cytosol. The Krebs cycle and the electron transport chain reactions occur within the mitochondria. ① During glycolysis, hydrogen atoms containing high-energy electrons are removed as each glucose molecule is broken down into two molecules of pyruvic acid. ② The pyruvic acid enters the mitochondrion, where Krebs cycle enzymes remove more hydrogen and decompose the pyruvic acid to carbon dioxide. During glycolysis and the Krebs cycle, small amounts of ATP are formed. ③ Chemical energy from glycolysis and the Krebs cycle, in the form of hydrogen atoms containing energy-rich electrons, is then transferred to the electron transport chain, which is built into the membrane of the cristae. The electron transport chain carries out oxidative phosphorylation, which produces most of the ATP generated by cellular respiration and finally unites the removed hydrogen with oxygen to form water.

mitochondrial cristae membranes (Figure 14.19). There the hydrogen atoms are split into hydrogen ions (H^+) and electrons (e^-). The electrons "fall down an energy hill" going from each carrier to a carrier of lower energy. They give off their "load" of energy in a series of steps in small enough amounts to enable the cell to attach phosphate to ADP and make ATP. Ultimately free oxygen is reduced (the electrons and hydrogen ions are united with molecular oxygen), forming water and a large amount of ATP. The beauty of this system is that, unlike the explosive reaction that happens when O_2 is combined with hydrogen, relatively small amounts of energy are lost as heat (and light).

Because glucose is the major fuel for making ATP, homeostasis of blood glucose levels is critically important. If there are excessively high levels of glucose in the blood (*hyperglycemia* [hi″per-gli-se′me-ah]), some of the excess is stored in body cells (particularly liver and muscle cells) as glycogen. If blood glucose levels are still too high, excesses are converted to fat. There is no question that eating large amounts of empty-calorie foods such as candy and other sugary sweets causes a rapid deposit of fat in the body's adipose tissues. When blood glucose levels are too low *(hypoglycemia),* the liver breaks down stored glycogen and releases glucose to the blood for cellular use.

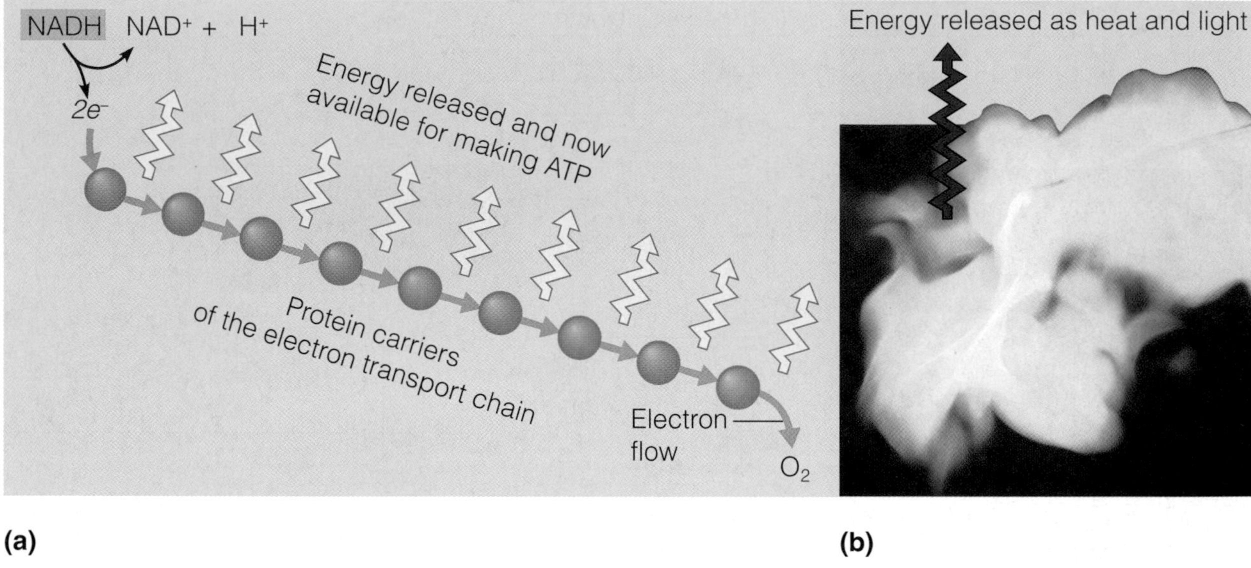

FIGURE 14.19 Electron transport chain versus one-step reduction of oxygen. (a) In cellular respiration, cascading electrons release energy in small steps and finally reduce O_2. The energy released is in quantities easily used to form ATP. (NADH is a niacin-containing coenzyme that delivers H^+ to the electron transport chain.) **(b)** When O_2 is reduced (combined with hydrogen) in one step, the result is an explosion.

These various fates of carbohydrates are shown in Figure 14.20a.

Fat Metabolism

As described shortly, the liver handles most lipid, or fat, metabolism that goes on in the body. The liver cells use some fats to make ATP for their own use; some to synthesize lipoproteins, thromboplastin (a clotting protein), and cholesterol; and then release the rest to the blood in the form of relatively small, fat-breakdown products. Body cells remove the fat products and cholesterol from the blood and build them into their membranes or steroid hormones as needed. Fats are also used to form myelin sheaths of neurons (see Chapter 7) and fatty cushions around body organs. In addition, stored fats are the body's most concentrated source of energy. (Catabolism of 1 gram of fat yields twice as much energy as the breakdown of 1 gram of carbohydrate or protein.)

For fat products to be used for ATP synthesis, they must first be broken down to acetic acid (Figure 14.20d). Within the mitochondria, the acetic acid (like the pyruvic acid product of carbohydrates) is then completely oxidized, and carbon dioxide, water, and ATP are formed. When there is not enough glucose to fuel the needs of the cells for energy, larger amounts of fats are used to produce ATP. Under such conditions, fat oxidation is fast but incomplete, and some of the intermediate products such as acetoacetic acid and acetone begin to accumulate in the blood. These cause the blood to become acidic (a condition called *acidosis,* or *ketosis*), and the breath takes on a fruity odor as acetone diffuses from the lungs. Ketosis is a common consequence of "no-carbohydrate" diets, uncontrolled diabetes mellitus, and starvation in which the body is forced to rely almost totally on fats to fuel its energy needs. While fats are an important energy source, cholesterol is *never* used as a cellular fuel. Its importance lies in the functional molecules and in the structures it helps to form.

Excess fats are stored in fat depots such as the hips, abdomen, breasts, and subcutaneous tissues. Although the fat in subcutaneous tissue is important as insulation for the deeper body organs, excessive amounts restrict movement and place greater demands on the circulatory system. The metabolism and uses of fats are shown in Figure 14.20b.

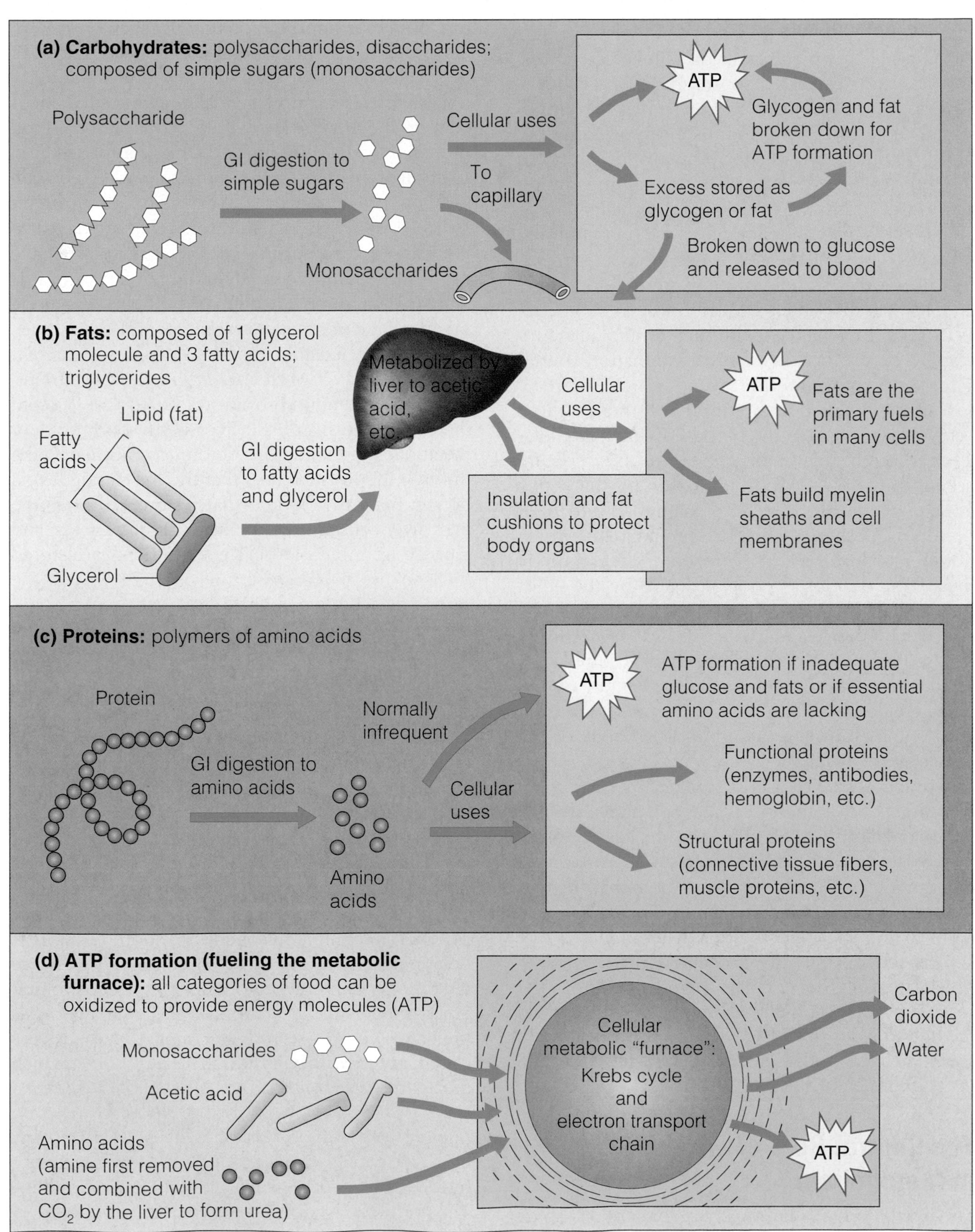

FIGURE 14.20 Metabolism by body cells. (a) Carbohydrate metabolism. **(b)** Fat metabolism. **(c)** Protein metabolism. **(d)** ATP formation.

Protein Metabolism

Proteins make up the bulk of cellular structures, and they are carefully conserved by body cells. Ingested proteins are broken down to amino acids. Once the liver has finished processing the blood draining the digestive tract and has taken its "fill" of amino acids, the remaining amino acids circulate to the body cells. The cells remove amino acids from the blood and use them to build proteins, both for their own use (enzymes, membranes, mitotic spindle proteins, muscle proteins) and for export (mucus, hormones, and others). Cells take few chances with their amino acid supply. They use ATP to actively transport amino acids into their interior even though in many cases they may contain more of those amino acids than there is in the blood flowing past them. Even though this may appear to be "cellular greed," there is an important reason for this active uptake of amino acids. Cells cannot build their proteins unless *all* the needed amino acids, which number around 20, are present. Since nine of these amino acids cannot be made by the cells, they are available to the cells only through the diet. Such amino acids are called *essential amino acids.* This helps explain the avid accumulation of amino acids, which ensures that all amino acids needed will be available for present and (at least some) future protein-building needs of the cells (Figure 14.20c).

Amino acids are used to make ATP only when proteins are overabundant and/or when carbohydrates and fats are not available. When it is necessary to oxidize amino acids for energy (Figure 14.20d), their amine groups are removed as *ammonia,* and the rest of the molecule enters the Krebs cycle pathways in the mitochondria. The ammonia that is released during this process is toxic to body cells, especially nerve cells. The liver comes to the rescue by combining the ammonia with carbon dioxide to form **urea** (u-re′ah). Urea, which is not harmful to the body cells, is then flushed from the body in urine.

The Central Role of the Liver in Metabolism

The liver is one of the most versatile and complex organs in the body. Without it we would die within 24 hours. Its role in digestion (that is, the manufacture of bile) is important to the digestive process to be sure, but it is only one of the many functions of liver cells. The liver cells detoxify drugs and alcohol, degrade hormones, make many substances vital to the body as a whole (cholesterol, blood proteins such as albumin and clotting proteins, and lipoproteins), and play a central role in metabolism as they process nearly every class of nutrient. Because of the liver's key roles, nature has provided us with a surplus of liver tissue. We have much more than we need, and even if part of it is damaged or removed, it is one of the few body organs that can regenerate rapidly and easily.

As described in Chapter 11, a unique circulation, the *hepatic portal circulation,* brings nutrient-rich blood draining from the digestive viscera directly to the liver. The liver is the body's major metabolic organ, and this detour that nutrients take through the liver ensures that the liver's needs will be met first. As blood circulates slowly through the liver, liver cells remove amino acids, fatty acids, and glucose from the blood. These nutrients are stored for later use or processed in various ways. At the same time, the liver's phagocytic cells remove and destroy bacteria that have managed to get through the walls of the digestive tract and into the blood.

General Metabolic Functions

The liver is vitally important in helping to maintain blood glucose levels within normal range (around 100 mg glucose/100 ml of blood). After a carbohydrate-rich meal, thousands of glucose molecules are removed from the blood and combined to form the large polysaccharide molecules called **glycogen** (gli′ko-jen), which are then stored in the liver. This process is **glycogenesis** (gli″ko-jen′ĕ-sis), literally, "glycogen formation" (*genesis* = beginning). Later, as body cells continue to remove glucose from the blood to meet their needs, blood glucose levels begin to drop. At this time, liver cells break down the stored glycogen, by a process called **glycogenolysis** (gli″ko-jen-ol′ĭ-sis), which means "glycogen splitting." The liver cells then release glucose bit by bit to the blood to maintain homeostasis of blood glucose levels. If necessary, the liver can also make glucose from noncarbohydrate substances such as fats and proteins. This process is **gluconeogenesis** (glu″ko-ne″o-jen′ĕ-sis), which means "formation of new sugar" (Figure 14.21). As described in Chapter 9, hormones

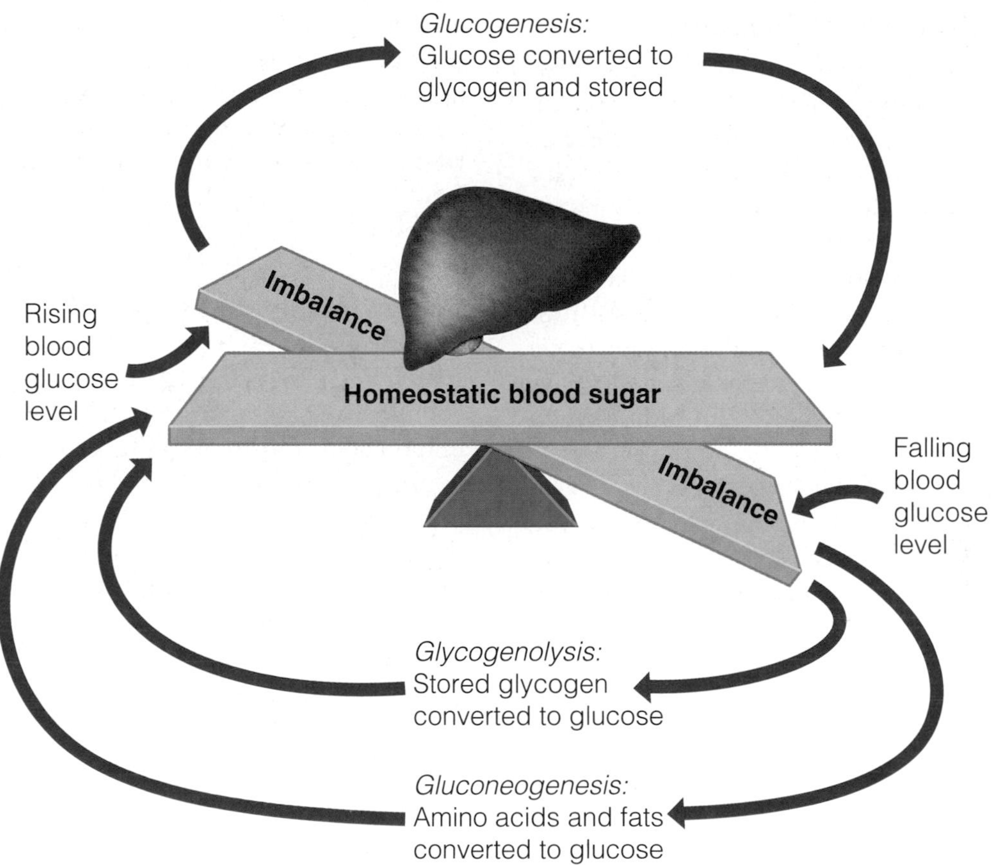

FIGURE 14.21 Metabolic events occurring in the liver as blood glucose levels rise and fall. When the blood glucose level is rising, the liver removes glucose from the blood and stores it as glycogen (glycogenesis). When the blood glucose level falls, the liver breaks down stored glycogen (glycogenolysis) and makes new glucose from amino acids and fats (gluconeogenesis). The glucose is then released to the blood to restore homeostasis of blood sugar.

such as thyroxine, insulin, and glucagon are vitally important in controlling the blood sugar levels and the handling of glucose in all body cells.

Some of the fats and fatty acids picked up by the liver cells are oxidized for energy (to make ATP) for use by the liver cells themselves. The rest are broken down to simpler substances such as *acetic acid* and *acetoacetic acid* (two acetic acids linked together) and released into the blood, or stored as fat reserves in the liver. The liver also makes cholesterol and secretes cholesterol's breakdown products in bile.

All blood proteins made by the liver are built from the amino acids its cells pick up from the blood. The completed proteins are then released back into the blood to travel throughout the circulation. *Albumin,* the most abundant protein in blood, holds fluids in the bloodstream. When insufficient albumin is present in blood, fluid leaves the bloodstream and accumulates in the tissue spaces, causing edema. The role of the protective *clotting proteins* made by the liver was discussed in Chapter 10. Liver cells also synthesize nonessential amino acids and, as mentioned earlier, detoxify ammonia (produced when amino acids are oxidized for energy) by converting it to urea.

Nutrients not needed by the liver cells, as well as the products of liver metabolism, are released into the blood and drain from the liver in the hepatic vein to enter the systemic circulation, where they become available to other body cells.

Cholesterol Metabolism and Transport

Although it's a very important lipid in the diet, **cholesterol** is not used as an energy fuel. Instead, it serves as the structural basis of steroid hormones and vitamin D and is a major building block of plasma membranes. Because we hear so much about "cutting down our cholesterol intake" in the media, it is always surprising to learn that only about 15 percent of blood cholesterol comes from the diet. The other 85 percent or so is made by the liver. Cholesterol is lost from the body when it is broken down and secreted in bile salts, which eventually leave the body in feces.

Because of the important role they play in fat and cholesterol transport, the lipoproteins, one class of proteins made by the liver and known by the buzzwords *HDLs* and *LDLs,* deserve a bit more attention.

Since fatty acids, fats, and cholesterol are insoluble in water, they cannot circulate freely in the bloodstream. Instead they are transported bound to the small lipid-protein complexes called lipoproteins. Although the entire story is complex, the important thing to know is that the **low-density lipoproteins,** or **LDLs,** transport cholesterol and other lipids *to* body cells, where they are used in various ways. If large amounts of LDLs are circulating, the chance that fatty substances will be deposited on the arterial walls, initiating atherosclerosis, is high. Because of this possibility, the LDLs are unkindly tagged as "bad lipoproteins." By contrast, the lipoproteins that transport cholesterol *from* the tissue cells (or arteries) to the liver for disposal in bile are **high-density lipoproteins,** or **HDLs.** High HDL levels are considered "good" because the cholesterol is destined to be broken down and eliminated from the body. Obviously both LDLs and HDLs are "good and necessary"; it is just their relative ratio in the blood that determines whether or not potentially lethal cholesterol deposits are likely to be laid down in the artery walls. In general, aerobic exercise, a diet low in saturated fats and cholesterol, and abstaining from smoking and drinking coffee all appear to favor a desirable HDL/LDL ratio.

Body Energy Balance

When any fuel is burned, it consumes oxygen and liberates heat. The "burning" of food fuels by body cells is no exception. As stated in Chapter 2, energy cannot be created or destroyed—it can only be converted from one form to another. If we apply this principle to cellular metabolism, it means that a dynamic balance exists between the body's energy intake and its energy output:

Energy intake = total energy output
(heat + work + energy storage)

Energy intake is the energy liberated during food oxidation—that is, during the reactions of glycolysis, the Krebs cycle, and the electron transport chain. **Energy output** includes the energy we immediately lose as heat (about 60 percent of the total), plus that used to do work (driven by ATP), plus energy that is stored in the form of fat or glycogen. Energy storage is important only during periods of growth and during net fat deposit.

Regulation of Food Intake

When energy intake and energy outflow are balanced, body weight remains stable. When they are not, weight is either gained or lost. Since body weight in most people is surprisingly stable, mechanisms that control food intake or heat production or both must exist.

But how is food intake controlled? That is a difficult question, and one that is still not fully answered. Researchers believe that several factors—such as rising or falling blood levels of nutrients (glucose and amino acids), hormones (insulin, glucagon, and leptin), or body temperature (rising temperature is inhibitory), and psychological factors—have an effect on eating behavior through feedback signals to the brain. Indeed, psychological factors are believed to be a very important cause of obesity. However, even when psychological factors *are* the underlying cause of obesity, individuals do *not* continue to gain weight endlessly. It seems that their feeding controls still operate, but they act to maintain total body energy content at higher-than-normal levels.

Metabolic Rate and Body Heat Production

Basal Metabolic Rate When nutrients are broken down to produce cellular energy (ATP), they yield different amounts of energy. As mentioned earlier, the energy value of foods is measured in a unit called the *kilocalorie (kcal).* In general, carbohydrates and proteins each yield 4 kcal/gram, and fats yield 9 kcal/gram when they are broken down

TABLE 14.3 Factors Determining the Basal Metabolic Rate (BMR)

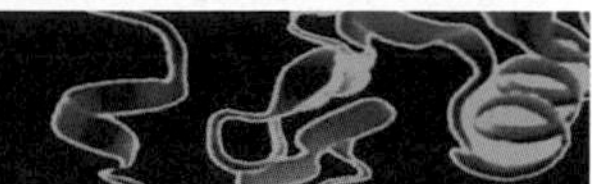

Factor	Variation	Effect on BMR
Surface area	Large surface area in relation to body volume, as in thin, small individuals	Increased
	Small surface area in relation to body volume, as in large, heavy individuals	Decreased
Sex	Male Female	Increased Decreased
Thyroxine production	Increased Decreased	Increased Decreased
Age	Young, rapid growth Aging, elderly	Increased Decreased
Strong emotions (anger or fear) and infections		Increased

for energy production. Most meals, and even many individual foods, are mixtures of carbohydrates, fats, and proteins. To determine the caloric value of a meal, we must know how many grams of each type of foodstuff it contains. For most of us, this is a difficult chore indeed, but approximations can easily be made with the help of a simple, calorie-values guide available in most drugstores.

The amount of energy used by the body is also measured in kilocalories. The **basal metabolic rate (BMR)** is the amount of heat produced by the body per unit of time when it is under basal conditions—that is, at rest. It reflects the energy supply a person's body needs just to perform essential life activities such as breathing, maintaining the heartbeat, and kidney function. An average 70-kg (154-pound) adult has a BMR of about 60 to 72 kcal/hour.

Many factors influence BMR, including surface area and gender. As shown in Table 14.3, small, thin males tend to have a higher BMR than large, obese females. Age is also important; children and adolescents require large amounts of energy for growth and have relatively high BMRs. In old age, the BMR decreases dramatically as the muscles begin to atrophy.

The amount of **thyroxine** produced by the thyroid gland is probably the most important factor in determining a person's BMR; hence, thyroxine has been dubbed the "metabolic hormone." The more thyroxine produced, the higher the oxygen consumption and ATP use, and the higher the metabolic rate. In the past, most BMR tests were done to determine whether sufficient thyroxine was being made. Today, thyroid activity is more easily assessed by blood tests.

Homeostatic Imbalance

Hyperthyroidism causes a host of effects due to the excessive metabolic rate it produces. The body catabolizes stored fats and tissue proteins, and despite increased hunger and food intake, the person often loses weight. Bones weaken and body muscles, including the heart, atrophy. In contrast, *hypothyroidism* results in slowed metabolism, obesity, and diminished thought processes. ▲

Total Metabolic Rate When we are active, more glucose must be oxidized to provide energy for the additional activities. Digesting food and even modest physical activity increase the body's caloric requirements dramatically. These additional fuel requirements are above and beyond the energy required to maintain the body in the basal state. **Total metabolic rate (TMR)** refers to the total amount of kilocalories the body must consume to fuel all ongoing activities. Muscular work is the major body activity that increases the TMR. Even slight increases in skeletal muscle activity cause remarkable leaps in metabolic rate. When a well-trained athlete exercises vigorously for several minutes the TMR may increase to 15 to 20 times normal, and it remains elevated for several hours afterward.

When the total amount of calories consumed is equal to the TMR, homeostasis is maintained, and our weight remains constant. However, if we eat more than we need to sustain our activities, excess calories appear in the form of fat deposits. Conversely, if we are extremely active and do not properly feed the "metabolic furnace," we begin to break down fat reserves and even tissue proteins to satisfy our TMR. This principle is used in every good weight-loss diet. (The total calories needed are calculated on the basis of body size and age. Then, 20 percent or more of the requirements are cut from the daily diet.) If the dieting person exercises regularly, weight drops off even more quickly because the TMR increases above the person's former rate.

Body Temperature Regulation

Although we have been emphasizing that foods are "burned" to produce ATP, remember that ATP is not the only product of cell catabolism. Most of the energy released as foods are oxidized escapes as heat. Less than 40 percent of available food energy is actually captured to form ATP. The heat released warms the tissues and, more importantly, the blood, which circulates to all body tissues, keeping them at homeostatic temperatures, which allows metabolism to occur efficiently.

Body temperature reflects the balance between heat production and heat loss. The body's thermostat is in the *hypothalamus* of the brain. Through autonomic nervous system pathways, the hypothalamus continually regulates body temperature around a set point of 35.6° to 37.8°C (96° to 100°F) by initiating heat-loss or heat-promoting mechanisms (Figure 14.22).

Heat-Promoting Mechanisms When the environmental temperature is cold (or the temperature of circulating blood falls), body heat must be conserved (increased). Short-term means of accomplishing this are **vasoconstriction** of blood vessels of the skin and **shivering.**

When the skin vasculature constricts, the skin is temporarily bypassed by the blood, and blood is rerouted to the deeper, more vital, body organs. When this happens, the temperature of the exposed skin drops to that of the external environment.

Homeostatic Imbalance

Restriction of blood delivery to the skin is no problem for brief periods of time. But if it is extended, the skin cells, chilled by internal ice crystals and deprived of oxygen and nutrients, begin to die. This condition, called *frostbite,* is extremely serious. ▲

When the *core* body temperature (the temperature of the deep organs) drops to the point beyond which simple constriction of skin capillaries can handle the situation, shivering begins. Shivering, involuntary shudderlike contractions of the voluntary muscles, is very effective in increasing the body temperature, because skeletal muscle activity produces large amounts of heat.

Homeostatic Imbalance

Extremely low body temperature resulting from prolonged exposure to cold is *hypothermia.* In hypothermia, the individual's vital signs (respiratory rate, blood pressure, heart rate) decrease. The person becomes drowsy and oddly comfortable, even though previously he or she felt extremely cold. Uncorrected, the situation progresses to coma and finally death as metabolic processes grind to a stop. ▲

Heat-Loss Mechanisms Just as the body must be protected from becoming too cold, it must also be protected from excessively high temperatures. Most heat loss occurs through the skin via **radiation** or **evaporation.** When body temperature increases above what is desirable, the blood vessels serving the skin dilate and capillary beds in the skin become flushed with warm blood. As a result, heat radiates from the skin surface. However, if the external environment is as hot as or hotter than the body, heat cannot be lost by radiation, and the only means of getting rid of excess heat is by the evaporation of perspiration off the skin surface. This is an efficient means of body-heat loss as long as the air is dry. If it is humid, evaporation occurs at a much

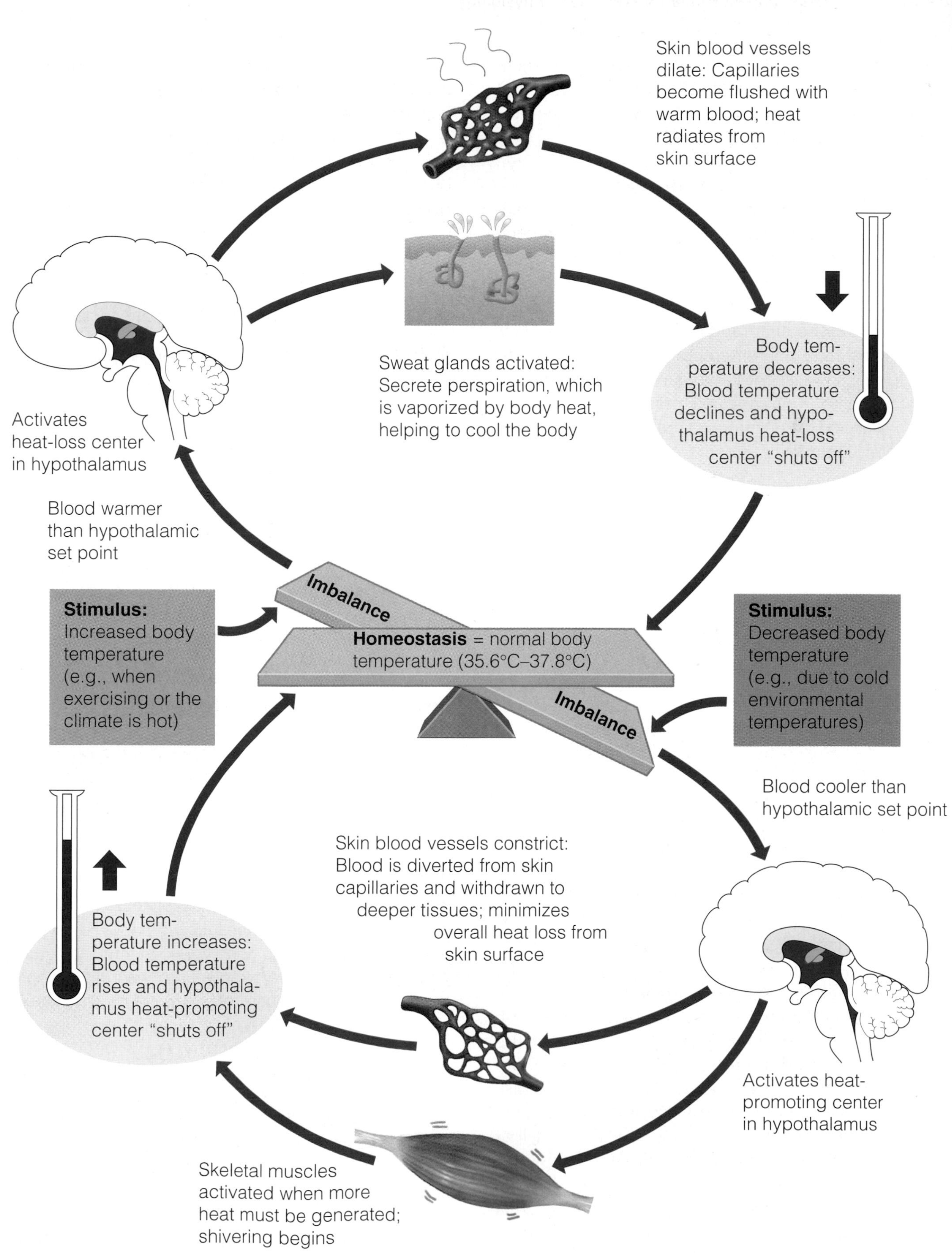

FIGURE 14.22 Mechanisms of body temperature regulation.

slower rate. In such cases, our heat-liberating mechanisms don't work well, and we feel miserable and irritable. This is why the hot, humid days of August are often called the "dog days."

Homeostatic Imbalance

When normal heat loss processes become ineffective, the *hyperthermia,* or elevated body temperature, that results depresses the hypothalamus. As a result, a vicious positive feedback cycle occurs: Soaring body temperature increases the metabolic rate, which in turn increases heat production. The skin becomes hot and dry; and, as the temperature continues to spiral upward, permanent brain damage becomes a distinct possibility. This condition, called **heat stroke,** can be fatal unless rapid corrective measures are taken immediately (immersion in cool water and administration of fluids).

Heat exhaustion is the term used to describe the heat-associated collapse of an individual during or following vigorous physical activity. Heat exhaustion results from excessive loss of body fluids (dehydration) and is evidenced by low blood pressure, a rapid heartbeat, and cool, clammy skin. In contrast to heat stroke, heat-loss mechanisms still operate in heat exhaustion. ▲

Fever is *controlled hyperthermia.* Most often, it results from infection somewhere in the body, but it may be caused by other conditions (cancer, allergic reactions, and CNS injuries). Macrophages, white blood cells, and injured tissue cells release chemical substances called *pyrogens* (pi′ro-jenz) that act directly on the hypothalamus, causing its thermostat to be set to a higher temperature (*pyro* = fire). After the thermostat resetting, heat-promoting mechanisms are initiated. Because of vasoconstriction, the skin becomes cool, and shivering begins to generate heat. This situation, called "the chills," is a sure sign that body temperature is rising. Body temperature is allowed to rise until it reaches the new setting. Then, it is maintained at the "fever setting" until natural body defense processes or antibiotics reverse the disease process. At that point, the thermostat is reset again to a lower (or normal) level, causing heat-loss mechanisms to swing into action—the individual begins to sweat, and the skin becomes flushed and warm. Physicians have long recognized that these signs signaled a turn for the better in their patients and said that the patient had "passed the crisis" because body temperature was falling.

As explained in Chapter 12, fever, by increasing the metabolic rate, helps speed the various healing processes, and it also appears to inhibit bacterial growth. The danger of fever is that if the body thermostat is set too high, body proteins may be denatured, and permanent brain damage may occur.

PART III: DEVELOPMENTAL ASPECTS OF THE DIGESTIVE SYSTEM AND METABOLISM

The very young embryo is flat and pancake-shaped. However, it soon folds to form a cylindrical body, and its internal cavity becomes the cavity of the alimentary canal. By the fifth week of development, the alimentary canal is a continuous tube-like structure extending from the mouth to the anus. Shortly after, the digestive glands (salivary glands, liver, and pancreas) bud out from the mucosa of the alimentary tube. These glands retain their connections (ducts) and can easily empty their secretions into the alimentary canal to promote its digestive functions.

Homeostatic Imbalance

The digestive system is susceptible to many congenital defects that interfere with feeding. The most common is the *cleft palate/cleft lip* defect. Of the two, cleft palate is more serious because the child is unable to suck properly. Another relatively common congenital defect is a *tracheoesophageal fistula* (tra″ke-o-ĕ-sof′ah-je-al fis′tu-lah). In this condition, there is a connection between the esophagus and the trachea. In addition, the esophagus often (but not always) ends in a blind sac and does not connect to the stomach. The baby chokes, drools, and becomes cyanotic during feedings because food is entering the respiratory passageways. All three defects can be corrected surgically.

There are many types of inborn errors of metabolism (genetically based problems that interfere with metabolism), but perhaps the two most common are *cystic fibrosis (CF)* and *phenylketonuria* (fen″il-ke′to-nu″re-ah) *(PKU).* CF primarily affects the lungs, but it also significantly impairs the activity of the pancreas. In CF, huge amounts of mucus are produced, which block the passages of involved organs. Blockage of the pancreatic duct prevents pancreatic fluid from reaching the small intestine. As a result, fats and fat-soluble vitamins are not digested or absorbed, and bulky, fat-laden stools result. This condition is usually handled by administering pancreatic enzymes with meals.

PKU involves an inability of tissue cells to use phenylalanine (fen"il-al'ah-nin), an amino acid present in all protein foods. In such cases, brain damage and retardation occur unless a special diet low in phenylalanine is prescribed. ▲

The developing infant receives all its nutrients through the placenta, and at least at this period of life, obtaining and processing nutrients is no problem (assuming the mother is adequately nourished). Obtaining nutrition is the most important activity of the newborn baby, and several reflexes present at this time help in this activity. For example, the *rooting reflex* helps the infant find the nipple (mother's or bottle), and the *sucking reflex* helps him or her to hold on to the nipple and swallow. The stomach of a newborn infant is very small, so feeding must be frequent (every three to four hours). Peristalsis is rather inefficient at this time, and vomiting is not at all unusual.

Teething begins around age 6 months and continues until about the age of 2 years. During this interval, the infant progresses to more and more solid foods and usually is eating an adult diet by toddlerhood. Appetite decreases in the elementary school–aged child and then increases again during the rapid growth of adolescence. (Parents of adolescents usually bewail their high grocery bills!)

All through childhood and into adulthood, the digestive system operates with relatively few problems unless there are abnormal interferences such as contaminated food or extremely spicy or irritating foods (which may cause inflammation of the gastrointestinal tract, or *gastroenteritis* [gas"tro-en-ter-i'tis]). Inflammation of the appendix, *appendicitis,* is particularly common in teenagers for some unknown reason. Between middle age and early old age, the metabolic rate decreases by 5 to 8 percent in every 10-year period. This is the time of life when the weight seems to creep up, and obesity often becomes a fact of life. To maintain desired weight, we must be aware of this gradual change and be prepared to reduce caloric intake. Two distinctly middle-age digestive problems are *ulcers* (see the "A Closer Look" box on pp. 472–473) and *gallbladder problems* (inflammation of the gallbladder or gallstones).

During old age, activity of the GI tract declines. Fewer digestive juices are produced, and peristalsis slows. Taste and smell become less acute, and periodontal disease often develops. Many elderly individuals live alone or on a reduced income. These factors, along with increasing physical disability, tend to make eating less appealing, and nutrition is inadequate in many of our elderly citizens. Diverticulosis and cancer of the gastrointestinal tract are fairly common problems in the elderly. Cancer of the stomach and colon rarely have early signs and often progress to an inoperable stage (that is, spread to distant parts of the body as well) before the individual seeks medical attention. However, when detected early, both diseases are treatable, and it has been suggested that diets high in plant fiber and low in fat might help to decrease the incidence of colon cancer. Additionally, since most colorectal cancers derive from initially benign mucosal tumors called *polyps,* and the incidence of polyp formation increases with age, a yearly colon examination should be a health priority in everyone over the age of 50.

Prove It Yourself

Calculate Your Basal Metabolic Rate

How much energy does your body need just to perform essential life activities? You can calculate your BMR as follows:

1. Calculate your weight in kilograms:
   Divide the number of pounds by 2.2 = ________
2. If you are male:
   Multiply your weight × 1.0 = ________
3. If you are female:
   Multiply your weight × 0.9 = ________
4. This number approximates the number of kilocalories your body consumes per hour. To estimate how many kilocalories you need each day to support basic metabolic functions, multiply this number by 24 = ________

For example, a 200-pound man (91 kg) has a BMR of 91 kcal/hour; his metabolic activities consume approximately 2,184 kilocalories per day. A 130-pound woman (59 kg) has a BMR of 53 kcal/hour, and her metabolic activities consume only 1,272 kilocalories per day. Of course, other factors besides gender and body weight can influence BMR; see Table 14.3.

Obesity: Magical Solution Wanted

How fat is too fat? The bathroom scale is an inaccurate guide because body weight tells nothing of body composition. A skilled dancer with dense bones and well-developed muscles may weigh several pounds more than an inactive person of the same relative size.

The most common view of obesity is that it is a condition of excessive triglyceride storage. Although we bewail our inability to rid ourselves of fat, the real problem is that we keep refilling the storehouses by consuming too many calories. A body fat content of 18 to 22 percent of body weight (males and females, respectively) is considered normal for adults. Anything over that is defined as obesity.

The official medical measure of obesity and body fatness is called the *body mass index (BMI)*, an index of a person's weight relative to height. To estimate BMI, multiply weight in pounds by 705 and then divide by your height in inches squared:

$$\text{BMI} = \frac{\text{wt [lb]} \times 705}{(\text{height [inches]})^2}$$

Overweight is defined by a BMI between 25 and 30; *obese* is a BMI greater than 30.

However it's defined, obesity is a perplexing and poorly understood disease. The term *disease* is appropriate because all forms of obesity involve some imbalance in food intake control mechanisms. Despite its well-known adverse effects on health (the obese have a higher incidence of arteriosclerosis, hypertension, coronary artery disease, and diabetes mellitus), it is the most common health problem in the United States. One out of three of us is obese. Not only are U.S. kids getting fatter, but, because they are opting for home video games and nachos instead of touch football and apples, their general cardiovascular fitness is declining as well.

Besides the health problems mentioned, the obese may store excessive levels of fat-soluble toxic chemicals, such as the insecticide DDT and PCB (a cancer-causing chemical) in their bodies. DDT interferes with the liver's ability to rid the body of other toxins, so these effects may be very far reaching. As if this were not enough, the social stigma and economic disadvantages of obesity are legendary. A fat person pays higher insurance premiums, is discriminated against in the job market, has fewer clothing choices, and is frequently humiliated during both childhood and adulthood. With all its attendant problems, it's a pretty fair bet that few people choose to be obese. So what causes obesity? Let's look at three of the more recent theories.

Settling Point Theory

Some believe that overeating behaviors develop early in life (the "clean your plate" syndrome) and set the stage for adult obesity by increasing the number of fat cells formed during childhood. During adulthood, increases in adipose tissue mass occur because more fat is deposited in the existing cells. Thus, the more cells there are, the more fat can be stored.

Signals delivered by bloodborne nutrients or so-called satiety chemicals (hormones and others) should prevent massive fat deposit, but it appears that systems controlling hunger and satiety respond too slowly to stop a high-fat intake before the body has had too much. There are also hints that the fat cells themselves may stimulate overeating. Supporting this idea is the observation that when yo-yo dieters lose weight, their metabolic rate falls sharply. But when they subsequently gain weight, their metabolic rate increases like a furnace being stoked. Each successive weight loss occurs more slowly, but lost weight is regained three times as fast. Thus, it appears that people, like laboratory animals subjected to alternating "feasts and fasts," become increasingly food efficient, and their metabolic rates adjust to prevent any deviation from their weight "settling point." The instrument used to solve the problem—dieting (again and again)—becomes self-defeating.

Fuel Efficiency Theory

Obese people are more fuel efficient and more effective "fat storers." Although it is often assumed that obese people eat more than other people, this is not necessarily true—many actually eat less than those of normal weight.

Fat, the nutrient, is the obese person's worst enemy. Fats pack more wallop per calorie (are more fattening) than proteins or carbohydrates because very little energy is used in their processing. For example, when someone ingests 100 excess *fat* calories, only 3 calories are "burned" and the rest (97) are stored.

These facts apply to everyone, but when you are obese the picture is even bleaker. For example, fat cells of overweight people "sprout" more alpha receptors (the kind that favors fat accumulation), and their lipoprotein lipase enzyme, which unloads fat from the blood (usually to fat cells), is exceptionally efficient.

Genetic Predisposition Theory

Morbid obesity is the fate of those inheriting two obesity genes. However, a true genetic predisposition for "fatness" appears to account for only about 5 percent of obese people in the United States. These people, given excess calories, will always deposit them as fat, as opposed to those who lay down more muscle with some of the excess calories.

False and Risky Cures

Rumors and poor choices for dealing with obesity abound. Some of the most unfortunate strategies used for coping with obesity are listed here.

- *"Water pills."* Diuretics prompt the kidneys to excrete more water. At best, these may cause a few pounds of weight loss for a few hours; they can also cause serious electrolyte imbalance and dehydration.
- *Diet pills.* Some obese people use amphetamines (such as Dexedrine and Benzedrine ["speed"]) to reduce appetite and elevate the metabolic rate. These work, but only temporarily (until tolerance develops) and can cause a dangerous dependency. Furthermore, diet aids that provide fiber to prevent absorption of nutrients can cause serious malnutrition. In the early 1990s, a new combination drug therapy regimen, the fen-phen program, was conducted with mixed success. Fenfluramine (the "fen") increases levels of serotonin, which causes satiety, and phentermine ("phen") increases metabolic rate. These drugs were shown to promote weight loss of about 10 percent; but in September 1997, spooked by reports of heart valve abnormalities, the makers of these drugs withdrew them from the market. More recently, another diet drug, sibutramine (Merida), has come under fire for suspected cardiovascular side effects.
- *Fad diets.* Many magazines print at least one new diet regimen yearly, and diet products sell well. However, many of these diets are nutritionally unhealthy. For example, some of the liquid high-protein diets contain such poor-quality (incomplete) protein that they are actually dangerous. (The worst are those that contain collagen protein instead of a milk or soybean protein source.)
- *Surgery.* Sometimes sheer desperation prompts surgical solutions, such as having the jaws wired shut or the stomach stapled, intestinal bypass surgery, and biliopancreatic diversion (BPD). BPD "rearranges" the digestive tract: two-thirds of

Obesity: Magical Solution Wanted *(continued)*

the stomach is removed; the small intestine is cut in half and one 8-foot-long portion is sutured into the stomach opening. Since pancreatic juice and bile are diverted away from this "new intestine," fewer nutrients (and no fats) are digested and absorbed. Although the patients can eat anything they want without gaining weight, BPD is major surgery and carries all of its risks. Liposuction, the removal of adipose tissue by suction to reshape the body, does remove fat, but unless the patient changes his or her eating habits, the remaining fat deposits in other parts of the body overfill.

Unfortunately, there is no magical solution for obesity, and no current explanation of weight regulation leaves much room for voluntary control. Either your genes are helpful and your regulatory peptides are interacting with receptors that are responding as they should—or they're not! Nonetheless, without pharmacological help, the only way for most of us to lose weight is to take in fewer fat calories and increase physical activity. Fidgeting and resistance exercise help to increase muscle mass (muscle consumes more energy at rest than does fat). Low activity levels actually stimulate eating, while physical exercise depresses food intake and increases metabolic rate not only during activity, but also for some time after. The only way to keep the weight off is to make these dietary and exercise changes lifelong habits.

Systems in Sync

Homeostatic Relationships between the Digestive System and Other Body Systems

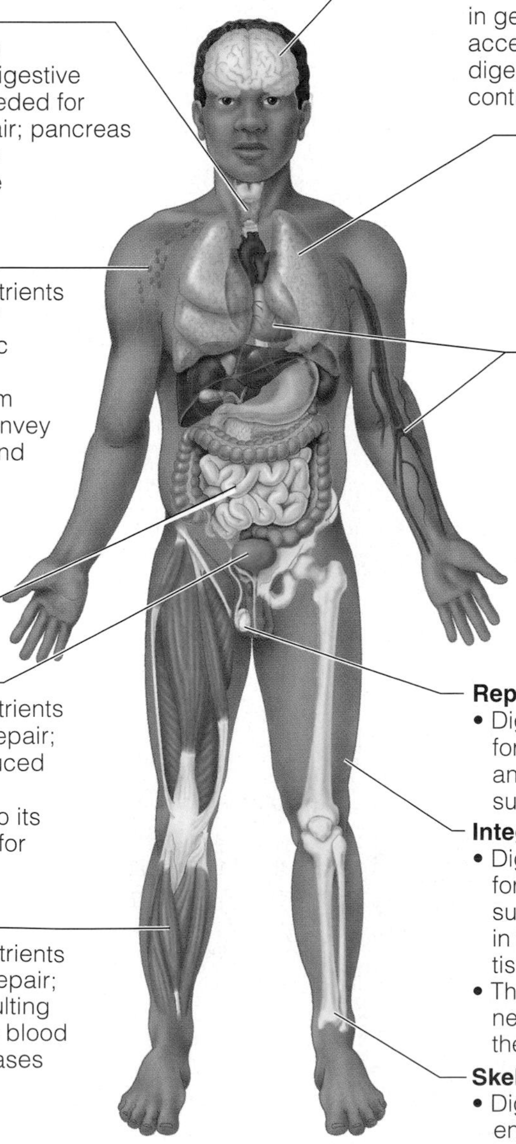

Nervous System
- Digestive system provides nutrients for normal neural functioning
- Neural controls of digestive function; in general, parasympathetic fibers accelerate and sympathetic fibers inhibit digestive activity; reflex and voluntary controls of defecation

Endocrine System
- Liver removes hormones from blood, ending their activity; digestive system provides nutrients needed for energy fuel, growth, and repair; pancreas has hormone-producing cells
- Local hormones help regulate digestive function

Respiratory System
- Digestive system provides nutrients for energy metabolism, growth, and repair
- Respiratory system provides oxygen and carries away carbon dioxide produced by digestive system organs

Lymphatic System/Immunity
- Digestive system provides nutrients for normal functioning; HCl of stomach provides nonspecific protection against bacteria
- Lacteals drain fatty lymph from digestive tract organs and convey it to blood; Peyer's patches and lymphoid tissue in mesentery house macrophages and immune cells that protect digestive tract organs against infection

Cardiovascular System
- Digestive system provides nutrients to heart and blood vessels; absorbs iron needed for hemoglobin synthesis; absorbs water necessary for normal blood volume
- Cardiovascular system transports nutrients absorbed by alimentary canal to all tissues of the body; distributes hormones of the digestive tract

Digestive System

Urinary System
- Digestive system provides nutrients for energy fuel, growth, and repair; excretes some bilirubin produced by the liver
- Kidneys transform vitamin D to its active form, which is needed for calcium absorption

Reproductive System
- Digestive system provides nutrients for energy fuel, growth, and repair and extra nutrition needed to support fetal growth

Integumentary System
- Digestive system provides nutrients for energy fuel, growth, and repair; supplies fats that provide insulation in the dermal and subcutaneous tissues
- The skin synthesizes vitamin D needed for calcium absorption from the intestine; protects by enclosure

Muscular System
- Digestive system provides nutrients for energy fuel, growth, and repair; liver removes lactic acid, resulting from muscle activity, from the blood
- Skeletal muscle activity increases motility of GI tract

Skeletal System
- Digestive system provides nutrients for energy fuel, growth, and repair; absorbs calcium needed for bone salts
- Skeletal system protects some digestive organs by bone; cavities store some nutrients (e.g., calcium, fats)

SUMMARY

Media study tools that could provide you with additional help in reviewing specific key topics of Chapter 14 are referenced below.

IP = Interactive Physiology;

WEB = A&P Place website.

PART I: ANATOMY AND PHYSIOLOGY OF THE DIGESTIVE SYSTEM (pp. 454–478)

ANATOMY OF THE DIGESTIVE SYSTEM (pp. 454–467)

1. The digestive system consists of the alimentary canal (a hollow tube extending from mouth to anus) and several accessory digestive organs. The wall of the alimentary canal has four main tissue layers—mucosa, submucosa, muscularis externa, serosa. The serosa (visceral peritoneum) is continuous with the parietal peritoneum, which lines the abdominal cavity wall.

   WEB Exercise: Chapter 14, Digestive System; Basic Structure of the Alimentary Wall.

2. Organs of the alimentary canal:
   a. The mouth, or oral cavity, contains teeth and tongue and is bounded by lips, cheeks, and palate. Tonsils guard its posterior margin.
   b. The pharynx is a muscular tube that provides a passageway for food and air.
   c. The esophagus is a muscular tube that completes the passageway from the pharynx to the stomach.
   d. The stomach is a C-shaped organ located on the left side of the abdomen beneath the diaphragm. Food enters it through the cardioesophageal sphincter and leaves it to enter the small intestine through the pyloric sphincter. The stomach has a third oblique layer of muscle in its wall that allows it to perform mixing or churning movements. Gastric glands produce hydrochloric acid, pepsin, rennin, mucus, gastrin, and intrinsic factor. Mucus protects the stomach itself from being digested.
   e. The tubelike small intestine is suspended from the posterior body wall by the mesentery. Its subdivisions are the duodenum, jejunum, and ileum. Food digestion and absorption are completed here. Pancreatic juice and bile enter the duodenum through a sphincter at the distal end of the bile duct. Microvilli, villi, and circular folds increase the surface area of the small intestine for enhanced absorption.
   f. The large intestine frames the small intestine. Subdivisions are the cecum; appendix; ascending, transverse, and descending colon; sigmoid colon; rectum; and anal canal. The large intestine delivers undigested food residue (feces) to the body exterior.

3. Salivary glands (three pairs—parotid, submandibular, and sublingual) secrete saliva into the oral cavity. Saliva contains mucus and serous fluids. The serous component contains salivary amylase.

4. Two sets of teeth are formed. The first set consists of 20 deciduous teeth that begin to appear at 6 months and are lost by 12 years of age. Permanent teeth (32) begin to replace deciduous teeth around 7 years. A typical tooth consists of crown covered with enamel and root covered with cementum. Most of the tooth is bonelike dentin. The pulp cavity contains blood vessels and nerves.

5. Several accessory organs duct substances into the alimentary tube.
   a. The pancreas is a soft gland lying in the mesentery between the stomach and small intestine. Pancreatic juice contains enzymes (which digest all categories of foods) in an alkaline fluid.
   b. The liver is a four-lobed organ overlying the stomach. Its digestive function is to produce bile, which it ducts into the small intestine.
   c. The gallbladder is a muscular sac that stores and concentrates bile. When fat digestion is not occurring, the continuously made bile backs up the cystic duct and enters the gallbladder.

FUNCTIONS OF THE DIGESTIVE SYSTEM (pp. 467–478)

1. Foods must be broken down to their building blocks to be absorbed. Building blocks of carbohydrates are simple sugars, or monosaccharides. Building blocks of proteins are amino acids. Building blocks of fats, or lipids, are fatty acids and glycerol.

   WEB Exercise: Chapter 14, Gastrointestinal Tract Activities.

2. Both mechanical (chewing) and chemical food breakdown begin in the mouth. Saliva contains

mucus, which helps bind food together into a bolus, and salivary amylase, which begins the chemical breakdown of starch. Saliva is secreted in response to food in the mouth, mechanical pressure, and psychic stimuli. Essentially no food absorption occurs in the mouth.

3. Swallowing has two phases: The buccal phase is voluntary; the tongue pushes the bolus into the pharynx. The involuntary pharyngeal-esophageal phase involves the closing off of nasal and respiratory passages and the conduction of food to the stomach by peristalsis.

4. When food enters the stomach, gastric secretion is stimulated by vagus nerves and by gastrin (a local hormone). Hydrochloric acid activates the protein-digesting enzyme pepsin, and chemical digestion of proteins begins. Food is also mechanically broken down by the churning activity of stomach muscles. Movement of chyme into the small intestine is controlled by the enterogastric reflex.

5. Chemical digestion of fats, proteins, and carbohydrates is completed in the small intestine by intestinal enzymes and, more importantly, pancreatic enzymes. Alkaline pancreatic juice neutralizes acid chyme and provides the proper environment for the operation of enzymes. Both pancreatic juice (the only source of lipases) and bile (formed by the liver) are necessary for normal fat breakdown and absorption. Bile acts as a fat emulsifier. Secretin and cholecystokinin, hormones produced by the small intestine, stimulate release of bile and pancreatic juice. Segmental movements mix foods; peristaltic movements move foodstuffs along the small intestine. Most nutrient absorption occurs by active transport into the capillary blood of the villi. Fats are absorbed by diffusion into both capillary blood and lacteals in the villi.

6. The large intestine receives bacteria-laden indigestible food residue. Activities of the large intestine are absorption of water and salts and of vitamins made by resident bacteria. When feces are delivered to the rectum by peristalsis and mass peristalsis, the defecation reflex is initiated.

PART II: NUTRITION AND METABOLISM (pp. 478–490)

NUTRITION (pp. 478–480)

1. Most foods are used as fuels to form ATP; the exception is proteins.

2. A nutrient is a substance in food used to promote growth, maintenance, and repair of the body.

3. The major nutrients are carbohydrates, lipids, and proteins. Vitamins and minerals are required in minute amounts.

4. Dietary carbohydrates (sugars and starch) are included in fruits and vegetables (plant products).

5. Dietary lipids are found in meats, dairy products, and vegetable oils.

6. Eggs, milk, meats, poultry, and fish are rich sources of protein.

7. Vitamins, found mainly in fruits, vegetables, and milk, function mainly as coenzymes in the body.

8. Minerals, most plentiful in vegetables and legumes, are mainly important for enzyme activity. Calcium is important for building bone, blood clotting, and secretory activities.

METABOLISM (pp. 480–490)

1. Metabolism includes all chemical breakdown (catabolic) and building (anabolic) reactions needed to maintain life.

2. Carbohydrates, most importantly glucose, are the body's major energy fuel. As glucose is oxidized, carbon dioxide, water, and ATP are formed. The sequential pathways of glucose catabolism are glycolysis, which occurs in the cytosol, and the Krebs cycle and electron transport chain (which function in the mitochondria). During hyperglycemia, glucose is stored as glycogen or converted to fat. In hypoglycemia, glycogenolysis, gluconeogenesis, and fat breakdown occur to restore normal blood glucose levels.

   WEB Exercise: Chapter 14, Overview of Cellular Respiration.

3. Fats insulate the body, protect organs, build some cell structures (membranes and myelin sheaths), and provide reserve energy. When carbohydrates are not available, more fats are oxidized to produce ATP. Excessive fat breakdown causes blood to become acidic. Excess dietary fat is stored in subcutaneous tissue and other fat depots.

4. Proteins form the bulk of cell structure and most functional molecules. They are carefully conserved by body cells. Amino acids are actively taken up from blood by tissue cells; those that cannot be made by body cells are called *essential amino acids.* Amino acids are oxidized to form ATP mainly when other fuel sources are not available. Ammonia, released as amino acids are catabolized, is detoxified by liver cells that combine it with carbon dioxide to form urea.

5. The liver is the body's key metabolic organ. Its cells remove nutrients from hepatic portal blood. It performs glycogenesis, glycogenolysis, and gluconeogenesis to maintain homeostasis of blood glucose levels. Its cells make blood proteins and other substances and release them to blood. Fats are burned by liver cells to provide some of their energy (ATP); excesses are stored or released to blood in simpler forms that can be used by other tissue cells. Phagocytic cells remove bacteria from hepatic portal blood. Most cholesterol is made by the liver; cholesterol breakdown products are secreted in bile. Fats and cholesterol are transported in the blood by lipoproteins. LDLs transport cholesterol to body cells; HDLs carry cholesterol to the liver for degradation. Cholesterol is used to make functional molecules and for some structural purposes; it is not used for energy.

6. A dynamic balance exists between energy intake and total energy output (heat + work + energy storage). Interference with this balance results in obesity or malnutrition leading to body wasting.

7. When the three major types of foods are oxidized for energy, they yield different amounts of energy. Carbohydrates and proteins yield 4 kcal/gram; fats yield 9 kcal/gram. Basal metabolic rate (BMR) is the total amount of energy used by the body when one is in a basal (resting) state. Age, sex, body surface area, and amount of thyroxine produced influence BMR.

8. Total metabolic rate (TMR) is number of calories used by the body to accomplish all ongoing daily activities. It increases dramatically as muscle activity increases. When TMR equals total caloric intake, weight remains constant.

9. As foods are catabolized to form ATP, more than 60 percent of energy released escapes as heat, warming the body. The hypothalamus initiates heat-loss processes (radiation of heat from skin and evaporation of sweat) or heat-promoting processes (vasoconstriction of skin blood vessels and shivering) as necessary to maintain body temperature within normal limits. Fever (hyperthermia) represents body temperature regulated at higher-than-normal levels.

PART III: DEVELOPMENTAL ASPECTS OF THE DIGESTIVE SYSTEM AND METABOLISM (pp. 490–491)

1. The alimentary tract forms as a hollow tube. Accessory glands form as outpocketings from this tube.

2. Common congenital defects include cleft palate, cleft lip, and tracheoesophageal fistula, all of which interfere with normal nutrition. Common inborn errors of metabolism are phenylketonuria (PKU) and cystic fibrosis (CF).

3. Various inflammatory conditions plague the digestive system throughout life. Appendicitis is common in adolescents, gastroenteritis and food poisoning can occur at any time (given the proper irritating factors), ulcers and gallbladder problems increase in middle age. Obesity and diabetes mellitus are bothersome during later middle age.

4. Efficiency of all digestive system processes decreases in the elderly. Gastrointestinal cancers, such as stomach and colon cancer, appear with increasing frequency in an aging population.

WEB Exercise: Chapter 14, Gastrointestinal Case Study.

REVIEW QUESTIONS

MULTIPLE CHOICE

More than one choice may apply.

1. Which of the following terms are synonyms?
   a. Gastrointestinal tract c. Digestive tract
   b. Digestive system d. Alimentary canal

2. A digestive organ that is *not* part of the alimentary canal is the:
   a. stomach d. large intestine
   b. liver e. pharynx
   c. small intestine

3. The GI tube tissue layer responsible for the actions of segmentation and peristalsis is the:
   a. serosa c. muscularis externa
   b. mucosa d. submucosa

4. The closure of which valve is assisted by the diaphragm?
   a. Ileocecal c. Cardioesophageal
   b. Pyloric d. Upper esophageal

5. Where in the stomach do the strongest peristaltic waves occur?
   a. Body c. Fundus
   b. Cardiac region d. Pylorus

6. Which of these organs lies in the right hypochondriac region of the abdomen?
   a. Stomach c. Cecum
   b. Spleen d. Liver

7. Release of CCK leads to:
   a. contraction of smooth muscle in the duodenal papilla
   b. increased activity of liver cells
   c. contraction of the gallbladder wall
   d. release of enzymes by the pancreas
8. The pH of chyme entering the duodenum is adjusted by:
   a. bile
   b. intestinal juice
   c. enzyme secretions from the pancreas
   d. bicarbonate-rich secretions from the pancreas
9. A 3-year-old girl is rewarded with a hug because she is now completely toilet trained. Which muscle is one that she has learned to control?
   a. Levator ani
   b. Internal anal sphincter
   c. Internal and external obliques
   d. External anal sphincter
10. Hormones that act to decrease blood glucose level include:
    a. insulin c. epinephrine
    b. glucagon d. growth hormone
11. Which events occur shortly after eating?
    a. Use of amino acids as a major source of energy
    b. Lipogenesis (and fat deposit)
    c. Breakdown of fat reserves
    d. Increased uptake of glucose by skeletal muscles and other body tissues
12. The material that forms the bulk of a tooth is:
    a. cementum c. enamel
    b. dentin d. pulp
13. Complete this statement. In glycolysis ___________ is oxidized and ___________ is reduced.
    a. vitamin-containing coenzyme; glucose
    b. ATP; ADP
    c. glucose; oxygen
    d. glucose; vitamin-containing coenzyme

SHORT ANSWER ESSAY

1. Make a simple line drawing of the organs of the alimentary tube, and label each organ.
2. Add three labels to your drawing—salivary glands, liver, and pancreas—and use arrows to show where each of these organs empties its secretion into the alimentary tube.
3. Name the layers of the alimentary tube wall from the lumen outward.
4. What is the mesentery? The peritoneum?
5. Name the subdivisions of the small intestine in a proximal to distal direction. Do the same for the subdivisions of the large intestine.
6. The digestive system has many structural modifications. Describe the structure and function of villi.
7. What is the normal number of permanent teeth? Of deciduous teeth? What substance covers the tooth crown? What substance makes up the bulk of a tooth? What is pulp, and where is it?
8. Name the three pairs of salivary glands. Name two functions of saliva.
9. Assume you have been chewing a piece of bread for five or six minutes. How would you expect its taste to change during this time? Why?
10. Name two regions of the digestive tract where mechanical food breakdown occurs, and explain how it is accomplished in those regions.
11. Name the organ where protein digestion begins.
12. Why is it necessary for the stomach contents to be so acidic? How does the stomach protect itself from being digested?
13. Only one organ produces enzymes capable of digesting all groups of foodstuffs. What organ is this?
14. Explain why fatty stools result from the absence of bile and/or pancreatic juice.
15. Define *emulsify*.
16. What is the function of gastrin? of secretin?
17. Describe the two phases of swallowing.
18. How do segmental and peristaltic movements differ?
19. A cream cheese and jelly sandwich contains proteins, carbohydrates, and fats. Describe what happens to the sandwich when you eat it relative to events occurring in ingestion, digestion, absorption, and elimination.
20. Where does most nutrient absorption occur?
21. What substances are absorbed in the large intestine?
22. What is the composition of feces?
23. Define *defecation reflex, constipation,* and *diarrhea.*
24. Define *metabolism, anabolism,* and *catabolism.*
25. Define *gluconeogenesis, glycogenolysis,* and *glycogenesis.*

26. Which food group is most important as a fuel source (that is, for ATP production)? Which is most important for building cell structures?

27. What is the harmful result when excessive amounts of fats are burned to produce ATP? Name two conditions that might lead to this result.

28. Define *BMR,* and name two factors that are important in determining an individual's BMR.

29. If your total caloric intake exceeds your TMR, what can you expect to happen?

30. How many calories are produced when 1 gram of carbohydrate is oxidized? 1 gram of protein? 1 gram of fat? If you just ate 100 grams of food that was 20 percent protein, 30 percent carbohydrate, and 10 percent fat, how many calories did you consume?

31. Some of the energy released as foods are oxidized is captured to make ATP. What happens to the rest of it?

32. Where is the body's thermostat?

33. Name two ways in which heat is lost from the body. Name two ways in which heat is retained or generated.

34. What is fever? What does it indicate?

35. Name three digestive system problems common to middle-aged adults. Name one common to teenagers. Name three common in elderly persons.

AT THE CLINIC

1. After chopping wood for about two hours on a hot but breezy afternoon, John stumbled into the house and then fainted. His T-shirt was wringing wet with perspiration, and his pulse was faint and rapid. Was he suffering from heat stroke or heat exhaustion? Explain your reasoning, and note what you should do to help John recover.

2. Harry is hospitalized with bacterial pneumonia. When you visit him, his teeth are chattering, his skin is cool and clammy to the touch, and he complains of feeling cold even though his room is quite warm. Explain his symptoms.

3. A young woman is put through an extensive battery of tests to determine the cause of her stomach pains. She is diagnosed with gastric ulcers. An antihistamine drug is prescribed, and she is sent home. What is the mechanism of her medication? What life-threatening problems can result from a poorly managed ulcer? Why did the clinic doctor warn the woman not to take aspirin?

4. Continuing from the previous question, the woman's ulcer got worse. She started complaining of back pain. The physician discovered that the back pain occurred because the pancreas was now damaged. Use logic to deduce how a perforating gastric ulcer could come to damage the pancreas.

5. Benny, a 6-year-old child who is allergic to milk, has extremely bowed legs. What condition do you suspect, and what is the connection to not drinking milk?

6. A new mother is worried about her week-old infant. The baby has begun to turn blue whenever she is fed and chokes during each feeding. What developmental abnormality do you suspect, and how will it be corrected?

7. An anorexic girl shows high levels of acetone in her blood. What is this condition called, and what has caused it?

8. Every year dozens of elderly people are found dead in their unheated apartments—victims of hypothermia. What is hypothermia, and how does it kill?

9. Mr. Harolde has had nonstop diarrhea all day and is severely weakened. Why is his nurse concerned?

15

The Urinary System

YOUR GOALS

After completing this chapter, you will have a working knowledge of the functions of the urinary system and will have mastered the objectives listed below.

FUNCTION PREVIEW

- The urinary system rids the body of nitrogenous wastes while regulating water, electrolyte, and acid-base balance of the blood.

OBJECTIVE CHECKLIST

KIDNEYS (pp. 502–511)

- Describe the location of the kidneys in the body.
- Identify the following regions of a kidney (longitudinal section): hilus, cortex, medulla, medullary pyramids, calyces, pelvis, and renal columns.
- Recognize that the nephron is the structural and functional unit of the kidney, and describe its anatomy.
- Describe the process of urine formation, identifying the areas of the nephron that are responsible for filtration, reabsorption, and secretion.
- Describe the function of the kidneys in excretion of nitrogen-containing wastes.
- Define *polyuria, anuria, oliguria,* and *diuresis.*
- Describe the composition of normal urine.
- List abnormal urinary components.

URETERS, URINARY BLADDER, AND URETHRA (pp. 512–514)

- Describe the general structure and function of the ureters, bladder, and urethra.
- Compare the course and length of the male urethra to that of the female.
- Define *micturition.*

>>

(Objective Checklist, continued)

- Describe the difference in control of the external and internal urethral sphincters.
- Name three common urinary tract problems.

FLUID, ELECTROLYTE, AND ACID-BASE BALANCE (pp. 514–520)

- Name and localize the three main fluid compartments of the body.
- Explain the role of antidiuretic hormone (ADH) in the regulation of water balance by the kidney.
- Explain the role of aldosterone in sodium and potassium balance of the blood.
- Compare and contrast the relative speed of buffers, the respiratory system, and the kidneys in maintaining the acid-base balance of the blood.

DEVELOPMENTAL ASPECTS OF THE URINARY SYSTEM (pp. 520–521)

- Describe three common congenital problems of the urinary system.
- Describe the effect of aging on urinary system functioning.

The kidneys, which maintain the purity and constancy of our internal fluids, are perfect examples of homeostatic organs. Much like sanitation workers who keep a city's water supply drinkable and dispose of its waste, the kidneys are usually unappreciated until there is a malfunction and "internal garbage" piles up. Every day, the kidneys filter gallons of fluid from the bloodstream. They then process this filtrate, allowing wastes and excess ions to leave the body in urine while returning needed substances to the blood in just the right proportions. Although the lungs and the skin also play roles in excretion, the kidneys bear the major responsibility for eliminating nitrogenous (nitrogen-containing) wastes, toxins, and drugs from the body.

Disposing of wastes and excess ions is only one part of the work of the kidneys. As they perform these excretory functions, they also regulate the blood's volume and chemical makeup so that the proper balance between water and salts and between acids and bases is maintained. Frankly, this would be tricky work for a chemical engineer, but the kidneys do it efficiently most of the time.

The kidneys have other regulatory functions as well: By producing the enzyme *renin* (re'nin), they help regulate blood pressure, and their hormone *erythropoietin* stimulates red blood cell production in bone marrow (see Chapter 10). Kidney cells also convert vitamin D to its active form.

The kidneys alone perform the functions just described and manufacture urine in the process. The other organs of the **urinary system**—the paired ureters and the single urinary bladder and urethra (Figure 15.1)—provide temporary storage reservoirs for urine or serve as transportation channels to carry it from one body region to another.

Kidneys

Location and Structure

Although many believe that the **kidneys** are located in the lower back, this is *not* their location. Instead, these small, dark red organs with a kidney-bean shape lie against the dorsal body wall in a *retroperitoneal* position (beneath the parietal peritoneum) in the *superior* lumbar region. The kidneys extend from the T_{12} to the L_3 vertebra; thus they receive some protection from the lower part of the rib cage. Because it is crowded by the liver, the right kidney is positioned slightly lower than the left. An adult kidney is about 12 cm (5 inches)

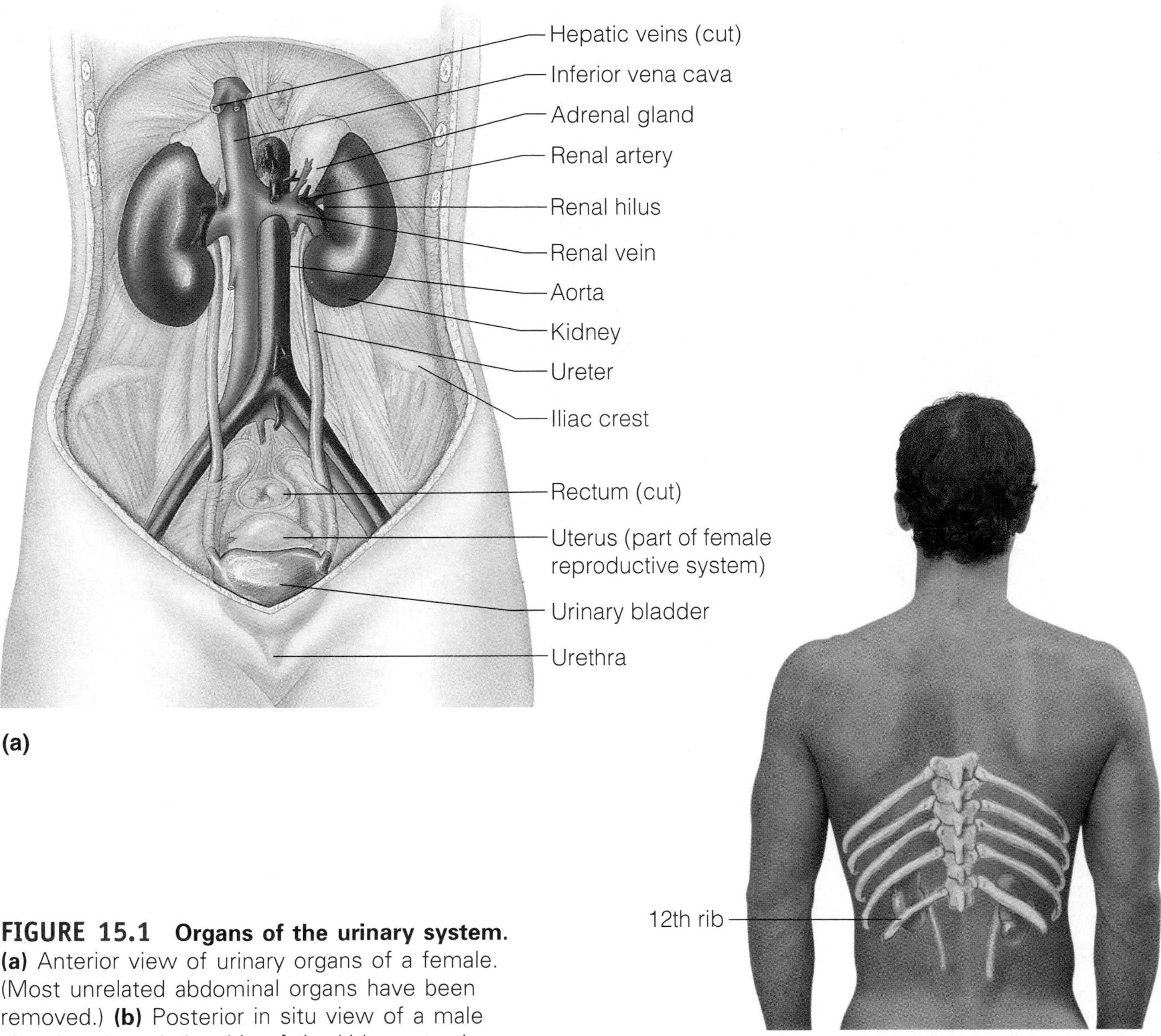

FIGURE 15.1 Organs of the urinary system. **(a)** Anterior view of urinary organs of a female. (Most unrelated abdominal organs have been removed.) **(b)** Posterior in situ view of a male showing the relationship of the kidneys to the 12th rib pair.

long, 6 cm (2.5 inches) wide, and 3 cm (1 inch) thick, about the size of a large bar of soap. It is convex laterally and has a medial indentation called the *renal hilus*. Several structures, including the ureters, the renal blood vessels, and nerves, enter or exit the kidney at the hilus (see Figures 15.1 and 15.2). Atop each kidney is an *adrenal gland,* which is part of the endocrine system and is a distinctly separate organ functionally.

A fibrous, transparent **renal capsule** encloses each kidney and gives a fresh kidney a glistening appearance. A fatty mass, the *adipose capsule,* surrounds each kidney and helps hold it in place against the muscles of the trunk wall.

Homeostatic Imbalance

The fat surrounding the kidneys is extremely important in holding them in their normal body position. If the amount of fatty tissue dwindles (as with rapid weight loss), the kidneys may drop to a lower position, a condition called *ptosis* (to′sis; "a fall"). Ptosis creates problems if the ureters, which drain urine from the kidneys, become kinked. When this happens, urine that can no longer pass through the ureters

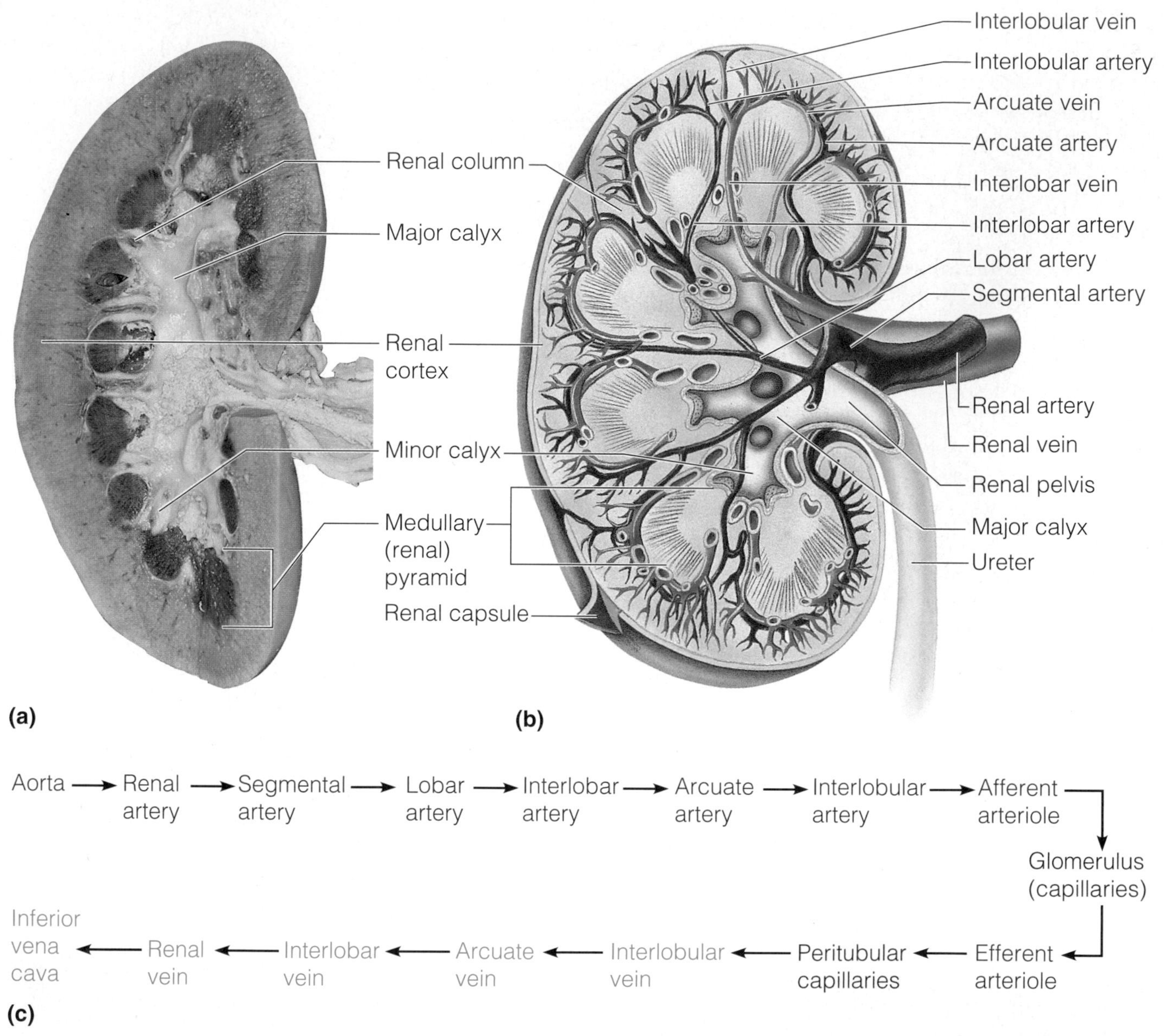

FIGURE 15.2 Internal anatomy of the kidney. (a) Photograph of a coronally sectioned kidney. **(b)** Diagrammatic view of a coronally sectioned kidney, illustrating major blood vessels. **(c)** Summary of the pathway of renal blood vessels.

backs up and exerts pressure on the kidney tissue. This condition, called *hydronephrosis* (hi″dro-nĕ-fro′sis), can severely damage the kidney. ▲

When a kidney is cut lengthwise, three distinct regions become apparent, as can be seen in Figure 15.2. The outer region, which is light in color, is the **renal cortex.** (The word *cortex* comes from the Latin word meaning "bark.") Deep to the cortex is a darker reddish-brown area, the **renal medulla.** The medulla has many basically triangular regions with a striped appearance, the **medullary** (med′u-lar″e) **pyramids.** The broader *base* of each pyramid faces toward the cortex; its tip, the *apex,* points toward the inner region of the kidney. The pyramids are separated by extensions of cortex-like tissue, the **renal columns.**

Medial to the hilus is a flat, basinlike cavity, the **renal pelvis.** As Figure 15.2b shows, the pelvis is continuous with the ureter leaving the hilus. Extensions of the pelvis, **calyces** (kal′ĭ-sēz; singular *calyx*), form cup-shaped areas that enclose the tips of the pyramids. The calyces collect urine, which continuously drains from the tips of the pyramids into the renal pelvis. Urine then flows from the pelvis into the ureter, which transports it to the bladder for temporary storage.

Blood Supply

The kidneys continuously cleanse the blood and adjust its composition, so it is not surprising that they have a very rich blood supply (see Figure 15.2b and c). Approximately one-quarter of the total blood supply of the body passes through the kidneys each minute. The arterial supply of each kidney is the **renal artery.** As the renal artery approaches the hilus, it divides into **segmental arteries.** Once inside the pelvis, the segmental arteries break up into **lobar arteries,** each of which gives off several branches called **interlobar arteries,** which travel through the renal columns to reach the cortex. At the cortex-medulla junction, interlobar arteries give off the **arcuate** (ar′ku-at) **arteries,** which curve over the medullary pyramids. Small **interlobular arteries** then branch off the arcuate arteries and run outward to supply the cortical tissue. Venous blood draining from the kidney flows through veins that trace the pathway of the arterial supply but in a reverse direction—**interlobular veins** to **arcuate veins** to **interlobar veins** to the **renal vein,** which emerges from the kidney hilus. (There are no lobar or segmental veins.)

Nephrons and Urine Formation

Nephrons

Each kidney contains over a million tiny structures called **nephrons** (nef′ronz). Nephrons are the structural and functional units of the kidneys and, as such, are responsible for forming urine. Figure 15.3 shows the anatomy and relative positioning of nephrons in each kidney.

Each nephron consists of two main structures: a **glomerulus** (glo-mer′u-lus), which is a knot of capillaries, and a **renal tubule.** The closed end of the renal tubule is enlarged and cup-shaped and completely surrounds the glomerulus. This portion of the renal tubule is called the **glomerular** (*glom* = little ball), or **Bowman's, capsule.** The inner (visceral) layer of the capsule is made up of highly modified octopus-like cells called **podocytes** (pod′o-sītz). Podocytes have long branching processes called *foot processes* that intertwine with one another and cling to the glomerulus. Because openings, the so-called *filtration slits,* exist between their extensions, the podocytes form a porous, or "holey," membrane around the glomerulus (Figure 15.3c and d).

The rest of the tubule is about 3 cm (approximately 1.25 inches) long. As it extends from the glomerular capsule, it coils and twists before forming a hairpin loop and then again becomes coiled and twisted before entering a collecting tubule called the **collecting duct.** These different regions of the tubule have specific names (see Figure 15.3); in order from the glomerular capsule they are the **proximal convoluted tubule (PCT),** the **loop of Henle** (hen′le), and the **distal convoluted tubule (DCT).** The lumen surfaces (surface exposed to the filtrate) of the tubule cells in the proximal convoluted tubules are covered with dense microvilli, which increases their surface area tremendously. Microvilli also occur on the tubule cells in other parts of the tubule but in much reduced numbers.

Most nephrons are called **cortical nephrons** because they are located almost entirely within the cortex. In a few cases, the nephrons are called **juxtamedullary nephrons** because they are situated close to the cortex-medulla junction, and their loops of Henle dip deep into the medulla (see Figure 15.3a). The **collecting ducts,** each of which receives urine from many nephrons, run downward through the medullary pyramids, giving them their striped appearance. They deliver the final urine product into the calyces and renal pelvis.

Each and every nephron is associated with two capillary beds—the glomerulus (mentioned earlier) and the *peritubular* (per″ĭ-tu′bu-lar) *capillary bed.* The glomerulus is both fed and drained by *arterioles.* The **afferent arteriole,** which arises from an *interlobular artery,* is the "feeder vessel," and the **efferent arteriole** receives blood that has passed through the glomerulus. The glomerulus, specialized for filtration, is unlike any other capillary bed in the entire body. Because it is both fed *and* drained by arterioles, which are high-resistance vessels, and the afferent arteriole has a larger diameter than the efferent, blood pressure in the glomerular capillaries is much higher than in other capillary beds. This extremely high pressure forces fluid and solutes (smaller than proteins) out of the blood into the glomerular capsule. Most of this filtrate (99 percent) is eventually reclaimed by the

Q *What path would a creatinine molecule in the glomerular blood take to reach the renal pelvis?*

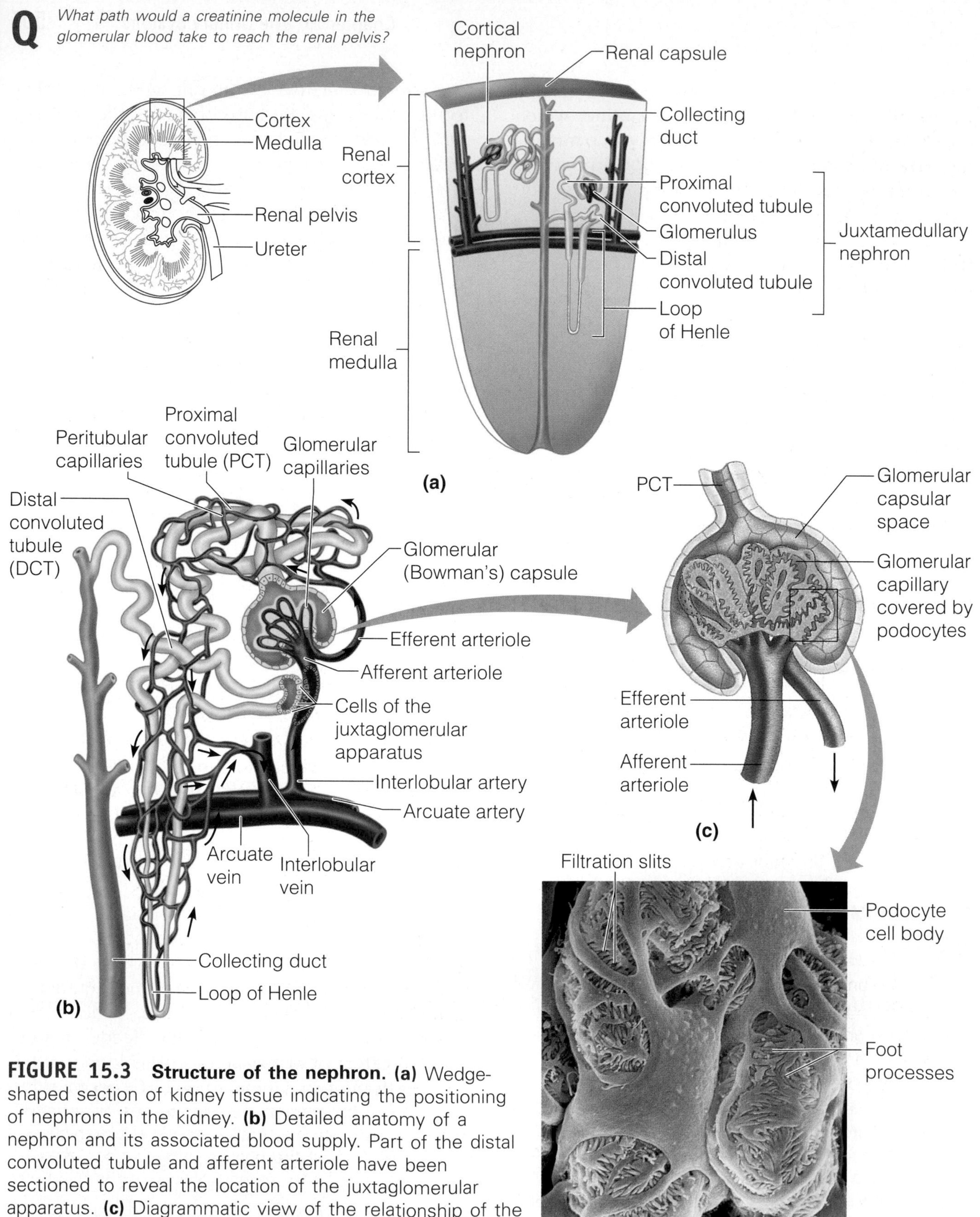

FIGURE 15.3 Structure of the nephron. (a) Wedge-shaped section of kidney tissue indicating the positioning of nephrons in the kidney. **(b)** Detailed anatomy of a nephron and its associated blood supply. Part of the distal convoluted tubule and afferent arteriole have been sectioned to reveal the location of the juxtaglomerular apparatus. **(c)** Diagrammatic view of the relationship of the visceral layer of the glomerular capsule to the glomerular capillaries. **(d)** Scanning electron micrograph of podocytes clinging to the glomerular capillaries.

A *From the glomerular blood into the glomerular capsular space, then the proximal convoluted tubule to the loop of Henle, the distal convoluted tubule, the collecting duct through the renal medulla to a calyx, and finally the renal pelvis.*

renal tubule cells and returned to the blood in the peritubular capillary beds.

The second capillary bed, the **peritubular capillaries,** arises from the efferent arteriole that drains the glomerulus. Unlike the high-pressure glomerulus, these capillaries are low-pressure, porous vessels that are adapted for absorption instead of filtration. They cling closely to the whole length of the renal tubule, where they are in an ideal position to receive solutes and water from the tubule cells as these substances are reabsorbed from the filtrate percolating through the tubule. The peritubular capillaries drain into interlobular veins leaving the cortex.

Urine Formation

Urine formation is a result of three processes—*filtration, tubular reabsorption,* and *tubular secretion*. Each of these processes is illustrated in Figure 15.4 and described in more detail next.

Filtration As just described, the glomerulus acts as a filter. **Filtration** is a nonselective, passive process. The filtrate that is formed is essentially blood plasma without blood proteins. Both proteins and blood cells are normally too large to pass through the filtration membrane, and when either of these appear in the urine, it is a pretty fair bet that there is some problem with the glomerular filters. As long as the systemic blood pressure is normal, filtrate will be formed. If arterial blood pressure drops too low, the glomerular pressure becomes inadequate to force substances out of the blood into the tubules, and filtrate formation stops.

Homeostatic Imbalance

An abnormally low urinary output is called *oliguria* (ol″i-gu′re-ah) if it is between 100 and 400 ml/day, and *anuria* (ah-nu′re-ah) if it is less than 100 ml/day. Low urinary output usually indicates that glomerular blood pressure is too low to cause filtration, but anuria may also result from transfusion reactions and acute inflammation or from crush injuries of the kidneys. ▲

Tubular Reabsorption Besides wastes and excess ions that must be removed from the blood, the filtrate contains many useful substances (including water, glucose, amino acids, and ions), which must be reclaimed from the filtrate and returned to the blood. **Tubular reabsorption** begins as soon as the filtrate enters the proximal convoluted tubule

Q *How would liver disease, in which the liver is unable to make many of the blood proteins, affect process a? (Reviewing Chapter 10 might help.)*

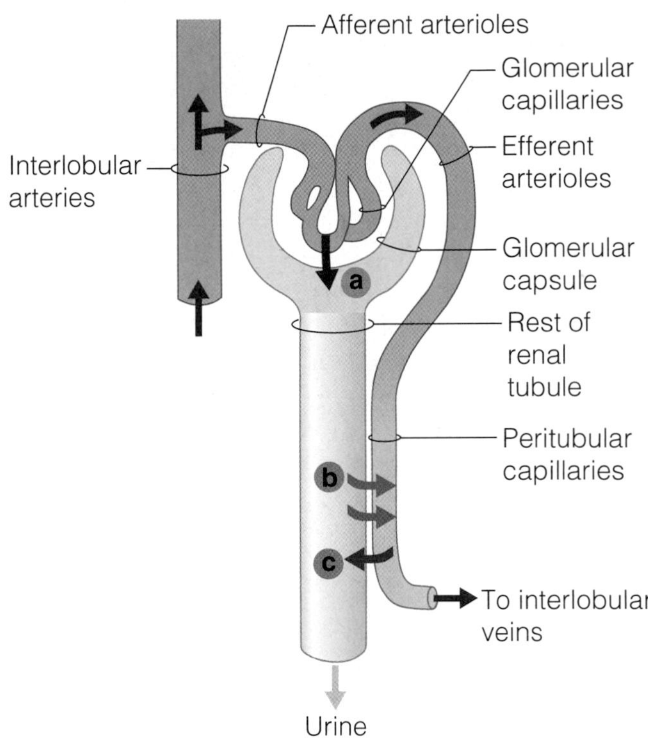

KEY:

a → **Filtration:** Water and solutes smaller than proteins are forced through the capillary walls and pores of the glomerular capsule into the renal tubule.

b → **Tubular Reabsorption:** Water, glucose, amino acids, and needed ions are transported out of the filtrate into the tubule cells and then enter the capillary blood.

c → **Tubular Secretion:** H^+, K^+, creatinine, and drugs are removed from the peritubular blood and secreted by the tubule cells into the filtrate.

FIGURE 15.4 The kidney depicted schematically as a single large, uncoiled nephron. A kidney actually has millions of nephrons acting in parallel. The three processes by which the kidneys adjust the composition of plasma are **(a)** filtration, **(b)** tubular reabsorption, and **(c)** tubular secretion.

The amount of renal filtrate formed is a function of filtration (blood) pressure and blood osmotic pressure (exerted largely by blood proteins). Normally the osmotic pressure is a constant; but, in the situation described, more filtrate than normal will be formed because the blood pressure is opposed to a lesser extent by osmotic pressure of the blood.

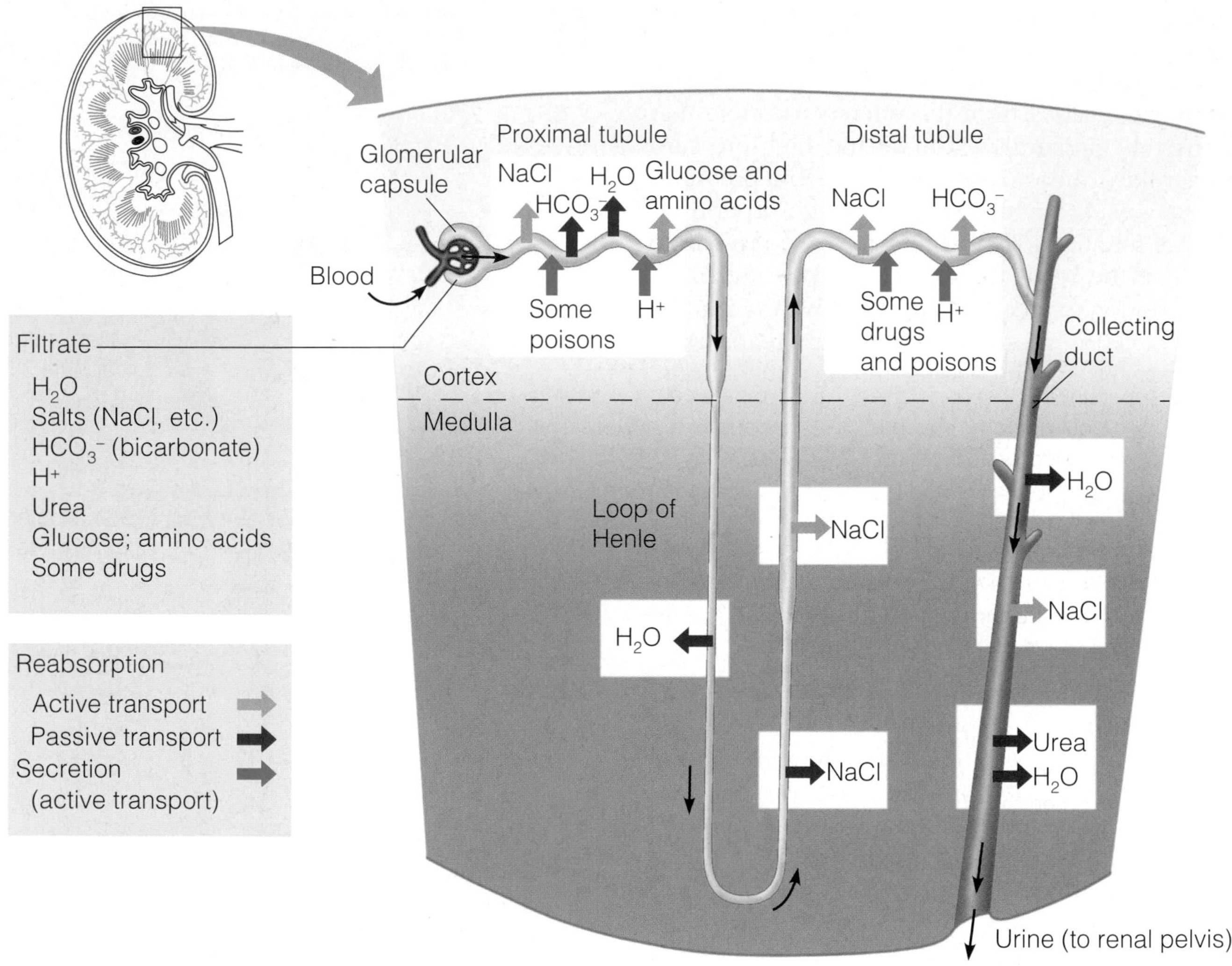

FIGURE 15.5 Sites of filtration, reabsorption, and secretion in a nephron.

(Figure 15.5). The tubule cells are "transporters," taking up needed substances from the filtrate and then passing them out their posterior aspect into the extracellular space, from which they are absorbed into peritubular capillary blood. Some reabsorption is done passively (for example, water passes by osmosis), but the reabsorption of most substances depends on active transport processes, which use membrane carriers and are very selective. There is an abundance of carriers for substances that need to be retained, and few or no carriers for substances of no use to the body. Needed substances (for example, glucose and amino acids) are usually entirely removed from the filtrate. **Nitrogenous waste products** are poorly reabsorbed, if at all. These include **urea** (u-re′ah), formed by the liver as an end product of protein breakdown when amino acids are used to produce energy; **uric acid,** released when nucleic acids are metabolized; and **creatinine** (kre-at′ĭ-nin), associated with creatine (kre′ah-tin) metabolism in muscle tissue. Because tubule cells have few membrane carriers to reabsorb these substances, they tend to remain in the filtrate and are found in high concentrations in urine excreted from the body. Various ions are reabsorbed or allowed to go out in the urine, according to what is needed at a particular time to maintain the proper pH and electrolyte composition of the blood. Most reabsorption occurs in the proximal convoluted tubules, but the distal convoluted tubule and the collecting duct are also active.

A Closer Look

Renal Failure and the Artificial Kidney

LIKE master chemists, the kidneys continually maintain the purity of our internal environment. Without their continual efforts, body fluids quickly become contaminated with nitrogen-containing wastes, blood pH tumbles into the acidic range, and *uremia* sets in, totally disrupting life processes. Signs and symptoms of uncontrolled uremia include diarrhea, vomiting, labored breathing, irregular heartbeat, convulsions, coma, and finally death.

Although not common, renal failure may occur when the number of functioning units (nephrons) becomes too low to carry out the normal kidney functions. The leading cause of renal failure is diabetes mellitus (approximately 33% of new cases each year) with hypertension a close second. (Notice that hypertension is both a cause and a symptom.) Other possible causes of renal failure include:

- Repeated damaging infections of the kidneys
- Physical trauma to the kidneys (crush injury and others)
- Chemical poisoning of the tubule cells by heavy metals (mercury or lead) or organic solvents (dry-cleaning fluids, paint thinner)
- Inadequate blood delivery to the tubule cells (as sometimes happens with arteriosclerosis)
- Prolonged pressure on skeletal muscles (causes release of myoglobin, a muscle pigment that can clog renal tubules)
- Chronic hypertension or immune system problems (see glomerulonephritis, p. 521)

In renal failure, filtrate formation decreases or stops completely. Because toxic wastes accumulate quickly in the blood when the kidney tubule cells are not working, *dialysis* (*dialys* = separate) is necessary to cleanse the blood while the kidneys are shut down. In *hemodialysis,* which uses an "artificial kidney" apparatus (see illustration), the patient's blood is passed through a membrane tubing that is permeable only to selected substances, and the tubing is immersed in a bathing solution that differs slightly from normal "cleansed" plasma. As blood circulates through the tubing, substances such as nitrogenous wastes and potassium ions (K^+) present in the blood (but not in the bath) diffuse out of the blood through the tubing into the surrounding solution, and substances to be added to the blood (mainly buffers for hydrogen ions [H^+] and glucose for malnourished patients) move from the bathing solution into the blood. In this way, needed substances are retained in the blood or added to it, while wastes and ion excesses are removed. Hemodialysis is routinely done three times weekly, and each session takes four to eight hours. Serious problems occasionally encountered by hemodialysis patients are thrombosis, infection, and

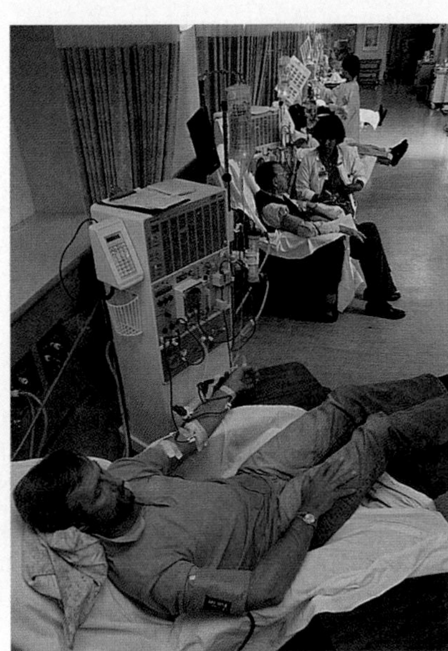

A renal patient undergoing hemodialysis.

Renal Failure and the Artificial Kidney *(continued)*

ischemia at the shunt site. Hemorrhage is an added risk, because the blood must be heparinized to prevent clotting during hemodialysis. (*Heparin* is an anticoagulant.)

A less efficient but more convenient procedure for patients who are not hospitalized is *continuous ambulatory peritoneal dialysis (CAPD)*. CAPD uses the patient's own peritoneal membrane as the dialyzing membrane. Fluid that is equal in chemical content to normal plasma and interstitial fluid is introduced into the patient's peritoneal cavity with a catheter and left to equilibrate there for 15 to 60 minutes. Then the dialysis fluid is retrieved from the peritoneal cavity and replaced with fresh dialysis fluid. The procedure is repeated until the patient's blood chemistry reaches normal. Because some ambulatory patients may be inattentive to cloudy or bloody dialysis fluid, infection is more common in CAPD than in hemodialysis.

When renal damage is nonreversible, as in chronic, slowly progressing renal failure, the kidneys become totally unable to process plasma or concentrate urine, and a kidney transplant is the only answer. Unhappily, the signs and symptoms of this life-threatening problem become obvious only after about 75 percent of renal function has been lost. The end-stage of renal failure, *uremia,* occurs when about 90 percent of the nephrons have been lost.

Tubular Secretion **Tubular secretion** is essentially reabsorption in reverse. Some substances, such as hydrogen and potassium ions and creatinine, also move from the blood of the peritubular capillaries through the tubule cells or from the tubule cells themselves into the filtrate to be eliminated in urine. This process seems to be important for getting rid of substances not already in the filtrate, such as certain drugs, or as an additional means for controlling blood pH (Figure 15.5).

Characteristics of Urine

In 24 hours, the marvelously complex kidneys filter some 150 to 180 liters of blood plasma through their glomeruli into the tubules, which process the filtrate by taking substances out of it (reabsorption) and adding substances to it (secretion). In the same 24 hours, only about 1.0 to 1.8 liters of urine are produced. Obviously, urine and filtrate are quite different. Filtrate contains everything that blood plasma does (except proteins), but by the time it reaches the collecting ducts, the filtrate has lost most of its water and just about all of its nutrients and necessary ions. What remains, **urine,** contains nitrogenous wastes and unneeded substances. Assuming we are healthy, our kidneys can keep our blood composition fairly constant despite wide variations in diet and cell activity.

Freshly voided urine is generally clear and pale to deep yellow. The normal yellow color is due to *urochrome* (u′ro-krōm), a pigment that results from the body's destruction of hemoglobin. The

TABLE 15.1 Abnormal Urinary Constituents

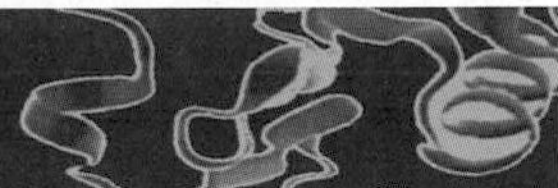

Substance	Name of condition	Possible causes
Glucose	Glycosuria (gli″ko-su′re-ah)	Nonpathological: Excessive intake of sugary foods Pathological: Diabetes mellitus
Proteins	Proteinuria (pro″te-ĭ-nu′re-ah) (also called albuminuria)	Nonpathological: Physical exertion, pregnancy Pathological: Glomerulonephritis, hypertension
Pus (WBCs and bacteria)	Pyuria (pi-u′re-ah)	Urinary tract infection
RBCs	Hematuria (he″mah-tu′re-ah)	Bleeding in the urinary tract (due to trauma, kidney stones, infection)
Hemoglobin	Hemoglobinuria (he″mo-glo-bĭ-nu′re-ah)	Various: Transfusion reaction, hemolytic anemia
Bile pigment	Bilirubinuria (bil″ĭ-roo-bĭ-nu′re-ah)	Liver disease (hepatitis)

more solutes are in the urine, the deeper yellow its color; on the other hand, dilute urine is a pale, straw color. At times, urine may be a color other than yellow. This might be a result of eating certain foods (beets, for example) or the presence of bile or blood in the urine.

When formed, urine is sterile, and its odor is slightly aromatic. If it is allowed to stand, it takes on an ammonia odor caused by the action of bacteria on the urine solutes. Some drugs, vegetables (such as asparagus), and various diseases (such as diabetes mellitus) alter the usual odor of urine.

Urine pH is usually slightly acid (around 6), but changes in body metabolism and certain foods may cause it to be much more acidic or basic. For example, a diet that contains large amounts of protein (eggs and cheese) and whole-wheat products causes urine to become quite acid; thus, such foods are called *acid-ash foods*. Conversely, a vegetarian diet is called an *alkaline-ash diet* because it makes urine quite alkaline as the kidneys excrete the excess bases. Bacterial infection of the urinary tract also may cause the urine to be alkaline.

Since urine is water plus solutes, urine weighs more, or is more dense, than distilled water. The term used to compare how *much* heavier urine is than distilled water is **specific gravity.** Whereas the specific gravity of pure water is 1.0, the specific gravity of urine usually ranges from 1.001 to 1.035 (dilute to concentrated urine, respectively). Urine is generally dilute (that is, it has a low specific gravity) when a person drinks excessive fluids, uses diuretics (drugs that increase urine output), or has chronic renal failure (a condition in which the kidney loses its ability to concentrate urine). Conditions that produce urine with a high specific gravity include inadequate fluid intake, fever, and a kidney inflammation called *pyelonephritis* (pi″ĕ-lo-nĕ-fri′tis).

Solutes normally found in urine include sodium and potassium ions, urea, uric acid, creatinine, ammonia, bicarbonate ions, and various other ions, depending on blood composition. With certain diseases, urine composition can change dramatically, and the presence of abnormal substances in urine is often helpful in diagnosing the problem. This is why a routine urinalysis should always be part of any good physical examination.

Substances *not* normally found in urine are glucose, blood proteins, red blood cells, hemoglobin, white blood cells (pus), and bile. Names and possible causes of conditions in which abnormal urinary constituents and volumes might be seen are given in Table 15.1.

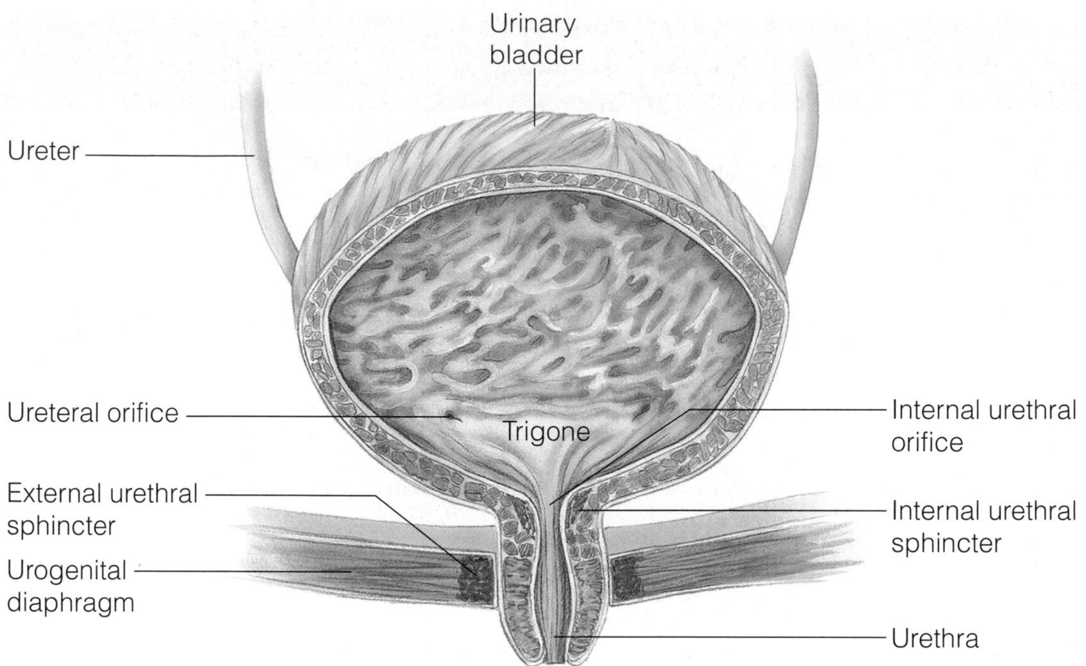

FIGURE 15.6 Basic structure of the urinary bladder and urethra of the female. The urethra of the male, which runs through the length of the penis, is substantially longer than that of the female.

Ureters, Urinary Bladder, and Urethra

Ureters

The **ureters** (u-re'terz) are slender tubes each 25 to 30 cm (10 to 12 inches) long and 6 mm (¼ inch) in diameter. Each ureter runs behind the peritoneum from the renal hilus to the posterior aspect of the bladder, which it enters at a slight angle (see Figures 15.1 and 15.6). The superior end of each ureter is continuous with the pelvis of the kidney, and its mucosal lining is continuous with that lining the renal pelvis and the bladder below.

Essentially, the ureters are passageways that carry urine from the kidneys to the bladder. Although it might appear that urine could simply drain to the bladder below by gravity, the ureters *do* play an active role in urine transport. Smooth muscle layers in their walls contract to propel urine into the bladder by peristalsis. Once urine has entered the bladder, it is prevented from flowing back into the ureters by small valvelike folds of bladder mucosa that flap over the ureter openings.

Homeostatic Imbalance

When urine becomes extremely concentrated, solutes such as uric acid salts form crystals that precipitate in the renal pelvis. These crystals are called **renal calculi** (kal'kyoo-li; *calculus* = little stone), or kidney stones. Excruciating pain that radiates to the flank occurs when the ureter walls close in on the sharp calculi as they are being eased through the ureter by peristalsis or when the calculi become wedged in a ureter. Frequent bacterial infections of the urinary tract, urinary retention, and alkaline urine all favor calculi formation. Surgery has been the treatment of choice, but a newer noninvasive procedure *(lithotripsy)* that uses ultrasound waves to shatter the calculi is becoming more popular. The pulverized, sandlike remnants of the calculi are painlessly eliminated in the urine. ▲

Urinary Bladder

The **urinary bladder** is a smooth, collapsible, muscular sac that stores urine temporarily. It is located retroperitoneally in the pelvis just posterior to the pubic symphysis. If the interior of the bladder is scanned, three openings are seen—the two ureter openings *(ureteral orifices)* and the single opening of the **urethra** (u-re′thrah) (the *internal urethral orifice*), which drains the bladder (Figure 15.6). The smooth triangular region of the bladder base outlined by these three openings is called the **trigone** (tri′gon). The trigone is important clinically because infections tend to persist in this region. In males, the *prostate gland* (part of the male reproductive system) surrounds the neck of the bladder where it empties into the urethra.

The bladder wall contains three layers of smooth muscle, collectively called the *detrusor muscle,* and its mucosa is a special type of epithelium, *transitional epithelium* (see p. 90). Both of these structural features make the bladder uniquely suited for its function of urine storage. When the bladder is empty, it is collapsed, 5 to 7.5 cm (2 to 3 inches) long at most, and its walls are thick and thrown into folds. As urine accumulates, the bladder expands and rises superiorly in the abdominal cavity (Figure 15.7). Its muscular wall stretches, and the transitional epithelial layer thins, allowing the bladder to store more urine without substantially increasing its internal pressure. A moderately full bladder is about 12.5 cm (5 inches) long and holds about 500 ml (1 pint) of urine, but it is capable of holding more than twice that amount. When the bladder is really distended, or stretched by urine, it becomes firm and pear-shaped and may be felt just above the pubic symphysis. Although urine is formed continuously by the kidneys, it is usually stored in the bladder until its release is convenient.

Urethra

The **urethra** is a thin-walled tube that carries urine by peristalsis from the bladder to the outside of the body. At the bladder-urethra junction, a thickening of the smooth muscle forms the **internal urethral sphincter** (see Figure 15.6), an involuntary sphincter that keeps the urethra closed when urine is not being passed. A second sphincter, the **external urethral sphincter,** is fashioned by skeletal muscle as the urethra passes through the pelvic floor. This sphincter is voluntarily controlled.

The length and relative function of the urethra differ in the two sexes. In females, it is about 3 to 4 cm (1½ inches) long, and its external orifice, or opening, lies anteriorly to the vaginal opening (see also Figure 16.8a, p. 537). Its function is to conduct urine to the body exterior.

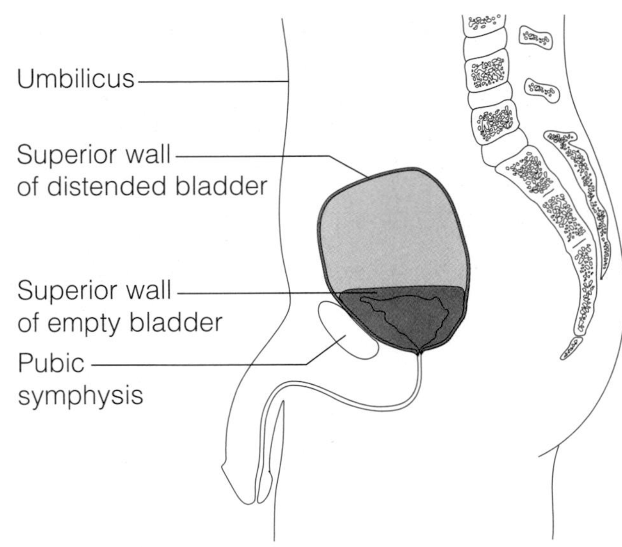

FIGURE 15.7 Position and shape of a distended and an empty urinary bladder in an adult male.

Homeostatic Imbalance

Since the female urinary orifice is so close to the anal opening, and feces contain a good deal of bacteria, improper toileting habits (that is, wiping from back to front rather than from front to back) can easily carry bacteria into the urethra. Moreover, since the mucosa of the urethra is continuous with that of the rest of the urinary tract organs, an inflammation of the urethra, or *urethritis* (u″re-thri′tis), can easily ascend the tract to cause bladder inflammation *(cystitis)* or even kidney inflammation (*pyelonephritis,* or *pyelitis*). Symptoms of urinary tract infection include *dysuria* (painful urination), urinary *urgency* and *frequency,* fever, and sometimes cloudy or blood-tinged urine. When the kidneys are involved, back pain and a severe headache are common. ▲

In males, the urethra is approximately 20 cm (8 inches) long and has three named regions (see Figure 16.2, p. 530), the *prostatic, membranous,* and *spongy* (or *penile*) *urethrae.* It opens at the tip of the penis after traveling down its length. The urethra of the male has a double function. It carries urine out of the body, and it provides the passageway through which sperm is ejected from the body. Thus, in males, the urethra is part of both the urinary and reproductive systems.

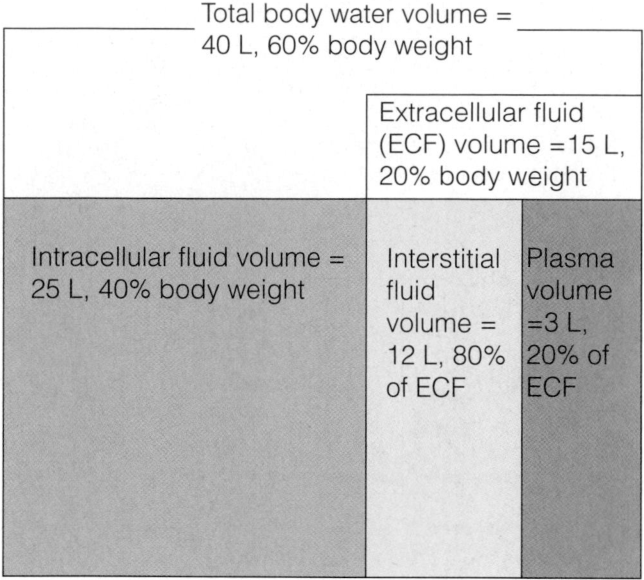

FIGURE 15.8 The major fluid compartments of the body. Approximate values are noted for a 70-kg (154-pound) male.

Micturition

Micturition (mik″tu-rish′un), or **voiding,** is the act of emptying the bladder. Two sphincters, or valves, the internal urethral sphincter (more superiorly located) and the external urethral sphincter (more inferiorly located) control the flow of urine from the bladder (see Figure 15.6). Ordinarily, the bladder continues to collect urine until about 200 ml have accumulated. At about this point, stretching of the bladder wall activates stretch receptors. Impulses transmitted to the sacral region of the spinal cord and then back to the bladder via the *pelvic splanchnic nerves* cause the bladder to go into reflex contractions. As the contractions become stronger, stored urine is forced past the internal urethral sphincter (the smooth muscle, involuntary sphincter) into the upper part of the urethra. It is then that a person feels the urge to void. Because the lower external sphincter is skeletal muscle and is voluntarily controlled, we can choose to keep it closed and postpone bladder emptying temporarily. On the other hand, if it is convenient, the external sphincter can be relaxed so that urine is flushed from the body. When one chooses not to void, the reflex contractions of the bladder will stop within a minute or so, and urine will continue to accumulate in the bladder. After 200 to 300 ml more have been collected, the micturition reflex occurs again. Eventually, micturition occurs whether one wills it or not.

Homeostatic Imbalance

Incontinence (in-kon′tĭ-nens) occurs when we are unable to voluntarily control the external sphincter. Incontinence is normal in children 2 years old or younger, because they have not yet gained control over their voluntary sphincter. It may also occur in older children who sleep so soundly that they are not awakened by the stimulus. However, after the toddler years, incontinence is usually a result of emotional problems, pressure (as in pregnancy), or nervous system problems (stroke or spinal cord injury).

Urinary retention is essentially the opposite of incontinence. It is a condition in which the bladder is unable to expel its contained urine. There are various causes for urinary retention. It often occurs after surgery in which general anesthesia has been given because it takes a little time for the smooth muscles to regain their activity. Another cause of urinary retention, occurring primarily in elderly men, is enlargement, or *hyperplasia,* of the prostate gland, which surrounds the neck of the bladder. As it enlarges, it narrows the urethra, making it very difficult to void. When urinary retention is prolonged, a slender flexible drainage tube called a *catheter* (kath′ĭ-ter) must be inserted through the urethra to drain the urine and prevent bladder trauma from excessive stretching. ▲

Fluid, Electrolyte, and Acid-Base Balance

Blood composition depends on three major factors: diet, cellular metabolism, and urine output. In general, the kidneys have four major roles to play, which help keep the blood composition relatively constant. These are (1) excretion of nitrogen-containing wastes, maintaining (2) water and (3) electrolyte balance of the blood, and (4) ensuring proper blood pH. Excretion of nitrogenous wastes has already been considered; roles 2 through 4 are discussed briefly next.

Maintaining Water and Electrolyte Balance of Blood

Body Fluids and Fluid Compartments

If you are a healthy young adult, water probably accounts for half or more of your body weight—

50 percent in females and about 60 percent in males. These differences reflect the fact that females have relatively less muscle and a larger amount of body fat (and of all body tissues, fat contains the least water). Babies, with little fat and low bone mass, are about 75 percent water, but total body water content declines through life and accounts for only about 45 percent of body weight in old age. The importance of water to the functioning of the body and its cells is described in Chapter 2. That information will not be repeated here except to say that water is the universal body solvent within which all solutes (including the very important electrolytes) are dissolved.

Water occupies three main locations within the body, referred to as *fluid compartments* (Figure 15.8). About two-thirds of body fluid, the so-called **intracellular fluid (ICF),** is contained within the living cells. The remainder, called **extracellular fluid (ECF),** includes all body fluids located outside the cells. Although ECF most importantly includes blood plasma and interstitial (or tissue) fluid, it also accounts for cerebrospinal and serous fluids, the humors of the eye, lymph, and others.

The Link between Water and Salt

Water certainly accounts for nearly the entire volume of body fluids, regardless of type, and all body fluids are similar, but there is more to *fluid balance* than just water. The types and amounts of solutes in body fluids, especially electrolytes such as sodium, potassium, and calcium ions, are also vitally important to overall body homeostasis, and water and electrolyte balance are tightly linked as the kidneys continuously process the blood. (Recall from Chapter 2 that electrolytes are charged particles [ions] that conduct an electrical current in an aqueous solution.) Very small changes in **electrolyte balance,** the solute concentrations in the various fluid compartments, cause water to move from one compartment to another. Not only does this alter blood volume and blood pressure, but it can also severely impair the activity of irritable cells like nerve and muscle cells. For example, a deficit of sodium ions (Na^+) in the ECF results in water loss from the bloodstream into the tissue spaces (edema) and muscular weakness.

If the body is to remain properly hydrated, we cannot lose more water than we take in. Most water intake is a result of fluids and foods we ingest in our diet. However, a small amount (about 10 percent) is produced during cellular metabolism, as explained in Chapter 14 and indicated in Figure 15.9. There are several routes for water to leave the body. Some water vaporizes out of the lungs, some is lost in perspiration, and some leaves

Q *How would the values shown here be affected by (1) drinking a six-pack of beer? (2) fasting (only water is ingested)?*

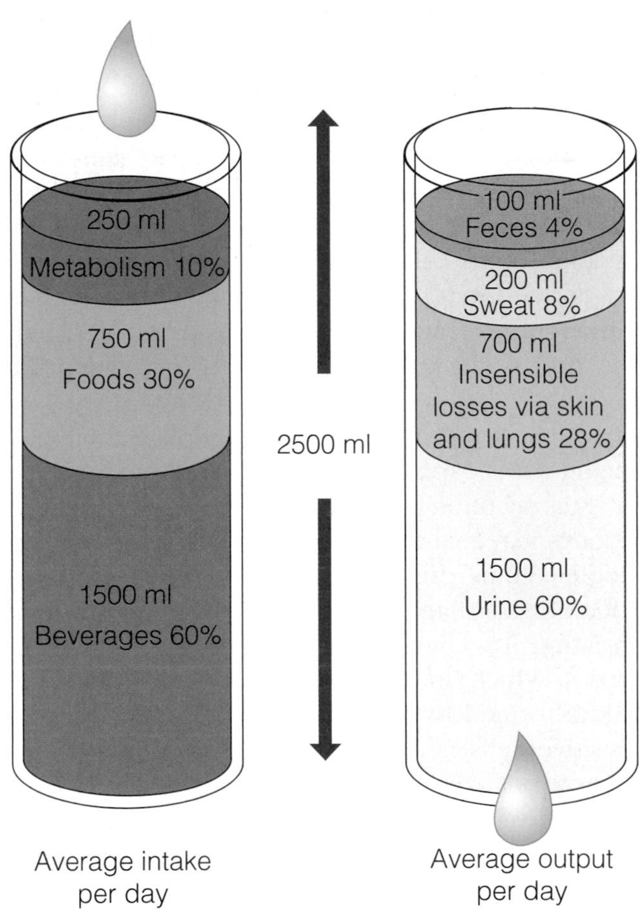

FIGURE 15.9 Water intake and output. Major sources of body water and routes of water loss from the body are shown. When intake and output are in balance, the body is adequately hydrated.

In case #1, the fluid ingested would greatly increase as would the amount of urinary output. In case #2, there would be no fluid intake from ingested foods and less fluid resulting from metabolism. Fluid output would be severely curtailed—no output in feces and less in urine to conserve body water. Insensible losses would not change. Loss of water in perspiration would depend on body and ambient temperatures.

the body in the stool. The job of the kidneys is like that of a juggler. If large amounts of water are lost in other ways, they compensate by putting out less urine to conserve body water. On the other hand, when water intake is excessive, the kidneys excrete generous amounts of urine, and the anguish of a too-full bladder becomes very real.

Likewise, the proper concentrations of the various electrolytes must be present in both intracellular and extracellular fluids. Most electrolytes enter the body in foods and "hard" (mineral-rich) water. Although very small amounts are lost in perspiration and in feces, the major factor regulating the electrolyte composition of body fluids is the kidneys. Just how the kidneys accomplish their balancing act is explained in more detail next.

Reabsorption of water and electrolytes by the kidneys is regulated primarily by hormones. When blood volume drops for any reason (for example, due to hemorrhage or excessive water loss through sweating or diarrhea), arterial blood pressure drops, which in turn decreases the amount of filtrate formed by the kidneys. In addition, highly sensitive cells in the hypothalamus called **osmoreceptors** (oz″mo-re-sep′torz) react to the change in blood composition (that is, less water and more solutes) by becoming more active. The result is that nerve impulses are sent to the posterior pituitary (Figure 15.10), which then releases **antidiuretic hormone (ADH).** (The term *antidiuretic* is derived from *diuresis* [di″u-re′sis], which means "flow of urine from the kidney," and *anti,* which means "against.") As one might guess, this hormone prevents excessive water loss in the urine. ADH travels in the blood to its main target, the kidney's collecting ducts, where it causes the duct cells to reabsorb more water. As more water is returned to the bloodstream, blood volume and blood pressure increase to normal levels, and only a small amount of very concentrated urine is formed. ADH is released more or less continuously unless the solute concentration of the blood drops too low. When this happens, the osmoreceptors become "quiet," and excess water is allowed to leave the body in the urine.

Homeostatic Imbalance

When ADH is *not* released (perhaps because of injury or destruction of the hypothalamus or posterior pituitary gland), huge amounts of very dilute urine (up to 25 liters/day) flush from the body day after day. This condition, *diabetes insipidus* (in-sip′ĭ-dus), can lead to severe dehydration and electrolyte imbalances. Affected individuals are always thirsty and have to drink fluids almost continuously to maintain normal fluid balance. ▲

A second hormone that helps to regulate blood composition and blood volume by acting on the kidney is **aldosterone** (al″dos′ter-on). Aldosterone is the major factor regulating sodium ion content of the ECF and in the process helps regulate the concentration of other ions (Cl^-, K^+, and Mg^{2+} [magnesium]) as well. Sodium ion (Na^+) is the electrolyte most responsible for osmotic water flows. When too little sodium is in the blood, the blood becomes too dilute. Consequently, water leaves the bloodstream and flows out into the tissue spaces, causing edema and possibly a shutdown of the circulatory system. Whether aldosterone is present or not, about 80 percent of the sodium in the filtrate is reabsorbed in the proximal convoluted tubules of the kidneys. When aldosterone concentrations are high, most of the remaining sodium ions are actively reabsorbed in the distal convoluted tubules and the collecting ducts. Generally speaking, for each sodium ion reabsorbed, a chloride ion follows and a potassium ion is secreted into the filtrate. Thus, as the sodium content of the blood increases, potassium concentration decreases, bringing these two ions back to their normal balance in the blood. Still another effect of aldosterone is to increase water reabsorption by the tubule cells, because as sodium is reclaimed, water follows it passively back into the blood. A little rule to keep in mind here is: *water follows salt.*

Recall that aldosterone is produced by the adrenal cortex. Although rising potassium levels or falling sodium levels in the ECF directly stimulate the adrenal cells to release aldosterone, the most important trigger for aldosterone release is the **renin-angiotensin mechanism** (see Figure 15.10) mediated by the *juxtaglomerular (JG) apparatus* of the renal tubules. The juxtaglomerular apparatus (see Figure 15.3b) consists of a complex of modified smooth muscle cells (JG cells) in the afferent arteriole plus some modified epithelial cells forming part of the distal convoluted tubule. The naming of this cell cluster reflects its location close by

Q *Which pathway would also stimulate thirst?*

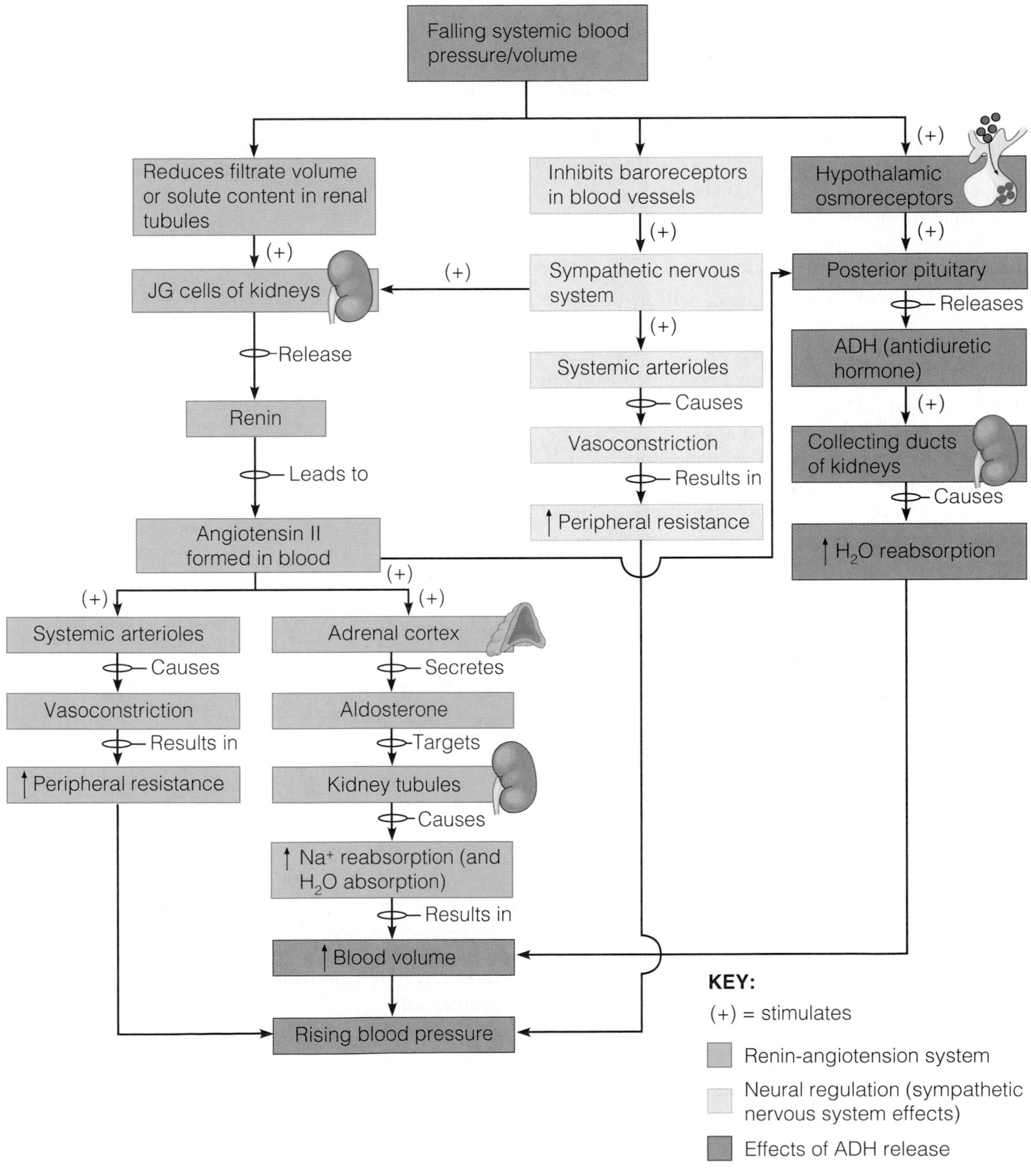

FIGURE 15.10 Flow chart of mechanisms regulating sodium and water balance to help maintain blood pressure homeostasis.

A *The pathway through the hypothalamus, which not only causes water retention, via the effect of ADH on the kidneys, but also stimulates thirst to increase water ingestion.*

(juxta) the glomerulus. When the cells of the JG apparatus are stimulated by low blood pressure in the afferent arteriole or changes in solute content of the filtrate, they respond by releasing the enzyme **renin** into the blood. (Notice the different spelling of this enzyme from *rennin,* an enzyme secreted by stomach glands.) Renin catalyzes the series of reactions that produce angiotensin II, which in turn acts directly on the blood vessels to cause vasoconstriction (and an increase in peripheral resistance) and on the adrenal cortical cells to promote aldosterone release. As a result, blood volume and blood pressure increase (see Figure 15.10). The renin-angiotensin mechanism is extremely important for regulating blood pressure.

The pressure drop also excites baroreceptors in the larger blood vessels. These baroreceptors alert sympathetic nervous system centers of the brain to cause vasoconstriction (via release of epinephrine and norepinephrine), which increases the peripheral resistance (see Figure 15.10). However, this neural mechanism's major focus is blood pressure regulation, not water and electrolyte balance.

Homeostatic Imbalance

People with Addison's disease (hypoaldosteronism) have *polyuria* (excrete large volumes of urine) and lose tremendous amounts of salt and water to urine. As long as adequate amounts of salt and fluids are ingested, people with this condition can avoid problems, but they are constantly teetering on the brink of dehydration. ▲

Maintaining Acid-Base Balance of Blood

For the cells of the body to function properly, blood pH must be maintained between 7.35 and 7.45, a very narrow range. Whenever the pH of arterial blood rises above 7.45, a person is said to have **alkalosis** (al″kah-lo′sis). A drop in arterial pH to below 7.35 results in **acidosis** (as″ĭ-do′sis). Because a pH of 7.0 is neutral, 7.35 is not acidic, chemically speaking; however, it represents a higher-than-optimal hydrogen ion concentration for the functioning of most body cells. Therefore, any arterial pH between 7.35 and 7.0 is called **physiological acidosis.**

Although small amounts of acidic substances enter the body in ingested foods, most hydrogen ions originate as by-products of cellular metabolism, which continuously adds substances to the blood that tend to disturb its **acid-base balance.** Many different acids are produced (for example, phosphoric acid, lactic acid, and many types of fatty acids). In addition, carbon dioxide, which is released during energy production, forms carbonic acid. Ammonia and other basic substances are also released to the blood as cells go about their usual "business." Although the chemical buffers in the blood can temporarily "tie up" excess acids and bases, and the lungs have the chief responsibility for eliminating carbon dioxide from the body, the kidneys assume most of the load for maintaining acid-base balance of the blood. Before describing how the kidneys function in acid-base balance, let's take a look at how each of our other two pH-controlling systems, blood buffers and the respiratory system, works.

Prove It Yourself

Demonstrate the Water-Retaining Power of Salt

The amount of water your body retains depends more on the amount of salt you consume than on how much water you drink. Here's how to demonstrate this.

On a normal day between meals, empty your bladder. Wait half an hour, and then urinate again. Measure the volume of urine you produced the second time. This is your baseline rate of urine production. Now quickly drink a quart of water and measure your urine output for four more consecutive half-hour periods. Subtract the baseline amount of urine from each time period to find out how much water you excreted in two hours.

Repeat the experiment on another day. This time dissolve 5 grams (approximately ¾ teaspoon) of salt in the water first. You should find that you excrete much less water during the next two hours than you did the first time.

People with certain health conditions, such as hypertension or congestive heart failure, may need to retain less water. Generally, they are advised to restrict their intake of salt, not water. Can you see why?

Blood Buffers

Chemical buffers are systems of one or two molecules that act to prevent dramatic changes in hydrogen ion (H^+) concentration when acids or bases are added. They do this by binding to hydrogen ions whenever the pH drops and by releasing hydrogen ions when the pH rises. Since the chemical buffers act within a fraction of a second, they are the first line of defense in resisting pH changes.

To better understand how a chemical buffer system works, let's review the definitions of strong and weak acids and bases. Recall that acids are proton (H^+) donors, and that the acidity of a solution reflects only the *free* hydrogen ions, not those still bound to anions. *Strong acids* dissociate completely and liberate all their H^+ in water. Consequently they can cause large changes in pH. By contrast, *weak acids* like carbonic acid dissociate only partially and so have a much slighter effect on a solution's pH (Figure 15.11). However, weak acids are very effective at preventing pH changes since they are forced to dissociate and release more H^+ when the pH rises over the desirable pH range. This feature allows them to play a very important role in the chemical buffer systems.

Also recall that bases are proton or hydrogen ion acceptors. *Strong bases* like hydroxides dissociate easily in water and quickly tie up H^+, but *weak bases* like bicarbonate ion (HCO_3^-) and ammonia (NH_3) are slower to accept H^+. However, as pH drops, the weak bases become "stronger" and begin to tie up more hydrogen ions. Thus, like weak acids, they are valuable members of the chemical buffer systems.

The three major chemical buffer systems of the body are the *bicarbonate, phosphate,* and *protein buffer systems,* each of which helps to maintain the pH in one or more of the fluid compartments. They all work together, and anything that causes a shift in H^+ concentration in one compartment also causes changes in the others. Thus, drifts in pH are resisted by the entire buffering system. Since all three systems operate in a similar way, examining just one, the bicarbonate buffer system, which is so important in preventing changes in blood pH, should be sufficient.

The **bicarbonate buffer system** is a mixture of *carbonic acid* (H_2CO_3) and its salt, *sodium bicarbonate* ($NaHCO_3$). Since carbonic acid is a weak acid, it does not dissociate much in neutral or acidic solutions. Thus, when a strong acid, such as hydrochloric acid (HCl) is added, most of the

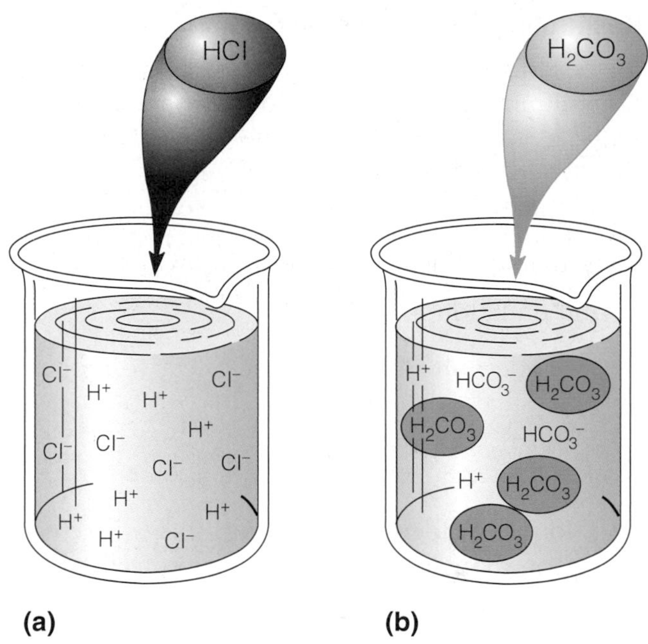

FIGURE 15.11 Dissociation of strong and weak acids. (a) When HCl, a strong acid, is added to water it dissociates completely into its ions (H^+ and Cl^-). **(b)** By contrast, dissociation of H_2CO_3, a weak acid, is very incomplete, and some molecules of H_2CO_3 remain undissociated (symbols shown in green ovals) in solution.

carbonic acid remains intact. However, the *bicarbonate ions* (HCO_3^-) of the salt act as bases to tie up the H^+ released by the stronger acid, forming more carbonic acid:

$$\underset{\text{strong acid}}{HCl} + \underset{\text{weak base}}{NaHCO_3} \longrightarrow \underset{\text{weak acid}}{H_2CO_3} + \underset{\text{salt}}{NaCl}$$

Because the strong acid is (effectively) changed to a weak one, it lowers the pH of the solution only very slightly.

Similarly, if a strong base like sodium hydroxide (NaOH) is added to a solution containing the bicarbonate buffer system, $NaHCO_3$ will not dissociate further under such alkaline conditions. However, carbonic acid will be forced to dissociate further by the presence of the strong base—liberating more H^+ to bind with the OH^- released by NaOH.

$$\underset{\text{strong base}}{NaOH} + \underset{\text{weak acid}}{H_2CO_3} \longrightarrow \underset{\text{weak base}}{NaHCO_3} + \underset{\text{water}}{H_2O}$$

The net result is replacement of a strong base by a weak one, so that the pH of the solution rises very little.

Respiratory System Controls

As described in Chapter 13, the respiratory system eliminates carbon dioxide from the blood while it "loads" oxygen into the blood. Remember that when carbon dioxide (CO_2) enters the blood from the tissue cells, most of it enters the red blood cells where it is converted to bicarbonate ion (HCO_3^-) for transport in the plasma as shown by the equation

$$\underset{\text{Carbon dioxide}}{CO_2} + \underset{\text{water}}{H_2O} \xrightleftharpoons{\text{Carbonic anhydrase}} \underset{\text{carbonic acid}}{H_2CO_3} \rightleftharpoons \underset{\text{hydrogen ion}}{H^+} + \underset{\text{bicarbonate ion}}{HCO_3^-}$$

The double-headed arrows reveal that an increase in carbon dioxide pushes the reaction to the right, producing more carbonic acid. Likewise, an increase in hydrogen ions pushes the equation to the left, producing more carbonic acid. In healthy people, carbon dioxide is expelled from the lungs at the same rate as it is formed in the tissues. Thus the H^+ released when carbon dioxide is loaded into the blood is not allowed to accumulate because it is tied up in water when CO_2 is unloaded in the lungs. So, under normal conditions, the hydrogen ions produced by carbon dioxide transport have essentially no effect on blood pH. However, when CO_2 accumulates in the blood (for example, during restricted breathing) or more H^+ is released to the blood by metabolic processes, the chemoreceptors in the respiratory control centers of the brain (or in peripheral blood vessels) are activated. As a result, breathing rate and depth increase, and the excess H^+ is "blown off" as more CO_2 is removed from the blood.

On the other hand, when blood pH begins to rise (alkalosis), the respiratory center is depressed. Consequently, the respiratory rate and depth fall, allowing carbon dioxide (hence, H^+) to accumulate in the blood. Again blood pH is restored to the normal range. Generally, these respiratory system corrections of blood pH (via regulation of CO_2 content of the blood) are accomplished within a minute or so.

Renal Mechanisms

Chemical buffers can tie up excess acids or bases temporarily, but they cannot eliminate them from the body. And while the lungs can dispose of carbonic acid by eliminating carbon dioxide, only the kidneys can rid the body of other acids generated during metabolism. Additionally, only the kidneys have the power to regulate blood levels of alkaline substances. Thus, although the kidneys act slowly and require hours or days to bring about changes in blood pH, they are the most potent of the mechanisms for regulating blood pH.

The most important means by which the kidneys maintain acid-base balance of the blood are by (1) excreting bicarbonate ions and (2) conserving (reabsorbing) or generating new bicarbonate ions. Look back again at the equation showing how the carbonic acid–bicarbonate buffer system operates. Notice that losing a HCO_3^- from the body has the same effect as gaining an H^+ because it pushes the equation to the right (that is, it leaves a free hydrogen ion). By the same token, reabsorbing or generating new HCO_3^- is the same as losing H^+ because it tends to combine with a hydrogen ion and pushes the equation to the left. Renal mechanisms undertake these adjustments: As blood pH rises, bicarbonate ions are excreted and hydrogen ions are retained by the tubule cells. Conversely, when blood pH falls, bicarbonate is reabsorbed and hydrogen ions are secreted. Urine pH varies from 4.5 to 8.0, which reflects the ability of the renal tubules to excrete basic or acid ions to maintain blood pH homeostasis.

Developmental Aspects of the Urinary System

When you trace the development of the kidneys in a young embryo, it almost seems as if they can't "make up their mind" about whether to come or go. The first tubule system forms and then begins to degenerate as a second, lower, set appears. The second set, in turn, degenerates as a third set makes its appearance. This third set develops into the functional kidneys, which are excreting urine by the third month of fetal life. It is important to remember that the fetal kidneys do not work nearly as hard as they will after birth, because exchanges with the mother's blood through the placenta allow her system to clear many of the undesirable substances from the fetal blood.

Homeostatic Imbalance

There are many congenital abnormalities of this system. Two of the most common are polycystic kidney and hypospadias.

Polycystic (pol″e-sis′tik) *kidney* is a degenerative condition that appears to run in families. In this disease, one or both kidneys are enlarged and have many blisterlike sacs (cysts) containing urine. These cysts interfere with renal function by obstructing urine drainage. Currently, not too much can be done for this condition except to prevent further kidney damage by avoiding infection. Renal failure is the eventual outcome, but kidney transplants improve chances for survival.

Hypospadias (hi″po-spa′de-as) is a condition found in male babies only. It occurs when the urethral orifice is located on the ventral surface of the penis. Corrective surgery is generally done when the child is around 12 months old. ▲

Because the bladder is very small and the kidneys are unable to concentrate urine for the first two months, a newborn baby voids from 5 to 40 times per day, depending on fluid intake. By 2 months of age, the infant is voiding approximately 400 ml/day, and the amount steadily increases until adolescence, when adult urine output (about 1500 ml/day) is achieved.

Control of the voluntary urethral sphincter goes hand in hand with nervous system development. By 15 months, most toddlers are aware when they have voided. By 18 months they can hold urine in their bladder for about two hours, which is the first sign that toilet training (for voiding) can begin. Daytime control usually occurs well before nighttime control is achieved. It is generally unrealistic to expect that complete nighttime control will occur before the child is 4 years old.

During childhood and through late middle age, most urinary system problems are infectious, or inflammatory, conditions. Many types of bacteria may invade the urinary tract to cause urethritis, cystitis, or pyelonephritis. *Escherichia coli* (esh″er-i′ke-ah ko′li) are normal residents of the digestive tract and generally cause no problems there, but these bacteria act as pathogens (disease-causers) in the sterile environment of the urinary tract. Bacteria and viruses responsible for *sexually transmitted diseases (STDs),* which are primarily reproductive tract infections, may also invade and cause inflammation in the urinary tract, which leads to the clogging of some of its ducts.

Homeostatic Imbalance

Childhood streptococcal (strep″to-kok′al) infections, such as strep throat and scarlet fever, may cause inflammatory damage to the kidneys if the original infections are not treated promptly and properly. A common sequel to untreated childhood strep infections is *glomerulonephritis* (glo-mer″ŭ-lo-ne-fri′tis), in which the glomerular filters become clogged with antigen-antibody complexes resulting from the strep infection. ▲

As we age, there is a progressive decline in kidney function. By age 70, the rate of filtrate formation is only about half that of the middle-aged adult. This is believed to result from impaired renal circulation due to atherosclerosis, which affects the entire circulatory system of the aging person. In addition to a decrease in the number of functional nephrons, the tubule cells become less efficient in their ability to concentrate urine.

Another consequence of aging is bladder shrinkage and loss of bladder tone, causing many elderly individuals to experience *urgency* (a feeling that it is necessary to void) and *frequency* (frequent voiding of small amounts of urine). *Nocturia* (nok-tu′re-ah), the need to get up during the night to urinate, plagues almost two-thirds of this population. In many, incontinence is the final outcome of the aging process. This loss of control is a tremendous blow to the pride of many aging people. Urinary retention is another common problem; most often it is a result of hypertrophy of the prostate gland in males. Some of the problems of incontinence and retention can be avoided by a regular regimen of activity that keeps the body as a whole in optimum condition and promotes alertness to elimination signals.

Systems in Sync

Homeostatic Relationships between the Urinary System and Other Body Systems

Endocrine System
- Kidneys dispose of nitrogenous wastes; maintain fluid, electrolyte, and acid-base balance of blood; produce the hormone erythropoietin; renal regulation of Na^+ and water balance essential for blood pressure homeostasis and hormone transport in the blood
- ADH, aldosterone, ANP, and other hormones help regulate renal reabsorption of water and electrolytes

Lymphatic System/Immunity
- Kidneys dispose of nitrogenous wastes; maintain fluid, electrolyte, and acid-base balance of blood
- By returning leaked plasma fluid to cardiovascular system, lymphatic vessels help ensure normal systemic blood pressure needed for kidney function; immune cells protect urinary organs from infection, cancer, and other foreign substances

Digestive System
- Kidneys dispose of nitrogenous wastes; maintain fluid, electrolyte, and acid-base balance of blood; also, metabolize vitamin D to the active form needed for calcium absorption
- Digestive organs provide nutrients needed for kidney cell health; liver synthesizes most urea, a nitrogenous waste that must be excreted by the kidneys

Urinary System

Muscular System
- Kidneys dispose of nitrogenous wastes; maintain fluid, electrolyte, and acid-base balance of blood; renal regulation of Na^+, K^+, and Ca^{2+} content in ECF crucial for muscle activity
- Muscles of pelvic diaphragm and external urethral sphincter function in voluntary control of micturition; creatinine is a nitrogenous waste product that must be excreted by the kidneys

Nervous System
- Kidneys dispose of nitrogenous wastes; maintain fluid, electrolyte, and acid-base balance of blood; renal regulation of Na^+, K^+, and Ca^{2+} content in ECF essential for normal neural function
- Neural controls involved in micturition; sympathetic nervous system activity triggers the renin-angiotensin mechanism

Respiratory System
- Kidneys dispose of nitrogenous wastes; maintain fluid, electrolyte, and acid-base balance of blood
- Respiratory system provides oxygen required by kidney cells; disposes of carbon dioxide; cells in the lungs convert angiotensin I to angiotensin II

Cardiovascular System
- Kidneys dispose of nitrogenous wastes; maintain fluid, electrolyte, and acid-base balance of blood; renal regulation of Na^+ and water balance essential for blood pressure homeostasis. Na^+, K^+, and Ca^{2+} regulation help maintain normal heart function
- Systemic arterial blood pressure is the driving force for glomerular filtration; heart secretes atrial natriuretic peptide; blood vessels transport nutrients, oxygen, etc. to urinary organs

Reproductive System
- Kidneys dispose of nitrogenous wastes; maintain fluid, electrolyte, and acid-base balance of blood

Integumentary System
- Kidneys dispose of nitrogenous wastes; maintain fluid, electrolyte, and acid-base balance of blood
- Skin provides external protective barrier; serves as site for water loss (via perspiration); vitamin D synthesis site

Skeletal System
- Kidneys dispose of nitrogenous wastes; maintain fluid, electrolyte, and acid-base balance of blood
- Bones of rib cage provide some protection to kidneys

SUMMARY

Media study tools that could provide you with additional help in reviewing specific key topics of Chapter 15 are referenced below.

IP = Interactive Physiology;

WEB = A&P Place website.

KIDNEYS (pp. 502–511)

1. The paired kidneys are retroperitoneal in the superior lumbar region. Each kidney has a medial indentation (hilus), where the renal artery, renal vein, and ureter are seen. Each kidney is enclosed in a tough fibrous capsule. A fatty cushion holds the kidneys against the body trunk wall.

   IP Urinary System Topic: Anatomy Review, page 4.

2. A longitudinal section of a kidney reveals an outer cortex, deeper medulla, and medial pelvis. Extensions of the pelvis (calyces) surround the tips of medullary pyramids and collect urine draining from them.

3. The renal artery, which enters the kidney, breaks up into segmental, lobar, and then interlobar arteries that travel outward through the medulla. Interlobar arteries split into arcuate arteries, which branch to produce interlobular arteries, which serve the cortex.

   IP Urinary System Topic: Anatomy Review, pages 5–6.

4. Nephrons are structural and functional units of the kidneys. Each consists of a glomerulus and a renal tubule. Subdivisions of the renal tubule (from the glomerulus onward) are glomerular capsule, proximal convoluted tubule, loop of Henle, and distal convoluted tubule. A second (peritubular) capillary bed is also associated with each nephron.

   WEB Exercise: Chapter 15, Anatomy and Function of the Nephron.

5. Nephron functions include filtration, reabsorption, and secretion. Filtrate formation is the role of the high-pressure glomerulus. Filtrate is essentially plasma without blood proteins. In reabsorption, done by tubule cells, needed substances are removed from filtrate (amino acids, glucose, water, some ions) and returned to blood. The tubule cells also secrete additional substances into filtrate. Secretion is important to rid the body of drugs and excess ions and to maintain acid-base balance of blood.

   WEB Exercise: Chapter 15, Nephron Activity; Early Filtrate Processing.

   IP Urinary System Topic: Anatomy Review, pages 7–9.

6. Urine is clear, yellow, and usually slightly acidic, but its pH value varies widely. Substances normally found in urine are nitrogenous wastes, water, various ions (always including sodium and potassium). Substances normally absent from urine include glucose, blood proteins, blood, pus (WBCs), bile.

URETERS, URINARY BLADDER, AND URETHRA (pp. 512–514)

1. The ureters are slender tubes running from each kidney to the bladder. They conduct urine by peristalsis from kidney to bladder.

   IP Urinary System Topic: Anatomy Review, page 3.

2. The bladder is a muscular sac posterior to the pubic symphysis. It has two inlets (ureters) and one outlet (urethra). In males, the prostate gland surrounds its outlet. The function of the bladder is to store urine.

3. The urethra is a tube that leads urine from the bladder to the body exterior. In females, it is 3 to 4 cm long and conducts only urine. In males, it is 20 cm long and conducts both urine and sperm. The internal sphincter of smooth muscle is at the bladder-urethra junction. The external sphincter of skeletal muscle is located more inferiorly.

4. Micturition is emptying of the bladder. The micturition reflex causes the involuntary internal sphincter to open when stretch receptors in the bladder wall are stimulated. Since the external sphincter is voluntarily controlled, micturition can ordinarily be temporarily delayed. Incontinence is the inability to control micturition.

FLUID, ELECTROLYTE, AND ACID-BASE BALANCE (pp. 514–520)

1. Blood composition depends on diet, cellular metabolism, and urinary output. To maintain blood composition, the kidneys must:

   a. Allow nitrogen-containing wastes (urea, ammonia, creatinine, uric acid) to leave the body in the urine.

   b. Maintain water and electrolyte balance by absorbing more or less water and reclaiming ions in response to hormonal signals. ADH increases water reabsorption and conserves body water. Aldosterone increases reabsorption of sodium and water and decreases potassium reabsorption.

c. Maintain acid-base balance by actively secreting bicarbonate ions (and retaining H^+) and by absorbing bicarbonate ions (and secreting H^+). Chemical buffers tie up excess H^+ or bases temporarily; respiratory centers modify blood pH by retaining CO_2 (decreases pH) or by eliminating more CO_2 from the blood (increases blood pH). Only the kidney can remove metabolic acids and excess bases from the body.

IP Fluids and Electrolytes Topic: Introduction to Body Fluids, pages 3, 8–14, 18–22.

DEVELOPMENTAL ASPECTS OF THE URINARY SYSTEM (pp. 520–521)

1. The kidneys begin to develop in the first few weeks of embryonic life and are excreting urine by the third month.
2. Common congenital abnormalities include polycystic kidney and hypospadias.
3. Common urinary system problems in children and young to middle-aged adults are infections caused by fecal microorganisms, sexually transmitted disease–causing microorganisms, and *Streptococcus.*
4. Renal failure is an uncommon, but extremely serious, problem in which kidneys are unable to concentrate urine, and dialysis must be done to maintain chemical homeostasis of blood.
5. With age, filtration rate decreases and tubule cells become less efficient at concentrating urine, leading to urgency, frequency, and incontinence. In males, urinary retention is another common problem.

REVIEW QUESTIONS

MULTIPLE CHOICE

More than one choice may apply.

1. Microscopic examination of a section of the kidney shows a thick-walled vessel with glomeruli scattered in the tissue on one side of the vessel but not on the other side. What vessel is this?
   a. Arcuate vein
   b. Interlobular artery
   c. Interlobular vein
   d. Arcuate artery
2. What is the glomerulus?
   a. The same as the renal tubule
   b. The same as Bowman's capsule
   c. The same as the nephron
   d. Capillaries
3. Urine passes through the ureters by which mechanism?
   a. Ciliary action
   b. Peristalsis
   c. Gravity alone
   d. Suction
4. Effects of aldosterone include:
   a. increase in sodium ion excretion
   b. increase in water retention
   c. increase in potassium ion concentration in the urine
   d. higher blood pressure
5. Which of the following is dependent on tubular secretion?
   a. Clearing penicillin from the blood
   b. Removal of nitrogenous wastes that have been reabsorbed
   c. Removal of excess potassium ions from the blood
   d. Control of blood pH
6. Which is a normal value for percent of body weight that is water for a young to middle-aged man?
   a. 73%
   b. 50%
   c. 45%
   d. 60%
7. The smallest fluid compartment is the:
   a. intracellular fluid
   b. extracellular fluid
   c. plasma
   d. interstitial fluid
8. In the carbonic acid–bicarbonate buffer system, strong acids are buffered by:
   a. carbonic acid
   b. water
   c. bicarbonate ion
   d. the salt of the strong acid

SHORT ANSWER ESSAY

1. Name the organs of the urinary system, and describe the general function of each organ.
2. Describe the location of the kidneys in the body.

3. Make a diagram of a longitudinal section of a kidney. Identify and label the cortex, medulla, medullary pyramids, renal columns, and pelvis.
4. Name the structural and functional unit of the kidney.
5. Trace the pathway a uric acid molecule takes from a glomerulus to the urethra. Name every gross or microscopic structure it passes through on its journey.
6. What is the function of the glomerulus? What two functions do the renal tubules perform?
7. Besides ridding the body of wastes formed during cell metabolism, the kidney adjusts blood chemistry in other ways. What are these three other ways?
8. Explain the difference between filtrate and urine.
9. How does aldosterone modify the chemical composition of urine?
10. What hormone's name means "against urine flow"? What condition happens if it is not secreted?
11. Name three substances normally found in blood that are not normally found in urine. Name the condition when each named substance *is* found in urine.
12. Why is urinalysis a routine part of any good physical examination?
13. How do the internal and external urethral sphincters differ structurally and functionally?
14. Define *micturition,* and then describe the micturition reflex.
15. What sometimes happens when urine becomes too concentrated or remains too long in the bladder?
16. Define *incontinence.*
17. How is the female urethra different from that of the male in structure and function?
18. Why is cystitis more common in females?
19. What type of problem most commonly affects the urinary system organs?
20. Describe the changes that occur in kidney and bladder function in old age.

AT THE CLINIC

1. A 55-year-old woman is awakened by an excruciating pain that radiates from her right abdomen to her flank on the same side. The pain is not continuous, but it recurs every 3 to 4 minutes. Diagnose this patient's problem and cite factors that might favor its occurrence. Explain why this woman's pain comes in "waves."
2. Mitchell's parents bring him to the clinic because his "wee-wee" (urine) is tinged with blood. Two days before he was diagnosed with strep throat. His face and hands are swollen. What is the probable cause of Mitchell's current kidney problem?
3. A young woman has come to the clinic with dysuria and frequent urination. What is the most likely diagnosis?
4. Mrs. Gates exhibits excessive thirst and polyuria, with excessive blood levels of sodium. Will you check the pituitary or the adrenal glands? What hormone do you suspect is involved, and is hyposecretion or hypersecretion more likely to be occurring? What is the name of this condition?
5. B.J. has been diagnosed with a hypersecretory tumor of a parathyroid gland. What effect will this have on his renal tubule function?
6. What happens to the rate of RBC production in a patient on dialysis with total renal failure? What could be given to the patient to counteract such a problem? (Hint: You might want to check Table 9.2 on page 321 for help with this one.)
7. Two physiology students are having a disagreement about renal function. Dan says that the kidneys work harder when you eat a high salt diet while Peter says that they work harder when you drink lots of water. Who is right, and why?
8. Mr. Jessup, a 55-year-old man, is operated on for a cerebral tumor. About one month later, he is complaining of excessive thirst and that he has been voiding almost continuously. A urine sample is collected and its specific gravity is 1.001. What is your diagnosis of Mr. Jessup's problem, and what might its connection be to his previous surgery?

16 The Reproductive System

YOUR GOALS

After completing this chapter, you will have a working knowledge of the functions of the reproductive system and will have mastered the objectives below.

FUNCTION PREVIEW

- The reproductive system ensures continuity of the species by producing offspring.

OBJECTIVE CHECKLIST

ANATOMY OF THE MALE REPRODUCTIVE SYSTEM (pp. 529–533)

- Discuss the common purpose of the reproductive system organs.
- When provided with a model or diagram, identify the organs of the male reproductive system, and discuss the general function of each.
- Name the endocrine and exocrine products of the testes.
- Discuss the composition of semen, and name the glands that produce it.
- Trace the pathway followed by a sperm from the testis to the body exterior.
- Define *erection, ejaculation,* and *circumcision.*

MALE REPRODUCTIVE FUNCTIONS (pp. 533–535)

- Define *meiosis* and *spermatogenesis.*
- Describe the structure of a sperm, and relate its structure to its function.
- Describe the effect of FSH and LH on testis functioning.

ANATOMY OF THE FEMALE REPRODUCTIVE SYSTEM (pp. 535–539)

- When provided with an appropriate model or diagram, identify the organs of the female reproductive system, and discuss the general function of each.
- Describe the functions of the vesicular follicle and corpus luteum of the ovary.

>>

(Objective Checklist, continued)

- Define *endometrium, myometrium,* and *ovulation.*
- Indicate the location of the following regions of the female uterus: cervix, fundus, body.

FEMALE REPRODUCTIVE FUNCTIONS AND CYCLES (pp. 539–543)

- Define *oogenesis.*
- Describe the influence of FSH and LH on ovarian function.
- Describe the phases and controls of the menstrual cycle.

MAMMARY GLANDS (pp. 543–544)

- Describe the structure and function of the mammary glands.

SURVEY OF PREGNANCY AND EMBRYONIC DEVELOPMENT (pp. 545–557)

- Define *fertilization* and *zygote.*
- Describe implantation.
- Distinguish between an embryo and a fetus.
- List the major functions of the placenta.
- Indicate several ways that pregnancy alters or modifies the functioning of the mother's body.
- Describe how labor is initiated, and briefly discuss the three stages of labor.
- List several agents that can interfere with normal fetal development.

DEVELOPMENTAL ASPECTS OF THE REPRODUCTIVE SYSTEM (pp. 557–560)

- Describe the importance of the presence/absence of testosterone during embryonic development of the reproductive system organs.
- Define *menarche* and *menopause.*
- List common reproductive system problems seen in adult and aging males and females.

Most organ systems of the body function almost continuously to maintain the well-being of the individual. The reproductive system, however, appears to "slumber" until puberty. The **gonads** (go′nadz; "seeds"), or **primary sex organs,** are the *testes* in males and the *ovaries* in females. The gonads produce sex cells, or **gametes** (gam′ēts; "spouses"), and secrete sex hormones. The remaining reproductive system structures are **accessory reproductive organs.** Although male and female **reproductive systems** are quite different, their joint purpose is to produce offspring.

The reproductive role of the male is to manufacture male gametes called **sperm** and deliver them to the female reproductive tract. The female, in turn, produces female gametes, called **ova** (singular, *ovum*), or *eggs.* If the time is suitable, the sperm and egg fuse to produce a fertilized egg, which is the first cell of a new individual. Once fertilization has occurred, the female uterus provides a protective environment in which the *embryo,* later called the *fetus,* develops until birth.

The sex hormones play vital roles both in the development and function of the reproductive organs and in sexual behavior and drives. These gonadal hormones also influence the growth and development of many other organs and tissues of the body.

Anatomy of the Male Reproductive System

As already noted, the primary reproductive organs of the male are the **testes** (tes′tēz), or *male gonads*, which have both an exocrine (sperm-producing) function and an endocrine (testosterone-producing) function. The accessory reproductive structures are ducts or glands that aid in the delivery of sperm to the body exterior or to the female reproductive tract.

Testes

Each olive-sized testis is approximately 4 cm (1½ inches) long and 2.5 cm (1 inch) wide. A fibrous connective tissue capsule, the *tunica albuginea* (tu′nĭ-kah al″bu-jin′e-ah; "white coat") surrounds each testis. Extensions of this capsule *(septa)* plunge into the testis and divide it into a large number of *lobules*. Each lobule contains one to four tightly coiled **seminiferous** (sem″in-if′er-us) **tubules,** the actual "sperm-forming factories" (Figure 16.1). Seminiferous tubules of each lobe empty sperm into another set of tubules, the *rete* (re′te) *testis,* located at one side of the testis. Sperm travel through the rete testis to enter the first part of the duct system, the *epididymis* (ep″ĭ-did′ĭ-mis), which hugs the external surface of the testis.

Lying in the soft connective tissue surrounding the seminiferous tubules are the *interstitial* (in″ter-stish′al) *cells,* functionally distinct cells that produce androgens—most importantly, *testosterone.* Thus, the sperm-producing and hormone-producing functions of the testes are carried out by completely different cell populations.

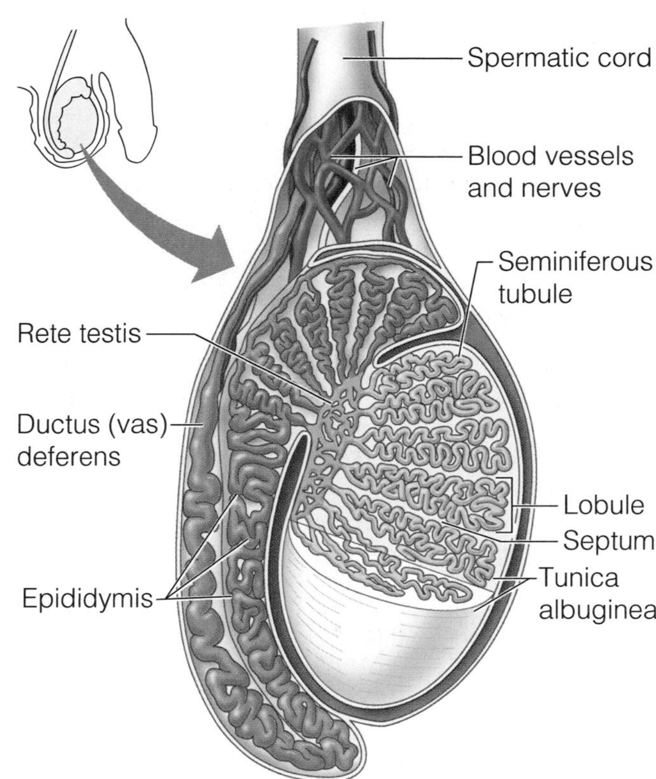

FIGURE 16.1 Sagittal section of the testis and associated epididymis.

Duct System

The accessory organs forming the male duct system, which transports sperm from the body, are the *epididymis, ductus deferens,* and *urethra* (Figure 16.2).

Epididymis

The comma-shaped **epididymis** is a highly coiled tube about 6 m (20 feet) long that caps the superior part of the testis and then runs down its posterolateral side (see Figure 16.1). The epididymis is the first part of the male duct system and provides a temporary storage site for the immature sperm that enter it from the testis. While the sperm make their way along the snaking course of the epididymis (a trip that takes about 20 days), they mature, gaining the ability to swim. When a male is sexually stimulated, the walls of the epididymis contract to expel the sperm into the next part of the duct system, the ductus deferens.

Ductus Deferens

The **ductus deferens** (duk′tus def′er-enz; "carrying away"), or **vas deferens,** runs upward from the epididymis through the inguinal canal into the pelvic cavity and arches over the superior aspect of the urinary bladder. This tube is enclosed, along with blood vessels and nerves, in a connective tissue sheath called the **spermatic cord** (see Figure 16.1). The end of the ductus deferens empties into the **ejaculatory** (e-jak′u-lah-to″re) **duct,** which passes through the prostate gland to merge with the urethra. The main function of the ductus deferens is to propel live sperm from their storage

Q *Looking at this diagram, it is easy to see how hypertrophy of the prostate gland might lead to urinary system problems. How so?*

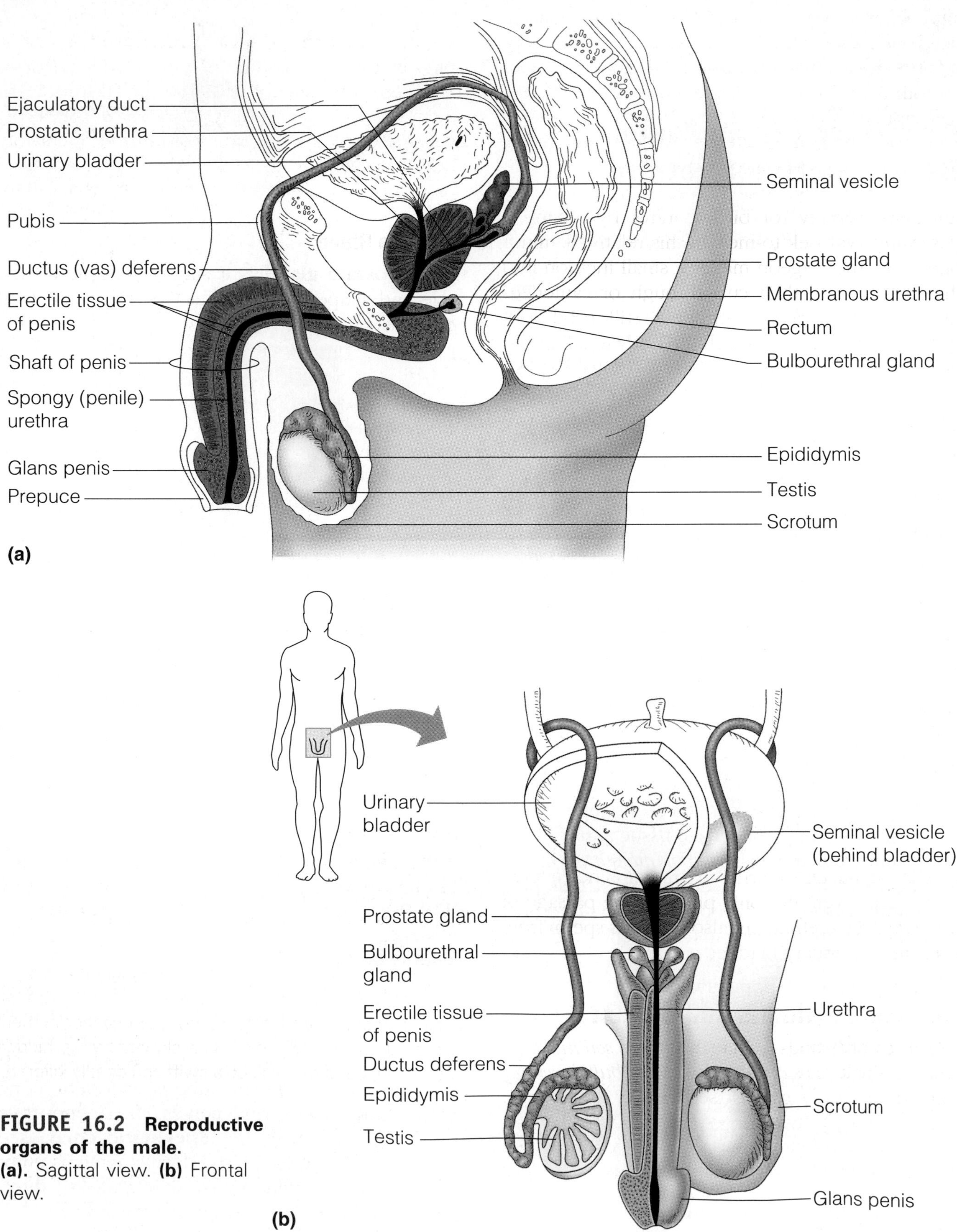

FIGURE 16.2 Reproductive organs of the male. **(a).** Sagittal view. **(b)** Frontal view.

A *Because the prostate gland encircles the superior part of the urethra, prostate hypertrophy would constrict the urethra in that region, impairing urination.*

sites, the epididymis and distal part of the ductus deferens, into the urethra. At the moment of ejaculation (*ejac* = to shoot forth), the thick layers of smooth muscle in its walls create peristaltic waves that rapidly squeeze the sperm forward.

As Figure 16.2 illustrates, part of the ductus deferens lies in the scrotum, which hangs outside the body cavity. Some men voluntarily opt to take full responsibility for birth control by having a *vasectomy* (vah-sek′to-me). In this relatively minor operation, the surgeon makes a small incision into the scrotum and then cuts through or cauterizes the ductus deferens. Sperm are still produced, but they can no longer reach the body exterior and eventually they deteriorate and are reabsorbed. A man is sterile after this procedure, but because testosterone is still produced, the sex drive and secondary sex characteristics are retained.

Urethra

The **urethra,** which extends from the base of the urinary bladder to the tip of the penis, is the terminal part of the male duct system. It has three named regions: (1) the **prostatic urethra,** surrounded by the prostate gland; (2) the **membranous urethra,** spanning the distance from the prostatic urethra to the penis; and (3) the **spongy (penile) urethra,** running within the length of the penis. As mentioned in Chapter 15, the male urethra carries both urine and sperm to the body exterior; thus, it serves two masters, the urinary and reproductive systems. However, urine and sperm never pass at the same time. When ejaculation occurs and sperm enter the prostatic urethra from the ejaculatory ducts, the bladder sphincter (internal urethral sphincter) constricts. This event not only prevents the passage of urine into the urethra, but also prevents sperm from entering the urinary bladder.

Accessory Glands and Semen

The accessory glands include the paired *seminal vesicles,* the single *prostate gland,* and the *bulbourethral* (bul-bo-u-re′thral) *glands* (see Figure 16.2). These glands produce the bulk of *semen* (se′men), the sperm-containing fluid that is propelled out of the male's reproductive tract during **ejaculation.**

Seminal Vesicles

The **seminal** (sem′ĭ-nul) **vesicles,** located at the base of the bladder, produce about 60 percent of the fluid volume of semen. Their thick, yellowish secretion is rich in sugar (fructose), vitamin C, prostaglandins, and other substances, which nourish and activate the sperm passing through the tract. The duct of each seminal vesicle joins that of the ductus deferens on the same side to form the ejaculatory duct (see Figure 16.2). Thus, sperm and seminal fluid enter the urethra together during ejaculation.

Prostate Gland

The **prostate gland** is a single gland about the size and shape of a chestnut (see Figure 16.2). It encircles the upper (prostatic) part of the urethra just below the urinary bladder. Prostate gland secretion is a milky fluid that plays a role in activating sperm. During ejaculation, it enters the urethra through several small ducts. Since the prostate is located immediately anterior to the rectum, its size and texture can be palpated (felt) by digital (finger) examination through the anterior rectal wall.

Homeostatic Imbalance

The prostate gland has a reputation as a health destroyer. Hypertrophy of the prostate gland, which affects nearly every elderly male, strangles the urethra. This troublesome condition makes urination difficult and enhances the risk of bladder infections *(cystitis)* and kidney damage. Traditional treatment has been surgical, but some newer options are becoming more popular. These include

- Using drugs (finasteride) or microwaves to shrink the prostate
- Inserting a small inflatable balloon to compress the prostate tissue away from the prostatic urethra
- Inserting a tiny needle that emits bursts of radiofrequency radiation, which incinerate excess prostate tissue

Inflammation of the prostate is the single most common reason for a man to consult a urologist, and prostatic cancer is the third most prevalent cancer in men. As a rule, prostatic cancer is a slow-growing, hidden condition, but it can also be a swift and deadly killer. ▲

Bulbourethral Glands

The **bulbourethral glands** (formerly called *Cowper's glands*) are tiny, pea-sized glands inferior to the prostate gland. They produce a thick, clear mucus that drains into the penile urethra. This secretion is the first to pass down the urethra when a man becomes sexually excited. It is believed to

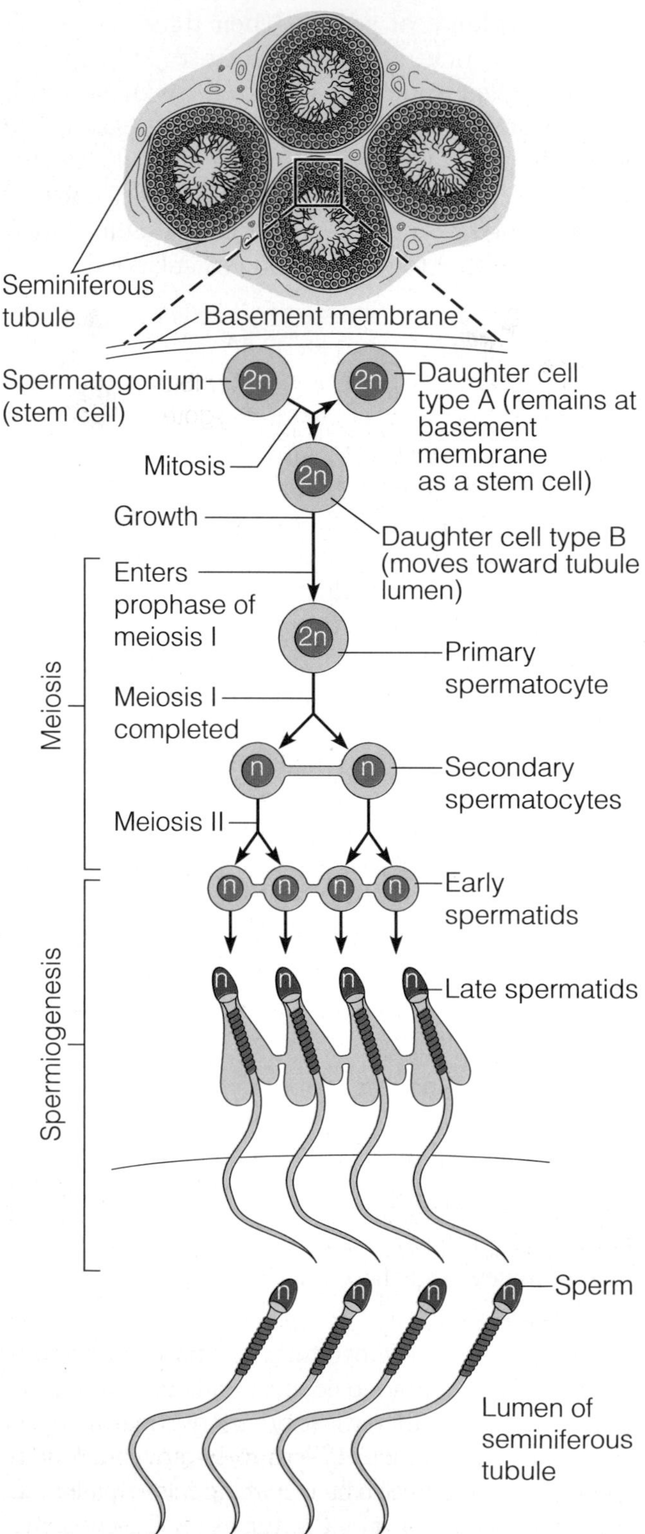

FIGURE 16.3 Spermatogenesis. Flow chart showing the relative position of the spermatogenic cells in the wall of the seminiferous tubule. Although the stem cells and the primary spermatocytes have the same number of chromosomes (46, designated as 2*n*) as other body cells, the products of meiosis (spermatids and sperm) have only half as many (23, designated as *n*).

cleanse the urethra of traces of acidic urine, and it serves as a lubricant during sexual intercourse.

Semen

Semen is a milky white, somewhat sticky mixture of sperm and accessory gland secretions. The liquid provides a transport medium and nutrients and contains chemicals that protect the sperm and aid their movement. Mature sperm cells are streamlined cellular "missiles" containing little cytoplasm or stored nutrients. The fructose in the seminal vesicle secretion provides essentially all of their energy fuel. The relative alkalinity of semen as a whole (pH 7.2–7.6) helps neutralize the acid environment (pH 3.5–4) of the female's vagina, protecting the delicate sperm. Sperm are very sluggish under acidic conditions (below pH 6). Semen also contains *seminalplasmin* (an antibiotic chemical that inhibits bacterial multiplication), the hormone relaxin, and certain enzymes that enhance sperm motility.

Semen also dilutes sperm; without such dilution, sperm motility is severely impaired. The amount of semen propelled out of the male duct system during ejaculation is relatively small, only 2 to 5 ml (about a teaspoonful), but there are between 50 and 130 million sperm in each milliliter.

Homeostatic Imbalance

Male infertility may be caused by obstructions of the duct system, hormonal imbalances, environmental estrogens, pesticides, excessive alcohol, and many other factors. One of the first series of tests done when a couple has been unable to conceive is *semen analysis.* Factors analyzed include sperm count, motility and morphology (shape and maturity), and semen volume, pH, and fructose content. A sperm count lower than 20 million per milliliter makes impregnation improbable. ▲

External Genitalia

The **external genitalia** (jen″i-tal′e-ah) of the male include the *scrotum* and the *penis* (see Figure 16.2). The **scrotum** (skro′tum; "pouch") is a divided sac of skin that hangs outside the abdominal cavity, between the legs and at the root of the penis. Under normal conditions, the scrotum hangs loosely from its attachments, providing the testes with a temperature that is below body temperature. This is a rather exposed location for a man's testes, which contain his entire genetic heritage, but apparently viable sperm cannot be produced at normal body temperature. The scrotum, which provides a temperature about 3°C (5.4°F) lower, is necessary for the production of healthy sperm. When the external temperature is very cold, the scrotum becomes heavily wrinkled as it pulls the testes closer to the warmth of the body wall. Thus, changes in scrotal surface area can maintain a temperature that favors viable sperm production.

The **penis** (pe′nis; "tail") is designed to deliver sperm into the female reproductive tract. The skin-covered penis consists of a **shaft,** which ends in an enlarged tip, the **glans penis.** The skin covering the penis is loose, and it folds downward to form a cuff of skin, the **prepuce** (pre′pus), or **foreskin,** around the proximal end of the glans. Frequently, the foreskin is surgically removed shortly after birth, by a procedure called *circumcision.*

Internally, the spongy urethra (see Figure 16.2) is surrounded by three elongated areas of *erectile tissue,* a spongy tissue that fills with blood during sexual excitement. This causes the penis to enlarge and become rigid. This event, called **erection,** helps the penis serve as a penetrating organ to deliver the semen into the female's reproductive tract.

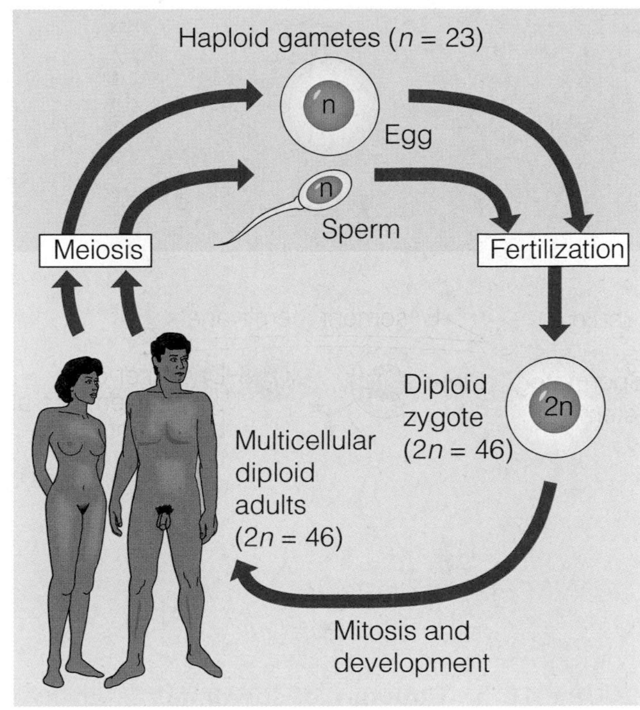

FIGURE 16.4 The human life cycle.

Male Reproductive Functions

The chief role of the male in the reproductive process is to produce sperm and the hormone testosterone. These processes are described next.

Spermatogenesis

Sperm production, or **spermatogenesis** (sper″mah-to-jen′ĕ-sis), begins during puberty and continues throughout life. Every day a man makes millions of sperm. Since only one sperm fertilizes an egg, it seems that nature has made sure that the human species will not be endangered for lack of sperm.

Sperm formation occurs in the seminiferous tubules of the testis, as noted earlier. As shown in Figure 16.3, the process is begun by primitive stem cells called **spermatogonia** (sper″mah-to-go′ne-ah), found in the outer edge, or periphery, of each tubule. Spermatogonia go through rapid mitotic divisions to build up the stem cell line. From birth until puberty, all such divisions simply produce more stem cells. During puberty, however, *follicle-stimulating hormone (FSH)* is secreted in increasing amounts by the anterior pituitary gland, and, from this time on, each division of a spermatogonium produces one stem cell (a *type A daughter cell*) and another cell called a *type B daughter cell.* The type A cell remains at the tubule periphery to maintain the stem cell population. The type B cell gets pushed toward the tubule lumen, where it becomes a **primary spermatocyte,** destined to undergo *meiosis* (mi-o′sis) and form four sperm. **Meiosis** is a special type of nuclear division that occurs for the most part only in the gonads (testes and ovaries). It differs from *mitosis* (described in Chapter 3) in two major ways. Meiosis consists of two successive divisions of the nucleus (called meiosis I and II) and results in four (instead of two) daughter cells, or more precisely, four *gametes.* In spermatogenesis, the gametes are called **spermatids** (sper′mah-tidz). Spermatids have only half as much genetic material as other body cells. In humans, this is 23 chromosomes (or the so-called n number of chromosomes) rather than the usual 46 ($2n$). Then, when the sperm and the egg (which also has 23 chromosomes) unite, forming the fertilized egg, or zygote, the normal $2n$ number of 46 chromosomes is reestablished and is maintained in subsequent body cells by the process of mitosis (Figure 16.4).

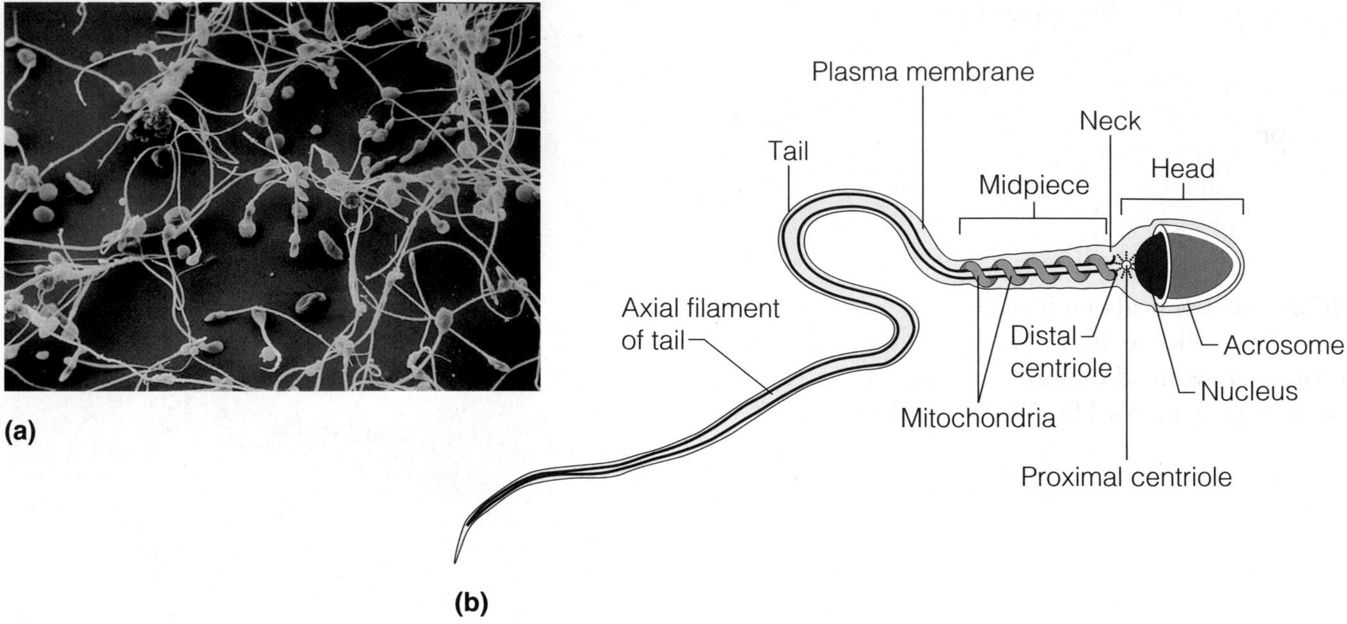

FIGURE 16.5 Structure of sperm. (a) Scanning electron micrograph of mature sperm (4303×). **(b)** Diagrammatic view of a sperm.

As meiosis occurs, the dividing cells (primary and then secondary spermatocytes) are pushed toward the lumen of the tubule. Thus, the progress of meiosis can be followed from the tubule periphery to the lumen. The spermatids, which are the products of meiosis, are *not* functional sperm. They are nonmotile cells and have too much excess baggage to function well in reproduction. They must undergo further changes, in which their excess cytoplasm is stripped away and a tail is formed (see Figure 16.3). In this last stage of sperm development, called **spermiogenesis** (sper″me-o-gen′ĕ-sis), all the excess cytoplasm is sloughed off, and what remains is compacted into the three regions of the mature sperm—the *head, midpiece,* and *tail* (Figure 16.5). The mature sperm is a greatly streamlined cell equipped with a high rate of metabolism and a means of propelling itself, enabling it to move long distances in a short time to get to the egg. It is a prime example of the fit between form and function.

The sperm head contains DNA, the genetic material. Essentially, it *is* the nucleus of the spermatid. Anterior to the nucleus is the helmetlike **acrosome** (ak′ro-sōm), which is similar to a large lysosome. When a sperm comes into close contact with an egg (or more precisely, an *oocyte*), the acrosomal membrane breaks down and releases enzymes that help the sperm penetrate through the follicle cells that surround the egg. *Filaments,* which form the tail, arise from centrioles in the midpiece. *Mitochondria* wrapped tightly around these filaments provide the ATP needed for the whiplike movements of the tail that propel the sperm.

The entire process of spermatogenesis, from the formation of a primary spermatocyte to release of immature sperm in the tubule lumen, takes 64 to 72 days. Sperm in the lumen are unable to "swim" and incapable of fertilizing an egg. They are moved by peristalsis through the tubules of the testes into the epididymis. There they undergo further maturation, which results in increased motility and fertilizing power.

Homeostatic Imbalance

Environmental threats can alter the normal process of sperm formation. For example, some common antibiotics, such as penicillin and tetracycline, may suppress sperm formation. Radiation, lead, certain pesticides, marijuana, tobacco, and excessive alcohol can cause production of abnormal sperm (two-headed, multiple-tailed, and so on). ▲

Testosterone Production

As noted earlier, the interstitial cells produce **testosterone** (tes-tos'tĕ-rōn), the most important hormonal product of the testes. During puberty, as the seminiferous tubules are being prodded to produce sperm by FSH, the interstitial cells are being activated by **luteinizing hormone (LH),** sometimes called *interstitial cell–stimulating hormone (ICSH),* which is also released by the anterior pituitary gland (Figure 16.6). From this time on, testosterone is produced continuously (more or less) for the rest of a man's life. The rising blood level of testosterone in the young male stimulates the adolescent growth spurt, prods his reproductive organs to develop to their adult size, underlies the sex drive, and causes the secondary male sex characteristics to appear. **Secondary sex characteristics** typical of males include:

- Deepening of the voice due to enlargement of the larynx
- Increased hair growth all over the body, and particularly in the axillary and pubic regions and the face (the beard and mustache)
- Enlargement of skeletal muscles to produce the heavier muscle mass typical of the male physique
- Increased heaviness of the skeleton due to thickening of the bones

Because testosterone is responsible for the appearance of these typical masculine characteristics, it is often referred to as the "masculinizing" hormone.

Homeostatic Imbalance

If testosterone is not produced, the secondary sex characteristics never appear in the young man, and his other reproductive organs remain childlike. This is *sexual infantilism.* Castration of the adult male (or the inability of his interstitial cells to produce testosterone) results in a decrease in the size and function of his reproductive organs, as well as a decrease in his sex drive. *Sterility* also occurs because testosterone is necessary for the final stages of sperm production. ▲

Q *What is the effect of negative feedback of testosterone on anterior pituitary and hypothalamic cells?*

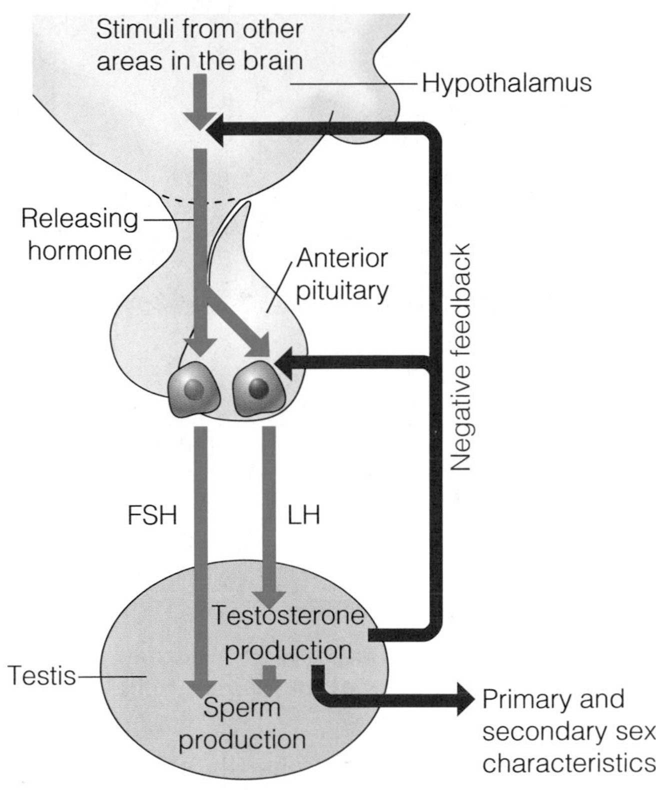

FIGURE 16.6 Hormonal control of the testis.

Anatomy of the Female Reproductive System

The reproductive role of the female is much more complex than that of the male. Not only must she produce the female gametes (ova), but her body must also nurture and protect a developing fetus during nine months of pregnancy. **Ovaries** are the primary reproductive organs of a female. Like the testes of a male, ovaries produce both an exocrine product (eggs, or *ova*) and endocrine products (estrogens and progesterone). The other organs of the female reproductive system serve as accessory structures to transport, nurture, or otherwise serve the needs of the reproductive cells and/or the developing fetus.

Ovaries

The paired *ovaries* (o'vah-rēz) are pretty much the size and shape of almonds. An internal view of an ovary reveals many tiny saclike structures called

A *It "turns off" the hypothalamic releasing factor that prompts ICSH secretion; hence, ICSH secretion also falls.*

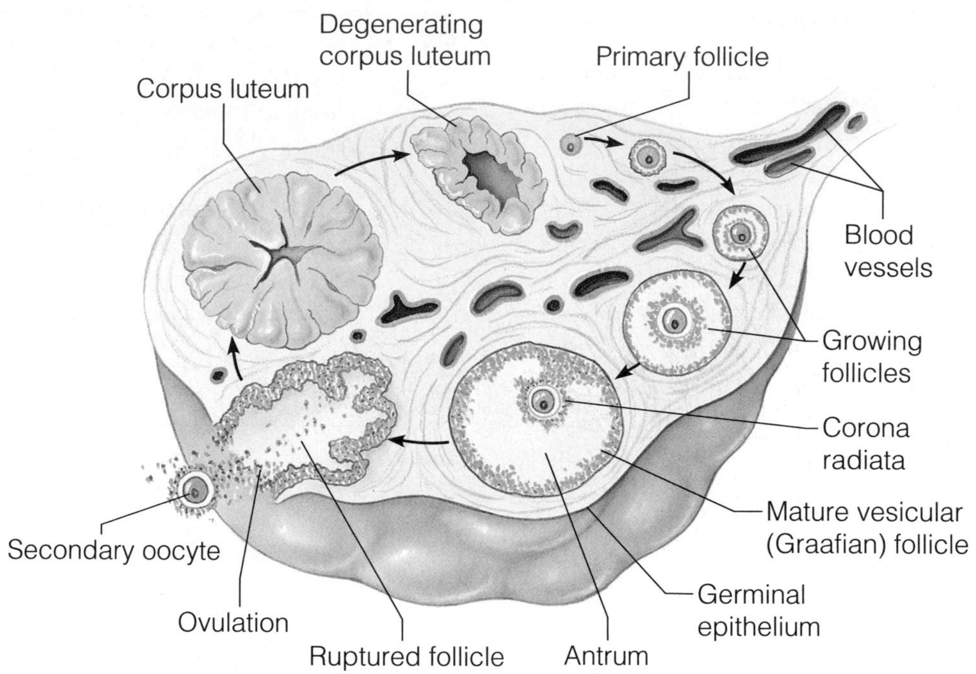

FIGURE 16.7 Sagittal view of a human ovary showing the developmental stages of an ovarian follicle.

ovarian follicles (Figure 16.7). Each follicle consists of an immature egg, called an **oocyte** (o′o-sīt), surrounded by one or more layers of very different cells called **follicle cells.** As a developing egg within a follicle begins to ripen or mature, the follicle enlarges and develops a fluid-filled central region called an *antrum.* At this stage, the follicle, called a **vesicular,** or **Graafian** (graf′e-an), **follicle,** is mature, and the developing egg is ready to be ejected from the ovary, an event called **ovulation.** After ovulation, the ruptured follicle is transformed into a very different-looking structure called a **corpus luteum** (kor′pus lu′te-um; "yellow body"), which eventually degenerates. Ovulation generally occurs every 28 days, but it can occur more or less frequently in some women. In older women, the surfaces of the ovaries are scarred and pitted, which attests to the fact that many eggs have been released.

The ovaries are secured to the lateral walls of the pelvis by the *suspensory ligaments.* They flank the uterus laterally and anchor to it medially by the *ovarian ligaments* (Figure 16.8). In between, they are enclosed and held in place by a fold of peritoneum, the *broad ligament.*

Duct System

The *uterine tubes, uterus,* and *vagina* form the duct system of the female reproductive tract (Figure 16.8).

Uterine (Fallopian) Tubes

The **uterine** (u′ter-in), or **fallopian** (fal-lo′pe-an), **tubes** form the initial part of the duct system. They receive the ovulated oocyte and provide a site where fertilization can occur. Each of the uterine tubes is about 10 cm (4 inches) long and extends medially from an ovary to empty into the superior region of the uterus. Like the ovaries, the uterine tubes are enclosed and supported by the broad ligament. Unlike in the male duct system, which is continuous with the tubule system of the testes, there is little or no actual contact between the uterine tubes and the ovaries. The distal end of each uterine tube expands as the funnel-shaped *infundibulum,* which has fingerlike projections called **fimbriae** (fim′bre-e) that partially surround the ovary. As an oocyte is expelled from an ovary during ovulation, the waving fimbriae create fluid currents that act to carry the oocyte into the uterine tube, where it begins its journey toward the uterus. (Obviously, however,

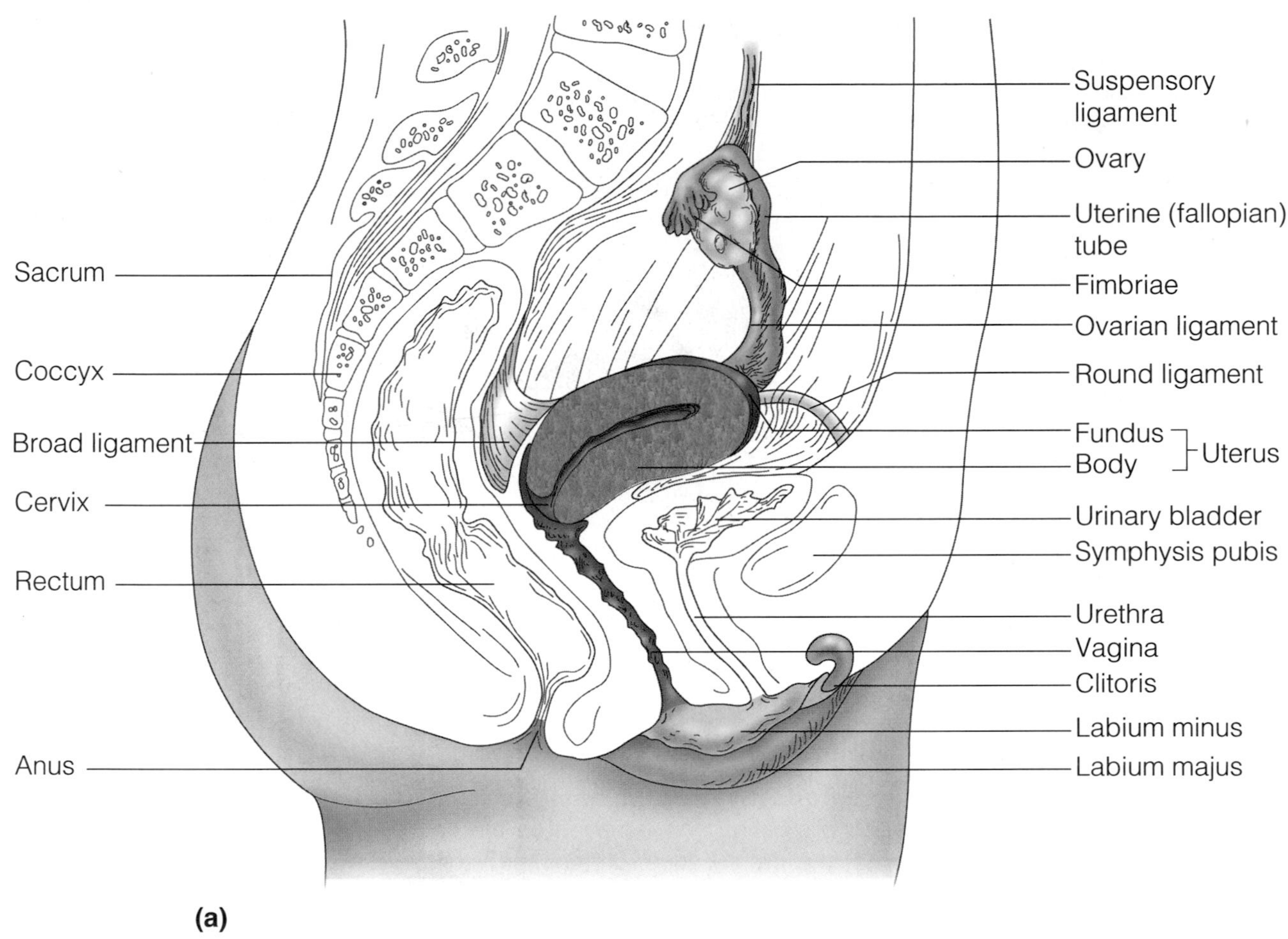

FIGURE 16.8 The human female reproductive organs. (a) Sagittal section. (The plural of *labium minus* and *majus* is *labia minora* and *majora,* respectively.)

many potential eggs are lost in the peritoneal cavity.) The oocyte is carried toward the uterus by a combination of peristalsis and the rhythmic beating of *cilia.* Because the journey to the uterus takes 3 to 4 days and the oocyte is viable for up to 24 hours after ovulation, the usual site of fertilization is the uterine tube. To reach the oocyte, the sperm must swim upward through the vagina and uterus to reach the uterine tubes. Because they must swim against the downward current created by the cilia, it is rather like swimming against the tide!

Homeostatic Imbalance

The fact that the uterine tubes are not continuous distally with the ovaries places women at risk for infections spreading into the peritoneal cavity from the reproductive tract. *Gonorrhea* (gon″o-re′ah) and other sexually transmitted bacteria sometimes infect the peritoneal cavity in this way, causing an extremely severe inflammation called *pelvic inflammatory disease (PID).* Unless treated promptly, PID can cause scarring and closure of the narrow uterine tubes, which is one of the major causes of female infertility. ▲

Uterus

The **uterus** (u′ter-us; "womb"), located in the pelvis between the urinary bladder and rectum, is a hollow organ that functions to receive, retain, and nourish a fertilized egg. In a woman who has never been pregnant, it is about the size and shape of a pear. (During pregnancy, the uterus increases tremendously in size and can be felt well above the umbilicus during the latter part of pregnancy.) The uterus is suspended in the pelvis by the broad ligament and anchored anteriorly and posteriorly by the *round* and *uterosacral ligaments,* respectively (see Figure 16.8).

The major portion of the uterus is referred to as the **body.** Its superior rounded region above the

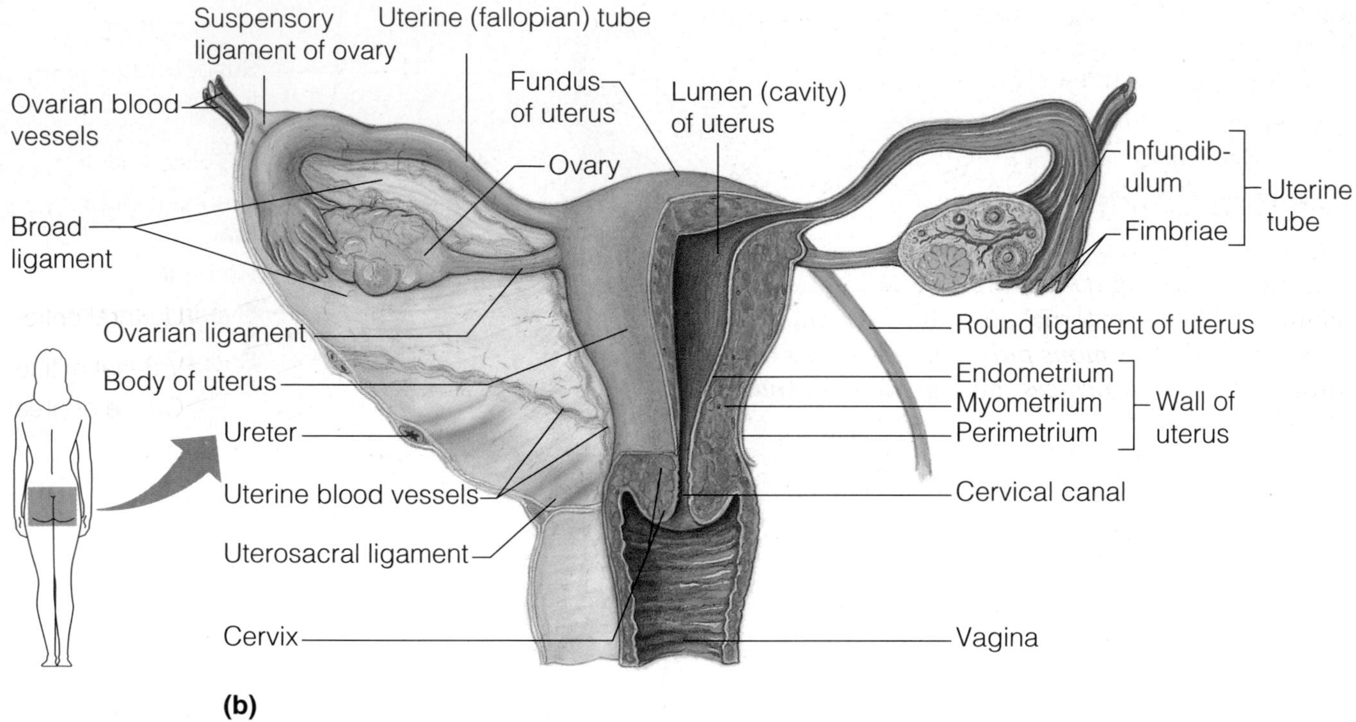

FIGURE 16.8 The human female reproductive organs (*continued*) **(b)** Posterior view. The posterior organ walls have been removed on the right side to reveal the shape of the lumen of the uterine tube, uterus, and vagina.

entrance of the uterine tubes is the **fundus,** and its narrow outlet, which protrudes into the vagina below, is the **cervix.**

The wall of the uterus is thick and composed of three layers. The inner layer or mucosa is the **endometrium** (en-do-me′tre-um). If fertilization occurs, the fertilized egg (actually the young embryo by the time it reaches the uterus) burrows into the endometrium (this process is called **implantation**) and resides there for the rest of its development. When a woman is not pregnant, the endometrial lining sloughs off periodically, usually about every 28 days, in response to changes in the levels of ovarian hormones in the blood. This process, called menstruation or *menses,* is discussed on pp. 541–543.

Homeostatic Imbalance

Cancer of the cervix is common among women between the ages of 30 and 50. Risk factors include frequent cervical inflammation, sexually transmitted diseases, multiple pregnancies, and many sexual partners. A yearly *Pap smear* is the single most important diagnostic test for detecting this slow-growing cancer. ▲

The **myometrium** (mi-o-me′tre-um) is the bulky middle layer of the uterus (see Figure 16.8b). It is composed of interlacing bundles of smooth muscle. The myometrium plays an active role during the delivery of a baby, when it contracts rhythmically to force the baby out of the mother's body. The outermost serous layer of the uterus is the *perimetrium* (per-ĭ-me′tre-um), or the visceral peritoneum.

Vagina

The **vagina** (vah-ji′nah) is a thin-walled tube 8 to 10 cm (3 to 4 inches) long. It lies between the bladder and rectum and extends from the cervix to the body exterior (see Figure 16.8). Often called the *birth canal,* the vagina provides a passageway for the delivery of an infant and for the menstrual flow to leave the body. Since it receives the penis (and semen) during sexual intercourse, it is the female organ of copulation.

The distal end of the vagina is partially closed by a thin fold of the mucosa called the **hymen** (hi′men). The hymen is very vascular and tends to bleed when it is ruptured during the first sexual intercourse. However, its durability varies. In some females, it is torn during a sports activity, tampon insertion, or pelvic examination. Occasionally, it is so tough that it must be ruptured surgically if intercourse is to occur.

External Genitalia

The female reproductive structures that are located external to the vagina are the **external genitalia** (Figure 16.9). The external genitalia, also called the **vulva,** include the *mons pubis, labia, clitoris, urethral* and *vaginal orifices,* and *greater vestibular glands.*

The **mons pubis** ("mountain on the pubis") is a fatty, rounded area overlying the pubic symphysis. After puberty, this area is covered with pubic hair. Running posteriorly from the mons pubis are two elongated hair-covered skin folds, the **labia majora** (la′be-ah ma-jo′ra), which enclose two delicate hair-free folds, the **labia minora.** The labia majora enclose a region called the **vestibule,** which contains the external openings of the urethra,* followed posteriorly by that of the vagina. A pair of mucus-producing glands, the **greater vestibular glands** *(Bartholin's glands),* flank the vagina, one on each side. Their secretion lubricates the distal end of the vagina during intercourse.

Just anterior to the vestibule is the **clitoris** (kli′to-ris; "hill"), a small, protruding structure that corresponds to the male penis. Like the penis, it is hooded by a prepuce and is composed of sensitive erectile tissue that becomes swollen with blood during sexual excitement. The clitoris differs from the penis in that it lacks a reproductive duct. The diamond-shaped region between the anterior end of the labial folds, the anus posteriorly, and the ischial tuberosities laterally is the **perineum** (per″ĭ-ne′um).

Female Reproductive Functions and Cycles

Oogenesis and the Ovarian Cycle

As described earlier, sperm production in males begins at puberty and generally continues throughout life. The situation is quite different in females.

*The male urethra carries both urine and semen, but the female urethra has no reproductive function—it is strictly a passageway for urine.

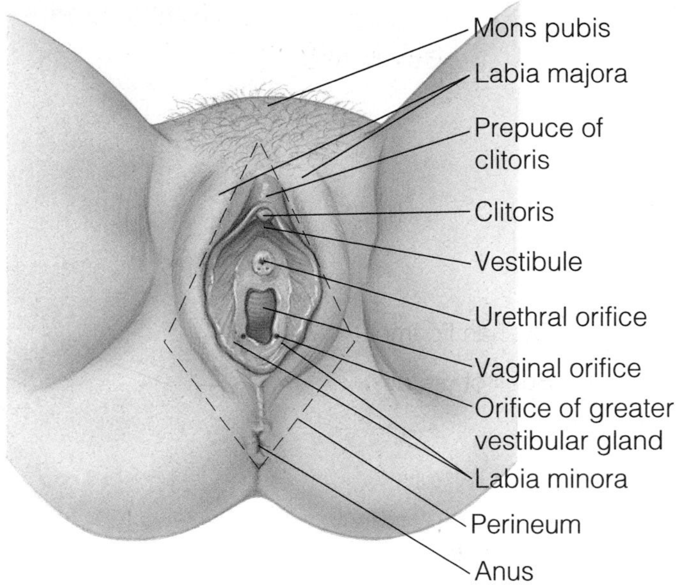

FIGURE 16.9 External genitalia of the human female.

The total supply of eggs that a female can release is already determined by the time she is born. In addition, a female's reproductive ability (that is, her ability to release eggs) usually begins during puberty and ends in her fifties or before. The period in which a woman's reproductive capability gradually declines and then finally ends is called *menopause* (see pp. 558–560).

Meiosis, the special kind of cell division that occurs in the male testes to produce sperm, also occurs in the female ovaries. But in this case, female gametes, or sex cells, are produced, and the process is called **oogenesis** (o″o-jen′ĕ-sis; "the beginning of an egg"). This process is shown in Figure 16.10 and described in more detail next.

In the developing female fetus, **oogonia** (o″o-go′ne-ah), the female stem cells, multiply rapidly to increase their number, and then their daughter cells, **primary oocytes,** push into the ovary connective tissue, where they become surrounded by a single layer of cells to form the *primary follicles.* By birth, the oogonia no longer exist, and a female's lifetime supply of primary oocytes (approximately 2 million of them) is already in place in the ovarian follicles, awaiting the chance to undergo meiosis to produce functional eggs. Since the primary oocytes remain in this state of suspended animation all through childhood, their wait is a long one—10 to 14 years at the very least.

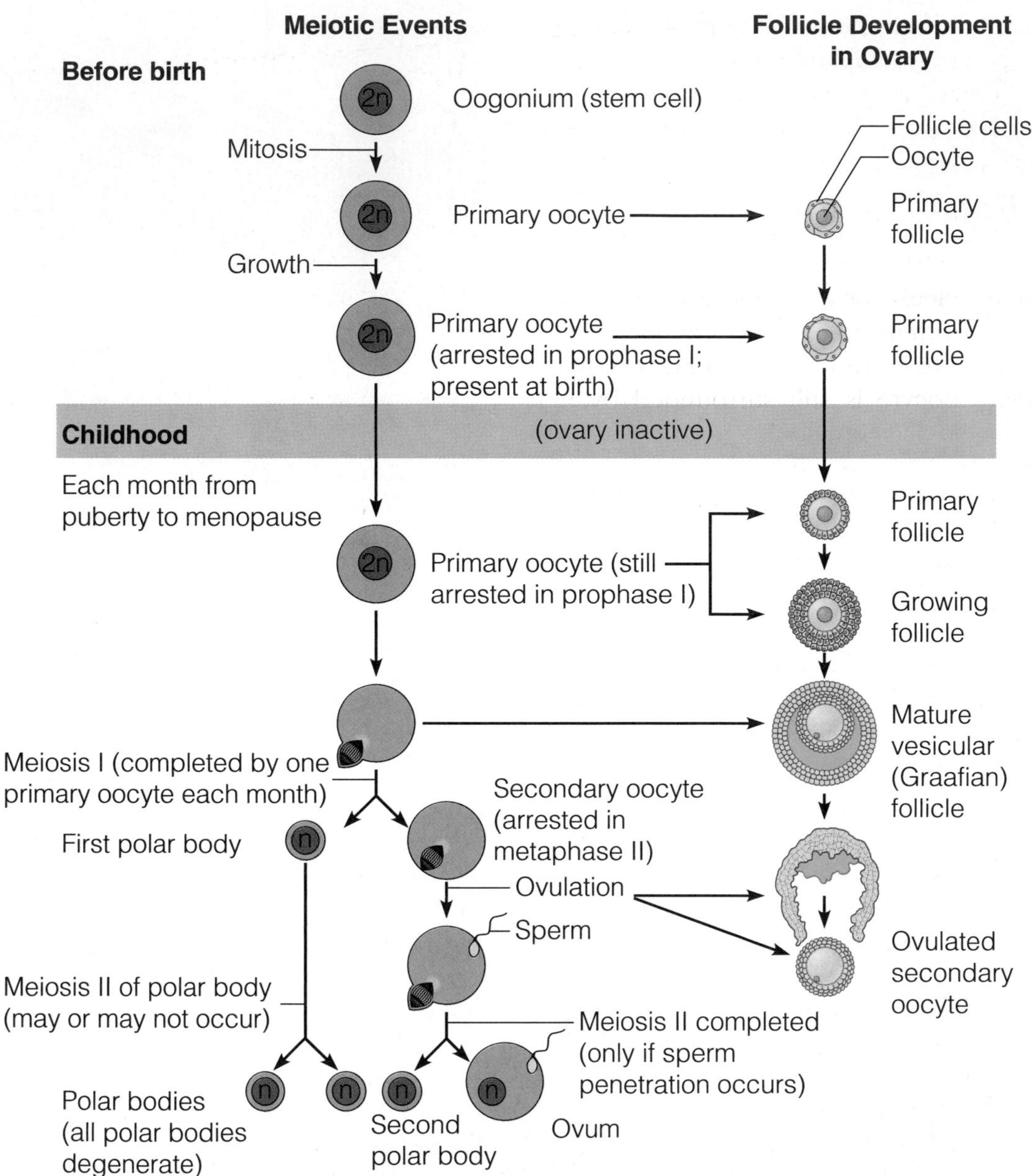

FIGURE 16.10 Events of oogenesis. Left, flow chart of meiotic events. Right, correlation with follicular development and ovulation in the ovary.

At puberty, the anterior pituitary gland begins to release *follicle-stimulating hormone (FSH),* which stimulates a small number of primary follicles to grow and mature each month, and ovulation begins to occur each month. These cyclic changes that occur monthly in the ovary constitute the **ovarian cycle.** At puberty, perhaps 400,000 oocytes remain; and, beginning at this time, a small number of oocytes are activated each month. Since the reproductive life of a female is at best about 40 years (from the age of 11 to approximately 51) and there is typically only one ovulation per month, fewer than 500 ova out of her potential of 400,000 are released during a woman's lifetime. Again, nature has provided us with a generous oversupply of sex cells.

As a follicle prodded by FSH grows larger, it accumulates fluid in the central chamber called the *antrum* (see Figure 16.7), and the primary oocyte it contains replicates its chromosomes and begins meiosis. The first meiotic division produces two cells that are very dissimilar in size (see Figure 16.10).

The larger cell is a **secondary oocyte** and the other, very tiny cell is a **polar body.** By the time a follicle has ripened to the mature *(vesicular follicle)* stage, it contains a secondary oocyte and protrudes like an angry boil from the external surface of the ovary. Follicle development to this stage takes about 14 days, and ovulation (of a secondary oocyte) occurs at just about that time in response to the burstlike release of a second anterior pituitary hormone, *luteinizing hormone (LH).* As shown in Figures 16.7, 16.10, and 16.11, the ovulated secondary oocyte is still surrounded by its follicle-cell capsule, now called the *corona radiata* ("radiating crown"). Some women experience a twinge of abdominal pain in the lower abdomen when ovulation occurs. This phenomenon, called *mittelschmerz* (mit′el-shmārts; German for "middle pain"), is caused by the intense stretching of the ovarian wall during ovulation.

Generally speaking, one of the developing follicles outstrips the others each month to become the dominant follicle. Just how this follicle is selected or selects itself is not understood, but the follicle that is at the proper stage of maturity when the LH stimulus occurs will rupture and release its oocyte into the peritoneal cavity. The mature follicles that are not ovulated soon become overripe and deteriorate. In addition to triggering ovulation, LH also causes the ruptured follicle to change into a very different glandular structure, the *corpus luteum.* (Both the maturing follicles and the corpus luteum produce hormones, as will be described later.)

If the ovulated secondary oocyte is penetrated by a sperm, it undergoes the second meiotic division that produces another polar body and the **ovum.** Once the ovum is formed, its 23 chromosomes are combined with those of the sperm to form the fertilized egg, which is the first cell of the yet-to-be offspring. However, if the secondary oocyte is not penetrated by a sperm, it simply deteriorates without ever completing meiosis to form a functional egg. Although meiosis in males results in four functional sperm, meiosis in females yields only one functional ovum and three tiny polar bodies. Since the polar bodies have essentially no cytoplasm, they deteriorate and die quickly.

Another major difference between males and females concerns the size and structure of their sex cells. Sperm are tiny and equipped with tails for locomotion. They have little nutrient-containing cytoplasm; thus, the nutrients in seminal fluid are vital to their survival. In contrast, the egg is a large, nonmotile cell, well stocked with nutrient reserves that nourish the developing embryo until it can take up residence in the uterus.

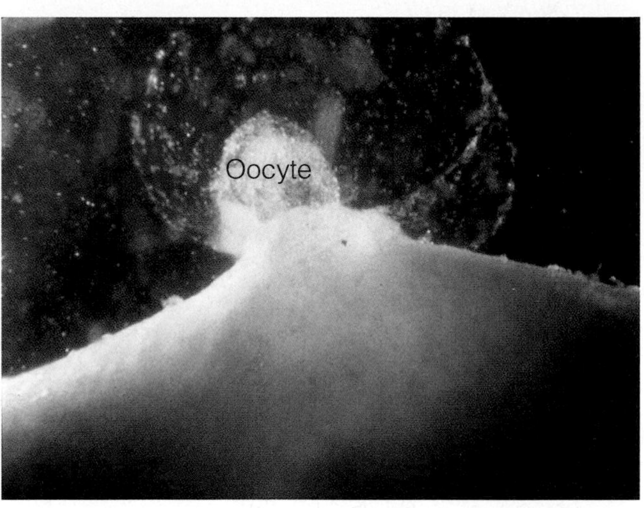

FIGURE 16.11 Ovulation. A secondary oocyte is released from a follicle at the surface of the ovary. The orange mass below the ejected oocyte is part of the ovary. The "halo" of follicle cells around the secondary oocyte is the *corona radiata.*

Uterine (Menstrual) Cycle

Although the uterus is the receptacle in which the young embryo implants and develops, it is receptive to implantation only for a very short period each month. Not surprisingly, this brief interval coincides exactly with the time when a fertilized egg would begin to implant, approximately seven days after ovulation. The events of the **uterine,** or **menstrual, cycle** are the cyclic changes that the endometrium, or mucosa of the uterus, goes through month after month as it responds to changes in the levels of ovarian hormones in the blood.

Since the cyclic production of estrogens and progesterone by the ovaries is, in turn, regulated by the anterior pituitary gonadotropic hormones, FSH and LH, it is important to understand how these "hormonal pieces" fit together. Generally speaking, both female cycles are about 28 days long (a period commonly called a *lunar month*), with ovulation typically occurring midway in the cycles, on or about day 14. Figure 16.12 illustrates the events occurring both in the ovary (the ovarian cycle) and in the uterus (menstrual cycle) at the

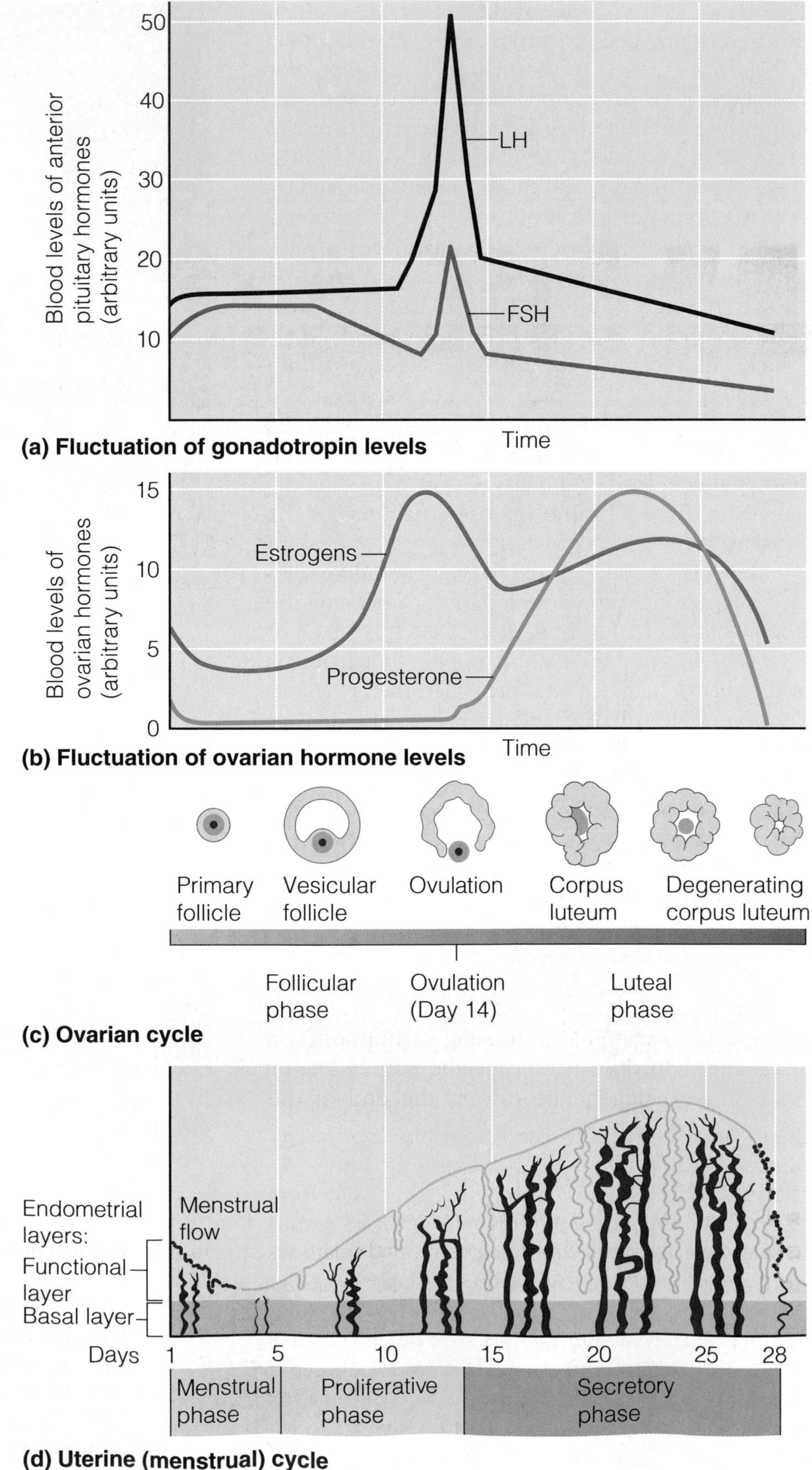

FIGURE 16.12 Hormonal interactions of the female cycles. Relative levels of anterior pituitary gonadotropins correlated with hormonal and follicular changes of the ovary and with the menstrual cycle.

same time. The three stages of the menstrual cycle are described next.

- **Days 1–5: Menstrual phase.** During this interval, the superficial *functional layer* of the thick endometrial lining of the uterus is sloughing off, or becoming detached, from the uterine wall. This is accompanied by bleeding for three to five days. The detached tissues and blood pass through the vagina as the menstrual flow. The average blood loss during this period is 50 to 150 ml (or about ¼ to ½ cup). By day 5, growing ovarian follicles are beginning to produce more estrogen.
- **Days 6–14: Proliferative phase.** Stimulated by rising estrogen levels produced by the growing follicles of the ovaries, the basal layer of the endometrium regenerates the functional layer, glands are formed in it, and the endometrial blood supply is increased. The endometrium once again becomes velvety, thick, and well vascularized. (Ovulation occurs in the ovary at the end of this stage, in response to the sudden surge of LH in the blood.)
- **Days 15–28: Secretory phase.** Rising levels of progesterone production by the corpus luteum of the ovary act on the estrogen-primed endometrium and increase its blood supply even more. Progesterone also causes the endometrial glands to increase in size and to begin secreting nutrients into the uterine cavity. These nutrients will sustain a developing embryo (if one is present) until it has implanted. If fertilization does occur, the embryo produces a hormone very similar to LH, which causes the corpus luteum to continue producing its hormones. If fertilization does not occur, the corpus luteum begins to degenerate toward the end of this period as LH blood levels decline. Lack of ovarian hormones in the blood causes the blood vessels supplying the functional layer of the endometrium to go into spasms and kink. When deprived of oxygen and nutrients, those endometrial cells begin to die, which sets the stage for menses to begin again on day 28.

Although this explanation assumes a classic 28-day cycle, the length of the menstrual cycle is quite variable. It can be as short as 21 days or as long as 40 days. Only one interval is fairly constant in all females; the time from ovulation to the beginning of menses is almost always 14 or 15 days.

Hormone Production by the Ovaries

As the ovaries become active at puberty and start to produce ova, production of ovarian hormones also begins. The follicle cells of the growing and mature follicles produce **estrogens,*** which cause the appearance of the *secondary sex characteristics* in the young woman. Such changes include

- Enlargement of the accessory organs of the female reproductive system (uterine tubes, uterus, vagina, external genitals)
- Development of the breasts
- Appearance of axillary and pubic hair
- Increased deposits of fat beneath the skin in general, and particularly in the hips and breasts
- Widening and lightening of the pelvis
- Onset of menses, or the menstrual cycle

The second ovarian hormone, **progesterone,** is produced by the glandular *corpus luteum* (see Figure 16.7). As mentioned earlier, after ovulation occurs the ruptured follicle is converted to the corpus luteum, which looks and acts completely different from the growing and mature follicle. Once formed, the corpus luteum produces progesterone (and some estrogen) as long as LH is still present in the blood. Generally speaking, the corpus luteum has stopped producing hormones by 10 to 14 days after ovulation. Except for working with estrogen to establish the menstrual cycle, progesterone does not contribute to the appearance of the secondary sex characteristics. Its other major effects are exerted during pregnancy, when it helps maintain the pregnancy and prepare the breasts for milk production. (However, the source of progesterone during pregnancy is the placenta, not the ovaries.)

Mammary Glands

The **mammary glands** are present in both sexes, but they normally function only in females. Since the biological role of the mammary glands is to produce milk to nourish a newborn baby, they are actually important only when reproduction has already been accomplished. Stimulation by female sex hormones,

---

*Although the ovaries produce several different estrogens, the most important are *estradiol, estrone,* and *estriol.* Of these, estradiol is the most abundant and is most responsible for mediating estrogenic effects.

Q *Flat-chested women are perfectly able to nurse their newborn infants; hence, it is not glandular tissue that accounts for the bulk of the breast tissue. So, what does?*

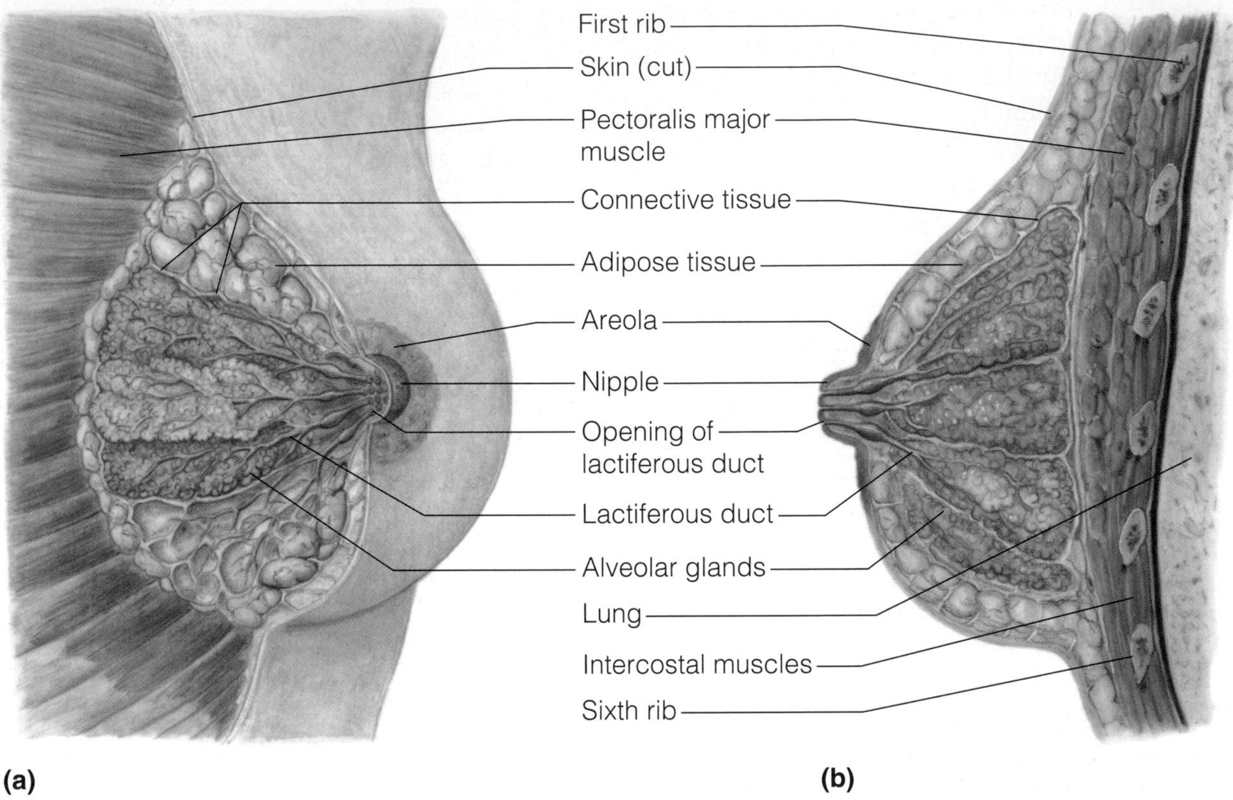

FIGURE 16.13 Female mammary glands. (a) Anterior view. **(b)** Sagittal section.

especially estrogens, causes the female mammary glands to increase in size at puberty.

Developmentally, the mammary glands are modified *sweat glands* that are actually part of the skin. Each mammary gland is contained within a rounded skin-covered breast anterior to the pectoral muscles of the thorax. Slightly below the center of each breast is a pigmented area, the **areola** (ah-re′o-lah), which surrounds a central protruding **nipple** (Figure 16.13).

Internally, each mammary gland consists of 15 to 25 *lobes,* which radiate around the nipple. The lobes are padded and separated from each other by connective tissue and fat. Within each lobe are smaller chambers called *lobules,* which contain clusters of **alveolar glands** that produce the milk when a woman is **lactating** (producing milk). The alveolar glands of each lobule pass the milk into the **lactiferous** (lak-tif′er-us) **ducts,** which open to the outside at the nipple.

Homeostatic Imbalance

Cancer of the breast is the second most common cause of death in American women. One woman in eight will develop this condition. Breast cancer is often signaled by a change in skin texture, puckering, or leakage from the nipple. Early detection by breast self-examination and mammography is unquestionably the best way to increase one's chances of surviving breast cancer. Since most breast lumps are discovered by women themselves in routine monthly breast exams, this simple examination should be a priority in every woman's life. Currently the American Cancer Society recommends scheduling **mammography**—X-ray examination that detects breast cancers too small to feel (less than 1 cm)—every 2 years for women between 40 and 49 years old and yearly thereafter (Figure 16.14). ▲

A Adipose tissue.

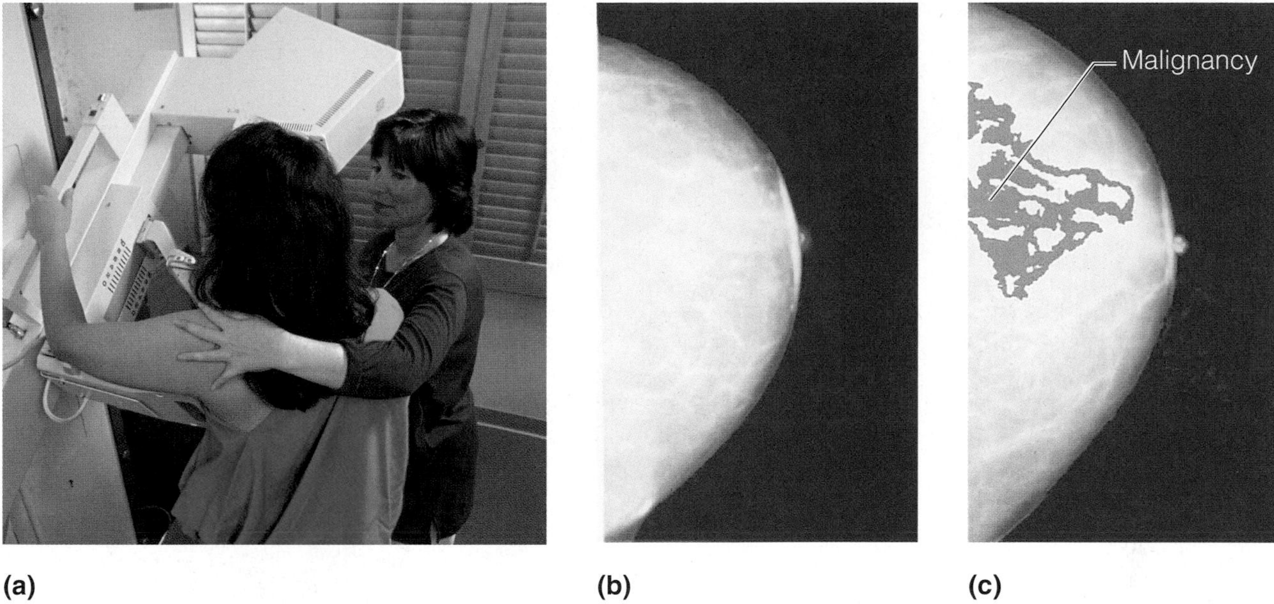

FIGURE 16.14 **Mammograms. (a)** Photograph of woman undergoing mammography. **(b)** Normal breast. **(c)** Breast with tumor.

Survey of Pregnancy and Embryonic Development

Because the birth of a baby is such a familiar event, we tend to lose sight of the wonder of this accomplishment. In every instance it begins with a single cell, the fertilized egg, and ends with an extremely complex human being consisting of trillions of cells. The development of an embryo is very complex, and the details of this process can fill a good-sized book. Our intention here is simply to outline the important events of pregnancy and embryonic development.

Accomplishing Fertilization

Before fertilization can occur, the sperm must reach the ovulated secondary oocyte. The oocyte is viable for 12 to 24 hours after it is cast out of the ovary, and sperm generally retain their fertilizing power within the female reproductive tract for 12 to 48 hours after ejaculation. Some "super sperm," however, are viable for 72 hours. Consequently, for fertilization to occur, sexual intercourse must occur no more than 72 hours before ovulation and no later than 24 hours after, at which point the oocyte is approximately one-third of the way down the length of the uterine tube. Remember that sperm are motile cells that can propel themselves by lashing movements of their tails. If sperm are deposited in a female's vagina at the approximate time of ovulation, they are attracted to the oocyte by chemicals that act as "homing devices," allowing them to locate the oocyte. It takes one to two hours for sperm to complete the journey up the female duct system to the end of the uterine tubes, and if an oocyte is en route in the tube, fertilization is a distinct possibility.

When the swarming sperm reach the oocyte, hundreds of their acrosomes rupture, releasing enzymes that break down the "cement" that holds the follicle cells of the corona radiata together around the oocyte. Once a path has been cleared and a single sperm makes contact with the oocyte's membrane receptors, its head (nucleus) is pulled into the oocyte cytoplasm. This is one case that does not bear out the adage "The early bird catches the worm." A sperm that comes along later, after hundreds of sperm have undergone acrosomal reactions to expose the oocyte membrane, is in the best position to be *the* fertilizing sperm. Once a single sperm has penetrated the oocyte, the oocyte nucleus completes the second meiotic division, forming the ovum and a polar body.

After sperm entry, changes occur in the fertilized egg to prevent other sperm from gaining entry. Of the millions of sperm ejaculated by a male,

Q *Why is the multicellular blastocyst only slightly larger than the single-cell zygote?*

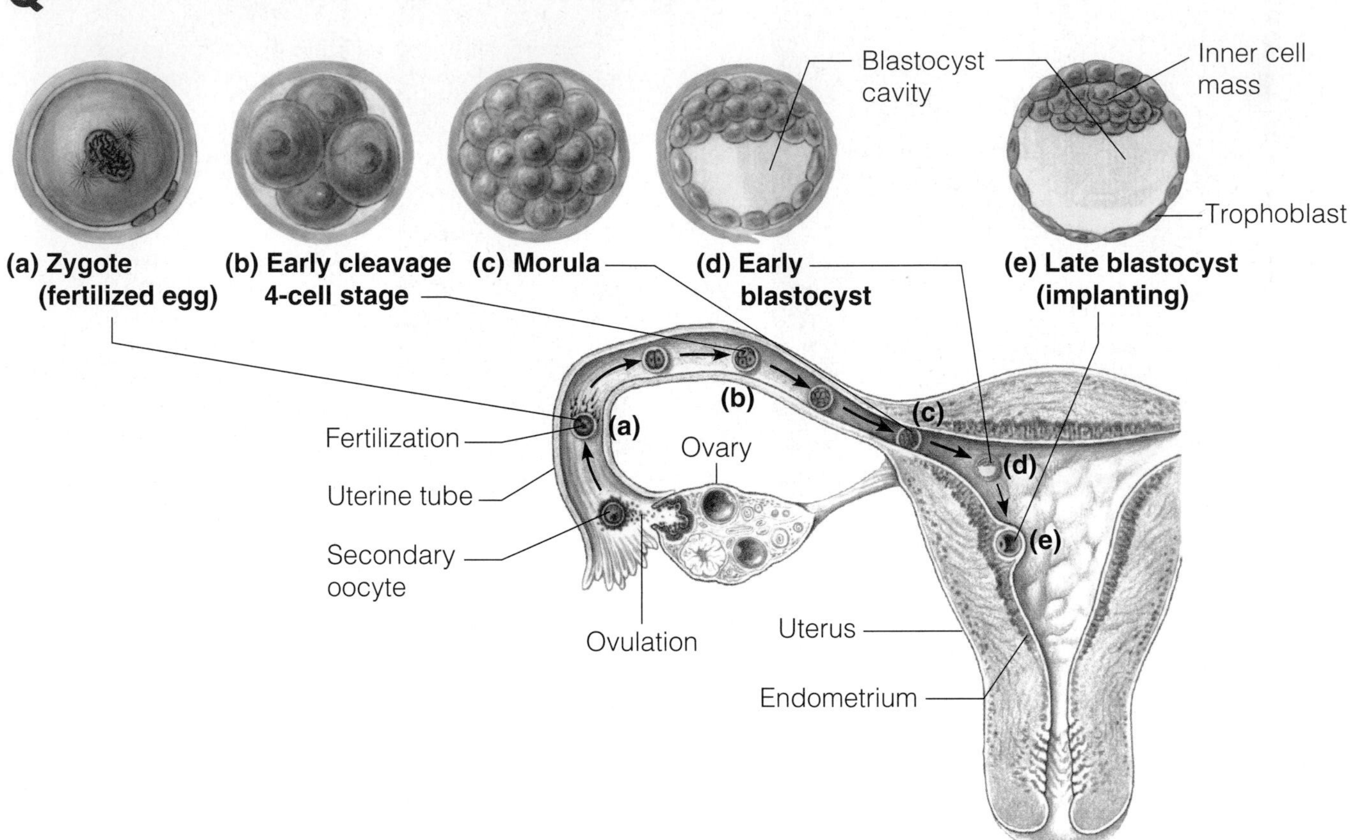

FIGURE 16.15 Cleavage is a rapid series of mitotic divisions that begins with the zygote and ends with the blastocyst. The zygote begins to divide about 24 hours after fertilization and continues to divide rapidly (undergo cleavage) as it travels down the uterine tube. Three to four days after ovulation, the preembryo reaches the uterus and floats freely for two to three days, nourished by secretions of the endometrial glands. At the late blastocyst stage, the embryo is implanting into the endometrium; this begins at about day 7 after ovulation. **(a)** Zygote (fertilized egg). **(b)** Four-cell stage. **(c)** Morula, a solid ball of blastomeres. **(d)** Early blastocyst; the morula hollows out, and fills with fluid. **(e)** Late blastocyst, composed of an outer sphere of trophoblast cells and an off-center cell cluster called the inner cell mass.

only *one* can penetrate an oocyte. **Fertilization** occurs at the moment the genetic material of a sperm combines with that of an ovum to form a fertilized egg, or **zygote** (zi′gōt). The zygote represents the first cell of the new individual.

A *Because as the zygote and then its descendants divide, little or no time is provided for growth between subsequent division cycles. As a result, the cells get smaller and smaller and the size of the cell mass stays approximately the same size as the initial zygote.*

Events of Embryonic and Fetal Development

As the zygote journeys down the uterine tube (propelled by peristalsis and cilia), it begins to undergo rapid mitotic cell divisions—forming first two cells, then four, and so on. This early stage of embryonic development, called **cleavage,** is shown in Figure 16.15. Since there is not much time for cell growth between divisions, the daughter cells become

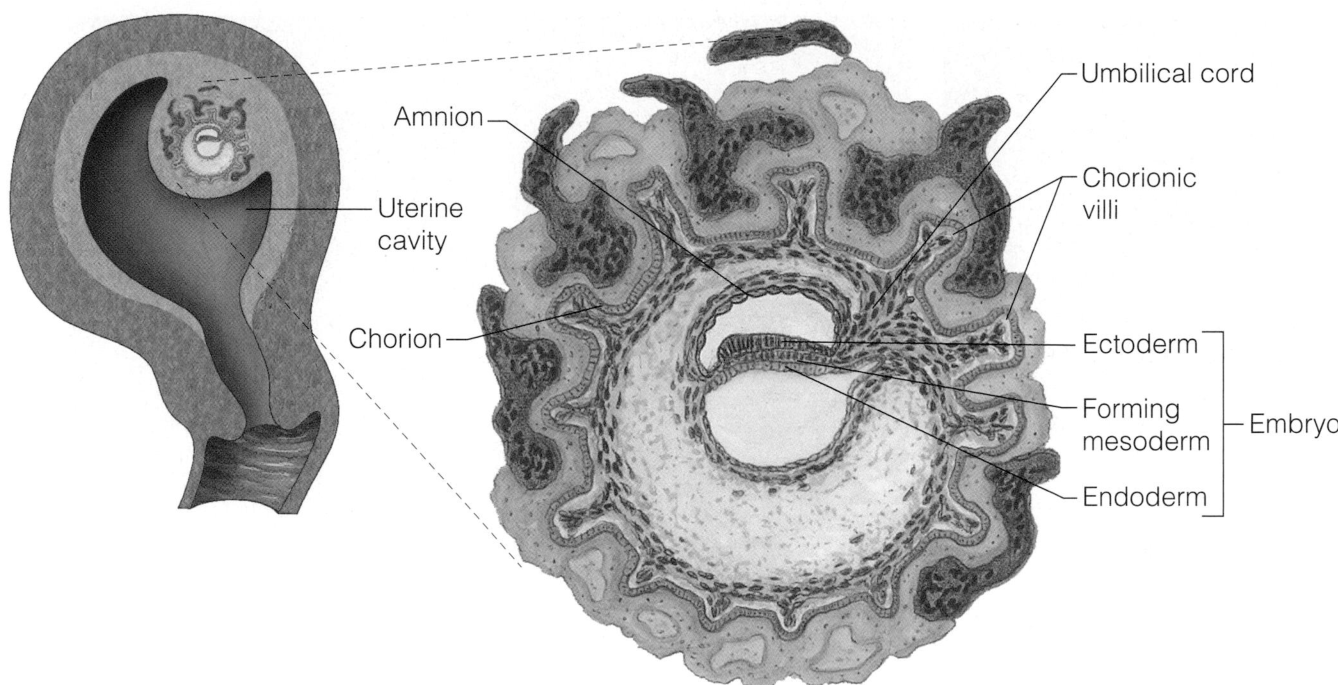

FIGURE 16.16 Embryo of approximately 22 days. Embryonic membranes present.

smaller and smaller. Cleavage provides a large number of cells to serve as building blocks for constructing the **embryo** (developmental stage until the ninth week). Consider for a moment how difficult it would be to construct a building from one huge block of granite. If you now consider how much easier your task would be if you could use hundreds of brick-size granite blocks, you will quickly grasp the importance of cleavage. By the time the developing embryo reaches the uterus (about 3 days after ovulation, or on day 17 of the woman's cycle), it is a *morula,* a tiny ball of 16 cells that looks like a microscopic raspberry. The uterine endometrium is still not fully prepared to receive the embryo at this point, so the embryo floats free in the uterine cavity, temporarily using the uterine secretions for nutrition. While still unattached, the embryo continues to develop until it has about 100 cells, and then it hollows out to form a ball-like structure, a **blastocyst** (blas′to-sist) or **chorionic** (ko″re-on′ik) **vesicle.** At the same time, it secretes an LH-like hormone called **human chorionic gonadotropin (hCG),** which prods the corpus luteum of the ovary to continue producing its hormones. (If this were not the case, the functional layer of the endometrium would be sloughing off shortly in menses.) It is hCG that many home pregnancy tests assay for in a woman's urine.

The blastocyst has two important functional areas: the *trophoblast,* which forms the large fluid-filled sphere, and the *inner cell mass,* a small cluster of cells displaced to one side (see Figure 16.15e). By day 7 after ovulation, the blastocyst has attached to the endometrium and has eroded away the lining in a small area, embedding itself in the thick velvety mucosa. All of this is occurring even while development is continuing and the three primary germ layers are being formed from the inner cell mass (Figure 16.16). The *primary germ layers* are the **ectoderm** (which gives rise to the nervous system and the epidermis of the skin), the **endoderm** (which forms mucosae and associated glands), and the **mesoderm** (which gives rise to virtually everything else). Implantation has usually been completed and the uterine mucosa has grown over the burrowed-in embryo by day 14 after ovulation—the day the woman would ordinarily be expecting to start menses. After it is securely implanted, the trophoblast part of the blastocyst develops elaborate projections, called **chorionic**

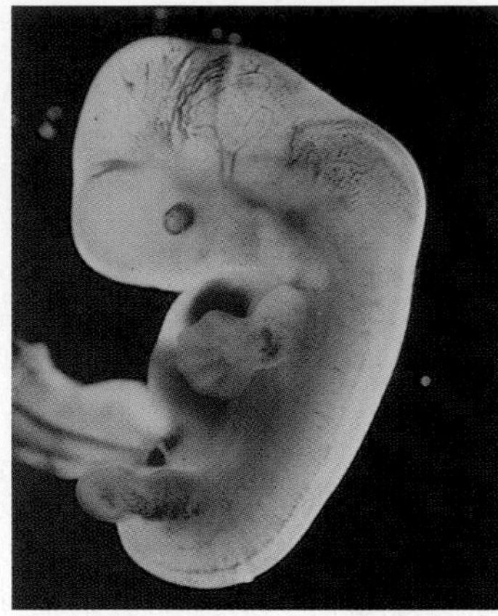

(a) **5 weeks.** Limb buds, eyes, the heart, the liver, and rudiments of all other organs have started to develop in the embryo, which is only about 1 cm long.

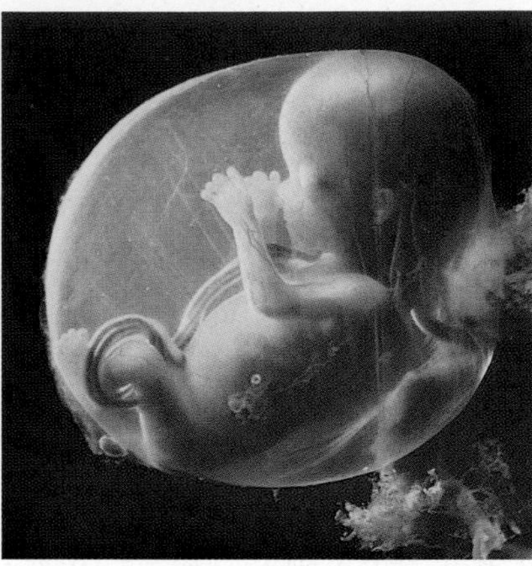

(b) **14 weeks.** Growth and development of the offspring, now called a fetus, continue during the second trimester. This fetus is about 6 cm long.

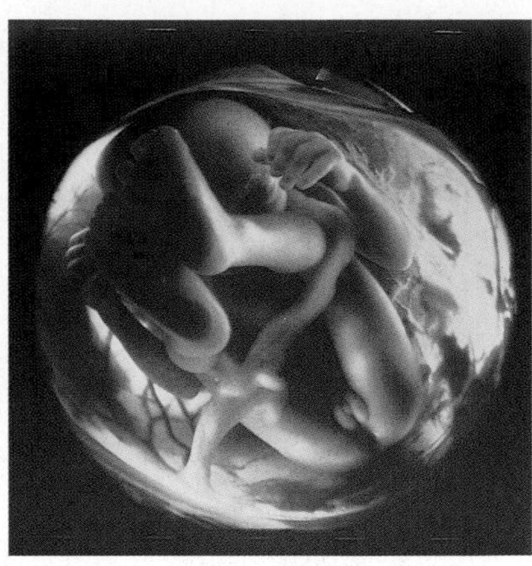

(c) **20 weeks.** By the end of the second trimester (at 24 weeks), the fetus grows to about 30 cm in length.

FIGURE 16.17 Human fetal development.

villi, which cooperate with the tissues of the mother's uterus to form the **placenta** (plah-sen′tah). Once the placenta has formed, the platelike embryonic body, now surrounded by a fluid-filled sac called the **amnion** (am′ne-on), is attached to the placenta by a blood vessel–containing stalk of tissue, the **umbilical cord** (Figure 16.16). (The special features of the umbilical blood vessels and fetal circulation are discussed on pp. 370–372.) Generally by the third week, the placenta is functioning to deliver nutrients and oxygen to and remove wastes from the embryonic blood. All exchanges are made through the placental barrier. By the end of the second month of pregnancy, the placenta has also become an endocrine organ and is producing estrogen, progesterone, and other hormones that help to maintain the pregnancy. At this time, the corpus luteum of the ovary becomes inactive.

By the eighth week of embryonic development, all the groundwork has been completed. All the organ systems have been laid down, at least in rudimentary form, and the embryo looks distinctly human. Beginning in the ninth week of development, the embryo is referred to as a **fetus.** From this point on, the major activities are growth and organ specialization, accompanied by changes in body proportions. During the fetal period, the developing fetus grows from a crown-to-rump length of about 3 cm (slightly more than 1 inch) and a weight of approximately 1 g (0.03 ounce) to about 36 cm (14 inches) and 2.7 to 4.1 kg (6–10 pounds) or more. (Total body length at birth is about 55 cm, or 22 inches.) As you might expect with such tremendous growth, the changes in fetal appearance are quite dramatic (Figure 16.17). The most significant of these changes are summarized in Table 16.1. By approximately 270 days after fertilization (the end of the tenth lunar month), the fetus is said to be "full-term" and is ready to be born.

Effects of Pregnancy on the Mother

Pregnancy (the period from conception to the birth of her baby) can be a difficult time for the mother. Not only are there obvious anatomical changes, but striking changes occur in her physiology as well.

Anatomical Changes

The ability of the uterus to enlarge during pregnancy is nothing less than remarkable. Starting as a fist-sized organ, the uterus grows to occupy most

Focus on Careers

Doula

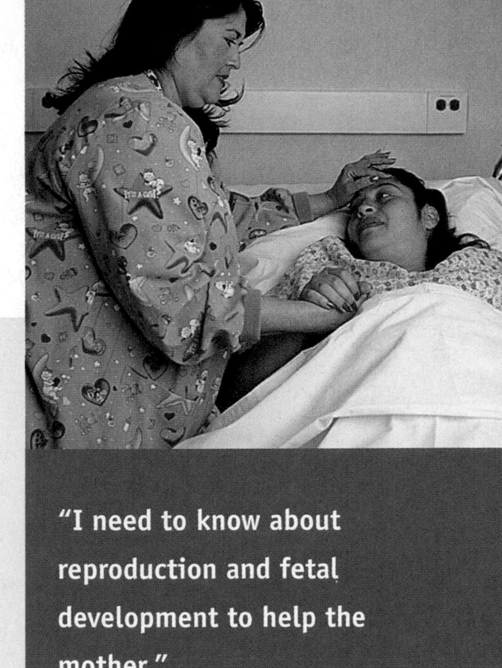

Anatomy and physiology classes are central in the training of doulas, or labor support professionals.

"WHY do I want to be a doula? Because I know how much one could have helped me when I gave birth," says Morgan Tharan. "I had a long labor, and it affected my baby's breathing and heart rate, so my midwife had to focus all her efforts on helping him. He improved quickly and turned out to be fine, but it was a really anxious time. I was exhausted, and my partner was wiped out. It made me appreciate the importance of emotional support during labor and delivery."

Today Tharan is studying to become a doula, a professional who provides emotional and physical support to mothers during pregnancy, labor, birth, and the hours immediately afterward. "Doulas focus on the mother by supplying reassurance and comfort measures," explains Tharan. "We are there to make sure the mother's and partner's emotional needs are being met."

Sometimes called labor support professionals, doulas (from the Greek word for "servant") are a relatively new profession. While training requirements vary, anatomy and physiology classes are always central. "I need to know about reproduction and the process of fetal development in order to help the mother and answer questions," explains Tharan. "We study how the body changes during pregnancy. We have a whole section on labor pain as well as the process of delivery. We learn the mechanics of breastfeeding and how the breast works, too."

Tharan loves the variety in her job and the rapport she develops with clients. "Like me, most doulas are 'birth junkies,'" she smiles. "We love participating in this wonderful life experience." Many doulas work on a freelance basis and contract directly with the expectant parents, meeting at least once to discuss the mother's expectations and concerns about labor and delivery. Notes Tharan, "Sometimes I schedule a second prenatal visit if there are difficult issues, like health problems or a past miscarriage. We can also be a big help to the partner/father, because they often have a lot of questions and fears of their own. We facilitate communication and help them help the mother." Depending on the situation, a doula may answer questions about medical procedures and offer suggestions about nutrition, exercise, and other ways to make pregnancy more comfortable.

When her client goes into labor, Tharan gets an urgent phone call. "Even if it's the middle of the night, if the mother wants me, I'm there," she says. "I meet her at the hospital or at home, depending on where she wants to give birth, and I stay through labor. I may give a massage, help her change positions, coach the partner—whatever helps the mother the most. After birth, I can help her begin breastfeeding. By the time I leave, everyone is settled and happy."

A national association, Doulas of North America (DONA), certifies many labor support professionals. Certification steps include:

- Complete one of the following: training in childbirth education or midwifery; nursing work experience in labor and delivery; or observing a childbirth preparation series (not as an expectant parent)
- Complete an approved training course of 14 or more hours
- Read four books from the DONA reading list
- Provide continuous labor support at three births
- Obtain positive evaluations from healthcare professionals and clients
- Write an essay on the purpose and value of labor support
- Sign the DONA Code of Ethics and Standards of Practice

Emotional support pays off physically: According to DONA, mothers who have doulas spend less time in labor, request less pain medication, and undergo 50 percent fewer cesarean sections than the national average.

For more information on doulas, contact DONA:
P.O. Box 626
Jasper, IN 47547
(888) 788-3662
http://www.dona.org

"I need to know about reproduction and fetal development to help the mother."

For additional information on this career, click the Focus on Careers link at www.anatomyandphysiology.com

TABLE 16.1 Development of the Human Fetus

Time	Changes/accomplishments
8 weeks (end of embryonic period)	Head nearly as large as body; all major brain regions present
	Liver disproportionately large and begins to form blood cells
	Limbs present; though initially webbed, fingers and toes are free by the end of this interval
	Bone formation begun
	Heart has been pumping blood since the fourth week
8 weeks	All body systems present in at least rudimentary form
	Approximate crown-to-rump length: 30 mm (3 cm; 1.2 inches); weight: 1 gram (0.03 ounces)
9–12 weeks (third month)	Head still dominant, but body elongating; brain continues to enlarge
	Facial features present in crude form
	Walls of hollow visceral organs gaining smooth muscle
	Blood cell formation begins in bone marrow
12 weeks	Bone formation accelerating
	Sex readily detected from the genitals
	Approximate crown-to-rump length at end of interval: 90 mm (9 cm)
13–16 weeks (fourth month)	General sensory organs are present; eyes and ears assume characteristic position and shape; blinking of eyes and sucking motions of lips occur
	Face looks human and body beginning to outgrow head
	Kidneys attain typical structure
16 weeks	Most bones are distinct and joint cavities apparent
	Approximate crown-to-rump length at end of interval: 140 mm (14 cm)

of the pelvic cavity by 16 weeks. As pregnancy continues, the uterus pushes higher and higher into the abdominal cavity. As birth nears, the uterus reaches the level of the xiphoid process and occupies the bulk of the abdominal cavity. The crowded abdominal organs press superiorly against the diaphragm, which intrudes on the thoracic cavity. As a result, the ribs flare, causing the thorax to widen.

The increasing bulkiness of the abdomen changes the woman's center of gravity, and many women develop an accentuated lumbar curvature (lordosis), often accompanied by backaches, during the last few months of pregnancy. Placental production of the hormone **relaxin** causes pelvic ligaments and the pubic symphysis to relax, widen, and become more flexible. This increased motility eases birth passage, but it may also result in a waddling gait during pregnancy.

Obviously, good maternal nutrition is necessary throughout pregnancy if the developing fetus is to have all the building materials (proteins, calcium, iron, and the like) it needs to form its tissues and organs. The old expression "A pregnant woman is eating for two" has encouraged many women to eat *twice* the amount of food actually

TABLE 16.1 *(continued)*

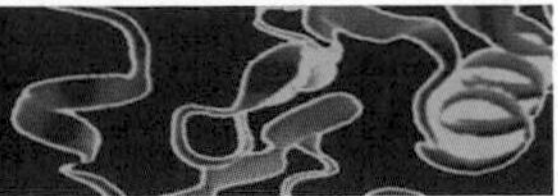

Time	Changes/accomplishments
17–20 weeks (fifth month)	Vernix caseosa (fatty secretions of sebaceous glands) covers body; silklike hair (lanugo) covers skin
	Fetal position (body flexed anteriorly) assumed because of space restrictions
	Limbs achieve near-final proportions
	Quickening occurs (mother feels spontaneous muscular activity of fetus)
	Approximate crown-to-rump length at end of interval: 190 mm (19 cm)
21–30 weeks (sixth and seventh months)	Substantial increase in weight (may survive if born prematurely at 27–28 weeks, but hypothalamus still too immature to regulate body temperature, and surfactant production by the lungs is still inadequate)
	Myelination of spinal cord begins; eyes are open
	Skin is wrinkled and red; fingernails and toenails are present
	Body is lean and well proportioned
	Bone marrow becomes sole site of blood cell formation
	Testes enter scrotum in seventh month (in males)
At birth	Approximate crown-to-rump length at end of interval: 280 mm (28 cm)
30–40 weeks (term) (eighth and ninth months)	Skin whitish pink; fat laid down in subcutaneous tissue
	Approximate crown-to-rump length at end of interval: 350–400 mm (35–40 cm; 14–16 inches); weight: 2.7–4.1 kg (6–10 pounds)

needed during pregnancy, which, of course, leads to excessive weight gain. Actually, a pregnant woman needs only about 300 additional calories daily to sustain proper fetal growth. The emphasis should be on high-quality food, not just more food.

Homeostatic Imbalance

Since many potentially harmful substances can cross through the placental barrier into the fetal blood, the pregnant woman should be very much aware of what she is taking into her body. Substances that may cause life-threatening birth defects (and even fetal death) include alcohol, nicotine, and many types of drugs (anticoagulants, antihypertensives, sedatives, and some antibiotics). Maternal infections, particularly German measles (rubella), may also cause severe fetal damage. Termination of a pregnancy by loss of a fetus during the first 20 weeks of pregnancy is called **abortion.** ▲

Physiological Changes

Gastrointestinal System Many women suffer nausea, commonly called *morning sickness,* during the first few months of pregnancy, until their system adjusts to the elevated levels of progesterone and estrogens. *Heartburn* is common because the esophagus is displaced and the stomach is crowded by the growing uterus, which favors reflux of stomach acid into the esophagus. Another problem is constipation, because motility of the digestive tract declines during pregnancy.

Urinary System The kidneys have the additional burden of disposing of fetal metabolic wastes, and they produce more urine during pregnancy. Because the uterus compresses the bladder, urination becomes more frequent, more urgent, and sometimes uncontrollable. (The last condition is called *stress incontinence.*)

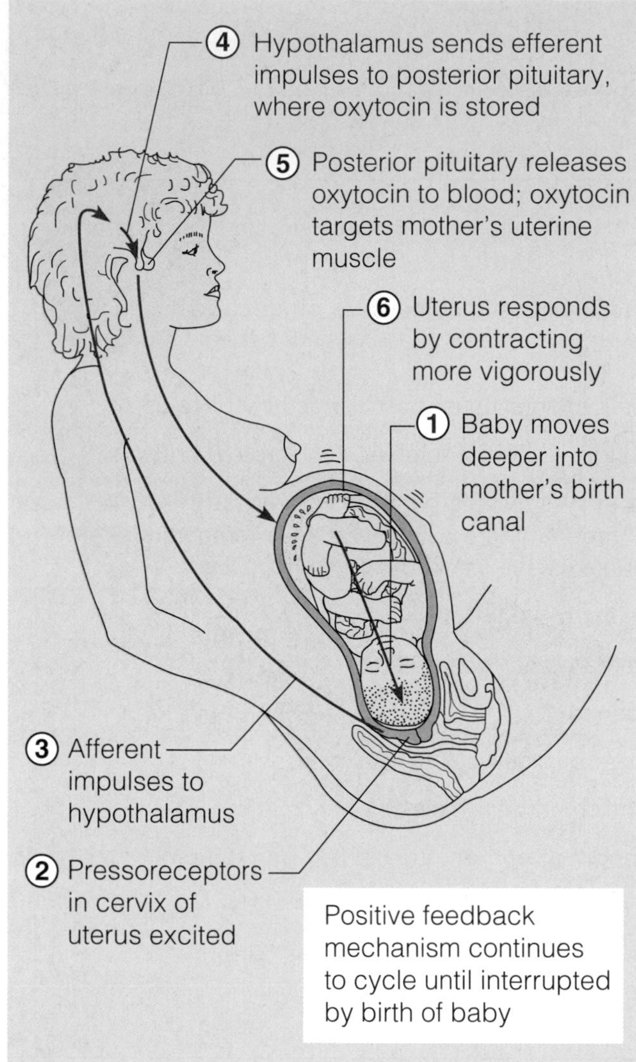

FIGURE 16.18 The positive feedback mechanism by which oxytocin promotes labor contractions during birth.

Respiratory System The nasal mucosa responds to estrogen by becoming swollen and congested; thus, nasal stuffiness and occasional nosebleeds may occur. Vital capacity and respiratory rate increase during pregnancy, but residual volume declines, and many women exhibit *dyspnea* (difficult breathing) during the later stages of pregnancy.

Cardiovascular System Perhaps the most dramatic physiological changes occur in the cardiovascular system. Total body water rises and blood volume increases by 25 to 40 percent to accommodate the additional needs of the fetus. The rise in blood volume also acts as a safeguard against blood loss during birth. Blood pressure and pulse typically rise and increase cardiac output by 20 to 40 percent; this helps propel the greater blood volume around the body. Because the uterus presses on the pelvic blood vessels, venous return from the lower limbs may be impaired somewhat, resulting in varicose veins.

Childbirth

Childbirth, also called **parturition** (par″tu-rish′un; "bringing forth young"), is the culmination of pregnancy. It usually occurs within 15 days of the calculated due date (which is 280 days from the last menstrual period). The series of events that expel the infant from the uterus are referred to collectively as **labor.**

Initiation of Labor

The precise trigger for labor is not clear, but several events appear to be interlocked in this process. During the last few weeks of pregnancy, estrogens reach their highest levels in the mother's blood. This has two important consequences: it causes the myometrium to form abundant *oxytocin* receptors (so that it becomes more sensitive to the hormone oxytocin), and it interferes with progesterone's quieting influence on the uterine muscle. As a result, weak, irregular uterine contractions begin to occur. These contractions, called *Braxton Hicks contractions,* have caused many women to go to the hospital, only to be told that they were in **false labor** and sent home.

As birth nears, two more chemical signals cooperate to convert these false labor pains into the real thing. Certain cells of the fetus begin to produce oxytocin, which in turn stimulates the placenta to release *prostaglandins.* Both hormones stimulate more frequent and powerful contractions of the uterus. At this point, the increasing emotional and physical stresses activate the mother's hypothalamus, which signals for oxytocin release by the posterior pituitary. The combined effects of rising levels of oxytocin and prostaglandins initiate the rhythmic, expulsive contractions of true labor. Once the hypothalamus is involved, a positive feedback mechanism is propelled into action: Stronger contractions cause the release of more oxytocin, which causes even more vigorous contractions, forcing the baby ever deeper into the mother's pelvis, and so on (Figure 16.18).

Since both oxytocin and prostaglandins are needed to initiate labor in humans, anything that interferes with production of either of these hormones will hinder the onset of labor. For example, antiprostaglandin drugs such as aspirin and ibuprofen can inhibit labor at the early stages, and such drugs are used occasionally to prevent premature births.

Stages of Labor

The process of labor is commonly divided into three stages (Figure 16.19). These stages are described next.

Stage 1: Dilation Stage The **dilation stage** is the time from the appearance of true contractions until the cervix is fully dilated by the baby's head (about 10 cm in diameter). As labor starts, regular but weak uterine contractions begin in the upper part of the uterus and move downward toward the vagina. Gradually, the contractions become more vigorous and more rapid, and, as the infant's head is forced against the cervix with each contraction, the cervix begins to soften, becomes thinner *(effaces),* and dilates. Eventually, the amnion ruptures, releasing the amniotic fluid, an event commonly referred to as "breaking the water." The dilation stage is the longest part of labor and usually lasts for 6 to 12 hours (sometimes considerably more).

Stage 2: Expulsion Stage The **expulsion stage** is the period from full dilation to delivery of the infant. In this stage, the infant passes through the cervix and vagina to the outside of the body. During this stage, a mother experiencing natural childbirth (that is, undergoing labor without local anesthesia) has an increasing urge to push, or bear down, with the abdominal muscles. Although this phase can take as long as 2 hours, it is typically 50 minutes in a first birth and around 20 minutes in subsequent births.

When the infant is in the usual head-first *(vertex)* position, the skull (its largest diameter) acts as a wedge to dilate the cervix. The head-first presentation also allows the baby to be suctioned free of mucus and to breathe even before it has completely exited from the birth canal. Once the head has been delivered, the rest of the baby's body is delivered much more easily. After birth, the umbilical cord is clamped and cut. In *breech* (buttocks-first) presentations and other nonvertex

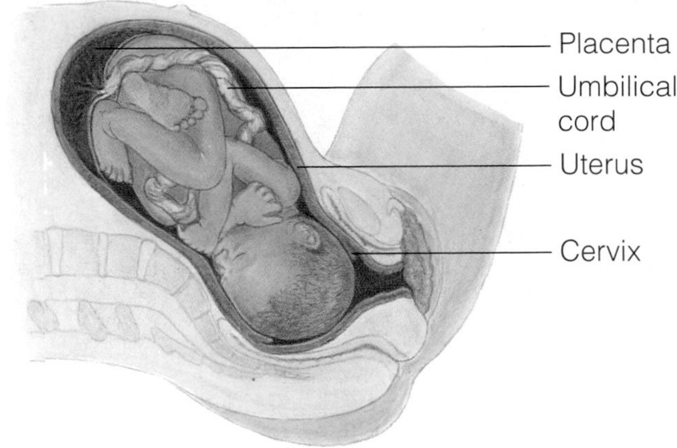

(1) Dilation of the cervix

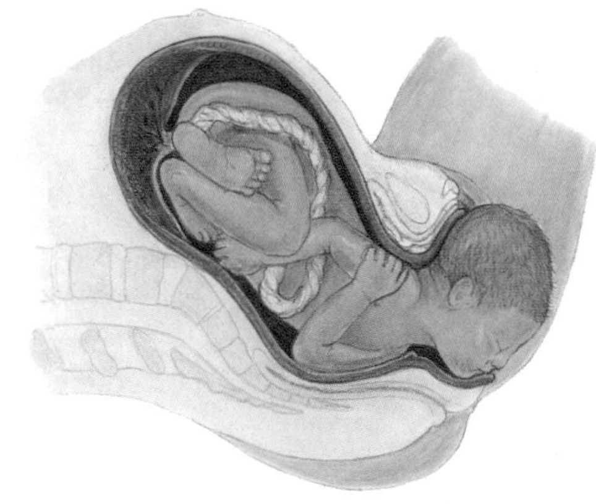

(2) Expulsion: delivery of the infant

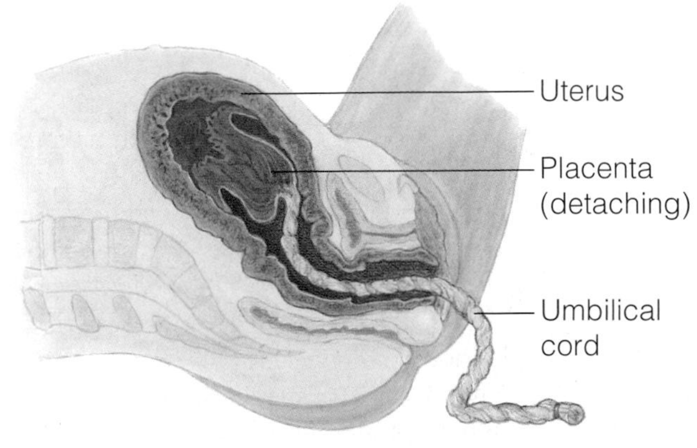

(3) Delivery of the placenta

FIGURE 16.19 The three stages of labor.

A Closer Look

Contraception: To Be or Not To Be

IN a society such as ours, where many women opt for professional careers or must work for economic reasons, **contraception** (*contra* = against; *cept* = taking), or *birth control,* is often seen as a necessity. Thus far, much of the burden for birth control has fallen on women's shoulders, and most birth control products are female-directed.

The key to birth control is dependability. As shown by the red arrows in the accompanying flow chart, the birth control techniques and products currently available have many sites of action for blocking the reproductive process. Let's examine the relative advantages of a few of these current methods more closely.

The most-used contraceptive product in the United States is the *birth control pill,* or simply, "the pill," a preparation taken daily that contains tiny amounts of estrogens and progestins (progesterone-like hormones), except that for the last seven days of the 28-day cycle the tablets are hormone-free. The pill tricks the hypothalamic-pituitary control system and "lulls it to sleep" because the relatively constant blood levels of ovarian hormones make it appear that the woman is pregnant (both estrogen and progesterone are produced throughout pregnancy). Ovarian follicles do not mature, and ovulation ceases. The endometrium does proliferate slightly and is sloughed off when the hormones are discontinued each month, but menstrual flow is much reduced. However, since hormonal balance in the body is precisely controlled, some women simply cannot tolerate the changes caused by the pill—they become nauseated and/or hypertensive.

For a while, the pill was suspected of increasing the incidence of breast and uterine cancer. Its influence on breast cancer is still a question. However, it appears that the new, very low dose preparations may actually help protect against some forms of cancer (ovarian and endometrial) and may also have reduced the incidence of serious cardiovascular side effects, such as strokes, heart attacks, and blood clots, that occurred (rarely) with earlier forms of the pill.

Presently, the pill is one of the most widely used drugs in the world; well over 50 million women use these drugs to prevent pregnancy. The incidence of failure is less than 6 pregnancies per 100 women per year.

A different hormone pill, the *morning-after pill (MAP),* is widely prescribed on college campuses and the therapy of choice for rape victims and is one of the best-kept secrets in the United States—most teenagers and women in their thirties and forties have never heard of them. Taken within 72 hours of unprotected intercourse, the concentrated estrogen-progesterone combination pills "mess up" the normal hormonal signals so much that fertilization is prevented altogether or a fertilized egg is prevented from implanting. Planned Parenthood clinics have begun to publicize the availability of MAPs for postcoital contraception, a use now approved by the U.S. Food and Drug Administration (FDA).

Other approaches include a device that is implanted just under the skin and injections of synthetic progesterone. The Norplant implant, six tiny silicone rods that release progestin over a 5-year period (see the photo, p. 556), was approved late in 1990 and quickly gained users. Its failure rate (0.05 percent) is even lower than that of the pill, but there have been some problems with rod removal. *Depo-Provera,* a synthetic progesterone developed 35 years before, was finally approved for use as an injectable contraceptive in October 1992. Administered in a 150-mg dose once every three months, Depo-Provera's failure rate is only 0.4 percent.

For several years, the second most used contraceptive method was the *intrauterine device (IUD),* a plastic or metal device inserted into the uterus that prevented implantation of the fertilized egg (see photo). The failure rate of the IUD was nearly as low as that of the pill. For a while, IUDs were taken off the market in the United States because of problems with occasional contraceptive failure, uterine perforation, or pelvic inflammatory disease (PID). New IUD products that deliver sustained doses of synthetic progesterone to the endometrium are currently being recommended for those who have given birth and for those with a lower risk of developing PID in monogamous relationships.

Some methods, such as *tubal ligation* and *vasectomy* (cutting or cauterizing the uterine tubes or ductus deferens, respectively), are nearly foolproof and are the choice of approximately 33 percent of couples of childbearing age in the United States. Both procedures can be done in the physician's office. These changes are usually permanent, however, so they are not for individuals who still plan to have children but want to choose the time.

Coitus interruptus, or withdrawal of the penis just before ejaculation, is simply against nature, and control

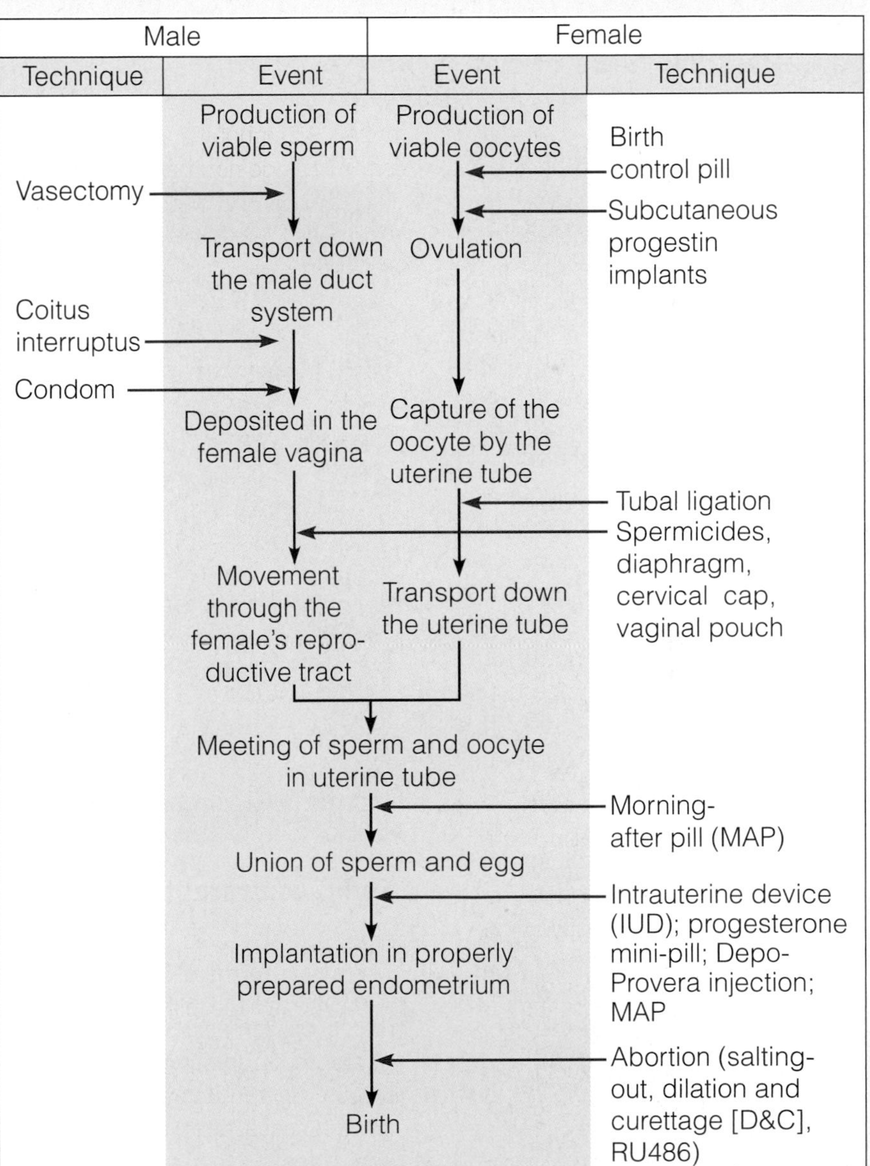

Flow chart of the events that must occur to produce a baby. Techniques or products that interfere with the process are indicated by red arrows at the site of interference; they act to prevent the next step.

Contraception: To Be or Not To Be *(continued)*

of ejaculation is never assured. *Rhythm,* or *fertility awareness, methods* depend on recognizing the period of ovulation or fertility and avoiding intercourse during those intervals. This may be accomplished by (1) recording daily basal body temperatures (body temperature drops slightly immediately prior to ovulation and then rises slightly after ovulation) or (2) recording changes in the consistency of vaginal mucus (the mucus first becomes sticky and then clear and stringy, much like egg white, during the fertile period). Rhythm techniques require accurate record-keeping for several cycles before they can be used with confidence. *Barrier methods,* such as diaphragms, cervical caps, condoms (see photo), spermicidal foams, gels, and sponges, are quite effective, especially when some agent is used by both partners. But many people avoid them because they can reduce the spontaneity of sexual encounters.

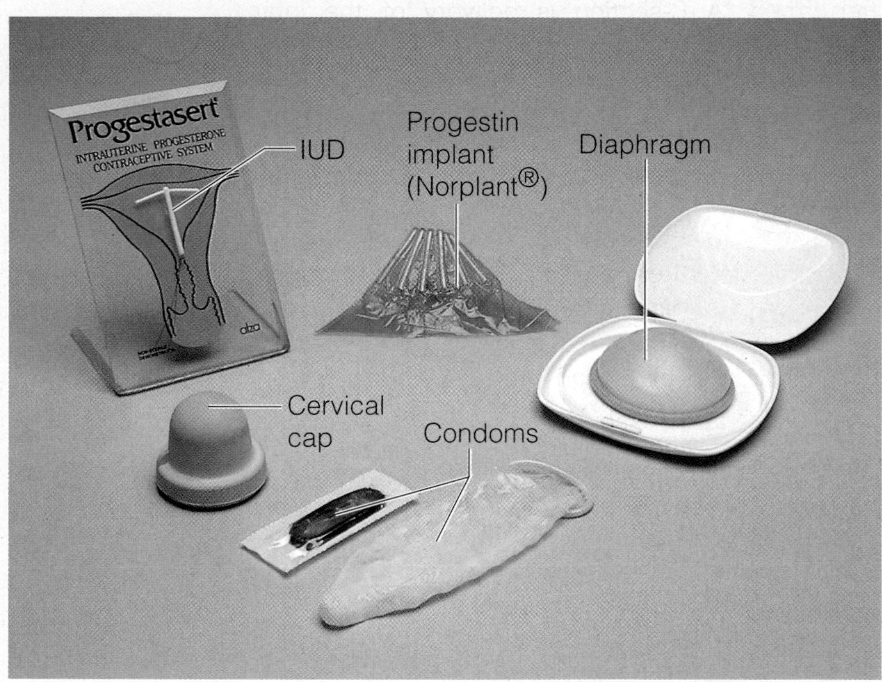

Some contraceptive devices.

RU486, the so-called *abortion pill* developed in France, may well be the next "big seller" in the United States. When taken during the first 7 weeks of pregnancy in conjunction with a tiny amount of prostaglandin to induce uterine contractions, it induces miscarriage by blocking progesterone's quieting effect on the uterus. RU486 has a 96 to 98 percent success rate with virtually no side effects. Now approved by the FDA, it has become a topic of bitter controversy among pro-choice and pro-life groups in the United States.

This summary has barely touched on the large number of experimental birth control drugs now awaiting clinical trials; and other methods are sure to be developed in the near future. In the final analysis, however, the only 100 percent effective means of birth control is the age-old one—*total abstinence.*

presentations, these advantages are lost and delivery is much more difficult, sometimes requiring the use of forceps or a vacuum extractor.

Homeostatic Imbalance

During an extremely prolonged or difficult stage 2, a condition called *dystocia* (dis-to'se-ah) may occur. In dystocia, oxygen delivery to the infant is inadequate, leading to fetal brain damage (resulting in cerebral palsy or epilepsy) and decreased viability of the infant. To prevent these outcomes, a *cesarean* (se-zayr'e-an) *section,* also called a *C-section,* may be performed. A C-section is delivery of the infant through a surgical incision made through the abdominal and uterine walls. ▲

Stage 3: Placental Stage The **placental stage,** or the delivery of the placenta, is usually accomplished within 15 minutes after birth of the infant. The strong uterine contractions that continue after birth compress uterine blood vessels, limit bleeding, and cause the placenta to detach from the uterine wall. The placenta and its attached fetal membranes, collectively called the **afterbirth,** are then easily removed by a slight tug on the umbilical cord. It is very important that all placental fragments be removed to prevent continued uterine bleeding after birth *(postpartum bleeding).*

Developmental Aspects of the Reproductive System

Although the sex of an individual is determined at the time of fertilization (males have X and Y sex chromosomes and females have two X sex chromosomes), the gonads do not begin to form until about the eighth week of embryonic development. Prior to this time, the embryonic reproductive structures of males and females are identical and are said to be in the *indifferent stage.* After the gonads have formed, development of the accessory structures and external genitalia begins. Whether male or female structures will form depends entirely on whether testosterone is present or absent. The usual case is that, once formed, the embryonic testes produce testosterone, and the development of the male duct system and external genitalia follows. When testosterone is not produced, as is the case in female embryos that form ovaries, the female ducts and external genitalia result.

Homeostatic Imbalance

Any interference with the normal pattern of sex hormone production in the embryo results in bizarre abnormalities. For example, if the embryonic testes fail to produce testosterone, a genetic male develops the female accessory structures and external genitalia. On the other hand, if a genetic female is exposed to testosterone (as might happen if the mother has an androgen-producing tumor of her adrenal gland), the embryo has ovaries but develops male accessory ducts and glands, as well as a penis and an empty scrotum. Individuals having accessory reproductive structures that do not "match" their gonads are called *pseudohermaphrodites* (su"do-her-maf'ro-dītz) to distinguish them from true *hermaphrodites,* rare individuals who possess both ovarian and testicular tissues. In recent years, many pseudohermaphrodites have sought sex change operations to match their outer selves (external genitalia) with their inner selves (gonads). ▲

The male testes, formed in the abdominal cavity at approximately the same location as the female ovaries, descend to enter the scrotum about one month before birth. Failure of the testes to make their normal descent leads to a condition

Prove It Yourself

The Stages of Mitosis and Meiosis Are Similar but Different

Here is a low-tech, but effective, way to compare the stages of mitosis and meiosis.

You'll need two pens of one color and two of another color, each with removable caps. Each pair of pens of a given color can represent the duplicated sister chromatids of a single chromosome (2 sister chromatids = 1 chromosome). Together, the four pens represent the sister chromatids of a pair of corresponding chromosomes (4 sister chromatids = a pair of sister chromosomes).

Refer to Figure 3.15, and shift your pens to follow the movements of the sister chromatids through the stages of mitosis. Now look at Figure 16.3, and move the pens through the stages of meiosis. To represent crossing-over, switch pen caps. How do the events of mitosis and meiosis differ?

called *cryptorchidism* (krip-tor'kĭ-dizm). Because this condition results in sterility of a male (and also puts him at risk for cancer of the testes), surgery is usually performed during childhood to rectify this problem.

Homeostatic Imbalance

Abnormal separation of chromosomes during meiosis can lead to congenital defects of this system. For example, males who have an extra female sex chromosome have the normal male accessory structures, but their testes atrophy, causing them to be sterile. Other abnormalities occur when a child has only one sex chromosome. An XO female appears normal but lacks ovaries; YO males die during development. Other much less serious conditions affect males primarily; these include *phimosis* (fi-mo'sis), which essentially is a narrowing of the foreskin of the penis, and misplaced urethral openings. ▲

Since the reproductive system organs do not function until puberty, there are few problems with this system during childhood. **Puberty** is the period of life, generally between the ages of 10 and 15 years, when the reproductive organs grow to their adult size and become functional under the influence of rising levels of gonadal hormones (testosterone in males and estrogen in females). After this time, reproductive capability continues until old age in males and menopause in females. Since the secondary sex characteristics and major events of puberty were described earlier, these details will not be repeated here. It is important to remember, however, that puberty represents the earliest period of reproductive system activity.

The events of puberty occur in the same sequence in all individuals, but the age at which they occur varies widely. In males, the event that signals puberty's onset is enlargement of the testes and scrotum, around the age of 13 years, followed by the appearance of pubic, axillary, and facial hair. Growth of the penis goes on over the next 2 years, and sexual maturation is indicated by the presence of mature sperm in the semen. In the meantime, the young man has unexpected, and often embarrassing, erections and frequent nocturnal emissions ("wet dreams") as his hormones surge and hormonal controls struggle to achieve a normal balance.

The first sign of puberty in females is budding breasts, often apparent by the age of 11 years. The first menstrual period, called **menarche** (mĕ-nar'ke), usually occurs about two years later. Dependable ovulation and fertility are deferred until the hormonal controls mature, an event that takes nearly two more years.

Homeostatic Imbalance

In adults, the most common reproductive system problems are infections. Vaginal infections are more common in young and elderly women and in those whose resistance is low. Common infections include those caused by *Escherichia coli* (spread from the digestive tract); sexually transmitted microorganisms (such as gonorrhea, syphilis, and herpesvirus); and yeasts (a type of fungus). Untreated vaginal infections may spread throughout the female reproductive tract, causing pelvic inflammatory disease and sterility. Problems involving painful or abnormal menses may result from infection or hormone imbalance.

The most common inflammatory conditions in males are *urethritis, prostatitis,* and *epididymitis* (ep″ĭ-did-ĭ-mi'tis), all of which may follow sexual contacts in which sexually transmitted disease (STD) microorganisms are transmitted. *Orchiditis* (or″kĭ-di'tis), inflammation of the testes, is rather uncommon but is serious because it can cause sterility. Orchiditis most commonly follows STD or mumps (in an adult male).

As noted earlier, neoplasms represent a major threat to reproductive system organs. Tumors of the breast and cervix are the most common reproductive cancers in adult females, and prostate cancer (a common sequel to prostatic hypertrophy) is a widespread problem in adult males. ▲

Most women reach peak reproductive abilities in their late twenties. After that, a natural decrease in ovarian function occurs. As estrogen production declines, ovulation becomes irregular and menstrual periods become scanty and shorter in length. Eventually, ovulation and menses cease entirely, ending childbearing ability. This event, called **menopause,** is considered to have occurred when a whole year has passed without menstruation.

Although estrogen production continues for a while after menopause, the ovaries finally stop functioning as endocrine organs. When deprived of the stimulatory effects of estrogen, the reproductive organs and breasts begin to atrophy. The vagina becomes dry; intercourse may become painful (particularly if infrequent), and vaginal infections become increasingly common. Other consequences

Systems in Sync

Homeostatic Relationships between the Reproductive System and Other Body Systems

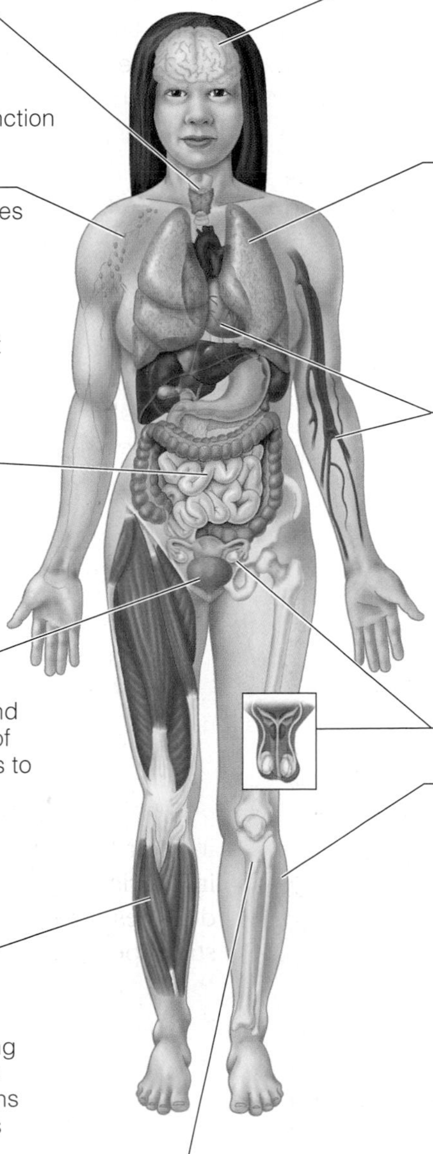

Endocrine System
- Gonadal hormones exert feedback effects on hypothalamic-pituitary axis; placental hormones help to maintain pregnancy
- Gonadotropins help regulate function of gonads

Lymphatic System/Immunity
- Developing embryo/fetus escapes immune surveillance (not rejected)
- Lymphatic vessels drain leaked tissue fluids; transport sex hormones; immune cells protect reproductive organs from disease; IgA is present in breast milk

Digestive System
- Digestive organs crowded by developing fetus; heartburn, constipation common during pregnancy
- Digestive system provides nutrients needed for health

Urinary System
- Hypertrophy of the prostate gland inhibits urination; compression of bladder during pregnancy leads to urinary frequency and urgency
- Kidneys dispose of nitrogenous wastes and maintain acid-base balance of blood of mother and fetus; semen exits the body through the urethra of the male

Muscular System
- Androgens promote increased muscle mass
- Abdominal muscles active during childbirth; muscles of the pelvic floor support reproductive organs and aid erection of penis/clitoris

Nervous System
- Sex hormones masculinize or feminize the brain and influence sex drive
- Hypothalamus regulates timing of puberty; neural reflexes regulate sexual response

Respiratory System
- Pregnancy impairs descent of the diaphragm, causing difficult breathing
- Respiratory system provides oxygen; disposes of carbon dioxide; vital capacity and respiratory rate increase during pregnancy

Cardiovascular System
- Estrogens lower blood cholesterol levels and promote cardiovascular health in premenopausal women; pregnancy increases workload of the cardiovascular system
- Cardiovascular system transports needed substances to organs of reproductive system; local vasodilation involved in erection; blood transports sex hormones

Reproductive System

Integumentary System (Skin)
- Male sex hormones (androgens) activate oil glands, which lubricate skin and hair; gonadal hormones stimulate characteristic fat distribution and appearance of pubic and axillary hair; estrogen increases skin hydration; enhances facial skin pigmentation during pregnancy
- Skin protects all body organs by enclosing them externally; mammary gland secretions (milk) nourish the infant

Skeletal System
- Androgens masculinize the skeleton and increase bone density; estrogen feminizes skeleton and maintains bone mass in females
- The bony pelvis encloses some reproductive organs; if narrow, the bony pelvis may hinder vaginal delivery of an infant

of estrogen deficit include irritability and other mood changes (depression in some); intense vasodilation of the skin's blood vessels, which causes uncomfortable sweat-drenching "hot flashes"; gradual thinning of the skin and loss of bone mass; and slowly rising blood cholesterol levels, which place postmenopausal women at risk for cardiovascular disorders. Some physicians prescribe low-dose estrogen-progestin preparations to help women through this often difficult period and to prevent the skeletal and cardiovascular complications.

There is no equivalent of menopause in males. Although aging men exhibit a steady decline in testosterone secretion, their reproductive capability seems unending. Healthy men well into their eighties and beyond are able to father offspring.

SUMMARY

Media study tools that could provide you with additional help in reviewing specific key topics of Chapter 16 are referenced below.

IP = Interactive Physiology;

WEB = A&P Place website.

ANATOMY OF THE MALE REPRODUCTIVE SYSTEM (pp. 529–533)

1. The paired testes, the male gonads, reside in the scrotum outside the abdominopelvic cavity. Testes have both an exocrine (sperm-producing) and an endocrine (testosterone-producing) function.
2. The male duct system includes the epididymis, ductus deferens, and urethra. Sperm maturation occurs in the epididymis. When ejaculation occurs, sperm are propelled through duct passageways to the body exterior.
3. Accessory glands of the male include the seminal vesicles, prostate gland, and bulbourethral glands. Collectively, these glands produce a fluid that activates and nourishes sperm.
4. External genitalia:
   a. Scrotum—a skin sac that hangs outside the abdominopelvic cavity and provides the proper temperature for producing viable sperm.
   b. Penis—consists of three columns of erectile tissue surrounding the urethra. Erectile tissue provides a way for the penis to become rigid so it may better serve as a penetrating device during sexual intercourse.

   WEB Exercise: Chapter 16, Male Reproductive Anatomy: Sagittal View.

MALE REPRODUCTIVE FUNCTIONS (pp. 533–535)

1. Spermatogenesis (sperm production) begins at puberty in seminiferous tubules in response to FSH. Spermatogenesis involves meiosis, a special nuclear division that halves the chromosomal number in resulting spermatids. An additional process that strips excess cytoplasm from the spermatid, called spermiogenesis, is necessary for production of functional, motile sperm.
2. Testosterone production begins in puberty in response to LH. Testosterone is produced by interstitial cells of the testes. Testosterone causes the appearance of male secondary sex characteristics and is necessary for sperm maturation.

ANATOMY OF THE FEMALE REPRODUCTIVE SYSTEM (pp. 535–539)

1. The ovaries, the female gonads, are located against the lateral walls of the pelvis. They produce female sex cells (exocrine function) and hormones (endocrine function).
2. The duct system:
   a. Uterine (fallopian) tubes extend from the vicinity of an ovary to the uterus. Ends are fringed and "wave" to direct ovulated oocytes into uterine tubes, which conduct the oocyte (embryo) to the uterus by peristalsis and ciliary action.
   b. The uterus is a pear-shaped organ in which the embryo implants and develops. Its mucosa (endometrium) sloughs off each month in menses unless an embryo has become embedded in it. The myometrium contracts rhythmically during the birth of a baby.
   c. The vagina is a passageway between the uterus and the body exterior that allows a baby or the menstrual flow to leave the body. It also receives the penis and semen during sexual intercourse.
3. External genitalia of the female include labia majora and minora (skin folds), clitoris, and urethral and vaginal openings.

FEMALE REPRODUCTIVE FUNCTIONS AND CYCLES (pp. 539–543)

1. Oogenesis (production of female sex cells) occurs in ovarian follicles, which are activated at puberty by FSH and LH to mature and eject oocytes (ovulation) on a cyclic basis. The female egg (ovum) is formed only if sperm penetrates the secondary oocyte. In females, meiosis produces only one functional ovum (plus three nonfunctional polar bodies), as opposed to the four functional sperm per meiosis produced by males.
2. Hormone production: Estrogens are produced by ovarian follicles in response to FSH. Progesterone, produced in response to LH, is the main hormonal product of the corpus luteum. Estrogens stimulate development of female secondary sex characteristics.
3. The menstrual cycle concerns changes in the endometrium in response to fluctuating blood levels of ovarian hormones. The phases are
   a. Menstrual phase. Endometrium sloughs off and bleeding occurs. Ovarian hormones are at their lowest levels.
   b. Proliferative phase. Endometrium is repaired, thickens, and becomes well vascularized in response to increasing levels of estrogens.
   c. Secretory phase. Endometrial glands begin to secrete nutrients, and lining becomes more vascular in response to increasing levels of progesterone.

   If fertilization does not occur, the phases are repeated about every 28 days.

   WEB Exercise: Chapter 16, The Female Menstrual Cycle.

MAMMARY GLANDS (pp. 543–544)

1. Mammary glands are milk-producing glands found in the breasts. After the birth of a baby, they produce milk in response to hormonal stimulation.

SURVEY OF PREGNANCY AND EMBRYONIC DEVELOPMENT (pp. 545–557)

1. An oocyte is fertilizable for up to 24 hours after release; sperm are viable within the female reproductive tract for up to 72 hours. Hundreds of sperm must release their acrosomal enzymes to break down the egg's corona radiata.
2. Following sperm penetration, the secondary oocyte completes meiosis II. Then ovum and sperm nuclei fuse (fertilization), forming a zygote.
3. If fertilization occurs, embryonic development begins immediately. Cleavage, a rapid series of mitotic divisions without intervening growth, begins with the zygote and ends with a blastocyst.
4. By day 14 after ovulation, the young embryo (blastocyst) has implanted in the endometrium, and the placenta is being formed. Human chorionic gonadotropin (hCG) released by the blastocyst maintains hormone production of the corpus luteum, preventing menses, until the placenta assumes its endocrine role.
5. The placenta serves respiratory, nutritive, and excretory needs of the embryo and produces hormones of pregnancy.
6. All major organ systems have been laid down by eight weeks, and at nine weeks the embryo is called a fetus. Growth and tissue/organ specialization are the major events of the fetal period.
7. A pregnant woman has increased respiratory, circulatory, and urinary demands placed on her system by the developing fetus. Good nutrition is necessary to produce a healthy baby.
8. Childbirth (parturition) includes a series of events called labor. It is initiated by several factors but most importantly by rising levels of oxytocin and prostaglandins, which promote vigorous uterine contractions. The three stages of labor are dilation, expulsion, placental.

DEVELOPMENTAL ASPECTS OF THE REPRODUCTIVE SYSTEM (pp. 557–560)

1. Reproductive system structures of males and females are identical during early development. Gonads begin to develop in the eighth week. The presence or absence of testosterone determines whether male or female accessory reproductive organs are formed.
2. Important congenital defects result from abnormal separation of sex chromosomes during sex cell formation.
3. The reproductive system is inactive during childhood. Reproductive organs mature and become functional for childbearing at puberty.
4. Common reproductive problems during young adulthood are infections of the reproductive tract. Neoplasms of breast and cervix are major threats to females. Prostate cancer is the most common reproductive system cancer seen in males.

   WEB Exercise: Chapter 16, Reproductive Case Study.

5. During menopause, female reproductive capabilities end, and reproductive organs begin to atrophy. Hot flashes and mood changes may occur. Reproductive capacity does not appear to decline significantly in aging males.

REVIEW QUESTIONS

MULTIPLE CHOICE

More than one choice may apply.

1. Which of the following are accessory sex structures in the male?
   a. Gonads
   b. Gametes
   c. Broad shoulders
   d. Seminal vesicles

2. Which of these is mismatched?
   a. Vagina—penis
   b. Testis—ovary
   c. Labia majora—scrotum
   d. Oviduct—ductus deferens

3. The myometrium is the muscular layer of the uterus, and the endometrium is the __________ layer.
   a. serosa
   b. adventitia
   c. submucosa
   d. mucosa

4. All of the following are true of the gonadotropins except that they are:
   a. secreted by the pituitary gland
   b. LH and FSH
   c. hormones with important functions in both males and females
   d. the sex hormones secreted by the gonads

5. The approximate area between the anus and clitoris in the female is the:
   a. peritoneum
   b. perineum
   c. vulva
   d. labia

6. Which of the following attach to the ovary?
   a. Fimbriae
   b. Ovarian ligament
   c. Suspensory ligaments
   d. Broad ligament

7. Human ova and sperm are similar in that:
   a. about the same number of each is produced per month
   b. they have the same degree of motility
   c. they are about the same size
   d. they have the same number of chromosomes

8. Select the false statement about the cervix of the uterus.
   a. It is the superiormost part of the uterus.
   b. It projects into the vagina.
   c. Its cervical glands secrete mucus.
   d. It contains the cervical canal.

9. Each month, only one:
   a. primary follicle is stimulated
   b. follicle secretes estrogen
   c. vesicular follicle undergoes ovulation
   d. ovary is stimulated

10. After ovulation, the ruptured follicle:
    a. degenerates
    b. becomes a corpus luteum
    c. sloughs off as waste material
    d. mends and produces another oocyte

11. The outer layer of the blastocyst, which attaches to the uterine wall, is the:
    a. yolk sac
    b. inner cell mass
    c. amnion
    d. trophoblast

12. The usual and most desirable presentation for birth is:
    a. vertex
    b. breech
    c. nonvertex
    d. head first

13. During human embryonic development, organogenesis occurs:
    a. during the first trimester
    b. during the second trimester
    c. during the third trimester
    d. during blastula formation

SHORT ANSWER ESSAY

1. What are the primary sex organs, or gonads, of males? What are their two major functions?

2. Name the organs forming the male duct system, as listed from the male gonad to the body exterior.

3. What is the function of seminal fluid? Name the three types of glands that help produce it.

4. The penis contains erectile tissue that becomes engorged with blood during sexual excitement. What term is used to describe this event?

5. Define *ejaculation*.

6. Why are the male gonads not found in the abdominal cavity? Where are they found?

7. How does enlargement of the prostate gland interfere with a male's reproductive function?

8. What structures in the testes form the sex cells? When does spermatogenesis begin? What causes it to begin?

9. The process of spermatogenesis actually forms cells called spermatids. How are the spermatids converted to functional sperm?

10. Testosterone causes the male secondary sex characteristics to appear at puberty. Name three examples of male secondary sex characteristics.

11. Explain why a male's sexual responsiveness and secondary sex characteristics generally remain unchanged after a vasectomy.

12. Name the female gonad, and describe its two major functions.

13. Why is the term *urogenital system* more applicable to males than females?

14. Name the structures of the female duct system, and describe the important functions of each.

15. Since the uterine tubes are not continuous with the ovaries, how can you explain the fact that all ovulated "eggs" do not end up in the female's peritoneal cavity?

16. What anterior pituitary hormones cause follicle development and ovulation to occur in the ovary? What is a follicle? What is ovulation?

17. The female cell that is ovulated is not a mature sex cell (ovum). When or under what conditions does it become mature?

18. What hormone can be called the "feminizing" hormone? What ovarian structures produce this hormone? Name the second hormone produced by the same structure.

19. List and describe the events of the menstrual cycle. Why is the menstrual cycle so important?

20. Define *menopause*. What does this mean to a female?

21. What is the role of the mammary glands?

22. Define *fertilization*. Where does fertilization usually occur? Describe the process of implantation.

23. What are the functions of the placenta?

24. How is a pregnant woman's body functioning altered by her pregnancy?

25. Compare the effects of aging on the male and female reproductive systems.

AT THE CLINIC

1. A pregnant woman in substantial pain called her doctor and explained (between sobs) that she was about to have her baby "right away." The doctor calmed her and asked how she had come to that conclusion. She said that her water had broken and that her husband could see the baby's head. **(a)** Was she right to believe that birth was imminent? If so, what stage of labor was she in? **(b)** Do you think she had time to make it to the hospital 60 miles away? Why or why not?

2. Lucy had both her left ovary and her right uterine tube removed surgically at age 17 because of a cyst and a tumor in these organs. Now, at age 32, she remains healthy and is expecting her second child. How could Lucy conceive a child with just one ovary and one uterine tube, widely separated on opposite sides of the pelvis like this?

3. Adolf, a 68-year-old man, has trouble urinating and is given a rectal exam. What is Adolf's most probable condition, and what is the purpose of the rectal exam?

4. Some anatomy students were saying that the bulbourethral glands of males act like city workers who come around and clear parked cars from the street before a parade. What did they mean by this analogy?

5. A young woman visited her doctor for a premarital gynecological examination and requested information on contraception. She plans to have a large family but "not just yet." She reveals that her faith prevents her from using any drugs or mechanical means of birth control. What are the only options for this young woman?

6. Mrs. Montez is experiencing Braxton Hicks contractions in the sixth month of her pregnancy. Why did her doctor prescribe ibuprofen (with what mechanism does this drug interfere)?

7. In medieval times, young boys with beautiful soprano voices were sometimes castrated (had their testes removed) so that their singing voices would not change. Explain the cause and effect here.

Appendix A

Word Roots, Prefixes, and Suffixes

WORD ROOTS

Word element	Meaning
Circulatory system	
cardio	heart
angio, vaso	vessel
hem, hema, hemato	blood
vena, phlebo	vein
arteria	artery
lympho	lymph
thrombo	clot (of blood)
embolus	moving clot
Digestive system	
bucca	cheek
os, stomato	mouth
gingiva	gum
glossa	tongue
pharyngo	pharynx
esophago	esophagus
gastro	stomach
hepato	liver
cholecyst	gallbladder
pancreas	pancreas
entero	intestines
duodeno	duodenum
jejuno	jejunum
ileo	ileum
appendeco	appendix
colo	colon
recto	rectum
ano, procto	anus
Skeletal system	
skeleto	skeleton
Respiratory system	
naso, rhino	nose
tonsillo	tonsil
laryngo	larynx
tracheo	trachea
bronchus, broncho	bronchus (pl. bronchi)
pulmo, pneuma, pneum	lung (sac with air)
Nervous system	
neuro	nerve
cerebro	brain; cerebrum
oculo, ophthalmo	eye
oto	ear
psych, psycho	mind
Urinary system	
urethro	urethra
cysto	bladder
uretero	ureter
reni, reno, nephro	kidney
pyelo	pelvis of kidney
uro	urine
Female reproductive system	
vulvo	vulva
perineo	perineum
labio	labium (pl. labia)
vagino, colpo	vagina
cervico	cervix
utero	womb; uterus
tubo, salpingo	fallopian tube
ovario, oophoro	ovary
Male reproductive system	
orchido	testes
Regions of the body	
crani, cephalo	head
cervico, tracheo	neck
thoraco	chest
abdomino	abdomen
dorsum	back
Tissues	
cutis, dermato	skin
lipo	fat
musculo, myo	muscle
osteo	bone
myelo	marrow
chondro	cartilage
Miscellaneous	
cyto	cell
gene	formation, origin
gram	tracing or mark
graph	writing, description
kinesis	motion
lapar	through the abdominal wall
meter	measure
oligo	small, few
phobia	fear
photo	light
pyo	pus
roentgen	X ray
scope	instrument for visual examination

Word Roots, Prefixes, and Suffixes (*continued*)

Word element	Meaning
PREFIXES	
a, an, ar	without or not
ab	away from
acro	extremities
ad	toward, to
adeno	glandular
aero	air
ambi	around, on both sides
amyl	starch
ante	before, forward
anti	against, counteracting
bi	double
bili	bile
bio	life
bis	two
brachio	arm
brady	slow
broncho	bronchus (pl. bronchi)
cardio	heart
cervico	neck
chole	gall or bile
cholecysto	gallbladder
circum	around
co	together
contra	against, opposite
costo	ribs
cysto	bladder
cyto	cell
demi	half
derma	skin
dis	from
dorso	back
dys	abnormal, difficult
electro	electric
en	into, in, within
encephal	brain
entero	intestine
equi	equal
eryth	red
ex	out, out of, away from
extra	outside of, in addition to
ferro	iron
fibro	fiber
fore	before, in front of
gastro	stomach
glosso	tongue
glyco	sugar
hemi	half
hemo	blood
hepa, hepato	liver
histo	tissue
homo	same
hydro	water
hygro	moisture
hyper	too much, high
hypo	under, decreased
hyster	uterus
ileo	ileum
in	in, within, into
inter	between
intra	within
intro	in, within, into
juxta	near, close to
lapar	abdomen
laryngo	larynx
latero	side
leuk	white
macro	large, big
mal	bad, poor
mast	breast
medio	middle
mega, megalo	large, great
meno	menses
mono	single
multi	many
myelo	bone marrow, spinal cord
myo	muscle
neo	new
nephro	kidney
neuro	nerve
nitro	nitrogen
noct	night
non	not
ob	against, in front of
oculo	eye
odonto	tooth
ophthalmo	eye
ortho	straight, normal
os	mouth, bone
osteo	bone
oto	ear
pan	all
para	beside, accessory to
path	disease
ped	child, foot
per	by, through
peri	around
pharyngo	pharynx
phlebo	vein
photo	light
phren	diaphragm,mind
pneumo	air, lungs
pod	foot
poly	many, much
post	after
pre	before
proct	rectum
pseudo	false
psych	mind
pyel	pelvis of the kidney
pyo	pus
pyro	fever, heat
quadri	four
radio	radiation
re	back, again
reno	kidney
retro	backward
rhin	nose
sacro	sacrum
salpingo	fallopian tube
sarco	flesh
sclero	hard, hardening
semi	half
sex	six
skeleto	skeleton
steno	narrowing, constriction
sub	under
super	above, excess
supra	above
syn	together
tachy	fast
thyro	thyroid, gland
trache	trachea
trans	across, over
tri	three
ultra	beyond
un	not, back reversal
uni	one
uretero	ureter
urethro	urethra
uro	urine, urinary organs
vaso	vessel

continued

Word Roots, Prefixes, and Suffixes (*continued*)

Word element	Meaning
SUFFIXES	
able	able to
algia	pain
cele	tumor, swelling
centesis	surgical puncture to remove fluid
cide	killing, destructive
cule	little
cyte	cell
ectasia	dilating, stretching
ectomy	excision, surgical removal of
emia	blood
esis	action
form	shaped like
genesis, genetic	formation, origin
gram	tracing, mark
graph	writing
ism	condition
itis	inflammation
ize	to treat
lith	stone, calculus
lithiasis	presence of stones
lysis	disintegration
megaly	enlargement
meter	instrument that measures
oid	likeness, resemblance
oma	tumor
opathy	disease of
orrhaphy	surgical repair
osis	disease, condition of
ostomy	to form an opening or outlet
otomy	to incise
pexy	fixation
phage	ingesting
phobia	fear
plasty	plastic surgery
plegia	paralysis
rhage	to burst forth
rhea	excessive discharge
rhexis	rupture
scope	lighted instrument for visual examination
scopy	to examine visually
stomy	to form an opening
tomy	incision into
uria	urine
zyme	ferment

Sources: Courtesy of Margaret Ling, Director of Vocational Nursing, Santa Rosa Junior College, Santa Rosa, Calif.; Kozier, B., G. Erb, A. J. Berman, and K. Burke: *Fundamentals of Nursing: Concepts and Procedures,* sixth ed., Prentice Hall, 2000. Used by permission.

Appendix B

Periodic Table of the Elements

Representative (main group) elements | Transition metals | Representative (main group) elements

Periodic Table of the Elements

Period	IA	IIA	IIIB	IVB	VB	VIB	VIIB	VIIIB	VIIIB	VIIIB	IB	IIB	IIIA	IVA	VA	VIA	VIIA	VIIIA
1	1 **H** 1.0079																	2 **He** 4.003
2	3 **Li** 6.941	4 **Be** 9.012											5 **B** 10.811	6 **C** 12.011	7 **N** 14.007	8 **O** 15.999	9 **F** 18.998	10 **Ne** 20.180
3	11 **Na** 22.990	12 **Mg** 24.305											13 **Al** 26.982	14 **Si** 28.086	15 **P** 30.974	16 **S** 32.066	17 **Cl** 35.453	18 **Ar** 39.948
4	19 **K** 39.098	20 **Ca** 40.078	21 **Sc** 44.956	22 **Ti** 47.88	23 **V** 50.942	24 **Cr** 51.996	25 **Mn** 54.938	26 **Fe** 55.845	27 **Co** 58.933	28 **Ni** 58.69	29 **Cu** 63.546	30 **Zn** 65.39	31 **Ga** 69.723	32 **Ge** 72.61	33 **As** 74.922	34 **Se** 78.96	35 **Br** 79.904	36 **Kr** 83.8
5	37 **Rb** 85.468	38 **Sr** 87.62	39 **Y** 88.906	40 **Zr** 91.224	41 **Nb** 92.906	42 **Mo** 95.94	43 **Tc** 98	44 **Ru** 101.07	45 **Rh** 102.906	46 **Pd** 106.42	47 **Ag** 107.868	48 **Cd** 112.411	49 **In** 114.82	50 **Sn** 118.71	51 **Sb** 121.76	52 **Te** 127.60	53 **I** 126.905	54 **Xe** 131.29
6	55 **Cs** 132.905	56 **Ba** 137.327	57 **La** 138.906	72 **Hf** 178.49	73 **Ta** 180.948	74 **W** 183.84	75 **Re** 186.207	76 **Os** 190.23	77 **Ir** 192.22	78 **Pt** 195.08	79 **Au** 196.967	80 **Hg** 200.59	81 **Tl** 204.383	82 **Pb** 207.2	83 **Bi** 208.980	84 **Po** 209	85 **At** 210	86 **Rn** 222
7	87 **Fr** 223	88 **Ra** 226.025	89 **Ac** 227.028	104 **Rf** 261	105 **Db** 262	106 **Sg** 263	107 **Bh** 264	108 **Hs** 265	109 **Mt** 266	110 **Uun** 269	111 **Uuu** 272	112 **Uub** 277		114		116		118

Rare earth elements

Lanthanides	58 **Ce** 140.115	59 **Pr** 140.908	60 **Nd** 144.24	61 **Pm** 145	62 **Sm** 150.36	63 **Eu** 151.964	64 **Gd** 157.25	65 **Tb** 158.925	66 **Dy** 162.5	67 **Ho** 164.93	68 **Er** 167.26	69 **Tm** 168.934	70 **Yb** 173.04	71 **Lu** 174.967
Actinides	90 **Th** 232.038	91 **Pa** 231.036	92 **U** 238.029	93 **Np** 237.048	94 **Pu** 244	95 **Am** 243	96 **Cm** 247	97 **Bk** 247	98 **Cf** 251	99 **Es** 252	100 **Fm** 257	101 **Md** 258	102 **No** 259	103 **Lr** 262

The periodic table arranges elements according to atomic number and atomic weight into horizontal rows called *periods* and 18 vertical columns called *groups* or *families*. The elements in the groups are classified as being in either A or B classes.

Elements of each group of the A series have similar chemical and physical properties. This reflects the fact that members of a particular group have the same number of valence shell electrons, which is indicated by the number of the group. For example, group IA elements have one valence shell electron, group IIA elements have two, and group VA elements have five. In contrast, as you progress across a period from left to right, the properties of the elements change in discrete steps, varying gradually from the very metallic properties of groups IA and IIA elements to the nonmetallic properties seen in group VIIA (chlorine and others), and finally to the inert elements (noble gases) in group VIIIA. This change reflects the continual increase in the number of valence shell electrons seen in elements (from left to right) within a period.

Class B elements are referred to as *transition elements*. All transition elements are metals, and in most cases they have one or two valence shell electrons. (In these elements, some electrons occupy more distant electron shells before the deeper shells are filled.)

In this periodic table, the colors are used to convey information about the phase (solid, liquid, or gas) in which a pure element exists under standard conditions (25 degrees centigrade and 1 atmosphere of pressure). If the element's symbol is solid black, then the element exists as a solid. If its symbol is red, then it exists as a gas. If its symbol is dark blue, then it is a liquid. If the element's symbol is green, the element does not exist in nature and must be created by some type of nuclear reaction.

Appendix C

Key Information About Vitamins and Many Essential Minerals

Vitamin	Major dietary sources	Major functions	Signs of severe, prolonged deficiency	Signs of extreme excess
Fat-Soluble				
A	Fortified milk, cheese, butter, eggs, liver, deep orange fruits and vegetables, dark green leafy vegetables.	Promotes healthy eyes, skin, and bones; hormone synthesis.	Night blindness; keratinization of epithelial tissues including the cornea of the eye (xerophthalmia) causing permanent blindness; dry, scaling skin; increased susceptibility to infection.	Joint pain, dry and itchy skin, cracked lips, nausea and vomiting, weight loss.
D	Fortified milk, eggs, liver (diet often not as important as sunlight exposure).	Promotes absorption and use of calcium and phosphorus (strong bones and teeth).	Rickets (bone deformities) in children; osteomalacia (bone softening) in adults.	Calcification of soft tissues, kidney stones, fatigue.
E	Vegetable oils, egg yolk, whole grain products, green leafy vegetables.	Antioxidant to prevent cell membrane damage.	Possible anemia and neurological effects.	Generally nontoxic; dizziness, weakness, may worsen clotting defect in vitamin K deficiency.
K	Green leafy vegetables, cabbage, cauliflower, liver.	Aids in formation of certain proteins, especially those for blood clotting.	Defective blood coagulation causing severe bleeding on injury.	Liver damage and anemia, interference with anticlotting medications.
Water-Soluble				
Thiamin (B_1)	Pork, organ meats, legumes, peanuts, enriched or whole-grain products.	Coenzyme used in energy metabolism.	Nerve changes, sometimes edema, heart damage; beriberi.	Rapid pulse, weakness, headache, insomnia.
Riboflavin (B_2)	Dairy products, whole and enriched grain products, green leafy vegetables.	Coenzyme used in energy metabolism.	Skin rash, eye sensitivity to the sun.	Generally nontoxic.
Niacin	Nuts, legumes, enriched and whole-grain products, meats, fish, all protein foods.	Coenzyme used in energy metabolism.	Weakness, dizziness, loss of appetite, confusion, diarrhea, rash.	Flushing of face, neck, hands; potential liver damage, swollen tongue, low blood pressure.
B_6	High-protein foods in general.	Coenzyme used in amino-acid metabolism.	Nervous, skin, and muscular disorders; anemia.	Damaged nerves, weakness, bloating.
Folate	Green leafy vegetables, orange juice, nuts, legumes, grain products.	Coenzyme used in DNA and RNA metabolism.	Megaloblastic anemia (large, immature red blood cells); digestive disturbances.	Masks vitamin B_{12} deficiency; interferes with drugs used to control epilepsy, diarrhea, insomnia.
B_{12}	Meat, fish, poultry, milk, cheese, eggs.	Coenzyme used in DNA and RNA metabolism; helps maintain nerve cells.	Megaloblastic anemia, fatigue.	Thought to be nontoxic.
Pantothenic acid	Meat, eggs; widely distributed in many foods.	Coenzyme used in energy metabolism.	Fatigue, vomiting.	Generally nontoxic; occasionally causes diarrhea.
Biotin	Liver, egg yolk, green vegetables, widely distributed in foods.	Coenzyme used in energy metabolism; fat synthesis.	Scaly dermatitis, loss of appetite, nausea.	Thought to be nontoxic.

Key Information About Vitamins and Many Essential Minerals (*continued*)

Vitamin	Major dietary sources	Major functions	Signs of severe, prolonged deficiency	Signs of extreme excess
C (ascorbic acid)	Citrus fruits and dark greem leafy vegetables, broccoli, cabbage, cauliflower, strawberries, melons, tomatoes, potatoes.	Functions in synthesis of collagen; is an antioxidant; aids in detoxification; improves iron absorption.	Scurvy; petechiae (minute hemorrhages); weakness, delayed wound healing; impaired immune response.	GI upsets, fatigue; confounds certain lab tests.
Calcium	Milk products, dark green leafy vegetables, broccoli, shrimp, salmon, clams, fortified orange juice, legumes, tofu.	Bone and tooth formation; blood clotting; muscle function; nerve transmission; blood pressure.	Stunted growth in children; bone loss (osteoporosis) in adults.	Extra calcium usually excreted; possible depressed absorption of some other minerals and kidney damage.
Phosphorus	Milk products, meat, poultry, eggs, legumes, nuts.	Bone and tooth formation; acid-base balance; component of coenzymes.	Weakness; demineralization of bone.	Depressed absorption of some minerals. such as calcium
Magnesium	Whole grains, nuts, legumes, dark green leafy vegetables.	Component of enzymes; bone and tooth formation.	Weakness, growth failure in children, confusion.	Unknown toxicity, but extreme amount can cause diarrhea.
Sulfur	Sulfur-containing amino acids in dietary proteins.	Component of cartilage, tendons, and proteins.	(Related to protein deficiency.)	Excess sulfur-containing amino acid intake leads to poor growth; liver damage.
Sodium	Salt in most foods: soy sauce, cured meats, pickles, canned soups, processed cheese.	Fluid balance, nerve impulse transmission.	Muscle cramps; reduced appetite, weakness.	High blood pressure in some people.
Potassium	Meat, milk, many fruits and vegetables, legumes, bananas.	Fluid balance; nerve and muscle function.	Muscular weakness; fatigue, confusion.	Muscular weakness; cardiac arrest.
Chloride	(Same as sodium)	Plays a role in acid-base balance; formation of gastric juice.	Muscle cramps; reduced appetite; poor growth in children.	Vomiting.
Iron	Red meat, organ meat, egg yolk, legumes, enriched cereals and breads, green leafy vegetables, dried fruits.	Components of hemoglobin and enzymes.	Anemia, weakness, infections, fatigue, pale eye membranes.	Acute: shock, death. Chronic: liver damage, cardiac failure.
Iodine	Seafood, iodized salt.	Component of thyroid hormones.	Enlarged thyroid, weight gain; retardation of infant both mentally and physically.	Iodide goiter.
Fluoride	Flouridated water and toothpastes; seafood.	Formation of bones and teeth.	Higher frequency of tooth decay.	Acute: Digestive distress. Chronic: mottling of teeth; skeletal deformation.
Zinc	Meat, fish, poultry, whole grain cereals and breads, legumes, nuts.	Component of enzymes; growth factor, production of insulin; making of sperm.	Growth failure; scaly dermatitis; reproductive failure; impaired immune function; taste and smell deficits.	Acute: nausea; vomiting; diarrhea. Chronic: Adversely affects copper metabolism, anemia, and immune function.
Selenium	Seafood, meat, whole grains.	Antixoidant properties; functions in close association with vitamin E.	Heart disease.	Nervous and digestive system disorders; hair loss.
Copper	Shellfish, nuts, legumes, organ meats.	Component of enzymes, production of hemoglobin.	Anemia; bone and cardiovascular changes.	Vomiting, diarrhea.
Chromium	Meat, vegetable oil, whole grain cereal.	Involved in glucose and energy metabolism; associated with hormone insulin.	Impaired glucose metabolism.	Lung and kidney damage (occupational exposures only).
Manganese	Nuts, whole grains, vegetables, and fruits.	Component of enzymes.		Central nervous system damage (occupational exposures).
Molybdenum	Organ meat, legumes, cereals.	Component of enzymes.	Disorder in nitrogen excretion.	Inhibition of enzymes; adversely affects copper metabolism.

Sources: Adapted from *www.wvda.org/nutrient/*

Photographic Credits

Chapter 1 A Closer Look a, 1.6 (middle left and middle right): © Howard Sochurek. A Closer Look b: Custom Medical Stock Photography. Focus on Careers: © Elena Dorfman/Addison Wesley Longman. 1.1: © Corbis. 1.6 (top): © Jenny Thomas/Addison Wesley Longman. (middle center): © Petit Format/Photo Researchers, Inc.
Chapter 2 A Closer Look: © Geoff Tomkinson/SPL/Photo Researchers. 2.8b: © George Bernard/Animals, Animals. Focus on Careers: © Benjamin Cummings.
Chapter 3 A Closer Look: © David M. Philips/Visuals Unlimited. Focus on Careers: © Elena Dorfman/Addison Wesley Longman. 3.7a: Courtesy of Drs. J.V. Small and G. Rinnerthaler, Austrian Academy of Sciences. 3.12b: © Dr. Birgit H. Stair, Department of Anatomy and Structural Biology, Albert Einstein College of Medicine. 3.18a: © G.W. Willis/Visuals Unlimited. 3.18b, e, f, 3.19c, g, 3.20c: © Allen Bell, University of New England/Benjamin Cummings 3.18c: © Cabisco/Visuals Unlimited 3.18d: © R. Calentine/Visuals Unlimited. 3.19a, b, e, h, 3.20b: © Ed Reschke. 3.19d: © Ed Reschke/Peter Arnold. 3.19f: © Carolina Biological/Visuals Unlimited. 3.20a: © Eric Graves/Photo Researchers. 3:21: © Biophoto Associates/Photo Researchers.
Chapter 4 A Closer Look: © Rachel Epstein/Stuart Kenter Associates. 4.3: © From Gray's Anatomy/Henry Gray. Churchill Livingstone, UK. 4.5: From Kozier, Erb, and Olivieri, *Fundamentals of Nursing,* 4th edition (Redwood City, CA: Benjamin Cummings, 1991) © Addison Wesley Publishing. 4.6: © Cabisco/Visuals Unlimited. 4.8: © CNRI/Science Photo Library/Photo Researchers, Inc. 4.10a: © Rotker/Phototake. 4.10b,c: © NMSB/Custom Medical Photography. 4.11b (top): © Myrleen Ferguson/Photo Edit. 4.11b (middle), 4.12b: © Dr. P. Marazzi/Science Photo Library/Photo Researchers, Inc. 4.11b (bottom): © Jon Meyer/Custom Medical Stock Photography. All rights reserved. 4.12a: © Bart's Medical Library/Phototake. 4.12c: © Zeva Oelbaum/Peter Arnold.
Chapter 5 A Closer Look 1: © John Wilson White/Addison Wesley/Addison Wesley Longman. 5.30: © CNRI/Science Photo Library/Photo Researchers. 5.31: © Carolina Biological Supply. 5.33: © P. Motta, Department of Anatomy, University "La Sapienza," Rome/Science Photo Library/Photo Researchers.
Chapter 6 A Closer Look, 1: © Superstock. Focus on Careers: © Elena Dorfman/Addison Wesley Longman. Table 1 (left), 6.4b: © Eric Graves/Photo Researchers, Inc. Table 1 (center): © Ed Reschke. Table 1 (right): © Allen Bell, University of New England/Benjamin Cummings. 6.11a: © B. Daemmrich/The Image Works 6.11b: © Vandystadt Angence De Presse/Photo Researchers, Inc. 6.13: © John Wilson White/Addison Wesley/Addison Wesley Longman.
Chapter 7 A Closer Look 1: © Associated Press Photo/Toby Talbot. A Closer Look 2: © Alexander Tsiara/Photo Researchers, Inc. 7.4b: © Benjamin Cummings Publishing Company. Photo by Victor Eroschenko, University of Idaho. 7.11d: © Jim Selby/Science Photo Library/Custom Medical Stock Photo. 7.13b, 7.16b: From *A Stereoscopic Atlas of Human Anatomy* by David Bassett.
Chapter 8 8.3b: From *A Stereoscopic Atlas of Human Anatomy* by David Bassett. 8.7: © NMSB/Custom Medical Photography. 8.8: © 1990/Custom Medical Stock Photography. All rights reserved. 8.15b: © John Wilson White/Addison Wesley/Addison Wesley Longman.
Chapter 9 A Closer Look: © Chuck Savage/CORBIS. 9.6b: © Ed Reschke. 9.7: © Photo Researchers. 9.08: © Photo Researchers. 9.11: Courtesy of Dr. Charles B. Wilson, Neurological Surgery, University of California Medical Center, San Francisco. 9.13a: © Victor Eroschencko, Benjamin Cummings.
Chapter 10 A Closer Look: © Ed Eckstein/Phototake NYC. Focus on Careers: © Elena Dorfman/Addison Wesley Longman. 10.2: © Benjamin Cummings Publishing Company. Photo by Victor Eroschenko, University of Idaho. 10.3: © Stan Flegler/Visuals Unlimited. 10.7: © Boehringer Ingelheim International GmbH. Photo: Lennart Nilsson. 10.8: Courtesy of Jack Scanlon, Holyoke Community College.
Chapter 11 A Closer Look 2: a. © GA/CNRI/Phototake NYC. b. Custom Medical. Focus on Careers: © Elena Dorfman/Addison Wesley Longman. 11.8a: From R.G. Kessel and R.H. Kardon, *Tissues and Organs.* ©1979 W.H. Freeman.
Chapter 12 A Closer Look: Carroll H. Weiss/Camera M.D. Studios. 12.7a: Boehringer Ingelheim Internaional GmbH. Photo: Lennart Nilsson, *The Body Victorious,* Dell Publishing Company. 12.15a: Computer Modeling and Photography, A.J.Olson © The Scripps Institute.
Chapter 13 A Closer Look: © Martin M. Rotker. Focus on Careers: © Elena Dorfman/Addison Wesley Longman. 13.3: © Science Photo Library/Photo Researchers, Inc. 13.4a: © Benjamin Cummings Publishing Company. Photo by Richard Tauber. 13.5: CNRI/Science Photo Library/Photo Researchers, Inc.
Chapter 14 A Closer Look 1: © Javier Domingo/Phototake A Closer Look 2: © Peter Arnold 14.4: From *Color Atlas of Histology* by Leslie P. Garner and James L. Hiatt © 1990 Williams and Wilkins. 14.5: From *A Stereoscopic Atlas of Human Anatomy,* by David L. Bassett. 14.19b: © Stephen Frisch.
Chapter 15 A Closer Look: © Hank Morgan/Science Source/Photo Researchers, Inc. 15.1b: © Benjamin/Cummings Publishing Inc., Photo by Richard Tauber. 15.2a, 15.3d: From *A Stereoscopic Atlas of Human Anatomy,* by David L. Bassett.
Chapter 16 A Closer Look 2: © Anthony Loveday/Benjamin/Cummings. Contraceptive devices courtesy of Planned Parenthood, Inc. Focus on Careers: © Elena Dorfman/Addison Wesley Longman. 16.5a: © Manfred Kage/Peter Arnold, Inc. 16.11: © C. Edleman/La Vilette/Photo Researchers, Inc. 16.14a: © Spencer Grant/Stock Boston. 16.14 b and c: Richard D'Amico. M.D./Custom Medical Stock Photography. 16.17: From Lennart Nilsson, *A Child is Born,* © Dell Publishing.

Art Credits

Illustrations by Imagineering unless otherwise noted.

Chapter 3 3.2: Carla Simmons/Kristin Mount 3.3: Tomo Narashima. 3.7: Tomo Narashima. 3:15: Adapted from Campbell, Mitchell, Reece, Taylor *Biology: Concepts and (Connections),* 3rd Edition, Figure 8.6, © 2000, Benjamin Cummings.
Chapter 4 4.1: Precision Graphics/Imagineering. 4.2: Kristin Mount/ Precision Graphics. 4.4: Tomo Narashima. 4.6: Precision Graphics/ Imagineering.
Chapter 5 A Closer Look: Adapted from Ranier I. Kotz, Reinhard Windhager, et al., *Nature,* Fig. 1, Vol. 406:143, 13 July 2000, © Macmillan Publishers, Ltd. 5.5, 5.28: Barbara Cousins/Kristin Mount. 5.6: Raychel Ciemmna/Kristin Mount. 5.7, 5.8, 5.9, 5.11, 5.17: Kristin Mount. 5.16, 5.19, 5.22, 5.24. 5.25: Laurie O'Keefe/Kristin Mount.
Chapter 6 6.8: Precision Graphics/Imagineering. 6.14: Adapted from Martini, *Fundamentals of Anatomy and Physiology,* 4e, Fig. 11.1, Upper Saddle River, NJ: Prentice Hall, © Frederic H. Martini, 1998. 6.15, 6.16, 6.20, 6.21: Raychel Ciemmna/Kristin Mount. 6.17, 6.18, 6.19: Kristin Mount.
Chapter 7 7.18, 7.19: Laurie O'Keefe/Kristin Mount. 7.20: Charles Hoffman/Kristin Mount. 7.21: Precision Graphics/Imagineering.
Chapter 8 8.14a: Adapted from Campbell, Mitchell, Reece, Taylor *Biology: Concepts and Connections,* 4th Edition, Figure 29.9D. © 2003, Benjamin Cummings.
Chapter 9 9.10: Precision Graphics/Imagineering.
Chapter 11 11.4: Barbara Cousins/Imagineering. 11.5: Barbara Cousins/ Kristin Mount. 11.8b: Adapted from Tortora and Grabowski, *Principles of Anatomy and Physiology* 9th Edition, Figure 21.1, page 671 (New York: John Wiley & Sons, Inc. 2000) © 2000 Biological Textbooks, Inc. and Sandra Reynolds Grabowski. 11.13, 11.14, 11.15: Kristin Mount.
Chapter 14 14.3: Adapted from Seeley, Stephens, Tate *Anatomy & Physiology* 4th Edition, figure 24.2. page 778 (New York, WCB/McGraw Hill, 1998) © 1998 The McGraw Hill Companies, Inc. 14.4, 14.5: Kristin Otwell/Kristin Mount. 14.19: From Campbell, Mitchell, Reece, Taylor *Biology: Concepts and Connections,* 3rd Edition, © 2000, Benjamin Cummings.
Chapter 15 15.6: Linda McVay/Imagineering.
Chapter 16 16.3, 16.5: Precision Graphics/Imagineering. 16.8: Martha Blake/Kristin Mount.

Glossary

Pronunciations in the text and this glossary use the following rules:

1. Accent marks follow stressed syllables. The primary stress is shown by ′ and the secondary stress by ″.
2. Unless otherwise noted, assume that vowels at the ends of syllables are long and vowels followed by consonants are short. Exceptions to this rule are indicated by a bar (¯) over the vowel, which indicates a long vowel, or a breve sign (˘) over the vowel, indicating that the vowel is short.

For example, the phonetic spelling of "thrombophlebitis" is *throm″bo-flĕ-bi′tis.* The next-to-last syllable (*bi′*) receives the greatest stress, and the first syllable (*throm″*) gets the secondary stress. The vowel in the second syllable comes at the end of the syllable and is long. The vowel that comes at the end of the third syllable is short because it has a breve sign.

Abdomen *ab′do-men* the portion of the body between the diaphragm and the pelvis.

Abduct *ab-dukt′* to move away from the midline of the body.

Abortion *ah-bor′shun* termination of a pregnancy before the embryo or fetus is viable outside the uterus.

Absorption *ab-sorp′shun* passage of a substance into or across a blood vessel or membrane.

Accommodation (1) adaptation in response to differences or changing needs; (2) adjustment of the eye for seeing objects at close range.

Acetabulum *as″ĕ-tab′u-lum* the cuplike cavity on the lateral surface of the hip bone that receives the head of the femur.

Acetylcholine *as″ĕ-til-ko′lēn* a chemical transmitter substance released by certain nerve endings.

Achilles tendon *ah-kil′ēz ten′don* the tendon that attaches the calf muscles to the calcaneus, or heel bone; also called the calcaneal tendon.

Acid a substance that liberates hydrogen ions when in an aqueous solution; compare with *base.*

Acidosis *as″ĭ-do′sis* a condition in which the blood has an excess hydrogen ion concentration and a decreased pH.

Acne inflammatory disease of the skin; infection of the sebaceous glands.

Acromion *ah-kro′me-on* the outer projection of the spine of the scapula; the highest point of the shoulder.

Acrosome *ak′ro-sōm* an enzyme-containing structure covering the nucleus of the sperm.

Actin *ak′tin* a contractile protein of muscle.

Action potential an electrical event occurring when a stimulus of sufficient intensity is applied to a neuron or muscle cell, allowing sodium ions to move into the cell and reverse the polarity.

Active immunity immunity produced by an encounter with an antigen; provides immunologic memory.

Active transport net movement of a substance across a membrane against a concentration or electrical gradient; requires release and use of cellular energy.

Adaptation (1) any change in structure or response to suit a new environment; (2) decline in the transmission of a sensory nerve when a receptor is stimulated continuously and without change in stimulus intensity.

Adduct *ah-dukt′* to move toward the midline of the body.

Adenosine triphosphate (ATP) *ah-den″o-sin tri-fos′fāt* the compound that is the important intracellular energy source; cellular energy.

Adipose *ad′ĭ-pōs* fatty.

Adrenal glands *ah-dre′nal glanz* hormone-producing glands located superior to the kidneys; each consists of a medulla and cortex areas.

Adrenergic fibers *ad″ren-er′jik* nerve fibers that release norepinephrine.

Aerobic *a-er-o′bik* requiring oxygen to live or grow.

Aerobic respiration respiration in which oxygen is consumed and glucose is broken down entirely; water, carbon dioxide, and large amounts of ATP are the final products.

Afferent *af′er-ent* carrying to or toward a center.

Afferent neurons *nu′ronz* nerve cells that carry impulses toward the central nervous system.

Agglutination *ah-gloo″tin-a′shun* clumping of (foreign) cells, induced by cross-linking of antigen-antibody complexes.

Agglutinins *ah-gloo′tĭ-ninz* antibodies in blood plasma that cause clumping of corpuscles or bacteria.

Agglutinogens *ag″loo-tin′o-jenz* (1) antigens that stimulate the formation of a specific agglutinin; (2) antigens found on red blood cells that are responsible for determining the ABO blood group classification.

Agonist *ag′o-nist* a muscle that bears the primary responsibility for causing a certain movement; a prime mover.

AIDS acquired immuno deficiency syndrome; caused by human immunodeficiency virus (HIV); symptoms include severe weight loss, night sweats, swollen lymph nodes, opportunistic infections.

Albumin *al-bu′min* a protein found in virtually all animals; the most abundant plasma protein.

Albuminuria *al″bu-mĭ-nu′re-ah* presence of albumin in the urine.

Alimentary *al″ĭ-men′tar-e* pertaining to the digestive organs.

Alkalosis *al″kah-lo′sis* a condition in which the blood has a lower hydrogen ion concentration than normal, and an increased pH.

Allergy *al′er-je* overzealous immune response to an otherwise harmless antigen; also called hypersensitivity.

Alopecia *al″o-pe′she-ah* baldness, condition of hair loss.

Alveolus *al-ve′o-lus* (1) a general term referring to a small cavity or depression; (2) an air sac in the lungs.

Amino acid *ah-me′no* an organic compound containing nitrogen, carbon, hydrogen, and oxygen; the building block of protein.

Amnion *am′ne-on* the fetal membrane that forms a fluid-filled sac around the embryo.

Amphiarthrosis *am″fe-ar-thro′sis* a slightly movable joint.

Anabolism *an-nab′o-lizm* the energy-requiring building phase of metabolism in which simpler substances are combined to form more complex substances.

Anaerobic *an-a′-er-ōb-ik* not requiring oxygen.

Anatomy the science of the structure of living organisms.

Anemia *ah-ne′me-ah* reduced oxygen-carrying capacity of the blood caused by a decreased number of erythrocytes or decreased percentage of hemoglobin in the blood.

Aneurysm *an′u-rizm* blood-filled sac in an artery wall caused by dilation or weakening of the wall.

Angina pectoris *an-ji′nah pek′tor-is* severe, suffocating chest pain caused by brief lack of oxygen supply to heart muscle.

Anorexia *an″o-rek′se-ah* loss of appetite or desire for food.

Anoxia *ah-nok′se-ah* a deficiency of oxygen.

Antagonists *an-tag′o-nists* muscles that act in opposition to an agonist or prime mover.

Antecubital anterior surface of elbow.

Anterior *an-ter′e-er* the front of an organism, organ, or part; the ventral surface.

Antibody *an″tĭ-bod′e* a specialized substance produced by the body that can provide immunity against a specific antigen.

Antigen *an′tĭ-jen* any substance—including toxins, foreign proteins, or bacteria—that, when introduced to the body, is recognized as foreign and activates the immune system.

Anus *a′nus* the distal end of the digestive tract; the outlet of the rectum.

Aorta *a-or′tah* the major systemic artery; arises from the left ventricle of the heart.

Aortic body a receptor in the aortic arch sensitive to changing oxygen, carbon dioxide, and pH levels of the blood.

Apocrine gland *ap' o-krin* the less numerous type of sweat gland. Produces a secretion containing water, salts, and proteins.

Aponeurosis *ap" o-nu-ro' sis* fibrous or membranous sheet connecting a muscle and the part it moves.

Appendicular skeleton *ap" en-dik' u-lar* bones of the limbs and limb girdles that are attached to the axial skeleton.

Appendix *ah-pen' diks* a wormlike extension of the large intestine.

Aqueous humor *a' kwe-us hu' mer* the watery fluid in the anterior chambers of the eye.

Arachnoid *ah-rak' noid* weblike; specifically, the weblike middle layer of the three meninges.

Areola *ah-re' o-lah* the circular, pigmented area surrounding the nipple.

Arrector pili *ah-rek' tor pi' li* tiny, smooth muscles attached to hair follicles, which cause the hair to stand upright when activated.

Arteriole *ar-tēr' e-ol* minute artery.

Arteriosclerosis *ar-tēr' e-o-skler-o' sis* any of a number of proliferative and degenerative changes in the arteries leading to their decreased elasticity.

Artery a vessel that carries blood away from the heart.

Arthritis *ar-thri' tis* inflammation of the joints.

Articulation joint; point where two bones meet.

Asthma *az' mah* disease or allergic response characterized by bronchial spasms and difficult breathing.

Astigmatism *ah-stig' mah-tizm* a visual defect resulting from irregularity in the lens or cornea of the eye causing the image to be out of focus.

Atherosclerosis *a' ther-o" skler-o' sis* changes in the walls of large arteries consisting of lipid deposits on the artery walls. The early stage of arteriosclerosis and increased rigidity.

Atlas the first cervical vertebra; articulates with the occipital bone of the skull and the second cervical vertebra (axis).

Atom *at' um* the smallest part of an element; indivisible by ordinary chemical means.

Atomic mass number the sum of the number of protons and neutrons in the nucleus of an atom.

Atomic number the number of protons in an atom.

Atomic symbol a one- or two-letter symbol indicating a particular element.

Atomic weight average of the mass numbers of all of the isotopes of an element.

Atrioventricular node (AV node) *a' tre-o-ven-trik' u-lar* a specialized mass of conducting cells located at the atrioventricular junction in the heart.

Atrium *a' tre-um* a chamber of the heart receiving blood from the veins; superior heart chamber.

Atrophy *at' ro-fe* a reduction in size or wasting away of an organ or cell resulting from disease or lack of use.

Auditory *aw' dĭ-to" re* pertaining to the sense of hearing.

Auditory ossicles *aw' sih-kulz* three tiny bones, the malleu, incus, and stapes, located within the middle ear that serve as transmitters of sound vibrations.

Auscultation *aws" kul-ta' shun* the act of examination by listening to body sounds.

Autoimmune response the production of antibodies or effector T cells that attack a person's own tissue.

Automaticity *aw-to" mah-tis' ĭ-te* the ability of a structure, organ, or system to initiate its own activity.

Autonomic *aw" to-nom' ik* self-directed; self-regulating; independent.

Autonomic nervous system the division of the nervous system that functions involuntarily; innervates cardiac muscle, smooth muscle, and glands.

Axial skeleton *ak' se-al* the bones of the skull, vertebral column, thorax, and sternum.

Axilla *ak-sih' lah* armpit.

Axis (1) the second cervical vertebra; has a vertical projection called the dens around which the atlas rotates; (2) the imaginary line about which a joint or structure revolves.

Axon *ak' son* neuron process that carries impulses away from the nerve cell body; efferent process; the conducting portion of a nerve cell.

B cells lymphocytes that oversee humoral immunity; their descendants differentiate into antibody-producing plasma cells; also called B lymphocytes.

Bacteria any of a large group of microorganisms, generally one-celled; found in humans and other animals, plants, soil, air, and water; have a broad range of functions.

Basal metabolic rate *met" ah-bol' ik* the rate at which energy is expended (heat produced) by the body per unit time under controlled (basal) conditions: 12 hours after a meal, at rest.

Basal nuclei *nu' kle-i* gray matter areas deep within the white matter of the cerebral hemispheres; also called basal ganglia.

Base a substance that accepts hydrogen ions; proton acceptor; compare with *acid.*

Basement membrane a thin layer of extracellular material to which epithelial cells are attached in mucosal surfaces.

Basophils *ba' zo-filz* white blood cells whose granules stain deep blue with basic dye; have a relatively pale nucleus and granular-appearing cytoplasm.

Benign *be-nīn'* not malignant.

Biceps *bi' seps* two-headed, especially applied to certain muscles.

Bicuspid *bi-kus' pid* having two points or cusps.

Bile a greenish-yellow or brownish fluid produced in and secreted by the liver, stored in the gallbladder, and released into the small intestine.

Biopsy *bi' ŏp-se* the removal and examination of live tissue; usually done to detect or rule out the presence of cancerous cells.

Blastocyst *blas' to-sist* a stage of early embryonic development.

Blood-brain barrier a mechanism that inhibits passage of materials from the blood into brain tissues.

Bolus *bo' lus* a rounded mass of food prepared by the mouth for swallowing.

Bony thorax *bōn' e tho' raks* bones of the thorax, including ribs, sternum, and thoracic vertebrae.

Brachial *bra' ke-al* pertaining to the arm.

Bradycardia *brad" e-kar' de-ah* slow heart beat, usually defined as a rate under 60 beats per minute.

Brain stem the portion of the brain consisting of the medulla, pons, and midbrain.

Bronchitis *brong-ki' tis* inflammation of the mucous membranes of the bronchi.

Bronchus *brong' kus* one of the two large branches of the trachea leading to the lungs.

Buccal *buk' al* pertaining to the cheek.

Buffer a substance or substances that help to stabilize the pH of a solution.

Bundle branch block a blocking of heart action resulting from damage to one of the bundle branches; delayed contraction of one ventricle.

Bursa *ber' sah* a small sac filled with fluid and located at friction points, especially joints.

Calculus *kal' ku-lus* a stone formed within various body parts.

Calorie *kal' o-re* unit of heat; the large calorie (spelled with a capital letter C) is the amount of heat required to raise 1 kg of water 1°C; also used in metabolic and nutrition studies as the unit to measure the energy value of foods.

Calyx *ka' liks* a cuplike extension of the pelvis of the kidney.

Canal a duct or passageway; a tubular structure.

Canaliculus *kan" ah-lik' u-lus* extremely small tubular passage or channel.

Cancer a malignant, invasive cellular neoplasm that has the capability of spreading throughout the body or body parts.

Capillary *kap' ĭ-lar" e* a minute blood vessel connecting arterioles with venules.

Carbohydrate *kar" bo-hi' drāt* organic compound composed of carbon, hydrogen, and oxygen; includes starches, sugars, cellulose.

Carcinogen *kar-sin' o-jen* cancer-causing agent.

Carcinoma *kar" sĭ-no' mah* cancer; a malignant growth of epithelial cells.

Cardiac *kar' de-ak* pertaining to the heart.

Cardiac cycle sequence of events encompassing one complete contraction and relaxation of the atria and ventricles of the heart.

Cardiac muscle specialized muscle of the heart.

Cardiac output the blood volume (in liters) ejected per minute by the left ventricle.

Cardioesophageal sphincter *kar" de-o-ĕ-sof" ah-je' al sfingk' ter* valve between the stomach and esophagus.

Cardiovascular system organ system that distributes blood to all parts of the body.

Carotid *kah-rot' id* the main artery in the neck.

Carotid body a receptor in the common carotid artery sensitive to changing oxygen, carbon dioxide, and pH levels of the blood.

Carotid sinus *si' nus* a dilation of a common carotid artery; involved in regulation of systemic blood pressure.

Carpal *kar' pal* one of the eight bones of the wrist.

Cartilage *kar' tĭ-lij* white, semiopaque connective tissue.

Cartilaginous joint *kar" ti-laj' ĭ-nus* bones united by cartilage; no joint cavity is present.

Catabolism *kah-tab' o-lizm* the process in which living cells break down substances into simpler substances; destructive metabolism.

Cataract *kat' ah-rakt* partial or complete loss of transparency of the crystalline lens of the eye.

Catecholamines *kat" ĕ-kol' ah-menz* epinephrine and norepinephrine.

Caudal *kaw' dal* toward the tail; in humans, the inferior portion of the anatomy.

Cecum *se' kum* the blind-end pouch at the beginning of the large intestine.

Cell the basic biological unit of living organisms, containing a nucleus and a variety of organelles enclosed by a limiting membrane.

Cell membrane see *plasma membrane.*

Cellular immunity *sel' u-lar ĭ-mu' nĭ-te* immunity conferred by lymphocytes called T cells; also called cell-mediated immunity.

Cellular respiration metabolic processes in which ATP is produced.

Cementum *se-men' tum* the bony connective tissue that covers the root of a tooth.

Central nervous system (CNS) the brain and the spinal cord.

Centriole *sen' tre-ōl* a minute body found near the nucleus of the cell composed of microtubules; active in cell division.

Cerebellum *ser" ĕ-bel' um* part of the hindbrain; involved in producing smoothly coordinated skeletal muscle activity.

Cerebral aqueduct *ser' ĕ-bral ak' we-dukt"* the slender cavity of the midbrain that connects the third and fourth ventricles; also called the aqueduct of Sylvius.

Cerebrospinal fluid the fluid produced by choroid plexi; fills the ventricles and surrounds the central nervous system.

Cerebrum *ser' ĕ-brum* the largest part of the brain; consists of right and left cerebral hemispheres.

Cerumen *sĕ-roo' men* earwax.

Cervical *ser' vĭ-kal* refers to the neck or the necklike portion of an organ or structure.

Cervix *ser' viks* the inferior necklike portion of the uterus leading to the vagina.

Chemical bond an energy relationship holding atoms together; involves the interaction of electrons.

Chemical energy energy form stored in chemical bonds.

Chemical reaction process in which molecules are formed, changed, or broken down.

Chemoreceptors *ke' mo-re-sep' torz* receptors sensitive to various chemicals in solution.

Chiasma *ki-as' mah* a crossing or intersection of two structures, such as the optic nerves.

Cholecystokinin *ko" le-sis" to-kin' in* an intestinal hormone that stimulates gallbladder contraction and pancreatic juice release.

Cholesterol *ko-les' ter-ol* a steroid found in animal fats as well as in most body tissues; made by the liver.

Cholinergic fibers *ko" lin-er' jik* nerve endings that, upon stimulation, release acetylcholine.

Chondrocyte *kon' dro-sīt* a mature cartilage cell.

Chorion *ko' re-on* the outermost fetal membrane; helps form the placenta.

Choroid *ko' roid* the pigmented nutritive layer of the eye.

Chromatin *kro' mah-tin* the structures in the nucleus that carry the hereditary factors (genes).

Chromosome *kro' mo-sōm* barlike body of tightly coiled chromatin; visible during cell division.

Chyme *kīm* the semifluid stomach contents consisting of partially digested food and gastric secretions.

Cilia *sil' e-ah* tiny, hairlike projections on cell surfaces that move in a wavelike manner.

Ciliary zonule suspensory ligament that attaches the lens to the ciliary body in the anterior eye.

Circle of Willis a union of arteries at the base of the brain.

Circumduction *ser' kum-duk' shun* circular movement of a body part.

Cirrhosis *sĭ-ro' sis* a chronic disease of the liver, characterized by an overgrowth of connective tissue or fibrosis.

Cleavage *klēv' ij* an early embryonic phase consisting of rapid cell divisions without intervening growth periods.

Clitoris *kli' to-ris* a small, erectile structure in the female, homologous to the penis in the male.

Clonal selection *klo' nul* the process during which a B cell or T cell becomes sensitized through binding contact with an antigen.

Clone descendants of a single cell.

Coagulation clotting (of blood).

Cochlea *kok' le-ah* a cavity of the inner ear resembling a snail shell; houses the hearing receptor.

Coitus *ko' ĭ-tus* sexual intercourse.

Coma *ko' mah* unconsciousness from which the person cannot be aroused.

Complement a group of plasma proteins that normally circulate in inactive forms; when activated by complement fixation, causes lysis of foreign cells and enhances phagocytosis and inflammation.

Compound substance composed of two or more different elements, the atoms of which are chemically united.

Concave *kon' kāv* having a curved or depressed surface.

Conductivity *kon" duk-tiv' ĭ-te* ability to transmit an electrical impulse.

Condyle *kon' dīl* a rounded projection at the end of a bone that articulates with another bone.

Cones one of the two types of photoreceptor cells in the retina of the eye. Provides for color vision.

Congenital *kon-jen' ĭ-tal* existing at birth.

Congestive heart failure (CHF) condition in which the pumping efficiency of the heart is depressed so that circulation is inadequate to meet tissue needs.

Conjunctiva *kon" junk-ti' vah* the thin, protective mucous membrane lining the eyelids and covering the anterior surface of the eyeball.

Connective tissue a primary tissue; form and function vary extensively. Functions include support, storage, and protection.

Contraception *kon" trah-sep' shun* the prevention of conception; birth control.

Contraction *kon-trak' shun* to shorten or develop tension, an ability highly developed in muscle cells.

Contralateral *kon" trah-lat' er-al* opposite; acting in unison with a similar part on the opposite side of the body.

Convergence *kon-ver' jens* turning toward a common point from different directions.

Convoluted *kon' vo-lūted* rolled, coiled, or twisted.

Cornea *kor' ne-ah* the transparent anterior portion of the eyeball.

Coronal plane see *frontal plane.*

Corpus *kor' pus* body; the major portion of an organ.

Cortex *kor' teks* the outer surface layer of an organ.

Costal *kos' tal* pertaining to the ribs.

Covalent bond *ko-va' lent* a bond involving the sharing of electrons between atoms.

Coxal pertaining to the hip.

Cramp painful, involuntary contraction of a muscle.

Cranial *kra' ne-al* pertaining to the skull.

Cranial nerves the 12 pairs of nerves that arise from the brain.

Crenation *kre-na' shun* the shriveling of a cell, for example an erythrocyte, resulting from loss of water.

Crural pertaining to the leg.

Cryptorchidism *krip-tor' kĭ-dizm* a developmental defect in which the testes fail (or one testis fails) to descend into the scrotum.

Cupula *ku' pu-lah* a domelike structure.

Cushing's syndrome *koosh' ingz sin' drōm* a disease produced by excess secretion of adrenocortical hormone; characterized by adipose tissue accumulation, weight gain, and osteoporosis.

Cutaneous *ku-ta' ne-us* pertaining to the skin.

Cutaneous membrane the skin; composed of epidermal and dermal layers.

Cyanosis *si" ah-no' sis* a bluish coloration of the mucous membranes and skin caused by deficient oxygenation of the blood.

Cystitis *sis-ti' tis* an inflammation of the urinary bladder.

Cytokinesis *si"to-kĭ-ne'sis* division of cytoplasm that occurs after the cell nucleus has divided.

Cytology *si-tol'o-je* the science concerned with the study of cells.

Cytoplasm *si'to-plazm"* the substance of a cell other than that of the nucleus.

Cytotoxic T cell see *killer T cell*.

Deciduous *de-sid'u-us* temporary.

Deciduous (milk) teeth the 20 temporary teeth replaced by permanent teeth; "baby" teeth.

Decomposition reaction a destructive chemical reaction in which complex substances are broken down into simpler ones.

Defecation *def"ih-ka'shun* the elimination of the contents of the bowels (feces).

Deglutition *deg"loo-tish'un* the act of swallowing.

Dehydration *de"hi-dra'shun* a condition resulting from excessive loss of water.

Dehydration synthesis process by which a larger molecule is synthesized from smaller ones by removal of a water molecule at each site of bond formation.

Dendrites *den'drīts* the branching extensions of neurons that carry electrical signals to the cell body; the receptive portion of a nerve cell.

Dentin *den'tin* the calcified tissue forming the major part of a tooth; deep to the enamel.

Deoxyribonucleic acid (DNA) *de-ok"se-ri"bo-nu"kla'ik* nucleic acid found in all living cells; carries the organism's hereditary information.

Depolarization *de-po"lar-i-za'shun* the loss of a state of polarity; the loss of a negative charge inside the plasma membrane.

Dermatitis *der"mah-ti'tis* an inflammation of the skin; nonspecific skin allergies.

Dermis *der'mis* the deep layer of the skin; composed of dense, irregular connective tissue.

Diabetes mellitus *di"ah-be'tēz mel-li'tus* a disease caused by deficient insulin release or inadequate responsiveness to insulin, leading to inability of the body cells to use carbohydrates at a normal rate.

Diapedesis *di"ah-pĕ-de'sis* the passage of blood cells through intact vessel walls into the tissues.

Diaphragm *di'ah-fram* (1) any partition or wall separating one area from another; (2) a muscle that separates the thoracic cavity from the abdominopelvic cavity.

Diaphysis *di-af'ĭ-sis* elongated shaft of a long bone.

Diarthrosis *di"ar-thro'sis* a freely movable joint.

Diastole *di-as'to-le* a period (between contractions) of relaxation of the heart during which it fills with blood.

Diencephalon *di"en-sef'ah-lon* that part of the forebrain between the cerebral hemispheres and the midbrain including the thalamus, the third ventricle, and the hypothalamus.

Diffusion *dĭ-fu'zhun* the spreading of particles in a gas or solution with a movement toward uniform distribution of particles.

Digestion *di-jest'jun* the bodily process of breaking down foods chemically and mechanically.

Digestive system system that processes food into absorbable units and eliminates indigestible wastes.

Digital pertaining to the digits; fingers, toes.

Dilate *di'lāt* to stretch; to open; to expand.

Disaccharide *di-sak'ĭ-rīd* literally, double sugar; examples include sucrose and lactose.

Distal *dis'tal* farthest from the point of attachment of a limb or origin of a structure.

Diverticulum *di"ver-tik'u-lum* a pouch or sac in the walls of a hollow organ or structure.

Dorsal *dor'sal* pertaining to the back; posterior.

Dorsiflexion up and down movement that includes lifting the foot so that its superior surface approaches the shin (standing on your heels).

Duct *dukt* a canal or passageway.

Duodenum *du"o-de'num* the first part of the small intestine.

Dura mater *du'rah ma'ter* the outermost and toughest of the three membranes (meninges) covering the brain and spinal cord.

Dynamic equilibrium sense that reports on angular or rotatory movements of the head in space.

Dyspnea *disp'ne-ah* labored, difficult breathing.

Ectopic *ek-top'ik* not in the normal place; for example, in an ectopic pregnancy the egg is implanted at a place other than the uterus.

Edema *ĕ-de'mah* an abnormal accumulation of fluid in body parts or tissues; causes swelling.

Effector *ef-fek'tor* an organ, gland, or muscle capable of being activated by nerve endings.

Efferent *ef'er-ent* carrying away or away from.

Efferent neurons neurons that conduct impulses away from the central nervous system.

Ejaculation *e-jak"u-la'shun* the sudden ejection of semen from the penis.

Electrical energy energy form resulting from the movement of charged particles.

Electrocardiogram (ECG) *e-lek"tro-kar'de-o-gram"* a graphic record of the electrical activity of the heart.

Electroencephalogram (EEG) *e-lek"tro-en-sef'ah-lo-gram"* a graphic record of the electrical activity of nerve cells in the brain.

Electrolyte *e-lek'tro-līt* a substance that breaks down into ions when in solution and is capable of conducting an electric current.

Electron negatively charged subatomic particle; orbits the atomic nucleus.

Electron transport chain metabolic pathway within the mitochondria in which energy harvested from high-energy hydrogen atoms is used to make ATP. Final delivery of H to molecular oxygen produces water.

Element *el'ĕ-ment* any of the building blocks of matter; oxygen, hydrogen, carbon, for example.

Embolism *em'bo-lizm* the obstruction of a blood vessel by an embolus (blood clot, bubble of air, or fatty mass floating in the blood).

Embryo *em'bre-o* an organism in its early stages of development; in humans, the first 2 months after conception.

Emesis *em'ĕ-sis* vomiting.

Emmetropia the eye that focuses images correctly on the retina is said to have this "harmonious vision."

Emphysema *em"fĭ-se'mah* a condition caused by overdistension of the pulmonary alveoli and fibrosis of lung tissue.

Enamel the hard, calcified substance that covers the crown of a tooth.

Endocarditis *en"do-kar-di'tis* an inflammation of the inner lining of the heart.

Endocardium *en"do-kar'de-um* the endothelial membrane lining the interior of the heart.

Endocrine glands *en'do-krin* ductless glands that empty their hormonal products directly into the blood.

Endocrine system body system that includes internal organs that secrete hormones.

Endometrium *en-do-me'tre-um* the mucous membrane lining of the uterus.

Endomysium *en"do-mis'e-um* the thin connective tissue surrounding each muscle cell.

Endoneurium a delicate connective tissue sheath that surrounds each fiber in a nerve.

Endoplasmic reticulum *en"do-plas'mik rĕ-tik'u-lum* a membranous network of tubular or saclike channels in the cytoplasm of a cell.

Endothelium *en"do-the'le-um* the single layer of simple squamous cells that line the walls of the heart and the vessels that carry blood and lymph.

Energy the ability to do work.

Enzyme *en'zīm* a substance formed by living cells that acts as a catalyst in bodily chemical reactions.

Eosinophils *e"o-sin'o-filz* granular white blood cells whose granules readily take up a stain called eosin.

Epicardium a serous membrane that tightly hugs the external surface of the heart and is actually part of the heart wall; also called visceral pericardium.

Epidermis *ep"ĭ-der'mis* the outer layers of the skin; epithelium.

Epididymis *ep"ĭ-did'ĭ-mis* that portion of the male duct system in which sperm mature. Empties into the ductus, or vas, deferens.

Epiglottis *ep"ĭ-glot'is* the elastic cartilage at the back of the throat; covers the glottis during swallowing.

Epimysium *ep"ĭ-mis'e-um* the sheath of fibrous connective tissue surrounding a muscle.

Epineurium a tough, fibrous sheath that binds together the fascicles in a nerve.

Epiphysis *ĕ-pif'ĭ-sis* the end of a long bone.

Epithelium *ep"ĭ-the'le-um* one of the primary tissues; covers the surface of the body and lines the body cavities, ducts, and vessels.

Equilibrium *ē′kwĭ-lib′ re-um* balance; a state when opposite reactions or forces counteract each other exactly.

Erythrocytes *ĕ-rith′ ro-sīts* red blood cells.

Erythropoiesis *ĕ-rith″ ro-poi-e′ sis* the process of erythrocyte formation.

Estrogens *es′ tro-jenz* hormones that stimulate female secondary sex characteristics; female sex hormones.

Eupnea *ūp-ne′ ah* easy, normal breathing.

Eversion special movement of the foot achieved by turning the sole laterally.

Exchange reaction a chemical reaction in which bonds are both made and broken; atoms become combined with different atoms.

Excretion *ek-skre′ shun* the elimination of waste products from the body.

Exocrine glands *ek′ so-krin* glands that have ducts through which their secretions are carried to a body surface (skin or mucosa).

Expiration *eks″ pĭ-ra′ shun* the act of expelling air from the lungs; exhalation.

Extension movement that increases the angle of a joint; e.g., straightening a flexed knee.

External respiration the actual exchange of gases between the alveoli and the blood (pulmonary gas exchange).

Extracellular *eks″ trah-sel′ u-lar* outside a cell.

Extracellular matrix nonliving material in connective tissue consisting of ground substance and fibers that separate the living cells.

Extracellular fluid fluid within the body but outside the cells.

Fallopian tube *fal-lo′ pe-an* see *uterine tube.*

Fascia *fash′ e-ah* layers of fibrous tissue covering and separating muscles.

Fascicle *fas′ ĭ-kul* a bundle of nerve or muscle fibers bound together by connective tissue.

Fatty acid a building block of fats.

Feces *fe′ sēz* material discharged from the bowel composed of food residue, secretions, and bacteria.

Femoral pertaining to the thigh.

Fertilization *fer′ tĭ-lĭ-za′ shun* fusion of the nuclear material of an egg and a sperm.

Fetus *fe′ tus* the unborn young; in humans the period from the third month of development until birth.

Fibrillation *fi-brĭ-la′ shun* irregular, uncoordinated contraction of muscle cells, particularly of the heart musculature.

Fibrin *fi′ brin* the fibrous insoluble protein formed during the clotting of blood.

Fibrinogen *fi-brin′ o-jen* a blood protein that is converted to fibrin during blood clotting.

Fibrous joint bones joined by fibrous tissue; no joint cavity is present.

Fibrous protein a strandlike protein that appears most often in body structures. They are very important in binding structures together and for providing strength in certain body tissues.

Fibular pertaining to the area of the fibula, the lateral bone of the leg.

Filtration *fil-tra′ shun* the passage of a solvent and dissolved substances through a membrane or filter.

Fissure *fis′ zher* (1) a groove or cleft; (2) the deepest depressions or inward folds on the brain.

Fixators *fiks-a′ torz* muscles acting to immobilize a joint or a bone; fixes the origin of a muscle so that muscle action can be exerted at the insertion.

Flaccid *flak′ sid* soft; flabby; relaxed.

Flagella *flah-jel′ ah* long, whiplike extensions of the cell membrane of some bacteria and of sperm; serve to propel the cell.

Flexion *flek′ shun* bending; the movement that decreases the angle between bones.

Focus *fo′ kus* creation of a sharp image by a lens.

Follicle *fol′ lĭ-kul* (1) structure in an ovary consisting of a developing egg surrounded by follicle cells; (2) colloid-containing structure in the thyroid gland.

Fontanels *fon″ tah-nelz′* the fibrous membranes in the skull where bone has not yet formed; babies' "soft spots."

Foramen *fo-ra′ men* a hole or opening in a bone or between body cavities.

Formed elements cellular portion of blood.

Fossa *fos′ ah* a depression; often an articular surface.

Fovea *fo-ve′ ah* a pit.

Frontal (coronal) plane a longitudinal plane that divides the body or an organ into anterior and posterior parts.

Fundus *fun′ dus* the base of an organ; that part farthest from the opening of the organ.

Gallbladder the sac beneath the right lobe of the liver used for bile storage.

Gallstones particles of hardened cholesterol or calcium salts that are occasionally formed in gallbladder and bile ducts.

Gamete *gam′ ēt* male or female sex cell (sperm/egg).

Gametogenesis *gam″ ē-to-jen′ ĕ-sis* the formation of gametes.

Ganglion *gang′ gle-on* a group of nerve cell bodies located in the peripheral nervous system.

Gastrin *gas′ trin* a hormone that stimulates gastric secretion, especially hydrochloric acid release.

Gene *jēn* biological unit of heredity located in chromatin; transmits hereditary information.

Genetics *je-net′ iks* the science of heredity.

Genitalia *jen″ ĭ-ta′ le-ah* the external sex organs.

Germ layers the initial or primary tissues formed in the embryo (ectoderm, mesoderm, endoderm) from which all body tissues arise.

Gingiva *jin-jī′ vah* the gums.

Gland an organ specialized to secrete or excrete substances for further use in the body or for elimination.

Glaucoma *glaw-ko′ mah* an abnormal increase of the pressure within the eye.

Glia *glī′ ah* see *neuroglia.*

Globular protein a protein whose functional structure is basically spherical. Also referred to as functional protein; includes hemoglobin enzymes and some hormones.

Glomerular capsule *glō-mer′ yoo-ler* double-walled cuplike end of a renal tubule; encloses a glomerulus; also called Bowman's capsule.

Glomerulus *glo-mer′ u-lus* a knot of coiled capillaries in the kidney; forms filtrate.

Glottis *glot′ is* the opening between the vocal cords in the larynx.

Glucose *gloo′ kōs* the principal sugar in the blood; a monosaccharide.

Glycerol *glis′ er-ol* a sugar alcohol; one of the building blocks of fats.

Glycogen *glī′ ko-jen* the main carbohydrate stored in animal cells; a polysaccharide.

Glycogenesis *glī″ ko-jen′ ĕ-sis* formation of glycogen from glucose.

Glycogenolysis *glī″ ko-jĕ-nol′ ĭ-sis* breakdown of glycogen to glucose.

Glycolysis *gli-kol′ ĭ-sis* breakdown of glucose to pyruvic acid; an anaerobic process.

Goblet cells individual cells (simple glands) that produce mucus.

Goiter *goi′ ter* a benign enlargement of the thyroid gland.

Gonads *go′ nadz* organs producing gametes; ovaries or testes.

Graafian follicle *graf′ e-an fol′ lĭ-kul* see *vesicular follicle.*

Graded potential a local change in membrane potential that varies directly with the strength of the stimulus, declines with distance.

Graded response a response that varies directly with the strength of the stimulus.

Gray matter the gray area of the central nervous system; contains unmyelinated nerve fibers and nerve cell bodies.

Groin the junction of the thigh and the trunk; the inguinal area.

Gustation *gus-ta′ shun* taste.

Gyrus *jī′ rus* an outward fold of the surface of the cerebral cortex.

Hamstring muscles the posterior thigh muscles: the biceps femoris, semimembranosus, and semitendinosus.

Haversian system *ha-ver′ shan* see *osteon.*

Heart block impaired transmission of impulses from atrium to ventricle.

Heart murmur an abnormal heart sound (usually resulting from valve problems).

Heat stroke the failure of the heat-regulating ability of an individual under heat stress.

Helper T cell the type of T lymphocyte that orchestrates cellular immunity by direct contact with other immune cells and by releasing chemicals called lymphokines; also helps to mediate the humoral response by interacting with B cells.

Hematocrit *he-mat′ o-krit* the percentage of erythrocytes to total blood volume.

Hematopoiesis *he″ mato-poi-e′ sis* formation of blood cells.

Hemiplegia *hem″ e-ple′ je-ah* paralysis of one side of the body.

Hemocytoblasts *he′ mo-si′ to-blastz* stem cells that give rise to all the formed elements of the blood.

Hemoglobin *he″mo-glo′bin* the oxygen-transporting pigment of erythrocytes.

Hemolysis *he-mol′ĭ-sis* the rupture of erythrocytes.

Hemophilia *he″mo-fil′e-ah* an inherited clotting defect caused by absence of a blood-clotting factor.

Hemorrhage *hem′or-ij* the loss of blood from the vessels by flow through ruptured walls; bleeding.

Hepatic portal system *hĕ-pat′ik* the circulation in which the hepatic portal vein carries dissolved nutrients from the digestive tract to the liver for processing.

Hepatitis *hep″ah-ti′tis* inflammation of the liver.

Hilum, hilus *hi′lum, hi′lus* a depressed area where vessels enter and leave an organ.

Histamine *his′tah-mēn* a substance that causes vasodilation and increased vascular permeability.

Histology *his-tol′o-je* the branch of anatomy dealing with the microscopic structure of tissues.

Homeostasis *ho″me-o-sta′sis* a state of body equilibrium or stable internal environment of the body.

Homologous *ho-mol′o-gus* parts or organs corresponding in structure but not necessarily in function.

Hormones *hor′mōnz* chemical messengers secreted by endocrine glands; responsible for specific regulatory effects on certain parts or organs.

Humoral immunity *hu′mor-al* immunity provided by antibodies released by sensitized B cells and their plasma cell progeny. Also called antibody-mediated immunity.

Hyaline *hi′ah-lin* glassy; transparent.

Hydrochloric acid *hi″dro-klo′rik* HCl; aids protein digestion in the stomach; produced by parietal cells.

Hydrogen bond weak bond in which a hydrogen atom forms a bridge between two electron-hungry atoms; an important intramolecular bond.

Hydrolysis *hi-drol′ĭ-sis* the process in which water is used to split a substance into smaller particles.

Hyperopia *hi″per-o′pe-ah* farsightedness.

Hypertension *hi″per-ten′shun* an abnormally high blood pressure.

Hypertonic *hi″per-ton′ik* excessive, above normal, tone or tension.

Hypertrophy *hi-per′tro-fe* an increase in the size of a tissue or organ independent of the body's general growth.

Hypotension low blood pressure.

Hypothalamus *hi″po-thal′ah-mus* the region of the diencephalon forming the floor of the third ventricle of the brain.

Hypothermia *hi″po-ther′me-ah* subnormal body temperature.

Hypotonic *hi-po-ton′ik* below normal tone or tension.

Hypoxia *hi-pok′se-ah* a condition in which inadequate oxygen is available to tissues.

Ileum *il′e-um* the terminal part of the small intestine; between the jejunum and the cecum of the large intestine.

Immune response antigen-specific defenses mounted by activated lymphocytes (T cells and B cells).

Immunity *ĭ-mu′nĭ-te* the ability of the body to resist many agents (both living and nonliving) that can cause disease; resistance to disease.

Immunocompetence *im″mu-no-kom′pĕ-tents* the ability of the body's immune cells to recognize (by binding) specific antigens; reflects the presence of plasma membrane–bound receptors.

Immunodeficiency disease *im″mu-no-de-fish′en-se* disease resulting from the deficient production or function of immune cells or certain molecules (complement, antibodies, and so on) required for normal immunity.

Immunoglobulin *im″mu-no-glob′u-lin* a protein molecule, released by plasma cells, that mediates humoral immunity; an antibody.

Infarct *in′farkt* a region of dead, deteriorating tissue resulting from a lack of blood supply.

Inferior (caudal) pertaining to a position near the tail end of the long axis of the body.

Inflammation *in″flah-ma′shun* a physiological response of the body to tissue injury; includes dilation of blood vessels and increased vessel permeability.

Inguinal *ing′gwĭ-nal* pertaining to the groin region.

Inner cell mass a cluster of cells in the blastocyst from which the embryo develops.

Innervation *in″er-va′shun* the supply of nerves to a body part.

Inorganic compound a compound that lacks carbon; for example, water.

Insertion *in-ser′shun* the movable attachment of a muscle as opposed to its origin.

Inspiration *in″spĭ-ra′shun* the drawing of air into the lungs; inhalation.

Integumentary system *in-teg-u-men′tar-e* the skin and its accessory organs.

Intercellular *in″ter-sel′u-lar* between the body cells.

Intercellular matrix *ma′triks* the material between adjoining cells; especially important in connective tissue.

Internal respiration the use of oxygen by body cells; also called *cellular respiration.*

Interneurons complete the pathway between afferent and efferent neurons; also called *association neurons.*

Interstitial fluid *in″ter-stish′al* the fluid between the cells.

Intervertebral discs *in″ter-ver′tĕ-bral* the discs of fibrocartilage between the vertebrae.

Intracellular *in″trah-sel′u-lar* within a cell.

Intracellular fluid fluid within a cell.

Intrinsic factor *in-trin′sik* a substance produced by the stomach that is required for vitamin B_{12} absorption.

Invert to turn inward.

Ion *i′on* an atom with a positive or negative electric charge.

Ionic bond bond formed by the complete transfer of electron(s) from one atom to another (or others). The resulting charged atoms, or ions, are oppositely charged and attract each other.

Ipsilateral *ip″sĭ-lat′er-al* situated on the same side.

Iris *i′ris* the pigmented, involuntary muscle that acts as the diaphragm of the eye.

Irritability *ir″ĭ-tah-bil′ĭ-te* ability to respond to a stimulus.

Ischemia *is-ke′me-ah* a local decrease in blood supply.

Isometric *i″so-met′rik* of the same length.

Isotonic *i″so-ton′ik* having a uniform tension; of the same tone.

Isotope *i′sĭ-tōp* different atomic form of the same element. Isotopes vary only in the number of neutrons they contain.

Jaundice *jawn′dis* an accumulation of bile pigments in the blood producing a yellow color of the skin.

Jejunum *je-joo′num* the part of the small intestine between the duodenum and the ileum.

Joint the junction of two or more bones; an articulation.

Keratin *ker′ah-tin* a tough, insoluble protein found in tissues such as hair, nails, and epidermis of the skin.

Killer T cell effector T cell that directly kills foreign cells; also called a cytotoxic T cell.

Kilocalories (kcal) unit used to measure the energy value of food.

Kinetic energy energy of motion.

Kinins *ki′ninz* group of polypeptides that dilate arterioles, increase vascular permeability, and induce pain.

Krebs cycle the aerobic pathway occurring within the mitochondria, in which energy is liberated during metabolism of carbohydrates, fats, and amino acids and CO_2 is produced.

Labia *la′be-ah* lips.

Labyrinth bony cavities and membranes of the inner ear that house the hearing and equilibrium receptors.

Lacrimal *lak′rĭ-mal* pertaining to tears.

Lactation *lak-ta′shun* the production and secretion of milk.

Lacteal *lak′te-al* special lymphatic capillaries of the small intestine that take up lipids.

Lactic acid *lak′tik* the product of anaerobic metabolism, especially in muscle.

Lacuna *lah-ku′nah* a little depression or space; in bone or cartilage, lacunae are occupied by cells.

Lamina *lam′ĭ-nah* (1) a thin layer or flat plate; (2) the portion of a vertebra between the transverse process and the spinous process.

Laryngitis *lar″in-ji′tis* an inflammation of the larynx.

Larynx *lar' inks* the cartilaginous organ located between the trachea and the pharynx; voice box.

Lateral *lat' er-al* away from the midline of the body.

Lens the elastic, doubly convex structure in the eye that focuses the light entering the eye on the retina.

Lesion *le' zhun* a tissue injury or wound.

Leukemia *lu-ke' me-ah* a cancerous condition in which there is an excessive production of immature leukocytes.

Leukocyte *lu' ko-sīt* white blood cell.

Ligament *lig' ah-ment* a cord of fibrous tissue that connects bones.

Lipid *lip' id* organic compound formed of carbon, hydrogen, and oxygen; examples are fats and cholesterol.

Lumbar *lum' bar* the portion of the back between the thorax and the pelvis.

Lumen *lu' men* the space inside a tube, blood vessel, or hollow organ.

Lymph *limf* the watery fluid in the lymph vessels collected from the tissue spaces.

Lymph node a mass of lymphatic tissue.

Lymphatic system *lim-fat' ik* a system of lymphatic vessels, lymph nodes, and other lymphoid organs and tissues.

Lymphocytes *lim' fo-sītz* agranular white blood cells formed in the bone marrow that mature in the lymphoid tissue.

Lymphoid organs refers to organs in the lymphatic system including lymphatic vessels, lymph nodes, spleen, and tonsils; see *lymphatic system.*

Lymphokines *lim' fo-kīnz* proteins involved in cell-mediated immune responses that enhance the immune and inflammatory responses.

Lysosomes *li' so-sōmz* organelles that originate from the Golgi apparatus and contain strong digestive enzymes.

Lysozyme *li' so-zīm* an enzyme found in sweat, saliva, and tears that is capable of destroying certain kinds of bacteria.

Macrophage *mak' ro-fāj"* cell particularly abundant in lymphatic and connective tissues; important in the immune response as an antigen-presenter to T cells and B cells.

Malignant *mah-lig' nant* life threatening; pertains to neoplasms that spread and lead to death, such as cancer.

Mammary glands *mam' mer-e* milk-producing glands of the breasts.

Mastication *mas" tĭ-ka' shun* the act of chewing.

Matter anything that occupies space and has mass.

Meatus *me-a' tus* the external opening of a canal.

Mechanical energy energy form directly involved in putting matter into motion.

Mechanoreceptors *mek' ah-no-rē-sep' terz* receptors sensitive to mechanical pressures such as touch, sound, or contractions.

Medial *me' de-al* toward the midline of the body.

Mediastinum *me' de-as-ti' num* the region of the thoracic cavity between the lungs.

Medulla *mĕ-dul' ah* the central portion of certain organs.

Meiosis *mi-o' sis* the two successive cell divisions in gamete formation producing nuclei with half the full number of chromosomes (haploid).

Melanin *mel' ah-nin* the dark pigment synthesized by melanocytes; responsible for skin color.

Melanocyte a cell that produces melanin.

Memory cell member of T cell and B cell clones that provides for immunologic memory.

Menarche *mĕ-nar' ke* establishment of menstrual function; the first menstrual period.

Meninges *mĕ-nin' jēz* the membranes that cover the brain and spinal cord.

Meningitis *men" in-ji' tis* inflammation of the meninges of the brain or spinal cord.

Menopause *men' o-pawz* the physiological end of menstrual cycles.

Menses *men' sēz* monthly discharge of blood from the uterus. Menstruation.

Menstruation *men" stroo-a' shun* the periodic, cyclic discharge of blood, secretions, tissue, and mucus from the mature female uterus in the absence of pregnancy.

Mesentery *mes' en-ter" e* the double-layered membrane of the peritoneum that supports most organs in the abdominal cavity.

Metabolic rate *met" ah-bol' ik* the energy expended by the body per unit time.

Metabolism *mĕ-tab' o-lizm* the sum total of the chemical reactions that occur in the body.

Metabolize *mĕ-tab' o-līz* to transform substances into energy or materials the body can use or store by means of anabolism or catabolism.

Metacarpal *met" ah-kar' pal* one of the five bones of the palm of the hand.

Metastasis *mĕ-tas' tah-sis* the spread of cancer from one body part or organ into another not directly connected to it.

Metatarsal *met" ah-tar' sal* one of the five bones between the tarsus and the phalanges of the foot.

Microvilli *mi" kro-vil' i* the tiny projections on the free surfaces of some epithelial cells; increase surface area for absorption.

Micturition *mik' tu-rish' un* urination, or voiding; emptying the bladder.

Midsagittal (median) section specific sagittal plane that lies exactly in the midline.

Minerals the inorganic chemical compounds found in nature.

Mitochondria *mi" to-kon' dre-ah* the rod-like cytoplasmic organelles responsible for ATP generation.

Mitosis *mi-to' sis* the division of the cell nucleus; often followed by division of the cytoplasm of a cell.

Mixed nerves nerves containing the processes of motor and sensory neurons; their impulses travel to and from the central nervous system.

Molecule *mol' ĕ-kyool* particle consisting of two or more atoms held together by chemical bonds.

Monoclonal antibodies *mon" o-klōn' ul* pure preparations of identical antibodies that exhibit specificity for a single antigen.

Monocyte *mon' o-sīt* large single-nucleus white blood cell; agranular leukocyte.

Monosaccharide *mon" o-sak' ĭ-rīd* literally, one sugar; the building block of carbohydrates; examples include glucose and fructose.

Mons pubis *monz pu' bis* the fatty eminence over the pubic symphysis in the female.

Motor unit a motor neuron and all the muscle cells it supplies.

Mucous membrane (mucosa) membrane that forms the linings of body cavities open to the exterior (digestive, respiratory, urinary, and reproductive tracts).

Mucus *myoo' kus* a sticky, thick fluid, secreted by mucous glands and mucous membranes, that keeps the free surface of membranes moist.

Muscle fibers muscle cells.

Muscle spindle encapsulated sensory receptor found in skeletal muscle that is sensitive to stretch.

Muscle tone sustained partial contraction of a muscle in response to stretch receptor inputs; keeps the muscle healthy and ready to react.

Muscle twitch a single rapid contraction of a muscle followed by relaxation.

Muscular dystrophy *dis' tro-fe* a progressive disorder marked by atrophy and stiffness of the muscles.

Muscular system organ system consisting of skeletal muscles and their connective tissue attachments.

Myelin *mi' ĕ-lin* a white, fatty lipid substance.

Myelinated fibers *mi' ĕ-lĭ-nāt" ed* axons (projections of a nerve cell) covered with myelin.

Myocardial infarction *mi" o-kar' de-al infark' shun* a condition characterized by dead tissue areas in the myocardium caused by interruption of blood supply to the area.

Myocardium *mi" o-kar' de-um* the cardiac muscle layer of the heart wall.

Myofibrils *mi" o-fi' brilz* contractile organelles found in the cytoplasm of muscle cells.

Myofilament *mi" o-fil' ah-ment* filaments composing the myofibrils. Of two types: actin and myosin.

Myometrium *mi" o-me' tri-um* the thick uterine musculature.

Myopia *mi-ō' pe-ah* nearsightedness.

Myosin *mi' o-sin* one of the principal contractile proteins found in muscle.

Nares *na' rēz* the nostrils.

Nasal pertaining to the nose.

Necrosis *nĕ-kro' sis* the death or disintegration of a cell or tissues caused by disease or injury.

Negative feedback feedback that causes the stimulus to decline or end.

Neoplasm *ne' o-plazm* an abnormal growth of cells; sometimes cancerous.

Nephrons *nef′ ronz* structural and functional units of the kidney.

Nerve bundle of neuronal processes (axons) outside the central nervous system.

Nerve fiber axon of a neuron.

Nerve impulse a self-propagating wave of depolarization; also called an *action potential.*

Nervous system fast-acting control system that employs nerve impulses to trigger muscle contraction or gland secretion.

Neuroglia *nu-rog′ le-ah* the nonneuronal tissue of the central nervous system that performs supportive and other functions; also called glia.

Neuromuscular junction *nu″ ro-mus′ ku-lar* the region where a motor neuron comes into close contact with a skeletal muscle cell.

Neurons *nu′ ronz* cells of the nervous system specialized to transmit messages throughout the body.

Neurotransmitter chemical released by neurons that may, upon binding to receptors of neurons or effector cells, stimulate or inhibit them.

Neutralization *nu″ tral-i-za′ shun* (1) a chemical reaction that occurs between an acid and a base; (2) blockage of the harmful effects of bacterial exotoxins or viruses by the binding of antibodies to their functional sites.

Neutron *nu′ tron* uncharged subatomic particle; found in the atomic nucleus.

Neutrophils *nu′ tro-filz* the most abundant of the white blood cells.

Nucleic acid *nu-kle′ ic* class of organic molecules that includes DNA and RNA.

Nucleoli *nu-kle′ o-li* small spherical bodies in the cell nucleus; function in ribosome synthesis.

Nucleotide *nu′ kle-o-tīd* building block of nucleic acids.

Nucleus *nu′ kle-us* a dense central body in most cells containing the genetic material of the cell.

Occipital *ok-sip′ ĭ-tal* pertaining to area at the back of the head.

Occlusion *ŏ-kloo′ zhun* closure or obstruction.

Olfaction *ol-fak′ shun* smell.

Oogenesis *o″ o-jen′ ĕ-sis* the process of formation of the ova.

Ophthalmic *of-thal′ mik* pertaining to the eye.

Opposition the action by which the thumb is used to touch the tips of the other fingers on the same hand. This unique action makes the human hand such a fine tool for grasping and manipulating things.

Optic *op′ tik* pertaining to the eye.

Optic chiasma *op′ tik ki-as′ mah* the partial crossover of fibers of the optic nerves.

Oral relating to the mouth.

Orbital eye area.

Organ a part of the body formed of two or more tissues that performs a specialized function.

Organ system a group of organs that work together to perform a vital body function; e.g., nervous system.

Organelles *or″ gan-elz′* specialized structures in a cell that perform specific metabolic functions.

Organic compound a compound containing carbon; examples include proteins, carbohydrates, and fats.

Organism an individual living thing.

Origin attachment of a muscle that remains relatively fixed during muscular contraction.

Osmoreceptor *oz″ mo-re-cep′ tor* a structure sensitive to osmotic pressure or concentration of a solution.

Osmosis *oz-mo′ sis* the diffusion of a solvent through a membrane from a dilute solution into a more concentrated one.

Ossicles *os′ sĭ-kulz* the three bones of the middle ear: hammer, anvil, and stirrup.

Osteoblasts *os′ te-o-blasts″* bone-forming cells.

Osteoclasts *os′ te-o-klasts″* large cells that resorb or break down bone matrix.

Osteocyte *os′ te-o-sīt″* a mature bone cell.

Osteon *os′ te-on* a system of interconnecting canals in the microscopic structure of adult compact bone; unit of bone.

Osteoporosis *os″ te-o-po-ro′ sis* an increased softening of the bone resulting from a gradual decrease in rate of bone formation; a common condition in older people.

Otic *o′ tik* pertaining to the ear.

Otolith *o′ to-lith* one of the small calcified masses in the utricle and saccule of the inner ear.

Ovarian cycle *o-va′ re-an* the monthly cycle of follicle development, ovulation, and corpus luteum formation in an ovary.

Ovary *o′ var-e* the female sex organ in which ova (eggs) are produced.

Ovulation *ov″ u-la′ shun* the release of an ovum (or oocyte) from the ovary.

Ovum *o′ vum* the female gamete (germ cell); an egg.

Oxidation *ok″ sĭ-da′ shun* the process of substances combining with oxygen or the removal of hydrogen.

Oxygen debt *ok′ sĭ-jen* the volume of oxygen required after exercise to oxidize the lactic acid formed during exercise.

Oxyhemoglobin *ok″ sĭ-he″ mo-glo′ bin* hemoglobin combined with oxygen.

Palate *pal′ at* roof of the mouth.

Palpation *pal-pa′ shun* examination by touch.

Pancreas *pan′ kre-as* gland posterior to the stomach, between the spleen and the duodenum; produces both endocrine and exocrine secretions.

Pancreatic juice *pan″ kre-at′ ik* a secretion of the pancreas containing enzymes for digestion of all food categories.

Papilla *pah-pil′ ah* small nipplelike projection.

Papillary muscles *pap′ ĭ-ler″ e* cone-shaped muscles found in the heart ventricles.

Paralysis *pah-ral′ ĭ-sis* the loss of muscle function.

Paraplegia *par″ ah-ple′ je-ah* paralysis of the lower limbs.

Parasympathetic division *par″ ah-sim″ pah-thet′ ik* a division of the autonomic nervous system; also referred to as the craniosacral division.

Parathyroid glands *par″ ah-thi′ roid* small endocrine glands located on the posterior aspect of the thyroid gland.

Parathyroid hormone (PTH) hormone released by the parathyroid glands that regulates blood calcium level.

Parietal *pah-ri′ ĕ-tal* pertaining to the walls of a cavity.

Parotid *pah-rot′ id* located near the ear.

Passive immunity short-lived immunity resulting from the introduction of "borrowed antibodies" obtained from an immune animal or human donor; immunological memory is not established.

Passive transport membrane transport processes that do not require cellular energy (ATP); e.g., diffusion, which is driven by kinetic energy.

Patella *pah-tel′ ah* the kneecap.

Pathogen disease-causing microorganism (e.g., some bacteria, fungi, viruses, etc).

Pathogenesis *path″ o-jen′ ĕ-sis* the development of a disease.

Pectoral *pek′ to-ral* pertaining to the chest.

Pectoral girdle see *shoulder girdle.*

Peduncle *pe-dung′ kul* a stalk of fibers, especially that connecting the cerebellum to the pons, midbrain, and medulla oblongata.

Pelvic girdle incomplete bony basin formed by the two coxal bones that secures the lower limbs to the sacrum of the axial skeleton.

Pelvis *pel′ vis* a basin-shaped structure; lower portion of the skeleton of the body trunk.

Penis *pe′ nis* the male organ of copulation and urination.

Pepsin an enzyme capable of digesting proteins in an acid pH.

Pericardium *per″ ĭ-kar′ de-um* the membranous sac enveloping the heart.

Perimysium *per″ ĭ-mis′ e-um* the connective tissue enveloping bundles of muscle fibers.

Perineum *per″ i-ne′ um* that region of the body extending from the anus to the scrotum in males and from the anus to the vulva in females.

Perineurium coarse connective tissue wrapping that binds groups of fibers in a nerve, forming fascicles, or fiber bundles.

Periosteum *per″ e-os′ te-um* double-layered connective tissue that covers and nourishes the bone.

Peripheral nervous system (PNS) *pĕ-rif′ er-al* a system of nerves that connects the outlying parts of the body with the central nervous system.

Peripheral resistance the resistance to blood flow offered by the systemic blood vessels; a measure of the amount of friction encountered by blood.

Peristalsis *per″ ĭ-stal′ sis* the waves of contraction seen in tubelike organs; propels substances along the tract.

Peritoneum *per″ĭ-to-ne′um* the serous membrane lining the interior of the abdominal cavity and covering the surfaces of the abdominal organs.

Peritonitis *per″ĭ-to-ni′tis* an inflammation of the peritoneum.

Permeability *per″me-ah-bil′ĭ-te* that property of membranes that permits passage of molecules and ions.

Peroneal *per″o-ne′al* pertaining to the lateral aspect of the leg.

pH the symbol for hydrogen ion concentration; a measure of the relative acidity or alkalinity of a solution.

Phagocyte *fag′o-sīt* a cell capable of engulfing and digesting particles or cells harmful to the body.

Phagocytosis *fag″o-si-to′sis* the ingestion of solid particles by cells.

Phalanges *fah-lan′jēz* the bones of the finger or toe.

Pharyngotympanic (auditory) tube tube that connects the middle ear and the pharynx; allows pressure to be equalized on both sides of the ear drum; also called the Eustachian tube.

Pharynx *far′inks* the muscular tube extending from the posterior of the nasal cavities to the esophagus.

Phospholipid *fos′fo-lip″id* a modified triglyceride containing phosphorus.

Photoreceptors *fo″to-re-sep′torz* specialized receptor cells that respond to light energy.

Physiology *fiz″e-ol′o-je* the science of the functioning of living organisms.

Pinocytosis *pi″no-si-to′sis* the engulfing of extracellular fluid by cells.

Pituitary gland *pĭ-tu′ĭ-tār″e* the neuroendocrine gland located beneath the brain that serves a variety of functions including regulation of the gonads, thyroid, adrenal cortex, water balance, and lactation.

Placenta *plah-sen′tah* the temporary organ that provides nutrients and oxygen to the developing fetus, carries away wastes, and produces the hormones of pregnancy.

Plantar *plan′tar* pertaining to the sole of the foot.

Plasma *plaz′mah* the fluid portion of the blood.

Plasma cell member of a B cell clone; specialized to produce and release antibodies.

Plasma membrane membrane that encloses cell contents; outer limiting membrane.

Platelet *plāt′let* one of the irregular cell fragments of blood; involved in clotting.

Pleura *ploor′ah* the serous membrane covering the lungs and lining the thoracic cavity.

Pleurisy *ploor′ĭ-se* inflammation of the pleurae, making breathing painful.

Plexus *plek′sus* a network of interlacing nerves, blood vessels, or lymphatics.

Plica *pli′kah* a fold.

Pneumothorax *nu″mo-tho′raks* the presence of air or gas in a pleural cavity.

Polar body a minute cell produced during meiosis in the ovary.

Polar molecules nonsymmetrical molecules that contain electrically unbalanced atoms.

Polarized *po′lar-īzd* the state of an unstimulated neuron or muscle cell in which the inside of the cell is relatively negative in comparison to the outside; the resting state.

Polycythemia *pol″e-si-the′me-ah* presence of an abnormally large number of erythrocytes in the blood.

Polypeptide *pol″e-pep′tīd* a chain of amino acids.

Polysaccharide *pol″e-sak′ĭ-rīd* literally, many sugars; a polymer of linked monosaccharides; examples include starch and glycogen.

Pons (1) any bridgelike structure or part; (2) the brain area connecting the medulla with the midbrain, providing linkage between upper and lower levels of the central nervous system.

Positive feedback feedback that tends to cause a variable to change in the same direction as the initial change; enhances the stimulus.

Postganglionic (postsynaptic) neuron *pōst″gang-gle-on′ik* a neuron of the autonomic nervous system having its cell body in a ganglion and its axon extending to an organ or tissue.

Precipitation formation of insoluble complexes that settle out of solution.

Preganglionic (presynaptic) neuron *pre″gang-gle-on′ik* a neuron of the autonomic nervous system having its cell body in the brain or spinal cord and its axon terminating in a ganglion.

Pressoreceptor *pres″o-re-sep′tor* a nerve ending in the wall of the carotid sinus and aortic arch sensitive to vessel stretching.

Pressure gradient difference in hydrostatic (fluid) pressure that drives filtration.

Primary (immune) response the initial response of the immune system to an antigen; involves clonal selection and establishes immunological memory.

Prime mover muscle whose contractions are primarily responsible for a particular movement; agonist.

Process (1) a prominence or projection; (2) a series of actions for a specific purpose.

Pronation *pro-na′shun* the inward rotation of the forearm causing the radius to cross diagonally over the ulna—palms face posteriorly.

Prone refers to a body lying horizontally with the face downward.

Proprioceptor *pro″pre-oh-sep′tor* a receptor located in a muscle or tendon; concerned with locomotion, posture, and muscle tone.

Protein *pro′tēn* a complex nitrogenous substance; the main building material of cells.

Proteinuria *pro″te-in-u′re-ah* the passage of proteins in the urine.

Proton *pro′ton* subatomic particle that bears a positive charge; located in the atomic nucleus.

Proximal *prok′sĭ-mal* toward the attached end of a limb or the origin of a structure.

Puberty *pu′ber-te* the period at which reproductive organs become functional.

Pubic pertaining to the genital region.

Pulmonary *pul′mo-ner″e* pertaining to the lungs.

Pulmonary circulation system of blood vessels that carry blood to and from the lungs for gas exchange.

Pulmonary edema *ĕ-de′mah* a leakage of fluid into the air sacs and tissue of the lungs.

Pulmonary ventilation breathing; consists of inspiration and expiration.

Pulse the rhythmic expansion and recoil of arteries resulting from heart contraction; can be felt from the outside of the body.

Pupil an opening in the center of the iris through which light enters the eye.

Purkinje fibers *pur-kin′je* the modified cardiac muscle fibers of the conduction system of the heart.

Pus the fluid product of inflammation composed of white blood cells, the debris of dead cells, and a thin fluid.

Pyelonephritis *pi″ĕ-lo-nĕ-fri′tis* an inflammation of the kidney pelvis and surrounding kidney tissues.

Pyloric region *pi-lor′ik* the final portion of the stomach; joins with the duodenum.

Pyramid any cone-shaped structure of an organ.

Pyrogen *pi′ro-jen* an agent or chemical substance that induces fever.

Quadriplegia *kwod″rĭ-ple′je-ah* the paralysis of all four limbs.

Radiant energy energy of the electromagnetic spectrum, which includes heat, light, ultraviolet waves, infrared waves, and other forms.

Radiate *ra′de-āt* diverge from a central point.

Radioactivity the process of spontaneous decay seen in some of the heavier isotopes, during which particles or energy are emitted from the atomic nucleus; results in the atom becoming more stable.

Radioisotope *ra″de-o-i′sĭ-tōp* isotope that exhibits radioactive behavior.

Ramus *ra′mus* a branch of a nerve, artery, vein, or bone.

Receptor *re-sep′tor* (1) a peripheral nerve ending specialized for response to particular types of stimuli; (2) molecule that binds specifically with other molecules, e.g., hormones and neurotransmitters.

Reduction restoring broken bone ends (or a dislocated bone) to its original position.

Reflex automatic reaction to a stimulus.

Reflex arc neural pathway for reflexes.

Refract bend; usually refers to light.

Refractory period *re-frak′to-re* the period of unresponsiveness to threshold stimulation.

Renal *re′nal* pertaining to the kidney.

Renal calculus *re′nal kal′ku-lus* a kidney stone.

Renin *re′nin* a substance released by the kidneys that is involved with raising blood pressure.

Repolarization restoration of the membrane potential to the initial resting (polarized) state.

Reproductive system organ system that functions to produce offspring.

Respiratory system organ system that carries out gas exchange; includes the nose, pharynx, larynx, trachea, bronchi, and lungs.

Responsiveness the ability to sense changes (stimuli) in the environment and then to react to them; see also *irritability.*

Reticulum *rĕ-tik' u-lum* a fine network.

Retina *ret' ĭ-nah* light sensitive layer of the eye; contains rods and cones.

Ribonucleic acid (RNA) *ri" bo-nu-kle' ic* the nucleic acid that contains ribose; acts in protein synthesis.

Ribosomes *ri' bo-sōmz* cytoplasmic organelles at which proteins are synthesized.

Rods one of the two types of photosensitive cells in the retina.

Rotate to turn about an axis.

Rugae *roo' je* elevations or ridges, as in the mucosa of the stomach.

Sacral *sa' kral* the lower portion of the back, just superior to the buttocks.

Sagittal section (plane) *sa' jih-tul* a longitudinal (vertical) plane that divides the body or any of its parts into right and left portions.

Saliva *sah-li' vah* the secretion of salivary gland which is ducted into the mouth.

Salt ionic compound that dissociates into charged particles (other than hydrogen or hydroxyl ions) when dissolved in water.

Sarcomere *sar' ko-mēr* the smallest contractile unit of muscle; extends from one Z disc to the next.

Sclera *skle' rah* the firm white fibrous outer layer of the eyeball; protects and maintains eyeball shape.

Sebaceous glands *se-ba' shus* glands that empty their sebum secretion into hair follicles.

Sebum *se' bum* the oily secretion of sebaceous glands.

Second messenger intracellular molecule generated by binding of a chemical to a membrane receptor; mediates intracellular responses.

Secondary (immune) response second and subsequent responses of the immune system to a previously met antigen; more rapid and more vigorous than the primary response.

Secondary sex characteristics anatomical features that develop under influence of sex hormones (male or female pattern of muscle development, bone growth, body hair, etc.) that are not directly involved in the reproductive process.

Secretion *se-kre' shun* (1) the passage of material formed by a cell to its exterior; (2) cell product that is transported to the cell exterior.

Semen *se' men* fluid mixture produced by male reproductive structures; contains sperm, nutrients, and mucus.

Semilunar valves *sem" e-lu' nar* valves that prevent blood return to the ventricles after contraction.

Seminiferous tubules *se" mĭ-nif' er-us* highly convoluted tubes within the testes that form sperm.

Sensory nerve a nerve that contains processes of sensory neurons and carries nerve impulses to the central nervous system.

Sensory nerve cell an initiator of nerve impulses following receptor stimulation.

Serous fluid *ser' us* a clear, watery fluid secreted by the cells of a serous membrane.

Serous membrane membrane that lines a cavity without an opening to the outside of the body (except for joint cavities); serosa.

Sex chromosomes chromosomes that determine genetic sex; the X and Y chromosomes.

Shoulder girdle composite of two bones, scapula and clavicle, that attach the upper limb to the axial skeleton; also called the pectoral girdle.

Sinoatrial node *si" no-a' tre-al* the mass of specialized myocardial cells in the wall of the right atrium; pacemaker of the heart.

Sinus *si' nus* (1) a mucous membrane–lined, air-filled cavity in certain cranial bones; (2) a dilated channel for passage of blood or lymph.

Skeletal muscle muscle composed of cylindrical multinucleate cells with obvious striations; the muscle(s) attached to the body's skeleton; also called *voluntary muscle.*

Skeletal system system of protection and support composed primarily of bone and cartilage.

Skull bony enclosure for the brain.

Smooth muscle muscle consisting of spindle-shaped, unstriped (nonstriated) muscle cells; involuntary muscle.

Solute *sol' yoot* the dissolved substance in a solution.

Solution a homogenous mixture of two or more components.

Somatic nervous system *so-mat' ik* a division of the peripheral nervous system; also called the voluntary nervous system.

Spermatogenesis *sper" mah-to-jen' ĕ-sis* the process of sperm production in the male; involves meiosis.

Sphincter *sfink' ter* a circular muscle surrounding an opening; acts as a valve.

Spinal nerves the 31 pairs of nerves that arise from the spinal cord.

Squamous *skwa' mus* (1) flat, scalelike; (2) pertaining to flat, thin cells that form the free surface of some epithelial tissues.

Stasis *sta' sis* (1) a decrease or stoppage of flow; (2) a state of nonchange.

Static equilibrium *stat' ik e" kwĭ-lib' re-um* balance concerned with changes in the position of the head.

Stenosis *stĕ-no' sis* abnormal constriction or narrowing.

Sternal breastbone area.

Steroids *stĕ' roidz* a specific group of chemical substances including certain hormones and cholesterol.

Stimulus *stim' u-lus* an excitant or irritant; a change in the environment producing a response.

Stratum *stra' tum* a layer.

Stressor any stimulus that directly or indirectly causes the hypothalamus to initiate stress-reducing responses, such as the fight-or-flight response.

Striated muscle *stri' āt-ed* muscle consisting of cross-striated (cross-striped) muscle fibers; includes cardiac and skeletal muscle.

Stroke a condition in which brain tissue is deprived of a blood supply, as in blockage of a cerebral blood vessel.

Stroke volume a volume of blood ejected by a ventricle during systole.

Subcutaneous *sub" ku-ta' ne-us* beneath the skin.

Sudoriferous glands *su' do-rif' er-us* see *sweat glands.*

Sulcus *sul' kus* a furrow on the brain, less deep than a fissure.

Summation *sum-ma' shun* the accumulation of effects, especially those of muscular, sensory, or mental stimuli.

Superficial (external) located close to or on the body surface.

Superior refers to the head or upper body regions.

Supination *su" pĭ-na' shun* the outward rotation of the forearm causing palms to face anteriorly.

Supine refers to a body lying with the face upward.

Suppressor T cells regulatory T lymphocytes that suppress the immune response.

Surfactant *sur-fak' tant* a chemical substance coating the pulmonary alveoli walls that reduces surface tension, thus preventing collapse of the alveoli after expiration.

Sutures *soo' churz* immovable fibrous joints that connect the bones of the adult skull.

Sweat glands the glands that produce a saline solution called sweat; also called sudoriferous glands.

Sympathetic division a division of the autonomic nervous system; opposes parasympathetic functions; called the fight-or-flight division.

Synapse *sin' aps* the region of communication between neurons.

Synaptic cleft *sĭ-nap' tik* the fluid-filled space at a synapse between neurons.

Synarthrosis *sin" ar-thro' sis* an immovable joint.

Synergists *sin' er-jists* muscles cooperating with another muscle or muscle group to produce a desired movement.

Synovial fluid *sĭ-no' ve-al* a fluid secreted by the synovial membrane; lubricates joint surfaces and nourishes articular cartilages.

Synovial joint freely movable joint exhibiting a joint cavity enclosed by an articular (fibrous) capsule lined with synovial membrane.

Synovial membrane membrane that lines the capsule of a synovial joint.

Synthesis reaction chemical reaction in which larger molecules are formed from simpler ones.

System a group of organs that function cooperatively to accomplish a common purpose; there are eleven major systems in the human body.

Systemic *sis-tem' ik* general; pertaining to the whole body.

Systemic circulation system of blood vessels that carries nutrient and oxygen-rich blood to all body organs.

Systemic edema *ĕ-de' mah* an accumulation of fluid in body organs or tissues.

Systole *sis' to-le* the contraction phase of heart activity.

Systolic pressure *sis-tol' ik* the pressure generated by the left ventricle during systole.

T cells lymphocytes that mediate cellular immunity; include helper, killer, suppressor, and memory cells. Also called T lymphocytes.

Tachycardia *tak" e-kar' de-ah* an abnormal, excessively rapid heart rate; over 100 beats per minute.

Tarsal *tahr' sal* one of the seven bones that form the ankle and heel.

Taste buds receptors for taste on the tongue, roof of mouth, pharynx, and larynx.

Tendon *ten' dun* cord of dense fibrous tissue attaching a muscle to a bone.

Testis *tes' tis* the male primary sex organ that produces sperm.

Testosterone *tes-tos' tĕ-rōn* male sex hormone produced by the testes; during puberty promotes virilization and is necessary for normal sperm production.

Tetanus *tet' ah-nus* (1) the tense, contracted state of a muscle; (2) an infectious disease.

Thalamus *tha' luh-mis* a mass of gray matter in the diencephalon of the brain.

Thermoreceptor *ther" mo-re-sep' ter* a receptor sensitive to temperature changes.

Thoracic *tho-ras' ik* refers to the chest.

Thorax *tho' raks* that portion of the body trunk above the diaphragm and below the neck.

Threshold stimulus the weakest stimulus capable of producing a response in an irritable tissue.

Thrombin *throm' bin* an enzyme that induces clotting by converting fibrinogen to fibrin.

Thrombophlebitis *throm" bo-fle-bi' tis* an inflammation of a vein associated with blood clot formation.

Thrombus *throm' bus* a fixed clot that develops and persists in an unbroken blood vessel.

Thymus gland *thi' mus* an endocrine gland active in the immune response.

Thyroid gland *thi' roid* one of the largest of the body's endocrine glands; straddles the anterior trachea.

Tidal volume amount of air inhaled or exhaled with a normal breath.

Tissue a group of similar cells specialized to perform a specific function; primary tissue types are epithelial, connective, muscle, and nervous tissues.

Trachea *tra' ke-ah* The windpipe; the respiratory tube extending from larynx to bronchi.

Tract a collection of nerve fibers in the CNS having the same origin, termination, and function.

Transverse plane plane that divides the body or its parts into superior and inferior portions; also called a cross section.

Trauma *traw' mah* an injury, wound, or shock; usually produced by external forces.

Triglycerides *tri-glis' er-īdz* compounds composed of fatty acids and glycerol; fats and oils; also called neutral fats.

Trochanter *tro-kan' ter* a large, somewhat blunt process.

Tropic hormone *trōp' ik* a hormone that regulates the function of another endocrine organ.

Tubercle *tu' ber-kul* a nodule or small rounded process.

Tuberosity *tu" bĕ-ros' ĭ-te* a broad process, larger than a tubercle.

Tunica *too' nĭ-kah* a covering or tissue coat; layer.

Tympanic membrane *tim-pan' ik* the eardrum.

Ulcer *ul' ser* a lesion or erosion of the mucous membrane, such as gastric ulcer of stomach.

Umbilical cord *um-bĭ ' lĭ-kul* a structure bearing arteries and veins connecting the placenta and the fetus.

Umbilicus *um-bĭ ' lĭ-kus* navel; marks site of the umbilical cord in the fetal stage.

Urea *u-re' ah* the main nitrogen-containing waste excreted in the urine.

Ureters *u-re' terz* tubes that carry urine from kidney to bladder.

Urethra *u-re' thrah* the canal through which urine passes from the bladder to the outside of the body.

Urinary system system primarily responsible for water, electrolyte, and acid-base balance and the removal of nitrogenous wastes from the blood.

Urine filtrate containing waste and excess ions excreted by the kidneys.

Uterine tube the oviduct. The tube through which the ovum is transported to the uterus; also called fallopian tube.

Uvula *u' vu-lah* tissue tag hanging from soft palate.

Valence shell *va' lens* the outermost energy level of an atom that contains electrons; the electrons in the valence shell determine the bonding behavior of the atom.

Vas *vas* a duct; vessel.

Vascular *vas' ku-lar* pertaining to blood channels or vessels.

Vasoconstriction *vas" o-kon-strik' shun* narrowing of blood vessels.

Vasodilation *vas" o-di-la' shun* relaxation of the smooth muscles of blood vessels producing dilation.

Vasomotor nerve fibers *vas-o-mo' tor* the nerve fibers that regulate the constriction or dilation of blood vessels.

Vein *vān* a vessel carrying blood away from the tissues toward the heart.

Venae cavae two large blood vessels that drain oxygen-poor blood from the veins into the right side of the heart.

Ventral anterior or front.

Ventricles *ven' trĭ-kulz* (1) discharging chambers of the heart; (2) cavities within the brain.

Venule *ven' ūl* a small vein.

Vertebral column the spine, formed of a number of individual bones called vertebrae and two composite bones (sacrum and coccyx).

Vesicular follicle a mature ovarian follicle.

Villi *vil' i* fingerlike projections of the small intestinal mucosa that tremendously increase its surface area for absorption.

Viscera *vis' era* the internal organs.

Visceral *vis' er-al* pertaining to the internal part of a structure or the internal organs.

Viscosity *vis-kos' ĭ-te* the state of being sticky or thick.

Visual acuity *ah-ku' ĭ-te* the ability of the eye to distinguish detail.

Vital capacity the volume of air that can be expelled from the lungs by forcible expiration after the deepest inspiration; total exchangeable air.

Vitamins organic compounds required by the body in minute amounts for physiological maintenance and growth.

Vitreous humor a gel-like substance that helps prevent the eyeball from collapsing inward by reinforcing it internally.

Voluntary muscle muscle under control of the will; skeletal muscle.

Vulva *vul' va* female external genitalia.

White matter white substance of the central nervous system; the myelinated nerve fibers.

Zygote *zi' gōt* the fertilized ovum; produced by union of two gametes.

Index

NOTE: A *t* following a page number indicates tabular material, an *f* following a page number indicates a figure, and a *b* following a page number indicates a boxed feature.

continues

continues

continues